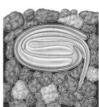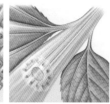

The Molecular Life of Plants

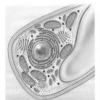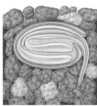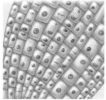

The Molecular Life of Plants

Russell Jones, Helen Ougham,
Howard Thomas and Susan Waaland

WILEY-BLACKWELL

A John Wiley & Sons, Ltd., Publication

Library of Congress Cataloging-in-Publication Data

The molecular life of plants / Russell Jones . . . [et al.].
   p. cm.
Includes bibliographical references and index.
ISBN 978-0-470-87012-9 (pbk.: alk. paper)—ISBN 978-0-470-87011-2 (hardcover: alk. paper)
1. Plant molecular biology—Textbooks. 2. Plant physiology—Textbooks. 3. Botany—Research—Textbooks.
I. Jones, Russell L.
QK728.M634 2012
572.8′2928—dc23

                                                                                                    2012007650

A catalogue record for this book is available from the British Library.

Brief contents

Part V Maturation

Part VI Renewal

Companion website

This book is accompanied by a companion website:

www.wiley.com/go/jones/molecularlifeofplants

The website includes:

- Powerpoints of all figures
- PDFs of all tables from the book for downloading
- PDFs of the table of contents and index

Contents

Part III
Emergence

8   Light perception and transduction 251

Part IV
Growth

10   Hormones and other signals 329

Part V
Maturation

13   Mineral nutrient acquisition and assimilation 455

Part VI
Renewal

17 Development and dormancy of resting structures 629

Companion website

This book is accompanied by a companion website:

www.wiley.com/go/jones/molecularlifeofplants

The website includes:

- Powerpoints of all figures
- PDFs of all tables from the book for downloading
- PDFs of the table of contents and index

Preface

The Molecular Life of Plants, a textbook designed to introduce undergraduate students to contemporary experimental plant biology, was inspired by *Biochemistry and Molecular Biology of Plants* published in 2000 by the American Society of Plant Biologists. *Biochemistry and Molecular Biology of Plants* was written with a graduate student readership in mind, its 24 chapters covering key topics at the forefront of plant biochemistry and molecular biology. In addition to presenting advances in plant science research, an outstanding feature of *Biochemistry and Molecular Biology of Plants* is the quality and richness of the artwork, many aspects of which have been emulated by other textbooks. *Biochemistry and Molecular Biology of Plants* has been translated into Chinese, Japanese and Italian and an English language version of the book has been produced for the Indian market.

The success of *Biochemistry and Molecular Biology of Plants* led the American Society of Plant Biologists to collaborate with Wiley-Blackwell to produce other publications in plant biology. The editors of *Biochemistry and Molecular Biology of Plants*, Bob Buchanan, Willi Gruissem and Russell Jones, developed the outline of *The Molecular Life of Plants*, an undergraduate textbook incorporating the strengths of *Biochemistry and Molecular Biology of Plants*, especially the outstanding illustrative material. The goal of this new text is to include broader aspects of contemporary experimental plant biology in a typical undergraduate plant physiology curriculum. Thus *The Molecular Life of Plants* is designed to show how the unifying influence of molecular advances in the fields of biochemistry, physiology, development, adaptation and evolution is revolutionizing the teaching of experimental plant biology.

The Molecular Life of Plants uses the life cycle of a seed plant as a framework to discuss the key aspects of plant function from seed to seed. Helen Ougham and Sid Thomas, both at Aberystwyth University, and Susan Waaland at the University of Washington, joined Russell Jones at the University of California at Berkeley to write this book. Russell Jones, Helen Ougham and Sid Thomas wrote the elements of the 18 chapters and Susan Waaland was the scientific editor ensuring that the whole book read uniformly and was factually accurate.

The teaching of functional plant biology has a long and illustrious history in Europe and North America. Many outstanding textbooks have been devoted to this topic. In the middle to late 19th century this field was dominated by books written in German by authors such as Haberlandt, Sachs and Pfeffer, whose texts were translated into English and were used in English-speaking countries worldwide. It was only in the 1930s that textbooks dealing with the mechanisms of plant growth and development written by North American authors began to be published, the first by Edwin Miller in 1931. Miller's Preface to *Plant Physiology* stated: 'The various texts by European investigators and teachers, while summarizing the work that has been done on the continent, have failed to cover adequately the contributions of American and English plant physiologists.' From the mid-20th century, authorship of textbooks covering the field has had a decidedly North American bias, largely due to the impact that Land Grant Universities in the USA have had on the teaching of this subject so essential to agriculture. Publication of Miller's book was followed by the now famous text *Plant Physiology* by Bernard Meyer and Donald Anderson, first published in 1938 and surviving in various editions until the 1970s. Two other textbooks have since dominated the field, one by Frank Salisbury and Cleon Ross first published in 1969 and the most recent by Lincoln Taiz and Eduardo Zeiger, first published in 1991 and now in its fifth edition.

The Molecular Life of Plants differs from its predecessors in that it reflects the dramatic changes made possible in biology by the revolution in molecular genetics. The complete genome sequences of a large number of plant species have been published and the ability to generate mutants with unique phenotypes in *Arabidopsis*, in *Zea mays* and in many other species has allowed the dissection of biochemical pathways and cell processes and an enhanced understanding of the fundaments of plant growth and development. Following the lead of *Biochemistry and Molecular Biology of Plants*, we have incorporated many of these topics into *The Molecular Life of Plants*. We have also introduced 'Key points' as a tool to summarize material and facilitate student learning. Salient aspects within each section of a chapter are summarized in a 'Key points' text box that condenses the topic to its essence in 100–150 words.

This book is organized into six parts beginning with *Origins* which has four chapters, the first providing a primer on plant structure and reproduction, then Chapter 2 presenting the basics of cellular chemistry, followed by Chapter 3 on plant genomes, their organization and expression, and Chapter 4 on cell structure. Part II (*Germination*) has three chapters. Chapter 5 describes the cellular events crucial for germination including membrane transport and intracellular protein trafficking. The other two chapters in this part discuss germination and the mobilization of stored food reserves (Chapter 6), and how these reserves are metabolized to provide energy and carbon skeletons for the developing plant (Chapter 7).

Part III (*Emergence*) deals with the roles of light in seedling growth and development. Chapter 8 discusses light perception and the developmental consequences of this, while Chapter 9 addresses photosynthesis and photorespiration. Part IV (*Growth*) covers hormone synthesis and action (Chapter 10); the cell cycle and meristems (Chapter 11); and cell elongation, embryogenesis and vegetative development (Chapter 12).

Maturation (Part V) and *Renewal* (Part VI) complete the functional aspects of the plant life cycle. In Part V, Chapter 13 discusses nutrient acquisition, Chapter 14 covers the topics of long-distance transport with a focus on the mature plant, and Chapter 15 deals with interactions of the plant with its environment. In the final part, Chapter 16 describes the development of flowers, seeds and fruits, while Chapter 17 discusses the development of resting structures and dormancy mechanisms. Events in the plant life cycle are completed in Chapter 18 with a detailed treatment of senescence, ripening and death in the final stages in the life of a plant.

A comprehensive list of credits and permissions for the use of the many figures and tables is provided at the end of the book. Special thanks are due to several members of the editorial team at Wiley-Blackwell. Celia Carden our Development Editor deserves particular recognition for moving the project forward and for her good humor under all conditions. In addition to her deep inside knowledge of the publication business, Celia demonstrated her broad knowledge of plant biology, helpful in selecting the anonymous reviewers of the manuscript in its various stages of development. We are indebted to these reviewers. Celia was instrumental in hiring Debbie Maizels as the illustrator for the book. Debbie is an extraordinarily talented artist with the bonus of having a sound background in the biological sciences. We owe Debbie special gratitude for her work on this book. Fiona Seymour, Senior Project Editor at Wiley-Blackwell provided excellent support during the production process. Fiona's knowledge of the intricacies of textbook production was invaluable in ensuring the overall very high quality of *The Molecular Life of Plants*. Jane Andrew, Project Manager, has helped immensely with the details of production including copy-editing, liaising with the typesetters, proofreading and indexing. Last, but not least thanks are due to Andy Slade at Wiley-Blackwell and Nancy Winchester at ASPB headquarters in Rockville, Maryland. Andy and Nancy were instrumental in the launching of the joint ASPB–Wiley publication venture and they were both very supportive of *The Molecular Life of Plants*, cheering from the sidelines when needed and making sure that the project did indeed come to fruition.

Russell Jones
Helen Ougham
Howard Thomas
Susan Waaland
2012

Part I
Origins

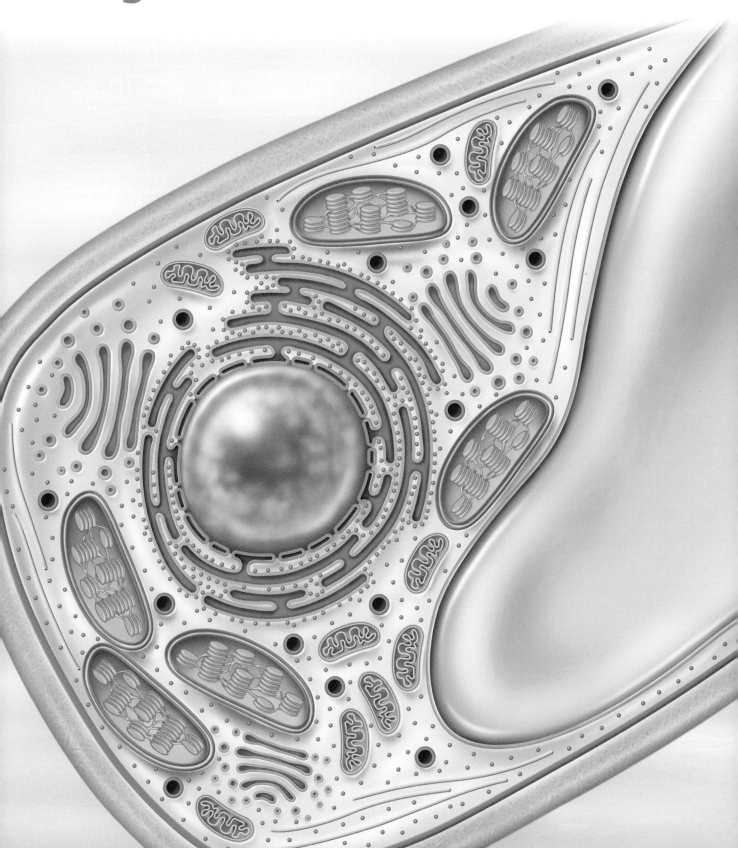

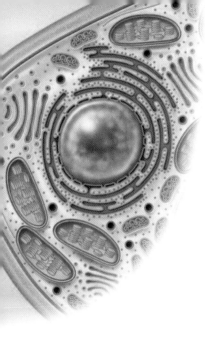

Chapter 1

Plant life: a primer

1.1 An introduction to plant biology

We begin our investigation of how genes, proteins, metabolites and environmental signals interact in living plants by recognizing that readers may approach this subject from different backgrounds. To provide a common knowledge base, we have developed this chapter as a plant biology primer. For readers well versed in the evolution, development, anatomy and morphology of plants, this chapter will review familiar topics. For those not yet exposed to these disciplines, the chapter provides grounding in the biology of whole plants and introduces the plant life cycle on which this textbook is structured. Many of the terms and concepts introduced here will be revisited as later chapters delve into the processes and mechanisms that underlie each stage of plant life, describing the intricate network of cellular, molecular, biochemical and physiological events through which plants make life on land possible. We will be discussing the types of evidence used to develop modern classification schemes, and the evolutionary history and relationships among the groups of green plants alive today. These will provide a basis for the discussion of the fundamentals of plant anatomy, development and reproductive biology.

1.2 Plant systematics

What makes a plant a plant? This seemingly simple question has challenged biologists for centuries. The science of **systematics** seeks to identify organisms and order them in hierarchical classification schemes based on their evolutionary (**phylogenetic**) relationships. The levels of classification range from the **domain**, the most inclusive group, to the **species**, the most exclusive group (Table 1.1). Such schemes have predictive value, making it easier to distinguish individual organisms by name and to recognize groups of close or distant relatives. Members of a group of species at any level of classification are sometimes referred to as a **taxon** (plural: **taxa**), and the science of classification is called **taxonomy**.

1.2.1 Each species has a unique scientific name that reflects its phylogeny

The scientific name of a plant includes its genus and species names. Carolus Linnaeus developed the genus/species binomial in 1753 as a shorthand version of the long polynomial name he gave each plant in his major taxonomic work, *Species Plantarum*. Linnaeus's polynomials have fallen out of use, but the binomial system has survived as the cornerstone of all biological classification schemes.

Latin binomials are italicized, with the first letter of the genus name capitalized and the first letter of the species epithet in lower case. Often a specific attribution is added to the binomial. In the case of domesticated barley, *Hordeum vulgare*, this binomial was used first by Linnaeus, so the abbreviation 'L.' is appended in Roman typeface: *Hordeum vulgare* L. (Figure 1.1). After first use of the full binomial in a document or in a discussion of

The Molecular Life of Plants, First Edition. Russell Jones, Helen Ougham, Howard Thomas and Susan Waaland.
© 2013 John Wiley & Sons, Ltd. Published 2013 by John Wiley & Sons, Ltd.

Table 1.1 The ranks used in the classification of plants, as illustrated for domesticated barley (*Hordeum vulgare*).

Domain	Eukarya
Kingdom	Viridoplantae (green plant)
Phylum (Division)	Magnoliophyta (flowering plant)
Class	Liliopsida (monocot)
Order	Poales
Family	Poaceae (grass family)
Genus	*Hordeum* (barley)
Species	*vulgare*

Figure 1.1 The Latin binomial for this barley plant is *Hordeum vulgare* L. 'Galena'.

several barley species, the shortened form *H. vulgare* can be used.

The value of coupling the binomial system to phylogeny-based classification becomes apparent when considering the muddle of common botanical names. Take 'beans', for example. The many plants that are referred to as beans do not belong to the same genus, the same family or even the same order (Figure 1.2). The common edible beans you might find on a dinner plate belong to the bean family, Fabaceae, but the plant that produces the castor bean is *Ricinus communis* in the family Euphorbiaceae, and the coffee bean comes from *Coffea arabica*, a member of the Rubiaceae. To make the situation more complex, the beans in the family Fabaceae belong to a number of different genera and often have many different common names. One example is *Phaseolus vulgaris*, a single species whose cultivated varieties produce adzuki, dry, French, green, pinto, runner, snap and wax beans. Other species in the same genus include lima beans (*P. limensis*) and butter beans (*P. lunatus*). Another genus in Fabaceae, *Vicia*, has 160

separate species, including *Vicia faba*. As you might guess from the specific epithet *faba*, *Vicia faba* is the fava bean, but this species is also called the broad, English, field, horse, pigeon, tick or Windsor bean (Figure 1.2).

Plant scientists often encounter the term **cultivar**, which is used to describe the cultivated varieties that plant breeders produce from wild species. When the cultivar is known, it is denoted by single quotation marks and/or the abbreviation 'cv.' and follows the Latin binomial. Cultivar names are in a language other than Latin. They are not italicized, and first letter(s) are capitalized: *Hordeum vulgare* L. 'Golden Promise' is the current convention, but the forms *Hordeum vulgare* L. cv. 'Golden Promise' and *Hordeum vulgare* L. cv. Golden Promise have also been used.

> **Key points** Common names have limited usefulness in identifying plants. The broad bean in the UK is the fava bean in the USA, but this plant is also referred to as the faba, field or horse bean, among other names. Plant biologists use the Latin binomial system devised by Linnaeus to describe species and the binomial for fava bean is *Vicia faba*. Binomials are written with a strict set of rules. The first epithet, as in *Vicia*, is the organism's genus and the second, *faba*, is the species. The binomial is often followed by an abbreviated attribution that identifies the person who gave the organism its binomial. In the case of *V. faba*, it is followed by L., identifying Linnaeus as the authority. Binomials are abbreviated when they are used repetitively as above, *V. faba*, the genus denoted by the first letter followed by the full species name. Binomials are also displayed in italic font whereas the authority is typed plain font.

1.2.2 Modern classification schemes attempt to establish evolutionary relationships

Classification schemes based on phylogeny attempt to construct taxa that are **monophyletic**, that is, groups that include an ancestral species, all of its descendants and only its descendants. Modern methods of phylogenetic analysis are known as **cladistics**, a term derived from the word clade. A **clade** is a monophyletic taxon. In this chapter we will be using evolutionary trees called **cladograms** (Figure 1.3) to illustrate current hypotheses about the evolutionary history of plants.

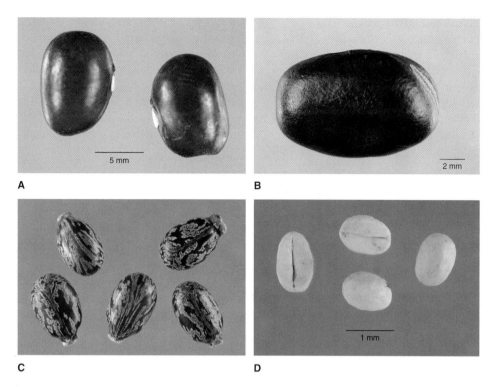

Figure 1.2   The common name 'bean' is used to refer to plants belonging to many different families and genera. These images show four 'beans' belonging to three different families and four different genera. This illustrates the importance of using the Latin binomial when identifying plants. (A) French bean (Fabaceae, *Phaseolus vulgaris*). (B) Fava bean (Fabaceae, *Vicia faba*). (C) Castor bean (Euphorbiaceae, *Ricinus communis*). (D) Coffee bean (Rubiaceae, *Coffea arabica*).

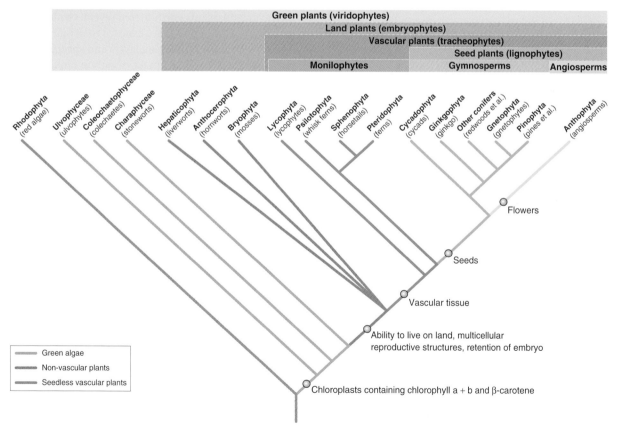

Figure 1.3   Cladogram illustrating the evolution of green plants.

To construct cladograms, systematists and evolutionary biologists use morphological, anatomical and metabolic traits as well as biochemical and molecular-genetic data. The tree is rooted using an **outgroup**, a relative of the taxon under investigation (the **ingroup**). The ingroup and the outgroup share certain **primitive traits**; members of the ingroup have acquired new **derived traits** that distinguish them from the outgroup. For example, in the cladogram in Figure 1.3, red algae are the outgroup for the green plant clade. Red algae and green plants share certain primitive traits, such as having cell walls composed of cellulose (see Chapter 4) and chloroplasts enclosed by two membranes. Chloroplasts of green plants, unlike those of red algae, contain both chlorophyll a and chlorophyll b; the presence of chlorophyll b is a major character that separates green plants from red algae. The remainder of the cladogram is constructed in a similar fashion. At each branch point, the clade above the branch point has derived traits that separate it from the taxa below the branch.

1.3 The origin of land plants

To understand the evolution of the land plants, it is useful to look at their evolutionary origins. Life originated on Earth more than 3.5 bybp (billion years before present) and it is believed that the fundamental design of the photosynthetic apparatus was established early in evolution. Chloroplasts, the subcellular structures that carry out photosynthesis in plants, are derived from photosynthetic bacteria (cyanobacteria) that, according to most estimates, entered into an endosymbiotic association with early animal-like unicellular organisms (protozoa) around 1.5 bybp. The primary endosymbiotic event soon gave rise to two evolutionary lineages, the so-called glaucophytes (a small group of alga-like unicellular freshwater organisms) and the red algal/green plant branch. Other photosynthetic protists acquired their chloroplasts by secondary endosymbiosis during which they engulfed unicellular red or green algae. Here we examine the further diversification of the green plant group leading to the land plants, the subject of this book.

1.3.1 The green plant clade, viridophytes, includes the green algae and land plants

The ancestors of land plants are widely believed to be the green algae. These two groups make up the green plant clade, the **viridoplantae**. The viridophytes are photosynthetic organisms whose chloroplasts are enclosed by two membranes, contain chlorophylls a and b and store starch. Cytological and molecular data support the division of the viridophytes into two clades that diverged more than 1 bybp: the **chlorophytes**, composed of most green algae, and the **streptophytes**, which include **charophycean green algae**, such as *Chara* and *Coleochaete* (Figure 1.4), and the **land plants**. Land plants, the **embryophytes**, have multicellular reproductive organs and produce embryos that are protected and nourished by the parent plant.

1.3.2 Unlike their green algal ancestors, embryophytes have evolved adaptations to life on land

Evolution of land plants has been driven by problems associated with living on dry land. Plants require carbon dioxide, water and light for photosynthesis. In addition, they need access to O_2 and to inorganic ions containing a number of different elements, including nitrogen, phosphorus, potassium, sulfur, etc. (see Chapter 13). Green algal ancestors of land plants were relatively simple, aquatic organisms (Figure 1.5). To survive and thrive on land, plants encountered several challenges. Organisms living on land are exposed to dry air and therefore need **waterproofing** layers and **specialized pores, stomata**, to allow uptake of carbon dioxide. Air does not provide buoyant support; this led to the evolution of support tissues to help keep plants upright. **Resources** essential for plant growth are usually **spatially separated**: light is above ground and water and minerals are in the soil. This led to the evolution of specialized aerial **photosynthetic organs** (leaves), **underground organs** (roots) specialized for uptake of water and inorganic ions and **connecting organs** (stems) with efficient transport systems to move sugar from the photosynthetic tissues to the roots and water and minerals upwards from the soil to the leaves.

The transition from aquatic living to life on land also impacted on reproduction. To understand this impact, we need to look at sexual reproduction in plants. The pattern of sexual reproduction in land plants is quite different from that of animals (Figure 1.6). In animals, including humans, the adult body is made of **diploid (2n)** cells. Meiosis produces **haploid (n) gametes**, eggs and sperm; these are the only haploid cells in the animal life cycle. Egg and sperm fuse at fertilization to produce a diploid zygote that gives rise to the next diploid generation. In contrast, sexual reproduction in land plants involves an **alternation** between a **diploid**

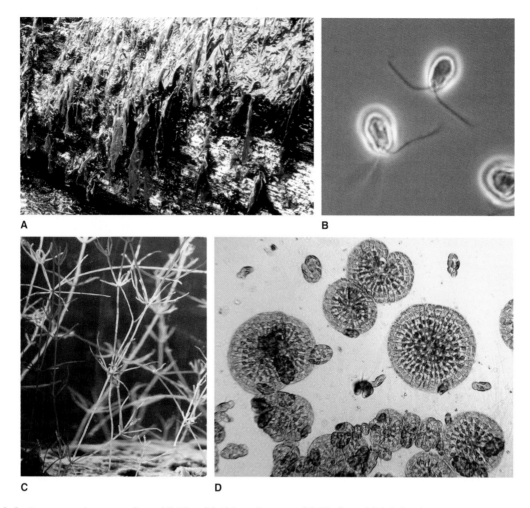

Figure 1.4 Representative green algae: (A) *Ulva*, (B) *Chlamydomonas*, (C) *Nitella* and (D) *Coleochaete*.

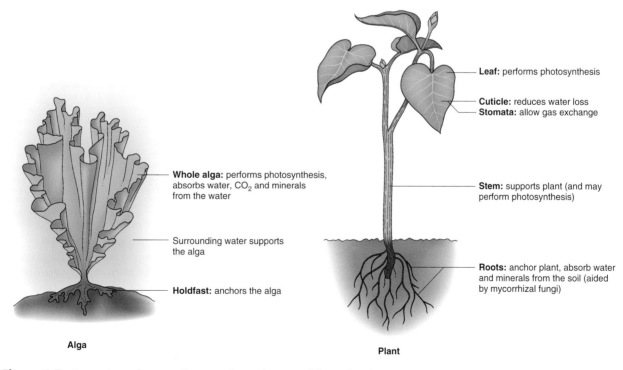

Figure 1.5 Comparison of an aquatic green alga and a terrestrial vascular plant.

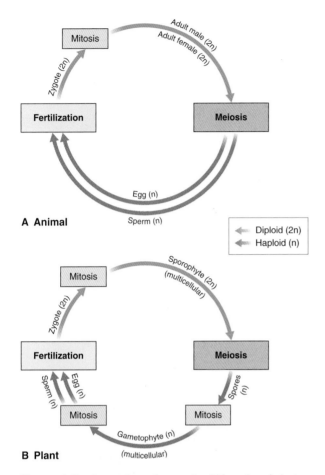

A Animal

Diploid (2n)
Haploid (n)

B Plant

Figure 1.6 Comparison of generalized life cycles of plants and animals. (A) The animal life cycle features one multicellular generation, which is diploid. The only haploid cells are the gametes produced by meiosis. (B) The plant life cycle has an alternation of multicellular generations; one is haploid, the other diploid. Gametes are produced by the haploid generation via mitosis. Meiosis in the diploid generation generates spores that germinate to yield the haploid generation.

generation, the **sporophyte**, and a **haploid generation**, the **gametophyte**. Furthermore, in land plants gametes are produced by mitosis. The diploid sporophyte produces multicellular **sporangia** in which **meiosis** occurs to produce **haploid spores**. These spores divide by mitosis to produce a multicellular haploid gametophyte. The gametophyte produces multicellular **gametangia** in which **gametes** are produced by **mitosis**. Gametes fuse to form a diploid **zygote** that in turn gives rise to a new sporophyte generation. Alternation of generations in seed plants is not obvious because the spores develop into gametophytes within the sporangium. The gametophytes are quite small, depend

on the sporophyte for nutrition and, in the case of female gametophytes, remain physically associated with the sporophyte.

In land plants, the zygote is retained in the female **gametangium**, where it develops into an **embryo** that is protected and nourished within tissues of the parent gametophyte. For land plants, both the gametophyte and sporophyte generation must be able to survive and reproduce on land. In water, gametes and spores could swim or float; on land, new mechanisms for bringing gametes together and for dispersing offspring were needed. Using Figure 1.3 as a guide, we will start with the most primitive clade of land plants, the bryophytes, and follow the evolution of land plants noting the adaptations that have evolved to allow plants to conquer land.

Key points Evolutionary trees or cladograms are used to establish evolutionary relationships among groups of organisms. The goal of this branch of biology, called cladistics, is to identify ancestral organisms and all of their descendants using key features that distinguish them from all other organisms. Organisms belonging to a unique group are referred to as a clade or a taxon. Plants can be divided into two distinct clades, ancestral green algae and land plants. Land plants are distinguished from algae by retention of the fertilized egg and protection and nourishment of the embryo by the parent plant. The retention of the embryo was among several features that allowed the transition from an aquatic existence to life on land. Among the other key features that evolved were a protective outer waxy layer that limited evaporation of water, closable pores for gas exchange, water transporting tissue and mechanisms of reproduction that eventually eliminated the need for liquid water.

1.4 Bryophytes

The **bryophytes** are the most primitive group of embryophytes. This group is **paraphyletic** (i.e. its members do not share a common ancestor); it includes three monophyletic clades, the **hornworts, liverworts** and **mosses** (see Figures 1.3 and 1.7). About 24 000 bryophyte species exist today. The first bryophytes are

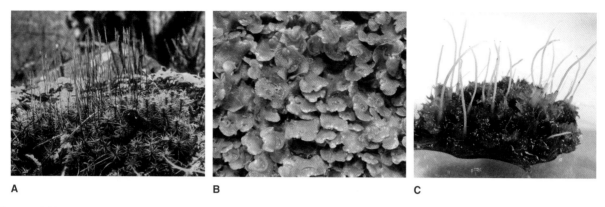

A   B   C

Figure 1.7   Representative bryophytes: (A) moss, (B) liverwort and (C) hornwort.

thought to have diverged from the vascular plant lineage more than 450 million years before present (mybp).

Key points Reproduction in plants is fundamentally different from that in animals. While in animals eggs and sperm are produced by meiosis, plants produce eggs and sperm by mitosis. Haploid spores are produced by meiosis by the diploid spore-making sporophyte plants. Spores germinate to produce a haploid gamete-making plant called the gametophyte that produces eggs and sperm by mitosis. This so-called alternation of generations between sporophyte and gametophyte is well illustrated in ferns. Large leafy ferns are sporophytes and spores produced by meiosis are found on the underside of leaves. Spores are shed and germinate to produce a gametophyte, rarely more that a few millimeters in diameter. Eggs and motile sperm are produced by the gametophyte plant and after fertilization the resulting zygote grows to the mature sporophyte plant.

1.4.1 Bryophytes have adapted to a range of environments and show a limited degree of differentiation into tissues and organs

Bryophytes are usually associated with damp, even wet, environments such as acidic peat bogs in which *Sphagnum* moss is abundant, but they also occur in environments that are seasonally dry and hot. For example, mosses are widely distributed on surfaces such as roofing tiles, slate and rocks. In these environments, they survive dry periods in a desiccated state and rehydrate and grow when moisture is available. Bryophytes are also abundant above the tree line and cover large areas in the Arctic regions.

The photosynthetic plants that we recognize as mosses, liverworts or hornworts, are haploid gametophytes (Figure 1.7). In hornworts and many liverwort species, the gametophyte takes the form of a **thallus**, a flattened body that hugs the ground. On its underside it produces **rhizoids**, elongate single cells that anchor the gametophyte to the substrate. The gametophytes of leafy liverworts and mosses are differentiated into leaf-like and stem-like organs (Figures 1.7 and 1.8) that may be covered in a cuticle. Often, the central part of the moss 'stem' contains elongate cells that are thought to participate in the transport of water and solutes, but these cells are different in structure and origin to the cells of the water-conducting tissue of vascular plants. Multicellular, filamentous rhizoids anchor the moss gametophyte to its substrate. Diploid sporophytes are less conspicuous and are nutritionally dependent on gametophytes.

1.4.2 Gametophytes dominate the bryophyte life cycle

The life cycle of a typical moss is shown in Figure 1.8. The gametophyte produces gametes in multicellular gametangia. The egg remains within the female gametangium (**archegonium**); biflagellate sperm are released from the male gametangium (**antheridium**) and swim through a film of water to the egg. **Fertilization** takes place in the archegonium.

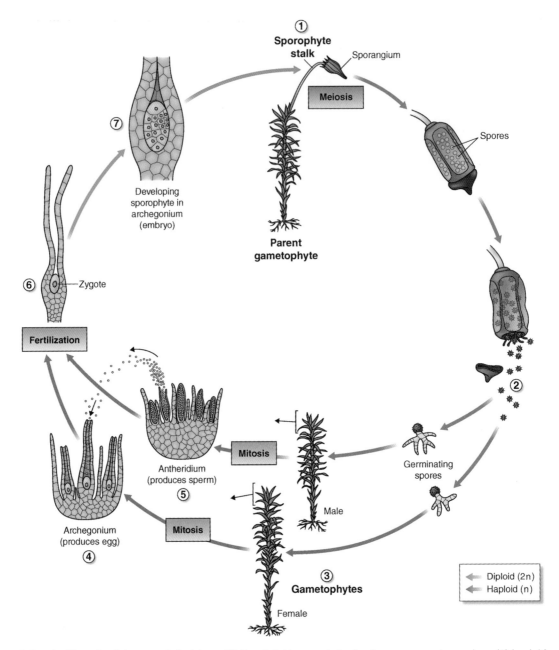

Figure 1.8 The life cycle of the moss *Polytrichum*. (1) The diploid sporophyte develops a sporangium, where (2) haploid spores are produced by meiosis. (3) These germinate to give female and male gametophyte plants. (4) Mitosis produces a haploid egg in the archegonium of the female plant and (5) haploid motile sperm in the antheridium of the male plant. (6) Motile sperm swim down the neck of the archegonium, fertilizing the egg, and producing (7) a diploid zygote that develops into the diploid sporophyte.

The resulting diploid **zygote** divides in situ, to produce a diploid **embryo**. The embryo develops into the **sporophyte** that remains attached to and dependent on the gametophyte.

A mature sporophyte has a stalk (**seta**) and **sporangium** (**capsule**). In the stalk of the sporophyte, a central cylinder of differentiated cells may be the functional progenitor of the vascular system of vascular plants; **stomata** may also be present in the sporophyte. Within the sporangium, diploid cells divide by **meiosis** to produce **haploid spores**. Bryophyte spores are enclosed in a thick cell wall that contains **sporopollenin**, a polymer known to reduce water loss and resist degradation. Spores are dispersed by wind. When spores germinate, they develop into a new gametophyte generation.

Key points Bryophytes are the most primitive land plants. Reproduction in bryophytes requires liquid water for sperm to reach the egg. The resulting zygote develops into an embryo surrounded by maternal tissue. The sporophyte remains attached to and is nourished by the female gametophyte. Haploid spores, produced by meiosis, are wind dispersed. The free-living gametophytes of bryophytes lack many adaptations for living on dry land, including a waterproof cuticle and a well-developed vascular system, but are thought to be the progenitors of vascular plants.

1.4.3 Many features of bryophytes suggest a link to the vascular plants

Sporophytes of bryophytes show several adaptations to life on dry land, including the presence of **cuticle, stomates** and **transport cells** and the production of **desiccation-resistant**, wind-dispersed **spores**. However, because they are dependent on gametophytes for nutrition, sporophytes are limited to environments that are at least seasonally wet, where gametophytes can thrive. While gametophytes have some cuticle, their tissues are subject to desiccation; absorption of water and minerals takes place over the surface of the whole plant. They lack stiffening tissue and so remain relatively small. **Sexual reproduction** requires a **film of water** through which **motile sperm** can **swim** to **non-motile eggs**. However, the presence of multicellular sex organs and the fact that the diploid embryo is protected and nourished by the gametophyte represent adaptations to life on land.

1.5 Vascular plants

Fossil evidence indicates that the first vascular plants (**tracheophytes**) were simple, **dichotomously branched** organisms about 5–10 cm in height. In the earliest vascular plants, the gametophyte and sporophyte are believed to have been free-living and approximately equal in size. These early land plants had specialized **vascular tissue** for conducting water, sugar and minerals (see Sections 1.8.2–1.8.4). Morphological and molecular evidence supports the division of living vascular plants

into three clades: **lycophytes, ferns and allied taxa** (**monilophytes**) and **seed plants** (**lignophytes**) (see Figure 1.3). Within the vascular plant clade, we shall see a progressive reduction in the size and life span of the gametophyte generation and an increase in the size and importance of the sporophyte generation. Plants that are recognized as ferns, pine trees and flowering plants are diploid sporophytes. We will first examine the **vascular spore plants**, the lycophytes and the monilophytes. In these two groups, although sporophytes are more prominent, both sporophytes and gametophytes are free-living, independent plants. In both groups, sporophytes produce spores that are wind dispersed.

Key points Vascular plants are characterized by the presence of a well-developed transport system, epidermis with a cuticle and stomates on leaves and stems, and their sporophytes are well adapted to life on land. Three clades of vascular plants are extant, the lycophytes, ferns and allied taxa, and seed plants. Lycophytes and ferns and their allies all have alternation of free-living sporophyte and gametophyte generations, with sporophytes being the dominant generation and best adapted to land. Their gametophytes are much reduced in size and the motile sperm they produce require water for fertilization. Extant lycophytes such as the quillworts and *Selaginella* are relatively small and inconspicuous and live in well-watered environments. In the Carboniferous period, members of this clade formed forests with trees in excess of 30 m in height that gave rise to today's coal deposits. Ferns were also abundant in the Carboniferous; these plants were larger than extant ferns.

1.5.1 Lycophytes were among the first tracheophytes to evolve

Extant **lycophytes** form a distinct clade that split from other tracheophytes about 400 mybp. Today they are few in number, about 1200 species, but during the **Carboniferous** period, diverse lycophytes dominated the landscape and gave rise to **coal** deposits. Extinct members of this clade include large trees that produced woody trunks by processes similar to those found in woody seed plants. Living lycophytes are represented by the **club mosses** (e.g. *Lycopodium*), the **spike mosses**,

Figure 1.9   Representative members of the Lycophyta: (A) *Lycopodium* (club moss), (B) *Selaginella* (spike moss) and (C) *Isoetes* (quillwort).

commonly found in the tropics (e.g. *Selaginella*), and the **quillworts** (e.g. *Isoetes*) (Figure 1.9). Unlike bryophytes, lycophyte sporophytes have true roots, stems and leaves. Their leaves are small and have a single vein. Sporangia, borne on specialized leaves, produce wind-dispersed spores. Gametophytes of lycophytes are independent plants that have a simple morphology and lack vascular tissue. They produce flagellate sperm that swim through a film of water to reach eggs.

1.5.2 Ferns, horsetails and whisk ferns constitute a single monophyletic clade, the monilophytes

The **monilophytes**, which arose more than 360 mybp, include the **ferns** (Ophioglossaceae, Marattiales and Polypodiales), **horsetails** (Equisetales) and **whisk ferns** (Psilotaceae) (Figure 1.10). Along with the lycophytes, the ferns and allied groups contributed to the formation of coal during the Carboniferous period. Today, although they have limited economic value, they play important ecological roles. Ferns are numerous (exceeding 11 000 species) and abundant, especially in the tropics.

The phylogenetic relationships among monilophytes have been inferred from gene sequences. When only morphological characters were available for comparison, systematists did not consider these plants monophyletic. For example, the leafless, rootless whisk ferns (e.g. ***Psilotum***) were thought to be more primitive than the lycophytes, but molecular phylogeny places them firmly in the monilophyte clade.

1.5.3 Although adapted to land, ferns require water for reproduction

Although some fern orders (e.g. Salviniales) are aquatic, most ferns are terrestrial. As in lycophytes, the fern life cycle includes a large sporophyte generation and an inconspicuous, though independent, gametophyte (Figure 1.11). Fern sporophytes are well adapted to land; their above-ground parts are covered by a cuticle and an epidermis in which stomata are embedded. Leaves are generally large, multiveined and may be highly divided. They are often the only above-ground part of the fern plant. Most ferns have a horizontal underground stem (a **rhizome**) and a complex root system. Fern sporophytes produce wind-dispersed, haploid spores in sporangia borne on the undersides of leaves. Spores germinate to form haploid gametophytes that may have male and female gametangia on the same plant or on separate gametophytes. In a few species, sporophytes produce two types of sporangia: **megasporangia** that make large spores (**megaspores**) and **microsporangia** that make small spores (**microspores**). Megaspores develop into female gametophytes and microspores develop into male gametophytes.

Fern gametophytes are photosynthetic. They are small, rarely more than 1 cm in diameter, and usually only one to two cells thick. Gametophytes lack cuticle, vascular tissue and true organs; simple unicellular rhizoids anchor them to the substrate. They can only survive in damp areas. Gametophytes produce eggs in archegonia and flagellated sperm in antheridia. The sperm swim through a layer of water to reach the egg.

A   B

C   D

Figure 1.10 Diverse morphologies of some members of the fern clade (monilophytes), which do not demonstrate their membership in a monophyletic clade. These relationships were established using DNA sequence data. (A) *Equisetum* (horsetails). (B) *Psilotum* (whisk fern). (C, D) The true ferns: (C) *Polystichum* and (D) *Cyathea*, a tree fern.

After fertilization, the diploid zygote develops into an embryonic sporophyte within an archegonium of the gametophyte. As the young sporophyte produces leaves, stems and roots, it overgrows the parent gametophyte.

Sporophytes of vascular spore-producing plants (lycophytes, ferns and allied groups) are well adapted to life on dry land. However, their gametophytes remain tied to wet areas by their anatomy and the fact that fertilization requires water. Because sporophytes begin life attached to gametophytes, they must start life in moist areas. The problem of a vulnerable gametophyte was overcome when the seed plants evolved.

1.5.4 Seed plants are successful conquerors of land

Living seed plants are divided two monophyletic clades, the **gymnosperms**, including five major lineages—the **cycads**, the **pine family, other conifers, gnetophytes** and *Ginkgo*—and the **angiosperms** (the flowering plants) (see Figure 1.3). The gymnosperm lineages contain more than 800 extant species. Flower-bearing angiosperms are by far the largest seed plant lineage, including more than 254 000 species. This number is probably an under-

estimate because it is likely that many members of this clade await discovery.

Seed plants provide many key resources, including food, lumber, fiber and fuel. Given their significance in our lives and their centrality to this book, we will describe in detail their reproduction, structure and development. First, we will examine the phylogeny and reproductive biology of gymnosperms and angiosperms separately. Later, in Sections 1.7–1.10, we discuss the anatomy and development of angiosperms noting similarities and differences with gymnosperms.

1.5.5 Seeds encase the embryo and its food, facilitating dispersal of the new sporophyte generation

Seed plant sporophytes produce two kinds of sporangia: **ovules** (**megasporangia**) and **pollen sacs** (**microsporangia**). Spores produced in these sporangia are not released, but divide in situ to produce gametophytes. Ovules enclose female gametophytes; pollen sacs contain male gametophytes, called **pollen**.

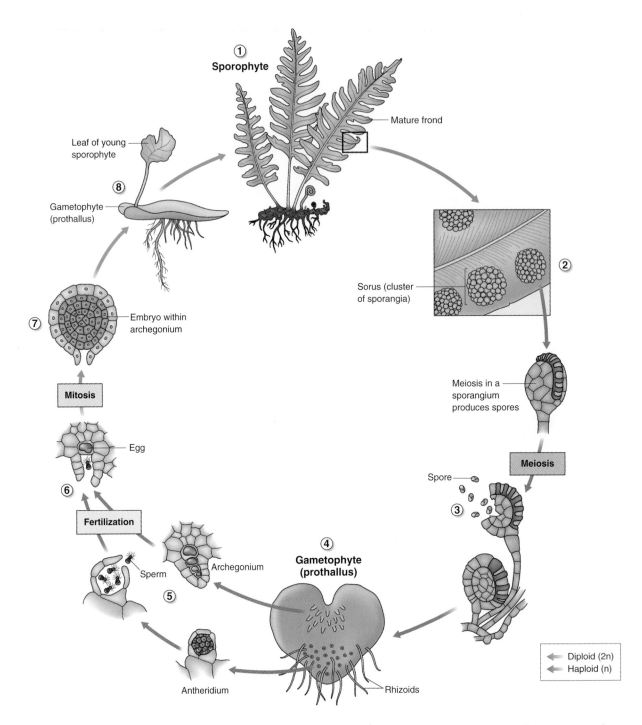

Figure 1.11   The life cycle of a fern. (1) The large sporophyte bears (2) sporangia in groups, each group called a sorus, on the underside of fronds where (3) meiosis occurs, giving rise to haploid spores. (4) These spores germinate to produce a small haploid gametophyte, rarely more than a few millimeters in size. The gametophyte produces (5) motile sperm in the antheridia and eggs in the archegonia and (6) the motile sperm swim in a layer of water down the neck of the archegonium to fertilize the egg, producing (7, 8) a diploid zygote that develops in situ into a diploid sporophyte.

Three major reproductive advances in adapting to life on land are found in seed plants. First, haploid gametophytes are reduced in size and are protected within sporangia on the parent sporophyte. Second, seed plants no longer require a film of water for fertilization. The male gametophyte, a pollen grain, develops a protective coat and is delivered to the vicinity of the female gametophyte by wind or by animal pollinators.

Finally, a new dispersal stage, the **seed** has evolved. A seed, with its protective seed coat, contains a new sporophyte with a source of food. We shall see below how each of these features is incorporated into the life cycles of first a gymnosperm and then an angiosperm.

> **Key points** Seed plants contain two distinct clades, gymnosperms and angiosperms. Seed plants successfully colonized land because a mode of reproduction evolved that made sexual reproduction independent of liquid water. They were also successful because the seed was relatively long-lived and provided seedlings with stored food for survival after germination. The sporophyte is the dominant part of the seed plant life cycle. The gametophytes are retained, protected and nourished by the sporophyte. Pollen evolved as the vehicle to deliver sperm to the egg without a film of water.

1.6 Gymnosperm phylogeny and reproduction

The term gymnosperm is derived from Greek: *gymnos*, bare or naked, and *sperm*, seed. The name refers to the gymnosperm seed, which is not enclosed in a protective structure as it is in the angiosperm ovary (*ang(os)* = vessel (Greek)). Gymnosperms were the dominant land plants in the Cretaceous and Jurassic periods, the age of dinosaurs. There are about 800 species of gymnosperms in the present-day flora.

1.6.1 Gymnosperm phylogeny reveals five lineages

Gymnosperms arose more than 320 mybp. Previously, Gymnosperms had been classified into four groups: the **cycads**; the **ginkgophytes**, containing one extant species *Ginkgo biloba*; the **gnetophytes**, including *Ephedra, Gnetum* and *Welwitschia*; and the **conifers** (Figure 1.12A–D). Analyses of DNA sequences of chloroplast, mitochondrial and nuclear genes, however, have shown that the conifers are divided into two lineages, the Pinaceae (pine family) and the rest of the

conifers (see Figure 1.3). Pinaceae consists of nine genera including *Pinus* (pine), *Abies* (fir), *Picea* (spruce), *Cedrus* (cedar), *Tsuga* (hemlock) and *Pseudotsuga* (Douglas fir). The second conifer lineage includes five families: Araucariaceae (*Araucaria* family), Cephalotaxaceae (plum yew family), Podocarpaceae (*Podocarpus* family), Taxaceae (yew family) and Cupressaceae (Cypress family, including *Sequoia, Sequoiadendron* (Figure 1.12E), *Chamaecyparis, Thuja* and *Juniperus*.

1.6.2 Conifers constitute an important natural resource

The best-known and most diverse gymnosperms are found in the pine and other conifer lineages. Sporophytes in these lineages are among the largest and oldest living organisms in the biosphere. They are well adapted for life on dry land. Many have needle-like leaves that resist water loss. The needles are circular in cross section giving them a low surface to volume ratio; the epidermis is covered by a thick **cuticle**. Pines and other conifers produce large woody stems (see Section 1.10). The coastal redwood (*Sequoia sempervirens*) can exceed 100 m in height and the giant sequoia (*Sequoiadendron giganteum*) 8 m in diameter. Some bristlecone pine trees (*Pinus longaeva*) have lived to an age of more than 4900 years.

In contrast to the other gymnosperm lineages, the pine family and other conifers have great economic value and broad ecological significance. They provide lumber for building materials and pulp for paper. The northern coniferous forest (taiga) is among Earth's largest terrestrial biomes, and pines and other conifer species are also common in many temperate forests at lower latitudes. Given their significance as a commodity and as habitat, we focus our discussion of gymnosperm reproduction in the genus *Pinus*.

1.6.3 Sporangia and gametophytes of pines and other conifers are produced in cones

The life cycle of *Pinus* is illustrated in Figure 1.13. As is the case in most gymnosperms, **ovules** are produced in **seed cones** (female) and **pollen sacs** are produced in **pollen cones** (male) (Figure 1.14). The pollen cone consists of cone scales, each of which bears two pollen sacs. Inside a pollen sac are many diploid **microspore mother cells** that undergo meiosis, forming haploid

Figure 1.12   Representatives of the five gymnosperm lineages: (A) a cycad, (B) *Ephedra* (a gnetophyte), (C) *Ginkgo biloba*, (D) *Pinus* and (E) *Sequoiadendron giganteum* (giant sequoia).

microspores. Each microspore divides by mitosis to produce a **male gametophyte (pollen grain)**. The mature pollen grain is enclosed in a thick cell wall; it has a **generative cell** that will give rise to the sperm, a **tube cell** that functions in sperm delivery, and two **prothallial** cells that typically degenerate. When mature, pollen grains are released from the male cone and are dispersed by the wind to female cones.

In the female cone, ovules are produced on ovuliferous scales; each scale bears two ovules. An ovule is enclosed in a protective integument that has an opening at one end that allows the pollen tube to enter. A cell at the center of the ovule, a **megaspore mother cell**, divides by meiosis to produce four haploid **megaspores**, three of which degenerate. The remaining megaspore divides by mitosis to produce a **female gametophyte**. The female gametophyte is retained within and nourished by the ovule.

1.6.4 Pine reproduction is characterized by a long delay between pollination and fertilization

In pine, **pollination**—the transfer of pollen from a male cone to an ovule in a female cone—occurs a few weeks before megaspores are formed within the ovule. The outer surface of the female cone exudes sticky secretions between the cone scales called **pollination droplets**. Wind-borne pollen grains stick to the droplets; as these dry, they draw the pollen grains toward the ovules. About 3 months after pollination, the pollen grains germinate and produce pollen tubes that will carry sperm to the female gametophyte within the ovule. It can take up to a year for the pollen tube to reach the female gametophyte.

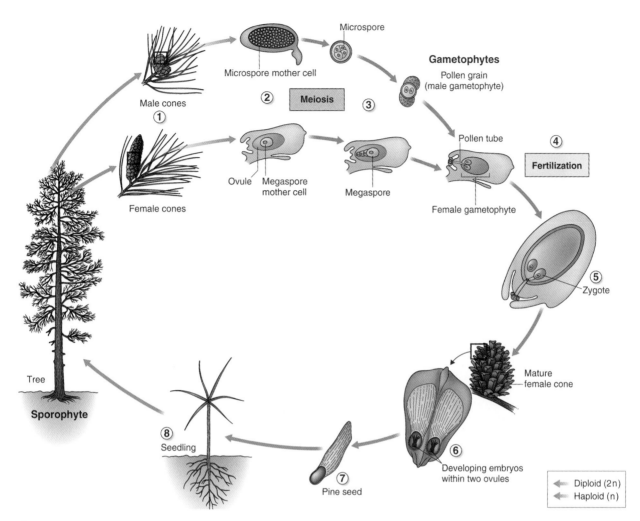

Figure 1.13 Life cycle of a pine (*Pinus* sp.). (1) Male and female cones have scales bearing two sporangia, (2) pollen sacs or microsporangia in the male and ovules or megasporangia in the female. (3) In each pollen sac many microspore mother cells divide by meiosis to produce haploid microspores that develop into pollen grains (male gametophytes). In the female cone, in each ovule one megaspore mother cell divides by meiosis to produce a single megaspore which develops into a female gametophyte that bears archegonia with eggs. (4) Pollen is dispersed by wind and lands on an ovule within a female cone and germinates to produce a pollen tube containing two sperm, one of which will fertilize the egg. (5) The resulting diploid zygote develops into an embryo. (6) A single female cone scale bears two ovules, each of which contains an embryo sporophyte and develops into (7) a winged seed. (8) The seeds germinate to produce a new seedling sporophyte. Reproduction in *Pinus* from the initiation of cone development to the release of seeds takes about 3 years.

Meanwhile, shortly after pollination, female gametophyte development begins. After 6 months to 1 year, a mature female gametophyte produces several archegonia, each containing one egg. At about this time the pollen tube reaches the female gametophyte. Its generative cell divides once to produce a sterile cell and a **spermatogenous cell**. As the pollen tube approaches an archegonium, the spermatogenous cell divides to produce two sperm that lack flagella. The pollen tube fuses with the plasma membrane of the egg cell and discharges both sperm into the egg cytoplasm. One

sperm fertilizes the egg, producing a diploid zygote, and the other degenerates. Many pollen tubes may grow into a single ovule, and several eggs can be fertilized. Typically, only one zygote develops into an embryo.

1.6.5 Pine seeds contain both diploid and haploid tissues

After fertilization, as embryo development proceeds, the inner tissue of the ovule may increase in size. The

A   B

C   D

Figure 1.14 Male and female cones in *Pinus*. (A) Young male cones releasing pollen. (B–D) Female cones: (B) spring of year one when pollination occurs; (C) spring of year two when fertilization occurs; and (D) fall of year two when seeds are released.

enveloping integument becomes the seed coat. The mature pine seed (Figure 1.15) contains tissues from three genetically distinct generations: (1) the seed coat, from the parent sporophyte, that forms a hard protective layer around the seed; (2) the haploid female gametophyte that contains stored food that will be used to nourish the embryo; and (3) the diploid embryo. Edible pine nuts found in grocery stores are pine female gametophytes plus embedded embryos. Seeds of pines and other conifers are often winged to aid in their dispersal.

1.7 Angiosperm phylogeny and reproduction

An extensive fossil record indicates that flowering plants arose more than 140 mybp. Traditionally, the angiosperms were divided into two clearly recognizable groups, the **eudicots** and **monocots**. Most of the eudicots have **two cotyledons** (**seed leaves**), **broad leaves** with **branched venation**, and a **taproot**; many eudicots can form **wood**. The monocots have **one**

Female
gametophyte   Embryo

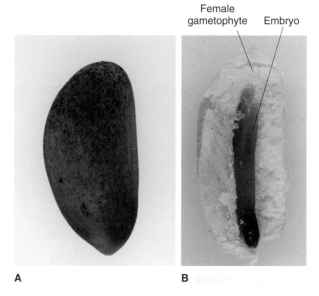

A   B

Figure 1.15 Seed of *Pinus edulis*. (A) Intact seed; the seed coat develops from ovule tissue (2n) of the parent sporophyte generation. (B) Bisected seed; the embryo (2n) (new sporophyte generation) is embedded in female gametophyte tissue (1n) which will serve as a food source for the embryo during seed germination.

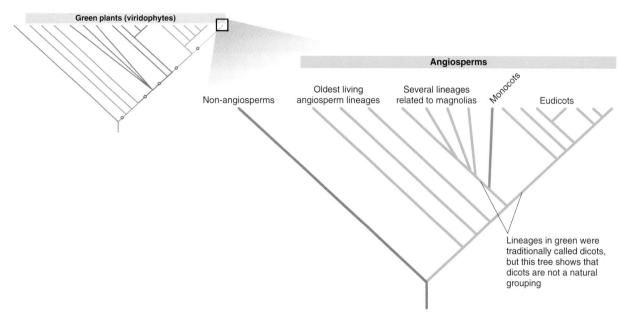

Figure 1.16 Cladogram illustrating the evolution of the angiosperms.

Key points The gymnosperms are defined by the presence of an ovule that is not enclosed within an ovary as is the case in angiosperms. The female gametophyte of *Pinus* is located within the ovule located on the surface of a female cone scale. Meiosis in the ovule gives rise to a megaspore that divides by mitosis to produce a haploid female gametophyte. Pollen develops from microspores that are the product of meiosis in male cones. In *Pinus* each pollen grain is a male gametophyte and has four cells. Pollen is wind dispersed in gymnosperms and is produced in vast amounts. Pollen grains land on the surface of mature female cones and are drawn into the cone by a drying pollination droplet. The pollen grain produces a pollen tube which grows through the ovule to deliver non-motile sperm to the female gametophyte. The pollen tube fuses with the egg cell membrane, discharging the sperm, and fertilization ensues. The zygote develops into an embryo in situ. Outer layers of the ovule form the seed coat enclosing the embryo and female gametophyte tissue which functions as a food reserve for the seedling.

cotyledon and **narrow leaves** with **parallel venation**; they generally **lack wood**.

More recent phylogenetic analyses, especially those incorporating data from morphology and gene sequences, provide a more complex view of angiosperm phylogeny. Figure 1.16 shows how angiosperms can be divided into a number of major groups based on gene sequences and the number of pores in the pollen cell wall.

1.7.1 The flower is the defining feature of angiosperms

Flowers are found only in angiosperms. The flower contains the reproductive organs of the angiosperm sporophyte: ovules and pollen sacs and associated sterile organs. A **complete flower** consists of **four whorls** (concentric rings) of organs (Figure 1.17). The **sepals** make up the outermost whorl, known also as the **calyx**; sepals are usually green and protect the internal floral organs in a flower bud. Moving inward, the next whorl, the **corolla**, is made up of **petals** which are often brightly colored and may serve to attract animal pollinators. In contrast to wind pollination, the use of animal pollinators allows more precise targeting of pollen delivery and allows pollen to be transported over longer distances. **Stamens** form the next whorl, the **androecium**. Each stamen consists of a stalk-like **filament** and an anther made up of **pollen sacs**. The innermost whorl, the **gynoecium** consists of one to many **carpels**. Each vase-shaped carpel has a **stigma** upon which pollen lands, a neck-like **style** through which pollen tubes grow, and an **ovary** that contains one to many **ovules**. The carpel is unique to angiosperms; it protects the ovules and, as we shall see below, develops into a fruit.

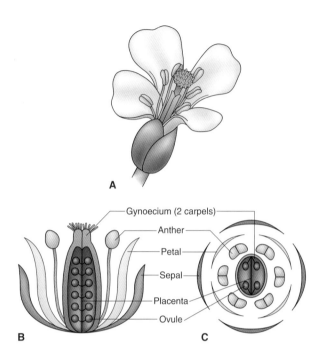

A

Gynoecium (2 carpels)

Anther

Petal

Sepal

Placenta

Ovule

B   **C**

Figure 1.17   (A) Sketch of an *Arabidopsis* flower with (B) longitudinal and (C) transverse sections shown. The outermost whorl contains four sepals, the second four petals. The third whorl contains six stamens, each with a filament and a pollen-bearing anther. The fourth (innermost) whorl contains the central gynoecium with two carpels which house the ovules.

1.7.2 Gametophytes of angiosperms are much smaller than those of gymnosperms

The flower is the site of reproduction in angiosperms; all the events from meiosis, to development of gametophytes by mitosis, to fertilization and production of seeds take place here (Figure 1.18). The production of male gametophytes takes place within the pollen sacs of the anthers. Here diploid cells undergo meiosis to form haploid microspores. Each microspore divides by mitosis to produce a male gametophyte, a pollen grain that has two cells: a **tube cell** and a **generative cell**. In some angiosperms the generative cell divides immediately to form two **non-motile sperm**; in the remaining genera, division of the generative cell takes place after pollination. In either case, the pollen grain develops a thick, waterproof cell wall containing sporopollenin. The wall of pollen grains is often highly sculptured and decay-resistant, a feature that can aid in identifying the species of plant that produced it.

As in gymnosperms, the production of female gametophytes in angiosperms takes place within the ovules. In this case, however, the ovules are enclosed in the ovary of a carpel. Each ovule has protective

integuments with a small opening, the **micropyle**, at one end. Within the ovule, one cell divides by meiosis to produce four haploid megaspores of which three degenerate. The surviving megaspore divides by mitosis to produce a female gametophyte, the **embryo sac**. Three rounds of mitosis produce seven cells: three uninucleate cells at each end of the gametophyte and one large **central cell** with two polar nuclei. Of the trio of cells at the micropylar end of the embryo sac, the middle cell is the egg; the two adjacent cells are the **synergids**, one of which determines the point of entry of the pollen tube into the embryo sac. The three cells at the opposite end of the gametophyte are called the **antipodals**. The mature angiosperm female gametophyte has only seven cells in contrast to the female gametophyte of gymnosperms that typically has thousands.

1.7.3 Double fertilization in angiosperms leads to the formation of a diploid embryo and polyploid endosperm

Pollination occurs when pollen is transferred from a pollen sac to the stigma of a carpel. Mechanisms exist that prevent pollen germination or growth unless the pollen and stigma/style are compatible (see Chapter 16). When pollen germinates, it produces a pollen tube containing two sperm. The pollen tube grows down the style of the carpel carrying the two sperm near its tip. It enters an ovule through the micropyle. When it reaches the embryo sac, the pollen tube fuses with one of the synergid cells and releases the sperm. **One sperm** fuses with the egg to produce a diploid **zygote**. The **second sperm** fuses with the two polar nuclei and forms a triploid, **primary endosperm cell**. These events are called **double fertilization**; the second fertilization event is unique to angiosperms.

After double fertilization, the zygote divides to produce the new embryo and the primary endosperm cell divides to form a unique nutritive tissue, the **endosperm**. The endosperm of angiosperm seeds takes the place of the large, multicellular female gametophyte that nourishes the developing embryo in all other embryophytes. Development of the embryo and endosperm starts almost immediately after fertilization. The patterns of early embryonic development, **embryogeny**, are strikingly similar in all angiosperms (Figure 1.19). The first division of the zygote is transverse and produces two cells, an apical cell that gives rise to most of the embryo and a basal cell that gives rise to part of the embryonic root apex and to the **suspensor**. Continued cell division in the embryo gives rise to a ball of cells referred to as the **globular stage**. During this

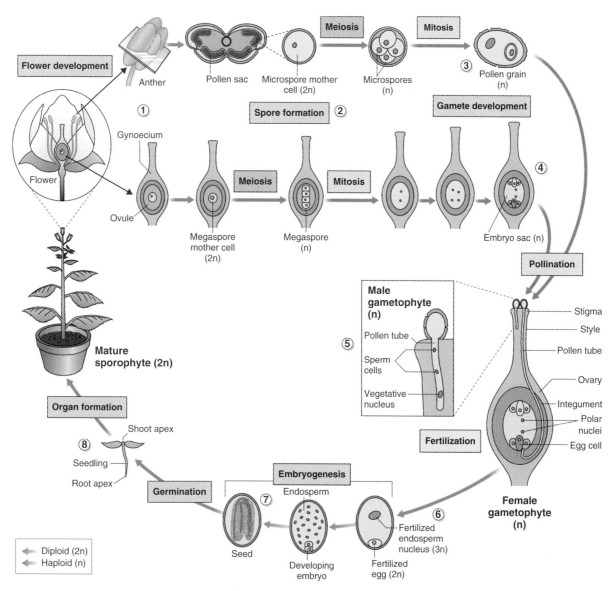

Figure 1.18 The life cycle of a flowering plant. Note the alternation of multicellular haploid (gametophytic) and diploid (sporophytic) generations. (1) Flowers have anthers and a gynoecium that (2) produce spores by meiosis. (3) Microspores divide by mitosis to produce a male gametophyte, a pollen grain that has two cells: a tube cell and a generative cell. (4) In the ovule one megaspore divides by mitosis to give an embryo sac containing an egg cell and two polar nuclei. (5)The germinating pollen grain produces a pollen tube that grows down the style towards the embryo sac and the generative cell divides by mitosis to produce two non-motile sperm. When the pollen tube reaches the embryo sac, its tip bursts. (6) One sperm fertilizes the egg, giving rise to the diploid embryo and the other fuses with two polar nuclei giving rise to the triploid endosperm nucleus. (7) The embryo and endosperm grow by mitotic division and in the example shown in this figure the endosperm is absorbed by the developing embryo and the mature seed consists of an embryo axis and two cotyledons surrounded by seed coats. (8) After the seed germinates, the embryo develops into a seedling and ultimately the mature sporophyte plant.

stage, tissue precursors begin to form. In primitive dicots and eudicots, cell division coupled with differing rates of cell expansion result in the formation of two **cotyledons** (**seed leaves**) that become evident at the **heart stage** of embryogeny. The shoot apical meristem arises in the notch between the two developing cotyledons. In monocots, a single cotyledon is formed. In this case, the shoot apical meristem forms on one side of the embryo,

near the base of the cotyledon (Figure 1.20). A mature embryo usually consists of an embryonic root, the **radicle**; one or two **cotyledons**; and an embryonic shoot axis, including a **hypocotyl** (*hypo* = below, *cotyl* = cotyledons), an **epicotyl** (*epi* = above) and a **shoot apical bud**. The embryos of cereals and grasses are more complex than those of most monocots and eudicots (see Chapter 6).

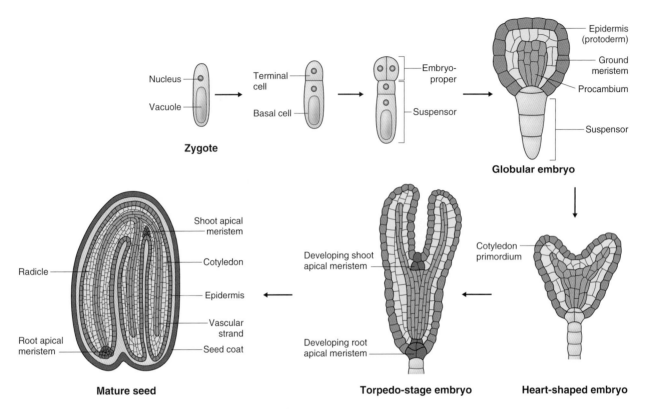

Figure 1.19   Embryo development in *Arabidopsis*, a eudicot.

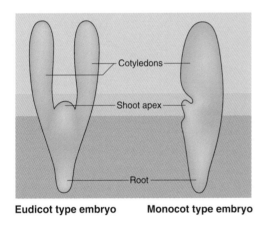

Figure 1.20   Comparison of basic monocot and eudicot embryos.

Endosperm development typically begins before embryogenesis starts. The primary endosperm cell may undergo many rounds of free-nuclear division in which cell walls are not laid down. Walls are eventually laid down between nuclei to form a fully cellularized tissue. In the coconut palm, cellularization is not completed until the time of seed germination. As a result, coconuts contain both cellularized, solid endosperm, known as copra, and the free-nuclear liquid endosperm familiar as coconut water (Figure 1.21). The mature angiosperm seed consists of an **embryo** enclosed by a protective coat,

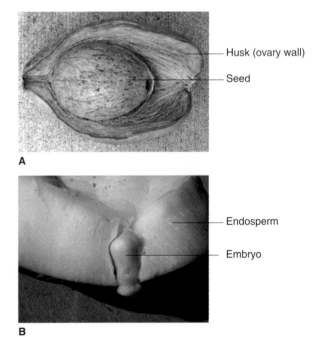

Figure 1.21   Structure of coconut fruit. (A) Fruit halved to show seed. (B) Portion of coconut seed showing endosperm and embryo.

the **testa**, derived from the outer tissues of the **ovule** (see Chapter 6). In some cases the mature seed contains endosperm while in other seeds the endosperm is consumed during the later stages of embryo development. In Chapter 6 we describe the structure and function of seed tissues in more detail and examine their roles in dormancy and germination.

1.7.4 In angiosperms, fruits promote seed dispersal

While seeds are maturing within the ovary, the ovary wall enlarges and develops into a **fruit**. Fruits are ripened carpels containing seeds and are unique to angiosperms. In some species with single-seeded fruits, e.g. cereal grains, the wall of the ovary fuses with the seed coat; therefore the grain is technically a fruit not a seed.

Fruits are usually modified to enhance seed dispersal (Figure 1.22). In fleshy fruits, such as tomatoes and cherries, the ovary develops into tasty flesh. These fruits attract animals that eat them and deposit the indigestible seeds at some distance from the parent plant. In other plants, such as maple and elm, the ovary wall develops into a wing that aids in wind dispersal of seeds. Some fruits have fluffy parachutes that float seeds in the wind. In cocklebur, the ovary wall produces hooks that snag onto fur and feathers allowing the fruit and its enclosed seeds to hitchhike on animals. In coconut, the ovary wall becomes buoyant allowing the fruit to float.

When environmental conditions are favorable, seeds germinate and a new sporophyte seedling begins to grow. The process of germination is discussed in detail in Chapter 6. Below we will examine the organization and structure of a seed plant sporophyte using angiosperms as an example. Then we will discuss the basic principles of how this plant increases in size and produces new organs throughout its life.

1.8 The seed plant body plan I. Epidermis, ground tissue and vascular system

The body plan in angiosperms and gymnosperms is based on three types of vegetative (non-reproductive)

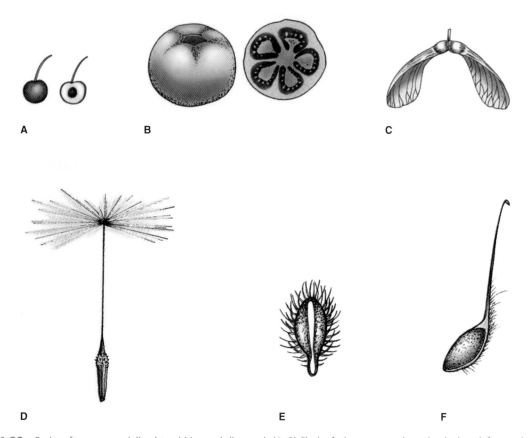

Figure 1.22 Fruits often are specialized to aid in seed dispersal. (A, B) Fleshy fruits are eaten by animals that defecate the seeds at some distance from the parent plant. (C, D) Fruits with parachutes or wings are dispersed by wind. (E, F) Fruits with spines and barbs attach to fur or feathers or clothing and hitchhike to new locales.

Key points Angiosperms were originally thought to consist of two major groups, dicots and monocots, but more recent analysis shows that angiosperms are made up of at least 15 different lineages. The evolution of the flower with a carpel enclosing ovules was a major step in the evolution of plants. In addition to protecting the ovules, the carpel develops into a fruit that enhances seed dispersal. Another major evolutionary advance in angiosperms was the formation of endosperm as a result of double fertilization. The gametophyte phase of the angiosperm life cycle is even more reduced than that found in gymnosperms. The female gametophyte consists of an embryo sac containing seven cells, one of which is an egg with two polar nuclei. Pollen is delivered to the stigma of the carpel; pollen germination and pollen tube growth down the style to the ovule begins immediately. The pollen tube carries two non-motile sperm and delivers them to the embryo sac. One sperm fuses with the embryo to produce a zygote and the other sperm fuses with two polar nuclei to give a triploid endosperm. Embryo and endosperm grow by mitotic divisions to produce the seed; the ovule forms a protective seed coat.

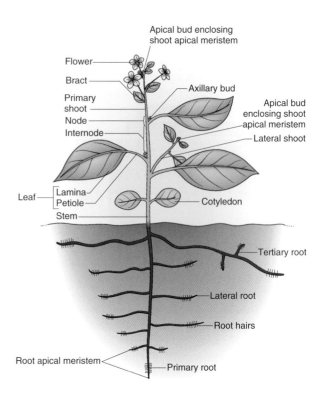

Figure 1.23   The organization of the shoot and root systems of a representative angiosperm. The shoot system includes two non-reproductive organs, leaves and stems, in addition to flowers; the roots of a plant make up the root system.

organs: **roots, stems** and **leaves**. All roots of a plant comprise its **root system**, while its stems and leaves make up its **shoot system** (Figure 1.23). Each organ is made up of three tissue systems: **dermal, ground** and **vascular tissue systems**. The arrangement of these tissue systems within a plant in shown in Figure 1.24. As we shall see below each tissue system has a different function and is made up of a number of different cell types (Table 1.2).

1.8.1 Epidermal tissue covers the outside of a plant while ground tissue makes up the bulk of a plant

The **dermal tissue system**, composed of **epidermis**, covers and protects the internal tissues of each organ. The epidermis of aerial organs secretes a waterproof layer called **cuticle** to slow water loss. In regions of the root that take up water and minerals, cuticle is absent, rather epidermal cells produce long tubular extensions called **root hairs** that increase the surface area available for uptake. The **epidermis** contains several specialized cell

types including **gland cells, hair cells** (also called **trichomes**) and **guard cells** that surround the **stomata**, pores in the epidermis that allow gas exchange (Figures 1.25 and 1.26). In contrast to other epidermal cells, stomatal guard cells contain chloroplasts; in addition they have specialized wall thickenings that allow them to regulate the size of the stomatal pore. Two types of guard cells are found in plants, the dumbbell-shaped guard cells characteristic of grasses and palms, and the sausage-shaped guard cells found in most eudicots (see Figure 14.29). Both types of guard cell have differentially thickened cell walls. The role of guard cells in opening and closing stomata is discussed in more detail in Chapter 14.

The **ground tissue system** makes up most of the volume of a plant. Most of the metabolic activities of the plant occur in this tissue. It is largely composed of **parenchyma cells** (Figure 1.27) with thin flexible primary cell walls (see Chapter 4). They are relatively unspecialized and are involved in the production and storage of carbohydrate and other organic molecules. **Collenchyma** and **sclerenchyma tissues** can also be found in the ground tissue system (Figure 1.27). **Collenchyma cells** have primary cell walls with uneven

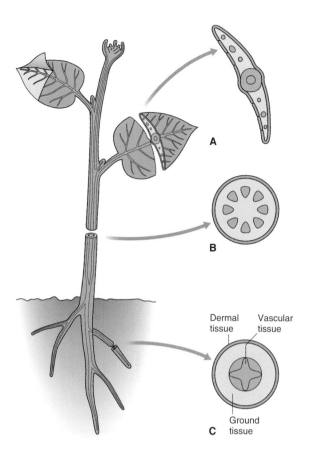

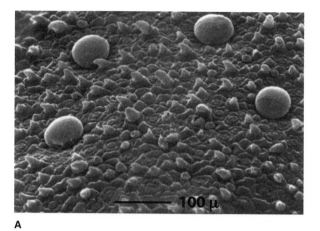

A

Figure 1.24 Arrangement of three tissue systems, vascular, ground and dermal, in a vascular plant in longitudinal and cross-section views. Shown are cross-sections of (A) a leaf, (B) a shoot and (C) a root.

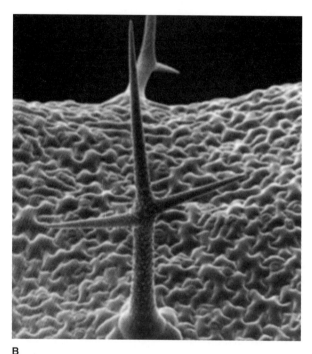

B

Figure 1.25 Scanning electron micrographs of some of the cell types found in the dermal tissue system. (A) Gland cells that secrete oils. (B) Trichomes that may discourage predators or reflect excess light.

Table 1.2 Tissue systems, tissues and cell types found in plant organs.

Tissue systems	Tissues	Types of cells
Dermal	Epidermis	Epidermal cells, guard cells, gland cells, hair cells
Ground	Ground tissue	Parenchyma cells
	Sclerenchyma	Fibers, sclerids
	Collenchyma	Collenchyma cells
Vascular	Xylem tissue	Tracheary elements (vessel elements and tracheids), fiber cells, parenchyma cells
	Phloem tissue	Sieve tube elements (members), companion cells, fiber cells, parenchyma cells

thickenings, usually at the corners of the cells. These cells usually occur in bundles, e.g. in the ribs of celery stalks, and provide flexible support to petioles and the main veins of leaves. **Sclerenchyma cells**—fibers and sclereids—have thick, highly lignified, secondary cell walls and may be dead at maturity. **Sclereids**, often called stone cells, are cube-shaped and tend to occur in clusters. They are best known for the gritty texture they give to fruits such as pears. **Fibers** are elongate cells, found individually or in groups, that add strength to tissues, providing rigid support.

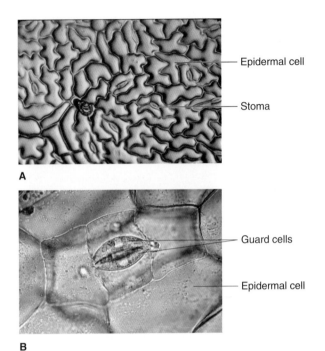

A

B

Figure 1.26   Leaf epidermis. (A) Replica of leaf epidermis from the cherry laurel (*Prunus laurocerasus*), showing stomata and irregularly shaped epidermal cells. (B) Epidermal peel from *Tradescantia* sp. Note the stomatal guard cells with chloroplasts.

Key points The body plan of seed plants is simple and composed of a shoot and root system. The shoot includes two organs, stems and leaves; roots are the organs of the root system. Each of these organs is made up of three primary tissue types, all derived from the root and shoot apical meristems. The apical meristem gives rise to dermal, ground and meristem tissue systems. Dermal tissue covering the entire plant is epidermis, a tissue specialized for protection and water conservation in stems and leaves and for water uptake in young roots. Epidermis encloses ground tissue within which vascular tissue is embedded. Ground tissue is comprised of primarily parenchyma that has different functions depending on its location in the plant. The vascular tissue system consists of two complex tissues, the phloem and the xylem.

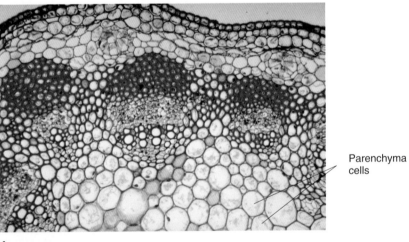

A

Parenchyma cells

Figure 1.27   Cell types found in the ground tissue system. (A) Parenchyma cells in the pith of a *Helianthus* stem. (B) Confocal image of collenchyma cells from a celery petiole (stalk); red asterisks indicate cell wall thickening at the corners of cells. (C) Sclereids (S) in *Trochodendron* leaf cells.

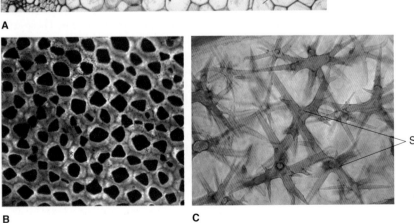

B

C

S

1.8.2 Vascular tissues are specialized for long-distance transport

The **vascular tissue system** is composed of two complex, long-distance transport tissues: **xylem** and **phloem** (see Table 1.2). Xylem tissue contains **tracheary elements** through which long-distance water transport occurs. It also contains xylem parenchyma cells and fibers. Phloem tissue contains sugar-conducting **sieve tubes** and their companion cells. Phloem parenchyma cells and fibers are also found in phloem.

1.8.3 Long-distance transport of water occurs in tracheary elements

Tracheary elements comprise two cell types, **tracheids** and **vessel elements** (Figure 1.28); they have thick, highly lignified, secondary cell walls and are dead and hollow at functional maturity. The pattern of secondary cell wall deposition in tracheary elements varies. Secondary cell walls of tracheary elements that differentiate in tissue that is still elongating are laid down in rings or in spiral bands. In regions of the plant where elongation has ceased, secondary walls of tracheary elements form a solid sheet with gaps called **pits**. Columns of vessel elements or tracheids stacked end to end form the pipes in which water and dissolved minerals are transported in **xylem**; these columns extend from the deepest roots to the tips of the tallest stems. Columns of vessel elements are called **vessels**.

  Tracheids are found in the xylem of all vascular plants, while vessels are found almost exclusively in angiosperms. Tracheids are long and narrow with tapered ends. They have pits in their end walls and side walls. Pits can be simple pores or more complex structures, e.g. the bordered pits found in tracheary elements of some gymnosperms (Figure 1.28). **Vessel elements** are shorter and wider than tracheids; they may be up to 700 μm in diameter. There are pits in the side walls of vessel elements; however their end walls have large perforations that allow water to flow unimpeded from cell to cell. In addition to water transport, tracheary elements provide structural support, allowing some vascular plants to reach heights of more than 100 m.

1.8.4 Long-distance transport of organic solutes occurs in sieve tubes

In all vascular plants except angiosperms, long-distance transport of sucrose and other organic molecules takes

Key points The evolution of vascular tissue that efficiently transport water and organic solutes was key to establishing plant life on land. Water is transported upward in the xylem. In angiosperms water-conducting cells in the xylem consist of tracheids and vessel elements, but in almost all gymnosperms xylem has only tracheids. Whereas tracheids are elongated cells connected to neighboring tracheids by pores called pits, vessels form a stacked series of dead cells that form unobstructed tubes for water movement through perforated end walls. Organic solutes are transported bidirectionally in the phloem. In angiosperms, phloem contains living sieve tube elements that lack a nucleus but have a nucleated companion cell associated with them. Sieve tube elements have highly perforated end walls called sieve plates and large cytoplasmic connections so that stacked elements form continuous tubes. Phloem transport, called translocation, requires living sieve cells.

place in tubes formed by stacks of elongate **sieve cells**. These cells, in contrast to tracheary elements, have thin primary cell walls and are alive at functional maturity. In angiosperms, organic solute transport occurs through stacked **sieve tube members** also known as **sieve elements**. These cells lack a nucleus, vacuole and several other organelles, but they have a functional plasma membrane and therefore are alive (Figure 1.29). Cytoplasmic strands connect adjacent sieve tube members in a stack. The perforated end walls of these cells are called **sieve plates**. Adjacent to each sieve tube member and connected to it by numerous plasmodesmata (see Chapters 5 and 14) is a **companion cell** containing a nucleus and the other organelles usually found in living plant cells. Companion cells play a key role in the metabolism of sieve tube members and participate in loading sucrose into sieve tube members for long-distance transport (see Chapter 14).

1.9 The seed plant body plan II. Form and function of organ systems

We have seen that the cells of a plant are organized in tissues and tissue systems, each of which serves a different function. Now we will examine how these cells and tissues are arranged in roots, stems and leaves and how they contribute to the functions of each organ.

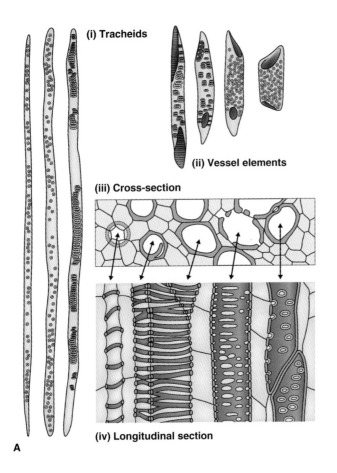

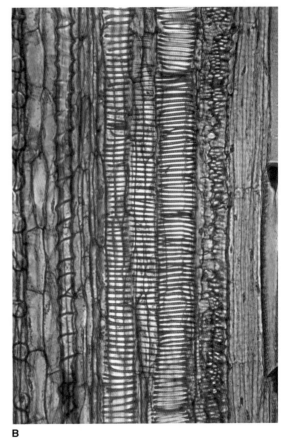

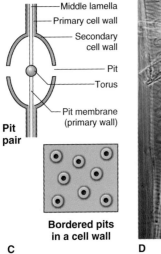

Figure 1.28 Tracheary elements in the xylem. (A) Drawing of (i) tracheids and (ii) vessel elements in a (iii) cross-section and (iv) longitudinal section of xylem. Note the thickened secondary walls and empty cell lumens. (B) Light micrograph of a longitudinal section of vessel elements and tracheids in a *Curcubita* stem. (C) Diagram of a bordered pit from a gymnosperm tracheid; if air enters the tracheid on one side of the pit, it pushes the torus to one side sealing off the air bubble. (D) Light micrograph of tracheids from *Pinus* wood showing bordered pits.

1.9.1 The root system acquires water and minerals

The root system is highly branched and generally has a larger surface area than the shoot system. The primary functions of roots are to absorb water and minerals from the soil, firmly anchor the plant and store carbohydrate. In most gymnosperms and eudicots, the radicle of the embryo becomes the primary root forming a **taproot**. The production of branched **lateral roots** further increases root surface area. Taproots can penetrate to

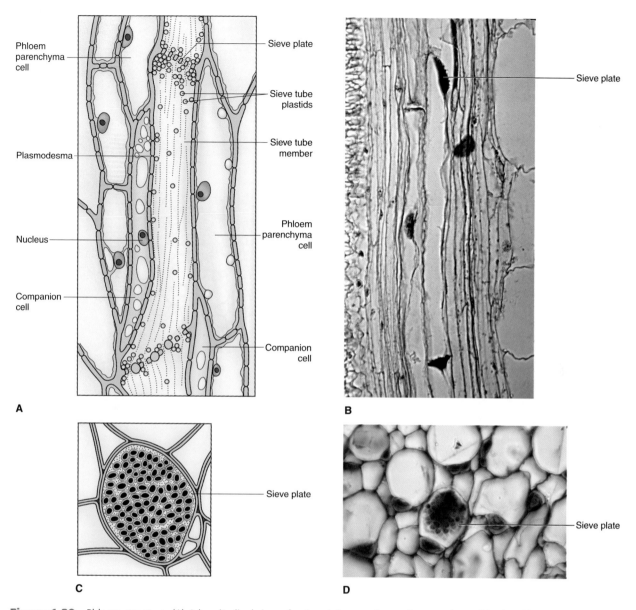

Figure 1.29 Phloem structure. (A) A longitudinal view of a sieve tube member and companion cell in the phloem. (B) Light micrograph of a longitudinal section of phloem from *Cucurbita pepo*. (C) A face view of a sieve plate; dark areas are cytoplasmic connections passing through holes in the sieve plate. (D) Light micrograph of a cross-section through phloem in *C. pepo* showing sieve tube elements, including a face view of a sieve plate, and companion cells.

great depths in the soil (Figure 1.30). Roots of mesquite, a desert shrub, can reach depths of 100 m. Taproots of carrots and beets are modified for increased food storage. In monocots, such as cereals and other grasses, the seedling primary root usually dies and is replaced by **adventitious roots**, produced from the base of the stem. These roots form a **fibrous root system** that spreads laterally in the soil and efficiently absorbs water and minerals from surface layers of soil. The ecological differences between these two root systems can be seen in an unwatered lawn during warm summer months. Grasses, with fibrous root systems, often turn brown while the leaves of eudicot weeds such as dandelion, that have deep taproots, remain green.

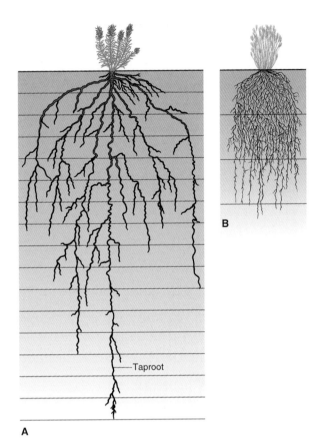

A

Figure 1.30 Two types of root systems. (A) A taproot system extends deeply into the soil. (B) A fibrous root system extends laterally near the soil surface rather than to great depths. The vertical distance between lines represents about 30.5 cm.

1.9.2 Primary tissues of the root consist of the central stele surrounded by the cortex and epidermis

The primary anatomy of a mature root of a eudicot is shown in Figure 1.31. At the center of the root is a core of vascular tissue, called the **vascular cylinder** or **stele**. In the roots of cone-bearing gymnosperms and most eudicots, primary xylem occupies the center of the stele; transverse sections through the stele show that the xylem resembles a star with three, four or five points. Primary phloem is found in notches between the points of the star. In monocot roots, on the other hand, a core of ground tissue called **pith** is located at the center of the stele. Concentric rings of primary xylem (next to the pith) and primary phloem surround the pith. In all cases, a layer of unspecialized cells, the **pericycle**, forms the outer boundary of the vascular cylinder. Between the epidermis and vascular cylinder is ground tissue called **cortex**. Cells of the cortex store reserves, mostly starch.

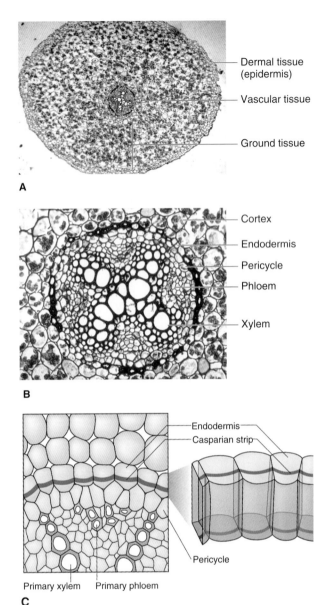

A

B

C

Figure 1.31 Organization of primary tissues in a young eudicot root (*Ranunculus*, buttercup). Light micrograph of cross (transverse) sections of (A) the whole root, and (B) its vascular cylinder showing the three primary tissue systems. (C) Diagram illustrating the endodermis including the Casparian strip.

The innermost layer of cortex cells is the **endodermis**. Periclinal walls (arranged at right angles to the root surface) of endodermal cells are impregnated with a waxy polymer, **suberin**; this layer is called the **Casparian strip** (Figure 1.31C). The plasma membrane of endodermal cells is firmly attached to the Casparian strip, creating a barrier that prevents diffusion of water and solutes along endodermal cell walls. Solutes must cross the plasma membranes of endodermal cells before they can reach the vascular system (see Chapter 13).

1.9.3 The shoot system is organized into repeating modules

The shoot system consists of two organs, **leaves** and **stems**; it also produces reproductive organs (see Figure 1.23). The shoot system is organized into repeating modules called **phytomers**. Each phytomer consists of a **node**, where leaves and **axillary buds** attach to the stem, and an **internode**, the stem between nodes. Leaves display a variety of forms; they are major photosynthetic organs converting water, carbon dioxide and light energy to carbohydrate (Figure 1.32). Stems connect underground roots with aerial leaves, conducting water and minerals from the roots to the leaves, and photosynthate from the leaves to the roots. Some stems, such as potato (*Solanum tuberosum*) tubers, iris (*Iris* spp.) rhizomes and crocus (*Crocus* spp.) corms, are modified for food storage, while others, such as cactus stems, are modified for water storage (Figure 1.33).

1.9.4 The tissues of an angiosperm leaf consist of an epidermis with stomata, photosynthetic mesophyll cells and veins

The arrangement of tissues in a generalized leaf is shown in Figure 1.34. An epidermis with cuticle surrounds the inner tissues of leaves. Stomata occur in the epidermis and their location and density varies from species to species (see Chapter 14). The ground tissue in angiosperm leaves is made up mostly of parenchyma cells that are specialized for photosynthesis and contain many chloroplasts. Photosynthetic parenchyma cells in leaves are often called **mesophyll** (*meso* = middle, *phyll* = leaf). In many angiosperms, there are two distinct types of parenchyma: tightly packed, elongate **palisade parenchyma**, located toward the adaxial surface, and loosely packed **spongy parenchyma**, found

Figure 1.32 Variations in leaf morphology. Simple leaves (C–H) have an entire blade, while in compound leaves (A, B, I) the blade is divided into leaflets. Blades of leaves range in size and shape; some are broad (e.g. C and E) whereas others are narrow (H) or even needle-like (J). There is also variation in petiole length; some are long (e.g. C, E, and K) while others are short (e.g. G) and some leaves lack petioles (D and F).

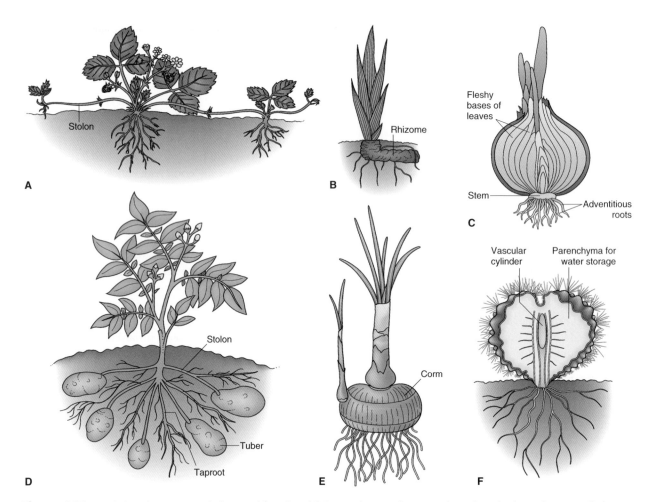

Figure 1.33 Variations in stem morphology and function. (A) Some plants such as strawberry have horizontal stems called stolons that run along the surface of the soil. (B) Iris is an example of a plant with a horizontal underground stem called a rhizome. (C) Onion bulbs are made up of the swollen bases of leaves; the leaves are attached to a small stem. (D) Potato plants have both above-ground stems and underground stolons; in this case the tips of stolons swell to form tubers. (E) Crocus corms are enlarged underground stems packed with reserves. (F) Cactus stems are rounded or flattened, and are the primary photosynthetic organ of the plant. The spines are modified leaves.

abaxially. The air spaces between spongy parenchyma cells favor rapid gas exchange to palisade cells. Collenchyma and sclerenchyma may also be found in leaf ground tissue, usually associated with major veins.

Vascular tissue in leaves is found in **veins** embedded in the ground tissue. The interconnected system of veins in a leaf is highly branched so that no cell in a leaf is far from a vein. In each vein, xylem occupies the adaxial half of the vein and phloem is found in the abaxial half. Veins are often surrounded by a **bundle sheath** of parenchyma cells. In most angiosperms, bundle sheath cells do not have chloroplasts but bundle sheath cells of plants with C_4 photosynthesis, such as *Zea mays* (maize), are packed with chloroplasts (see Chapter 9). In the leaves of most

eudicots and primitive dicots, veins occur in a reticulate (netted) pattern with a main vein that forms the midrib of a leaf and branches into progressively smaller veins. In most monocot leaves, veins run parallel to each other along the length of a leaf.

1.9.5 Primary tissues of the stem are organized differently in monocots and eudicots

The organization of tissues in typical monocot and eudicot stems is shown in Figure 1.35. The outmost layer

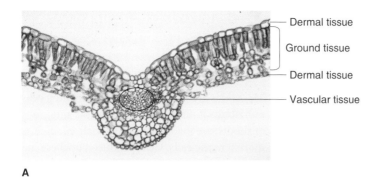

Dermal tissue
Ground tissue
Dermal tissue
Vascular tissue

A

Figure 1.34   Organization of tissue systems in a eudicot leaf. (A) Light micrograph of a transverse section of the main vein of a leaf showing vascular tissue (main vein), ground tissue, and upper and lower epidermis. In this leaf stomata are found in the lower (abaxial) epidermis. (B) Light micrograph of a transverse section of a leaf blade. In this leaf, the ground tissue is divided into palisade and spongy parenchyma. A small vein containing xylem and phloem can be seen.

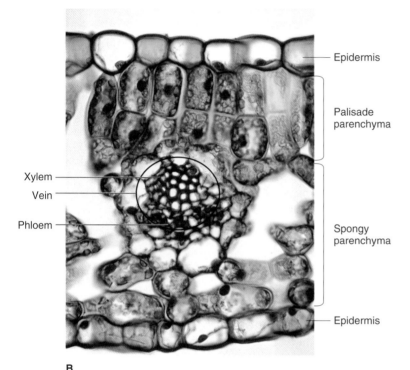

Epidermis
Palisade parenchyma
Xylem
Vein
Phloem
Spongy parenchyma
Epidermis

B

of cells is the epidermis with a waxy cuticle. In green stems the epidermis may contain stomata as well as hairs and glands. In transverse section, vascular tissue of stems occurs in **vascular bundles**. Each bundle is divided into **xylem** and **phloem**. Usually xylem is located on the inner half of a bundle and phloem in the outer half. In the stems of most eudicots and gymnosperms that have only undergone primary growth, a ring of vascular bundles divides the ground tissue into two regions, an outer **cortex** between the epidermis and vascular bundles, and an inner **pith**, ringed by vascular bundles. This differs from the arrangement of vascular bundles in monocots, such as maize, in which vascular bundles are scattered through the ground tissue. In green stems, the cells in the outer layers of cortex contain chloroplasts and are photosynthetic.

1.10 The seed plant body plan III. Growth and development of new organs

Now that we have considered the organization of the plant body, we can examine how this body grows and develops. In contrast to animals, a plant continues to grow and produce organs repeatedly throughout its life. New organs are initiated in localized centers of cell division called **meristems**. Primary growth occurs at the tips of roots and stems and leads to increase in length. Secondary growth may occur in mature tissues of roots and stems; this results in an increase in diameter.

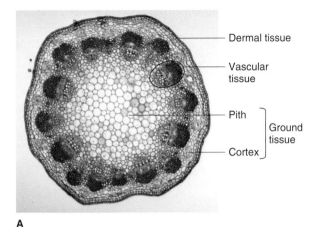

A

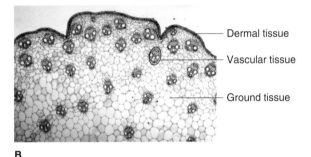

B

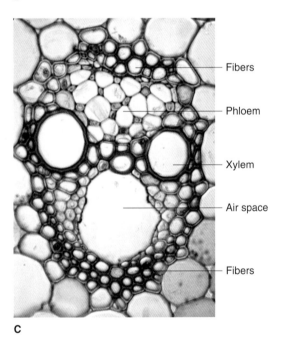

C

Figure 1.35 Light micrographs showing the organization of primary tissue systems in young stems. (A) In *Helianthus*, a eudicot, the stem contains a ring of vascular bundles (vascular tissue) that separates the ground tissue into cortex and pith. (B) The vascular bundles of monocots, such as maize (*Zea mays*) are dispersed throughout the ground tissue of the stem. (C) Higher magnification of one *Z. mays* vascular bundle showing the positions of phloem and xylem.

1.10.1 Apical meristems produce the primary plant body

Meristems are localized groups of dividing cells. **Apical meristems**, found at the tips of roots and growing points of shoots, are made up of unspecialized, dividing **meristematic cells**. Cells produced by these meristems allow continuous growth of the root and stem. In addition, the shoot apical meristem produces cells that will become new organs: leaves and axillary buds, potential branches. Thus plants have indeterminate growth, increasing in size and producing new organs throughout their lives. Cells derived from apical meristems form the three **primary meristems** of roots and shoots: the **protoderm**, the **ground meristem** and the **procambium**. As we have seen, each of these primary meristems produces one of the three tissue systems that make up a plant, respectively the **dermal, ground** and **vascular tissue systems**. Tissues that derive from primary meristems are called **primary tissues** and the increase in length that results from their action is called **primary growth**.

Key points Plants differ from animals in possessing groups of cells that in principle can grow and divide indefinitely. Meristems are defined as zones of undifferentiated cells that can undergo mitosis to produce new cells that differentiate into mature tissues. Apical meristems at the tips of stems and roots are responsible for primary growth, and long-lived species such as bristlecone pines (*Pinus longaeva*) owe their longevity in part to apical meristems. In woody plants lateral meristems are responsible for producing secondary tissues: secondary phloem and secondary xylem as well as cork in secondary growth. Lateral meristems are found in both stems and roots, in fact woody shoots and roots are almost impossible to distinguish from each other after several years of secondary growth.

1.10.2 The root apex consists of the meristem covered by the root cap, and lateral roots originate as primordia in the pericycle

There are many important differences in primary growth and the production of new organs in roots and shoots. We will look at these processes first in roots and then in shoots. The organization of the root apex is shown in Figure 1.36. In roots, the apical meristem is covered by a **root cap** that protects the root as it pushes through soil. As the root grows, root cap cells are sloughed off and are continually replaced by the apical meristem. The outermost cells of the root cap produce a polysaccharide slime that lubricates the tip as it moves through the soil. Roots may produce a prodigious amount of slime; it has been estimated that a large plant, with thousands of roots, can invest as much as 10% of all carbon produced in photosynthesis in slime production.

In most roots cell elongation is confined to the apical 10 mm behind the apical meristem. In the zone of cell elongation cells may increase in length 20–50 times. Some cell differentiation occurs in this zone, however most differentiation, including the production of root hairs, takes place in the zone of cell maturation after elongation has ceased (Figure 1.36).

The initiation of **branch** (= **lateral**) **roots** occurs in mature parts of a root (Figure 1.37). Cells of the pericycle are triggered to divide to form a group of cells, a **branch root primordium**. The primordium develops into a new root, with its own apical meristem; it burrows through the cortex and epidermis of the parent root until it reaches the outside. As vascular tissues in the basal end of the lateral root mature, they connect to the vascular cylinder of the main root.

1.10.3 The shoot apical bud is the source of leaves, axillary buds and floral organs

The organization of the shoot apex, the **shoot apical bud**, differs from that of the root tip (Figure 1.38). In addition to adding cells to primary meristems, the shoot apical meristem produces cells that will become new leaves and axillary buds or, under certain conditions, floral organs. **Leaf primordia** arise as groups of cells on the sides of the apical meristem. As cell division continues young leaves increase in size and arch over the apical meristem. In the **axil** of each leaf, cells derived from the apical meristem divide to form **axillary buds**, miniature, dormant (non-growing) replicas of the shoot apical bud. Cell elongation in the stem is somewhat delayed so the apical bud of a stem contains a number of nodes with very short internodes. When axillary buds start to grow they become lateral branches. If axillary buds grow soon after they are formed, a plant will have a 'bushy' appearance as in the ornamental plant *Coleus*. If axillary buds remain dormant, the plant will have a monopodial growth habit and lack branches (e.g. in tobacco, sunflower, etc.) (Figure 1.39).

1.10.4 Secondary growth is for the long haul, up to thousands of years

Many plants, such as monocots and herbaceous eudicots, increase in size only by primary growth, all cells being added by primary meristems. In gymnosperms and woody angiosperms, stems and roots increase in diameter as well as in height by **secondary growth**. During secondary growth, most of the tissues of the plant body are produced not by apical meristems, but by lateral meristems. Lateral meristems give rise to secondary tissues, and the consequent increase in plant size is termed secondary growth. Secondary growth is often associated with extreme size and longevity, as in trees such as the bristlecone pine and giant sequoia (see Section 1.6.2).

1.10.5 Lateral meristems allow for expansion in girth

In some species, secondary growth may increase the diameter of the stem from 1–2 cm to 10–15 m, and woody trees can reach heights in excess of 100 m. Growth in girth, sometimes referred to as secondary thickening, is initiated in parts of the root and stem that are no longer increasing in length. Two lateral meristems are involved in secondary growth: the **vascular cambium**, which produces **secondary xylem** and **secondary phloem**, and **cork cambium**, which produces **periderm**, a waterproof, protective outer coating of cells. Although most monocots lack secondary meristems, many, notably various species of palm, can increase in circumference

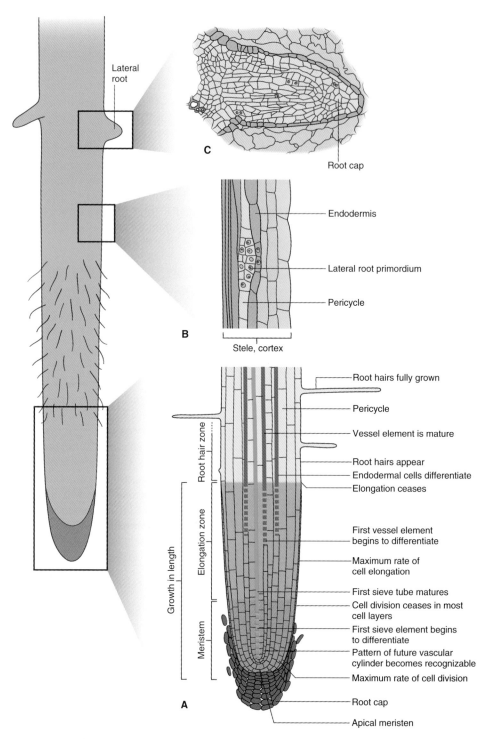

Lateral
root

C

Root cap

B

Endodermis

Lateral root primordium

Pericycle

Stele, cortex

Root hairs fully grown

Pericycle

Vessel element is mature

Root hairs appear
Endodermal cells differentiate
Elongation ceases

First vessel element
begins to differentiate

Maximum rate of
cell elongation

First sieve tube matures

Cell division ceases in most
cell layers

First sieve element begins
to differentiate

Pattern of future vascular
cylinder becomes recognizable

Maximum rate of cell division

Root cap

Apical meristen

Root hair zone

Elongation zone

Meristem

Growth in length

A

Figure 1.36 Primary growth in the root. (A) Elongating primary roots have an apical meristem that produces cells in two directions: acropetally to replenish the protective root cap; and basipetally to add cells to the elongation zone, where cells increase in length 20–50-fold. Cell differentiation primarily occurs after cell elongation has ceased. (B) At the edge of the vascular cylinder, in mature tissue, division of the pericycle produces lateral root primordia. (C) Developing lateral roots burrow through the root cortex, and eventually burst through the epidermis and grow into the soil.

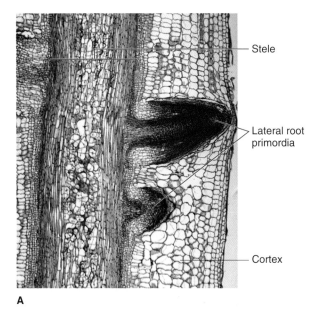

A

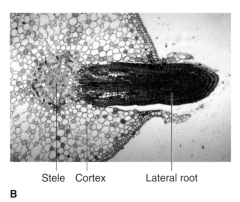

Stele   Cortex   Lateral root

B

Figure 1.37   Light micrographs showing lateral root formation. (A) Longitudinal section of a *Lupinus* root showing two stages of lateral root formation. (B) Transverse section of lateral root formation in willow (*Salix*). Lateral roots arise by division of pericycle cells. The root primordium burrows through the cortex and epidermis of the parent root to reach the soil.

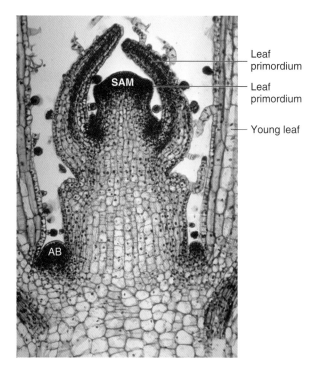

Figure 1.38   Light micrograph showing a longitudinal section of the shoot tip of *Coleus*. Note the shoot apical meristem (SAM) and successively older leaf primordia. Axillary bud primordia (AB) develop in the axils of young leaves. As they mature, these primordia form axillary buds, each with its own apical meristem and leaf primordial.

because structures such as leaf bases become lignified, die and persist, providing support for the stem.

Within the stem, a continuous ring of **vascular cambium** arises from undifferentiated remnants of procambium that are located between the phloem and xylem of vascular bundles and from **interfascicular cambium** produced by the dedifferentiation of parenchyma cells located between the vascular bundles (Figure 1.40). In the root, vascular cambium forms in the stele from residual procambium that surrounds the primary xylem. In both stems and roots, cells in the ring of vascular cambium divide periclinally, producing tiers of cells in both radial directions. Secondary xylem is

produced toward the center of a stem or root, while secondary phloem is produced toward the organ surface. At intervals, the vascular cambium produces files of parenchyma cells called **rays** both centripetally (toward the center of the organ) and centrifugally (towards the outside). Ray cells store starch and permit radial transport of water, inorganic ions and organic solutes across the woody stem and root. Cells of the vascular cambium also divide anticlinally (perpendicular to the surface of the organ) increasing the diameter of the cambium.

As growth proceeds from year to year, layers of secondary xylem accumulate forming **wood**. In an older tree trunk, one can see a distinction between darkly colored **heartwood**, at the center of the trunk, and paler, younger **sapwood** (Figure 1.41). In this case, water transport only occurs in sapwood. The tracheary elements of heartwood are filled with various polymers thought to be waste products that are deposited in the xylem for permanent storage.

Tissues outside the vascular cambium, including the primary and secondary phloem, the cortex and the epidermis cannot accommodate the increase in circumference. Periodically, outer layers split and are

A   B

Figure 1.39   Apical dominance affects the overall morphology of a plant. (A) Some plants, such as *Coleus*, have weak apical dominance resulting in a low, highly branched morphology. (B) In species such *Hibiscus*, apical dominance is much stronger, producing a plant with a single main stem with few to no branches.

sloughed off. To replace the protective epidermis, another lateral meristem, the **cork cambium**, initially arises in the outer layers of the cortex in the stem and from the pericycle cells of the root. The division of cells in the cork cambium produces **cork** (**phellem**) to the outside and parenchyma called **phelloderm** toward the inside. Cork cells are dead at functional maturity. Their cell walls contain lignin and suberin, which act to waterproof the outer parts of the organ (Figure 1.42). Slit-like openings in the cork, called **lenticels**, form at sites where the cork cambium generates a spongy tissue that, unlike cork, has many air spaces between the cells. The lenticels allow for gas exchange between the atmosphere and the living cells in the woody stem.

The **bark** of a tree consists of all the tissue outside the vascular cambium; the inner bark is made up of functional secondary phloem, and the outer bark contains alternating layers of cork and non-functional, older secondary phloem. A tree accumulates all the xylem tissue it produces, but only a few years worth of functional secondary phloem are present. Removing a ring of bark around the circumference of a tree, a practice called 'ringing' or 'girdling', removes phloem thus interrupting the flow of photosynthate to the roots and the tree will die.

1.10.6 Wood morphology is influenced by environmental and endogenous factors

The activity of the vascular cambium can be regulated by day length. In trees that grow in temperate latitudes where there are seasonal changes in day length, the cambium ceases to divide with the onset of the short days of fall. The last cells produced by the cambium in the fall have small diameters and thick walls. Tracheary elements, produced when cambial division is reinitiated in spring, are larger and have thinner walls than those produced in late summer. Annual growth rings in the xylem are delineated by the contrast between the late summer wood of one year and spring wood of the next year (Figure 1.43). The amount of xylem produced in a given year is influenced by a number of factors including temperature, water availability and nutrient availability. The measurement of changes in annual rings has given rise to the science of **dendrochronology** in which the pattern of annual rings can be linked to specific dates. In this way dendrochronologists can accurately determine the age of a tree and can date when a specific piece of wood was formed. Annual rings can be used to estimate

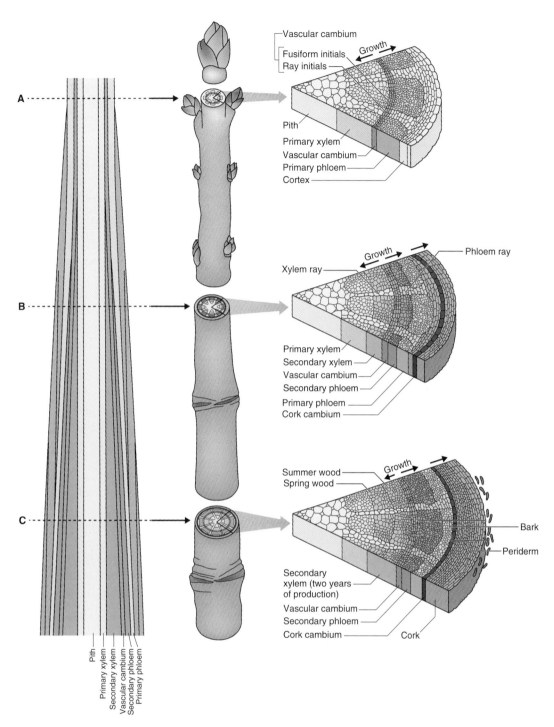

Figure 1.40 Secondary growth in the shoot. The activity of two lateral meristems—the vascular cambium and the cork cambium—increase stem circumference in many gymnosperms and non-monocot angiosperms. The vascular cambium produces secondary phloem outward and secondary xylem (wood) inward. The cork cambium produces layers of cork that form a portion of the outer bark.

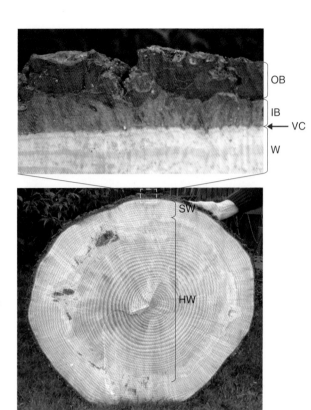

Figure 1.41   A cross-section of a woody stem and an enlargement of a portion of the bark and underlying wood (W), showing the corky outer bark (OB), functional phloem (inner bark, IB), vascular cambium (VC), water-conducting sapwood (SW) and non-conducting heartwood (HW).

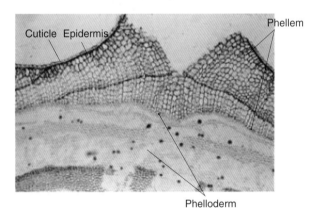

Figure 1.42   Outer bark of a woody plant. The periderm produced by the cork cambium replaces the epidermis, which splits and dies as cell divisions in the vascular cambium increase stem girth. The outer layer of periderm consists of dead cork cells.

the age of a tree and the climatic history of the area in which the tree is growing (see Section 17.2.3). Trees that grow in tropical latitudes, characterized by little or no seasonal change in day length, often do not show growth rings. However, in areas with wet and dry seasons, similar growth rings are produced when cambial growth stops in the dry season and restarts in the wet season.

Key points   Many common and technical names have been used to describe tissue in mature trees and these include bark and cork for the outer layers, and sapwood and heartwood, and soft wood and hard wood for secondary xylem. Bark is a non-technical term for all tissue in a stem or root that lies outside the vascular cambium and includes the secondary phloem and cork. Removing the bark from a tree is often referred to as ring girdling. It will kill the tree because all transport in the phloem is interrupted. Sapwood and heartwood refer to two zones in secondary xylem that can be distinguished by their appearance. Heartwood is located at the center of the stem and is darker in color as a result of the deposit of a range of natural, antimicrobial products. The outer layers of wood are called sapwood and are lighter in color because deposits in these cells are absent; this tissue is where water transport occurs. Heartwood is generally more resistant to decay by fungi and other microorganisms and is therefore more highly sought after in construction. Soft wood and hard wood are terms that refer to the physical properties of wood that result from the presence or absence of vessels and fibers. Soft wood is most generally found in gymnosperms; it lacks vessels and is a tissue of uniform consistency and is relatively easy to work, i.e. chisel, plane or saw. Hard wood is found in angiosperms and consists of a mixture of tracheids and vessels, making it harder to work.

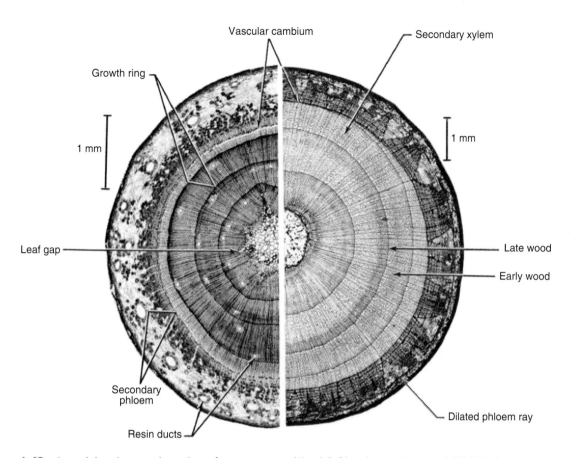

Figure 1.43 Annual rings in secondary xylem of a gymnosperm (*Pinus*) (left) and an angiosperm (*Tilia*) (right). Growth rings are formed because of differences in sizes of cells produced in spring (early wood, large cells) and summer/fall (late wood, very small cells). The very small cells of late wood form a distinct demarcation with the cells of the following spring. The easiest way to count annual rings is to count the number of lines that late wood makes in the secondary xylem.

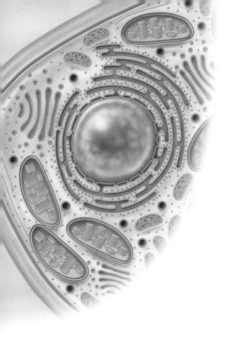

Chapter 2

Molecules, metabolism and energy

2.1 Introduction to biological chemistry and energetics

This book is about the internal workings of plants. At the outset we need to review some of the basic principles of **biochemistry** and **bioenergetics**. All organisms share basic requirements of a **supply of chemical elements** to produce biological molecules (**biomolecules**) and an **energy source** to fuel the chemical reactions that sustain life. Here we look at the structures of biomolecules, beginning with a survey of chemical bonding and representations of molecular structures. The major informational, catalytic and structural components of cells are **macromolecules**, large molecules made of repeated units. This chapter considers the chemical characteristics of the major classes of macromolecule: carbohydrates, lipids, proteins and nucleic acids.

Living organisms take in energy and use it to power biochemical processes. An understanding of the basics of **thermodynamics**, the study of energy change, helps make clear why and how particular biochemical reactions occur. Energy flows through living systems in the form of oxidation/reduction reactions and transformations of high-energy compounds, usually organic molecules carrying phosphate groups. Even when thermodynamically favored, many biochemical reactions require the participation of a **catalyst** to proceed at a biologically useful rate. A catalyst is a substance that makes a chemical reaction go faster, without itself being altered. **Enzymes** are catalytic protein molecules that facilitate chemical changes in living cells. In this chapter we examine general features of enzyme catalysis and its control.

2.2 Biological molecules

In this section we review the basic concepts of molecular structure as they apply to the molecules that characterize living cells. **Water** is essential as the medium within which biochemical reactions are carried out. We will discuss the chemical and physical properties of water that are critical for life. Carbon-based biomolecules are responsible for building and sustaining the cytoplasm, membranes and walls of living plant cells. We will look at each of the four major groups of biomolecules: carbohydrates, lipids, proteins and nucleic acids, reviewing their structures and functions.

2.2.1 Molecules consist of atoms linked by chemical bonds

Living cells are made from a relatively narrow range of different elements. Carbon, hydrogen and oxygen are the main constituents, followed by nitrogen, phosphorus,

The Molecular Life of Plants, First Edition. Russell Jones, Helen Ougham, Howard Thomas and Susan Waaland.
© 2013 John Wiley & Sons, Ltd. Published 2013 by John Wiley & Sons, Ltd.

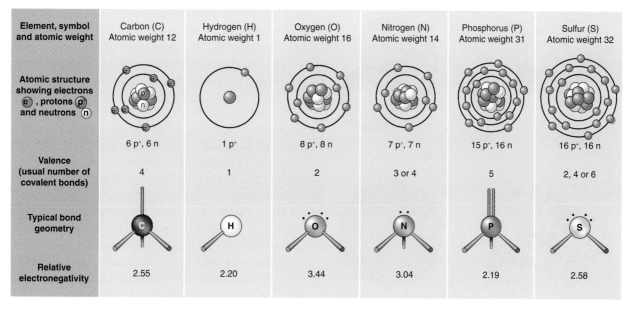

Element, symbol and atomic weight	Carbon (C) Atomic weight 12	Hydrogen (H) Atomic weight 1	Oxygen (O) Atomic weight 16	Nitrogen (N) Atomic weight 14	Phosphorus (P) Atomic weight 31	Sulfur (S) Atomic weight 32
Atomic structure showing electrons e , protons p and neutrons n	$6\,p^+$, $6\,n$	$1\,p^+$	$8\,p^+$, $8\,n$	$7\,p^+$, $7\,n$	$15\,p^+$, $16\,n$	$16\,p^+$, $16\,n$
Valence (usual number of covalent bonds)	4	1	2	3 or 4	5	2, 4 or 6
Typical bond geometry	C	H	O	N	P	S
Relative electronegativity	2.55	2.20	3.44	3.04	2.19	2.58

Figure 2.1 Physico-chemical properties of the most abundant elements of biomolecules. Atomic nuclei are shown as clusters of protons (red) and neutrons (white). Nuclei and electrons are not to scale. The hydrogen ion, H^+, arising from a hydrogen atom losing its single electron, is often referred to as a proton.

sulfur (Figure 2.1), a few metals and a number of minor elements (see Chapter 13). Metabolism is the making and breaking of the chemical bonds between these elements. In general, there are two ways in which atoms may bond together: covalently and non-covalently. In a **covalent bond** two atoms (A and B in this example) share a pair of electrons A:B. The number of chemical bonds that an atom of a particular element can form is its **valence** or **valency**. The valence of C is 4, H is 1, O is 2, N is 3 or 4, P is 5 and S is 2, 4 or 6 (Figure 2.1).

If the pair of electrons in a covalent bond is equally shared between two atoms, the molecule is said to be **non-polar**. Carbon–carbon bonds and carbon–hydrogen bonds are non-polar. If the shared electrons in a covalent bond spend more time with one atom than the other, that atom is said to be more **electronegative**. This unequal sharing results in a **polar bond** in which the more electronegative partner has a slight negative charge (δ^-) and the less electronegative partner has a slight positive charge (δ^+). Oxygen and nitrogen are more electronegative than carbon and hydrogen (Figure 2.1) and so bonds between these elements tend to be polar. The nature and implications of polar bonding are considered further in connection with the properties of water in Section 2.2.3. In broad terms, a polar molecule is likely to be soluble in water and insoluble in oils or organic solvents, and is referred to as **hydrophilic** ('water-loving') or **lipophobic** ('oil-hating'). Conversely, a **non-polar** molecule tends to be **hydrophobic** ('water-hating') or **lipophilic** ('oil-loving').

In addition to covalent bonds, atoms may form several other types of bonds including **hydrogen bonds** (see Section 2.2.3) and **ionic bonds**. Ionic bonds result from the electrostatic attraction between positive and negative ions. Ions are formed when, instead of sharing electrons, one or more electrons transfer from one atom (A) to another (B). This results in a net negative charge on B, which becomes an **anion**, B^-, and a net positive charge on A, which becomes a **cation**, A^+. A and B then form an **ionic bond**, A^+B^-. Like polar molecules, ions are hydrophilic.

2.2.2 Chemical structures are represented in a variety of ways

Chemical structures are represented in different ways depending on the complexity of the molecule and the information to be conveyed. Small molecules are conveniently written as chemical formulas, for example water as H_2O, methanol as CH_3OH. Sometimes it is useful to include the bond structure, as in triple-bonded dinitrogen gas $N\equiv N$ and double-bonded ethylene $H_2C=CH_2$. More complex molecules may be represented by abbreviations of chemical names, for instance adenosine triphosphate, ATP, or coenzyme A, CoA. Models of chemical structure in three dimensions, representing the relative sizes of atoms and the angles of the bonds between them, may take the form of ball-and-stick (Figure 2.2A) or space-filling (Figure 2.2B)

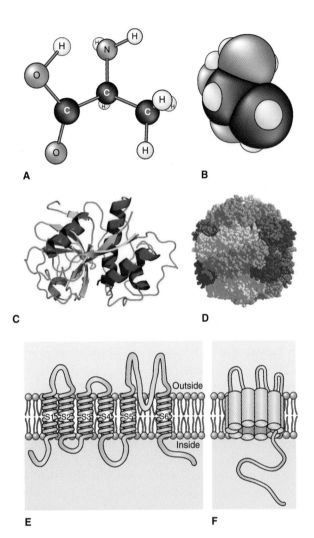

may mention **X-ray crystallography** as a particularly powerful technique for macromolecules like proteins and nucleic acids. To picture how this technique works, imagine an aquarium with fish of different kinds suspended at different depths. Light shone through the glass tank makes a pattern of shadows when projected onto a flat surface. By analyzing the pattern using the appropriate mathematical techniques one can recreate a three-dimensional picture of the sizes and neighbor-to-neighbor relationships of the aquarium's inhabitants. The same principle is used to determine molecular structures, looking at atoms in a crystal instead of fish in a tank, and using X-rays rather than visible light to project the **diffraction pattern**. Knowledge of the crystal structure of an enzyme or complex allows secondary and higher-order structures to be verified and catalytic mechanisms to be understood, pictured and modified. Many examples of molecular structures determined by crystallography occur in this book.

Figure 2.2 Representations of molecular structures. (A) Ball-and-stick model of the amino acid alanine. (B) Space-filling model of alanine (carbon atoms are shown in dark gray, hydrogen light blue, nitrogen dark blue, and oxygen in red). (C) Ribbon representation of secondary structure of the polypeptide chain of the enzyme papain showing helices in red and sheets as yellow. (D) Space-filling contour model of the tertiary structure of the multisubunit CO_2-fixing enzyme rubisco. (E) Membrane-spanning regions of a plant potassium channel protein represented as helices S1–S6 in relation to the inside and outside of the lipid bilayer. (F) Membrane-spanning regions of a plasma membrane H^+-ATPase shown as cylinders.

> **Key points** Biological molecules consist of C, H and O plus N, P, S and some minor elements, linked mainly by covalent, ionic or hydrogen bonds. Metabolism is the process of making and breaking chemical bonds within and between molecules. The number of bonds an atom can make is determined by its valence. The terms polar or hydrophilic or lipophobic describe any molecule in which the distribution of electrons in constituent bonds results in solubility in water and insolubility in oils. Oil-soluble molecules are referred to as non-polar, hydrophobic or lipophilic. Molecular structures are represented symbolically in different ways, including chemical formulas, abbreviations of chemical names, two- and three-dimensional structural drawings and computer-generated models.

images. Large molecules such as proteins are often represented as ribbon structures (Figure 2.2C) to show how they fold, or as contoured shapes (Figure 2.2D) to highlight features such as binding sites and subunit interactions. Membrane-spanning regions of membrane-associated proteins can be computed (see Section 2.2.8) and are often represented as helices or cylinders (Figure 2.2E, F).

Among the many tools for determining structures by identifying atoms and their bonding relationships, we

2.2.3 Water is an essential constituent of living cells

Water is the most abundant chemical constituent of living organisms. It can account for as much as 70–90% of the mass of non-woody plants. Water is more than simply the medium within which the biochemical reactions of cells are carried out. Its distinctive physico-chemical properties are critical for life in general and many aspects of plant physiology in particular.

Water is an example of a polar molecule. The water molecule consists of a single atom of oxygen covalently bonded to two hydrogen atoms (Figure 2.3A). The

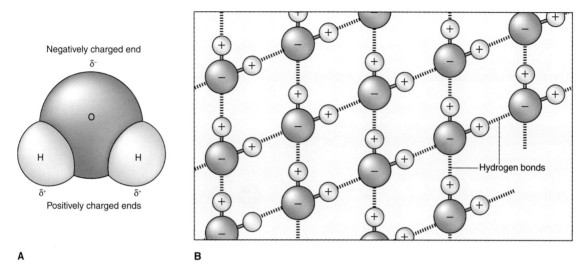

Figure 2.3　The molecular structure of water. (A) The water molecule is a dipole. (B) Water molecules form hydrogen bonds with each other.

oxygen atom, with its nucleus of eight positively charged **protons**, is strongly electronegative relative to hydrogen, the nucleus of which has just a single proton (see Figure 2.1). As a consequence, the shared electrons (negatively charged) tend to be more strongly attracted to the oxygen 'end' of the molecule, giving it a partial negative charge, while the hydrogens are partially positive (Figure 2.3A). Each water molecule therefore behaves as a tiny magnet (a **dipole**) and will form **hydrogen bonds** with other water molecules (Figure 2.3B). The strength of the hydrogen bond is around 5–10% of that of a covalent bond. Hydrogen bonds in liquid water are continuously made and unmade, the life span of an individual bond being of the order of 10^{-11} seconds.

The cohesive but labile nature of hydrogen bonding gives water its unusual and life-supporting properties. It accounts for water's anomalously high boiling and melting points (100 °C and 0 °C, respectively) compared with compounds of similar molecular structure (for example hydrogen sulfide, H_2S, melts at −86 °C and boils at −61 °C). The liquid state of water at ambient temperatures and pressures sustains the diffusion and transport processes essential for life. The 'stickiness' conferred on water molecules by hydrogen bonding accounts for the **surface tension** seen in the skin (meniscus) covering a water drop, and for the tendency of drops to fuse with each other. It also explains how water is able to form unbroken columns and to move through **capillaries** (narrow tubes) against the force of gravity, an effect essential for **transpirational flow** in vascular plants (see Chapter 14). Another consequence of water's tendency to form intermolecular hydrogen bonds is its unusually high **specific heat** (the heat energy required to increase the temperature of a unit quantity of a substance by a unit degree), which has important

buffering effects against temperature fluctuations and therefore greatly extends the range of environments where life is possible.

Water plays a critical role in the chemistry and structure of living cells. It acts as a solvent, dissolving polar molecules by forming **hydration shells** around electrically charged chemical groups or ions. Water excludes non-polar molecules such as lipids which, in aqueous media, coalesce into clusters sustained by hydrophobic associations and, more extensively, into the complex multimolecular assemblages of cell membrane systems.

Water is also a reactant, accepting or donating protons (hydrogen ions) and electrons in the critical biochemical and bioenergetic processes that sustain life. Although water is a covalent molecule, it will undergo spontaneous ionization to a small degree (about one molecule in 10 million at ambient temperature and pressure), to form a positively charged proton (H^+) and negatively charged hydroxide ion (OH^-). The proton exists in hydrated form as the **hydronium ion**, H_3O^+ (Equation 2.1).

Equation 2.1 Ionization of water

$$H_2O \rightleftharpoons OH^- + H^+$$

$$H^+ + H_2O \rightleftharpoons H_3O^+$$

$$2H_2O \rightleftharpoons OH^- + H_3O^+$$

The common measure of the acidity or alkalinity of an aqueous solution is its **pH** which is equal to minus the decimal logarithm of the hydrogen ion concentration ($-\log_{10}[H^+]$) of the solution at 25 °C. Pure water, which is neutral (that is, neither acid nor alkaline) has a pH of 7. An **acid** has a pH in the range <7 to 0. The pH of an alkali (also called a **base**) is between >7 and 14. Lemon juice has a pH of around 2 as a consequence of the accumulation of organic acids by cells in the fruit. The sap in the central vacuole of a typical plant cell has a pH of about 5. Cytosol is generally neutral or slightly basic. The ionization of water is critical for, among other things, the generation of reductant and oxygen during photosynthesis, solute transport across membranes, all aspects of energy metabolism, and the making and breaking of many kinds of chemical bonds during biosynthesis and biodegradation.

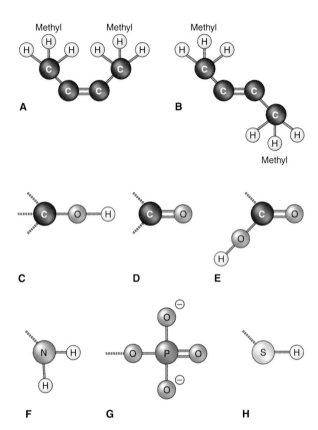

Key points The chemical properties of water are critical for life. Water molecules are polar and hydrogen bond to each other and to other polar molecules. Hydrogen bonding in liquid water underlies many important biological processes, including capillarity, buffering against temperature extremes and solute transport. Water is the universal solvent for metabolism, and a reactant in many biosynthetic and degradative pathways. It is the source of reductant and oxygen during photosynthesis. Ionization of water is the basis of the pH scale.

2.2.4 Biological molecules have carbon–carbon backbones

All biomolecules contain carbon atoms. Carbon atoms can form bonds with four other atoms including other carbon atoms. This results in an almost infinite range of chemical structures. Carbon can form single (C–C), double (C=C) and triple (C≡C) bonds. Carbon atoms joined by a single bond can rotate freely about the bond axis, but the carbon–carbon double bond does not allow rotation and so the atoms or substituent groups attached to a pair of double-bonded carbons will adopt one of two different configurations, **cis-** and **trans-**, depending on whether the groups are orientated in the same or opposite direction, respectively (Figure 2.4A, B). Compounds with identical chemical formulas but different structures, such as trans- and cis-forms, are called **isomers**. Carbons in the backbone of organic molecules may be attached to a number of different **functional groups**. Carbon bonds with hydrogen to form hydrocarbon structures such as the methyl group, –CH_3. Hydrocarbon groups tend to be highly non-polar. Groups with hydrogen and oxygen include **alcohols** C–OH, **carbonyls** C=O and **carboxyls** (**carboxylic acids**) COOH (Figure 2.4C–E). Carbon also links to

Figure 2.4 Structures of chemical groups. (A) Double-bonded carbon atoms with methyl groups in the cis position. (B) The trans isomer of (A). (C) Alcohol group. (D) Carbonyl group. (E) Carboxyl group. (F) Amine group. (G) Phosphate group. (H) Thiol group. Carbon atoms are shown in dark gray, hydrogen light blue, oxygen red, nitrogen blue, sulfur yellow, and phosphorus in purple.

nitrogen, commonly in the form of the **amine group** –NH_2 (Figure 2.4F), and to phosphorus and sulfur, mostly as **phosphates** –PO_4 and **thiols** –SH (Figure 2.4G, H). The presence of O or N in a group makes an organic molecule more polar (hydrophilic).

If a carbon atom is bonded to four different atoms or groups it is said to be an **asymmetrical** or **chiral** carbon. The groups attached to an asymmetrical carbon can be arranged in two mirror image configurations that cannot be superimposed (Figure 2.5). These are called **stereoisomers** or **enantiomers** and are designated D- and L-forms. They have identical chemical properties; however, enzymes can distinguish one form from another.

2.2.5 Monomers are linked to form macromolecules

Living organisms are able to make a wide range of large, structurally and functionally complex biomolecules

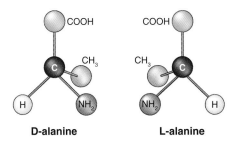

Figure 2.5 Stereoisomerism illustrated by the mirror image D- and L-enantiomers of the amino acid alanine.

(macromolecules). Macromolecules are generally made up of small repeating units, **monomers**. Molecules with a defined and relatively short chain of monomers are sometimes referred to as **oligomers**. Many of the most important macromolecules are **polymers**, that is, they are made of a potentially unlimited number of monomers. For example, the cellulose molecule is built from individual glucose monomers joined together to make a chain which might extend to many hundreds of units in length (see Chapter 4). DNA, RNA and proteins are polymers but in each case their monomeric units are of different kinds that must be put together in the right order to build structures able to work correctly. The precise **sequence** of the different monomeric units in nucleic acids represents **genetic information**, which in turn specifies the ordering of the monomers in proteins through the processes of gene **transcription** and **translation** (see Chapter 3). The molecular mass (weight) of a biomolecule is often given in units of Daltons (**Da**). The mass of a macromolecule with a molecular weight in the thousands will be expressed in kilo-Daltons (**kDa**).

The reaction between monomers during the biosynthesis of oligomers and polymers usually involves bond formation by **dehydration**, the loss of water (Equation 2.2).

Equation 2.2 Condensation of two molecules by dehydration

$$R^1\text{–OH} + R^2\text{H} \rightarrow R^1\text{–}R^2 + H_2O$$

An example of a dehydration reaction is that between a carboxylic acid and an alcohol. The product is an **ester** and the reaction is called **esterification** (Equation 2.3). The reverse reaction, **hydrolysis**, breaks down polymers to oligomeric and monomeric subunits, and also cleaves ester linkages and other bonds formed by dehydration.

Equation 2.3 Production of an ester by dehydration

$$R^1\text{–C–OH} + R^2\text{–OH} \rightarrow R^1\text{–C–O–}R^2 + H_2O$$

Having seen the general principles of how biomolecules are constructed, we will now examine each class of biological molecules in more detail. For each group of biomolecules, we will examine its monomeric units and see how they are linked together to form polymers. This material is important for understanding the structure and function both of individual plant cells and of the plant as a whole.

Key points Biomolecules are made of carbon atoms, which can bond to each other and to atoms of other elements to make chains, branched structures and rings. Carbon has a valence of 4 and can form single, double or triple bonds with other atoms. Two molecules with the same chemical formula but different bonding configurations are referred to as isomers. Hydrocarbon structures such as the methyl functional group CH_3 are non-polar. Other functional groups such as carbonyls (C=O) and amines (NH_2) are polar in nature. Complex carbohydrates, proteins, DNA and other large and complex biomolecules are polymers, macromolecules made up of repeating monomer units. In general, polymeric biomolecules are made by forming bonds between monomers through dehydration reactions.

2.2.6 Carbohydrates include simple sugars and complex polysaccharides

Carbohydrates are the group of biomolecules that includes sugars, starches and celluloses. The monomers of complex carbohydrates are simple sugars, **monosaccharides**. These are made of carbon, hydrogen and oxygen, according to the general empirical formula $(CH_2O)_n$ where n = 3–7. As a rule, carbohydrate names take the suffix -ose. A **triose** is a sugar with n = 3. **Tetroses, pentoses** and **hexoses** correspond to sugars with n = 4, 5 and 6, respectively. Where n is not specified, the general term **glycose** is used. The structure of glucose is shown in Figure 2.6. While all monosaccharides have stereoisomers (D- and L-forms), usually only one stereoisomer is present in cells. For example, glucose exists almost always as the D-isomer. All monosaccharides can assume a straight-chain (acyclic) conformation. Sugars with four or more carbons can also rearrange to form **heterocyclic rings**, rings containing more than one kind of atom (Figures 2.6 and 2.7). The six-membered configuration

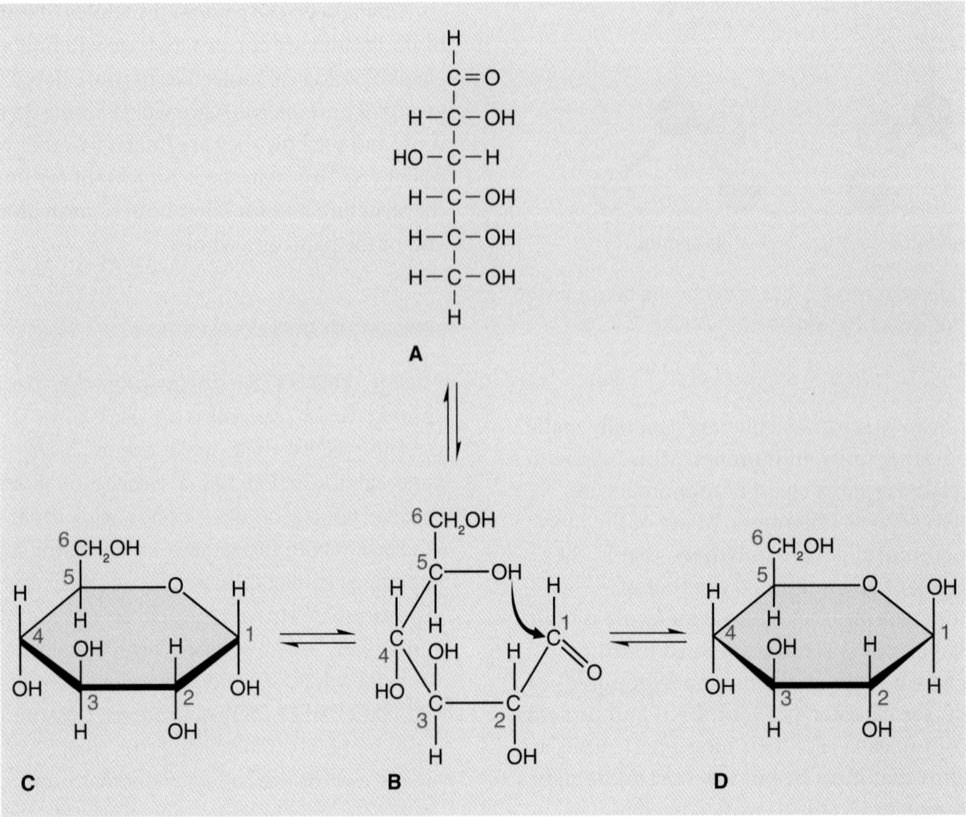

Figure 2.6 Structure of glucose. (A) Acyclic form of D-glucose. (B) Mechanism of cyclization to generate (C) α-D-glucopyranose or (D) β-D-glucopyranose.

(five carbons and an oxygen) is called the **pyranose** form. The five-membered ring (four carbons and an oxygen) is called the **furanose** form. Figure 2.6 shows the hexose glucose as a straight chain and a pyranose ring structure (sometimes called glucopyranose). Fructose is an **isomer** of glucose; it has the same empirical formula ($C_6H_{12}O_6$) but a different configuration. Straight-chain and fructofuranose depictions for fructose are given in Figure 2.7A and B. Figure 2.7C and D are examples of straight-chain and furanose structures of a pentose, in this case ribose.

In the ring form of a sugar, the carbon to which two oxygen atoms are attached is called the **anomeric** carbon: C-1 in glucose or C-2 in fructose. When a straight-chain sugar cyclizes, the –OH group attached to the anomeric carbon can adopt one of two orientations, α and β (see Figure 2.6). In solution a sugar switches between α- and β-forms as the ring opens and closes. A carbohydrate with a free anomeric group is able to react with oxidizing agents such as Cu^{2+} or Ag^{2+} ions and is said to be a **reducing sugar**. If the anomeric carbon participates in an intermolecular bond, it is locked into either the α- or β-form and the sugar is **non-reducing**.

Individual pentoses or hexoses may link together via **glycosidic bonds** to form oligomers (**oligosaccharides**)

and polymers (**polysaccharides**). **Sucrose**, table sugar, is an example of a disaccharide, consisting of a glucose residue linked α(1 → 2) to a fructose unit (Figure 2.8A). Note that sucrose is a non-reducing sugar, because the anomeric carbons of both monosaccharide components participate in the α(1 → 2) linkage. It is an important biomolecule that is unique to plants, where it functions as the principal molecule for transporting carbon and energy. Some plants also use sucrose for energy storage.

Stereoisomers are optically active. When **polarized light** (that is, light in which the waves are aligned in a single plane) is passed through a solution of a single stereoisomeric form of a compound, the polarization plane is rotated. A compound that rotates polarized light to the right is called **dextrorotatory** and one that rotates it to the left is **levorotatory**. Note that (confusingly) the designation D- and L-, as applied to monosaccharides for example, refers only to the *geometric configuration* of the molecule about the chiral center and not to whether the molecule is dextrorotatory or levorotatory.

D-glucopyranose is dextrorotatory, rotating polarized light +52.7°. D-fructofuranose, on the other hand, is strongly levorotatory, rotating polarized light −92° (Equation 2.4). Hydrolysis of sucrose (dextrorotatory, +66.5°) yields an equimolar mixture of glucose (as

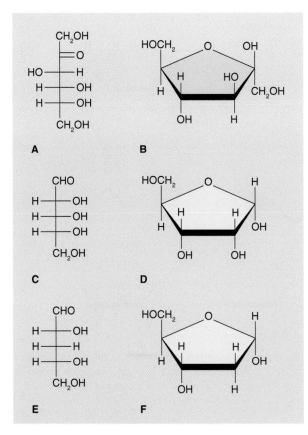

Figure 2.7 Structures of some monosaccharides: (A) acyclic form of D-fructose, (B) α-D-fructofuranose, (C) D-ribose, (D) α-D-ribofuranose, (E) 2-deoxyribose and (F) α-D-2-deoxyribofuranose.

(see Figure 6.10). **Cellulose**, a major structural component of cell walls, is the corresponding polymer based on the β(1 → 4) linkage of cellobiose (see Figure 4.4). The α(1 → 4) linkages in amylose give the molecule a spiral shape; while the β(1 → 4)-linked cellulose molecule is a straight rod. These differences in molecular shape are crucial in determining the very different functions of these molecules.

In describing glycosidic bonds, one must specify the configuration of the anomeric carbon involved in the bond (α or β), the numbers of the two carbons that participate in the bond, and the steroisomer (D or L) of each sugar. Thus the full, chemically explicit names for maltose, cellobiose and sucrose are, respectively: α-D-glucopyranosyl-(1 → 4)-D-glucose; β-D-glucopyranosyl-(1 → 4)-D-glucose; and β-D-fructofuranosyl-(2 → 1)-α-D-glucopyranoside.

Polysaccharides are named after the principal sugars that constitute them and often have the suffix -an. Amylose and cellulose are examples of **glucans**. Other examples include **fructan** (also called **levan**), a storage carbohydrate, which consists of a chain of fructose residues extended from sucrose, and **xylans** (polymers of the pentose sugar xylose). Branched polysaccharides, or polysaccharides decorated with side-chains of various lengths, are common in plants. Starch consists of amylose together with the highly branched glucan **amylopectin**, which has a backbone of α(1 → 4)-linked D-glucopyranose residues with additional α(1 → 6) bonds occurring every 24 to 30 glucose units (see Figure 6.10). As we shall see in Chapter 4, cell walls are particularly rich in branched polysaccharides, most of which, like **xyloglucans** for example, have more than one kind of monomer unit.

Sugar residues in polymeric molecules may be chemically modified. Perhaps the most important example is **deoxyribose** (2-deoxy-D-ribose; see Figure 2.7F), the pentose component of the nucleoside monomer unit which gives DNA (*deoxyribo*nucleic acid) its name. The corresponding sugar is D-ribose in RNA (*ribo*nucleic acid; see Figure 2.7D). L-fucose and L-rhamnose are common deoxy-sugars found in the structural polysaccharides of plant cell walls. Also prominent among the modified sugars of cell walls are **uronic acids**, formed by oxidation of the C_6 primary alcohol of hexoses to a carboxylic acid. D-mannuronic acid, D-galacturonic acid and D-glucuronic acid are the

D-glucopyranose) and fructose (as D-fructofuranose) and this mixture has a net rotation of −39.3°. The change from dextro- to levorotation during sucrose hydrolysis is referred to as **inversion** (Equation 2.4) and gives the enzyme that catalyzes the reaction, **invertase**, its name. The hydrolysis product is often called invert sugar.

Maltose and **cellobiose** are isomeric disaccharides; both are reducing sugars. Each consists of two glucose molecules linked C-1 → C-4, but the bond in maltose is in the α-configuration whereas that of cellobiose is β (Figure 2.8B, C). These two compounds have very different chemical and biological properties. The polysaccharide built from the chain of α(1 → 4)-linked glucose monomers based on maltose is **amylose**, a component of starch, used by plants to store energy

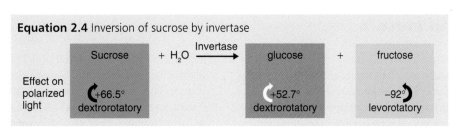

Equation 2.4 Inversion of sucrose by invertase

	Sucrose	+ H_2O	→ (Invertase)	glucose	+	fructose
Effect on polarized light	+66.5° dextrorotatory			+52.7° dextrorotatory		−92° levorotatory

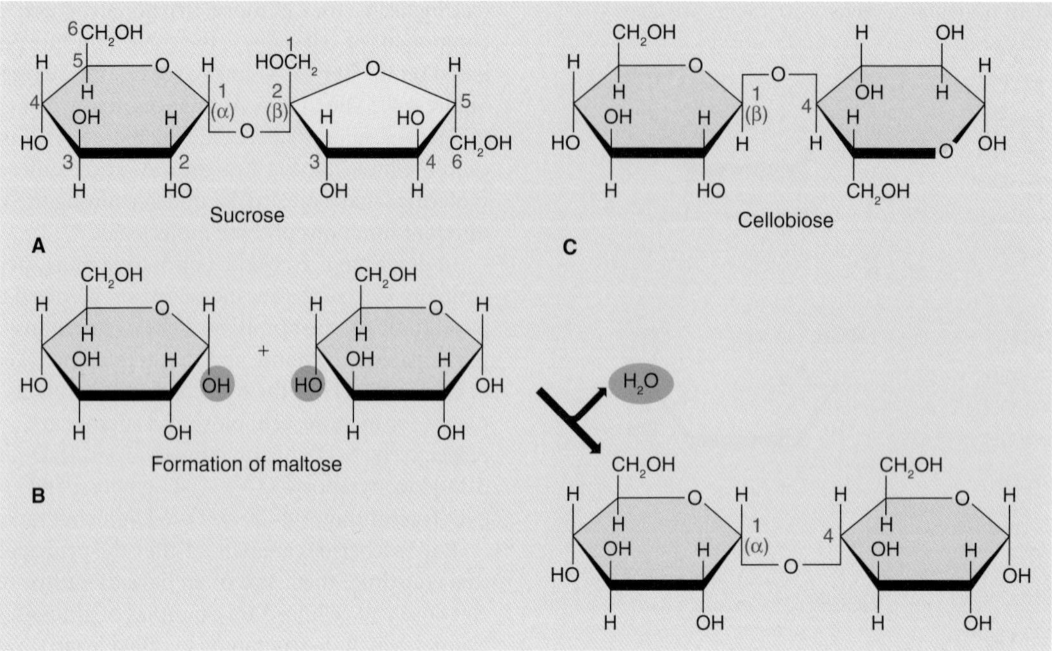

Figure 2.8 Structures and formation of disaccharides. (A) Sucrose consists of D-glucopyranose linked α 1 to the β 2 carbon of D-fructofuranose. (B) Dehydration reaction between two glucose molecules to make the disaccharide maltose. Maltose is a cleavage product of the amylose polysaccharide component of starch and is α(1 → 4) linked. (C) In cellobiose, a product of cellulose breakdown, the 1 → 4 glycosidic bond is in the β configuration, resulting in one glucose residue being flipped over relative to the other.

uronic acids derived from mannose, galactose and glucose, respectively. Other notable types of sugar derivative include **sugar alcohols** such as sorbitol (a transport and storage compound in some plants, derived from glucose) and **amino sugars**, for instance glucosamine.

As well as linking to one another, sugars and modified sugars can be attached to a range of molecules to form **glycosides**. Plants typically accumulate a vast array of environmental chemicals (**xenobiotics**) or secondary compounds, many of which are stored, secreted or detoxified as glycosides. An example of such a glycoside is salicin, a β-D-glucoside from willow, closely related to the drug aspirin. Sugars and oligosaccharides may form glycosidic bonds with lipids (see Section 2.2.7) and also with proteins, particularly those destined for secretion across the plasma membrane or into lytic or storage vacuoles.

2.2.7 Lipids include oils, fats, waxes and sterols

Lipids are a structurally diverse group of hydrophobic molecules, including oils, fats, waxes and sterols, that are preferentially soluble in non-aqueous solvents such as chloroform. Lipids have many functions. They are major components of biological membranes, determining

Key points Sugars, starches and celluloses are carbohydrates with the general empirical formula $(CH_2O)_n$. Trioses, tetroses, pentoses and hexoses are simple sugars where $n = 3$, 4, 5 or 6 respectively. Ribose is an example of a pentose, and glucose and fructose are hexoses. Sugars can form straight-chain and ring structures and exist in a range of isomeric forms. Di-, tri-, oligo- and polysaccharides are chains of two or more simple sugar monomer units. Sucrose (table sugar) is an example of a disaccharide, comprising a molecule of glucose linked to one of fructose. Cellulose is a polysaccharide made up of long chains of glucose monomers linked by bonds in the so-called β configuration, whereas starch is a polysaccharide of α-linked glucose units. The difference in bond configuration results in dramatic differences in the structure and functions of these two molecules.

essential biophysical and functional properties of the plasma membrane, organelles and endomembrane system. They act as reserves of carbon and free energy (yielding up to twice as much ATP as the equivalent mass of carbohydrate). They are used as storage materials, primarily in pollen and seeds. Fatty acids are also precursors for the synthesis of important biocompounds such as surface **waxes**, which form a protective skin for the plant epidermis, and regulatory substances including the plant hormone **jasmonic acid** and signaling molecules derived from **phosphatidylinositol**.

Most lipids are derivatives of **fatty acids**, which are carboxylic acids with long hydrocarbon tails. More than 200 different fatty acids have been identified in plants. Stearic acid, which has 18 carbons, is an example of a fully **saturated fatty acid**, that is, the hydrocarbon chain has no double bonds. In accordance with nomenclature conventions, it is referred to as octadecanoic acid (that is, the carboxylic acid related to the 18-carbon molecule octadecane) or, in short form notation, 18:0 (Figure 2.9A). The carbon atoms of fatty acids are numbered beginning with the carboxylic carbon as C-1. Oleic acid is an unsaturated 18-carbon fatty acid with one double bond that occurs at position C-9. Its short form designation is $18:1^{\Delta 9}$ (Figure 2.9B). The hydrogens on carbon atoms linked by a double bond can adopt one of two isomeric configurations, either cis or trans (see Figure 2.4A, B). Bonds in unsaturated fatty acids are almost exclusively cis isomers. The chemical name of oleic acid is thus cis-9-octadecenoic acid. A fatty acid such as oleic acid with one double bond is termed **monounsaturated**. A major fraction of the fatty acids in plants is **polyunsaturated**. Linoleic acid ($18:2^{\Delta 9,12}$) and α-linolenic acid ($18:3^{\Delta 9,12,15}$) are abundant in green tissues, where they constitute a major fraction of the lipids of photosynthetic membranes. The relative benefits and adverse effects of unsaturated compared with polyunsaturated fats in the diet is a much debated subject in human nutrition and healthcare.

Note that fatty acids, like other organic acids of living cells, are normally ionized at physiological pH and so biochemists frequently refer to them by the names of their respective anions. Thus oleic acid and oleate, for example, are used interchangeably, as are acetic acid and acetate, cinnamic acid and cinnamate, fumaric acid and fumarate and so on. The same convention applies to amino acids: for instance glutamic acid is often referred to as glutamate.

The major lipids of membranes and storage tissues are **glycerolipids**, consisting of fatty acids esterified to derivatives of the three-carbon sugar alcohol **glycerol**. The four principal types of glycerolipid found in plants are **triacylglycerol, phospholipid, galactolipid** and **sulfolipid**. Triacylglycerols are reserves of energy and carbon, found mostly in seeds and pollen. They are strongly non-polar and for this reason are sometimes referred to as neutral lipids. As the name suggests, a triacylglycerol molecule consists of a glycerol molecule with one fatty acid molecule esterified to each of its three carbons (Figure 2.10A). Phospholipids are **diglycerides**. Carbons 1 and 2 of glycerol are esterified with fatty acids. Carbon 3 carries a phosphate group that in turn is esterified with an alcohol such as choline (Figure 2.10B) or inositol. Phospholipids are essential components of membrane structure, and also function as messenger molecules in regulatory signaling pathways. An important property of the phospholipid molecule is its **amphipathic** character, that is, it consists of a hydrophilic (polar) phosphoester head group at one end and a hydrophobic (non-polar) fatty acid region at the other. In hydrophilic environments such as the cytosol, amphipathic lipids are able to form bilayers in which non-polar tails associate with each other while polar heads are orientated towards the aqueous surroundings (see Section 4.3).

Galactolipids and sulfolipids are major classes of diglycerides found in plastid membranes. These are **glycolipids** that have a sugar or sugar derivative on C-3 of glycerol. In sulfolipid (sulfoquinovosyldiacylglycerol) the substituent is sulfoquinovose, a sulfonated derivative of glucose bearing an SO_3^- group on C-6 (Figure 2.11A). Galactolipids include monogalactosyldiacylglycerol, in which C-3 of glycerol is linked to the hexose galactose (Figure 2.11B) and digalactosyldiacylglycerol in which the galactose residue of monogalactosyldiacylglycerol is linked 6 → 1 to a second galactose (Figure 2.11C). Galacto- and sulfolipids contain high concentrations of

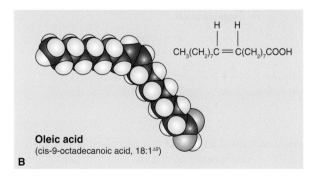

$CH_3(CH_2)_{12}CH_2CH_2CH_2CH_2COOH$

Stearic acid
(n-octadecanoic acid, 18:0)

A

$$CH_3(CH_2)_7\overset{\overset{\displaystyle H}{|}}{C}=\overset{\overset{\displaystyle H}{|}}{C}(CH_2)_7COOH$$

Oleic acid
(cis-9-octadecanoic acid, $18:1^{\Delta 9}$)

B

Figure 2.9 Structures of (A) a saturated and (B) an unsaturated C_{18} fatty acid.

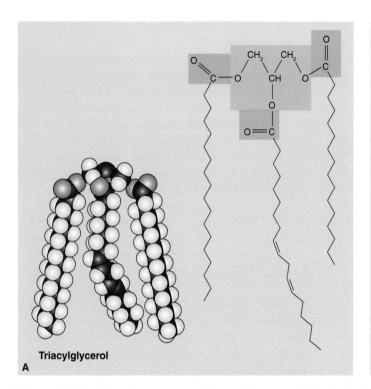
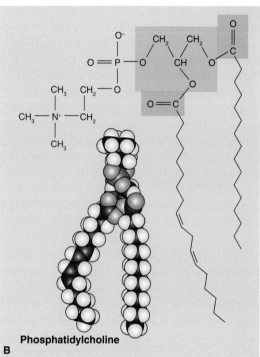

Figure 2.10 Molecular structures of (A) triacylglycerol and (B) phosphatidylcholine.

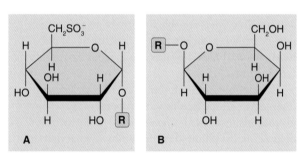
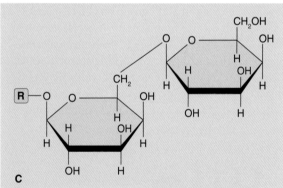

Figure 2.11 Sugar substituents of plant glycolipids: (A) sulfoquinovose, (B) galactose and (C) digalactose. R indicates the diacylglyceryl moiety.

polyunsaturated fatty acids and, like phospholipids, are markedly amphipathic in nature.

Among other plant constituents classified as lipids are **sphingolipids**, fatty acid esters of long-chain amino-

alcohols that are common in plasma membranes. **Waxes** are highly hydrophobic lipids containing a wide variety of long-chain hydrocarbon substituents, esters, polyesters and hydroxy esters of long-chain primary alcohols and fatty acids. Many **volatiles**, including essential oils, fragrances, insect attractants and anti-feedants, are lipid derivatives. Also grouped with the lipids are hydrophobic pigments (**carotenoids, chlorophylls**), sterols, resins, quinines and a vast array of secondary compounds that are metabolically unrelated to fatty acids.

Key points Lipids are highly hydrophobic molecules with major roles in membrane structure, storage of carbon and energy reserves, plant protection and cell signaling. Most are derivatives of fatty acids with long hydrocarbon chains that may be saturated, or may have one or more C=C double bonds, in which case they are referred to as unsaturated. Major structural and reserve lipids in plants consist of two or three fatty acids linked to glycerol. Membrane lipids include phospholipids (glycerol linked to two fatty acids and a third, phosphate-containing side group), glycolipids (two fatty acids and a carbohydrate side group) and sphingolipids (fatty acids linked to a long-chain amino-alcohol). Other important hydrophobic compounds classified with lipids are waxes, volatiles and most photosynthetic pigments.

Side-chain properties	Amino acid	Three-letter symbol	Single-letter symbol	Structure
Polar and fully charged Hydrophilic side chains act as acids or bases which tend to be fully charged (+ or −) under physiological conditions. Side-chains form ionic bonds and are chemically reactive	L-aspartic acid	Asp	D	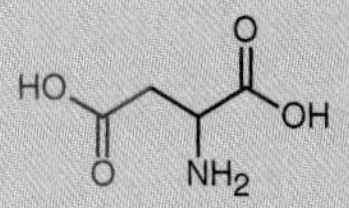
	L-glutamic acid	Glu	E	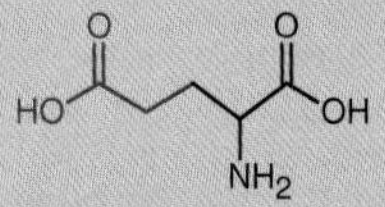
	L-lysine	Lys	K	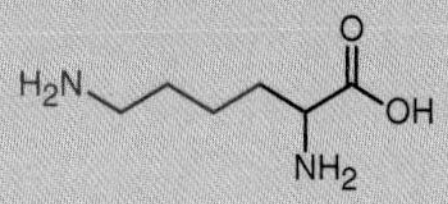
	L-arginine	Arg	R	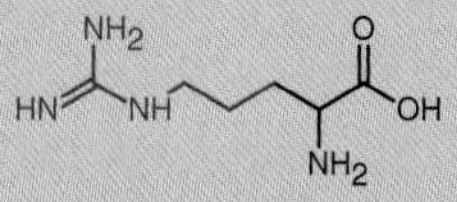
	L-histidine	His	H	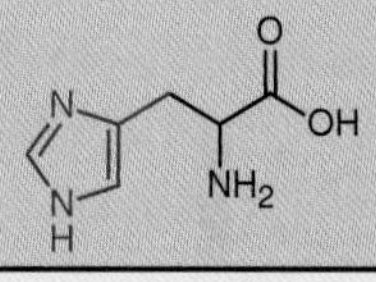
Polar and uncharged or partially charged Hydrophilic side-chains. When partially + or − charged, they are chemically reactive, form H bonds and associate with water	L-serine	Ser	S	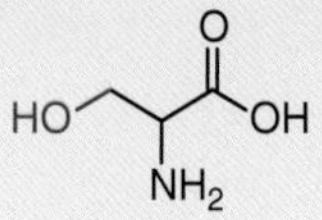
	L-threonine	Thre	T	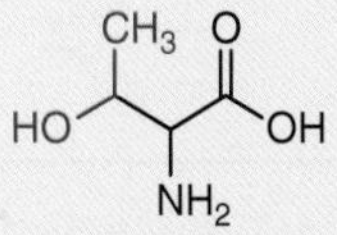
	L-glutamine	Gln	Q	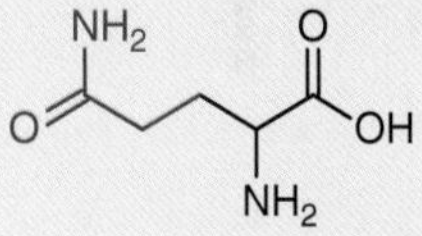
	L-asparagine	Asn	N	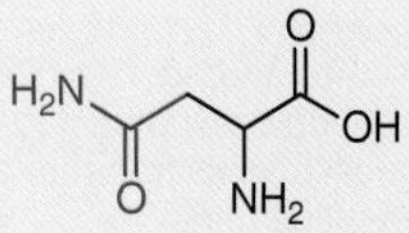
	L-tyrosine	Tyr	Y	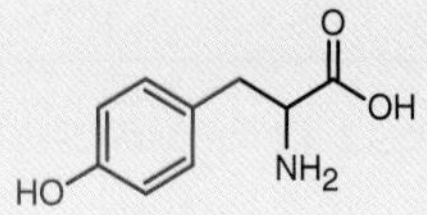

Figure 2.12 Molecular structures of the 20 amino acids most commonly found in proteins, classified according to the chemical properties of their side-chains.

Nonpolar				
	L-alanine	Ala	A	
Hydrophobic side-chains, consisting almost entirely of C and H atoms. In soluble proteins these amino acids tend to be buried inside tertiary structure. In hydrophobic proteins they tend to be on the outside, where they facilitate association with membrane lipids	L-valine	Val	V	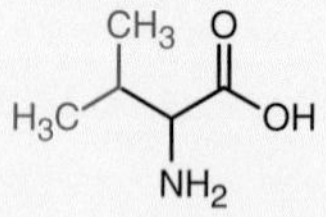
	L-leucine	Leu	L	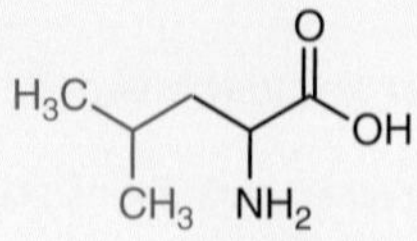
	L-isoleucine	Ile	I	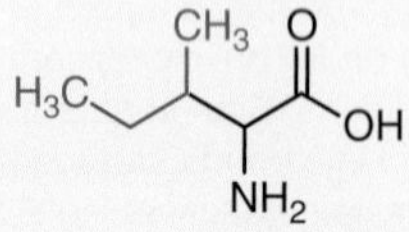
	L-methionine	Met	M	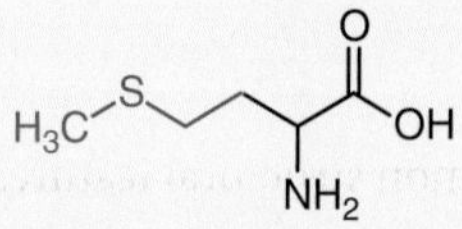
	L-phenylalanine	Phe	F	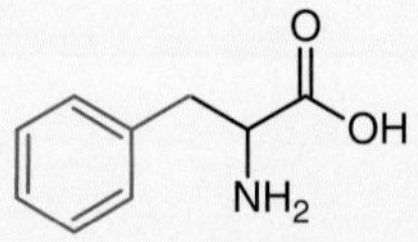
	L-tryptophan	Try	W	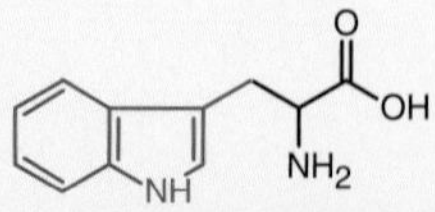
Side-chains with unique properties	Glycine	Gly	G	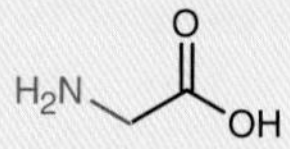
Gly can fit into either a hydrophobic or a hydrophilic environment. Cys can form an -S-S- with another Cys. Pro is a hydrophobic imino acid that introduces kinks into the polypeptide chain	L-cysteine	Cys	C	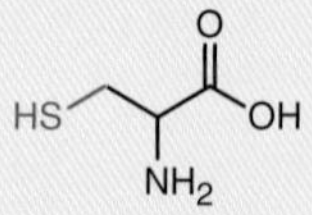
	L-proline	Pro	P	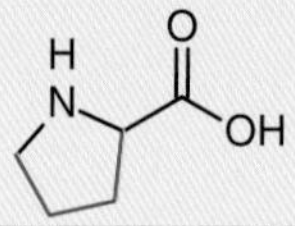

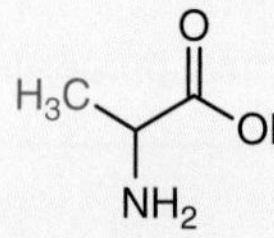

Figure 2.12 *(continued)*.

2.2.8 Proteins function as catalysts, structural and mechanical entities, and signaling molecules

Proteins are central to life. Enzymes, which facilitate the chemical reactions on which life depends, are proteins.

Proteins are also structural, mechanical and signaling molecules, essential for the form and function of cells and their component parts. Proteins are linear polymers built from a series of up to 20 different amino acids (Figure 2.12). There are 20^n possible combinations of a sequence of n amino acids. Thus for a modest-sized protein 100 amino acids in length there are 20^{100} potential primary sequences—an astronomical

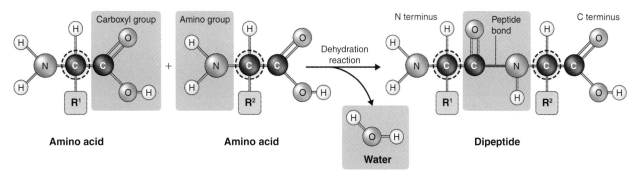

Figure 2.13 Amino acid structure and peptide bond formation. The α-carbons of two amino acids, identified by dotted circles, show the attached amino and carboxylic acid groups. The peptide bond is formed by condensation of the amino group of one amino acid with the carboxylic group of another. The resulting dipeptide has a carboxylic acid group at one end (the carboxy or C terminus) and an amino group at the other (the amino or N terminus).

number, giving a glimpse of the extreme variety of structures and dependent functions possible for a macromolecule built in this way. Virtually all of the different amino acid monomers of proteins have common structural features, notably an α-carbon to which an amino group, a carboxyl group and a variable side-chain, often designated as 'R', are bonded (Figures 2.12 and 2.13). An exception is the amino acid proline, in which the amine group is cyclized (Figure 2.12). In all amino acids other than glycine the α-carbon is asymmetrical and therefore each exists as a D- and an L-isomer form. The amino acids of proteins are exclusively in the L-configuration.

The dehydration reaction between the carboxylic acid group of one amino acid and the amine group of another results in formation of the **peptide bond** (Figure 2.13). **Oligopeptides** and **polypeptides** are chains of peptide-linked amino acids. There is an uncondensed amino group at one end (the amino or **N terminus**) of an oligo- or polypeptide and an uncondensed carboxylic acid group at the other (carboxy or **C terminus**). Proteins are polypeptides. Their higher-order structures and reactivity are determined by the variety of chemical structures and properties of the amino acids from which they are built.

The **primary structure** of a protein is the precise sequence of amino acids between the N terminus and C terminus. Each amino acid is represented by a three-letter abbreviation or, more concisely and usually, a single letter symbol (see Figure 2.12). So a protein's primary structure can be represented as a chain of letters, or diagrammatically in the form of a string of beads, each bead connected to the next through a peptide bond (Figure 2.14A). The configuration of peptide bonds is such that they readily form hydrogen bonds with each other to make loops, helices and other localized arrangements that together represent the **secondary structure** of a protein. These interactions give rise to

particular three-dimensional motifs, notably the **α-helix** (Figure 2.14B) and the pleated (β) **sheet** (Figure 2.14B). Figure 2.2C is a model of the plant protein-degrading enzyme papain in which β-sheet regions are shown in yellow and α-helices in red. In its active state a protein has a **tertiary structure** defined by the coordinates (spatial positions) of all its constituent atoms and maintained by interactions between different parts of the polypeptide chain (Figure 2.14C). Finally, folded proteins in many cases adopt **quaternary structures** by associating in multisubunit complexes (Figure 2.14D; the carbon-fixing enzyme rubisco shown in Figure 2.2D is an example). Complexes of two, three, four, etc. subunits are referred to, respectively, as dimers, trimers, tetramers and so on. A complex is **homomeric** if it is made of identical subunits and **heteromeric** if the subunits are different.

The side-chains of the different protein amino acids have different degrees of hydrophobicity (see Figure 2.12). Each amino acid has a relative hydrophobicity 'score'; the higher the score, the more hydrophobic the amino acid. If this index is represented in graphical form from one end of the primary structure to the other, the result is a **hydropathy plot** that identifies whole regions of the protein that are more or less hydrophobic in character. Figure 2.15A is a typical example of the primary structure of a protein, consisting of 275 amino acids listed from the N-terminal methionine (M) to the C-terminal asparagine (N). Figure 2.15B is the corresponding hydropathy plot. In this example, there are three peaks, starting at around amino acid 130 through to the C terminus, where the value exceeds the threshold beyond which a polypeptide region is sufficiently hydrophobic to span a membrane bilayer. According to hydropathy analysis the protein, which is located in the thylakoid membranes of chloroplasts, has three hydrophobic regions that are long enough to traverse the membrane in the form of

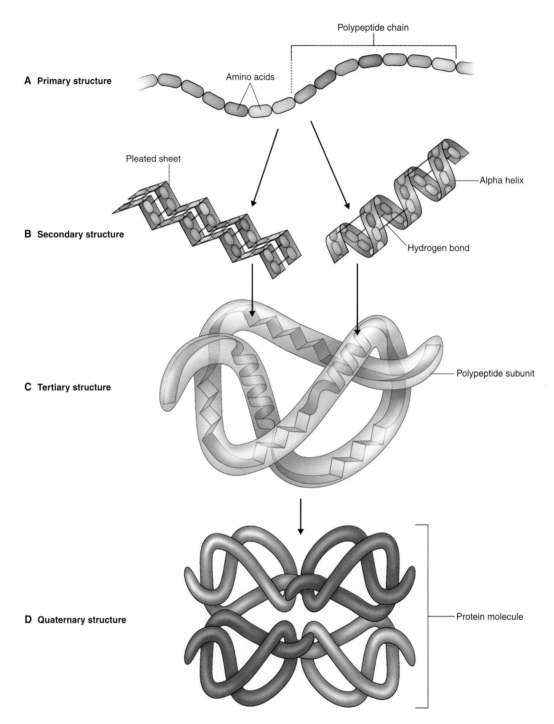

Figure 2.14   The four levels of protein structure. (A) Primary (sequence of amino acids in the polypeptide chain). (B) Secondary (local arrangement into helices and sheets). (C) Tertiary (folding of the entire polypeptide). (D) Quaternary (assembly into a multisubunit complex).

α-helices. Figure 2.15C is a computer prediction of the organization of the protein in relation to the membrane and the hydrophilic environment outside it. The structure corresponds well to experimental observations of the configuration of such proteins in vivo. Figure 2.2E and F shows further examples of proteins whose membrane-spanning helices have been predicted by hydropathy analysis and computer modeling.

```
MAATTAVAASYFSGTRTQYTKQNPGKIQALFGFGTKKSPPPPPPKKSSPKQFEDRLVWFPGASPPEWL
DGTMVGDRGFDPFALGKPAEYLQFDLDSLDQNLAKNLAGDVIGVRVDATEVKPTPFQPYSEVFGLQR
FRECELIHGRWAMLGTLGAIAVEALTGVAWQDAGKVELIEGSSYLGQPLPFSLTTLIWIEVIVVGYIEFQ
RNAELDPEKRLYPGGYFDPLGLASDPEKIENLQLAEIKHARLAMVAFLIFGIQAAFTGKGPISFVATFNN
```

A

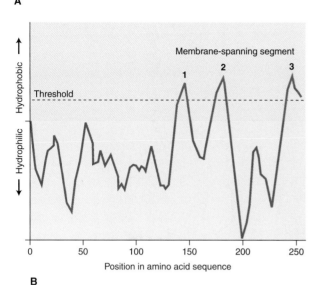

B

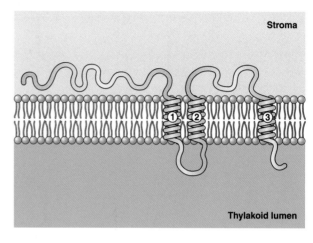

C

Figure 2.15 Structural features of a chlorophyll-binding protein from the photosynthetic membrane of poplar chloroplasts. (A) Primary structure, showing the sequence of amino acids (single-letter codes; see Figure 2.12) from the N-terminal methionine (M) to asparagine (N) at the C terminus. (B) Hydropathy plot, which scans the primary sequence from N to C terminus and assigns a hydrophobicity score based on the side-chain properties of the amino acids in each region. Peaks that cross the indicated threshold value are sufficiently hydrophobic to span a lipid membrane. The example shows three such membrane-spanning regions. (C) Structural model of this protein associated with the thylakoid membrane, based on hydropathy predictions.

Key points Proteins include enzymes, structural materials and signaling molecules. A protein is a polypeptide, a macromolecular polymer consisting of a chain of up to many hundred amino acid monomers linked by peptide bonds. The order in which the 20 different amino acids occur in the polypeptide chain is referred to as the protein's primary structure. Localized loops and helices in the chain represent the secondary structure, and the entire polypeptide is folded up to make the tertiary structure. Folded proteins often form quaternary structures by combining into multisubunit complexes. Membrane-associated proteins have regions of secondary structure, consisting of groups of hydrophobic amino acids, that allow them to embed in the lipid bilayer.

2.2.9 Nucleic acids contain the genetic information of an organism

We now consider nucleotides and nucleic acids. The structure of DNA and the story of how its famous **double helix** configuration was discovered have been written about so often that we will give only a brief summary of nucleic acid chemistry here. Chapter 3 has a more detailed account of DNA and RNA in relation to their functions in genetics. The monomer unit of nucleic acids is the **nucleotide**, which in turn consists of a pentose phosphate linked to a **nitrogenous heterocyclic base** (Figure 2.16A). The pentose is D-ribose in RNA and 2-deoxyribose in DNA (see Figure 2.7). The carbon atoms of the sugar moiety in the nucleotide molecule are numbered 1′ to 5′ (one prime to five prime). The phosphate group of the nucleotide monomer is attached

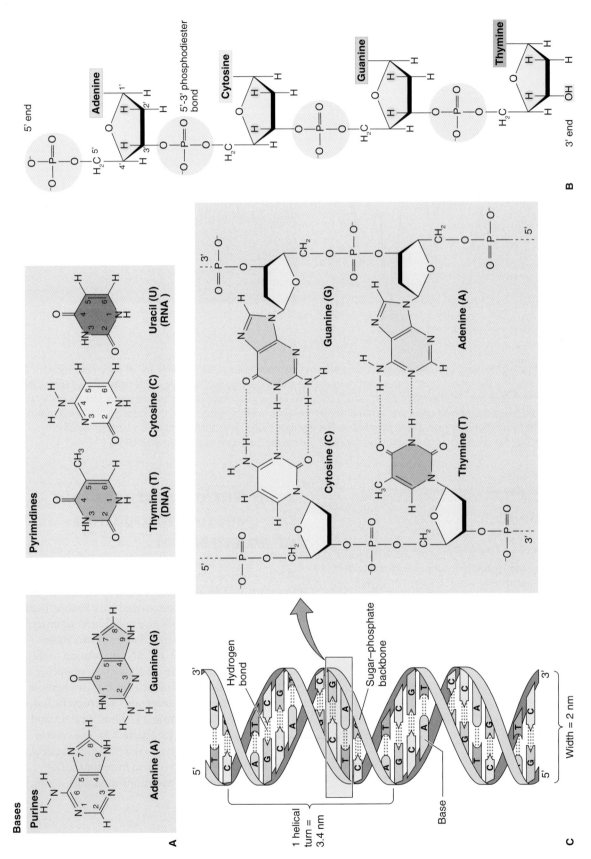

Figure 2.16   Structure of nucleic acids. (A) Purine and pyrimidine bases. (B) Primary structure of the polynucleotide chain. (C) Base-pairing and the double helix.

to C-5′ and the nitrogenous base to C-1′ of the pentose. A nucleotide without a phosphate is called a **nucleoside**. There are two kinds of heterocyclic bases: **pyrimidines**, which consist of a single six-membered ring; and **purines**, made of a pyrimidine ring fused to a five-membered imidazole ring. The purines **adenine** and **guanine** and the pyrimidine **cytosine** are found in both DNA and RNA. DNA also has the pyrimidine **thymine** whereas in RNA it is replaced by **uracil** (Figure 2.16A). The bases in nucleic acids are abbreviated A, G, C, T and U.

The polynucleotide chain of DNA and RNA consists of a sugar–phosphate backbone, pentoses alternating with phosphates that link the 3′ of one sugar with the 5′ of the next (Figure 2.16B). Nucleic acid bases on complementary strands form hydrogen-bonded pairs, A with T or U, C with G (Figure 2.16C). The specificity of pairing means that two polynucleotide strands with the appropriate sequence of purine and pyrimidine bases, running in opposite directions, will align precisely, each chain carrying all the information necessary to determine the base sequence of the other. Paired strands of DNA adopt the iconic double-helix configuration (Figure 2.16C). Except in some viruses and regulatory structures in cells, RNA is generally single-stranded. However, it may form helical **hairpin loop** structures where local stretches of base pairing occur. Among the essential cell components consisting of nucleic acids complexed with proteins are chromosomes and ribosomes (see Chapter 3).

2.3 Energy

Living cells require a source of energy to maintain their organization to grow and to reproduce. In this section we lay the groundwork for understanding energy transformations within cells. **Energy** is the term that describes the amount of work that can be carried out by the action of a force. The unit of energy is the **joule** (J). There are many, freely interchangeable, forms of energy; these can be divided into two major categories: **kinetic energy**, the energy of motion, and **potential** or stored **energy**. Thermal energy, radiant energy and electrical energy are examples of kinetic energy. In living cells, potential energy is found in chemical bonds. Gradients in concentration of solutes across cellular membranes also represent potential energy as does charge separation across membranes.

Living organisms can be divided into large groups based on how they obtain energy. **Autotrophs** (including photosynthetic bacteria and green plants) use CO_2 from the atmosphere as their sole source of carbon to make small precursor molecules and assemble them into macromolecules (Figure 2.17). Some bacteria, known as **chemoautotrophs**, obtain energy by the oxidation of inorganic molecules such as ferrous iron. In this book, we will focus on the algae and land plants, **photoautotrophic** organisms that are capable of using light as a source of energy. Organisms like ourselves, that must obtain their energy and carbon from the

Key points Nucleic acids are polymeric macromolecules consisting of chains of nucleotide monomers. A nucleotide is a pentose phosphate linked to a nitrogenous base. In DNA the pentose is deoxyribose and the base in a given nucleotide monomer is represented by the single letter A, C, G or T. The pentose in the case of RNA is ribose, and the bases are A, C, G or U. The sequence of bases represents genetic information. C specifically pairs with G by hydrogen bonding. Similarly A pairs with T (DNA) or U (RNA). The double helix of DNA consists of two sugar–phosphate chains running in opposite directions and held together by A-T and C-G base pairing. RNA is usually a single chain of nucleotides but often contains regions of internal base pairing that form loop structures.

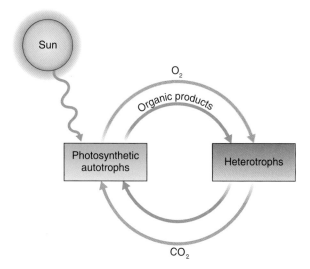

Figure 2.17 The capture of solar energy by photosynthetic autotrophs and the cycling of organic compounds, CO_2 and O_2 between autotrophic and heterotrophic domains in the biosphere.

degradation of organic molecules such as glucose, are categorized as **heterotrophs**. They ultimately depend on autotrophic organisms for their livelihood.

2.3.1 Biological systems obey the laws of thermodynamics

The study of energy transformations is called **thermodynamics**. Knowledge of the basics of thermodynamics is helpful for understanding how energy and chemistry are interrelated in living cells. The principles of thermodynamics enable us to predict whether a particular reaction will proceed spontaneously. The term 'thermodynamics' suggests something to do with heat ('thermo-') and change ('-dynamics'). Change refers to the interconversion of different forms of energy and is represented by the Greek symbol delta (Δ). The basic principles of thermodynamics were established through studies on the equivalence between mechanical work and heat; but these relationships apply to all types of energy, including those that drive the processes of living cells. When it changes from one form to another within a defined system, energy obeys the **laws of thermodynamics**, the first of which states that *during the change from one form to another, energy can be neither created nor destroyed*. This is sometimes referred to as the law of **conservation of energy**.

The energy change in a system under constant pressure and volume, as is the case for the vast majority of biochemical reactions in living tissues, is expressed as ΔH, where H refers to a quantity called the **enthalpy**. Thermodynamics is also concerned with the degree of disorder within the system, referred to as the **entropy** (S). ΔS is the change in entropy. The **second law of thermodynamics** states that *any closed system of its own accord will always undergo change in such a way as to increase the entropy*. Note that living cells, within which metabolism acts to decrease entropy and create order, must continuously exchange energy with the environment. They are therefore not closed systems but components of a larger system within which total entropy increases. In the words of the physicist Erwin Schrödinger, life 'feeds upon negative entropy', and thus does not violate the Second Law.

The part of the enthalpy change that is available to do useful work is called **Gibbs free energy**, ΔG; ΔG is dependent on temperature (T) and pressure (P). When P is 1 atmosphere (101.325 mega-Pascals, MPa) and T is 25 °C (298.15 degrees Kelvin, °K), the **standard free energy change** for a reaction involving 1 mole of a substance is designated $\Delta G°$. For such a reaction under standard conditions at a given pH (usually 7), the free energy change is represented by $\Delta G°'$. Since most biochemical reactions take place at or close to standard conditions of P, T and pH, in this book we will generally present free energy changes as $\Delta G°'$. Similarly, enthalpy and entropy changes under standard reaction conditions are designated $\Delta H°$, $\Delta H°'$, $\Delta S°$ and $\Delta S°'$ respectively.

Equation 2.5, the **Gibbs–Helmholtz equation**, shows the relationship between free energy, enthalpy and entropy. The units of ΔG and ΔH are Joules per mol (J mol^{-1}), and of ΔS are J °K^{-1} mol^{-1}. The term $T\Delta S°$ represents the quantity of energy used to create disorder in the system. It is the basis of the **third law of thermodynamics**: *a system may only achieve zero entropy at absolute zero*, that is, when T approaches 0 °K.

Equation 2.5 Gibbs–Helmholtz equation

$$\Delta G° = \Delta H° - T\Delta S°$$

2.3.2 Change in free energy can be used to predict the direction of a chemical reaction

Using the Gibbs–Helmholtz equation, we can classify chemical reactions according to their likelihood of running in the direction of forming products from reactants (Table 2.1). When the Gibbs free energy content of the reactants is greater than that of the products, $\Delta G°$ is negative and the reaction favors product formation. Such a reaction system is able to do work on the surroundings and is termed **exergonic**. Exergonic reactions are sometimes said to be **spontaneous**. Cellular respiration (sugars + $O_2 \rightarrow CO_2 + H_2O$ + chemical energy; see Chapter 7) is an example of an exergonic reaction. If $\Delta G°$ is positive then product formation is the unfavorable direction and the reaction is termed **endergonic**. Photosynthesis ($CO_2 + H_2O$ + light energy $\rightarrow$ sugars + O_2; see Chapter 9) is such a reaction. An endergonic reaction will not occur spontaneously, but can be driven if coupled to an exergonic reaction, provided that the combined change in free energy is negative overall. In the example of photosynthesis, the energy required to drive the reaction is supplied by light. Many biochemical reactions are endergonic, and energy metabolism in living systems is directed towards providing the high-energy substrates and exergonic reactions to enable such individually unfavorable reactions to occur.

In addition to classifying a reaction in terms of a change in free energy, we can also consider whether or

Table 2.1 The likelihood of a reaction proceeding in the direction of product formation predicted by the Gibbs–Helmholtz equation ($\Delta G^\circ = \Delta H^\circ - T\Delta S^\circ$) for different combinations of ΔG, ΔH and ΔS.

ΔH°	ΔS°	ΔG°	Spontaneous formation of product
Negative (exothermic)	Positive	Negative (exergonic)	Favorable
Negative (exothermic)	Negative and $T\Delta S^\circ < \Delta H^\circ$	Negative (exergonic)	Favorable
Positive (endothermic)	Positive and $T\Delta S^\circ > \Delta H^\circ$	Negative (exergonic)	Favorable
Positive (endothermic)	Negative	Positive (endergonic)	Unfavorable

not it produces heat. When ΔH° is negative, the reaction gives out energy in the form of heat to its surroundings and it is said to be **exothermic**. A reaction with a positive ΔH° takes heat energy from its surroundings and is **endothermic**. Note the distinction between endergonic/exergonic on the one hand, terms referring to *free energy change*, and endothermic/exothermic on the other, which are functions of *enthalpy change*. Table 2.1 summarizes the feasibility of reactions in terms of the direction of change in G, H and S. Note also that a reaction in which heat energy is taken up can proceed in the direction of product formation if the increase in entropy is large enough.

Key points Plants are photoautotrophs, obtaining energy from light. Heterotrophic organisms such as animals get their energy by oxidizing the biomolecules made by autotrophs. Energy transformations obey the laws of thermodynamics, which state that energy can be neither created nor destroyed and that disorder (entropy, S) tends to increase, except at a temperature of absolute zero. Enthalpy (H) is the total energy of a system under standard conditions and the Gibbs free energy, G, is the energy available to do work. If the change (represented by the symbol delta Δ) in G of a chemical reaction is negative, formation of product is favored and the reaction is said to be exergonic. An endergonic reaction has a positive ΔG and will not proceed spontaneously to product formation, but can be driven by coupling with an exergonic reaction of larger negative ΔG. Reactions with positive or negative ΔH are classified as endothermic or exothermic, respectively.

2.3.3 Electrons are transferred in oxidation/reduction reactions

One way in which energy is transferred in cellular reactions is by moving electrons from one molecule to another. The electron transport chains of cellular respiration (see Chapter 7) and photosynthesis (see Chapter 9) are two examples of this phenomenon. When a molecule gives up electrons (acts as an **electron donor**) it becomes **oxidized** and the molecule that gains electrons (**electron acceptor**) is **reduced**. Such reactions are called **redox reactions** for short. Electron donors are also called reducing agents or **reductants**; likewise electron acceptors are called oxidizing agents or **oxidants**. Both a reductant and an oxidant are needed for the transfer of electrons in a redox reaction.

A flow of electrons is, by definition, an electric current and is the result of the voltage difference between the electron donor and the acceptor. Accordingly, electron flow in a redox reaction occurs by virtue of the energy levels (called **redox potentials**) of reactants. Similar to the terms used in thermodynamics (see Section 2.3.1), the standard redox potential at 1 atmosphere, 25 °C, and pH 7 is designated $E^{\circ\prime}$. Redox potential (units: volts, V) is calculated in relation to the hydrogen electrode ($H_2 \rightarrow 2H^+ + 2e^-$) which is 0.000 V at pH 7. Electron donors (reductants) with more energetic (more negative) potential donate their electrons to acceptors (oxidants) with less energetic (less negative, more positive) potential.

The change in free energy ($\Delta G^{\circ\prime}$) associated with a redox reaction can be calculated using Equation 2.6, where n = number of electrons transferred from

Table 2.2 Standard reduction potentials and corresponding free energy change values for some biologically important half-reactions at pH 7.0 and 25 °C.

Half-reaction	$E^{\circ\prime}$ (V)	$\Delta G^{\circ\prime}$ (kJ mol^{-1})
$\frac{1}{2}O_2 + 2H^+ + 2e^- \rightarrow H_2O$	0.816	−157.5
$NO_3^- + 2H^+ + e^- \rightarrow NO_2^- + H_2O$	0.421	−81.2
Cytochrome f (Fe^{3+}) $+ e^- \rightarrow$ cytochrome f (Fe^{2+})	0.365	−35.2
Cytochrome a (Fe^{3+}) $+ e^- \rightarrow$ cytochrome a (Fe^{2+})	0.29	−28.0
Cytochrome c (Fe^{3+}) $+ e^- \rightarrow$ cytochrome c (Fe^{2+})	0.254	−24.5
Cytochrome b (Fe^{3+}) $+ e^- \rightarrow$ cytochrome b (Fe^{2+})	0.077	−7.4
Fumarate$^{2-} + 2H^+ + 2e^- \rightarrow$ succinate^{2-}	0.031	−6.0
Oxaloacetate$^{2-} + 2H^+ + 2e^- \rightarrow$ malate^{2-}	−0.166	32.0
Pyruvate$^- + 2H^+ + 2e^- \rightarrow$ lactate$^-$	−0.185	35.7
Glutathione$_{ox} + 2H^+ + 2e^- \rightarrow$ 2glutathione$_{red}$	−0.23	44.4
$NAD^+ + H^+ + 2e^- \rightarrow$ NADH	−0.32	61.7
$NADP^+ + H^+ + 2e^- \rightarrow$ NADPH	−0.324	62.5
Ferredoxin (Fe^{3+}) $+ e^- \rightarrow$ ferredoxin (Fe^{2+})	−0.432	41.7

reductant to oxidant and F = Faraday constant (96.48 kJ V^{-1} mol^{-1}).

Equation 2.6 Relationship between free energy and redox potential

$$\Delta G^{\circ\prime} = -nFE^{\circ\prime}$$

Table 2.2 shows the reduction potentials and $\Delta G^{\circ\prime}$ values of some biologically important redox half-reactions. Electrons flow from entries lower in the table to components nearer the top (for example NADH to oxaloacetate). In this process the free energy released is typically used to drive energetically unfavorable reactions. Note that the **sign** of the reduction potential or free energy change for the reverse of a given reaction is also **reversed**. We can illustrate this with the example of the **tricarboxylic acid cycle** (TCA) enzyme malate dehydrogenase (see Chapter 7). Equation 2.7 shows the partial reactions catalyzed by malate dehydrogenase. It also gives the $E^{\circ\prime}$ and derived $\Delta G^{\circ\prime}$ values for these reactions (Table 2.2) and the predicted $\Delta G^{\circ\prime}$ for the overall reaction. The reduction of oxaloacetate to malate is an endergonic reaction, while the oxidation of NADH is strongly exergonic. If these two reactions are coupled, the overall reaction is exergonic and therefore thermodynamically feasible. The value for the $\Delta G^{\circ\prime}$ of malate dehydrogenase has been determined experimentally, from equilibrium measurements under

standard conditions, to be about −26 kJ mol^{-1}, which is close to prediction.

Equation 2.7 Redox and Gibbs free energy relations of malate dehydrogenase

Oxaloacetate$^{2-} + 2H^+ + 2e^- \rightarrow$ malate^{2-}
$E^{\circ\prime} = -0.166$ $\Delta G^{\circ\prime} = +32.0$

NADH $\rightarrow NAD^+ + H^+ + 2e^-$
$E^{\circ\prime} = +0.32$ $\Delta G^{\circ\prime} = -61.7$
___
Oxaloacetate^{2-} + NADH + $H^+ \rightarrow$ malate^{2-} + NAD^+
$\Delta G^{\circ\prime} = -29.7$

2.3.4 Energy in cells flows through phosphorylated intermediates

The flow of energy through metabolism is principally in the form of **phosphorylated** metabolites, organic molecules to which one or more phosphate groups are attached. Phosphate, often designated P$_i$ (inorganic phosphate, $PO_4^{3-} \rightleftharpoons HPO_4^{2-} \rightleftharpoons H_2PO_4^-$) is attached via one of its oxygens to a carbon atom (Figure 2.18A, C). At physiological pH, two of the three unlinked oxygen atoms are negatively charged and associated with protons (hydrogen ions, H^+) or metal ions such as Mg^{2+}. One of

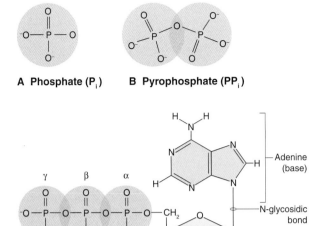

A Phosphate (P$_i$)   B Pyrophosphate (PP$_i$)

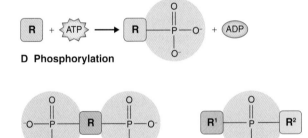

C Adenosine mono-, di-, tri-phosphate

D Phosphorylation

E Bisphosphate   F Phosphodiester

Figure 2.18 Forms in which phosphate participates in energy metabolism.

the three uncommitted oxygens of an organic phosphate may form a **phosphoanhydride** bond with another phosphate group to create a **diphosphate**. The second phosphate in turn is available to form a **triphosphate** via another phosphoanhydride bond (Figure 2.18C). A triphosphate may be hydrolyzed to a monophosphate with the release of **pyrophosphate**, PP$_i$ (Figure 2.18B). Note the distinction between a diphosphate (one phosphate linked to another) and a **bisphosphate** (two phosphate groups linked to two separate carbon atoms of the organic acceptor) (Figure 2.18E). A phosphate group may also bridge two organic molecules, forming a **phosphodiester** bond (Figure 2.18F). DNA and RNA are polymers made of phosphodiester-linked units (see Section 2.2.9).

Table 2.3 Standard free energies of hydrolysis (pH 7) of some important phosphorylated metabolites.

Phosphorylated compound	$\Delta G^{\circ\prime}$ (kJ mol^{-1})
Phosphoenolpyruvate	−54.4
1,2-bisphosphoglycerate	−49.4
Acetyl phosphate	−43.9
Creatine phosphate	−37.7
ATP → ADP + P$_i$	−30.5
Phosphodiesters (e.g. PP$_i$ → P$_i$)	−25.1
Glucose-1-phosphate	−20.9
AMP → adenosine + P$_i$	−13.8
Glucose-6-phosphate	−13.8
Glycerol-1-phosphate	−9.6

A considerable amount of energy is needed to drive the reaction that creates phosphoanhydride bonds. Conversely, release of P$_i$ or PP$_i$ by hydrolysis liberates chemical energy. Table 2.3 shows the standard free energy of hydrolysis ($\Delta G^{\circ\prime}$) of a number of phosphorylated metabolites. A negative value indicates that the reaction is exergonic. The energy relations of phosphorylated intermediates are often described in terms of the cleavage of a particular 'high-energy bond'. But it is important to be aware that the energy status of a molecule is a property of its structure as a whole and not of any particular bond.

2.3.5 ATP is the central player in cellular energy flow

Adenosine triphosphate (see Figure 2.18C), **ATP**, serves as the universal energy currency of living cells. The hydrolysis of ATP to adenosine diphosphate (ADP) and P$_i$ releases energy that is used to drive a myriad of energetically unfavorable biochemical reactions that sustain the life of the cell. In many cases the efficient use of ATP requires Mg^{2+} binding to its phosphate groups. The two phosphoanhydride bonds yield energy when they are either removed by hydrolysis (conversion to ADP + P$_i$ and AMP + PP$_i$) or transferred to an acceptor metabolite (see Figure 2.18D). In the latter case, the energy status of the phosphorylated metabolite changes so that previously unfavorable reactions become more favorable.

For example (Equation 2.8), the first step in glucose breakdown involves phosphorylation of glucose to produce glucose-6-phosphate (see Chapter 7), an endergonic reaction. In this case a P_i is removed from ATP, an exergonic reaction, and transferred to glucose to form glucose-6-phosphate making the total reaction energetically favorable.

Equation 2.8 Energy relations of glucose phosphorylation

ATP $\rightarrow$ ADP + P_i	$\Delta G^{\circ\prime} = -31.0$
Glucose + P_i $\rightarrow$ glucose-6-phosphate	$\Delta G^{\circ\prime} = +13.8$
ATP + glucose $\rightarrow$ glucose-6-phosphate + ADP	$\Delta G^{\circ\prime} = -17.2$

Because the cleavage of the phosphoester bond of AMP releases much less energy than those of ADP and ATP (see Table 2.3), AMP is not used to drive unfavorable reactions.

2.3.6 Synthesis of ATP occurs by two distinct mechanisms

ATP is synthesized by two distinct mechanisms. First is by **substrate-level phosphorylation**, the direct transfer of phosphate from a high-energy phosphorylated donor to ADP. In this case, the donor has a more negative $\Delta G^{\circ\prime}$ than that for the ATP $\rightarrow$ ADP + P_i reaction; phosphoenol pyruvate and creatine phosphate are examples (see Table 2.3). The second mechanism is **chemiosmotic coupling**, described in 1961 by the Nobel Laureate Peter Mitchell. This is essentially the reverse of ATP hydrolysis. The reaction is driven in the highly unfavorable direction of synthesis by the energy released when protons diffuse down a steep concentration gradient across a membrane, such as the thylakoid of chloroplasts or the inner mitochondrial membrane. There is no high-energy intermediate in this mode of ATP synthesis, which instead uses electron transport and movement of protons from one side of a membrane to the other (chemiosmosis) to bring about phosphorylation of ADP (see Chapters 7 and 9).

2.4 Enzymes

Enzymes are the catalysts that make biochemical processes possible. Like all catalysts, an enzyme can participate in a chemical reaction over and over again without getting used up. Unlike chemical catalysts (such as the metal-coated ceramics in the catalytic converters of car exhaust systems) enzymes are easily **denatured** by heat, pH or heavy metal ions. This is because enzymes are proteins and therefore sensitive to inactivation as a consequence of the loss of secondary and tertiary

---

Key points Energy transfer in metabolism may occur by electron transfer between donor and acceptor molecules. The electron donor becomes oxidized and the acceptor is reduced in a so-called redox reaction. Redox potential (E) is the voltage difference between donor and acceptor and is proportional to the Gibbs free energy change ΔG. Redox reactions with positive or negative values of E are, respectively, exergonic or endergonic. A related mechanism of energy flow is by the transfer of phosphate groups. Formation of the phosphate bond is strongly endothermic, while hydrolysis releases chemical energy. ATP is the universal phosphorylated high-energy compound in cells. It is synthesized either by direct transfer of phosphate from a phosphorylated donor with a more negative ΔG than that for the ATP hydrolysis reaction, or by a chemiosmotic process involving the movement of H^+ ions across a biological membrane.

---

structure. A further essential feature of enzymes that distinguishes them from ordinary chemical catalysts is their **specificity**: generally a particular enzyme catalyzes one type of biochemical reaction only.

Aside from some specialized RNAs (ribozymes) that catalyze their own splicing, and ribosomal RNAs that participate in protein synthesis, protein enzymes are the universal facilitators of the biochemical reactions on which living cells depend. Enzymes are very efficient catalysts and are highly specific. Cells contain thousands of enzymes, each converting one or more substrates to one or more products at a high reaction rate and under mild conditions. The different types of enzyme reactions are shown in Table 2.4, which follows the classification system introduced by the Enzyme Commission (EC) of the International Union of Biochemistry and Molecular Biology. Most of the reactions catalyzed by enzymes involve the transfer of electrons, atoms or functional groups. Enzymes are, therefore, classified according to the type of transfer reaction, the group donor and the group acceptor.

2.4.1 Enzymes often require cofactors

Enzymes often require the presence of non-protein **cofactors**. A cofactor may be a metal atom, an ion or an organic group. Cofactors that are organic molecules are called **coenzymes**; many coenzymes are derivatives of vitamins. When a cofactor is bound to a protein, the complex is called a **holoenzyme**. In a holoenzyme, the

Table 2.4 Enzymes and their reactions classified according to the Enzyme Commission (EC) system. Each enzyme is identified by a four-element EC number (for example, hexokinase [EC 2.7.1.2]), which places it in a unique reaction category within one of the six major groups described in this table.

EC 1 Oxidoreductases

Typical reactions: $AH + B \rightarrow A + BH$ (reduced)

$A + O \rightarrow AO$ (oxidized)

Typical enzyme names: dehydrogenase, oxidase

Example: alcohol dehydrogenase [EC 1.1.1.1]

Ethanol $+ NAD^+ \rightarrow$ acetaldehyde $+ NADH + H^+$

EC 2 Transferases

Typical reaction: $AB + C \rightarrow A + BC$

Typical enzyme names: transaminase, kinase

Example: hexokinase [EC 2.7.1.2]

D-glucose $+ ATP \rightarrow$ D-glucose-6-P $+ ADP$

EC 3 Hydrolases

Typical reaction: $AB + H_2O \rightarrow AOH + BH$

Typical enzyme names: lipase, amylase, peptidase

Example: papain [EC 3.4.22.2]

$[aa\text{-}aa]_n + H_2O \rightarrow [aa\text{-}aa]_m + [aa\text{-}aa]_{n-m}$

where aa = amino acid

EC 4 Lyases

Typical reactions: $RCOCOOH \rightarrow RCOH + CO_2$

$[A\text{--}X\text{--}Y\text{--}B] \rightarrow [A\text{--}B] + [X\text{--}Y]$

Typical enzyme name: decarboxylase

Example: pyruvate decarboxylase [EC 4.1.1.1]

Pyruvate $\rightarrow$ acetaldehyde $+ CO_2$

EC 5 Isomerases

Typical reaction: $AB \rightarrow BA$

Typical enzyme names: isomerase, mutase

Example: phosphoglucomutase [EC 5.4.2.2]

D-glucose-1-P $\rightarrow$ D-glucose-6-P

EC 6 Ligases

Typical reaction: $X + Y + ATP \rightarrow XY + ADP + P_i$

Typical enzyme names: synthetase, ligase

Example: Glutamate–cysteine ligase [EC 6.3.2.2]

L-cysteine $+$ L-glutamic acid $+ ATP \rightarrow$
L-γ-glutamyl-L-cysteine $+ ADP + P_i$

cofactor is called a **prosthetic group** and the protein an **apoprotein**. Figures 2.19–2.21 present the structures of some of the major coenzymes involved in plant metabolism.

Some coenzymes are derivatives of nucleotides; these coenzymes function as **electron carriers** in oxidation/reduction reactions in cells. They move from one enzyme to another during the process. Structures of the oxidized and reduced forms of **nicotinamide adenine dinucleotide**, respectively **NAD$^+$** and **NADH**, are shown in Figure 2.19. On reduction, the oxidized coenzyme accepts two electrons and one proton; the second proton dissociates and is released in the medium at physiological pH. Also note that in **nicotinamide adenine dinucleotide phosphate** (**NADP$^+$**) a phosphate group is introduced on the C-2 position of the ribose attached to adenine. NAD$^+$ is typically used in reactions that require or release energy, whereas NADPH is usually associated with reactions of biosynthesis. Structures of oxidized and reduced **flavin adenine dinucleotide** (**FAD**) and **flavin mononucleotide** (**FMN**) are shown in Figure 2.20. Like NAD$^+$ and NADP$^+$, the flavin coenzymes function in oxidation/reduction reactions. Unlike nicotinamide cofactors, the flavin electron carriers generally remain bound to their apoproteins and are not mobile.

Coenzyme A (CoA; Figure 2.21A) is a strategically important metabolite that stands at the point where the pathways of sugar and lipid breakdown meet (see Chapter 7). Before a fatty acid can be metabolized in cells, it usually must be activated by linking to the terminal thiol (–SH) moiety of CoA to form an acyl-CoA. Acetyl-CoA, CoA covalently bound to a two-carbon acetate, is also the product that links glycolysis with the TCA cycle (see Chapter 7). The thiol ester bond makes acyl-CoA a high-energy metabolite $(\Delta G^{\circ\prime} -59 \text{ kJ mol}^{-1})$, amply able to drive ATP synthesis. The reduced, unesterified form of coenzyme A, which is often abbreviated as CoASH, has a complex structure built from several components (Figure 2.21A). The ADP portion of this molecule serves as a recognition site that increases the affinity of CoA binding to enzymes. Plants and microorganisms are competent to biosynthesize pantothenic acid, a precursor of CoA, but animals are not and must obtain it, in the form of vitamin B$_5$, in their diet.

Figure 2.21 shows two other coenzymes with important roles in cellular respiration. **Thiamine pyrophosphate** (Figure 2.21B) is the active form of vitamin B$_1$ and is a prosthetic group in a number of enzymes including pyruvate dehydrogenase. **Lipoic acid** (α-lipoic acid, thioctic acid; Figure 2.21C) is a cyclic disulfide with strong antioxidant properties. It is a bound cofactor for several mitochondrial enzymes.

A number of important cofactors consist of one or more metal atoms held in critical positions for reactivity

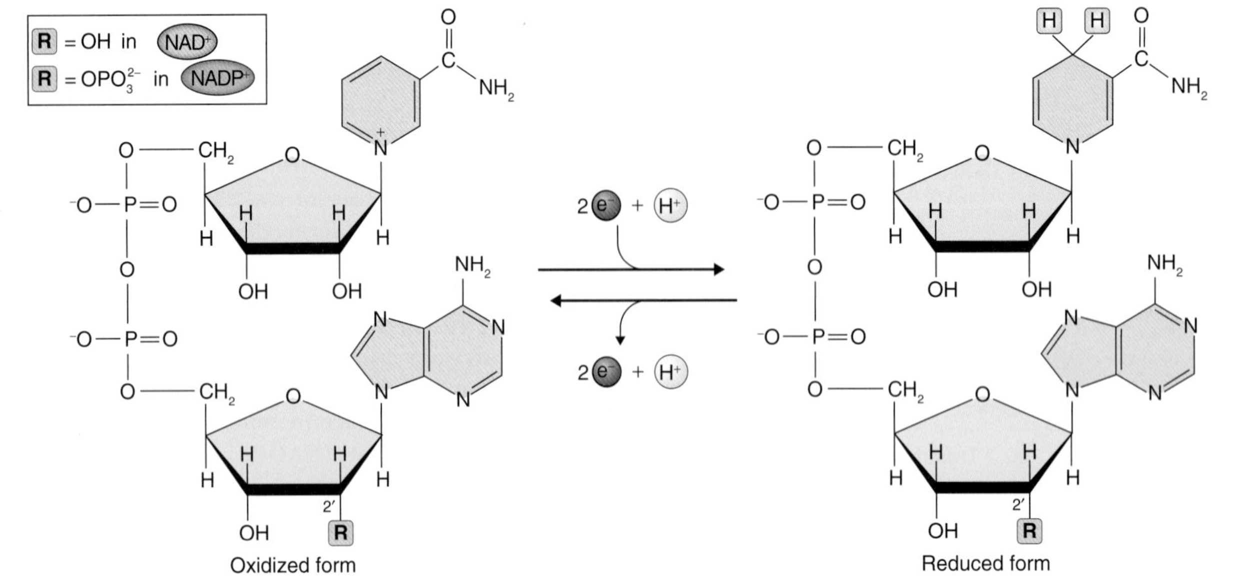

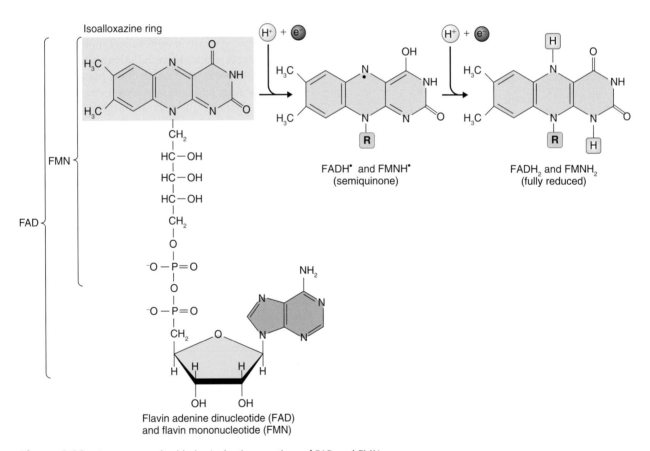

Figure 2.19 Structures and oxidation/reduction reactions of NAD and NADP.

Figure 2.20 Structures and oxidation/reduction reactions of FAD and FMN.

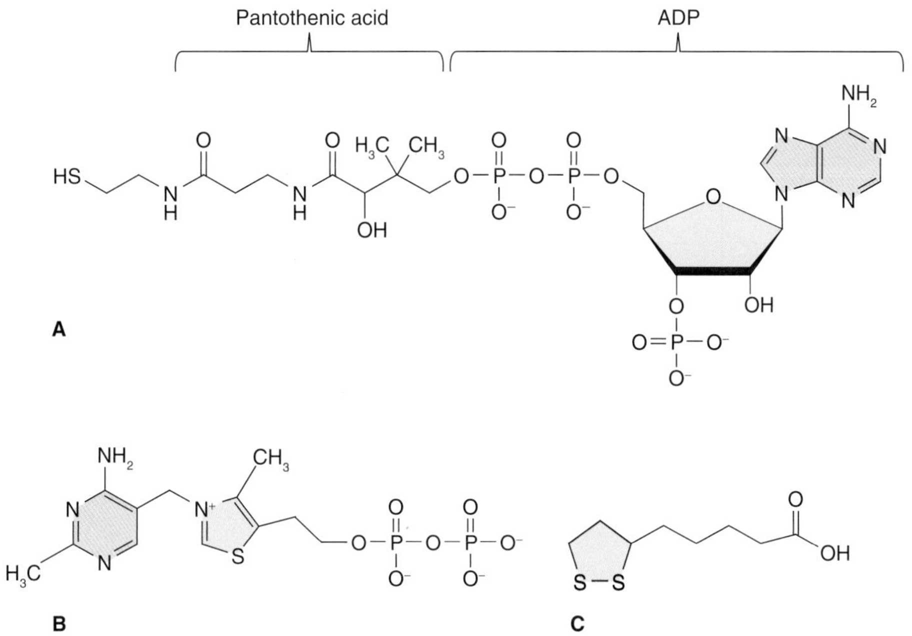

Figure 2.21 Coenzyme structures: (A) coenzyme A, (B) thiamine pyrophosphate and (C) lipoic acid.

by amino acid side-chains or coordinating prosthetic groups or both. Examples include the iron–sulfur centers of redox enzymes and electron transport components that function in respiration and photosynthesis (see Chapters 7 and 9), iron, magnesium and cobalt-containing tetrapyrroles (see Chapter 8) and the molybdenum cofactors involved in nitrogen and sulfur reduction (see Chapter 13).

> **Key points** Enzymes are highly specific catalytic proteins that bring about the transfer of electrons, atoms or functional groups during metabolism. Catalytic function often requires the presence of a non-protein cofactor such as a metal atom, an ion or an organic group. Coenzymes are organic cofactors, many of which are structurally related to vitamins, and include nucleotide derivatives that function as electron carriers in redox reactions. Important coenzymes include NAD^+, $NADP^+$, FAD, FMN and coenzyme A.

2.4.2 Catalysis greatly increases the rates of thermo-dynamically feasible reactions by reducing energy barriers

Even though a chemical reaction is energetically favorable, it may not occur under relatively mild conditions. Coal contains lots of stored energy; but that energy cannot be released without the application of a flame; then it becomes a self-sustaining net source of combustion energy. Likewise, sucrose is relatively stable at room temperature, but energy must be put in before sucrose can be broken down to glucose and fructose with the net release of energy. The energy that must be added to make these reactions occur is called the **activation energy**. The relationship between the activation energy and the free energy of a reaction can best be shown graphically. In Figure 2.22, the x coordinate shows the extent to which a reaction has taken place, and the y coordinate shows the free energy change for an idealized exergonic, energetically favorable reaction. At the maximum of the curve connecting the reactions and the products lies the **transition state** that has the needed amount of energy and the correct arrangement of atoms to form products.

Activation energy can be supplied by thermal energy such as the flame in our example of coal. Such conditions would be fatal to cells. Catalysts in general, and enzymes in particular, work by lowering the activation energy (Figure 2.22, dotted line), thereby increasing the reaction rate, in some cases by many factors of 10. Thus reactions can take place under normal cellular conditions. Although greatly accelerating the rate of reaction, a catalyst does not affect $\Delta G^{\circ\prime}$. This value is the same whether the reaction is catalyzed or uncatalyzed.

In achieving the transition state, the enzyme forms a molecular association with a substrate (**enzyme–substrate complex**) that is subsequently converted to free enzyme and product. The location on the enzyme where the substrate(s) binds is called the **active site** (Figure 2.23A). The interaction of enzyme and

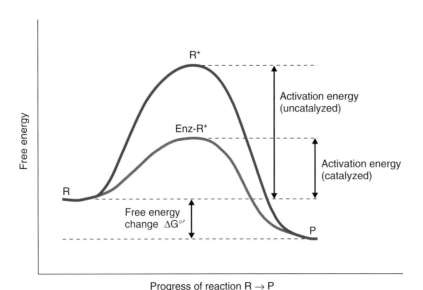

Figure 2.22 Energy transitions during enzyme catalysis of the conversion of a reactant (R) to a product (P). The enzyme increases the rate of reaction by forming a transition complex with substrate (Enz-R*), thereby reducing activation energy. The net free energy change of the reaction, $\Delta G^{\circ\prime}$, is unchanged by enzyme catalysis.

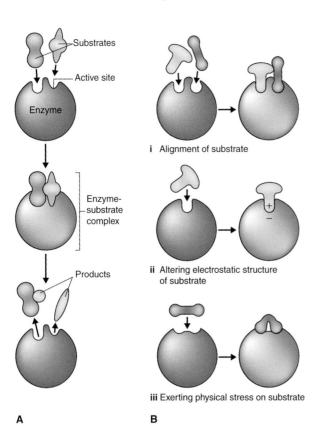

Figure 2.23 Interaction between enzyme and substrate. (A) Formation of an enzyme–substrate complex. (B) Examples of mechanisms by which an enzyme accelerates the rate of reaction.

substrate(s) brings substrate molecules close together in a precise orientation; it may also cause the bonds within the substrate to deform (Figure 2.23B). The shape of the enzyme also changes. This interaction is highly specific and allows enzymes to distinguish between different

molecules and even between stereoisomers of the same molecule. When the reaction is complete, products are released from the enzyme's active site regenerating an unchanged enzyme ready to catalyze another reaction.

2.4.3 A number of factors determine the rate of enzyme-catalyzed reactions

The rate of an enzyme-catalyzed reaction under defined conditions depends on the concentrations of the substrate(s), the product(s) and of the enzyme itself. Enzymes display different affinities for their substrates and also show differences in catalytic capability—that is, the rate at which substrate occupies and product leaves the enzyme's active site, a factor called the **turnover number**. One common way to measure substrate affinity is to plot the initial reaction rate as a function of substrate concentration and to calculate the **Michaelis–Menten constant**, K_m.

The participants in a simple enzyme reaction that obeys Michaelis–Menten kinetics consist of the enzyme E, its substrate S, and the reaction product P. Enzyme and substrate interact to form the enzyme–substrate complex, E-S. The enzyme catalyzes the change of S into P. P is released and E is recycled for another round of catalysis. Michaelis–Menten behavior assumes that E-S is formed from E and S by a reversible reaction and rapidly reaches a **steady state**, whereas product formation from E-S is irreversible. This is represented in Equation 2.9.

Equation 2.9 Enzymatic reaction obeying Michaelis–Menten kinetics

$$E + S \rightleftharpoons E - S \rightarrow E + P$$

The initial rate of the reaction (V) is determined by substrate concentration, [S]. At very low [S], availability of substrate to the enzyme is the rate-limiting factor and the reaction is said to be **first order** with respect to S. If [S] is greatly in excess of enzyme concentration, the reaction will proceed at maximal rate (V_{max}) and adding more substrate will not increase V further. Under these conditions the reaction is **zero order** with respect to S. A graph of V against [S] takes the form of the curve illustrated in Figure 2.24A. The Michaelis–Menten equation relates [S], V and V_{max} according to Equation 2.10A. As shown in Equation 2.10B, the K_m of an enzyme is the concentration of substrate that supports the reaction at one-half maximal rate ($V_{max}/2$) (Figure 2.24A). The affinity of an enzyme for its substrate is $1/K_m$. Thus a low K_m means an enzyme has a high substrate affinity and vice versa.

Equation 2.10 The Michaelis–Menten equation

2.10A

$$V = \frac{V_{max}[S]}{K_m + [S]}$$

where K_m is the Michaelis constant

2.10B
Rearranging:

$$K_m = \frac{[S](V_{max} - V)}{V}$$

$$\text{if } V = V_{max}/2$$

$$K_m = [S]$$

K_m and V_{max} may be determined by measuring the initial rate of the enzymic reaction at a range of [S] and plotting a graph like that in Figure 2.24A. In practice it is

often difficult to determine the zero-order region of such a curve with certainty. Rearrangements of the Michaelis–Menten equation provide more accurate ways of plotting V/[S] data for the determination of K_m and V_{max}. A commonly used approach is one that takes the **reciprocal** of both sides of the Michaelis–Menten equation. The double-reciprocal plot of $1/V$ against $1/[S]$ gives a straight line in which the intercept with the vertical axis gives $1/V_{max}$ and the horizontal axis intercept is equal to $-1/K_m$ (Figure 2.24B).

The K_m of an enzyme can be an important clue as to the enzyme's biological function. For example, plants contain at least two enzymes potentially able to assimilate ammonium ion, NH_4^+, into amino acids: glutamate dehydrogenase (Equation 2.11) and glutamine synthetase (Equation 2.12).

Equation 2.11 Glutamate dehydrogenase

$$NH_4^+ + \alpha\text{-ketoglutarate} + NAD(P)H$$

$$\rightleftharpoons glutamate + NAD(P)^+$$

Equation 2.12 Glutamine synthetase

$$Glutamate + NH_4^+ + ATP \rightarrow glutamine + ADP + P_i$$

When the aquatic plant *Lemna* (duckweed) is grown on a liquid medium containing nitrate or low concentrations of ammonium, the tissue concentration of NH_4^+ is in the range 0.2–1.0 mm. The K_m for *Lemna* glutamate dehydrogenase with respect to NH_4^+ is around 30 mm (a very high value, indicating low affinity), whereas the value for glutamine synthetase is no more than 0.015 mm. Tissue ammonium concentrations

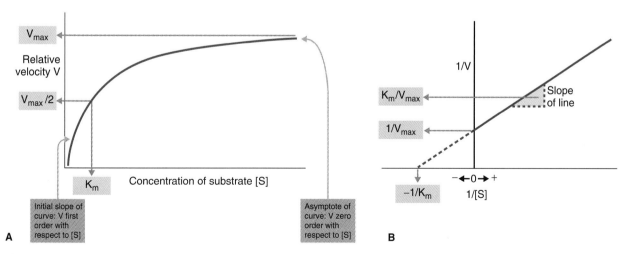

Figure 2.24   Relationship of velocity to substrate concentration for an enzyme reaction. (A) Graph of initial reaction velocity against substrate concentration for an enzyme reaction obeying Michaelis–Menten kinetics. (B) Double-reciprocal plot of $1/V$ against $1/[S]$, showing the relation between intercepts on the vertical and horizontal axes and the reciprocals of V_{max} and K_m respectively.

are less than one-thirtieth of those necessary for glutamate dehydrogenase to be operating at even a half of V_{max}, whereas the high affinity (low K_m) of glutamine synthetase for NH_4^+ means that it is working at close to maximal efficiency. This is strong evidence that ammonium assimilation in *Lemna* is via glutamine synthetase. As discussed further in Chapters 13 and 18, the function of glutamate dehydrogenase is predominantly to release NH_4^+ from glutamate (K_m with respect to glutamate for the *Lemna* enzyme is 2.5 mM) by the reverse of the assimilatory reaction.

2.4.4 Enzyme activity is under tight regulation

Organisms have methods of regulating enzyme activities at many levels. They may, for example, express the gene coding for an enzyme more or less strongly depending on the amount of enzyme needed; or they may break down and remove excess enzyme. However, one important method of controlling enzyme activity is **substrate-level regulation**. Enzymes can typically be inhibited or activated by certain compounds that interact with catalytic or regulatory sites on the protein molecule in a structure-specific manner. Substrate-level regulation, in which enzyme activity is modulated by different types of activation and inhibition, is essential to keep complex and sometimes competing reactions in balance. In this section we will first look at enzyme inhibition, and then move on to activation.

Enzyme inhibition can under some circumstances be **irreversible**. Here, an inhibitor molecule binds to the enzyme in such a way that it is difficult or impossible to remove, inactivating the enzyme in the process. In animals, many poisons work this way, as do some antibiotics, such as penicillin. Penicillin kills potentially harmful bacteria by preventing them from forming normal cell walls, and it does so by binding irreversibly to the bacterial enzyme that forms cross-links in the cell walls. Plants also use irreversible inhibitors as part of their range of defenses against predators and pathogens. An example is mimosine, an alkaloid originally isolated from the sensitive plant, *Mimosa pudica*. It is toxic to browsing animals because it irreversibly inhibits enzymes of DNA synthesis in susceptible cells.

However, irreversible inhibition is not generally a method by which organisms regulate the activity of their own enzymes, since control can be exerted more sensitively and less wastefully by **reversible enzyme inhibition**. Reversible inhibition falls into four main classes: **competitive**, **non-competitive**, **mixed** and **uncompetitive**.

A compound that is similar in structure to the enzyme's normal substrate or substrates may be able to bind to the active site and prevent the substrate from doing so—it is then acting as a **competitive inhibitor**. An example of such an inhibitor occupying the active site of a plant enzyme is the tripeptide leupeptin, which is made by some fungi. This compound binds in the catalytic cleft of the protease papain and related hydrolases (Figure 2.25). Because a substrate and a competitive inhibitor contend for the same site, if sufficient substrate is present, it displaces the inhibitor and overcomes the inhibition. Competitive inhibitors do not alter the maximum rate of catalysis (V_{max}) but, because they interfere with substrate binding, they do increase the K_m (that is, in the presence of the inhibitor the affinity of the enzyme for its substrate is reduced) (Figure 2.26A).

The other classes of inhibitor bind to the enzyme at a point other than the substrate-binding area in the active site. **Non-competitive inhibitors** reduce the maximum rate of the chemical reaction (V_{max}) without changing K_m, the binding affinity of the enzyme for its substrate (Figure 2.26B). Non-competitive inhibition cannot be overcome by increasing substrate concentration.

Uncompetitive inhibitors, which are rare, bind only to the enzyme–substrate complex, not to the free enzyme. Like mixed inhibitors, they reduce both V_{max} and substrate binding (hence they increase K_m); unlike mixed inhibitors, they are unaffected by increased substrate concentration (Figure 2.26C).

In **mixed inhibition**, the inhibitor binds to the enzyme at the same time as the substrate; although binding is at a site different from the substrate-binding site, the inhibitor still reduces the affinity of the enzyme for its substrate, so that K_m is increased and V_{max} is reduced. Increasing substrate concentration can partly, but not completely, overcome mixed inhibition.

The four types of reversible inhibition described above classify inhibitors according to how they affect the maximum catalytic rate (V_{max}) and the substrate-binding affinity (K_m) of the enzyme. They say little about how this inhibition is brought about. However, a common mode of regulation—activation or inhibition—by metabolites is the mechanism referred to as **allosteric or noncovalent regulation**. The metabolites involved are called **allosteric modifiers or effectors**. Allosteric regulation normally affects enzymes with more than one subunit, and causes a change in the conformation of the enzyme molecule so that the ability of the active site to bind substrate is increased or reduced. Most mixed inhibitors fall into this category. Figure 2.27 indicates how the conformation of an oligomeric enzyme may be switched between forms with

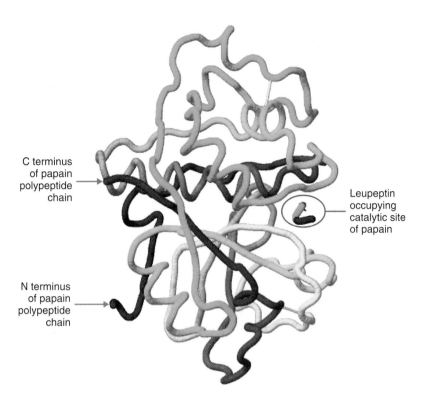

Figure 2.25 Molecular structure, derived by X-ray crystallographic analysis, of the complex between the plant protease papain and the competitive inhibitor leupeptin. The polypeptide chain is shown in N → C rainbow format, where the amino (N) terminus of the protein is blue, the carboxy (C) terminus is red and the polypeptide between them follows a spectral rainbow sequence. Note that leupeptin is a tripeptide and so also appears as a rainbow structure.

C terminus of papain polypeptide chain

Leupeptin occupying catalytic site of papain

N terminus of papain polypeptide chain

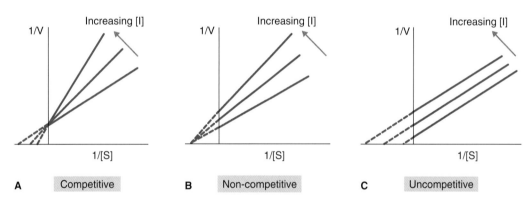

Figure 2.26 Double-reciprocal plots (see Figure 2.24B) illustrating the effects of inhibitors on enzyme kinetics. (A) Competitive inhibitor changes K_m but has no effect on V_{max}. (B) Non-competitive inhibitor decreases V_{max} but K_m remains the same. (C) Uncompetitive inhibitor changes both K_m and V_{max}. Arrows indicate increasing concentration of inhibitor [I].

differing affinities for substrates and products in response to the presence of allosteric effectors. Where allosteric binding causes inhibition of enzyme activity, it is most often non-competitive inhibition.

A common biochemical mechanism for fine-tuning the rates of enzyme-catalyzed pathways is to use an intermediate or final compound in the pathway as an activator or inhibitor of a key step. Terms used to describe this mode of control include **feedback, feed-forward** or **end-product regulation**. Galactose dehydrogenase is one of many examples from plants; it catalyzes a reaction in the pathway of ascorbic acid (vitamin C) biosynthesis (Equation 2.13).

Equation 2.13 Galactose dehydrogenase

L-galactose + NAD$^+$

$\rightleftharpoons$ L-galactono-1,4-lactone + NADH + H$^+$

Ascorbic acid is a classic competitive inhibitor of galactose dehydrogenase, as the double-reciprocal plot of Figure 2.28 demonstrates. The sensitivity of this enzyme to end-product inhibition prevents ascorbic acid from building up to excessive levels.

Enzymes can also be regulated by covalent modification. In a particularly important type of covalent

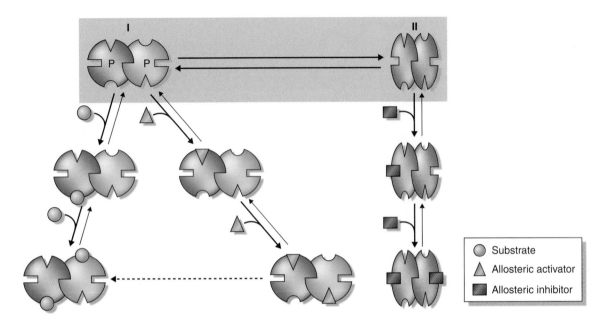

Figure 2.27 Allosteric regulation of the conformation-dependent activity of an enzyme consisting of two identical subunits P, each with substrate-, activator- and inhibitor-binding sites. The enzyme can adopt two conformations, I (catalytically active) and II (inactive). Binding substrate or allosteric activator favors conformation I and the cooperative binding of further substrate or activator molecules. Occupation of the inhibitor-binding site favors state II, in which activator and substrate sites are non-binding.

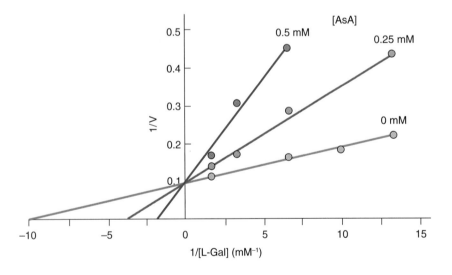

Figure 2.28 Feedback inhibition of galactose dehydrogenase by ascorbic acid. The double-reciprocal plot of L-galactose concentration $(1/[L\text{-}Gal])$ against initial reaction velocity $(1/V)$, measured at different concentrations of ascorbic acid $([AsA])$, shows competitive inhibition.

regulation, serine or tyrosine residues in the protein's polypeptide chain are phosphorylated by reaction with ATP that is catalyzed by specific **protein kinases**. Fine control of protein function can be exerted through the balance between kinase-mediated phosphorylation and **phosphatase**-catalyzed dephosphorylation. Another example of covalent modification that often plays a role in metabolic regulation is reduction, for example conversion of the bridge linking two cysteine residues from S–S to –SH HS–, which changes the protein's higher-order conformation.

By responding to these different types of regulation, enzymes located at critical points control the rates of pathways. In certain cases, the enzymes of opposing pathways respond in an opposite manner to the same type of regulation. When undergoing catalysis or regulation, the conformation of an enzyme usually changes dramatically, thereby affecting the catalytic rate.

Key points Catalysts, including enzymes, increase the rates of chemical reactions by reducing activation energies, but have no effect on ΔG. Catalysis requires formation of a transition-state structure which in biochemical reactions is the enzyme–substrate complex, consisting of an active site on the protein molecule occupied by the reactants. Conformational changes in enzyme and substrate facilitate the formation of reaction products, releasing the enzyme for a further cycle of catalysis. Many enzymes obey Michaelis–Menten kinetics, in which the parameters K_m and V_{max} define the affinity for substrate and maximal catalytic rate, respectively. Enzyme inhibitors and allosteric effectors are chemicals that interact with the enzyme or enzyme–substrate complex and alter K_m or V_{max} or both. Enzyme activity may also be altered by covalent modifications such as phosphorylation.

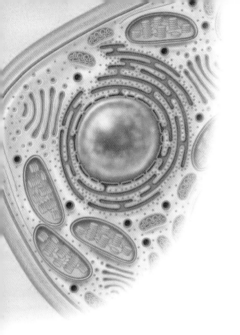

Chapter 3

Genome organization and expression

3.1 Introduction to genes and genomes

Like all living organisms except certain viruses, plants contain the information required for their growth and functioning in the form of DNA. This molecule contains the genes that direct the synthesis of proteins via RNA intermediates, as well as intervening regions of non-gene DNA; the total DNA content of a cell is known as its **genome**. The term is perhaps best known in the context of the Human Genome Project, an ambitious undertaking that resulted in the publication of the essentially complete genome sequence of a human being in the year 2000. Developments in technology for determining DNA sequences mean that it is becoming progressively easier and cheaper to analyze the genomes of other organisms.

A plant's genome is more-or-less constant throughout its life span and in all its tissues. The growth, shape and behavior of the plant are determined by which parts of the DNA are acted upon at any one time; that is, which genes are being expressed. This chapter begins with an overview of the plant genome (or rather, the three plant genomes, those of the mitochondrion, the plastid and the nucleus). The structure and features of plant chromosomes are described, followed by a section on the features of typical plant genes. The remainder of the chapter addresses how plants regulate the expression of their genes, and the

processes of transcription and translation that result in the synthesis of proteins.

3.2 Organization of plant genomes I. Plastid, mitochondrial and nuclear genomes

The genome of a eukaryotic organism consists of the nuclear genome, where the majority of the genes are found, and the genomes of organelles. Mitochondria have their own genome which is found in all eukaryotic organisms including plants. However, photosynthetic eukaryotes like plants are unique in that they possess three genomes in total. In addition to the nuclear and the mitochondrial genome, they also have a plastid genome. As discussed in Chapter 4, plastids and mitochondria are derived from **prokaryotic endosymbionts**.

3.2.1 Plastid genomes do not contain all the genes required for plastid function

Genome sizes are described in terms of the number of pairs of DNA bases they contain. The most commonly

The Molecular Life of Plants, First Edition. Russell Jones, Helen Ougham, Howard Thomas and Susan Waaland.
© 2013 John Wiley & Sons, Ltd. Published 2013 by John Wiley & Sons, Ltd.

Table 3.1 Terms used to describe the length of DNA.

bp	one pair of nucleotide bases
kb	1 kilobase (kilobase pairs) = 1000 bp
MB	1 megabase (megabase pairs) = 1 000 000 bp

used unit is the kilobase (kb), meaning 1000 base pairs (bp); other terms are listed in Table 3.1. The size of the plastid genome varies between plants, typically containing 120–160 kb; the largest known plastid genome, at approximately 400 kb, is that of the giant unicellular green alga *Acetabularia*. In illustrations the plastid genome is usually represented as a circle, but in living plastids it can adopt several different conformations including linear and branched molecules as well as circles. The organization of the plastid genome also varies between species, but it is most commonly made up of four sections (Figure 3.1). There are two regions, one large and one small, of **single-copy genes** (**LSC** and **SSC** regions, respectively). These regions are separated by two copies of an **inverted repeat** (**IR$_A$** and **IR$_B$**), though these are absent in plastid genomes of some conifers, algae and legume species. It is believed that IR regions were present in the common ancestor of higher plant plastids but have been lost in certain groups during the course of evolution. Where IR regions are present, they account for most of the variation in plastid genome size, since they can range from just 0.5 kb (500 bp) up to 76 kb.

All of the non-reproductive cells in a single organism have the same set of nuclear genes but express those genes in different combinations according to cell position, environment and other factors. Likewise all plastids in a single plant contain identical DNA but may differ in the way it is expressed according to developmental stage and metabolic activity. Now that plastid genome sequences are available for almost 300 plant and algal species, it is possible to make some generalizations about the genes that they contain. Most plastid genomes contain all the genes encoding plastid ribosomal RNAs and transfer RNAs, which are different from those encoded by the nuclear genome. They also include about 100 single-copy protein-coding genes, most of which are known to encode proteins required for photosynthetic functions (Table 3.2). There are still a few sections of the plastid genome which are predicted to encode proteins, but whose proposed products have as yet unknown functions. Plastid genomes in algae are generally larger than those of land plants and contain additional protein-coding genes not found in plant plastids. For example, the plastid genome of the red alga *Porphyra purpurea* encodes 70 proteins that in land plants are encoded by the nuclear genome. In contrast,

the plastid genomes of non-photosynthetic plants, such as the parasitic plants *Cuscuta* and *Epifagus*, are small (50–73 kb), having lost many of the genes that they no longer need for photosynthesis.

Plastids contain by no means all the genes that they require for their own functions. Many genes that were present in the ancestral organelle soon after symbiotic assimilation were probably transferred to the nucleus some time in the course of evolution. A good example is the enzyme that is required for CO_2 fixation, **ribulose-1,5-bisphosphate carboxylase/oxygenase** (**rubisco**; see Chapter 9). This enzyme, the most abundant protein in photosynthetic organisms, is a multiprotein complex with two types of subunits. The gene for the larger subunit (**rbcL**) is present in the plastid genome, but the gene for the smaller subunit (**RBCS**) is found in the nucleus. Many of the protein complexes involved in photosynthesis (see Chapter 9) are similarly encoded by a mixture of nuclear and plastid genes. However, there are also important biochemical pathways in the chloroplast and other types of plastids (see Figure 4.23) for which all of the enzymes are encoded in the nuclear genome, synthesized on cytosolic ribosomes and transported into the plastids.

3.2.2 Plant mitochondrial genomes vary greatly in size between different plant species

Plant **mitochondrial genomes** show enormous variation in size, from about 200 kb in *Oenothera* (evening primrose and its relatives) and *Brassica* species to 2600 kb in muskmelon (*Cucumis melo*). This is in contrast to animal mitochondrial genomes, which, for those sequenced to date, are only about 16 kb. Most of the difference between plant and animal mitochondrial genomes can be accounted for by variations in the number of non-coding sequences in the regions between genes (Figure 3.2). In the model plant *Arabidopsis thaliana*, for example, coding regions make up less than 10% of the 367 kb mitochondrial DNA. In contrast to much of the non-coding DNA found between genes in plant nuclear genomes, plant mitochondrial intergenic regions are not made up of repetitive DNA sequences.

Unlike the plastid genome, it does not appear that the plant mitochondrial genome always exists as a single DNA molecule. Instead, as illustrated in Figure 3.3 for *Zea mays* (maize), mitochondrial genomes sometimes exist as circular DNA molecules of variable size, known as **subgenomic circles**. The combined DNA sequence content of the subgenomic circles can account for the entire mitochondrial genome. The largest possible

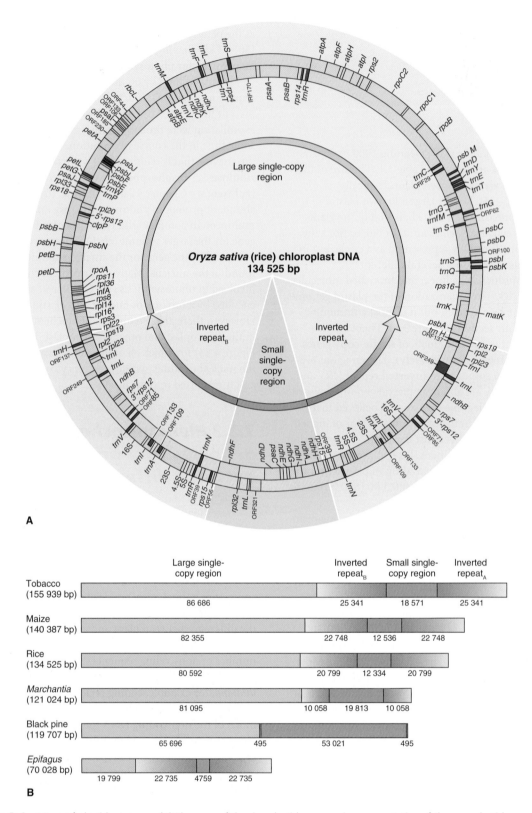

Figure 3.1 Maps of plastid genomes. (A) The map of the rice plastid genome is representative of the way plastid genes are organized in most flowering plants. (B) Schematic representations of several plant plastid genomes, showing conserved features: two regions of single-copy genes and two inverted repeats. The single-copy regions in the plastid genome of the non-photosynthetic parasitic plant *Epifagus* are quite small because most of the genes encoding photosynthetic proteins have been eliminated. The number beneath each DNA segment refers to its length in base pairs.

Table 3.2 Genes identified in complete plastid genome sequences.

Gene products	Gene acronym	Land plants		Algae	
		Photosynthetic plants	*Epifagus*[a]	*Euglena*	*Porphyra*[b]
Number of genes		101–150	40	82	182
rRNA	*rrn*	4	4	3	3
tRNA	*trn*	30–32	17	27	35
Ribosomal protein	*rps, rpl*	20–21	15	21	46
Rubisco and complexes of the thylakoid membrane system	e.g. *rbcL, psa, psb, pet, atp*	29–30	0	26	40
NADH dehydrogenase[c]	*ndh*	11	0	0	0
Biosynthesis, gene expression and miscellaneous functions		6–11	4	5	58
Number of introns		18–21	6	155	0

[a] *Epifagus* (beechdrops) is a non-photosynthetic, parasitic flowering plant.
[b] *Porphyra* is a red alga.
[c] The plastid genome of black pine does not contain genes for NADH dehydrogenase.

Table 3.3 Types of genes identified in the *Zea mays* mitochondrial genome.

Gene products	Gene abbreviations	Function
rRNAs	*rrn18, rrn26, rrn5*	Protein synthesis
tRNAs	*trn*	Protein synthesis
Ribosomal proteins	*rps, rpl*	Protein synthesis
NADH dehydrogenase	*nad*	Respiratory electron transport
Cytochrome c oxidase	*cox*	Respiratory electron transport
Apocytochrome	*cob*	Respiratory electron transport
F_0F_1-ATPase proteins	*atp*	ATP synthesis

Z. mays mitochondrial circular DNA molecule, called the **master circle**, which would, in theory, encode the complete set of mitochondrial genes, has never been isolated. The formation of subgenomic DNA circles has been observed in living plant cells, but it is not clear whether some or all subgenomic DNA circles replicate independently or whether they can be generated only from the hypothetical master circle. Not all plants form subgenomic DNA circles; for example, the liverwort *Marchantia* and white mustard (*Brassica hirta*) have homogeneous circular mitochondrial genomes, and the alga *Chlamydomonas* has a linear mitochondrial genome. The reasons for the diversity of organization in organellar genomes are not yet well understood.

Complete DNA sequences are now available for many plant and algal mitochondrial genomes, including those of sugar beet (*Beta vulgaris*), *Arabidopsis*, *Marchantia*, the green alga *Prototheca*, and the red alga *Chondrus*. Although plant mitochondrial DNAs are very variable in size, they all contain essentially the same genetic information. They do not contain many genes; most of the enzymes required for mitochondrial DNA replication and transcription are encoded by the nucleus. The products of the genes that are encoded in the mitochondrial genome are mainly ribosomal and transfer RNAs required for protein synthesis, or enzymes with roles in oxidative respiration (see Chapter 7) and ATP synthesis (Table 3.3 summarizes these products for *Z. mays*).

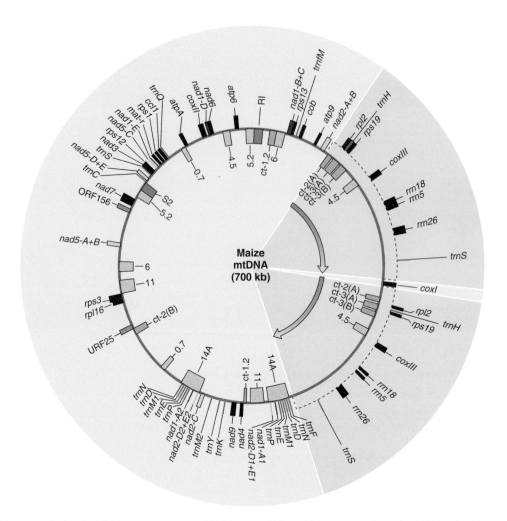

Figure 3.2 Maps of mitochondrial genomes are usually drawn as circles which represent hypothetical circular DNA molecules containing all of the mitochondrial genes. The mitochondrial DNA in *Zea mays*, shown here, is considerably larger than the chloroplast DNA, but it contains fewer genes. There are several inverted and direct repeat DNA sequences (shown as blue, green and magenta boxes on the inner circle) which participate in recombination events, producing small, subgenomic circular DNA molecules, as shown in Figure 3.3.

3.2.3 Some plant nuclear genomes are much larger than the human genome, others are much smaller

Like the **nuclear genomes** of other eukaryotic organisms, those of plants cover a wide range of sizes. Size may be expressed either as the number of nucleotide base pairs per genome or by the weight (in picograms, pg) of the DNA in one nucleus. The number of base pairs in a genome is determined by sequencing the genome. The model organism *Arabidopsis thaliana* has one of the smallest known plant genomes. *Arabidopsis* is a diploid species, with five pairs of chromosomes. Its haploid DNA content or **C value**—that is, the amount of DNA in one of the sets of five chromosomes—is 1.35×10^8 bp. In

contrast, the genome of the lily *Fritillaria assyriaca* is one of the largest, with 1×10^{11} bp. The genomes of most major crop species are intermediate in size between these two extremes; rice (*Oryza sativa*), *Z. mays*, and wheat (*Triticum aestivum*), for example, have genome sizes of 5×10^8, 6.6×10^9 and 1.6×10^{10} bp, respectively. For comparison, the human genome lies in the middle of this range with 3×10^9 bp (Figure 3.4).

Where sequence data are not available, plant genome size can still be determined quite accurately by flow cytometry, a technique for measuring the nuclear DNA content of individual cells in picograms. The number of base pairs can then be estimated from the weight in picograms, since 1 pg corresponds to 978×10^6 bp, meaning that *Arabidopsis* has a haploid DNA content of 0.138 pg, whereas *F. assyriaca* has a haploid DNA content of 102.25 pg. The Plant DNA C Value Database maintained by the Royal Botanical Gardens in Kew, UK

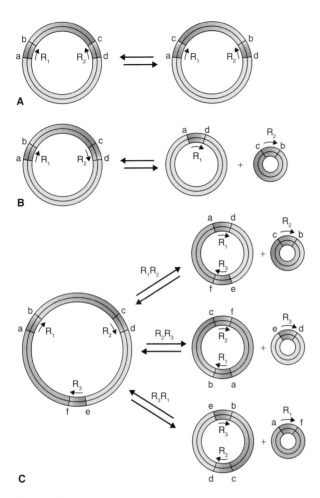

A

B

C

Figure 3.3 Formation of subgenomic circles in plant mitochondrial DNA. Plant mitochondrial DNA contains regions of repeats, which can be either direct or inverted. The presence of these repeats makes it possible for the mitochondrial DNA to recombine into a variety of different circular conformations and generate subgenomic circles containing only part of the total mitochondrial genome. Three examples are shown here, based on three hypothetical 'master circle' DNA structures. (A) Recombination between one pair of inverted repeats produces two forms of the master circle DNA. (B) Recombination between one pair of direct repeats yields two subgenomic circles. (C) When a master circle contains three copies of a direct repeat, there are three different possible recombination events, each of which produces a different pair of subgenomic circles.

(http://data.kew.org/cvalues) is a good source of information, mostly obtained by flow cytometry, about genome sizes in plants for which one knows the Latin name (binomial). It also lists the haploid chromosome number for many species.

Why do plant genomes vary so widely in size? The number of genes encoding proteins is not greatly different from one plant to another; estimates range from about 30 000 up to around 60 000, depending on species; the human genome is considered to contain

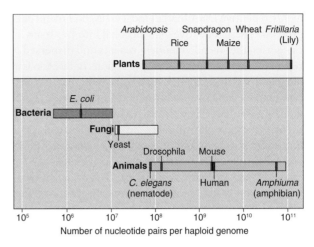

Figure 3.4 C values (numbers of DNA base pairs in the haploid genome) from different organisms. Most eukaryotes have C values between 10^7 and 10^{11} base pairs.

between 20 000 and 25 000 genes. With plant sequence data available, we now know about many of the features of different classes of DNA that contribute to this huge size variation, although its significance for the growth and performance of the plant itself is still poorly understood. At least some of the size variation can be attributed to one or more rounds of genome duplication followed by gene rearrangements which have occurred at some point in the evolution of the species, and to the expansion of families of repetitive elements (see Section 3.2.4), as well as to polyploidy (see Section 3.3.7).

> **Key points** Plants have three genomes: plastid, mitochondrial and nuclear. The nuclear genome contains the majority of the genes necessary for plant development and function. The plastid and mitochondrial genomes each have a subset of the genes required for their own functions. Many of the genes that encode chloroplast and mitochondrial proteins are found in the nuclear genome. Plant nuclear genomes vary greatly in size; some are much larger than the human genome, while others are much smaller. The larger plant genomes are dominated by repetitive DNA regions.

3.2.4 Repetitive DNA makes up much of the genome in many plants

One reason for the large variation in the sizes of plant nuclear genomes is that some genomes have regions of highly repetitive DNA that, in most cases, do not encode proteins. These repetitive sequences fall into two main categories, the so-called **tandem repeats** and **dispersed repeats**. Tandem repeats are short regions of DNA, a few nucleotides or tens of nucleotides in length, that are repeated over and over to form blocks of the same sequence element. They are often associated with particular structural features of chromosomes, such as centromeres or telomeres. In many organisms, including yeast (*Saccharomyces cerevisiae*) and most animals, tandem repeats are commonly richer in the nucleotides A and T than in the genome as a whole. In plants, on the other hand, tandem repeats tend to be GC-rich. **Dispersed repeats**, found scattered throughout the genome, are often, but not always, derived from transposable elements (see below) that have propagated themselves throughout the chromosomes before becoming inactive. Large-genome species such as *Z. mays* contain many different families of such sequences. Figure 3.5 shows the region of the *Z. mays* genome that surrounds the alcohol dehydrogenase gene *Adh1*. All the repetitive sequences in this figure are inactive versions of former transposable elements; some are related to each other. In every case there are many more copies of the element elsewhere in the genome.

Transposable elements (**TEs**) are sections of DNA that move, or **transpose**, from one site in the genome to another. These mobile DNA elements carry genetic information with them as they transpose, making them important features of genome structure. They were first discovered in *Z. mays* by Nobel Laureate Barbara McClintock in the 1940s. Since then, mobile elements have been identified in most higher organisms, including yeast, insects and mammals as well as plants. When a TE inserts itself into the coding region of a gene or its regulatory elements, it disrupts gene function. The best characterized TEs in plants have been those of *Z. mays*

and *Antirrhinum*, where pigmented kernels and flowers, respectively, have made it possible to trace the activity of transposable elements as they 'jump' in and out of genes encoding enzymes of pigment biosynthesis. This 'jumping' takes place continually so that even adjacent cells in the same organ can have the same gene with and without a TE insertion (Figure 3.6).

Transposable elements are classified into two main types, according to the mechanism by which they transpose (Figure 3.7). **Class I elements**, known as **retrotransposons**, synthesize an RNA intermediate as part of the transposition process. The RNA is then transcribed into DNA by a **reverse transcriptase** (RNA → DNA) and the DNA is inserted into another location in the genome. This process is similar to the way in which a retrovirus replicates itself throughout the genome of its plant or animal host. Therefore, these elements are believed to have originated as viruses. Retrotransposons, and degenerate, inactive (non-motile) elements derived from them, can make up a large proportion of the total genome in a large-genome plant species. In some cereals, and in many species of iris (*Iris* spp.) and lily (*Lilium* spp.), 50–90% of the genome can consist of retrotransposon-derived sequences. Sometimes these sequences are organized in a complex manner making it possible to infer the sequence of events that gave rise to the present-day genome. Figure 3.8 illustrates a group of nested transposable elements in the *Z. mays* genome. It was possible to determine the relative ages of different elements by sequencing the whole region and comparing sequence similarities between different members of this array of insertions.

Class II elements, often called **transposons**, move via a DNA rather than an RNA intermediate (see Figure 3.7). An element is cut out of one site; it is then inserted into another site. Each active Class II TE encodes one or two gene products that are needed for transposition. They

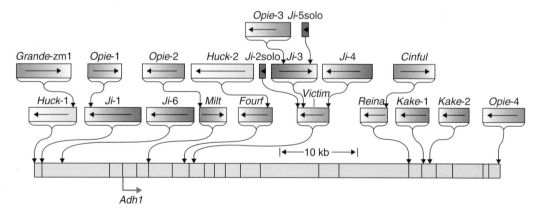

Figure 3.5 The region of the *Zea mays* genome around the *Adh1* (alcohol dehydrogenase 1) gene, showing the large number of repetitive sequences surrounding the gene. These sequences are mostly retrotransposons, and have been color-coded according to the family of related sequences to which they belong (e.g. yellow for the *Huck* family).

Most plant genomes contain only a small number of active TEs that have the potential to move around. Over time, a much large number of TEs have undergone mutations, including DNA deletions, which have rendered them inactive. These inactive elements contribute to the size and organization of the plant genome, but are not known to have other effects.

The transposition of active elements can cause mutations if they insert into the protein-coding or regulatory regions of genes, thereby inactivating those genes or causing them to be incorrectly regulated. An important difference between Class I and Class II elements is that the original Class I element remains in position, whereas a Class II element is removed from one place and transposed to another. Thus while Class I elements cause stable mutations, Class II elements generally produce unstable mutations which can be partially or fully reversed when the transposable element is excised. However, since a portion of duplicated sequence remains after excision, its position will determine whether the mutation is reversible. If it inserts in a regulatory or a coding region portion of the gene, the gene may remain non-functional. Many of the variegated patterns in flowers and in *Z. mays* kernels (see Figure 3.6B) result from excision events involving Class II transposable elements.

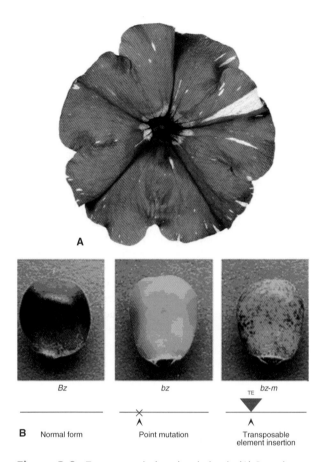

A

Bz   *bz*   TE *bz-m*

B   Normal form   Point mutation   Transposable element insertion

Figure 3.6   Transposon-induced variation in (A) *Petunia* flower and (B) *Zea mays* kernel pigmentation. (A) A transposon insertion into the *difF* gene of this *Petunia* plant results in the production of reddish flowers. The insertion blocks the expression of cytochrome b$_5$ which in turn causes reduction in the activity of an enzyme that normally modifies precursors of the pink and red pigments (anthocyanins), rendering them colorless. Where the transposon spontaneously excises from the gene, cytochrome b$_5$ expression is restored, and pale, unpigmented spots and sectors are formed. (B) In these kernels, a point mutation has inactivated the *Bronze1* (*Bz1*) gene, which encodes an enzyme, UDP-glucose:flavonoid-3-*O*-glucosyl transferase, required for pigment biosynthesis; this inactivation leads to kernels which are orange (*bz*) rather than dark (*Bz*). If a transposon inserts into the gene close to the point mutation (*bz-m*), and then jumps out again in some cells of the kernel, the normal function of the gene may be restored in these cells. This results in kernels with dark spots on the orange background.

also have inverted repeats, about 10 bp long, flanking their coding sequence. These repeats are recognized by **transposase**, an enzyme that binds to them and integrates the transposon into its target site in the genome. When a Class II element is excised from one position, a portion of the element remains in the original location.

3.2.5 Related plant species show conserved organization of gene content and order

Related species of plants often have similar gene content and order across large segments of their chromosomes. This phenomenon is known as **synteny**. The first, and still the most striking, example of synteny is found in cereal and other grass species—wheat, rye (*Secale cereale*), rice, forage grasses and others all show a remarkable conservation of gene order (Figure 3.9). As more genomes are sequenced, it is becoming clear that similar relationships occur in other plant groups, including the family Solanaceae (which includes tomato (*Solanum lycopersicum*), potato (*S. tuberosum*), bell pepper (*Capsicum annuum*), eggplant (*S. melongena*) and so on) and the legumes soybean (*Glycine max*), common bean (*Phaseolus vulgaris*), alfalfa (*Medicago sativa*) and clover (*Trifolium* spp.), etc. The lengths of genome over which synteny extends decrease as progressively more distantly-related species are compared, but even between species that diverged 50 million years ago it is still possible to detect short blocks of synteny (**microsynteny**).

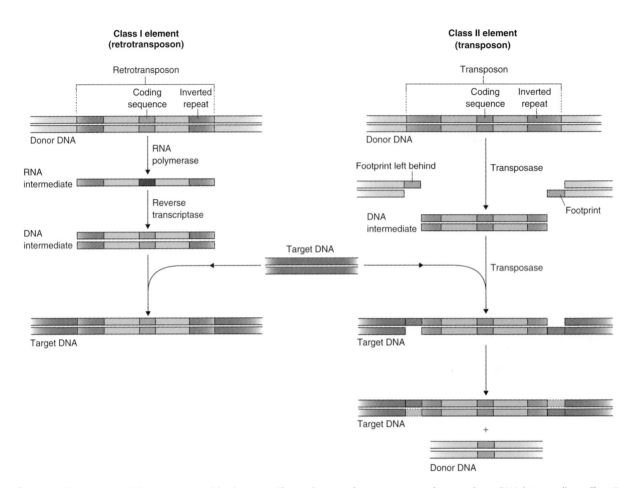

Figure 3.7 Class I and Class II transposable elements. Class I elements (retrotransposons) move via an RNA intermediate. Class II elements transpose via a DNA intermediate, and leave a small insertion at the point in the genome where they have excised.

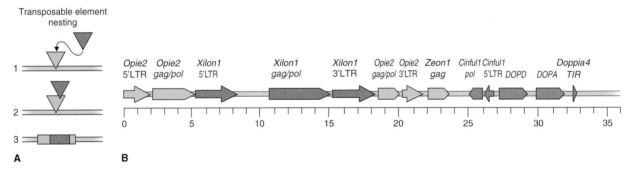

Figure 3.8 Nested transposable elements. (A) One transposable element can insert within another, leading to 'nesting' of elements. (B) A region from the *Zea mays* genome in lines McC and B73, showing a group of *Xilon* family repeats nested within a group of *Opie* repeats, adjacent to repeats from other families. The ruler shows distance (kb) from the start of the region.

3.3 Organization of plant genomes II. Chromosomes and chromatin

Since the early 20th century, chromosomes have been known to carry the hereditary material in eukaryotes; by the middle of the century, the hereditary material had been identified as DNA. It is now known that each chromosome contains just one double-stranded DNA molecule that may be many centimeters, in some cases even meters, in length. This molecule and its associated proteins can be packaged so compactly that chromosomes are only microns in length during mitosis (Figure 3.10). Highly condensed chromosomes, visible

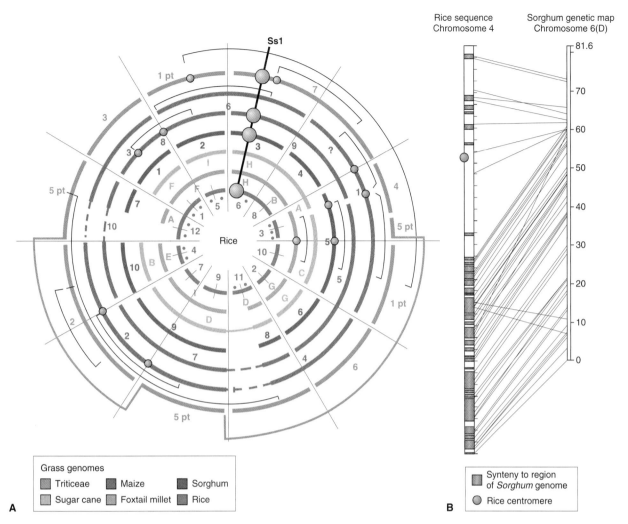

Figure 3.9 Synteny in grass family genomes. (A) The whole genomes of cereals and other grasses can be aligned in a circular form. Starting at a given gene (in this case *Ss1*) in rice (*Oryza sativa*), and moving outwards, it is possible to predict the location of this gene and the genes flanking it in the other species. The circle representing each genome is labeled with colored numbers or letters corresponding to the chromosome set of that species; the convention for identifying chromosomes is not consistent across all organisms. (B) Similarities in DNA sequences between chromosome 4 of rice and chromosome 6 of *Sorghum*. The vertical lines represent the chromosomes. Lines drawn between the two chromosomes indicate a high level of similarity in DNA sequence. In most cases the order of DNA sequences, and therefore of genes, is conserved, but where the lines cross, genes are in a different place in sorghum compared with rice.

during nuclear division, have characteristic morphological features including a **centromere** where sister chromatids are joined, **arms** that extend out from the centromere in both directions and a **telomere** at the end of each arm (Figure 3.11). While these features have been observed for years, we can now characterize them in terms of their DNA sequence.

3.3.1 Chromosome arms are gene-rich

Chromosome arms stretch from the telomeres to close to the regions near the centromeres (the pericentromeric

regions) and are where the majority of plant genes are found. In a species such as *Arabidopsis thaliana*, which has a small genome, chromosome arms consist of genes and little else. There is an average of one gene every 4.5 kb in the *Arabidopsis* genome, with each pair of genes being separated by, on average, 2 kb of DNA that does not encode a gene product but usually contains regulatory regions. Most plant species have genomes larger than that of *Arabidopsis*. Since these large genomes do not have many more genes, it follows that the average gene density is lower for most plants. In *Z. mays*, for example, about 80% of the genome consists of repetitive DNA, most of which encodes retrotransposons. The *Z. mays* genome can be thought of as consisting of gene

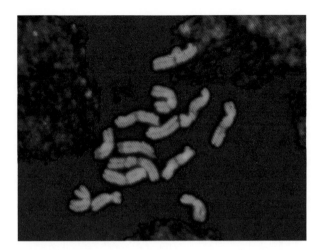

Figure 3.10   The chromosome set of the diploid grass species *Lolium perenne* (2n = 2x = 14) during mitosis, visualized using a fluorescent probe.

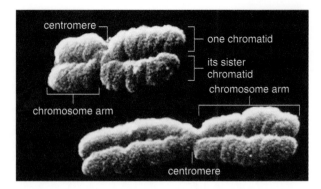

Figure 3.11   Scanning electron micrograph of replicated chromosomes showing the position of centromeres (visible as constrictions in chromosomes), chromosome arms and chromatids. Each chromatid terminates in a telomere.

> **Key points**   Repetitive regions in the genome may be tandem repeats—regions of DNA, a few nucleotides or tens of nucleotides in length—that are repeated over and over to form blocks of the same sequence element. Alternatively, they may be dispersed repeats, often derived from transposable elements, that can move around the genome. When a transposable element inserts itself into a gene, the function of that gene is disrupted. Transposable elements are classified into two types. Class I elements, also known as retrotransposons, synthesize an RNA intermediate, which is then transcribed into DNA by a reverse transcriptase and the DNA is inserted into another place in the genome. This normally causes a stable mutation. Class II elements, often called transposons, do not synthesize an RNA intermediate. Instead, a Class II element is cut out of one site, leaving a portion of itself behind, and inserted into another site in the genome by an enzyme called transposase. The portion left behind may continue to disrupt gene function, but in some cases a mutation caused by Class II element insertion is reversed when the element is excised. In related plant species, the organization of gene content and order are conserved.

islands, each containing at most seven genes, in oceans of retrotransposons which can extend for hundreds of kilobases.

In all plants so far studied, the density of genes is greater towards the ends of the chromosome arms than closer to the centromeres. This is more striking in the large-genome species, but is also observed in *Arabidopsis*. Figure 3.12 shows the distribution of known genes in *Arabidopsis* and in rice.

3.3.2 Each chromosome arm terminates in a telomere

At the end of each chromosome arm is a **telomere**, a specialized structure that protects the chromosome end, ensures that it is replicated accurately, and prevents what would otherwise be a natural tendency to shorten at each successive round of DNA synthesis (because the mechanism of DNA replication results in the loss of 50–100 nucleotides from the 5′ end of the sequence). Telomeres also seem to have a role in maintaining 'non-sticky' ends on the chromosome. When chromosomes break, which can happen, for example, in response to radiation, the broken ends are very sticky, readily joining with any other available DNA fragment. In intact chromosomes, telomeres prevent this. Finally, telomeres appear to play a role in the organization of chromosomes in the nucleus, attaching the chromosome to the inner surface of the nuclear envelope.

Telomeres are composed of multiple repeats of a short DNA sequence. In most plants so far studied, this is TTTAGGG, though in species in the order Asparagales (which includes asparagus, onions, agave, yucca and orchids) it is TTAGGG, identical to the version found in humans and other vertebrates. The overall length of each telomeric region depends on the species and genotype. In *Arabidopsis*, for example, there are 2–5 kb regions of perfect repeats of the TTTAGGG motif at the extreme end of each chromosome, followed by several kilobases more of degenerate (imperfect) versions of the motif. Tobacco (*Nicotiana tabacum*), in contrast, has telomeric regions about 150 kb long.

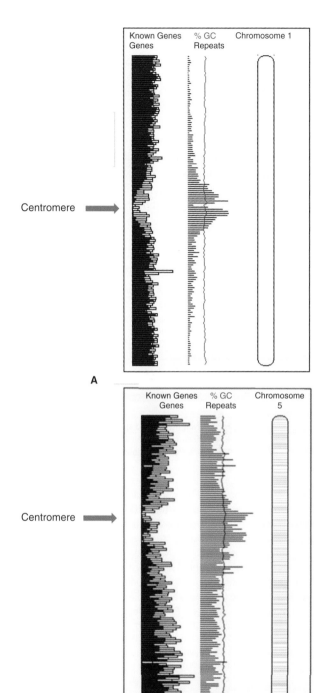

Figure 3.12 The distribution of known (red) and presumed (gray) genes on (A) chromosome 1 of *Arabidopsis thaliana* and (B) chromosome 5 of *Oryza sativa*. In both cases the genes are more abundant towards the chromosome ends. The distribution of regions containing GC-rich repetitive DNA is also shown; in general, such regions are most abundant where there are few genes, especially around the centromere.

3.3.3 The centromere is a complex structure visible as a constriction in the chromosome

Like all eukaryotic chromosomes, each plant chromosome contains a **centromere**. Centromeres are constrictions in chromosomes, which are clearly visible when the chromosomes condense during the cell division cycle (see Figure 3.11). As discussed in more detail in Chapter 11, the spindle fibers attach to centromeres to facilitate the separation of replicated chromatids in mitosis and meiosis. Despite the name, centromeres are not necessarily found at the physical centers of chromosomes. Figure 3.13 illustrates examples of plant chromosomes that have their centromeres located in different positions. Variation in centromere location between different chromosomes in a species can be a useful cytogenetic tool for distinguishing one chromosome from another.

Plant centromeres are large and complex structures. In *Arabidopsis*, for example, centromeres mainly consist of tandem arrays of 180 bp DNA sequence, repeated over and over, while *Z. mays* centromeres contain a different, 150 bp, repeat. There is considerable variation in the number of copies of a given repeat sequence among different chromosomes within a plant, and even for the same chromosome in different varieties within a species. In addition to these tandem repeats, centromeres often contain retrotransposon elements that are usually centromere-specific. The amount of repetitive DNA in centromeres has made sequencing in and around this region technically difficult, so the exact size of most plant centromeres has not yet been established. Rice is an exception; two of its 12 centromeres have been completely sequenced and analyzed. They have been

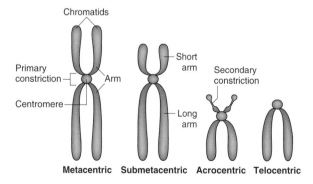

Figure 3.13 Plant chromosomes with their centromeres in different positions, from the nearly symmetrical metacentric to the telocentric, which effectively lacks a second arm.

shown to be between 1 and 2 megabases (Mb) in length. The centromeres in *Arabidopsis* have been estimated to be 1 Mb each, although there is also evidence to suggest that they may be larger.

3.3.4 Chromosomes have other distinctive structural features

In addition to the major morphological landmarks on chromosomes, there are characteristic regions that can be distinguished by their DNA sequences. These include **pericentromeric regions** found on all chromosomes and **nucleolar organizer regions** (**NORs**) and **heterochromatic knobs** that are located on specific chromosomes. Adjacent to each centromere are two **pericentromeric regions** that are many megabases long and contain few genes. As discussed further in Chapter 11, these normally show low rates of genetic recombination during meiosis. They generally contain large numbers of transposons and retrotransposons.

Nucleolar organizer regions are the sites where the genes for ribosomal RNA are found. The **nucleolus** forms around the NORs and cytosolic ribosomes are assembled here. Ribosomes are needed in very large quantities for protein synthesis in active cells, and it would be impossible for one or a few copies of a gene to keep up with the demand for ribosomal RNA synthesis. Each subcellular compartment in which protein synthesis takes place—the cytosol, plastid and mitochondrion—has its own ribosomes with characteristic S values (Table 3.4). Ribosomes and their individual subunits are designated using the sedimentation coefficient S (Svedberg, named for the Swedish chemist Theodor Svedberg) value, which indicates the speed at which they would sediment during centrifugation in a gradient of increasing sucrose concentration. The function of ribosomes is discussed in Section 3.6.2.

Cytosolic ribosomes contain four RNA species. The NORs consist of tandem repeats of a sequence about 10 kb long that encodes the **28S, 18S** and **5.8S RNA** components of the ribosome (Figure 3.14). The coding sequences of these ribosomal genes are highly conserved across species, even across the boundary between the plant and animal kingdoms. However these coding regions are separated by the **intergenic spacer** (**IGS**), which is much more variable. It has been used for exploring relationships between species by analyzing the extent of IGS sequence divergence. The number and chromosomal location(s) of NORs have been

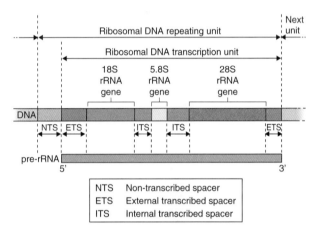

Figure 3.14 Organization of genes encoding ribosomal RNAs. Most plant eukaryotic ribosomal genes lie within a repeating unit, which can be from 7800 to 185 000 base pairs long. It consists of rRNA genes, whose sequences are highly conserved, separated by short stretches of spacer sequences that are not transcribed into RNA. There are four rRNA genes, of sizes 18S, 5.8S, 28S and 5S. The first three of these are contained on a repeating unit that is initially transcribed as a pre-RNA unit; the 5S gene is encoded elsewhere in the genome. The four different rRNA molecules encoded by these genes together make up the RNA part of the ribosome.

Table 3.4 Summary of the composition and properties of various ribosome types.

	Svedberg units (S)		
Ribosome	**Subunits**	**rRNAs**	**Number of proteins**
Plant cytosolic, 80S	Small 40	18	32
	Large 60	28, 5.8, 5	48
Plant plastids, 70S	30	16	~25
	50	23, 5, 4.5	~33
Plant mitochondria, ~70S	30	18	>25
	50	26, 5	>30
Prokaryotic, 70S	30	16	21
	50	23, 5	31

determined in many plant species. In *Arabidopsis*, for example, there are two NORs, each about 4 Mb in size, one on chromosome 2 and one on chromosome 4. In addition to the 28S, 18S and 5.8S RNA components, each ribosome contains a **5S RNA**. This RNA is also encoded in tandem arrays that are found at other locations in the genome from the 28S, 18S and 5.8S RNA cluster, often on different chromosomes altogether.

Heterochromatic knobs are regions of highly condensed chromatin found in locations separate from the centromeres. They are commonest in plants with large genomes, and have been well-characterized in *Z. mays* (Figure 3.15). *Arabidopsis* has a single heterochromatic knob in its haploid genome, on chromosome 4. Sequencing has shown that it consists mainly of retrotransposons.

3.3.5 DNA in the nucleus is packaged with histones to form chromatin

So far, chromosomes have been described solely in terms of their DNA composition. But plant chromosomes, like those of other eukaryotes, consist of more than just DNA. Chromosomal DNA is associated with a number of different proteins. These proteins are important both in maintaining the structure of a chromosome and in regulating gene expression. DNA and its associated proteins are called **chromatin**. A large portion of the proteins in chromatin are **histones**. Histones are very basic proteins (that is, they carry a positive charge), and this facilitates their interaction with the acidic DNA molecule. During interphase, regions of chromatin, called **heterochromatin**, are tightly coiled and stain darkly with dyes; heterochromatin is usually considered to be relatively 'inert' DNA in which genes are undergoing little or no transcription. In contrast other regions, designated **euchromatin**, are more loosely packed, stain lightly and often contain genes that are undergoing active transcription.

Using scanning electron microscopy at the highest levels of resolution, chromatin resembles a 'beads on a string' structure (Figure 3.16), about 10 nm in diameter. The 'beads' represent a **nucleosome** array. A nucleosome consists of DNA wrapped two full turns (166 bp) around a globular cluster of eight histone proteins—two tetramers, each consisting of H2A, H2B, H3 and H4 (Figure 3.17). One additional histone, H1, binds outside the nucleosome core; one of its functions is to stabilize both the nucleosome array and higher-order chromatin structures (Figure 3.18). The core histones H2A, H2B, H3 and H4 can all undergo covalent modifications to their protein structures. These modifications may include, among others, methylation, acetylation and

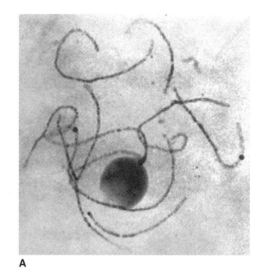

A

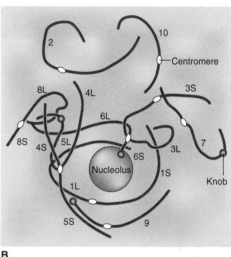

B

Figure 3.15 The ten chromosomes of a *Zea mays* nucleus, shown at a stage in the cell division cycle when they are visible as distinct, elongated strands. (A) Electron micrograph. (B) Schematic drawing of (A), identifying the individual chromosomes and highlighting their centromeres and a heterochromatic knob. S and L refer to the short and long arms of each chromosome respectively.

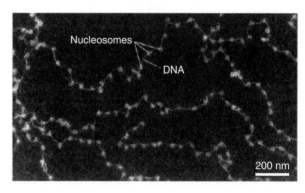

Figure 3.16 Micrograph showing nucleosomes along a strand of DNA—the 'beads on a string' structure, which is approximately 10 nm in diameter.

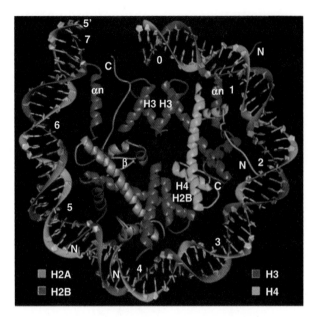

Figure 3.17 Three-dimensional representation of a nucleosome. The core nucleosome particle, the 'bead' in Figure 3.16, consists of 146 base pairs of DNA wrapped around a complex of eight histone proteins. This octamer contains two tetramers, each made up of histones H2A, H2B, H3 and H4.

Nucleosome structure

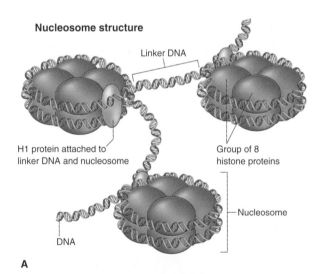

In some cases, nucleosomes may be grouped into 30-nanometer fibers

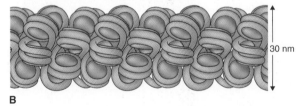

Figure 3.18 The 10 nm nucleosome array. The array (A) is usually condensed into a more densely-packed 30 nm fiber, the 'solenoid' (B). This structure is stabilized by histone H1, which links coils that consist of six nucleosomes.

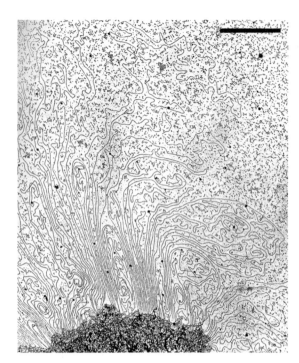

Figure 3.19 Attachment of chromatin to nuclear scaffold proteins results in the formation of variably-sized loops, as shown in this electron micrograph.

phosphorylation. These alterations affect interactions between histones and DNA, and in some cases they are involved in the regulation of gene expression.

The 'beads on a string' assembly in turn undergoes higher-order coiling, producing the so-called '**solenoid**' structure, 30 nm in diameter, so that the genome adopts a very compact conformation in the nucleus. The solenoid structure is further compressed when it attaches to **nuclear scaffold proteins** resulting in the formation of variably-sized looped domains of chromatin (Figure 3.19). These various levels of compression are essential, because while the higher plant nucleus is only a few micrometers in diameter, the total length of DNA in, for example, a single *Z. mays* nucleus is about 4 m. Thus DNA needs to be packaged so that it fits into the nucleus and does not suffer breakage or become tangled.

3.3.6 Each species has a characteristic chromosome number

The number of chromosomes per nucleus is characteristic of a species. As was the case with genome size, there is huge variation between species in the number and sizes of chromosomes. At the lower end of the range, three species (*Brachycome dichromosomatica*, *Zingeria biebersteiniana* and *Haplopappus gracilis*) have

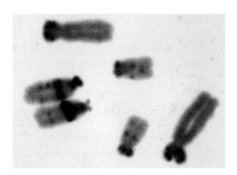

Figure 3.20 Micrograph showing the three pairs of chromosomes of *Crepis capillaris*.

haploid chromosome numbers (n) of two; at the other extreme, some plants of the genus *Equisetum* (horsetails) have more than 100 chromosomes in their haploid set. There is no correlation between the number of chromosomes and the total DNA content. For example, the soft rush *Juncus effusus*, which is used commercially in Japan for the weaving of tatami mats, has 23 tiny chromosomes in its haploid complement, but a nuclear DNA content of 0.3 pg, less than twice that of *Arabidopsis* (n = 5). In contrast, *Crepis capillaris*, a member of the daisy family, packs its 2.1 pg of DNA into just three chromosomes (Figure 3.20).

3.3.7 Polyploidy and genome duplication are common in plants

Humans, like most mammals, are diploid—they have just two full sets of chromosomes, one set derived from each parent. In contrast plant species with more than two sets of chromosome are commonly found; this condition

is called **polyploidy**. Tetraploid and hexaploid species, with four and six full sets of chromosomes respectively, are common, and octaploids also exist.

Polyploids may arise in one of two ways. If the entire chromosome complement of a diploid plant is doubled the result is an **autotetraploid**: a plant that has four sets of chromosomes, two chromosome sets from each parent. Autotetraploids may arise spontaneously or may be induced by using drugs such as colchicine, which disrupt the mitotic spindle. Formally, if the parent plant had 2n = 6, the autotetraploid would have 12 chromosomes and would be designated 2n = 4x = 12 where n = the new haploid number of chromosomes and x = the original haploid number. If the chromosome set were doubled again, the resulting plant would be an octoploid: 2n = 8x = 24 (Figure 3.21).

As more plant genome sequences become available, there is increasing evidence that there have been many **genome duplication** events throughout the history of plant evolution. The timing of these duplications— anything from a few million years ago to more than 100 million years in the past—can be inferred from the extent to which the duplicated sequences have become different from one another. Protein-coding sequences are used for these analyses, since the non-coding regions of the genome are under less strong selective pressure, and often rapidly diverge to the point at which their common origin is no longer recognizable. Where coding sequences of two genes, in the same or different plant species, are still identifiably related, the genes are called **homologs**. More specifically, when homologous genes are found in different plant species they are known as **orthologs**. The term given to two or more homologous genes in the same species is **paralogs**.

Following genome duplication events, individual duplicate genes can be lost or undergo changes in gene function. These changes can occur because, following

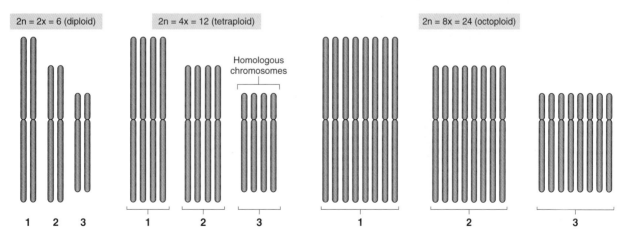

Figure 3.21 Formation of an autopolyploid converts a hypothetical diploid plant with three pairs of chromosomes (2n = 2x = 6) to a tetraploid (2n = 4x = 12) and then to an octoploid (2n = 8x = 24).

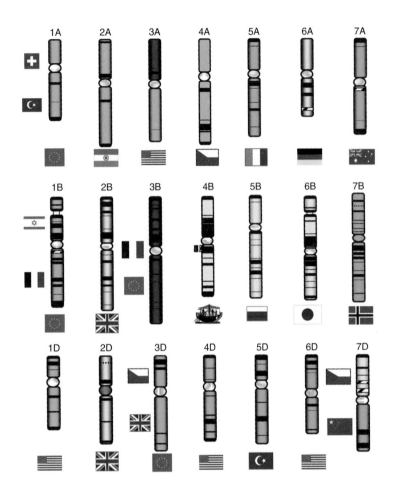

Figure 3.22 The chromosome complement of bread wheat (*Triticum aestivum*). Bread wheat is a hexaploid (2n = 6x = 42), formed as a result of hybridization between three different diploid ancestral species, probably during domestication of wheat as a food crop. A, B and D refer to the ancestral species that contributed the corresponding chromosome. Flags indicate countries working on different chromosomes in the wheat genome sequencing project.

duplication, at least one of the duplicate genes is free to diverge or evolve without needing to maintain its original function that might have been essential for an organism's survival. The end result may be that one of the duplicated genes takes on a different role than that of its original parent. Alternatively, the two duplicates together may cover the roles that were originally fulfilled by the parent gene; for example, one may be expressed in leaves and the other in roots; or one during early development and one later on. In both cases, these changes can affect either the gene's protein-coding region or its regulatory regions.

The other class of polyploids, **allopolyploids**, arises from events in which the genomes of two or more ancestral species are combined into a single new species, without reduction in the total number of chromosomes. Such **interspecific hybridization events** are infrequent in nature and the progeny are often sterile. However, genetic analysis has shown that a large proportion of modern plant species underwent interspecific hybridization at some point in their evolutionary history. Many important crop species are allopolyploids, probably as a result of early human attempts to improve yield and quality by bringing together the best traits of wild species. For example, bread wheat (*Triticum*

aestivum) has three sets of chromosomes, derived from three ancestral diploid species. Its genome is described as 2n = 6x = 42 (Figure 3.22).

Key points In the nuclear genome, the hereditary material, DNA, is carried on chromosomes. Each chromosome has two arms, which terminate in telomeres and are separated by a complex structure called the centromere. Chromosomes have other distinctive structural features including nucleolar organizer regions and heterochromatic knobs. The chromosome arms are gene-rich, whereas the region immediately around the centromere has few genes. Nuclear DNA is packaged with histone proteins to form chromatin, which consists of a series of nucleosomes in a 'beads on a string' conformation; the nucleosomes undergo further coiling to produce a compact 'solenoid' structure. Each plant species has a characteristic number of pairs of chromosomes: this may be anything from two to over 100 pairs in the nucleus of every cell. Polyploidy and genome duplication are common events in plants.

3.4 Expression of the plant genome I. Transcription of DNA to RNA

For a gene to be expressed, the DNA of which it is composed must be transcribed into RNA. In most cases the RNA is then translated into protein. However, the products of transcription include not only mRNA, which is the template for protein synthesis, but also a number of RNA molecules that either are components of the machinery of protein synthesis or play regulatory roles (Table 3.5). With very few exceptions, only one strand (the **coding strand**) of any region of the DNA double helix actually encodes a protein, and it may be either strand. Thus protein-coding genes on a given section of a chromosome may be transcribed to RNA in either direction. This process is under the control of regulatory elements that determine whether, and to what extent, a gene is expressed at a given time and under given conditions. We will consider the processes of transcription and translation of nuclear genes and examine ways in which gene expression is regulated.

3.4.1 Plant nuclear genes have complex structures

We have discussed the general structure and composition of plant genomes. Now let us look at the organization of individual plant genes. Plant nuclear genes are organized like the genes of other eukaryotes. On average, the part of a gene that codes for its gene product stretches over about 1300 DNA bases (1.3 kb). However, in most cases, these bases are interspersed and surrounded by other non-coding sequences, so that the

Table 3.5 Different types of RNA are transcribed by different types of RNA polymerase.

Type of polymerase	Types of genes transcribed/RNAs synthesized
Pol I	18S, 5.8S and 25S ribosomal RNAs
Pol II	Messenger RNA, micro RNAs
Pol III	5S ribosomal RNA, transfer RNAs
Pol IV	Small interfering RNAs (siRNAs)

typical eukaryotic gene covers 4 kb or more. The physical location of a gene in a chromosome is knows as its **locus** (plural: loci).

The sections of a protein-coding gene that encode the gene product are called **exons**, while the intervening stretches of non-coding DNA are known as **introns**. Initially, when a protein-coding gene is transcribed into RNA, the whole exon–intron region is transcribed to produce a **primary transcript**. Subsequently, the sections corresponding to the introns are excised, in a process known as **splicing** (see Section 3.4.11) to produce the **messenger RNA** (**mRNA**) that is translated into the protein product.

The non-coding sequence that flanks the 5′ end of the gene (**5′ leader sequence**) usually contains DNA elements that have roles in the regulation of gene transcription. The 3′ flanking region contains sequence elements for modifying mRNA and a transcription stop site, and may contain additional regulatory elements (Figure 3.23). Sometimes determining where the 3′ flanking region of one gene 'ends' and the 5′ flanking region of the next gene 'begins' is not easy. These intergenic regions contain much of the DNA that makes some plant genomes so large, including tandem repeats and active or inactivated transposable elements.

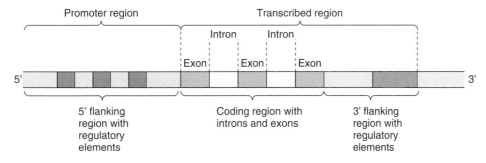

Figure 3.23 The structure of a typical plant gene, containing a regulatory region (promoter) and a region that is transcribed to produce an RNA copy that encodes the gene's protein product. This coding region also contains regulatory information. Introns are sequences present in the coding region that do not contribute to the mature messenger RNA used to produce the protein; only the exons (sequences that do encode the protein) appear in the final mRNA.

3.4.2 Histones and chromatin organization play important roles in gene expression

Before a nuclear gene can be transcribed, chromatin must be remodeled to make the gene accessible to the transcription machinery. The chromatin surrounding a gene that is being, or is about to be, transcribed is less condensed than genomic chromatin as a whole. These regions of the genome can be recognized because they are hypersensitive to digestion by the enzyme DNase I.

In addition to their role in chromatin structure, two of the histones, **H3** and **H4**, can affect gene transcription if they have undergone post-translational modification by the addition of components to their polypeptide chains after they have been synthesized. These changes can affect gene transcription both by altering chromatin structure and by promoting or reducing transcription factor binding. The two major forms of modification that histones H3 and H4 undergo are **acetylation** and **methylation**. During acetylation, an acetyl group is transferred from acetyl coenzyme A to one or more of five lysine residues in either of the two histones. The enzymes that carry out this process are known as **histone acetyl transferases** (**HATs**). Acetylated histones form nucleosomes that are less compact and generally more accessible to transcription factors (Figure 3.24). Methylation of histones H3 and H4 also affects lysine residues. It is carried out by **histone methyl transferases** (**HMTases**) that transfer methyl groups from *S*-adenosylmethionine or *S*-adenosylhomocysteine to the histones. Depending which lysine residues are acetylated,

methylation can result in either increased gene expression or heterochromatin formation and silencing of genes. It is becoming increasingly clear that histone modifications and chromatin structure play important roles in gene expression in plants and other eukaryotes, but our understanding of these processes is far from complete.

3.4.3 Higher-order chromatin structure also regulates gene expression

The chromosomes of higher eukaryotes are attached at many points to a nuclear scaffold structure, forming variably-sized looped domains of chromatin (see Figure 3.19). The sequences where attachment occurs, known as **matrix attachment regions** (**MARs**), are AT-rich DNA sequence motifs which are 200–1000 bp long. MARs are believed to facilitate transcription of a gene or a group of genes by causing a reduction in chromatin condensation (Figure 3.25). The loops created

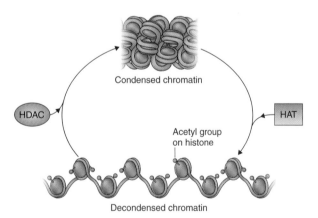

Figure 3.24 Histone modification. Histones within the nucleosome, and the linker histone H1, can all be modified at the N terminal by acetylation, carried out by acetyltransferases. When an acetyl group is present on a histone, it creates a local region of decondensation of the chromatin. This makes the DNA in this area more accessible for transcription. When acetyl groups are removed by histone deacetylases (HDAC), the histones interact more tightly with the DNA and the structure is more condensed.

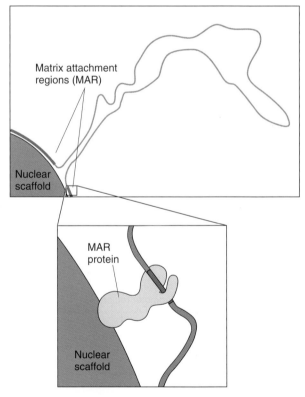

Figure 3.25 Matrix attachment regions. MARs in chromatin can form attachments to the nuclear matrix, creating looped, supercoiled structures. MARs are believed to influence gene expression by facilitating differential levels of condensation of the genome in different regions. There are also several classes of specialized proteins associated with MARs, which may function in tethering the MAR DNA sequences to the nuclear matrix.

by different MARs are independent of one another and show various degrees of supercoiling, which is known to influence gene expression.

Several plant genes have been identified as having associated MARs which appear to influence gene structure or function; examples are known from pea (*Pisum sativum*), bean (*P. vulgaris*), soybean (*Glycine max*), *Nicotiana tabacum*, sugarcane (*Saccharum officinarum*) and *Zea mays*. MARs are often found flanking the coding regions of genes, and in many cases they are associated with regulatory elements. For example, in *P. vulgaris*, the 5′ flanking region of the β-phaseolin gene contains a MAR that acts together with a transcriptional enhancer sequence to promote transcription of the gene.

Key points Plant nuclear genes have complex structures. The sections of a protein-coding gene that encode the product are called exons; they are separated by intervening non-coding regions, the introns. The regions that flank the protein-coding portion of the gene include promoters and other regulatory regions that control the timing and extent of gene expression. For a gene to be expressed, an RNA copy of its DNA must be made, in a process called transcription. In most cases, the RNA then undergoes translation into protein. However, some products of transcription are RNA molecules that are components of the machinery of protein synthesis, or that play regulatory roles. With very few exceptions, only one strand (the coding strand) of any region of the DNA double helix actually encodes an RNA product, and either strand may be the coding strand. The expression of DNA is regulated at many levels: histones, chromatin organization and higher-order chromatin structure all play important roles in gene expression.

3.4.4 Promoters and other regulatory elements control the timing and extent of gene transcription

The transcription of DNA to RNA is carried out by a class of enzymes called **RNA polymerases**. For transcription to occur, a polymerase must bind to a specific promoter sequence upstream (i.e. some distance before the 5′ region) of the gene to be transcribed. In a broad sense, a **gene promoter** includes a variety of sequence elements that control the recruitment of proteins called

transcription factors which are required for, or enhance, transcription of a gene. A gene's promoter is often considered to be the 1–2 kb of sequence immediately upstream of the gene itself. It consists of **core elements** and **promoter regulatory elements**. Core elements are sequences that are required for binding of the RNA polymerases and therefore for the transcription of genes; they are usually within 50 bp upstream (to the 5′ side of) or downstream (to the 3′ side) of the site at which transcription is initiated. Promoter regulatory elements are responsible for the activation, repression or modulation of gene expression. Transcription factors bind to these regulatory elements and influence the rate at which the genes are transcribed by the appropriate RNA polymerase. Promoter regulatory elements are upstream of, and close to, the transcription initiation site and core elements. However, other regulatory elements may also be found in a gene's introns, downstream of the coding region, or at distances of many kilobases from the gene, on the same or a different chromosome. Regulatory elements that are located in the same chromosome and on the same DNA strand as the coding region of the gene they influence, and in close proximity to it, are known as **cis elements**; whereas proteins (or other molecules) encoded elsewhere in the genome that interact with the promoter or promoter-associated proteins are called **trans-acting factors**.

The promoters of some plant genes contain DNA sequences that confer specificity of gene expression in response to factors such as nutrient status, hormone concentration or environmental stress. These **responsive elements** (also often called **response elements**) generally have a short, highly-conserved core DNA sequence, which is essential for response to a particular factor, surrounded by a less highly-conserved region. For example, the abscisic acid responsive element (ABRE), which is found in over 2000 genes in both *Arabidopsis* and rice, contains the core sequence AGCT. The presence of this sequence in a gene's promoter region means that expression of the gene can be induced by abscisic acid. Elements responsive to ethylene, iron deficiency and dehydration have also been well characterized in plants.

3.4.5 RNA polymerases catalyze transcription

Plants have four different nuclear RNA polymerases, each of which is involved in transcribing different classes of genes in the nucleus. There are additional RNA polymerases for the transcription of mitochondrial and plastid genes. The types of genes transcribed by each of the nuclear polymerases are summarized in Table 3.5. RNA polymerases I, II and III are multisubunit enzymes that are evolutionarily related to each other, and are

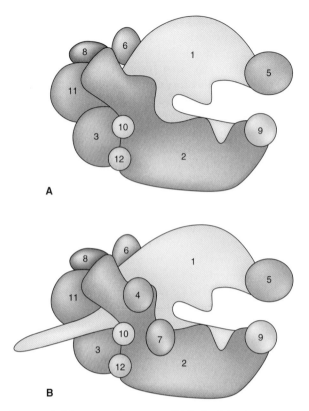

A

B

Figure 3.26   RNA polymerases. (A) Each of the three eukaryotic RNA polymerases has five core subunits (1, 2, 3, 6 and 11) and five common subunits (5, 8, 9, 10 and 12). (B) The polymerases all also have unique features, as shown here for RNA polymerase II. In addition to the five core and five common subunits, it has two unique subunits (Rpb4 and Rpb7, labeled 4 and 7 on the diagram and a 'tail', the carboxy-terminal domain, which is required for initiating transcription and RNA processing.

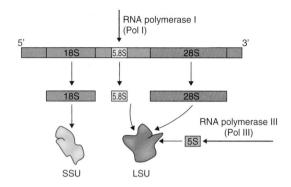

Figure 3.27   RNA polymerase I (Pol I) transcribes the group of ribosomal genes (18S, 5.8S and 28S sequences) to produce a 45S ribosomal RNA precursor. This is then cleaved to give the three individual subunits. The 18S rRNA is a component of the small ribosomal subunit (SSU) while the 5.8S and 28S rRNAs are part of the large ribosomal subunit (LSU).

found in all eukaryotes. Each consists of two large subunits and 10–15 smaller subunits (Figure 3.26). Some of the subunits are common to two or to all three of the other polymerases. Each of the four polymerases has its own characteristic core promoter type or types.

RNA polymerase I (**Pol I**) is involved solely in the transcription of the group of ribosomal genes (18S, 5.8S and 25S sequences; see Section 3.3.4) that are transcribed in tandem to produce a **45S ribosomal RNA precursor** (Figure 3.27). This transcript is subsequently processed to excise spacer sequences and produce the individual rRNA components.

The Pol I **core promoter** is found about 50 bp upstream of the ribosomal genes. This core promoter normally contains a highly conserved region around the transcription start site (TATAT<u>A</u>(A/G) GGG) but otherwise it is not yet known what the critical regulatory elements are. Even though transcription by Pol I of the multiple RNA genes in growing cells can represent up to 80% of all RNA synthesis, there is still little

understanding of how this transcription is regulated. The fourth RNA component of cytosolic ribosomes, 5S RNA, is transcribed by Pol III.

RNA polymerase II (**Pol II**) is responsible for the transcription of all protein-coding genes in the nucleus. Pol II promoters have been studied in more detail than other promoters. The Pol II promoter associated with a gene plays an important part in regulating when, in which cell types, and in response to what environmental conditions the gene is expressed.

Many Pol II core promoters contain a region called a **TATA box**. The TATA box is the DNA sequence TATAA, usually followed by three more A-T base pairs and surrounded by GC-rich regions, and it is found about 25 bp upstream of the starting point for RNA synthesis. When Pol II promoters were first identified, it was believed that the TATA box was present and essential for them all, whether from plants, animals or fungi. Now that whole-genome sequences are becoming more widely available, it turns out that the TATA box occurs only in about one Pol II promoter in three. The genes that have this feature are mainly ones that are highly expressed, at least under some conditions. There are whole classes of genes—for example, those involved in photosynthesis—for which there is no TATA box. These genes usually have an alternative region known as the **downstream promoter element** located 3′ of the transcription factor IIB (**TFIIB**) binding site.

Initiation of transcription by Pol II requires the interaction of a number of factors to ensure accurate transcription of a protein-coding gene. Generalized transcription factors (TFIIs), such as TFIIB, work together with the **TATA-binding protein, TBP**, to ensure the correct binding of RNA polymerase to the core promoter of every protein-coding gene (Figure 3.28A).

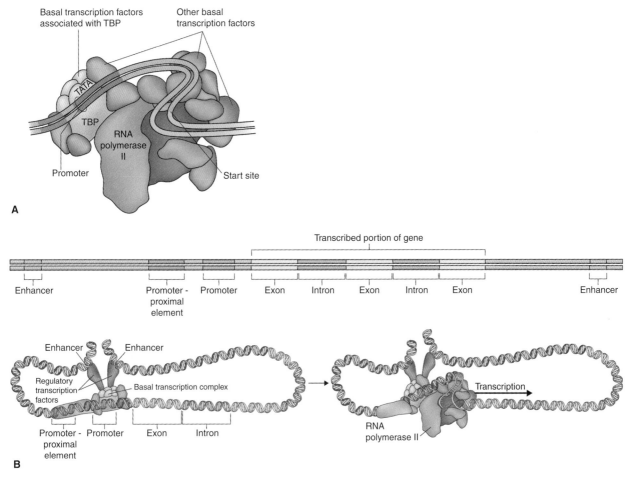

Figure 3.28 (A) The complex formed between DNA, transcription factors and other regulatory proteins, and RNA Pol II when transcription is initiated. TBP, TATA-binding protein. (B) Enhancer elements may be located at a distance either upstream or downstream of the genes whose expression they regulate.

Specific transcription factors that are involved in the transcription of particular genes or groups of related genes interact with the basal transcription complex to regulate the initiation of transcription (see Section 3.4.6). Additional TBP-associated factors mediate the interactions between the transcription complex and specific transcription factors, thus playing a part in regulating the way different genes are expressed.

Regulatory elements bind transcription factors; they may be found both in the promoter itself and far upstream from the TATA box, in the untranslated 5′ leader sequence of the gene and even in its introns. These regulatory elements often act as **enhancers**, increasing the efficiency of RNA Pol II in initiating gene transcription (Figure 3.28B). Enhancer sequences can be located at a considerable distance from the coding region of the gene. However, as cis elements, they must by definition be on the same DNA strand as the gene(s) they influence. Enhancers may have roles in regulating the specificity of gene expression, controlling whether, and to what extent, a gene is expressed in a particular tissue

or in response to a certain environmental factor such as light or pathogen attack. The term **silencers** is used for elements similar to enhancers that act to downregulate, rather than increase, gene expression.

The remaining two RNA polymerases, **Pol III** and **Pol IV**, are less well studied. Pol III is the most complex of the RNA polymerases: it is composed of 17 subunits, five of which have no equivalents in Pol I and Pol II. It transcribes genes encoding a number of small RNAs, including tRNAs, 5S ribosomal RNAs and others. Pol IV is completely distinct from Pol I–III. It was discovered relatively recently and appears to be unique to plants. Pol IV synthesizes **small interfering RNAs** (**siRNAs**). These are RNA molecules, typically 20–25 nucleotides long, that play a variety of roles in eukaryotic cells. They were first discovered in plants, in the late 1990s, but have since been identified in insects, mammals and other organisms. One function of siRNAs is RNA silencing. In this process an individual siRNA can suppress expression of a gene to which it has sequence homology. They also play a part in resistance to certain viruses.

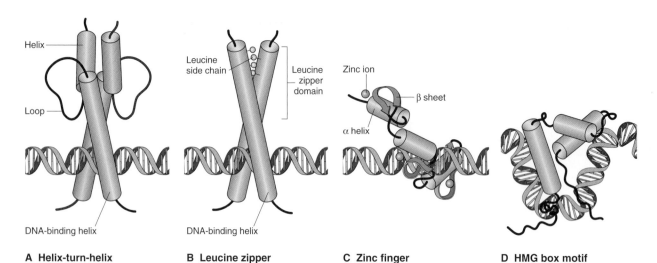

Helix
Loop
DNA-binding helix

A Helix-turn-helix

Leucine side chain
Leucine zipper domain
DNA-binding helix

B Leucine zipper

Zinc ion
β sheet
α helix

C Zinc finger

D HMG box motif

Figure 3.29   Transcription factor motifs. (A) Helix-turn-helix proteins consist of two α-helical regions separated by a loop. This type of factor always operates as a dimer; that is, two molecules of the protein are required for it to function. One α-helix in each monomer functions to keep the dimer together by interacting with its counterpart in the other monomer. The other pair of α-helices forms a structure like a pair of scissors in which the two 'blades' fit into the adjacent major grooves in the DNA molecule. (B) Basic leucine zipper proteins also form dimers that bind in the major groove of the DNA molecule. In this case, however, each monomer has a single α-helix, in which every seventh amino acid is a leucine. The leucines face each other in the dimer, and their orientation 'zips' the two helices together into a coiled coil. This zipped coil can then hold the basic regions of the two monomers in place within the major groove of the DNA. (C) Zinc finger proteins, which generally act as monomers, create their own sets of projections to insert into the major groove of the DNA molecule. These projections, or 'fingers', consist of a zinc ion coordinated with four amino acids in the protein (four cysteines, or a combination of histidines and cysteines). A zinc finger protein can have from two to nine of these projections, which insert into successive turns of the major grooves along the DNA molecule. (D) HMG-box motif proteins are named after a class of chromatin-associated proteins from which they were first identified. Like helix-turn-helix and leucine zipper transcription factors, they typically bind to DNA as a dimer. Each of the two protein molecules in the dimer contains three α-helices. Two of these bind to a gene's promoter, creating a structure that distorts the DNA. This distortion brings other regulatory sequences, outside the HMG recognition motif itself, sufficiently close to allow their joint interaction with additional transcription factors.

3.4.6 Transcription factors bind to DNA regulatory sequences

For proper plant development and function, it is essential that the correct genes are expressed at the right time and in response to the right stimulus. As we have seen, for protein-coding genes, this control is achieved by interaction between the promoter regulatory cis elements and trans-acting DNA-binding proteins (transcription factors). This interaction is required to ensure that RNA Pol II transcribes the gene appropriately.

Most transcription factors have at least two domains that are essential for their function. One of these is required for recognition and binding of the cis element target sequence associated with the gene itself; the other domain functions in organizing additional proteins involved in activating transcription (Figure 3.28). The majority of the known transcription factors can be classified into groups based on their structural motifs. They fall into four major categories: **helix-turn-helix** motifs, **basic leucine zippers, zinc fingers** and **high-mobility group box** (**HMG-box**) motifs

(Figure 3.29). These motifs are conserved between species, and can be found either in the DNA-binding domain or in the protein-binding domain of the transcription factor.

Transcription factors are often encoded by families of genes that code for a set of related proteins. Members of a family may be involved in the transcription of closely related genes, or of the same genes in different tissues or environments. Those transcription factors that act as heterodimers, formed when two different members of a transcription factor family interact, allow for further fine tuning of the DNA binding specificity.

3.4.7 Homeobox proteins are important in regulating development and determining cell fate

One class of transcription factors deserves a special mention, because of their importance in regulating development and in determining the fate of cells and

Figure 3.30   The effect of a mutation in the *antennapedia* homeobox gene of a fruit fly (*Drosophila melanogaster*). The fly on the left is normal. The one on the right carries the mutation, which results in the development of additional legs where its antennae should be.

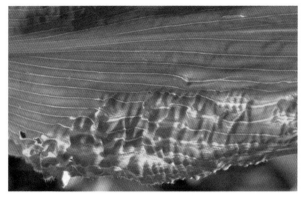

Figure 3.31   Abnormal expression of the *Knotted1* (*Kn1*) gene in a *Zea mays* leaf. *Kn1* is another example of a homeobox gene. When it is expressed at the wrong stage of development, the vascular cells of the leaf divide outside the normal plane, forming 'knots' of tissue.

organs. They are known as **homeobox proteins** (or **homeodomain proteins**) because genes that encode them all share a common sequence, approximately 180 bp long, called the 'homeobox'. Homeobox genes were first discovered in the fruit fly *Drosophila melanogaster*. In the fly, mutations in homeobox genes produce dramatic developmental effects such as the replacement of an antenna by a leg (Figure 3.30). Subsequently, homeobox genes have been found in all multicellular eukaryotes investigated to date, including plants. They are believed to control the expression of other regulatory proteins, including transcription factors, and thereby act as 'master genes'. Genes, like homeobox genes, that regulate tissue or organ identity are called **homeotic** genes.

The transcription factors encoded by homeobox genes are all of the helix-turn-helix class. Their DNA-binding domains are encoded by the highly conserved 180 bp homeobox. They regulate their target genes in a precise, coordinated pattern in time and space. Despite the high degree of conservation of the DNA-binding domain itself, homeobox genes are able to achieve specificity by differential associations with other regulatory proteins or cofactors, or by variations in the DNA flanking the homeobox which alter binding efficiency. Outside the homeobox domain, the DNA sequence is not conserved between homeobox genes among species, or even within species. In *Zea mays*, for example, two families of homeobox genes, *KNOX*, named for the *Knotted1* (*Kn1*) gene, and *ZMH1/ZMH2*, have been identified. The amino acid sequence of the homeodomains is highly conserved between *Kn1* and the *ZMHs* (57 of 64 amino acids in the two proteins are identical), but the remaining sequences are very different.

The *Z. mays Kn1* gene was the first plant homeobox gene to be identified. Mutations in the gene alter leaf structure by causing vascular tissue to be produced outside its normal position (Figure 3.31). Since the discovery of *Kn1*, additional homeotic genes have been isolated from *Z. mays*, *Arabidopsis* and many other plants. Because the homeodomains from plants and animals, including humans, are so similar, it is likely the homeobox genes arose as a way of regulating development before the stage in evolution at which animals and plants diverged.

3.4.8 The MADS-box family includes homeotic genes and regulators of flowering time

The **MADS-box genes** arose early in the development of eukaryotic organisms, and are found in species ranging from yeast and flies to humans and plants. The name is derived from four members of the gene family: *MCM1* from budding yeast, *Saccharomyces cerevisiae*, *AGAMOUS* from *Arabidopsis thaliana*, *DEFICIENS* from the snapdragon *Antirrhinum majus* and *SRF* (serum response factor) from *Homo sapiens*. The MADS-box is a highly conserved region of about 56 amino acids towards the N terminus of the protein, which is a DNA-binding domain. Plants contain many MADS-box genes—for example, there are at least 30 in *Arabidopsis*—and they have evolved to perform a range of functions. Several are homeotic genes, which specify the identity of different parts of flowers; thus a mutation in the MADS-box gene *APETALA3*, for example, results in flowers with petals transformed to sepals and stamens converted to carpels. Others genes in this group are not homeotic, but are important in ensuring the correct timing of flowering in response to environmental factors such as day length and temperature.

3.4.9 Many genes are named after mutant phenotypes

Analysis of **mutants** and other genetic variants has been essential for identifying plant genes and understanding their regulation. Different versions of the same gene, with one or more variations at the DNA sequence level, are called **alleles**, and some allelic variations give rise to mutant phenotypes. Mutants are often given descriptive names (in some cases not entirely serious!) which, usually in abbreviated form, become attached to the corresponding malfunctioning genes when they have been isolated and sequenced. The convention is to notate the name of the mutant gene in *lower case*, and the **wild-type** (functional) variant in *UPPER CASE*. Gene names are usually written in *italics*, and the corresponding protein translation product is non-italicized. Regulatory pathways are represented as interacting networks of functioning versions of mutant genes. The inhibitory influence of one gene, or gene product, on another is usually represented by a bar-ended line; a positive interaction is shown by an arrow. For example, *Arabidopsis* mutants identified by the production of disorganized groups of leaves and shoots were given the name *wuschel*, meaning 'tousled hair' in German. Another kind of mutant, in which meristems grow as a band or a ring instead of as a point, was called *clavata* (from the Latin for 'club-shaped'). The corresponding wild-type genes are *WUSCHEL* (*WUS*) and *CLAVATA* (*CLV*), encoding the proteins WUS (a homeobox transcription factor) and CLV (a receptor kinase), respectively. Figure 3.32 shows a **regulatory model**, based on analysis of mutants, in which *WUS* and *CLV* are components of opposing pathways that interact to regulate the patterning of meristem growth (see Section 12.3.3).

3.4.10 Transcription proceeds via initiation, elongation and termination

With the transcription initiation complex containing the necessary transcription factors in place, the process of transcription can begin. RNA Pol II separates the two strands of the DNA molecule, maintaining a region of about ten nucleotides over which the strands are held

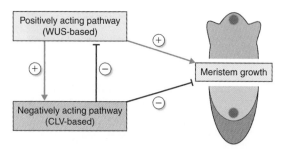

Figure 3.32 Example of a regulatory interaction between positively and negatively acting pathways. The pathway based on the regulatory factor WUSCHEL (WUS) simultaneously promotes meristem growth and stimulates the negatively acting pathway based on the gene *CLAVATA* (*CLV*). CLV suppresses meristem growth through its negative influence on the WUS pathway, and may also directly inhibit the meristem.

Key points Transcription of DNA to RNA is carried out by RNA polymerases. The nuclear genome is transcribed by four different RNA polymerases, each of which is responsible for transcribing a different class of gene. Protein-encoding genes are transcribed to messenger RNA by RNA polymerase II. Transcription requires the binding of proteins called transcription factors to promoter regions in the DNA. There are several classes of transcription factors, which are classified according to their structural motifs. Transcription factors are often encoded by families of genes, and members of a family may be involved in the transcription of structurally-related genes, or of the same genes in different environments or parts of the plant. Homeobox proteins are a special class of transcription factor, important in the regulation of development and in determining cell fate. Genes of the MADS-box family are widespread in plants, and play regulatory roles in the development of plant parts and in the control of processes such as flowering in response to environmental cues.

apart. Ribonucleotides complementary to the template (non-coding) strand of this region are added one at a time by RNA Pol II to produce an RNA molecule that is a copy of the DNA-coding strand, except that each thymine residue in the DNA is represented by a uracil in the RNA. When transcription is complete, RNA Pol II dissociates from the DNA and the transcript of the gene is released. Multiple RNA Pol II molecules can associate with a single DNA template at any one time, producing many copies of the gene transcript.

3.4.11 Messenger RNA molecules undergo post-transcriptional modifications

The initial product of transcription of a protein-coding gene is a 1:1 RNA copy of the gene's DNA sequence, including the non-coding introns as well as the coding exons. This primary transcript must undergo considerable modification to form a mature messenger RNA that will be translated into protein. These modifications include the addition of a 5′ structure known as a **cap**, **splicing** to remove introns and the addition of a chain of adenylic acid residues at its 3′ end (the **poly(A) tail**). Figure 3.33 shows the relationship between primary transcript and mature mRNA.

The 5′ cap is added to the growing mRNA molecule, just after it exits the RNA polymerase complex. A guanosine residue is first attached to the existing 5′ nucleotide by a triphosphate bridge. The guanine then undergoes methylation at the N-7 position. The 5′ cap has an important role during initiation of mRNA translation. It is also believed to help protect mRNA from being degraded while it is being synthesized because mRNA is normally very susceptible to attack by nucleases.

Once transcription is complete, introns are excised from the primary RNA transcript and the exons are joined together to yield the mature RNA sequence in a process called RNA splicing. **Small nuclear ribonucleoproteins (snRNPs)** recognize and bind to intron–exon boundaries. The bound snRNPs then aggregate to form a **spliceosome** that catalyzes the removal of the exon and the joining of introns to form the sequence that will be used to synthesize a protein.

Before most nuclear-encoded mRNAs move from the nucleus to the cytoplasm for translation, a sequence of between 25 and 250 adenosine residues is added to their 3′ ends. This poly(A) tail is not encoded in the gene but is added post-transcriptionally. It facilitates the process of mRNA export from the nucleus and, like the 5′ cap, it plays a role in stabilizing the mRNA against degradation by nucleases and in the initiation of translation.

Unlike nuclear-encoded mRNAs, those encoded by plastid and mitochondrial genomes lack the 5′ cap and long poly(A) tail. A small proportion of plastid mRNAs do have a short poly(A) segment, or an A-rich region, at the 3′ end. In this case these regions appear to regulate degradation of the mRNA; they do not function in the initiation of translation.

Key points Transcription of DNA to messenger RNA proceeds in three phases: initiation, elongation and termination. A transcription initiation complex, containing transcription factors and other regulatory proteins, and RNA polymerase II, is assembled on the DNA to be transcribed. RNA polymerase II holds the two DNA strands apart over a short region, and adds ribonucleotides complementary to the template (non-coding) strand of this region one at a time, to produce an RNA molecule that is a copy of the corresponding DNA coding strand, except that each thymine residue in the DNA is represented by a uracil in the RNA. At the termination stage, RNA polymerase II dissociates from the DNA, releasing the RNA transcript of the gene. Messenger RNA molecules undergo post-transcriptional modifications that include the addition of a 5′ structure known as a cap; removal of the introns by splicing to leave only the coding portion of the RNA; and addition of the poly(A) tail, a chain of adenylic acid residues at the 3′ end. Gene expression can be regulated at the post-transcriptional level by interaction between micro RNAs and messenger RNAs. A typical plant micro RNA is 21–23 nucleotides long, and contains a sequence complementary to part of the protein-coding region of one or more messenger RNAs. This sequence allows the micro RNA to form a section of double-stranded RNA with its mRNA target. The double-stranded region promotes destruction of the messenger RNA by enzymatic cleavage.

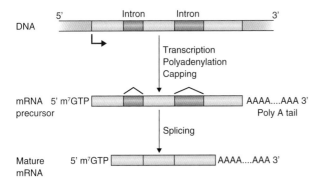

Figure 3.33 The relationship between a gene, the initial RNA product of transcription, and the mature mRNA following the addition of a 5′ cap and a poly(A) tail and the removal of introns by splicing. It is the mature mRNA which is translated to give the protein that the gene encodes.

3.4.12 Micro RNAs are regulators of gene expression at the post-transcriptional level

Micro RNAs (**miRNAs**) are one of several classes of small RNA molecules found in eukaryotes that have roles in regulating gene expression; another of these classes, siRNAs, was described in Section 3.4.5. A typical miRNA is 21–23 nucleotides long and is single-stranded. A ribonuclease called Dicer produces miRNAs by cleaving longer RNA precursors. In plants a miRNA usually has a sequence that is complementary (or nearly complementary) to part of the protein-coding region of one or more messenger RNAs. Therefore a miRNA can form a section of double-stranded RNA with its target mRNA; this double-stranded region promotes enzymatic cleavage of the mRNA. Thus miRNAs regulate gene expression at a point after transcription has occurred, but before the message can be translated (see Section 3.5). They have been shown to have a wide range of roles in plants, both during normal development and in response to abiotic stresses and to bacterial and fungal pathogens. The way miRNAs and siRNAs function is known as RNA interference (RNAi).

3.5 Expression of the plant genome II. Epigenetic regulation of gene expression

Traditionally, scientists believed that all the heritable traits that determine whether and when genes are expressed, and hence the appearance and behavior of an individual, are encoded in the primary sequence of its DNA, its genome. However, for decades there has been evidence that there can be changes in gene expression that are stably passed on from cell to cell (mitotically), and sometimes from generation to generation (meiotically), but that do not seem to involve any alterations in the organism's **primary DNA sequence**. These changes may reveal themselves through an alteration such as loss of pigmentation in flowers or cereal grains. Other, less apparent, changes are now recognized as being more widespread than previously thought. These changes are said to be **epigenetic**; a new field of research, **epigenetics**, has developed that studies these changes. There are many ways in which epigenetic changes can arise in plants; some of the best understood are described below.

3.5.1 DNA methylation is an important mediator of epigenetic regulation of gene expression

DNA methylation is the best studied, though not the only, mechanism that accounts for epigenetic changes which are inherited both mitotically (within the cells of a given organism) and meiotically (from one generation to the next). Cytidine groups in genomic DNA undergo methylation; in plants this normally occurs where there is a CpG dinucleotide (the p represents a phosphodiester bond) or a CpNpG site, where N can be any nucleotide. When the parent strand of a newly-replicated region of DNA contains methylated CpG or CpNpG sites, the new strand is a strong substrate for **DNA methyltransferase** which ensures that pre-existing methylation patterns are passed on to both the daughter chromosomes that result from a cell division. DNA methylation can affect chromatin structure which, in turn, can affect patterns of gene transcription.

3.5.2 Epigenetic changes through paramutation can be passed on from one generation to the next

Paramutation is a type of epigenetic change that is meiotically heritable—that is, it can be passed on from one generation to the next. The term describes an interaction between two alleles in which the expression of one allele in a heterozygote is altered by the presence of the other. Figure 3.34 illustrates one example from *Zea mays*, in which a gene controlling pigmentation can be present in one of two forms. One of these, the allele *B-I* of the gene *b* (*booster*), is strongly expressed and gives intense coloration, so plants with two copies are intensely colored. The *B'* allele is weakly expressed, and plants with two copies of this allele are less intensely colored. However, in heterozygous plants with one copy of each allele (*B-I/B'*) the color intensity is low, and only *B'* alleles are transmitted sexually from these heterozygous plants via pollen and seed to the next generation. The two *b* alleles, *B'* and *B-I*, have identical DNA sequences, yet their rates of transcription differ by 10–20-fold.

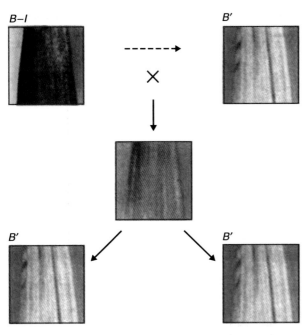

Figure 3.34 The phenomenon of paramutation in *Zea mays*, showing an example affecting coloration of the ear husks. Parental, F1, and the resulting segregant phenotypes (F2) are represented in successive rows of the figure. Weakly expressed paramutagenic *B'* states can arise spontaneously from strongly expressed paramutable *B-I* alleles (dashed arrow). *B-I* alleles change exclusively to *B'* when exposed to *B'* in the F1 heterozygote.

3.5.3 Transgenes can silence a plant's own genes by cosuppression

In some cases, the suppression of gene expression without any change in the DNA sequence of the gene (**epigenetic gene silencing**) does not result in heritable changes of gene activity. When it first became possible to make transgenic plants, researchers made the surprising discovery that inserting an engineered gene (a **transgene**), that was similar or identical in DNA sequence to a gene that was already present in the plant, caused the endogenous gene to be silenced. This phenomenon is known as **cosuppression**. One of the best-known examples comes from an attempt to intensify flower color in *Petunia* by introducing a transgene encoding an enzyme in the pigment biosynthetic pathway (see Chapter 15). The transgene not only did not result in increased pigment synthesis, it actually suppressed activity of the corresponding endogenous gene by causing degradation of the gene's mRNA transcript. This meant that the pigment was not

synthesized at all, resulting in colorless flowers (Figure 3.35). Several transgenic studies, in *Arabidopsis* and *Nicotiana tabacum* as well as *Petunia*, have shown a correlation between repetition of genes at a given position in the genome and silencing. The more copies of the transgene that are present at one position, the greater the chance that expression from those linked transgenes will be silenced, and the greater their ability to silence related genes, whether transgenes or part of the plant's own DNA complement, elsewhere in the genome. Cosuppression does not lead to a permanently heritable change because endogenous gene activity is restored once the transgene is separated from the endogenous gene by recombination during meiosis to form gametes. It may take several rounds of sexual reproduction to accomplish this separation.

3.5.4 Imprinting occurs only at certain stages in plant development

Imprinting is an example of epigenetic change that operates only during certain stages of plant development. It happens when the expression of certain alleles differs, depending on whether the allele came from the male or the female parent. Examples of imprinting are known in mammals and fungi as well as in plants.

Imprinting is particularly evident during some phases of reproductive development, and has been intensively studied in the endosperm of the seed in flowering plants. Figure 3.36 shows imprinting in the case of the red gene (*r*) that affects endosperm color. Certain alleles of the gene (the *R* alleles) are strongly expressed, giving a solid seed color, when transmitted through the female egg but weakly expressed (variegated seed color) when transmitted through the male pollen. This reflects a male-specific epigenetic change in *r* gene activity.

3.6 Expression of the plant genome III. Translation of RNA to protein

The process of cytoplasmic and mitochondrial protein synthesis in plants is similar to that in other eukaryotic organisms, except that translation of some mRNAs can be regulated by light. Protein synthesis in plastids is unique to plant cells. In photosynthetically active tissue,

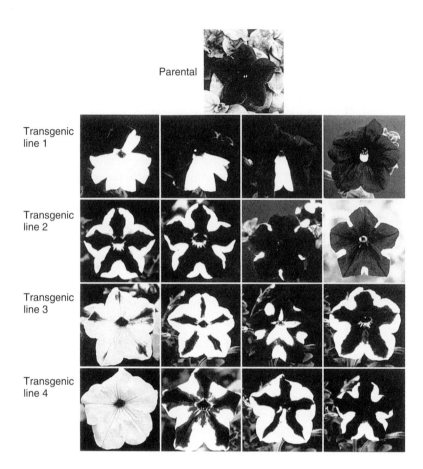

Parental

Transgenic line 1

Transgenic line 2

Transgenic line 3

Transgenic line 4

Figure 3.35 Cosuppression of the plant's own gene by the introduction of a transgene. The flowers of the parental petunia plants are deep purple because of the presence of anthocyanin pigments. The insertion of a chalcone synthase transgene suppresses the expression of the native chalcone synthase gene, reducing or preventing anthocyanin biosynthesis. The most extreme examples have white, unpigmented flowers. Intermediate examples result where the native gene is less effectively suppressed, or suppressed only in certain groups of cells. Each row shows four representative flowers for an individual transgenic line.

Key points Some changes in gene expression can be passed on from cell to cell in an organism, and in certain cases from one generation to the next, but nevertheless do not involve changes to the primary DNA sequence. These changes are described as epigenetic. DNA methylation at cytidine residues, affecting chromatin structure, is one important mechanism for epigenetic regulation of gene expression. Paramutation is a type of epigenetic change in which the expression of one allele in a heterozygous individual is altered by the presence of the other allele; this class of alteration can be passed on to subsequent generations. Cosuppression is a form of epigenetic gene silencing in which overexpression of a transgene results in suppression of the expression of the plant's own copy of the gene. In imprinting, the expression of certain alleles differs depending on whether they came from the male or the female parent.

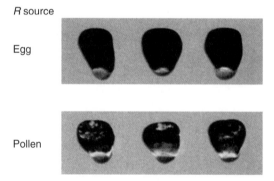

R source

Egg

Pollen

Figure 3.36 Imprinting is a type of epigenetic regulation of gene expression in which the expression of a gene varies according to whether it came from the male or female parent. In this example, *R* alleles are strongly expressed when transmitted from the female parent, giving darkly colored *Zea mays* kernels. However, they are more weakly expressed when derived from the male parent, so that the kernels are less intensely colored.

about 75% of the total protein synthesis takes place in the cytosol, using mRNAs transcribed in the nucleus as templates. The products can total more than 25 000 different proteins, whereas chloroplasts and mitochondria together account for, at most, 300 different proteins.

Once mature mRNA has been moved out of the nucleus into the cytosol it can be translated to produce a protein. **Translation** is the mechanism by which the sequence of nucleotides in mRNA is decoded and used as

Amino acid	3-Letter code	1-Letter code	Codons
Alanine	Ala	A	GCC, GCU, GCG, GCA
Arginine	Arg	R	CGC, CGG, CGU, CGA, AGA, AGG
Asparagine	Asn	N	AAU, AAC
Aspartic acid	Asp	D	GAU, GAC
Cysteine	Cys	C	UGU, UGC
Glutamic acid	Glu	E	GAA, GAG
Glutamine	Gln	Q	CAA, CAG
Glycine	Gly	G	GGU, GGC, GGA, GGG
Histidine	His	H	CAU, CAC
Isoleucine	Ile	I	AUU, AUC, AUA
Leucine	Leu	L	UUA, UUG, CUA, CUG, CUU, CUC
Lysine	Lys	K	AAA, AAG
Methionine	Met	M	AUG
Phenylalanine	Phe	F	UUC, UUU
Proline	Pro	P	CCU, CCC, CCA, CCG
Serine	Ser	S	UCU, UCC, UCA, UCG, AGU, AGC
Threonine	Thr	T	ACU, ACC, ACA, ACG
Tyrosine	Tyr	Y	UAU, UAC
Tryptophan	Trp	W	UGG
Valine	Val	V	GUU, GUC, GUA, GUG
"Stop"	—	—	UAA, UAG, UGA

Figure 3.37 The genetic code, which relates the codons (the triplets of nucleotides in mRNA) to the amino acids that each triplet encodes in the protein. Amino acids can be represented as three-letter or one-letter abbreviations (see Figure 2.12).

a template that specifies the order of amino acids in a protein. The **genetic code** specifies the relationship between the sequence of nucleotides in the mRNA and the corresponding sequence of amino acids in the protein. The mRNA is translated three nucleotides at a time, and each trinucleotide sequence, or **codon**, encodes one amino acid (Figure 3.37). The genetic code is said to be degenerate, because there is more than one codon for most of the 20 amino acids that are found in proteins. There are 64 possible mRNA codons; three of them are reserved for use as **stop codons**, which identify the position where translation of the mRNA terminates. It is important that the protein synthesis machinery initiates translation at the correct point in the sequence; in the example shown in Figure 3.38 if translation were initiated at the U instead of the A in the AUG, a protein with an entirely different—and probably non-functional—amino acid sequence would be produced.

The process of translation requires **transfer RNA (tRNA)**, an adaptor molecule that recognizes codons in mRNA and the amino acids for which they code. Translation also needs **ribosomes**, the structures that will catalyze the formation of peptide bonds between amino acids. In the following sections we examine these participants in more detail and then discuss the process of protein synthesis and the production of functional proteins.

3.6.1 Transfer RNAs are the link between mRNA codons and amino acids

Transfer RNAs are small RNA molecules, generally 70–90 nucleotides long. Each tRNA has a position to which a specific amino acid can be attached and a three-nucleotide **anticodon** sequence that base-pairs only with the mRNA triplet that encodes the bound amino acid. Base-pairing within the tRNA sequence results in an L-shaped three-dimensional conformation that exposes the anticodon and enables base-pairing with the complementary codon in the mRNA sequence (Figure 3.39). The attachment of specific amino acids to the 3′ ends of tRNA molecules is catalyzed by specific aminoacyl-tRNA synthetases. A tRNA with an attached amino acid is said to be **aminoacylated** or 'charged' and is represented as **aa-tRNA^{aa}** (e.g. Phe-tRNA^{Phe}). Together the ribosome and aa-tRNAs participate in decoding the information carried in the nucleotide sequence of the mRNA.

While there are 61 codons that specify amino acids, there are only 30–40 different tRNAs, depending on species. This is possible because some of the tRNAs can recognize more than one codon for their particular amino acid; for example, both the codons 5′-UUU-3′

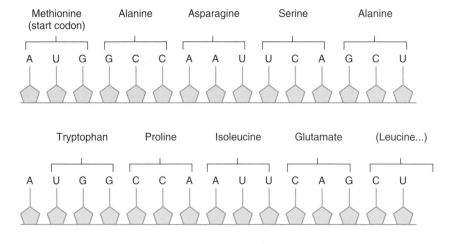

Figure 3.38 Translation of the sequence of five mRNA codons shown here begins at AUG, the start codon. If the initiation of translation were to be displaced by one nucleotide, and begin at UGG, the protein product would have a completely different amino acid sequence.

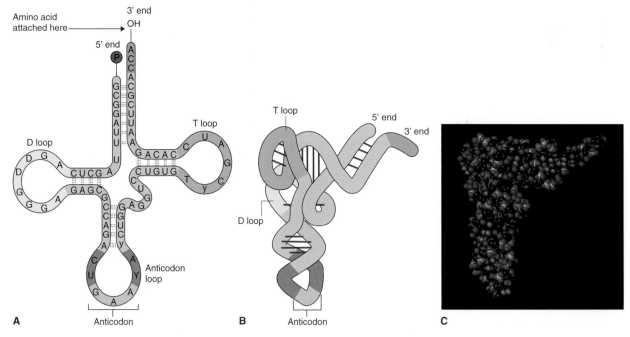

Figure 3.39 The structure of a typical tRNA. (A) The 'cloverleaf' secondary structure of tRNA results from the pattern of pairing of certain bases, forming loops and stems. (B) The tRNA undergoes further folding in three dimensions. (C) This space-filling model shows the compact form of a tRNA molecule.

and 5′-UUC-3′ are translated to the same amino acid: phenylalanine. This is possible because the 5′-GAA-3′ anticodon of a single phenylalanine tRNA can pair with either of the two phenylalanine codons in mRNA (Figure 3.40). One is an exact match; the other has a mismatch at the third nucleotide of the codon, resulting in what is known as **wobble-pairing**. This phenomenon allows 30–40 different tRNAs to read the 61 codons that specify amino acids in the genetic code. Wobble-pairing can only occur with certain combinations of anticodon–codon, which is why more than 30, not just 20, different tRNAs are required to translate the 61 possible codons into amino acids.

3.6.2 Protein biosynthesis takes place on ribosomes

Protein synthesis takes place on structures called ribosomes, which contain ribosomal RNA (rRNA) and proteins (see Table 3.4). The ribosomes found in the cytosol are larger (80S) than those found in mitochondria and plastids (70S); organellar ribosomes resemble those found in prokaryotic cells. Each ribosome consists of two **subunits**, referred to as the large and small subunits, based on their relative S values. Each subunit in turn contains one or more rRNAs, which fold

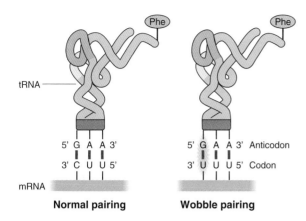

Figure 3.40 Wobble pairing means that a single tRNA can read more than one RNA codon. In this example, tRNA^{Phe}, which carries the amino acid phenylalanine and has the anticodon 5'-GAA-3', pairs with the complementary codon UUC but also with UUU. Thus both UUC and UUU in mRNA translate to phenylalanine in the protein.

A ribosome reads the mRNA-coding sequence in a 5' to 3' direction, directing the addition of the amino acid corresponding to each codon and extending the growing polypeptide chain from the amino (or N) terminus to the carboxy (or C) terminus (Figure 3.41). Proteins are synthesized on ribosomes in a process comprising three distinct phases: **initiation, elongation** and **termination**. Each of these phases requires the participation of specific protein factors named according to the phase in which they are involved: respectively **eukaryotic initiation factors (eIFs), eukaryotic elongation factors (eEFs)** and **eukaryotic release factors (eRFs)**. In the next several sections we will explore each of these phases and the roles that protein factors play in each.

3.6.3 Protein synthesis is initiated from the 5' end of the mRNA

into highly ordered structures and associate with proteins. Plant cytosolic ribosomes can contain up to 80 proteins that are highly conserved between species. They are mostly basic (positively charged), which facilitates their interaction with negatively charged rRNA. However, cytosolic ribosomes also contain a small family of acidic proteins, called the **P-proteins**. One of those found in plants, P3, is not found in other eukaryotes.

Protein synthesis begins at the universal **start codon AUG** that encodes the amino acid **methionine**. Thus a methionine residue or a derivative is at the N terminus of every nascent (growing) polypeptide. Often the N-terminal methionine is removed during post-translational processing. There are two different tRNAs that carry the anticodon that specifies methionine. One of these is a **specific initiator, tRNA_i^{Met}**, which only functions to deliver the N-terminal

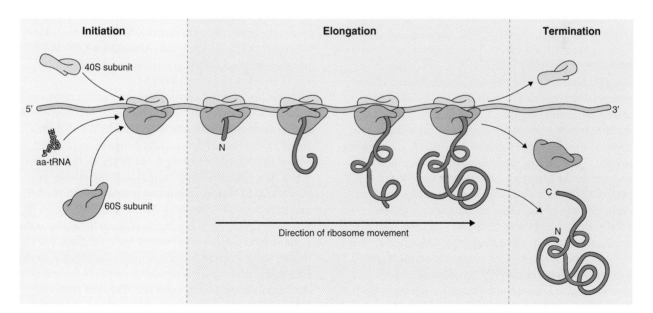

Figure 3.41 Protein synthesis. The mRNA is read from the 5' end to the 3' end, and the polypeptide chain is synthesized, starting from the amino terminus, by sequential addition of amino acids. The process of translation can be divided into three stages: initiation, elongation and termination.

Translation of messenger RNA to the protein it encodes is dependent on the triplet code, in which each group of three nucleotides, or codon, in the DNA of the gene encodes an amino acid in the protein product. There are 64 possible trinucleotide codons; three of them are used as stop codons, identifying positions in the DNA where transcription of the mRNA terminates, and the other 61 encode amino acids. The genetic code is said to be degenerate because in most cases a given amino acid can be specified by more than one codon. Transfer RNAs are the adaptor molecules that recognize codons in mRNA and ensure that they are translated into the correct amino acids. Protein biosynthesis takes place on ribosomes, structures that contain ribosomal RNAs and proteins.

methionine to the start codon. The other Met-tRNA^{Met} species carries the rest of the methionines incorporated into the elongating polypeptide chain. Initiation at the start codon establishes the **reading frame** (the phase in which the triplets in the mRNA are read).

During the initiation phase of protein synthesis, a 40S ribosomal subunit with an attached Met-tRNA$_i^{Met}$ binds to the 5′ end of an mRNA. It then scans along the mRNA until it encounters a start codon. This process is complex and requires the participation of a number of auxiliary proteins referred to as eukaryotic initiation factors (Figure 3.42). First, the factor eIF2 interacts with initiator Met-tRNA^{Met} and GTP to form an RNA–protein **ternary complex**. Next, the ternary complex binds to a free 40S ribosomal subunit to form a **preinitiation complex**, a process that is facilitated by several other eIFs (Figure 3.42). Members of the eIF4 initiator family assemble around the 5′ cap of an mRNA, allowing the ribosomal subunit–Met-tRNA-eIF complex to bind. The complex scans down the mRNA until it encounters the Met start codon where the initiator Met-tRNA^{Met} base-pairs with the start codon. At this point, the eIFs of the initiation complex are released. Finally, the large ribosomal subunit associates with the small subunit, positioning the initiator Met-tRNA^{Met} in the P site (peptidyl-tRNA-binding site) on the large subunit. Members of the eIF5 family participate in positioning the large subunit and in the release of GDP, the product of GTP hydrolysis (Figure 3.42). Now everything is in place and protein synthesis can begin. **Translation-level regulation** of gene expression in eukaryotes occurs primarily at the initiation step. In this respect, the participation of eIF2 and GTP and the release of GDP are important control points.

3.6.4 Polypeptide chain elongation occurs by the sequential addition of amino acid residues to the growing polypeptide chain

At the end of the initiation phase there is an initiator tRNA^{Met} in the P site of the ribosome; in subsequent steps in the synthesis cycle this site will be occupied by the aa-tRNA that holds the growing polypeptide chain. During the elongation phase of protein synthesis, a ribosome holds mRNA and tRNAs loaded with amino acids in position and the **peptidyl transferase** site of the ribosome catalyzes the formation of peptide bonds between successive amino acids to build up the polypeptide chain (Figure 3.43). The addition of amino acids to a growing polypeptide chain requires three sites on the fully assembled ribosome, known as the **A, P** and **E sites**. These three sites are used in turn, in a cycle that takes as little as a twentieth of a second to add an amino acid to the chain. The process of chain elongation requires three elongation factors: eEF1A, eEF1B and eEF2.

Elongation factor eEF1A binds to aa-tRNAs and GTP delivers them to the A site (aminoacyl-tRNA-binding site; also known as the decoding site) on the ribosome where the next codon on the mRNA is exposed. If the aa-tRNA anticodon is complementary to the mRNA codon in the A site, GTP is hydrolyzed, causing a conformational change in the aa-tRNA in the A site. This brings the amino acid on that tRNA closer to the amino acid on the tRNA in the P site. Also eEF1A-GDP is released during this step; it is subsequently cleaved by eEF1B to allow recycling of eEF1A.

Now, the peptidyl transferase center of the ribosome catalyzes the formation of a peptide bond between the amino acids attached to the tRNAs in the A and P sites. During this process, the growing polypeptide chain is passed from the tRNA in the P site to the aa-tRNA in the A site. The ribosome then moves along the mRNA in a process called **translocation** that is catalyzed by eEF2 and requires GTP hydrolysis. As the ribosome moves the now deacylated tRNA is transferred from the P site to the E site (exit site) from which it will leave the ribosome during the next cycle. Simultaneously, the tRNA with the attached peptide chain is repositioned to the P site and a new codon is exposed at the A site. The next cycle of amino acid addition to the peptide chain can begin.

A single 80S ribosome occupies about 30–35 nucleotides on an mRNA. Once a ribosome has started to read an mRNA, it moves downstream. This exposes the start codon, and allows a second ribosome to initiate translation of the same mRNA. Thus, when an mRNA is

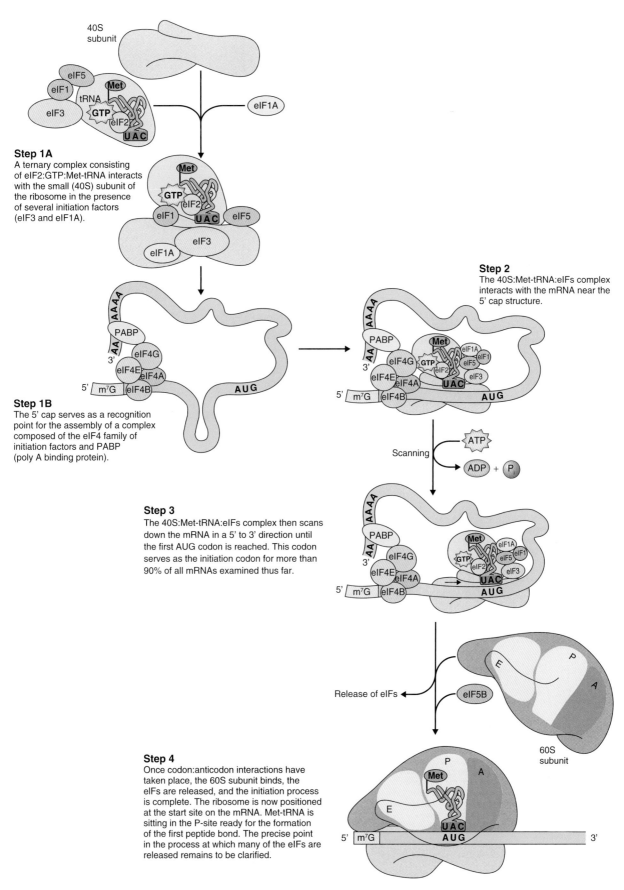

Step 1A
A ternary complex consisting of eIF2:GTP:Met-tRNA interacts with the small (40S) subunit of the ribosome in the presence of several initiation factors (eIF3 and eIF1A).

Step 1B
The 5' cap serves as a recognition point for the assembly of a complex composed of the eIF4 family of initiation factors and PABP (poly A binding protein).

Step 2
The 40S:Met-tRNA:eIFs complex interacts with the mRNA near the 5' cap structure.

Step 3
The 40S:Met-tRNA:eIFs complex then scans down the mRNA in a 5' to 3' direction until the first AUG codon is reached. This codon serves as the initiation codon for more than 90% of all mRNAs examined thus far.

Step 4
Once codon:anticodon interactions have taken place, the 60S subunit binds, the eIFs are released, and the initiation process is complete. The ribosome is now positioned at the start site on the mRNA. Met-tRNA is sitting in the P-site ready for the formation of the first peptide bond. The precise point in the process at which many of the eIFs are released remains to be clarified.

Figure 3.42   The mechanism of initiation of polypeptide chain synthesis in the cytosol.

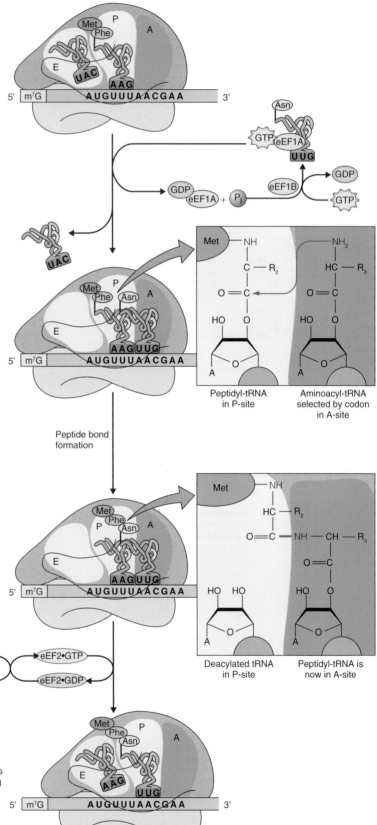

Step 1
A growing polypeptide chain is covalently attached to the tRNA in the P-site. The A-site is empty, exposing the next codon in the mRNA. The E-site is occupied by the uncharged tRNA from the previous cycle.

Step 2
A charged tRNA binds to the ribosomal A-site only if its anticodon matches the exposed codon on the mRNA. When this occurs, the tRNA in the E-site is ejected from the ribosome. A required elongation factor, eEF1A, forms a ternary complex with GTP and the charged tRNA and promotes the binding of the charged tRNA to the A-site of the ribosome.

Step 3
The ribosome uses the peptidyl transferase center to catalyze the formation of a peptide bond between the growing polypeptide chain and the new amino acid. Recent experiments suggest that rRNA plays a particularly important role as a ribozyme in this catalytic step. The net result of this process is that the nascent polypeptide has been transferred from the tRNA in the P-site to the new amino acid attached to its tRNA in the A-site. The polypeptide is one residue longer, the peptidyl-tRNA now occupies the A-site, and the tRNA in the P-site is free of its amino acid. This tRNA is said to be deacylated.

Step 4
The complex must rearrange, exposing the next triplet. This process, called translocation, involves three rearrangements: The deacylated tRNA in the P-site moves into the vacant E-site; the peptidyl-tRNA in the A-site moves into the P-site; and the ribosome moves relative to the mRNA by exactly three nucleotides (one codon), exposing a new triplet in the A-site. Moving one or two nucleotides too few or too many would initiate a new reading frame, probably resulting in an inactive polypeptide.

Figure 3.43 The elongation phase in cytosolic protein synthesis.

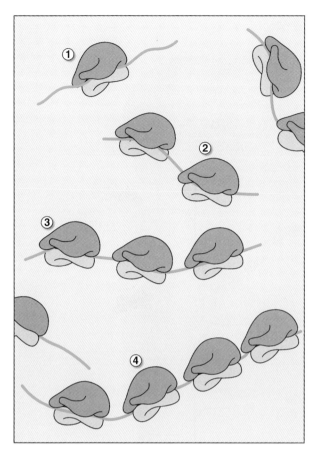

Figure 3.44 Polysomes are messenger RNA molecules bound to two or more ribosomes, as indicated by numbers in the diagram.

being actively translated, it is usually associated with several ribosomes, separated by as few as 80–100 nucleotides, forming a **polyribosome** or **polysome** (Figure 3.44). Depending on the proteins they are synthesizing, polyribosomes may be free in the cytosol or attached to the endoplasmic reticulum (ER).

3.6.5 Protein synthesis terminates when a stop codon is reached

When a ribosome reaches one of three stop codons on the mRNA (UAA, UAG or UGA), this is a signal that polypeptide chain elongation should cease (Figure 3.45). For protein synthesis to terminate, two specific proteins, known as release factors, eRF1 and eRF3, must bind to the A site. eRF1 is a protein that structurally mimics tRNAs and recognizes all three stop codons. It binds eRF3, which has a bound GTP. Interaction between the RFs and the ribosome sets off a series of events.

Hydrolysis of GTP by RF3 catalyzes the cleavage of the bond between the completed polypeptide chain and the final tRNA molecule at the P site, releasing the polypeptide from the ribosome. It also stimulates the release of the RFs from the ribosome. At this point the tRNAs and the ribosome are also released from the mRNA and become available to participate in another cycle of translation. The ribosome may dissociate into its small and large subunits at this point, a process promoted by eIF6. However, this may not occur until small subunits are required for further translation, so intact ribosomes lacking an mRNA often accumulate when initiation of translation is limited, such as under conditions of stress (for example, hypoxia, heat shock or dehydration stress).

3.6.6 Most proteins undergo post-transcriptional modifications

A newly synthesized polypeptide frequently undergoes further modifications to reach its final form as a mature, functional protein. First, some of the amino acids may undergo covalent modification by the addition of acetyl, phosphate or other groups. Second, the polypeptide must be folded into its correct three-dimensional structure. Examples of three-dimensional protein structures are shown elsewhere in this book. Third, many proteins consist of two or more subunits, which must assemble to form a functional protein. For example, the

Key points The synthesis of a protein is initiated from the 5′ end of the mRNA, at the start codon AUG. Initiation requires several proteins known as initiation factors, in addition to mRNA and ribosomes. Elongation of the polypeptide chain occurs by the sequential addition of amino acid residues. A ribosome holds the mRNA and transfer RNAs loaded with the appropriate amino acids in position so that the peptidyl transferase site of the ribosome can catalyze the formation of peptide bonds between successive amino acids. Elongation continues until a stop codon is reached; then the new polypeptide is released and the ribosome and mRNA dissociate. Newly-synthesized polypeptides often undergo post-transcriptional modifications to become mature, functional proteins. These modifications include covalent additions of acetyl, phosphate or other groups, and folding of the polypeptide into the correct secondary, tertiary and, where appropriate, quaternary structure.

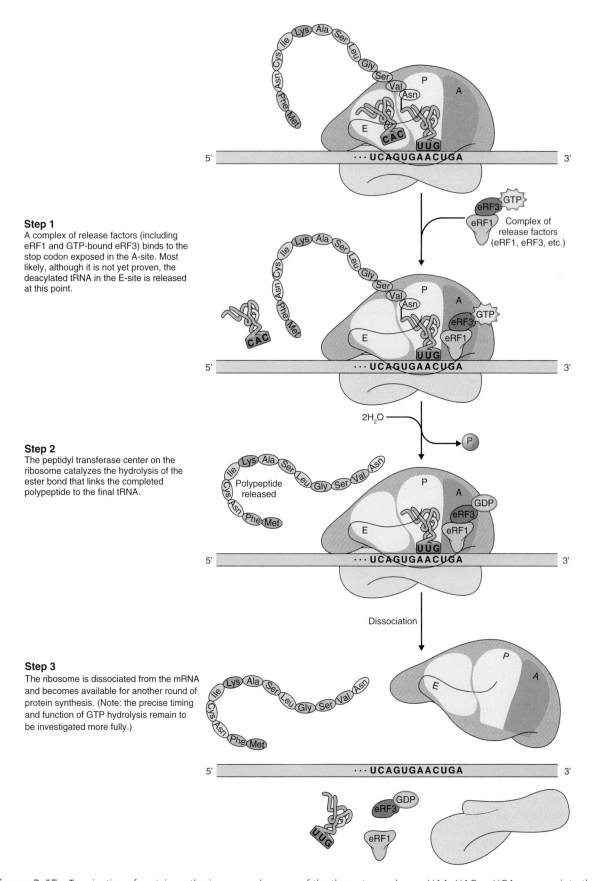

Step 1
A complex of release factors (including eRF1 and GTP-bound eRF3) binds to the stop codon exposed in the A-site. Most likely, although it is not yet proven, the deacylated tRNA in the E-site is released at this point.

Step 2
The peptidyl transferase center on the ribosome catalyzes the hydrolysis of the ester bond that links the completed polypeptide to the final tRNA.

Step 3
The ribosome is dissociated from the mRNA and becomes available for another round of protein synthesis. (Note: the precise timing and function of GTP hydrolysis remain to be investigated more fully.)

Figure 3.45 Termination of protein synthesis occurs when one of the three stop codons—UAA, UAG or UGA—moves into the A site of the ribosome.

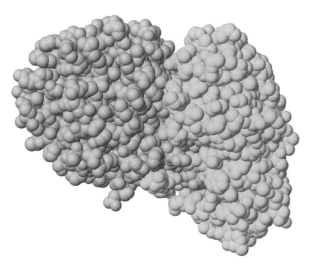

Figure 3.46   The structure of a lectin from mistletoe (*Viscum album*). The protein consists of two different subunits, of 254 and 263 amino acids, colored blue and green in the figure.

major lectin of mistletoe (*Viscum album*), which is highly toxic to humans and other animals, is a heterodimer (Figure 3.46), meaning that it has two non-identical subunits. Its 1A subunit is 254 amino acids in length, while the 1B subunit is 263 amino acids long. Lectins are carbohydrate-binding molecules, and in the mistletoe lectin, the 1B subunit binds to the sugar-containing backbone of the 28S ribosomal RNA subunit while the 1A subunit cleaves an adenine base from the ribosomal RNA, blocking protein synthesis at the elongation stage and leading to cell death.

3.7 Expression of organellar genes

So far we have focused on the transcription and translation of nuclear genes. As described in Section 3.2, chloroplasts and mitochondria also have genomes that encode RNAs and proteins. The expression of chloroplast and mitochondrial genes is in many important respects like that of nuclear genes. It consists of DNA transcription by RNA polymerase to produce RNA, followed, in the case of protein-coding genes, by translation of mRNA on ribosomes to make polypeptides. However, the details of the gene expression process in chloroplasts and mitochondria more closely resemble what takes place in bacteria, reflecting the

origins of these organelles as prokaryotic endosymbionts. Chloroplast gene expression has been much more intensively studied than plant mitochondrial gene expression. Mitochondrial gene expression in plants is similar to that in other eukaryotes, and will not be further discussed here.

3.7.1 The machinery of chloroplast gene expression resembles that of bacteria more than that of nuclear genes

The main RNA polymerase of chloroplasts is similar to that of bacteria, consisting of a core of four subunits (two of them identical) and one of the several regulatory subunits encoded by the plastid genome. This polymerase recognizes genes with promoter regions similar to the bacterial −10 and −35 sequences, so called because they lie approximately 10 and 35 nucleotides, respectively, upstream from the transcription initiation site. However, not all chloroplast genes have such promoter sequences. Plants also have a nuclear-encoded RNA polymerase that is imported into the plastid and can recognize alternative promoter sequences.

The ribosomes, messenger RNAs, enzymes and other components needed for the synthesis of chloroplast-encoded proteins are different from those used for the synthesis of nuclear-encoded proteins. The numbers and functions of specific initiation factors in the plastid are considerably different from those of the cytosol. Bioinformatic analysis has shown that plastids have homologs to the **bacterial initiation factors** IF1, IF2 and IF3. Similarly, the elongation factors, the termination factors and the **ribosome recycling factor** of plastids have sequence homology to the bacterial equivalents rather than the eukaryotic ones.

Plastid ribosomes contain homologs of almost all the ribosomal proteins found in bacteria, but there are also **plastid-specific ribosomal proteins** (**PSRPs**) that have no bacterial equivalents. The number of these PSRPs varies between different plant species; for example, spinach chloroplast ribosomes have six of them, four in the small subunit and two in the large subunit. It has been suggested that some of the PSRPs may have roles in the regulation of protein synthesis by light, an important property of chloroplasts but not of mitochondria or the majority of bacteria.

3.7.2 Transcripts encoded by the plastid genome are often polycistronic and are translated by prokaryotic-type mechanisms

One of the key differences between the translation systems of the chloroplast and cytosol is the way in which the initiation signal on the mRNA is selected. While cytosolic mRNAs have a 5′ cap, which flags the 5′ end for initiation factors, plastid mRNAs are not capped. Like bacterial mRNAs, they are often transcribed in the first instance as **polycistronic messages** with the potential to be translated to several proteins.

In bacterial systems, the correct AUG codon for initiation is selected by the 16S rRNA in the small (30S) subunit, which can base-pair with a short sequence found just upstream of the initiation AUG. This sequence, which is called the **Shine/Dalgarno sequence** after its discoverers, directs selection of the start codon by the 30S subunit. The secondary structure of the mRNA is also important in determining the efficiency of translational initiation.

The plastids of some plants use a mechanism similar to the bacterial system. In such species, the chloroplast mRNAs contain a Shine/Dalgarno sequence upstream from the start codon (Figure 3.47). This sequence,

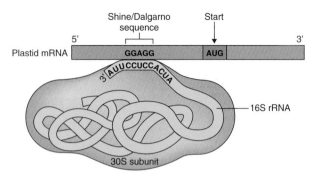

Figure 3.47 The Shine/Dalgarno sequence and its participation in selection of the start codon in a plastid gene. Hydrogen bonding takes place between a polypurine sequence in the Shine/Dalgarno sequence and a polypyrimidine sequence near the 3′ end of the 16S rRNA.

together with the secondary structure of the mRNA surrounding the start codon, plays an important role in initiation of translation. In contrast, plastid mRNAs of other plant species do not contain a Shine/Dalgarno sequence within 20 nucleotides of the start codon. In these mRNAs the translational start site is specified by the presence of an AUG region of the mRNA with little or no secondary structure (Figure 3.48) and evidence suggests that in many of these cases a nuclear-encoded protein is also required to facilitate initiation.

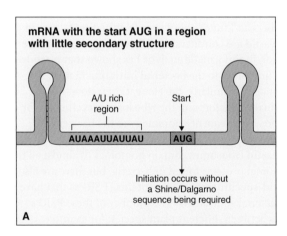
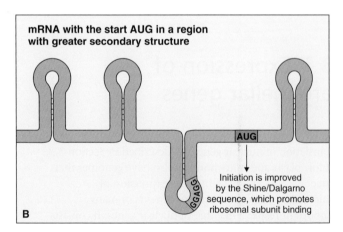

Figure 3.48 Selection of the start site on chloroplast mRNAs proceeds by one of two routes, depending on the mRNA sequence and its secondary structure. (A) The start codon may be in a region with little secondary structure, requiring a trans-acting factor for effective initiation. (B) In the second route, a Shine/Dalgarno sequence is located near the start codon, facilitating its selection for initiation.

Key points The expression of chloroplast and mitochondrial genes follows the central dogma of molecular biology: DNA → RNA → protein. Organellar DNA is transcribed by RNA polymerase to produce RNA and, in the case of protein-coding genes, RNA is then translated on ribosomes to make polypeptides. The details of ribosome structure, initiation factors and gene expression in chloroplasts and mitochondria are more similar to those of eubacteria than to those of the nucleus and cytoplasm, reflecting the endosymbiotic origins of plastids and mitochondria. Mitochondrial gene expression in plants is similar to that in other eukaryotes. The main RNA polymerase of chloroplasts resembles that of bacteria, and can recognize promoter regions similar to bacterial promoter sequences. Because some chloroplasts lack such promoter regions, chloroplasts also import a nuclear-derived RNA polymerase that can recognize alternative promoters. Selection of the correct site for translation initiation on plastid RNAs is achieved by one of two mechanisms. The first is similar to that used in bacteria, whereas the other appears specific to plastids.

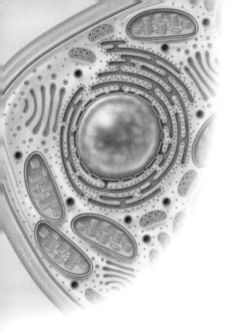

Chapter 4
Cell architecture

4.1 Introduction to cell structure

Living things consist of cells, distinct structural units that house the machinery that makes life possible. Plants have eukaryotic cells; a generalized plant cell is shown in Figure 4.1. Unlike animal cells, plant cells are surrounded by an extracellular coat, the **cell wall**. Within the cell wall lies the living **protoplast**, whose outer boundary is the **plasma membrane** that contains a variety of membrane-bound **organelles**, compartments with specialized functions.

Advances in our understanding of cell structure have come from developments in microscopy as well as in chemistry and molecular biology. Although electron microscopes were widely used to study cell structure in the mid-20th century, recent developments in tissue preparation have greatly expanded the scope of **electron microscopy** (**EM**). These advances include the use of **rapid freezing** and **freeze-substitution** methods that preserve cell structure at conditions nearer to those found in vivo because harsh chemical fixation procedures are avoided. Freeze substitution also preserves the antigenicity of proteins, allowing the use of antibodies to detect proteins at the EM level.

There have been recent dramatic advances in **light microscopy** as well, especially with **laser scanning confocal microscopy** which allows details of structure of living cells to be viewed at the three-dimensional level. The power of light microscopy has also been enhanced by the development of reporter molecules that allow the detection of a range of biologically important ions and macromolecules. These reporters include fluorescent dyes that are specific for a variety of molecules such as Ca^{2+} and H^+ as well as fluorescent proteins such as **green fluorescent protein** (**GFP**) whose expression can be engineered in cells after fusion to an appropriate promoter or target gene.

Advances in chemistry have also been crucial in enhancing our understanding of plant cell structure, especially as it pertains to the cell wall. The plant cell wall is a complex of **polysaccharide, protein** and **lignin** polymers whose structures are being resolved, especially through advances in **mass spectrometry**. Mass spectrometry allows the identification of cell wall sugars and phenolic compounds and this information is beginning to reveal how polysaccharides and lignins are assembled.

In this chapter we will address the cellular structures that are present in most living plant cells. The chemical compositions of many of these structures, a topic integral both to plant biology and to animal and human nutrition, are discussed in Chapters 2 and 5.

4.2 The cell wall

A strong, flexible primary cell wall composed of complex polysaccharides and proteins surrounds each plant cell. The shape of a plant cell and eventually of the plant itself is dictated by its cell wall, and this shape is important for the cell's function (Figure 4.2). The cell wall prevents

The Molecular Life of Plants, First Edition. Russell Jones, Helen Ougham, Howard Thomas and Susan Waaland.
© 2013 John Wiley & Sons, Ltd. Published 2013 by John Wiley & Sons, Ltd.

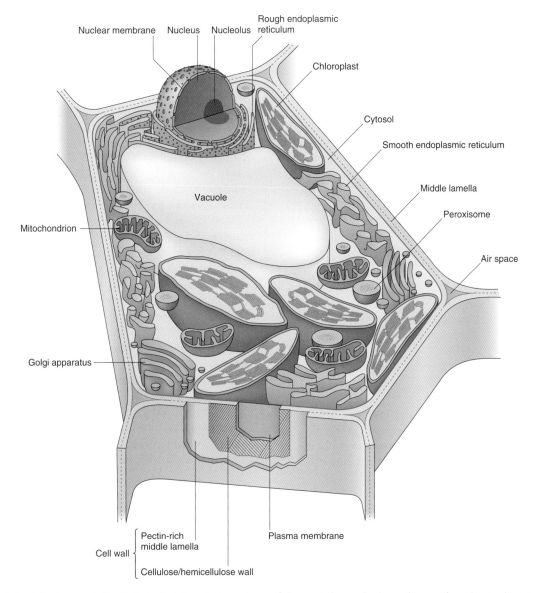

Figure 4.1   A leaf mesophyll cell. Note that the relative volume of the central vacuole shown here underestimates its actual size in a mature mesophyll cell. The plasmodesmata that connect the protoplasts of neighboring cells are not shown.

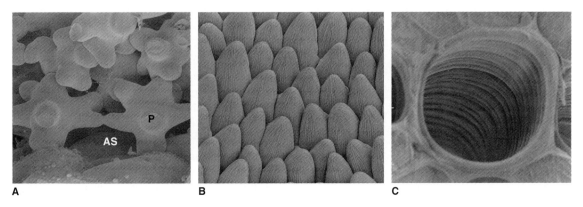

Figure 4.2   Scanning electron micrographs showing cell wall architecture of a selection of developing plant cells. (A) Spongy parenchyma (P) of *Zinnia* leaves showing large airspaces (AS) between cells that facilitate gas exchange. (B) Epidermal cells of snapdragon (*Antirrhinum majus*) petals that reflect light efficiently to enhance brightness. (C) Xylem tracheid with radial thickenings that strengthen the wall in these water-conducting cells which function under extreme tension.

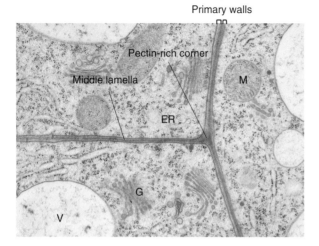

Primary walls

Pectin-rich corner

Middle lamella

ER

M

G

V

Figure 4.3   Transmission electron microscopy (TEM) image of the corners of three abutting cells. The primary walls of these cells are glued together by the middle lamella, which is formed during cell division and grows with the primary wall during cell expansion. The cell corners are often filled with pectin-rich polysaccharides; in older cells, the material in the cell corners is sometimes degraded and an airspace forms (not shown). ER, endoplasmic reticulum; G, Golgi apparatus; M, mitochondrion; V, vacuole.

excess water uptake by living cells, allowing plants to survive in fresh water. Adjacent plant cells are glued together along their abutting walls by a layer called the **middle lamella** so that the cells in a plant do not normally move relative to each other (Figure 4.3). Together, all the cell walls form a network, much like a skeleton, that helps hold the plant upright and gives each plant its characteristic shape or **morphology**. The cell wall is dynamic, changing its structure and composition throughout the life of the cell and rapidly increasing in surface area during cell expansion. In living cells, the cell wall constrains the rate and direction of growth and exerts a profound influence on plant development and morphology. Finally, during differentiation, many cells manufacture a more rigid **secondary cell wall**, producing complex structures uniquely suited to the cell's function.

4.2.1 Cellulose is a major component of the fundamental framework of primary cell walls

The primary cell wall is made up of several structurally independent but interacting networks. The fundamental framework of the wall is polysaccharide, consisting of **cellulose** rods connected by **cross-linking glycans**. This network may be embedded in a matrix of **pectic polysaccharides**. A third independent,

non-polysaccharide network can consist of **structural proteins** or **phenylpropanoids**. We will describe each of these networks.

Cellulose forms the basic scaffold of cell walls. Cellulose is the most abundant plant polysaccharide, accounting for 15–30% of the dry mass of all primary cell walls and an even larger percentage of secondary walls. Cellulose molecules are linear chains of 2000–20 000 (1 → 4)-β-D-linked glucose units (see Chapter 2). In cell walls, cellulose exists as **microfibrils**, paracrystalline arrays of many cellulose molecules, hydrogen bonded to one another along their lengths (Figure 4.4). On average, each microfibril is 36 individual cellulose chains in diameter. Because the individual cellulose chains begin and end at different places within a microfibril, a microfibril can contain thousands of chains and be hundreds of micrometers long.

Callose, another polymer of glucose, is a major component of the walls of certain types of cells and at certain times during development. It differs from cellulose because its glucose units have a (1→3)-β-D-linkage, which gives the molecule a helical rather than a linear shape. As a result, callose molecules do not form microfibrils because they do not hydrogen bond along their length as do linear cellulose chains. Callose is produced in the cell walls of pollen grains, in elongating pollen tubes, in the cell plates of dividing cells, and in cells that have been wounded mechanically or attacked by pathogenic fungi.

4.2.2 Cross-linking glycans interlock the cellulosic scaffold

Cellulose microfibrils form a mesh that is held together by a class of cell wall polysaccharides called **cross-linking glycans**, which form hydrogen bonds with cellulose and with each other. They coat the cellulose microfibrils, link to other glycans, and span the distance between microfibrils to form the fundamental structural network of the cell wall. In the past, cross-linking glycans were called 'hemicelluloses', a term that refers to cell wall material that can be extracted with strong alkali.

The two major cross-linking glycans of all primary cell walls of flowering plants are **xyloglucans** (XyGs) and **glucuronoarabinoxylans** (GAXs). Xyloglucans consist of linear chains of (1→4)-β-D-glucose with numerous **α-D-xylose** units linked at regular intervals at carbon 6 of the glucose units (Figure 4.5A). They may also have side linkages to other sugars. Glucuronoarabinoxylans are linear chains of (1→4)-β-D-xylose with side linkages to **glucuronic acid** and **arabinose** (Figure 4.5B). In addition to these major classes of cross-linking glycans, there are other less abundant polysaccharides, such as **glucomannans, galactomannans** and

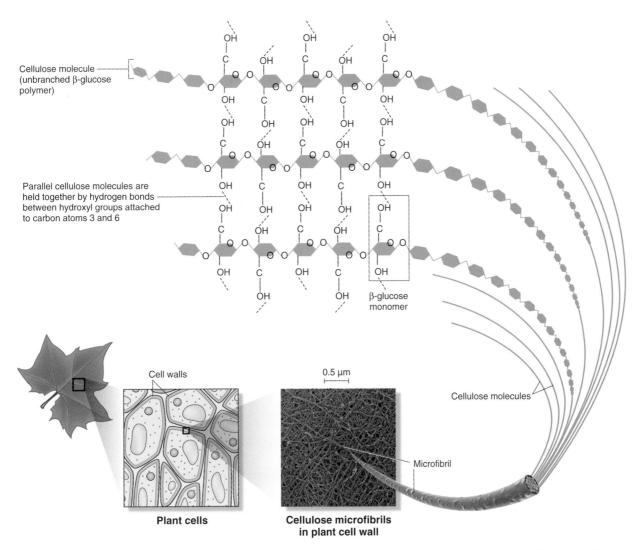

Figure 4.4 Arrangement of cellulose in cell walls. Linear cellulose molecules hydrogen bond to each other to form microfibrils. A microfibril is usually made up of about 36 cellulose molecules. Note that only the hydroxyl groups at positions 3 and 6 are shown on the glucose molecules in this figure.

galactoglucomannans, which may interlock the microfibrils in some primary walls. These mannans are found in virtually all angiosperms examined.

Angiosperm primary cell walls are classified into two groups based on the chemical composition of the non-cellulosic components of these walls (Table 4.1). **Type I cell walls** are found in all eudicot species and in about one-half of the monocots. Xyloglucans are the major cross-linking glycans in Type I walls; these walls have about an equal amount of cellulose and xyloglucan (Figure 4.6). **Type II cell walls** are found in commelinoid monocots (including bromeliads, palms, sedges and grasses). In Type II walls the major cross-linking glycans are glucuronoarabinoxylans (Figure 4.6). Within this latter group, the grasses, including the cereals, have additional cross-linking glycans, **β-glucans**, which are 'mixed-linkage' $(1\rightarrow 3)$, $(1\rightarrow 4)$-β-D-glucans

(Figure 4.5C). Mixed linkage glycans are especially abundant in the cell walls of cereal grain endosperm and make up the bulk of '**soluble fiber**'. The differences in the cross-linking glycans present in Type I and Type II cell walls may be reflected in differences in the responses of these cell walls to pH and other wall-loosening factors (see Chapter 12).

4.2.3 Pectin matrix polymers can form a second network in primary cell walls

Pectins, a heterogeneous mixture of branched, and highly hydrated, polysaccharides that are rich in

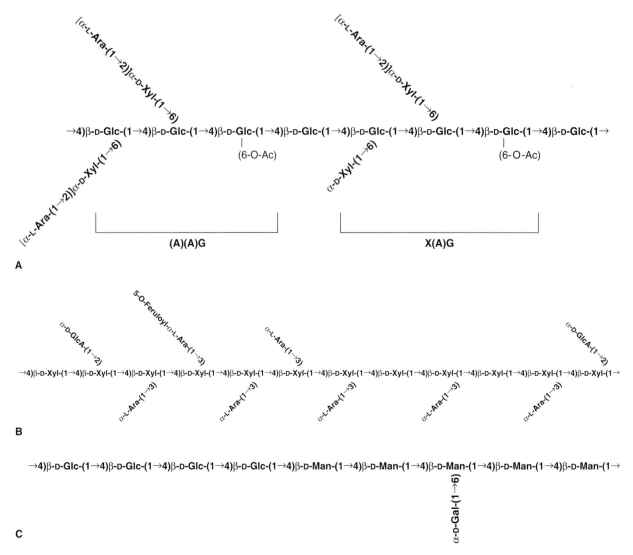

Figure 4.5 Examples of cross-linking glycans from primary cell walls. (A) Solanaceous arabinoxyloglucan, a xyloglycan, with a backbone of β(1→4)-linked glucose and side-chains of xylose and arabinose. (B) A commelinoid glucuronoarabinoxylan with a backbone of β(1→4)-linked xylose and side-chains made up mostly of arabinose and galacturonic acid. (C) The mixed link glucan of grasses made up entirely of glucose linked by either β(1→3) or β(1→4) linkages.

D-galacturonic acid, form a second network in primary cell walls. They bind Ca^{2+} and can be extracted from the cell wall by calcium chelators. Two fundamental constituents of pectins are **homogalacturonan** (**HGA**), a linear chain of galacturonic acid, and **rhamnogalacturonan**, a linear chain of alternating residues of galacturonic acid and **rhamnose**. Homogalacturonan chains can link with one another via Ca^{2+} bridges in which negatively charged galacturonic acid residues from two different HGA molecules associate with the same calcium ion.

Pectins affect the porosity of walls, modulate wall pH and ion balance, and act as recognition molecules to help plant cells detect the presence of symbiotic organisms, pathogens and insects. In addition, pectins in the **middle**

lamella glue adjacent cells to one another (see Figure 4.3). Type I cell walls contain much more pectin than Type II walls but in both types the middle lamella is mostly pectin. Cell walls of fruits are very rich in pectins, which can be extracted (e.g. by boiling) and used as gelling agents for making jams and jellies.

4.2.4 Non-polysaccharide constituents form a third structural network in primary cell walls

Primary cell walls can also contain a third network of polymers. In Type I walls, this network is made of

Table 4.1 Comparison of the composition of Type I and Type II primary cell walls in different taxa.

	Type I	Type II	
	Eudicots, most monocots	Commelinoid monocots (except grasses)	Members of the grass family (includes cereals)
Major cross-linking glycans	Xyloglucan (1:1 cellulose:XyG)	Glucuronoarabinoxylans	Glucuronoarabinoxylans β-Glucans
Pectin matrix	Yes	No	No
Protein network	Yes	Limited	Limited
Phenylpropanoid network	No*	Yes	Yes

*Except Chenopodiaceae.

structural proteins. It plays a role in the control of cell wall extensibility and in pollen–stigma interactions (see Chapters 12 and 16). Five classes of these structural proteins have been identified: **hydroxyproline-rich glycoproteins, proline-rich glycoproteins, glycine-rich proteins, threonine-rich proteins** and **arabinogalactan proteins**. The synthesis and deposition of these proteins is developmentally regulated, their relative amounts varying among tissues and species. The best-studied hydroxyproline-rich glycoprotein is **extensin**, a rod-shaped protein (see Figure 4.6) that confers structural rigidity to the wall by cross-linking to other cell wall polymers. The **proline-rich proteins** are similar in composition and shape to extensin. The **glycine-rich proteins** are thought to form a plate-like structure at the plasma membrane–cell wall interface. The fourth group, **arabinogalactan proteins**, may be more than 95% sugar by mass; these proteins are anchored to the plasma membrane via a **phosphatidylinositol** anchor and are thought to play a role in cell signaling and incompatibility responses.

Type II walls contain less structural protein than Type I walls and lack extensin, but threonine-rich proteins that are structurally similar to extensin are found in Type II walls together with phenolic compounds, shown as phenylpropanoids in Figure 4.6. Phenolic compounds,

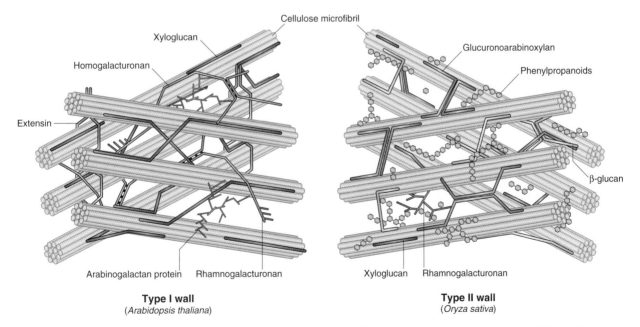

Type I wall
(*Arabidopsis thaliana*)

Type II wall
(*Oryza sativa*)

Figure 4.6 Three-dimensional models of a Type I primary cell wall (*Arabidopsis*) and a Type II primary cell wall (*Oryza*). Type I walls are found in eudicots and many monocots, while commelinoid monocots (including bromeliads, palms, sedges and grasses) have Type II walls. The Type II wall of rice, illustrated here, contains β-glucans, which is typical of grasses. These glucose-based cross-linking glycans have stretches of cellulose-like linkages (cellodextrins) that allow them to hydrogen bond together, as well as segments in which short runs of cellulose-like linkages (cellotriosyl- and cellotetraosyl-rich) are interspersed with callose-like linkages.

primarily **hydroxycinnamic acids**, link polysaccharides to proteins or to one another. These networks render the walls rigid, locking them in place after growth has ceased.

Key points Plant cells are surrounded by a primary cell wall made up of cellulose, cross-linking glycans, pectic polymers and proteins. The structure of the primary cell wall is similar to that of steel-reinforced concrete. Reinforcing cellulose microfibrils (analogous to reinforcing steel bars in concrete) are embedded in a matrix of amorphous polymers rich in cross-linking glycans, pectins and proteins (the concrete). Matrix polysaccharides and proteins form non-covalent linkages with cellulose. While the cell walls of all plants contain cellulose, the composition of the matrix polymers varies between species. Differences in matrix polymers can be used to divide angiosperms into two groups: those with Type I walls and those with Type II walls.

4.2.5 Biosynthesis and assembly of primary cell walls occurs during cell expansion

New cell walls are formed during cell division. Golgi vesicles containing glycoproteins and non-cellulosic wall polysaccharides are guided to the mid-point of the spindle where they fuse to form a plate-like, membranous organelle, the **phragmosome** (Figure 4.7). The phragmosome grows outward toward the lateral walls of the parent cell where it fuses with the plasma membrane to separate the contents of the daughter cells. Additional non-cellulosic components of the developing

wall are delivered by secretory vesicles derived from the Golgi apparatus (see Section 4.5.5). Cellulose microfibrils are synthesized by **cellulose synthase**, an enzyme that is embedded in the plasma membrane. As cellulose microfibrils are synthesized, they are extruded from the outer surface of the plasma membrane into the cell wall (Figure 4.8).

Plant cells expand 20–50-fold in length during their development. Primary cell walls must also increase in size during cell expansion. Existing wall must be loosened and new wall material inserted in a coordinated process that does not compromise the structural integrity of the existing wall. This process is discussed in detail in Chapter 12.

4.2.6 Secondary walls are produced after growth of the primary wall has stopped

When cells stop enlarging, the primary wall is cross-linked into its final shape. At this point, deposition of the **secondary wall** on the inner surface of the primary wall may begin (Figure 4.9). The composition and pattern of deposition of the secondary wall vary from cell to cell (Figure 4.10). For example, the secondary walls of cotton fiber cells are nearly 98% cellulose, whereas the secondary walls in some seed tissues are made of mostly non-cellulosic polysaccharides. The secondary walls of cereal endosperm cells are composed of β-glucans, whereas those of the date endosperm are mostly **mannans**, polymers of mannose. In these cases, secondary wall polysaccharides are used as a food source by the embryo during seed germination and early seedling growth. The β-glucans in cereal endosperm are important in human diets as well. They provide 'soluble fiber', thought to be important in lowering blood cholesterol levels. Because the yeast used in

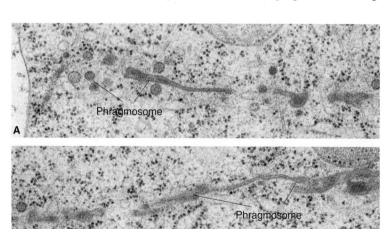

Figure 4.7 High-pressure frozen and freeze-substituted TEM image of a phragmosome from *Eucalyptus sieberti* root tip cells. (A) The phragmosome consists of a series of flattened membranous sacs that form near the equator of two dividing cells. (B) The beginning of membrane fusion that eventually completes the process of cytokinesis and separates the two daughter cells, each of which will have its own plasma membrane.

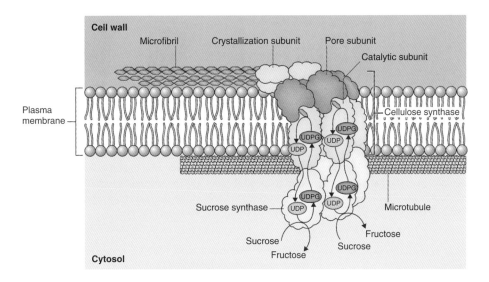

Figure 4.8 Model of the cellulose synthase complex. Cellulose synthase is embedded in the plasma membrane and extrudes cellulose microfibrils to the exterior surface of the cell. The positioning of the cellulose synthase complex, and therefore the orientation of the microfibril, is determined by microtubules. Glucose units for cellulose synthesis are supplied by sucrose synthase which is associated with the cellulose-synthesizing complex.

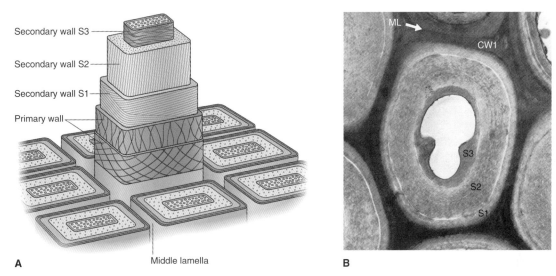

Figure 4.9   Secondary cell walls. Once they have achieved their final size and shape, some cells elaborate a multilayered secondary wall. In the diagram (A), a living protoplast has generated several distinct layers of secondary walls (S1 through S3, with the innermost S3 being laid down last), whereas the micrograph (B) shows secondary wall layers surrounding the empty lumen of a dead fiber cell in the young stem of a locust tree. In both panels, the original primary wall (CW1) and the middle lamella (ML) form the outermost layers of the wall.

beer making does not easily breakdown these glycans, they accumulate in beer causing a cloudy 'haze' (see Chapter 6).

Many secondary walls contain **lignins**, complex networks of aromatic compounds called **phenylpropanoids**. In the water-conducting cells of the xylem, lignified secondary walls are deposited in distinctive patterns, such as annular or helical coils or reticulate and pitted sheets (see Figure 1.28). Lignins make the wall rigid and waterproof. In trees, 15–25% of the dry mass of wood is lignin; lignified secondary walls give wood its strength and make up the skeleton that keeps 100 m tree trunks upright. When wood is processed to make paper, the lignins must be extracted by harsh chemical treatments.

Secondary deposition of suberin and cutin can render cell walls impermeable to water. **Suberin** is found in cell walls in specific cell types, notably stem and older root epidermal cells, cork cells of bark, the surfaces of wounded cells and parts of endodermal and bundle sheath cell walls. The core of suberin is lignin-like, and the attachment of the long-chain hydrocarbons imparts a strongly hydrophobic characteristic to suberin that prevents water movement. **Cutin**, a polymer of ester-linked fatty acids, and its associated waxes are also found on leaf and stem surfaces, providing a barrier to the diffusion of water vapor. Waxes are generally esters of long-chain fatty acids and alcohols that are synthesized in the ER.

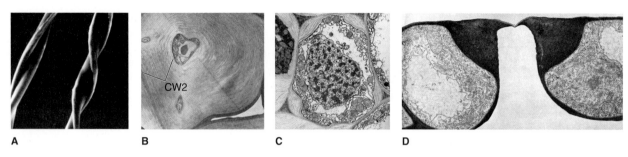

A   B   C   D

Figure 4.10   Secondary walls may differ in composition in different cell types. They serve a number of different functions. (A) The thick secondary wall of almost pure cellulose gives the flattened helical form to a cotton fiber. (B) The secondary wall (CW2) nearly fills the lumen of a pear fruit stone cell. (C) A collenchyma cell has reinforced thickenings only in the cell corners. (D) The thickened and elaborated inner wall of a guard cell pair provides the physical form needed to control formation of the stoma.

Key points   In expanding cells, the primary cell wall must be softened so that growth can occur. The linkages between cellulose and the matrix components are easily broken, for example by changes in cell wall pH, permitting growth to occur. In certain types of cells, when cell expansion is complete, a secondary cell wall is laid down. Secondary walls are rich in cellulose and lignin, a compound made up of phenylpropanoid units. Lignin forms a rigid network, stiffening the wall and thus blocking continued cell expansion. Secondary walls provide rigid support for plant organs.

4.3 Membranes of the cell

Inside the cell wall is the living **protoplast** enclosed by the **plasma membrane**. The plasma membrane creates and maintains an electrochemical environment within the cell that is different from the outside environment (see Chapter 5). Within the protoplast are many organelles whose membranes maintain distinctive electrochemical environments within them. The functioning of these organelles depends on the integrity of their membranes and their embedded transporters. Plant cells contain at least 14 distinct membrane types (Table 4.2). We will discuss the features common to all cellular membranes and then look specifically at the plasma membrane.

4.3.1 Biological membranes have common structural and functional properties

Almost all biological membranes consist of a bilayer of polar lipid molecules in association with a variety of proteins. Most cellular membranes have equal masses of

Table 4.2 Membrane types found in plant cells.

Endomembrane system	Other membranes
Plasma membrane	Peroxisomal membrane
Nuclear membrane	Plastid envelope membranes (inner and outer)
Endoplasmic reticulum	Thylakoid membranes of the plastid
Golgi cisternae (cis, medial and trans types)	Mitochondrial membranes (inner and outer)
Trans-Golgi network/ partially-coated reticulum	
Clathrin- and COP-coated transport vesicles	
Endocytic vesicle membrane	
Endosomal membrane	
Multivesicular body/autophagic vacuole membranes	
Tonoplast (vacuole membrane)	

COP, coat protein.

protein and lipid components. Virtually all membrane molecules are able to diffuse freely within the planes of the membrane, permitting membranes to change shape and membrane molecules to rearrange rapidly.

The primary components of the polar lipid bilayer of cellular membranes are **phospholipids** and **glycolipids** (see Section 2.2.7). These molecules are **amphipathic** because they have polar or charged **hydrophilic** ('water-loving') heads and **hydrophobic** ('water-hating') tails that consist of two fatty acids that contain between 14 and 24 carbon atoms. At least one of the fatty acids

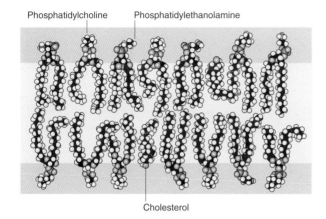

Phosphatidylcholine   Phosphatidylethanolamine

Cholesterol

Figure 4.11 Organization of amphipathic lipid molecules in a membrane bilayer. The hydrophilic heads of the molecules interact with water; their hydrophobic tails repel water and interact with each other in the center of the bilayer.

has one or more double bonds that form a kink in the otherwise straight fatty acid chain (see Figures 2.9 and 2.10). The hydrophilic head of phospholipids contains a phosphate group to which other polar molecules may be attached, while the hydrophilic head of glycolipids contains a sugar. In water, these molecules spontaneously form bilayers so that their hydrophilic heads interact with the water and their hydrophobic tails exclude water and interact with each other (Figure 4.11). Another class of membrane lipids is that of **sterols**; these molecules

have a hydrocarbon skeleton that is hydrophobic, and a single, hydrophilic hydroxyl group. Cholesterol is one sterol commonly found in membranes. The ratio of lipid classes varies between different membranes within a cell.

Membrane proteins may be associated with the lipid bilayer in a number of different ways (Figure 4.12). **Integral proteins** are embedded in the bilayer and, like membrane lipids, have both hydrophilic and hydrophobic domains. Many integral membrane proteins are decorated with chains of sugars. Water-soluble **peripheral proteins** interact with membrane lipids or proteins through salt bridges, hydrogen bonds, electrostatic reactions or a combination thereof. Unlike integral proteins, they can be removed by using treatments that do not disrupt the membrane itself. **Anchored proteins** attach to the bilayer via lipid anchors that are covalently bound to the proteins.

The lipid bilayer of cellular membranes serves as a permeability barrier to many molecules. While non-polar molecules and gases can readily diffuse across the hydrophobic core of the bilayer, hydrophilic polar molecules and ions cannot. An important exception to this rule is water, which, although polar, can freely cross the lipid bilayer. Membrane proteins serve as highly specific transporters allowing cells to take up or excrete ions and molecules (see Chapter 5). Membrane proteins perform a wide array of other functions including signal transduction, enzymatic catalysis and structural roles.

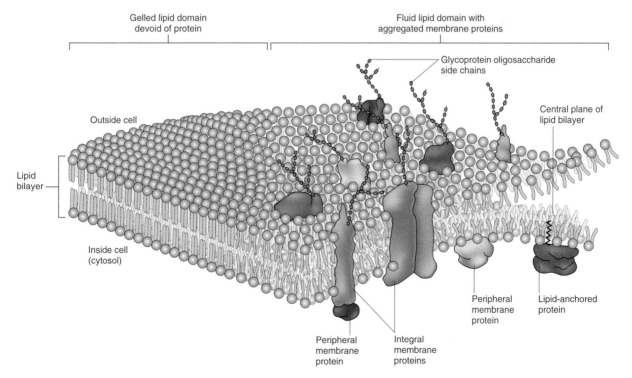

Gelled lipid domain devoid of protein

Fluid lipid domain with aggregated membrane proteins

Glycoprotein oligosaccharide side chains

Central plane of lipid bilayer

Outside cell

Lipid bilayer

Inside cell (cytosol)

Peripheral membrane protein

Integral membrane proteins

Peripheral membrane protein

Lipid-anchored protein

Figure 4.12 The fluid–mosaic membrane model. Integral, peripheral and lipid-anchored membrane proteins are shown. Note that the drawing is not to scale.

What's crucial to remember is that membrane proteins define the specificity of each membrane system.

4.3.2 The plasma membrane is the boundary between living protoplast and the external environment

All living cells have a plasma membrane. The lipids, proteins and their associated carbohydrate side-chains that make up plant plasma membranes occur in a molecular ratio of about 2:2:1. The most prevalent lipids of plasma membranes may be phospholipids or sterols. The ratio of different lipids may vary in the plasma membranes of cells in different organs or of cells in the same organ type in different species. Proteins in the plasma membrane participate in transmembrane transport and signaling, catalyze reactions such as the synthesis of cellulose (see Section 4.2.5), and connect both the membrane and its attached cytoskeletal elements to the cell wall.

Most cells in a plant are connected by numerous narrow (40–50 nm) cytoplasmic channels lined by plasma membrane. These **plasmodesmata** (singular = plasmodesma) span adjoining cell walls (Figure 4.13). Thus most cells in a plant share a physically continuous plasma membrane. Notable exceptions to this rule are stomatal guard cells and cells of the embryo sac. The structure of plasmodesmata is fairly complex (see Figures 4.13 and 14.7); ions and small molecules up to about 800 Da in mass can freely diffuse from cell to cell through plasmodesmata. Some cells have the ability to increase the size of plasmodesmata so that larger molecules may pass through; in certain cases, molecules up to 10 kDa have been shown to move via this pathway. The network of interconnected protoplasts in a plant is called the **symplasm** and the parts of the cell outside the plasma membrane, the cell wall and airspaces, collectively are called the **apoplast** (see Chapter 14).

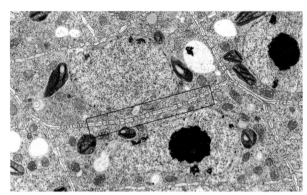

A

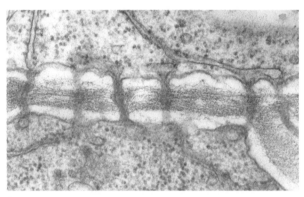

B

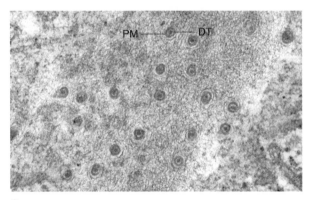

C

Figure 4.13 Plasmodesmata. (A) TEM image of *Azolla* root tip cells showing the many plasmodesmata (box) interconnecting neighboring cells. (B) Higher magnification image showing a side view of plasmodesmata from a *Vicia faba* extrafloral nectary. (C) TEM image of a face view of plasmodesmata in an *Abutilon* nectary trichome. Note that a plasmodesma is bounded by the plasma membrane (PM) and the core is filled with a piece of tightly rolled endoplasmic reticulum that forms the desmotubule (DT).

4.4 The nucleus

A prominent organelle in most living plant cells is the **nucleus**. One exception is the sugar-conducting sieve tube elements of the phloem, which lack nuclei at maturity. Nuclei are 3–10 μm in diameter, contain most of the cell's genetic information, and serve as the center of regulatory activity (Figure 4.14). The nucleus is bounded by a **nuclear envelope** that consists of two bilayer membranes separated by a lumen, the **perinuclear space**. This envelope contains complex **nuclear pores** through which RNA and ribosome subunits, synthesized in the nucleus, are exported to the cytosol, and proteins, synthesized in the cytosol, are imported into the nucleus (Figure 4.15). Molecules smaller than 40 kDa can pass through nuclear pores by diffusion but those greater than 40 kDa move by energy-dependent transporters. The outer membrane of the nuclear envelope has ribosomes on its cytoplasmic face. It is continuous with the membranes of the endoplasmic reticulum and therefore may be considered to be part of the endomembrane system (see below).

Chromosomes are made up of DNA–protein complexes called **chromatin** (see Chapter 3). Chromatin forms a network in the **nucleoplasm** during interphase when gene transcription and DNA synthesis are occurring. Individual chromosomes occupy discrete nuclear domains throughout this part of the cell cycle.

A typical interphase nucleus contains one to several **nucleoli** (singular = **nucleolus**), distinctive regions that stain densely in transmission electron micrographs (see Figure 4.14). These are the sites where ribosomal RNA is synthesized, processed and assembled with ribosomal proteins to form ribosomal subunits. The subunits then move through nuclear pores to the cytosol.

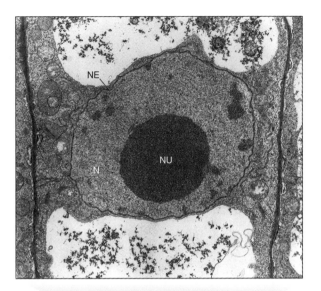
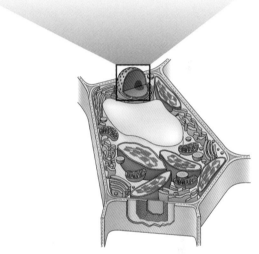

Figure 4.14   TEM image showing the nucleus (N) of a bean (*Phaseolus vulgaris*) root tip cell. Note the large, darkly stained central nucleolus (NU); the nuclear envelope (NE) consists of two membranes.

Key points The nucleus is one of the defining organelles of eukaryotes. The nucleus contains nucleoplasm where genetic material, chromatin, is found. The nuclear envelope is made up of two phospholipid bilayers punctuated by many nuclear pores. Nuclear pores allow traffic between the cytosol and nucleoplasm. This includes movement of various forms of RNA and ribosomes from the nucleus to cytosol and the passage of regulatory molecules such as transcription factors from cytosol to the nucleus. Movement through nuclear pores can occur by diffusion, a process that can accommodate molecules ranging in size from ions to macromolecules of about 40 kDa. Movement of molecules larger than 40 kDa through nuclear pores requires the expenditure of cellular energy.

4.5 The endomembrane system

The **endomembrane system** of a plant cell is an extensive, interconnected series of organelles that are responsible for the synthesis, processing and storage of a wide variety of macromolecules. Ten of the 14 cellular membrane types listed in Table 4.2 are part of the endomembrane system. Extensive vesicle traffic between these compartments not only transports secreted molecules to the cell surface and vacuolar proteins to the storage vacuoles, but also distributes membrane proteins and membrane lipids from their sites of synthesis to

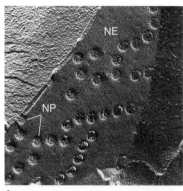

A

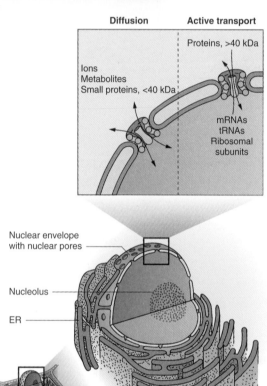

B

Figure 4.15   (A) TEM image of a freeze-fractured nuclear envelope (NE) with nuclear pores (NP). The continuity of the inner and outer membranes becomes apparent when the membranes are seen in cross-section. (B) Transport of small molecules into the nucleus via the nuclear pores takes place by diffusion, whereas import or export of larger molecules requires active (energy-consuming) transport. ER, endoplasmic reticulum.

other parts of the endomembrane system (Figure 4.16). A large number of sorting, targeting and retrieval systems regulate traffic between different compartments, ensuring delivery of molecules to the correct organelle and the maintenance of organelle identity (see Chapter 5).

4.5.1 The endoplasmic reticulum is a membrane system that is continuous with the nuclear envelope

The **endoplasmic reticulum** (**ER**) is the most extensive, versatile and adaptable organelle in eukaryotic cells. It consists of a three-dimensional network of continuous tubules and flattened sacs that connect to the nuclear envelope, course through the cytoplasm and underlie the plasma membrane (Figures 4.16 and 4.17). In plants, the principal functions of ER include the synthesis, processing and sorting of proteins that will be targeted to various membranes, to vacuoles or to the secretory pathway. Also, a diverse array of lipid molecules is synthesized within the ER. In addition to its synthetic functions, the ER provides anchoring sites for actin filament bundles that drive cytoplasmic streaming. It also plays a critical role in regulating the cytosolic concentration of calcium that in turn influences many other cellular activities. Three types of ER membranes are recognized: **rough ER**, which is studded with ribosomes and is a site of protein synthesis (Figure 4.17B), **smooth ER** (Figure 4.17C), which is free of ribosomes and is associated with lipid synthesis, and the **nuclear envelope**. In addition, many more morphologically distinct subdomains of ER that perform various different functions are found in plant cells (see Figure 4.16).

4.5.2 Many proteins are synthesized on the rough endoplasmic reticulum

Translation of all nuclear messenger RNAs (**mRNAs**) begins in the cytosol on free ribosomes. The synthesis of proteins that are destined to be secreted or are targeted to organelles of the endomembrane system is completed on the ER. As soluble proteins are synthesized, they are translocated across the ER membrane into the lumen; integral membrane proteins are incorporated directly into the ER membrane (see Chapter 5). Many of these proteins undergo further processing within the ER, modifications that affect their final three-dimensional conformation. For example, monomeric proteins may

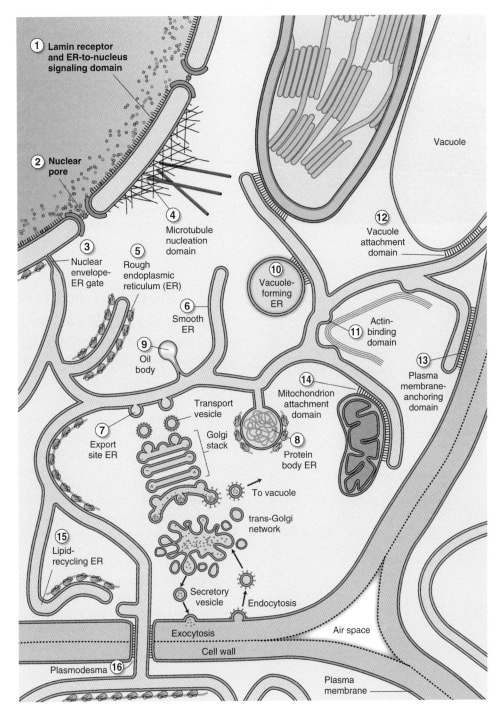

Figure 4.16 Diagram of the plant endomembrane system. Note the 16 different domains of the ER that are separately numbered in this diagram.

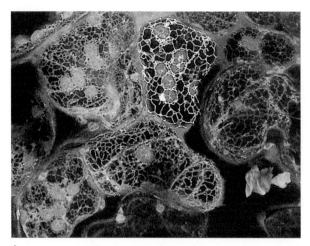

A

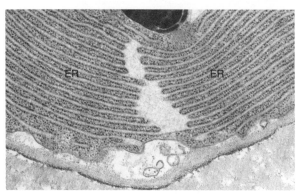

B

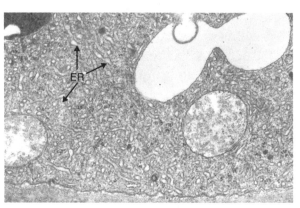

C

Figure 4.17   Confocal (A) and TEM images (B, C) of endoplasmic reticulum (ER). (A) Fluorescently labeled ER forming a net within *Nicotiana tabacum* (tobacco) leaf mesophyll cells. (B) Rough ER in a *Coleus* gland cell and (C) smooth tubular ER in a cap cell of a *Primula malacoides* farina gland (a surface structure that secretes a floury white wax).

assemble into polymeric protein complexes. In addition, some proteins are glycosylated by enzymes that add branched sugar chains to the amide nitrogen of asparagine residues. Once ER-localized protein processing is complete, most proteins are transported to the **Golgi apparatus** in vesicles that bud from the ER and fuse with the Golgi cisternae (see Figure 4.16 and Section 4.5.4). Budding and transport of vesicles from the ER is aided by **coat proteins** (**COPs**).

Some storage proteins that are synthesized during seed development are notable exceptions to the general rule of ER-to-Golgi transport. **Prolamins**, the alcohol-soluble storage proteins of cereal grains, are synthesized in specific ER regions and either aggregate in the lumen of the ER or are transported directly to vacuoles by vesicles that by-pass the Golgi (see Figure 4.16 and Chapter 6).

4.5.3 Smooth endoplasmic reticulum participates in fatty acid modification, lipid synthesis and the production of oil bodies

The smooth ER membrane is the site of many reactions involved in lipid synthesis. Some of the 16- and 18-carbon fatty acids that are synthesized in the plastids are transported to the ER. There they are combined with glycerol-3-phosphate to make phosphatidic acid, which is further modified to make the phospholipid molecules from which most cellular membranes are made. **Triacylglycerols** (oils) that are produced in developing seeds and pollen grains (see Chapter 6) are also synthesized in the ER. As they are synthesized, triacylglycerols accumulate between the inner and outer leaflets of the lipid bilayer of the ER membrane in regions of the membrane that contain proteins called **oleosins**. Mature **oil bodies** (**oleosomes**) are therefore enclosed by a 'half-unit' membrane, a lipid monolayer that contains oleosins and other proteins that are important for the integrity of this organelle (Figure 4.18).

Finally, the monomers that make up the waterproof waxes that coat the above-ground surfaces of plants (see Section 4.2.6) are synthesized by the ER of epidermal cells (Figure 4.19). Elongase enzymes located in ER membranes elongate fatty acids that are produced in the chloroplast. The resulting very-long-chain fatty acids are the monomers of waxes and cutin. It is not known how these monomers are transported to the exterior of the cell, although one hypothesis is that vesicles transport monomers to the plasma membrane where 'flippase' enzymes bring about their relocation to the cell's outer surface (Figure 4.19).

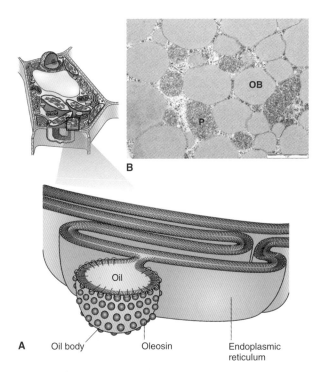

A   Oil body   Oleosin   Endoplasmic reticulum

Figure 4.18 Formation of oil bodies. (A) Triglycerides (oil) accumulate between the two lipid monolayers of the ER membrane and bud to form oil bodies at sites defined by the presence of molecules of the protein oleosin. (B) TEM iamge of oil bodies (OB) showing their close proximity to peroxisomes (P).

Key points The endoplasmic reticulum (ER) forms an extensive intracellular network of flattened sacs and tubules. The ER is continuous with the outer nuclear envelope. ER that is coated with ribosomes is called rough ER; smooth ER lacks ribosomes. Rough ER is involved in the synthesis of proteins that enter its lumen. Many of these proteins are destined for transport to other organelles or to the cell's exterior. Smooth ER is involved in the synthesis of lipids and a range of other molecules including hormones and secondary metabolites. The synthesis of triglycerides by the ER results in the formation of a unique organelle, the oleosome, in which oil accumulates between the inner and outer leaflet of the membrane bilayer.

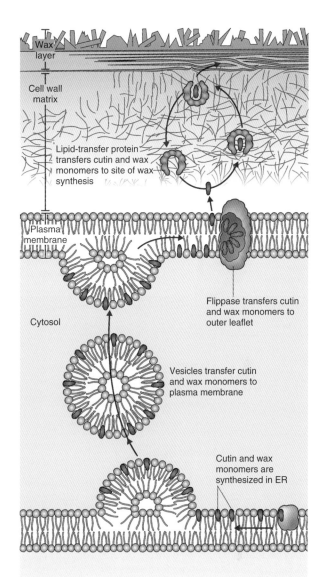

Figure 4.19 Hypothetical scheme for the transport of fatty acid precursors for cutin and wax synthesis to the site of polymerization in the cell wall. It is proposed that fatty acids move from the endoplasmic reticulum (ER) to the plasma membrane in the membranes of vesicles. A 'flippase' transfers cutin monomers from the inner side of the plasma membrane to the outer leaflet, where they can be extracted from the membrane by a water-soluble, lipid-transfer protein. The lipid-transfer protein diffuses in the solution phase of the cell wall matrix and transfers cutin monomers to enzymes that polymerize them into waxes near the surface of epidermal cell walls.

4.5.4 The Golgi apparatus processes and packages newly synthesized macromolecules

The **Golgi apparatus** occupies a central position in the secretory pathway. It is responsible for the synthesis,

modification, packaging and sorting of a variety of macromolecules destined for export to the cell wall or for delivery to other organelles. The Golgi apparatus receives newly synthesized glycoproteins and glycolipids from the ER and modifies their sugar side-chains, creating a wide variety of oligosaccharide groups. The sugar groups on proteins may specify plasma membrane–cell wall

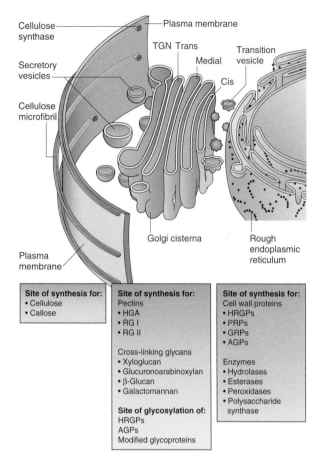

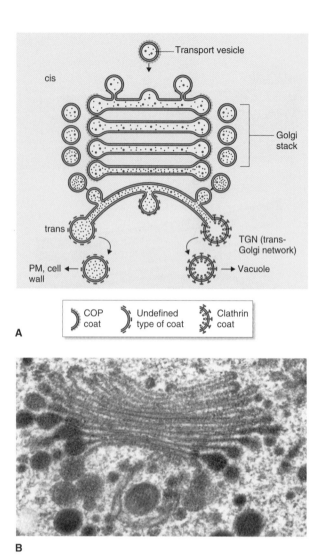

Figure 4.20 Cell wall biosynthesis requires the coordination of multiple compartments of the endomembrane system. Cellulose microfibrils are synthesized at the plasma membrane surface (see Figure 4.8).Structural proteins and wall-modifying enzymes are translated on the rough ER and transported to the Golgi apparatus, where some are glycosylated before being packaged in secretory vesicles for delivery to the cell exterior (AGP, arabinogalactan protein; GRP, glycine-rich protein; HRGP, hydroxyproline-rich protein; PRP, proline-rich protein). The Golgi also synthesizes and packages for secretion the various non-cellulose cell wall polysaccharides (HGA, homogalacturonan; RG I and RG I, rhamnogalacturonan I and II). At the cell surface, vesicle contents are integrated with the newly synthesized microfibrils. Assembly of the new wall is estimated to begin when the cellulose chains are no more than ten glucose residues in length. TGN, trans-Golgi network.

Figure 4.21 Structural features of the Golgi apparatus. (A) The spatial relationship of a Golgi stack with its associated trans-Golgi network (TGN). Two types of vesicles bud from the TGN: vesicles coated with clathrin protein are targeted to the vacuoles, while a different type of protein-coated vesicle moves products to the plasma membrane (PM) or cell wall. (B) TEM image of a Golgi stack and its TGN in a green alga, *Micrasterias*. Note the fuzzy protein coats on vesicles on the cis face (upper) and around the margins of the cis-cisternae and around the TGN cell.

interactions, prevent the premature activation of lectins, or contribute to protein folding or the assembly of multiprotein complexes. In addition, the Golgi is responsible for the synthesis of cross-linking glycans and other non-cellulosic cell wall polysaccharides (Figure 4.20).

The Golgi apparatus is a membranous organelle that has several components: the **Golgi stack**, about 1 μm in diameter and comprised of five to eight membrane-enclosed flattened sacs (**cisternae**); the

trans-Golgi network (**TGN**), a tubular structure associated with one side of the Golgi stack; and the **Golgi matrix**, a filamentous cage that surrounds the Golgi stack and its TGN (Figure 4.21). The number of Golgi stacks per cell varies widely depending on cell type and function. For example, small, shoot apical meristem cells of willow-herb (*Epilobium*) contain 20 Golgi stacks, the larger onion (*Allium cepa*) root apical meristem cells have about 400 stacks, and the giant fiber cells of cotton (*Gossypium hirsutum*) contain more than 10 000.

4.5.5 Transport through the Golgi is directional

Each Golgi stack is asymmetrical. Transport vesicles from the ER deliver their contents by fusing with the cis face of the Golgi stack. ER-derived products move sequentially via transport vesicles through the cis-, medial-, and trans-cisternae of the Golgi stack and then to the TGN. In addition, different compartments of the Golgi apparatus participate in the synthesis of the various classes of cell wall polysaccharides. Vesicles containing the non-cellulosic polysaccharide components of the cell wall bud from the TGN and move to the plasma membrane for delivery to the cell wall. Vesicles involved in intra-Golgi and Golgi-to-TGN transport bud from cisternae with the aid of a COP.

Within the TGN, materials are sorted for distribution to vacuoles or the plasma membrane, or for transport to the cell exterior. Proteins destined for vacuoles contain specific amino acid sequences that target them for transport to this organelle and are transported via the **multivesicular body** (**MVB**) (see Chapter 5). All other proteins are packaged for transport to the plasma membrane. Vesicles from the Golgi destined for vacuoles are coated by a protein called **clathrin**, which differs from the coat protein of vesicles bound for the plasma membrane (Figure 4.21).

4.5.6 Vesicles exchange materials with the cell exterior by exocytosis and endocytosis

When a transport vesicle from the Golgi apparatus reaches the plasma membrane, the vesicle membrane fuses with the plasma membrane and its contents are discharged into the cell wall, a process called **exocytosis**. Exocytosis allows the delivery of proteins and polysaccharides to the cell wall and thus the cell exterior. In addition, it allows the addition of new membrane proteins and lipids to the plasma membrane.

A second mechanism, **endocytosis**, allows plant cells to retrieve plasma membrane components for recycling or degradation. This process also may allow a plant cell to take up extracellular molecules. Endocytosis involves the import of plasma membrane-derived vesicles that are coated in clathrin. Models for endocytosis in plants propose that imported materials move from the clathrin-coated vesicles through the MVB before arriving at vacuoles, where they are broken down and their component parts are recycled into new molecules (see Figures 4.16 and 5.30).

Key points The Golgi apparatus and multivesicular body (MVB) sit at the crossroads of the secretory pathway. With the exception of some seed storage proteins, all proteins that leave the ER enter the Golgi apparatus where they can be modified, for example, by the addition of glycan chains to glycoproteins, or by modification of some of their amino acids. Proteins directed to the cell exterior leave the Golgi apparatus in secretory vesicles that fuse with the plasma membrane by exocytosis. Proteins destined for the vacuole are transported to the MVB in clathrin-coated vesicles. Clathrin-coated vesicles are produced by the Golgi, for example in the case of seed storage proteins or vacuolar enzymes, or by endocytosis. Proteins and other macromolecules brought into the cell by endocytosis are delivered to the vacuole to be broken down or recycled.

4.5.7 Vacuoles are multifunctional compartments

Vacuoles, fluid-filled compartments enclosed in a single membrane, the **tonoplast**, can occupy 30–90% of the volume of a plant cell (see Figures 4.1 and 4.22). Large, mature plant cells have a large central vacuole, while meristematic cells often contain numerous small vacuoles (Figure 4.22C) that coalesce into one or a few larger vacuoles as the cell matures and expands. Vacuoles typically originate as provacuoles that bud from the ER.

Vacuoles serve several functions. In mature plant cells, the large central vacuole occupies most of the volume of a cell and confines the metabolically-active cytoplasm to a thin layer, increasing its surface area and maximizing gas and nutrient exchange. The vacuole also allows plants to produce large cells without having to produce very much protein-rich cytoplasm.

Plant cells also use vacuoles for short- and long-term storage of a variety of ions and molecules. Vacuoles act as reservoirs of protons, organic acids and metabolically important ions, such as calcium; they also serve as storage organelles for polysaccharides called fructans in grasses (see Chapter 17). In addition vacuoles may contain lytic enzymes that can be used in digestion of stored reserves as well as in the breakdown and recycling of cellular components. Vacuoles may also store heavy metals, and plant defense compounds such as tannins, oxalic acid crystals, proteases and nucleases, and, in the case of one marine alga, sulfuric acid. In some cells, water-soluble pigments, such as the anthocyanins that

A

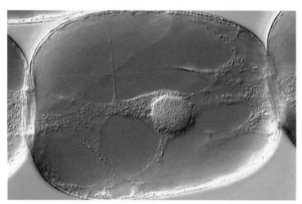

B

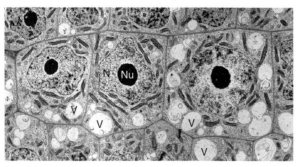

C

Figure 4.22   Vacuoles. (A) Flower of *Tradescantia*. Note the purple stamen hairs on the filaments of the stamens; each hair is a chain of cells. (B) Light micrograph of a single stamen hair cell. Vacuole contents are purple due to the presence of anthocyanin pigment. Note a very thin layer of cytoplasm around the large central vacuole; strands of cytoplasm cross the vacuole, and one strand contains the nucleus. (C) TEM iamge of *Hydrocharis* root tip cells showing early stages of vacuolation. N, nucleus; Nu, nucleolus; V, vacuole.

give leaves and flowers a red, pink or blue color (see Chapter 15), are found in vacuoles (Figure 4.22A, B). In seeds, there are specialized, protein-storing vacuoles.

Key points Vacuoles are dynamic organelles. They are key regulators of growth in plants where the vacuole may occupy more than 95% of the cell's volume. Since plant cells grow by an increase in volume, the bulk of this volume change occurs in the vacuole. Vacuoles are important storage reservoirs for minerals, metabolites, polysaccharides and proteins and are a repository for toxic foreign compounds, e.g. synthetic herbicides. Digestion of cellular components also occurs in vacuoles; this role is particularly important during cell death when cellular proteins are broken down to amino acids for redistribution to other parts of the plant. Vacuoles also play an important role as the final resting place for breakdown products of molecules such as chlorophyll. Finally, storage proteins in seeds are usually contained in protein storage vacuoles and protein bodies where they form an important source of dietary proteins for humans and animals.

4.6 Plastids

Plastids are found in the cells of all autotrophic eukaryotes; in plants they are present in almost all cells. They produce sugar in photosynthesis (see Chapter 9) and store carbohydrate as starch. They are involved in the biosynthesis of other essential metabolites, including **chlorophylls, carotenoids, purines, pyrimidines, amino acids, fatty acids** and hormones, e.g. **gibberellins** and **abscisic acid.** They also reduce **nitrite** and **sulfate**, facilitating the incorporation of nitrogen and sulfur into organic compounds.

4.6.1 Plastids are bounded by two membranes and possess prokaryotic-type genomes and protein synthesis machineries

Plastids are enclosed by two membranes, the **outer** and **inner membranes,** and contain internal membrane systems that vary in complexity according to the plastid's developmental stage and function. The outer membrane contains non-specific protein pores that allow ions and molecules up to 10 kDa in mass to diffuse into the intermembrane space between the two membranes. The enzymes involved in galactolipid metabolism are localized in the outer membrane. The inner membrane, which encloses an aqueous compartment called the **stroma**, is less permeable. Gases and small, uncharged molecules can cross the inner membrane unassisted, but most metabolites enter or leave the stroma by way of specific transporters. The inner membrane invaginates to produce the internal **thylakoid** membrane system of the plastid. In plants, de novo synthesis of fatty acids takes place exclusively in the stroma of plastids.

Plastids contain the machinery required to synthesize some of their own proteins, including a naked DNA genome (discussed in more detail in Chapter 3) and ribosomes that resemble those of prokaryotes more closely than they do the linear nuclear chromosomes, and cytoplasmic ribosomes of eukaryotic cells. Morphological, biochemical and nucleic acid sequence evidence indicates that plastids originated as prokaryotic endosymbionts related to cyanobacteria. However, plastids have since lost many genes and must import about 75% of their proteins, which are encoded in the nucleus and transcribed on free ribosomes in the cytosol.

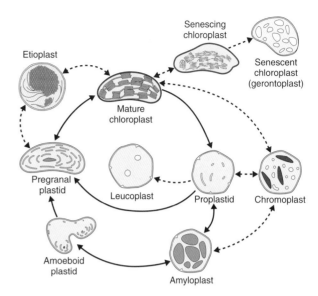

Figure 4.23 Plastid developmental cycle and interconversion of various plastid types. Solid arrows depict normal steps of chloroplast development; dotted arrows show plastid interconversions that occur under certain environmental or developmental conditions. The gray structures seen in the amyloplast are starch granules.

4.6.2 Different types of plastids are developmentally related to one another

All plastid types are related to the undifferentiated **proplastids** that are present in meristematic cells. Plastids possess a remarkable capacity to differentiate, dedifferentiate and redifferentiate (Figure 4.23); differentiated plastids can convert from one type to another in response to developmental or environmental conditions.

Proplastids are spherical or ovoid and range in diameter from about 0.2 to 1.0 μm. A proplastid contains few ribosomes, and its internal membrane system is poorly developed, consisting only of a few invaginations of the inner envelope and a small number of flattened sacs called lamellae (Figure 4.24A).

Chloroplasts are the site of photosynthesis and contain the green pigment chlorophyll. They are commonly hemispherical or lens-shaped in vascular plants (Figure 4.25A). They range in diameter from 3 to 10 μm and from 0.5 to 1.0 μm in thickness. Within each chloroplast the **thylakoids** form an extensive continuous network that encloses a single, branched chamber, the **thylakoid lumen**. The thylakoid network contains two types of membrane domains: stacks of thylakoids called

grana (singular = **granum**) are connected by unstacked **stroma thylakoids** (Figure 4.25B). The components of the light-energy-capture reactions, including the photosynthetic pigments, are associated with the thylakoid membranes (see Chapter 9). In a mature chloroplast the number of thylakoids in a granum may vary from several to 40 or more depending on the plant species and light conditions. For example, plants grown in shade generally have not only more numerous grana, but also more thylakoids per granum, than individuals of the same species growing in bright light.

Chloroplasts contain numerous ribosomes that can account for up to half the ribosomes present in photosynthetic tissues. The chloroplast enzyme, **rubisco**, makes up about half the soluble protein in leaves and may be Earth's most abundant protein. Rubisco's large subunit is encoded by chloroplast DNA and is synthesized on chloroplast ribosomes. In contrast, the small subunit of rubisco is encoded by nuclear DNA and is synthesized on cytoplasmic ribosomes (see Chapter 9). This is an example of the importance of coordinated nuclear and plastid gene expression during plastid development.

Etioplasts are plastids whose development from **proplastids** to chloroplasts has been arrested by the absence of light or by very low light; they are 2–3 μm in length. Etioplasts lack chlorophyll, but produce **protochlorophyllide**, a precursor of chlorophyll. A prominent structure in the etioplast is the **prolamellar body** (Figure 4.26), a lattice-like inclusion that consists

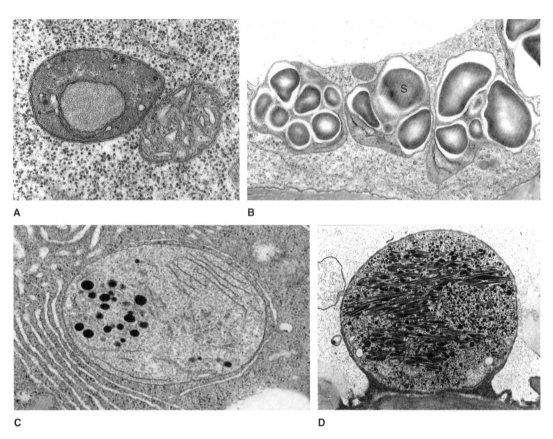

A   B   C   D

Figure 4.24   A range of plastid types found in plant cells. (A) TEM image showing a proplastid (left) adjacent to a mitochondrion in a bean (*Phaseolus vulgaris*) root cell. The electron-dense particles near the proplastid's periphery are lipid droplets. Beneath the large, round protein deposit in the middle of the proplastid can be seen the flattened lamella of an internal membrane. (B) TEM image of amyloplasts in a soybean (*Glycine max*) root cap columella cell containing many starch granules (S). (C) TEM image showing a leucoplast in an actively secreting glandular trichome of peppermint (*Mentha* spp.). The black, electron-dense globules are lipid droplets. Note the presence of extensive tubular smooth endoplasmic reticulum cisternae in the surrounding cytosol. (D) TEM image of a chromoplast in a cell of a ripe Jerusalem cherry (*Solanum pseudocapsicum*) fruit. The electron-dense bodies in the plastid contain the carotenoids.

of membrane lipids, protochlorophyllide and the light-requiring enzyme protochlorophyllide oxidoreductase (see Chapter 8). Protochlorophyllide is converted to chlorophyll, chlorophyll–protein stable complexes are produced, and lipids from the prolamellar body are incorporated into developing thylakoid membranes.

Amyloplasts (see Figure 4.24B) are unpigmented plastids that synthesize and store **starch**; they are about the same size as chloroplasts. Their name is derived from **amylose**, a major component of starch (see Chapter 6). Amyloplasts are especially common organelles in storage tissue of roots and stems and the endosperm of cereal grains. In potato tubers, starch grains commonly fill the entire plastid except for a thin layer of stroma lining the inner envelope. Amyloplasts occasionally redifferentiate into chloroplasts, as happens when potato tubers are left in the light. Note that amyloplasts in the cereal

endosperm cannot redifferentiate as this tissue is dead at functional maturity.

Leucoplasts (see Figure 4.24C) are colorless plastids that synthesize and store **monoterpenes**, volatile compounds contained in **essential oils** (see Chapter 15). They have few internal membranes or ribosomes and contain lipid droplets. They are found in specialized secretory gland cells associated with leaf and stem trichomes (hairs), the nectaries of flowers, and secretory cavities such as those found in citrus peel.

Chromoplasts (see Figure 4.24D) are plastids that store **carotenoids**; they may be yellow, orange or red, depending on the particular combination of **carotenes** and **xanthophylls** present. They are responsible for the colors of many fruits (e.g. tomatoes, oranges), flowers (e.g. buttercups, marigolds), roots (e.g. carrots) and tubers (e.g. sweet potatoes). Chromoplasts have an irregular shape ranging from round to

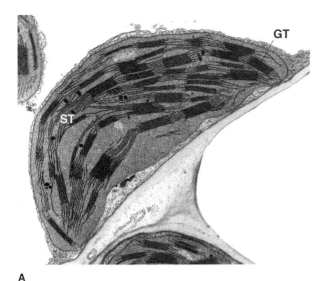

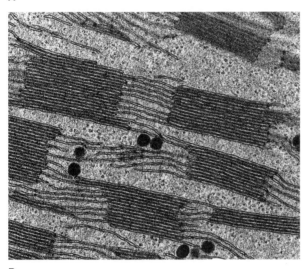

Figure 4.25   Chloroplast ultrastructure. (A) TEM image showing a chloroplast with stacked grana (GT) and unstacked stroma (ST) thylakoids in a maize leaf mesophyll cell. The dark granules are lipid droplets. (B) High-magnification TEM iamge of stacked grana and unstacked stroma thylakoids in a chloroplast from a leaf of timothy grass (*Phleum pratense*). The membranes, which are stacked in one area and unstacked in another, enclose a single lumen.

amoeboid to elongate; in some cases they contain large crystals. Chromoplasts can develop directly from proplastids or by the dedifferentiation of chloroplasts, as happens in ripening tomato fruits. The plastids of senescing leaves are sometimes called **gerontoplasts** and resemble chromoplasts in accumulating carotenoids that account for the bright yellow and orange colors of foliage in fall (see Chapter 18). Occasionally chromoplasts will redifferentiate into chloroplasts, as they do in some yellow and orange citrus fruits that regreen under the appropriate conditions, or in carrot roots exposed to light.

Key points Plastids characterize plant cells; they are thought to have evolved from cyanobacteria that were engulfed by ancestral eukaryotic cells. Chloroplasts are the sole sites of the reactions of photosynthesis in eukaryotes. Plastids also are responsible for many of the bright colors of leaves and stems. Chloroplasts and other plastids differentiate from small, colorless proplastids. Chromoplasts of senescing leaves contain accessory pigments such as carotenoids and xanthophylls that contribute to the bright colors of fall. Mature fruits such as apples, peppers and tomatoes contain chromoplasts. Plastids also perform an important storage function. The starch found in cells of storage organs and in the endosperm of cereal grains is stored in amyloplasts.

4.6.3 Plastids reproduce by division of existing plastids and are inherited differently in angiosperms and gymnosperms

Plastids arise by **binary fission** of existing plastids (see Figure 4.26). The division of proplastids, etioplasts and young chloroplasts is most common, but fully developed chloroplasts can also divide. In root and shoot meristems, proplastid division keeps pace with cell division, so that mother and daughter cells both possess approximately the same number of plastids—about 20. As cells expand, the number of plastids per cell increases due to continued plastid division.

Cytoplasmic organelles can be passed from one generation to the next by way of the egg (**maternal inheritance**), the sperm (**paternal inheritance**) or both (**biparental inheritance**). Genetic evidence indicates that in most angiosperms, the plastids and mitochondria are inherited maternally. These organelles are either excluded from angiosperm sperm cells or are degraded during male gametophyte development or after fertilization. This is discussed further in Chapter 16. In gymnosperms, plastids are usually passed to the next generation through sperm. Biparental inheritance of plastids and mitochondria has been documented both cytologically and genetically in a few flowering plant genera. Whether from egg, or sperm, plastids are in the proplastid stage at the time they are inherited.

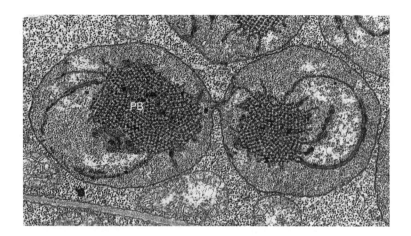

Figure 4.26   TEM image of an etioplast from *Phaseolus vulgaris* undergoing binary fission. Note the prominent prolamellar body (PB) in each etioplast and the constriction between the two daughter plastids.

4.7 Mitochondria and peroxisomes

Mitochondria, often called the powerhouses of the cell, are found in nearly all plant cells. Mitochondria are the site of **aerobic respiration** (see Chapter 7) and participate in **gluconeogenesis** (see Chapter 6) and **photorespiration** (see Chapter 9). Electron transport in mitochondria is a key source of ATP, and mitochondrial metabolic pathways supply important intermediates such as organic acids and amino acids that are used in biosynthesis elsewhere in the cell. Like plastids, mitochondria have their own DNA and prokaryote-like ribosomes, and divide by **fission**. Sequence data from mitochondrial ribosomal DNA indicates that mitochondria originated as endosymbionts related to the α-proteobacteria.

In electron micrographs, plant mitochondria appear spherical to ellipsoid, generally about 1 μm wide and 1–3 μm long (Figure 4.27). Viewed in living cells, however, these organelles change shape frequently from round to oval to dumbbell-shaped and back. A plant cell may contain hundreds or thousands of mitochondria, depending on the cell type and its stage of development. The cells of the maize root cap have about 200 mitochondria when young and 2000–3000 when fully enlarged and mature.

Mitochondria are bounded by two membranes, which are separated by the intermembrane space. The **outer membrane** contains few enzymes. The **inner membrane** encloses the **mitochondrial matrix**, which contains soluble enzymes, mitochondrial DNA and ribosomes. The inner membrane is highly infolded producing long, tubular **cristae** extending into the matrix (Figure 4.27). The inner membrane (including cristae) is the site of the respiratory electron transport chain and contains ATP synthase that makes ATP.

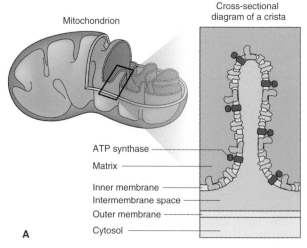

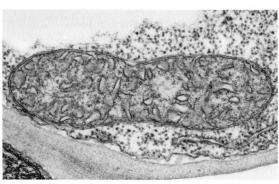

Figure 4.27   Mitochondrion morphology and structure. (A) The three-dimensional structure of a mitochondrion and the distribution of ATP synthase molecules in the inner membrane. (B) TEM image of a mitochondrion in an *Avena* (oat) leaf mesophyll cell.

As in plastids, the outer membrane of a mitochondrion is highly permeable to small molecules, whereas the inner membrane has a very low permeability. Protein complexes, known as **porins**, mediate the transport of small molecules across the outer membrane. Each complex creates an aqueous channel through which ions and molecules move freely. In contrast, solute-specific transporters control solute movement across the inner membrane.

Peroxisomes are roughly spherical organelles, about 0.2–2.0 μm in diameter, surrounded by a single membrane (Figure 4.28). These organelles grow by importing lipids, membrane proteins and matrix proteins from the cytosol, and periodically divide. Unlike plastids and mitochondria, peroxisomes contain neither DNA nor ribosomes. All proteins in these organelles are encoded by nuclear genes, translated on free polysomes in the cytosol and targeted to the organelle.

While their function varies in different tissues and organs, peroxisomes contain **oxidases** that generate **hydrogen peroxide**, a toxic by-product of metabolism, and **catalase**, an enzyme that breaks down hydrogen peroxide to form oxygen and water. In leaves, **peroxisomes** participate in the **glycolate pathway**, also known as **photorespiration** (see Chapter 9). In nitrogen-fixing root nodules of some legumes, peroxisomes are involved in the conversion of fixed nitrogen into nitrogen-rich organic compounds. Peroxisomes are the site of the **β-oxidation** of fatty acids and the conversion of fats to sugars (see Chapter 6). Peroxisomes are especially prominent is seeds that store fats and oils; in this case they have often been called **glyoxysomes**. Peroxisomes are implicated in the production of the plant hormone jasmonic acid.

4.8 The cytoskeleton

Spatial organization within the eukaryotic cell and directed movements of the cell or its contents are mediated by the **cytoskeleton**, a network of proteinaceous fibers that permeates the cytosol (Figure 4.29). The cytoskeleton includes three major families of proteins: **intermediate filaments, actin** and **tubulin**. Associated with these fibers are accessory proteins that link, move and modify the network. The cytoskeleton provides structural stability to cytoplasm, anchoring proteins and other macromolecules, and supporting organelles during and after their assembly (Figure 4.29A). The cytoskeleton moves cellular components during cytoplasmic streaming and plays a key role in cell wall deposition as well as in the orientation of the new cell wall during cell division. It can confer polarity to a cell, for example, by determining the direction in which organelles move (Figure 4.29B–D).

The cytoskeleton evolved before plants diverged from animals and the main features of the cytoskeleton have been conserved in both. However, the plant cytoskeleton

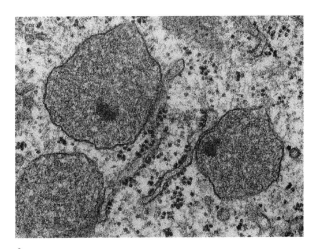

A

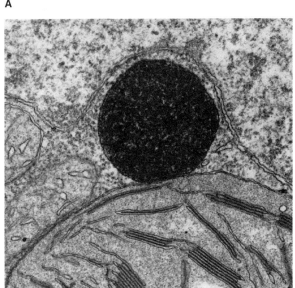

B

Figure 4.28 Peroxisomes. (A) TEM image showing three unspecialized peroxisomes closely associated with endoplasmic reticulum in a bean (*Phaseolus vulgaris*) root cell. (B) TEM image of a peroxisome adjacent to a chloroplast in a tobacco (*Nicotiana tabacum*) leaf cell that has been stained with diaminobenzidine, which becomes electron-dense where the enzyme catalase is active. In photosynthetic tissues, peroxisomes engage in photorespiration.

has also evolved unique functions and many of these are related to properties of the cell wall and the way in which the plant cell divides.

4.8.1 The cytoskeleton consists of a network of fibrous proteins

In plant cells, three types of cytoskeletal filaments have been identified: **intermediate filaments, actin filaments**

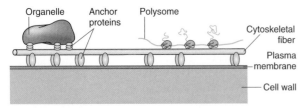

A Anchorage

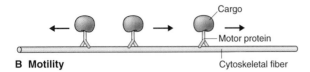

B Motility

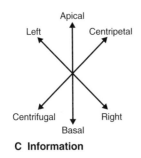

C Information

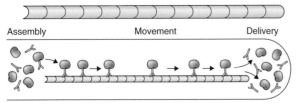

D Polarity

Figure 4.29 Different functions of the cytoskeleton: (A) anchorage, (B) motility, (C) information and (D) polarity.

or **microfilaments**, and **microtubules**. Each type of filament is composed of self-assembling, asymmetrical protein subunits. Non-covalent bonds link these subunits; therefore they assemble and disassemble in response to factors that affect protein–protein interactions, such as ionic strength, pH and temperature.

Intermediate filaments are 10–15 nm diameter protein filaments that are thinner than microtubules but thicker than actin filaments. Plants lack the various classes of intermediate filament proteins found in other eukaryotes, e.g. lamins, keratins, vimentins and neurofilaments. Plants also lack a **nuclear lamina**, a sheet-like structure lying just inside the nuclear envelope that stabilizes the nuclear envelope and maintains the shape of the organelle. Although homologs of the intermediate filament proteins of other eukaryotes are absent from plants, several groups of plant proteins that form **coiled-coil interactions** typical of intermediate filament proteins have been identified. Among these are

a group of proteins known as **filament-like proteins** (**FLPs**). FLPs show coiled-coil interactions with partner filaments that allow the formation of rope-like, 10–15 nm long intermediate filaments (Figure 4.30). The coiled-coil proteins identified in plants may be functional homologs of the intermediate filament proteins found in other eukaryotes. These proteins diverged over evolutionary time and now share limited sequence homology.

Microtubules are made of heterodimeric subunits of the globular proteins α- and β-**tubulin** (Figure 4.31). The dimers associate to form a hollow structure 25 nm in diameter. Tubulin dimers bind together at their sides and at their ends. The dimers lie in straight columns called **protofilaments**. Most microtubules have 13 protofilaments, but the number can vary from 11 to 16.

Actin filaments or microfilaments are made of the protein actin. **Soluble actin** is a globular protein of 375 amino acids (sometimes termed '**G-actin**' in contrast to the **filamentous** form called '**F-actin**'). Subunits polymerize into a tightly helical filament, about 8 nm wide (Figure 4.32A).

Both actin and tubulin occur in all eukaryotes. Actin and tubulin are highly conserved proteins that appear to have arisen from single-copy genes present before

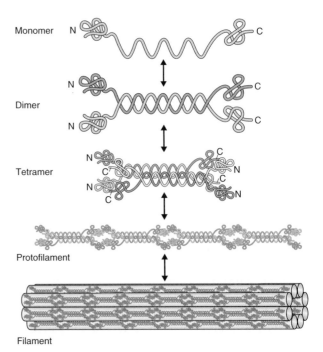

Figure 4.30 Assembly of intermediate filaments. Identical monomers pair through coiled-coil interactions to form a dimer. Dimers associate laterally to form tetramers. Tetrameters polymerize to form protofilaments. Finally, protofilaments associate laterally to form a rope-like filament.

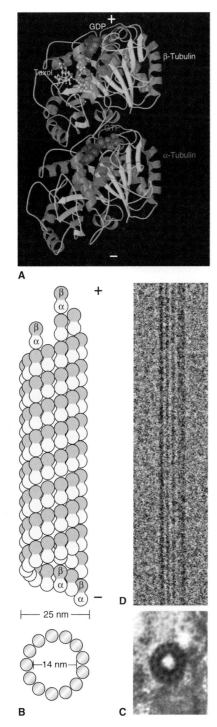

Figure 4.31 Structure of microtubules. (A) Ribbon diagram of the tubulin dimer within a microtubule. Regions of β-sheet structure are shown in green and the α-helix in blue. (B) Diagrams of a microtubule in cross-section and side view. (C) Electron micrograph of a cross-section of microtubule from a plant cell showing 13 protofilaments. (D) TEM image of a microtubule assembled from purified tubulin and viewed from the side. Unlike (C), this side view was prepared using cryotechniques, without fixation or staining with heavy metals.

multicellular eukaryotes diverged. In plants, α- and β-tubulin genes are present in similar numbers (four to nine copies), whereas the number of actin genes present varies widely, e.g. *Arabidopsis* contains ten actin genes whereas as *Petunia* has more than 100.

Key points The cytoskeleton in plant cells consists of a number of different proteinaceous filaments. These include actin microfilaments, microtubules assembled from tubulin monomers, and intermediate filaments composed of a number of different, but related, intermediate filament proteins. The cytoskeleton forms a constantly changing internal scaffold to which organelles are anchored and along which they move. The cytoskeleton can assemble into a large and complex machine, as in the case of the spindle apparatus that anchors and moves chromatids and chromosomes during meiosis and mitosis. Energy in the form of nucleoside triphosphates is used to assemble and power cytoskeletal machines.

4.8.2 Microtubules and actin filaments have an intrinsic polarity

Actin filaments and microtubules are polar structures because their protein subunits are asymmetrical. In polymer assembly, the asymmetrical subunits line up end to end with a uniform orientation (Figure 4.32B). The polarity thus conferred to the polymer means that each end has a different biochemical character and therefore each end of the polymer may have different rate constants for assembly and disassembly reactions. In addition, each end of the polymer can be recognized specifically so that cells may build polymer arrays with uniform polarity. The more dynamic end of the polymer is designated 'plus' and the less active end 'minus'.

4.8.3 Spontaneous assembly of cytoskeletal components occurs in three steps

Actin filaments and microtubules assemble by similar processes that share several characteristics (Figure 4.33). The first stage of assembly is called **nucleation** and results in the formation of a template. During the second stage, **elongation**, templates formed during nucleation grow by the endwise addition of subunits. The rate of

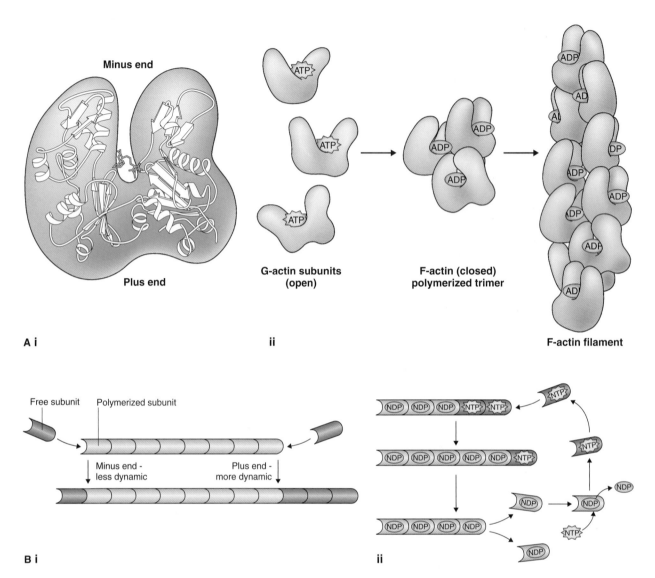

Figure 4.32   Structure and assembly of actin filaments. (A) (i) Ribbon diagram of G-actin monomers in the open configuration with bound ATP shown in yellow. (ii) Sequence showing open G-actin, an F-actin trimer in the closed configuration following hydrolysis of ATP, and a growing actin filament showing staggering of the monomers that give the filament its helical configuration. (B) (i) Actin filaments and microtubules form by the addition of polarized subunits to the plus and minus ends. (ii) Actin and tubulin monomers bind nucleoside triphosphates (NTPs) and the NTP is hydrolyzed when the monomer becomes incorporated into the growing polymer.

elongation represents the difference between the rate of subunit addition and the rate of subunit loss. The third stage is a **steady state** in which a constant length of fiber is maintained over time. The steady state is dynamic, the fiber gaining and losing subunits constantly. However, at the steady state, the rate of subunit addition is balanced exactly by the rate of subunit loss.

The assembly of microtubules and microfilaments requires the binding and hydrolysis of **nucleoside triphosphate** (NTP) by their respective protein subunits. G-actin binds ATP and α- and β-tubulin bind guanosine triphosphate (GTP). The γ-phosphate bond of the NTP is hydrolyzed only after a subunit has bound to the

growing structure. Nucleotide hydrolysis has different consequences for the assembly of actin and tubulin. In the case of actin, the presence of ATP at the plus end of the filament results in an assembly rate far exceeding that at the minus end, where the actin subunits contain ADP. This difference underlies a type of dynamic behavior called **treadmilling**, which occurs when the concentration of free subunits supports growth at the plus end that is balanced by shrinkage at the minus end. The net rate of change in polymer length can be zero, but subunits incorporated at the plus end will 'treadmill' through the polymer and are eventually released from the minus end (Figure 4.34). Because the difference in

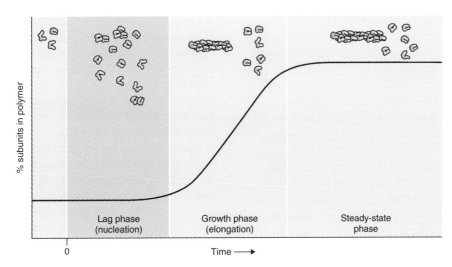

Figure 4.33 Graph showing the process and kinetics of polymerization. At time zero, a solution of individual subunits is induced to polymerize. During nucleation (pink panel), subunits must associate to create a stable template that serves as a platform for further elongation. In this example, using actin filaments, templates are formed when three subunits associate, generating a trimer. During elongation (growth phase, blue panel), subunits are rapidly added to the ends of growing polymers. A steady-state phase (yellow panel) is reached when the rate of subunit addition is balanced exactly by the rate of subunit loss.

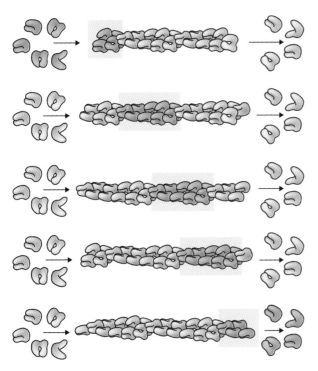

Figure 4.34 Treadmilling, a dynamic behavior observed in cytoskeletal polymers. Subunits (dark pink) added at one end (top row) 'treadmill' through the polymer and are released from the other end (bottom row). Although the diagram illustrates an actin filament, treadmilling also occurs in microtubules.

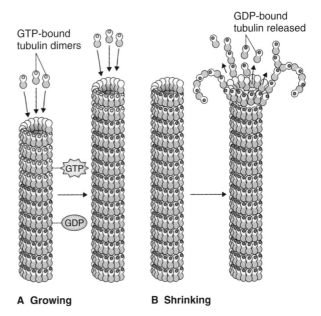

A Growing　　　**B Shrinking**

Figure 4.35 Dynamic instability of microtubules. (A) Diagram of growing microtubules (GTP is shown by yellow dots and GDP by green dots). Since hydrolysis of GTP usually lags behind polymerization of new subunits, the growing ends of microtubules are rich in subunits in which β-tubulin monomers bind GTP. Such microtubules are said to have a 'GTP cap'. (B) Hydrolysis of GTP causes a conformational change that tends to bend protofilaments outward, weakening lateral contacts between dimers shown with green dots in adjacent protofilaments.

assembly rate at the plus and minus ends is much larger for actin than for microtubules, treadmilling is presumed to be more common among actin assemblies. However, both polymers demonstrate this behavior.

Nucleotide hydrolysis also gives rise to a different kind of behavior in microtubules, called **dynamic instability**. Here, energy released by GTP hydrolysis accelerates the rate of depolymerization. Growing microtubules add GTP-bound subunits. Because hydrolysis lags slightly behind assembly, the growing ends of a microtubule will have a cap of subunits with bound GTP. This cap stabilizes the structure and favors further growth (Figure 4.35A, B). However, when the supply of tubulin dimers runs low, or when the rate of GTP hydrolysis accelerates, the end of the microtubule may contain a majority of **GDP-bound** subunits. When freed of the GTP cap, the protofilaments peel apart and initiate a catastrophic disassembly, hundreds of times faster than the rate of growth.

Key points Actin and tubulin assemble into microfilaments and microtubules, respectively. These filaments are distinguished by their intrinsic polarity and are both made up of asymmetrical subunits. Polarity of these filaments confers different biological properties upon their ends. The end that is more dynamic in terms of assembly or disassembly is called the plus end and the less dynamic end is the minus end. Actin microfilaments grow by the addition of G-actin subunits after ATP has bound. Microtubules grow by the addition of α- and β-tubulin after they have bound GTP. The attachment of subunits to the growing cytoskeletal filament requires the hydrolysis of the nucleoside triphosphate.

4.8.4 Accessory proteins regulate the assembly and function of the cytoskeleton

Accessory proteins interact with and regulate the different functions of both actin and tubulin. These accessory proteins fall into two general classes, those that influence the assembly and organization of cytoskeletal proteins, and those, called **motor proteins**, that convert chemical energy into motion.

Motor proteins fall into three families: **myosins**, **dyneins** and **kinesins**. All motor proteins contain a globular, force-producing 'head' domain that binds a cytoskeletal polymer and a rod-shaped 'tail' domain that binds cargo (Figure 4.36). Each also uses ATP as a source of energy. Myosin powers many types of cellular motility involving actin filaments (Figure 4.37A), whereas dynein and kinesin interact with microtubules (Figure 4.37B). Dynein transports cargo to the minus end of a microtubule while kinesin moves cargo towards the plus end. Kinesins are implicated in **vesicle traffic** and the formation of **mitotic spindles**. All three families of accessory proteins are found in plants but their relative abundance differs from that in animals. In plants, **axonemal dynein** is only found in those phyla that have motile sperm; angiosperms appear to lack this motor protein. In contrast, plants have more kinds of kinesin than animals, including forms that are plant-specific.

In addition to motor proteins, several other types of accessory proteins affect cytoskeleton assembly and function (Figure 4.38). Some, called **'cross-linking' or 'bundling' proteins**, form bonds between cytoskeletal polymers of the same type. Some cytoskeletal accessory proteins stabilize polymers; others sever or weaken them. Some proteins bind specifically to one of the polymer ends. Others, including the pollen allergen **profilin** (see Chapter 16), bind soluble actin, lowering the concentration of subunits available for polymerization. Finally, some proteins cross-link the cytoskeleton to other components of the cell, such as membranes, biosynthetic enzymes or signal transduction components.

Key points Assembly and function of the cytoskeleton requires activity of the motor proteins dynein, kinesin and myosin. Motor proteins have a globular head and a tail. The globular head of the motor protein attaches to the cytoskeleton and performs work, whereas the tail portion attaches to cargo. Myosin powers cellular motility involving actin filaments, whereas dynein and kinesin interact with microtubules. Dynein transports cargo to the minus end of a microtubule while kinesins move cargo toward the plus end. Kinesins are implicated in vesicle traffic and the formation of mitotic spindles.

4.8.5 Cytoplasmic streaming and movement and anchoring of organelles require actin

In many plant cells, organelles such as mitochondria and plastids flow through the cytoplasm; this movement is

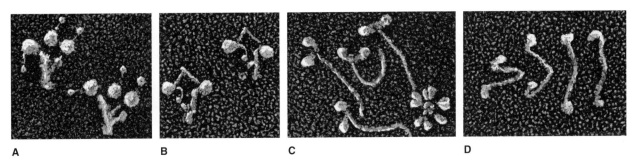

Figure 4.36 TEM images of highly purified motor proteins. (A) The flagellar form of dynein has three spherical head domains. (B) The cytoplasmic form of dynein has two head domains. (C) Myosin II; the asterisk-shaped structure in the lower right of the image is an IgM antibody used to purify the myosin. (D) Kinesin, showing two small motor domains at one end of the rod-shaped protein while the light chains at the other end are not distinguishable from each other.

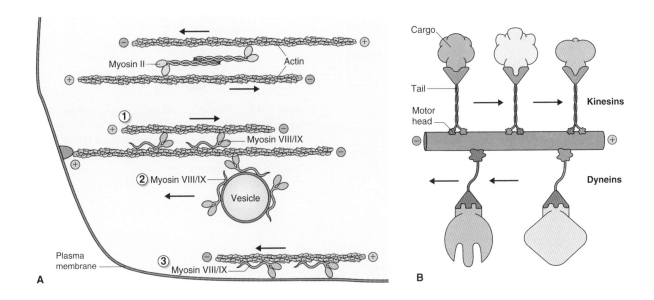

Figure 4.37 Motor proteins and their interactions with actin filaments. (A) Myosin proteins interact with actin filaments, moving from the minus (pointed) end to the plus (barbed) end of the polymer. Myosin II proteins, such as those in muscle, have long rod-like domains that promote assembly into bipolar (head-to-tail) filaments. Myosin VIII/IX motors have short tail domains that can bind an actin filament or a membrane. Force transduction by the myosin VIII/IX head may then (1) move one actin filament relative to another, (2) move a vesicle along an actin filament, or (3) move an actin filament along a membrane. (B) Diagram of the microtubule-associated motor proteins dynein and kinesin. Dyneins move cargo toward the minus end of a microtubule, whereas most kinesins move cargo toward the plus end.

called **cytoplasmic streaming**. Cytoplasmic streaming is thought to involve an interaction between myosin that is bound to the surfaces of organelles and actin filaments in the cytoplasm. The cytoskeleton also moves organelles to specific locations and anchors them at specific positions within the cell. For example in photosynthetic cells, actin filaments are involved in repositioning chloroplasts for optimal light absorption in response to changes in light intensity (Figure 4.39). Also actin filaments and microtubules can collaborate to reposition and anchor the nucleus.

4.8.6 Actin filaments participate in secretion

Actin filaments play a pivotal role in exocytosis in tubular cells that undergo 'tip growth' (also discussed in Chapter 12); in these cells exocytosis is localized to the extending apical end of the cell. In tip-growing cells (e.g. pollen tubes, root hairs), polarized actin filaments are arranged perpendicular to the tip and deliver secretory vesicles to the tip. The vesicles have myosin in their membranes and associate with actin filaments

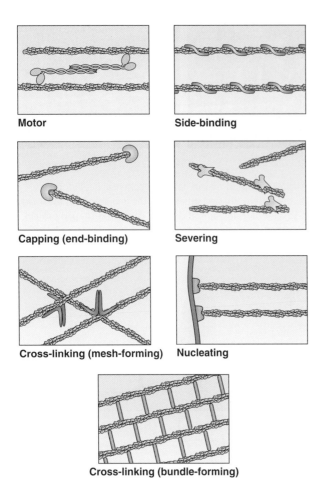

Motor

Side-binding

Capping (end-binding)

Severing

Cross-linking (mesh-forming)   **Nucleating**

Cross-linking (bundle-forming)

Figure 4.38   Various functions of accessory proteins. Proteins that associate with the cytoskeleton have various functions, as illustrated here with actin filaments. Although these proteins have been studied mainly in animal cells, some similar and possibly homologous proteins have been identified in plant cells.

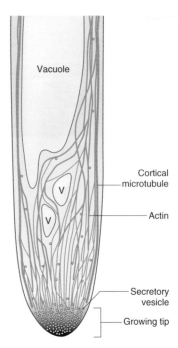

Figure 4.40   Schematic, cross-sectional view of the movement of vesicles to the tip of a growing pollen tube. Cortical microtubules are shown in green, actin filaments and cables in orange (V, vacuole). The high cytoplasmic Ca^{2+} gradient in the tip region is represented by gray shading with black representing the highest Ca^{2+} concentration. Blue dots represent secretory vesicles.

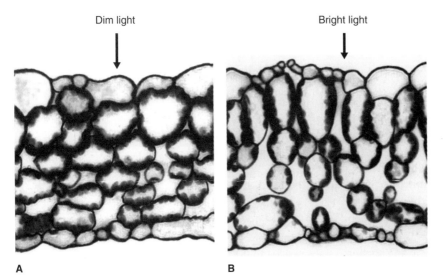

Figure 4.39   Reorientation of chloroplasts in an *Arabidopsis* leaf. (A) Under dim light, chloroplasts move to walls parallel to the leaf surface, thus maximizing light absorption for photosynthesis. (B) In bright light, chloroplasts migrate to cell walls perpendicular to the leaf surface, thus minimizing light absorption and photodamage.

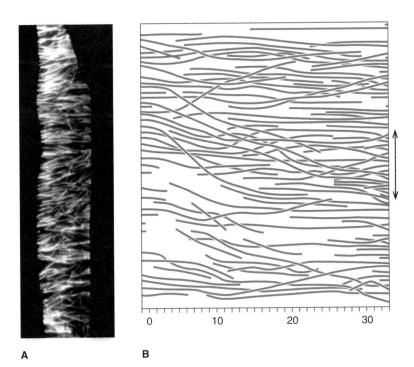

A   **B**

Figure 4.41   Orientation of microtubules in an elongating *Arabidopsis* root cell. (A) Fluorescence micrograph of a cell from the root of *Arabidopsis thaliana*, showing the cortical array of microtubules. Microtubules are orientated mainly with their axes transverse to the long axis of the cell (and root). (B) Region of the cortical array from a cell in the root of *Azolla pinnata*. The double-headed arrow represents the direction of maximum cell expansion rate.

(Figure 4.40). The high rates of relative expansion sustained by tip-growing cells presumably requires the delivery of a large quantity of wall precursors without interruption. Whether other cell types use actin-dependent exocytosis remains to be determined.

4.8.7 Cortical microtubules help orientate cell expansion by aligning cellulose microfibrils

An array of microtubules usually lies just inside the plasma membrane of plant cells and perpendicular to the direction of most rapid expansion. In single cells, or cylindrical organs such as stems and roots, the direction of maximum expansion parallels the long axis of the organ. Hence, the cortical microtubules, orientated perpendicular to that axis, are commonly described as 'transverse' (Figure 4.41). The principal function of the cortical array is to influence the direction of cellulose microfibril deposition, which, in turn, specifies the direction of maximal cell expansion in growing cells.

Part II
Germination

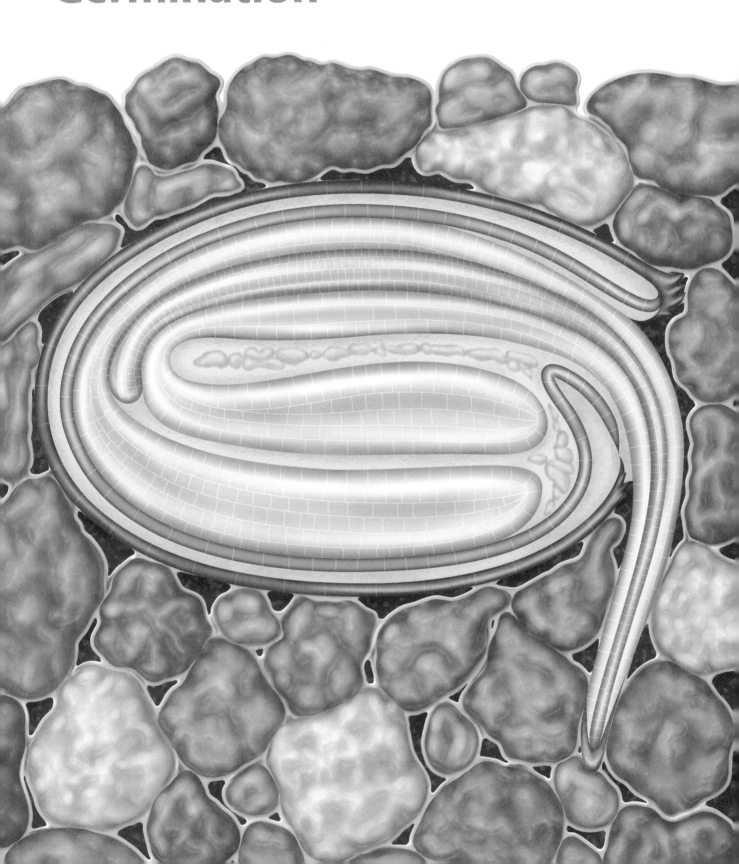

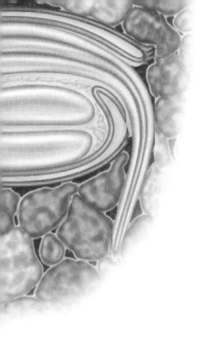

Chapter 5

Membrane transport and intracellular protein trafficking

5.1 Introduction to the movement of solutes and macromolecules

In this chapter we will discuss how solutes move into and out of plant cells and organelles. We will also examine how macromolecules are transported within cells. Minerals, sugars, metabolites and other small molecules move across the limiting membranes of cells and organelles, whereas larger macromolecules, such as proteins and glycoproteins, are packaged within vesicles for transport to their specific destinations. Membranes form barriers that insulate the contents of cells from the harsh external environment and allow the partitioning of the cytoplasm into compartments that carry out specific metabolic functions. As evolution progressed, the partitioning of the cytoplasm into membranous organelles allowed eukaryotic cells to increase in size and adopt highly specialized functions. In contrast to prokaryotes, eukaryotic cells contain many specialized organelles responsible for diverse biosynthetic, catabolic and storage functions (Figure 5.1; see Chapter 4).

Compartmentation of solutes and macromolecules within membrane-bound organelles not only concentrates reactants and catalysts, but also segregates incompatible processes.

Movements of water and solutes are among the first changes that occur when dormant or quiescent seeds are activated. Mature seeds generally contain less than 10% water and the hydration of cells and tissues is one of the first events that occurs when seeds are imbibed. The movement of solutes across membranes is facilitated by transport proteins embedded within membranes (Figure 5.2). These transporters are selective for the compound being transported, and their activities are often highly regulated. Vesicular transport of macromolecules is also selective, distributing cargo to various destinations both inside and outside the cell. The selectivity of vesicular traffic can lie in information contained on the surface of the vesicle as well as in the cargo contained in the vesicle.

The field of membrane transport has benefited enormously from advances in molecular cloning and genome biology. The nucleotide and amino acid sequences of many transport proteins have been determined and, as you will see in this chapter, detailed

The Molecular Life of Plants, First Edition. Russell Jones, Helen Ougham, Howard Thomas and Susan Waaland.
© 2013 John Wiley & Sons, Ltd. Published 2013 by John Wiley & Sons, Ltd.

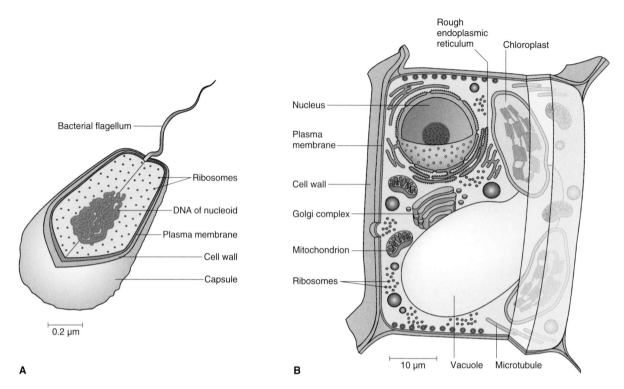

Figure 5.1   Bacterial (A) and plant (B) cells. The plant cell has many membrane-enclosed compartments while the bacterial cell does not.

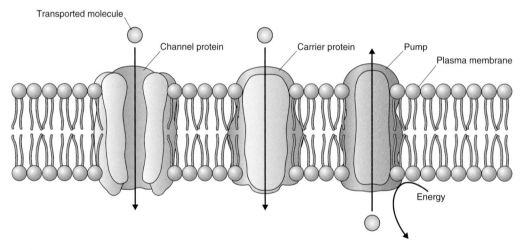

Figure 5.2   Channels, carriers and pumps are the principal classes of membrane transporters. Channels and carriers transport solutes down a favorable electrochemical gradient whereas pumps can move solutes against an electrochemical gradient by expending energy, usually in the form of ATP.

information is now available on the molecular structure of many of these proteins. Another important advance in the field of solute transport came about with the development of the patch clamp technique more than 30 years ago. This method allows the behavior of single transport proteins to be studied in the membranes in which they are located, and for the cell physiologist it represents the ultimate molecular approach as the function of the molecule can be analyzed in near in vivo conditions.

Membrane and vesicular transport must be sufficiently precise to ensure that solutes are properly distributed within the cell and that macromolecules reach their final destinations. Differences in the concentrations of ions and small molecules across membranes can vary by several orders of magnitude and these differences in concentration are achieved in part by the activities of membrane transporters. Differences are also found in the distribution of macromolecules in cells. Soluble enzymes are present in all subcellular compartments, whereas membrane-bound proteins occur in the more than 14 different lipid bilayers that delimit these compartments (see Table 4.2). It is estimated that as many as 30% of all cellular proteins are found in membranes. Some proteins are unique to a particular structure, compartment or membrane; other very similar proteins with comparable amino acid sequences, structures and functions occur in more than one compartment. Clearly, the specificity of cellular function requires that the transport machinery in cells be highly specific and carefully regulated.

In this chapter we will first discuss the principles that govern the movement of solutes across cellular membranes. We will then look at the mechanisms by which this transport occurs. Finally, we will discuss intracellular sorting and the transport of macromolecules.

5.2 Physical principles

The driving force for the movement of a molecule either in solution or in the gas phase is a **gradient** in its **potential energy**. Except at absolute zero, molecules are in constant motion and will distribute themselves uniformly in the space that they occupy, as illustrated in Figure 5.3. This movement is called **diffusion**.

5.2.1 Diffusion is a spontaneous process and obeys Fick's law

Diffusion occurs spontaneously; no additional energy must be provided. Diffusion occurs in biological systems, for example from one point in a cell to another or across a membrane. The rate at which a molecule, s, diffuses is related to the size of the molecule, the magnitude of its concentration gradient, the viscosity of the medium and temperature. This relationship is shown in Equation 5.1, known as **Fick's law**.

Equation 5.1 Fick's law

$$J_s = -D_s(\Delta C_s/\Delta x)$$

The rate of diffusion (J_s) of species s is determined by its diffusion coefficient (D_s) and concentration gradient ($\Delta C_s/\Delta x$), i.e. the concentration difference (ΔC_s) between two points (Δx). J_s is expressed as moles per area per time. Small molecules in the gas phase diffuse rapidly, whereas large molecules, such as proteins or polysaccharides, in solution diffuse slowly. The negative sign in Equation 5.1 shows that substances move by diffusion down a concentration gradient.

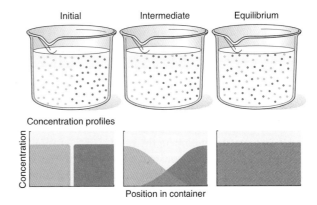

Figure 5.3 Random thermal motion of molecules dissipates concentration gradients, eventually leading to complete mixing. Initially the blue and yellow molecules are entirely separate; at equilibrium they are randomly distributed and therefore completely mixed. The diffusion of the molecules of each species is driven by its own concentration gradient. Diffusion is most rapid in the gas phase, slower in the liquid phase and much less rapid in the solid phase.

The importance of Fick's law is that one can calculate the time that it takes a molecule to diffuse a given distance. The time that it takes substance s to move distance L is given by the expression L^2/D_s. Thus the time it takes for s to diffuse increases with the square of the distance. The value of D_s for ions is around 10^{-9} m^2 s^{-1} and for larger biological molecules D_s is around 10^{-11} to 10^{-10} m^2 s^{-1}. A molecule with a D_s of 10^{-9} m^2 s^{-1} will take 2.5 s to move 50 μm, the diameter of an average plant cell, but to move a distance of 1 m would take 32 years. Thus diffusion may be an effective way for molecules and ions to move cellular distances, but it is ineffective for long-distance movement. As we shall see in Chapter 14, other mechanisms operate in plants to move solutes over long distances, e.g. from organ to organ.

5.2.2 The chemical potential of a solute is expressed as free energy per mole

Fick's law describes the flux of an uncharged solute; it only considers its concentration. When discussing transport of solutes in biological systems, it is important to consider the **sum of the physical forces** that act upon solute s. The **chemical potential** (μ_s) of a solute is its free energy per mole. It is affected by the concentration of the solute, its electrical charge and hydrostatic pressure (Equation 5.2).

Equation 5.2 Chemical potential of solute **s**

$$\mu_s = \mu_s{}^* + RT\ln C_s + z_s FE + V_s P$$

In this equation μ_s^* is the chemical potential of s under standard conditions; it provides the reference point for the measurement of the chemical potential of s. The term $RTlnC_s$, the **concentration component**, includes the concentration C of s in moles per liter at standard pressure, R, and absolute temperature T. The term z_sFE represents the **electrical component**; z is the charge on s (zero for an uncharged molecule and +1, +2 or −1, −2, etc. depending on the valence of the cation or anion), F is Faraday's constant and is equal to the amount of charge in one mole of electrons, and E is the electrical potential of the solution with respect to ground. The **hydrostatic pressure component**, V_sP, includes the partial molal volume (V) of s and pressure (P).

When considering the chemical potential of biological molecules at the cellular level, the contributions of V and P are generally small compared with the contributions of the chemical concentration and electrical properties of s. Therefore, for an ion or a charged molecule, Equation 5.2 can be simplified to Equation 5.3, which contains only the concentration and charge components.

Equation 5.3 Chemical potential of an ion or charged molecule

$$\mu_s = \mu_s^* + RTlnC_s + z_sFE$$

The chemical potential of a charged species is often called its **electrochemical potential**. For an uncharged solute the electrical component of the equation equals zero and the equation for its chemical potential becomes:

Equation 5.4 Chemical potential of an uncharged solute

$$\mu_s = \mu_s^* + RTlnC_s$$

Although the hydrostatic component of the chemical potential of solutes in cells is very small, in Chapter 14 we will see that pressure is an important factor when considering the chemical potential of water.

5.2.3 Differences in chemical potential drive solute movement

The driving force for diffusion of a solute into or out of a cell is the difference between its chemical potential (μ_s) inside the cell and that in the external solution (Figure 5.4). Therefore it is important to be able to describe this difference quantitatively. For uncharged solutes the difference in chemical potential ($\Delta\mu_s$) between the inside (i) and the outside (o) of a cell can be

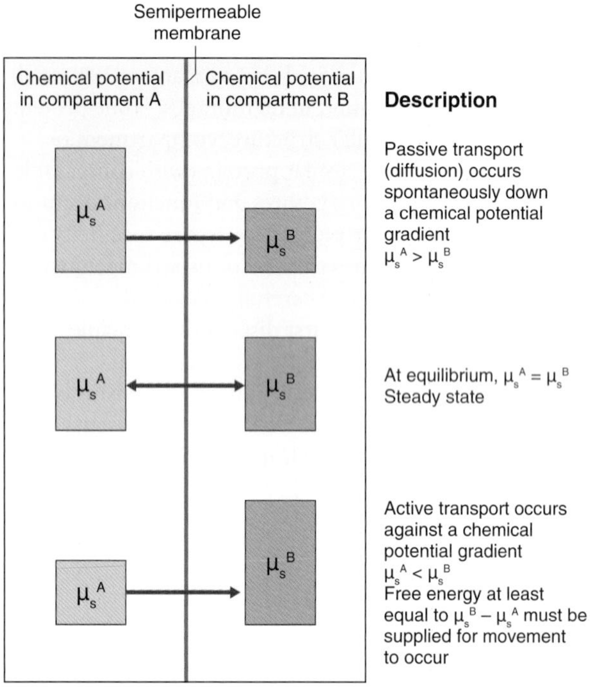

Figure 5.4 The magnitude of the chemical potential gradient drives the movement of solutes across membranes. The height of the boxes represents the chemical potential of solute s in each compartment; the membrane is permeable to solute s. When a difference in the concentration of s occurs across the membrane, s will flow passively down its chemical potential gradient (top) until equilibrium is reach (middle). To move s against its chemical potential gradient, free energy must be supplied (bottom), a process called active transport.

described by an expanded expression of Equation 5.4 (Equation 5.5).

Equation 5.5 Difference in chemical potential between cell interior and exterior

$$\Delta\mu_s = (\mu_s^* + RTlnC_s^i) - (\mu_s^* + RTlnC_s^o)$$

This equation reduces to:

Equation 5.6

$$\Delta\mu_s = RTln(C_s^i/C_s^o)$$

The conclusion that can be drawn from Equation 5.6 is that the driving force for the diffusion of uncharged solute s is the magnitude of its concentration difference across the membrane. The sign of the product of this equation indicates the net direction of movement of s; a negative sign shows that s will tend to diffuse into the cell down its chemical potential gradient. The other factor that will determine the movement of s across a membrane is the membrane's permeability to s, a topic that we will explore in more detail below.

For charged molecules and ions, the electrical component as well as the concentration component of Equation 5.2 must be taken into account when calculating the gradient in electrochemical potential across the membrane, as shown in Equation 5.7.

Equation 5.7 Gradient in electrochemical potential for a charged solute

$$\Delta\mu_s = (\mu_s{}^* + RT\ln C_s{}^i + z_s FE^i) - (\mu_s^* + RT\ln C_s{}^o + z_s FE^o)$$

This equation reduces to:

Equation 5.8

$$\Delta\mu_s = RT\ln(C_s{}^i/C_s{}^o) + zF(E^i - E^o)$$

The term $E^i - E^o$ in Equation 5.8 is often called the **membrane potential** or **membrane voltage** and is abbreviated $\mathbf{V_m}$. We can conclude from Equation 5.8 that movement of a charged solute responds to two independent forces, namely the difference in the concentration of s across the membrane and the membrane potential. While only solute *s* can contribute to changes in its own concentration, any charged solute can change the membrane potential. This means that charged solutes can diffuse into or out of cells against their own concentration gradient if the membrane potential is favorable. We will discuss this idea in a little more detail below.

Key points When dealing with transport of solutes in biological systems, it is important to consider all physical forces that act on the solute. These include a solute's concentration, its electrical charge and the effects of temperature and pressure. For solute movement into and out of cells, changes in pressure are relatively unimportant, but for the movement of solutes in trees effects of pressure are considerable. The sum of physical forces is expressed as the electrochemical potential of the solute. For the biologist what is critical to know is the difference in solute electrochemical potential, as this information dictates whether a solute moves and the direction in which movement occurs as all solutes move down their electrochemical potential gradient.

5.2.4 Unequal distributions of charged solutes across membranes give rise to a membrane potential

The rate at which different charged molecules and ions cross membranes is often not equal and even small differences in their distribution across the membrane can give rise to a membrane potential. The development of a membrane potential can be illustrated by the example shown in Figure 5.5; here two KCl solutions of different concentration are separated by a membrane that is more permeable to K^+ than Cl^-. In this illustration, both K^+ and Cl^- diffuse down their concentration gradients from compartment A to compartment B, but K^+ diffuses through the membrane at a slightly higher rate than Cl^-, giving rise to a difference in charge across the membrane. If electrodes were placed in compartments A and B, the difference in concentration of K^+ and Cl^- could be measured as a difference in voltage. Very small differences in the distribution of charge can give rise to significant differences in membrane potential. For example, it takes only a concentration difference of 0.001% in anions over cations across the membrane to produce a membrane potential of -100 mV. The membrane potential of most plant cells falls in the range -100 to -250 mV.

5.2.5 The Nernst equation predicts internal and external ion concentrations for a given membrane potential

Solutes moving into and out of cells by diffusion should reach equilibrium. When equilibrium between the inside and outside of a cell is reached for a given solute, $\Delta\mu_s$ is zero. If we rearrange Equation 5.8:

Equation 5.9 Rearrangement of Equation 5.8 for the case where $\Delta\mu_s = 0$

$$RT\ln(C_s{}^i/C_s{}^o) = -zF(E^i - E^o)$$

Equation 5.9 can then be further rearranged to show the difference in electrical potential, ΔE, between the inside and outside of the cell as shown in Equation 5.10:

Equation 5.10 Difference in electrochemical potential between cell interior and exterior

$$\Delta E = (RT\ln(C_s{}^o/C_s{}^i))/(z_s F)$$

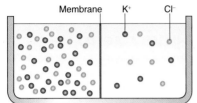

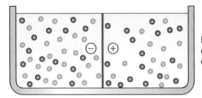

Initial conditions:
$[KCl]_A > [KCl]_B$

Diffusion potential exists until chemical equilibrium is reached

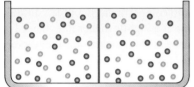

Equilibrium conditions:
$[KCl]_A = [KCl]_B$

At chemical equilibrium, diffusion potential equals zero

Figure 5.5 A membrane potential can develop when a membrane is differentially permeable to ions such as K^+ and Cl^-. Upper panel: K^+ and Cl^- will diffuse across the membrane from compartment A to compartment B because of the electrochemical potential difference for each ion. Initially, because the membrane is more permeable to K^+ than to Cl^-, a difference in charge—a membrane potential—will develop (middle panel). At equilibrium (bottom panel), the chemical concentrations of K^+ and Cl^- are equal on both sides of the membrane and the membrane potential is zero.

For a monovalent cation (s) at 25° C, this equation can be simplified to Equation 5.11 (note that the logarithm term is now given as log(base 10) rather than natural logarithm (ln).

Equation 5.11 Nernst equation

$$\Delta E_s = 59mV \, \log(C_s^{\,o}/C_s^{\,i})$$

Equation 5.11 is known as the **Nernst equation** and the expression ΔE for a specific ion is known as the **Nernst potential**. The Nernst equation has great predictive value for cell physiologists. It can be used to determine whether or not ions are accumulated against their electrochemical potential gradient. If the membrane potential of a cell can be measured and if the concentration of solutes outside ($C_s^{\,o}$) and inside ($C_s^{\,i}$) the cell is known, it becomes easy to establish whether a solute is taken up by active transport or whether it moves down its electrochemical gradient. Table 5.1 shows data obtained from pea roots where the concentration of ions inside root cells and in the soil solution were determined

Table 5.1 Comparison of predicted and observed ion concentrations in pea root tissue, where the membrane potential was measured as −110 mV.

Ion	External concentration (mmol L^{-1})	Internal concentration (mmol L^{-1})	
		Predicted	Observed
K^+	1	74	75
Na^+	1	74	8
Mg^{2+}	0.25	1340	3
Ca^{2+}	1	5360	2
NO_3^-	2	0.0272	28
Cl^-	1	0.0136	7
$H_2PO_4^-$	1	0.0136	21
SO_4^{2-}	0.25	0.00005	19

at a membrane potential of −110 mV. What this table clearly demonstrates is that while K^+ is at equilibrium in pea roots relative to soil solution, ions such as Cl^-, $H_2PO_4^-$ and NO_3^- are present inside pea root cells at concentrations much higher than those predicted by the Nernst equation. Nitrate, for example, is accumulated to a concentration 1000-fold greater than predicted by the Nernst equation. From this it can be concluded that energy must be expended for nitrate to accumulate inside cells. Ions such as Ca^{2+} and Mg^{2+}, on the other hand, are present inside pea root cells at far lower concentrations than those predicted by the Nernst equation, indicating that there are mechanisms that maintain these ions at a lower concentration, either by preventing their entry into cells or by pumping them out.

5.3 Regulation of solute movement by membranes and their associated transporters

Biological membranes are selectively permeable. Lipophilic/hydrophobic molecules such as O_2, CO_2, N_2, NH_3 and H_2O_2 move fairly freely through the lipid bilayer (Figure 5.6). The lipid bilayer is even permeable to water, as well as to somewhat larger polar molecules such as urea. As long as favorable chemical potential gradients exist for these molecules they will diffuse through the membrane until there is no difference in

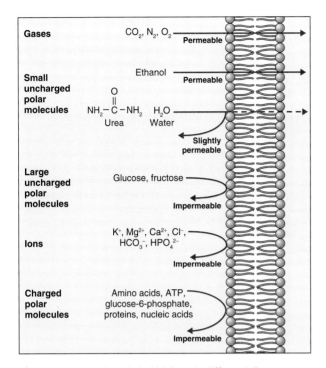

Figure 5.6 A phospholipid bilayer is differentially permeable. It is freely permeable to gases such as CO_2, O_2 and N_2 and to small, uncharged polar molecules such as ethanol. Other small, uncharged polar molecules such as urea and water have limited permeability. Membranes are impermeable to larger polar molecules, ions and macromolecules.

their chemical potentials on either side of the membrane. Charged solutes and larger polar molecules such as nucleotides and sugars do not diffuse easily across membranes. Specific transport proteins are therefore required to facilitate their movement into and out of cells. Transport proteins are embedded in the membrane (see Figure 5.2). They generally consist of alternating blocks of hydrophobic and hydrophilic amino acids that allow them to span the phospholipid bilayer of the membrane and, at the same time, have hydrophilic domains exposed to the cytosol, the lumen of an organelle or the cell exterior (Figure 5.7).

Solutes may cross membranes by diffusion, moving *down* their electrochemical potential gradients (Figure 5.8A). Solutes that freely cross the lipid bilayer are said to move by **simple diffusion**. Those that require the presence of a membrane transporter to diffuse across a membrane move by **facilitated diffusion**. There are two types of transporters that facilitate diffusion across membranes: **channels** and **carriers**.

Protein transporters may also move solutes *against* their electrochemical gradients; this process consumes cellular energy and is called **active transport** (Figure 5.8B). Active transport can be subdivided into two categories: **primary active transport** and **secondary active transport**. During primary active transport, ATP or pyrophosphate is hydrolyzed to provide the energy required to establish ion gradients. Transporters that perform primary active transport are called **pumps**.

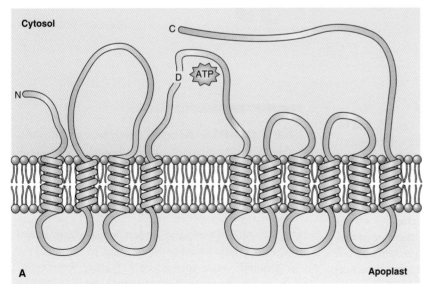

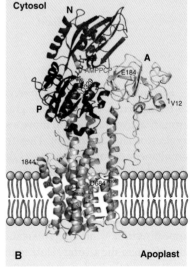

Figure 5.7 Configuration of membrane-associated ATPases. (A) Membrane disposition of the plasma membrane ATPase showing ten transmembrane domains and a long autoinhibitory C-terminal domain. The ATP-binding aspartate residue D is shown between loops 4 and 5. (B) Ribbon diagram of the H^+-ATPase based on X-ray crystallographic data of the active form of the pump lacking the long C-terminal tail. The ten transmembrane domains are shown in orange, green and brown, the ATP-binding domain is labeled N, and bound Mg is labeled AMPPCP.

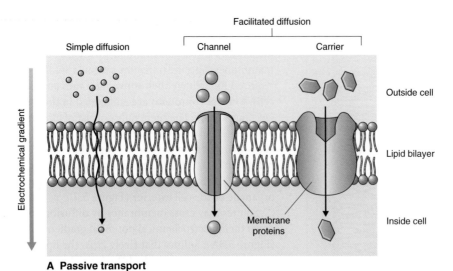

Facilitated diffusion

Simple diffusion | Channel | Carrier

Outside cell

Lipid bilayer

Membrane proteins

Inside cell

Electrochemical gradient

A Passive transport

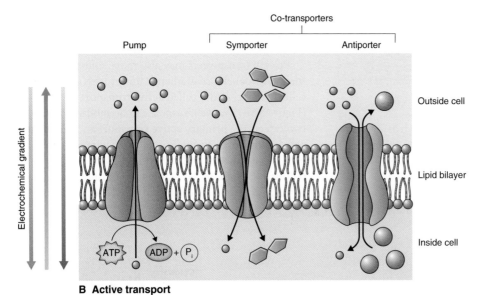

Co-transporters

Pump | Symporter | Antiporter

Outside cell

Lipid bilayer

Inside cell

ATP $ADP + P_i$

Electrochemical gradient

B Active transport

Figure 5.8 Molecules can move across membranes by passive (A) or active (B) transport. Passive transport occurs down an electrochemical gradient by simple diffusion through the lipid bilayer or through proteinaceous channels or carriers. Active transport occurs against an electrochemical gradient and can occur through pumps, symporters or antiporters, all of which are transmembrane proteins.

    Secondary active transporters, called **co-transporters,** exploit the ion gradients established by primary active transport to move a second solute against its electrochemical gradient. They transport two solutes simultaneously; one solute moves down its electrochemical gradient while the second moves up its electrochemical gradient. Co-transporters can be divided into two groups: **symporters** and **antiporters.** Symporters move two solutes across a membrane in the same direction, while antiporters move one solute in one direction and the second solute in the opposite direction. In plant cells, secondary active transport usually exploits H^+ gradients established by proton pumps. Let us examine the characteristics of each of these types of transport protein.

Key points Biological membranes are impermeable to a wide range of biologically important solutes; for these to cross membranes specific transporters are required. These fall into two general classes depending on whether they move solutes downhill, that is down an electrochemical potential gradient, or uphill against such a gradient. Channels and carriers move solutes downhill by diffusion whereas primary and secondary active transporters move solutes uphill. Primary active transporters use the energy released from the hydrolysis of ATP or pyrophosphate to establish ion gradients. Secondary active transporters use these ion gradients to drive the accumulation of other solutes.

5.4 Pumps

Pumps move solutes against their electrochemical gradients. They bind to the solute to be transported and hydrolyze ATP or pyrophosphate as a source of energy. Pumps transport solutes at a rate of tens to hundreds of molecules or ions per second. They are located in most cellular membranes (Figure 5.9). Pumps that hydrolyze ATP can be divided into three categories based on their structure, inhibitor sensitivity and mechanism of action (Figure 5.10). **F-type ATPases** (ATP synthases), such as those found in chloroplasts and mitochondria, have multiple subunits and are insensitive to the inhibitor vanadate. The **V-ATPase** found in vacuole and other cellular membranes is related to F-type ATPases, having multiple subunits, and is also vanadate-insensitive (see Section 5.4.4). **P-type ATPases** have a simpler structure than the other types of ATPase. They form phosphorylated intermediates after hydrolyzing ATP

(Figure 5.11) and are inhibited by vanadate. Members of the **ABC superfamily of transporters** are P-type ATPases. These transporters consist of four core domains, two that span the membrane and two that are cytosolic. Their cytosolic domains bind and hydrolyze ATP and the energy released drives solute movement across the membrane.

Proton (H$^+$)ATPases play a very important role because protons constitute one of the major energy currencies of the plant cell. At the inner mitochondrial membrane and at the thylakoid membrane, a transmembrane H$^+$ gradient is generated using light or chemical energy; the gradient is then used to energize the synthesis of ATP by H$^+$ pumps (ATP synthases). At all other membranes in the cell, **H$^+$ pumps** hydrolyze ATP or pyrophosphate to power the transport of protons out of the cytosol. The electrochemical gradient in protons is called a **proton motive force** (**pmf**). The pmf can be used in secondary active transport to power the transport of other ions and solutes across the membranes (see Figure 5.8B).

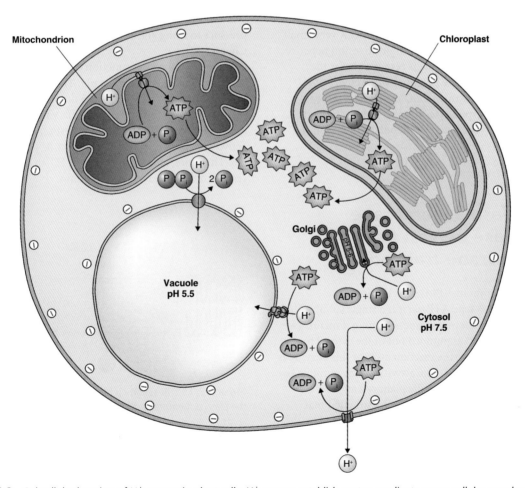

Figure 5.9   Subcellular location of H$^+$ pumps in plant cells. H$^+$ pumps establish proton gradients across cellular membranes that are utilized in the synthesis of ATP by the F-type enzyme, or in powering the transport of solutes in the case of the P-type and V-type pumps and the proton-pumping pyrophosphatase. The pyrophosphatase and V-ATPase are found on membranes other than the tonoplast, including the plasma membrane, Golgi apparatus, trans-Golgi network and prevacuolar compartment.

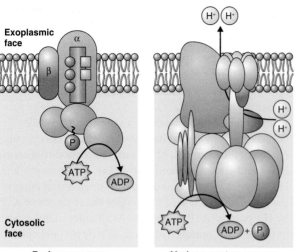

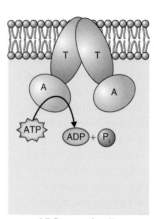

P-class pumps   **V-class proton pumps**   **ABC superfamily**

Figure 5.10 Three types of ATP-hydrolyzing pumps are found in membranes. P-type and V-type pumps transport H^+, and ABC-type pumps transport a variety of solutes. Note that in plants some P-type pumps such as the plasma membrane H^+-ATPase and the Ca^{2+}-ATPase, lack a β-subunit. A, ATP-binding site; T, transmembrane domain.

The transport of protons by H^+ pumps is not balanced by the movement of anions. Therefore the movement of protons establishes a charge gradient that changes the membrane potential. Pumps that create a charge gradient are said to be **electrogenic**. A very small difference in the actual number of positive and negative charges across the membrane is responsible for the change in the membrane potential V_m. Activity of H^+ pumps in the plasma membrane makes its V_m more negative, i.e. **hyperpolarizes** the V_m; if the activity of these pumps decreases, the V_m **depolarizes**, becoming less negative. As we shall see these changes have an important role in the uptake of ions.

5.4.1 Plasma membrane H^+-ATPase plays a key role in membrane transport

The plasma membrane (PM) H^+-ATPase is a Mg-dependent, P-type ATPase; it is a single subunit, 100 kDa protein with ten membrane-spanning domains (see Figure 5.7). One H^+ is pumped out of the cytosol for each ATP hydrolyzed; the reaction cycle of this enzyme is illustrated in Figure 5.11. The activity of this pump is key to the transport activities of the plant cell and it is a major consumer of cellular ATP. It is estimated that 30–50% of ATP produced by a cell is hydrolyzed by the PM H^+-ATPase. This pump establishes the cell's pmf, powering the activity of symporters and antiporters (see Figure 5.8B). It also influences the V_m, thus affecting the movement of ions through channels. Acidification of the

cell wall by the H^+ pump also influences growth and development. Auxin activates the PM H^+ pump within minutes, and the resulting acidification of the cell wall increases its extensibility, allowing turgor pressure to bring about cell expansion (see Chapter 12).

Another important function of the PM H^+ pump is the maintenance of cytosolic pH. The pH of the cytosol is maintained in the range 7.3–7.5 despite the fact that many of the reactions of intermediary metabolism generate an excess of H^+. An interesting aspect of the activity of the PM H^+-ATPase is that its pH optimum is 6.6. Therefore, an accumulation of H^+ in the cytosol results in the activation of the pump.

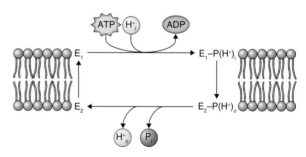

Figure 5.11 Reaction cycle of P-type ATPase in the plane of the plasma membrane. Phosphorylation of the ATPase (E_1) leads to binding of H^+ on one side of the membrane, leading to formation of E_2. E_2 has a lower affinity for H^+ leading to the release of protons and inorganic phosphate on the other side of the membrane. Hydrolysis of inorganic phosphate from E_2 returns it to the E_1 state.

5.4.2 Plasma membrane H^+-ATPase is regulated predominantly through enzyme activity rather than gene expression

Given the central role of the plasma membrane H^+ pump in plant cell physiology, the regulation of its activity has been intensively studied. This pump is encoded by a large multigene family; 11 genes have been identified in *Arabidopsis*. Two of these genes are expressed in all *Arabidopsis* cells and there is evidence that individual cells can express multiple isoforms of the gene.

Despite the physiological importance of the PM proton pump, its expression is not strongly regulated at the level of transcription or translation. Instead, evidence points to the regulation of the pump at the level of the enzyme's activity. One way in which the activity of this pump might be regulated is through the action of a 14-3-3 protein. 14-3-3 proteins are a family of conserved proteins found in all eukaryotic cells that bind a range of target proteins and modify their activity. The name 14-3-3 comes from the protein separation pattern after ion exchange chromatography and gel electrophoresis. The PM H^+ pump has a C-terminal regulatory, autoinhibitory domain (Figure 5.12) that contains several phosphorylation sites. When a conserved threonine residue in the autoinhibitory domain of the pump is phosphorylated, a 14-3-3 protein binds to the C-terminal domain and forms an active pump —14-3-3 complex (Figure 5.12). Dephosphorylation of the conserved threonine releases the14-3-3 protein and deactivates the pump.

The identification of 14-3-3 proteins has helped to explain the mechanism of action of the fungal toxin, fusicoccin. Work in the 1970s showed that fusicoccin promotes the acidification of the cell walls of cells in various tissues. It was proposed that fusicoccin caused wall acidification by increasing the activity of the plasma membrane H^+ pump. We now know that fusicoccin stimulates H^+ pumping activity by binding to a site that is formed by the interaction of 14-3-3 proteins and the plasma membrane H^+ pump (Figure 5.12).

5.4.3 A Ca^{2+} pumping ATPase on endomembranes regulates cytosolic Ca^2 concentrations

Calcium is an important regulator of cellular activities; because it forms insoluble salts with phosphate, its concentration in the cytosol is maintained in the low nanomolar range. The concentration of Ca^{2+} in the cytosol is regulated by a number of transporters, among them a Ca^{2+}-ATPase that is found on almost all cellular membranes including the membranes of chloroplasts and mitochondria. Like the plasma membrane H^+-ATPase, the Ca^{2+} pump belongs to the P-type family of ATPases. It is a single polypeptide of about 110 kDa with an extended N-terminal domain (Figure 5.13A). ATP hydrolysis by this enzyme is accompanied by the transport of two Ca^{2+} ions across the membrane (Figure 5.13B).

The role of Ca^{2+} pumps is to maintain cytosolic Ca^{2+} in the range of 50–200 nM by pumping Ca^{2+} out of the cell and into the vacuole and other organelles. Two broad classes of Ca^{2+}-ATPases have been identified in plant cells based on their ability to bind calmodulin, a cellular Ca^{2+} sensor. The calmodulin-binding class of Ca^{2+}-ATPases has an N-terminal, autoinhibitory domain that binds calmodulin in a Ca^{2+}-dependent manner. When calmodulin binds to the N terminus of the Ca^{2+}-ATPase, it inhibits the pump and thus raises the cytosolic Ca^{2+} concentration. This is a good example of a negative feedback loop that maintains cytosolic Ca^{2+} homeostasis.

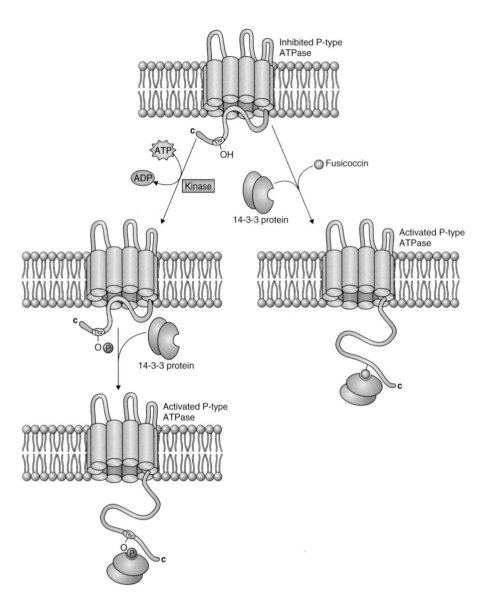

Figure 5.12 Activation of the plasma membrane H^+-ATPase by 14-3-3 proteins and the fungal toxin fusicoccin. The autoinhibitory C terminus of P-type ATPase can bind two 14-3-3 molecules at a phosphorylated threonine residue (Thr), leading to relief from autoinhibition. Fusicoccin can also activate the enzyme by binding with a single 14-3-3 molecule at the C terminus.

Key points Ca^{2+} pumping ATPases are found on almost all endomembranes including the plasma membrane. Their role is to maintain cytosolic Ca^{2+} at concentrations in the range of 50–200 nM. Cytosolic Ca^{2+} must be kept low for the maintenance of many metabolic reactions and for Ca^{2+} to function as a signaling molecule. For example, if levels of cytosolic Ca^{2+} are too high, calcium phosphate will precipitate, depleting the cytosolic phosphate pool. This pool is crucial for supporting a rapid turnover of ATP in cellular metabolism. In addition, if Ca^{2+} is to function as an effective signaling molecule there must be mechanisms for quickly changing its concentration in the cytosol. Whereas Ca^{2+} channels quickly raise the concentration of Ca^{2+} in the cytosol, it is Ca^{2+} pumps that modulate the Ca^{2+} signal by rapidly reducing its concentration.

5.4.4 V-type H^+-ATPases in plants are related to F-type ATPases

Vacuolar H^+-ATPases (V-ATPases) are found in almost all living cells; they share a common origin with F-type ATPases. V-ATPases have a complex multisubunit structure that is reminiscent of F-type ATPases (Figure 5.14). In *Arabidopsis* V-ATPases are made up of 13 subunits, each of which can be encoded by any one of 27 genes. If subunit isoforms can combine randomly, it is clear that an enormous number of different V-ATPase complexes could be assembled. V-ATPases pump about three H^+ per ATP hydrolyzed. They are electrogenic and contribute to the pmf and membrane potential (V_m) of the vacuole membrane, the **tonoplast**.

Although described as vacuolar enzymes, V-ATPases are ubiquitously distributed among the endomembranes

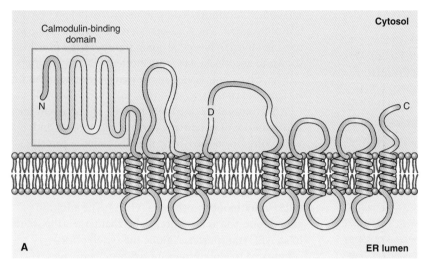

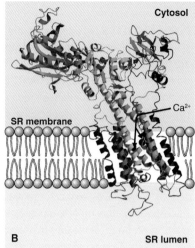

Figure 5.13 Configuration of membrane-associated Ca^{2+}-ATPase. (A) Membrane disposition of the vacuolar Ca^{2+}-ATPase showing ten transmembrane domains and an extended N-terminal calmodulin-binding domain. D, ATP-binding domain. (B) Ribbon diagram of the Ca^{2+}-ATPase based on X-ray crystallographic data, showing the ATP-hydrolyzing domain of a Ca^{2+} pump from an animal sarcoplasmic reticulum (SR; equivalent to the endoplasmic reticulum (ER)) and probable route of Ca^{2+} transport.

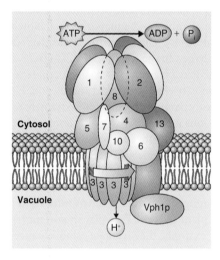

Figure 5.14 Model of V-ATPase from the yeast *Saccharomyces cerevisiae*. The multiple subunits of this complex enzyme are numbered 1 to 13 with an additional subunit labeled Vph1p.

5.4.5 Two types of H^+ pumping pyrophosphatase are found in plants

Plants, some protozoa and many species of archaebacteria and eubacteria possess an H^+ pumping pyrophosphatase (H^+-PPase), a protein of about 80 kDa. This pump is abundant on the tonoplast membrane but it is also located in the Golgi, trans-Golgi network (TGN), multivesicular body (MVB) and plasma membrane. The functional H^+-PPase is a homodimer. Two types of H^+-PPase are found in plants. Type I pumps are activated by cytosolic K^+ and inhibited by Ca^{2+}, whereas Type II H^+-PPases are strongly activated by Ca^{2+} and are insensitive to K^+. H^+-PPases have a remarkable degree of homology across phyla; all H^+-PPases from plants and bacteria share >85% sequence homology at the amino acid level.

Physiologists have puzzled over the existence of two types of H^+ pumps on the tonoplast of plant cells. One explanation proposes that the H^+-PPases evolved to exploit the abundant supply of pyrophosphate in plant cells, where its concentration can reach millimolar levels. Pyrophosphate is produced during several reactions of carbohydrate metabolism, e.g. during the synthesis of ADP-glucose and uridine diphosphate (UDP) glucose. The concerted actions of the V-ATPase and H^+-PPase acidify the vacuole and maintain a tonoplast V_m of about 20–30 mV, with the vacuole being positive relative to the cytosol.

of plant cells, but are absent from the membranes of chloroplasts and mitochondria. The V-ATPases on the membranes of the Golgi apparatus are thought to be involved in exocytosis and vesicle trafficking in the secretory pathway (see Section 5.8). The H^+ pumping activity of the V-ATPase at the tonoplast causes vacuolar acidification. The net positive charge of the vacuole lumen is balanced by the accumulation of organic anions such as malate and oxalate and inorganic anions such as chloride. The pH of the vacuole in most plant cells is about 5.5 but more specialized vacuoles, such as those of lime fruits, have a pH around 1.7, and those of *Oxalis* spp. are between 1.9 and 2.6.

Key points In addition to the plasma membrane H⁺-ATPase, plant cells contain two other H⁺ pumps. They are the vacuolar H⁺-ATPase, found on the tonoplast and other cellular membranes, and the H⁺ pumping pyrophosphatase that is also abundant on the tonoplast. It has been hypothesized that H⁺ pyrophosphatase evolved in plant cells to exploit the large pool of pyrophosphate generated during carbohydrate metabolism. Synthesis of ADP-glucose (for starch formation) and UDP-glucose (for cellulose formation) generate pyrophosphate, which in turn acts as a negative feedback regulator to slow synthesis, especially that of ADP-glucose. By using pyrophosphate to pump H⁺ at the tonoplast and other membranes, pyrophosphate concentrations are maintained at levels that do not adversely affect carbohydrate metabolism.

5.4.6 ABC transporters are P-type ATPases that facilitate solute transport

ABC (ATP-binding cassette) transporters make up a large family of transporters that hydrolyze ATP and transport a variety of organic molecules. They belong to the P-type class of ATPases. More than 120 ABC transporter genes have been identified in *Arabidopsis* and rice. All ABC transporters are made up of two basic structural elements: integral, membrane-spanning domains and cytoplasmically orientated, nucleotide-binding folds that are involved in ATP hydrolysis (Figure 5.15). ABC transporters transport uncharged solutes out of the cytosol and are therefore not electrogenic.

Although they were first discovered on vacuolar membranes, ABC transporters are also found on the plasma membrane. Among the compounds transported by ABC transporters are anthocyanins, chlorophyll catabolites, antifungal compounds and the plant hormone auxin. The *BR2* gene of maize, which is known to be involved in basipetal auxin transport, has now been shown to encode an ABC transporter. ABC transporters also have been implicated in the transport of waxes to the surfaces of leaf cells.

ABC transporters are thought to be important in the sequestration of potentially harmful metabolites and xenobiotic (from the Greek: literally, *xenos* = foreign, *bios* = to life) molecules in the vacuole. Many xenobiotic molecules are first modified by the addition of the tripeptide **glutathione**, catalyzed by a family of enzymes called glutathione transferases (GSTs). The addition of glutathione acts as an address label that targets the solute to an ABC transporter. Among the xenobiotics conjugated to glutathione are synthetic herbicides.

5.5 Channels

Two types of **channels** have been well characterized in plants, ion channels and aquaporins that allow water movement. An ion channel forms a selective, water-filled pore in the membrane (Figure 5.16). Unlike carrier-mediated transport (see below), there is little interaction between the channel and the ion. Fluxes through ion channels are therefore extremely rapid with up to 10^8 ions moving per second, compared to carriers and co-transporters that move between 10^2 and 10^4 ions per second. Ion channels are therefore ideally suited to permit the transport of large numbers of ions or molecules very rapidly.

Ion channels are ubiquitously distributed in the membranes of plant cells and they play important regulatory roles. For example, they regulate osmotic concentration by allowing the flux of potassium into and out of cells, and set the concentration of cytosolic Ca^{2+}, an ion that is important in cellular signaling. Movement through ion channels is passive and is dictated by the electrochemical potential for the particular ion. The diffusion of ions through channels is strongly affected by the

Figure 5.15 Model of a vacuolar ATP-binding cassette (ABC) transporter from *Arabidopsis*. Two nucleotide-binding folds, NBF1 and NBF2, are separated by transmembrane domains containing multiple, integral-membrane helices.

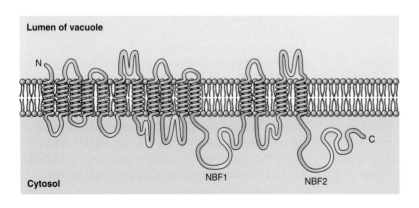

Lumen of vacuole

N

Cytosol

NBF1

NBF2

C

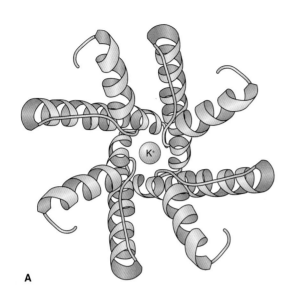

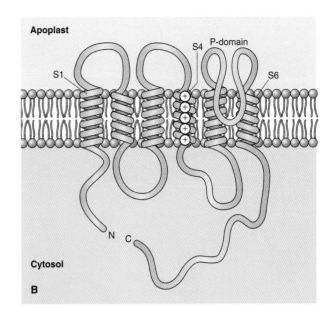

Figure 5.16 A plant K$^+$ channel. (A) Top view, based on ribbon structures showing four channel subunits that come together to form the central K$^+$ translocating pore. (B) Side view, showing disposition of the six transmembrane domains of one subunit of *Arabidopsis* AKT1. The P-domain between transmembrane loops 5 and 6 contributes to pore formation, whereas the charged amino acid residues in domain S4 contribute to the voltage sensitivity of the channel.

membrane potential (see Equation 5.8). Because the V_m of the plasma membrane is usually negative, cations tend to diffuse into the cytosol and anions tend to diffuse out.

Most ion channels show strong selectivity for either anions or cations. The selectivity filter of channels operates on the basis of size exclusion. **Cation channels** can be divided into those that are selective for K$^+$ over other monovalent cations, others that are relatively non-selective among monovalent cations, and those that are selective for Ca^{2+}. Most plasma membrane **anion channels** allow the passage of a wide range of anions, including Cl$^-$, NO$_3^-$ and organic acids. There are anion channels in the vacuolar membrane that select specifically for malate.

5.5.1 Ion channel activity is studied using patch clamping

The patch clamp technique can resolve the activity of a single ion channel molecule as it catalyzes ion translocation. The technique allows the detection of small electric currents carried by ions as they move through channels, made possible by advances in solid-state electronics that allowed accurate measurement of picoampere (10^{-12} A) currents. A high-resistance seal is produced when a blunt-tipped glass micropipette is pressed against a biological membrane with simultaneous application of suction. The high resistance of this seal forces currents flowing

into or out of the pipette to go through the membrane patch covering the pipette tip (Figure 5.17). This configuration is called the **cell-attached mode**; it can record the activities of individual ionic channels in the membrane patch at the tip of the micropipette. Pulling the pipette away from the rest of the membrane generates an **inside-out patch**, in which the cytosolic face of the membrane is exposed to the bathing medium (Figure 5.17). As with cell-attached recording, the activities of single ionic channels can be assessed, but the solution composition on both sides of the membrane is defined. Alternatively, the membrane patch that covers the pipette tip can be disrupted with a high-voltage pulse or suction, thereby giving electrical access to the inside of the cell, a configuration called **whole-cell mode** of recording. The relatively large volume of medium in the pipette exchanges rapidly with the cell contents, and becomes the defined intracellular solution. Finally, if the pipette is pulled away from the cell after attainment of the whole-cell mode, a membrane bleb is also pulled away and reseals itself across the pipette tip as an **outside-out patch**. This recording mode is particularly useful for testing the effects of putative cytosolic regulators on channel activity.

A major advantage of the patch clamp technique is that transport activity can be analyzed in relatively small cells. Thus most if not all types of plant cells are amenable to this approach. In contrast, classic microelectrode impalement techniques can irreparably damage small cells and thus are mostly limited to very large cells.

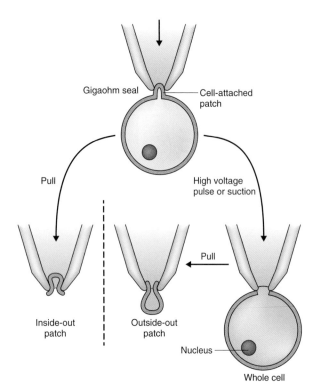

Figure 5.17 Recording modes used in the patch clamp technique. The initial, electrically-tight gigaohm seal yields a cell-attached mode of recording. When the pipette is pulled away from the rest of the membrane, an inside-out patch is formed in which the cytosolic face is exposed to the bathing medium. Alternatively, if the membrane patch that covers the pipette tip in the attached cell mode is disrupted by a high-voltage pulse or suction, a whole-cell mode of recording is produced. If the pipette is pulled away from the cell after attainment of the whole-cell mode, an outside-out membrane patch is formed.

5.5.2 The movement of ions through channels results in current flow

The movement of ions through channels results in the flow of current. This current can be used as a measure of the activity of a channel. Current flow is a function of membrane potential and the resistance that the membrane offers. Ohm's law ($I = V/R$, where current, I, is the membrane potential, V, divided by the resistance of the membrane, R) can therefore be used to model the movement of ions across the membrane (Figure 5.18). The resistance of a membrane to a particular ion depends on its selectivity and the number of channels. No current will cross a completely impermeable membrane, whereas membranes that are freely permeable will behave as a resistance-free circuit where $I = V$.

An example of the currents that can be measured using the whole-cell patch clamp technique is shown in

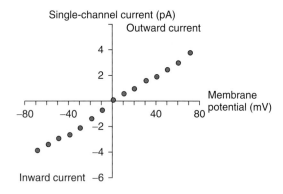

Figure 5.18 Current–voltage relationship of a single ion channel. This graph depicts the current that flows through the channel at different membrane potentials. Current–voltage plots allow determination of reversal potential, E_{rev}, and can help define the permeability of the membrane to specific ions.

Figure 5.19. Figure 5.19A shows current flow through the plasma membrane when different voltages ranging from -100 to $+100$ mV are applied to the membrane. What is apparent is that the membrane of this cell has channels that respond to voltage and open at either a positive or a negative V_m. Thus, with an increase in either negative or positive voltage, there is a concomitant increase in current flowing through channels in the membrane. An I–V plot of the data from the experiment illustrated in Figure 5.19A is presented in Figure 5.19B; the curve shows that activation of these channels occurs only after a given threshold voltage has been achieved. This is illustrated most clearly when positive voltage is applied, as channel activity is only activated when a considerable threshold is reached.

5.5.3 Opening and closing of channels is tightly regulated

Because ion channels allow the movement of large numbers of ions per unit time, their activities must be tightly regulated. The regulation of channel opening and closing is often referred to as **gating**. Many ion channels function as part of a feedback loop where membrane potential is one of the key regulators of channel opening and closing. Voltage-gated K^+ channels are important in maintaining the cell's membrane potential, because a shift in V_m can be compensated for by altering the opening and closing of channels. The plasma membrane has channels that allow either inward or outward movement of K^+. These operate like valves and are said to rectify. Those channels that allow inward movement of K^+ are referred to as **inwardly rectifying channels** whereas those that allow outward movement of K^+ are called **outwardly rectifying channels**. By opening and closing in response to voltage these channels can keep

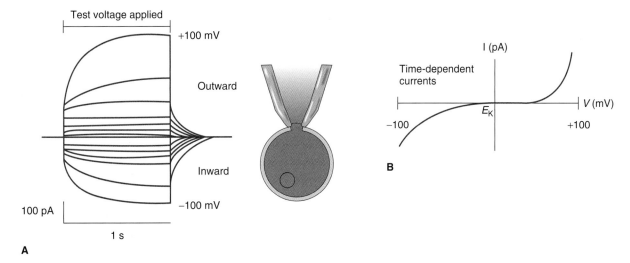

Figure 5.19 Whole-cell ion currents in a plant cell as measured by the patch clamp technique. (A) A pipette is sealed to the plasma membrane in the whole-cell configuration. In response to a series of 1-second voltage pulses, current flows through channels in the plasma membrane. The 13 current traces, corresponding to 13 different voltages between −100 and +100 mV, are superimposed in this plot. (B) Current (I)–voltage (V) relationship for the time-dependent component of currents in (A). Note that this whole-cell I–V curve is decidedly non-linear.

V_m constant. On the other hand, changes in V_m can be used to regulate osmotic concentration by altering the electrochemical potential gradient of ions. Ion channels can also be gated by ligands (chemicals that bind to the channel), such as hormones, Ca^{2+}, **G-proteins** and pH. Other channels are stretch-sensitive and are gated by changes in the turgor pressure of a cell.

Complex regulatory circuits can be built up by the interaction of these various regulators of ion channel activity and the way in which these regulators interact is nicely illustrated by what happens at the tonoplast of guard cells when stomatal closure is induced by **abscisic acid** (**ABA**) (Figure 5.20). The tonoplast has at least four Ca^{2+} channels, one regulated by voltage, one by Ca^{2+} and the other two by other ligands. The presence of multiple Ca^{2+} channels gated by different signals allows for dynamic changes in cytosolic Ca^{2+} in response to a variety of stimuli.

Key points Channels that transport ions and water across biological membranes are well suited for their roles in the rapid responses of cells to signals. For example, the response of stomatal guard cells to changes in the water content of soil or the atmosphere are a result of the ability of channels to quickly change concentration of solutes and water in the cell. Whereas primary and secondary active transporters move ion solutes at a rate of 10^2 to 10^4 species per second, channels allow up to 10^8 species per second to be transported. Because channels transport solute into or out of the cell at a very high rate they permit a rapid cascade of events to occur that lead, for example, to rapid stomatal closure.

5.5.4 Aquaporins are a class of channels facilitating water movement

It is somewhat counterintuitive that the lipid bilayer of biological membranes is highly permeable to water. It is perhaps even more surprising that membranes contain specialized water channels called **aquaporins**. Aquaporins in plant cells were initially found on the tonoplast but they are also localized to the plasma membrane as well as some other endomembranes including the endoplasmic reticulum. They belong to relatively large gene families in plants; for example *Arabidopsis* has 35 aquaporin genes, maize has 36 and rice 33. By definition, aquaporins act as water channels but there is growing evidence that they may facilitate the movement of CO_2, NH_3, H_2O_2, boron and silicon.

Aquaporins fall into four subgroups (Figure 5.21); there is high sequence homology (>97%) among members of each subgroup. The TIPs and PIPs are tonoplast and plasma membrane intrinsic proteins, respectively. Nodulin intrinsic proteins (NIPs) are found in the peribacteroid membranes of symbiotic nitrogen-fixing nodules and are present in non-leguminous plants as well. The fourth class of aquaporins, the SIPs (small basic intrinsic proteins), which are encoded by a small family of two to three genes, are found in the endoplasmic reticulum. The four subgroups of aquaporins are found in all land plants, from mosses to the angiosperms.

Aquaporins are relatively small proteins, around 30 kDa in size, and have six transmembrane domains

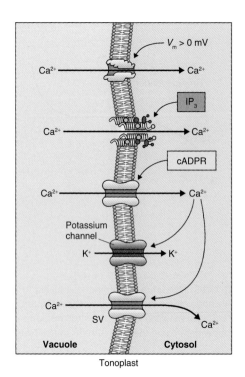

Figure 5.20 Ca²⁺ and K⁺ fluxes through channels in the tonoplast of a stomatal guard cell during abscisic acid-induced stomatal closure. Shown are: a K⁺ channel that is regulated by Ca²⁺; several Ca²⁺ channels that are regulated by voltage (V_m), inositol trisphosphate (IP3) or cyclic-ADP-ribose (cADPR); and the slow vacuolar Ca²⁺ channel (SV).

(Figure 5.22). The C- and N-terminal sequences of all aquaporins localize to the cytosol as do connecting loops II and IV. The I, III and V connecting loops of PIPs face the apoplast, whereas those of TIPs and SIPs face the lumen of the vacuole and endoplasmic reticulum, respectively. Four aquaporin monomers make a functional complex but it is noteworthy that each subunit of the tetramer forms a water channel (Figure 5.23).

5.5.5 Flux of water through aquaporins is regulated by many factors

Aquaporin activity is regulated at various levels. Many aquaporins are regulated by changes in the abundance of their mRNAs. Indeed, there is evidence for the coordinated downregulation of aquaporins in response to nutrient and water stress. They are also subject to regulation at the mRNA level by hormones and other regulatory molecules.

Aquaporins are extensively modified post-translationally by phosphorylation and methylation of amino acids at the N and C termini. Phosphorylation occurs at multiple amino acids at the C terminus of most

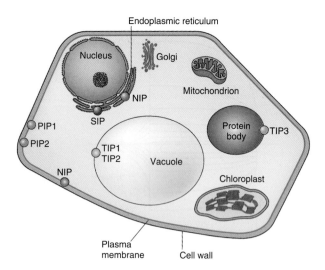

Figure 5.21 Diagram showing the subcellular localization of aquaporins. Plasma membrane intrinsic proteins, PIP1 and PIP2, and nodulin intrinsic protein, NIP, localize to the plasma membrane. Tonoplast intrinsic proteins, TIP1 and TIP2, localize to vacuoles, TIP3 to protein storage vacuoles, and NIP and small basic intrinsic protein (SIP) are found on the endoplasmic reticulum.

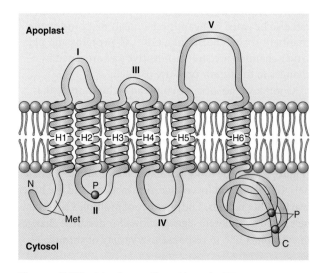

Figure 5.22 Membrane disposition of a PIP-type aquaporin. Six transmembrane helices are shown, forming five amino acid loops with I, III and V in the apoplast and II and IV facing the cytoplasm. Several phosphorylation sites are located in the long C terminus of the protein and in loop II, whereas regulatory sites that are methylated (Met) are shown at the N terminus.

aquaporins, and methylation of amino acids at the N terminus has been shown for PIPs (see Figure 5.22). Phosphorylation of aquaporins has been shown to regulate the opening of aquaporins, and the observation that Ca²⁺-dependent protein kinases may catalyze this phosphorylation suggests a link between Ca²⁺ signaling and the regulation of water movement between cells.

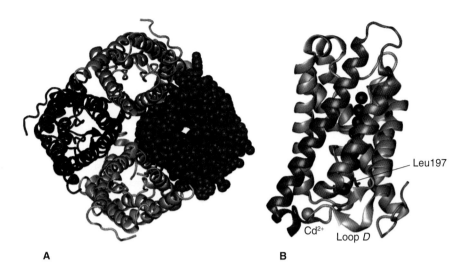

Figure 5.23 Ribbon diagrams of aquaporins based on crystallographic data. (A) A view from above the membrane of four aquaporin subunits; each subunit forms a water-transporting pore; one subunit is shown in a space-filling model to illustrate its pore. (B) A side view of one subunit showing the orientation of the amino acid chain in the plane of the membrane; water molecules (blue spheres) move through the pore.

Leu197

Cd²⁺   Loop *D*

A   B

Another interesting aspect of aquaporin regulation comes from observations that changes in cytoplasmic pH alter water movement. When *Arabidopsis* roots are subjected to anoxia, the cytoplasm becomes highly acidic (see Chapters 7 and 15) and water transport is dramatically reduced. The effect of anoxia on water movement can be mimicked experimentally by acidifying the cytoplasm of root cells, suggesting that it is the pH of the cytoplasm that is affecting water flux.

Key points At first glance the presence of aquaporins in membranes may appear to be an anomaly. The lipid bilayer is permeable to water so why do aquaporins exist? For example, a large portion of the area of membranes is occupied by transport proteins and macromolecules that adhere to their surface; this leaves a limited area of free bilayer for water diffusion. In fact, plant aquaporins were first isolated from tonoplast because they are the most abundant proteins in this membrane. The vacuole may make up >95% of the volume of the plant cell and water channels allow for rapid and regulated movement of water across the tonoplast in response to changing external solute concentrations.

5.6 Carriers and co-transporters, mediators of diffusion and secondary active transport

Carriers and **co-transporters** belong to a large and diverse class of proteins that move a wide range of ions and metabolites across cellular membranes. They are ubiquitously distributed in cellular membranes. In addition to their association with the plasma membrane, they are found in the endomembrane system as well as the envelopes of chloroplasts and mitochondria, where they are especially abundant. The array of ions and molecules translocated by these transporters is vast. They play roles in the uptake of inorganic nutrients, including NH_4^+, NO_3^-, SO_4^{2-} and $H_2PO_4^-$, and are important in loading sugar into the phloem for long-distance transport. Co-transporters and carriers are involved in the exchange of metabolites across mitochondrial and chloroplast membranes. Co-transporters on the tonoplast have a role in the storage of ions and organic solutes in the vacuole.

In transport mediated by carriers and co-transporters, the solute being transported binds to the transporter and causes a conformational change (Figure 5.24). This conformational change is thought to lead to the movement of the solute across the membrane. Because these transport proteins interact with solutes, their properties are reminiscent of those of enzymes. Carriers and co-transporters show saturable kinetics with increasing substrate concentration from which a Michaelis–Menten constant, K_m, can be derived (see Figure 2.24A). The interaction between transporter and solute confers very high selectivity and these proteins can discriminate between stereoisomers of sugars or amino acids.

Carriers facilitate the diffusion of organic solutes across membranes. For example, vacuolar uptake of glucose and some amino acids is driven by concentration gradients in these molecules. **Co-transporters** are involved in secondary active transport, coupling the downhill movement of ions, such as H^+, to the uphill movement of inorganic ions and organic solutes. The H^+-ATPases of the plasma membrane and tonoplast pump protons out of the cytosol, creating proton gradients across these membranes. Therefore protons

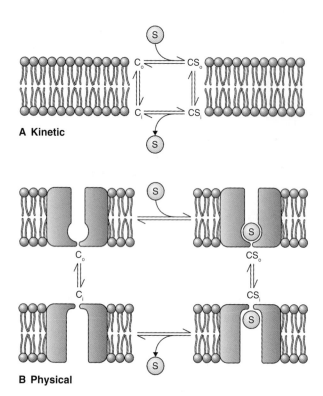

A Kinetic

B Physical

Figure 5.24   Models of the reaction cycle of a solute carrier, C. (A) Kinetic modeling of C as it interacts with solute S on each side of a membrane (C_o and C_i). (B) Model of changes in the structure of C as it interacts with S. Differences in C_o and C_i lead to binding and release of S.

tend to move energetically downhill across these membranes in the direction of the cytosol. Transporters that catalyze solute flux in the same direction as H^+ flux are known as **symporters** (see Figure 5.8B). Symporters typically drive uptake of solutes into the cytosol, either from the external medium or from intracellular compartments. Proton-coupled symporters include the H^+/sucrose symporter that is involved in loading sucrose into the phloem, several different H^+/anion symporters and a number of H^+/amino acid symporters. Conversely, secretion of solutes from the cytosol can be accomplished by **antiporters**, which exchange solutes for protons. Antiporters are present at the plasma membrane and at endomembranes. Proton-coupled antiporters include the H^+/Ca^{2+} antiporter on the tonoplast and the Na^+/H^+ antiporter on the plasma membrane. Both symporters and antiporters tend to dissipate the proton gradient (pmf).

In addition to co-transporters that are powered by a proton gradient, there are those that utilize gradients of other ions to energize transport of their substrates. For example, the most abundant protein in the chloroplast envelope is a phosphate translocator that exchanges inorganic phosphate for triose phosphate (see Chapter 9). Because there are many types of carriers and

co-transporters, it is difficult to provide a generalized model for their structure or regulation.

5.7 Intracellular transport of proteins

Plant genomes encode tens of thousands of proteins. Nearly all of a cell's proteins are encoded by nuclear DNA and synthesized by cytosolic ribosomes, which may or may not be attached to the endoplasmic reticulum. Proteins encoded by mitochondrial and chloroplast DNA (around 100 or so) are synthesized by ribosomes within these organelles and incorporated directly into the organelle compartments. However proteins synthesized in the cytoplasm or endoplasmic reticulum must be targeted to the compartment or membrane in which they will be active (Figure 5.25). To ensure that a protein reaches its proper location, it needs labels that designate

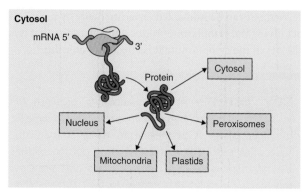

A Free ribosomes in cytosol

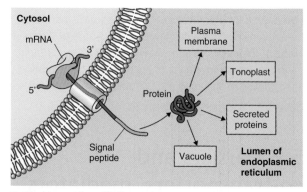

B Membrane-bound ribosomes

Figure 5.25   Protein synthesis occurs on 80S ribosomes that are (A) free in the cytoplasm or (B) bound to the surface of the endoplasmic reticulum (ER). Proteins synthesized on free ribosomes are destined for the cytoplasm, nucleus, plastids, mitochondria and peroxisomes, whereas those synthesized on ER-attached ribosomes enter the secretory pathway.

that location. We will examine the nature of these address labels and the machinery that reads those labels. In addition, these proteins must often pass into or through various cellular membranes. We shall see that this also requires special cellular machinery. Finally, we will examine how proteins that are destined for the cell wall and several internal organelles are processed, packaged and targeted for transport.

5.7.1 Protein transport requires peptide address labels and protein-sorting machinery

Proteins that are synthesized on cytosolic ribosomes and destined for transport to other compartments or to cellular membranes have one or more targeting domains that act as address labels, specifying their final destination. Targeting domains are usually short peptides or amino acid motifs that can be located anywhere in the protein's amino acid sequence. Specific cellular machinery interacts with this information to translocate the protein into, or retain it within, the proper compartment. Each compartment and membrane system requires a different targeting domain and sorting machinery. Although the targeting domain is essential for protein transport, it may not be part of the active protein. Proteases in the target location often remove the targeting domain to create a functional, mature polypeptide.

For those proteins that remain in the cytoplasm or are targeted to chloroplasts, mitochondria, nuclei and peroxisomes, synthesis of the protein is generally completed on free ribosomes (Figure 5.25A). Proteins destined for the secretory pathway are synthesized on ribosomes that are attached to the endoplasmic reticulum (Figure 5.25B). Whether mRNAs are translated by free or membrane-bound ribosomes is determined by the presence of a signal peptide, a short amino acid sequence located at the N terminus of the protein. When the growing polypeptide chain contains a signal peptide, this targets the protein-synthesizing machinery to the surface of the endoplasmic reticulum.

Although there are major differences in the mechanisms that underlie targeting to simple organelles such as peroxisomes or to more complex organelles such as chloroplasts and mitochondria, the underlying principles are similar. **Targeting domains** are found on most proteins destined for cellular organelles. The targeting domain is recognized by a receptor at the surface of the organelle membrane. The specificity of both the targeting domain and the receptor ensures that proteins reach the correct destination.

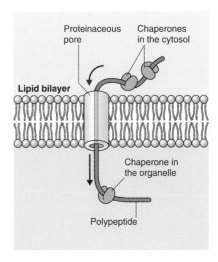

Figure 5.26 Proteins that cross the lipid bilayer of membranes are targeted to receptors embedded in the membrane; they pass through a proteinaceous pore and require the activity of molecular chaperones on both sides of the membrane. Chaperones keep proteins in the unfolded state to facilitate transport through the pore and catalyze the folding of protein when transport is completed.

As the polypeptide elongates, cytosolic chaperones keep the chain in the unfolded state in the cytosol. The movement of proteins through pores in the organelle membrane is facilitated by several polypeptides that make up a translocation motor, including those that hydrolyze nucleoside triphosphates (NTPs). As the translocated protein enters the lumen of the organelle, it interacts with another set of chaperones that catalyze protein folding (Figure 5.26).

5.7.2 To reach its destination, a protein often crosses at least one membrane

To enter target organelles, proteins must pass through at least one membrane. Most proteins have hydrophilic surfaces and therefore do not readily pass through the hydrophobic core of cellular membranes. Translocation through a membrane involves a proteinaceous pore or channel through which a protein must pass in an extended or unfolded configuration rather than in its final, folded form (Figures 5.26 and 5.27). As a polypeptide passes through a pore, it is assisted by molecular **chaperones**, proteins that bind the polypeptide during the folding and assembly process. Some cytosolic chaperones interact with newly synthesized proteins, keeping them unfolded so they can pass through a protein pore to an appropriate compartment or membrane. Other chaperones bind to

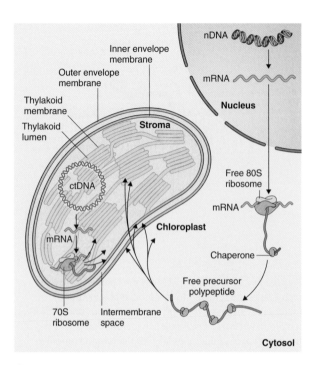

Figure 5.27 labels:
- Inner envelope membrane
- Outer envelope membrane
- Thylakoid membrane
- Thylakoid lumen
- nDNA
- mRNA
- Nucleus
- Stroma
- ctDNA
- Free 80S ribosome
- mRNA
- mRNA
- Chloroplast
- Chaperone
- Free precursor polypeptide
- 70S ribosome
- Intermembrane space
- Cytosol

Figure 5.27 Proteins synthesized on cytosolic ribosomes are imported into the chloroplast and can be targeted to six distinct destinations: the two outer membranes of plastid, intermembrane space, stroma, thylakoid membrane and thylakoid lumen. Proteins synthesized in the plastid stroma on 70S ribosomes either remain in the stroma or are targeted to the thylakoid membrane, stroma or inner envelope of the plastid.

the amino acid chain as it emerges from the membrane and facilitate folding. Still others function as repair stations to correct minor misfolding. The many members of three families of chaperones—the **60 kDa, 70 kDa and 90 kDa heat shock proteins** (Hsp60, Hsp70 and Hsp90, respectively; see Chapter 15)—fulfill all these roles, interacting with a wide spectrum of proteins. Hsp70 chaperones are present in all cells. The synthesis of many heat shock proteins is upregulated in response to heat stress, perhaps to prevent misfolding and to repair misfolded proteins (see Chapter 15).

5.7.3 Transport into chloroplasts and mitochondria involves translocation through several membrane barriers

Chloroplasts and mitochondria are surrounded by two unit membranes (see Figures 4.25 and 4.27). In chloroplasts an additional set of membranes called the thylakoid membranes form from an infolding of the inner chloroplast membrane. These membrane barriers

in turn enclose three aqueous compartments, the intermembrane space between the outer and inner chloroplast membrane, the stroma and the thylakoid lumen. Proteins must be targeted to each of the membranes and the compartments in these organelles (see Figure 5.27).

Although the principles involved in peptide targeting to chloroplasts and mitochondria are similar, the components of the molecular machinery involved are very different. Here we will describe the principles involved in targeting to chloroplasts. As indicated in Section 5.7.1, membrane receptors, peptide targeting domains, molecular chaperones and translocation proteins are required for proteins to enter the various compartments and membranes of chloroplasts. Figure 5.28 illustrates the processes involved in the targeting of proteins to the chloroplast stroma, thylakoid membrane and thylakoid lumen. Although the vast majority of chloroplast proteins are synthesized on free cytosolic ribosomes, some proteins are imported into the chloroplast after synthesis on endoplasmic reticulum-bound ribosomes and transported to the chloroplast via Golgi-derived vesicles.

Most chloroplast proteins synthesized on free cytosolic ribosomes interact with molecular chaperones of the Hsp70 class before they are imported into the stroma. Two distinct transport complexes denoted TOC (translocon at the outer chloroplast envelope) and TIC (translocon at the inner chloroplast envelope) participate in protein import (Figure 5.28). Chloroplast-bound proteins have a short amino terminal transit peptide that targets them to the outer surface of the chloroplast envelope where they interact with a receptor in the TOC complex. Interaction of imported proteins with TOC and translocation through the outer membrane requires ATP and GTP hydrolysis whereas translocation through TIC requires only ATP. Unlike protein translocation into the mitochondrion, import into the stroma of the chloroplast does not involve transmembrane pmf. Although transport through TOC and TIC can be viewed as biochemically separate events they are thought to occur almost simultaneously at points where the outer and inner membranes of the plastid make close contact.

Once the protein enters the stroma, the transit peptide is cleaved by a protease. This is the final destination of a stromal protein. It interacts with chaperones of the Hsp60 and Hsp70 class and assumes its final conformation. The interaction of the protein with molecular chaperones and the formation of the mature folded protein require the expenditure of energy in the form of ATP (Figure 5.28).

Proteins targeted to the thylakoid membrane and lumen arrive at their destinations by different routes and use different mechanisms. Proteins destined for the thylakoid membrane are targeted there by one of two

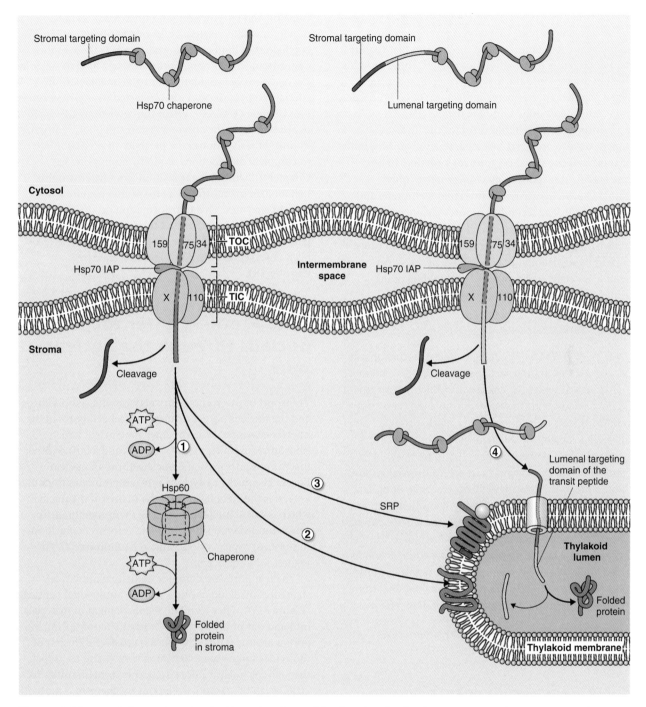

Figure 5.28 Protein import into the plastid requires the participation of molecular chaperones, targeting sequences and energy. Stromal transit peptides target unfolded proteins to the outer plastid envelope and transport across the membrane is facilitated by transport complexes TOC and TIC. The transit peptide is cleaved by a protease and (1) stromal proteins interact with Hsp60/70 chaperones and assume their final conformation. Proteins destined for the thylakoid membrane are targeted there by one of two routes: (2) by a mechanism that requires neither targeting information nor energy, or (3) by using a signal peptide that interacts with a signal recognition particle (SRP) in the stroma and a receptor on thylakoid membrane. Those proteins that are inserted into the thylakoid membrane via route (3) require energy in the form of GTP to become incorporated into the membrane. (4) Transport of proteins destined for the thylakoid lumen have signal peptides to target them to transport machinery in the thylakoid membrane; the signal peptides are cleaved after transit through the membrane.

routes, either by a non-cleavable signal sequence similar to the cleavable peptide that targets proteins to the endoplasmic reticulum, or by a mechanism that requires neither targeting information nor energy. Those proteins that are inserted into the thylakoid membrane using a signal peptide require energy in the form of GTP in order to become incorporated into the membrane.

Transport of proteins destined for the thylakoid lumen can also occur via two separate mechanisms, both of which require signal peptides for targeting. In one case specific transporting machinery moves only unfolded proteins across the thylakoid membrane and requires ATP. The second transport machinery translocates folded and unfolded proteins and utilizes the transmembrane pmf to power transport. In both cases, once in the thylakoid lumen, cleavage of the signal peptide is catalyzed by proteases and folding of the protein requires molecular chaperones.

Key points Targeting proteins to chloroplasts is complicated by the fact that there are six different destinations within a chloroplast: three membranes (outer, inner and thylakoid membranes) and three compartments (intermembrane space, stroma and thylakoid lumen). In addition, while some chloroplast proteins are synthesized on 70S ribosomes in the stroma of the chloroplast, most are synthesized on 80S ribosomes in the cytosol. With the exception of proteins destined for the outer membrane and intermembrane space, these proteins are first targeted to the stroma where a stromal targeting sequence is removed. Further targeting is required for proteins destined for the inner membrane, thylakoid membrane and thylakoid lumen. Passage through and into each membrane and space usually requires specific protein translocation systems.

5.7.4 Passage across a single membrane is required for proteins to enter peroxisomes

Peroxisomes differ from chloroplasts and mitochondria in several important respects. First, they are devoid of nucleic acids and the machinery for protein synthesis, and they are surrounded by a single unit membrane. These organelles, however, need to import proteins. First, they have the capacity to increase in size and to divide by binary fission (simply the pinching of the organelle into two daughter organelles). In addition, the function and enzyme content of peroxisomes are developmentally related. Following germination, cotyledons of lipid-storing seeds, such as those of the castor bean (*Ricinus communis*), convert lipid to sugar by the reactions of β-oxidation and gluconeogenesis in specialized peroxisomes often referred to as glyoxysomes (see Chapters 6 and 7). When the cotyledons are depleted of lipid they become green and photosynthesize. Photosynthesis requires the presence of peroxisomes to complete the reactions of photorespiration (see Chapter 9). During the transition from heterotrophic metabolism to autotrophy peroxisomes acquire a new suite of enzymes. This transformation occurs through the import of membrane and luminal proteins into the peroxisome. Redifferentiation of peroxisomes has also been observed to occur during senescence (see Chapter 18).

5.7.5 Proteins enter the nucleus through the nuclear pore

Transport of macromolecules between the cytoplasm and the nucleus occurs through specialized pores in the nuclear envelope. Proteins enter the nucleus and ribosomal subunits, mRNAs, tRNAs and a host of other macromolecules leave the nucleus through **nuclear pores**. The nuclear envelope is two membranes thick and separates the cytoplasm from the matrix of the nucleus, which is called **nucleoplasm**. The outer membrane is often studded with ribosomes and it forms connections to sheets of rough endoplasmic reticulum (see Figures 4.14 and 4.15).

Nuclear pores are complex, radially symmetrical structures made up of more than 100 individual peptides (Figure 5.29). They facilitate the movement of materials into and out of the nucleus. The diameter of the nuclear pore is around 9 nm. This permits passive diffusion of molecules up to about 40 kDa in size. Active transport through the nuclear pore also occurs; proteins of up to 20 kDa can be transported by this mechanism. This process involves a targeting sequence called the **nuclear localization signal** (**NLS**), receptor proteins in the nuclear pore complex, soluble cytoplasmic factors and energy in the form of GTP. The process can be divided into two steps. First the NLS on the protein binds to receptors in the nuclear pore. This step requires both cytosolic factors and GTP. Then the protein is translocated through the pore; this step also requires GTP to supply additional energy.

Transport through the nuclear pore is regulated by a number of factors, including those that effectively mask the NLS on targeted proteins, and environmental cues,

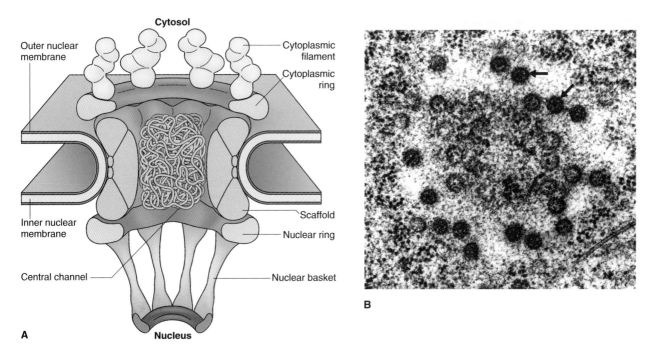

Figure 5.29 The nuclear pore complex. (A) Model showing the organization of proteins in the nuclear pore complex. (B) Electron micrograph of a tangential section through several nuclear pore complexes. Part (A) shows a cross-section of membrane, whereas (B) shows a surface view of the nuclear membranes with embedded pore complexes (arrows).

such as light. For example, the effect of light on photomorphogenesis is affected by a protein, COP1, which shuttles between the cytoplasm and nucleus in response to light. Note that this protein is different from the COPI coat protein that is involved in vesicle transport (see Section 5.8.3). The COP1 protein is a repressor of photomorphogenesis. COP1 normally resides in the nucleus in dark-grown plants, but it moves into the cytoplasm when plants are exposed to light. Movement of COP1 into the cytoplasm relieves repression of photomorphogenesis genes (see Chapter 8).

Key points Most cellular proteins are synthesized on cytosolic ribosomes. Whether they are synthesized on ribosomes attached to the endoplasmic reticulum and destined for the secretory system or on ribosomes that are free in the cytosol and targeted to organelles such as chloroplasts, mitochondria or peroxisomes, these proteins must cross at least one lipid bilayer. While seemingly complex, the principles that underlie transport of proteins across membranes are fundamentally similar and follow similar rules. First, proteins are targeted to the correct membrane using a specific amino acid sequence; second, molecular chaperones maintain them in the unfolded state; and third, they cross the membrane through a proteinaceous pore.

5.8 The protein secretory pathway

The **secretory pathway** forms an extensive membranous system, the **endomembrane system**, which branches throughout the cell. It consists of **endoplasmic reticulum** (**ER**), the **Golgi apparatus** and **trans-Golgi network** (**TGN**), **multivesicular bodies** (**MVB**) and various classes of vesicles and vacuoles (Figure 5.30). Although the synthesis of all proteins begins on free ribosomes, proteins destined for the endomembrane system or for export to the cell exterior have a signal peptide that targets the ribosome–protein complex to the surface of the ER where synthesis of the protein is completed (Figure 5.31). ER with attached ribosomes is called **rough ER**. It is particularly abundant in cells that specialize in protein secretion (e.g. cereal aleurone cells during seed germination; see Chapter 6) or in the storage of proteins in vacuoles (e.g. storage parenchyma cells in developing seeds).

5.8.1 Signal peptides target proteins to the endoplasmic reticulum

The mRNAs that encode proteins destined to function outside the cell, in the cell wall or inside the organelles of

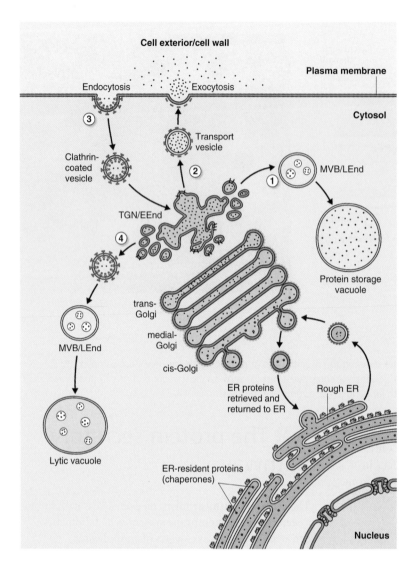

Figure 5.30 The secretory pathway. Proteins synthesized on 80S ribosomes attached to ER are transported via the Golgi apparatus to the trans-Golgi network (TGN). The TGN plays a central role in directing vesicle traffic. (1) Proteins destined for the vacuole are transported from the TGN via multivesicular bodies (MVB) to protein storage vacuoles. (2) Proteins destined for the exterior are transported in vesicles to the plasma membrane for exocytosis. (3) Proteins imported into the cell by endocytosis move in clathrin-coated vesicles to the TGN (4) where they are sent via a MVB to lytic vacuoles. During this process, the TGN is functionally equivalent to an early endosome (EEnd) in animal cells and the MVB functions as a late endosome (LEnd).

the endomembrane system, contain a signal sequence about 48–90 nucleotides in length that is located near the start codon where translation begins. Translation of this mRNA sequence results in the synthesis of a signal peptide of about 16–30 amino acids at the amino (N) terminus of the encoded protein (Figure 5.31). As the mRNA is translated and the N terminus of the resulting protein emerges from the ribosome, the **signal peptide** is recognized by the **signal recognition particle** (SRP), an RNA–protein complex in the cytosol that mediates the docking of the ribosome onto a protein complex within the ER membrane. Although the amino acid sequence of signal peptides vary enormously, almost all of them can function interchangeably to target proteins to the ER. For example, plant signal peptides can target animal proteins to the ER of animal cells and vice versa. Furthermore, addition of a signal peptide to the N terminus of a protein that is not normally secreted can target that protein to the secretory pathway.

When an SRP binds to a signal peptide, protein synthesis is temporarily arrested. The ribosome–nascent protein–SRP complex is recognized by an **SRP receptor** on the surface of the ER that allows docking of the complex on the membrane (Figure 5.31). When docking has occurred, the SRP and several other components of the SRP receptor dissociate using energy from the hydrolysis of GTP. Synthesis of the anchored protein is reinitiated, and the nascent protein is guided through a transmembrane channel in the ER. GTP hydrolysis also provides the energy required to move the protein through the membrane.

Several classes of proteins that are synthesized on rough ER do not completely cross the ER membrane; these include integral membrane proteins of the ER and other parts of endomembrane system. Their transport through the membrane is arrested by the presence of a hydrophobic sequence of amino acids called a **stop transfer sequence**. It is easy to imagine how a protein

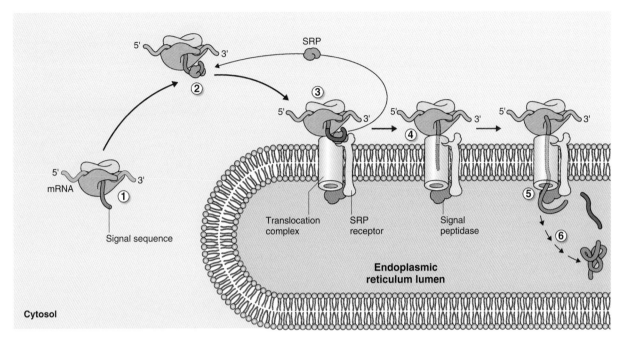

Figure 5.31 Targeting of proteins to the endoplasmic reticulum lumen involves (1) a signal sequence, (2) a signal recognition particle (SRP), (3) an SRP receptor, (4) a translocation complex, (5) signal peptidase and (6) molecular chaperones. These work in a coordinated fashion so that a folded protein synthesized in the cytosol resides in the lumen of the ER.

with an N-terminal signal peptide and a stop transfer sequence located half-way along its length can become stuck in the membrane, as illustrated in Figure 5.32A. These proteins, designated Type I, have their N terminus in the lumen of the ER and the C terminus in the cytosol. Type II proteins, whose C terminus lies in the lumen of the ER, do not use a signal peptide to target the ER; rather they use other hydrophobic amino acid sequences that allow their orientation (Figure 5.32B). Proteins that have multiple membrane-spanning domains, such as the membrane transporters described in earlier pages of this chapter, possess a series of alternating signal peptides and stop transfer sequences that work together to thread the protein into and out of the membrane (Figure 5.32C).

Soluble proteins enter the ER lumen in their entirety. In this case, the N-terminal signal peptide is removed by the action of a signal peptidase. The polypeptide is then folded with the aid of chaperones of the Hsp class as well as BiP (binding protein), calnexin and calreticulin, on the inner face of the ER membrane. Chaperone-catalyzed protein folding requires GTP. Metalloproteins bind metals as folding occurs. For example, the activity of cereal α-amylase is dependent on Ca^{2+} binding. That this binding occurs in the ER lumen is shown by the fact that α-amylase isolated from the ER is enzymatically active. Misfolded proteins are targeted for degradation in the lumen of the ER or else are transported out of the ER and degraded in the cytoplasm by the 26S **proteasome** (see Section 5.9).

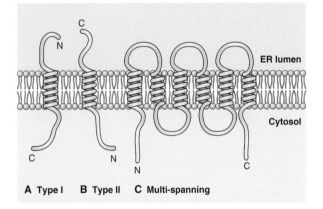

A Type I   **B Type II**   **C Multi-spanning**

Figure 5.32 Orientation of N- and C-terminal domains of membrane-spanning proteins synthesized on ribosomes bound to the endoplasmic reticulum. (A) Type I proteins have cytosolic C-terminal domains, whereas (B) Type II proteins have their N terminus in the cytosol. (C) Proteins that have several membrane-spanning domains have both their N and C termini in the cytosol.

5.8.2 Post-translational modification of proteins begins in the endoplasmic reticulum

The process of post-translational modification of proteins begins in the ER lumen. Informational amino acid sequences in these proteins determine whether and how they will be modified and whether they will progress further along the secretory pathway. Post-translational modifications that occur in the ER include the addition of glycan chains to glycoproteins, the modification of amino acids (e.g. conversion of proline to hydroxyproline), the formation of disulfide bonds catalyzed by protein disulfide isomerases, and the assembly of multimeric complexes, a process that requires chaperones.

Many soluble plant proteins, destined for the secretory pathway, are glycosylated by enzymes that add sugar chains. Glycans may be attached to the amide nitrogen of asparagine residues (N-linked) or to the OH groups of serine or hydroxyproline residues (O-linked). Glycosylation begins in the ER but may be completed in the Golgi. Proteins with N-linked glycans, including many legume storage proteins, have a characteristic amino acid sequence, Asn-X-Ser/Thr, where X can be any amino acid except Pro. Proteins with O-linked glycans also carry a specific amino acid motif. The hydroxyproline-rich cell wall protein **extensin** is a soluble protein with O-linked oligoarabinan chains attached to hydroxyprolines. There is no codon for hydroxyproline: it is produced post-transcriptionally by the hydroxylation of proline catalyzed by the enzyme prolyl hydroxylase. Thus the sequence of events in extensin modification in the ER is first the hydroxylation of prolines followed by the addition of arabinose. Specific amino acid signals dictate which proline residues are hydroxylated as well as which of these residues carry the arabinose chain.

Some soluble proteins are retained in the lumen of the ER for proper functioning. These include chaperones such as BiP, calnexin and calreticulin and enzymes such as protein disulfide isomerase and prolyl hydroxylase. Such proteins have a C-terminal ER retention signal, the amino acid sequence Lys-Asp-Glu-Leu, also referred to by the single letter amino acid code KDEL, or variants thereof such as HDEL (for amino acid abbreviations, see Figure 2.12). The ER retention signal is used to retrieve these proteins if they move out of the ER to the Golgi.

5.8.3 Coat proteins govern the shuttling of vesicles between the endoplasmic reticulum and Golgi

The transfer of proteins from the ER to other destinations in the secretory pathway occurs by the movement of vesicles initially from the ER to the Golgi where proteins may undergo further modification and sorting. The movement of vesicles from the ER to the Golgi is referred

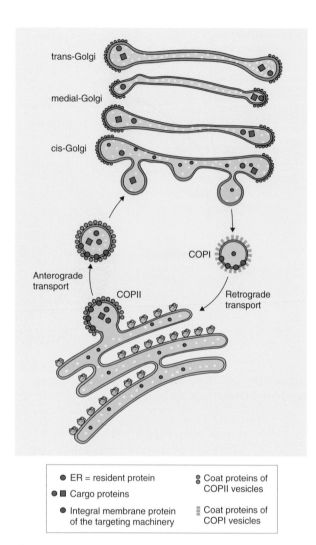

Legend:
- ER = resident protein
- Cargo proteins
- Integral membrane protein of the targeting machinery
- Coat proteins of COPII vesicles
- Coat proteins of COPI vesicles

Figure 5.33   Proteins can be recycled between the endoplasmic reticulum (ER) and Golgi apparatus. The movement of proteins from the ER to the Golgi is referred to as anterograde transport and occurs in vesicles having COPII coats, whereas recycling from the Golgi to the ER is known as retrograde transport and occurs in vesicles with COPI coats. Retrograde transport returns both soluble and membrane ER-resident proteins.

to as **anterograde transport** (Figure 5.33). It occurs in vesicles that have a specific class of **coat proteins** called **COPII** on their surface. COPII provides the information that dictates the direction of transport.

The retrieval of ER-localized, soluble proteins and membrane proteins from the Golgi occurs by a process called **retrograde transport** (Figure 5.33). For soluble proteins a KDEL receptor, ERD2, recognizes the KDEL motif on the protein. ERD2 is embedded in membranes of ER vesicles that shuttle proteins between the Golgi and ER. Membrane proteins are also recycled from the Golgi to ER but the amino acid motifs and receptors for these have yet to be identified in plants. In both cases receptors act as retrieval mechanisms for resident ER proteins. These proteins are returned to the ER in vesicles that have a second type of coat protein, **COPI**, on their surfaces.

5.8.4 Proteins are transported from the Golgi to a range of destinations

From the Golgi, proteins are transported to other downstream destinations such as the TGN, MVB, vacuoles or cell exterior (Figure 5.33). Movement of vesicles from the TGN to downstream destinations such as vacuoles or the plasma membrane occurs in **clathrin-coated** vesicles (see Chapter 4). Soluble secretory proteins that lack sequence information, which acts as specific address labels, reach the cell surface via the so-called **default pathway**. They are packaged into vesicles at the trans-Golgi that then move to the TGN (Figure 5.34). From there vesicles containing these proteins move to and fuse with the plasma membrane, releasing their contents outside the cell. This process is called **exocytosis**. Cereal aleurone α-amylase is an example of this type of protein.

In contrast to the default pathway, protein trafficking to vacuoles requires specific sorting signals. In addition to the central vacuole that is present in all mature plant cells, cells may contain at least three other kinds of vacuoles: **protein storage vacuoles**, acidic **lytic vacuoles** and **vacuoles with a neutral pH**. Proteins move to these compartments along different routes that are determined by distinct vacuolar sorting sequences. Proteins, such as **sporamin**, a soluble storage protein from sweet potato tubers, that are destined for lytic vacuoles contain an N-terminal propeptide signal (NTPP). At the TGN, the NTPP is recognized by a vacuolar sorting receptor (VSR) that directs sporamin into clathrin-coated vesicles that

move to the MVB from which the clathrin-coated vesicles move to lytic vacuoles. Proteins having C-terminal propeptide signals, on the other hand, are targeted to vacuoles that have a neutral pH along a route that has not been precisely defined. As we will discuss in Chapter 6, there are still more routes that vacuolar proteins can take after their synthesis on the ER. For example, pumpkin albumin, a vacuolar storage protein, is transported in precursor accumulating vesicles (PACs) from the ER to a protein storage vacuole bypassing the Golgi, TGN and MVB (see Figure 6.18).

In addition to the secretory pathway, proteins may be trafficked through the **endocytotic pathway** (see Figure 5.30). Vesicles produced by **endocytosis**, the invagination of the plasma membrane, have clathrin coats and are transported to the TGN. Here both membrane and soluble proteins are sorted and transported to vacuoles via the MVB. The TGN is thought to be the functional equivalent of the early endosome in animal cells and the MVB to be equivalent to the late endosome.

> **Key points** Unlike their animal counterparts, plant cells have hundreds to thousands of small Golgi distributed throughout the cytosol. The large central vacuole causes the cytoplasm of plant cells to be spread out over a larger area. Products originating in plant Golgi have a large number of destinations, including the cell wall and a variety of different types of vacuoles, some large and involved in growth and development, others dedicated to the storage of proteins such as those found in cotyledons and endosperm of seeds. In plant cells macromolecular traffic in the secretory and endocytotic systems is regulated by the trans-Golgi network associated with each separate Golgi apparatus and by the prevacuolar compartment.

5.9 Protein turnover and the role of the ubiquitin–proteasome system

Protein degradation is required for normal cellular function, not only to break down mis-synthesized or misfolded proteins, but also to remove developmentally regulated proteins. Degradation of these proteins must

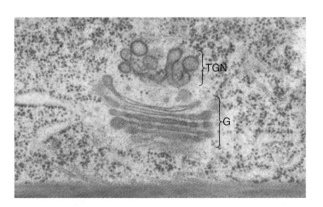

Figure 5.34 Electron micrograph of a *Eucalyptus siebertii* cell showing a Golgi apparatus (G) and an adjacent trans-Golgi network (TGN).

be highly selective to prevent the loss of other, needed proteins. Proteins targeted for removal must be specifically tagged before they are broken down. Cells use the **ubiquitin–proteasome system** (**UbPS**) as one of the principal pathways of protein degradation. Proteins are first tagged using **ubiquitin** (**Ub**), a small 76-amino acid protein, and are then broken down by a complex proteolytic machine called the **26S proteasome**; the designation 26S refers to the size of the proteasome in Svedberg units. The importance of protein removal by the UbPS in plants can be appreciated from genome analyses in *Arabidopsis* that show more than 1300 genes to be implicated in this pathway. In the next sections, we will first look at the process of ubiquitin addition to target proteins and then discuss the structure and function of the proteasome.

5.9.1 Ubiquitin targets proteins for degradation

Ubiquitin addition to target proteins occurs in an ATP-dependent cascade of reactions that proceeds in three distinct steps (Figure 5.35) catalyzed sequentially by Ub-activating enzymes (**E1**), followed by Ub-conjugating enzymes (**E2**) and finally by Ub-ligating enzymes (**E3**). E1 activates Ub using ATP to catalyze the formation of an E1-Ub intermediate. Ub is then transferred to E2. Finally, a Ub ligase (E3) binds both E2-Ub and the target protein and then catalyzes transfer of ubiquitin to the target. After transfer of Ub to the target substrate, additional Ubs are added to form a polyubiquitin chain, often up to a total of seven Ub molecules. The process of ubiquitination is highly specific; specificity is determined largely by the E3 ligation process. There are more than 1200 E3 genes in *Arabidopsis* but only two genes encoding E1, and 37 encoding E2.

The E3 ligase complex has been classified into two groups based on the mechanism of Ub transfer to target

proteins. **HECT** E3 ligases accept Ub from E2 and then transfer it to target proteins (Figure 5.36A). **RING** E3 ligases bind E2-Ub and the target protein and facilitate the direct transfer of Ub from E2 to the target. This latter group includes simple monomeric RING E3 ligases, which bind both E2-Ub and the target protein (Figure 5.36B), and complex multimeric RING E3 ligases such as the **CULLIN/RING** ligases. The CULLIN/RING ligases are complex assemblages of proteins, each of which includes: a **RING finger domain** protein that binds E2-Ub; a variable component, an adaptor that recognizes and binds the target protein; and a **CULLIN**-type protein that forms a scaffold for the rest of the complex (Figure 5.36C). Eleven different CULLIN proteins and hundreds of different adaptors have been identified in plants. The most complex of the CULLIN/RING E3 ligases are the APCs (anaphase-promoting complexes) that contain more than ten different subunits (see Chapter 11).

5.9.2 The 26S proteasome is a molecular machine that breaks down ubiquitinated proteins

Ubiquitinated proteins are recognized by the **26S proteasome** and undergo ATP-dependent proteolysis. The proteasome is a large complex made up of a 20S **core protease** (**CP**) and a 19S **regulatory particle** (**RP**) (Figure 5.37). The CP can be visualized as an open-ended barrel capped at both ends by the RP. The RP confers specificity to proteolysis by recognizing ubiquitinated proteins and allowing those that are to be degraded to enter the CP where proteolysis occurs. The RP also catalyzes removal of the ubiquitin tag and the unfolding of target proteins to allow for rapid proteolysis by the CP.

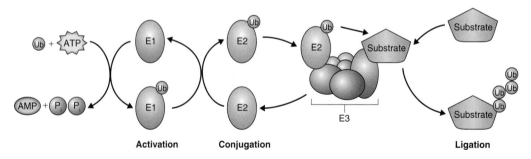

Figure 5.35 Protein degradation via the ubiquitin–proteasome system. Ub-activating enzyme E1 catalyzes the ATP-dependent addition of Ub. Ub is transferred from E1 to Ub-conjugating enzyme E2. The Ub ligase E3 binds both E2 and the target protein to be ubiquitinated facilitating specific transfer of up to seven Ub molecules to the target.

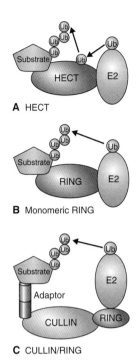

Figure 5.36 Two types of E3-Ub ligases are known. (A) In HECT-domain ligases Ub is transferred from the conjugating enzyme E2 to the HECT domain protein and it in turn transfers Ub to the target protein that is also bound to the HECT domain. (B, C) In the second type, Ub is transferred directly from E2 to the target protein. (B) RING/U-box domain E3 ligases are monomeric and directly bind both E2 and the target protein. (C) CULLIN/RING domain ligases are made up of a number of proteins. A CULLIN platform anchors the other proteins, which include a RING-domain protein that binds E2 and adaptor proteins that bind the target.

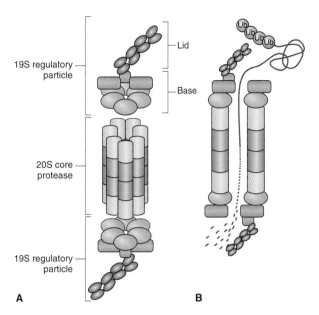

Figure 5.37 Model of the 26S proteasome showing key features of this protein-degrading machine. (A) The core protease consists of a 20S particle that is capped at each end by a 19S regulatory particle made up of a lid and base. The association of the regulatory particle with the core protease is ATP-dependent. (B) Ubiquitinated proteins are recognized by the regulatory particle and protein chaperones in the base unfold the target protein for proteolysis by the core protease.

5.9.3 Cytosolic and endoplasmic reticulum localized proteins are degraded by the UbPS

Proteolysis by the ubiquitin–proteasome system occurs in the cytosol and nucleus where it brings about regulated proteolysis of proteins synthesized on non--membrane-bound ribosomes. Surprisingly, defective proteins synthesized on the ER are also degraded by the cytosolic UbPS. As many as 30% of total cellular proteins are synthesized on the ER and it is estimated that a third of these newly synthesized proteins are degraded soon after synthesis. Because of the enclosed nature of the ER lumen there must be efficient mechanisms to degrade defective proteins synthesized in this compartment. The mechanism by which this is done is called **ER-associated degradation** (**ERAD**), a highly regulated process.

Key points Defective proteins and those involved in cellular regulation such as transcriptional activators and repressors must be identified and degraded. This is accomplished by the ubiquitin–26S proteasome system, which is found in both the nucleus and cytosol. Proteins targeted for degradation are ubiquitinated by a protein complex called E3 ligase, which identifies proteins to be degraded and facilitates transfer of as many as seven ubiquitin molecules to target proteins. Ubiquitination requires two other proteins, a Ub-activating enzyme E1 to which Ub is added in an ATP-dependent reaction and a Ub-conjugating enzyme E2 that transfers Ub to the target either indirectly via HECT E3 ligases or directly via RING ligases. Addition of Ub to target proteins allows them to be identified by the 26S proteasome. Ubiquitinated proteins are recognized by the regulatory particle of the proteasome and protein chaperones in its base unfold the target protein for proteolysis by the core protease. In ERAD, defective proteins in the ER lumen are translocated to the cytosol for degradation by the UbPS.

In ERAD, ER-localized chaperones recognize defective proteins in the ER lumen, unfold them and permit their transport to the cytosol to be degraded by the UbPS. These chaperones include the ER proteins, calnexin and calreticulin, and Hsp chaperones. Stresses that increase the proportion of defective proteins synthesized by the ER also increase the synthesis of ER-localized chaperones. The transport of aberrant proteins from the ER to the cytosol occurs via a translocation pore in the ER membrane; this process is called **retrotranslocation**. An interesting aspect of ERAD is that the 26S proteasome is enriched at the surface of the ER.

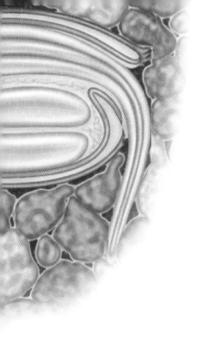

Chapter 6

Seed to seedling: germination and mobilization of food reserves

6.1 Introduction to seeds and their germination

The life of a plant begins with a seed and is renewed when the seed germinates and the embryo grows and develops into a seedling. The seed initially develops in a hydrated state but loses water and dries during maturation. In the dry (resting or quiescent) state, the seed can survive for extended periods, often years, in some species tens or even hundreds of years, before it takes up water and germinates. Some seeds are dormant when they mature; this dormancy must be overcome before germination can occur.

The visible sign that germination is complete is the penetration of the seed coat by the embryonic root, the **radicle**. As the embryo grows and develops, the reserves within the storage tissues of the seed are mobilized to support seedling growth and development. Germination typically occurs below the surface of the soil and the young seedling is often dependent on the stored reserves in the seed. In this regard seedling growth can be entirely **heterotrophic** and independent of photosynthesis.

Seeds are an important source of food for humans and livestock and it has been estimated that as much as 50% of human caloric intake and 47% of protein comes from seeds. Seventeen plant species provide 90% of human food and, of these, cereal grains make up the bulk of edible dry matter consumed. Wheat, maize, rice and barley grains make up about 70% of the food and feed consumed worldwide. The prevalence of cereal grains in the human diet is of relatively recent origin. Agriculture had its origins in the Fertile Crescent, an area encompassing what is now Turkey, Iraq, Israel, Palestine and Jordan, less that 10 000 years ago—a relatively short time in the context of human evolution. A consequence is that humans are not particularly well adapted to a diet

The Molecular Life of Plants, First Edition. Russell Jones, Helen Ougham, Howard Thomas and Susan Waaland.
© 2013 John Wiley & Sons, Ltd. Published 2013 by John Wiley & Sons, Ltd.

rich in starch and seed proteins. Many human diseases are recognized as being linked to a diet that relies on cereal grains.

The impact of plant breeding on agricultural production has been profound. Despite Malthusian claims that population growth will outstrip the food supply, conventional plant breeders have succeeded in increasing the yield of cereals on an almost annual basis. The high-yielding wheat and rice varieties that were the basis of the Green Revolution are testament to the breeders' skills. The Nobel Peace Prize was awarded to Norman Borlaug for his work in developing many of these new wheat varieties.

Molecular approaches are now becoming more widely used to develop new and improved varieties of seed plants. New maize varieties carrying resistance to various pests, and soybeans that are herbicide-resistant, have been introduced, and rice varieties with enhanced vitamin D synthesis and elevated iron levels have been successfully developed using recombinant DNA methods. There have also been dramatic improvements in the genetic modification of oil seeds; for example, the development of canola plants having very low levels (<2%) of the toxic fatty acid erucic acid (see Chapter 17). Molecular engineering has also led to improvements in the protein quality of various seeds. Perhaps the biggest hurdle faced by biotechnologists will be the acceptance of genetically modified foods by consumers.

In this chapter we will examine seed structure in more detail and describe the types of reserves that are stored in

seeds. Then we will look at the processes involved in seed germination, and examine how stored reserves are mobilized during early seedling growth.

6.2 Seed structure

The mature seed consists of an **embryo** enclosed by a protective seed coat (Figure 6.1). In addition, endosperm is present in the seeds of all monocots and many eudicots; these are **endospermous seeds**. The eudicot endosperm can vary in size from a single layer of cells as in *Arabidopsis* seeds (Figure 6.2), to the large, oil-storing endosperm of castor bean (*Ricinus communis*; Figure 6.1B) and the starch-storing endosperm of cereals such as maize (*Zea mays*; Figure 6.1D). Seeds of some **eudicots** such as the green bean (*Phaseolus vulgaris*; Figure 6.1A) are **non-endospermous**, i.e. they lack endosperm at maturity. We will now examine the parts of a seed in more detail.

6.2.1 Seeds contain an embryonic plant

With few exceptions the eudicot embryo has an axis consisting of an embryonic root, the **radicle**, and an embryonic shoot, the **plumule**, which bears a pair of

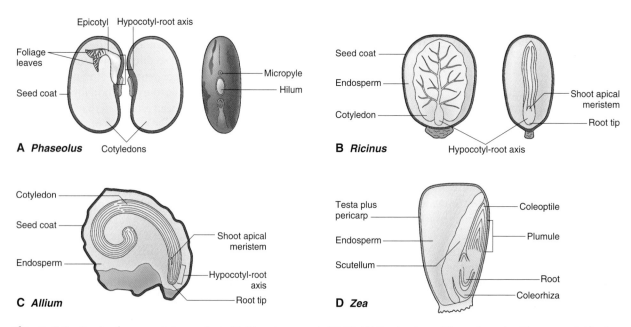

Figure 6.1 Seeds of representative eudicots (A, B) and monocots (C, D). (A) Garden bean (*Phaseolus vulgaris*) open and side view; the embryo has two fleshy food-storing cotyledons and no endosperm. (B) Castor bean (*Ricinus communis*) open and side view; note the large endosperm and thin, leaf-like cotyledons. (C) Onion (*Allium cepa*) with the large endosperm and shoot apex at the base and to the side of the single cotyledon. (D) Maize (*Zea mays*). Note the large endosperm, complex embryo with fleshy scutellum, shoot apical bud (plumule) encased by the coleoptile, and root surrounded by the coleorhiza.

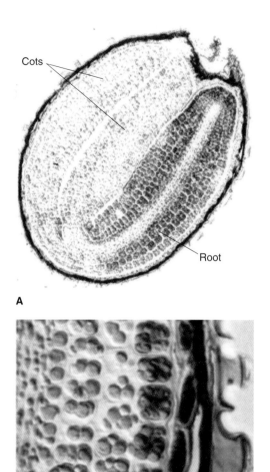

Figure 6.2 Light micrographs of an *Arabidopsis* seed. (A) Section through a whole seed showing two cotyledons (Cots) and root surrounded by a layer of aleurone and the testa. (B) Higher magnification showing the testa (seed coat), single-cell-layered aleurone (Al) and edge of one cotyledon.

6.2.2 Seed coats, made of layers of dead cells, protect the embryo

The testa, derived from the integuments of the ovule (see Figure 1.18), consists of several layers of dead and compressed cells (Figure 6.2). It is often impregnated with phenolic compounds that impart a red to brown color to many seeds and grains (Figure 6.3A). These phenolic compounds protect the seed from attack by herbivores and pathogens, filter out harmful short wavelength light, and can impose dormancy by making the seed coat rigid or impermeable to water or oxygen. **Tannins** and other **polyphenols** in seed coats contribute to the astringency of many fruits and seeds and deter feeding by herbivores. The effectiveness of polyphenolic compounds in protecting embryos from the damaging effects of ultraviolet light becomes clear from their structure (Figure 6.3B). The presence of a large number of double bonds in polyphenols such as the **proanthocyanidins** makes these compounds effective in absorbing short wavelength light, protecting the embryo from the mutagenic effects of these wavelengths.

The testa often has an exterior mucilage layer that hydrates quickly when seeds are exposed to water (Figure 6.4). In *Arabidopsis* seeds, mucilage is composed largely of esterified pectins made up of polygalacturonic acid polymers. This mucilage is deposited on the outer surface of the testa as the seed matures. Its presence is thought to be important in the uptake and retention of water by the imbibing seed, allowing for uniform water absorption over its surface.

The coats of many seeds are much more complex because the testa is fused to the **pericarp**. The pericarp arises from the ovary wall; therefore, seeds enclosed by a testa and pericarp are technically **fruits**. Examples of fruits in which a fused testa and pericarp surround a single embryo are the caryopses of grasses (Figure 6.1D) and the achenes of strawberry, dandelion and lettuce (Figure 6.5). Bracts called the **palea** and **lemma**, which are modified leaves, also enclose the grass caryopsis and are often referred to as **hulls** (Figure 6.6). Hulls are a major component of the **chaff** that is produced when grain has been threshed.

Seed coats are believed to be beneficial for human health when included as part of a normal diet. They may provide dietary fiber. Because seed coat lignins and polysaccharides are largely indigestible they may aid in the passage of food through the gut. They may also be a rich source of antioxidants. Among the compounds found in seed coats, the proanthocyanidins (Figure 6.3) are known to be strong antioxidants. The hulls of barley grains play an important role in traditional beer brewing

seedling leaves, the **cotyledons**. Monocot embryos show more variation in their structure. They vary from embryos of onion that are similar in many respects to eudicot embryos except that they have only one cotyledon (Figure 6.1C), to the embryos of grasses with their relatively more complex structure (Figure 6.1D). In grass embryos the single cotyledon has been modified to form the **scutellum**, an absorptive organ that lies adjacent to the endosperm. In addition, a sheath called a coleoptile surrounds the plumule of the grass embryo and a second sheath, the coleorhiza, surrounds the radicle.

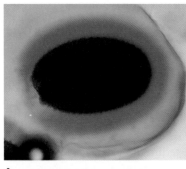

Red sorghum

Proanthocyanidin (*n* = 1–30)

Figure 6.3 Testa color in red sorghum. (A) Heads of red sorghum grain. (B) Molecular structure of proanthocyanidin, one of many polyphenols present in numerous species, including sorghum, that protect grains and seeds from damaging ultraviolet light.

A

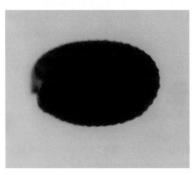

B

Figure 6.4 Mucilage surrounding an *Arabidopsis* seed is made up of two layers that facilitate the uptake of water. (A) Wild-type seed placed in the dye ruthenium red showing two mucilage layers, an outer cloudy layer and an inner intensely staining layer. (B) Seed of a mutant lacking mucilage and stained with ruthenium red.

methods, where they are used as filters in the beer-making process (Figure 6.7).

> **Key points** The seed consists of an embryo and food reserves surrounded by a seed coat of varying complexity; food may be stored in the embryo or in specialized, polyploid tissue called endosperm. The eudicot embryo axis includes the seedling root, stem and leaves, in the form of the radicle, plumule and cotyledons, respectively. Monocot embryos either have one cotyledon, as in the case of the onion embryo, or lack a true cotyledon. In grasses the scutellum is a modified cotyledon that absorbs nutrients from the endosperm. The seed coat or testa surrounds the embryo. It can consist of one to many layers of dead cells in the mature seed and is often impregnated with phenolic compounds.

6.2.3 Endosperm, a tissue unique to angiosperms, contains stored food

Almost all seeds contain stored food, which will be used to fuel the growth of the seedling after germination. The dust-like seeds of orchids are an exception as these lack stored food reserves and germination requires the participation of symbiotic endomycorrhizal fungi. In non-endospermous seeds, such as those of many

A   **B**   **C**

Figure 6.5   Single-seeded fruits. (A) Strawberry (*Fragaria* spp.), where fruits sit on a fleshy red receptacle. (B) Lettuce (*Lactuca sativa*). (C) Dandelion (*Taraxacum* sp.). The seed coat (testa) of the seeds of these species is fused with the pericarp to form a simple fruit called an achene.

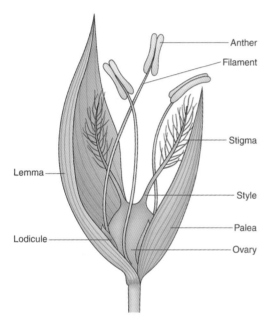

Figure 6.6   A stylized cereal flower showing a spike or head of grain—an individual floret. A lemma and palea enclose the stamens and ovary. Scale-like lodicules are shown and these are thought to be the remnants of sepals and petals. The lemma and palea form the hulls that surround a mature grain.

legumes, food is stored in the fleshy cotyledons of the embryo. In most endospermous seeds, the endosperm is specialized for storage of food reserves. However, in some species, it plays only a minor role in food storage. In *Arabidopsis* seeds, for example, the endosperm consists of only a single layer of cells that have relatively thick cell walls (see Figure 6.2). As we shall see later, the endosperm and testa of *Arabidopsis* impose dormancy by restricting growth of the embryo. These more specialized

endosperm cells also store protein and lipid reserves, but the bulk of food reserves in *Arabidopsis* are found in cotyledons.

The endosperm of many seeds is often not uniform; it may be differentiated into areas of varying thickness. In grasses the endosperm differentiates into two structurally and functionally distinct **tissues**: the **starchy endosperm** and the **aleurone layer** (Figure 6.8). Although they contain enzymes that function during germination, starchy endosperm cells are dead at maturity; these cells store protein as well as starch. The aleurone layer is alive in the mature seed. After germination, it functions as a digestive tissue that synthesizes and secretes large quantities of enzymes such as α-amylase to mobilize food reserves that are located in the starchy endosperm.

Endospermous cereals and non-endospermous legumes make up the bulk of the human diet. Recent figures show that more than 75% of dry matter consumed by humans comes from cereal grains and about 6% from legume seeds. Table 6.1 shows the tonnage of various seed crops produced worldwide and illustrates how cereal grains represent by far the largest proportion of seed crops grown. Indeed, the diet recommended by the United State Department of Agriculture (USDA) advocates 6–11 servings per day of cereals or cereal products, and USDA's food plate includes cereals as a large portion of the human diet (Figure 6.9).

The prominent role of cereals in the human diet is relatively recent in the context of human evolution and is a result of the rise of agriculture. The domestication of cereals began about 10 000 years ago and it is likely that cereals only became a staple in the human diet from around 3000 to 4000 years ago. Because human metabolism evolved for at least 100 000 years largely on a

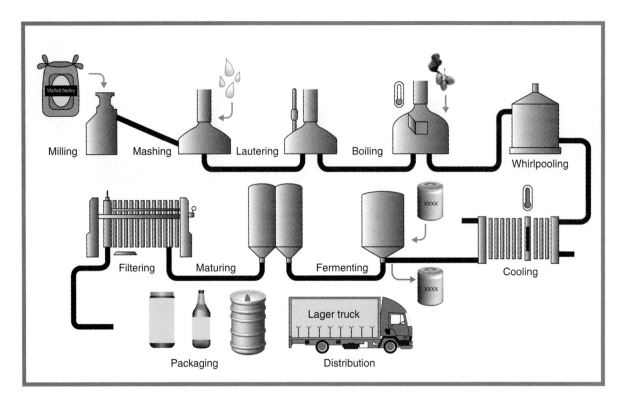

Figure 6.7   Malting and brewing of beer. During malting, mature barley or wheat grains are germinated under controlled conditions and then quickly dried. During germination, enzymes are produced that will eventually break down the stored endosperm starch and proteins, but rapid drying of the germinated grain halts endosperm breakdown. The dried product is malted grain. Malted grain is ground (milled) and added to a tank (mash tun) where it is mixed with water. During mashing, enzymes convert stored starch and proteins to sugars and amino acids to produce a solution called wort. Wort is strained through the residual grain hulls that lie at the bottom of the mash tun (the process is called lautering). The clarified wort is transferred to the brew kettle where it is combined with hops to provide flavor and the mixture is boiled. After cooling, yeast is added and the wort is transferred to a fermentation tank where the yeast converts sugars to alcohol. After filtration, the liquid, now called beer, is cooled in a conditioning tank and stored for weeks or months depending on the type of beer being brewed. Draft beer may be consumed directly after conditioning. Bottled beer is often pasteurized (heated to 60 °C for several minutes) to destroy harmful microorganisms and increase shelf life. In Germany there are strict laws that govern beer brewing, called Reinheitsgebot. These laws stipulate that only malted grain, yeast, hops and water may be used in the making of beer. In the USA and many other countries, grains other than malted barley (or wheat) are used in beer making to supplement the starch and proteins of malted barley. Commonly, milled endosperm from maize, rice and wheat may be used. These additives affect beer flavor and they are often used to make a less intensely flavored product.

meat-based, hunter-gatherer diet (in which starchy roots and tubers may have played an important role; see Chapter 17), it is perhaps not surprising that we are relatively poorly adapted to cereal-based food. Numerous diseases, including wheat allergy and celiac disease, can be traced to the preponderance of cereals in the human diet. The latter disease is caused by intolerance to gluten, a storage protein found in some cereal grains. Other less well-recognized illnesses are also associated with a high intake of cereals, including deficiencies of certain essential amino acids (lysine, methionine, tryptophan) and minerals (calcium, iron, zinc) (see Section 6.3.5).

6.3 Use of seed storage reserves by the germinating embryo

Seeds store polysaccharides, lipids, proteins and minerals as well as a wide range of other molecules that will be used by the embryo when it starts to grow (Tables 6.2 and 6.3). As mentioned earlier, these reserves may be stored in the endosperm, the cotyledons and in other tissues that make up the embryo (Table 6.4). For

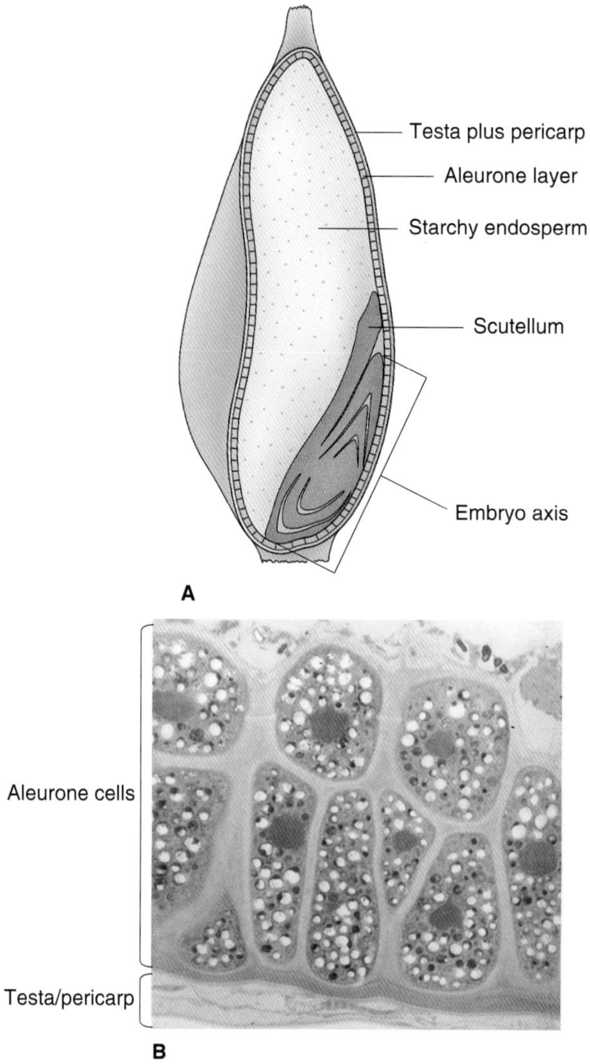

Figure 6.8 Details of the morphology of a barley grain. (A) Diagram of a longitudinal section through a barley grain showing the outer pericarp, aleurone layer, starchy endosperm, scutellum and embryo. (B) Micrograph of the aleurone layer of barley showing the densely filled cytoplasm of the aleurone cell surrounded by thick cell walls.

example, in cereals the scutellum is a major site of reserve storage and corn oil is obtained largely from this organ. While the starchy endosperm is a major site for the storage of protein and starch, the aleurone layer, scutellum and the rest of the embryo also contain stored protein and lipid. Small-grain cereals such as rice (*Oryza sativa*) are processed so that the oil of the aleurone layer can be harvested. In oil palm (*Elaeis* spp.) and castor

Table 6.1 Annual production of seed crops worldwide.

Crop	Production (kg $\times 10^9$)
Cereals	
Wheat (*Triticum* spp.)	468
Maize (*Zea mays*)	429
Rice (*Oryza sativa*)	330
Barley (*Hordeum* spp.)	160
Sorghum spp.	60
Oats (*Avena sativa*)	43
Rye (*Secale cereale*)	29
Millets and other small-grained cereals	26
Legumes	
Soybean (*Glycine max*)	88
Pea (*Pisum sativum*)	12
Bean	14
Peanut (*Arachis hypogaea*)	13
Other	
Rapeseed (*Brassica napus*)	19
Sunflower (*Helianthus annuus*)	10
Cottonseed (*Gossypium* sp.)	5

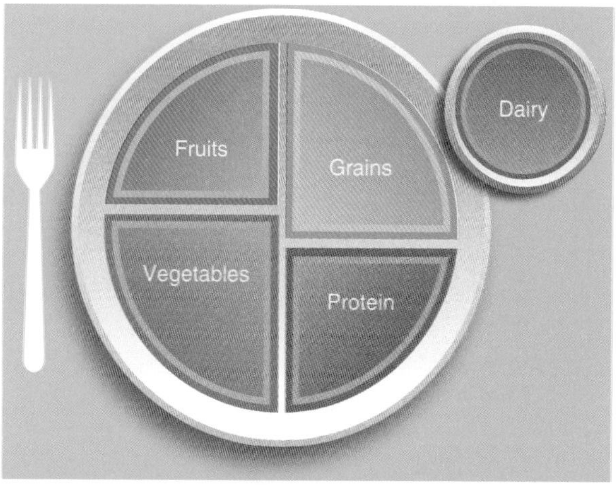

Figure 6.9 The food plate devised by the US Department of Agriculture to define an ideal diet for the American population. Note the recommended servings of cereal grains, vegetables and fruits versus proteins. The trend by nutritionists today is to advocate a balanced diet without reference to particular serving sizes of various foods.

Table 6.2 Vitamin and mineral content of the unprocessed grains of five cereals (100 g samples).

	Wheat (*Triticum* spp.)	Maize (*Zea mays*)	Rice (*Oryza sativa*)	Barley (*Hordeum*)	Oats (*Avena sativa*)
Vitamins					
B_1 (mg)	0.38	0.39	0.40	0.65	0.76
	(35%)	(35%)	(36%)	(59%)	(69%)
B_2 (mg)	0.12	0.20	0.09	0.29	0.14
	(9%)	(15%)	(7%)	(22%)	(11%)
B_3 (mg)	5.47	0.20	5.09	4.60	0.96
	(36%)	(24%)	(34%)	(31%)	(6%)
B_6 (mg)	0.30	0.62	0.51	0.32	0.12
	(21%)	(39%)	(32%)	(20%)	(7%)
Folate (mg)	38.2	19.0	19.5	19.0	56
	(21%)	(11%)	(11%)	(11%)	(31%)
Pantothenic acid (mg)	0.95	0.42	1.49	0.28	1.35
	(17%)	(8%)	(27%)	(5%)	(24%)
Minerals					
Potassium (mg)	363	287	223	452	429
	(18%)	(14%)	(11%)	(23%)	(21%)
Calcium (mg)	29.0	7.0	23.0	33.0	53.9
	(4%)	(1%)	(3%)	(4%)	(7%)
Phosphorus (mg)	228	210	333	264	523
	(36%)	(26%)	(42%)	(33%)	(65%)
Magnesium (mg)	126	127	143	133	177
	(45%)	(45%)	(51%)	(48%)	(63%)
Iron (mg)	3.19	2.71	1.47	3.60	4.72
	(21%)	(18%)	(10%)	(24%)	(31%)
Zinc (mg)	2.65	2.21	0.27	0.50	0.63
	(22%)	(14%)	(12%)	(22%)	(28%)

Values in (parentheses) represent recommended daily allowance percentages. No detectable amounts of vitamins A, C, D, B_{12} are found in any grain.

bean (*Ricinus communis*), oil is stored in the endosperm. In brazil nut (*Bertholletia excelsa*), on the other hand, the fleshy portion of the seed that is eaten by humans is the swollen hypocotyl/radicle of the embryo. In pine seeds (*Pinus* spp.), reserves are stored in cells of the megagametophyte (see Figure 1.15).

6.3.1 Starch is the major carbohydrate reserve of plants

Starch is the major carbohydrate reserve in most plants. It forms the bulk of the storage carbohydrate in many

Table 6.3 The food reserves of some important crop species.

	Average composition (%)			
	Protein	Oil	Carbohydrate[a]	Major storage site
Cereals				
Barley (*Hordeum* spp.)	12	3[b]	76	Endosperm
Maize (*Zea mays*)	10	5	80	Endosperm
Oats (*Avena sativa*)	13	8	66	Endosperm
Rye (*Secale cereale*)	12	2	76	Endosperm
Wheat (*Triticum* spp.)	12	2	75	Endosperm
Legumes				
Broad bean (*Vicia fava*)	23	1	56	Cotyledons
Garden pea (*Pisum sativum*)	25	6	52	Cotyledons
Peanut (*Arachis hypogaea*)	31	48	12	Cotyledons
Soybean (*Glycine max*)	37	17	26	Cotyledons
Other				
Castor bean (*Ricinus communis*)	18	64	Negligible	Endosperm
Oil palm (*Elaeis* sp.)	9	49	28	Endosperm
Pine (*Pinus* spp.)	35	48	6	Megagametophyte
Rape (*Brassica napus*)	21	48	19	Cotyledons

[a] Mainly starch.
[b] In cereals, oils are stored within the scutellum, an embryonic tissue, and aleurone.

Table 6.4 Percent composition of stored reserves in different parts of a maize (*Zea mays* cv. Iowa 939) kernel.

Reserve	Whole grain	Endosperm (starchy and aleurone layer)	Embryo (including scutellum)
Starch	74	88	9
Oil	4	<1	31
Protein	8	7	19

seeds and grains, especially cereals, but also in legumes such as garden pea (*Pisum sativum*) and broad bean (*Vicia faba*) (Table 6.3).

Starch is made up of unbranched **amylose** chains and branched **amylopectin** (Figure 6.10). Amylose is a polymer of up to about 300–3000 glucose residues joined

Key points Seed storage reserves function to nourish the embryo; they may be found in a variety of tissues. In pine seeds, protein and oil are stored in megagametophyte tissue. In angiosperms, seed reserves are stored in the endosperm and/or in the embryo itself. In cereal grains, the endosperm consists of dead cells that contain stored starch and proteins whereas oil is found mostly in the scutellum of the embryo. In other monocots such as coconut, the endosperm in a mature seed contains living cells that store starch, carbohydrate and lipids. In many eudicot seeds, the endosperm is reduced to a single layer of cells that surrounds the embryo and the bulk of reserves are found in the embryonic cotyledons. Some endospermous eudicot seeds, for example castor bean, have a fleshy endosperm that stores reserves such as oil.

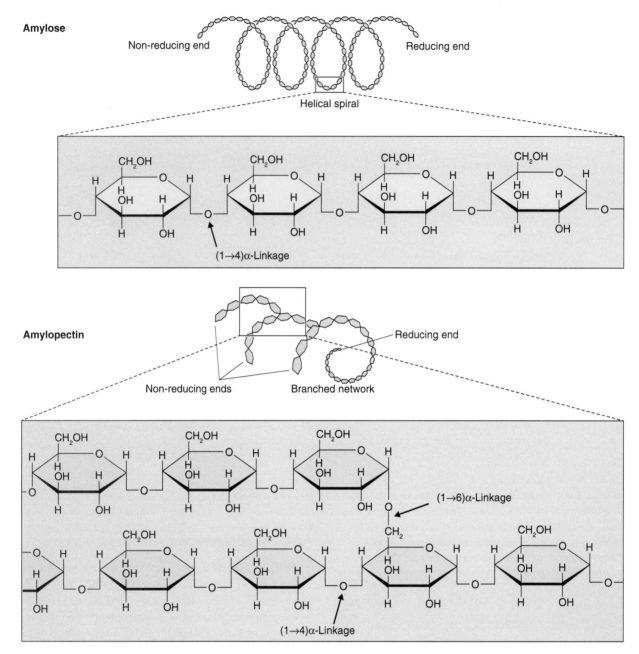

Figure 6.10   Structure of starch. Starch contains amylose, constructed of long, unbranched chains of $\alpha(1 \rightarrow 4)$-linked glucose units, and amylopectin, which consists of amylose chains linked by $\alpha(1 \rightarrow 6)$ linkages. Branch points in amylopectin are separated by an average of 20–30 glucose residues. Diagrammatic representations of the two molecules show how $\alpha(1 \rightarrow 4)$ and $\alpha(1 \rightarrow 6)$ linkages influence polymer organization. Note that amylose has a pronounced helical structure that is absent in amylopectin. The helical structure of amylose does not allow hydrogen bonds to readily form between adjacent amylose chains giving starch a 'floury' texture.

by $\alpha(1 \rightarrow 4)$ linkages. The backbone of the amylopectin polymer is much longer than that of amylose. It can be up to 10^5 glucose residues joined by $\alpha(1 \rightarrow 4)$ linkages with side-chains of about 20–25 $\alpha(1 \rightarrow 4)$-linked glucose monomers that are attached by $\alpha(1 \rightarrow 6)$ linkages to the backbone (Figure 6.10).

Amylose can be distinguished from amylopectin by its reaction with an iodine and potassium iodide (I_2KI) mixture. The $\alpha(1 \rightarrow 4)$ linkages in amylose give the molecule a helical structure whereas amylopectin, because it is branched, does not form helices. Iodine is able to intercalate into the helical amylose molecule and,

Table 6.5 Amylose content and granule size of starch grains from various species.

Starch source	Amylose (%)	Granule size range (μm)	Average size (μm)
Maize (*Zea mays*)	28	5–25	14
High amylose maize	70	4–20	10
Waxy maize	<2	5–25	14
Waxy rice[a] (*Oryza sativa*)	0	2–15	6
Waxy *Sorghum*[a]	0	—	—
Wheat (*Triticum* spp.)	26	3–35	7 and 20
Sweet potato (*Ipomoea batatas*)	18	4–40	19
Cassava (*Manihot esculenta*)	17	3–30	14
White potato (*Solanum tuberosum*)	20	10–100	36

[a]Starch is all amylopectin.

when it does, it changes color from golden brown to blue-black. Because the amylopectin molecule is not a helix, iodine does not bind to it and its color remains unchanged.

The degree of branching in amylopectin can vary considerably and this changes the property of the starch that is synthesized. The ratio of amylose to amylopectin also varies according to species; in most cereals starch is on average 30% amylose and 70% amylopectin. Table 6.5 shows the extremes in amylose content in some commercial crops. In waxy cereals, such as waxy mutants of maize, starch is composed entirely of amylopectin, whereas in sweet maize mutants deficient in enzymes catalyzing the conversion of sugar to starch, amylose content can be as high as 70%.

Sugars imported from the parent plant into developing seeds and grains are converted into starch within the **amyloplast**, a non-green plastid that is modified for starch synthesis and accumulation (see Figure 4.24). Starch accumulates in starch grains that can vary in size from 2 μm to more than 50 μm, such as those found in the tubers of white potato (*Solanum tuberosum*) (Table 6.5). The starch grains of cereal endosperm show considerable variation in size even within one cell, as is shown in the electron micrograph in Figure 6.11A. Starch is added to grains in layers, which remain distinct in the growing grain and appear as concentric rings when examined by electron microscopy (Figure 6.11B).

Starch has many uses beyond its predominance as a basic food source for humans and livestock. For example, it is widely used in the making of paper, in the printing industry as a bonding agent or glue, as a stiffening agent in the laundering of clothes, in the molding of candy and as a thickening agent in a wide variety of food stuffs.

6.3.2 Cell walls are also an important store of polysaccharides in many seeds

Although starch is clearly an important source of stored polysaccharide, cell walls may be the predominant site for carbohydrate storage in many seeds. The cell walls of almost all tissues of a seed can be modified for polysaccharide storage. Cell walls of the endosperm are especially important storage sites. In seeds such as date (*Phoenix dactylifera*) and fenugreek (*Trigonella foenum-graecum*), the cell walls are the principal reserve of mannan polysaccharides. Endosperm walls in these species become extensively thickened so that they almost totally eliminate the cytoplasm (Figure 6.12).

Several types of polysaccharide may function as storage polymers in cell walls. In the cell walls of many legume seeds, as well as those of date and fenugreek, polymers with a mannose backbone are the principal storage polymers. Pure **mannans** form a hard, almost crystalline, matrix similar to that of cellulose; in **galactomannans** galactose side-chains change the structural properties of the polymer (Figure 6.13). **Glucomannans**, in which the backbone of the polymer is made up of varying proportions of mannose and glucose, are found in the seeds of many lilies (*Lilium* spp.), whereas the seeds of tamarind (*Tamarindus indica*) have polysaccharides that are galactose-rich **xyloglucans**.

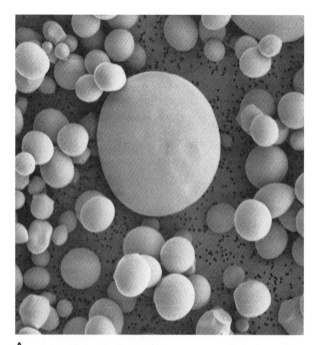

A

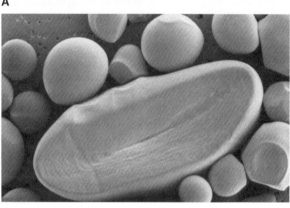

B

Figure 6.11 Scanning electron microscopy of starch granules isolated from wheat (*Triticum aestivum*). (A) Starch granules vary widely in size and in wheat endosperm there are two size classes of starch granules, large and small. (B) Detail of a starch granule showing that starch is deposited in concentric rings.

The cell walls of the cereal endosperm have a characteristic polysaccharide known as mixed-link glucan or β-glucan (Figure 6.14). This polymer consists of alternating runs of cellotriose (three glucose units) and cellotetraose (four glucose units) linked by $\beta(1 \rightarrow 3)$ and $\beta(1 \rightarrow 4)$ linkages. Barley and oat endosperm can contain up to 4–5% β-glucan. It forms a highly viscous polymer and as such creates problems for brewers, causing haze in beer at cold temperatures as well as interfering with the filtering of finished beer. Cereal β-glucans are thought to benefit human health by lowering serum cholesterol levels. They are frequently marketed as 'soluble fiber'.

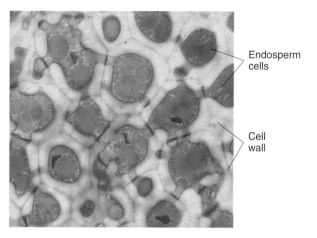

Figure 6.12 Light micrograph of endosperm cells of date (*Phoenix dactylifera*). Note the extremely thick mannan-containing walls and large protein storage vacuoles in the cytoplasm. Sections were stained with toluidine blue O; this dye stains proteins but not cell wall mannans.

6.3.3 Storage proteins in eudicot seeds include globulins and albumins

Storage proteins make up as much as 40% of the dry weight of legume seeds but comprise only 10–12% of cereal grains (see Table 6.3). Despite this difference in protein content, cereal storage proteins represent a far higher proportion of human and livestock protein intake (about 2×10^5 kg annually) than do those of legumes (about 7×10^4 kg) (see Table 6.1). Work at the beginning of the 20th century led to the classification of plant proteins according to their solubility in water, saline solution, dilute alkali and alcohol (Table 6.6). Water-soluble proteins are called **albumins**, those soluble in dilute salt and alkali solutions are **globulins** and **glutelins**, respectively, and those proteins soluble in 70% ethanol are **prolamins**.

Salt-soluble globulins and water-soluble albumins make up a large fraction of the storage proteins of eudicot seeds. In seeds of alfalfa (*Medicago sativa*), for example, globulins make up around 40% of seed storage proteins and albumins make up about 20%. The globulins of legumes are made up of two classes of oligomeric (multisubunit) proteins, the **legumins** and the **vicilins**. Legumins and vicilins are distinguished based on their sedimentation coefficients in Svedberg units (S). Proteins of the legumin class are between 11S and 14S in size whereas vicilins are 7S–8S. Both classes of proteins are made up of variable numbers of subunits, up to eight in vicilin and seven in legumin of *Pisum sativum*. Other eudicot seeds may contain legumin-like and/or vicilin-like proteins. *Arabidopsis* seeds lack globulins of

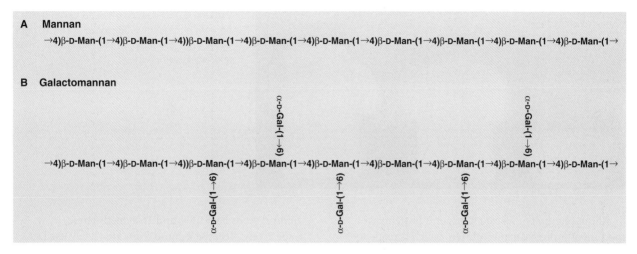

Figure 6.13 Cross-linking glycans that contain mannose. (A) Adjacent chains of pure linear mannans can hydrogen bond to each other to form paracrystalline arrays that are similar in structure to cellulose. (B) Galactomannans have backbones composed exclusively of β(1 → 4)-D-mannose with α-D-galactose units added at the 6 positions. These side-chains prevent the formation of ordered paracrystalline arrays.

→4)β-D-Glc-(1→4)β-D-Glc-(1→3)β-D-Glc-(1→4)β-D-Glc-(1→4)β-D-Glc-(1→3)β-D-Glc-(1→4)β-D-Glc-(1→4)β-D-Glc-(1→3)β-D-Glc-(1→4)β-D-Glc-(1→

Figure 6.14 A mixed-link glucan is a characteristic polysaccharide in cereal endosperm. The glucan chain consists of a mixture of β(1 → 3) and β(1 → 4) linkages.

Table 6.6 Classification of plant proteins.

Protein group	Solubility in solvent
Albumin	Water
Globulin	Dilute salt (0.1 M NaCl)
Glutelin	Dilute alkali (0.1 M NaOH)
Prolamin	70% aqueous ethanol

Table 6.7 The approximate protein composition (%) of some cereal grains and the common names of some storage proteins.

Cereal	Albumin	Globulin	Prolamin	Glutelin
Wheat (*Triticum* spp.)	9	5	40 (gliadin)	46 (glutenin)
Maize (*Zea mays*)	4	2	55 (zein)	39
Barley (*Hordeum*)	13	12	52 (hordein)	23 (hordenin)
Oats (*Avena sativa*)	11	56	9 (avenin)	23
Rice (*Oryza sativa*)	5	10	5 (oryzin)	80 (oryzenin)
Sorghum spp.	6	10	46 (kafirin)	38

the 7S vicilin class and about 85% of its storage proteins are made up of 12S legumin and 2S albumin.

6.3.4 Storage proteins in cereal grains differ from those found in eudicot seeds

The storage protein composition of cereal grains is markedly different from that of eudicot seeds. Cereal grains contain high concentrations of alcohol-soluble prolamins and alkali-soluble glutelins, proteins that are essentially absent from the seeds of eudicot plants (Table 6.7). Despite these differences in overall protein composition, the storage proteins of cereal grains and eudicot seeds share some interesting homologies. The globulins of cereals have sedimentation coefficients of about 7S and show sequence homologies with the 7S vicilins of legumes. Similarly, the globulins found in the starchy endosperm and aleurone of cereals show sequence similarities with the 11S family of legumins.

Figure 6.15   Loaves of bread baked from flour with different sulfur content. Left to right: flours from low sulfur wheat, high sulfur wheat and excess sulfur wheat.

About one-half of wheat endosperm protein is a mixture of proteins, named **gluten**, which is made up largely of prolamins and glutelins. These proteins are crucial for the properties of foodstuffs such as bread and pasta. The baking properties of bread are strongly influenced by the presence of disulfide bonds (S–S) between adjacent gluten chains; the elasticity of the dough is a function of the oxidation status of gluten. During bread-making, dough is kneaded to evenly distribute the gluten in the flour and to oxidize the SH groups, enabling the formation of S-S bonds between adjacent gluten molecules. This forms a protein network that traps the carbon dioxide produced by the added fermenting yeast and allows the bread to rise. Flour that is low in the sulfur-containing amino acid cysteine produces bread that does not rise, probably because it forms fewer S-S bonds. On the other hand, excess cysteine in the flour also adversely affects bread quality (Figure 6.15). Agents such as potassium bromate that oxidize cysteine and promote the formation of the disulfide network are often used to enhance dough quality.

6.3.5 The amino acid content of seed proteins affects their nutritional value for humans and livestock

Globulins, abundant in legume seeds, have high levels of the amino acids arginine, asparagine and glutamine and are therefore an important source of amino-nitrogen for the developing seedling. These proteins are nutritionally poor for humans and livestock as they are relatively depleted in several essential amino acids, including the sulfur-containing amino acids cysteine and methionine, as well as lysine and tryptophan. In contrast several seed proteins of the albumin class are especially rich in methionine; for example, the 2S albumins from Brazil nut (*Bertholletia excelsa*) and sunflower (*Helianthus*

annuus) seeds have 18% and 16% methionine, respectively. Because legume crops are widely used as animal feed, attempts have been made to increase their methionine content. To this end, 2S albumins from sunflower have been successfully introduced into narrow-leafed lupine (*Lupinus angustifolius*), a major grain legume. Transgenic lupines carrying the sunflower albumin protein contain higher methionine levels and are nutritionally superior to wild-type lupines when fed to test animals. A cautionary note, however, should be added regarding attempts to improve the nutritional quality of seeds by introducing genes that encode proteins of a desired amino acid content. Experiments to improve the nutritional qualities of soybean (*Glycine max*) by introducing a gene from Brazil nut that encodes a 2S albumin inadvertently introduced a strong allergen into transgenic plants. Indeed, it is well known that among the 2S albumins of seeds are inhibitors of the amylases and proteases that are necessary for the digestion of seed storage products as well as proteins that are highly allergenic. The antinutritional and allergenic properties of these albumins are usually retained when transgenes encoding them are introduced into other species.

Seed storage proteins of the legumin, vicilin and prolamin classes are localized in specialized **protein storage vacuoles**, also called **protein bodies**, and **aleurone grains** (Figures 6.16 and 6.17). All seed storage proteins are synthesized on rough endoplasmic reticulum (ER) but they can follow distinct paths from the ER to the vacuole. The globulins of dicot and monocot seeds are transported from the ER to vacuoles via dense vesicles produced by the Golgi apparatus (Figure 6.18). Proteins such as pumpkin seed albumin and the prolamins of maize and rice do not pass through the Golgi apparatus. In the case of pumpkin (*Cucurbita pepo*), albumin synthesized in the ER is transported directly to the vacuole by precursor accumulating vesicles (Figure 6.18). On the other hand, cereal prolamins accumulate in the lumen of the ER to form an ER-derived protein storage vacuole (Figure 6.17).

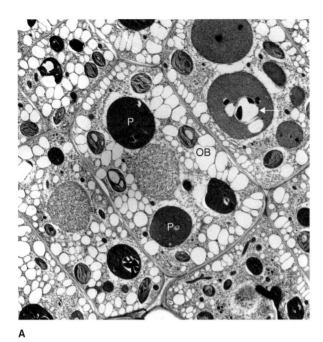

A

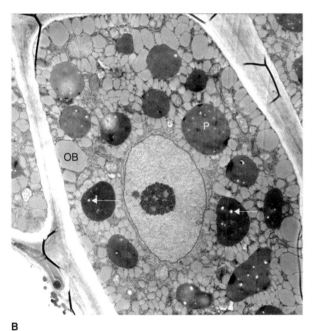

B

Figure 6.16   Protein storage vacuoles in *Arabidopsis* seeds. Electron micrographs of (A) cotyledon and (B) aleurone cells. Large, darkly-staining protein storage vacuoles (P) contain crystalline inclusions of phytin that are much larger in the cells of cotyledons. Crystalline phytin is often lost from the vacuoles during the preparation of cells for electron microscopy, leaving an empty space in the protein matrix of the vacuole (white arrows). Note also the abundance of oil bodies (OB) in both cell types and the much thicker cell walls in the aleurone.

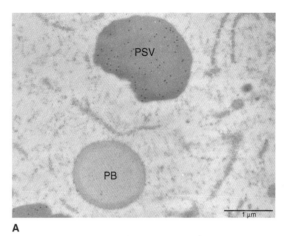

A

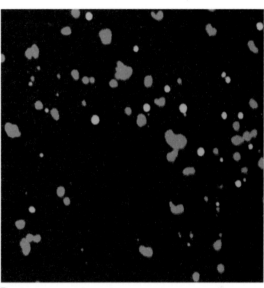

B

Figure 6.17   Protein bodies in rice. (A) Electron micrograph of the major rice storage protein bodies with immunocytochemical localization of glutelin. Gold particles (fine black dots) mark the location of glutelin in a protein storage vacuole (PSV). Below the PSV is a prolamin protein body (PB) within the rough endoplasmic reticulum. (B) Fluorescence micrograph of isolated rice protein bodies labeled and visualized using double indirect immunofluorescence techniques. The green-labeled protein bodies are prolamin protein bodies while the red ones are glutelin protein storage vacuoles.

6.3.6 Seed storage proteins may act as antinutrients

Certain seed storage proteins possess undesirable properties and can be the cause of many diseases. Among these proteins are lectins, enzyme inhibitors and thionins. Enzyme inhibitors fall into two general classes, the α-amylase inhibitors and the protease inhibitors.

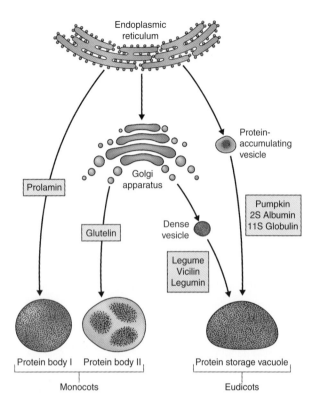

Figure 6.18   Diagram illustrating how storage proteins are deposited in endomembranes. Glutelins, legumin and vicilin are synthesized on the endoplasmic reticulum (ER), transferred to the Golgi apparatus and transported to protein storage vacuoles via vesicles produced by the Golgi. Storage proteins such as 11S globulin and the 2S albumin of pumpkin seeds are synthesized on the ER and transported directly to the protein storage vacuoles by precursor accumulating vesicles. Other proteins, such as cereal prolamins, accumulate in the lumen of the ER to form ER-derived protein bodies.

Seeds contain a wide range of **amylase inhibitors** that have specificity for α-amylases in a wide range of animals, from insects to humans. They are especially abundant proteins in grasses that belong to the tribe Triticeae, which includes barley, rye and wheat. Amylase inhibitors make up a large proportion (up to 80%) of the water-soluble albumin fraction of seed storage proteins. They function by binding near the active site of α-amylase, preventing hydrolysis of the glucosyl linkages in the amylose chain. It is thought that these proteins evolved as part of the plant's defense mechanism against insects. Certainly many seed amylase inhibitors show high specificity toward insect α-amylases, but cereal amylase inhibitors also inhibit cereal α-amylases as well as human salivary and pancreatic α-amylases.

These properties of amylase inhibitors have been exploited in several ways. Seeds have been engineered to synthesize amylase inhibitors as a means of reducing losses to insects, which can be very large in some parts of the world. These inhibitors have also been used as dietary

supplements in humans for weight reduction. They have been shown to effectively reduce starch digestion and thus lead to weight loss. Amylase inhibitors are therefore commonly used to reduce obesity. They can also be administered in order to decrease sugar production in people with diabetes.

In addition, **protease inhibitors** are found in seed storage proteins; they act to inhibit protein breakdown by binding to proteases. Seeds contain a spectrum of these inhibitors with a range of specificities and, like amylase inhibitors, they are thought to have evolved as part of the plant's defense mechanism. Various protease inhibitors have been expressed in new plant hosts as a means of elevating defense against pests, especially insects.

Lectins include a large group of carbohydrate-binding proteins. Like enzyme inhibitors, lectins can affect the properties of carbohydrates and glycosylated proteins. They may bind to enzymes, to structural proteins and to molecules that are taken up as food. Wheat germ agglutinin (WGA) is a widely studied lectin because it has diverse effects on human cells and tissues. WGA binds to surface glycans on the cells that border the digestive system and it interferes with uptake from the gut and alters the gut microflora. WGA is also very stable, surviving heat during cooking as well as passage through the digestive tract. WGA is a strong immunogen, and because it survives in the gut and can be taken up into the blood stream, it can elicit a wide range of deleterious effects.

Thionins are small basic proteins common in seeds but also found in vegetative tissues. Most thionins contain 45–50 amino acids and possess several conserved cysteine residues. The structure of γ-thionins shows that they resemble toxins produced by arthropods, e.g. insect defensins and scorpion toxin. Thionins are known to be toxic to animals as well as plant pathogens and their mode of action is varied; they can function as inhibitors of enzyme synthesis and activity and affect transport systems in cells.

6.3.7 Unlike most plant tissues, seeds often contain storage lipids

While most plant tissues store only carbohydrates, many seeds store lipids, in the form of **triacylglycerols** (see Figures 2.10A and 6.16, and Tables 6.3, 6.4 and 6.8). In *Macadamia* and pecan (*Carya illinoinensis*), oils can make up more than 70% of seed dry weight; seeds of many other species have oil contents in excess of 50%. Oils that are of commercial importance are derived from a relatively small number of crops including coconut (*Cocos nucifera*), maize (*Zea mays*), cotton (*Gossypium* sp.), flax (*Linum usitatissimum*), olive (*Olea europaea*), palm (*Elaeis* sp.), rapeseed/canola (*Brassica napus*), soybean (*Glycine max*) and sunflower (*Helianthus annuus*) (Table 6.8). Although oil makes up only 4.5% of maize kernel dry weight, it can comprise up to 50% of the weight of the embryo including the scutellum where most corn oil is stored. The embryo, called the germ, is removed from mature maize kernels during processing for animal feed and breakfast cereals. The germ, including the large scutellum, is pressed for oil production whereas the endosperm is used for feed, ethanol production and related uses. Because maize is such a widely grown crop in the USA, it is one of the major sources of vegetable oil.

Grains of rice, the dominant cereal grown in Asia, are also used as a source of oil. Almost all rice for human consumption is white rice, produced by **pearling** whole grains. Pearling is a process that removes the embryo, pericarp, testa and aleurone; you may be familiar with pearled barley, barley that is almost like white rice. The layers removed by pearling are often called **bran**. Bran can be pressed to produce rice bran oil that is thought to be nutritionally beneficial because it is rich in vitamin E. Rice bran oil also has a high smoking point making it excellent for frying food.

Because fatty acid molecules are more reduced than carbohydrates, lipid reserves yield more energy mole for mole than polysaccharides. Hydration of starch grains and fructans reduces their energy density, whereas triacylglycerols are largely hydrophobic and represent a more concentrated form of potential energy. As a consequence of their physicochemical properties, the

ATP yield from catabolism to carbon dioxide and water is approximately twice as high for triacylglycerols as for an equivalent mass of carbohydrates (Figure 6.19). Storing carbon and energy reserves in the form of lipid rather than carbohydrate may be advantageous where a compact seed mass is a requirement for ecological fitness. The triacylglycerol reserves of seeds and fruits typically contain C_{16} and C_{18} saturated and unsaturated fatty acids, though some oils, such as that of palm kernels, are rich in fatty acids with shorter chain lengths, and *Arabidopsis* seeds have high levels of monounsaturated C_{20} gondoic acid (see Table 17.1). The polyunsaturated fatty acids linoleic acid ($18:2^{\Delta 9,12}$) and α-linolenic acid ($18:3^{\Delta 9,12,15}$) are the major types found in plant membrane lipids (see Chapter 2).

6.3.8 The fatty acid content of seed oil is important for human uses

The fatty acid composition of seed oil is key to its end use. Table 6.9 shows the fatty acids commonly found in plants and Table 6.8 shows the distribution of these fatty acids in seeds. **Saturated fatty acids**, i.e. those without double bonds, are abundant in palm and coconut, where **palmitic acid** makes up 85–90% of the seed oil. Saturated fatty acids are also plentiful in Brazil nut, cotton seed and peanut, in which **stearic** acid comprises 17–24% of the fatty acid content of the seed oil. The most abundant **unsaturated fatty acids** in seeds are **linoleic** (18:2), **linolenic** (18:3) and **oleic** (18:1).

Nutritionists have identified a category of essential fatty acids that are required components of the human diet. Essential fatty acids, like essential amino acids, are those that cannot be synthesized by humans. They are called omega-3 and omega-6 fatty acids; among those made by plants are linoleic and α- and γ-linolenic acids (Table 6.10). The omega-3 and -6 terminology is based on the position of the double bond in the fatty acyl chain relative to the terminal carbon atom that is farthest away from the carboxyl end of the molecule. Omega-3 signifies that the third carbon atom from the terminal CH_3 group has a double bond; in omega-6 fatty acids a double bond is found on the sixth carbon atom. Plants are a good source of essential fatty acids in the human diet and α-linolenic and linoleic acids are examples of omega-3 and omega-6 fatty acids, respectively.

The fatty acid composition and degree of saturation in vegetable oils vary (see Table 6.8), and have been subject to chemical and biotechnological interventions on a large scale to tailor them to particular culinary and industrial uses. Catalytic hydrogenation reduces double bonds, increases the ratio of saturated to unsaturated fatty acids and raises the oil's melting point, but is implicated in the

Table 6.8 Oil content and fatty acid composition of seed oils. (Note that content and composition can vary widely with variety and growing conditions. Where sources vary, average values are given).

Seed	Oil content (%)	Fatty acids (% total oil)[a]				
		$18:3^{\Delta9,12,15}$ α-linolenic	$18:2^{\Delta9,12}$ linoleic	$18:1^{\Delta9}$ oleic	$18:0$ stearic	$16:0$ palmitic
Almond (*Prunus dulcis*)	54		17	78	5	
Arabidopsis thaliana[b]	40	18	29	16	4	9
Brazil nut (*Bertholletia excelsa*)	67		24	48	24	
Coconut (*Cocos nucifera*)	35		3	6		91
Maize (*Zea mays*)	4		59	24	17	
Cottonseed (*Gossypium* sp.)	40		50	21	25	
Flax (linseed; *Linum usitatissimum*)	35	58	14	19	4	5
Macadamia sp.	72		10	71	12	
Peanut (*A. hypogaea*)	48		29	47	18	
Pecan (*Carya illinoinensis*)	71		20	63	7	
Pistachio (*Pistacia vera*)	54		19	65	9	
Rape (canola; *Brassica napus*)	30	7	30	54	7	
Sesame (*Sesamum indicum*)	49		45	42	13	
Soybean (*Glycine max*)	18	7	50	26	6	9
Wheatgerm (*Triticum* spp.)	11	5	50	25	18	

[a] In the fatty acid name, e.g. $18:3^{\Delta9,12,15}$, 18 indicates its number of carbons, :3 is the number of double bonds, and $\Delta9,12,15$ indicates the positions of the double bonds relative to the terminal carboxyl group of the fatty acid. The structure of these fatty acids is shown in Table 6.9.
[b] *Arabidopsis* seeds also contain the unusual C_{20} monounsaturated fatty acid gondoic acid (22% of total oil).

ATP/g from sugar

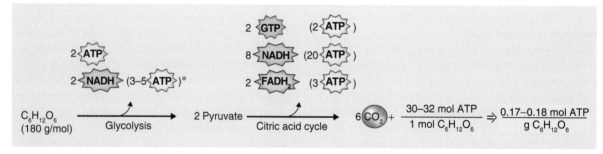

ATP/g from fatty acid

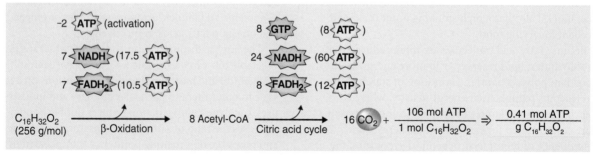

Figure 6.19   Comparison of ATP yields from the catabolism of sugar and fatty acid.

Table 6.9 Selected fatty acids present in plants.

Common name	Systematic name	Structure	Abbreviation[a]
Saturated fatty acids			
Lauric acid	n-Tetradecanoic acid	$CH_3(CH_2)_{10}COOH$	12 : 0
Palmitic acid [a]	n-Hexadecanoic acid	$CH_3(CH_2)_{12}CH_2CH_2COOH$	16 : 0
Stearic acid [a]	n-Octadecanoic acid	$CH_3(CH_2)_{12}CH_2CH_2CH_2CH_2COOH$	18 : 0
Arachidic acid	n-Eicosanoic acid	$CH_3(CH_2)_{12}CH_2CH_2CH_2CH_2CH_2CH_2COOH$	20 : 0
Behenic acid	n-Docosanoic acid	$CH_3(CH_2)_{12}CH_2CH_2CH_2CH_2CH_2CH_2CH_2CH_2COOH$	22 : 0
Lignoceric acid	n-Tetracosanoic acid	$CH_3(CH_2)_{12}CH_2CH_2CH_2CH_2CH_2CH_2CH_2CH_2CH_2CH_2COOH$	24 : 0
Unsaturated fatty acids			
Oleic acid [a]	cis-9-Octadecenoic acid	$CH_3(CH_2)_7CH{=}CH(CH_2)_7COOH$	$18:1^{\Delta 9}$
Petroselenic acid	cis-6-Octadecenoic acid	$CH_3(CH_2)_{10}CH{=}CH(CH_2)_4COOH$	$18:1^{\Delta 6}$
Linoleic acid [a]	cis,cis-9,12-Octadecatrienoic acid	$CH_3(CH_2)_4CH{=}CH{-}CH_2{-}CH{=}CH(CH_2)_7COOH$	$18:2^{\Delta 9,12}$
α-linolenic acid [a]	all-cis-9,12,15-Octadecatrienoic acid	$CH_3CH_2CH{=}CH{-}CH_2{-}CH{=}CH{-}CH_2{-}CH{=}CH(CH_2)_7COOH$	$18:3^{\Delta 9,12,15}$
γ-linolenic acid	all-cis-6,9,12-Octadecatrienoic acid	$CH_3(CH_2)_4CH{=}CH{-}CH_2{-}CH{=}CH{-}CH_2{-}CH{=}CH(CH_2)_4COOH$	$18:3^{\Delta 6,9,12}$
Roughanic acid	all-cis-7,10,13-Hexadecatrienoic acid	$CH_3CH_2CH{=}CH{-}CH_2{-}CH{=}CH{-}CH_2{-}CH{=}CH(CH_2)_5COOH$	$16:3^{\Delta 7,10,13}$
Erucic acid	cis-13-Eicosenoic acid	$CH_3(CH_2)_7CH{=}CH(CH_2)_{11}COOH$	$22:1^{\Delta 13}$
Some unusual fatty acids			
Ricinoleic acid	12-Hydroxyoctadeca-9-enoic acid	$CH_3(CH_2)_5{-}\overset{OH}{\underset{H}{C}}{-}CH_2{-}CH{=}CH(CH_2)_7COOH$	$12\text{-}OH\text{-}18:1^{\Delta 9}$
Vernolic acid	12,13-Epoxyoctadeca-9-enoic acid	$CH_3(CH_2)_4{-}\overset{O}{\overset{\diagup\!\!\diagdown}{CH{-}CH}}{-}CH_2{-}CH{=}CH(CH_2)_7COOH$	
Gondoic acid	cis-11-Eicosenoic acid	$CH_3(CH_2)_7CH{=}CH(CH_2)_9COOH$	$20:1^{\Delta 11}$

[a] Fatty acids commonly found in plant membrane lipids.

formation of so-called trans-fats which are under suspicion as unhealthy components of the human diet. In many species there is substantial natural variation in oil composition which can be exploited by plant breeders. For example, the oleic acid ($18:1^{\Delta 9}$) composition of canola oil may be anything from 43% to 78% of total fatty acids. Canola is a cultivar of *Brassica napus* (oilseed rape) and takes its name from Canadian oil low acid. Rapeseed oil naturally contains high levels of erucic acid ($22:1^{\Delta 13}$). Accumulation of this long-chain fatty acid has been reduced in canola by naturally-occurring knockout mutation of a gene encoding an elongase that specifically extends the chains

of C_{18} fatty acids. There are concerns that dietary erucic acid is harmful to heart tissue. Mutagenesis to increase the range of genetic variation in oil composition has been successfully applied to a number of species, including flax and sunflower. In recent years the modification of oil properties by recombinant DNA technology has been directed, with varying degrees of success, towards reducing relative levels of saturates, decreasing the tendency of plant fatty acids to become oxidized to potentially hazardous products during cooking or storage in air, increasing dietary antioxidants, reducing trans-acids and the production of essential long-chain fatty acids. The contribution of such approaches to

Table 6.10 Examples of omega-3 and omega-6[a] fatty acids.

Common name	Lipid name[b]	Systemic name
Omega-3 fatty acids		
Roughanic acid	$16:3^{\Delta7,10,13}$	all-cis-7,10,13-hexadecatrienoic acid
α-Linolenic acid (ALA)	$18:3^{\Delta9,12,15}$	all-cis-9,12,15-octadecatrienoic acid
Stearidonic acid (STD)	$18:4^{\Delta6,9,12,15}$	all-cis-6,9,12,15-octadecatetraenoic acid
Eicosatrienoic acid (ETE)	$20:3^{\Delta11,14,17}$	all-cis-11,14,17-eicosatrienoic acid
Eicosatetraenoic acid (ETA)	$20:4^{\Delta8,11,14,17}$	all-cis-8,11,14,17-eicosatetraenoic acid
Eicosapentaenoic acid (EPA)	$20:5^{\Delta5,8,11,14,17}$	all-cis-5,8,11,14,17-eicosapentaenoic acid
Docosapentaenoic acid (DPA), clupanodonic acid	$22:5^{\Delta7,10,13,16,19}$	all-cis-7,10,13,16,19-docosapentaenoic acid
Docosahexaenoic acid (DHA)	$22:6^{\Delta4,7,10,13,16,19}$	all-cis-4,7,10,13,16,19-docosahexaenoic acid
Tetracosapentaenoic acid	$24:5^{\Delta9,12,15,18,21}$	all-cis-9,12,15,18,21-docosahexaenoic acid
Tetracosahexaenoic acid (nisinic acid)	$24:6^{\Delta6,9,12,15,18,21}$	all-cis-6,9,12,15,18,21-tetracosenoic acid
Omega-6 fatty acids		
Linoleic acid	$18:2^{\Delta9,12}$	all-cis-9,12-octadecadienoic acid
γ-Linolenic acid	$18:3^{\Delta6,9,12}$	all-cis-6,9,12-octadecatrienoic acid
Eicosadienoic acid	$20:2^{\Delta11,14}$	all-cis-11,14-eicosadienoic acid
Dihomo-γ-linolenic acid	$20:3^{\Delta8,11,14}$	all-cis-8,11,14-eicosatrienoic acid
Arachidonic acid	$20:4^{\Delta5,8,11,14}$	all-cis-5,8,11,14-eicosatetraenoic acid
Docosadienoic acid	$22:2^{\Delta13,16}$	all-cis-13,16-docosadienoic acid
Adrenic acid	$22:4^{\Delta7,10,13,16}$	all-cis-7,10,13,16-docosatetraenoic acid
Docosapentaenoic acid	$22:5^{\Delta4,7,10,13,16}$	all-cis-4,7,10,13,16-docosapentaenoic acid

[a]The designations 3 and 6 indicate the position of the first double bond in the fatty acid relative to the terminal methyl group.
[b]In the lipid name, e.g. $16:3^{\Delta XX}$, 16 indicates its number of carbons, :3 is the number of double bonds, and $^{\Delta XX}$ indicates the positions of the double bonds counting from the carboxyl end of the molecule (see Table 6.9 for the structure of representative fatty acids).

human nutrition will depend on the safety and acceptability of foodstuffs from genetically modified crop sources. In addition to their importance as sources of feed and food, there are many industrial uses of seed oils and these include their use in the manufacture of cosmetics, detergents, lubricants and paints (Table 6.11).

6.3.9 Seeds store the bulk of mineral elements in a complexed form

Seeds store the bulk of mineral elements in a complex with **phytic acid**, myo-inositol hexaphosphoric acid (IP_6) (Figure 6.20). More than 90% of the phosphorus and major cations including K^+, Mg^{2+}, Ca^{2+}, Zn^{2+} and

Fe^{3+} in seeds are found as salts of phytic acid collectively called **phytate** (Table 6.12). Phytic acid is synthesized from glucose-6-phosphate; myo-inositol monophosphate synthase converts glucose-6-phosphate (G6P) to inositol-1-P and by successive phosphorylation reactions, with ATP as the phosphate (P_i) donor, to IP_3. Two inositol phosphate kinases convert IP_3 to IP_6. The subcellular location of IP_6 synthesis is not known although it is likely that it occurs in the cytosol. Phytate accumulates in protein storage vacuoles as a crystalline deposit known as a **globoid** surrounded by an envelope. Globoid crystals can vary greatly in size even in tissues of the same species. Figure 6.16 shows electron micrographs of the cotyledon and endosperm cells of *Arabidopsis* seeds. Both cell types contain protein storage vacuoles containing phytate but the globoid crystals in

Table 6.11 Some non-food uses of plant fatty acids.

Lipid type	Example	Major sources	Major uses	Approximate US market 1989 (10^6 \$)
Medium chain	Lauric acid (12:0)	Coconut (*Cocos nucifera*), palm kernel (*Elaeis* sp.)	Soaps, detergents, surfactants	350
Long chain	Erucic acid (22:1)	Rapeseed (*Brassica napus*)	Lubricants, slip agents	100
Epoxy	Vernolic acid	Epoxidized soybean oil (*Glycine max*), *Vernonia*	Plasticizers, coatings, paints	70
Hydroxy	Ricinoleic acid	Castor bean (*Ricinus communis*)	Coatings, lubricants, polymers	50
Trienoic	Linolenic acid (18:3)	Flax (*Linum*)	Paints, varnishes, coatings	45
Wax esters	Jojoba oil	Jojoba (*Simmondsia chinensis*)	Lubricants, cosmetics	10

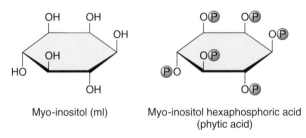

Myo-inositol (ml)   Myo-inositol hexaphosphoric acid (phytic acid)

Figure 6.20 Structural formulas of myo-inositol and its phytic acid derivative. Specific kinases use ATP to form the phosphoester bonds of phytic acid.

Key points Oils, polysaccharides and proteins make up the bulk of storage reserves in seeds. Mineral elements are generally stored as a chelated complex with inositol hexaphosphate in a compound called phytate. Starch grains stored in amyloplasts and cell wall carbohydrates make up the bulk of storage polysaccharides. In some seeds, such as date and fenugreek, cell wall polymers make up the bulk of storage polysaccharide. Seed storage proteins are complex and are classified according to their solubility in water, dilute salt or dilute alcohol solutions. They are found in protein storage vacuoles and protein bodies. Oils, as triglycerides, are stored in specialized oil bodies (oleosomes). They are also chemically diverse, varying both in the length of the fatty acid chain and in their degree of saturation.

cotyledonary cells are much larger than those found in the endosperm.

Recent molecular genetic experiments show that phytate is not required for the production of viable seed.

Arabidopsis double mutants that lack both of the inositol phosphate kinases that catalyze the last two steps in IP_6 synthesis, produce seeds that lack phytate and have soluble phosphate concentrations that are almost ten times higher than wild-type (Table 6.13). Seed yield is not affected by the absence of phytate and germination of phytate-free seeds is normal.

6.3.10 Phytate is another antinutrient in seeds

Phytate is a potent antinutrient because of its ability to chelate cations such as Ca^{2+}, Fe^{3+} and Zn^{2+}. Diets rich in seeds, such as those in countries where a mixture of beans and corn form the staple diet, may lead to defects in bone mineralization because Ca^{2+} is sequestered by phytate and is therefore not taken up by the digestive system. Other diseases that result from diets high in phytate are anemia from reduced iron availability, and hypogonadal dwarfism from zinc deficiency. Seeds also act as antinutrients in animal feed, especially for monogastric animals with single stomachs (as opposed to bovines and other ruminants with multiple stomachs) that lack the intestinal flora to hydrolyze phytate and mobilize nutrients. Birds, especially turkeys, that are fed on high cereal diets suffer from weakened limb bones. In pigs, much of the phytate that is ingested is lost via feces, leading to eutrophication of rivers and streams because of increased phosphorus levels. Treatment of seed-based livestock feed with the enzyme phytase can alleviate some of the problems associated with high phytate content, as can cooking methods that hydrolyze phytate. Another strategy being tested to reduce the adverse effects of phytate in human and livestock diets is to develop crop plants that are deficient in phytate; this approach has

Table 6.12 The content of the main inorganic elements of phytate in various species of seeds, expressed on a percent dry weight basis.

Species	Mg	Ca	K	P	Fe	Mn	Cu
Barley (*Hordeum* spp.)	0.16	0.03	0.56	0.43			
Oats (*Avena sativa*)	0.4	0.19	1.1	0.96	0.035	0.008	0.005
Soybean (*Glycine max*)	0.22	0.13	2.18	0.71			
	0.4	0.13	2.18	0.79	0.059	0.003	0.005
Sunflower (*Helianthus annuus*)	0.4	0.2	1	1.01			

Table 6.13 Inositol phosphate, free phosphate and seed yield analysis in *Arabidopsis*.

Genotype[a]	IP$_4$ (nmol mg^{-1})	IP$_5$ (nmol mg^{-1})	IP$_6$ (nmol mg^{-1})	P$_i$ (nmol mg^{-1})	Weight of 200 seeds (mg)
Wild-type	ND	ND	22.3 ± 0.4	13.7 ± 5.3	3.3 ± 0.1
atipk1-1	3.3 ± 0.3	29.7 ±	3.9 ± 0.9	23.9 ± 0.3	3.0 ± 0.4
atipk2β -1	2.5 ± 0.9	15.3 ± 1.4	14.5 ± 0.2	23.7 ± 0.9	3.3 ± 0.3
atipk1-1 atipk2β-1	<1	6.7 ± 1.1	<1	127.3 ± 6.0	3.1 ± 0.2

[a]atipk1-1 and atipk2β-1 are kinases required for the synthesis of IP6. *atipk1-1 atipk2β -1* is a double mutant. Data are the average ± SD if two or three independent samples were assayed in triplicate.
IP, inositol phosphate; ND, none detected; P$_i$, inorganic phosphate.

been successful in the model plant *Arabidopsis* (Table 6.13).

6.3.11 Seed maturation produces seeds that can survive for long periods

Mature seeds have a water content below 10%, a condition that is doubtless required for seeds to remain viable during extended periods of dormancy or quiescence. Seeds with higher water content are known to have a higher metabolic rate and will oxidize their food reserves more rapidly than do quiescent or dormant seeds. In addition to providing the reserves required to support growth of a new seedling, the polymerization of stored reserves is required to ensure water loss as the seed matures. Glucose is polymerized to starch where, as noted above, the degree of polymerization can easily exceed 10^5 glucose residues. The organization of starch into the complex starch grain ensures its insolubility. Similarly, incorporation of sugars into cell wall polymers ensures that the osmotic properties of the cytoplasm are unaffected by the accumulation of polysaccharide.

Storage proteins with a large molecular mass are packaged into protein storage vacuoles where they form deposits that are often soluble only in saline solution or aqueous alcohol. Triacylglycerols are neutral, water-insoluble oils that are partitioned from the cytoplasm by the half-unit membrane of the **oleosome** (**oil body**; see Chapter 4). The ultimate sequestration of mineral elements as crystalline phytate within vacuoles further ensures that the water potential of the cytosol will allow water to be easily lost from the seed by evaporation. In the next sections we will describe how the process of reserve deposition is reversed when a seed germinates and the seedling starts to grow. These processes will support survival of the seedling as a heterotrophic organism until its photosynthetic machinery is developed and becomes active.

6.4 Germination and early seedling growth

Germination is defined as the penetration of the seed coat by the elongating embryonic root (Figure 6.21). In addition to the pericarp, testa and endosperm as in the

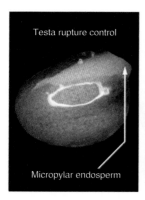 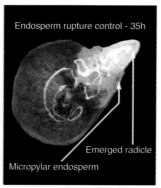

Figure 6.21 Radicle protrusion in *Lepidium* seeds. After imbibition of non-dormant seeds the testa ruptures followed by protrusion of the radicle through the enveloping endosperm.

example of *Lepidium* shown in this figure, the radicle of a cereal embryo must also penetrate the coleorhiza and often the hulls (palea and lemma) (see Figure 6.6). After germination, the embryo starts to grow and the seedling emerges from the seed coats. The elongation of the embryonic stem may take place in the **epicotyl** (above the cotyledons) or **hypocotyl** (below the cotyledons), depending on whether the cotyledon(s) of the germinating seedling remain below ground or above ground (Figure 6.22). In **hypogeal** seedlings such as garden pea (*Pisum sativum*) the cotyledon(s) remain below ground after germination and the first true leaves of the seedling are borne on an epicotyl. In peas, the epicotyl forms a hook that pulls the shoot apical bud upward, while in cereal seedlings, the coleoptile protects the plumule of the embryo as it is pushed upward through the soil. In **epigeal** seedlings such as garden bean (*Phaseolus vulgaris*), castor bean (*Ricinus communis*) and onion (*Allium cepa*) the cotyledons are propelled upwards into the air by growth of the hypocotyl (Figure 6.22). In onion and castor bean seedlings, the cotyledons serve not only as a source of stored food, but also as the first photosynthetic organs of the seedling.

6.4.1 Imbibition of water is necessary for seed germination

As noted above, mature seeds have a low water content, often below 10% of seed dry weight. The water content of seeds will vary with the water content of the atmosphere in which they are stored as well as with the type of reserves stored within the seed. In general, oil seeds have a lower water content (about 5%) than starchy seeds (about 10%) stored at the same relative humidity. When seeds are placed in water, they rapidly

rehydrate, a process known as **imbibition**. Imbibition takes place in both living and dead seeds; it is a physical process resulting from hydration of seed storage polymers. A **micropyle** is present in the testa of most seeds (see Figure 6.1A); this may be a thinner area of tissue or a microscopic hole in the testa. It is a remnant of the gap between the integuments of the ovule where the pollen tube entered the ovule (see Figure 1.18). Hydration of the seed often begins by water uptake through the micropyle. Uniform water uptake by seeds, especially under conditions of limited water availability, is aided by the presence of mucilage layers on the surface of seed coats (see Figure 6.4). The mucilage hydrates very quickly, often within seconds of exposure to water. Very large forces, up to 10–20 MPa (equal to 100–200 atmospheres of pressure, sufficient to break rocks!) can be generated by imbibitional water uptake. Splitting of the testa and pericarp often occurs during imbibition, but this may not result in germination.

6.4.2 Dormant seeds do not germinate after imbibition

Viable seeds that do not germinate under normally favorable conditions of water and oxygen availability are said to be in a state of **dormancy**. Like other dormant structures—resting buds, for example—dormant seeds require additional stimuli to commence and sustain growth. Different types of dormancy in different organs have distinctive features but also share some common physiological and regulatory mechanisms; these aspects are described in Chapter 17. Here we focus on the major ways in which dormancy inhibits seed germination.

Dormancy is a complex characteristic. Seeds that have **primary dormancy** are dormant when they are shed from the mature plant, but seeds can acquire **secondary dormancy** in response to environmental conditions that they experience after shedding. Secondary dormancy is most commonly imposed on non-dormant seeds by extremes of temperature but the presence or absence of light may also induce secondary dormancy. Primary dormancy ensures that seeds do not germinate prematurely while they are still attached to the parent plant. Secondary dormancy delays germination until environmental conditions favor successful growth of the emerging seedling.

Dormancy may be a property of the seed coats or of the embryo. **Seed coat-enhanced dormancy** is found in a wide variety of species including dormant varieties of *Arabidopsis*, barley, lettuce, rice and wild oats. A basic characteristic of seed coat dormancy is that when the embryo from a dormant seed is separated from the

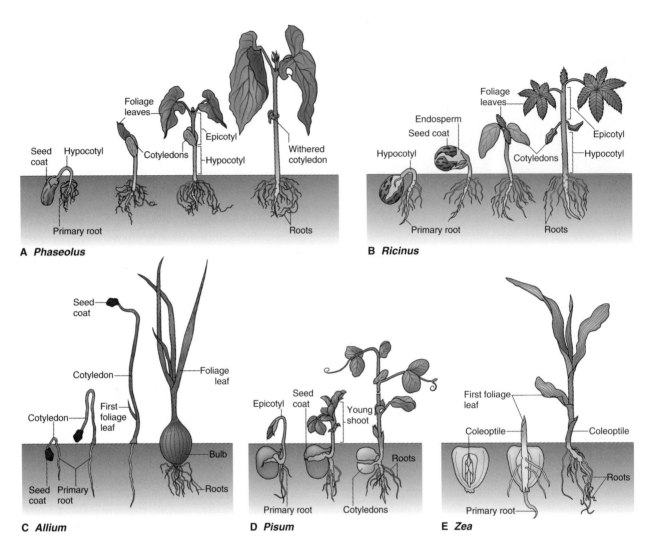

Figure 6.22 Patterns of seedling development in selected monocots and eudicots. Germination in (A) garden bean (*Phaseolus vulgaris*), (B) castor bean (*Ricinus communis*) and (C) onion (*Allium cepa*) is epigeal; the cotyledons of these seedlings are propelled above ground by growth of the hypocotyl. Germination in (D) pea (*Pisum sativum*) and (E) maize (*Zea mays*) is hypogeal; the cotyledons of these seedlings remain below ground and the elongating epicotyl (mesocotyl in cereals) carries the shoot apex (plumule) above ground. These seedlings illustrate different strategies that have evolved to protect the plumule as it moves through soil. In eudicots (A, B, D) the hypocotyl or epicotyl is hooked and in some monocots, such as onion (C), the cotyledon is also hooked. In the cereals, such as maize (E), the plumule is protected by a tubular coleoptile. In many species (e.g. corn, pea, garden bean) the cotyledons function only as storage organs, but in some (e.g. castor bean, onion) they are also the first photosynthetic organs of the seedling.

enveloping seed coats and is incubated in water, it is able to grow and develop into a seedling (Figure 6.23). The growth of embryos removed from dormant seeds is indistinguishable from that of embryos in non-dormant seeds. As discussed further in Chapter 17 (see Sections 17.2.1 and 17.4.1), mechanisms proposed to explain seed coat dormancy include: mechanical restraint of the embryo; impermeability of seed coats to the uptake of water or gases; the coat acting as a barrier preventing inhibitory compounds such as the plant hormone

abscisic acid (**ABA**) from being removed by leaching; or a combination of these factors.

Embryos that do not grow when seed coats have been removed are said to show **embryo dormancy**. An example of embryo dormancy is found in freshly harvested onion (*Allium*) seeds that must be stored for several months before they will germinate. Less is known about why such embryos do not grow but investigators suspect that metabolic or developmental deficiencies in the embryo play a role (see Section 17.4.2).

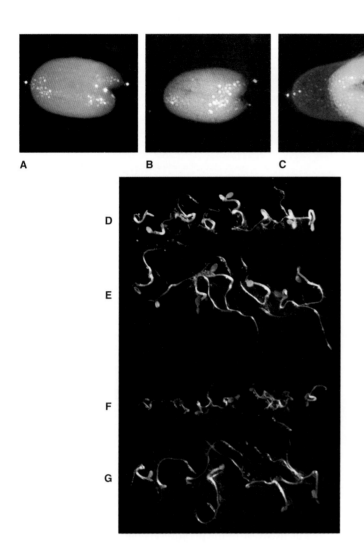

A   B   C

D

E

F

G

Figure 6.23 Embryos from dormant *Arabidopsis* seeds germinate after removal of the testa and aleurone. Top: (A, B) dormant seeds with the testa removed and an intact aleurone layer do not germinate after 3 days or 28 days, respectively. (C) If the testa is removed and the aleurone layer disrupted, embryos germinate in 3 days. Bottom: germination of (D) non-dormant Columbia ecotype (Co), (E) moderately dormant C24, and highly dormant (F) Cvi and (G) Kas2 *Arabidopsis* seeds after removal of the testa and aleurone. Embryos were incubated in water and photographs taken after 3 days. Note that intact seeds of the Cvi and Kas2 ecotypes will remain dormant for several months when sown on water.

6.4.3 Environmental signals may trigger the breaking of dormancy

Seeds that are dormant eventually germinate, and some of the signals that bring about germination are now well understood. Dormant seeds normally germinate in response to environmental signals—usually temperature or light—or as a result of prolonged storage. As Table 6.14 shows, seeds often respond to more than one germination-triggering signal. Grains of many cereals are among the seeds that are dormant when freshly harvested but germinate after storage under dry conditions, a process called **after-ripening**. A requirement for after-ripening in cereals means that freshly harvested grain cannot be sown immediately but must be stored under suitable conditions, often for several months, before dormancy is completely broken.

Many seeds lose dormancy after storage under moist, cool conditions, a treatment called **stratification** (Table 6.14). Most seeds that respond to chilling are from species that grow in temperate regions; this suggests that stratification may have adaptive advantages. Because seeds shed in late summer or early fall must receive an over-wintering period so that dormancy can be lost, these seeds germinate only in the following spring or summer when conditions are more favorable for growth. Dormancy of vegetative resting buds and vascular cambium is also attuned to winter conditions, as discussed further in Chapter 17 (see Sections 17.4.2 and 17.4.5).

6.4.4 Light can be an important trigger for germination

Light is clearly an important trigger for germination; light of specific wavelengths, intensity or duration may

Table 6.14 Termination of dormancy by various factors.

Species	After-ripening	Chilling (stratification)	Light
Colonial bent grass (*Agrostis tenuis*)	+		+
Wild oat (*Avena fatua*)	+	+	
Barley (*Hordeum* spp.)	+	+	
Annual bluegrass (*Poa annua*)		+	+
Wheat (*Triticum aestivum*)	+	+	
Sycamore (*Acer pseudoplatanus*)	+	+	
White birch (*Betula pubescens*)	+	+	+
Lamb's quarters (*Chenopodium album*)	+		+
Hazel nut (*Corylus avellana*)	+	+	
Lettuce (*Lactuca sativa*; some cultivars)	+	+	+
Tobacco (*Nicotiana tabacum*)			+
Plum (*Prunus domestica*)	+	+	
Apple (*Malus domestica*)	+	+	(+)?
Scotch pine (*Pinus sylvestris*)		+	+

+, effective in dormancy breaking.

Table 6.15 Illumination conditions required for the breaking of dormancy.

Illumination conditions	Examples
Seconds or minutes	Colonial bentgrass (*Agrostis tenuis*) Lamb's quarters (*Chenopodium album*) Lettuce (*Lactuca sativa* cv. Grand Rapids) Tobacco (*Nicotiana tabacum*)
Several hours	Pig nut (*Hyptis suaveolens*) Purple loosestrife (*Lythrum salicaria*)
Days	*Epilobium cephalostigma* *Kalanchoe blossfeldiana*
Long days	*Begonia evansiana* White birch (*Betula pubescens*) (at 15 °C) Jerusalem oak (*Chenopodium botrys*) (at 30 °C)
Short days	Jerusalem oak (>30 °C) Eastern hemlock (*Tsuga canadensis*) White birch (>15 °C)

be the effective stimulus (Table 6.15). For example, brief pulses of red or blue light can trigger germination, as can long periods of light or light given as a series of either short days or long days, the equivalent of winter and summer day lengths, respectively.

The discovery that red light at around 660 nm could break dormancy in lettuce (*Lactuca sativa*) seeds led to a landmark discovery in the 1950s. Photodormant lettuce seeds were shown to germinate after exposure to a brief pulse of red light, an experiment that eventually led to the discovery of the photoreversible pigment, **phytochrome** (see Chapter 8). A key experiment showed that a brief pulse of far-red (FR) light (approximately 710 nm) could reverse the promotive effects of a brief pulse of red (R) light (about 660 nm) on germination (Figure 6.24). Indeed, it was shown that when alternating pulses of R and FR light were given to dormant lettuce, seeds remained dormant or germinated depending on the wavelength of light that was given last (Table 6.16). These experiments were interpreted as showing that a photoreversible pigment was present in lettuce seeds and that the form of the pigment present in seeds after R light treatment favored germination. It was further hypothesized that exposure of this form of the pigment to FR light caused its photoconversion to a form that would block germination. We now know this pigment to be phytochrome (P). The relationship between the P_R and P_{FR} forms of phytochrome that exist after FR and R light exposure is shown in the simplified diagram in Figure 6.25. You will learn more about phytochrome and its role in plant growth and development in Chapter 8 and about the adaptive and ecological significance of dormancy photoregulation in Chapter 17 (see Section 17.4.3 and 17.4.5).

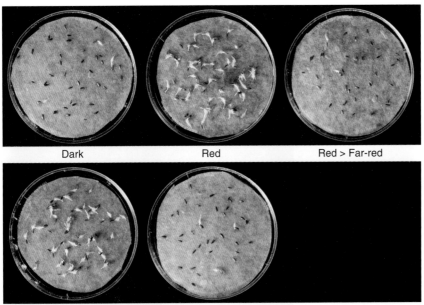

Red > Far-red > Red   Red > Far-red > Red > Far-red

Figure 6.24 Photodormant seeds of many species, including lettuce shown here, require a light trigger for germination. The photoreceptor for this response is phytochrome. Lettuce seed germination is stimulated by red light, but not by far-red light. As can be seen in this experiment, when seeds are exposed to red light followed by far-red light, the stimulatory effect of the red light is cancelled and the seeds remain dormant. If seeds are given alternating exposures to red and far-red light, their ability to germinate depends on the last light treatment received.

Table 6.16 Phytochrome photoreversibility and the breaking of dormancy.[a]

Irradiation sequence	Germinated (%)
None (darkness)	4
R	98
FR	3
R, FR	2
R, FR, R	97
R, FR, R, FR	0
R, FR, R, FR, R	95

[a] Seeds of the Grand Rapids cultivar of lettuce were imbibed in darkness and then exposed to red light (640–680 nm) (R) for 1.5 min and far-red light (>710 nm) (FR) for 4 minutes in the sequence shown. After irradiation, they were returned to darkness for 24 hours before germinated seeds were counted.

Figure 6.25 Diagram illustrating the photoreversibility of phytochrome from the red-light-absorbing form (P_R) to the far-red-light-absorbing form (P_{FR}). P_{FR} can also undergo enzyme-catalyzed reversion in darkness as indicated by the dotted line.

6.4.5 Plant hormones play important roles in the maintenance and breaking of seed dormancy

Although a range of environmental signals can trigger germination of dormant seeds, it is likely that the underlying molecular mechanisms for breaking dormancy are similar. Plant growth hormones have been implicated in the regulation of dormancy. Genetic and physiological experiments point to important roles for

ABA and gibberellins (GAs). Mutants that overproduce GAs, or that have a constitutive GA response in the absence of elevated GA levels, lack dormancy. Similar behavior is shown by mutants that lack ABA, have reduced ABA levels or have a compromised response to ABA. A useful rule-of-thumb is that in most seeds ABA imposes dormancy and a reduction in its concentration and an increase in GA concentration lead to breaking of dormancy.

In barley, the ABA concentration is high in dormant and non-dormant dry grain but following imbibition its level declines much more dramatically in embryos of non-dormant grain than in dormant grain (Figure 6.26A). GA concentration in barley grains also changes during the breaking of dormancy, increasing rapidly in non-dormant embryos while remaining low in dormant embryos (Figure 6.26B). Taken together, the reciprocal changes in ABA and GA concentrations in dormant and non-dormant embryos result in the breaking of dormancy. In *Arabidopsis* seeds, red light acting via phytochrome can stimulate the breaking of dormancy. In this case, light triggers the expression of

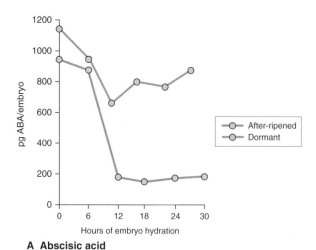

A Abscisic acid

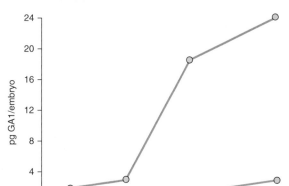

B Gibberellin

Figure 6.26   Changes in (A) abscisic acid (ABA) and (B) gibberellin (GA) concentrations in dormant and non-dormant (after-ripened) barley embryos. Embryos were dissected from dormant and after-ripened seeds after various periods of imbibition on water and the levels of ABA and GA measured.

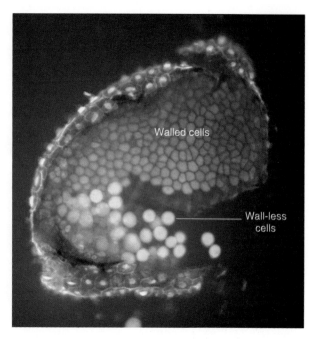

Figure 6.27   *Arabidopsis* aleurone cell walls can undergo dramatic wall weakening. Cells of aleurone dissected from seeds of the *Spy-1* mutant of *Arabidopsis* become rounded as a result of the loss of the rigid cell wall. Cell wall dissolution in the *Spy-1* mutant is extreme, causing cells to separate from one another and become rounded. *Spy-1* mutant seeds are not dormant and germinate very readily on water.

have a constitutive GA response show extreme cell wall weakening in the endosperm adjacent to the radicle (Figure 6.27).

Most general models of the molecular regulation of dormancy include ABA–GA interaction as a central feature. This is discussed further in Section 17.4.5.

GA biosynthetic genes and this in turn leads to an increase in bioactive GAs in the seed.

How are changes in the concentration of ABA and GA translated into radicle emergence in dormant seeds? There is widespread agreement that seed coat dormancy is lost when the coat can no longer restrain growth of the embryo axis. In dormant seeds of *Arabidopsis*, lettuce (*Lactuca*), peppergrass (*Lepidium*) and tobacco (*Nicotiana*), the seed coat and especially the endosperm do not yield in response to pressure from the growing embryo and they thereby prevent radicle emergence. When the seeds of these species lose dormancy, the radicle is able to penetrate the weakened walls of the endosperm. In *Arabidopsis* and *Lactuca*, hormonal treatments affect the strength of endosperm cell walls. In GA-treated seeds, endosperm cells have reduced cell wall thickness while added ABA prevents the weakening of these cell walls. Furthermore, *Arabidopsis* mutants that

Key points Not all seeds germinate under favorable conditions of moisture and temperature. Seeds that fail to germinate are dormant. Dormancy is an important property because it prevents precocious germination of seeds under conditions that may not support seedling growth. Many environmental signals and endogenously produced compounds break dormancy. Among the environmental triggers are alterations in temperature, especially cold periods, the presence or absence of light, and concentration of soil anions, in particular nitrate and nitrite. The hormones abscisic acid (ABA), gibberellins (GAs) and nitric oxide (NO) all play a role in regulating dormancy. Genetic and biochemical evidence point to a key role of ABA in establishing dormancy, whereas GA and NO are known to break dormancy.

6.5 Mobilization of stored reserves to support seedling growth

Germination usually takes place underground. The growing seedling must depend on food stored in the seed to support its growth until it reaches light and can become photosynthetic. As we have seen, seeds contain reserves of protein, carbohydrate, lipid and minerals. The growing seedling will use the minerals as a source of elements, such as P, K, Mg and Fe, and the organic molecules as sources of energy and carbon skeletons, all of which are needed to make new macromolecules. Before the growing seedling can use its stored reserves, they must be broken down into smaller subunits that can be transported throughout the developing seedling. We will look at how each of the major storage molecules is mobilized and used by the growing seedling.

6.5.1 Mobilization of protein involves the enzymatic breakdown of proteins to amino acids

All seeds store some protein. During early seedling development, these proteins are broken down to yield amino acids that are required for growth. The amino acids formed during proteolysis may be reused directly for new protein synthesis or, following loss of the amino group, they may be oxidized to release energy. Hydrolysis of stored proteins to their constituent amino acids requires the action of a variety of proteases. Some of these catalyze the total hydrolysis of protein to amino acids whereas others break down proteins to small polypeptides that must be degraded further by peptidases. A large number of proteolytic enzymes have been identified and characterized in seeds. All of these appear to be members of the metallo-, cysteine and serine protease families.

The proteases that are active during seed germination can be grouped by the way in which they hydrolyze (break) peptide bonds of their substrates (Figure 6.28):

- **Endopeptidases**: cleave internal peptide bonds to yield smaller peptides.
- **Aminopeptidases**: sequentially cleave the terminal amino acids from the free amino acid end of the polypeptide chain.
- **Carboxypeptidases**: resemble aminopeptidases except amino acids are cleaved from the carboxyl end of the chain.
- **Peptidases**: degrade small peptides to amino acids.

Recently, it has been shown that thioredoxin, a regulatory disulfide protein, facilitates the mobilization of both protein and carbohydrate reserves in developing seedlings. In the cereal endosperm, thioredoxin is reduced by NADPH and the flavin enzyme NADP-thioredoxin reductase (NTR) (Equation 6.1).

Equation 6.1 Role of NTR in proteolysis

6.1A

$$\text{NADPH} + \underset{\text{(S–S)}}{\text{thioredoxin}_{ox}} \rightarrow \underset{\text{(–SH HS–)}}{\text{thioredoxin}_{red}} + \text{NADP}^+$$

6.1B

$$\underset{\text{(–SH HS–)}}{\text{Thioredoxin}_{red}} + \underset{\text{(S–S)}}{\text{protein}_{ox}} \rightarrow \underset{\text{(–SH HS–)}}{\text{protein}_{red}} + \underset{\text{(S–S)}}{\text{thioredoxin}_{ox}}$$

Reduced thioredoxin then reduces disulfide (S–S) groups of a number of target seed proteins. For example, when storage proteins of the seed are reduced by thioredoxin, their solubility increases and they are more susceptible to proteolysis. In addition, thioredoxin increases the activity of selected enzymes, either directly by reducing them or indirectly by inactivating disulfide inhibitor proteins. One enzyme that is inhibited by a disulfide protein is starch debranching enzyme, which is involved in starch breakdown (see Equation 6.3). Thioredoxin has been shown to affect enzyme activity in cereal grains and legume seeds.

Transgenic experiments in which expression of thioredoxin was altered have confirmed the central role it plays in germination. Cereal grains that overexpress thioredoxin were found to germinate more rapidly than control seeds, and grains in which the expression of thioredoxin was downregulated germinated more slowly. Development of transgenic wheat cultivars with reduced thioredoxin levels may lead to the solution of a long-standing agricultural problem. In very wet seasons, wheat grains may begin to germinate while still attached to the parent plant, a condition known as preharvest sprouting. Wheat cultivars with lowered thioredoxin levels show reduced preharvest sprouting (Figure 6.29).

Some of the proteases that mobilize storage proteins are synthesized de novo but there is evidence that others may already be present in the seed in an inactive form. In cereals, for example, the living aleurone layer that surrounds the starchy endosperm synthesizes and secretes a wide range of acid hydrolases, i.e. hydrolytic

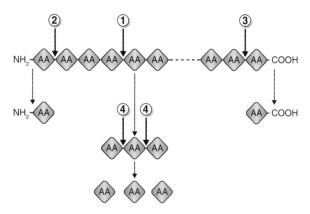

Figure 6.28   The differences in the specificity of different proteases ensure that stored proteins are completely degraded; endopeptidases (1), aminopeptidases (2), carboxypeptidases (3) and peptidases (4) cleave proteins at specific points in the polypeptide chain.

Key points   The mobilization of stored reserves in seeds is brought about by enzymes. In the dead starchy endosperm of cereals, starch and proteins are accessible to enzymes synthesized and secreted from the aleurone layer. In storage cotyledons of eudicots, reserves are located in living cells, proteins in protein storage vacuoles (PSVs) and starch in amyloplasts. The synthesis of enzymes that break down storage protein in eudicot cotyledons occurs at the ER. These enzymes are then delivered to vacuoles by Golgi-derived vesicles. A similar process occurs for protein degradation in the PSVs of cereal aleurone. Synthesis of starch-degrading enzymes occurs on free ribosomes; these enzymes are taken up into the amyloplast via specific transporters in the outer and inner plastid membranes.

enzymes that have pH optima between 4 and 5. Among these acid hydrolases are a number of proteases that break down storage proteins in the endosperm. These secreted enzymes complement those present in the starchy endosperm of dry grain.

6.5.2 Stored protein mobilization in eudicots takes place in living cells

The digestion of proteins present in the cotyledons and endosperm of eudicot seeds is different from that in the cereal endosperm. Cells in these tissues are alive and their stored protein reserves are localized in protein storage vacuoles. There is evidence that both newly synthesized and pre-existing proteases break down protein reserves in vacuoles. One hypothesis suggests that a newly synthesized protease is transported to the vacuole where it initiates a cascade of proteolytic activities that bring about the breakdown of stored proteins. According to this hypothesis, the role of the newly synthesized protease is to activate preformed but inactive proteases in the lumen of the protein storage vacuole. The pH of the protein storage vacuole may also play an important role in regulating the degradation of storage proteins. In *Arabidopsis* a vacuolar processing enzyme (VPE) has been identified that participates in the mobilization of storage proteins by initiating a proteolytic cascade. VPE is synthesized in the ER as an inactive precursor; it is self-activated at an acidic pH, presumably in the vacuole. VPE then activates other proteases. Because proteases that degrade storage protein have acidic pH optima, it is easy to see how acidification of the vacuole lumen can govern the rate at which these proteases break down their substrates. The protein mobilization process in seed

Figure 6.29   Transgenic wheat (T) in which the thioredoxin gene is downregulated shows suppressed preharvest sprouting compared to non-transformed control plants (C). Preharvest sprouting causes grain to germinate in the ear and is a major cause of crop loss in cool temperate regions, for example in some regions of China 10–20% of the wheat crop is lost due to preharvest sprouting.

germination has many points of similarity with the proteolysis that occurs in senescence and cell death, as discussed further in Chapter 18.

Key points Proteases that hydrolyze storage proteins to peptides and amino acids fall into four broad categories, depending on how the enzyme attacks the polypeptide chain. Aminopeptidases digest from the amino terminal end of the polypeptide, carboxypeptidases from the carboxyl terminus, and endopeptidases at random points between the amino and carboxyl ends. Peptidases cleave the small peptides produced by the action of endopeptidases to amino acids. Amino acids formed as a result of proteolysis of stored protein can be used in the synthesis of new enzymes (e.g. the α-amylase produced by the cereal aleurone) or transported to the embryo to support new protein synthesis in the developing seedling.

6.5.3 Mobilization of stored starch may be catalyzed by phosphorolytic enzymes

Starch is the major carbohydrate reserve in most plants. It consists of two molecules, amylose and amylopectin (Section 6.3.1). Starch may be cleaved into smaller polymers or monomers by either of two classes of enzymes: **phosphorolytic enzymes** that use P_i to break glycosidic (glucose to glucose) bonds and **hydrolytic enzymes** that use water to break these bonds (Figure 6.30).

At least three enzymes contribute to **phosphorolytic starch degradation: starch phosphorylase, debranching enzyme** and **glucosyltransferase** (Equations 6.2–6.4).

Equation 6.2 Starch phosphorylase

$$\alpha\text{-Glucan}_{(n)} + P_i \rightarrow \alpha\text{-glucan}_{(n-1)} + \text{glucose-1-phosphate}$$

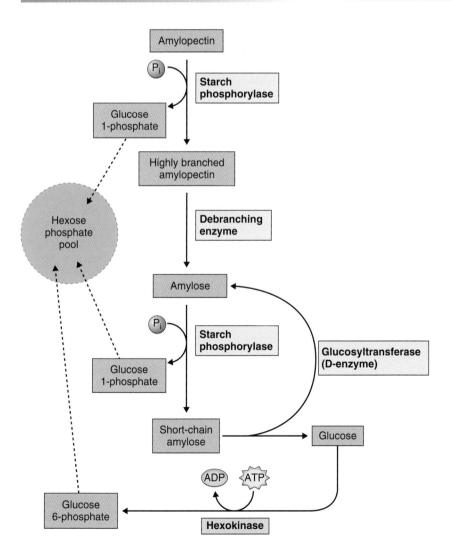

Figure 6.30 Phosphorolytic starch degradation. Starch phosphorylase initiates the breakdown of starch (amylopectin), but complete break-down requires the subsequent participation of debranching enzyme and glucosyltransferase (D-enzyme).

Equation 6.3 Debranching enzyme (R-enzyme)

$$\text{Branched } \alpha(1 \rightarrow 6)\text{-}\alpha\text{-}(1 \rightarrow 4)\text{-glucan}$$

$$\rightarrow \text{linear } \alpha\text{-}1,4\text{-glucans}$$

Equation 6.4 Glucosyltransferase (D-enzyme)

$$\alpha\text{-Glucan}_{(m)} + \alpha\text{-glucan}_{(n)} \rightarrow \alpha\text{-glucan}_{(m+n-1)} + \text{glucose}$$

Starch phosphorylase cleaves individual glucose residues from the non-reducing end of the starch molecule, generating glucose-1-phosphate. Starch phosphorylase can attack only bonds located at least four glucose residues from a branch point. Continued phosphorolytic attack on the branched starch molecule is made possible by a debranching enzyme (also called **pullulanase** or **R-enzyme**) that cleaves the $\alpha(1 \rightarrow 6)$ bonds, releasing linear chains on which the starch phosphorylase can act. In addition, the **D-enzyme**, or glucosyltransferase, can condense short glucan polymers to produce new substrate for the phosphorylase (Figure 6.30).

Although starch phosphorylase plays a central role in carbohydrate metabolism, how the activity of this enzyme is regulated in plants has not been discovered. One hypothesis is that it may be regulated by the availability of inorganic phosphate, which fluctuates in response to changes in growth conditions including different stresses. This pattern of regulation of plant starch phosphorylase is different from that of its glycogen phosphorylase counterpart in animals. The latter enzyme is regulated by a sophisticated protein phosphorylation/dephosphorylation network.

6.5.4 Amylases also play a role in starch breakdown

Starch can also be cleaved by a group of hydrolytic enzymes known as **amylases** (Equations 6.5–6.7) that break down linear $\alpha(1 \rightarrow 4)$-glucans. α-Amylase catalyzes the cleavage of internal glucosyl bonds, giving rise to short glucans called **dextrins**. The smallest unit formed by α-amylase is the disaccharide maltose. β-Amylase hydrolyzes starch by cleaving maltose residues from the non-reducing end of the starch molecule. The maltose and short glucan molecules that are the products of amylase digestion are further degraded to glucose molecules by the action of α-**glucosidase**. Complete digestion of amylopectin requires the action of

debranching enzyme (Equation 6.3) as well as the amylases and glucosidase (Figure 6.31).

Equation 6.5 α-Amylase

$$1, 4\text{-}\alpha\text{-D-glucan}_{(n)} \rightarrow 1, 4\text{-}\alpha\text{-D-glucan}_{(x)}$$

$$+ 1, 4\text{-}\alpha\text{-D-glucan}_{(y)}$$

$$(n \geq 3; x + y = n)$$

Equation 6.6 β-Amylase

$$1, 4\text{-}\alpha\text{-D-glucan}_{(n)} \rightarrow 1, 4\text{-}\alpha\text{-D-glucan}_{(n-2)} + \text{maltose}$$

Equation 6.7 α-Glucosidase

$$1, 4\text{-}\alpha\text{-D-glucose}_{(n)} \rightarrow 1, 4\text{-}\alpha\text{-D-glucose}_{(n-1)} + \text{D-glucose}$$

The enzymes involved in hydrolytic starch cleavage are particularly active in seed germination. During germination of cereals, there is rapid breakdown of starch in the endosperm. The most studied of the starch-hydrolyzing enzymes is α-amylase, which is synthesized in the scutellum and the aleurone layer and secreted into the starchy endosperm (Figure 6.32). The de novo synthesis of α-amylase by the aleurone is promoted by GA and repressed by ABA. All secretory proteins are synthesized on the ER and transported to the cell exterior by vesicles produced by the Golgi apparatus. Like the proteases secreted by the aleurone layer and scutellum, α-amylases have an acidic pH optimum and show maximum activity at around pH 4.

β-Amylase of cereal grains is not synthesized de novo by the aleurone layer. In this case, a precursor of the enzyme is present in bound form in the starchy endosperm before germination. It is released during germination and is activated either by removal of a small peptide from the carboxyl terminus of the enzyme or by the action of sulfhydryl agents such as thioredoxin (see Section 6.5.1). Some α-glucosidase is present in the dry seed, but during germination large amounts of this enzyme are synthesized by the aleurone in response to increased GA levels. Seed-localized debranching enzyme and α-amylase activities are further regulated by small, specific disulfide proteins that act as inhibitors. The hydrolytic enzymes associated with starch breakdown in storage tissues such as cotyledons also appear to play a role in the mobilization of starch stored temporarily in chloroplasts in foliage leaves. In this case hydrolytic enzymes are synthesized on free ribosomes and targeted to the chloroplast.

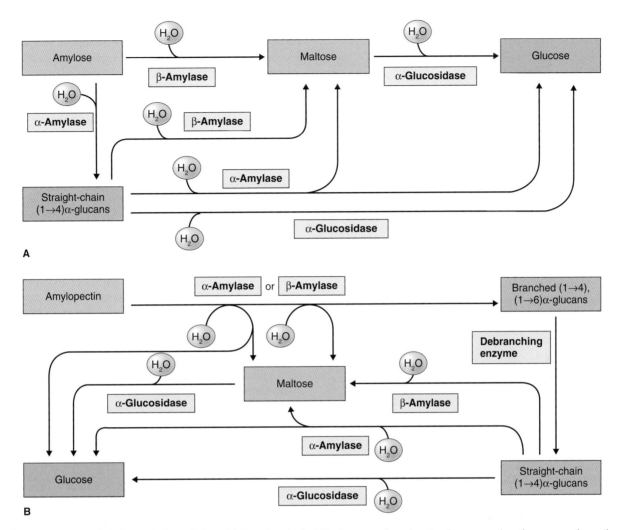

Figure 6.31 Hydrolytic starch degradation. (A) Complete hydrolytic cleavage of amylose to glucose requires the cooperative action of several hydrolytic enzymes. (B) In the case of amylopectin, debranching enzyme is also required to hydrolyze $\alpha(1 \to 6)$ bonds.

6.5.5 Cell walls are another source of carbohydrates

Cell walls are a major source of stored polysaccharide in many seeds, and enzymes secreted by living cells also mobilize these. In the case of eudicot seeds that have endosperm, cell wall-digesting enzymes—such as mannanases that hydrolyze mannans and galactomannans—are synthesized by endosperm cells and secreted into the cell wall. These enzymes break down the endosperm cell walls. In the case of seeds such as those of *Arabidopsis*, *Lactuca* and *Lepidium*, the weakening of the cell walls by digestive enzymes breaks dormancy and allows the radicle to emerge from the seed. The aleurone and scutellum of cereals also secrete glucanases that digest the walls of starchy endosperm cells. In eudicots, the cell walls of storage cotyledons are also extensively degraded. It is thought that the cell wall-degrading enzymes in cotyledons are synthesized de novo in the cytosol and secreted into the cell wall.

6.5.6 Mobilization of stored lipids involves breakdown of triacylglycerols

During germination and early seedling growth, triacylglycerols, stored in seeds, are degraded and converted into glucose and a wide variety of other essential metabolites. Within the cells of storage tissue, triacylglycerols may be converted to sucrose for transport to other tissues in the growing seedling. In seeds, the mobilization of triacylglycerols involves three organelles: (i) the oil body (oleosome), where stored triacylglycerols are converted to glycerol and fatty acids; (ii) the specialized **peroxisome**, often called a **glyoxysome**, where fatty acids are converted by **β-oxidation** to

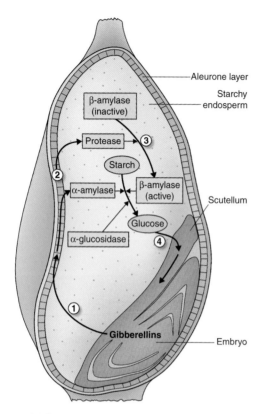

Figure 6.32   Role of gibberellins in mobilizing carbohydrate reserves of germinating cereal grains. Gibberellins released by the embryo (1) promote the synthesis of α-amylase and proteases (2). Proteases activate β-amylase (3) which together with α-amylase convert starch to the disaccharide maltose, which is converted to glucose by α-glucosidase. The scutellum then takes up glucose (4) for transport to the embryo axis.

Key points   Storage polysaccharides are broken down by a wide range of carbohydrases, enzymes that cleave glycosyl bonds in carbohydrates. Starch is broken down by a family of enzymes that include debranching enzyme, α- and β-amylases, and α-glucosidase. Debranching enzyme attacks the β(1 → 6) linkages in amylopectin producing linear chains of β(1 → 4)-linked amylose. Amylose is attacked by α-amylase at random points producing short oligomeric dextrins. β-Amylase attacks amylose chains and dextrins from the non-reducing end to form maltose, a disaccharide. α-Glucosidase breaks down maltose to glucose. Cell wall polysaccharides are broken down by specific carbohydrases, for example xylans are hydrolyzed by xylanases, β-glucans by β-glucanases, etc.

acetyl-coenzyme A (CoA) and then to succinate by the **glyoxylate cycle**; and (iii) the **mitochondrion**, where succinate is either respired or converted to malate which is transported to the cytosol for final conversion to hexoses by the reactions of **gluconeogenesis**. Here we will discuss the breakdown of triacylglycerols to glycerol and fatty acids and of fatty acids to acetyl-CoA. Subsequent conversion of acetyl-CoA to hexoses will be addressed in Chapter 7.

Triacylglycerols are stored in unusual organelles called oil bodies (see Chapter 4). Because of their insolubility in water, triacylglycerols must be hydrolyzed to **fatty acids** and **glycerol** by lipases before they are available for metabolism (Figure 6.33). The glycerol released in the lipase reaction is converted in the cytosol to triose phosphates, which, in turn, are used to synthesize sucrose via gluconeogenesis (see Chapter 7). In contrast, the fatty acids are transferred to **peroxisomes** for the next stage in the mobilization process. The mechanism by which this transfer is accomplished is not known although oil bodies are usually found adjacent to peroxisomes (see Figure 6.33B).

In plants, the oxidation of fatty acids takes place in peroxisomes of germinating seeds and leaves (Figure 6.33). This is in contrast to animal cells in which the major site of fatty oxidation is the mitochondrial matrix. Fatty acids are broken down to two carbon units (C2) during a cyclic process called **β-oxidation** because the number 2 (or β) carbon of the fatty acid is the carbon that undergoes oxidation (Figure 6.34). Fatty acids are first activated by the addition of coenzyme A using an enzyme called **fatty acid-CoA synthase**. They are then oxidized in a four-step process shown in Figure 6.34. The energy released in the first oxidative step of fatty acid breakdown is dissipated as heat, because **acyl-CoA oxidase** transfers electrons directly to oxygen, producing H_2O_2. The hydrogen peroxide produced is a strong and potentially damaging oxidant; it is immediately cleaved by **catalase** to water and oxygen (Equation 6.8).

Equation 6.8 Catalase

$$2H_2O_2 \rightarrow 2H_2O + O_2$$

In contrast, animal cell β-oxidation occurs in mitochondria and the electrons removed in the first oxidation step pass through the respiratory chain to oxygen, a process that is accompanied by the capture of energy by means of ATP synthesis (see Chapter 7). As we will see in Chapter 7, reactions similar to those catalyzed by **enoyl-CoA hydratase** and **β-hydroxyacyl-CoA dehydrogenase** in β-oxidation are found in the citric

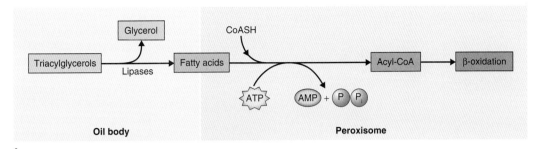

A

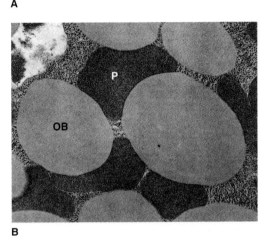

B

Figure 6.33   The breakdown of stored triacylglycerols in oil-storing seeds takes place in the oil body and peroxisome. (A) In the oil body, triacylglycerols are degraded by lipase to glycerol and fatty acids. Fatty acids are transported to the peroxisome where they are further degraded to acetyl-CoA by the β-oxidation pathway. (B) An electron micrograph showing a peroxisome (P) and adjacent oil bodies (OB).

acid cycle phase of aerobic respiration (e.g. fumarase and malate dehydrogenase) in mitochondria.

6.5.7 Stored minerals are mobilized by breaking down phytic acid

As discussed in Section 6.3.9, mineral elements are stored in seeds in a complex with phytic acid, myo-inositol hexaphosphoric acid (IP_6). During germination IP_6 is hydrolyzed by the enzyme **phytase**, a phosphatase that releases phosphate and chelated cations. Phytase levels increase following seed germination making phosphate and other minerals available to support embryo growth and development. The hydrolysis of IP_6 leads to the formation of a series of inositol phosphates ranging from inositol-1-phosphate to IP_6. It is well known that certain inositol phosphates, especially inositol-1,4,5-trisphosphate, are important regulatory molecules in eukaryotic cells. There is as yet no evidence that IP_6

hydrolysis products play a regulatory role in seeds; indeed *Arabidopsis* mutants lacking phytate germinate normally. The presence of phytate makes the seed and young seedling independent of an external supply of essential minerals for at least the first several days after germination.

We have seen how protein, starch and lipid reserves are mobilized and converted to, respectively, amino acids, simple sugars and acetyl-CoA. These compounds (or their derivatives) are transported from storage tissues to growing tissues. In the cereal grain the scutellum plays a prominent role in the transfer of solutes from the endosperm to the growing shoot and root systems. These solutes may be used by the embryo either as a source of carbon skeletons or as a source of energy to power metabolism. Mobilization of lipid reserves is more complex since acetyl-CoA and glycerol must be converted to sucrose or succinate for transport and further utilization. The processes involved in this conversion, including the tricarboxylic acid and glyoxylate cycles and gluconeogenesis, are discussed in detail in Chapter 7.

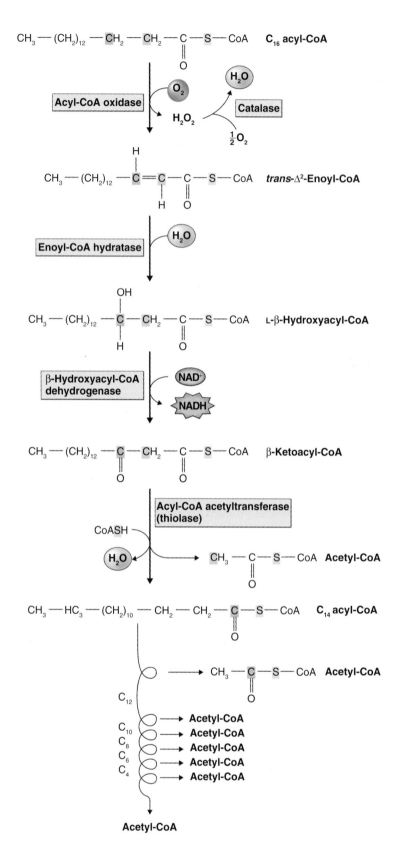

Figure 6.34 β-Oxidation pathway in peroxisomes. Fatty acids are activated by the addition of coenzyme A (CoA) to form fatty acyl-CoA molecules. The fatty acyl-CoA is broken down to two carbon acetyl units in a cyclic process. In each pass through the sequence, one acetyl residue is removed in the form of acetyl-CoA from the carboxyl end of an acyl-CoA. Seven passes through the cycle are required to oxidize a C_{16} fatty acid to eight molecules of acetyl-CoA.

Key points Triglycerides are broken down to glycerol and fatty acyl chains in the oil body by the action of lipase. Glycerol can be further metabolized by glycolysis. Fatty acids are converted to acetyl-CoA by the reactions of β-oxidation and then oxidized in a four-step series of reactions in the peroxisome to form the two carbon molecule acetyl-CoA. The first step in this process is catalyzed by acyl-CoA oxidase, an enzyme that yields H_2O_2 and heat but no usable energy. The H_2O_2 is broken down to H_2O and O_2 by the enzyme catalase that is abundant in all peroxisomes. The acetyl-CoA produced in the peroxisome is further metabolized in mitochondria and cytoplasm to eventually yield sugars, reactions called gluconeogenesis. Minerals such as K, Mg and Fe are stored in the vacuoles of seed tissues as insoluble salts of phytic acid. During germination the enzyme phytase hydrolyzes phytate and the resulting phosphate salts ionize. releasing minerals for transport to growing seedling tissues.

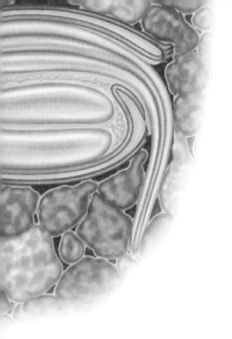

Chapter 7

Metabolism of reserves: respiration and gluconeogenesis

7.1 Introduction to catabolism and anabolism

All living tissues of the plant need to obtain **energy** to sustain vital processes, and **carbon skeletons** to build the organic molecules required for cell structure and function. The main starting point is sugars, the direct or indirect products of photosynthesis. In Chapter 6 we followed the early stages of seed germination and saw how seed reserves are mobilized to produce smaller molecules. Focusing on the metabolic pathways by which cells of the seedling use glucose obtained from the break-down of stored starch, we now describe the generation of energy and creation of raw materials to make the new macromolecules necessary for growth, adaptation and reproduction. The hundreds of different enzyme-catalyzed reactions that participate in this process are remarkably similar in different forms of life. Another source of carbon skeletons during germination is reserve lipid; fats (acylglycerols) stored in the seed are broken down to produce fatty acids. In this chapter we will examine the pathways by which cells can use fatty acids to synthesize carbohydrates. Aspects of carbon and energy metabolism in the seedling establishment phase

of development recur during senescence and are described in Chapter 18.

Metabolism comprises two types of pathway: one engaged in **anabolism**, the energy-requiring activities that biosynthesize cell components from smaller precursors; and the other in **catabolism**, the phase of intermediary metabolism concerned with the degradation of nutrients and complex molecules to release energy, or to salvage materials, or both (Figure 7.1). Because it is quite usual for synthesis and breakdown of a given metabolite to occur simultaneously, net accumulation or decrease of such a metabolic product will reflect the difference between how fast it is made and its rate of degradation. The amount of a metabolite in the cell at a particular time is its **pool size** and the rate at which molecules join and leave the pool is the **turnover rate** or **flux**. A pool that remains constant in size is said to be at the **steady state**. In a phase of rapid growth and development, such as seedling establishment, anabolic processes building the structural and functional machineries of new tissues and organs predominate over catabolic activities. The reverse is true of phases such as senescence or mobilization of reserves in seed storage tissues. Regulation of the anabolism–catabolism balance, exercised through developmental and environmental

The Molecular Life of Plants, First Edition. Russell Jones, Helen Ougham, Howard Thomas and Susan Waaland.
© 2013 John Wiley & Sons, Ltd. Published 2013 by John Wiley & Sons, Ltd.

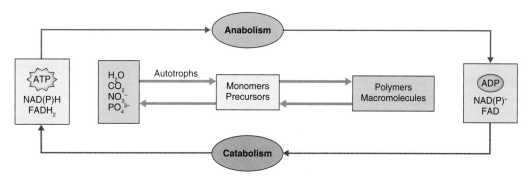

Figure 7.1   Relationship between anabolism and catabolism. Anabolic pathways convert small precursor molecules into cell macromolecules that provide chemical energy when processed by catabolic pathways. The overall biological cycle is driven by the assimilation of inorganic molecules by autotrophic organisms.

control of the expression of genes encoding biosynthetic and degradative activities, is a central feature of progression through the life cycles of plants and their parts. High-energy metabolites, coenzymes, oxidation and reduction reactions, and enzymes—the leading players in the catabolic processes taking place in seedlings and adult plants—are described in Chapter 2.

Key points   Respiratory breakdown of sugars provides energy and carbon skeletons for biological structures and functions. Anabolism, the biosynthetic reactions that build cell components from smaller precursors, uses the energy provided by the breakdown processes collectively called catabolism. Anabolism and catabolism contribute to metabolic turnover. During metabolism, molecules enter and exit the pools of biochemical intermediates, which are in a state of continuous flux. At the steady state, the pool size of a metabolite is constant. Anabolism is dominant over catabolism during growth and development. The reverse is true for tissues undergoing senescence or mobilization of reserves.

7.2 Anaerobic phase of carbohydrate breakdown

One of the central features of catabolic metabolism in plants and other organisms is the breakdown of energy-rich molecules such as glucose to yield energy. In this process, called **cellular respiration**, glucose is

oxidized to carbon dioxide and water. The energy released during this oxidation ($\Delta G^{\circ\prime} - 2870$ kJ mol^{-1}) drives the synthesis of ATP (Equation 7.1) which is used primarily for biosynthetic processes, but also for powering ion pumps such as the plasma membrane H$^+$-ATPase and for movement of organelles and cytosol by the cytoskeleton. The pathways of cellular respiration also serve as a source of the **precursors** of proteins, carbohydrates, nucleic acids and lipids that are needed for growth and for the replacement of molecules as they turn over in fully developed tissues. Cellular respiration takes place in all tissues of a plant both in the light and in the dark. It has been estimated that 30–60% of the sugar produced in photosynthesis is used in cellular respiration.

Equation 7.1 Overview of aerobic respiration

$$C_6H_{12}O_6 + 6O_2 + 32-34(ADP + P_i)$$
$$\rightarrow 6CO_2 + 6H_2O + 32-34ATP$$

Cellular respiration consists of three major pathways: **glycolysis**, the **tricarboxylic acid (TCA) cycle** and **electron transport/oxidative phosphorylation**. Each of these pathways takes place in a different subcellular compartment. The enzymes of glycolysis are located in the cytosol. The TCA cycle uses enzymes of both the matrix and cristae of mitochondria, and electron transport/oxidative phosphorylation take place on the inner mitochondrial membrane (Figure 7.2). The initial steps in the breakdown of carbohydrates in aerobic cells do not require the presence of oxygen (**anaerobic phase**). In the second stage (described in Sections 7.3 and 7.4), the organic acid products of glucose catabolism in the anaerobic phase are oxidized fully to carbon dioxide and water. Energy is released, albeit in different amounts, at each of these stages. We will follow a molecule of glucose through these pathways, keeping in

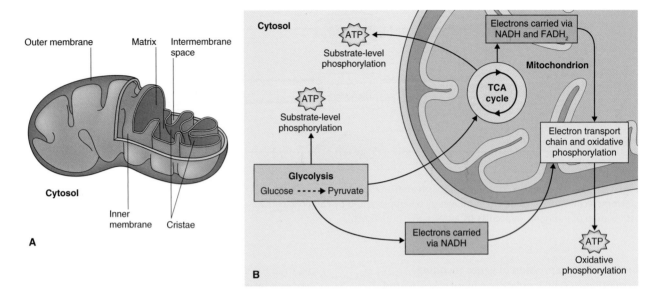

Figure 7.2   Mitochondrion structure and metabolism. (A) Structure of a mitochondrion showing membranes, matrix and intermembrane space. The outer membrane contains channels (porins) to allow metabolites and ions to traverse the barrier. The inner membrane contains numerous enzymes, the respiratory complex and specific transporters for metabolites. (B) Function of the mitochondrion in respiration, showing the TCA cycle accepting pyruvate from glycolysis in the cytosol and the membrane-associated electron transport chain producing ATP by oxidative phosphorylation.

mind that some intermediates in these reaction sequences may be diverted to provide carbon skeletons for other cellular compounds. We also describe two other reaction sequences important for carbon and energy metabolism under some circumstances: the **oxidative pentose phosphate pathway** (see Section 7.5) and the **glyoxylate cycle** (see Section 7.6).

7.2.1 Glycolysis converts glucose to pyruvate

Glycolysis brings about the conversion of one six-carbon (glucose) to two three-carbon (pyruvate) molecules in reactions that are independent of oxygen. Glycolysis is a series of ten enzyme-catalyzed reactions that take place in the cytosol. The names of the glycolytic enzymes and intermediates are given in Figure 7.3. During glycolysis, two molecules of ATP are invested per molecule of glucose, two molecules of NAD^+ are reduced to NADH, and four molecules of ATP are produced (Equation 7.2). The overall free energy change for glycolysis is -95.5 kJ mol^{-1}.

Equation 7.2 Overview of glycolysis

Glucose + 2ATP + 2NAD$^+$ + 4ADP + 4P$_i$ →

   2 pyruvate + 2NADH + 2H$^+$ + 2ADP + 4ATP + 2H$_2$O

Glycolysis can be divided into two phases: an **investment or preparatory** phase and a **payoff phase** (Figure 7.3). During the preparatory phase, two molecules of ATP are used to phosphorylate glucose, first to glucose-6-phosphate and then to fructose-1, 6-bisphosphate (Figure 7.3A). Fructose-1,6-bisphosphate is then cleaved to form two interconvertible triose phosphate molecules, glyceraldehyde-3-phosphate and dihydroxyacetone phosphate. Notice how the two halves of the fructose-1,6-bisphosphate molecule are related, carbon-by-carbon and substituent-by-substituent group, to the structures of the two triose phosphate cleavage products. This reaction and its reverse are key steps not just in glycolysis but also in anabolism, where they link the supply of three-carbon compounds, such as those produced during gluconeogenesis (see Section 7.6) or photosynthetic carbon fixation, to the formation of hexoses.

The **payoff phase** of glycolysis (Figure 7.3B) begins when glyceraldehyde-3-phosphate undergoes an oxidation reaction in which NAD^+ is reduced (to give NADH) and P$_i$ is directly incorporated into glyceraldehyde-3-phosphate (**substrate-level phosphorylation**) to give 1,3-bisphosphoglycerate. The phosphate of glyceraldehyde-3-phosphate is transferred to ADP via substrate phosphorylation to produce ATP, and 1,3-bisphosphoglycerate is converted to 3-phosphoglycerate.

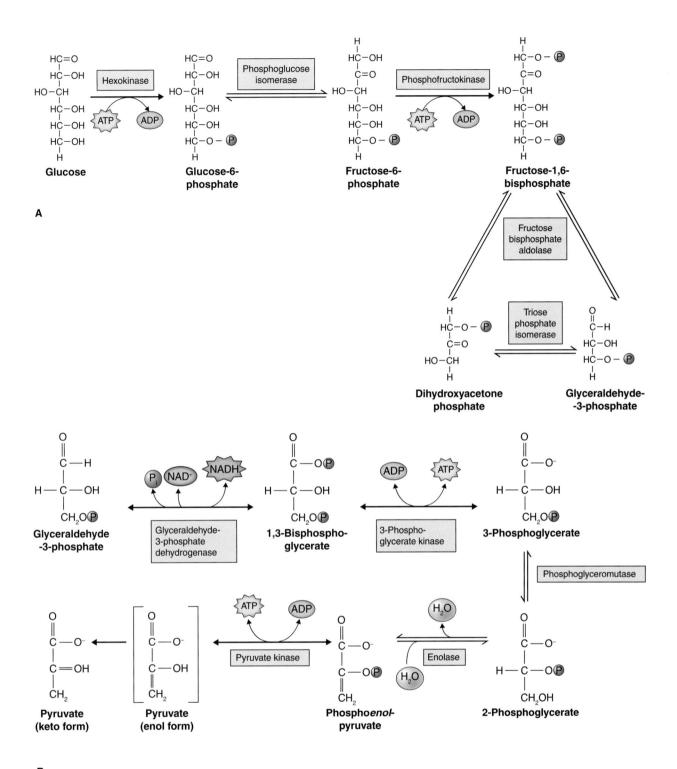

Figure 7.3 Glycolysis. For each molecule of glucose that is processed through the preparatory phase (A), two molecules of glyceraldehyde-3-phosphate are formed; both pass through the payoff phase (B). Pyruvate is the end-product of the second phase of glycolysis.

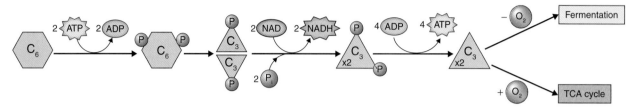

Figure 7.4   Stoichiometry of glycolysis. For each glucose molecule, two ATPs are consumed in the preparatory phase and four ATPs plus two NADHs are produced in the payoff phase, giving a net yield of two ATPs per molecule of glucose converted to pyruvate.

Next, conversion of the 3-phosphoglycerate to 2-phosphoglycerate is followed by formation of phosphoenolpyruvate (PEP) and, in the final glycolytic step, PEP and ADP are converted to pyruvate plus ATP (Figure 7.3B). Note that Mg^{2+} participates in each of the reactions that involve transfer of a phosphate group. The substrates used in these reactions are complexes between Mg^{2+} and the individual phosphorylated intermediates.

For every glucose molecule that begins the investment phase of glycolysis, two glyceraldehyde-3-phosphate molecules enter the payoff phase, and for every glyceraldehyde-3-phosphate molecule that passes through the payoff phase of glycolysis, two molecules of ATP and one NADH are produced (Figures 7.3 and 7.4). Thus, the two ATPs that were invested in the preparatory phase are returned, and an additional two ATPs plus two NADHs are formed as net products of glycolysis.

When oxygen is present, the pyruvate, produced by glycolysis, enters the mitochondrion where, as described in Section 7.3, it is further oxidized via the TCA cycle, and the energy released is used to drive the synthesis of ATP during oxidative phosphorylation. In addition, NADH is oxidized to NAD^+ within the mitochondrion, where its electrons are transferred to the electron transport chain. The NAD^+ produced is shuttled back to the cytosol and reused during glycolysis.

7.2.2 Alcoholic fermentation allows glycolysis to continue in the absence of oxygen

In the absence of O_2, for example in the cells of roots in flooded soil, further oxidation of pyruvate within the mitochondrion stops. If NADH were to remain reduced, NAD^+ would no longer be available to be used in the preparatory phase of glycolysis. Under such anaerobic conditions, the NADH generated in glycolysis may be reoxidized during the process of **alcoholic fermentation** (Figure 7.5). Here, pyruvate is decarboxylated to yield

acetaldehyde and CO_2 and acetaldehyde is then reduced to yield ethanol and NAD^+, recycling the NADH generated in glycolysis. In this way, the cell can continue to harvest at least a portion of the energy available in the glucose molecule, even under anaerobic conditions. Not only does glycolysis continue in the absence of O_2, its rate often increases, a response known as the **Pasteur effect** after its discovery by the illustrious French microbiologist Louis Pasteur. Through elevated levels of glycolytic intermediates and increased expression of genes encoding glycolytic and fermentative enzymes, the plant is able to compensate for the low energetic efficiency of fermentation. Since ethanol can be toxic to cells, alcoholic fermentation is usually limited to situations such as flooding, where alcohol can diffuse away from the plant.

Key points   Cellular respiration is the breakdown of glucose to CO_2 and H_2O, with the release of energy in the form of ATP. Glycolysis is a respiratory pathway located in the cytosol. The breakdown of carbohydrates by glycolysis in aerobic cells does not require the presence of O_2. Glycolysis converts each C_6 molecule of glucose into two C_3 pyruvate molecules in a series of ten enzyme reactions. Two molecules of ATP are utilized in the investment phase of glycolysis and four produced in the payoff phase, a net gain of two ATP. In addition two molecules of NAD^+ are reduced to NADH. Under aerobic conditions, pyruvate is further metabolized by the TCA cycle in the mitochondrion. In the absence of O_2, pyruvate is converted to acetaldehyde and then to ethanol by the process of alcoholic fermentation. This oxidizes the NADH generated in glycolysis, recycling the NAD^+. Stimulation of glycolysis by anaerobic conditions is a response known as the Pasteur effect. Fermentation has been exploited by humans from the earliest times for baking and brewing.

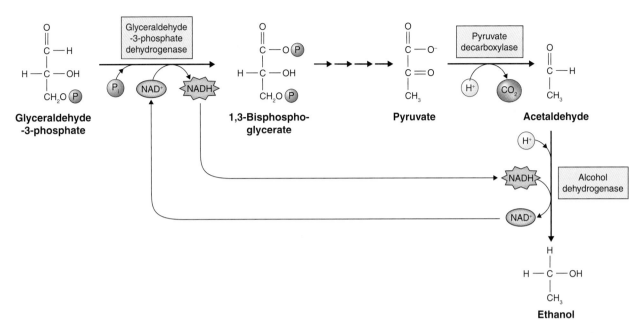

Figure 7.5 Fermentation, the anaerobic conversion of pyruvate to ethanol. The NADH generated during glycolysis is used to reduce the product of pyruvate decarboxylation, acetaldehyde, to ethanol, enabling NAD^+ to be recycled.

Fermentation has been exploited for countless centuries by humans, in activities ranging from the baking of bread to the production of beer and wine (see Figure 6.7). More recently, fermentation is being used to supply ethanol as a replacement for gasoline in transport fuels. In each of these cases, the process is carried out by yeast (or related microorganisms) that employ the same glycolytic enzymes discussed above for plants. While efforts to improve fermentation biotechnology for the food and beverage industries have been underway for many years, the eventual need to replace petroleum as a source of fuel has prompted a renaissance in this field of research. The substrate for much of the current production of biofuel is starch derived from corn, a source of both food and feed. Intensive research is now focused on improving the production of ethanol from cellulose—a major plant product that cannot be used as a human food source. It is predicted that the production of biofuels will emerge as a major industry worldwide in the years to come. As such, it will have an impact on both agriculture and industry.

7.3 The tricarboxylic acid cycle

The remainder of the reactions of cellular respiration take place within the mitochondrion (see Figure 7.2A). With the exception of one membrane-bound component, enzymes of the TCA cycle (also called the Krebs or citric acid cycle) are soluble and located in the mitochondrial matrix, whereas electron transport/oxidative phosphorylation take place completely on the inner mitochondrial membrane. Substrates for these reactions, including pyruvate, NADH, ADP and P_i, must cross both the outer and inner mitochondrial membranes. The outer mitochondrial membrane is permeable to certain smaller molecules (of molecular mass up to 1 kDa), but many others are transported by a **porin**, a protein that acts as a transport channel. Passage across the inner mitochondrial membrane requires substrate-specific transporters. Paramount among these are transporters for pyruvate, ADP/ATP, P_i and citrate and malate (Figure 7.6).

7.3.1 Pyruvate is converted to acetyl-CoA in preparation for entry to the TCA cycle

During the second stage of cellular respiration, pyruvate from glycolysis is oxidized to CO_2 within the mitochondrial matrix; its electrons are transferred to the electron carriers NAD^+ and FAD (see Figures 2.19 and 2.20). The first step in this stage is the **preparatory reaction** that converts pyruvate into a form that can enter the TCA cycle. This reaction is catalyzed by the **pyruvate dehydrogenase enzyme complex**, composed of three separate enzymes and bound cofactors: flavin adenine dinucleotide (FAD), thiamine pyrophosphate and lipoic acid (see Figures 2.20 and 2.21). The

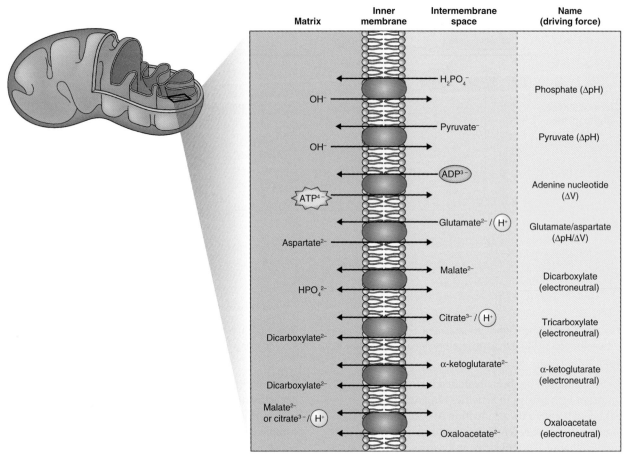

Figure 7.6   Metabolite carriers of the inner mitochondrial membrane.

substrates for this reaction are pyruvate, NAD^+ and coenzyme A (CoASH; see Figure 2.21). Pyruvate is decarboxylated, releasing CO_2, and then oxidized and NAD^+ is reduced. The resulting C_2 compound, acetate, is attached to CoASH to form **acetyl-CoA** (Equation 7.3). The bond between acetate and CoA is unstable and reactive, conserving some of the energy in the original pyruvate molecule. Acetyl-CoA is now ready to enter the TCA cycle, the enzymes and intermediates of which are given in Figure 7.7.

Equation 7.3 Pyruvate dehydrogenase complex

$$Pyruvate + CoASH + NAD^+ \rightarrow acetyl\text{-}CoA + NADH + CO_2$$

7.3.2 The TCA cycle completes the breakdown of pyruvate to carbon dioxide and reduced electron carriers

As the TCA cycle begins, acetyl-CoA and oxaloacetate, a four-carbon organic acid, condense to form the C_6 acid,

citrate, and free CoASH (Figure 7.7). Citrate is then isomerized to isocitrate. The next two steps of the cycle involve oxidative decarboxylations that each produce CO_2 and results in transfer of electrons to NAD^+. First, isocitrate is decarboxylated to form CO_2, NADH and the five-carbon organic acid α-ketoglutarate. Then the α-ketoglutarate product is oxidized in a reaction catalyzed by the **α-ketoglutarate dehydrogenase enzyme complex** to produce succinyl-CoA, CO_2 and NADH. The structure of this enzyme complex is similar to that of the pyruvate dehydrogenase complex that converts pyruvate to acetyl-CoA (Section 7.3.1). The reactions catalyzed by these two complexes are chemically analogous and both have the coenzymes thiamine pyrophosphate and lipoic acid as prosthetic groups. Although the reaction mechanisms of these two enzyme complexes are very similar, pyruvate dehydrogenase activity is regulated by reversible phosphorylation (see Section 7.7.1) while α-ketoglutarate dehydrogenase activity does not appear to be.

Succinyl-CoA is converted to succinate, with the concomitant phosphorylation of ADP to ATP; this is the only TCA cycle reaction that directly produces ATP by substrate-level phosphorylation. The oxidation of

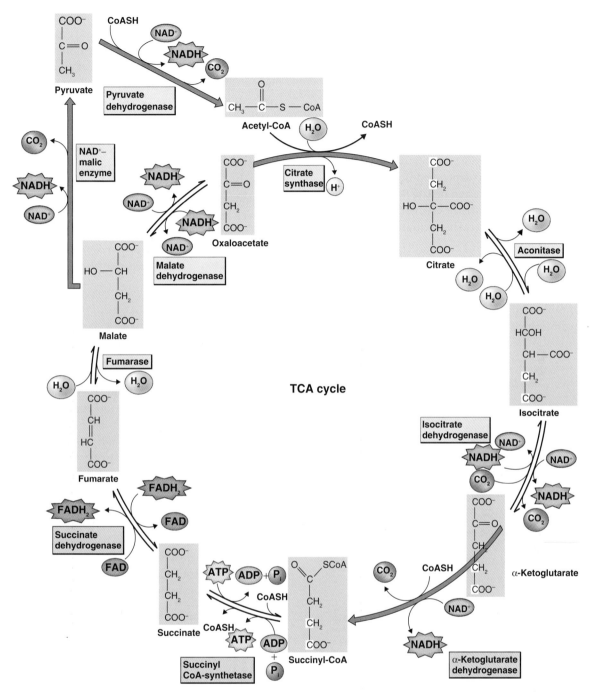

Figure 7.7 The TCA cycle. In a reaction catalyzed by citrate synthase, acetyl-CoA from the preparatory reaction combines with oxaloacetate to produce the tricarboxylic C_6 compound, citrate. The irreversible reactions of the cycle are identified with solid brown arrows. C atoms derived from the acetyl moiety of acetyl-CoA are highlighted in white.

succinate to fumarate is catalyzed by the only membrane-associated enzyme of the TCA cycle, **succinate dehydrogenase**. It transfers electrons from succinate to covalently bound FAD and is the catalytic component of Complex II of the respiratory electron transport chain (see Section 7.4.2).

Next, fumarate is reversibly hydrated to form malate. The reaction is catalyzed by **fumarase**, which is unique to

the mitochondrion and therefore is a convenient marker for the mitochondrial matrix. The final step of the TCA cycle is the oxidation of malate to oxaloacetate, producing NADH. This reaction is freely reversible. In vitro the equilibrium of this reaction strongly favors the reduction of oxaloacetate to malate. In vivo, however, the equilibrium is shifted towards the formation of oxaloacetate because the products of the oxidation reaction are consumed rapidly.

Overall, during the preparatory reaction and one subsequent turn of the TCA cycle proper, the three carbon atoms of pyruvate are released as CO_2, one molecule of ATP is formed directly, and four NADH and one $FADH_2$ molecules are produced (Figure 7.8). The ATP, NADH and $FADH_2$ constitute major sources of energy for the cell.

Whereas most TCA cycle enzymes use NAD^+ as an electron acceptor, **NADP-dependent** isoforms of **isocitrate dehydrogenase** and **malate dehydrogenase** are present in other parts of the cell. For example, peroxisomes contain an NADP isocitrate dehydrogenase. The NADPH produced by such enzymes has many possible fates. It can be oxidized directly by the electron transport chain of plant mitochondria. Moreover, NADPH is an electron donor for several mitochondrial reactions including: the reduction of dihydrofolate to tetrahydrofolate, a substrate for the C_2 photorespiratory

cycle (see Chapter 9); the reduction of oxidized glutathione to protect against reactive O_2 species (see Chapters 15 and 18) generated during mitochondrial electron transport; and the reduction of mitochondrial thioredoxins for the activation of the alternative oxidase (see Section 7.7.2). The cytosolic form of NADP-dependent isocitrate dehydrogenase is abundant in leaf cells, but knocking out the activity of the gene that encodes it has only a small effect on growth and the profile of metabolites of carbon and nitrogen metabolism. It is thought that the enzyme has functions in amino acid metabolism and redox signaling linked to pathogen responses.

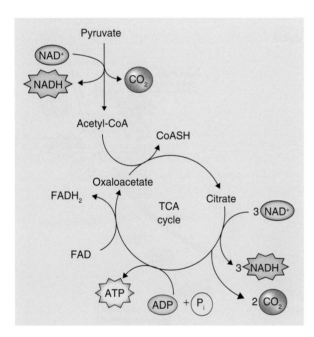

Figure 7.8   Summary of products formed by the TCA cycle. In the overall cycle, citrate is oxidized to produce two molecules of CO_2 in a series of reactions that lead to the formation of one oxaloacetate, three NADH, one $FADH_2$ and one ATP. The resulting oxaloacetate reacts with another molecule of acetyl-CoA to continue the cycle. The oxidative decarboxylation of pyruvate yields an additional CO_2 and NADH. Thus the TCA cycle brings about the complete oxidation of pyruvate to $3CO_2$ plus $10e^-$, which are stored temporarily as 4NADH and $1FADH_2$. In addition, one molecule of ATP is synthesized directly from ADP and P_i by substrate-level phosphorylation. Malate, which can be an alternative product of glycolysis in plants, is converted to pyruvate, CO_2 and NADH by the action of NAD-malic enzyme.

Key points   Pyruvate from glycolysis undergoes a preparatory reaction before it enters the tricarboxylic acid cycle. Pyruvate dehydrogenase, a multiprotein enzyme complex, catalyzes the reaction of pyruvate with NAD^+ and coenzyme A to form acetyl-CoA and NADH, with the release of CO_2. One turn of the TCA cycle, catalyzed by a sequence of eight mitochondrial enzymes, releases the two acetyl carbon atoms of acetyl-CoA as two molecules of CO_2, and produces one molecule of ATP, three of NADH and one of $FADH_2$. One molecule of oxaloacetate is regenerated to start the next cycle. Seven of the enzymes are soluble, located in the mitochondrial matrix. One, succinate dehydrogenase, is membrane-bound and is also part of one of the complexes of the respiratory electron transport chain. Most TCA enzymes have isoenzymic forms located in cell compartments other than mitochondria. Fumarase is an exception, being exclusively mitochondrial, and is often used as a specific marker for the organelle.

7.3.3 Amino acids and acylglycerols are oxidized by glycolysis and the TCA cycle

While glucose is the primary substrate for glycolysis, the breakdown products of lipids and amino acids can also feed into the pathway. Furthermore, catabolism of lipids and certain amino acids, including glutamate, yields acetyl-CoA or intermediates metabolized by the TCA cycle (Figure 7.9). In germinating seeds that have a large amount of stored protein—those of legume species, for example—amino acids such as glutamate are a major source of energy. Catabolism of amino acids in tissues

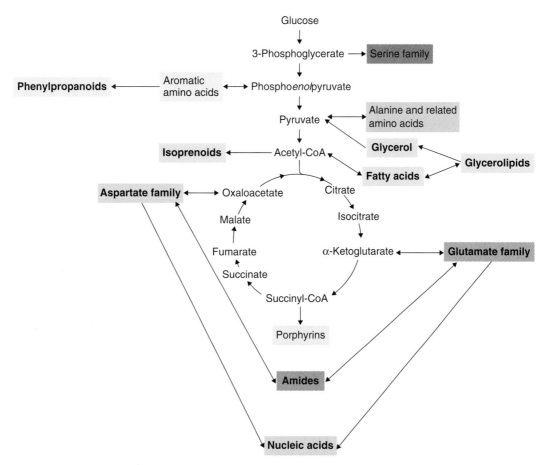

Figure 7.9 The dual catabolic and anabolic functions of glycolysis and the TCA cycle. As well as metabolizing sugars, the pathway breaks down carbon skeletons released by lipid hydrolysis and the de- and trans-amination of amino acids. It is also a source of precursors for the biosynthesis of a range of metabolites, including amino acids, porphyrins, nucleic acids, isoprenoids, fatty acids and phenylpropanoid secondary compounds.

where nitrogen is being remobilized from stored protein (seeds or senescing leaves, for example) is often linked to the synthesis of amides such as glutamine and asparagine, which are the major organic molecules in which nitrogen is translocated over long distances within the plant. A summary of the substrates that are degraded by glycolysis and the TCA cycle is shown in Figure 7.9.

7.3.4 The TCA cycle and glycolysis provide carbon skeletons for biosynthesis

In plants, the TCA cycle functions in both catabolism (to generate energy) and anabolism (for biosynthesis). The cycle provides precursors for many biosynthetic pathways (Figure 7.9). Especially prominent are the use of acetyl-CoA for the synthesis of fatty acids, and of oxaloacetate and α-ketoglutarate which serve as the respective precursors of the amino acids aspartate and glutamate. In addition to being a component of proteins,

glutamate is a precursor of nucleic acids. Succinyl-CoA is used for porphyrin biosynthesis. Glycolytic intermediates also serve precursor functions for biosynthesis: 3-phosphoglycerate for serine, phosphoenolpyruvate for aromatic amino acids and pyruvate for alanine. Thus, both glycolysis and the TCA cycle not only provide energy, but also supply the basic structural elements to meet the biosynthetic needs of the cell.

As intermediates of the TCA cycle are removed to satisfy the needs of biosynthesis, the lost TCA components are restocked by **anaplerotic reactions** (anaplerotic means 'replenishing'), in which CO_2 is added to glycolytic intermediates. **Phosphoenolpyruvate carboxylase** (Equation 7.4) and **malic enzyme** (Equation 7.5) are the main enzymes that fulfill this function in plants.

Equation 7.4 PEP carboxylase

Phosphoenolpyruvate + CO_2 → oxaloacetate + P_i

Equation 7.5 Malic enzyme

Pyruvate + NADPH + $CO_2 \rightarrow$ malate + $NADP^+$

Key points As well as glucose, breakdown products of lipids and amino acids feed into the TCA cycle via acetyl-CoA. Synthesis of amides, for example in geminating seeds, is often linked to amino acid catabolism and production of TCA cycle intermediates such as α-ketoglutarate. Glycolysis and the TCA cycle also provide precursors for biosynthesis. Among the products derived from respiratory intermediates are fatty acids, amino acids, amides, nucleotides, porphyrins and phenylpropanoids. TCA cycle metabolites are replenished in anaplerotic reactions, catalyzed by the CO_2-fixing enzymes malic enzyme and phosphoenolpyruvate carboxylase.

7.4 Mitochondrial electron transport and ATP synthesis

The metabolic events discussed so far have taken place without using O_2. However, the further processing of the products of these reactions is strictly O_2-dependent. The next phase of metabolism centers on the transport of electrons from carriers reduced in glycolysis and the TCA cycle (NADH, $FADH_2$) to O_2. The formation of ATP is directly linked to electron transport through an ordered series of reactions taking place on the mitochondrial inner membrane (see Figure 7.2A).

7.4.1 Mitochondrial electron transport and oxidative phosphorylation generate ATP

The operation of glycolysis and the TCA cycle generates large fluxes of NADH and $FADH_2$. The final step in cellular respiration is the extraction of the energy stored in these reduced coenzymes and the use of that energy to drive the synthesis of ATP from ADP and P_i. To do this, electrons from NADH and $FADH_2$ are passed down a chain of **electron carriers** to O_2. In terms of bioenergetics, movement of these electrons is downhill, and is accompanied by the transport of **protons** (hydrogen ions, H^+) from the mitochondrial matrix to the inner membrane space. This results in a gradient in protons across the inner membrane. When this proton gradient is discharged, the energy stored in it is used to drive the chemiosmotic synthesis of ATP from ADP and P_i. In this way, the cell is able to conserve a large portion of the energy released when glucose is oxidized to CO_2.

The respiratory electron transport chain consists of a series of membrane-bound **redox centers** that catalyze a multistep transfer of electrons from NADH and $FADH_2$ to O_2, forming water and translocating protons from the **matrix** to the **intermembrane space**. This **endergonic** (energy-consuming) proton pumping is driven by the **exergonic** (energy-releasing) transfer of electrons from strong reducing agents to a strong oxidant. The reduction of $^1/_2 O_2$ by $2e^-$ from NADH involves a redox potential difference ($\Delta E^{\circ\prime}$) of 1.14 V (Figure 7.10). This translates to 220 kJ of free energy, equivalent to approximately seven ATPs released for every mole of NADH oxidized. The relationships between redox potential, free energy and ATP are described in Chapter 2. If NADH or $FADH_2$ reduced O_2 directly, the energy would be released as heat and would not be biologically useful. It would, in fact, kill the cell. The electron transport chain facilitates several modestly exergonic redox reactions, rather than a single explosive one, and conserves energy through proton translocation. The $F_0 F_1$-**ATP synthase** enzyme provides a path for controlled proton diffusion from the intermembrane space back into the matrix and uses the free energy released to drive the phosphorylation of ADP. The structures of **ubiquinone**, a **heme** and an **iron–sulfur cluster**, the major types of prosthetic groups of the proteins that make up the electron transport chain, are shown in Figure 7.11.

7.4.2 The electron transport chain moves electrons from reduced electron carriers to oxygen

The typical mitochondrial electron transport chain, the **cytochrome pathway**, consists of four multisubunit protein complexes, commonly referred to as Complexes I through IV (Figure 7.12). The $F_0 F_1$-ATP synthase, which does not have electron transfer activity, is occasionally referred to as Complex V. One of the keys to elucidating the electron transport chain in mitochondria was the use of chemicals that block the pathway by specifically inhibiting the functions of individual complexes (Figure 7.13). Note that some well-known poisons, such as cyanide and carbon monoxide, work by blocking mitochondrial electron transport and oxidative phosphorylation.

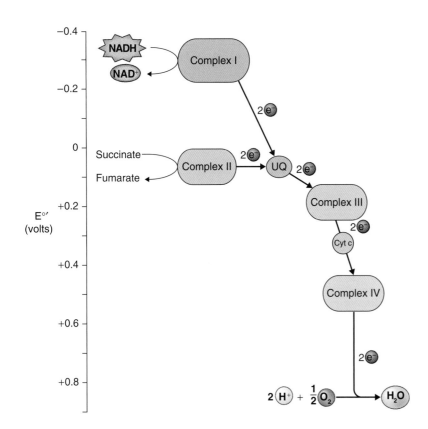

Figure 7.10 The flow of electrons along the thermodynamic gradient during the operation of the mitochondrial respiratory chain. Components of the electron transport system are arranged according to their approximate positions on the redox potential energy scale. Release of energy drives proton translocation at three sites on the chain: between Complex I and ubiquinone (UQ); between UQ and cytochrome c; and between cytochrome c and Complex IV.

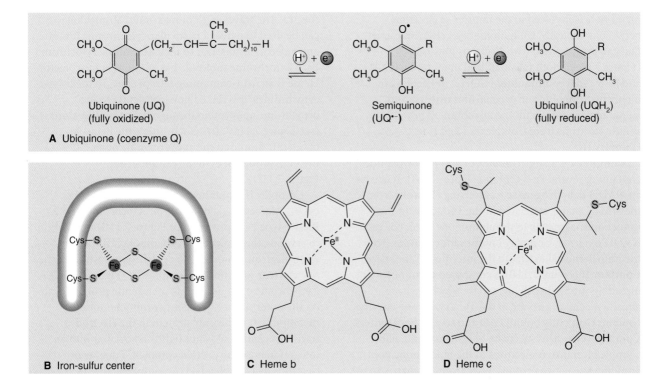

Figure 7.11 Structures of prosthetic groups of carriers of the mitochondrial electron transport chain. (A) Ubiquinone. Note the partly reduced (semiquinone) and fully reduced forms. (B) An iron–sulfur center of the type found in the electron transport chain; in Complex II for example. (C) Heme b, prosthetic group of b-type cytochromes. (D) Heme c, the prosthetic group of cytochrome c, which differs from heme b in its covalent bonding with cysteine groups in the cytochrome apoprotein.

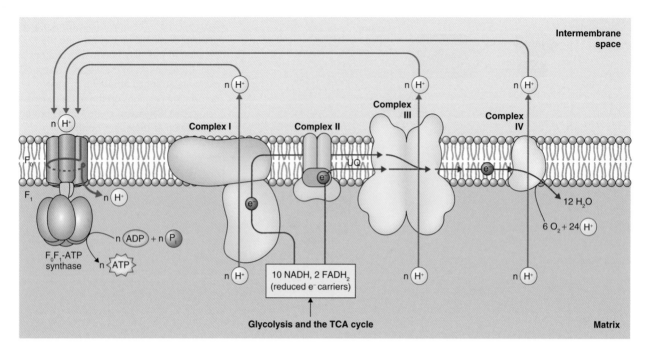

Figure 7.12 Organization of the plant mitochondrial electron transfer chain in the inner mitochondrial membrane, showing electron transfer Complexes I–IV and F_0F_1-ATP synthase (Complex V). Red lines indicate electron transfer pathways. Blue arrows show the direction of proton translocation at Complexes I, III and IV and the transmembrane proton movement that generates an electrochemical gradient driving ATP synthesis via F_0F_1-ATP synthase.

A pool of **ubiquinone** within the inner mitochondrial membrane is involved in the transfer of electrons between complexes. Ubiquinone is an electron carrier that can accept one or two electrons plus H^+ (Figure 7.11A). Both fully oxidized ubiquinone (UQ) and fully reduced **ubiquinol** (UQH_2) molecules are highly hydrophobic and capable of transverse and lateral movement within the membrane. They shuttle electrons from Complexes I and II to Complex III.

Complex I is a large complex of 30–40 polypeptides containing several Fe-S centers (Figure 7.13A). It acts as an **NADH dehydrogenase** transferring electrons from the NADH that is generated in the TCA cycle in the mitochondrial matrix to the ubiquinone pool. For each NADH oxidized, Complex I pumps four protons from the mitochondrial matrix to the intermembrane space. **Complex II** (Figure 7.13B) is composed of four proteins, including succinate dehydrogenase which we have already met as an enzyme of the TCA cycle. The enzyme contains covalently bound FAD and several Fe-S centers. Complex II oxidizes succinate to fumarate and, like Complex I, transfers electrons to the ubiquinone pool. Simultaneously two protons move from the matrix to the ubiquinone pool.

Complex III (Figure 7.13C) is also known as the **cytochrome bc_1 complex**. The **cytochromes** are a group of proteins that function in redox reactions throughout

the cell. Cytochromes have characteristic absorption spectra. The wavelength of maximum absorption can be used to detect a particular cytochrome; it is often appended as a subscript to the name of the cytochrome. Each cytochrome molecule has a bound heme cofactor. Heme is an iron-containing tetrapyrrole. In the b-type cytochromes, it is linked non-covalently to the cytochrome apoprotein through the central Fe atom (see Figure 7.11C). In cytochrome c the heme is covalently bound to the apoprotein through the thiol groups of cysteine residues (see Figure 7.11D).

Complex III contains two b-type cytochromes, cyt b_{566} and cyt b_{560}, one c_1-type cytochrome, and a Rieske-type Fe-S protein (see Figure 7.11B) in addition to several other polypeptides. Complex III transfers electrons from ubiquinol to cytochrome c which carries one electron at a time to Complex IV (Figure 7.13C). Cytochrome c is a peripheral membrane protein located on the outer surface of the inner mitochondrial membrane. It is the only protein in the electron transport chain that is not tightly associated with an integral membrane protein complex. Transfer of electrons by Complex III is accompanied by the pumping of 4 protons per 2 electrons by a mechanism known as the **Q cycle** (see below).

Complex IV, also called **cytochrome c oxidase** (Figure 7.13D), is made up of seven to nine polypeptides

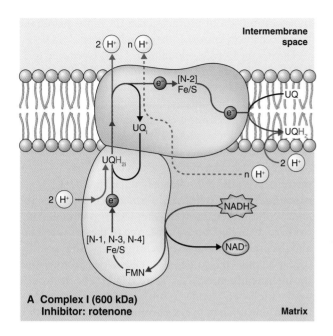

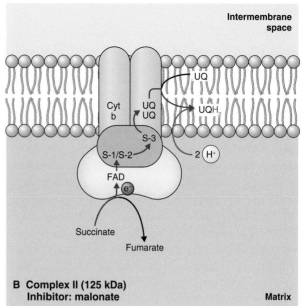

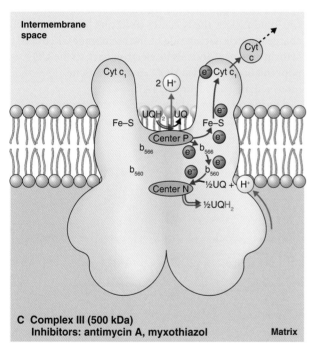

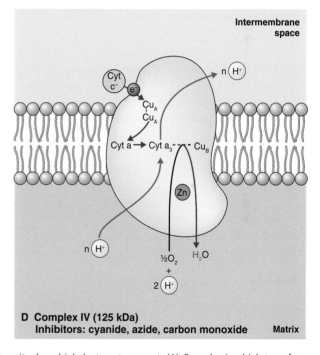

Figure 7.13 Functions and inhibitors of individual complexes in mitochondrial electron transport. (A) Complex I, which transfers electrons from NADH to ubiquinone (UQ). (B) Complex II, which oxidizes succinate to fumarate and transfers electrons to UQ. (C) Complex III supports operation of the proton motive Q cycle as it oxidizes UQH_2 and reduces UQ in two steps and passes electrons to cytochrome c. (D) Complex IV is cytochrome c oxidase, transferring electrons to the terminal acceptor O_2.

and includes two hemes and two Cu centers. It is the terminal electron carrier in the electron transport chain. For every four electrons transferred from cytochrome c (the equivalent of two molecules of NADH or $FADH_2$), one molecule of O_2 is reduced to two molecules of water. This complex pumps two protons per two electrons transferred.

7.4.3 Proton pumping at Complex III occurs via the Q cycle

Protons are pumped across the inner membrane from matrix to intermembrane space at Complexes I,

The energy stored in the reduced cofactors NADH and FADH$_2$ produced during glycolysis and the operation of the TCA cycle is released through the mitochondrial electron chain linked to ATP synthesis and reduction of O$_2$ to H$_2$O. The chain includes four multisubunit protein complexes associated with the mitochondrial membrane. Electrons are transferred between complexes by ubiquinone (UQ) and its reduction product ubiquinol (UQH$_2$). Complex I consists of multiple polypeptides and acts as an NADH dehydrogenase, transferring electrons from NADH to UQ. Complex II is succinate dehydrogenase and contains covalently bound FAD. Complex III is the cytochrome bc$_1$ complex, containing bound heme cofactors. Complexes I, II and III all contain iron–sulfur centers. Complex III transfers electrons from UQH$_2$ to cytochrome c. Complex IV, cytochrome oxidase, receives electrons from cytochrome c and reduces O$_2$ to H$_2$O.

translocation of four protons across the inner membrane by a mechanism known as the Q cycle (Figure 7.14).

During the Q cycle, ubiquinone (UQ) is reduced to UQH$_2$ (see Figure 7.11A) on the matrix side of the inner membrane by electrons from Complex I and Complex II (Figure 7.14A). In both cases, the reduction of UQ to UQH$_2$ by two electrons is accompanied by the uptake of two protons from the mitochondrial matrix. UQH$_2$ then diffuses to the outer face of the membrane and binds to the P center (so named because this face of the membrane is positively charged), a specific UQH$_2$ oxidation site on Complex III formed by cytochrome b and the Rieske Fe-S protein. After UQH$_2$ is bound, it is oxidized. One electron is transferred to the Fe-S protein and the other is passed to cytochrome b$_{566}$. The oxidation of UQH$_2$ to UQ also releases two protons into the intermembrane space (Figure 7.14A). The electron on the Rieske Fe-S center is subsequently transferred to cytochrome c$_1$, then to cytochrome c and eventually to O$_2$. Reduced cytochrome b$_{566}$ transfers its electron to cytochrome b$_{560}$, which is associated with a second UQ-binding site, the N center (referring to the negatively charged matrix-facing side of the membrane). Reduced cytochrome b$_{560}$ donates its electron back to the UQ pool (Figure 7.14B) via UQ$^{\bullet-}$, a semiquinone intermediate (see Figure 7.11A).

During operation of the Q cycle, two protons are translocated across the inner membrane for each

III and IV. The mechanisms of proton pumping at Complexes I and IV are not well understood. At Complex III, the transfer of one electron pair from ubiquinol (UQH$_2$) to cytochrome c is accompanied by the

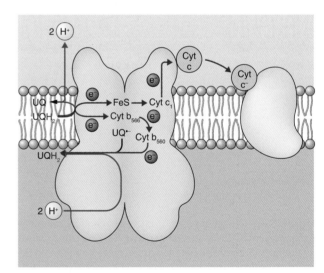

A First UQH$_2$ oxidized

B Second UQH$_2$ oxidized

Figure 7.14 The Q cycle at Complex III of the mitochondrial electron transport chain. (A) The first molecule of UQH$_2$ binds at a site on the intermembrane space side of Complex III and is oxidized to UQ, releasing two protons. One electron is transferred to the Rieske Fe-S protein and thence to cytochrome c$_1$ and cytochrome c. The second electron reduces the high-potential cytochrome b$_{560}$ via b$_{566}$. A molecule of UQ binds at a site on the matrix-facing side of Complex III, accepts an electron from cytochrome b$_{560}$ and is reduced to the semiquinone. (B) Another molecule of UQH$_2$ binds and in a second turn of the redox cycle, two protons are released, one electron passes to cytochrome c and one to cytochrome b$_{560}$. The latter reduces UQ$^{\bullet-}$ to UQH$_2$ with the uptake of two protons from the matrix. The net outcome of the operation of the Q cycle is the transfer of four protons from the matrix to the intermembrane space for every two electrons transferred from Complex III to cytochrome c.

electron that reduces cytochrome c. This stoichiometry occurs because for every two molecules of UQH_2 oxidized by Complex III, four protons are released into the intermembrane space, but only two of the four resulting electrons proceed down the electron transport chain to ultimately reduce O_2. The other two electrons are recycled through the b-type cytochromes to re-reduce one UQ to UQH_2 (Figure 7.14B). A similar Q cycle operates at the cytochrome b_6f complex of the chloroplast thylakoid membrane during photosynthetic electron transport (see Chapter 9).

7.4.4 The F_0F_1-ATP synthase complex couples proton gradient to ATP formation

The transport of electrons down the electron transport chain results in an accumulation of protons in the intermembrane space (see Figure 7.12). The discharge of this gradient leads to the synthesis of ATP from ADP and P_i. The enzyme that catalyzes this synthesis is F_0F_1-ATP synthase. This enzyme is a multisubunit complex that spans the inner membrane; it has two major components, F_1 and F_0 (Figure 7.15). The integral membrane protein complex, F_0, functions as a **proton channel** through the inner membrane. The peripheral membrane protein complex, F_1, extends into the matrix space on a stalk anchored at its base to the F_0 complex. The F_1 complex contains at least five separate polypeptides, α through ϵ. The catalytic sites for ATP synthesis from ADP and P_i are localized primarily on the β-subunits. The F_1 stalk includes the γ-polypeptide and several other subunits, one of which is the target for the respiration inhibitor oligomycin.

The free energy input from proton diffusion is believed to induce a conformational change in the F_1 complex that releases ATP from one of the three active sites, designated L (loose), T (tight) and O (open). ADP and P_i initially bind to an unoccupied O site, while ATP occupies the T site, and ADP and P_i the L site (Figure 7.16, stage 1). Driven by energy released by proton movement through the F_0 channel, the γ-subunit rotates, simultaneously changing the conformations of each of the three nucleotide-binding sites (Figure 7.16, stage 2). Conversion of the T site to O releases the bound ATP. The L site containing ADP and P_i becomes a tight-binding hydrophobic pocket that facilitates ATP synthesis, while the O site that bound ADP and P_i in step 1 adopts the L conformation. Conversion of the ADP and P_i to ATP in the T site does not require additional energy input or conformational change (Figure 17.16, step 3). The CF_0CF_1-ATP synthase of chloroplasts makes ATP by a closely similar three-stage mechanism (see Chapter 9).

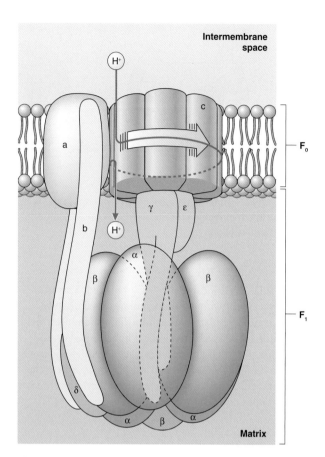

Figure 7.15 Proposed structure and membrane topography of F_0F_1-ATP synthase. The catalytic sites for conversion of ADP and P_i to ATP are located principally on the β-subunits of F_1. The F_0 complex is an integral membrane-spanning complex that acts as a proton channel, providing the pathway for movement of protons across the inner membrane into the matrix. Oligomycin B, a complex cyclic antibiotic made by the bacterium *Streptomyces*, is an inhibitor of ATP synthase.

7.4.5 An overall energy balance sheet for oxidative phosphorylation can be worked out from moles of NADH in and ATP out

We can now calculate the yield of energy from oxidative phosphorylation. The transport of two electrons from NADH to O_2 is accompanied by the pumping of ten protons from the mitochondrial matrix to the intermembrane space. Taking into consideration the energy cost of transporting ADP and P_i across the inner membrane into the matrix, about one ATP is produced for every four protons moved by electron transport. Therefore, the ratio of ADP:O for the oxidation of NADH should be 2.5, i.e. 1ADP phosphorylated per $4H^+$

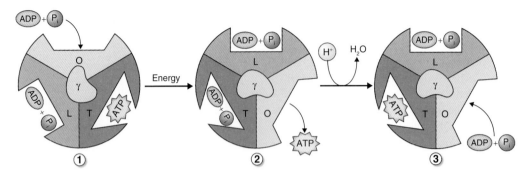

Figure 7.16 Three-step model of ATP synthesis in the F_1 complex of ATP synthase. The movement of protons from the intermembrane space to the matrix drives conformational changes in the L, T and O sites and ADP phosphorylation.

of the $10H^+$ translocated per $1/2O_2$ reduced. This is in good agreement with experimental measurements of ADP:O ratio in isolated mitochondria, which often yield a value of 2.4–2.7. The theoretical figure based on the reduction potential difference between NADH and $1/2O_2$ is approximately 7 (see Section 7.4.1), which makes the **energetic efficiency** of oxidative phosphorylation in vivo 2.5/7 or about 30%.

Oxidation of the bound FADH in Complex II yields theoretical and experimentally observed ADP:O ratios of 1.5, because transport of electrons starting at Complex II involves only two of the three proton translocation sites (see Figure 7.12). ATP formation at each stage from glycolysis and the TCA cycle through to direct substrate phosphorylation, and indirectly via electron transport in the mitochondria, is summarized in Table 7.1.

Key points During electron transport, Complexes I, II and IV pump protons across the inner membrane from the matrix to intermembrane space and this drives oxidative phosphorylation, the synthesis of ATP from ADP and P_i. The transfer of protons across the membrane at Complex III occurs by the Q cycle mechanism. The full Q cycle consists of two turns, during which two molecules of UQH_2 are oxidized, four protons are transferred, two electrons are sent to complex IV and a molecule of UQ is reduced. The protons accumulated in the intermembrane space during electron transport drive ATP synthesis by the multisubunit F_0F_1-ATP synthase complex. The F_0 part is buried in the membrane and functions as a proton channel. F_1 is peripherally located and synthesizes ATP by a conformational change mechanism involving three substrate-binding sites. Overall, ten protons are pumped for every two electrons transported along the chain from NADH to O_2, with a yield of about 2.5 molecules of ATP. The ATP yield for oxidation of the bound FADH in Complex II is about 1.5.

7.4.6 Bypass dehydrogenases are associated with mitochondrial Complex I

In addition to the NADH dehydrogenase activity associated with Complex I, plant mitochondria have a number of dehydrogenases that can oxidize NADH or NADPH (Figure 7.17). Unlike Complex I, these dehydrogenases do not pump protons. There are both NADH and NADPH dehydrogenases localized on the outer face of the inner membrane. These dehydrogenases oxidize NADH and NADPH produced in the cytosol, (e.g. NADH from glycolysis), the plastid (e.g. NADPH from the oxidative pentose phosphate pathway) and the peroxisome (e.g. NADPH from fatty acid degradation; see Chapter 6). In addition, matrix-orientated bypass NAD(P)H dehydrogenases compete with Complex I for NADH produced by enzymes in the mitochondrial matrix, for example those of the TCA cycle. This bypass is insensitive to the Complex I inhibitor rotenone. It has substantially less affinity for NADH than does Complex I

Table 7.1 ATP yield from the conversion of glucose to CO_2 and H_2O via glycolysis, the TCA cycle and oxidative phosphorylation.

Metabolic pathway	Substrates	Products	ATP yield
Glycolysis: cytosol	1 Glucose	2 Pyruvate	
	$2ADP + 2P_i$	2ATP	2
	$2NAD^+$ (cytosolic)	2NADH (cytosolic)	
TCA cycle: mitochondria	2 Pyruvate	$6CO_2$	
	$2ADP + 2P_i$	2ATP	2
	$8NAD^+$	8NADH	
	2FAD (bound)	$2FADH_2$ (bound)	
Oxidative phosphorylation: mitochondria	$6O_2$	$12H_2O$	
	$ADP + P_i$		
	2NADH (cytosolic)	2NAD+ (cytosolic)	3[a]
	8NADH (mitochondrial)	8NAD+ (mitochondrial)	20[b]
	$2FADH_2$ (bound)	2FAD (bound)	3
Cumulative ATP yield			30

[a] Assumes 1.5ATP/NADH (cytosolic, formed in glycolysis).
[b] Assumes 2.5ATP/NADH (mitochondrial).

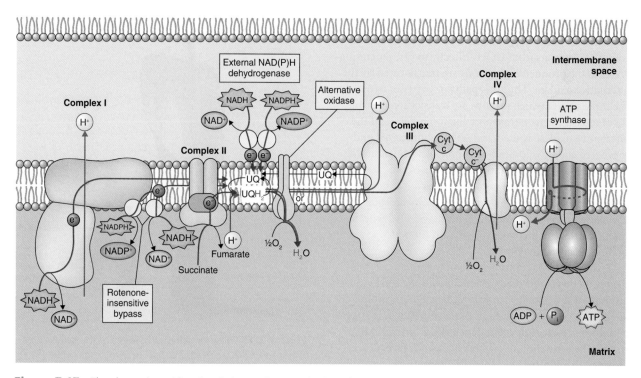

Figure 7.17   The alternative oxidase in relation to the organization of the plant mitochondrial electron transfer chain in the inner mitochondrial membrane. The alternative oxidase and two additional rotenone-resistant NAD(P)H dehydrogenases are shown, together with an external NAD(P)H dehydrogenase, Complexes I–IV and ATP synthase.

and consequently operates only when matrix concentrations of NADH are high. The ADP:O ratio for NADH and NADPH oxidized by supplementary dehydrogenases is about 1.5 because the electrons donated by these enzymes are routed around the Complex I proton pumping site and passed directly to the UQ pool.

7.4.7 Plant mitochondria have an alternative oxidase that transfers electrons to oxygen

Besides having supplementary pathways for the oxidation of NAD(P)H, plant mitochondria contain a pathway for the transfer of electrons from UQ to O_2 that bypasses cytochrome c oxidase (Complex IV) (Figure 7.17). This **alternative oxidase** is found in land plants, many algae and fungi, and some protozoa. Electron flow through the alternative oxidase is insensitive to classic inhibitors of cytochrome c oxidase (cyanide and carbon monoxide) and Complex III (antimycin-A). However, it can be specifically inhibited by salicylhydroxamic acid (SHAM) and *n*-propylgallate. The alternative oxidase is tightly bound to the inner membrane and diverts electrons from the standard electron transport chain (commonly referred to as the **cytochrome pathway**) at the level of the UQ pool. It transfers electrons from UQH_2 directly to O_2, generating water as the product. No protons are translocated by the alternative oxidase. Thus this pathway bypasses two of the three sites at which protons are translocated during the standard electron transport chain. Therefore, if the alternative oxidase is used, less ATP is synthesized and more free energy is lost as heat. Some plants exploit the **thermogenic** (heat-generating) property of the alternative oxidase, most notably species of the genus *Arum* such as eastern skunk cabbage (*Symplocarpus renifolius*; Figure 7.18), in which heat generated within maturing flowers increases the volatilization of pollinator-attracting odors.

Tissues of most plants demonstrate some degree of alternative oxidase activity, but the total amount of this activity can vary widely. For example, alternative oxidase synthesis is often stimulated by stress conditions, such as nutrient deficiency or drought, that induce the formation of reactive O_2 species (see Chapter 15). It is believed that the alternative oxidase pathway may be a way to prevent over-reduction of the quinone pool and thereby reduce the formation of reactive O_2 species.

Key points Plant mitochondria have a number of NAD(P)H dehydrogenases, in addition to Complex I, that generally utilize NAD(P)H produced outside the mitochondrion. A matrix-orientated NAD(P)H dehydrogenase is insensitive to the Complex I inhibitor rotenone. By routing electrons around a proton pumping site, this so-called bypass dehydrogenase reduces the ATP yield of NAD(P)H oxidation to about 1.5. Plant mitochondria also have an alternative pathway that transfers electrons from UQ to O_2, bypassing Complex IV. The alternative oxidase is tightly bound to the inner membrane and is insensitive to inhibitors of Complexes III and IV. By diverting electrons around two of the three proton translocation sites, the alternative oxidase reduces ATP yield and increases loss of energy as heat. This is harnessed by some so-called thermogenic species to increase the attractiveness of mature flowers to pollinators. A more general function of the alternative oxidase is to control the formation of harmful reactive oxygen species.

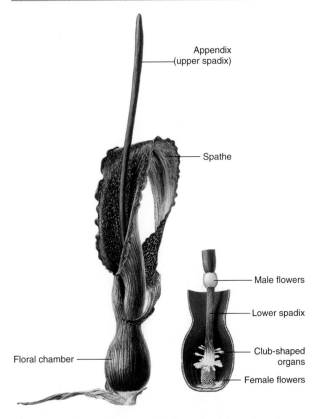

Appendix (upper spadix)

Spathe

Male flowers

Lower spadix

Club-shaped organs

Female flowers

Floral chamber

Figure 7.18   Thermogenesis in flowers of the genus *Arum*. The spadix of some aroids, such as eastern skunk cabbage shown here, can be up to 30° C above a cool ambient air temperature. This helps to vaporize the stench of the flowers and stimulate the activity of pollinator insect visitors. The site of thermogenesis is the upper spadix.

7.5 The oxidative pentose phosphate pathway

In addition to the glycolytic pathway, there are other ways in which cells can breakdown carbohydrates. Of particular significance in plants is the **oxidative pentose phosphate pathway** (Figure 7.19). This should not be confused with the Calvin–Benson cycle (see Chapter 9) which is sometimes referred to as the **reductive pentose phosphate pathway**. The oxidative pathway oxidizes and decarboxylates glucose-6-phosphate via the pentose intermediate ribulose-5-phosphate, producing CO_2; here $NADP^+$ rather than NAD^+ is the electron acceptor (Equation 7.6).

Equation 7.6 Overview of oxidative pentose phosphate pathway

$$6 \text{ Glucose-6-phosphate} + 7H_2O + 12NADP^+$$
$$\rightarrow 5 \text{ glucose-6-phosphate} + 6CO_2 + P_i$$
$$+ 12NADPH + 12H^+$$

The pathway is present both in plastids and in the cytosol, with the plastid pathway generally predominating. The plastid pathway operates primarily at night when its product, NADPH, is not available via the reactions of photosynthesis (see Chapter 9).

7.5.1 The pentose phosphate pathway has oxidative and regenerative phases

The oxidative pentose phosphate pathway has two stages: an **oxidative phase**, in which glucose-6-phosphate is converted, sequentially, to 6-phosphogluconate+ NADPH then to ribulose-5-phosphate, CO_2 and NADPH (Figure 7.19A); and a **regenerative (non-oxidative) phase** in which glucose-6-phosphate is regenerated from ribulose-5-phosphate (Figure 7.19B). The stoichiometry of the pathway (Figure 7.20), assuming none of the intermediates is diverted to other biosynthetic destinations, shows that one molecule of glucose-6-phosphate recruited to the C_6 pool is completely oxidized to CO_2 in six turns of the cycle. In other words, out of every six C_6 participants, one is emitted in the form of six molecules of CO_2 and the remaining five C_6 skeletons are regenerated though a process of shuffling between C_5, C_4, C_7 and C_3 intermediates in the regenerative phase. These, with

newly recruited glucose-6-phosphates, become available to enter the oxidative phase again.

Note that there is no ATP generation by substrate-level phosphorylation during the operation of the pentose phosphate pathway; all the energy in glucose-6-phosphate is conserved in NADPH. The NADPH can be used either for biosynthesis or to produce ATP in the mitochondria when NADPH is oxidized by the mitochondrial respiratory electron transport chain.

7.5.2 The pentose phosphate pathway is a source of intermediates for a number of biosynthetic pathways

The intermediates of the regenerative phase of this pathway may be used for the biosynthesis of cellular components. For example ribose-5-phosphate is used in the synthesis of RNA, DNA, ATP and several coenzymes. In plants, erythrose-4-phosphate, a four-carbon intermediate of the pathway, is a precursor in **aromatic amino acid** synthesis; it is also used in the synthesis of plant **natural product** derivatives, e.g. lignin and flavonoids (see Chapter 15).

Key points The oxidative pentose phosphate pathway, localized in plastids and the cytosol, oxidizes and decarboxylates glucose-6-phosphate to produce CO_2, with $NADP^+$ as the electron acceptor. The three reactions of the oxidative phase convert glucose-6-phosphate to ribulose-5-phosphate and CO_2 with the reduction of $2NADP^+$. The regenerative (non-oxidative) phase regenerates glucose-6-phosphate from ribulose-5-phosphate by way of six enzymic reactions. Out of every six molecules of glucose-6-phosphate, one is completely oxidized to CO_2 in six turns of the cycle. Regeneration produces five hexose phosphates (C_6) via phosphorylated C_5, C_4, C_7 and C_3 intermediates. All energy released by the pentose phosphate pathway is captured as NADPH; no ATP is synthesized by substrate-level phosphorylation. NADPH feeds into mitochondrial oxidative phosphorylation or it may, along with pathway intermediates, have other metabolic fates including coenzyme, nucleic acid and natural product biosynthesis, and interaction with the Calvin–Benson cycle of photosynthesis.

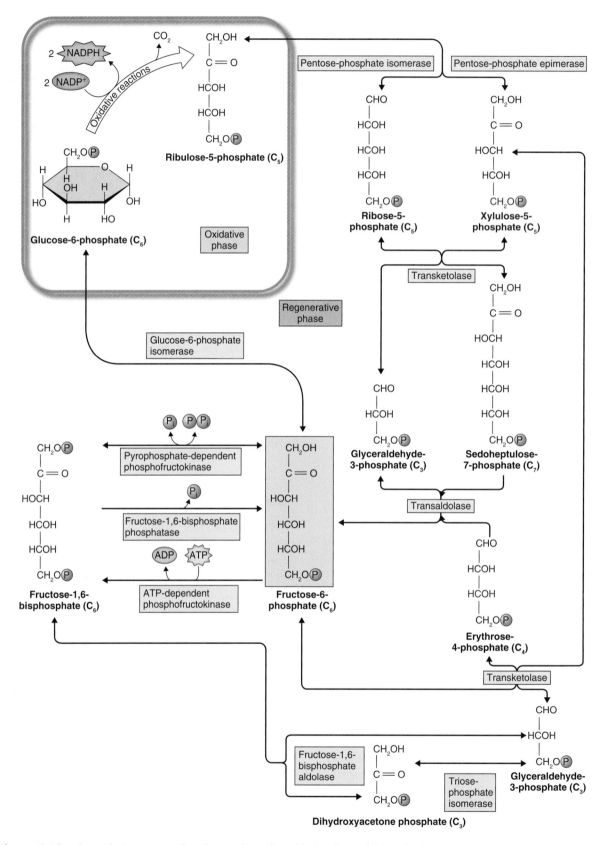

Figure 7.19   The oxidative pentose phosphate cycle. In the oxidative phase of the cycle, glucose-6-phosphate is oxidized to ribulose-5-phosphate, yielding NADPH and CO_2. Glucose-6-phosphate is regenerated from ribose-5-phosphate in the regenerative or non-oxidative phase of the cycle. Aside from transaldolase, enzymes of the regenerative phase are discussed in connection with the Calvin–Benson cycle in Chapter 9. Reactions of the oxidative phase are irreversible, while those of the regenerative phase are freely reversible.

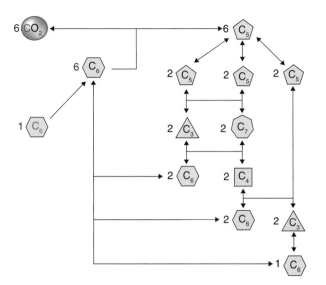

Figure 7.20 Stoichiometry of the oxidative pentose phosphate pathway. Each C_6 sugar joining the C_6 pool is completely oxidized to CO_2 by six turns of the cycle. In the process, five C_6 sugar phosphates are regenerated from six C_5 precursors.

The intermediates and most enzymes of the regenerative phase of the pentose phosphate cycle are also participants in the Calvin–Benson cycle and will be discussed in Chapter 9. In plastids undergoing **greening**, intermediates of the oxidative pentose phosphate pathway can also be used as a source of intermediates of the Calvin–Benson cycle until the capability for photosynthesis is fully developed.

7.6 Lipid breakdown linked to carbohydrate biosynthesis

As we have seen in Chapter 6, the breakdown of triacylglycerols to acetyl-CoA in seeds involves two organelles: the **oil body** (**oleosome**), where the stored lipids are converted to fatty acids; and the peroxisome, also known as the glyoxysome, where fatty acids are converted by **β-oxidation** to acetyl-CoA (see Figure 6.33). Other acyl lipids, such as the glyco- and phospholipids of membranes, are catabolized similarly during metabolic turnover or net breakdown. Acetyl-CoA may be respired through the mitochondrial TCA cycle in the normal way. Alternatively it may be converted to succinate in the glyoxylate cycle and used as a substrate for the synthesis of hexoses in the reactions of **gluconeogenesis** (Figure 7.21). Succinate is converted to malate in the mitochondrion via reactions of the TCA cycle. Most of the malate produced from acetyl-CoA is

transported to the cytosol for final conversion to hexoses that can be used in the synthesis of other carbohydrates or can be used to make sucrose for export to other cells of the plant.

7.6.1 The glyoxylate cycle converts acetyl-CoA to succinate

Before it can be used for synthesizing carbohydrates, acetyl-CoA, produced by β-oxidation within the peroxisome, is converted by the enzymes of the glyoxylate cycle into succinate, a C_4 compound (Equation 7.7).

Equation 7.7 Overview of the glyoxylate cycle

$$2\text{Acetyl-CoA} + \text{NAD}^+ + 2\text{H}_2\text{O} \rightarrow$$
$$\text{succinate} + 2\text{CoASH} + \text{NADH} + \text{H}^+$$

Acetyl-CoA first condenses with oxaloacetate to form citrate, which is then converted to isocitrate. Isocitrate is broken down into glyoxylate and succinate by **isocitrate lyase**. The succinate is released as a net product and the glyoxylate combines with another molecule of acetyl-CoA to form malate in a reaction catalyzed by **malate synthase**. Malate is converted to oxaloacetate to continue the glyoxylate cycle. The succinate, formed in the isocitrate lyase reaction, moves out of the peroxisome and into the mitochondrion (Figure 7.21).

By converting acetyl-CoA to isocitrate and then to succinate in the glyoxylate cycle, instead of using the TCA cycle, the two CO_2-emitting reactions of the TCA cycle are bypassed. The carbon thus retained is available ultimately to be used for the net synthesis of sugars. Because animals lack the two critical enzymes of the glyoxylate cycle, isocitrate lyase and malate synthase, they are unable to convert lipids to carbohydrates. This is a fundamental biochemical difference between plants and animals.

7.6.2 Mitochondria convert succinate to malate, a precursor of carbohydrates

Before succinate which is produced by glyoxysomes can be used to make carbohydrate, it must be converted to malate. Succinate is transported into the mitochondrial matrix. Here it is converted to malate by the enzymes of the TCA cycle (see Figure 7.7). Malate is then transported to the cytosol where it is converted to

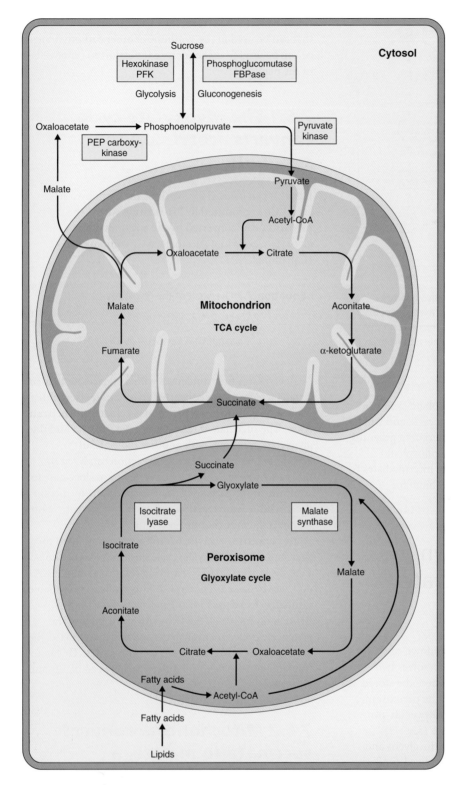

Figure 7.21 The glyoxylate cycle and gluconeogenesis. Isocitrate lyase and malate synthase are unique to the peroxisome. Opposing pathways of glycolysis and gluconeogenesis (sucrose synthesis) are shown. The bypass reactions of gluconeogenesis are: phosphoenolpyruvate (PEP) carboxykinase, which bypasses the irreversible glycolytic enzyme pyruvate kinase; fructose-1,6-bisphosphatase (FBPase), which bypasses phosphofructokinase (PFK); and phosphoglucomutase, which bypasses hexokinase.

oxaloacetate by malate dehydrogenase. Oxaloacetate is converted to phosphoenolpyruvate by **PEP carboxykinase** (Equation 7.8). Phosphoenolpyruvate is the substrate for synthesis of hexoses via gluconeogenesis (see Figure 7.21).

Equation 7.8 PEP carboxykinase

$$\text{Oxaloacetate} + \text{ATP} \rightarrow$$

$$\text{phosphoenolpyruvate} + \text{ADP} + \text{CO}_2$$

We have seen that there are two possible fates for malate in the mitochondrial matrix. It may be exported to the cytosol and be used to synthesize carbohydrate or it can be used in the TCA cycle where it will be oxidized to CO_2 and H_2O. What determines the pathway that malate will take? There is evidence that isocitrate dehydrogenase plays a critical role in specifying which way carbon flows in mitochondria. When the enzyme is inhibited (e.g. by intermediates of the TCA cycle), malate may be exported from the mitochondrion; alternatively, when the enzyme is active, malate flows through the cycle and is oxidized.

7.6.3 Gluconeogenesis converts phosphoenolpyruvate to hexoses

Gluconeogenesis is the final set of reactions in the conversion of lipids to carbohydrates. It begins with phosphoenolpyruvate and ends with the production of fructose-6-phosphate and glucose-6-phosphate, which can be used to synthesize sucrose or other products. Seven of the ten reactions of this pathway are the reverse of the glycolytic reactions (see Figure 7.3). However three of the glycolytic reactions are essentially irreversible in vivo and represent roadblocks on the route from phosphoenolpyruvate to hexose phosphate. These are the reactions catalyzed by **hexokinase** (conversion of glucose to glucose-6-phosphate), **phosphofructokinase** (PFK; phosphorylation of fructose-6-phosphate to fructose-1,6-bisphosphate) and **pyruvate kinase** (conversion of phosphoenolpyruvate to pyruvate) (see Figure 7.21).

During gluconeogenesis, the pyruvate kinase reaction is bypassed by PEP carboxykinase (Equation 7.8). The conversion of fructose-1,6-bisphosphate (Figure 7.22A) to fructose-6-phosphate is bypassed by **fructose-1,6-bisphosphatase** (FBPase), which catalyzes the essentially irreversible hydrolysis of the C-1 phosphate (Equation 7.9).

Equation 7.9 Fructose-1,6-bisphosphatase (FBPase)

Fructose-1,6-bisphosphate + H_2O →

fructose-6-phosphate + P_i

This enzyme has a chloroplast counterpart in the Calvin–Benson cycle, but the two enzymes are regulated in very different ways. Like the enzyme from animals, plant cytosolic FBPase is negatively regulated by the allosteric metabolite AMP, and a specific regulatory metabolite, **fructose-2,6-bisphosphate** (Figure 7.22B).

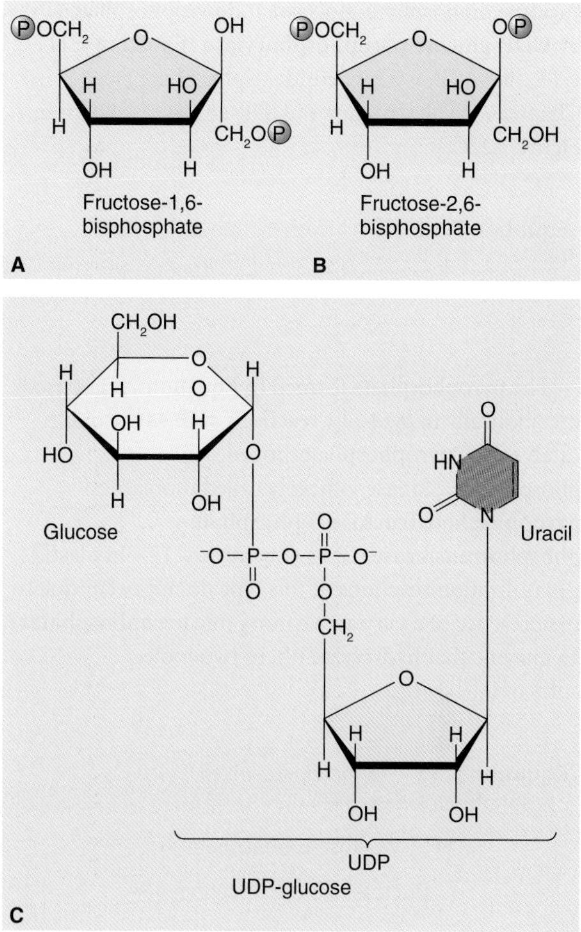

Figure 7.22   Structures of: (A) fructose-1,6-bisphosphate, (B) fructose-2,6-bisphosphate and (C) uridine diphosphate-glucose (UDP-glucose).

The concentration of fructose-2,6-bisphosphate is determined by the activities of the kinase and phosphatase enzymes that synthesize and degrade it and which, in turn, are sensitive to levels of regulatory intermediates (see Chapter 9). Chloroplast FBPase, by contrast, is regulated by the ferredoxin/thioredoxin system in conjunction with a light-induced change in pH and Mg^{2+} concentration (see Chapter 15).

In seeds the irreversible hexokinase reaction of glycolysis is bypassed by converting glucose-6-phosphate to glucose-1-phosphate with **phosphoglucomutase** (Equation 7.10).

Equation 7.10 Phosphoglucomutase

Glucose-6-phosphate → glucose-1-phosphate

In the presence of uridine triphosphate (UTP), glucose-1-phosphate is then converted to **UDP-glucose** (uridine diphosphate-glucose) and pyrophosphate (PP_i) by **UDP-glucose pyrophosphorylase** (Equation 7.11). UTP, like ATP, is a nucleoside triphosphate (see Chapter 2). The structure of UDP-glucose is shown in Figure 7.22C.

Equation 7.11 UDP-glucose pyrophosphorylase

Glucose-1-phosphate + UTP → UDP-glucose + PP_i

The pyrophosphate formed in Equation 7.11 is used metabolically in cytosolic reactions such as the one catalyzed by pyrophosphate-linked phosphofructokinase, officially called inorganic **pyrophosphate fructose-6-phosphate phosphotransferase** (PFP) (Equation 7.12). In plastids PP_i utilization reactions of this type do not occur due to the presence of a very active inorganic **pyrophosphatase**, an enzyme that hydrolyzes PP_i to two molecules of P_i.

Equation 7.12 Pyrophosphate-linked phosphofructokinase (PFP)

PP_i + fructose-6-phosphate →

fructose-1,6-bisphosphate + P_i

In the final stages of sucrose synthesis, sucrose-6-phosphate is synthesized from UDP-glucose and fructose-6-phosphate by **sucrose-6-phosphate synthase** (Equation 7.13). Finally, sucrose-6-phosphate is hydrolyzed to sucrose and P_i by **sucrose-6-phosphate phosphatase** (Equation 7.14).

Equation 7.13 Sucrose-6-phosphate synthase

UDP-glucose + fructose-6-phosphate →

sucrose-6-phosphate + UDP

Equation 7.14 Sucrose-6-phosphate phosphatase

Sucrose-6-phosphate + H_2O → sucrose + P_i

The net result of lipid breakdown is the conversion of fats to sucrose, sugar that can be translocated to other parts of the plant. There it can be used for a variety of purposes, such as biosynthesis or, following breakdown via glycolysis and the TCA cycle, the synthesis of ATP.

Key points During acyl lipid breakdown, β-oxidation of fatty acids in the peroxisome results in the production of acetyl-CoA, which may be respired to CO_2 via the mitochondrial TCA cycle. Alternatively the carbon atoms of acetyl-CoA can be conserved through the operation of the glyoxylate cycle and used for gluconeogenesis, the synthesis of hexose sugars. Acetyl-CoA enters the peroxisome and is converted to isocitrate. Isocitrate lyase and malate synthase, the characteristic enzymes of the glyoxylate cycle, convert isocitrate to malate, which continues the glyoxylate cycle, and succinate. Succinate moves to the mitochondrion and joins the TCA cycle. During gluconeogenesis, malate derived from succinate is converted in the cytosol to oxaloacetate, which in turn is converted to phosphoenolpyruvate (PEP) by PEP carboxykinase. All enzymes of glycolysis between hexose and PEP except phosphofructokinase (PFK) and hexokinase are reversible. Fructose-1,6-bisphosphatase (FBPase) and phosphoglucomutase bypass these reactions, allowing PEP to be converted to sugars essentially by running glycolysis in reverse. FBPase is allosterically regulated by AMP and fructose-2,6-bisphosphate.

7.7 Control and integration of respiratory carbon metabolism

Respiration is the core of primary metabolism and must operate sensitively to ensure that physiological integrity is maintained. Metabolic control is the process of adjusting the flux of intermediates through a biochemical pathway by a combination of fine (moment-to-moment) and coarse (long-term) regulation. The amount of an enzyme has a direct effect on the rate of the reaction it catalyzes and, as described below, may be altered by up- or downregulating gene expression. But the relatively extended timescale required to bring about appreciable change in enzyme activity this way means that a mechanism based on changes in protein synthesis usually contributes to coarse control only. Regulation of respiration over the short term is generally achieved through the responsiveness of existing enzymes to metabolites that interact directly with the catalytic or regulatory sites of the protein and enhance or reduce its metabolic capacity. In this section we summarize mechanisms of fine and coarse control by which respiration is attuned to physiological requirements.

7.7.1 Fine control of respiration is exercised through metabolic regulation of enzyme activities

The flow of metabolites along the pathways of cellular respiration is ultimately regulated through adjustments to the rate of NADH reoxidation by the respiratory electron transport chain and the cellular rate of ATP utilization. The study of the flux of substrates and products through the enzymes of a biochemical pathway is called **metabolic control analysis** (**MCA**). In the very simplest case of a short, unbranched reaction sequence, reaction flux may be set by the rate-determining step, that is, the activity of the pace-setting enzyme for the whole pathway. In reality, metabolic pathways are complex networks and flux is a property of the whole interconnected system rather than being determined by any particular step. MCA quantifies the contribution of individual reactions to the flux behavior of the entire pathway and the responses of the system and its components to perturbations such as a change in the concentration of a metabolite, an environmental signal or a developmental event.

MCA has been applied to glycolysis, the TCA cycle and the mitochondrial electron transport chain in plants. Analysis of these pathways reveals flexible organization and alternative reactions that confound the identification of particular rate-determining steps and single points of flux regulation. As Figure 7.23 shows, however, in terms of contributions to metabolic control in the overall system of respiratory biochemistry, some reactions are points of particular sensitivity. Phosphofructokinase (PFK) control of glycolysis is an example of negative feedback by downstream intermediates. Conversion of fructose-6-phosphate to fructose-1,6-bisphosphate by PFK is activated by P_i and is strongly inhibited by PEP (Figure 7.24) and, to a lesser extent, by other metabolites further along the glycolytic pathway. Likewise, pyruvate kinase is activated by ADP and inhibited by some TCA cycle products (Figure 7.23). Metabolic control of the TCA cycle is best illustrated by regulation of the pyruvate dehydrogenase complex in the preparatory reaction. The pyruvate dehydrogenase component is inhibited when it is phosphorylated by an ATP-dependent enzyme, a **protein kinase**. Removal of the phosphate by a **phosphatase** reactivates the enzyme (Figure 7.25). The activity of the protein kinase in turn is modulated by several metabolites. Pyruvate inhibits the kinase, ensuring that the dehydrogenase is active when its substrate is plentiful. In addition, pyruvate dehydrogenase is subject to feedback inhibition by its products, acetyl-CoA and NADH. Other dehydrogenases of the TCA cycle (malate,

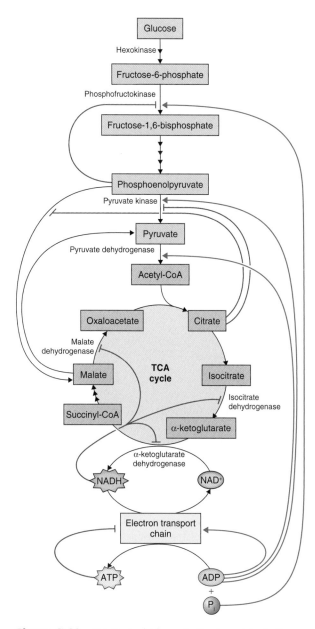

Figure 7.23 Points at which respiration is metabolically regulated by feedback control.

isocitrate, α-ketoglutarate dehydrogenases) are also inhibited by NADH and acetyl-CoA through negative feedback (Figure 7.23).

The rate of electron transport, and with it the rate of O_2 uptake, are regulated by the availability of ADP and P_i. This phenomenon is called **respiratory control** (Figure 7.26A). In the absence of ADP or P_i, the F_0 proton channel of the ATP synthase is blocked. As a consequence the proton gradient across the inner membrane builds until it exerts a chemiosmotic back pressure that restricts further proton translocation across the membrane. Because electron transport is obligately

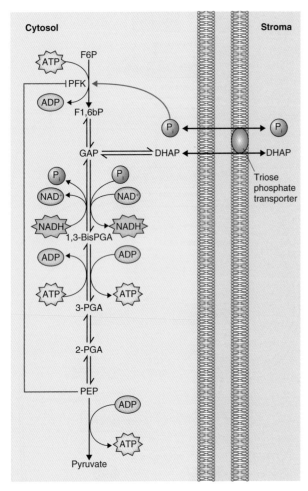

Figure 7.24 Feedback regulation of phosphofructokinase (PFK), and interaction between glycolysis and photosynthesis in the chloroplast via the plastid envelope triose phosphate transporter.

rapid; in the absence of ADP or P_i, or both, the protons leak slowly across the inner membrane.

The proton leak can be dramatically stimulated by certain compounds, called **uncouplers**, that act as **protonophores** or proton channels. Uncouplers, such as NH_4^+, allow protons to equilibrate across the inner membrane (Figure 7.26B). This leads to a collapse of the proton gradient and ATP synthesis stops; however, the rate of electron transport and O_2 consumption increase because the block in the proton channel is removed. In this way electron transport is uncoupled from ATP synthesis.

7.7.2 Respiration interacts with other carbon and redox pathways

In the light, green tissues assimilate CO_2 through the photosynthetic Calvin–Benson cycle, and simultaneously lose CO_2 through photorespiration (see Chapter 9) and cellular respiration. In darkness, leaves can no longer photosynthesize but continue to produce CO_2 and ATP through respiration. Differences in respiratory physiology under illumination and in the dark are the consequence of interactions with photosynthetic carbon and energy metabolism. Photosynthesis supplies respiratory substrate. An important point of contact between glycolysis and chloroplast carbon metabolism is the triose phosphate transporter (TPT) of the plastid envelope. By exchanging C_3 Calvin–Benson intermediates and P_i, the TPT can directly modulate PFK activity (see Figure 7.24).

Respiration is also an optimizing influence on photosynthesis. For example, respiration facilitates the export of excess reducing equivalents produced by photochemical reactions. The mitochondrial electron transport chain plays an essential role in dissipating such equivalents, thereby preventing over-reduction of chloroplast electron transport components and consequent oxidative damage to thylakoid membranes.

linked to proton translocation, a large proton buildup will also restrict the rate of O_2 consumption. In a steady state, the rate of electron transport is determined by the rate at which protons in the intermembrane space flow back into the matrix. When ADP and P_i are available, the backflow of protons via the ATP synthase is

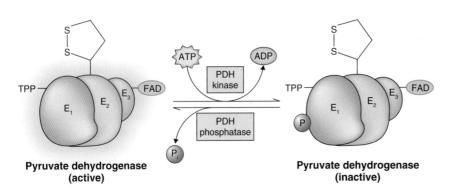

Pyruvate dehydrogenase (active)

Pyruvate dehydrogenase (inactive)

Figure 7.25 Regulation of the pyruvate dehydrogenase (PDH) complex by phosphorylation/dephosphorylation. The enzyme, which is inhibited by phosphorylation of a specific serine, is reactivated on subsequent dephosphorylation by a PDH phosphatase.

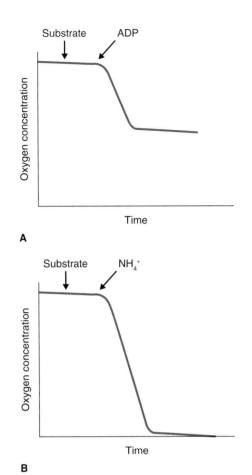

A

B

Figure 7.26   Electron flow in mitochondria, determined by measuring oxygen uptake from the medium. (A) Electron flow depends on ADP in addition to a substrate such as NADH, a phenomenon known as respiratory control, to discharge the proton gradient. Electron flow is said to be coupled to ATP formation. (B) Ammonium ion is an uncoupler, a compound that equilibrates the protons on the two sides of the membrane, thereby collapsing the gradient. Electron flow is then said to be uncoupled.

by covalent modification of protein thiol groups. Among the respiratory components shown to be activated by thiol reduction are cytosolic NAD-glyceraldehyde-3-phosphate dehydrogenase and aldolase, and mitochondrial citrate synthase. The activity of the alternative oxidase is known to be under redox control; formation of an intersubunit –S–S– bond inactivates the alternative oxidase. Reduced thioredoxin has been shown to break the bond and activate the protein.

> **Key points**   Metabolic control analysis quantifies the contribution of each enzyme in a pathway to the regulation of flux through the pathway as a whole. Among the steps in glycolysis that are most influential on respiratory flux are those involving PFK and pyruvate kinase, which are feedback-regulated by downstream intermediates. The pyruvate dehydrogenase complex is regulated by phosphorylation/dephosphorylation and allosteric feedback. Some TCA cycle enzymes are inhibited by NADH and acetyl-CoA. The rate of electron transport and O_2 uptake is determined by the sensitivity of the F_0 proton channel of the ATP synthase to ADP and P_i availability. The chloroplast triose phosphate transporter is a point at which photosynthesis regulates glycolysis. The mitochondrion influences photosynthesis by preventing oxidative damage to the thylakoid membranes, by oxidizing glycine produced during photorespiration, and by competing for precursors of sucrose biosynthesis. Redox control of the activity of some respiratory components is exerted through thioredoxin-mediated covalent modification of –S–S– bonds.

Another critical function of mitochondrial respiration during photosynthesis is oxidation of glycine produced during photorespiration (see Chapter 9). **Transgenic plants** with a reduced expression of genes encoding TCA cycle components such as aconitase or malate dehydrogenase have increased rates of photosynthesis and dry matter accumulation. This is interpreted to be the result of reduced competition from the TCA cycle for the precursors of sucrose biosynthesis.

It is well established that light directly regulates several photosynthetic enzymes through the redox state of **thioredoxin** linked, via ferredoxin, to electron transport flow in the chloroplast (see Chapter 15). The recent discovery of cytosol- and mitochondrion-specific thioredoxins is evidence for redox control of respiration

7.7.3 Coarse control of respiratory activity is exerted through regulation of gene expression

As well as exerting an effect through the interaction between photosynthetic and respiratory metabolism, light directly influences respiration through photoreceptor-mediated **transcriptional control** of mitochondrial proteins, particularly electron transport chain components (Table 7.2). Most TCA cycle enzymes are not light-regulated, but two malate dehydrogenases of *Arabidopsis* have been observed to be induced by light. In the case of one of these, the effect is sensitive to red

Table 7.2 Mitochondrial proteins whose genes are regulated by light and sugars.

Process	Protein	Light[a,b] INDUCED/ repressed	Sugar INDUCED/ repressed[b]
TCA cycle	Pyruvate dehydrogenase E1 subunit	w	sucrose
	Citrate synthase-4[c]	W	SUCROSE
	Citrate synthase-3[c]	fr	sucrose
	Aconitase-2	fr	SUCROSE
	Malate dehydrogenase-2[c]	FR, R	SUCROSE
	Malate dehydrogenase-5[c]	W	SUCROSE, GLUCOSE
Respiratory chain/ oxidative phosphorylation	Alternative oxidase	R	SUCROSE, GLUCOSE
	Rotenone-insensitive NAD(P)H dehydrogenase	B, FR, R	
	Cytochrome c	B, FR, R	GLUCOSE
	ADP/ATP carrier protein	r	sucrose
	Respiratory chain complex assembly protein	FR	

[a] Light quality influencing gene expression is indicated by: B/b, blue; FR/fr, far red; R/r, red; W/w, white.
[b] UPPER CASE indicates that the gene is upregulated; lower case indicates it is downregulated.
[c] Citrate synthase-3 and -4 are mitochondrial-targeted enzymes encoded by different genes. Likewise malate dehydrogenase-2 and -5.

and far-red light, indicating the involvement of **phytochrome** (see Chapter 8). Light also influences three enzyme steps at the point of carbon entry to the TCA cycle: pyruvate dehydrogenase, citrate synthase and aconitase (Table 7.2). About 10% of the 100 or so genes encoding components of the mitochondrial respiratory chain display light-related changes in expression. They include genes for cytochrome c (which shows sensitivity to blue as well as red and far-red light), ADP/ATP carrier protein, and a protein required for respiratory complex assembly (Table 7.2). The transcription of genes encoding the rotenone-insensitive NAD(P)H dehydrogenase (see Figure 7.17) is strongly light-dependent, mediated by phytochrome and the blue light receptor **cryptochrome**. The alternative oxidase capacity of mitochondria isolated from light-grown leaves is higher than that of organelles from dark-grown tissue. Both photosynthetically active light and low-intensity red light have been shown to be inductive (Table 7.2). These observations suggest that a specific, non-proton pumping, alternative respiratory chain may be activated in the light. This may be related to the need to deal with the high levels of NADH generated by

glycine decarboxylase during photorespiration (see Chapter 9). Figure 7.27 summarizes the light responses of respiratory genes, comparing the photosynthetic and photomorphogenic pathways in green tissues with those of tissues containing etioplasts (plastids that develop in darkness). The scheme includes the interaction between photoreceptors and the internal biological clock (see Chapter 8), which underlies the daily rhythm of respiratory activities.

Soluble sugars are the principal metabolic resources utilized by respiration. Sugars are also sensed by specific signaling pathways and exert a powerful influence on the expression of genes, not only for carbon metabolism but also for a very wide range of developmental pathways and environmental responses. Sucrose and the hexose products of sucrose hydrolysis (glucose and fructose) have inductive or repressive effects on the transcription of genes encoding a number of respiratory enzymes (Table 7.2). Plant cells have separate sensors for sucrose and glucose or fructose; changes in the sucrose:hexose ratio detected by these systems lead to different transduction pathways. Sugar-sensing systems are either growth promoting or growth inhibiting. Here we

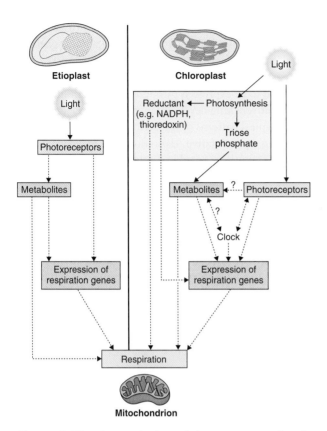

Figure 7.27 Photosynthetic- and photoreceptor-mediated regulation of respiration and expression of genes for respiratory enzymes. Metabolites, reductants and transcription are modulated by white, red, far-red or blue light in etiolated or green tissues.

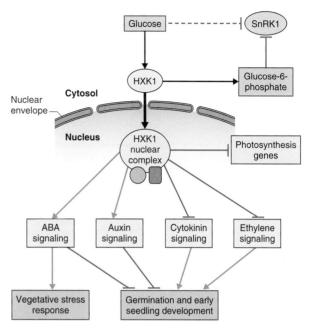

Figure 7.28 Hexokinase acts as a sugar sensor in addition to its enzymatic function in phosphorylating glucose. The scheme shows the regulatory network controlled by the protein encoded by the *HXK1* gene of *Arabidopsis* and its interaction with the signaling pathway mediated by SnRK1 (see Figure 7.29). HXK1 is shown diagrammatically in the nucleus in the form of a high molecular mass complex with other proteins. Upregulation is indicated by arrows, downregulation by bar-ended lines. ABA, abscisic acid.

describe the intracellular regulatory pathway centered on the enzyme that catalyzes the first step in glycolysis, **hexokinase** (**HXK**), as an example of a growth-promoting pathway, and on a protein kinase called **SnRK1** (**Snf1-Related Kinase1**), as an example of a component of an inhibitory network. The two systems interact through their sensitivity to sugar metabolism (Figure 7.28).

The enzymes encoded by the *Arabidopsis HXK1* and rice *HXK5* and *HXK6* genes function as glucose sensors. The glucose-sensing property of HXK resides mainly in the mitochondrion and is independent of its glycolytic role in converting glucose to glucose-6-phosphate. A fraction of HXK exists in the nucleus in high-molecular-weight complexes, which repress the expression of photosynthetic genes and promote proteasome-mediated degradation of transcription factors that function in plant hormone signaling pathways (Figure 7.28). *Arabidopsis hxk1* mutants are glucose-insensitive and show reduced shoot and root growth, delayed flowering and senescence and altered sensitivities to auxin and cytokinin. These results show

the importance of HXK in promoting both vegetative and reproductive growth.

SnRK1 is the plant homolog of Snf1, one of the main regulators of the shift from fermentation to aerobic metabolism in yeast. SnRK1 is a protein kinase which acts as both a post-translational inhibitor and a transcriptional activator with wide-ranging influence on development and environmental responses (Figure 7.29). It is activated by high cellular concentrations of sucrose or low glucose or both, by a period of darkness, and by nutrient starvation. SnRK1 inhibits several key metabolic reactions by phosphorylating the corresponding enzymes, including nitrate reductase and sucrose phosphate synthase. Activated SnRK1 also stimulates the transcription of genes that encode enzymes of carbon mobilization, including sucrose synthase and α-amylase. Plants in which SnRK1 expression has been experimentally downregulated display a number of developmental irregularities, including abnormal pollen, stunted roots and premature senescence. High sensitivity to salt stress and susceptibility to pathogen attack have also been observed in SnRK1-deficient plants. Such experiments support the idea that SnRK1 is a key player in the global regulation of metabolism and development.

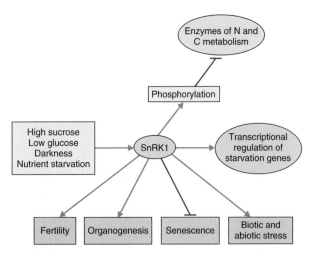

Figure 7.29 SnRK1 functions in the regulation of metabolism, development and stress responses. Activated SnRK1 inactivates several key metabolic enzymes by phosphorylating them, and is a transcriptional activator of a number of genes including those encoding sucrose synthase and α-amylase. SnRK1 also stimulates the signaling networks necessary for vegetative and reproductive development and stress responses, and delays senescence.

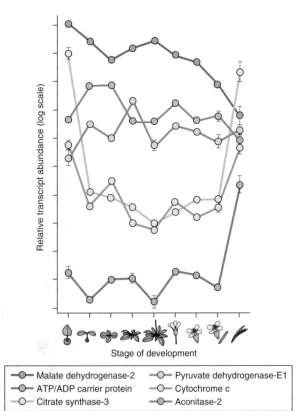

Figure 7.30 Expression of genes encoding respiratory enzymes of mitochondria at different stages of the *Arabidopsis* life cycle. The enzymes selected correspond to entries in Table 7.2. Transcript abundances were computed from a number of microarray experiments and are displayed using the Genevestigator meta-analysis tool (https://www.genevestigator.com/).

As might be expected of components of primary metabolic pathways that have to meet the fundamental requirements of virtually all plant tissues for energy and reductants, respiratory enzymes are encoded by genes that tend to be expressed throughout the plant life cycle, with relatively little developmental variation. Figure 7.30 shows typical patterns of transcript abundance for genes encoding representative respiration-related enzymes, selected from those listed in Table 7.2. The control of respiratory activity is a clear illustration of the hierarchical nature of metabolic regulation in plants, from the fine-scale moment-to-moment tuning at the level of substrate flux and feedback through the coarse adjustments of low amplitude and extended timeframe that result from gene activation and repression in response to environmental and developmental signals.

Key points Light directly influences the transcription of some respiratory genes, including those encoding pyruvate dehydrogenase, TCA cycle components, cytochrome c, ADP/ATP carrier protein, bypass NAD(P)H dehydrogenase and alternative oxidase. In many cases there is evidence for the involvement of the photoreceptors phytochrome and cryptochrome, working with the endogenous biological clock. Sugars are powerful regulators of gene expression. In addition to its enzymatic function in glycolysis, hexokinase is a sugar sensor that forms a multicomponent complex in the cell nucleus where it downregulates the expression of genes for photosynthesis and modulates hormone signaling pathways. SnRK1 is another sugar-sensing regulator. It is a kinase activated in response to limited availability of carbon, nutrient or light, and inhibits several key enzymes by phosphorylation. SnRK1 also transcriptionally upregulates carbon mobilization, and is required for normal development and stress resistance. There is relatively little variation during development in the expression of genes that encode respiratory enzymes.

Part III
Emergence

Chapter 8

Light perception and transduction

8.1 Introduction to light and life

Light is the ultimate source of energy for most of life on Earth. Plants and other photoautotrophs capture light energy in the process of photosynthesis (see Chapter 9). But light is more than the driver of energy-demanding metabolic activities; it also reports on the state of the environment through its quality (the balance of photons of different wavelengths), intensity (energy flux) and interactions with other environmental factors.

Plants are acutely sensitive to seasonal, daily and moment-to-moment variations in solar radiation. Light has a dramatic effect on growth and development. When seedlings germinate and begin to grow underground, their stems elongate rapidly and their cotyledons and/or leaves do not expand. They are pale in color because they lack chlorophyll. In eudicot seedlings, the apical portion of the stem is hooked so that the shoot apex faces

downwards. In cereals and other grasses, the coleoptile encloses the shoot apex and young leaves. Seedlings grown in the dark are said to be **etiolated**. This type of growth is called **skotomorphogenesis** (*skoto* = dark). In the light, **photomorphogenesis** begins. Photomorphogenesis is defined as the developmental response of an organism to information in light, which may be its quantity, quality (i.e. wavelengths present) and direction or the relative length of day and night (**photoperiod**). On exposure to light, stem elongation slows; cotyledons and/or leaves expand and become green. In eudicots, the apical hook straightens and in grasses coleoptile growth decelerates and stops as expanding leaves pierce its tip (Figure 8.1).

In order to respond to light, organisms must possess **photoreceptors**, molecules that absorb light and set in motion a cascade of events leading to biological responses. In this chapter we will review the general properties of light, discuss the way plants detect light, and consider how light perception leads to biological responses.

The Molecular Life of Plants, First Edition. Russell Jones, Helen Ougham, Howard Thomas and Susan Waaland.
© 2013 John Wiley & Sons, Ltd. Published 2013 by John Wiley & Sons, Ltd.

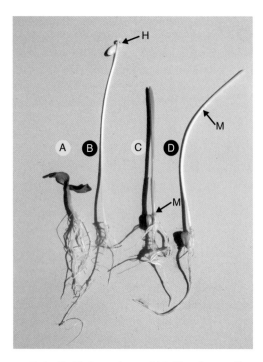

Figure 8.1 (A, B) Cucumber and (C, D) maize seedlings grown for 1 week in the light (A, C) or in the dark (B, D). H, apical hook; M, mesocotyl–coleoptile border.

Table 8.1 The fate of solar power reaching Earth.

Global solar power balance	Amount in terawatts[a]
Solar power input[b]	178 000
Reflected to space immediately	53 000
Absorbed and then reflected as heat	82 000
Used to evaporate water (weather)	40 000
Captured by photosynthesis (net primary productivity)[c]	100
Total power used by human society:	
In 2005	13
Projected use in 2100	46
Total used for food	0.6

[a]The **watt** is the unit of **power** and is related to the **joule**, the unit of **energy**, by watts = joules per unit time. A terawatt is 10^{12} watts and is equal to 10^{12} joules s^{-1}.
[b]Total solar energy input per year = 5.62×10^{12} terawatts (5.62×10^{24} joules).
[c]Total solar energy per year captured by photosynthetic organisms = 3.16×10^9 terawatts (3.16×10^{21} joules).

8.1.1 Visible light is part of the electromagnetic spectrum

The sun is the ultimate source of all non-nuclear energy on Earth. **Solar energy** is the product of the sun's nuclear fusion reaction that converts protons into helium nuclei at a rate equivalent to almost 10^{17} kg of TNT per second. The total annual solar energy absorbed by the Earth's atmosphere, oceans and landmasses amounts to some 5.62×10^{24} joules, of which photosynthesis captures 3.16×10^{21} joules per year (Table 8.1).

Visible light is just one part of the electromagnetic spectrum, which stretches from γ- and X-rays at one extreme through to radiowaves at the other (Figure 8.2). Light has the properties simultaneously of a wave and a particle. In simple terms it can be thought of as individual packets of energy or **quanta** that move in waves. A quantum of light energy is called a **photon**. The **wavelength** (λ, Greek letter lambda) of visible light is usually expressed in nanometers (nm). The **visible spectrum**, which we see as the colors of the rainbow, runs from a wavelength of about 380 nm (violet) through to 760 nm (far red). Equation 8.1 expresses the relationship between wavelength (in meters) and **frequency** (υ, Greek letter nu; units = s^{-1}) and **speed** of light (c, units = m s^{-1}).

Equation 8.1 Relationship between wavelength, frequency and speed of light

$$c = \upsilon\lambda$$

The energy of a quantum is directly related to its frequency and inversely related to its wavelength, as given in Equation 8.2. The proportionality constant **h** is called Planck's constant. E, for photons, is expressed in units of **electron-volts** (eV): 1 eV is equal to 1.6×10^{-19} joules. The term **hυ** is often used to represent a photon.

Equation 8.2 Relationship between wavelength or frequency of electromagnetic radiation and energy

$$E = hc/\lambda = h\upsilon$$

where c = speed of light (approximately 300×10^6 m s^{-1}) and h = Planck's constant (4.14×10^{-15} eV · s). It follows that the energy of a given wavelength of light (in the commonly expressed unit of nm) is:

$$E = 1240/\lambda_{nm}$$

Gamma- and X-rays are at the short wavelength end of the electromagnetic spectrum and are very energetic

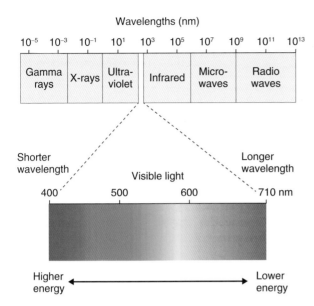

Figure 8.2 The electromagnetic spectrum with the portion from 400 to 710 nm expanded to show the colors of the visible wavelengths. The limits of human perception can extend beyond this range, as far as 380 nm at the blue end and 760 nm at the red end.

$(E > 10^5$ eV). Radiowaves have long wavelengths and are comparatively low energy $(< 10^{-6}$ eV). The photon energy of visible light at the short wavelength (blue-violet) end of the spectrum is higher than that at long wavelengths (red)—about 3.3 eV compared with 1.6 eV. **Ultraviolet** (UV) light is generally subdivided into UVA (wavelength range 400–320 nm, photon energy 3.1–3.94 eV), UVB (320–280 nm, 3.94–4.43 eV) and UVC (280–100 nm, 4.43–12.4 eV).

8.1.2 Light interacts with matter in accordance with the principles of quantum physics

The interaction of light with living organisms can be characterized by the principles of quantum physics. A photon colliding with an atom may transfer its energy to an electron. The subsequent fate of the electron and photon will depend on the energy level of the photon and the nature of the atom. Short wavelength photons such as γ- or X-rays are energetic enough to give the electron sufficient kinetic energy to break free of the atom altogether. The short wavelength portion of the electromagnetic spectrum is therefore often referred to as the source of **ionizing radiation**. Life on Earth is possible because solar energy reaching the planet's surface (Figure 8.3) has been filtered by the atmosphere to remove most of the ionizing radiation (wavelengths <295 nm) that would otherwise be hazardous to living

matter. Even so, enough short wavelength photons (principally UVB) reach the biosphere to make it necessary for organisms to equip themselves with antioxidant defenses, repair mechanisms and sunblockers (discussed in detail in Chapter 15).

A molecule that can interact with photons within the visible part of the electromagnetic spectrum is called a **pigment**. Pigments are colored to the human eye because they selectively absorb certain parts of the visible wavelength range. When light of visible wavelengths impinges upon such a pigment molecule, it is not energetic enough to cause ionization but may make an electron jump to a higher energy level within the atom (Figure 8.4). When the atom of a pigment absorbs a light quantum with an energy that matches the energy difference between the molecule's non-excited (ground) state (E_g) and excited state (E_e), one of its electrons is shifted from a lower-energy to a higher-energy molecular orbital (Equation 8.3).

Equation 8.3 Relationship between energy of absorbed photon, ground state and excited state

$$E_e - E_g = hc/\lambda$$

Because a pigment molecule consists of many different atoms and electrons, each with its own E_e and E_g, it will absorb light across broad wavelength ranges (for example, one wave band in the blue and one in the red spectral region in the case of chlorophyll) rather than the sharp bands characteristic of individual atoms (Figure 8.4). An energized electron (sometimes called an **exciton**) in this molecular environment may have one of a number of possible fates. As Figure 8.4 shows, it may immediately fall back to its original energy level, re-emitting energy as light (**fluorescence**) and/or heat (**infra-red** energy). It may remain in the high-energy state for a relatively prolonged period until it gives up its energy and reverts to the low-energy state (phosphorescence). It may be transferred to another molecule, leaving the donor pigment molecule positively charged and the acceptor negatively charged (a process called **charge separation**, of central importance in photosynthesis; see Chapter 9). The energy levels of infra-red and other long wavelength photons are too low to make electrons jump, but their absorption by molecules can translate into **vibrational** energy in bonding systems. Absorption of infra-red radiation by 'greenhouse gases' such as carbon dioxide and methane traps heat in the atmosphere and is of international concern as the cause of global warming and **climate change**. In all cases the energy transfers from photons to atoms strictly obey the laws of thermodynamics, as shown by the higher entropy levels of the system and the

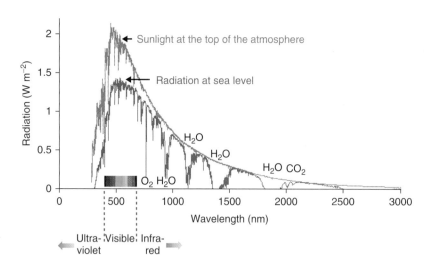

Figure 8.3 The spectrum of solar energy reaching the Earth. The red and blue lines are the solar spectra, respectively, before and after filtering by the Earth's atmosphere. The visible spectrum, ultraviolet and infra-red regions, and bands of radiation absorbed by atmospheric O_2, H_2O and CO_2 are identified.

increased wavelength (decreased energy) of photons re-emitted when excitons revert to the low-energy state (Figure 8.4).

In vivo, if pigment molecules absorb light energy in excess of the capacity of normal energy metabolism to use it, **photosensitization** may occur. In this case the energized pigment brings about chemical alteration of another molecule in the system. Photosensitization is

frequently harmful and makes it necessary for defenses against damage by excess light to be built into cell structure and function. This is particularly important in plants, whose primary energy source is light. Chapter 9 describes how photosynthesis is organized to deal with the hazards of photosensitivity, and defenses against toxic chemical products of excess energy are discussed in Chapter 15.

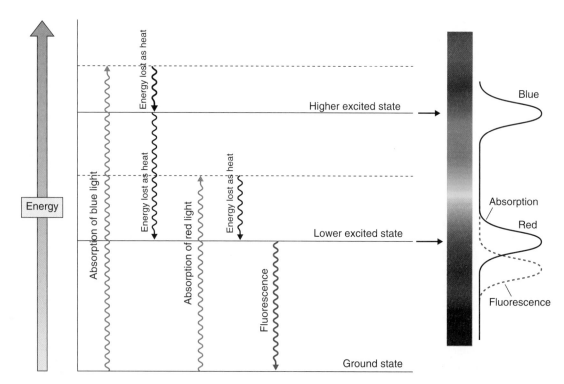

Figure 8.4 Energy levels within a pigment molecule interacting with light. The example shows a molecule such as chlorophyll that absorbs light in both the blue and red regions of the spectrum. During transitions between excited states, energy is lost as heat, in accordance with the second law of thermodynamics. Fluorescence is the re-emission of light from the lowest excited state. The wavelength maximum of photons emitted through fluorescence is longer (i.e. they are of lower energy) than the maxima of photons absorbed by the pigment molecule. The spectra for fluorescence and absorption are shown at the right of the figure. The short wavelength absorption band corresponds to a transition to the higher excited state, and the long wavelength absorption band corresponds to a transition to the lower excited state.

Key points Sunlight drives photosynthesis and is also the source of environmental signals that regulate plant development. Skotomorphogenesis is development in darkness. Photomorphogenesis is development in response to light and is mediated by photoreceptors. Visible light, a small region of the electromagnetic spectrum, covers the wavelength range 380 nm (violet) to 760 nm (far red). The energy of a photon (a quantum of light) is inversely related to its wavelength. A photon may interact with an atom by transferring its energy to an electron, which then jumps from the ground to the energized state. Highly energetic photons, with wavelengths less than 295 nm, cause the ionization of atoms by completely displacing electrons. Photoreceptor pigments are molecules that absorb visible light. The energized electron of a pigment atom can revert to the ground state while re-emitting light energy (fluorescence) and/or heat (infra-red energy), or it may transfer to an acceptor molecule (charge separation). Charge separation provides the energy for photosynthesis. Photon energy at infra-red and longer wavelengths is too low to move electrons from the ground state but can increase molecular bond vibrational energy. In this way, greenhouse gases such as carbon dioxide absorb infra-red radiation, with consequences for global climate change.

8.1.3 Photobiology is the study of the interaction of light with living organisms

The study of the interaction between light and life is called **photobiology**. Light-absorbing **photoreceptor** molecules allow an organism to monitor environmental rhythms and fluctuations and to adjust its physiology accordingly. In humans and other animals, rhodopsin is one of the photoreceptors for vision. Plants have a number of different photoreceptors. These include the photosynthetic pigments (see Section 8.4 and Chapter 9), phytochromes (see Section 8.2), and cryptochromes and phototropins (see Section 8.3). Each photoreceptor has a characteristic **absorption spectrum**. The wavelengths of light that are absorbed by a photoreceptor activate specific responses. Plotting the intensity of a particular physiological response against the wavelengths that trigger it produces an **action spectrum**. Measuring the action spectrum for a photoresponse helps to identify the photoreceptor for the response. Examples of action spectra will be introduced in Section 8.2.3.

Most aspects of vegetative and reproductive growth and development are attuned to the light environment, as illustrated in Figure 8.5. For instance buried, light-sensitive seeds remain dormant until the soil is disturbed and they are exposed to light (see Chapter 17). Many plant species will not switch from the vegetative to reproductive state unless particular daylength conditions

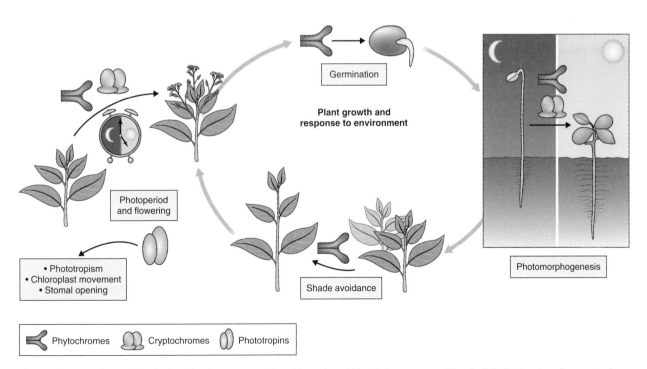

Figure 8.5 Light-regulated plant development mediated by red and blue light receptors. The clock indicates circadian control.

for flower induction are met. Seasonal behavior, like bud dormancy and the shedding of parts such as leaves, is also determined by changing light conditions over the annual cycle. Furthermore, plants respond to the direction of light. Charles Darwin and his son Francis provided one of the earliest scientific accounts of **phototropism**, the growth of plants and their parts towards or away from a light source (tropism = directional growth). In the following sections we will look at the major photoreceptors and present examples of how absorption of light by the photoreceptor leads to a biological response.

> **Key points** Photobiology studies the response of living organisms to light. The major photoreceptors of plants are chlorophyll and other photosynthetic pigments, phytochromes, cryptochromes and phototropins. Each photoreceptor interacts with photons at particular wavelengths and therefore has a characteristic absorption spectrum. Measuring a photomorphogenic process at different wavelengths produces an action spectrum, which is often directly related to the absorption spectrum of the mediating photoreceptor. Photoreceptors enable plants to coordinate phases of vegetative and reproductive development, including growth towards or away from sources of illumination (phototropism), with the state of the light environment during the life cycle.

8.2 Phytochrome

Germination of light-sensitive lettuce seeds is promoted by exposure to **red light** (R; λ around 660 nm). If the red light stimulus is followed by exposure to **far-red light** (**FR**; λ 730 nm), seeds do not germinate (see Figure 6.24). In turn, the effect of far-red light can be reversed by red light. The property of R/FR photoreversibility allowed scientists to use spectrophotometry to isolate and identify the photoreceptor pigment **phytochrome**. Phytochrome has been purified and the genes that encode it have been identified. Developmental processes regulated by phytochrome include: induction of germination in light-sensitive seeds; de-etiolation and increased chlorophyll synthesis (see Figure 8.1); decreased rate of stem elongation; promotion of leaf expansion; shade avoidance; photoperiodism; and flowering (see Figure 8.5).

8.2.1 Light acts through isomerization of the phytochrome chromophore

Phytochrome is a **biliprotein**, consisting of a 120 kDa apoprotein with a photoreactive prosthetic group (**chromophore**). The chromophore is a straight-chain tetrapyrrole, a bilin designated (**3E**)-**phytochromobilin** or PΦB (3E refers to the isomeric form of the photoactive molecule). Phytochromobilin is structurally related to, and shares its biosynthetic origin with, cyclic tetrapyrroles such as heme and chlorophyll (Section 8.4). The four pyrrole rings are designated A, B, C and D (Figure 8.6). Phytochromobilin adopts two molecular configurations, $P\Phi B_R$ and $P\Phi B_{FR}$. $P\Phi B_R$, the red-absorbing form, has a peak of absorbance at a wavelength of 660 nm. On exposure to red light, $P\Phi B_R$ is converted to $P\Phi B_{FR}$ and the absorbance peak is shifted to 730 nm in the far-red region of the spectrum. In far-red light, $P\Phi B_{FR}$ changes back to $P\Phi B_R$. The phytochrome biliprotein with the chromophore in the $P\Phi B_R$ configuration is the **red-absorbing form** and is referred to as P_R; the **far-red-absorbing form** is P_{FR}, with the bilin chromophore in the $P\Phi B_{FR}$ configuration. The light absorption profiles of P_R and P_{FR} are illustrated in Figure 8.7A and the corresponding difference in color between P_R and P_{FR} in solution is shown in Figure 8.7B. P_{FR} has considerable absorbance at red wavelengths. For this reason, under red light the population of phytochrome molecules consists of a maximum of 85% P_{FR}. Conversely, the proportion of P_R under far-red light can reach 95%. Photoreversible $P\Phi B_R - P\Phi B_{FR}$ interconversion is the chemical basis of the phytochrome effect in photomorphogenesis.

The A pyrrole ring of the chromophore is linked to a cysteine residue in the polypeptide chain of the phytochrome apoprotein. In $P\Phi B_R$ the C and D pyrroles are attached to each other through a **methine bridge** (–CH=) in the cis configuration (Figure 8.6). $P\Phi B_{FR}$ is the trans isomer. Red light flips the cis bond into the trans conformation and far-red light drives the reverse reaction. High-resolution spectroscopy has identified transient intermediate states of the chromophore between $P\Phi B_R$ and $P\Phi B_{FR}$, known as **lumi-** forms, arising from the effects of protein conformation changes. In addition, $P\Phi B_{FR}$ slowly converts to $P\Phi B_R$ in the absence of light in a process referred to as **dark reversion**. $P\Phi B_{FR}$ is the form that triggers photomorphogenesis. The biological outputs from the phytochrome cycle are determined by the ratio of $P\Phi B_R$ and $P\Phi B_{FR}$ forms, which in turn reflects the relative levels of red and far red in the light environment and the

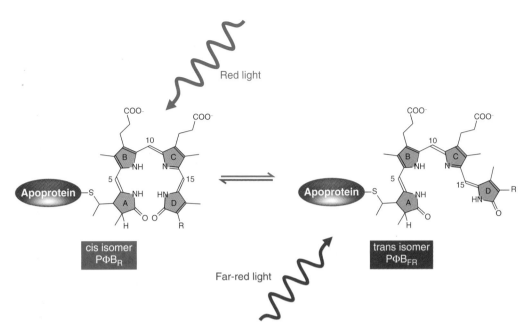

Figure 8.6 Cis–trans isomerization of the phytochrome chromophore (PΦB) following exposure to red or far-red light. Note the change in configuration about the double bond between carbon atom 15 and the D-ring.

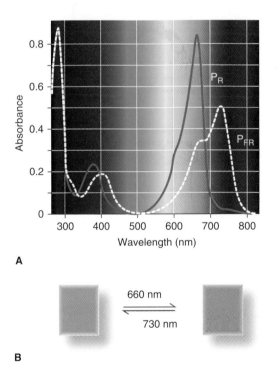

A

B

Figure 8.7 Light perception is mediated by the photoreversible receptor protein phytochrome. (A) Absorption of light in the visible wavelength range by the two molecular forms of phytochrome. (B) Reversible photoconversion of the phytochrome chromophore is associated with a visible change in the color of the photoreceptor in solution.

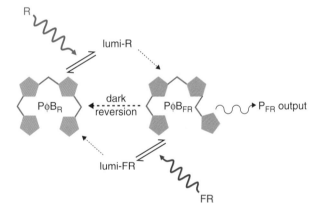

Figure 8.8 The phytochrome photocycle driven by red and far-red light. Lumi-R and lumi-FR are transient intermediate states of the chromophore during its conversion from one configuration to the other.

rate of dark reversion. The phytochrome photocycle is summarized in Figure 8.8.

8.2.2 Phytochrome protein has a complex multidomain structure

Rapid advances in our understanding of phytochrome structure and mode of action have come from

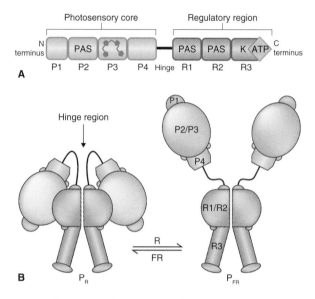

Figure 8.9 Domain structure and conformations of phytochrome. (A) Domain configuration, showing the P1–P4 regions of the photosensory domain, the PAS motif repeated in the regulatory region of the molecule, and the C-terminal histidine kinase module. The chromophore is shown attached to P3. (B) Dimerization of phytochrome and conformations of P_R and P_{FR}.

comparative studies of photoreceptive biliproteins across a range of plants and microorganisms, from seed plants to fungi to cyanobacteria. Alignments of protein sequences and modeling of three-dimensional structures based on X-ray crystallography data have identified functional and regulatory motifs within the phytochrome molecule (Figure 8.9). The phytochrome polypeptide consists of an N-terminal **photosensory** region and a C-terminal **regulatory** domain, with a **hinge** between them. The protein has a modular architecture within which a number of structural motifs are recognized. The photosensory part has four domains (Figure 8.9A). P1, at the N terminus, and P4 function in the inhibition of P_{FR} to P_R reversion in the dark. P2 is a motif with the so-called PAS fold, a conserved structural feature that is found in a wide range of proteins in prokaryotes and eukaryotes and which is involved in signal sensing. P3 is the binding site of phytochromobilin. The three-dimensional model of phytochrome shows that the chromophore occupies a pocket formed by the folding of the P3 domain. The regulatory fragment of phytochrome consists of two PAS-type domains and a C-terminal kinase region (Figure 8.9A).

In cells phytochrome exists as a **dimer** (Figure 8.9B). In the red-absorbing form of the holoprotein, P_R, both subunits are folded at the hinge so that the N-terminal photosensory and C-terminal regulatory regions are brought into contact with each other. Photoconversion

of the chromophore by red light results in disruption of the interaction between the N- and C-terminal domains, opening up the structure and exposing buried surfaces (Figure 8.9B) in the far-red-absorbing form of phytochrome, P_{FR}. This makes these surfaces accessible to **phytochrome-interacting factors** (PIFs) in the signal transduction pathway. The relationship between P_{FR} and such PIFs is described in Section 8.2.4. The conformational change also allows for **light-dependent phosphorylation** of exposed serine and threonine residues which may result from autophosphorylation by the R3 domain kinase. A P_{FR}-specific **phosphatase** has been identified that removes phosphate groups from amino acids in the hinge region; this increases their affinity for PIFs and susceptibility to proteolysis. Phytochrome has also been shown to phosphorylate other sensory photoreceptors, including cryptochromes (see Section 8.3) and proteins of the auxin signaling pathways (see Chapter 10), which can promote **cross-talk** between light and other regulators of plant development. Another consequence of the conformational difference between P_R and P_{FR} is the difference in stability of the two forms in vivo. The **half-life** of P_R is about 1 week, whereas that of P_{FR} is 1–2 hours. In part this reflects the difference in compactness of the protein which reveals peptide linkages in P_{FR} that are susceptible to hydrolysis by proteolytic enzymes. Thus the combination of red/far-red light modulation,

Key points The photoreceptor phytochrome exists in two photoreversible forms. P_R absorbs red light (wavelength around 660 nm) and is converted to P_{FR}, which reverts to P_R when exposed to far-red light (about 730 nm). Red/far-red reversibility of a developmental process, for example germination of light-sensitive lettuce seeds, is diagnostic of mediation by phytochrome. Phytochrome protein has a photoreactive prosthetic group (phytochromobilin) that flips between two molecular forms in response to red or far-red light. The prosthetic group is bound to the protein in the N-terminal photosensory domain, which is connected to a C-terminal regulatory domain by a hinge region. Active phytochrome is a dimer. In the P_R form, the N- and C-terminal regions of each subunit are in contact. Photoconversion to P_{FR} opens the hinge, makes buried surfaces accessible to phytochrome-interacting factors (PIFs), phosphorylation and proteolytic enzymes, and initiates photomorphogenesis. As a consequence of the difference in protein conformation, the half-life of P_R is about 1 week and that of P_{FR} is 1–2 hours.

phosphorylation/dephosphorylation and differential proteolytic stability allows phytochrome action to be subject to a fine degree of control.

8.2.3 Different forms of phytochrome are encoded by multiple genes

In the discussion of phytochrome structure to this point, we have assumed that there is just one type of phytochrome molecule. In fact, molecular analysis has shown that there is a family of phytochrome genes. For example, *Arabidopsis* has five genes, designated *PHYA* to *PHYE*, that encode phytochrome. Rice has three *PHY* genes (*PHYA* to *PHYC*). Most conifers and mosses have four *PHY* genes, cycads three, ferns two and lycopods one. Analysis of phylogenetic relationships, based on comparison of gene sequences, groups *PHYA* with *PHYC* and puts *PHYB* in a divergent group with *PHYD* and *PHYE* (Figure 8.10A). The various phytochromes encoded by these genes have distinctive molecular properties. For example, PHYA is unstable in the light and consequently is most abundant in etiolated and newly-emerged tissues, whereas all the other phytochromes are light-stable and predominate in light-grown plants (Figure 8.10B). *Arabidopsis* PHYA forms only **homodimers**, whereas the other *Arabidopsis* PHYs can form **heterodimers** with each other, though the physiological significance of such contrasting dimerization behavior is not clear.

In addition to variation in expression between different tissues during the life cycle (Figure 8.10B), the

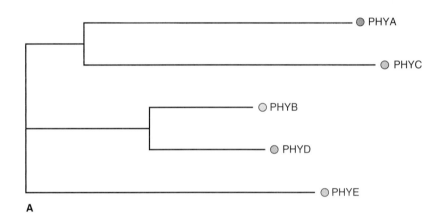

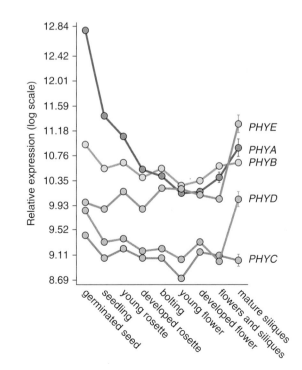

Figure 8.10 (A) Relationships between members of the *Arabidopsis* PHY family based on protein sequence alignment. PHYA and PHYC have about 50% homology to each other. PHYB and PHYD are closely related with about 80% homology. PHYE has less than 50% homology with the PHYA/C group and more than 50% with PHYB/D.
(B) Expression of each *PHY* gene in different tissues during the *Arabidopsis* life cycle.

diverse forms of phytochrome differ in their sensitivities to light and in the responses they control. Phytochrome-mediated responses can be classified into the high irradiance reaction (**HIR**) and low fluence responses. **Fluence** is defined as the number of photons impinging on a unit surface area and has units of mol m^{-2}. **Irradiance** is fluence rate, that is, fluence per unit time, with units of mol m^{-2} s^{-1}. A characteristic of low fluence responses (but not HIR) is that they obey the principle of **reciprocity**: they may be invoked by long exposure to dim light or short exposure to higher irradiance. Low fluence responses are further subdivided into the low fluence red/far-red reversible reaction (**LFR**) and the very low fluence response (**VLFR**). Figure 8.11 summarizes the fluence-response relationships of VLFR,

LFR and HIR and gives examples of photomorphogenic events associated with each class.

Mutations in one or other of the phytochrome genes have been identified in a number of plant species and these have allowed individual members of the phytochrome gene family to be associated with particular fluence responses. For example, the *lh* mutant of cucumber (Figure 8.12) and the *hy3* mutant of *Arabidopsis*, which are deficient in PHYB, grow taller than wild-type plants in white light. This phenotype mimics the effect of shading and is consistent with the ecological role of phytochrome in proximity-sensing and competition for light in plant communities. When the PHYA-deficient *aurea* mutant of tomato is grown in white light, it has yellow-green leaves, elongated

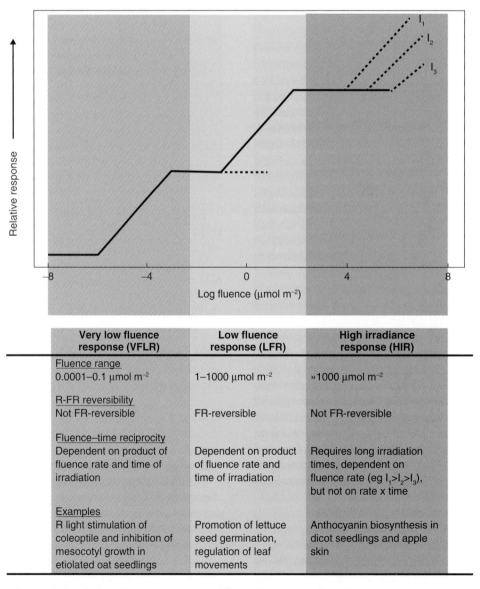

	Very low fluence response (VFLR)	Low fluence response (LFR)	High irradiance response (HIR)
Fluence range	0.0001–0.1 µmol m^{-2}	1–1000 µmol m^{-2}	»1000 µmol m^{-2}
R-FR reversibility	Not FR-reversible	FR-reversible	Not FR-reversible
Fluence–time reciprocity	Dependent on product of fluence rate and time of irradiation	Dependent on product of fluence rate and time of irradiation	Requires long irradiation times, dependent on fluence rate (eg I$_1$>I$_2$>I$_3$), but not on rate x time
Examples	R light stimulation of coleoptile and inhibition of mesocotyl growth in etiolated oat seedlings	Promotion of lettuce seed germination, regulation of leaf movements	Anthocyanin biosynthesis in dicot seedlings and apple skin

Figure 8.11 Characteristics of phytochrome responses at different fluences. FR, far red; R, red.

Figure 8.12 Growth of wild-type and *lh* mutants of cucumber. Plants were grown for 20 days on a 14/10-hour light/dark cycle and exposed to 20 minutes of far-red light (FR) or darkness (control treatment, D) at the end of each day period.

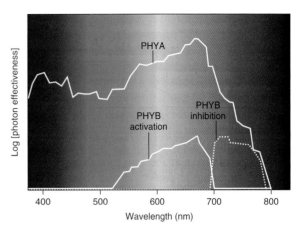

Figure 8.13 Phytochrome action spectra measured as the capacity of various wavelengths of light to induce seed germination or to reverse red light-dependent promotion of germination in *Arabidopsis*. The PHYA action spectrum is determined on seed of a PHYB-less mutant, imbibed in water for 2 days in darkness. The action spectrum for PHYB is determined on seed of a PHYA mutant imbibed in water for 3 hours in darkness. PHYB activation of seed germination by red light is reversible by exposure to far-red light while the PHYA effect is not. Germination response is determined as photon effectiveness (fluence required for the induction or inhibition) and is expressed on a log scale.

hypocotyls and reduced anthocyanin content. The evidence from such studies identifies PHYB and PHYA as the prime regulators of LFR and VLFR, respectively. The different functions of PHYA and PHYB are reflected in their action spectra. Figure 8.13 shows the action spectrum of PHYA, determined for the VLFR of *Arabidopsis* seed imbibed in water for 2 days in darkness, and the PHYB action spectrum determined in seed imbibed for very short periods. The PHYA-related reaction is not far-red reversible. By contrast, the PHYB response is a typical LFR (see Figure 8.11) in being reversed by exposure to far-red light.

The three other phytochromes of *Arabidopsis* have a diversity of functions, some of which overlap (Table 8.2). Although phytochromes clearly are involved in HIR, evidence indicates that other photoreceptors that absorb UV or blue light (see Section 8.3) contribute to this control. In relation to the different fluence responses, it seems that the various phytochromes within or between species share similar functions but have diverged to adopt particular roles, reflecting evolutionary separation and particular ecophysiological adaptations.

8.2.4 Phytochrome regulates gene expression by interacting with a number of proteins

Induction of photomorphogenesis by phytochrome involves large changes in the patterns of gene expression. If phytochrome is to affect gene expression, a number of events must take place. First P_R, which is found in the cytosol, must be converted to P_{FR}, which then moves into the nucleus. Next P_{FR} interacts with one of a number of **phytochrome-interacting factors** (**PIFs**). The first PIF was discovered by using yeast **two-hybrid screening**, a technique that allows the identification of molecules that bind to a protein of interest (Figure 8.14). When PHYB was used as bait in this technique, a PIF called PIF3 was discovered. PIF3 is a nuclear-localized basic helix-loop-helix (bHLH) transcription factor (see Chapter 3) that binds to PHYB in preference to PHYA, at the C-terminal domain. Subsequently more members of the PIF3 bHLH protein family have been identified; these include PIF1, -4, -5 and -6. PIF3 and related factors promote the transcription of skotomorphogenic genes and suppress photomorphogenesis. After P_{FR} complexes with PIF3, the interaction leads to the phosphorylation of PIF3, which is followed by its ubiquitination and degradation via the proteasome (Figure 8.15; see Section 5.9). Thus P_{FR} removes the transcription factors

Table 8.2 Examples of functions influenced by phytochromes A–E, determined by analysis of *Arabidopsis* mutants.

Function	PHYA	PHYB	PHYC	PHYD	PHYE
Promotion of seed germination	+	+			+
Regulation of seedling de-etiolation	+	+	+	+	+
Stimulation of chlorophyll synthesis	+	+			
Regulation of root gravitropic curvature		+			
Suppression of root hair growth		+			
Regulation of leaf architecture	+	+	+	+	+
Suppression of internode elongation	+	+			+
Suppression of shade avoidance		+		+	+
Regulation of stomate:epidermal cell ratio		+			
Circadian clock entrainment	+	+		+	+
Photoperiodic perception	+		+		
Repression of flowering		+	+	+	+

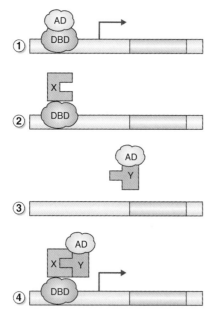

Figure 8.14 Yeast two-hybrid screening. (1) The transcription factor that regulates a yeast gene encoding a selectable marker is engineered so that it is separated into a DNA-binding domain (DBD) and a transcription activation domain (AD). The DBD or AD alone is unable to activate transcription. (2) If one fragment binds a protein of interest, X, usually referred to as the 'bait' (phytochrome in this case), and (3) the other fragment binds an interacting protein, Y, the 'prey' (a PIF, for example), the transcription factor is reconstituted (4) and the yeast reporter gene is transcribed.

that promote skotomorphogenesis and rebalances transcription in favor of photomorphogenesis.

There are many other types of molecule that interact with phytochrome. **FHY1** (Far-red elongated Hypocotyl 1) and **FHL** (FHY1-like) are two PIFs that interact with the P_{FR} form of PHYA but not PHYB and facilitate the transport of PHYA to the nucleus. Some PIFs influence phytochrome activity by stabilizing P_{FR}. **ARR4** (*Arabidopsis* response regulator 4) is an example of this type of PIF; it stabilizes PHYB but not PHYA. ARR4 is a member of the type A family of *Arabidopsis* response regulators, which act as negative regulators of cytokinin signaling (see Section 10.4.3). ARR4 protein accumulates in red light, in a PHYB-dependent manner. It binds equally to the P_R and P_{FR} forms of PHYB and inhibits dark reversion of P_{FR} to P_R. By stabilizing the P_{FR} form of PHYB, ARR4 binding enhances phytochrome activity.

Another PIF that regulates the output activity of phytochrome is **COP1** (Constitutive Photomorphogenic 1). COP1 is the master repressor of photomorphogenesis; *cop1* mutants undergo constitutive photomorphogenesis, even in the dark. COP1 is an E3 ubiquitin ligase (see Section 5.9). It promotes skotomorphogenesis by targeting a number of the transcription factors that enhance expression of photoresponsive genes to the 26S proteasome for degradation. The cellular localization of COP1—cytosolic in the light, nuclear in the dark—is regulated by PHYA and PHYB (and also cryptochrome). COP1 bound to P_{FR} is excluded from the nucleus, thus stabilizing positive regulators of light-regulated genes

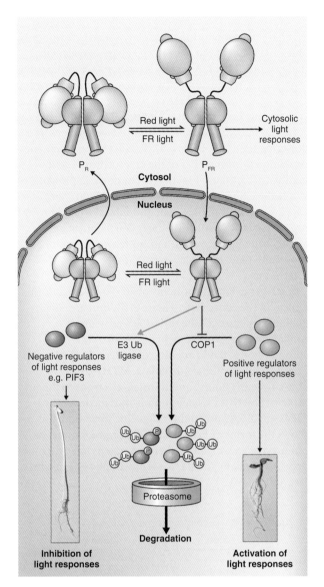

Figure 8.15 The effect of phytochrome on photomorphogenesis. P_{FR} promotes photomorphogenesis by stimulating the degradation of negative regulators of photoresponses via the ubiquitin–proteasome system and by preventing the degradation of positive regulators. COP1 is an E3 ubiquitin ligase that specifically targets positive regulators of light responses.

protein turnover via the ubiquitin–proteasome system (UbPS). The phytochrome regulatory network is extensive. At least 15 different proteins with roles in phytochrome signal transduction have been identified as interacting with P_{FR}, and over a dozen more are known from genomic analyses and protein-binding screens. We are only just beginning to explore the mechanisms and interactions of the pathways in which these components operate, but it is already clear that phytochrome-mediated light perception is of central importance in the control of plant growth and development.

Key points Phytochrome in angiosperms is encoded by a family of up to five genes (*PHYA–E*). The DNA sequence of the *PHYB/D/E* group is divergent from that of *PHYA/C*. Unlike the other PHYs, PHYA is unstable in the light (therefore abundant in etiolated tissues) and exists exclusively as a homodimer (other PHYs can form heterodimers with each other). PHY-mediated responses are divided into the high irradiance reaction (HIR; for example, anthocyanin biosynthesis), low fluence response (LFR; germination of light-sensitive seeds, for example) and very low fluence response (VLFR; for instance, red light stimulation of coleoptile growth). HIR requires participation of other photoreceptors in addition to PHYs. LFR is regulated primarily by PHYB, and VLFR by PHYA. The other PHYs have diverse, sometimes overlapping, functions. P_R is located in the cytosol, P_{FR} in the nucleus. P_{FR} interacts with the transcription factor PIF3 and targets it for degradation by the ubiquitin system. PIF3 and related PIFs are suppressors of photomorphogenesis. Other PIFs regulate PHY stability, subcellular localization and hormone signaling interactions. P_{FR} also blocks the activity of COP1 (Constitutive Photomorphogenic 1), which promotes ubiquitin-mediated degradation of activators of photomorphogenesis. COP1 binding to PHYA increases ubiquitination and degradation of PHYA in the light.

and permitting photomorphogenesis (Figure 8.15). COP1 also appears to play a role in the light-triggered degradation of PHYA. Binding to COP1 increases ubiquitination and degradation of PHYA; conversely, PHYA is stabilized in the *cop1* mutant.

These examples of phytochrome interactions show that light influences morphogenesis in a very precise way through photoconversion of the phytochrome chromophore, changes in protein conformation, phosphorylation/dephosphorylation, complex formation with PIFs, subcellular compartmentation and differential

8.3 Physiological responses to blue and ultraviolet light

Phytochrome provides the plant with morphogenetic information primarily from the red end of the light spectrum. Plants are also sensitive to wavelengths in the

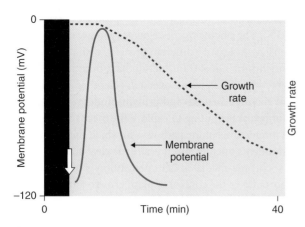

Figure 8.16 Blue light effect on plasma membrane potential and growth of etiolated hypocotyls. Individual cells of etiolated hypocotyls undergo a pronounced depolarization of the membrane potential when irradiated with blue light (white arrow). This is followed by a marked decline in their rate of growth.

range 320–500 nm. Among the reactions to blue light are changes in plant cell plasma membrane potential (Figure 8.16), inhibition of hypocotyl elongation, phototropism, stomatal opening, circadian clock setting and anthocyanin production. Biochemical and molecular responses include changes in redox reactions, electron transport and expression of blue light-regulated genes (Table 8.3). Plants are equipped with at least three classes of photoreceptors sensitive to UVA and blue wavelengths of light. **Cryptochromes** and **phototropins** are well-established receptors that mediate the effects of UVA/blue light. More recently a third class of blue light receptor, known as **zeitlupe** from the German for 'slow motion', has been identified. Zeitlupe proteins will be discussed in more detail in Section 8.3.3. All three groups of blue light photoreceptors contain **flavins** as chromophores. There is incomplete but growing knowledge of the processes regulated by these photoreceptors (Table 8.3) and the underlying mechanisms of photoreceptor activation and signal transduction.

8.3.1 Cryptochromes are responsible for regulating several blue light responses, including photo-morphogenesis and flowering

The *Arabidopsis* genome contains three genes, designated **CRY1–3**, that encode cryptochromes. *CRY1* was isolated

by screening mutagenized populations for individuals that have abnormally long hypocotyls when grown under blue or UVA light but have normal hypocotyl length under R or FR light. A second *Arabidopsis* cryptochrome gene, *CRY2*, was identified by sequence homology with *CRY1*. CRY1 and CRY2 are responsible for regulating various blue light responses, including hypocotyl and cotyledon growth, operation of the circadian clock and flowering time (see Table 8.3). The *Arabidopsis* genome contains a third member of the cryptochrome gene family, *CRY3*, whose function is unknown at present. While the CRY1 protein is stable in blue light, CRY2 is unstable in blue light and subject to rapid proteolysis. CRY2 and CRY1 are 58% identical in amino acid sequence at the amino-terminal region, but show only 14% sequence identity at the C terminus.

CRY genes encode proteins of 70–80 kDa that form dimers in the active state. Figure 8.17 compares the domain structures of the three CRY proteins of *Arabidopsis*. Cryptochromes contain two cofactor chromophores, **flavin adenine dinucleotide** (**FAD**) and **pterin** (methenyltetrahydrofolate, **MTHF**). FAD is non-covalently bound and transfers electrons to the protein moiety via a semiquinone intermediate when CRY1 and CRY2 are in their signaling state. Equation 8.4 shows the forms of FAD during the cryptochrome **photocycle**, in which the interconversions are driven not only by blue light and darkness but also by green light (see Section 8.3.5).

Equation 8.4 The cryptochrome photocycle

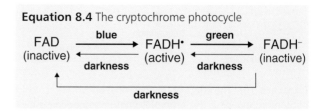

The cofactor-binding region of the protein is called the PHR (photolyase-related) domain because of its structural similarity to the bacterial repair enzyme DNA-photolyase (although plant CRYs have no photolyase enzyme activity). The PHR domain also mediates **dimerization**, which is essential for photoreceptor function. A conserved motif, referred to as DAS, occurs in the C-terminal domains of CRY1 and CRY2. The C-terminal region of CRY2 has also been shown to contain a **nuclear localization signal**. Although CRY1 lacks this sequence, it also has been shown to move between cytosol and nucleus. In CRY3 the DAS motif found between the PHR domain and an N-terminal extension (absent from CRY1 and CRY2) is required for the import of the protein into chloroplasts and mitochondria.

Table 8.3 Physiological responses to blue light and the corresponding photoreceptors.

| Response | Cryptochrome | | Phototropin | | Zeitlupe |
	CRY1	CRY2	PHOT1	PHOT2	(ZTL/ADO)
Inhibition of hypocotyl/stem elongation	High fluence rates	Low fluence rates	Rapid response		+
Cotyledon/leaf expansion	+		+	+	
Gross effects on gene expression	+	+			
Stimulation of chlorophyll synthesis	+	+			
Circadian clock entrainment	+	+			
Circadian clock function					+
Flowering time	+	+			+
Phototropism			Low intensity	High intensity	
Chloroplast movement: accumulation			+	+	
Chloroplast movement: avoidance				+	
Stomatal opening			+	+	
Increase in intracellular [Ca^{2+}]	+		+	+	
Increased proton efflux	+		+	+	

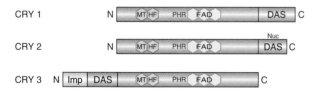

Figure 8.17 Domain structures of the three cryptochromes of *Arabidopsis*. FAD, flavin adenine dinucleotide cofactor; MTHF, methenyltetrahydrofolate (pterin) cofactor; PHR, photolyase-related domain (binds cofactors). DAS is a highly conserved motif found in cryptochromes from a wide range of organisms. Nuc shows the nuclear localization motif near the C terminus of CRY2. Imp indicates the organelle import signal at the N terminus of CRY3.

Several aspects of cryptochrome function in signal transduction resemble those of phytochrome. We have already seen that CRY2, like PHYA, is unstable in light, and both cryptochrome and phytochrome are active in the nucleus. Blue light stimulation of CRY1 and CRY2 results in a rapid **phosphorylation** that is at least partially self-mediated. There is evidence that blue light leads to a change in CRY protein conformation, exposing the C terminus to physical interaction with signaling factors. One of these is COP1, the repressor of photomorphogenesis in darkness that also interacts with phytochrome. Experiments have shown that CRY1 and CRY2 bind to phytochrome (see Section 8.2.4), and there are also reports that phytochrome can phosphorylate CRYs. The physiological significance of such CRY–PHY interactions is not clear, but it is likely that they mediate cross-talk between photoreceptor signaling pathways. CRY1 is known to be required for PHYA signaling to the circadian clock (see Section 8.5).

8.3.2 Phototropins are blue light receptors that contribute to optimizing growth, tropic responses and plastid orientation

As the name suggests, the **phototropin** class of UV/blue light receptors was discovered through studies of phototropism. The search for the blue light photoreceptor for phototropism led to identification of a plasma membrane-associated protein from pea epicotyls that becomes autophosphorylated on exposure to blue light. Subsequently a homologous protein was identified in *Arabidopsis* through analysis of a **non-phototropic mutant**, *nph1*. Two phototropin genes (*PHOT1* and *PHOT2*), which have partially overlapping functions, are currently known to occur in *Arabidopsis* (see Table 8.3). PHOT1 alone modulates cotyledon and hypocotyl growth, whereas both PHOT1 and PHOT2 are implicated in phototropism, chloroplast movement, stomatal opening and ion transport. In some cases, however, the two phototropins operate over different fluence ranges.

The phototropin protein (Figure 8.18A) consists of two regions: an N-terminal **photosensory domain** and a C-terminal **serine/threonine kinase domain** that are connected by a hinge. Within the photosensory region are two similar motifs of approximately 110 amino acids, called **LOV1** and **LOV2**, that bind the light-sensing

cofactor **FMN** (**flavin mononucleotide**). The acronym LOV refers to light, oxygen and voltage—the external signals that regulate proteins of the diverse superfamily to which phototropins belong. In the dark, the two domains of the molecule are appressed and the kinase is inactive. The FMN photocycle is shown in Figure 8.19. When FMN absorbs blue light it binds covalently to the LOV domain via a conserved cysteine residue. Intermediates in the photoconversion process are identifiable by their absorbance maxima. In the dark the LOV2 region interacts with a conserved helical region on the C-terminal side of the photosensory domain, the **Jα-helix** (Figure 8.18B). Autophosphorylation, a consequence of the disruption of this association by photostimulation, leads to activation of the C-terminal kinase. There is evidence that phosphorylation may be necessary for relocating the proteins from the plasma membrane to the Golgi apparatus, a rapid subcellular response to irradiation.

How the signal is transmitted from phototropins to downstream targets is not clear. For example, blue light, absorbed by both PHOT1 and PHOT2, stimulates the opening of **stomata**. This response requires activation of the guard cell plasma membrane H^+-ATPase and extrusion of protons (see Figure 14.32). Although activation of the H^+-ATPase involves its phosphorylation, there is no evidence that phototropin is directly involved. Other factors are thought to mediate the interaction between phosphorylated PHOT and plasma membrane H^+-ATPase, allowing transduction of the blue light signal into stomatal opening.

Stems and grass seedling coleoptiles are positively phototropic, growing towards a unilateral light source. This response is mediated by PHOT1. Unilateral irradiation induces a **gradient** of PHOT1 autophosphorylation, which in turn leads to a gradient of

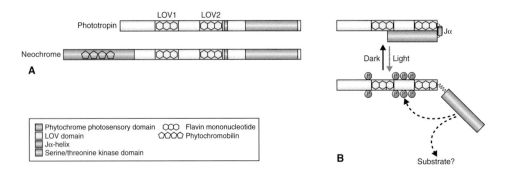

Figure 8.18   Phototropin and neochrome. (A) Domain structures of phototropin and neochrome. Neochrome, found in certain ferns and algae, is a chimeric protein consisting of a phytochrome photosensory domain fused to the N terminus of an entire phototropin receptor. (B) Mechanism of phototropin receptor activation by light. The unphosphorylated receptor is in the inactive ground state. Absorption of light by LOV2 causes the Jα-helix region to unfold. This activates the C-terminal kinase domain, which leads to autophosphorylation of the photoreceptor and (possibly) phosphorylation of other proteins.

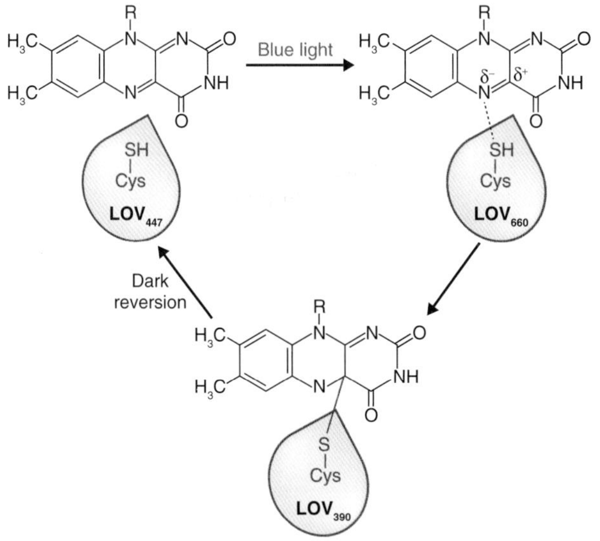

Figure 8.19 The FMN photocycle in the phototropin LOV domain. In darkness the FMN chromophore absorbs light maximally at a wavelength of 447 nm (LOV$_{447}$). Blue light drives covalent binding of FMN to a cysteine residue in the LOV domain forming LOV$_{390}$, via LOV$_{660}$, an intermediate absorbing maximally at 660 nm. The photoreaction process is fully reversible in darkness.

8.3.3 Other phototropin-like LOV receptor proteins act as photoreceptors in a wide range of species and plant processes

Proteins in the recently described zeitlupe class of blue light photoreceptors (also known as ZTL/ADO) are related to, but different from, phototropins. Like the phototropins, the zeitlupe proteins have a photoactive FMN-binding LOV region. However, they also have an **F-box** motif. This is a structural motif of approximately 50 amino acids that facilitates protein–protein interactions. F-box proteins are components of certain E3 ubiquitin ligases. E3 ligases with F-box proteins are associated with a range of cellular functions including signal transduction, hormone action and regulation of the cell cycle (see Chapter 10). Three *zeitlupe*-type genes have been identified in the *Arabidopsis* and rice genomes and putative homologs have been described in several other species, including poplar, maize and pine. The physiological functions of zeitlupe photoreceptors include the targeted proteolysis of components associated with the control of flowering time and the operation of the circadian clock (see Section 8.5).

Another type of photoreceptor with LOV domains is the **neochromes**. This family of receptors appears to have evolved independently in ferns and algae. Neochrome proteins have a phototropin-like protein sequence fused with a phytochrome chromophore-binding domain (see Figure 8.18A), making them dual red/blue light photoreceptors. The filamentous green alga *Mougeotia* has two neochromes and two phototropins. The single, large, ribbon-like chloroplast in each *Mougeotia* cell displays a striking light-avoidance response (Figure 8.20) under the control of these photoreceptors.

Rhodopsin-like and flavoprotein photoreceptors occur in many algae and also in animals. **Rhodopsin** is well known as a visual pigment in humans and other vertebrates; its chromophore is **retinal**, a vitamin A derivative. The flagellated, unicellular green alga *Chlamydomonas reinhardtii* has a pigmented eye-spot that detects light for **phototaxis** (movement in response to directional light). The photoreceptors in this alga are two rhodopsin-like molecules called channel rhodopsins (ChR1 and ChR2). The channel rhodopsins are light-gated proton channels that regulate phototaxis in response to blue-green light (absorption peak at 500 nm). In addition to rhodopsin-like photoreceptors, *C. reinhardtii* has a single cryptochrome and a single phototropin (but no phytochrome). In the photosynthetic protist *Euglena gracilis*, the photoreceptor

auxin concentration. Higher concentrations of auxin on the darker side of the organ result in more rapid growth on that side and the stem or coleoptile curves toward the light. Analysis of mutants with impaired phototropic responses has identified proteins that interact with PHOT1 and are thought to function in auxin redistribution. One of these is **phytochrome kinase substrate 1** (PKS1) protein, a PIF. PKS1 is a member of the PKS gene family, which has been shown by analysis of mutants in *Arabidopsis* to be required for normal phototropism under weak intensities of blue light. PKS1 is thus a link between the phototropin and phytochrome signaling pathways.

The movement of organelles within cells is also influenced by blue light acting through phototropin receptors. Under dim illumination the chloroplasts of mesophyll cells tend to orientate parallel to the leaf surface, thereby maximizing light capture, but in bright light they avoid possible photodamage by lining up along the cell walls perpendicular to the surface of the leaf (see Figure 4.39). Identification of *Arabidopsis* mutants lacking the chloroplast avoidance response has led to the identification of *CHUP1* (Chloroplast Unusual Positioning 1) and a family of *PMI* (Plastid Movement-Impaired) genes, the expression of which is necessary for linking phototropin excitation with organelle positioning within the cell.

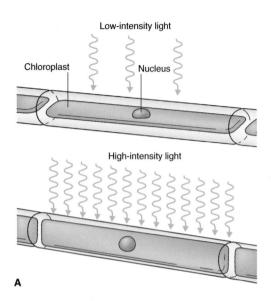

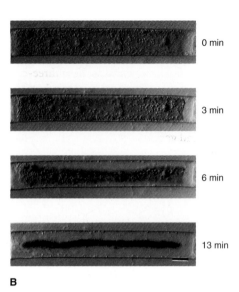

Figure 8.20 Neochrome- and phototropin-mediated movement of chloroplasts in cells of the filamentous green alga *Mougeotia*. (A) In dim light the flat side of the single ribbon-like chloroplast faces the light, while in bright light the edge faces the light. (B) Light micrographs of a rotating chloroplast taken at defined intervals. Scale bar = 10 μm.

Key points Plant responses to wavelengths at the blue-violet end of the visible spectrum are facilitated by cryptochrome (CRY), phototropin (PHOT) and zeitlupe (ZTL). In each case the photoreactive prosthetic group is a flavin. There are three *CRY* genes in *Arabidopsis*, *CRY1–3*. CRYs are dimeric proteins, each subunit of which includes a region that binds the chromophores FAD and pterin. CRY2 has a nuclear localization signal, and CRY3 a plastid/mitochondrion targeting motif. CRY1 (stable in blue light) and CRY2 (unstable) function in the control of growth, the circadian clock and flowering time. Blue light stimulation of CRY1 and CRY2 promotes phosphorylation and interaction with signaling components, including COP1 and PHY. PHOT proteins have an FMN-binding N-terminal photosensory domain connected by a hinge region to a C-terminal kinase domain. Blue light opens up the protein, resulting in phosphorylation and stimulation of stomatal opening, phototropic growth, organelle movements and interaction with the PHY network. Zeitlupes resemble PHOTs but also have an F-box domain that functions in the ubiquitin–proteasome system during floral induction and circadian regulation. Other less well-understood blue-sensitive photoreceptors in plants include neochromes, rhodopsins and flavoprotein adenyl cyclases.

for phototaxis is a photoactivated adenylyl cyclase. This flavoprotein complex responds to directional blue light information by increasing the levels of cytosolic cyclic AMP, a regulator of flagellar motility.

8.3.4 Wavelengths of light that are reflected or transmitted by leaves may be used to detect the presence of neighboring plants

Light absorbed by the major photoreceptors is predominantly at either extreme of the visible spectrum. Likewise, light capture by photosynthetic pigments takes place largely at red and blue wavelengths. Reflection and transmittance of non-absorbed light peaks at about 560 nm, which gives vegetation its characteristic green color (Figure 8.21). The **reflectance spectrum** of foliage

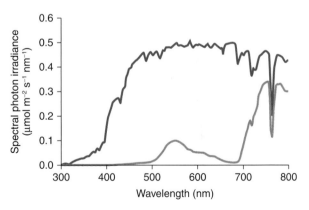

Figure 8.21 The spectral photon distributions of daylight (blue line) and light reflected from leaves of Japanese knotweed, *Fallopia japonica* (green line). Note that the light reflected from leaves is highly depleted in red quanta and relatively enriched in far-red quanta.

has been an important factor in the **coevolution** of plants and the visual systems of animals that interact with them. For example old-world primates, such as gorillas, chimpanzees and humans, have three-color (RGB) vision, whereas the photoreceptors in the eyes of new-world primates are sensitive to only two wavelength ranges, centered on green and blue. Comparative physiological and genetic studies indicate that the third photoreceptor of old-world primates (absorbance maximum at about 564 nm) evolved because it is **spectrally tuned** to the wavelengths reflected by leaves. This would have been an adaptive advantage, allowing the animal to pick out food and predators more easily against a background of foliage.

Light reflected from vegetation also has an important role in **plant–plant interactions**. A crop or a natural community is a population of individuals competing for resources, with foliage that forms **canopies**. Neighboring plants can sense each other by responding to shifts in the spectral quality of light resulting from canopy reflectance. For example, the light that is transmitted or reflected by green leaves has an increase in the ratio of far-red to red quanta (Figure 8.22). This change in R:FR ratio can be detected by phytochrome, leading to an increase in the proportion of phytochrome that is in the P_R (inactive) form. Shade-intolerant plants, like *Arabidopsis*, respond by increasing their rate of stem and petiole elongation. This response has been called **shade avoidance**, since rapid stem elongation may enable plants to escape shading by other plants. Mutations that abolish this mechanism for detecting changes in R:FR ratio cause plants to grow as though shaded even when they are placed in white light, as shown for the

PHYB-deficient *lh* mutant of cucumber in Figure 8.12. Shading also decreases blue light fluence rates that can be detected by blue/UV photoreceptors, and in some species phototropin may serve as a photoreceptor for the shade-avoidance response. This response is an essential strategy for survival and, because it directly influences light capture and photosynthetic productivity, is agriculturally important as a factor in crop yield.

8.3.5 Plants respond to wavelengths of light in addition to blue, red and far red

In laboratory studies of photophysiology, green is widely used as a 'safe' light, because it has minimal effects on the plants under investigation. There is increasing evidence, however, that plant morphogenesis is not completely insensitive to green light. A number of studies suggest that the green waveband has negative effects on aspects of extension and tropic growth, stomatal opening and the expression of genes in plastids. Many of these responses may be explained in terms of the absorbance spectra of known photoreceptors. For example, Figure 8.7 shows that phytochromes, particularly P_R, have appreciable absorbance in the 500–600 nm range, and the PHYA action spectrum for breaking seed dormancy extends well into the green waveband (see Figure 8.13).

Recent studies of cryptochrome show that green light can reverse the effect of blue and have led to a proposed

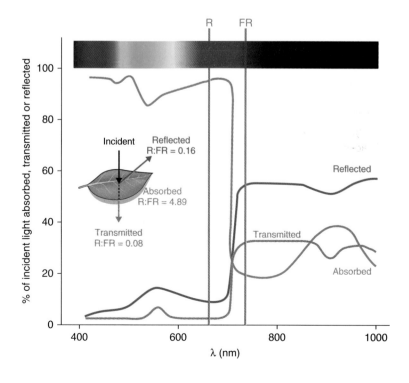

Figure 8.22 Spectral characteristics of typical plant leaves over the wavelength range 400–1000 nm. Leaves absorb more than 80% of incident light between 400 and 700 nm. There is a sharp fall in absorbance in the far-red and infra-red regions. Consequently the red:far-red (R:FR) ratio of absorbed light (4.89) is much higher than that of reflected (0.16) or transmitted (0.08) light. There are consequences for the state of the phytochrome, which will be converted to the P_R form in reflected and transmitted light.

model of the flavin photocycle in which oxidized and reduced forms interconvert in response to irradiation with blue or green light and are subject to dark reversion (see Equation 8.4). Nevertheless, some responses to green light are difficult to account for in terms of the known photoreceptors and the search for a fourth, green-sensitive, class has turned up some interesting candidates. Blue light stimulates stomatal opening. A short green light pulse eliminates the blue light response. The action spectrum for the green light effect, which shows a maximum at 540 nm and minor peaks at 490 and 580 nm, resembles the absorption spectrum of the carotenoid **zeaxanthin** red-shifted by about 50 nm. Stomata from a zeaxanthin-less *Arabidopsis* mutant lack a specific response to blue light. Zeaxanthin is also a central component in the **xanthophyll cycle**, a mechanism for dealing with light stress (see Chapters 9 and 15). Other pigments that could function as potential green-sensitive photoreceptors in seed plants have been described, including novel classes of flavoprotein and retinal-binding proteins.

As mentioned in Sections 8.1.1 and 8.1.2, UV light can be harmful to living organisms. In particular, nucleic acids and proteins absorb UVB and UVC strongly and may be damaged in the process. Ultraviolet damage is itself a photoperception event and triggers a range of signaling and biochemical responses. There is evidence, however, for the existence of specific **ultraviolet-sensitive photoreceptors** distinct from the known UV/blue receptor classes, one active in the range 280–300 nm and another at 300–320 nm. At present their chemistry and mode of action are unknown.

Key points Leaves appear green because they reflect and transmit wavelengths peaking at about 560 nm, which is close to the absorbance maximum of the long wavelength visual receptor of old-world primates (including humans) with three-color vision. Plants sense the red:far-red ratio of reflected and transmitted light from neighboring plants through phytochrome and use the information to trigger the shade-avoidance growth response. Blue light and phototropins may also play a role in shade avoidance. The carotenoid zeaxanthin has been identified as a possible photoreceptor responsible for stomatal responses to green light. Ultraviolet light in the range 400–320 nm (UVA) and 320–280 nm (UVB) also has photomorphogenetic, and sometimes harmful, effects and may be sensed by specific receptors.

8.4 Biosynthesis of chlorophyll and other tetrapyrroles

Seedlings germinated and grown in the dark are pale yellow, while those grown in the light are green. This change in color is due to the production of chlorophyll and a functional photosynthetic apparatus that only takes place in the light. In this section we will examine the process of chlorophyll synthesis and the basis for its dependence on light. Chlorophyll (**Chl**) is one of a group of molecules known as **tetrapyrroles**, so named because they take the form of four linked pyrrole groups. Phytochromobilin is also a tetrapyrrole. Another important tetrapyrrole is **heme**, best known as a component of the oxygen-carrying protein hemoglobin in blood. Leghemoglobin is a hemoprotein that is essential for the nitrogen-fixation process in plants because it sequesters oxygen to prevent inhibition of the nitrogen-fixing enzyme nitrogenase (see Section 12.4.4). Heme is also the cofactor of **cytochromes**, which are involved in electron transport in both respiration (see Chapter 7) and photosynthesis (see Chapter 9), and of certain enzymes such as **peroxidases**. The initial steps in the biosynthesis of all tetrapyrroles are identical; the pathway then branches according to the final product. In the case of Chl, all the biosynthetic enzymes are found in the plastid, although they are encoded by nuclear genes (Figure 8.23). Most of the genes needed for Chl biosynthesis are regulated in a light-dependent manner, and some of them are also expressed according to circadian rhythms. Environmental factors other than light also play a part in regulating Chl biosynthesis.

8.4.1 Aminolevulinic acid is the precursor of tetrapyrrole biosynthesis

Tetrapyrrole biosynthesis begins with the formation of aminolevulinic acid (ALA). The first step in ALA biosynthesis (Equation 8.5), catalyzed by **glutamyl tRNA reductase** (**GluTR**) in plants, is rate-limiting for the whole pathway. GluTR is encoded by the gene *HEMA1*. The photoreceptors that have been shown to act in light stimulation of Chl biosynthesis are PHYA, PHYB, CRY1 and CRY2. They achieve this effect mainly by regulating transcription of *HEMA1*.

Equation 8.5 Glu-tRNA reductase (GluTR)

Glutamate-1-semialdehyde + NADP$^+$ + tRNA(Glu)
$\rightleftharpoons$ glutamyl tRNA(Glu) + NADPH

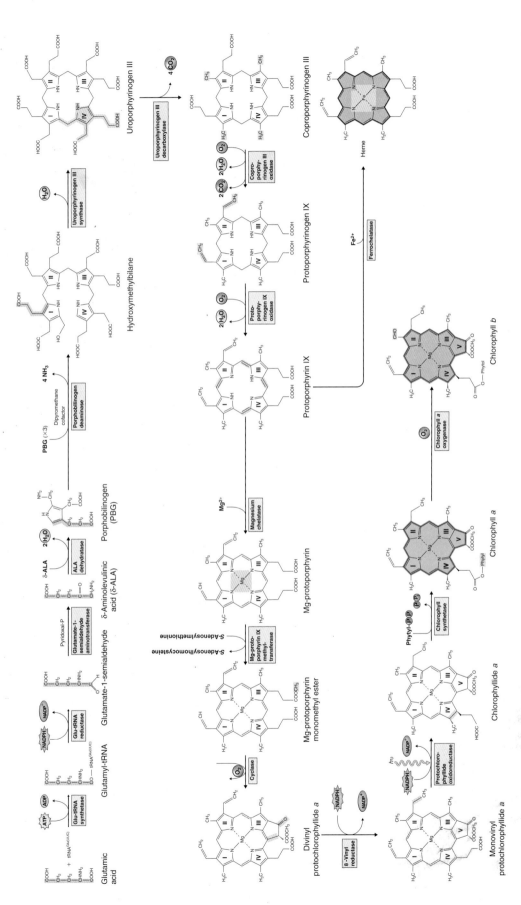

Figure 8.23 The tetrapyrrole biosynthesis pathway. Chlorophylls and hemes are biosynthesized by a branched pathway which begins with the synthesis of δ-aminolevulinic acid (ALA) from glutamic acid. In the next three steps eight molecules of ALA are combined, via the intermediate porphobilinogen (PBG), to form the ring structure of uroporphyrinogen III which undergoes further modifications to become protoporphyrin IX. After this point the heme and chlorophyll pathways diverge. Chlorophyll synthesis requires insertion of Mg into the protoporphyrin IX ring by magnesium chelatase, whereas heme contains Fe, which is inserted into the ring by a ferrochelatase. Chlorophyll synthesis involves further steps, the penultimate of which, in angiosperms, requires the light-dependent enzyme protochlorophyllide oxidoreductase. This produces chlorophyllide a, to which the phytol chain is attached to form chlorophyll a. Chlorophyll b is synthesized from chlorophyll a.

Heme, one of the major products of the pathway, acts as an allosteric inhibitor of GluTR in plants, so that if, for example, heme accumulates because its degradation is blocked, the biosynthesis of Chl precursors is also reduced. The synthesis of ALA is downregulated in darkness by the product of the **FLU** gene, which interacts physically with GluTR. Mutations in *FLU* result in plants that accumulate fluorescent Chl biosynthesis intermediates in the dark.

8.4.2 Cyclic intermediates in tetrapyrrole metabolism are potential photosensitizers

Equation 8.6 Synthesis of uroporphyrinogen from aminolevulinate
8.6A Porphobilinogen synthase

$$2 \text{ 5-Aminolevulinate} \rightarrow \text{porphobilinogen} + 2H_2O$$

8.6B Hydroxymethylbilane synthase

$$4 \text{ Porphobilinogen} + H_2O$$
$$\rightarrow \text{hydroxymethylbilane} + 4NH_3$$

8.6C Uroporphyrinogen III synthase

$$\text{Hydroxymethylbilane} \rightarrow \text{uroporphyrinogen III} + H_2O$$

The first cyclic tetrapyrrole in the biosynthesis pathway, uroporphyrinogen III, is formed by the condensation of eight molecules of ALA (Equation 8.6). It then undergoes a three-stage oxidative conversion to protoporphyrin, during which the molecules become increasingly photoreactive and potentially damaging to the plant. When Chl and other tetrapyrroles absorb light they can transfer the energy to other molecules, leading to the formation of reactive oxygen species, which in turn can damage proteins and lipid membranes. For this reason tetrapyrrole synthesis is normally tightly regulated and mutations or chemical treatments that cause the buildup of photoreactive intermediates cause plants to become very sensitive to light damage. Mutation of the *FLU* gene, described in Section 8.4.1, is only one of the possible causes for accumulation of these phototoxic compounds. Their properties make intermediates in tetrapyrrole biosynthesis useful in the treatment of diseases, especially cancerous tumors by **photodynamic therapy**. In this approach a photosensitizing molecule is targeted to the diseased tissue. Then high-intensity light is used to induce localized production of reactive oxygen species which kill cancerous cells. ALA, the precursor for all tetrapyrrole biosynthesis, has been widely used in

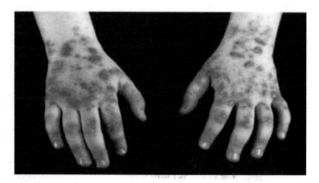

Figure 8.24   Skin lesions in a sufferer from porphyria, a disorder affecting the heme biosynthesis pathway, induced by accumulation of tetrapyrrole biosynthesis intermediates.

photodynamic therapy as have derivatives of bacterial chlorophyll. The early intermediates of **chlorophyll catabolism** (see Chapter 18) are also photoactive tetrapyrroles and one of them, pheophorbide, is another photosensitizer used therapeutically to cause lipid peroxidation and arrest of tumor growth.

Animals, like plants, are susceptible to damage if intermediates in tetrapyrrole biosynthesis accumulate abnormally. Several of the human diseases known as porphyrias are due to deficiencies of specific enzymes in the heme biosynthetic pathway. For example, uroporphyrinogen III synthase, coproporphyrinogen III oxidase, protoporphyrinogen oxidase and ferrochelatase deficiencies can all lead to illnesses with symptoms that include photosensitivity. Sufferers need to avoid bright light as otherwise they may develop painful skin lesions (Figure 8.24).

8.4.3 Protoporphyrin stands at the branch point leading to chlorophyll or heme

All the enzymes of the pathway illustrated in Figure 8.23 up to and including coproporphyrinogen III oxidase are found in the plastid, either in the stroma or attached loosely to the plastid membranes. However, the next enzyme in the pathway, protoporphyrinogen oxidase (PPX), is located in both mitochondrial and plastid membranes. At this stage in the pathway, if the tetrapyrrole is destined for conversion into a mitochondrial cytochrome it must be transported from the plastid to the mitochondrion, though the transporters responsible have not yet been identified.

Equation 8.7 Ferrochelatase

$$\text{Protoporphyrin} + Fe^{2+} \rightarrow \text{protoheme} + 2H^+$$

Equation 8.8 Magnesium chelatase

$$ATP + \text{protoporphyrin IX} + Mg^{2+} + H_2O$$

$$\rightarrow ADP + P_i + \text{Mg-protoporphyrin IX} + 2H^+$$

Protoporphyrin (**Proto**) can undergo two different fates (Equations 8.7 and 8.8). If Fe^{2+} is inserted into the center of the tetrapyrrole ring the molecule is destined to become **heme**; insertion of Mg^{2+} leads to the synthesis of **chlorophyllide** (**Chlide**) and thence Chl. In plants these processes are carried out by the metal **chelatase** enzymes ferrochelatase (**FeCh**) and magnesium chelatase (**MgCh**), respectively; when both enzymes are present in the same subcellular compartment they compete for their common substrate, Proto. FeCh is found both in plastids and in mitochondria, but there is evidence that in plants most heme biosynthesis takes place in the plastids. Plant MgCh consists of three subunits, CHLI, CHLD and CHLH. In angiosperms, the distribution of CHLH and CHLD between the plastid envelope membrane and the stroma is dependent on internal Mg^{2+} concentration. It is probable that the CHLI–CHLD complex provides magnesium ions, while CHLH carries the Proto substrate.

8.4.4 The chromophore of phytochrome is synthesized from heme

As discussed in Section 8.2.2, phytochromes are dimeric proteins. Each of the two monomers in a phytochrome molecule is covalently linked to the linear tetrapyrrole phytochromobilin (see Section 8.2.1 and Figure 8.6). This compound is synthesized from heme (Figure 8.25); the committed step in its synthesis is the cleavage of heme by **heme oxygenase**, producing biliverdin IXa, in a reaction that requires oxygen and the electron donor ferredoxin (Fd), and produces carbon monoxide and Fe^{2+} (Equation 8.9).

Equation 8.9 Heme oxygenase

$$Heme + 3Fd_{red} + 3O_2$$

$$\rightarrow biliverdin + Fe^{2+} + CO + 3Fd_{ox} + 3H_2O$$

The ferredoxin-dependent enzyme (3E)-phytochromobilin synthase then reduces biliverdin IXa to the so-called 3Z-isomer of phytochromobilin, which is finally isomerized by flipping a methyl group, resulting in (3E)-phytochromobilin, the active

configuration of the phytochrome chromophore (see Section 8.2.1). It is not known whether this isomerization step occurs spontaneously or requires enzymatic catalysis. The biosynthetic pathway shown in Figure 8.25 also produces other bilins, including phycocyanobilin and phycoerythrobilin. These compounds are chromophores of phycobiliprotein photosynthetic pigments in cyanobacteria and some algae (see Section 9.2.1).

Key points Chlorophyll, heme and phytochromobilin are tetrapyrroles with a common biosynthetic origin in the chloroplast. The rate-limiting reaction is catalyzed by glutamyl tRNA reductase (GluTR), encoded by the gene *HEMA1*, which is transcriptionally regulated by PHYA, PHYB, CRY1 and CRY2. Heme is an allosteric feedback inhibitor of GluTR activity. Formation of aminolevulinic acid, the product of GluTR and an aminotransferase, is downregulated in darkness by the FLU protein. Uroporphyrinogen III is the first cyclic tetrapyrrole intermediate and is converted via three oxidation reactions into the branch point metabolite protoporphyrin (Proto). Cyclic tetrapyrrole intermediates are potent photosensitizers and their accumulation to abnormal levels can cause harmful lesions in the light. Chlorophyll and photosynthetic hemes are made from Proto in the chloroplast. Proto transported to the mitochondrion is the precursor of the heme of respiratory cytochromes. The iron atom of heme is inserted by Fe chelatase, and the magnesium atom of chlorophyll by Mg chelatase. The phytochromobilin of phytochrome is synthesized from heme in two ferredoxin-dependent reactions, the first of which is catalyzed by heme oxygenase. The product of a second reduction isomerizes to produce the active chromophore.

8.4.5 Conversion of protochlorophyllide to chlorophyllide in seed plants is light-dependent

The last but one stage in the biosynthesis of chlorophyll a is the conversion of protochlorophyllide (Pchlide) to Chlide by reduction of the double bond between carbons C17 and C18 in ring D (see Figure 8.23). The enzyme catalyzing this reaction is **NADPH-Pchlide oxidoreductase** (**POR**; Equation 8.10).

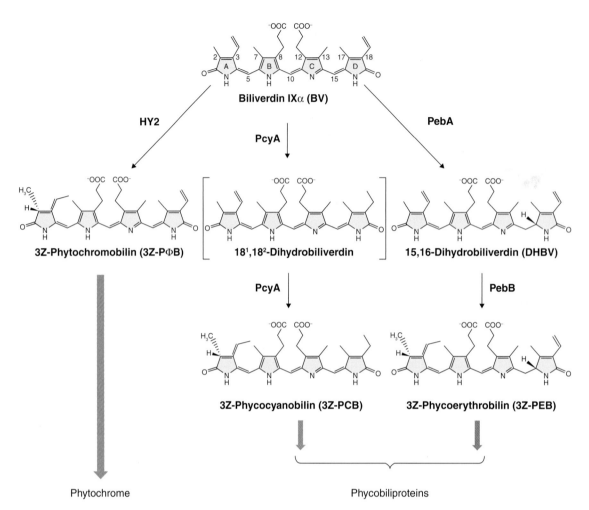

Figure 8.25   The biosynthesis of phytochromobilin and other plant bilins from the linear tetrapyrrole biliverdin IXα (BV). 3Z-phytochromobilin, the precursor of the 3E-phytochromobilin chromophore that is found in phytochrome, is produced by the action of phytochromobilin synthase (HY2). Phycocyanobilin:ferredoxin oxidoreductase (PcyA) is responsible for the synthesis of phycocyanobilin (PCB). Phycoerythrobilin (PEB) formation requires the sequential action of two enzymes: 15,16-dihydrobiliverdin:ferredoxin oxidoreductase (PebA), which reduces the C-15 methine bridge of BV; and phycoerythrobilin:ferredoxin oxidoreductase (PebB), which reduces the A-ring diene structure of 15,16-dihydrobiliverdin (DHBV).

Equation 8.10 NADPH-protochlorophyllide oxidoreductase (POR)

Protochlorophyllide + NADPH + H$^+$

→ chlorophyllide a + NADP$^+$

All algae, aerobic photosynthetic bacteria, liverworts and gymnosperms contain a three-subunit light-independent form of POR. However, at some point in evolution, angiosperms lost this enzyme, and they have only a light-dependent form, which is also present in all the other photosynthetic organisms except for some anaerobic bacteriochlorophyll-containing bacteria. This means that the final steps in Chl biosynthesis in angiosperms have an absolute requirement for light; in its absence, leaf plastids become **etioplasts** instead of chloroplasts (Figure 8.26).

Light-dependent POR, a key enzyme in Chl synthesis, is a monomeric enzyme of 35–38 kDa; in angiosperms it is usually encoded by small families of genes which are differentially regulated by light and at different developmental stages. In barley, for example, there are two *POR* genes, while *Arabidopsis* has three. *Arabidopsis PORA* is specifically expressed in the dark; phytochrome downregulates its transcription. *PORB* is constitutively expressed and *PORC* is induced by light. The lattice-like structure that occupies much of the etioplast, as shown in Figure 8.26, is largely composed of a complex of Pchlide, NADPH and POR. Pchlide acts as a

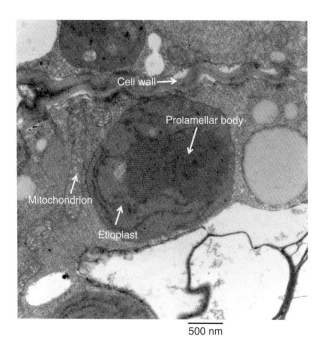

500 nm

Figure 8.26 Etioplast from *Arabidopsis* seedling grown in darkness.

Labels in figure: Cell wall →, Prolamellar body, Mitochondrion, Etioplast

photoreceptive prosthetic group in the complex. When tissue containing etioplasts is illuminated, POR rapidly converts Pchlide to Chlide and the internal structure of the etioplast is reorganized into the typical chloroplast configuration (see Section 4.6.2). The polypeptide prePORA, the precursor of the major 'dark' form of *Arabidopsis* POR, is imported into plastids from the site of synthesis in the cytosol in a substrate-dependent manner, so that sufficient Pchlide must be present in the plastid for the PORA precursor to enter.

PORA is one of a small number of plant proteins, also including phytochrome, which act as photoreceptors. In seedlings germinated in the dark, the etioplast complex of POR, Pchlide and NADPH described above is stable. However, upon illumination and the conversion of PChlide to Chlide, POR is released and the free enzyme is then subject to rapid **proteolysis**, so that most of the POR protein disappears in the light. Thus, the photoreceptor properties of POR are comparable with those of phytochrome: POR possesses a tetrapyrrole chromophore that triggers photomorphogenic changes in cells and tissues; and the light being received by the plant tissue regulates its form and abundance. The behavior of POR proteins is best understood from studies of the etioplast-to-chloroplast conversion. Other forms of POR, such as *Arabidopsis* PORC, which is responsible for chlorophyll synthesis in light-grown plants, still require light for activity, but the regulation of their turnover is much less well understood.

8.4.6 Phytol is added to chlorophyllide to make chlorophylls a and b

The product of the POR-catalyzed reaction is Chlide a. In the final step of chlorophyll a biosynthesis the 18-carbon **phytol** side-chain is added (Equation 8.11), a process that makes Chl, in contrast to Chlide, highly hydrophobic. The enzyme that carries out this step is **chlorophyll synthase**, which adds phytyl pyrophosphate to the C_{18} propionyl group of Chl in an esterification reaction.

Equation 8.11 Chlorophyll synthase

Chlorophyllide a + phytyl pyrophosphate

→ chlorophyll a + PP_i

Formation of chlorophyll b requires the oxidation of the C-7 methyl group of Chlide a to formyl as well as the addition of the phytol side-chain. This oxidation (Equation 8.12) is carried out by the enzyme **Chlide a oxygenase**, a monooxygenase that contains iron and sulfur, and the substrate is Chlide a rather than Chl a, producing Chlide b as a substrate for Chl synthase.

Equation 8.12 Chlorophyllide a oxygenase

7-Hydroxychlorophyllide a + O_2 + NADPH

→ chlorophyllide b + $2H_2O$ + $NADP^+$

Chlorophylls are at the core of the photosynthetic apparatus, and as this section shows, their biosynthesis as well as their function is light-dependent, reinforcing the central role of light in autotrophic plant growth and morphogenesis. The workings of the photosynthetic machinery will be described in Chapter 9.

8.5 Circadian and photoperiodic control

Early in evolution, living organisms evolved mechanisms to coordinate biological processes with the day–night cycle of light and temperature caused by the planet's rotation. It has been suggested that the original reason for this may have been to ensure that DNA was only replicated at night, to avoid the damaging effects of ultraviolet radiation. Present-day plants, animals, fungi and many other living organisms all show such cyclic regulation of biological activities. In some cases,

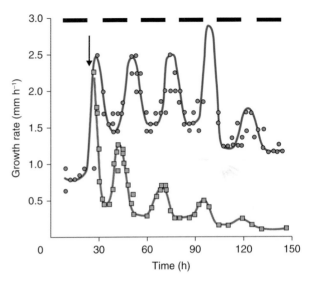

Figure 8.27 Circadian rhythm in leaf growth. Growth rates of the fourth leaf of *Lolium temulentum* seedlings at 20°C were measured in an 8/16-hour light/dark photoperiod (●) and after transfer to continuous darkness (■) at 24 hours (arrow).

Key points Chlorophyllide a, the immediate precursor of chlorophyll a, is the product of a light-requiring enzymatic reaction catalyzed by NADPH-protochlorophyllide oxidoreductase (POR). An additional light-independent form of POR present in algae, aerobic photosynthetic bacteria, liverworts and gymnosperms appears to have been lost during angiosperm evolution. The *POR* nuclear gene family of most angiosperms consists of two or three members. *Arabidopsis PORA* is specifically expressed in the dark and is downregulated by PHY. PORA forms large crystalline inclusions in etioplasts and is subject to proteolytic breakdown as the chloroplast develops on illumination. In *Arabidopsis, PORB* expression is constitutive and *PORC* is light-induced. POR behaves as a photoreceptor with protochlorophyllide acting as its chromophore. Chlorophyllide b is the product of a monooxygenase reaction of chlorophyllide a catalyzed by chlorophyllide a oxygenase. Chlorophyll a and b are synthesized from the corresponding chlorophyllides by esterification with the C_{18} alcohol phytol in a reaction catalyzed by chlorophyll synthase.

as described earlier in this chapter, light itself has a direct regulatory role, but organisms also possess **biological clocks** which schedule events to occur on a daily basis—the so-called **circadian rhythms**. Under normal light–dark cycles, circadian rhythms are **entrained** to these cycles; that is, the duration and phase of the circadian rhythm is matched to that of the light–dark sequence. But even when an organism is removed from the normal environment of day and night and placed, for example, in continuous light, circadian rhythms continue to operate, with a time period of about 24 hours, for many days or even longer. Figure 8.27 shows an example in which the timing of leaf growth in a grass species follows a circadian rhythm even when the plant is placed in darkness, though the amplitude of the rhythm gradually declines and the cycle lengthens relative to the original 24-hour period.

In regions near the equator, the cycle of light and darkness varies little throughout the year. In more northerly and southerly parts of the Earth, however, the planet's tilt on its axis means that the ratio of dark to light varies continuously (Figure 8.28). **Photoperiodic control** is the term given to the regulation of biological processes according to the relative lengths of day and night. As we shall see later, for photoperiodic control to operate properly a plant needs to 'compare' the external signal provided by light or its absence with its own endogenous circadian rhythm.

8.5.1 The circadian rhythm and day–night cycle must be synchronized in order to regulate biological functions correctly

Proper functioning of circadian rhythms is important for all organisms. In humans, for example, the phenomenon of jet-lag is caused when travel through a number of time zones means that the body's clocks are out of phase with the new day–night cycle under which the person is now living. Over a period of days, the clocks are **reset** thanks to exposure to the altered dark/light pattern, but different bodily processes are reset at different rates, leading to unpleasant symptoms that can include sleep disturbance, digestive problems, headache and disorientation. For plants too, research has shown that those whose circadian clock is correctly aligned with the current day–night cycle fix more carbon, grow faster and have a competitive advantage over plants of the same species with incorrect circadian clocks due to mutation. Selective pressure on all species must have favored the existence of accurate clocks.

Circadian rhythms are **cell autonomous**—that is, they are set largely independently in each cell of an organism. It has been shown that the clocks of individual cells in a leaf are weakly synchronized with one another, even in the prolonged absence of light, so that rhythmic behaviors such as stomatal opening occur together across the whole leaf. However, roots use only a subset of clock

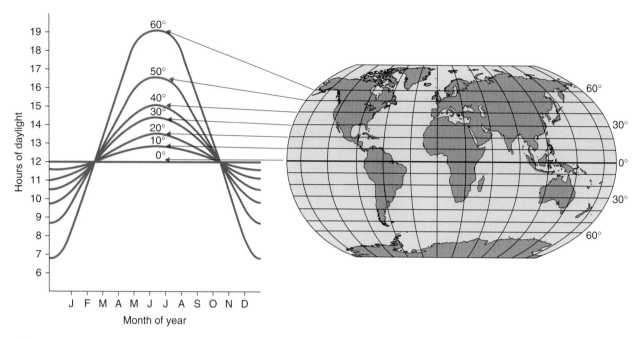

Figure 8.28   The relationship between day length throughout the year (measured on the 20th of each month) and latitude.

genes for circadian regulation. Because roots are not normally exposed to light it is believed that they depend on a signal from the leaves to maintain their clocks in synchrony with the day–light cycle.

Circadian clocks regulate a high proportion of plant genes—it has been estimated that 25–30% of the protein-coding genes in *Arabidopsis* are subject to circadian control. These genes participate in many essential processes, as illustrated in Figure 8.29. Plants also use the circadian clock to measure the **duration** of day and night, thus sensing the season and influencing processes such as the onset of flowering or the start of leaf senescence, which must occur at the right time of year if the plant is to flourish. We will first discuss

features of circadian clocks in plants and then examine **photoperiodism**, the response to seasonal changes in the lengths of day and night.

8.5.2 Genetically controlled interlocking feedback loops underlie the circadian clock mechanism

Although the way circadian clocks function appears similar in all organisms, the genes that control the clock

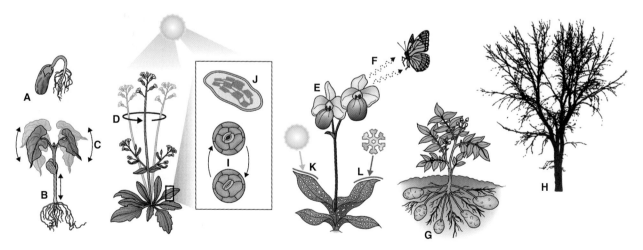

Figure 8.29   The circadian clock regulates many aspects of the plant life cycle: (A) germination; (B) elongation growth and shade avoidance; (C, D) leaf and flower movements; (E) flowering time and flower opening; (F) scent production; (G) tuberization; (H) winter dormancy; (I) stomatal opening; (J) photosynthesis; (K, L) protection from extremes of temperature.

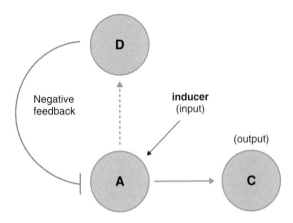

Figure 8.30 A feedback loop. Transcription of gene *A* is upregulated by an inducer which leads to the synthesis of a protein D, and to a pathway leading to a response C. Increasing concentration of D leads to downregulation of A. The net result is an oscillation in the abundance of D and of the response C.

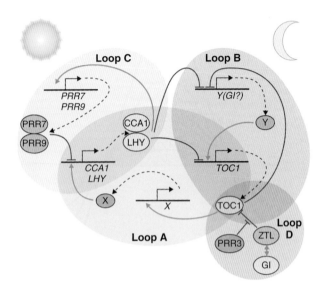

Figure 8.31 Model of the plant clock. Loop A consists of the dawn-phased factors CCA1 and LHY, negative regulators of *TOC1* expression. TOC1 directly or indirectly activates the postulated component X, which induces *CCA1* and *LHY* expression. Loop B comprises two or more evening-phased genes, an unknown factor Y (the activity of which may be partially provided by GI, the product of the gene *Gigantea*) and TOC1. Loop C consists of morning-phased *PRR7*, *PRR9*, *CCA1* and *LHY*. Loop D shows post-transcriptional regulatory interactions between ZTL, TOC1, GI and PRR3.

mechanisms in plants are very different from those in animals or fungi. Because possessing a clock confers several selective advantages, there were probably many independent evolutionary events that led to their presence in these different organisms. Studies at the molecular and genetic levels have shown that the synchronizing of clocks to the Earth's natural 24-hour cycle occurs via a series of **feedback loops** that control the expression of a fairly small number of so-called **clock genes**. Typically, in a simple feedback loop (Figure 8.30), a stimulus causes transcription of a gene to increase, leading to synthesis of the corresponding protein and other downstream effects. Increasing abundance of the protein causes transcription of the gene to be downregulated, usually after a delay, and the net effect is an **oscillation** in the abundance of the protein and in any processes controlled by it. More sophisticated regulation can be achieved when interlocking loops participate, as illustrated in Figure 8.31. All circadian clocks are regulated by such systems of interlocking feedback loops, and research on plant clocks is uncovering increasing degrees of subtlety. Figure 8.31 shows the current explanation for clock regulation in the leaves of *Arabidopsis* in which the proteins **CCA1** (Circadian Clock Associated 1), **LHY** (Late Elongated Hypocotyl) and **TOC1** (Timing of CAB Expression 1), along with other, as yet unidentified, factors, control a morning and an evening loop. The third, more recently discovered, morning loop also involves CCA1 and LHY together with PRR7 and PRR9 (Pseudo-Response Regulators 7 and 9). At dawn, LHY and CCA1 proteins are synthesized, and this upregulates the genes encoding their inhibitors, PRR7 and PPR9, which in turn downregulate expression of the *LHY* and *CCA1* genes in an example of negative

feedback. LHY and CCA1 also inhibit the expression of evening genes in the morning. The most important of these evening genes, *TOC1*, is expressed at dusk and it regulates expression of its own activator, a hypothetical protein encoded by gene *Y*, which has not yet been isolated but whose existence has been inferred from the properties of the clock. Another hypothetical gene, *X*, believed to be upregulated by TOC1, feeds back from the evening to the morning loop, activating expression of *LHY/CCA1* late in the night. Expression of the genes encoding LHY/CCA1, PRR9 and Y is stimulated by light, providing a means by which the clock can be maintained in synchrony with the length of the day.

While the best-understood components of the clock mechanism use transcriptional-level regulation, **post-transcriptional modifications** also play an important role, as in the case of the zeitlupe protein ZTL. ZTL negatively regulates abundance of the protein product of the key *TOC1* gene by binding to TOC1 and targeting it for degradation by the proteasome. Although it is clear that the relative abundances of the clock gene products at different times of day exert a regulatory effect over other plant genes, the mechanisms by which they do so—the outputs of the plant circadian clock—are still poorly understood.

Key points Biological processes are synchronized with day–night cycles of light and, in some cases, temperature. Daily physiological events scheduled by an organism's biological clock result in cell-autonomous circadian rhythms, which are usually entrained by the light–dark sequence. Roots, which normally grow in darkness, rely on a signal from leaves to synchronize their clocks. Up to a third of plant genes are regulated by circadian clocks. Circadian control underlies photoperiodic responses, in which growth and development are sensitive to day–night duration. Circadian clocks are synchronized to the daily cycle at the molecular level through a mechanism of interacting feedback loops that results in oscillations in the expression of clock genes. The mechanism in *Arabidopsis* comprises morning and evening loops in which some genes (for example *Circadian Clock Associated 1, CCA1*) are upregulated at dawn while expression of others (*Timing of CAB Expression 1, TOC1*, for instance) increases at dusk. Many clock genes are light-responsive, thereby staying in sync with day length. Some components of the clock, for example zeitlupe proteins, are post-transcriptional regulators of clock gene expression.

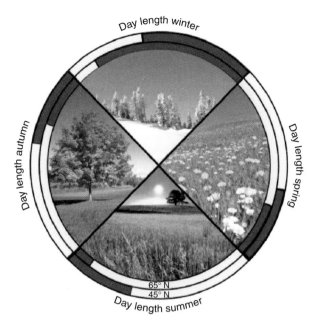

Figure 8.32 Variation of day length over latitude and season.; 65° and 45° north are compared (65°N is approximately the latitude of Reykjavík, Iceland and Anchorage, Alaska, USA while 45°N is the approximate latitude of Milan, Italy and Montreal, Canada). The start date of winter is taken to be January 1, spring May 1, summer June 20 and fall August 10.

8.5.3 Plants are classified as long-day, short-day or day-neutral according to their developmental responses to photoperiod

Many stages in plant reproductive development are under photoperiodic control. They include the initiation of flowering, onset of tuber development in species like potato, bulb initiation in onions and garlic, bud set in some trees, and entry into dormancy for many perennial species. Such regulation allows plants to anticipate and prepare for an oncoming season that may be unfavorable for growth. For example, shortening day lengths in mid to late summer are the most reliable predictor of the cold days of late fall and winter. Photoperiodic sensitivity also allows the plant to time its key developmental stages so that it can make the best possible use of light throughout the growing season to set seed, or produce vegetative storage structures, before the arrival of adverse growth conditions. Light availability can vary dramatically from season to season, particularly in far northern and southern latitudes (Figure 8.32). The best-studied

photoperiodic system is undoubtedly the control of flowering by day length.

The earliest experiments on environmental control of the switch from vegetative to reproductive development established that exposure to short days (long nights) promotes flowering in *Humulus* (hop) and *Cannabis* (hemp), whereas *Sempervivum funckii* (Funck's houseleek) flowers under long days (short nights). Subsequently, day/night duration (photoperiod) was found to regulate flowering and many other aspects of development in a wide variety of plants. Species in which flowering is promoted by long-night/short-day conditions are classified as **short-day plants** (**SDPs**); *S. funckii* is an example of a **long-day plant** (**LDP**) species (Figure 8.33). In general, LDPs bloom in spring or early summer as the days are lengthening, whereas SDPs blossom in late summer and early fall at a time of shortening days and lengthening nights. Flowering in many plant species is insensitive to photoperiod: such species are classified as **day-neutral**.

Not all photoperiodically sensitive plants can be neatly categorized in this way; some plants require a sequence of different photoperiods, while others require sequential temperature and photoperiod treatments. Flowering is frequently a developmental reaction to non-optimal environments, and many of the regulatory

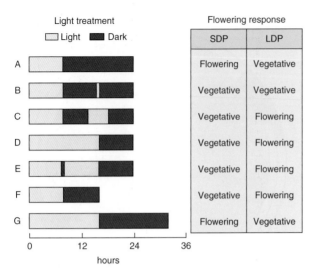

Figure 8.33 Typical responses of short-day plants (SDPs) and long-day plants (LDPs) to combinations of light and dark periods of different durations: (A–E), normal 24-hour day–night cycle; (F), artificially shortened day–night cycle; (G), artificially lengthened day–night cycle. SDPs flower in response to long nights (A), whether the light period is short (A) or long (G). Interrupting the long dark period with a short or long light break suppresses the flowering response (B, C). Most LDPs flower in response to short nights (D, E) even when the light period is also short (F) and remain vegetative under long nights with (B) or without (A) a short light break, even when the day is also long (G); but a longer light break in a long dark period will induce flowering (C). A dark break in a long light period does not modify the response of SDPs or LDPs (compare D and E).

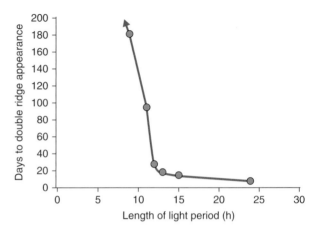

Figure 8.34 *Lolium temulentum* forma Ceres is a qualitative long-day plant, requiring one inductive photocycle for floral induction. Flowering occurs when the night is shorter than 10 hours, its critical night length. Double ridge formation describes the condition of the shoot apex in grasses at the stage when the apical meristem ceases to produce leaf primordia and switches instead to initiating flowers.

networks underlying stress responses (see Chapter 15) intersect with the pathways that control flowering.

Long-day and short-day plants are further divided into **qualitative daylength-sensitive**, which are absolutely dependent on the correct photocycle for floral induction, and **quantitative** or **facultative daylength-sensitive**, where flowering will eventually occur autonomously under non-inductive conditions but is hastened by exposure to inductive photoperiods. Qualitative types requiring a single long day or short day are experimentally useful because photoperiodic treatments can be applied with precision, allowing the light requirements for floral induction to be analyzed in detail. Morning glory (*Pharbitis nil*, also referred to as *Ipomoea nil*) and cocklebur (*Xanthium strumarium*) are examples of model qualitative SDPs in which flowering is promoted by exposure to a single inductive photocycle. Flowering of the so-called Ceres form of the annual grass *Lolium temulentum* (Figure 8.34) is induced by a single long day. Building on the physiological foundations established by work on these qualitative long day and short day models, molecular dissection of floral induction and development has focused on *Arabidopsis*, a facultative LDP species.

The terms LDP and SDP are a little misleading, because it is the length of the dark period rather than the length of the day that is critical in photoperiodic responses. In SDPs, short days only induce flowering if they are combined with long nights. When the plants are grown under artificial day–night cycles of less than 24 hours, in which the nights as well as the days are short, SDPs will not flower (Figure 8.33F). Moreover if a long night is interrupted with a short pulse of light, flowering is again prevented. For SDPs, therefore, it is a long night which initiates the transition from vegetative to reproductive growth. In LDPs, again the night length is critical. In this case a short night is required for induction, as shown for *L. temulentum* in Figure 8.34. If the night is sufficiently short, LDPs will flower even if the day is also short (Figure 8.33F), whereas a long day does not induce flowering in LDPs if the night is also long (Figure 8.33G). Thus SDPs are really long-night plants and LDPs are really short-night plants. Interrupting a non-inductive long night can induce flowering in many (but not all) LDPs, but only if the light break is long enough—a very short light break will not bring about flowering. This is in contrast to SDPs, where even a brief flash of light in the dark period is sufficient to prevent flowering.

How does a plant measure the length of its days and nights? The most widely-accepted explanation is that photoreceptors responding to the amount and quality of light in the environment control the stability of key proteins which are expressed according to circadian rhythms. These proteins in turn regulate the expression of other genes important in the process, e.g. bud set, growth cessation, tuberization and so on, which is under photoperiodic control. As for many other aspects of development, the best-understood experimental subject for analyzing the control of flowering time is *Arabidopsis*, a quantitative LDP in which the sensitivity to inductive floral signals increases with age.

8.5.4 The gene FT, which encodes a mobile floral inducer, is regulated by the transcription factor CO

Experiments in which different parts of the same plant are exposed to different photoperiods established that it is the leaf and not the shoot apex that perceives the inductive stimulus. Grafting a photoinduced leaf onto a vegetative shoot induces flowering. In studies of the SDP species *Perilla frutescens* (beefsteak plant) it has been shown that a single induced leaf can be grafted onto seven successive vegetative plants and will still sustain its inductive effect (Figure 8.35). These observations suggest that photoinduction causes a permanent change in the receptive leaf so that it becomes a continuous source of a mobile flowering stimulus. This stimulus was given the name **florigen** and its identification became the central objective of research on floral induction.

Mutants, particularly in *Arabidopsis*, have been essential tools for analyzing the genes and interactions controlling vegetative development (see Chapter 12). Similarly, studies of mutants that flower earlier or later than the wild-type have revealed the existence of networks that regulate the perception of a floral stimulus, transmission of the flowering signal, and transition from the vegetative to the reproductive condition in the shoot apex. In general, late-flowering genetic variants are defective in genes that promote flowering, whereas lines that flower early are indicative of genes that ordinarily repress flowering being inactivated or downregulated. Analysis of *Arabidopsis* late-flowering mutants led to identification of the regulatory gene **CONSTANS** (**CO**), which encodes a nuclear zinc-finger transcription factor. Under long days, CO induces transcription of the gene *FLOWERING LOCUS T* (**FT**).

FT is normally expressed in leaves and not in apices, but transgenic plants that overexpress *FT* in the shoot apex, under the control of a meristem-specific promoter, will flower under non-inductive conditions. When targeted to phloem, the CO protein stimulates both the synthesis of FT and the formation of a translocatable flowering signal. Such transgenic manipulations, together with grafting experiments, provide strong

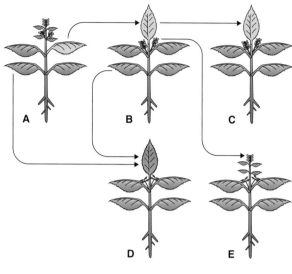

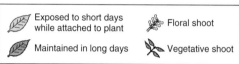

Figure 8.35 Permanent photoinduction of *Perilla* leaves. (A) A single leaf exposed to short days (SDs) induces flowering of an attached plant that has been exposed only to non-inductive long days (LDs). (B) Grafting the same SD leaf onto a second vegetative plant grown in LDs induces flowering. The same leaf can induce flowering on a third host plant (C) and so on. Neither LD leaves from plants in flower (D) nor a floral apex grafted onto a non-induced plant (E) can induce flowering in the host.

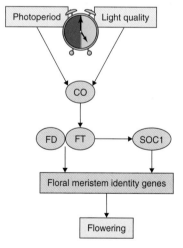

Figure 8.36 Gene networks connecting photoregulation and circadian control of flowering. Environmental signals upregulate the synthesis of the transcription factor CONSTANS (CO) which interacts with the floral integrator gene *FLOWERING LOCUS T* (*FT*), and, indirectly via *FT*, with *FLOWERING LOCUS D* (*FD*) and *SUPPRESSOR OF OVEREXPRESSION OF CONSTANS1* (*SOC1*). In turn, products of these genes interact with floral meristem identity genes leading to flowering.

evidence that the FT protein is the phloem-mobile florigenic factor that transmits the inductive signal from the leaf to shoot apex. In the shoot apex FT associates with the bZIP transcription regulator **FD** which is encoded by the gene *FLOWERING LOCUS D (FD)*. The FT–FD complex induces expression of *SUPPRESSOR OF OVEREXPRESSION OF CONSTANS1 (SOC1)*. FT–FD and SOC1 are the transcriptional integrators that activate floral identity genes (Figure 8.36; see Chapter 16).

8.5.5 The CO–FT system is regulated by the circadian clock, photoperiod and light quality

As we have seen, evidence suggests that FT, whose synthesis is promoted by CO, moves from the leaves to the shoot apical meristem where it initiates the changes in gene expression that result in a switch from **vegetative growth** to **flowering** (see Chapter 16). Figure 8.37 illustrates our current understanding of the regulation of CO expression in relation to day length in the LDP *Arabidopsis*. The circadian clock controls expression of the *CONSTANS* gene and the ability of phytochrome and cryptochrome to regulate the expression of *CO* or the stability of its protein product CO. When the day length is short the *CO* gene is expressed maximally during the night; however, the CO protein is degraded in darkness. In addition, during the early part of the day, PHYB promotes the degradation of CO. In short days CO never accumulates because it cannot be synthesized during the day and is degraded when it is made at night. On the other hand, when the day is longer a zeitlupe photoreceptor active near the end of the long day promotes the degradation of an inhibitor of *CO* expression. This leads to a peak of *CO* gene expression in the light when the CO protein that is synthesized is not subject to degradation. Additionally, phytochrome A and cryptochromes act to repress CO degradation at the end of the day, thus promoting CO stability in the light. Therefore CO accumulates and it can then activate *FT* expression leading to the induction of flowering. The mutants *phyA* and *cry2* are late flowering; in contrast, *phyB* mutant plants are early flowering.

Genes homologous to *CO* and *FT* that also have diurnally-regulated patterns of expression have been identified in other cases where photoperiod controls developmental processes. Overexpression of *FT* in rice and winter wheat, and in a number of dicots (including *Populus*, *Nicotiana* and *Pharbitis*), causes extreme early

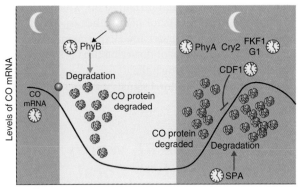

A Short days

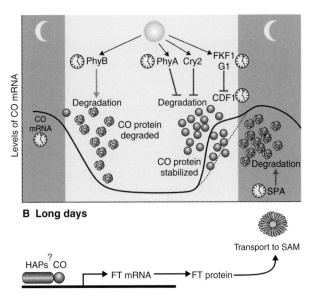

B Long days

Figure 8.37 Clock control of *CO* gene expression at the transcription and protein levels. Red spheres represent intact CO protein, red split spheres are degraded CO protein, and the black line shows the clock-controlled variation in levels of CO mRNA. The clock symbol indicates genes controlled by the circadian clock. (A) In short day conditions CO mRNA is mainly synthesized in the dark, and the resulting CO protein is degraded by proteins of the SPA family. Some CO protein is also produced in the morning and this too is degraded, in a process that depends on the presence of active PHYB. Inhibition of *CO* gene expression by CDF1 prevents synthesis of CO in the afternoon. (B) In long days CO produced in the morning is still degraded. However, in the light FKF1 and G1 prevent repression of *CO* mRNA synthesis by CDF1, so the *CO* gene is expressed in the afternoon. The resulting CO protein is stabilized in the light by the presence of PHYA and Cry2. It is believed that this stable CO protein can form a complex with HAP (heme activator protein) and the complex binds to the promoter of the *FT* gene, stimulating production of FT protein. FT is transported through the phloem to the shoot apical meristem, where it induces flowering.

flowering. Downregulating *FT* by mutation or RNA interference methods greatly delays flowering in *Arabidopsis* and rice. The evidence from such experiments strongly supports the case for *FT* as a universal flowering signal. The CO−FT module is highly conserved, not only in the network of genes regulating floral induction, but also in such photoperiodically determined processes as tuberization, dormancy and bud set (see Chapter 17).

An important practical implication of the conserved nature of the flowering process is that mechanistic knowledge derived from model systems can be directly applied to agricultural species, in which manipulation of reproductive development is a central objective for improving crop yields. The regulatory pathways centered on *FT* homologs have been studied in cereals. Flowering in *Oryza* (rice), an SDP, is regulated by members of the *HEADING DATE* (**Hd**) gene family. *Hd3a* encodes a phloem-mobile FT ortholog. *Hd1* is the rice equivalent of *CO* and mediates the photoperiodic control of *Hd3a* expression, though its functions are more complex than those of its *Arabidopsis* homolog, since not only does Hd1 stimulate *Hd3a* transcription under inductive conditions but it also represses it under long days. Genetic and environmental regulation of reproductive development is considered further in Chapter 16.

Key points The switch from vegetative to reproductive development, formation of resting structures, dormancy and other life-cycle events are under photoperiodic control. Plants are grouped into short-day (SD), long-day (LD) or day-neutral types according to the photoperiod that induces flowering. Daylength-sensitive plants may be qualitative (for example the SD plant *Pharbitis*, the LD plant *Lolium temulentum*) or quantitative (such as *Arabidopsis*, a LD plant). Usually it is the length of the dark phase of the photocycle that determines flowering in LD and SD plants, and responses to the light phase show the red/far-red reversibility indicative of phytochrome involvement. Photoperiod is perceived by leaves where the gene *CONSTANS* (*CO*) expresses an inducer of *FLOWERING LOCUS T* (*FT*). The abundance of CO is directly regulated by the circadian clock, and PHYA and CRY2 promote CO stability, whereas PHYB stimulates CO breakdown. FT protein moves from the induced leaf to the apex where it forms a complex with the bZIP transcription regulator FD and initiates the transition from vegetative to floral development by upregulating floral integrator and identity genes.

Chapter 9

Photosynthesis and photorespiration

9.1 Introduction to photosynthesis

The synthesis of organic molecules by nearly all living organisms is driven directly or indirectly by energy from the sun, through the process of **photosynthesis** (Figure 9.1). The figures on solar energy input and utilization given in Chapter 8 tell us that photosynthesis captures about one in every 1700 joules of photon energy arriving at the Earth's surface. The photosynthetic land plants, algae and prokaryotes are collectively known as **photoautotrophs** and are the primary producers of the organic material that in turn supports **heterotrophic** consumers through **food chains** and **food webs**. Conversion of inorganic carbon dioxide into the organic products of photosynthesis is often referred to as **carbon fixation** or **assimilation**. All of the **atmospheric oxygen** used in cellular respiration and other biological oxidation processes is a by-product of photosynthesis (Figure 9.1). Humans are dependent on current carbon fixation for food, feed, fiber and biomass, and on ancient photosynthesis for the **fossil fuels** that are at present essential to industrial economies. The formation of organic products and oxygen from carbon dioxide and water is a complex process in which light is absorbed and converted into biologically useable forms of energy and reducing power. In this chapter we describe the organization, function and regulation of different components of the photosynthetic machinery of green plants—light capture and electron transport, water splitting and gas exchange, CO_2 fixation and carbohydrate metabolism. We also discuss **photorespiration**, an oxidative activity that competes with CO_2 assimilation and limits net photosynthetic productivity.

9.1.1 Photosynthesis in green plants is a redox process with water as the electron donor and carbon dioxide as the electron acceptor

Photosynthesis is the process of using light energy to reduce CO_2 to carbohydrate (empirical formula $(CH_2O)_n$). The source of **reductant** in most photosynthetic organisms is water, but there are prokaryotic photoautotrophs that use other compounds: for example, hydrogen sulfide (H_2S) is used by **purple sulfur bacteria** (Figure 9.2). Photosynthesis may therefore be summarized as in Equation 9.1, where the reductant is represented in the generalized form H_2A.

Equation 9.1 Photosynthesis: generalized overall reaction

$$CO_2 + 2H_2A + light \rightarrow (CH_2O) + 2A + H_2O$$

The Molecular Life of Plants, First Edition. Russell Jones, Helen Ougham, Howard Thomas and Susan Waaland.
© 2013 John Wiley & Sons, Ltd. Published 2013 by John Wiley & Sons, Ltd.

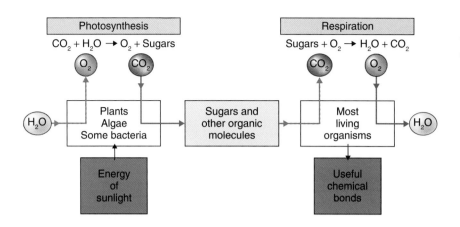

Figure 9.1 Photosynthesis provides organic molecules and oxygen for autotrophs and heterotrophs.

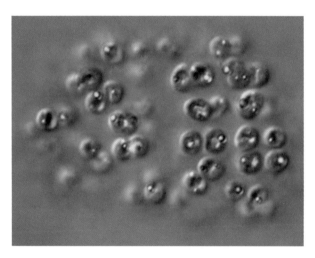

Figure 9.2   Purple sulfur bacteria from low-oxygen sediment of hot pools in Yellowstone National Park, USA.

During photosynthesis, purple sulfur bacteria convert H_2S into elemental sulfur, S. In green plants, algae and cyanobacteria, where the reductant H_2A is water, H_2O, it follows that the product A is oxygen. Photosynthesis that yields oxygen is referred to as **oxygenic**; photosynthesis in which the product A is other than oxygen, such as sulfur, is termed **anoxygenic**. The biochemistry of CO_2 fixation in anoxygenic photosynthetic prokaryotes is generally the same as that of green plants, and many of the structural and mechanistic features of light capture and energy transduction are also common to all photoautotrophs. Organisms such as purple sulfur bacteria have been important experimental models for understanding the molecular and biophysical fundamentals of photosynthesis.

Oxygenic photosynthesis is critical for maintaining oxygen in the Earth's atmosphere, and supplying energy and reduced carbon for organic synthesis (Figure 9.1). Oxygenic photosynthesis (Equation 9.2) is an **endergonic reaction** with a free energy change, $\Delta G^{\circ\prime}$, of $+2840$ kJ mol^{-1}.

Equation 9.2 Photosynthesis: overall reaction carried out by oxygenic photoautotrophs

$$CO_2 + 2H_2O + light \rightarrow (CH_2O) + O_2 + H_2O$$

9.1.2 Photosynthesis in green plants takes place in chloroplasts

Photosynthesis is carried out by leaves and other green tissues of multicellular plants. The cells of these tissues contain **chloroplasts**, specialized organelles in which all the biophysical and biochemical reactions of photosynthesis take place (Figure 9.3). Chloroplasts are thought to have originated in an **endosymbiotic** association of a protoeukaryotic cell and a cyanobacterium-like photosynthetic prokaryote more than 1500 million years ago. As described in Chapter 4, the chloroplast consists of a complex internal **thylakoid membrane** system embedded in an aqueous **stroma** matrix surrounded by a double-membrane **envelope**. In green plants the thylakoid system is organized into appressed (stacked) regions called **grana** (singular **granum**) and unstacked stroma membranes. Grana and stroma membranes are interconnected and enclose the **thylakoid lumen**, an internal space (Figure 9.3).

9.1.3 Thylakoids convert light energy to ATP and NADPH utilized in the stroma for carbon reduction

When carefully isolated, intact chloroplasts can carry out light-dependent conversion of CO_2 to carbohydrate. Photosynthesis can be divided into two phases (Figure 9.4): **light energy capture reactions** and **carbon**

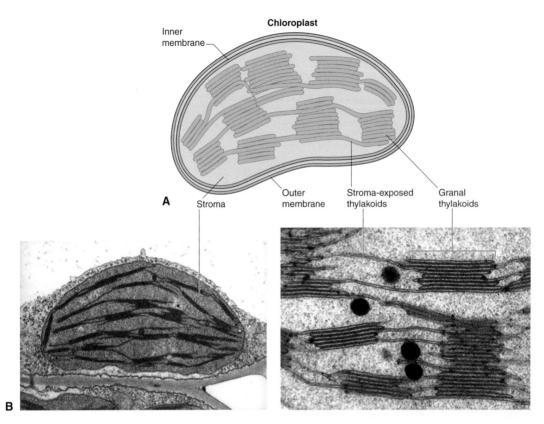

Figure 9.3 Chloroplast structure. (A) Schematic diagram of plant chloroplast, showing compartmentation of the organelle. In a typical higher plant chloroplast, the internal membranes (thylakoids) include stacked membrane regions (granal thylakoids) and unstacked membrane regions (stroma-exposed thylakoids). (B) Transmission electron micrographs of a plant chloroplast revealing its ultrastructure.

reduction reactions (**Calvin–Benson cycle**). In the first phase, light energy is converted into chemical energy. The products of this phase are O_2, ATP and NADPH. In the second phase, ATP and NADPH from the first phase reduce CO_2 to carbohydrate. In turn, the $NADP^+$ and $ADP + P_i$ products of carbon reduction are reused in the light energy capture reactions (Figure 9.4). It is important to note that the oxygen released during oxygenic photosynthesis comes from *water* and *not* from carbon dioxide. The light capture reactions take place exclusively in thylakoid membranes, while the reactions of the carbon reduction cycle are located in the chloroplast stroma. The two phases are sometimes referred to as the 'light' and 'dark' reactions, respectively, but this is misleading since they are coupled, i.e. the products of one phase are reactants in the other. In addition, light is required to activate some enzymes of the 'dark' phase. Both sets of reactions stop in the dark or in the absence of CO_2.

The photosynthetic pigments of green plants are components of two distinct multiprotein complexes in the thylakoid membrane, **photosystems I** and **II** (**PSI** and **PSII**). The water-splitting site is part of the PSII

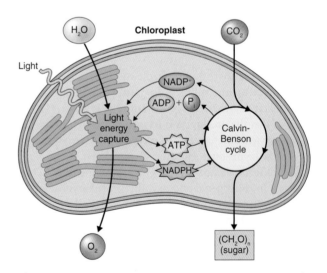

Figure 9.4 Relationship between light energy capture and carbon reduction reactions of photosynthesis. The light energy capture reactions take place in thylakoid membranes while the carbon reduction reactions take place in the stroma.

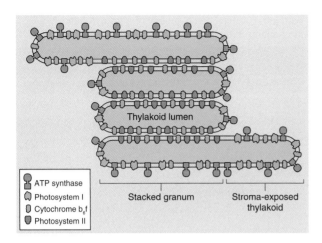

Figure 9.5   Organization of the thylakoid membrane and photosynthetic complexes. PSII is localized primarily in the stacked membrane regions of the thylakoids, whereas PSI and ATP synthase are almost exclusively localized in the unstacked membrane regions. The cytochrome b_6f complex is distributed evenly throughout the membrane regions. The separation of the photosystems necessitates mobile electron carriers such as plastoquinone and plastocyanin, which shuttle electrons between the spatially separated membrane complexes.

complex. PSI passes electrons to $NADP^+$ via the redox protein **ferredoxin**. The thylakoid also contains two other multiprotein structures: the transmembrane cytochrome complex (**cytochrome b_6f**) and chloroplast **ATP synthase**, a large stroma-exposed protein complex (Figure 9.5). The functional relationship between these structures is such that cytochrome b_6f receives electrons from PSII and passes them to PSI. Electrons are carried between PSII and cytochrome b_6f by a lipid-soluble quinone (**plastoquinone**), and between cytochrome b_6f and PSI by a water-soluble copper protein (**plastocyanin**). Photosynthetic electron transport involves the movement of electrons from water in the thylakoid lumen to $NADP^+$ in the stroma, where the carbon-linked Calvin–Benson cycle reactions occur. ATP synthase uses the proton gradient built up during

electron transport for the chemiosmotic synthesis of ATP. Photosynthetic electron transport can be summarized in the sequence shown in Figure 9.6.

An important feature of the thylakoid membrane is that the multiprotein complexes of the photosynthetic apparatus display **lateral heterogeneity**; that is, they are not uniformly distributed throughout the membrane (Table 9.1 and Figure 9.5). PSI is concentrated in the unstacked and stroma-exposed membranes, whereas PSII is found mostly in the grana stacks. ATP synthase is largely confined to the stroma-exposed membrane whereas the cytochrome b_6f complex is more or less evenly distributed throughout the thylakoid membrane. Plastocyanin is more abundant in the unstacked regions of the thylakoid lumen than in the grana. An implication of lateral heterogeneity is that there must be mechanisms for **long-range electron transfer** to allow electrons to move between distant thylakoid components during photosynthesis. The thylakoid membrane allows rapid lateral movement of the diffusible electron carriers plastoquinone and plastocyanin to accomplish this.

The polypeptides of the photosynthetic membrane complexes and the carbon-fixing stroma enzyme ribulose-1,5-bisphosphate carboxylase (rubisco) are products both of **nuclear genome** and of **chloroplast genome** transcription and translation. The correct assembly of these multicomponent structures requires a high degree of cooperation between the organelle and the nucleus (see Chapters 12 and 15).

9.2 Pigments and photosystems

As we have seen in Chapter 8, light energy must be absorbed by a pigment in order to have a biological effect. In vivo, photosynthetic pigments are components

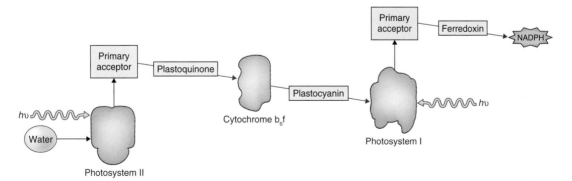

Figure 9.6   The photosynthetic electron transport chain. Photosystem II, cytochrome b_6f and photosystem I are multiprotein complexes located in the chloroplast thylakoid membrane. Plastoquinone, plastocyanin and ferredoxin are mobile electron carriers.

Key points Photosynthetic autotrophs convert CO_2 into the organic compounds that are the basis of the food and energy webs which support heterotrophs, including humans. Green plants carry out oxygenic photosynthesis, an endergonic process in which water is the electron donor and CO_2 the electron acceptor. Anoxygenic photosynthetic microorganisms such as purple sulfur bacteria are useful models for photosynthesis research. Light energy is captured in the thylakoid membranes of chloroplasts. ATP and NADPH are generated by the operation of a light-driven electron transport chain and phosphorylation mediated by a proton gradient. The carbon reduction (Calvin–Benson) cycle in the chloroplast stroma utilizes ATP and NADPH to assimilate CO_2 and produce organic metabolites.

Table 9.1 Lateral heterogeneity of photosynthetic components in thylakoid membranes.

Component	Thylakoids (%)	
	Grana stack	Stroma-exposed
PSII	85	15
PSI	10	90
Cytochrome b_6f complex	50	50
LHC-II (light-harvesting complex of PSII)	90	10
ATP synthase	0	100
Plastocyanin (located in thylakoid lumen)	40	60

of PSI and PSII, which are structurally, functionally and spatially distinct, but share features with all photosystems across the range of photosynthetic organisms. A photosystem contains an antenna array, also called a **light-harvesting complex**, whose pigments collect light energy and transfer it to a reaction center where light energy is converted to chemical energy. In most plants about 250 antenna chlorophyll molecules are associated with each reaction center. Antenna pigments are arranged so that energy is transferred between them, and ultimately to the reaction center, at high speed and with almost 100% efficiency.

9.2.1 Light energy used in photosynthesis is captured by chlorophylls, carotenoids and, in certain algae and cyanobacteria, phycobilins

In the plants, algae and cyanobacteria that carry out oxygenic photosynthesis, the primary pigment absorbing light energy is **chlorophyll**. A structurally related pigment called **bacteriochlorophyll** is the photoreceptor of anaerobic photosynthetic bacteria (Figure 9.7). Chlorophyll is a tetrapyrrole and is synthesized by a metabolic pathway that also provides the precursors of other tetrapyrroles, notably heme, phycobilins and the chromophore of phytochrome (see Chapter 8). The tetrapyrrole ring (sometimes called the **macrocycle**) of chlorophyll binds a central magnesium atom and is attached to **phytol**, a long (C_{20}) hydrophobic side-chain that renders the molecule extremely non-polar.

Different chemical side groups on the macrocycle give rise to different molecular variants of chlorophyll, referred to as chlorophylls a, b, c and d. As summarized in Table 9.2, all oxygen-evolving photosynthetic organisms contain **chlorophyll a** (Figure 9.7). In almost all cyanobacteria, chlorophyll a is the only form of chlorophyll. Viridophytes and certain cyanobacteria contain a second form, **chlorophyll b** (Table 9.2 and Figure 9.7). Chlorophyll b is synthesized from chlorophyll a by conversion of a methyl to a formyl side group (see Chapter 8). Algae of the brown, diatom and dinoflagellate groups contain **chlorophyll c** as well as chlorophyll a. In red algae, the second form of chlorophyll is **chlorophyll d** (Table 9.2).

The different side-chains and degrees of saturation of the macrocycle significantly alter the spectral absorption profiles of the various chlorophyll species, but all chlorophylls absorb light primarily in the blue and red wavelengths of the visible spectrum. The reflection of green light, which is weakly absorbed by chlorophylls a and b (Figure 9.8A), accounts for the color of the photosynthetic tissues of land plants.

All photosynthetic organisms contain **carotenoids**, and some also have **phycobiliproteins** (Table 9.2). These photosynthetic pigments absorb light at wavelengths where absorption by chlorophyll is low, thereby making more effective use of visible light energy than could be achieved by one pigment type alone (Figure 9.8B). In general, oxygenic organisms are able to use light of wavelengths 400–700 nm for photosynthesis. In many anoxygenic photosynthetic prokaryotes, which tend to occupy red-depleted aquatic environments, absorption by bacteriochlorophyll extends the waveband beyond 700 nm into the less energetic wavelengths of the near infra-red.

Carotenoids, orange-yellow pigments that absorb light between 400 and 500 nm (Figure 9.8B), are **tetraterpene** (C_{40}) molecules biosynthesized in

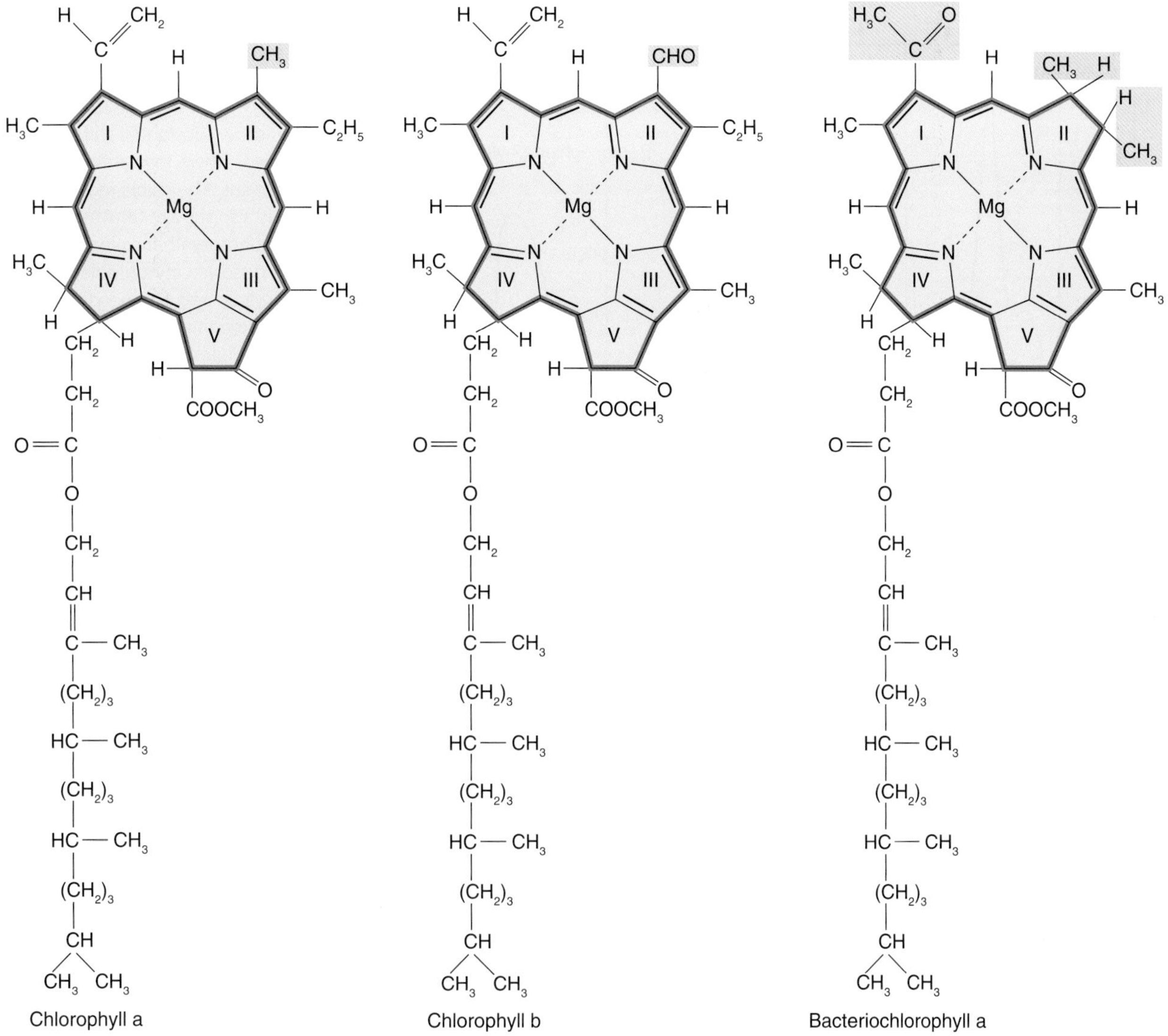

Chlorophyll a · Chlorophyll b · Bacteriochlorophyll a

Figure 9.7 Molecular structures of different chlorophylls. Note the highlighted constituents on the tetrapyrrole ring which differ between different types of chlorophyll.

Table 9.2 Pigments present in organisms capable of oxygenic photosynthesis.

	Chlorophyll					
Organism	a	b	c	d	Carotenoids	Phycobiliproteins
Land plants	+	+	−	−	+	−
Green algae	+	+	−	−	+	−
Diatoms	+	−	+	−	+	−
Dinoflagellates	+	−	+	−	+	−
Brown algae	+	−	+	−	+	−
Red algae	+	−	−	+	+	+
Cyanobacteria	+	−	−	−	+	+

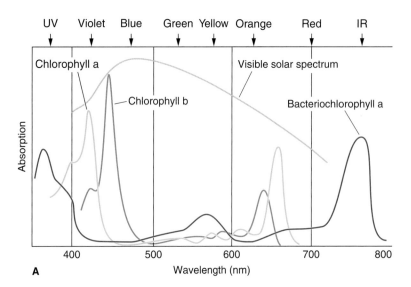

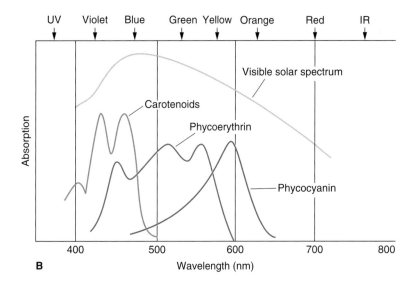

Figure 9.8 Light absorption spectra of photosynthetic pigments. (A) The absorption spectra are shown for chlorophylls a and b and bacteriochlorophyll a dissolved in non-polar solvents. Note that the spectra of these pigments show substantial shifts in absorbance in vivo, where they are associated with specific proteins. The visible region of the solar spectrum is also shown. (B) Absorption spectra of other photosynthetic pigments. The carotenoids are dissolved in non-polar solvents; phycocyanin and phycoerythrin spectra are measured in aqueous solution. IR, infra-red; UV, ultraviolet.

chloroplasts by the **isoprenoid pathway** (see Chapter 15). They comprise the **carotenes**, which contain a conjugated double-bond system (a chain of alternating single and double carbon-carbon bonds), and the **xanthophylls**, in which the terminal rings of carotenes are oxygenated in various ways. Carotenoids act as minor accessory pigments in photosynthesis, transferring light energy to chlorophyll molecules, but they have leading roles as essential structural components in the assembly of light-harvesting complexes, and in defending the photosynthetic apparatus against photooxidative damage (see Chapter 15).

Phycobilins, pigments that absorb light in the 500–650 nm range (Figure 9.8B), are found in red algae and in cyanobacteria (Table 9.2). They are water-soluble linear tetrapyrroles that are covalently linked to specific proteins: **phycoerythrin**, which binds the chromophore

phycoerythrobilin; and **phycocyanin** and **allophyco-cyanin**, which bind the chromophore **phycocyanobilin**. In red algae and cyanobacteria, phycobiliproteins are organized into complex structures called **phycobilisomes** which are tightly appressed to the thylakoid membrane. Phycobilisomes interfere with thylakoid stacking and consequently red algal chloroplasts lack grana.

9.2.2 Reaction centers are the sites of the primary photochemical events of photosynthesis

In Chapter 8, we saw that when a pigment absorbs a light quantum, one of its electrons jumps to an excited state.

The excited electron may be transferred to an acceptor molecule in a process called **charge separation**. Charge separation involving reduction of an electron acceptor by an excited pigment is the central photochemical event of photosynthesis (Equation 9.3). It takes place at the **reaction center** of each photosystem.

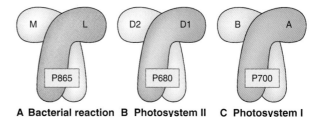

A Bacterial reaction B Photosystem II C Photosystem I
center

Figure 9.9   Diagrammatic representations of reaction center polypeptides and (bacterio)chlorophyll dimers, showing structural homology between the bacterial reaction center (A) and green plant PSII (B) and PSI (C).

Equation 9.3 Charge separation in photosynthesis

Pigment + acceptor + hν → pigment* + acceptor

→ pigment$^+$ + acceptor$^-$

where * denotes the excited state

Transfer of an excited electron to an acceptor molecule occurs about a thousand times more rapidly than fluorescence emission—of the order of 10^{-12} compared with 10^{-9} seconds. The efficiency of a photochemical event is expressed as **quantum yield**, the proportion of absorbed photons converted into chemical product. Under optimal conditions the measured quantum yield of photochemistry is approximately 1, which means that, providing there is an adequate supply of electron acceptors, de-excitation via the charge separation route will occur preferentially, photosynthesis will proceed at high efficiency and little fluorescence will be detected. If, however, there is an inadequate supply of oxidized electron acceptors, quantum yield will be <1 and fluorescence will increase. The measurement of **chlorophyll fluorescence** is the basis of highly sensitive methods for analyzing the mechanism, regulation and quantum yield of photosynthesis.

A critical advance in understanding of the molecular structures and interactions of reaction center components came with the resolution, by X-ray crystallography, of the three-dimensional configuration of a reaction center complex isolated from the photosynthetic bacterium *Rhodopseudomonas viridis*. The *R. viridis* reaction center provides the conceptual model for the structures and functions of PSI and PSII. It consists of a pair of polypeptides, L and M, in a pseudosymmetrical arrangement (Figure 9.9A) perpendicular to the photosynthetic membrane. Associated with L and M is a bacteriochlorophyll dimer, referred to as the **special pair**, the pigment that participates in primary charge separation. The molecular environment of the special pair gives it a characteristic absorption maximum, 865 nm, and hence it is commonly referred to as **P865**. There is close structural and functional homology between the *R. viridis* reaction center and those of PSI and PSII (Figure 9.9). In eukaryotic autotrophs, the reaction center consists of two homologous proteins, **D1** and **D2** in PSII (Figure 9.9B), and the C-terminal domains of PSI subunits A and B (Figure 9.9C). In each case the polypeptide chains are folded into five transmembrane α-helices, and the two subunits interlock to create a pseudo-symmetrical handshake motif. Associated with these proteins is a chlorophyll a dimer. The molecular environment of the dimer gives it a characteristic absorption maximum from which it is named. In PSII the dimer is referred to as **P680**; that of PSI is **P700** (Figure 9.9B, C). The reaction center chlorophyll absorbs a photon and subsequently transfers an electron to a series of acceptors (A_0, A_1, A_2). Table 9.3 identifies electron acceptors A_0, A_1 and A_2 for PSI and PSII. The symmetry observed in the transmembrane protein subunits is reflected in the arrangement of potential electron acceptors such that the entire reaction center appears to have arisen by **duplication** of a basic structure. Only one of the two symmetrical pathways is available for electron transfer. The second branch remains inactive for reasons, and by mechanisms, that are yet to be discovered. The genes encoding reaction center subunits for PSI are referred to by the symbol *psa*; those of PSII are named *psb* (Table 9.4). Synthesis and assembly of functional reaction center complexes requires cooperation between the genomes of the nucleus and the chloroplast.

Table 9.3 Electron transfer carriers in the PSI and PSII reaction center complexes.

Carrier	PSI	PSII
Reaction center chlorophyll	P700	P680
A_0	Chlorophyll a	Pheophytin a
A_1	Phylloquinone	Plastoquinone (Q_A)
A_2	F_x ([Fe_4S_4] center)	Plastoquinone (Q_B)

Table 9.4 Properties of proteins of reaction center chlorophyll–protein complexes.

Protein	Molecular mass of protein (kDa)	Gene encoding protein	Genome
Associated with PSI			
PsaA	18	psaA	Plastid
PsaB	18	psaB	Plastid
PsaN (plastocyanin dock)		psaN	Nuclear
PsaF	17	psaF	Nuclear
PsaC (redox protein)		psaC	Plastid
PsaD (binds ferredoxin)	18	psaD	Nuclear
PsaE (binds FNR)	10	psaE	Nuclear
Associated with PSII			
D1	32	psbA	Plastid
CP43	43	psbD	Plastid
CP47	51	psbC	Plastid
Cytochrome b_{559}		psbE, psbF	Plastid

9.2.3 Antenna pigments and their associated proteins form light-harvesting complexes in the thylakoid membrane

In all organisms with oxygenic photosynthesis, the reaction center chlorophyll is always chlorophyll a. Reaction center pigments present a small target for excitation by incoming photons. The remaining chlorophyll a plus the other photosynthetic pigments (see Table 9.2) are found in light-harvesting (antenna) complexes, which greatly increase the capacity for photon capture. In green plants, antenna complexes contain chlorophylls a and b plus carotenoids. Isolated chlorophyll a has a peak of absorption at around 660–670 nm (depending on the solvent). Chlorophyll b absorbs maximally at 640–650 nm (see Figure 9.8A). The absorption maxima of chlorophylls associated with the antenna proteins of photosystem light-harvesting complexes are shifted significantly towards the red (lower energy) end of the spectrum, closer to the longer wavelengths absorbed by reaction center chlorophylls. Reaction centers act as **traps** for light energy funneled to them from antennae.

Antenna pigments are bound to specific proteins in **light-harvesting complexes** (LHCs). A given LHC may contain from one to three different proteins. The

complexes associated with PSI and PSII are designated **LHC-I** and **LHC-II**, respectively, and the corresponding proteins are encoded by the **Lhca** and **Lhcb** families of nuclear genes. LHCs contain carotenoids as well as chlorophyll, generally in the ratio of 2:1 total chlorophyll:carotenoid. Table 9.5 summarizes size and compositional data for the major LHCs of PSI and PSII. While LHCs usually funnel light energy to their own reaction centers, under certain conditions the antenna of PSII may also transfer energy to the PSI reaction center (see Chapter 15). In red algae and cyanobacteria, phycobilisomes function as light-harvesting antennae for PSII, transferring their energy to the LHC-II embedded in the thylakoid membrane.

9.3 Photosystem II and the oxygen-evolving complex

PSII, one of the two photosystems found in all organisms that carry out oxygenic photosynthesis, functions as a **water-plastoquinone oxidoreductase**. The reaction center of the PSII of green plants closely follows the *Rhodopseudomonas viridis* model of structure and function (see Figure 9.9). The antenna complex of PSII

Table 9.5 Properties of light-harvesting chlorophyll–protein complexes.

Complex	Chlorophyll a:b ratio	Molecular mass of protein (kDa)	Nuclear gene encoding protein
Associated with PSI			
LHC-Ia	2.0–3.1	20.5	*Lhca3*
		18	*Lhca2*
LHC-Ib	2.2–4.4	20	*Lhca1*
		20	*Lhca4*
Associated with PSII			
LHC-IIa	4.0	29	*Lhcb4*
LHC-IIb	1.35	27–28	*Lhcb1*
		25–27	*Lhcb2*
		25	*Lhcb3*
LHC-IIc	2.9	26.5	*Lhcb5*
LHC-IId	1.51	24	*Lhcb6*

Key points The primary photosynthetic pigment of oxygenic autotrophs is chlorophyll a. In most cyanobacteria it is the only form of chlorophyll. Anaerobic bacteria have bacteriochlorophyll. Green algae and land plants have chlorophyll b in addition to chlorophyll a. Different algal groups have chlorophyll c or d. All photoautotrophs have carotenoid accessory pigments, and phycobilins are found in cyanobacteria and red algae. Light energy is captured by photosynthetic pigments organized into photosystems, each of which consists of light-harvesting antennae and a reaction center embedded in the thylakoid membrane. Photoexcitation causes charge separation, in which a pair of reactive chlorophyll a molecules in the reaction center transfer electrons from a donor to an acceptor. P680 and P700, the reaction center chlorophylls of the two photosystems of green plants, PSII and PSI respectively, are bound to two reaction center polypeptides in a pseudo-symmetrical structure homologous to the bacterial reaction center. Chlorophylls and accessory pigments are organized with membrane proteins into antenna complexes that transfer captured light energy to the reaction center.

consists of multiple light-harvesting units of at least four types built from combinations of six different chlorophyll a/b-binding proteins (Table 9.5). Also associated with PSII is the **oxygen-evolving complex**, the site that splits water and provides electrons and protons for the electron transport chain and ATP synthase. This section considers PSII structure, the mechanism by which water is cleaved to yield oxygen, and the nature of plastoquinone, the acceptor that conveys electrons from PSII to the cytochrome b_6f complex.

9.3.1 The PSII reaction center is an integral membrane multiprotein complex containing P680 and electron transport components

Each PSII reaction center contains two pheophytin molecules, an iron atom, and two plastoquinone molecules designated Q_A and Q_B, which act as the terminal electron acceptors. Electron carriers are organized in a symmetrical two-branch structure (Figure 9.10). P680, pheophytin and plastoquinone are the **prosthetic groups** of the two major reaction center proteins, D1 and D2. D1, which functions in the active branch of the reaction center, is a hydrophobic 32 kDa protein encoded by the plastid gene *psbA*. In the light, D1 is exposed to the extremely oxidizing environment created by excited P680. This leads to a high rate of D1 breakdown and simultaneous replacement by newly synthesized protein. The light-dependent **turnover** of D1 is discussed further in connection with environmental responses of photosynthesis in Chapter 15. D2, a 34 kDa

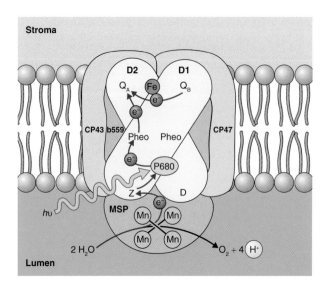

Figure 9.10   Photosystem II (PSII) and the oxygen-evolving complex. Components of photosystem II include: D1 and D2, PSII reaction center proteins; Z, a tyrosine residue in the D1 subunit; CP43 and CP47, chlorophyll a-binding proteins; pheophytin (Pheo); Q_A and Q_B, two plastoquinone molecules; and MSP, manganese-stabilizing protein of the oxygen-evolving complex.

protein encoded by the plastid gene *psbB*, represents the inactive branch of the reaction center (see Table 9.4).

CP43 (molecular mass 43 kDa) and **CP47** (51 kDa) are hydrophobic chlorophyll a-binding proteins associated with the PSII reaction center. They are encoded by the plastid genes *psbD* and *psbC* respectively (see Table 9.4). The reaction center also contains a number of small proteins whose specific functions are unclear but which are known to be required for correct assembly of PSII. Among these is **cytochrome b$_{559}$**, the subunits of which are encoded by plastid genes *psbE* and *psbF* (see Table 9.4).

9.3.2 The light-harvesting antenna complex of PSII accounts for half of total thylakoid protein

In green plants, the antenna complex of PSII consists of multiple light-harvesting units of at least four types built from combinations of six different chlorophyll a/b-binding proteins (see Table 9.5). LHC-II accounts for around 50% of the total protein in plant chloroplast membranes and, in the green tissues of many plants, is the second most abundant protein complex after rubisco (see Section 9.7.1). It exists as a **trimer** and binds half of all chlorophyll molecules in the chloroplast. Each monomeric LHC-II contains a single polypeptide of 230–250 amino acid residues (molecular mass 24–29 kDa), the product of one of the *Lhcb* family of nuclear genes (see Table 9.5). Non-covalently bound to the LHC-II protein are 13–15 chlorophyll a and

chlorophyll b molecules and two luteins. The complex also includes one tightly bound **phospholipid**. Hydropathy analysis (see Chapter 2) of the primary structure of LHC-II protein predicts a secondary structure consisting of three transmembrane α-helices (see Figure 2.15), designated A, B and C. This configuration is confirmed by **X-ray crystallography** of LHC-II from pea, *Pisum sativum* (Figure 9.11). The N-terminal segment of the protein, a sequence of 54 largely polar amino acids accessible to the stroma, leads to the C, B and A membrane-spanning helices (Figure 9.11A). The C-terminal region of the polypeptide on the lumen side of the thylakoid membrane includes helix D, a short helical run ten amino acids in length. A recent higher-resolution study of LHC-II from spinach reveals an almost identical higher-order structure to the *P. sativum* complex, with the addition of a fifth short helical region, helix E, between helices C and B.

The A and B helices form a distinctive X shape, braced by two molecules of the xanthophyll **lutein** (Figure 9.12A). Xanthophylls function as effective **accessory light-harvesting pigments**, absorbing light in the blue-green spectral region as a complement to the absorption of red light by chlorophylls a and b. They also have a role in protecting against the damaging effects of excessive light (see Chapter 15). Chlorophylls a and b in LHC-II are held in place by non-covalent **interactions** with backbone carbonyls and side-chains of histidine, glutamine, glutamate or asparagine ligands in the polypeptide chain, and with the phosphodiester group of a **phosphatidyl glycerol** molecule. The distances between the individual chlorophyll molecules are consistent with known energy transfer rates.

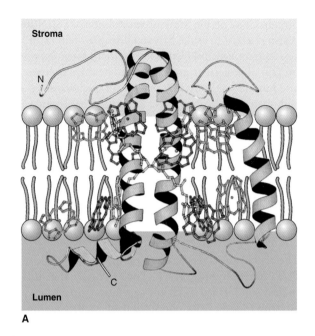

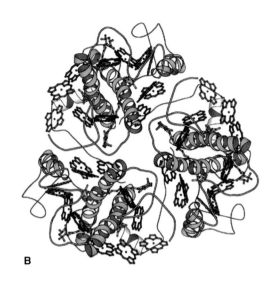

Figure 9.11 Light-harvesting complex of photosystem II. (A) View of the three transmembrane helices and the positions of the macrocycles of chlorophylls a and b. The stroma surface of the thylakoid is at the top, lumen at the bottom. (B) Model of the LHC-II trimer, viewed from the stroma side of the membrane.

Figure 9.12 Two xanthophylls associated with the light-harvesting complex of photosystem II: (A) lutein and (B) violaxanthin.

When protein and pigment components are present in the correct proportions under the appropriate conditions, LHC-II has been shown to **self-assemble** in vitro. Other xanthophylls may be substituted for lutein, though they are not so effective in supporting this assembly. Mutants of *Arabidopsis* lacking lutein have been observed to have normal levels of fully-functional light-harvesting complexes, in which the role of lutein is played by **violaxanthin** (Figure 9.12B). The presence of pigments facilitates folding of the protein of LHC-II into the correct functional form. This is significant for the biosynthesis and breakdown of the complex, which is attuned to pigment supply and removal. Phosphatidyl glycerol has an important structural role in stabilizing the hydrophobic interactions between LHC-II monomers that occur during assembly into trimers. The configuration of monomers in trimeric LHC-II is shown in Figure 9.11B.

Interactions between adjacent trimers are in turn mediated by the **galactolipids** of the thylakoid membrane, forming light-harvesting protein–pigment–lipid supercomplexes that further increase the cross-sectional area of the antenna and the photon capture capacity.

9.3.3 Oxidation of water and reduction of PSII electron acceptors requires four photons per molecule of oxygen released

Photosystem II removes electrons from water and passes them to plastoquinone (see Figure 9.10). The **oxygen-evolving complex** where water is split (Equation 9.4) is closely associated with PSII. The light-induced charge separation in the PSII reaction center produces P680$^+$, an oxidant strong enough to support the endergonic transfer of electrons from water and the release of oxygen (Figure 9.13). The cleavage of water involves a complex series of reactions on the oxidizing (lumen) side of PSII, during which four electrons are transferred for every molecule of oxygen produced. The oxygen-evolving complex contains **manganese, calcium** and (probably) **chloride**, and is electrically linked to P680 by **Z**, a redox-active **tyrosine** residue of the D1 protein of PSII (Figure 9.13).

Equation 9.4 The water-splitting reaction has an E°′ of +0.82 V

$$2H_2O \rightarrow O_2 + 4H^+ + 4e^-$$

Successive one-electron reductions of P680$^+$ are coupled to four-electron oxidations of water and the production of one molecule of O_2. This is thought to occur by the **S-state cycle**, a model in which single-photon events at P680 are stored as intermediate (S_1, S_2, S_3) states until S_4 is reached, the four electrons accumulated drive oxidation of water to O_2 and protons, and the system reverts to S_0 in readiness to repeat the cycle (Figure 9.14). High-resolution spectroscopy and X-ray crystallographic modeling of oxygen-evolving complexes indicate that the electron-acceptor components of the S-state cycle are four manganese atoms arranged with calcium in an oxygen-linked **cuboid** structure. The Z residue of the D1 subunit links the S system with P680.

Three proteins make up the oxygen-evolving complex (Table 9.6). The product of the nuclear gene *psbO* is a 33 kDa protein that is thought to bind to reaction center

protein CP47 on the **lumen-exposed** side of PSII. PsbO is also referred to as the Mn-stabilizing protein (MSP). The nuclear gene *psbP* encodes another lumen-exposed protein of molecular mass 23 kDa. The third protein, also luminal, is the 16 kDa product of nuclear gene *psbQ*. Although these proteins have structures that include various ion-binding and other motifs, details of their functions in water cleavage are not known yet.

9.3.4 Plastoquinone is the first stable acceptor of electrons from PSII

The primary electron acceptor on the reducing side of PSII is **pheophytin a** (Phe a). P680$^+$ Phe a$^-$, the initial radical pair formed by charge separation, drives forward electron transfer from Phe a$^-$ to the first plastoquinone acceptor Q_A (Equation 9.5A), and P680$^+$ is re-reduced by electron transfer from Z and the Mn$_4$Ca cluster of the oxygen-evolving complex (Figure 9.13). Q_A, the first stable electron acceptor, is tightly bound to the PSII reaction center and is rapidly reduced. It transfers two electrons in a slower reaction to the more loosely-bound secondary acceptor, Q_B (Equation 9.5B). Q_B^{2-} is

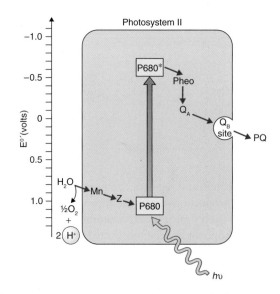

Figure 9.13 P680-mediated electron transfer from water to plastoquinone in photosystem II. Red arrows show the path taken by electrons. When a photon is absorbed by P680, an electron is transferred from P680 to pheophytin (Pheo), the first electron acceptor, and subsequently to two plastoquinone molecules, the first of which (Q_A) is bound tightly to the complex; the second, being mobile, is able to bind the Q_B site when oxidized (PQ), but not when fully reduced (PQH$_2$). P680$^+$ is reduced by Z, a tyrosine residue in the D1 subunit of the reaction center which, in turn, is reduced by electrons from water.

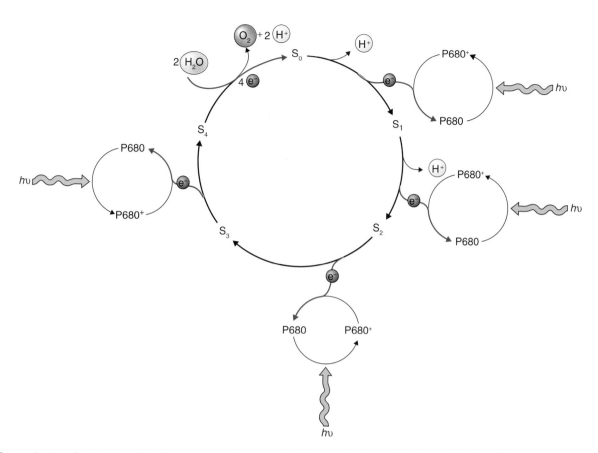

Figure 9.14 The S-state cycle in the oxygen-evolving complex. The complex is considered to exist in five different oxidation states (S_0 through S_4). The cycle is advanced sequentially by photons absorbed by PSII until the highly oxidized (positively charged) state (S_4) is produced. S_4 is the only state capable of water oxidation. Shown are the light-driven oxidation of P680 and its subsequent reduction by electrons that are derived ultimately from water. Four quanta, absorbed by PSII, are required to produce one O_2.

Table 9.6 Characteristics of proteins associated with the oxygen-evolving complex.

Protein (location)	Molecular mass (kDa)	Gene encoding protein	Genome
PsbO (Mn-stabilizing protein, MSP; binds to CP47)	33	*psbO*	Nuclear
PsbP (lumen-exposed protein)	23	*psbP*	Nuclear
PsbQ (lumen-exposed protein)	16	*psbQ*	Nuclear

Equation 9.5 The PSII electron transfer cycle

9.5A Formation of plastosemiquinone Q_A^-

$$P680^+ \text{ Phe a}^- + Q_A \rightarrow P680^+ \text{ Phe a} + Q_A^-$$

9.5B Two-step formation of fully reduced Q_B^{2-}

$$Q_A^- + Q_B \rightarrow Q_A + Q_B^-$$
$$Q_A^- + Q_B^- \rightarrow Q_A + Q_B^{2-}$$

9.5C Protonation of Q_B^{2-} to form plastoquinol

$$Q_B^{2-} + 2H^+ \rightarrow Q_BH_2$$

Figure 9.15A shows the structures of plastoquinone (PQ) and its reduced forms, plastosemiquinone ($PQ^{\bullet -}$) and plastoquinol (PQH_2). Compare these to the structure of ubiquinone, which is found in the respiratory electron transport chain (Figure 7.11). Such quinones are highly lipophilic. Their structure enables rapid lateral diffusion within the lipid bilayer of the thylakoid or inner mitochondrial membranes (Figure 9.15B).

protonated with two H^+ ions from the lumen (Equation 9.5C). It is displaced from its binding site on PSII by oxidized plastoquinone returning from the cytochrome b_6f complex.

Key points The electron donor to PSII is water, which is split by the oxygen-evolving complex consisting of three proteins, as well as manganese, calcium and chloride. Reduction of P680 and the production of one molecule of O_2 from water is accomplished by a four-photon S-state cycle. Plastoquinone accepts electrons in a series of cyclic transfers from reduced pheophytin and Q_A bound to the PSII reaction center and, in its protonated form plastoquinol, diffuses through the lipid bilayer to the cytochrome b_6f complex.

9.4 Electron transport through the cytochrome b₆f complex

The electrons of plastoquinol are transferred to cytochrome b_6f, an integral membrane protein complex

that functions as a **plastoquinol-plastocyanin oxidoreductase**. It resembles in structure and function the cytochrome bc_1 complex (Complex III) of the mitochondrial respiratory chain (see Chapter 7). For example, cytochrome f is a c-type cytochrome with a covalently bound heme cofactor (see Figure 7.11) and the redox and proton release functions of the complex are accomplished by the operation of a Q cycle (Figure 7.13C). The cytochrome b_6f complex is highly responsive to growth conditions and the developmental state of the plant, making it the predominant **control point** for matching photosynthetic electron flux to the metabolic demand for ATP and NADPH.

9.4.1 The cytochrome b₆f complex includes three electron carriers and a quinone-binding protein

Crystal structures have been determined for the cytochrome b_6f complexes of the filamentous cyanobacterium *Mastigocladus laminosus* and

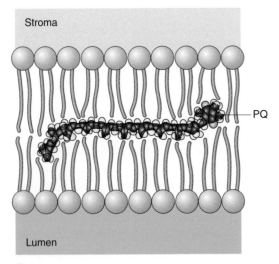

Figure 9.15 Plastoquinone. (A) Structure of plastoquinone and its reduction products. (B) Plastoquinone (PQ) in the lipid bilayer of a thylakoid membrane. Plastoquinone diffuses laterally in the membrane carrying electrons from PSII to the cytochrome b_6f complex.

the unicellular eukaryotic green alga *Chlamydomonas reinhardtii* (Figure 9.16A). Molecular and genetic studies have established that the structure and function of the complex are highly conserved from photosynthetic prokaryotes to seed plants. Cytochrome b_6f exists embedded in the thylakoid membrane as a **dimer** with a molecular mass of 220 kDa. The monomer consists of eight subunits: four redox components with comparatively high molecular mass (17–32 kDa); and four small polypeptides (<4 kDa) of uncertain function (Table 9.7). They are arranged as a bundle of 13 helices traversing the thylakoid membrane, with extensive regions of polypeptide chain also located within the lumen (Figure 9.16B).

Cytochrome f is a 32 kDa polypeptide that has a single transmembrane helical segment and a long N-terminal heme-binding region extending into the thylakoid lumen (Figure 9.16B). Cytochrome f is encoded by the *petA* gene (Table 9.7). The molecular mass of **cytochrome b_6** is 24 kDa. It has four membrane-spanning helices and non-covalently binds two b-type heme prosthetic groups (Figure 9.16B). The third redox-active subunit of the complex is a 19 kDa polypeptide located in the lumen and anchored to the thylakoid membrane by a hydrophobic region near the N terminus. The prosthetic group, a $[Fe_2S_2]$ **Rieske iron–sulfur center**, is bound to the protein by coordination with cysteine and histidine residues (Figure 9.16B). The cytochrome b_6f complex also includes a quinone-binding 17 kDa protein, **subunit IV** (Figure 9.16B); *petG*, *petL*, *petM* and *petN* encode small **peripheral proteins** of the complex, each consisting of a single membrane-spanning helix (Figure 9.16B). They do not seem to have redox functions, but are necessary for assembly and stability of the complex.

9.4.2 The cytochrome b_6f complex generates a proton gradient through the operation of a Q cycle

The cytochrome b_6f complex facilitates the exergonic transfer of an electron along an electrochemical gradient from plastoquinol to plastocyanin via the Rieske $[Fe_2S_2]$ subunit and cytochrome f. The $E^{o'}$ values for the components of this so-called **high potential** electron transport chain are shown in Table 9.8. A second electron, from the plastosemiquinone product of the high potential pathway, is transported along a **low potential** chain involving the hemes of cytochrome b_6. The operation of these two electron transport chains is linked to the net transfer of protons from the stroma side of the thylakoid membrane to the lumen (Figure 9.17) and resembles the **Q cycle** mechanism for electron and proton

Table 9.7 Characteristics of proteins of the cytochrome b_6f complex.

Protein	Molecular mass (kDa)	Gene encoding protein	Genome
Cytochrome f	32	*petA*	Plastid
Cytochrome b_6	24	*petB*	Plastid
Redox protein	19	*petC*	Nuclear
Quinone binding protein (subunit IV)	17	*petD*	Plastid
Peripheral proteins		*petG*, *petL*, *petM*, *petN*	Nuclear

transfer described for Complex III of the mitochondrial electron transport chain (see Chapter 7). Equation 9.6 shows the transfer of the electron in reduced Rieske $[Fe_2S_2]$ to plastocyanin, the electron carrier between cytochrome b_6f and PSI, via cytochrome f along the electrochemical gradient (Table 9.8) of the high potential pathway. Each turn of the cycle collects two protons from the stroma and releases them into the lumen. The protons can then drive **chemiosmotic ATP synthesis**.

Equation 9.6 Reduction of plastocyanin

$$[Fe_2S_2]_{red} + cyt\ f_{ox} \rightarrow [Fe_2S_2]_{ox} + cyt\ f_{red}$$

$$cyt\ f_{red} + PC_{ox} \rightarrow cyt\ f_{ox} + PC_{red}$$

9.4.3 Plastocyanin is a soluble protein that carries electrons from cytochrome b_6f to PSI

Plastocyanin is a small (11 kDa) **copper-containing** electron carrier protein. The plastocyanin pool is located in the aqueous phase of the thylakoid lumen. The rapidity with which it transfers electrons between the cytochrome b_6f complex and PSI indicates high mobility. Variations in amounts of plastocyanin have been observed to correlate with variations of photosynthetic electron transport activity in a number of seed plant and algal species. On the other hand, experiments in which levels were manipulated in *Arabidopsis* by down-regulating or overexpressing *PETE1* and *PETE2*, the two **nuclear genes** encoding plastocyanin in this species, have failed to demonstrate a role in limiting the rate of photosynthesis under optimal growth conditions.

Plastocyanin sequesters as much as 50% of the total copper in photosynthetic cells and thus has an important role in **copper nutrition and toxicity** (see Chapter 13). Copper availability regulates biosynthesis

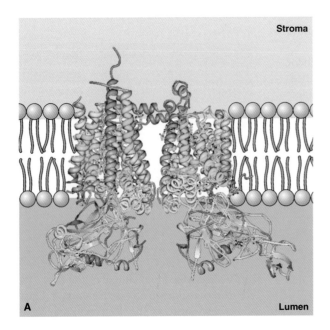

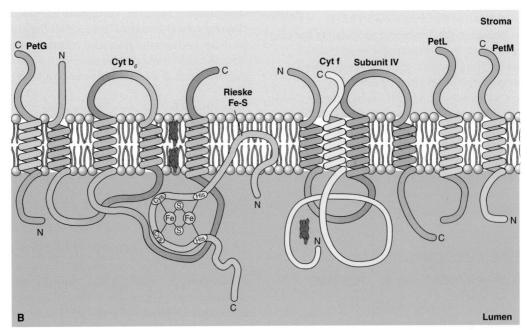

Figure 9.16 The cytochrome b$_6$f complex. (A) Model of the *Chlamydomonas* complex based on X-ray crystallographic studies. (B) Schematic representation of the arrangement of the constituent polypeptides in the thylakoid membrane. Several of the subunits, such as PetG, cyt f, PetL and PetM, are thought to contain only one membrane-spanning domain, whereas cyt b$_6$ and subunit IV contain several transmembrane helices. The cyt b$_6$ heme groups are shown on opposite sides of the membrane and are part of the quinol oxidation and reduction sites, respectively. The heme group of cyt f and the Fe-S center of the Rieske protein are localized within the thylakoid lumen. Heme groups are shown in red.

of plastocyanin in some algae and cyanobacteria. In seed plants under conditions of limited copper supply, plastocyanin gene expression and protein synthesis are maintained at the expense of other copper-containing proteins that are not essential for photoautotrophic growth, such as **copper/zinc superoxide dismutase**.

9.5 Photosystem I and the formation of NADPH

Photosystem I, which functions as a **plastocyanin-ferredoxin oxidoreductase**, has a core of

Key points The cytochrome b_6f complex is a plastoquinol-plastocyanin oxidoreductase and is structurally and functionally similar to the cytochrome bc_1 complex of the mitochondrial electron transport chain. It consists of eight subunits, of which cytochrome f, cytochrome b_6 and a $[Fe_2S_2]$ center protein have redox functions, and forms dimers in the thylakoid membrane. Electrons are transferred from the donor plastoquinol to the acceptor plastocyanin via high potential $[Fe_2S_2]$–cytochrome f and low potential cytochrome b_6 chains operating in a Q cycle similar to that of cytochrome bc_1 in respiration. The Q cycle also transfers protons from the stroma side of the membrane to the thylakoid lumen. Electrons are carried from cytochrome b_6f to PSI by plastocyanin, a mobile copper-containing protein.

16 protein subunits arranged in 45 transmembrane helices, together with 168 chlorophyll molecules, five carotenoids, three $[Fe_4S_4]$ clusters and two molecules of the electron carrier **phylloquinone** (vitamin K1; Figure 9.18). Chlorophyll a, phylloquinone and $[Fe_4S_4]$ function as the electron acceptors A_0, A_1 and A_2 respectively on the reducing side of the complex (see Table 9.3). The reaction center chlorophyll is P700. PSI is largely confined to the intergranal (stroma-exposed) regions of the thylakoid membrane (see Table 9.1 and Figure 9.5). The **crystal structure of PSI** from pea has been determined at atomic resolution (Figure 9.19). The nomenclature for PSI components parallels that for PSII; reaction center subunits are referred to by the symbol Psa, light-harvesting by Lhca.

Table 9.8 Redox potentials of components of the cytochrome b_6f complex.

Redox component	E°′ (V)
PQ/PQH$_2$	+0.10
Cytochrome b$_6$ (low potential electron transport chain)	
Heme b$_{n(ox/red)}$	−0.05 to −0.03
Heme b$_{p(ox/red)}$	−0.15 to −0.09
High potential electron transport chain	
Rieske [Fe$_2$S$_2$]$_{(ox/red)}$	+0.30
Cytochrome f$_{(ox/red)}$	+0.35
Plastocyanin$_{(ox/red)}$	+0.36

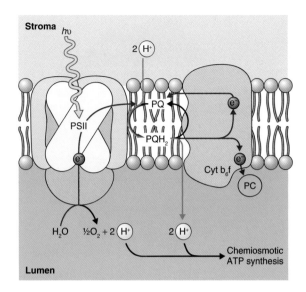

Figure 9.17 Proton movement across the thylakoid membrane during electron transfer between photosystem II (PSII) and the cytochrome b_6f complex. Red arrows show the path taken by an electron. Electrons carried from PSII by plastoquinone (PQ) are transported through the b_6f complex to plastocyanin (PC). Two electron transport chains, comprising a high potential pathway via the Rieske $[Fe_2S_2]$ center and cyt f and a low potential pathway via the hemes of cyt b_6, operate in a Q cycle that transfers protons from the stroma to the lumen side of the thylakoid membrane.

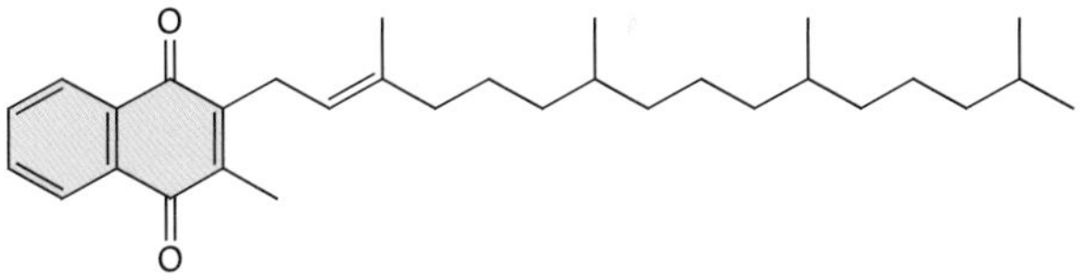

Figure 9.18 Structure of phylloquinone (vitamin K1), an electron carrier associated with photosystem I.

9.5.1 PSI reaction center subunits are associated with plastocyanin docking, P700 and primary and secondary electron acceptors

Photosystem I accepts electrons from plastoquinone and donates them to ferredoxin. It has been called 'an almost perfect Einstein photochemical machine', since it operates with a quantum yield close to 1.0 (see Section 9.2.2). Photoexcitation of P700 in the PSI reaction center oxidizes reduced plastocyanin arriving from the cytochrome b_6f complex. P700 transfers its electron to the first stable acceptor, F_x (A_2), via chlorophyll a (A_0) and phylloquinone (A_1), as shown in Figure 9.20. P700,

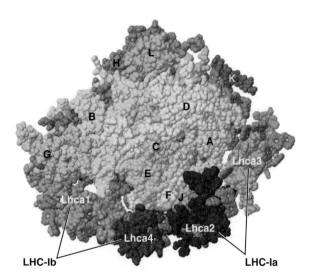

Figure 9.19 Structural model of plant photosystem I, seen from the stromal side of the thylakoid membrane. Crystal structure showing the positions of the major subunits: LHC, chlorophyll-protein subunits; Lhca, light-harvesting proteins; PsaA and PsaB, P700-binding reaction center polypeptides (see Figure 9.9). PsaN, the plastocyanin docking protein, is located on the lumen side of the complex, beneath Lhca2 and Lhca3. Capital letters correspond to different Psa proteins (for example, A is PsaA). Some properties of other polypeptides (PsaC–PsaF) are listed in Table 9.4.

A_0, A_1 and A_2 are bound to **PsaA** and **PsaB**, a pair of chloroplast genome-encoded polypeptides of molecular mass about 83 kDa. Plastocyanin is believed to dock with **PsaN**, a small luminal subunit that binds with **PsaF** (17 kDa; Figure 9.19).

The secondary electron acceptors F_A and F_B (Figure 9.20) are [Fe_4S_4] centers bound to a 9 kDa subunit, **PsaC. Ferredoxin** (Fd), which accepts electrons from F_B, in turn reduces $NADP^+$ in a reaction catalyzed by the enzyme **ferredoxin-$NADP^+$ reductase** (FNR; Equation 9.7). Subunit **PsaD** (18 kDa) binds to ferredoxin and PsaC; the 10 kDa **PsaE** subunit binds FNR. The PSI reaction center includes several other Psa polypeptides with a range of functions including the binding of Lhca subunits, photoregulation and structural stabilization.

Equation 9.7 Ferredoxin-$NADP^+$ reductase

$$2Fd_{red} + NADP^+ + H^+ \rightarrow 2Fd_{ox} + NADPH$$

9.5.2 The PSI antenna consists of four light-harvesting chlorophyll-binding proteins

The light-harvesting antenna of PSI consists of four peripheral chlorophyll-protein subunits arranged into

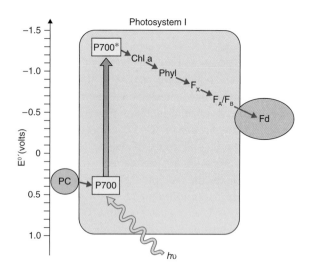

Figure 9.20 Photosystem I-mediated transfer of electrons (red arrows) from plastocyanin (PC) to ferredoxin (Fd). The pathway of electron transfer through the PSI complex consists of P700, a monomeric chlorophyll a (Chl a), a phylloquinone (Phyl) and a series of additional electron carriers that include three different Fe-S centers (F_X, F_A, F_B).

two dimers, **LHC-Ia** and **LHC-Ib**, on the PsaF side of the reaction center (Figure 9.19). **Lhca proteins** have molecular masses of 18–25 kDa and each is associated with 13 chlorophylls and two to three carotenoid molecules. LHC-Ib is tightly bound via the 11 kDa **PsaG** subunit. LHC-Ia is relatively weakly bound and amounts of the constituent Lhca proteins tend to be variable during growth and development depending on light intensity and nutrient availability.

9.5.3 Ferredoxin, the PSI electron acceptor, is a reductant in photosynthetic NADPH formation and many other redox reactions

Subunits PsaC, PsaD and PsaE, with the bound [Fe_4S_4] electron carriers F_A and F_B, facilitate the exergonic transfer of electrons from P700* to **ferredoxin** (Figure 9.20). Fd is a strong reductant ($E^{\circ\prime}$ −0.42 V). It functions as an **electron donor** in many important metabolic reactions, including reduction of $NADP^+$ to NADPH, nitrogen and sulfur assimilation (see Chapter 13), fatty acid synthesis (see Chapter 17), redox regulation via **thioredoxin** (see Chapter 15), catabolism of chlorophyll (see Chapter 18) and cyclic photophosphorylation (see Section 9.6.2).

Plant ferredoxins are low molecular mass (10–11 kDa) iron–sulfur proteins with a [Fe_2S_2]

prosthetic group. The *Arabidopsis* nuclear genome contains four genes that encode ferredoxins. *AtFd1* and *AtFd2* are expressed in green tissues. *AtFd3* encodes a ferredoxin localized in roots. The role of a fourth ferredoxin type, encoded by *AtFd4*, is currently not known. The primary amino acid sequence of all four ferredoxins includes an N-terminal **plastid-targeting region**, consistent with functions for these proteins in a range of redox reactions within chloroplasts and the plastids of non-green tissues.

The transfer of electrons from reduced ferredoxin to $NADP^+$ is catalyzed by ferredoxin-$NADP^+$ reductase (FNR) (see Equation 9.7). FNR is a **flavoprotein**, with **flavin adenine dinucleotide** (**FAD**) as the prosthetic group (see Figure 2.20). Fd_{red} binds to FNR and transfers a single electron reducing FAD to flavin semiquinone. A second transfer from Fd_{red} results in the formation of fully reduced $FADH_2$. FNR then transfers the two electrons to $NADP^+$.

9.6 Photophosphorylation

Chapter 2 describes how redox reactions may be characterized and ordered into an electrochemical series on the basis of **redox potential**, $E^{\circ\prime}$. Figure 9.21 places the components of the photosynthetic electron transport chain in relation to an $E^{\circ\prime}$ scale. The path taken by an electron between the water-splitting event on the oxidizing side of PSII and NADPH formation on the reducing side of PSI gives its name to the **Z scheme**. A large body of experimental evidence supports the Z

Key points Photosystem I is a light-dependent plastocyanin-ferredoxin oxidoreductase located in intergranal regions of the thylakoid membrane. The primary electron acceptors in the reaction center of PSI are chlorophyll a, phylloquinone and ferredoxin which, together with P700, are bound to PsaA (the active branch) and PsaB proteins. The PSI core consists of a total of 16 polypeptide subunits, including a plastocyanin-docking protein and three $[Fe_4S_4]$ centers that function as secondary electron acceptors. Two dimeric pigment protein complexes make up the PSI antenna. Ferredoxin, the electron acceptor from PSI, participates in redox reactions across a range of metabolic processes. The NADPH used in the carbon reduction reactions of photosynthesis is the product of electron transfer from reduced ferredoxin catalyzed by the flavoprotein ferredoxin-$NADP^+$ reductase.

scheme model of photosynthetic electron transport and has filled in a great deal of biophysical detail.

The products of primary photochemistry in the PSII reaction center are a strong oxidant ($P680^+$, with a high positive $E^{\circ\prime}$ of the order of +1.2 V, making it the most oxidizing species known in biology) and a weak, relatively stable reductant A_1 (Q_A^-, a plastosemiquinone). $P680^+$ has the oxidizing power to remove electrons from water, generating O_2 and protons. Q_A^- donates electrons to PSI via plastoquinone,

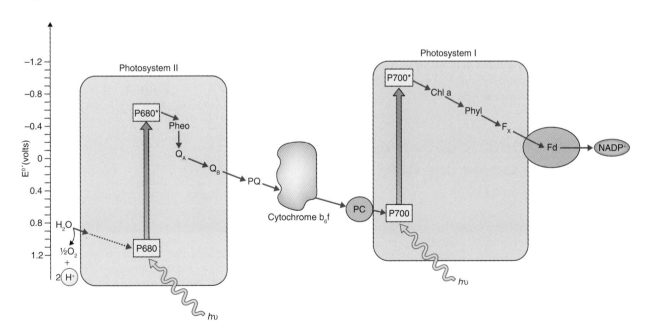

Figure 9.21   The Z scheme of photosynthetic electron transport. Red arrows show the path taken by an electron. The vertical placement of each electron carrier of the non-cyclic electron transfer chain corresponds to the midpoint of its redox potential ($E^{\circ\prime}$).

cytochrome b_6f complex and plastocyanin intermediates along an **electrochemical gradient**. Energy released by this **exergonic** electron transfer from PSII to PSI is used for the **chemiosmotic** synthesis of ATP. Charge separation in the PSI reaction center produces a strong, stable reductant (a reduced iron–sulfur center, F_X^- $E^{\circ\prime} -0.73V$) and $P700^+$, which is a relatively weak oxidant ($E^{\circ\prime} +0.49V$). Reduced ferredoxin, an intermediate product of P700 oxidation, reduces $NADP^+$ to NADPH (Figure 9.21).

By introducing redox reagents with particular $E^{\circ\prime}$ values, it is possible to donate electrons to, or intercept electrons from, the photosynthetic electron transport chain. Such chemicals have been used as research tools to dissect the Z scheme and examine specific sections of the pathway. Compounds that specifically inhibit the chloroplast electron transport chain have found widespread practical application as herbicides. Herbicides of the class exemplified by diuron and atrazine prevent the reduction of Q_B by binding to the D1 protein of the PSII reaction center. Paraquat-type herbicides act at the reducing side of PSI and inhibit ferredoxin reduction.

During photosynthetic electron transport, protons accumulate in the thylakoid lumen and their transfer to the stroma drives ATP synthesis by a multiprotein **ATP synthase** complex, often referred to as **coupling factor** (CF). Synthesis of ATP during photosynthesis is referred to as **photophosphorylation** and occurs by a chemiosmotic mechanism similar to that described for mitochondrial oxidative phosphorylation (see Chapter 7). In both cases ATP synthesis is usually **coupled** to electron transport, although by applying chemical **uncouplers** that collapse the proton gradient it is possible to obtain unimpeded electron flow in the absence of ATP formation. ATP synthesis during the transfer of electrons from water to NADPH is termed **non-cyclic phosphorylation. Cyclic phosphorylation** occurs by additional pathways that redirect electrons from PSI back to cytochrome b_6f.

9.6.1 The products of non-cyclic electron transport are ATP, oxygen and NADPH

Two regions of the photosynthetic electron transport chain liberate protons into the thylakoid lumen (see Figure 9.17). One is the oxygen-evolving complex of PSII, which cleaves water into electrons, O_2 and H^+. The other is on the oxidizing side of cytochrome b_6f, where plastoquinol transfers electrons to the Rieske $[Fe_2S_2]$ center and protons are pumped into the lumen by operation of the Q cycle. The degree of **coupling** of non-cyclic phosphorylation is generally in the range of

1.0–1.5 ATP molecules synthesized from ADP and P_i for every two electrons transferred from water to $NADP^+$.

9.6.2 ATP is the sole product of cyclic electron flow around PSI

In addition to non-cyclic phosphorylation, chloroplasts have a cyclic electron transport chain in which reduced ferredoxin from PSI donates electrons directly or indirectly to the cytochrome b_6f complex. Like non-cyclic photophosphorylation, the cyclic pathway pumps protons from the stroma to the thylakoid lumen; however water is not split, no oxygen is produced and there is no net NADPH production associated with cyclic electron flow. Increased understanding of the pathways of cyclic electron flow around PSI has come from recent genetic studies. Mutations in the gene ***ndhB*** of the cyanobacterium *Synechocystis* disrupt PSI cyclic electron transport activity. *Nicotiana tabacum* plants with a lesion in the homologous gene in the chloroplast genome have a similar phenotype. *ndhB* encodes a subunit of the chloroplast **NAD(P)H dehydrogenase** (**NDH**) complex that is the equivalent of mitochondrial **Complex I** which transfers electrons from NADH to ubiquinone (see Figure 7.13A). These observations suggest a pathway of cyclic electron flow from PSI in which $NADP^+$ is reduced by FNR, and NADPH in turn transfers electrons to plastoquinone via the thylakoid NDH complex (Figure 9.22). Mitochondrial Complex I pumps protons from one side of the membrane to the other; chloroplast NDH may do the same.

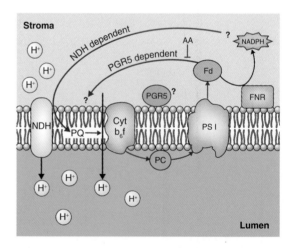

Figure 9.22 Cyclic electron transport. In vascular plants there are two partially redundant pathways: NDH-dependent and PGR5-dependent. The PGR5-dependent pathway is inhibited by antimycin A (AA). Neither the electron donor to the NDH complex nor the route taken by electrons in the PGR5-dependent pathway is known. FNR, ferredoxin-NADP$^+$ reductase; NDH, NAD(P)H dehydrogenase; PC, plastocyanin; PGR5, proton gradient regulation 5 protein.

The existence of a second non-cyclic electron chain is suggested by the characteristics of the *Arabidopsis* mutant **proton gradient regulation 5** (**pgr5**). This mutant is defective in cyclic electron transport but has normal NDH levels and functionality. The **PGR5** gene encodes a small (10 kDa), highly conserved thylakoid-associated protein. PGR5 is believed to be a regulator of ferredoxin-dependent plastoquinone reduction (Figure 9.22) but its detailed function remains to be determined. The PGR5 pathway is universally present in photosynthetic organisms. In contrast to the NDH route, it is blocked by **antimycin A**, an inhibitor of quinone redox reactions. A number of plant groups, including conifers and unicellular green algae, appear to lack *ndh* genes in their plastid genomes.

Cyclic phosphorylation has a number of physiological functions. The **balance** of non-cyclic to cyclic electron flow determines the relative supplies of ATP and NADPH, allowing flexibility in response to the demands of different metabolic pathways. Furthermore, adjustments to the balance prevent **over-reduction** of the stroma and disruptions to membrane transport processes. Over-reduction is also a cause of **photoinhibition** in PSI (see Chapter 15).

9.6.3 CF_0CF_1 is a multiprotein ATP synthase complex that uses the proton gradient across the thylakoid membrane to phosphorylate ADP

The ATP synthase of thylakoids is closely similar in structure and function to other ATP synthases such as the **F_0F_1-ATP synthase** of mitochondria (see Figure 7.15). The stromal and transmembrane domains of the chloroplast complex are known as **CF_1** and **CF_0**, respectively. Figure 9.23 shows the similarities in function and organization between the ATP synthases of chloroplasts and mitochondria. Like the other multiprotein complexes of the thylakoid membrane, assembly of CF_0CF_1-ATP synthase requires coordinated expression of chloroplast (six subunits) and nuclear (three subunits) genes (Table 9.9).

CF_0 has a molecular mass of around 170 kDa and consists of 17 polypeptides (Table 9.9): single subunits I, II and IV plus 14 copies of subunit III arranged in a ring of transmembrane helices. Five subunits (α, β, γ, δ, ε in the ratio 3:3:1:1:1) make up CF_1, which has a total molecular mass of 400 kDa (Table 9.9). The phosphorylation of ADP to form ATP is catalyzed by the α- and β-subunits in a three-stage **binding change mechanism** involving the operation of a proton-driven 'molecular motor'. The δ-subunit and subunits I, II and

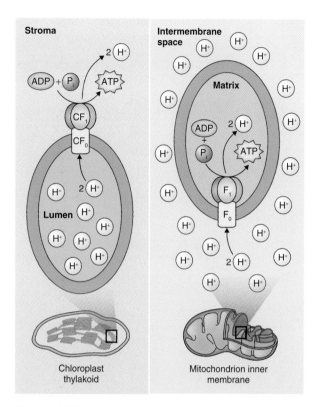

Figure 9.23 Organization of chloroplast and mitochondrial ATP synthases compared. ATP is produced on the stromal side of the chloroplast thylakoid and on the matrix side of the inner mitochondrial membrane.

IV function in binding CF_1 to CF_0, and subunit III is involved in proton translocation (Table 9.9). A distinctive feature of the ATP synthase of the chloroplast is that its activity is modulated by light. The ε-subunit plays an important role, blocking catalysis in the dark. The γ-subunit also participates in light activation and dark inactivation of ATP synthase mediated by the **ferredoxin/thioredoxin system** (see Chapter 15).

9.7 Carbon dioxide fixation and the photosynthetic carbon reduction cycle

NADPH and ATP, the products of the light energy capture reactions, are used during the carbon reduction phase of photosynthesis in the reactions of the **Calvin–Benson cycle**. This cycle, which comprises 13 enzymatic steps that take place in the stroma, is divided into three phases: (i) **carboxylation**, in which CO_2 is added to an acceptor molecule, ribulose-1,5-bisphosphate (**RuBP**), producing two molecules of a

Table 9.9 Polypeptide subunits of the ATP synthase complex.

Protein	Gene	Location of gene	Molecular mass (kDa)	Number of copies per complex	Function
CF_1					
α-Subunit	atpA	C	55	3	Catalytic
β-Subunit	atpB	C	54	3	Catalytic
γ-Subunit	atpC	N	36	1	Proton gating
δ-Subunit	atpD	N	20	1	Binding of CF_1 to CF_0
ε-Subunit	atpE	C	15	1	ATPase inhibition
CF_0					
I	atpF	C	17	1	Binding CF_0 to CF_1
II	atpG	N	16	1	Binding CF_0 to CF_1
III	atpH	C	8	14	Proton translocation
IV	atpI	C	27	1	Binding of CF_0 to CF1

C, chloroplast genome; N, nucleus.

Key points The shape described by the graphical illustration of electron flow along the electrochemical gradient from water to NADPH gives its name to the Z scheme. The energy released during operation of the Z scheme is captured in ATP. ATP synthesis associated with O_2 and NADPH formation is the product of non-cyclic photophosphorylation. ATP is also synthesized by cyclic photophosphorylation, in which reduced ferredoxin from PSI donates electrons to cytochrome b_6f via NAD(P)H dehydrogenase. The oxygen-evolving complex and cytochrome b_6f are sites at which protons are transferred into the thylakoid lumen. The proton gradient drives phosphorylation of ADP by the multiprotein CF_0CF_1-ATP synthase complex in a three-stage binding change mechanism similar to that occurring during ATP synthesis by the mitochondrial F_0F_1 complex.

three-carbon organic acid, 3-phosphoglycerate (**3-PGA**); (ii) **reduction**, which yields a three-carbon sugar, glyceraldehyde 3-phosphate (**GAP**) from each 3-PGA; and (iii) **regeneration**, which results in the formation of more acceptor molecules (Figure 9.24). Two molecules each of ATP and NADPH per atom of carbon fixed are

used in the reductive phase of the cycle. An additional ATP is consumed during the regeneration phase. The Calvin–Benson cycle is sometimes called the **reductive pentose phosphate pathway**; it shares enzyme activities and reaction intermediates with the oxidative pentose phosphate pathway of respiration (see Chapter 7).

The path taken by the carbon of CO_2 as it is fixed, reduced and distributed throughout the pools of metabolites in the chloroplasts of illuminated cells was established in a series of classic experiments with the green algae *Chlorella* and *Scenedesmus*. The first stable products of CO_2 fixation are glyceric acid mono- and bis-phosphates and phosphorylated trioses. These are all **three-carbon metabolites** and so the Calvin–Benson cycle is often referred to as the C_3 **pathway** of photosynthesis. As we shall see in Section 9.9, many plants, principally of tropical and subtropical environments, have additional CO_2-fixation mechanisms, notably C_4 and **CAM photosynthesis**.

Here we discuss the structure and function of the carboxylating enzyme, the net entry of carbon into metabolism through the reduction phase of the Calvin–Benson cycle and the interconversions leading to regeneration of the carboxylation acceptor. Figure 9.25 summarizes intermediates, reactions and enzymes in the pathway.

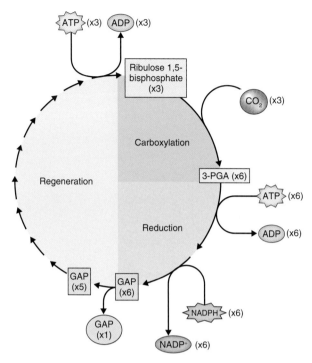

Figure 9.24 Phases of the carbon fixation cycle in C_3 photosynthesis. The cycle is divided into three phases: carboxylation, reduction and regeneration. Fixation of one molecule of CO_2 requires two molecules of NADPH and three of ATP. 3-PGA, 3-phosphoglyerate; GAP, glyceraldehyde 3-phosphate.

9.7.1 Rubisco catalyzes the first reaction in the Calvin–Benson cycle

The first reaction in the Calvin–Benson cycle is carboxylation of the five-carbon acceptor **ribulose-1,5-bisphosphate (RuBP)** by CO_2, a reaction catalyzed by the enzyme **RuBP carboxylase-oxygenase**, usually referred to as **rubisco**; the product of this reaction is **3-phosphoglycerate (3-PGA)** (Equation 9.8). As someone once said, it is perfectly appropriate that rubisco sounds like the brand name of a breakfast cereal, since it is ultimately responsible for just about all the carbohydrate in the human diet.

Equation 9.8 Carboxylation reaction catalyzed by rubisco

$$RuBP + CO_2 + H_2O \rightarrow 2\ 3\text{-PGA}$$

The reaction catalyzed by rubisco proceeds by a five-step mechanism, during which a transient **six-carbon intermediate** is formed and then hydrolyzed to yield two molecules of the 3-PGA product. This reaction is unique to the Calvin–Benson cycle and is the

route by which the vast majority of CO_2 enters into organic combination in the biosphere (an estimated 10^{14} kg annually). Rubisco is one of the most abundant enzymes on Earth. As the predominant protein of green tissues, accounting for more than half the total protein nitrogen of leaves, it is a major potential source of amino acid nitrogen for use elsewhere when its photosynthetic function comes to an end. Rubisco breakdown during senescence is discussed further in Chapter 18. The products of the rubisco reaction enter the **reduction** phase of the cycle, followed by a series of reactions that **regenerate** RuBP (Figure 9.25).

In addition to the carboxylation of RuBP, rubisco can catalyze a reaction between O_2 and a five-carbon transition state intermediate of RuBP, resulting in the formation of 3-PGA plus the two-carbon product **phosphoglycolate** (Figure 9.26). The **oxygenase** activity of rubisco is the basis of **photorespiration**, an important pathway that limits the efficiency of photosynthetic carbon fixation (see Section 9.8).

9.7.2 Rubisco is a complex enzyme with subunits encoded in both the nuclear and the plastid genomes

Green plants and practically all other eukaryotic photoautotrophs have **Type I rubisco**, made up of large (L) and small (S) subunits. The fully active Type I **holoenzyme** consists of four dimeric large and two tetrameric small subunits arranged as an $(L_2)_4(S_4)_2$ **hexadecamer** (Figure 9.27). The L subunit is a 55 kDa polypeptide, which is encoded by the chloroplast *rbcL* **gene**, and carries both the catalytic site of the enzyme and binding sites for effectors. The S subunit has no catalytic activity but is necessary for the active conformation of the Type I holoenzyme. It is encoded by members of the *rbcS* **gene family** in the nuclear genome. The S subunit is synthesized on cytosolic ribosomes as a 20 kDa precursor, and targeted to the chloroplast where it is processed to its final size of 14 kDa. Import and maturation of S polypeptides and assembly of S and L subunits into the holoenzyme requires a number of protein cofactors that act as **molecular chaperones**. Three other types of rubisco, each lacking S subunits, have been described in archea, proteobacteria and certain dinoflagellates.

Structure–function studies of rubisco show that the basic catalytic unit is a **dimer of L subunits** arranged head-to-tail. Each of the two active sites per dimer consists of amino acid residues primarily from the C terminus of one monomer, together with some residues from the N terminus of the second monomer. Rubisco

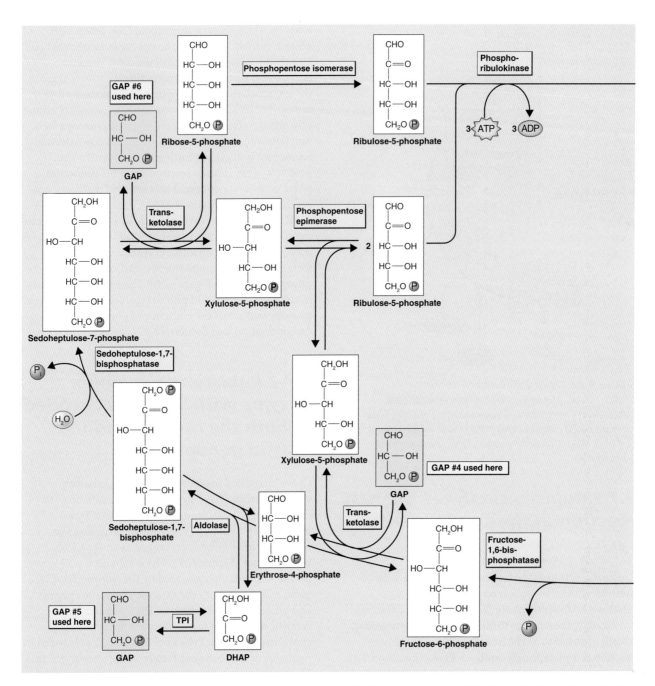

Figure 9.25 Individual steps in the photosynthetic carbon reduction (Calvin–Benson) cycle of C_3 plants. Carboxylation, reduction and regeneration phases are color-coded as in Figure 9.24. Figure continues on opposite page.

has a relatively slow **catalytic rate** (in the range of 3–10 reactions per second, compared with values for catalysis by a typical enzyme of 10^2 to 10^3 s^{-1}); the large amount of the protein present in the chloroplast compensates for its inefficiency. Rubisco activity is generally the **rate-determining step of CO_2 fixation** in bright light. The enzyme is regulated at the gross level through protein amount, which in turn is determined by the balance of subunit synthesis, assembly and protein breakdown.

Fine control is also exerted through modifications and interactions of the L subunit. Rubisco is enzymatically inactive until a positively charged lysine residue in the catalytic site is covalently modified by reaction with CO_2 to form a negatively charged **carbamylate**. Carbamylated enzyme forms a catalytically active **ternary complex** with Mg^{2+} (Equation 9.9A; Figure 9.27). Rubisco activity is further modulated in a light-sensitive manner by phosphorylated metabolites, including the inhibitor **2-carboxy-D-arabinitol-1-phosphate (CA1P;**

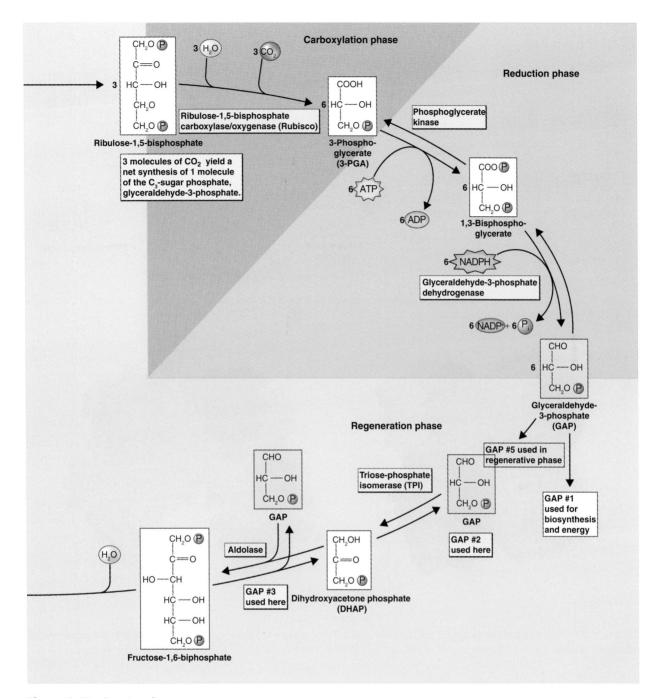

Figure 9.25 (continued).

Figure 9.28). CA1P is synthesized in the absence of light from 2-carboxy-D-arabinitol (CA), which in turn is derived from fructose-1,6-bisphosphate, one of the intermediates of the Calvin–Benson cycle. CA1P binds tightly to activated rubisco and renders it inactive (Equation 9.9B). In the light, CA1P is removed from the carbamylated rubisco–Mg^{2+} complex by **rubisco activase**, a redox-responsive protein factor that exists as two isoforms (molecular masses 43 and 46 kDa). The CA1P released by rubisco activase is converted back to CA by a phosphatase.

Equation 9.9 Rubisco (E-Lys-NH_2) inhibition and activation

9.9A

$$[E\text{-Lys-}NH_2]_{inactive} + CO_2 + Mg^{2+}$$
$$\rightarrow [E\text{-Lys-NH-COO}^- Mg^{2+}]_{active} + H^+$$

9.9B

$$[E\text{-Lys-COO}^- Mg^{2+}]_{active} + CA1P$$
$$\rightarrow [E\text{-Lys-COO}^- Mg^{2+} CA1P]_{inactive}$$

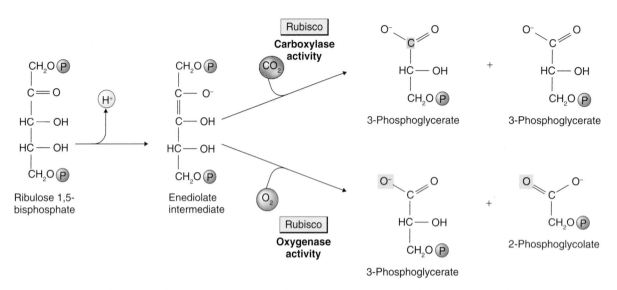

Figure 9.26 The carboxylase and oxygenase reactions catalyzed by rubisco. Oxygen and CO_2 compete at the enzyme's active site for reaction with the enediol intermediate derived from RuBP.

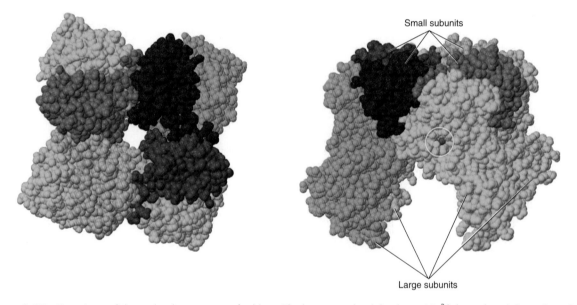

Figure 9.27 Two views of the molecular structure of rubisco. The image on the right shows Mg^{2+} (green) and the carbamylated side-chain of lysine (red) in the catalytic cleft of an L subunit (circled).

9.7.3 The two-step reduction phase of the Calvin–Benson cycle uses ATP and NADPH

The product of carboxylation catalyzed by rubisco, 3-PGA, is converted to the **triose phosphate** glyceraldehyde-3-phosphate (**GAP**) in two enzymatic steps (see Figure 9.25). **Phosphoglycerate kinase** catalyzes phosphorylation of 3-PGA to form 1,3-bisphosphoglycerate (Equation 9.10A), which in turn is reduced to GAP by **NADPH-dependent GAP dehydrogenase** (Equation 9.10B).

Equation 9.10 Enzyme reactions in the carbon reduction phase of the Calvin–Benson cycle

9.10A Phosphoglycerate kinase

$$3\text{-PGA} + \text{ATP} \rightleftharpoons 1,3\text{-bisphosphoglycerate} + \text{ADP}$$

9.10B Glyceraldehyde-3-phosphate dehydrogenase

$$1,3\text{-bisphosphoglycerate} + \text{NADPH} \rightleftharpoons \text{GAP} + \text{NADP}^+ + \text{P}_i$$

This phase of the cycle uses 1 mole each of ATP and NADPH, supplied by the light energy capture reactions, per CO_2 fixed. Notice the similarities (C_3 intermediates,

Key points Carbon dioxide is assimilated in the stroma by the Calvin–Benson pathway (also called the reductive pentose phosphate or C_3 pathway), a three-phase sequence of 13 enzymatic steps. The first phase, carboxylation catalyzed by ribulose-1,5-bisphosphate carboxylase-oxygenase (rubisco), is followed by reduction and regeneration. Rubisco, an abundant 16-subunit stromal enzyme made of two different polypeptides, large (plastid-encoded) and small (nuclear), catalyzes the carboxylation reaction between CO_2 and the C_5 acceptor ribulose-1,5-bisphosphate to form two molecules of the C_3 product 3-phosphoglycerate. Carboxylation requires Mg^{2+} for full activation and is regulated at the substrate level by light via 2-carboxy-D-arabinitol-1-phosphate and rubisco activase. The oxygenase function of rubisco catalyzes the formation of the C_2 product phosphoglycolate from ribulose-1,5-bisphosphate and O_2.

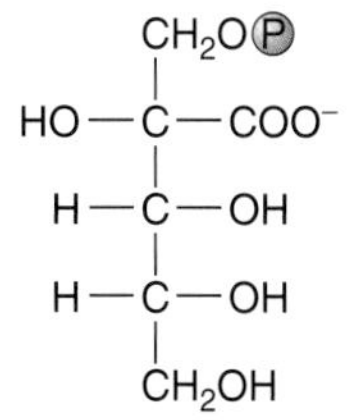

Figure 9.28 2-Carboxy-D-arabinitol-1-phosphate (CA1P), a metabolite responsible for the inhibition of rubisco activity in low light or darkness.

participation of ATP) and differences (use of NADP compared with NADH respectively) between the **carbon reduction phase** of photosynthesis and the first two reactions of the **payoff phase of glycolysis** working in reverse (see Figure 7.3B). GAP represents a major point of carbon removal from the cycle. It is a precursor of chloroplast starch synthesis (see Section 9.7.6). GAP may also be exchanged for phosphate across the chloroplast envelope by the activity of the **triose phosphate transporter** (see Section 9.7.5) and used as the precursor for the synthesis of **sucrose** and other products (see Section 9.7.6) or as an energy source.

9.7.4 During the regeneration phase of the Calvin–Benson cycle, ten enzyme reactions convert five 3-carbon to three 5-carbon intermediates

The regeneration phase comprises ten of the 13 enzymes of the Calvin–Benson cycle, including two that are unique to photosynthesis. **Sedoheptulose-1,7-bisphosphatase** dephosphorylates a C_7 diphosphosugar to yield a monophosphosugar. **Phosphoribulokinase** regenerates the initial C_5 acceptor of CO_2 by phosphorylating ribulose-5-phosphate to form RuBP (see Figure 9.25). Of every six triose phosphate molecules produced by the reduction phase, one represents the **net gain of carbon** to the plant and the remaining five regenerate three molecules of RuBP that re-enter the cycle.

The interchanges between intermediates in this cycle have similarities to those of the **oxidative pentose phosphate pathway** (see Figure 7.19) and are summarized in Figure 9.29. Two molecules of triose phosphate (C_3) combine to make fructose-1,6,-bisphosphate (C_6). After dephosphorylation, this C_6 sugar phosphate reacts with another triose phosphate to make one C_5 and one C_4 product. Reaction of the C_4 sugar phosphate with another triose phosphate molecule produces a C_7 sugar phosphate. After dephosphorylation, C_7 reacts with C_3 to yield two C_5 sugar phosphates. The final step in the regeneration cycle is the conversion of ribulose-5-phosphate to RuBP (see Figures 9.24 and 9.28).

The energetic requirements for the net synthesis of two trioses (one glucose equivalent) from six CO_2 molecules total 12 molecules of NADPH and 18 molecules of ATP. Measurements of the thermodynamics of the Calvin–Benson cycle under physiological conditions indicate its efficiency to be greater than 80%.

Key points The two-step conversion of 1 mole of 3-phosphoglycerate to triose phosphate during the reduction phase of the Calvin–Benson cycle utilizes 1 mole each of ATP and NADPH supplied by the light energy capture reactions. One out of every 6 moles of triose phosphate represents the net gain of carbon fixed in photosynthesis, and the remaining five C_3 intermediates are converted to three C_5 products during the regeneration phase of the cycle. The ten enzymes of this phase include sedoheptulose-1,7-bisphosphatase and phosphoribulokinase, which are unique to photosynthesis. The final reaction step which completes the carbon reduction cycle is ATP-dependent regeneration of the CO_2 acceptor ribulose-1,5-bisphosphate.

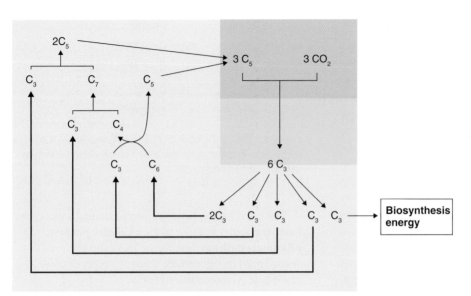

Figure 9.29
Interconversions during the Calvin–Benson cycle that regenerate three C$_5$ carboxylation acceptors from five C$_3$ precursors. The background colors correspond with the different phases of the Calvin–Benson cycle as shown in Figures 9.24 and 9.25.

9.7.5 Photosynthesis is dependent on the exchange of metabolites across the chloroplast envelope

For the products of photosynthesis to pass from the chloroplast to where they are needed elsewhere within the cell, and in remote cells, tissues and organs, they must first cross the two outer plastid membranes, together called the **chloroplast envelope**. Traffic across the envelope in the opposite direction, from the cytosol into the chloroplast, is essential to supply proteins, substrates and precursors required for the structure and

function of the photosynthetic apparatus. Figure 9.30 summarizes the transporters of the inner membrane with functions in the exchange of photosynthetic precursors and products. As well as photosynthesis-related import and export, metabolites are exchanged during nitrogen and sulfur assimilation, the synthesis of amino acids and fatty acids, carbohydrate and lipid storage, and the formation of secondary compounds. The plastid envelope, with its numerous **transporters**, represents the essential interface in the complex metabolic networks that link the chloroplast with other cell compartments.

Phosphate **antiport** systems, which exchange inorganic phosphate for various phosphorylated C$_3$, C$_5$ and C$_6$ compounds between the plastid and the cytosol, are of particular importance for sustaining

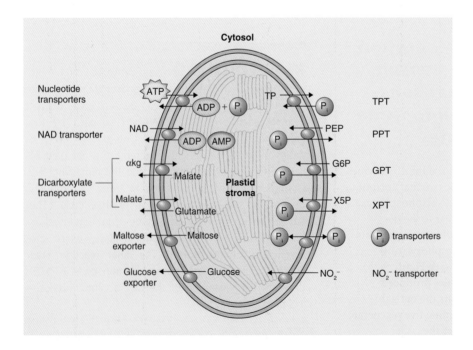

Figure 9.30   Solute transporters of the plastid inner membrane. αkg, α-ketoglutarate; G6P, glucose-6-phosphate; GPT, G6P translocator; PEP, phosphoenolpyruvate; PPT, PEP/phosphate translocator; TP, triose phosphate; TPT, triose phosphate translocator; X5P, xylulose-5-phosphate; XPT, X5P translocator.

photosynthesis and distributing the products of carbon assimilation. The best characterized of these is the triose phosphate translocator (**TPT**) of the chloroplast inner membrane. The TPT exchanges three-carbon Calvin–Benson cycle intermediates for inorganic phosphate in a strictly coupled reaction (Figure 9.31). The TPT is sometimes called the 'day path of carbon allocation' and in most plants the main uses of exported triose phosphate are for the synthesis of **sucrose** (1 mole of sucrose per 4 moles of exported triose phosphate) and for cellular respiration. Sucrose synthesis releases P_i which is transported from the cytosol back into the chloroplast and reused for photosynthetic phosphorylations. Another transporter of three-carbon compounds is **PPT**, the phosphoenolpyruvate (PEP)/phosphate translocator. This is of particular significance in C_4 **photosynthesis** and is discussed further in Section 9.9.

Another transporter supports the exchange of **free sugars** and represents the 'night path of carbon allocation' during which the assimilated carbon, stored in the chloroplast in the form of **starch**, is broken down to **maltose** and exported to the cytosol. **Hexose phosphates** are not generally exchanged across the chloroplast envelope, but non-green plastids such as **amyloplasts** possess a glucose-6-phosphate transporter (**GPT**) that provides substrate for the synthesis of starch and fatty acids or for metabolism via the plastid oxidative pentose phosphate pathway. Chloroplasts do, however, use a C_5 **sugar phosphate transporter, XTP**, to link the oxidative pentose phosphate pathway (some enzymatic steps of which are cytosolic) with the Calvin–Benson cycle (reductive pentose phosphate pathway).

9.7.6 The Calvin–Benson cycle provides the precursors of carbohydrates for translocation and storage

Triose phosphate transferred across the chloroplast envelope by the TPT enters metabolism in the cytosol where it supports the synthesis of a range of products of carbon fixation. Here we consider in particular the pathways leading to **sucrose**, the predominant form in which carbon is translocated in the phloem, and **starch**, the most common form in which carbon is stored. GAP exported from the chloroplast is converted to fructose-6-phosphate by a sequence of reactions similar to the first steps of the regeneration phase of the Calvin–Benson cycle, via dihydroxyacetone phosphate and fructose-1,6-bisphosphate (see Figure 9.25). The enzymes responsible are **cytosolic isoforms** of triose phosphate isomerase, aldolase and

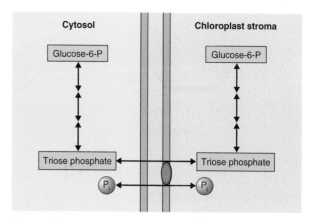

Figure 9.31   The triose phosphate translocator of the chloroplast inner membrane.

fructose-1,6-bisphosphatase (**F1,6BPase**). The reaction catalyzed by F1,6BPase is both the first irreversible step and an important control point in the committed pathway of sucrose synthesis.

Fructose-6-phosphate, glucose-6-phosphate and glucose-1-phosphate represent the **hexose phosphate pool** and are freely interconvertible through the activity of the reversible enzymes glucose-6-phosphate isomerase and phosphoglucomutase (Figure 9.32). The reactants in sucrose synthesis are fructose-6-phosphate and uridine diphosphate glucose (**UDP-glucose**; see Figure 7.22C); the latter is synthesized from glucose-1-phosphate by the enzyme **UDP-glucose pyrophosphorylase** (Equation 9.11).

Equation 9.11 UDP-glucose pyrophosphorylase

Glucose-1-phosphate + UTP $\rightleftharpoons$ UDP-glucose + PP_i

Sucrose-6-phosphate synthase (**SPS**) catalyzes the reaction between UDP-glucose and fructose-6-phosphate to yield sucrose-6-phosphate and UDP. Sucrose-6-phosphate is hydrolyzed to sucrose and P_i by **sucrose-6-phosphate phosphatase** (Figure 9.33). The negative free energy change for sucrose formation from glucose-1-phosphate ($\Delta G^{\circ\prime} -25 \text{ kJ mol}^{-1}$) indicates that the overall pathway is strongly exergonic and essentially irreversible. UDP, the product of the sucrose-6-phosphate synthase reaction, is recycled to UTP in an ATP-dependent reaction catalyzed by **nucleoside-diphosphate kinase**. P_i and PP_i released in the steps leading to sucrose may be transferred from the cytosol to the chloroplast by transporters, including the TPT, in the plastid envelope (see Figure 9.30).

Metabolic control over sucrose synthesis is exerted through the regulation of SPS and cytosolic F1,6BPase. F1,6BPase activity is modulated by the allosteric inhibitor **fructose-2,6-bisphosphate** (**F2,6BP**; see Figure 7.22B),

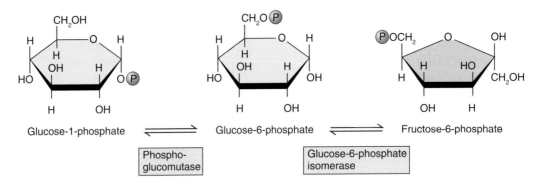

Figure 9.32   Interconversions of intermediates in the hexose phosphate pool leading to sucrose biosynthesis.

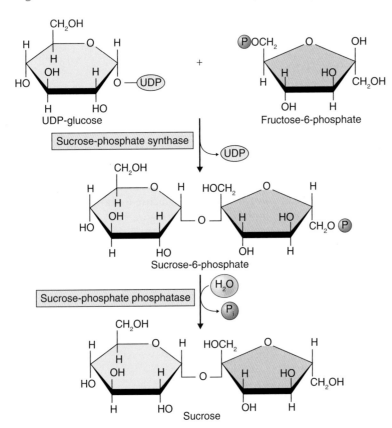

Figure 9.33   Enzymatic synthesis of sucrose from uridine diphosphate (UDP) glucose and fructose-6-phosphate.

the concentration of which is sensitive to levels of P_i and other metabolites. At high rates of photosynthesis, levels of F2,6BP are low, F1,6BPase is active, and fructose-1,6-bisphosphate is used in **sucrose synthesis**. When the supply of triose phosphate from photosynthesis decreases, F2,6BP builds up and inhibits F1,6BPase, and carbon is diverted away from sucrose synthesis towards glycolysis. SPS is responsive to the state of the hexose phosphate pool. Glucose-6-phosphate stimulates the enzyme by **allosteric interaction**, and P_i inhibits when it is attached to the enzyme by a specific **kinase**.

Sucrose is the principal form in which fixed carbon is transported from source tissues, through the vascular system, to **metabolic sinks**—tissues that require net import of photosynthate to meet the biosynthetic needs

of growth or storage (see Chapters 14 and 17). Generally speaking, if the supply of assimilate from current photosynthesis exceeds sink demand, fixed carbon is diverted away from sucrose and into chloroplast starch. In some plant groups, however, notably temperate pasture grass and cereal species and representatives of a few other families, sucrose can be accumulated in the vacuoles of the photosynthetic cells of leaves. High concentrations of sucrose in these species drive the synthesis of **fructans**, soluble fructose polymers which, depending on the species, may be linear or branched and have different monomer–monomer linkages (see Chapter 17). Fructans in the plants that accumulate them fulfill the short-term carbon storage role that in other species is played by chloroplast starch.

Starch, a polymeric glucan, represents a stable, osmotically inactive form of carbohydrate (see Figures 6.10 and 6.11). It occurs as a temporary storage compound in chloroplasts, and a long-term repository of mobilizable carbon in the non-green plastids (**amyloplasts**) of storage tissues in seeds, roots, tubers and other storage organs (see Chapters 6 and 17). The pathways of sucrose and starch synthesis are spatially separated and fed by different hexose phosphate pools (Figure 9.34). As in the reactions leading to sucrose, the precursor of starch is a nucleoside diphosphate glucose, but in this case the nucleoside base is adenine (A) rather than uracil (U). **ADP-glucose** is the product of the reaction between glucose-1-phosphate (derived from the Calvin–Benson cycle in chloroplasts or from the cytosol in amyloplasts) and ATP, catalyzed in the plastid by the enzyme **ADP-glucose pyrophosphorylase** (Figure 9.34).

Starch consists of two kinds of glucan: the linear $\alpha(1 \rightarrow 4)$-linked polymer **amylose**; and **amylopectin**, which is $\alpha(1 \rightarrow 4)$ with $\alpha(1 \rightarrow 6)$ branches (see Figure 6.10). The proportion of amylose to amylopectin varies with species, organ and environmental conditions, but is commonly in the region of about 1:3. Chloroplast starch is deposited as grains that often have a layered appearance, the result of periods of faster and slower growth (see Figure 6.11). Two enzymes are responsible for the synthesis of starch from ADP-glucose, namely **starch synthase**, which makes the $\alpha(1 \rightarrow 4)$ links, and **starch branching enzyme**, which introduces the $\alpha(1 \rightarrow 6)$ branches. Starch deposition in storage tissues is discussed in further detail in Chapter 17. ADP-glucose pyrophosphorylase is the major control point for the regulation of starch synthesis. This enzyme exists as chloroplast and cytosolic isoforms and in the active state is a **heterotetramer**, comprising two large and two small subunits. It is activated by 3-PGA and inhibited by P_i and is thereby directly responsive to the supply of fixed carbon and status of photophosphorylation (Figure 9.34).

The hexose phosphate pool, fed by the Calvin–Benson cycle, meets the needs of a number of pathways of carbon metabolism in addition to sucrose and starch synthesis (Figure 9.35). These include: the synthesis of **cell wall polysaccharides** derived from glucose-1-phosphate; **acyl lipid synthesis** from fructose-6-phosphate via triose phosphate and **acetyl-CoA** (Figure 9.36); and respiratory pathways supplying intermediates for the biosynthesis of other compounds such as amino acids.

9.8 Photorespiration

Photosynthesis by the C_3 pathway as described in Section 9.7 is sensitive to atmospheric oxygen. Doubling the

Key points Triose phosphate, representing the net gain from CO_2 fixation, is exported from the chloroplast by a specific phosphate antiporter in the chloroplast inner. Triose phosphate, which is converted to hexose phosphate in the cytosol, is exchanged for the inorganic phosphate essential for chloroplast phosphorylation reactions. The plastid envelope is also the site of a free sugar transporter that imports carbohydrate at night for the synthesis of storage starch from ADP-glucose, and a C_5 sugar phosphate transporter that links the Calvin–Benson and oxidative pentose phosphate pathways. The cytosolic hexose phosphate pool derived from triose phosphate exported from the chloroplast is the source of precursors for the synthesis of cell wall polysaccharides, acyl lipids, sucrose and fructans. Sucrose, the major long-range transport form of fixed carbon, is synthesized from UDP-glucose by the allosterically-regulated enzyme sucrose-6-phosphate synthase.

oxygen concentration from the normal ambient level of 21% can reduce CO_2 fixation rates by up to 50%. Conversely, decreasing oxygen concentration to <2% can result in a two-fold stimulation of carbon fixation. If photosynthesizing tissues are transferred to darkness, there is a **post-illumination burst** of CO_2 in direct proportion to the external oxygen concentration. These observations point to the existence of a light-stimulated process that competes with photosynthesis by consuming oxygen and evolving CO_2. The process is called **photorespiration** and is associated with characteristic metabolic pathways operating in different cell organelles. While photorespiration consumes O_2 and produces CO_2 like cellular respiration, we shall see that it is a very different process, one that consumes ATP rather than produces it.

9.8.1 The initial step of photorespiration is catalyzed by the oxygenase activity of rubisco

Oxygen, which is much more abundant than CO_2 in the atmosphere (molecular ratio of about 600:1), inhibits photosynthetic carbon fixation by **competing** with CO_2 for the same active site on rubisco. The oxygenase activity of rubisco results in the incorporation of O_2 into the carboxyl groups of 3-PGA and **2-phosphoglycolate**

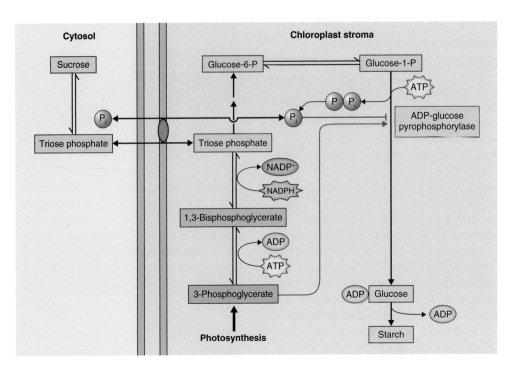

Figure 9.34 The pathway of starch synthesis in chloroplasts.

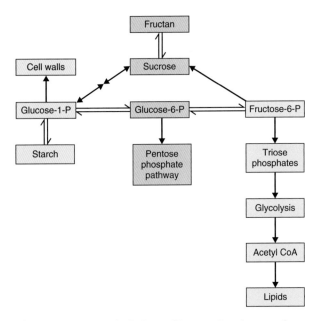

Figure 9.35 Metabolic fates of hexose phosphate products of photosynthesis include biosynthetic pathways and glycolysis.

(see Figure 9.26). The relative oxygenase and carboxylase activities of rubisco may be calculated according to Equation 9.12 from the maximal velocities (V) and Michaelis–Menten constants (K_m) of the two reactions (see Chapter 2). The ratio represented by $(V_c/K_c)/(V_o/K_o)$ is a comparative index of carboxylase to oxygenase capacity and is referred to as the **specificity factor**. Experimentally measured values of the specificity

factor for the rubisco of land plants are in the range 80 to 130. The oxygen and carbon dioxide used by rubisco are in solution: at 25 °C, the concentration of dissolved CO_2 in equilibrium with air is about 8 μM and the O_2 concentration is about 250 μM. Substituting these figures in Equation 9.12 gives a value of $v_c/v_o = 3.2$ for an enzyme with a specificity factor of 100. In other words, on average, the rate of carbon fixation by rubisco in terrestrial species is about three times that of the oxygenation reaction of photorespiration.

Equation 9.12 Rates of carboxylation and oxygenase activities of rubisco

$$v_c/v_o = ((V_c/K_c)/(V_o/K_o))[CO_2]/[O_2]$$

where subscripts c and o refer to carboxylase and oxygenase, respectively, and v = rate, $V = V_{max}$ and $K = K_m$

Oxygenation of RuBP is an intrinsic property of all known rubisco enzymes. This observation has been interpreted to mean that the ancestral rubisco originally evolved in an anaerobic atmosphere. Specificity factors determined for the rubisco of **anaerobic photosynthetic bacteria**, which may give a glimpse of the characteristics of the earliest enzyme, are as low as 15. The value for **cyanobacteria** is around 50–60. It seems that selective pressure over 1.5 billion years has gradually increased the specificity factor to the value of 100 and more seen in the most recently evolved plants. In Section 9.9 we will discuss some mechanisms that have evolved to shift the

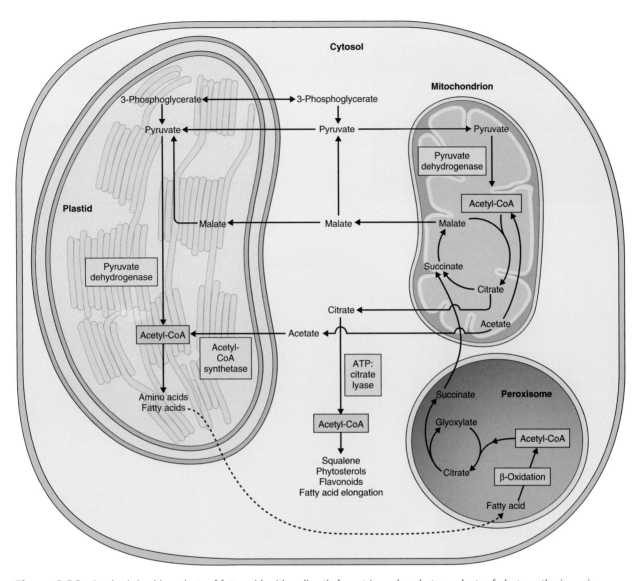

Figure 9.36 Synthesis in chloroplasts of fatty acids either directly from triose phosphate products of photosynthesis or via catabolic activities within mitochondria and peroxisomes. Note that the relative sizes of organelles are not to scale.

outcome of the carboxylase–oxygenase relationship towards carboxylase.

9.8.2 Enzymatic reactions of photorespiration are distributed between chloroplasts, peroxisomes and mitochondria

One of the products of the rubisco oxygenase reaction, 2-phosphoglycolate, is not used by the Calvin–Benson pathway. Instead it is metabolized by the **C$_2$ oxidative photosynthetic carbon cycle**, also known as the **photorespiratory carbon oxidation cycle**. These

reactions salvage most of the reduced carbon that is diverted to phosphoglycolate. The C$_2$ cycle is summarized in Figure 9.37. The pathway is notable for the way in which it operates across three subcellular organelles and integrates with both nitrogen metabolism and the Calvin–Benson cycle. 2-Phosphoglycolate is dephosphorylated to produce **glycolate**, which is transferred across the chloroplast envelope by a specific transporter and enters the **peroxisome** (Figure 9.37). The peroxisomal enzyme **glycolate oxidase** catalyzes the reaction of glycolate and O$_2$ to produce **glyoxylate** and **hydrogen peroxide** (H$_2$O$_2$). **Catalase** within the peroxisome (Figure 9.37A) converts H$_2$O$_2$ to water plus oxygen. **Glycine** is formed by the amination of glyoxylate catalyzed by two peroxisomal **aminotransferase** enzymes. The **amine donors** are serine and glutamate. One of these enzymes, glutamate:glyoxylate

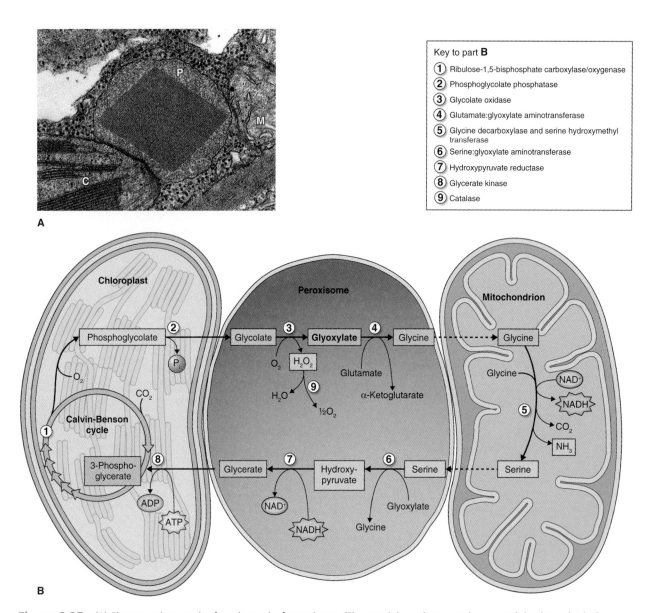

Figure 9.37 (A) Electron micrograph of a tobacco leaf peroxisome (P), containing a large catalase crystal, in close physical contact with both a chloroplast (C) and a mitochondrion (M). (B) The photorespiratory glycolate pathway, which involves enzyme systems located in peroxisomes, chloroplasts and mitochondria. Glycolate, derived from phosphoglycolate, the product of the oxygenase reaction catalyzed in the chloroplast by rubisco, is metabolized to glycine in the peroxisome. Glycine is decarboxylated in the mitochondrion with the production of serine, which is further metabolized in the peroxisome and returned to the chloroplast as glycerate. Organelles are not to scale.

aminotransferase, produces **α-ketoglutarate**, the fate of which is described in the next section. Both these aminotransferases are essential for the continued operation of the C_2 cycle.

Glycine next moves to the mitochondrion. Here, in a complicated reaction, the enzyme **glycine decarboxylase**, with its cofactor **tetrahydrofolate** (THF; Figure 9.38A), catalyzes the NAD^+-dependent oxidative conversion of a molecule of glycine to one each of CO_2, NH_3 and CH_2-THF (methylene tetrahydrofolate; Figure 9.38B). Glycine decarboxylase is an abundant

mitochondrial enzyme complex and the reaction it catalyzes is responsible for the production of CO_2 during photorespiration. The complex has four enzyme subunits, and pyridoxal phosphate, FAD and lipoamide cofactors in addition to THF. Reaction of CH_2-THF with a second glycine, catalyzed by **serine hydroxymethyltransferase**, regenerates THF and produces serine which returns to the peroxisome.

In the peroxisome, serine is the substrate for peroxisomal serine:glyoxylate aminotransferase (Figure 9.37). The amino group of serine

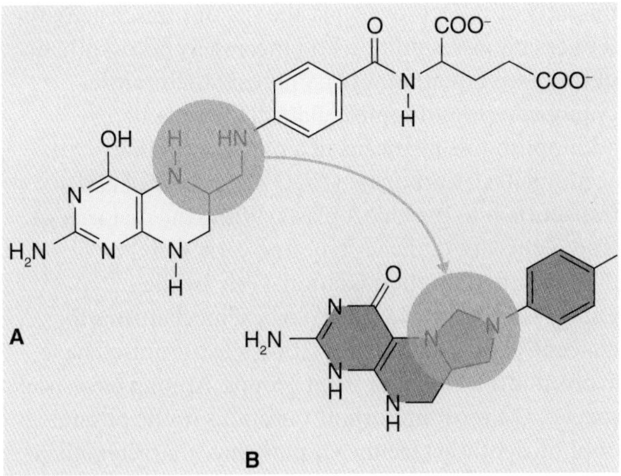

Figure 9.38 (A) Tetrahydrofolate (THF), cofactor of glycine decarboxylase, the mitochondrial enzyme responsible for CO_2 production during photorespiration. (B) The enzyme transfers a CH_2 fragment from glycine to THF, forming methylene-THF.

is transferred to glyoxylate producing **hydroxypyruvate**, which in turn is reduced to **glycerate**. Glycerate is transferred to the chloroplast and, after phosphorylation to 3-PGA, joins the Calvin–Benson cycle.

9.8.3 Ammonia produced during photorespiration is efficiently reassimilated

Ammonia is a product of glycine decarboxylation in the mitochondrion during photorespiration. **Reassimilation** of NH_3 (that is, return of the nitrogen to an organic combination) is necessary to avoid nitrogen loss or the buildup of toxic ammonia. This occurs in the chloroplast via the **glutamine synthetase–glutamate α-ketoglutarate aminotransferase (GS-GOGAT)** pathway (see Chapter 13). Equation 9.13 shows the net reaction catalyzed by GS and GOGAT acting in sequence. ATP and reduced ferredoxin are supplied by photosynthetic electron transport and photophosphorylation.

Equation 9.13 Photorespiratory ammonia reassimilation by the GS-GOGAT pathway

α-Ketoglutarate + NH_3 + ATP + 2Fd_{red}

→ glutamate + ADP + P_i + 2Fd_{ox}

The glutamate:α-ketoglutarate cycle is coordinated between the chloroplast and the peroxisome by the activities of GOGAT and glutamate:glyoxylate aminotransferase and transporters in the chloroplast envelope (see Figure 9.30). By linking the oxidative carbon cycle with nitrogen assimilation and peroxisomal

aminotransferase reactions, photorespiratory metabolism is kept in balance.

9.8.4 Energy costs and environmental sensitivities of photorespiration are significant for the impact of climate change on the biosphere

The net effect of the operation of the photorespiratory carbon cycle is to convert two molecules of RuBP into three of 3-PGA and one of CO_2 (Equation 9.14). In addition to losing one in ten carbon atoms from the carboxylation acceptor, the plant also expends NADPH and ATP, generated by the light energy capture reactions of photosynthesis, on reassimilating ammonia. In terms of gas exchange, three molecules of oxygen are taken up for every one of CO_2 liberated (Equation 9.14). In contrast, the molar ratio of CO_2 in to O_2 out during photosynthesis is 1:1. Theoretical calculations indicate that photorespiration increases the **energy cost of CO_2 fixation** and increases the number of photons required to assimilate one molecule of CO_2 from eight to nearly 14.

Equation 9.14 Overall stoichiometry of the C_2 cycle

2RuBP + 3O_2 + 2Fd_{red} + 2ATP

→ 3 3-PGA + CO_2 + 2Fd_{ox} + 2ADP + 2P_i

Equation 9.12 shows that photorespiration can be minimized by increasing the concentration of CO_2 relative to O_2 at the active site of rubisco. As we shall see in Section 9.9, the photosynthetic productivity of plants that have evolved metabolic mechanisms for minimizing photorespiration is correspondingly elevated. Enhancing CO_2 concentration within greenhouses to improve the yield of protected crops is a long-established horticultural technology.

At the center of contemporary political concern about global climate change is the rise, over the last 150 years or so, of **global atmospheric CO_2 concentration** from 260 to 380 parts per million (ppm), a trend that is likely to continue and even accelerate. Projections by the International Panel on Climate Change based on a range of scenarios anticipate a [CO_2] of 450–550 ppm by 2050. Higher [CO_2] should favor carboxylation relative to photorespiration and benefit **crop yields** through the 'CO_2 fertilization' effect. For increased photosynthetic potential under elevated CO_2 to pay off as higher crop yield there needs to be adequate **sink capacity** (for example, number and size of seeds per plant) to receive the additional fixed carbon. If productivity is **sink-limited** (as is often the case), feedback and adaptive processes will tend to counteract the CO_2 fertilization effect. A suggested example of such a developmental adjustment is a change in the density of stomata. Studies of herbarium specimens collected over the last 200 years show that, for a given species, there has been a clear decrease in the number of **stomata** per unit leaf area, which could lead to reduced CO_2 uptake as atmospheric CO_2 levels have increased. Potential yield increases in response to a rise in $[CO_2]$ from present levels to 550 ppm are estimated to be in the range of $11–32\%$ for wheat, rice and soybean. The true consequences of elevated CO_2, however, are likely to be significantly influenced by complex **interactions** with temperature and soil moisture as the global climate changes.

Key points The oxygenase activity of rubisco catalyzes the initial step of photorespiration, a pathway that limits photosynthetic productivity by competing with carbon fixation. The relative carboxylase and oxygenase activity of rubisco is expressed as the specificity factor and has increased during evolution from a low of 15 in primitive anaerobic autotrophs through 50—60 in cyanobacteria to more than 100 in land plants. The 2-phosphoglycolate product of the oxygenase reaction is metabolized by the C_2 oxidative photosynthetic carbon (photorespiratory carbon oxidation) cycle. Carbon moves in the form of glycolate from the chloroplast to the peroxisome where it is converted to glycine and passed to the mitochondrion. Mitochondrial glycine decarboxylase, an abundant multienzyme complex, is responsible for the photorespiratory release of CO_2 and also generates NH_3, which is reassimilated in the chloroplast. The loss of fixed CO_2 and the expenditure of reductant and ATP on N reassimilation during photorespiration increase the energy cost of photosynthesis. The balance between photosynthesis and photorespiration is responsive to atmospheric temperature and $[CO_2]$ and therefore critically sensitive to global climate change.

9.9 Variations in mechanisms of primary carbon dioxide fixation

As we have seen, rubisco can fix CO_2 leading to sugar synthesis, or O_2 leading to photorespiration. Increasing

the ratio of $[CO_2]$ to $[O_2]$ at the site of rubisco activity reduces photorespiration and increases photosynthetic output (see Equation 9.12). Photosynthetic algae, cyanobacteria and some aquatic higher plants, accomplish this by means of a mechanism that accumulates bicarbonate (HCO_3^-) in the cell and transports it to the chloroplasts. While the majority of land plants are of the C_3 type and carry out photosynthetic carbon fixation solely by the Calvin–Benson cycle, physiological mechanisms for concentrating CO_2 at the fixation site of rubisco have evolved in a number of plant groups. Among terrestrial species, the most important variations in the physiology of photosynthesis are the **C_4 pathway**, characteristic of species from tropical, subtropical and saline environments, and **crassulacean acid metabolism** (**CAM**), typical of plants such as succulents from arid environments. In these plant groups, initial CO_2 fixation is carried out by **phosphoenolpyruvate carboxylase** (**PEP carboxylase, PEPC**), an enzyme that does not fix O_2. In both groups, secondary CO_2 fixation and sugar synthesis occur via rubisco and the Calvin–Benson cycle.

9.9.1 C_4 plants have two distinct carbon dioxide-fixing enzymes and a specialized leaf anatomy

For terrestrial plants, a number of environmental conditions can lead to a decrease in the ratio of $[CO_2]$ to $[O_2]$ in leaf air spaces, and a consequent increase in photorespiration. For example, as temperature increases, the concentration of dissolved CO_2 decreases more rapidly than the concentration of O_2, effectively reducing $[CO_2]:[O_2]$. In addition, the oxygenase activity of rubisco increases with temperature faster than the carboxylation activity, favoring photorespiration at the expense of photosynthesis. Finally, conditions such as high temperature, low humidity and saline soil lead to rapid water loss by leaves or reduced water uptake by roots. Under these conditions plants partially close their stomata to conserve water (see Chapter 14). This in turn reduces the rates of diffusion of O_2 out of a leaf and CO_2 into a leaf, thereby lowering the ratio of CO_2 to O_2.

The **C_4 pathway of carbon fixation** evolved in some angiosperms that live in hot, sunny areas or in saline conditions. These include a number of **tropical grasses**, such as *Zea mays* and sugar cane (*Saccharum officinarum*), and a few eudicots (for example, species of the genus *Amaranthus*). Plants with the C_4 carbon fixation pathway have two distinguishing features, one biochemical and one anatomical. First, while 3-PGA is the first stable product of photosynthesis in C_3 species, in C_4 species large amounts of **four-carbon organic acids** are the earliest products of CO_2 fixation. This observation led to the discovery that primary initial CO_2 fixation in these plants is catalyzed by PEPC. Second, C_4 species have a distinctive leaf anatomy which allows them to concentrate CO_2 and decrease O_2 near the rubisco active site.

In leaves of **C_4 species**, chloroplast-containing tissues are differentiated into two cell types, referred to as

mesophyll cells and **bundle sheath cells** (Figure 9.39A). In contrast to C_3 leaves, mesophyll cells are not differentiated into palisade and spongy layers, but are tightly packed around prominent, chloroplast-containing bundle sheath cells that surround veins. A large number of plasmodesmata connect mesophyll and bundle sheath cells. This tissue organization is called **Kranz anatomy**, from the German word for 'wreath', describing the arrangement of bundle sheath cells.

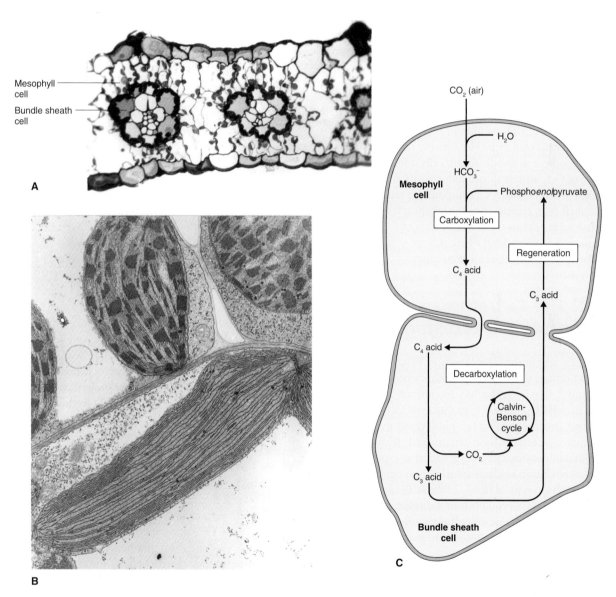

Figure 9.39   (A) Light microscope image of a leaf cross-section of a C_4 plant showing Kranz anatomy with mesophyll cells and large bundle sheath cells surrounding vascular bundles. (B) Electron micrograph contrasting chloroplast structure in bundle sheath (lower) and mesophyll (upper) cells. (C) Overview of the C_4 pathway. CO_2 enters the mesophyll cell and is converted to HCO_3^- in the aqueous environment of the cytosol. Bicarbonate ion reacts with PEP to form a C_4 acid (oxaloacetate), which is converted to a second C_4 acid and then transported to a neighboring bundle sheath cell. There, the C_4 acid is decarboxylated, and the CO_2 released is fixed by rubisco and the Calvin–Benson cycle. The C_3 acid product of decarboxylation is transported back to the mesophyll cell to regenerate PEP.

The spatial distribution of mesophyll and bundle sheath cells in Kranz anatomy is reflected in localization of the biochemical pathways that are critical for the organization and operation of C_4 photosynthesis. Carboxylation via PEPC occurs in mesophyll cells while rubisco activity is limited to bundle sheath cells. **Chloroplast ultrastructure** is correspondingly different in mesophyll and bundle sheath cells (Figure 9.39B). Thylakoid membranes in mesophyll cell chloroplasts are organized into grana stacks and stroma lamellae and contain PSI and PSII. PEPC synthesis is upregulated in mesophyll cells and rubisco is absent from mesophyll chloroplasts. The chloroplasts of bundle sheath cells contain rubisco but lack stacked thylakoid membranes and are depleted in PSII. There is little non-cyclic electron flow and thus little O_2 evolution in bundle sheath chloroplasts; ATP is generated predominantly by cyclic photophosphorylation.

The C_4 pathway begins with a C_4 organic acid, the initial product of CO_2 fixation in cells of the outer mesophyll catalyzed by PEPC. Transfer of the C_4 product from mesophyll to bundle sheath cells is followed by release of CO_2 and its subsequent refixation by the Calvin–Benson cycle. A C_3 compound is returned to the mesophyll cells, regenerating the acceptor for PEPC (Figure 9.39C). Thus the concentration of CO_2 is maintained at a high level where rubisco is active, and photorespiration is kept to a minimum.

9.9.2 The C_4 pathway minimizes photorespiration

In the first step of the C_4 pathway, PEPC carboxylates PEP to produce **oxaloacetate**, a four-carbon organic acid, and P_i (see Equation 7.4). In contrast to rubisco, PEPC uses **HCO_3^-** rather than dissolved CO_2 as its C_1 substrate. Because the HCO_3^--binding site does not recognize O_2, carboxylation can proceed efficiently without competition from oxygenation as experienced by rubisco. CO_2 in aqueous solution is in equilibrium with a small amount of HCO_3^- (about 1% at pH 7; Equation 9.15). This equilibrium is displaced in favor of bicarbonate by **carbonic anhydrase**, an important enzyme in C_4 and CAM photosynthesis, described further in Section 9.9.5.

Equation 9.15 Carbon dioxide and bicarbonate in aqueous solution

$$CO_2 + H_2O \rightleftharpoons HCO_3^- + H^+$$

Oxaloacetate, the initial product of carbon fixation in the mesophyll, may be further metabolized in one of three possible ways (Table 9.10). The **NADP$^+$-malic enzyme** route occurs in species such as sorghum (*Sorghum bicolor*) and *Zea mays* (Figure 9.40A). In this variant, oxaloacetate is converted to malate by NADP$^+$-malate dehydrogenase. Malate is transported from mesophyll to bundle sheath cells, where it is decarboxylated by NADP$^+$-malic enzyme. The CO_2 released is assimilated through rubisco and the Calvin–Benson cycle. Pyruvate, the product of malate decarboxylation, is returned to the mesophyll cell where it is phosphorylated by **pyruvate-orthophosphate dikinase** (PPDK) to regenerate PEP.

Proso millet (*Panicum miliaceum*) and switchgrass (*P. vergatum*) are examples of species that carry out **NAD$^+$-malic enzyme**-type C_4 photosynthesis (Figure 9.40B). In this case, carbon is transported between mesophyll and bundle sheath cells in the form of the transamination products of oxaloacetate and pyruvate, namely **aspartate** and **alanine** respectively (Table 9.10). In the third variant, exemplified by guinea grass (*Megathyrsus maximus*), transamination of aspartate transported from the mesophyll cell produces oxaloacetate, which is then decarboxylated by bundle sheath **PEP carboxykinase** to yield CO_2 and PEP (Figure 9.40C). Thus carbon is shuttled back to the mesophyll as PEP, alanine or pyruvate (Table 9.10).

The movement of C_4 and C_3 intermediates between cells takes place by diffusion, facilitated by the extensive system of **plasmodesmata** connecting mesophyll and bundle sheath cells. **ATP** consumed by the PPDK and (where active) PEP carboxykinase reactions is additional to that required by the operation of the Calvin–Benson cycle in the bundle sheath. Under dry, warm, high light conditions, this net energy penalty of C_4 photosynthesis is more than offset by the price that C_3 species pay for photorespiration. However, C_3 species will outcompete C_4 species when light is limiting or under cooler, wetter conditions where photorespiration is lower or stomata are wider open.

Mitochondria play important roles in NAD$^+$-malic enzyme and PEP carboxykinase-type C_4 plants. Photosynthetic carbon flux through the bundle sheath mitochondria in NAD$^+$-malic enzyme C_4 species is estimated to be many times greater than through the standard respiratory and C_3-type photorespiratory pathways. The PEP-phosphate transporter (PPT; see Section 9.7.5) of the chloroplast envelope is of particular significance for the movement of C_3 intermediates within mesophyll cells. PPDK, which converts pyruvate to PEP (Figure 9.40), is localized in the mesophyll cell chloroplast stroma, whereas PEPC is cytosolic. The PPT is responsible for transferring PEP across the chloroplast envelope.

The C_4 mechanism evolved in plants of hot, sunny or saline environments; it enhances photosynthetic efficiency and decreases water loss through improved **trapping of CO_2** under conditions of reduced stomatal

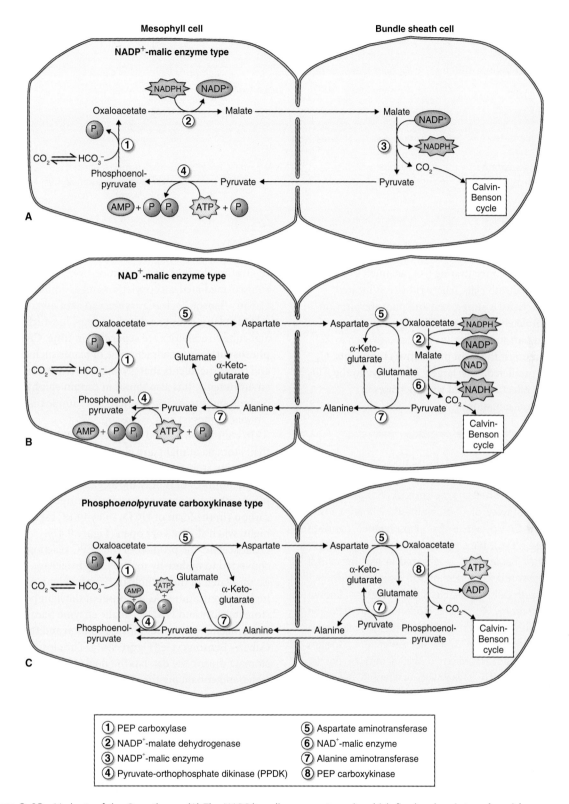

Figure 9.40   Variants of the C_4 pathway. (A) The $NADP^+$-malic enzyme type, in which fixed carbon is transferred from mesophyll to bundle sheath cells in the form of malate. (B) In the NAD^+-malic enzyme type, carbon is exchanged between the mesophyll and bundle sheath as the transamination products aspartate and alanine. (C) The PEP carboxykinase type, in which carbon is recycled from the bundle sheath as alanine and PEP.

Table 9.10 Variations in carbon transport and decarboxylation in C_4 species.

C_4 acid transported to bundle sheath cells	C_3 acid transported to mesophyll cells	Decarboxylase	Typical species
Malate	Pyruvate	$NADP^+$-malic enzyme	Maize (*Zea mays*), sorghum (*Sorghum bicolor*)
Aspartate	Alanine	NAD^+-malic enzyme	Switchgrass (*Panicum vergatum*), proso millet (*P. miliaceum*)
Aspartate	Alanine, PEP or pyruvate	PEP carboxykinase	Guinea grass (*Megathyrsus maximus*)

conductance. By concentrating CO_2 as much as 20-fold in the bundle sheath cells, where rubisco is located, oxygenase activity is suppressed and photorespiration almost completely eliminated. The C_4 adaptation accounts in part for the reason why species such as *Z. mays* are amongst the most productive crop plants. C_4 species are predicted to be largely insensitive to the CO_2 fertilization effect associated with increases in greenhouse gas emissions.

Key points Species from tropical, subtropical and saline environments have evolved the C_4 pathway, a mechanism for offsetting the costs of photorespiration to the C_3 pathway of photosynthesis. Primary carbon fixation in C_4 species is by phosphoenol pyruvate carboxylase (PEPC), the C_4 products of which give these species their name. PEPC is localized in leaf mesophyll cells whereas rubisco is confined to structurally distinct bundle sheath cells, a characteristic arrangement known as Kranz anatomy. C_4 organic acids transported to the bundle sheath are decarboxylated and the CO_2 released is fixed by rubisco and the Calvin–Benson cycle. C_4 species may be classified according to the identity of the C_4 products transported. By maintaining a high $[CO_2]$ at the rubisco active site, photorespiratory oxygenase activity is minimized in C_4 species.

9.9.3 In CAM plants, the processes of CO_2 capture and photosynthesis are separated in time

The C_4 adaptation addresses the problem of high temperatures and low water availability by **spatially**

separating the initial CO_2 fixation event from rubisco-mediated carboxylation and assimilation by the Calvin–Benson cycle. **Crassulacean acid metabolism (CAM)** plants adopt a related strategy, in which PEPC and rubisco activities are **separated in time. CAM photosynthesis** is characteristic of plants such as cacti and other succulents that grow in extremely arid environments. It is also found in certain epiphytic orchids and some crop species, for example pineapple (*Ananas comosus*) and *Agave* spp.

In contrast to C_3 and C_4 species, CAM plants open their stomata at night and close them during day. Restricting stomatal opening to the relatively cool and humid hours of darkness allows gas exchange to take place without excessive transpiration. Initial fixation of carbon (in the form of HCO_3^-) by PEPC takes place at night, when stomata are open (Figure 9.41). Oxaloacetate, the product of the PEPC reaction, is converted to **malate** by malate dehydrogenase. Malate is accumulated in the vacuole during the night, often to high concentrations. At the onset of the light period, the stomata close and $NADP^+$-malic enzyme releases CO_2 from malate for assimilation via rubisco and the Calvin–Benson cycle (Figure 9.41). The closure of stomata during the day has the dual effect of conserving water and enhancing the efficiency of rubisco as a result of the accumulation of high concentrations of CO_2.

The daily cycle of stomatal aperture is reflected in rhythms of enzyme activity and abundance, a pattern of behavior in which **circadian regulation** (see Chapter 8) plays an important role. Amounts of PEPC mRNA and protein vary from a maximum in darkness to a minimum during the day. By contrast, transcription of the gene for the PEP-phosphate transporter (see Section 9.7.5) occurs most actively during the light period. PEPC is also regulated at the **post-translational** level. It is activated in the dark by **phosphorylation**, catalyzed by a specific **kinase**. Synthesis and degradation of the kinase are controlled by the **circadian oscillator** (see

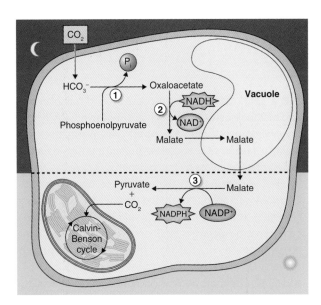

Figure 9.41 Crassulacean acid metabolism (CAM). At night, CAM plants open their stomata, allowing CO_2 to enter. PEP carboxylase (1) incorporates this CO_2 (as HCO_3^-) into the C_4 organic acid oxaloacetate, which is reduced to malate by malate dehydrogenase (2). Malate is stored in the vacuole overnight. In the light, stomata close, preventing water loss. The stored malate is decarboxylated by $NADP^+$-malic enzyme (3), and the resulting CO_2 is assimilated via the Calvin–Benson cycle.

Chapter 8). Furthermore, PEPC is sensitive to **feedback inhibition** by cytosolic malate. The endogenous rhythm in CAM photosynthesis may be overridden under conditions of extreme drought: stomata remain closed at night and intracellular recycling of CO_2 becomes a maintenance-only activity that does not contribute to net growth.

9.9.4 The transpiration ratio relates carbon dioxide fixation to water loss

CAM and C_4 adaptations allow plants to live in drier climates because they reduce the amount of water lost by transpiration compared to C_3 plants. The **transpiration ratio** can be used to compare how efficiently plants fix CO_2 relative to the amount of water lost by transpiration (see Chapter 14). The transpiration ratio is generally expressed as a ratio of the amount of water lost to dry weight gain. C_3 plants have a transpiration ratio of 450–950 g g^{-1}. In CAM, primary CO_2 capture into organic acids occurs at night when water loss through open stomata is minimized and these plants have the most favorable transpiration ratio in the range 18–125 g g^{-1}. C_4 plants have an intermediate transpiration ratio of

250–350 g g^{-1}. The two- to three-fold difference in transpiration ratio between C_3 and C_4 plants can be explained by differences in leaf anatomy and the mechanism of primary CO_2 capture. The mesophyll cells of C_4 leaves place chloroplasts with their more efficient CO_2-capturing enzymes adjacent to the large leaf air spaces. This allows internal CO_2 concentration of C_4 plants to be lower than that in the air spaces of leaves of C_3 plants. Lowering the CO_2 concentration in leaf air spaces creates a steeper gradient for CO_2 to diffuse into the leaf. This allows C_4 plants to maintain rapid uptake of CO_2 even when their stomata are partially closed to reduce water loss.

9.9.5 Clues to the evolutionary origins of C_4 and CAM photosynthesis come from studies of the enzyme carbonic anhydrase

C_4 photosynthesis occurs in about 7000 species of angiosperms distributed across 19 plant families. The pathway arose around 30 million years ago, first in grasses, subsequently in eudicots, and appears to have **evolved independently** in more than 40 lineages. CAM is more extensively distributed, occurring in at least 20 000 species from more than 30 families. It is an **ancient pathway**, found in primitive non-angiosperms, but in the flowering plants its origin and multilineage distribution are similar in scale to those of the C_4 syndrome pathway. Among the steps necessary for the acquisition of C_4 capacity are the development of Kranz anatomy, localization of glycine decarboxylase to the bundle sheath, upregulation of PEPC, and expression of other C_4 enzymes. For CAM photosynthesis, succulence and tight packing of mesophyll cells are the most significant structural innovations, and circadian control is the essential regulatory requirement.

A critical component of the PEPC route of initial carbon fixation is **carbonic anhydrase** (**CA**), the activity of which catalyzes the formation of the C_1 substrate HCO_3^- (see Equation 9.15). CA is a **metalloenzyme**, with an active site containing zinc. It is almost universally present in living organisms, from bacteria to humans. The enzymes of plants are members of the β-CA group. The evolution of CA in relation to C_4 has been studied in plants of the yellowtop (*Flaveria*) genus, which includes C_3, C_4 and intermediate species. β-CA in *Flaveria* is encoded by a small, multigene family. *CA3* encodes a **cytosolic form** which is highly expressed in leaves and is the enzyme responsible for supplying PEPC with bicarbonate. It has been suggested that chloroplast forms of CA have a role in regulating CO_2 levels and diffusion within the organelle in both C_3 and C_4 species. Comparison of CA gene structure and expression in *Flaveria* species, together with studies on the intracellular location of different isoforms of the enzyme, indicate that CA3, which catalyzes the first step in the C_4 pathway, arose by modifications to the *CA3* gene of the C_3 ancestor. During evolution, the **chloroplast transit peptide** was lost and expression of the gene became upregulated, leading to the high levels of cytosolic CA activity characteristic of C_4 mesophyll cells.

The evolution of elaborations on the basic C_3 mechanism of photosynthesis are of continuing practical interest. The crops on which world food supply depends are dominated by C_3-type cereals, especially wheat and rice. If photorespiratory losses in these species could be reduced by introducing or stimulating components of the C_4 mechanism, enhanced photosynthetic productivity and yield would be expected. Since evolution has independently given rise to C_4 behavior in C_3 lineages many times, perhaps plant breeding and biotechnology can pull off the same trick. In practice, success in this endeavor has been very limited so far, but it remains a **Grand Challenge** for agricultural research in the 21st century.

Key points Succulents have evolved crassulacean acid metabolism (CAM) to maintain photosynthesis in hot dry environments. The stomata of CAM plants open at night to take in CO_2 and close during the day, thereby minimizing water loss by transpiration. CO_2 fixed at night by PEPC accumulates in the central vacuole as malate. CO_2 released from malate during the day by $NADP^+$-malic enzyme is assimilated by rubisco and the Calvin–Benson cycle. Enzyme activities and gene expression in CAM plants are under circadian regulation. Carbonic anhydrase converts CO_2 to HCO_3^-, the substrate of PEPC in CAM and C_4 plants. Improving the photosynthetic productivity of crop plants by reducing photorespiration in imitation of the C_4 syndrome is an ongoing biotechnological objective.

Part IV
Growth

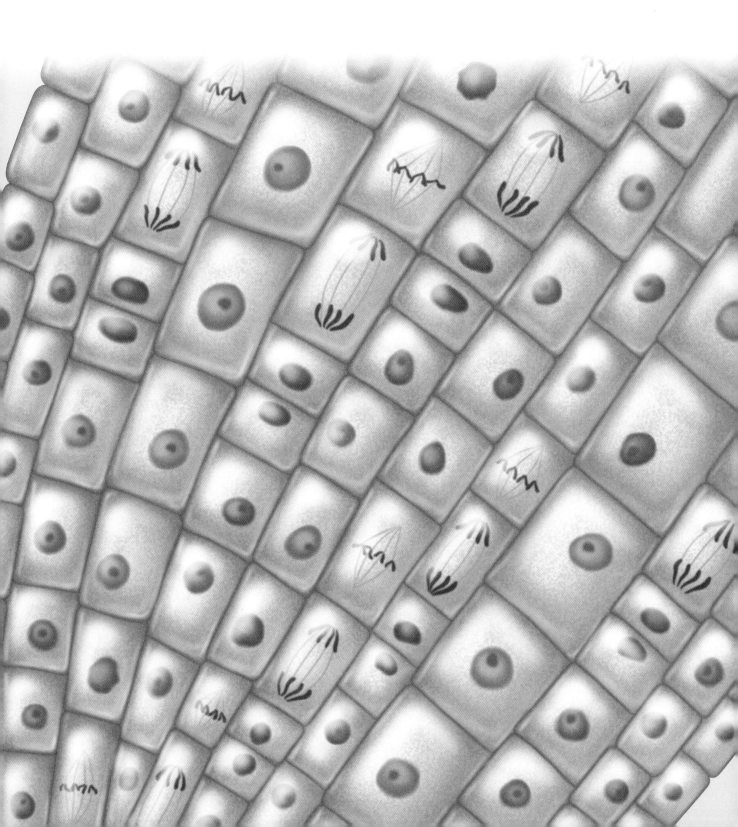

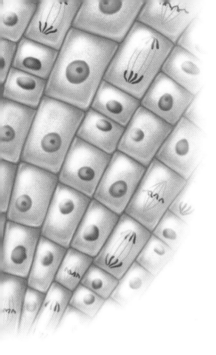

Chapter 10

Hormones and other signals

10.1 Introduction to plant hormones

Hormones are classically defined as chemical compounds produced in one part of an organ or organism and transported to another where they elicit responses. Charles Darwin and his son Francis were among the first to show that signals produced in one part of the plant affect growth in another. The Darwins exposed seedlings of various grasses and eudicots to unilateral sources of light to investigate growth responses; growth in response to a light source is called **phototropism** (see Chapter 8). The clearest evidence they obtained for the involvement of a transmissible signal was with canary grass (*Phalaris canariensis*), where exposing the tip of the seedling **coleoptile** to **unilateral light** caused the base to grow toward the source of illumination. In 1880 they wrote: 'We must, therefore, conclude that when seedlings are freely exposed to a lateral light, some influence is transmitted from the upper to the lower part, causing the latter to bend'.

Work in the early part of the 20th century helped establish that the signal produced by the tips of grass seedlings exposed to unilateral light was chemical in nature, and work by Frits Went in the 1920s led to the development of the first **bioassay** for plant hormones. Went showed a quantitative relationship between the amount of the chemical substance produced by coleoptile tips of grasses and the bending response in the lower part of the coleoptile. He called the chemical **auxin**. This bioassay allowed researchers to determine the amount of hormone in a plant extract by measuring the bending response of a de-tipped coleoptile. The coleoptile bioassay led to the identification of the hormone that caused coleoptile bending as **indole-3-acetic acid (IAA)** (Figure 10.1).

Bioassays proved to be key tools in early work identifying hormones from plants and animals and their use persisted until the 1960s and 1970s, when advances in chemical analysis allowed for the precise identification and quantification of hormones isolated from tissues. These advances included the development of sensitive gas chromatographs and mass spectrometers in the 1960s and 1970s, instruments that were later combined so that **GC-MS** (gas chromatography–mass spectrometry) and modern variants of these analytical instruments became the standard methods for analyzing plant natural products.

Representatives of the 11 major classes of plant hormones—**abscisic acid (ABA)**, **auxins**, **brassinosteroids (BRs)**, **cytokinins (CKs)**, **ethylene**, **gibberellins (GAs)**, **jasmonates (JAs)**, **polyamines (PAs)**, **salicylic acid (SA)**, **strigolactone (SL)** and **nitric oxide (NO)**—are shown in Figure 10.1. Although these hormones differ widely in their chemistry, many aspects of their synthesis and breakdown pathways are shared (Figure 10.2). The isoprenoid pathway serves as the source of the precursors of ABA, BRs, CKs, GAs and SL (see Chapter 15), whereas the auxins, ethylene and PAs

The Molecular Life of Plants, First Edition. Russell Jones, Helen Ougham, Howard Thomas and Susan Waaland.
© 2013 John Wiley & Sons, Ltd. Published 2013 by John Wiley & Sons, Ltd.

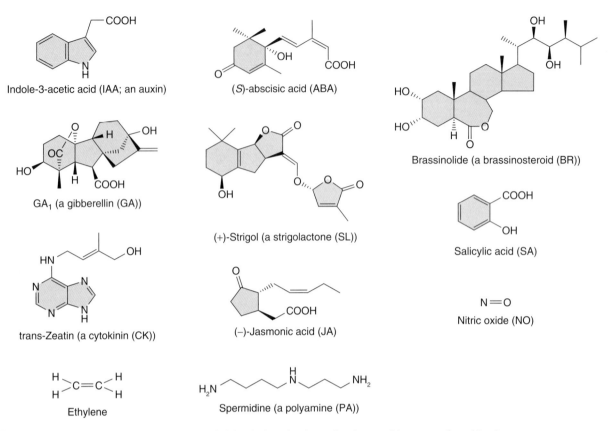

Figure 10.1   Chemical structures, names and abbreviations for the major classes of hormones found in plants.

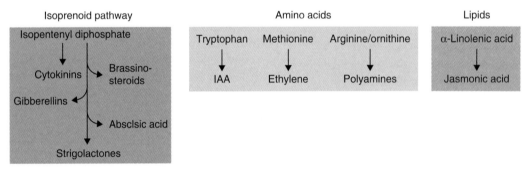

Figure 10.2   Plant hormones can be grouped into three general categories based on their biosynthetic origins: along the isoprenoid pathway (ABA, BR, CK. GA and SL), from amino acids (ethylene, IAA and polyamines)) and from lipids (JA).

arise from the amino acids tryptophan, methionine and arginine/ornithine, respectively. Lipids, specifically the fatty acid α-linolenic acid, are precursors of jasmonates. Two of the hormones, ethylene and NO, are gases and emerging evidence suggests that another gas, **hydrogen sulfide**, also produced by plants, can act as a potent signaling molecule.

The major plant hormones had been well characterized chemically by the end of the 1980s and a combination of chemical and biochemical approaches had elucidated the basic elements of their biosynthesis. The advent of molecular cloning led to major advances in understanding hormone biosynthesis, and the

powerful methods of molecular genetics have permitted the identification of the signal transduction pathways for each of the major classes of plant hormone. This has led to the identification of receptors for almost all of the major classes of hormones, progress that was crucially dependent on developments in molecular genetics. As we will see from this chapter, some of the most recent advances have resulted in what can be regarded as major surprises for those working in the field of hormone signaling. For example, we now know that strigolactones, in addition to auxin, regulate axillary bud dormancy. It is likely that more surprises await as we investigate the biology of mobile signals in plants.

Key points The development of specific bio-assays was an important step in the discovery and characterization of plant hormones. Bioassays are still used to identify hormones in plant extracts, but new analytical methods, especially mass spectrometry, allow the quantification of very small amounts of natural products in complex mixtures. Although plant hormones are represented by a diverse range of molecules, from simple gases such as ethylene and nitric oxide to terpenoid molecules having up to 27 carbon atoms as in the case of brassinosteroids, they can be categorized into three broad groups based on their biosynthetic origins. ABA, BRs, CKs, GAs and SL are all synthesized along the isoprenoid pathway, ethylene, IAA and PAs arise from amino acids and JA is synthesized from membrane lipids.

10.2 Auxins

In this chapter we define auxins as compounds that are active in auxin-specific bioassays. The most abundant auxin is IAA, but other indole ring compounds such as indole-3-butyric acid and 4-chlorindole-3-acetic acid, and at least one non-indole auxin, phenylacetic acid, are found in plants (Figure 10.3) and elicit responses in auxin bioassays. Auxins affect almost all aspects of plant development from pollination and fertilization, through vegetative development to flowering. Auxins are key signaling molecules in the tropic responses of plants to

gravity, light and touch, and they are known to play a role in the synthesis and action of other plant hormones, which means that their effects on plant growth and development are extensive.

The effects of auxins in promoting root initiation and inhibiting shoot branching are well-known phenomena that are widely exploited in agriculture and horticulture (Figure 10.4). Shoot cuttings can be induced to form adventitious roots by treating the cut ends of shoots with an auxin. In the case of the *Calycanthus* 'Venus' hybrid (Venus sweetshrub) shown in Figure 10.4A, the cut ends of cuttings were treated with the potassium salt of the auxin indole-3-butyric acid (**KIBA**). In the absence of auxin, adventitious roots do not develop, but as the concentration of KIBA is raised an increasing number of roots are formed. The growth of axillary buds is also controlled by auxin. In the case of the *P. sativum* plants shown in Figure 10.4B, axillary buds are normally suppressed, a phenomenon known as **apical dominance**, but when the **shoot** apical meristem is removed axillary buds grow. If the decapitated shoot is treated with IAA, axillary bud growth is again suppressed. The hormonal mechanisms underlying apical dominance are described in further detail in Chapter 12.

10.2.1 Both synthesis and catabolism of IAA are important in auxin signaling

When considering the effectiveness of plant hormones as regulatory molecules it is crucial to know how the pool of active hormone is regulated. For a hormone to be

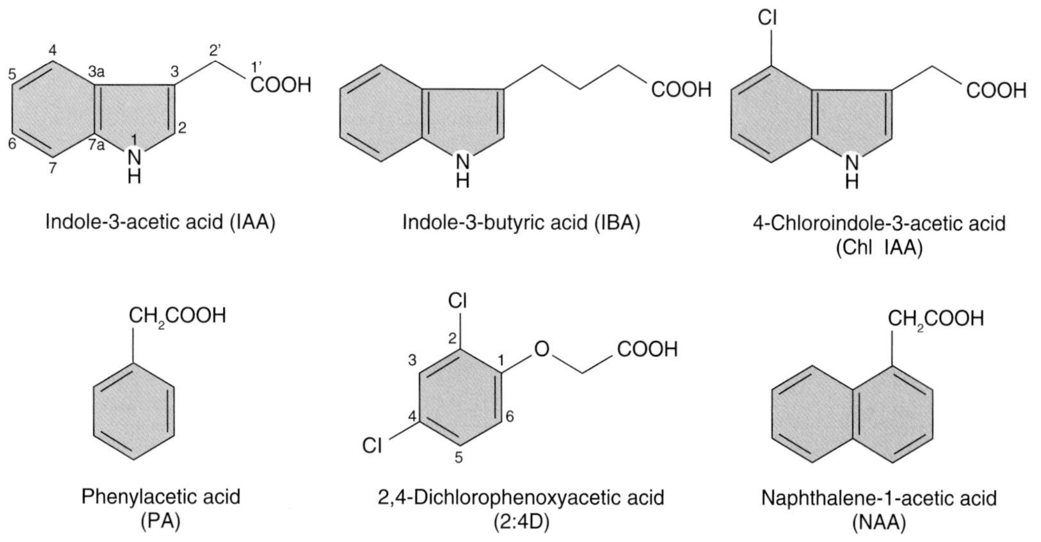

Figure 10.3 Structures of naturally occurring auxins (IAA, IBA, Chl IAA and PA) and two synthetic auxins, one widely used as a herbicide (2:4D) and the other as an inducer of root formation on stem cuttings (NAA).

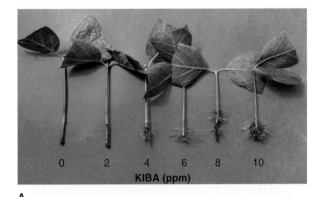

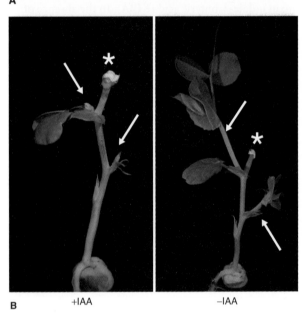

Figure 10.4 (A) Induction of lateral roots on shoot cuttings of Venus sweetshrub (*Calycanthus* 'Venus' hybrid) by increasing concentrations of the auxin KIBA, the potassium salt of indole-3-butyric acid. (B) Effect of indole-3-acetic acid (IAA) on apical dominance in peas (*Pisum sativum*). In the experiment shown here, pea seedlings were decapitated and their apices were replaced with lanolin paste (*) alone or with added IAA. When IAA is present (left) axillary buds (arrows) remain dormant and do not elongate; in absence of applied IAA (right), axillary buds are released from dormancy and develop into branches (arrows).

effective it must be synthesized and degraded. Simply synthesizing a hormone would not necessarily be sufficient for it act as a signal because in the absence of a mechanism to reduce its local concentration the signal would always be 'on'. Turning the signal 'off' is therefore as important as turning it on. For auxin, biosynthesis, conjugation, metabolic degradation and **polar transport** are all mechanisms that control its local concentration (Figures 10.5 and 10.6; see Section 10.2.2).

There are two principal pathways of IAA biosynthesis, one from the amino acid tryptophan and the other in a tryptophan-independent pathway (Figure 10.6). A common pathway proceeds from chorismate to indole; at this point it branches. One branch, the tryptophan-independent pathway, proceeds through indole-3-pyruvic acid to IAA. The other branch leads to the formation of tryptophan. At this point, analysis of mutants shows that there are at least four, possibly five, routes for synthesizing IAA from tryptophan, which has made dissection of these pathways difficult. Despite the multiplicity of biosynthetic routes from tryptophan to IAA, experimental evidence indicates that tryptophan is the predominant precursor of IAA in plants.

Catabolism of IAA can also follow several paths, one that results in the formation of conjugates, and others that bring about IAA degradation (Figure 10.5). The formation of some inactive IAA conjugates is reversible and their hydrolysis to active IAA is important in IAA homeostasis. Most of these reactions are under feedback regulation by IAA. Conjugation of IAA is enhanced by high levels of IAA; conversely, hydrolysis of conjugates is promoted by low levels of free IAA. Conjugation of IAA to L-amino acids and to glucose is widespread, and the formation of glucose esters is especially prevalent in seeds. While some IAA−amino acid conjugates can be remobilized and can therefore function as storage products, many of these conjugates are components of irreversible IAA degradation pathways. Degradation of free IAA and its amido conjugates can be carried out by enzymes of the peroxidase class.

10.2.2 Polar transport of auxins plays an important role in regulating development

The movement of auxins from cell to cell is directional. In the shoot and root, transport occurs **basipetally** (away from the organ apex). This feature, called polar transport, is unique to auxins and has an important role in many developmental processes including the formation and repair of tracheary elements, apical dominance, initiation of lateral roots (see Chapter 12) and growth responses to light (phototropism,; see Chapter 8) and gravity (gravitropism and the downward bending of the petiole termed **epinasty**; see Chapter 15).

Polar auxin transport is largely driven by gradients in protonated IAA (IAAH) in the apoplast, and anionic forms (IAA⁻) in the symplasm, which are maintained by specific auxin transporters in the plasma membrane (Figure 10.7). Gradients in IAAH and IAA⁻ are

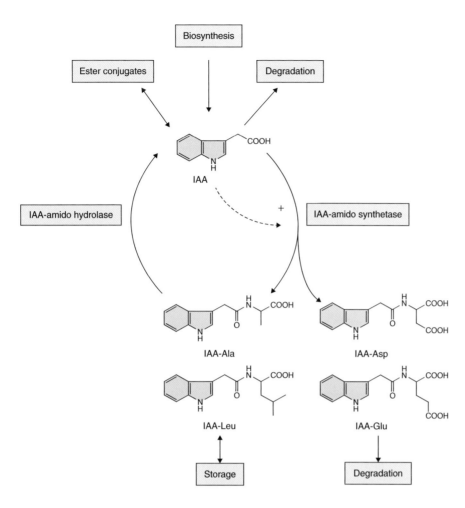

Figure 10.5 Routes for the degradation and storage of indole-3-acetic acid (IAA). IAA-amido synthase catalyzes the synthesis of amino acid conjugates. Some of these, such as IAA-Ala and IAA-Leu, can be stored and reconverted to active IAA by IAA-imido hydrolase. Others, such as IAA-Asp and IAA-Glu, are irreversibly degraded. Ester conjugates are another storage form of IAA.

generated by pH differences across the plant plasma membrane (see Chapter 5). IAA is a weak acid with a pKa of 4.7, therefore in the cytosol at a pH of about 7.0 it exists largely as IAA$^-$, whereas in the cell wall at a pH of about 5.5 it is predominantly IAAH. The plasma membrane is relatively permeable to uncharged IAAH but impermeable to IAA$^-$; therefore the ΔpH across the plasma membrane effectively traps IAA$^-$ in the cytosol.

The discovery that specific transporters play a role in polar auxin transport came from genetic analysis of *Arabidopsis* mutants. ***Auxin influx carrier* (*aux*)** mutants were discovered in *Arabidopsis* plants that are resistant to high concentrations of applied auxin. Sequence analysis suggests that AUX1 belongs to a family of proteins similar to prokaryotic amino acid permeases. Given that IAA and tryptophan are chemically related it seems reasonable to suggest that AUX1 is indeed involved in auxin transport, acting as a H$^+$/IAA$^-$ symporter, but direct evidence is lacking. Auxin influx transporters are asymmetrically localized in the plasma membrane at the apical end of the cell (Figure 10.7B).

Another class of asymmetrically distributed transporters called **PIN proteins** was first found in the *Arabidopsis pin1* mutant shown in Figure 10.8. This

mutant was called *pin* because of the needle-like shape of the inflorescence stem that lacks leaves, buds or flowers. The *pin* phenotype could be replicated by treatment of wild-type *Arabidopsis* with inhibitors of auxin transport; this suggested that PIN proteins participate in auxin efflux from cells. At least eight *PIN* genes have been isolated from *Arabidopsis*. A number of these have been implicated in tropic responses to light and gravity, responses that are known to involve auxin (see Chapters 8 and 15). Several of the PIN proteins are asymmetrically localized in the basal plasma membrane of cells where they are responsible for the polar efflux of auxin (Figure 10.7C).

A third set of auxin transporters is a group of the **ABC transporter** class of membrane proteins (see Chapter 5) that were discovered in loss-of-function mutants in *Arabidopsis*. Several members of the B subclass of the ABC transporter family have been shown to transport auxin, and they can function in IAA$^-$ efflux across the plasma membrane and tonoplast. Unlike PIN proteins they do not show asymmetrical distribution in cells so they are probably not involved in polar transport. Transporters of this class require energy in the form of ATP to transport IAA$^-$.

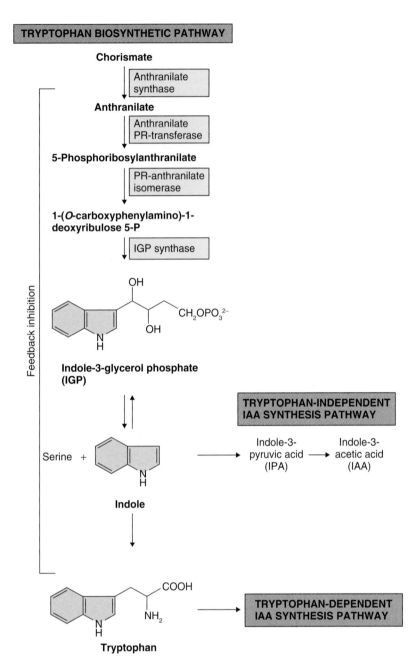

TRYPTOPHAN BIOSYNTHETIC PATHWAY

Figure 10.6 Two pathways for the synthesis of indole-3-acetic acid (IAA). Both pathways share a common set of reactions from chorismate to indole. At this point IAA may be synthesized from indole by the tryptophan-independent pathway via indole-pyruvic acid. Alternatively indole may be converted to tryptophan and subsequently to IAA via the tryptophan-dependent pathway.

The low pH of the apoplast, the permeability of the plasma membrane to IAAH and the dissociation of IAAH at the neutral pH of the cytosol lead to the trapping of IAA$^-$ in the cytosol. Polar transport of auxin results from the localization of AUX1 at the apical end of the cell, which brings about a directional influx of auxin, and the polar efflux of IAA$^-$ through PINs localized at the basal end of the cell. A significant driving force for the efflux of IAA$^-$ from the cell is the negative membrane potential of most cells, often between -200 and -300 mV.

10.2.3 The auxin receptor is a component of an E3 ubiquitin ligase

It takes less than 10 minutes for auxin to induce changes in growth of target tissue and some aspects of the mechanism by which auxin brings about such rapid changes are now well understood. At least two types of auxin receptor have been isolated, the auxin-binding

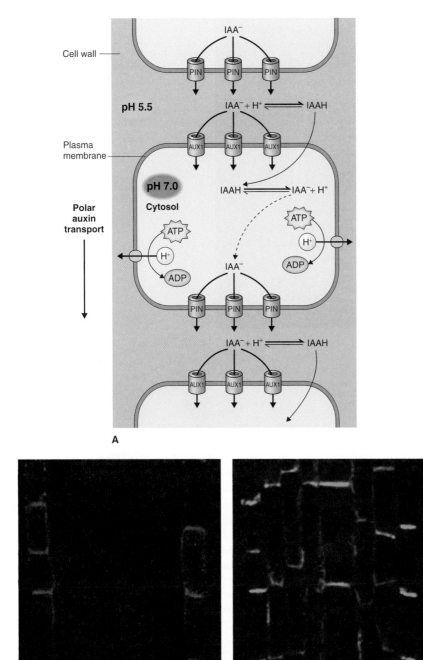

Cell wall

pH 5.5

Plasma membrane

pH 7.0

Cytosol

Polar auxin transport

A

B

C

Figure 10.7 Polar auxin transport. (A) Model showing membrane localized transporters that participate in the polar transport of auxin. The cell wall is relatively acidic at around pH 5.5 causing IAA to be protonated, but the cytoplasm has a pH of around 7 causing IAAH to dissociate into IAA$^-$ and H$^+$. IAAH can diffuse freely through the plasma membrane while IAA$^-$ requires a membrane transporter. The difference in pH between the wall and cytoplasm creates a gradient in IAA concentration that drives its movement into and out of the cell. Localized IAA$^-$ transporters establish auxin polar transport. The auxin influx carrier AUX1, located at the apical plasma membrane, allows transport of IAA$^-$ into the cell, whereas auxin efflux carriers of the PIN class that transport IAA$^-$ out of the cell are preferentially localized at the basal plasma membrane. (B, C) These carriers can be visualized in *Arabidopsis* root stele cells using immunofluorescence microscopy. This shows localization at the opposite ends of cells of the auxin influx carrier AUX1, stained red (B) and the auxin efflux carrier PIN1, stained green (C).

protein (**ABP**) and the F-box protein **TIR1** (Transport Inhibitor Response 1), a component of the AUX/IAA-SCF^{TIR1} E3 ubiquitin ligase complex that is described below. ABP has a KDEL/HDEL endoplasmic reticulum (ER) retention signal and is therefore a largely ER resident protein (see Chapter 5), although some ABP traffics to the plasma membrane. How ABP functions in either the ER or plasma membrane is not well understood, but it is known that the affinity of ABP for IAA is much stronger at an acidic pH. Since the lumen of the ER has a pH near neutrality and the extracellular face of the plasma membrane is about pH 5.5, it is likely that ABP functions as an IAA receptor only when it is transported to the plasma membrane.

Wild type

pin formed/pin 1

Figure 10.8 Phenotypes of the wild-type and *pin formed/pin1* double mutant of *Arabidopsis*. Note that the inflorescence stem of the mutant lacks discernible leaves or lateral buds, giving this mutant its pin-like phenotype. When this gene was cloned it was found to encode a transmembrane protein, PIN, that is now known to be an auxin efflux carrier.

Key points Auxins were the first class of hormones to be discovered in plants. With the exception of phenylacetic acid, all naturally occurring auxins such as IAA contain an indole ring, and they are synthesized from the amino acid tryptophan or from indole. As is the case for all hormones, mechanisms exist to break down auxins as well as to synthesize them. Polar transport is a fundamental property of auxins and it is another mechanism for regulating their concentration in cells. Differences in pH of the cell wall (pH 5.5) and cytoplasm (pH 7) are critical components of polar auxin transport. IAAH readily enters cells by diffusion, but IAA$^-$ requires specific membrane transporters to enter or leave cells. IAA$^-$ influx transporters are located at the apical ends of cells and efflux carriers are localized at the basal ends. Differences in membrane permeability to IAA and IAA$^-$ coupled with the polarized localization of influx and efflux transporters account for polar auxin transport.

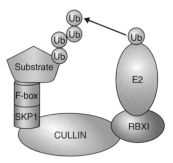

Figure 10.9 Model of a generic SKP1/CULLIN/F-box (SCF) E3 ligase. An SCF E3 ligase complex includes four proteins: CULLIN, SKP1, F-box and the RING finger protein RBX1. RBX1 binds an E2 ubiquitin conjugating enzyme. The F-box protein recognizes a target protein and positions it for ubiquitination by E2. Up to seven ubiquitin molecules (Ub) may be added to the target molecule, which is then degraded by the 26S proteasome. Target specificity is conferred by the F-box protein.

In contrast to ABP, a great deal is known about the role of the AUX/IAA-SCF^{TIR1} complex in auxin signaling. SCF complexes are members of the CULLIN/RING E3 ubiquitin ligase family of E3 ubiquitin ligases (see Figure 5.36C). These complexes are components of the ubiquitin–proteasome system (**UbPS**; see Chapter 5). **SCFs** are named for the three most important constituent subunits, SKP1, CULLIN and F-box protein (Figure 10.9) of the complex in

mammals; plants have homologous proteins. The F-box protein recognizes the substrate to be ubiquitinated. CULLIN forms a dimer with the RING domain protein RBX1 and CULLIN-RBX1 transfers ubiquitin to the target protein. The SCF complex involved in auxin signaling is localized in the nucleus and includes the SKP1 protein **ASK** (*Arabidopsis* SKP1-like), CULLIN and the F-box protein TIR1 (Figure 10.10).

Transcription of auxin-responsive genes is regulated by many factors that interact with their promoters, including transcription factors such as the auxin response factor (**ARF**) family of proteins. ARFs bind to specific auxin response elements in the promoters of

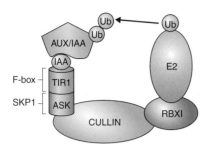

Figure 10.10   Model showing components of the AUX/IAA-SCF^{TIR1} E3 ligase complex involved in IAA perception and signaling. The F-box protein TIR1 is the IAA receptor and it forms a complex with the repressor proteins of the AUX/IAA class. ASK, the *Arabidopsis* SKP1 homolog, CULLIN and the RING domain protein RBX1 bring the AUX/IAA target proteins and E2 ligase together allowing AUX/IAA to be ubiquitinated (Ub) and subsequently broken down by the proteasome.

auxin-responsive genes and activate transcription of these genes. In the absence of auxin, the transcription of auxin-responsive genes is downregulated by transcriptional repressors called **AUX/IAAs**, which repress the activity of ARFs, and often by co-repressor proteins (Figure 10.11).

TIR1 functions as an IAA receptor. IAA initiates the expression of auxin-inducible genes by binding to TIR1. This allows TIR1 to bind AUX/IAAs and target them for ubiquitination by the SCF^{TIR1} complex and subsequent degradation by the 26S proteasome (Figure 10.11). Degradation of AUX/IAAs allows the transcription of auxin-responsive genes.

Key points   The auxin receptor TIR1 is a member of the F-box family of proteins and is a component of the ubiquitin E3 ligase SCF^{TIR1}. Auxin-responsive genes are regulated by transcription factors, including auxin response factors, whose activity is downregulated by transcriptional repressors called AUX/IAA. In the presence of auxin SCF^{TIR1} recognizes and binds AUX/IAA proteins allowing them to be ubiquitinated and subsequently degraded by the 26S proteasome. Removal of the AUX/IAA repressors allows auxin-responsive genes to be activated, leading to the auxin response.

10.3 Gibberellins

The gibberellins (GAs) belong to a large family of natural products produced by plants and fungi. They are tetracyclic diterpenes having 19 or 20 carbon atoms, and

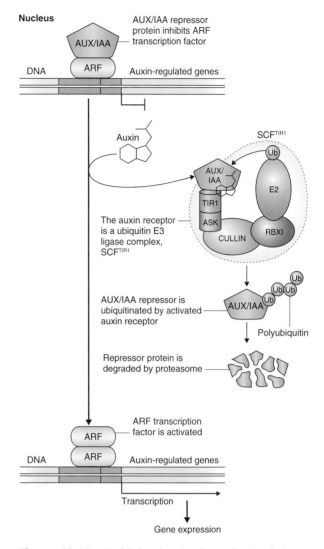

Figure 10.11   Model showing signal transduction during the response to auxin. In the absence of IAA, AUX/IAAs repress auxin-regulated genes by complexing with auxin response factors (ARFs). In the presence of IAA, AUX/IAA proteins become ubiquitinated by SCF^{TIR1} E3 ligase complex and are broken down by the proteasome. ARF can then function, often forming ARF–ARF dimers that allow the transcription of auxin-responsive genes.

to date 136 GAs have been isolated and characterized. The GAs are identified by numbers in the order of their discovery, i.e., GA$_1$, GA$_2$, etc. Not all GAs are biologically active in plants; the two most common active GAs are GA$_1$ and GA$_4$. Most other GAs isolated from plants and fungi are metabolic intermediates in the synthesis of active GAs or their degradation products.

Gibberellins were discovered in the 1920s by Japanese plant pathologists studying the 'foolish seedling' disease of rice (Figure 10.12). Foolish rice seedlings resulted from infection by the fungus *Gibberella fujikuroi* and

Figure 10.12 A rice (*Oryza sativa*) plant (white arrow) infected with the fungus *Gibberella fujikuroi* showing a highly elongated leaf and shoot.

they grew much taller than uninfected plants. It was quickly concluded that the infecting fungus produced a chemical that stimulated rice growth, and this was isolated and chemically characterized in the 1950s and called gibberellic acid (GA$_3$; see Figure 10.1).

The isolation of GA$_3$ led to work on its physiology, and results showing dramatic effects on plant growth generated immense interest in the GAs. The effects of GAs on shoot growth are clearly illustrated by the experiment shown in Figure 10.13A. *Arabidopsis* plants with mutation in the *GA1-3* gene have lower GA concentration and a very dwarf phenotype that can be

Figure 10.13 Mutations in gibberellin (GA) biosynthesis and signaling affect many aspects of plant growth and development. (A) The *ga1-3* mutant of *Arabidopsis* (center) lacks a key enzyme in GA biosynthesis and consequently has extremely low GA levels; mutant plants have a dwarf phenotype and delayed flowering. This mutant can be rescued by the application of gibberellic acid (GA$_3$) (right) and will grow to the same height as wild-type plants (WT) (left). (B) These wheat (*Triticum aestivum*) plants bear different alleles of the reduced height gene (*Rht*) that affects GA signaling. (C) This field of wheat shows the effect of wind and rain on plants of normal and reduced height: normal plants, on the left, have experienced lodging (flattening), while reduced height plants, on the right, are lodging-resistant.

A

B

C

rescued by spraying *ga1-3* plants with GA_3. In addition to having a profound effect on shoot growth, GAs control many developmental processes, including leaf growth, orientation and senescence, flowering, fruit and seed formation, seed germination and early seedling growth. Among the commercial uses of the GAs are the stimulation of fruit set and fruit growth in table grapes. Although the effects of GA on shoot growth suggested opportunities to increase plant biomass, technologies that reduce the GA content of plants have become far more valuable in increasing crop yield. Plants have been bred and engineered to reduce stem height and these new varieties have been instrumental in increasing grain yield, as seen in the **Rht** (reduced height) mutants of wheat (Figure 10.13B). Short straw wheat plants are often referred to as being part of the **Green Revolution**. These plants divert more photosynthate from stem growth to grain yield. They have the added trait of reduced lodging which occurs when wind causes wheat fields to be flattened and makes harvesting difficult or impossible (Figure 10.13C). The control of stature in cereal and other species is further discussed in Chapter 12.

10.3.1 The initial steps in gibberellin biosynthesis are similar to those for several other groups of hormones

Gibberellins are present in vegetative tissue at very low concentration, rarely more than 1 ng GA per gram of fresh tissue weight, which made unraveling their biosynthesis difficult. GAs are synthesized from isopentenyl diphosphate, which is synthesized in the chloroplast (see Chapter 15). Early steps in the synthesis of GAs are biochemically similar to those for abscisic acid, brassinosteroids and cytokinins (see Figure 10.2). The starting point for the synthesis of all these hormones is the same intermediate, namely the hemiterpene isopentenyl diphosphate (IPP) which isomerizes to dimethylallyl diphosphate (DMAPP; Figure 10.14). IPP and DMAPP are five-carbon compounds that can be synthesized by two distinct pathways using two different precursors. IPP destined for ABA and GA synthesis and DMAPP for the side-chain of CKs are synthesized in the plastid along the DOXP pathway (see Figure 15.23), whereas the IPP required for BRs is synthesized along the mevalonic acid (MVA) pathway in the cytosol (Figure 10.14).

The immediate diterpene precursor of GA is **geranylgeranyl diphosphate** which is converted to **ent-kaurene** in a two-step reaction also occurring in the plastid (Figure 10.15). The conversion of *ent*-kaurene to

kaurenoic acid and GA_{12} is catalyzed by two multifunctional enzymes of the cytochrome P450 class. It begins on the plastid envelope and is completed in the lumen of the ER. GA_{12} can follow one of two cytosolic pathways to bioactive GAs. In the **13-hydroxylation pathway** GA_{12} is converted to GA_{53}, then by a series of steps to bioactive GA_1. The **non-13-hydroxylation pathway** from GA_{12} leads to GA_4 (Figure 10.15). The immediate precursors of GA_1 and GA_4 are inactive GA_{20} and GA_9 respectively; both become active when hydroxylated at carbon 3 by the soluble dioxygenase enzyme GA-3 oxidase (GA3ox). Hydroxylation of GA_1 and GA_4 at carbon 2 is catalyzed by GA-2 oxidase (GA2ox), also a soluble dioxygenase, leading to inactivation of these hormones by forming GA_8 from GA_1 and GA_{34} from GA_4 (Figure 10.15).

10.3.2 Gibberellin concentration in tissues is subject to feedback and feed-forward control

Although GAs are present in plant tissues at extremely low levels, changes in their concentration must be strictly controlled if they are to be effective as regulatory hormones. The levels of GAs are subject to **feed-forward** and **feedback** mechanisms that bring about hydroxylation at the three- and two-carbon atoms of the GA molecule. In two examples of feedback regulation, the GA20ox and GA3ox that convert GA_{12} to GA_9, and GA_9 and GA_4, respectively, are inhibited by high concentrations of bioactive GAs (Figure 10.16). Sequential hydroxylation at carbons 20 and 3 of GA_{12} leads to the formation of active GA_4 and GA_1. When the levels of bioactive GAs are high, plants show a strong response to the presence of GA, and the synthesis of GA-20 and GA-3 oxidases is inhibited, thereby blocking the synthesis of additional active GAs. The dwarfing gene *le* studied by Gregor Mendel in *Pisum sativum* (see Figure 12.50) is now known to encode a GA3ox. Mutations in *Le* bring about a reduction in the amount of bioactive GA_1 and result in a dwarf phenotype, establishing a causal relationship between GA_1 concentrations and stem growth in this species.

It was thought that only GA_4 and GA_1, the principal bioactive GAs, were hydroxylated at carbon 2, but we now know that earlier GA precursors can also be hydroxylated at carbon 2. This converts them to inactive GAs that cannot become active GAs when further metabolized. An enzyme that oxidizes the double bond at carbon 16,17 also brings about inactivation of GAs. This reaction is an epoxidation and is restricted to those

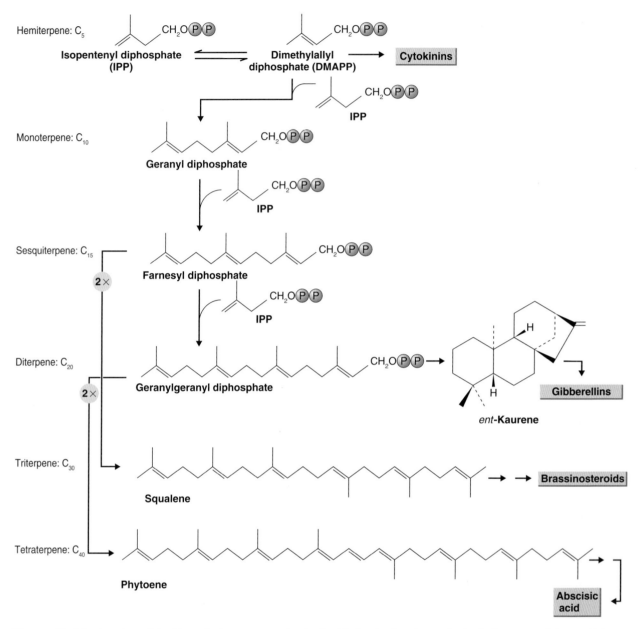

Figure 10.14 Overview of the biosynthetic steps occurring in plastids that lead to the synthesis of all or parts of the cytokinins, gibberrelins, brassinosteroids and abscisic acid.

GAs such as GA_9 and GA_4 that lack OH at carbon 13; 13-hydroxy GAs such as GA_{53}, GA_{20} and GA_1 are not inactivated by this pathway.

10.3.3 The gibberellin receptor GID1 is a soluble protein that promotes ubiquitination of repressor proteins

Molecular genetic approaches have identified the GA receptor in plants and the mechanism by which GAs regulate responses at the transcriptional level. The GA receptor was first identified in a rice mutant that was insensitive to GA. The mutant was called Gibberellin Insensitive Dwarf1 (GID1) and the gene responsible for this phenotype was called **GID1**. Only one copy of *GID1* is present in rice. Purification of the GID1 protein showed that it has a high affinity for biologically active GAs but not for inactive GA_8 and GA_{34}, providing strong evidence that it is a GA receptor. *GID1* is also found in *Arabidopsis* but in this species there are three homologs of the gene.

As is the case with auxin perception and signaling, the ubiquitin–proteasome system (UbPS) pathway is a key

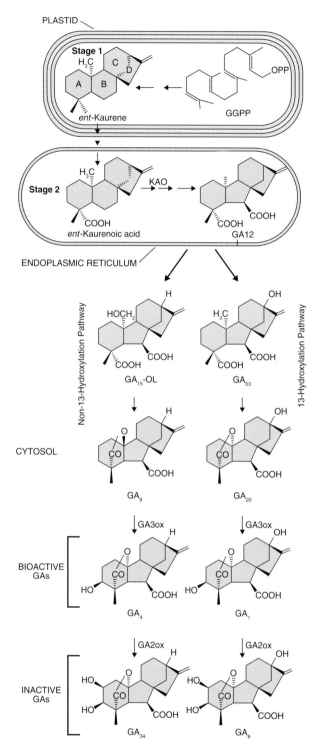

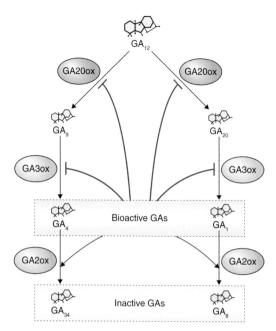

Figure 10.16 Biosynthesis of the bioactive gibberellins, GA_4 and GA_1, is regulated by feedback (⊢) and feed-forward (→) mechanisms. The activities of GA20ox and GA3ox are downregulated by bioactive GA_4 and GA_1, an example of a feedback-regulated step. On the other hand, the activity of GA2ox, which produces inactive GA_{34} and GA_8, is upregulated by these bioactive GAs, an example of feed-forward regulation.

Figure 10.15 Gibberellin biosynthesis showing the major steps in the pathways and their subcellular localization. The 13-hydroxylation pathway produces bioactive GA_1 and the non-13-hydroxylation pathway produces bioactive GA_4. GGPP, geranylgeranyl pyrophosphate); KAO, kaurenoic acid oxidase; PPO, pyrophosphate.

Key points Gibberellins belong to a large family of diterpenes synthesized by fungi and plants. The dramatic effects of GAs on stem growth initially stimulated agronomic interest in them but now there is more interest in mechanisms that bring about a reduction in the GA response; for example, in cereal plants with short stems that were key components of the Green Revolution. Although more than 130 GAs have been identified, only a very few, such as GA_1 and GA_4, are biologically active. Synthesis of bioactive GA is finely regulated by enzymes that are under feedback and feed-forward control. High concentrations of active GA_1 and GA_4 inhibit the enzymes that bring about their synthesis by feedback inhibition and simultaneously induce the synthesis of enzymes that bring about their inactivation. This combination of feedback inhibition and feed-forward induction of enzymes ensures that GAs turn over very rapidly and that GA signaling is turned off in the absence of continued synthesis of the hormone.

component of GA signal transduction. The basic elements of GA signaling via the receptor GID1 that lead to proteolysis of repressor proteins are illustrated in Figure 10.17. Unlike the auxin receptor TIR1, GID1 is a soluble, nuclear protein and is not part of the SCF E3 ligase complex. The F-box protein involved in GA signaling is called GID2 in rice and SLY1 in *Arabidopsis*. Thus the E3 ligase for GA signaling is SCF$^{SLY1/GID2}$. In the presence of bioactive GA and the receptor GID1, SLY1/GID2 recognizes a class of repressor proteins called **DELLA domain** proteins and this results in their ubiquitination and proteolysis by the proteasome (Figure 10.17).

Although DELLA domain proteins are referred to as repressor proteins, the transcription of some genes is promoted by DELLAs. Furthermore, DELLA domain proteins may not act directly as repressors, but function by interacting with other proteins to bring about repression or de-repression of genes indirectly. Examples of such interactions are shown in Figure 10.17. DELLA domain proteins have been shown to affect a suite of genes related to the synthesis of GA as well as those controlling growth. In addition to their involvement in GA responses, DELLA proteins are known to mediate a wide variety of plant responses to light and other abiotic factors including temperature. DELLA domain proteins interact with **PIF** (phytochrome-interacting factor) proteins leading to the regulation of transcription of both GA- and light-regulated genes (see Chapters 8 and 12). Treatment of plants with GA brings about degradation of DELLAs by the UbPS, which in turn leads to the expression or repression of genes which results in the promotion of a wide variety of GA responses including altered growth and development (see Figure 10.13), synthesis and secretion of hydrolytic enzymes by the cereal aleurone (see Figure 6.32), breaking of dormancy (see Chapters 6 and 17) and induction of flowering (see Chapter 16).

10.4 Cytokinins

Cytokinins (CKs) were discovered in the 1930s as compounds that, in concert with auxin, would support cell division and shoot development in plant tissue grown in sterile culture (Figure 10.18). When plant cells are grown on a medium that contain both CK and IAA, they form masses of undifferentiated cells called callus. If callus is transferred to a medium with high CK and low auxin concentration, it produces shoots. If transferred to a medium containing only auxin, callus produces roots. The first CK, called **kinetin**, was isolated from autoclaved herring sperm DNA. Subsequently several CKs have been isolated from plant tissue. Plant tissue culture became

Key points Gibberellin-responsive genes fall into two classes, those that are activated by GA and those repressed by the hormone. Activation and repression of GA-responsive genes is brought about by a family of transcriptional activator and repressor proteins, called DELLA domain proteins. The GA receptor GID1 is a soluble nuclear protein that binds GA and interacts with a SLY F-box protein, a component of an SCF^{SLY} complex. The SCF^{SLY}/GID1/GA complex recognizes and binds DELLA domain proteins leading to their ubiquitination and subsequent proteolysis by the proteasome. Degradation of DELLA proteins allows transcription of GA-responsive genes to be regulated.

the basis of the standard bioassay in the search for compounds that would stimulate shoot development, and compounds that were active in this bioassay were called cytokinins. CKs have a wide range of effects on plant growth and development. They play roles in shoot and root organogenesis, control lateral bud growth, and regulate senescence and seed germination.

Kinetin and benzyladenine, another synthetic cytokinin (Figure 10.19), do not occur naturally in plants but a range of natural CKs have now been isolated including isopentenyl adenine (**iPA**), trans-zeatin (**tZ**), cis-zeatin (**cZ**) and dihydrozeatin (**DZ**) (Figure 10.19). With the exception of diphenylurea, all native and synthetic CKs are derivatives of the purine base adenine.

The naturally occurring CKs fall into two groups based on the side-chain at the 6 position of the purine moiety. The most abundant naturally occurring CKs, including iPA, tZ, cZ and DZ, have an **isopentenyl** side-chain at this position; other less abundant CKs have an aromatic group (Figure 10.19). Of the isoprenoid CKs, iPA and tZ are more active and generally more abundant than cZ or DZ; however, the former are more susceptible to degradation by the enzyme cytokinin oxidase (see Section 10.4.2) than either cZ or DZ. The aromatic CKs are much less abundant than the isoprenoid CKs and less is known about their physiological role.

10.4.1 Cytokinin biosynthesis takes place in plastids

The biosynthesis of tZ is shown in Figure 10.20. The transfer of an isopentenyl side-chain from dimethylallyl diphosphate (DMAPP) to ATP or ADP is catalyzed by the enzyme **isopentenyl transferase** (**IPT**). This step of tZ biosynthesis occurs in the plastid because DMAPP is

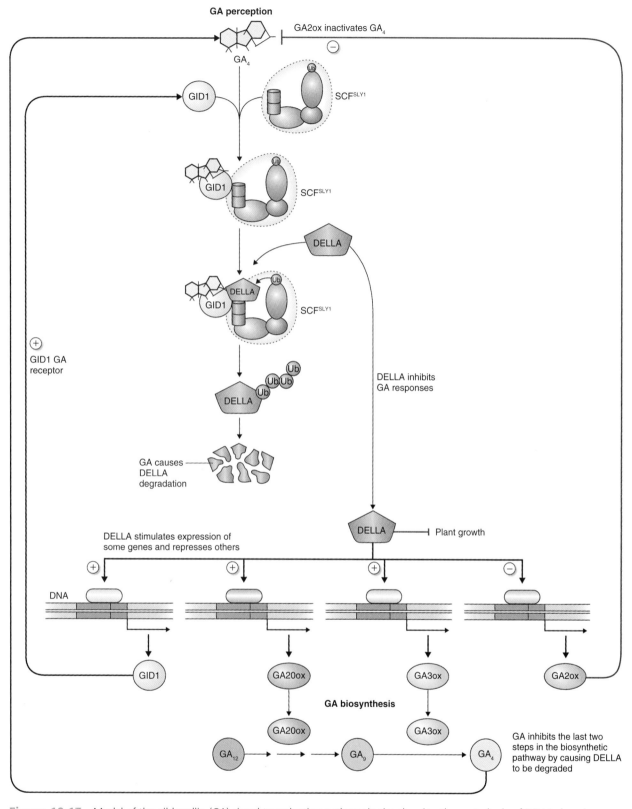

Figure 10.17 Model of the gibberellin (GA) signal transduction pathway in rice showing the central role of DELLA domain proteins as repressors of transcription of GA-regulated genes. DELLA proteins are degraded by the proteasome after they are ubiquitinated by the E3 ligase SCF^{SLY1}. GA binds to the GA receptor GID1. The hormone–receptor complex then interacts with SLY1, the F-box protein of SCF^{SLY1} and together they bind DELLA domain proteins allowing them to be ubiquitinated for subsequent proteolysis by the proteasome. Removal of DELLA proteins allows the transcription of GA-regulated genes to proceed. Among the genes that are upregulated by DELLA are those involved in GA perception (*GID1*) and biosynthesis (*GA20ox* and *GA3ox*). An example of a gene downregulated by DELLA domain proteins is *GA2ox*.

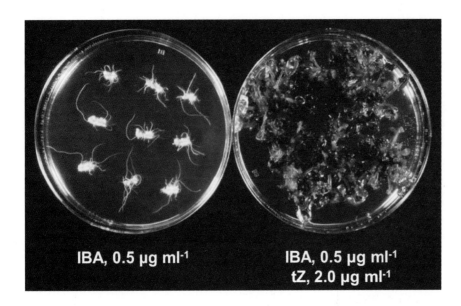

Figure 10.18 The production of roots and shoots by *Arabidopsis* callus tissue. Callus subcultured on a medium containing the auxin indole butyric acid (IBA) produces only roots (left); but on a medium containing a high ratio of the cytokinin trans-zeatin (tZ) to IBA it produces shoots (right).

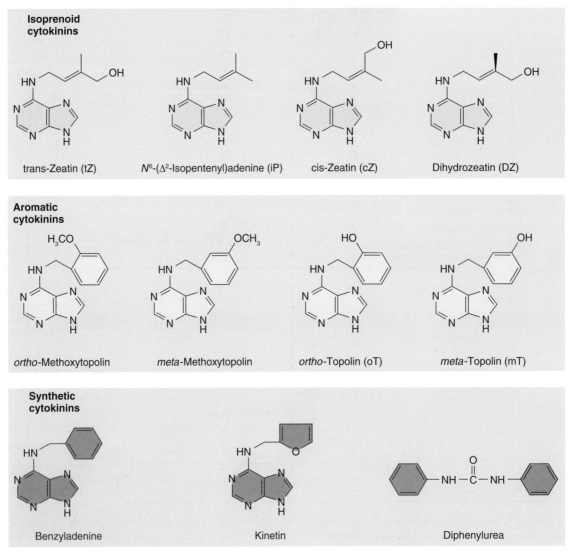

Isoprenoid cytokinins

trans-Zeatin (tZ) N^6-(Δ^2-Isopentenyl)adenine (iP)   cis-Zeatin (cZ)   Dihydrozeatin (DZ)

Aromatic cytokinins

ortho-Methoxytopolin   *meta*-Methoxytopolin   *ortho*-Topolin (oT)   *meta*-Topolin (mT)

Synthetic cytokinins

Benzyladenine   Kinetin   Diphenylurea

Figure 10.19 Structures of some naturally occurring and synthetic cytokinins. The naturally occuring cytokinins may be divided into two categories, isoprenoid cytokinins and aromatic cytokinins, based on the side-chain attached to the 6 position on the purine base adenine. Synthetic cytokinins such as benzyladenine are also 6-substituted purines, but diphenylurea lacks a purine ring structure.

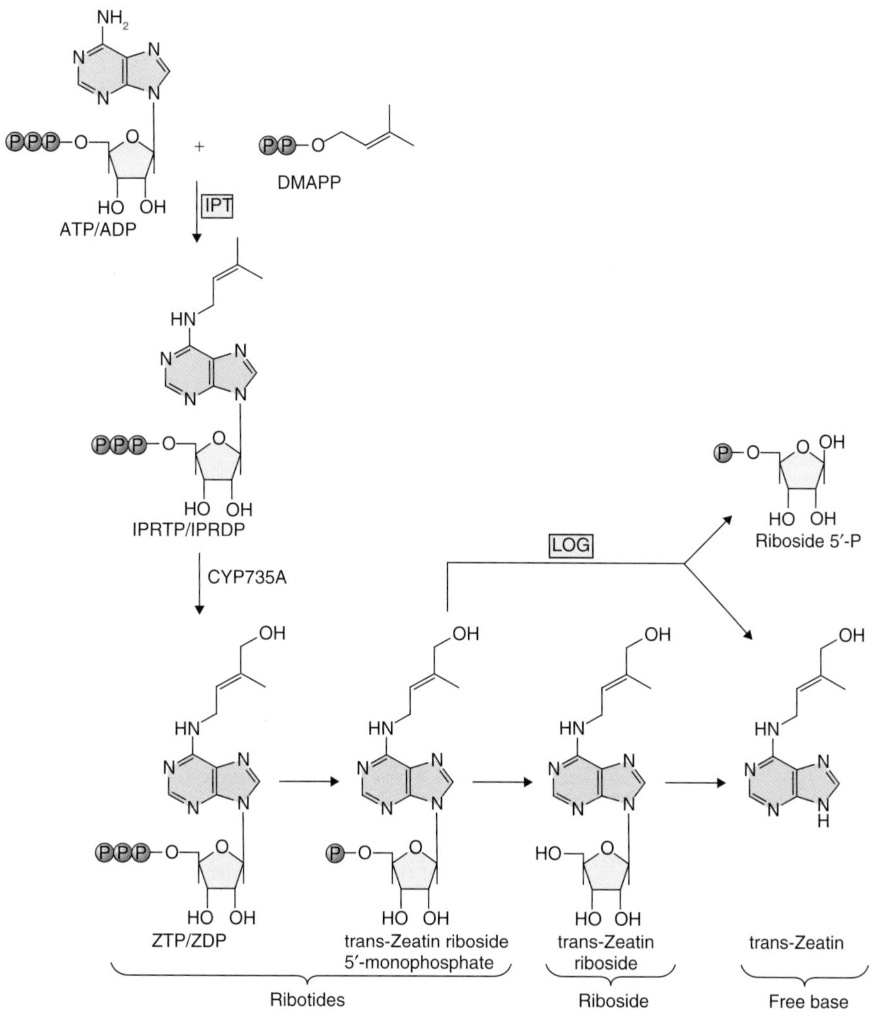

Figure 10.20 Major steps in the cytokinin biosynthetic pathway. CK biosynthesis begins with the addition of an isopentenyl side-chain from dimethylallyl diphosphate (DMAPP) to ATP or ADP. The product of this reaction is converted to zeatin triphosphate by a cytochrome P450 enzyme, followed by sequential removal of two additional phosphate groups and ribose to produce trans-zeatin. A single step formation of trans-zeatin can occur catalyzed by LOG, the product of the *LONELY GUY* gene in *Arabidopsis*.

synthesized in plastids (see Figure 10.14) and IPT enzymes are localized there. *IPT* genes constitute a small family (seven in *Arabidopsis* that are involved in CK synthesis). Mutations in *IPT* genes result in phenotypes with altered branching, enhanced root growth and reduced levels of iPA and tZ. The expression of some *IPT* genes is downregulated by CKs, suggesting a feedback loop in CK synthesis similar to that found in the GA biosynthesis pathway. The pattern of *IPT* gene expression in plant tissues indicates that the synthesis of CKs occurs in the apical region of shoots and roots as well as in developing seeds.

The next step in tZ biosynthesis is the hydroxylation of the isopentenyl side-chain by a member of the cytochrome P450 monooxygenase (**CYP**) family of enzymes. These reactions occur largely on the membranes of the ER and are regulated by several hormones, indicating that CYPs are points of cross-talk between the various hormone-related pathways. For example, CKs bring about the upregulation of *CYP* genes whereas IAA and ABA reduce the level of their

expression. CKs are biologically active only as the free base. The formation of the free base from the ribotide, trans-zeatin riboside-5′-monophosphate, can occur in a single step or in a two-step process. The single-step formation of a free base removes the phosphorylated ribose from the ribotide, releasing free tZ and is catalyzed by the product of the *LONELY GUY* gene in *Arabidopsis*. The two-step process requires hydrolytic cleavage of the phosphate from the ribotide by a phosphohydrolase to form the riboside followed by removal of ribose by a nucleosidase (Figure 10.20).

The possibility that CKs can be produced by the hydrolysis of tRNA has been widely explored. Specific IPT activities that can transfer isopentenyl side-chains to particular adenine residues in tRNA have been found in animal and plant cells. It has also been shown that if purified tRNA is hydrolyzed, the CK released shows biological activity. Whether tRNA contributes a physiologically important amount of CK to the endogenous pool of hormone has not been resolved.

10.4.2 There are two pathways for cytokinin breakdown

As with the other hormones discussed so far, regulation of the breakdown of CKs is key to their function as signaling molecules. There are two basic pathways by which CKs are catabolized. The first occurs by oxidation by **CKX** (**cytokinin oxidase**), a FAD oxidoreductase enzyme, which cleaves the isopentenyl side-chain of CK and releases adenine and 3-methylbutenal (Figure 10.21A). Evidence that a reduced activity of CKX causes developmental effects has been found in experiments on rice, where there is natural variation in CKX expression. Rice plants with reduced CKX expression show much larger panicle size and consequently higher grain yield (Figure 10.21B).

CK can also be inactivated by the addition of sugars or amino acids. Glucosylation can occur either at N-7 or N-9 positions of the adenine moiety, or on the OH group of the isopentenyl side-chain. N-glucosylation is

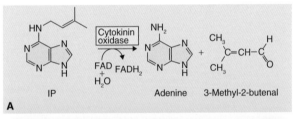

Figure 10.21   (A) Cytokinin oxidase is a flavin-containing oxidase that irreversibly degrades many cytokinins, including isopentenyladenine (IP) as shown here. (B) Wild-type rice (*Oryza sativa*) (left) and a mutant with lowered expression of cytokinin oxidase (right). Note that the mutant has a larger panicle size and consequently larger grain yield.

irreversible and results in inactivation of CK but O-glucosylation can be reversed. Alanine can be added to the N-9 position; this reaction is also reversible. Reversible modification of CK by the addition of sugars or amino acids may be an important mechanism for storage, since these modified CKs are located in the vacuole. Free CK can be released from the storage form, for example during seed germination.

10.4.3 The cytokinin receptor is related to bacterial two-component histidine kinases

Molecular genetic approaches have provided a comprehensive model of how CKs influence development at the cellular level, but several aspects of this pathway remain to be elucidated. The CK receptor is related to **bacterial two-component histidine kinase** family receptor proteins (Figure 10.22A). These receptors consist of a **histidine kinase sensor** with input and transmitter domains, and a **response regulator** with receiver and output domains. Ligand binding to the input domain of the sensor results in activation of its transmitter domain and autophosphorylation of the His residue. The sensor then transphosphorylates an Asp residue on the response regulator, and in bacteria this leads to activation of transcription.

In plants the CK receptor is more complex and is called a **phosphorelay two-component system** (Figure 10.22B). In this case a hybrid sensor has both sensor and receiver domains that contain His and Asp domains, respectively. Following autophosphorylation, the sensor transfers a phosphoryl group to a histidine phosphotransfer protein (Hpt), called **AHP** in *Arabidopsis*, which in turn transfers the phosphoryl group to an Asp residue on a response regulator, called **ARR** (*Arabidopsis* **response regulator**). Figure 10.23 shows the basic outlines of the CK signal transduction pathway. Three CK receptors have been isolated from *Arabidopsis*: CRE1 (Cytokinin Receptor1), AHK2 (*Arabidopsis* Hybrid sensor Kinase2) and AHK3. The CK receptor has a membrane-spanning domain and functions as a dimer. Although it has been proposed that this receptor is localized to the plasma membrane, more recent evidence suggests that it may be found in the ER. After CK binds to the receptor, a phosphoryl group is transferred from the receptor to one of many cytoplasmically localized AHP proteins. At least five authentic *AHP* genes have been identified in *Arabidopsis*

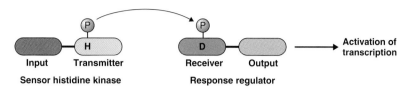

A Simple two-component signaling system

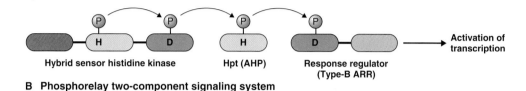

B Phosphorelay two-component signaling system

Figure 10.22   Models of simple and complex two-component histidine kinase systems. (A) In the simple two-component system the histidine kinase responds to the presence of a signal by autophosphorylation on a His (H) residue of its transmitter domain. The phosphoryl group is transferred to an Asp (D) residue on the receiver domain of the response regulator. (B) In the more complex phosphorelay model the histidine kinase receptor has both sensor and receiver domains that become sequentially phosphorylated. The phosphoryl group is then transferred from the histidine kinase to a histidine phosphotransferase (Hpt, called AHP in *Arabidopsis*), and finally to the response regulator ARR (*Arabidopsis* response regulator).

and they appear to play overlapping roles in CK signaling because deleting one or more of these genes does not entirely abolish responses to CK. Triple and quadruple mutants show a much reduced response to CK, suggesting that these His phosphotransfer proteins may function additively.

AHPs transfer their phosphoryl group to two potential targets, Type A and B ARRs, each of which can regulate subsets of the cell's response to CK; the ARRs are activated and stabilized by this phosphorylation. Eleven Type B ARRs have been identified in *Arabidopsis* and these are all transcriptional activators that affect CK homeostasis, shoot and root development, cell expansion and senescence. Type B ARRs also regulate the transcription of Type A ARRs.

Phosphorylated Type A ARRs negatively regulate their own phosphorylation by AHPs causing a feedback loop. Type A ARRs are represented by ten genes in *Arabidopsis*, and loss-of-function mutants have shown that eight of these ARRs negatively regulate CK signaling, although the precise mechanism by which they function is not known. *Arabidopsis* plants with mutations in Type A ARRs 3, 4, 5, 6, 8 and 9 show enhanced expression of CK primary response genes, indicating that these six ARRs play a role in suppressing CK signaling. At least two Type A ARRs act to positively regulate CK signaling, especially the effects of this hormone on circadian rhythms and phytochrome function.

Cytokinin response factors (**CRFs**) are another family of proteins that helps mediate the CK signal in cells (Figure 10.23). There are six *CRF* genes in *Arabidopsis* and the expression of three of these is enhanced by Type B ARRs. CRFs accumulate in the nucleus following CK treatment. Measurement of gene expression indicates that nuclear localization is required for CRF function, and deletion of these genes results in the reduced expression of a large number of CK-regulated genes.

Key points Cytokinins control cell division but also play a role in a range of other developmental processes including bud growth, senescence and seed germination. Most CKs are adenine derivatives with an isopentenyl group on carbon 6 of the purine moiety. Like all other hormones their synthesis and degradation are finely controlled by mechanisms including reversible inactivation by conjugation to sugars or amino acids and irreversible inactivation by the enzyme cytokinin oxidase. CK receptors are membrane-localized phosphorelay proteins. They initiate a phosphorelay in which the phosphoryl group on the receptor is transferred to a family of phosphotransfer proteins that in turn transfer the phosphoryl group to members of one of two groups of response regulator proteins called ARRs. Phosphorylated type A ARRs induce the transcription of cytokinin-responsive genes, among which are genes encoding type B ARRs. Type B ARRs in turn can cause feedback inhibition of the transcription induced by type- A ARRS as well as inducing their own set of CK-dependent processes.

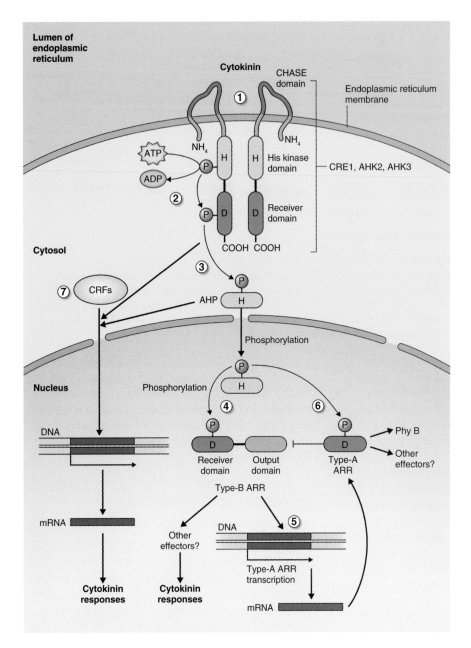

Figure 10.23 The cytokinin signal transduction pathway. (1) CK binds to the CHASE domain of its receptor (CRE1, AHK2, AHK3) at the ER, activating the histidine kinase domain of this protein (2). The phosphoryl group is transferred from the histidine kinase domain of CRE1 to the receiver domain and (3) it in turn transfers phosphate to histidine phosphotransfer protein AHP. (4, 6) AHP moves to the nucleus where it transfers its phosphoryl group to Type A and B *Arabidopsis* response regulators (ARRs). Phosphorylation of the type- B ARR leads to (5) transcription of type A **ARR** genes. ARRs bring about induction of the various CK responses. (7) Cytokinin response factors (CRFs) can also be activated by CRE1 or AHP. Activated CRFs move to the nucleus where they trigger transcription of cytokinin-responsive genes.

10.5 Ethylene

The role of ethylene in the regulation of plant growth and development was discovered in the early part of the 20th century when it was observed that street trees lost their leaves prematurely when they were growing near gas lamps, which release ethylene, a component of natural gas. Even though it was known as early as the mid-1930s that plants had the ability to produce ethylene, it was not until the advent of gas chromatography and the ability to measure low ethylene concentrations that plant biologists recognized that ethylene is an endogenous regulator of plant development.

Ethylene affects many aspects of plant development. In addition to its influence on abscission of leaves and other organs, it stimulates fruit ripening, promotes germination and affects cell expansion (see Chapters 12 and 18). Changes in cell expansion brought about by exposure to ethylene were described in the mid-19th century when the symptoms now known as the **triple response** were described (Figure 10.24). The response in seedlings includes bending and thickening of the shoot and pronounced curling of the apical hook. The triple response is a diagnostic of the response of plants to ethylene, a phenotype that has proven invaluable for selecting mutants in the ethylene signaling pathway.

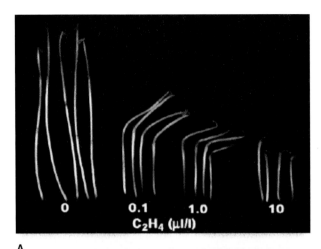

A

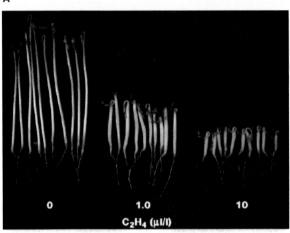

B

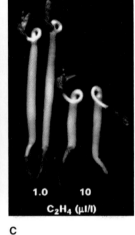

C

Figure 10.24 The triple response to ethylene of dark-grown pea (*Pisum sativum*) and mung bean (*Vigna radiata*) seedlings exposed to ethylene. (A) Control pea seedlings (0) and pea seedlings treated with increasing ethylene concentrations. Ethylene-treated plants have shorter, thicker epicotyls and, at intermediate ethylene concentration, the epicotyl grows horizontally rather than vertically. (B, C) Mung bean seedlings treated with ethylene show the classic triple response: inhibition of hypocotyl and root elongation, increase in hypocotyl width and extreme curvature of the apical hook.

10.5.1 Ethylene, a simple gas, is synthesized in three steps from methionine

Ethylene is a simple, gaseous, unsaturated hydrocarbon with the empirical formula C_2H_4 (see Figure 10.1). It is synthesized by almost all plant tissues, but its rate of synthesis is enhanced by wounding and other stresses and during fruit ripening, senescence and abscission (see Chapter 18). Not all ripening fruits produce large amounts of ethylene; indeed, this feature is limited to those fruits referred to as **climacteric** fruits. These fruits undergo a burst in respiration as they ripen, called the **respiratory climacteric**, which is stimulated by ethylene. Examples of fruits showing a respiratory climacteric are shown in Table 18.2. Tropical fruits, such as banana, mango and papaya, which are imported into countries in North America and Europe, are mostly climacteric fruits and can over-ripen during transportation from the tropics. As we discuss below, there has been considerable

research on strategies to postpone ripening, particularly focused on ethylene.

Ethylene is synthesized from methionine in three enzyme steps: **adenosyl methionine synthetase** which produces *S*-adenosyl methionine (**SAM**) from methionine; **ACC synthase** (**ACS**) which produces **ACC** (1-aminocyclopropane-1-carboxylic acid); and **ACC oxidase** (**ACO**) which produces ethylene as well as CO_2 and HCN from ACC (Figure 10.25). ACC synthase and ACC oxidase are highly regulated enzymes in vivo. The expression of ACS genes is stimulated by auxin, fruit ripening, wounding and several abiotic stresses, especially flooding, and ACO genes are strongly upregulated in ripening climacteric fruits.

ACS and *ACO* are represented by large multigene families. Tomato has at least ten *ACS* genes that can be subdivided into groups depending on whether their expression is regulated by auxin, ripening, wounding, abiotic stress, etc. ACS activity can also be regulated by alterations in its stability, for example by phosphorylation at the carboxyl terminus of the protein, which reduces its turnover by the 26S proteasome.

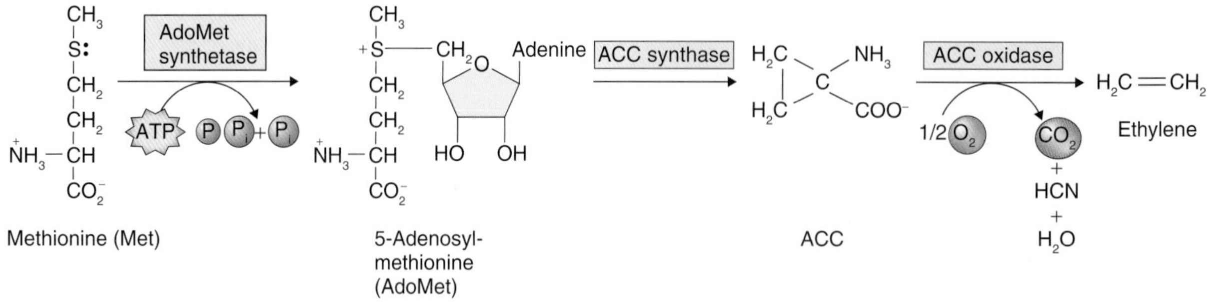

Figure 10.25 Biosynthesis of ethylene from the amino acid methionine. ACC, 1-aminocyclopropane-1-carboxylic acid.

Whereas specific mechanisms exist for the breakdown of the major plant hormones, ethylene is a notable exception. Ethylene quickly diffuses from sites of synthesis, meaning that specific catabolic reactions are not required to break it down. Nevertheless, catabolic reactions that can breakdown ethylene are found in plant tissues; however inhibition of these reactions does not affect the response of tissues to ethylene. These experiments support the conclusion that catabolism does not play a role in ethylene homeostasis. Rather, diffusion of the gas from its source is sufficient to remove the signal.

10.5.2 Ethylene receptors have some characteristics of histidine kinase response regulators

Ethylene receptors were identified in *Arabidopsis* plants screened for their insensitivity to the gas. A mutant isolated in this screen grew normally in the presence of ethylene and did not show the triple response; it was designated *etr1* (ethylene response1; Figure 10.26). Isolation and sequencing of the *etr1* gene showed it had similarities to the bacterial histidine kinase two-component response regulator; it was the first of this type of hormone receptor to be identified in plants. We now know that the CK receptor (see Figures 10.22 and 10.23) is also a member of this class of receptor. Other ethylene receptor genes have now been identified including *ERS2* (ethylene response sensor2), which is a homolog of *ETR1, ETR2* and *EIN4* (ethylene insensitive4). All of these ethylene receptors have an amino-terminal membrane-spanning domain that binds ethylene and a histidine kinase domain in the carboxyl half of the molecule. ETR1 and ERS1 have functional histidine kinase domains, whereas EIN4, ETR2 and ERS2 lack histidine kinase activity.

Ethylene receptors in *Arabidopsis* are thought to interact with each other in large complexes located in the membrane of the ER. The binding of **copper** is required for ethylene receptor function. In the absence of ethylene, ethylene receptors inhibit the ethylene response pathway. When ethylene receptors bind ethylene, they are inactivated and the response pathway is no longer inhibited. In other words ethylene receptors are 'on' in the absence of ethylene and 'off' in its presence.

10.5.3 Homologs of MAP kinases transmit the ethylene signal from the receptor to target proteins

Mitogen-activated protein kinases (**MAP kinases**) are a family of highly conserved eukaryotic proteins that transmit a cascade of signals, often from receptors, protein kinases or GTPases at the plasma membrane, to intracellular targets. While signals from receptors are short-lived, the phosphorylation cascade via MAP kinases is long-lived and widely broadcast throughout the cell by the many homologs of the proteins that belong to this protein kinase family. MAP kinases are serine/threonine protein kinases. *Arabidopsis* has 90 genes that encode this class of kinase. MAP kinases belong to three functionally distinct families that operate in series. **MAPKKK** (MAP kinase kinase kinase, for which there are 60 genes in *Arabidopsis*) becomes phosphorylated by upstream kinases such as the RAF-like kinase CTR1 in ethylene signaling; then MAPKKK phosphorylates **MAPKK** (MAP kinase kinase, ten *Arabidopsis* genes). MAPKK phosphorylates **MAPK** (MAP kinase, 20 *Arabidopsis* genes) on both threonine and tyrosine. MAPK is thought to be the last kinase in the phosphorylation cascade, as shown in Figure 10.27. Examples of the targets for this signaling cascade are transcription factors that are involved in stomatal guard cell functioning, the regulation of cell division, defense against pathogen attack and in responses to various abiotic stresses.

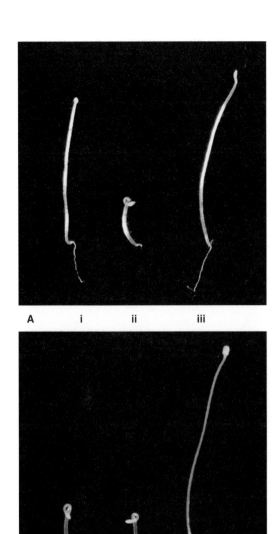

A   i   ii   iii

B   i   ii   iii

Figure 10.26 Wild-type and mutant *Arabidopsis* seedlings grown in the dark in the absence or presence of ethylene. (A) In the absence of ethylene, wild-type seedlings (i) and those of the ethylene-insensitive mutant *ein2-1* (iii) grow normally, but the *ctr1-1* mutant (ii) exhibits a constitutive triple response in the absence of ethylene, including reduced hypocotyl elongation, swelling of the hypocotyl and pronounced curling of the apical hook. (B) When seedlings of these three genotypes are grown in the presence of ethylene, both wild-type (i) and *ctr1-1* (ii) seedlings show the triple response but the ethylene insensitive *ein2-1* mutant (iii) grows as if ethylene were not present. Mutants such as these led to the identification of the ethylene receptor and the key steps in the ethylene signal transduction pathway.

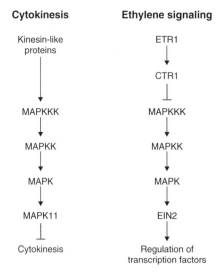

Figure 10.27 MAP kinase (MAPK) cascades are involved in regulating many aspects of plant growth and development, including cytokinesis and the response to ethylene in *Arabidopsis*. During regulation of cytokinesis (left) a kinesin-like protein is thought to be upstream of a MAPK cascade that targets protein kinases such as MAPK11, a kinase that inhibits cell division. For ethylene signaling (right), the ethylene receptor ETR1 and another kinase, CTR1, are thought to be upstream of the MAPK cascade and EIN2, a transmembrane protein in the ethylene signal transduction pathway, is downstream of this cascade. MAPKK, MAP kinase kinase; MAPKKK, MAP kinase kinase kinase.

MAP kinases are thought to function downstream of the ethylene receptor ETR1 (Figure 10.28) and a second protein kinase, called **CTR1** (Constitutive Triple Response1). The *ctr1* gene was identified in a genetic screen and, as the name suggests, plants carrying this mutation display a triple response in the absence of ethylene (see Figure 10.26). Because the *ctr1* mutation gave rise to the ethylene response phenotype in the absence of ethylene, the wild-type gene was proposed to be a negative regulator of ethylene responses. CTR1 is thought to interact directly with ETR1.

Ethylene brings about a change in the expression of a suite of genes that contain ethylene response elements, including those that regulate fruit ripening. It does this by affecting at least two sets of transcription factors, **EIN3** and **ERF1**. The link between CTR1 and the transcription factors that recognize ethylene response elements is **EIN2**, another protein with transmembrane motifs whose localization and function is unknown. Ethylene receptors are localized on the ER membrane with their receiver domains directed to the ER lumen. How information is transferred from CTR1 to EIN2, another membrane-localized protein, and then to transcription factors in the nucleus, remains to be resolved.

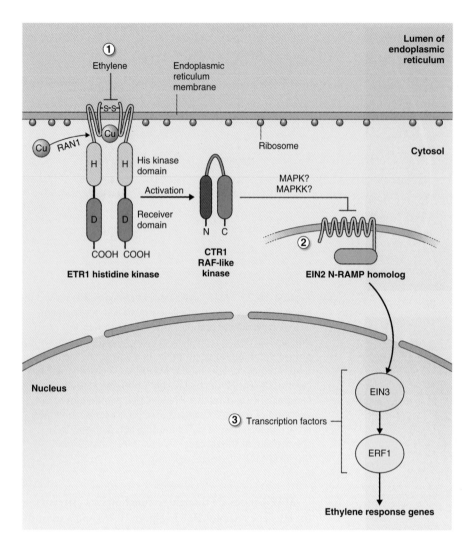

Figure 10.28 Ethylene signal transduction pathway in *Arabidopsis*. (1) Ethylene receptor ETR1 is localized in the ER membrane. In the absence of ethylene it activates a phosphorylation cascade that begins with CTR1, and includes members of the MAP kinase cascade. (2) EIN2 is a membrane-localized protein with 12 transmembrane spanning domains and is downstream of the MAP kinase cascade. (3) EIN2 brings about activation of a set of transcription factors, including EIN3 and ERF1, that regulate the expression of ethylene-responsive genes.

Key points Ethylene affects many aspects of plant growth and development including inhibition of cell elongation and promotion of leaf abscission and fruit ripening. Ethylene is synthesized from methionine via SAM and ACC. There is no requirement for ethylene breakdown in plants; rather, the ethylene signal is turned off by diffusion of the gas from its site of action. Ethylene receptors are ER-localized proteins related to bacterial histidine kinase two-component response regulators. In the absence of ethylene the receptor initiates a phosphorelay via protein kinases of the MAP kinase family resulting in the inactivation of the ethylene response pathway. In the presence of ethylene, the receptor is inactivated and a phosphorelay does not occur; this allows the membrane-localized protein EIN2 to become active. EIN2 activates transcription factors that regulate the expression of ethylene-responsive genes.

10.6 Brassinosteroids

Brassinosteroids (BRs), triterpene molecules with distinctive chemical structures containing from 27 to 29 carbon atoms, share biosynthetic origins with several other plant hormones (see Figure 10.14). **Brassinolide (BL)**, the most active of the native BRs in plants, has 28 carbon atoms; its structure with all its carbon atoms numbered is shown in Figure 10.29. The biological activity of BRs is dependent on the presence of hydroxyl groups at carbons 2 and 3 of ring A and positions 22 and 23 of the side-chain, and a seven-membered lactone B ring. BL was first isolated from germinating pollen of *Brassica rapa*. Later, castasterone, another BR, was isolated from insect galls of chestnut (*Castanea* spp.). BRs elicit a response in the rice lamina inclination bioassay, a test used to measure the biological activity of BR-like molecules (Figure 10.30).

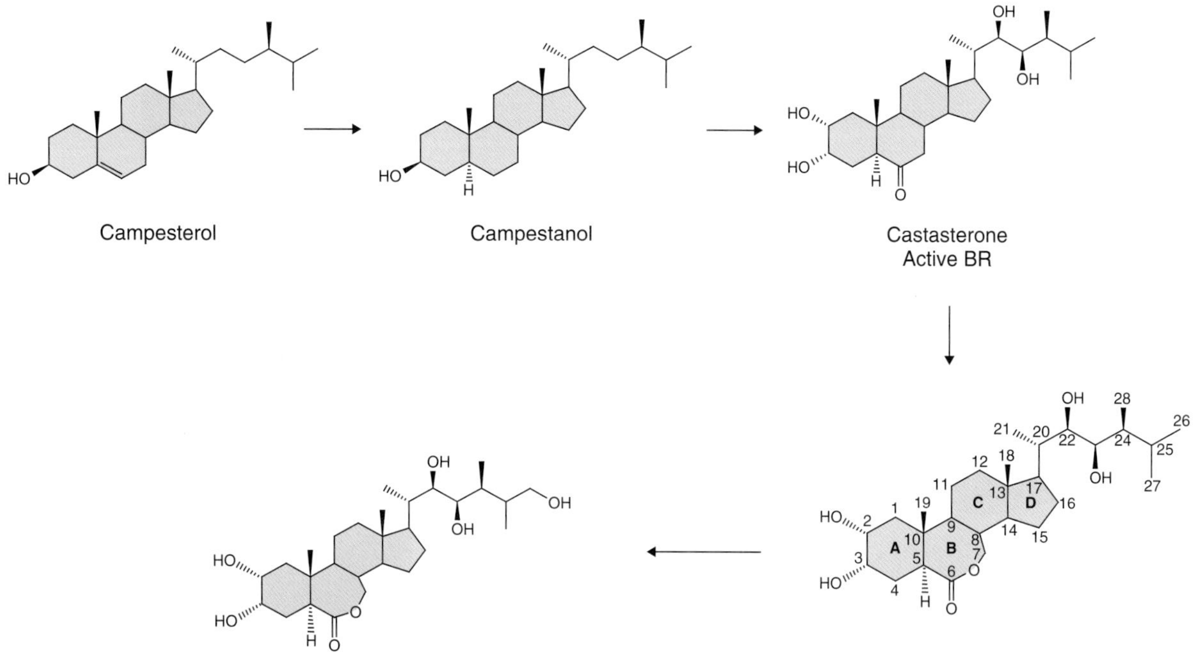

Figure 10.29   Highly simplified biosynthetic pathway of brassinosteroids (BRs). Campesterol and campestanol are inactive precursors of the biologically active compounds castasterone and brassinolide. Brassinolide is the most active naturally occurring BR and is metabolized to inactive 26-hydroxybrassinolide by a cytochrome P450 enzyme encoded by the gene *BAS1*. The biological activity of BRs is dependent on the presence of hydroxyl groups at carbons 2 and 3 of ring A and positions 22 and 23 of the side-chain, and a seven-membered lactone B ring.

Genetic analysis in *Arabidopsis* has shown that BRs are required for normal plant development. Mutations in BR biosynthesis and signaling pathways show a variety of phenotypes including extreme dwarfism as a result of ***Bri1*** (*BR insensitive1*), a mutation in the BR receptor, and de-etiolation in mutants such as ***det2*** (*de-etiolated2*) and ***cpd*** (*constitutive photomorphogenesis and dwarfism*). The de-etiolation phenotype is most pronounced in dark-grown *Arabidopsis* seedlings, where mutants exhibit short, fat hypocotyls, expanded cotyledons and the presence of anthocyanin pigments (Figure 10.31).

10.6.1 Brassinosteroid biosynthesis from campesterol is controlled by feedback loops

Brassinosteriods are synthesized from simpler sterol precursors such as sitosterol (C_{29}), stigmasterol (C_{29}), campesterol (C_{28}) and cholesterol (C_{26}) (see Figure 15.21). Although sitosterol is the most abundant sterol in plants, campesterol is the favored substrate for BR

biosynthesis. Campesterol is converted to castasterone through a series of reactions that involve oxidation at carbon 6, reduction at carbon 5 and hydroxylation by enzymes of the cytochrome P450 class that add hydroxyl groups at carbons 2, 3, 22 and 23 (Figure 10.29). There are many intermediates between campestanol and castasterone and research indicates that the order of hydroxylation reactions is random. Formation of a seven-carbon lactone B ring results in the conversion of castasterone to BL.

As with other hormones, maintenance of BL homeostasis is central to its effectiveness as a hormone. BL biosynthesis is regulated by feedback loops that transcriptionally regulate the expression of many of the genes that encode the cytochrome P450 enzymes involved in hydroxylation of campestanol. Thus when BL concentration is high, its biosynthesis is reduced. Catabolism of BL is less well understood but hydroxylation, epimerization, oxidation and sulfonation, as well as conjugation to sugars and lipids, are all known to lead to a reduction in BL activity. The best understood reaction is hydroxylation at carbon 26 by a cytochrome P450 enzyme encoded by the gene *BAS1* (Figure 10.29), which results in lowering of BL levels and produces dwarfism in *Arabidopsis*.

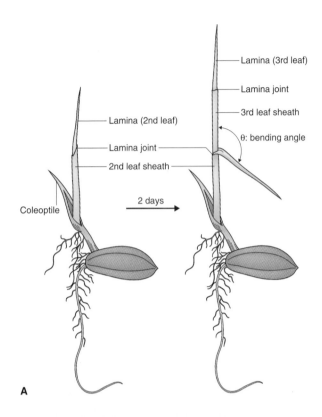

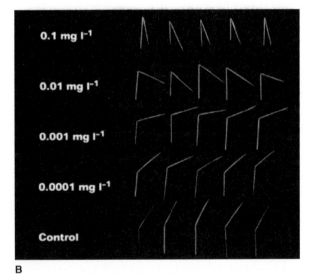

Figure 10.30   Brassinosteroid bioassay. (A) Diagram of rice seedling showing the lamina and leaf sheath. Brassinosteroids to be bioassayed are applied at the leaf sheath–lamina joint and the angle of the lamina is measured after 2 days of growth. (B) Photographs showing the effect of increasing BR on the lamina angle of rice seedlings.

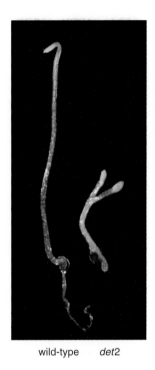

wild-type   *det2*

Figure 10.31   Effect of mutations in the brassinosteroid biosynthesis pathway on plant morphology, showing dark-grown wild-type and *de-etiolated2* (*det2*) mutant *Arabidopsis* plants defective in brassinosteroid synthesis.

10.6.2 The brassinosteroid receptor is a plasma membrane-localized LRR-receptor serine/threonine kinase

The BR receptor is a leucine-rich repeat (**LRR**) receptor serine/threonine kinase localized on the plasma membrane (Figure 10.32). This receptor was identified in a screen for *Arabidopsis* mutants that show normal root growth at high BL concentration. Mutants that grew normally in the presence of high BL were called *bri1* (*brassinolide insensitve1*). The extracellular domain of BRI1 consists of multiple LRR repeats and a BL-binding domain, while the cytoplasmic domain consists of a serine/threonine protein kinase domain. An inhibitor of BRI1 phosphorylation called BKI1 (BRI1 kinase inhibitor1) prevents phosphorylation of the cytoplasmic domain of BRI1 and this inhibition is relieved as a consequence of BL binding. BL attaches to the binding site located in the extracellular portion of the receptor and initiates BRI1 activation. Activation involves autophosphorylation at multiple sites and the formation

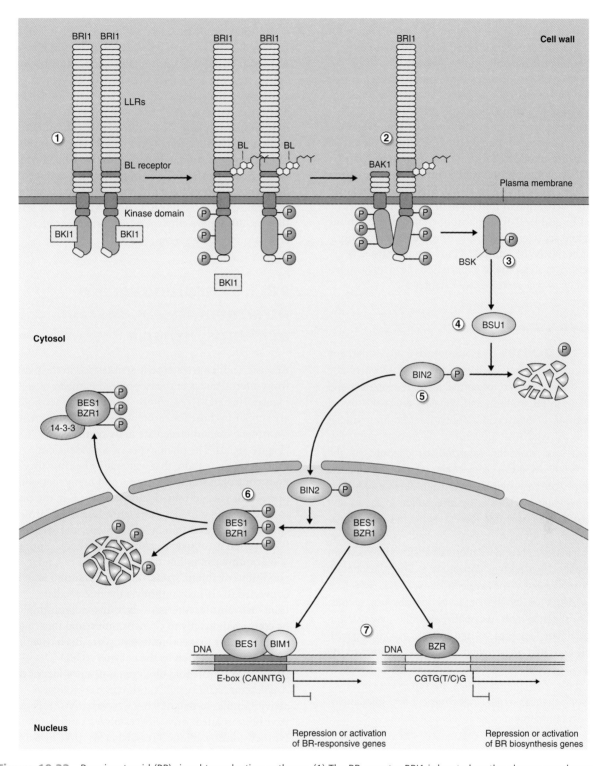

Figure 10.32 Brassinosteroid (BR) signal transduction pathway. (1) The BR receptor BRI1 is located on the plasma membrane where it forms homo-oligmers. Upon binding BR, BRI1 becomes phosphorylated and releases BRI1 kinase inhibitor (BKI1) into the cytosol. (2) Activated BRI1 forms a hetero-oligomer with BAK1, which becomes phosphorylated. (3) It in turn phosphorylates BSK proteins. (4) BSK then activates the protein phosphatase BSU. (5) BSU dephosphorylates BIN2, causing it to be degraded by the ubiquitin–proteasome system (UbPS). (6) In the absence of BR, phosphorylated BIN2 phosphorylates the transcriptional regulators BES1 and BZR1, thereby abolishing their activity and targeting them for degradation by the UbPS, or by enhancing their transport to the cytoplasm when they bind 14-3-3 proteins. (7) In the presence of BR, BES1 and BZR1 are not phosphorylated and are active as transcriptional regulators leading to brassinosteroid responses.

of homodimers. Activated BRI1 then forms heterodimers with BAK1, another LRR receptor kinase. Subsequently the activated heterodimer phosphorylates soluble cytoplasmic BR signaling kinases (BSKs).

The role of BSK proteins is to bind to and activate a protein phosphatase known as BSU1 (BRI1 suppressor1). Activated BSU1 dephosphorylates BIN2 (Br insensitive2), a repressor of BR-induced gene expression, leading to its degradation by the UbPS. In the absence of BL, the phosphorylated form of BIN2 phosphorylates two DNA-binding proteins BES and BZR1, which cannot bind DNA when phosphorylated. BIN2 may participate in either activation or inhibition of transcription, because once BES and BZR are dephosphorylated they can bind DNA and act to either activate or repress transcription. Phosphorylated BES and BZR1 are rapidly turned over by the 26S proteasome or exported to the cytoplasm with the aid of 14-3-3 proteins (see Chapter 5).

Key points Brassinosteroids are triterpene molecules, and brassinolide is the most active of the BRs. BRs are required for normal plant development and mutants in BR synthesis and signaling are characterized by a dwarf phenotype. BRs are synthesized from simpler sterols, of which campesterol is the favored precursor. A series of reactions catalyzed by enzymes of the cytochrome P450 class convert campesterol to BL via the intermediate castasterone. BL is inactivated by additional hydroxylation reactions or by conjugation to sugars and lipids. The BR receptor BRI1 is a plasma membrane-localized, leucine-rich repeat, serine/threonine kinase. BRI1 initiates a phosphorylation cascade that brings about activation of a protein phosphatase. The phosphatase dephosphorylates the protein kinase BIN2, which is then degraded by the UbPS. In the absence of BRs, BIN2 is active and phosphorylates the downstream transcription factors associated with BR-responsive genes. Phosphorylated transcription factors are exported to the cytosol or degraded by the UbPS. In the presence of BR, these transcription factors are not phosphorylated, allowing them to remain in the nucleus and stimulate or repress transcription of BR-responsive genes.

10.7 Abscisic acid

Abscisic acid (ABA) was isolated and its structure was elucidated in the early 1960s. At that time, one of the bioassays used to test for the presence of ABA in plant extracts measured abscission of petioles from cotton seedling explants in sealed containers, and this bioassay is the origin of the trivial name abscisic acid. It is now known that the effect of ABA on abscission in this assay was indirect. ABA stimulates ethylene production by cotton seedling explants, and enhanced ethylene synthesis in the sealed containers was responsible for the abscission observed. ABA has been shown to be involved in responses of plants to abiotic stress including drought, and one of its effects on plant function is to bring about rapid closure of stomatal guard cells (see Chapter 14). Other roles for ABA include imposing bud and seed dormancy, suppressing germination and accelerating leaf senescence (see Chapters 6, 17 and 18). As leaves senesce they produce more ethylene which brings about abscission in some species.

10.7.1 Carotenoids are intermediates in abscisic acid biosynthesis

Abscisic acid is a tetraterpene synthesized in the plastid from intermediates in the isoprenoid pathway (see Figures 10.2 and 10.14). Steps in the biosynthesis of ABA are similar to those for carotenoids, and the pigment β-carotene is an intermediate in its synthesis (Figure 10.33). The phenotypes of ABA-deficient mutants include those that wilt readily because they have poor control of stomatal closure, and many that exhibit precocious seed germination or vivipary as in the case of some *Zea mays* mutants (see Figure 17.20). β-Carotene is converted to zeaxanthin in several steps, and zeaxanthin is converted in a single reaction to trans-violaxanthin. Violaxanthin can follow two routes to ABA. Under conditions of abiotic stress, trans-violaxanthin is converted to cis-neoxanthin via the intermediate trans-neoxanthin. Alternatively, trans-violaxanthin can be converted directly to cis-neoxanthin and thence to xanthoxin, the first committed step in ABA biosynthesis, in a reaction catalyzed by the enzyme NCED (9-cis-epoxycarotenoid dioxygenase), a product of the *Arabidopsis NCED* gene. Xanthoxin has biological activity similar to that of ABA. Several *AtNCED* genes have been isolated; *AtNCED3* catalyzes xanthoxin synthesis during abiotic stress, whereas *AtNCED6* and *AtNCED9* catalyze xanthoxin synthesis in seeds.

Synthesis of ABA from xanthoxin occurs in the cytosol in two steps, beginning with the formation of ABA-aldehyde. In *Arabidopsis* the *ABA2* gene encodes an enzyme that catalyzes conversion of xanthoxin to ABA-aldehyde. The conversion of ABA-aldehyde to ABA is catalyzed by ABA-oxidase (AAO). Many mutants have been isolated that have defects in the conversion of ABA-aldehyde to ABA including *flacca* and *sitiens* in

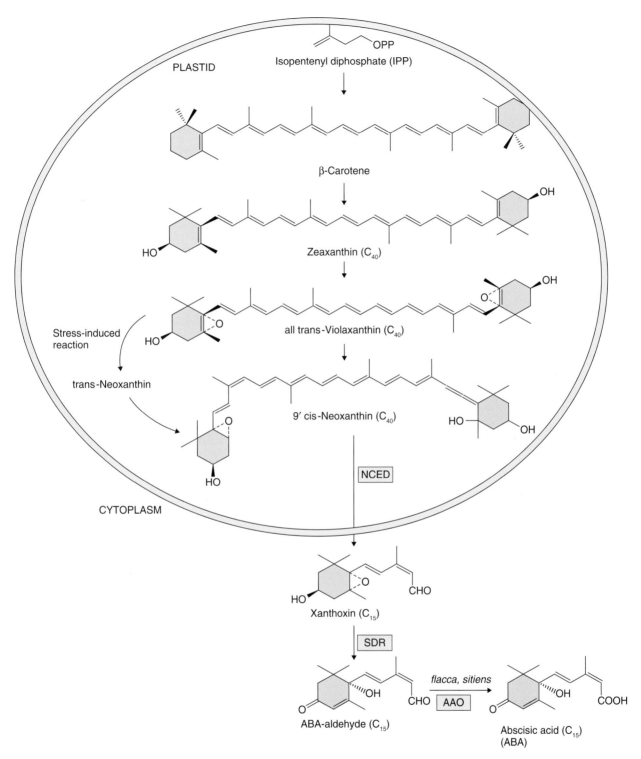

Figure 10.33 Abscisic acid (ABA) biosynthesis from isopentenyl diphosphate begins in the plastid. Xanthoxin conversion to ABA occurs in the cytosol. AAO, ABA-oxidase; NCED, 9-cis-epoxycarotenoid dioxygenase; SDR, short chain dehydrogenase/reductase.

tomato (*Solanum lycopersicum*), *nar2a* in barley (*Hordeum vulgare*) and *aba3* and *aao3* in *Arabidopsis*.

Catabolism of ABA and its conjugation to small molecules such as sugars are both mechanisms that ensure that ABA is quickly removed from cells, which

allows it to function as a signaling molecule. The main pathway for ABA catabolism is its oxidation to biologically inactive 8′-hydroxy-ABA, phaseic acid and dihydrophaseic acid (Figure 10.34). ABA 8′-hydroxylase catalyzes 8′-hydroxy-ABA synthesis and, like the

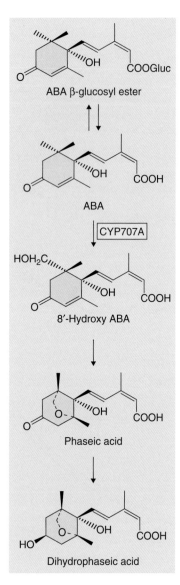

Figure 10.34 Steps in the inactivation of abscisic acid (ABA). ABA inactivation by esterification with glucose is reversible. ABA is irreversibly inactivated when converted to more oxidized derivatives such as 8'-hydroxy-ABA, phaseic acid and dihydrophaseic acid. The step from ABA to phaseic acid is catalyzed by a member of the CYP450 enzyme family.

enzymes that inactivate BRs and GAs, it is a cytochrome P450 monooxygenase, in this case **CYP707A**. Loss-of-function mutations in *CYP707A* cause a drastic reduction in germination of *Arabidopsis* seeds, emphasizing the importance of ABA in germination and of CYP707A in ABA inactivation (Figure 10.35). **Conjugation** of ABA to sugars is another route for removing active ABA from cells. ABA β-glucosyl ester is an important catabolite that is thought to be stored in the vacuole. Enzymes that can hydrolyze ABA conjugates have been isolated and it has been proposed that conjugates can provide a supply of active ABA.

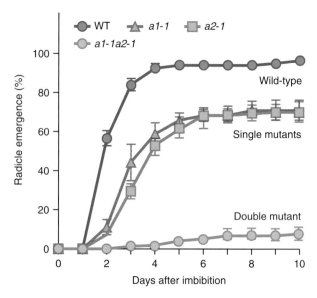

Figure 10.35 Mutations in the abscisic acid (ABA) 8'-hydroxylase gene *CYP707A* lead to an accumulation of active ABA, hence reduced germination of *Arabidopsis* seeds. All seeds of wild-type (WT) plants germinate by 4 days but single (*a1-1, a2-1*) and double (*a1-1a2-1*) mutants show decreasing levels of germination.

10.7.2 There are several classes of abscisic acid receptors

Abscisic acid elicits two distinct response types in plants: rapid responses such as the loss of turgor in guard cells to close stomata; and more gradual responses including effects on seed dormancy and germination and reactions to abiotic stresses. It was proposed that these relatively fast and slow responses use different receptors, making their isolation in a mutant screen more difficult. The likelihood that there are multiple ABA receptors is supported by evidence that indicates the existence of both soluble and membrane-bound receptors. Three classes of ABA receptors have been identified to date. These include the plasma membrane-localized **G proteins** GTG1 and GTG2; a plastid-localized enzyme that coordinates nucleus to plastid signaling; and cytosolic ligand-binding proteins of the START domain superfamily. Recent research shows that soluble START domain proteins are the principal ABA receptors that function in responses as diverse as stomatal closing, germination and abiotic stress.

The identification of START domain proteins as ABA receptors arose from two different experimental strategies. One employed **pyrabactin** (**PY**), a synthetic compound that mimics ABA, and searched for *Arabidopsis* mutants that were PY-insensitive. The gene that conferred insensitivity to PY was cloned and called *Pyrabactin Resistance1* (**PYR1**). PYR1, and its homologs,

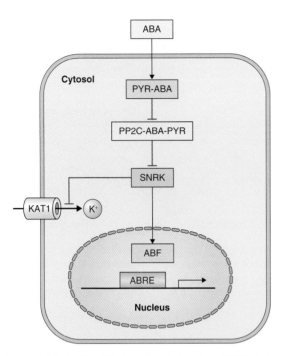

Figure 10.36 Simplified model of abscisic acid (ABA) signal transduction via the soluble PYR receptor. PYR binds ABA in the cytosol and then binds protein phosphatase PP2C, inhibiting its activity. Protein kinases of the SNRK class are downstream of PP2C. When phosphorylated, these kinases modulate the activity of ion channels, such as the potassium channel KAT1, and of transcription factors, such as the ABFs (ABRE binding factors) that regulate the activity of genes containing ABA response elements (ABREs).

which in *Arabidopsis* are called **PYL** (PYR-like), are START domain proteins and have been shown to be ABA-dependent inhibitors of one class of protein phosphatase type 2C (**PP2C**) proteins (Figure 10.36). What made this discovery exciting was the knowledge that the products of genes corresponding to several types of ABA-insensitive *Arabidopsis* mutants, including ***ABI1*** and ***ABI2***, were members of the PP2C class of proteins previously shown to be involved in ABA signaling. A different experimental approach, based on protein–protein interaction, also found a link between PYR1/PYL and PP2Cs. Using ABI1 and ABI2 as bait in the yeast two-hybrid system (see Figure 8.14) resulted in the isolation of additional PYR1/PYL homologs, called regulatory component of ABA receptors (**RCARs**).

Proteins of the PYR1/PYL/RCAR group form dimers, specifically bind ABA and function as ABA receptors. When one of these dimers binds ABA, it can, in turn, bind members of the PP2C class of protein phosphatase, inhibiting the activity of PP2Cs. The target of PP2Cs is a class of SNF1-related protein kinase 2 (**SNRK2**). Analysis of *Arabidopsis* mutants has shown that protein kinases of the SNRK class, such as OST (Open Stomata), are

positive regulators of the ABA signaling pathway. In other words, SNRK kinases are active in the presence of ABA and are required for ABA signal transduction. ABA exerts its physiological effects by regulating the activity of SNRK2 kinases, which in turn phosphorylate target proteins. In the absence of ABA, PP2Cs such as ABI1 and ABI2 are active and dephosphorylate SNRK2, inactivating these kinases. Conversely, in the presence of ABA, PP2C is inhibited and SNRK kinases are active. Many downstream targets of the PYR/PYL/RCAR-PP2C-SNRK2 pathway have now been identified. These include ion channels (e.g. the *Arabidopsis* potassium channel KAT1—a member of the same family of channels at AKT1 which is discussed in Chapter 13), the reactive oxygen-generating enzyme RbohF, an NADPH oxidase (see Chapter 18), and ABA-responsive transcription factors such as ABF2. Transcription factors of the ABF2 class interact with **ABA response elements** (ABREs) in the promoters of ABA-responsive genes (see Chapter 3).

Key points Abscisic acid is a tetraterpene whose synthesis from carotenoids begins in the plastid; the final steps in ABA formation occur in the cytosol. ABA regulates stomatal function, is involved in imposing dormancy in buds and seeds, and plays a role in senescence and cell death. ABA is inactivated by conjugation to sugars or by oxidation to phaseic acid by an enzyme of the cytochrome P450 class. ABA receptors are soluble cytoplasmic proteins that form dimers before binding ABA. The ABA–receptor complex interacts with PP2C protein phosphatases, inhibiting their activity. Inhibition of PP2C activity allows protein kinases of the SNRK class to phosphorylate and activate target proteins such as ion channels and transcription factors.

10.8 Strigolactones

Strigolactones (SLs) were initially isolated from plant root exudates as compounds that stimulate the germination of seeds of several parasitic plants from the family Orobanchaceae, such as *Orobanche* spp. (broom rape) and *Striga* spp. (witch weed) (Figure 10.37; see also Figure 15.42), and the branching of mycorrhizal hyphae. *Striga asiatica* is a common parasite of grasses and can cause total crop losses in some instances. *Orobanche* species, on the other hand, are highly specific parasites of many eudicot crops such as *Helianthus annuus*

Figure 10.37 *Orobanche hederae* (brown stalks) growing on ivy (*Hedera helix*).

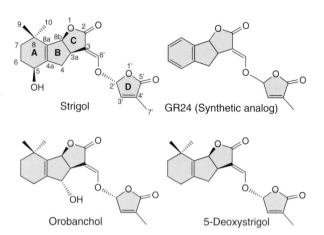

Figure 10.38 Chemical structures of strigolactones. Orobanchol and strigol are biologically active strigolactones isolated from plants, while 5-deoxystrigol is thought to be a precursor of strigol. GR24 is a synthetic strigol analog.

(sunflower). The severity of parasitism by *Striga* and *Orobanche* is increased by the fact that each plant produces thousands of very small seeds that persist in the soil.

In addition to their involvement in promoting invasion by harmful parasites, SLs have been shown to act as host recognition signals in arbuscular mycorrhizal fungi (see Chapter 12). These fungi form beneficial symbiotic associations with the roots of most (>80%) plants. SLs released by the roots of host plants serve to attract the arbuscular mycorrhiza (AM) to the root surface and are also thought to stimulate branching of the fungal hyphae. SLs are synthesized not only by plants that are hosts to AM but by non-host species too, such as *Arabidopsis* and white lupine (*Lupinus albus*). Taking these observations together with the fact that roots of plants that are not hosts to parasitic weeds also produce SLs, it became clear that these novel plant compounds were likely to play roles in normal plant growth and development.

10.8.1 Strigolactones were first isolated from cotton root exudates

Strigol was the first SL to be purified. It has a characteristic chemical structure that includes two lactone rings (rings C and D) linked via an enol ether bridge (Figure 10.38). Strigol was first isolated from root exudates of *Gossypium hirsutum* (cotton), a plant that is not parasitized by members of the Orobanchaceae. *G. hirsutum* is therefore one of many crops that is used as a '*Striga* trap', that is, it produces strigol which induces germination of *Striga* seeds in the absence of an

appropriate host plant, thus reducing the abundance of viable seeds in soil.

More than ten SLs have now been isolated from plants and all stimulate the germination of weedy parasites. **Orobanchol**, isolated from red clover (*Trifolium pratense*) root exudates, stimulates the germination of seeds of many *Orobanche* species. 5-Deoxystrigol has been found in root exudates of almost all plants that produce SLs, leading to the hypothesis that strigol, orobanchol and other SLs are products of its hydroxylation (Figure 10.38). SLs are present in plants at low concentration, and their chemical synthesis has only recently been achieved. Therefore, research on SL function has relied on the use of a synthetic SL analog known as GR24 (Figure 10.38). Note the similarity between GR24 and naturally occurring SLs, especially the presence of C and D rings and the enol ether bridge.

10.8.2 Strigolactones regulate lateral bud dormancy

As we have discussed earlier in this chapter, auxins and CKs have profound effects on branching in plants. The analysis of branching mutants of several species, especially *max* (*more axillary growth*) of *Arabidopsis, rms* (*ramosus*) of garden pea (*Pisum sativum*) and various dwarf (*d17* and *d10*) mutants of rice (*Oryza sativa*), led to the recognition that SLs are potent regulators of branching (Figure 10.39). MAX1, MAX3, MAX4, RMS1, RMS5, D10 and D17 are all members of the family of plastid localized enzymes known as carotenoid cleavage dioxygenases (**CCDs**). Mutations in CCD-encoding genes result both in increased branching, also called tillering in *O. sativa*, and in much lower endogenous SL

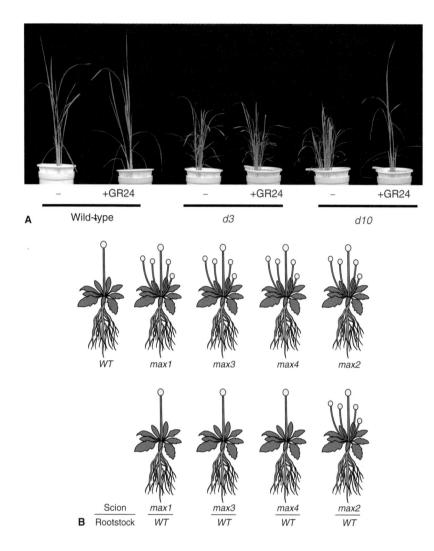

Figure 10.39 Branching mutants in *Arabidopsis* and rice indicate a role for strigolactones in regulating axillary bud growth. (A) Wild-type and *d3* and *d10* mutants of rice showing that the tillering (branch growth) phenotype of the mutant can be suppressed by treatment with the synthetic strigolactone GR24. (B) Drawings of wild-type (WT) and several *max* mutants of *Arabidopsis* (top row) and the corresponding mutant scion grafted to a WT rootstock (bottom row). Mutants in strigolactone synthesis (*max1*, *max3*, *max4*) can be rescued by grafting to WT but plants with a mutation in the strigolactone signaling gene *max2* cannot be rescued by grafting.

concentrations. These mutations can be rescued by the addition of the synthetic SL GR24 (Figure 10.39A). These and other results indicate that branching in *O. sativa* as well as in *Arabidopsis* and *P. sativum* is regulated by SL, whose synthesis is dependent on the activity of plastid-localized CCDs. Grafting experiments in *Arabidopsis* showed that SLs are produced by the root system and are translocated upward in the transpiration stream where they act to prevent the outgrowth of lateral branches (Figure 10.39B). In these experiments mutant scions were grafted on to wild-type (WT) rootstocks and in the case of *max3 and max4*, which lack enzymes required for SL biosynthesis, the mutant phenotype was rescued. The *max1* mutant of *Arabidopsis* encodes a CYP450 enzyme and can also be rescued by application of GR24 and by grafting the *max1* scions to WT rootstock. The results of these experiments suggest that SLs are produced in roots by a pathway that involves CCD and CYP450 enzymes. The *Arabidopsis max2* mutant cannot be rescued by root grafting to WT plants. In this case the mutation is in an F-box protein gene

that is involved in SL signaling, not synthesis, which means that no amount of added SL can rescue the phenotype.

Key points Strigolactones are terpenoid molecules synthesized from carotenoids in the plastid. SLs were originally isolated from cotton root exudates because roots of cotton and other plants are known to stimulate the germination of seeds of plants from the family Orobanchaceae. Members of this family of plants are noxious parasitic weeds that have deleterious effects on many crop plants including maize and sorghum. It is now known that SLs are involved in regulating plant growth and development, and experiments with several MAX mutants of *Arabidopsis* have established that SLs are important regulators of branching. Little is known about SL receptors or signal transduction except that SCF E3 ubiquitin ligases play a key role.

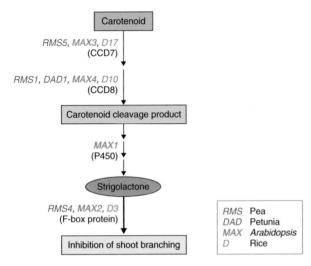

Figure 10.40   Steps in the synthesis and signal transduction pathway of strigolactones (SLs). SLs are synthesized in plastids from carotenoids and converted to active SLs in the cytosol by enzymes of the cytochrome P450 (CYP450) class. SL signaling involves the participation of the UbPS. Mutations affecting steps leading to SL synthesis in *Arabidopsis, Petunia, Pisum* and rice and an F-box protein in the SL signaling pathway are shown.

An SL biosynthetic scheme that incorporates work with the various branching mutants discussed above is shown in Figure 10.40. Carotenoids are cleaved by CCDs to a product that is converted to SL by the action of CYP450 enzymes. Signaling from SLs leading to inhibition of lateral bud growth involves an F-box protein, which is part of an SCF-type E3 ubiquitin ligase and is a product of the *MAX2* gene. The role of SLs in branching is discussed further in Chapter 12.

10.9 Jasmonates

Jasmonates are a group of naturally occurring compounds, similar to mammalian prostaglandins, that are synthesized from the fatty acid linolenic acid (Figure 10.41). Jasmonic acid (**JA**) and methyl jasmonate (**MeJA**) are the most commonly occurring jasmonates in plants. MeJA is also one of the principal components of fragrance produced by jasmine (*Jasminum grandiflorum*). The roles of jasmonates as plant hormones were discovered in the 1970s, when JA was reported to inhibit the growth of rice, wheat and lettuce seedlings. It was subsequently shown to inhibit seed and pollen germination, retard root growth and stimulate the curling of tendrils. JA also promotes tuber formation and the accumulation of storage proteins in seeds.

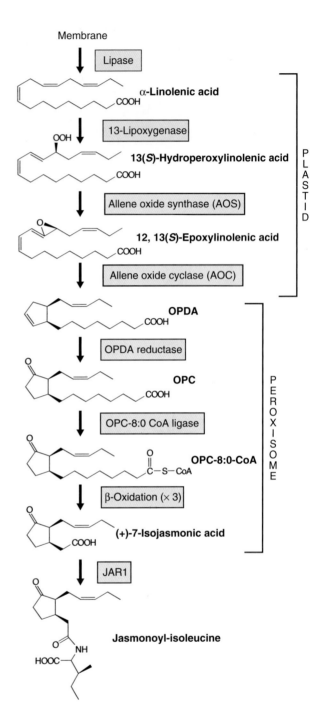

Figure 10.41   Biosynthesis of jasmonic acid. α-Linolenic acid released from membranes by lipase activity is converted to oxo-phytodienoic acid (OPDA) in the plastid followed by several steps, including three cycles of β-oxidation that occur in the peroxisome. JA conjugated to isoleucine is the active form of jasmonate in the cytosol. JAR1, product of the *JASMONATE RESISTANT 1* locus; OPC, oxo-cyclopentane octanoic acid.

Arabidopsis mutants that are defective in JA biosynthesis or response are male-sterile because anther filaments do not elongate to allow dispersal of pollen to the stigma. Furthermore, anther dehiscence does not occur when the flowers open, and the majority of mutant pollen grains are not viable. Fertility of these mutants can be rescued by JA treatment. Jasmonates also play an important role in plant defense as they are produced when herbivores damage plant tissue (see Chapter 15).

10.9.1 Jasmonate biosynthesis begins in the plastid and moves to the peroxisome

Two organelles, the plastid and peroxisome, are involved in the synthesis of JA from linolenic acid. The first step in JA biosynthesis is catalyzed by a lipase that releases α-linolenic acid, a C_{18} polyunsaturated fatty acid, from plastid membrane lipid. Linolenic acid is converted to oxo-phytodienoic acid (**OPDA**) in the plastid by three enzymes, lipoxygenase, allene oxide synthase and allene oxide cyclase. It is not understood how OPDA exits the plastid, but it is transported into the peroxisome by an ABC class transporter (see Chapter 5). OPDA is then converted to JA in the peroxisome in five further steps catalyzed by OPDA reductase and oxo-phytoenoic acid CoA ligase followed by three rounds of β-oxidation (Figure 10.41; see Chapter 6).

JA is conjugated via an amide link to isoleucine (Ile) and other amino acids to yield jasmonoyl-isoleucine (JA-Ile) and other jasmonoyl–amino acid conjugates, respectively (Figure 10.41). JA-Ile is thought to be the functionally active form of JA. In *Arabidopsis*, conjugation of JA to amino acids is catalyzed by an enzyme encoded by the *JAR1* (*JASMONATE RESISTANT 1*) locus. The *JAR1* gene was isolated in a screen for plants that were resistant to high JA concentrations, a phenotype that suggests that conversion of JA to JA-Ile is required to activate the JA signaling pathway. Unlike the synthesis pathways of other plant hormones, JA-Ile synthesis is under positive feedback regulation. JA treatment results in elevation of the synthesis of many JA biosynthetic genes including lipoxygenase, *AOS, AOC*, the gene encoding OPDA reductase (*OPR3*) and *JAR1*.

In addition to being converted to JA-Ile, JA is also the precursor of volatile methyl jasmonate, the product of a SAM-dependent carboxyl methyltransferase. Methyl jasmonate is biologically active and is thought to act as a volatile signal in plant defenses against herbivory (see Chapter 15). There are many other catabolites of JA that accumulate in plants and which have biological activity, including tuberonic acid which, as the name suggests, promotes tuber formation. Surprisingly little is known about the formation of inactive JA catabolites. As we have emphasized, mechanisms for removing hormones from cells are as important as those that bring about their synthesis. Perhaps, as is the case with ethylene, the volatility of JA and its derivatives makes chemical inactivation redundant.

10.9.2 The jasmonate receptor is an F-box protein that targets a repressor of jasmonate-responsive genes

Jasmonic acid regulates the transcription of a large number of genes, especially those that are related to defense responses in plants. The JA signaling pathway resembles the pathways of several other plant hormones in that the UbPS is involved in the degradation of a member of the **JAZ** family of transcriptional repressor proteins (Figure 10.42). JAZ functions to repress the activity of transcription factors that regulate JA-responsive genes. Proteolysis of JAZ results in relief of repression and JA-responsive genes are transcribed. JA targets JAZ proteins for ubiquitination by binding to the F-box protein COI1, a component of the SCF^{COI1} ubiquitin ligase complex.

JAZ proteins repress transcription by binding to **MYC2**, a transcription factor required for the expression of JA-responsive genes. When JA-Ile binds the SCF^{COI1}

Key points Jasmonates (JAs) are molecules that resemble mammalian prostaglandins; they have been implicated in the inhibition of seed germination and pollen tube growth as well as in orchestration of plant defense responses. JAs are synthesized from linolenic acid in a pathway that begins in the plastid and is completed in the peroxisome. JA is conjugated to amino acids such as isoleucine and JA-Ile is thought to be the active form. JA is also methylated to form volatile methyl jasmonate which is involved in signaling, especially in defense against herbivory. Formation of a volatile JA derivative may provide a mechanism to dissipate the signal after JA is synthesized. JA receptors are F-box proteins, components of the SCF^{COI1} E3 ubiquitin ligase that target the JAZ family of transcriptional repressors for degradation. JAZ proteins repress the expression of a suite of defense proteins and the proteolysis of JAZ by the proteasome relieves the inhibition of gene expression, allowing defense proteins to be transcribed.

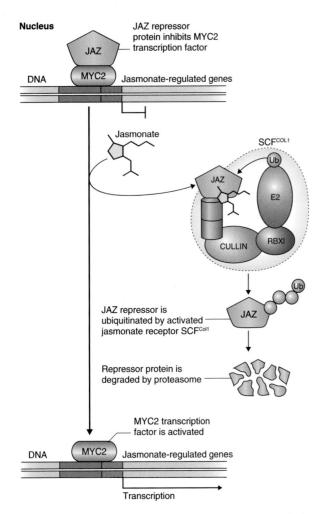

Figure 10.42 A simplified model of the jasmonic acid (JA) signal transduction pathway. Jasmonate forms a complex with the F-box protein COL1 that targets the transcription repressor JAZ for degradation by the UbPS. Release of inhibition of JAZ repression allows the transcription of JA-responsive genes by transcription factors such as MYC2. Transcribed genes include defense proteins.

complex it facilitates the binding of JAZ to the complex, which leads to its ubiquitination and eventual destruction by the proteasome. Repression of MYC2 is thus relieved and transcription of JA-responsive genes is initiated. Interestingly, genes that encode JAZ are among those upregulated by JA-Ile, giving rise to a negative feedback loop.

10.10 Polyamines

Polyamines (PAs) are small organic compounds that have two or more amino groups and range in molecular mass from 88 Da for putrescine to 202 Da for spermine. Putrescine, spermidine and spermine are the most

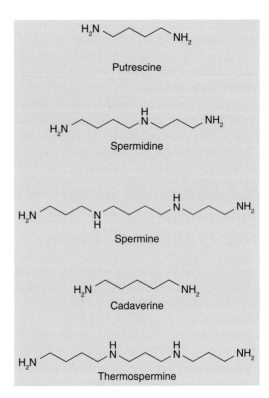

Figure 10.43 Chemical structures of polyamines (PAs) that occur naturally in plants. Putrescine, spermidine and spermine are the most common PAs but cadaverine is abundant in legumes.

abundant PAs in plants, but in legumes cadaverine is also abundant (Figure 10.43). Other less abundant PAs such as thermospermine are very important physiologically. Unlike the other hormones discussed so far, PAs are found in plants at millimolar concentrations. They elicit a range of responses, including cell differentiation and division, embryogenesis, flower development, fruit ripening, root initiation, abiotic stress tolerance and tuber formation.

PAs have a high affinity for anionic constituents such as DNA, RNA, phospholipids and acidic proteins, as well as anionic groups in membranes and cell walls, including hydroxycinnamic acids. In addition to being present as free amines, a significant fraction of the PA pool is found as amide conjugates with *p*-coumaric, ferulic and caffeic acids. Conjugates between PAs and hydroxycinnamic acids play roles in development including flowering and seed and fruit development, and in hypersensitive responses to viral and fungal infections.

10.10.1 Polyamines are synthesized from amino acids

Putrescine, the simplest of the PAs, can serve as the precursor of spermidine and spermine synthesis

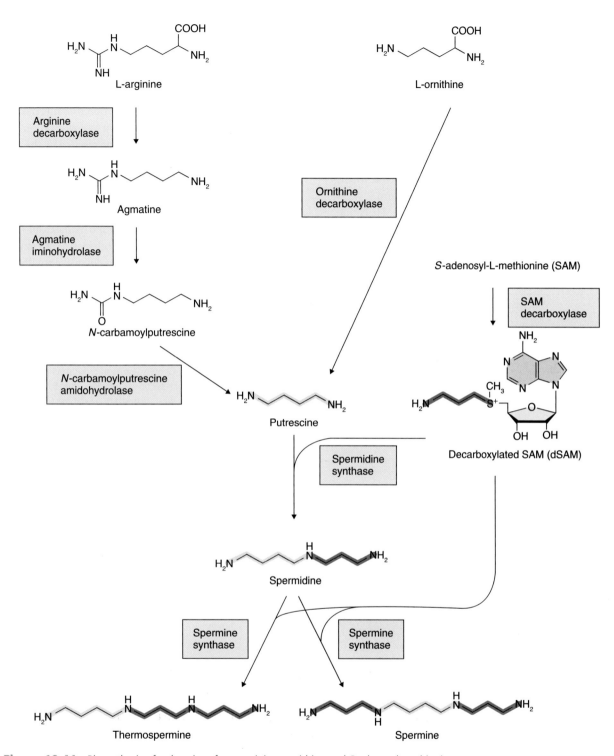

Figure 10.44 Biosynthesis of polyamines from arginine, ornithine and S-adenosyl methionine.

(Figure 10.44). In some species putrescine can be synthesized from L-ornithine (Orn) in one step catalyzed by Orn decarboxylase (**ODC**), or from L-arginine (Arg) in a two-step process initiated by Arg decarboxylase (**ADC**). ODC is usually less active than ADC, suggesting that Arg decarboxylation is the major pathway for

putrescine synthesis. Support for the importance of Arg in PA synthesis comes from the observation that *Arabidopsis* has two *ADC* genes but lacks an *ODC* gene. Furthermore, *Arabidopsis* mutants that lack both *ADC* genes produce abnormal seeds that fail to germinate. In many species putrescine production is increased by environmental stress as a consequence of upregulation of ADC activity. Exposing *Arabidopsis* to low temperature, high salt or dehydration, as well as ABA treatment, induces the expression of *ADC1* or *ADC2*.

SAM and putrescine are precursors of spermidine, spermine and thermospermine synthesis (Figure 10.44). As we have seen in Section 10.5.1, SAM is also a precursor of ethylene; consequently PAs and ethylene act as competitive inhibitors of each other's biosynthetic pathways. SAM is first decarboxylated to **dSAM**, which plays a role in the synthesis of spermidine from putrescine catalyzed by spermidine synthase. Spermine and thermospermine are synthesized from spermidine in reactions catalyzed by spermine synthases, in which dSAM also participates. The synthesis of dSAM is under

feed-forward and feedback control. Its synthesis increases as putrescine levels rise but it is inhibited by spermidine.

There are two *Arabidopsis* genes encoding spermine synthase, **SDMS** and **ACL5**. SDMS catalyzes the synthesis of spermine and ACL5 catalyzes the synthesis of thermospermine. *ACL5* loss-of-function mutants have defects in shoot xylem differentiation and elongation (Figure 10.45). The phenotype can be rescued by the application of thermospermine, but not spermine, indicating that thermospermine is required for normal stem elongation in *Arabidopsis*.

PA catabolism is catalyzed by two classes of enzymes, copper-containing diamine oxidases and flavin-linked polyamine oxidases. Both classes of enzyme generate H_2O_2 while PAs are broken down eventually

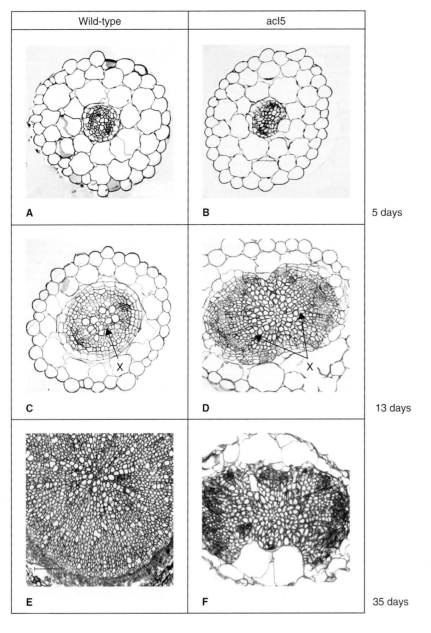

Figure 10.45 Mutations in the *acl5* gene, that encodes the enzyme for thermospermine synthesis from spermidine, cause defects in xylem (X) development in *Arabidopsis* seedling hypocotyls. Light micrographs of stem cross-sections sampled after 5, 13 and 35 days' growth of wild-type (A, C, E) and *acl5* (B, D, F) seedlings. Note the very aberrant xylem development in the *acl5* mutant.

to inactive compounds. Plants contain many forms of polyamine oxidase and some are localized to the apoplast where the H_2O_2 they release plays a role in lignification by cross-linking hydroxycinnamic acids. Three of the five polyamine oxidases of *Arabidopsis* are localized to peroxisomes, but H_2O_2 released by these enzymes is unlikely to be developmentally important since the abundant catalase in this organelle will convert H_2O_2 to H_2O.

10.10.2 Polyamines play roles in xylem differentiation

The effects of PAs on tracheary element differentiation have been extensively studied in *Arabidopsis* and in the model system *Zinnia elegans*. *Arabidopsis acl5* mutants have reduced thermospermine levels and defective xylem differentiation in hypocotyls (Figure 10.45). Compared with WT, vessel elements in *acl5* mutants are small, have only spiral thickenings and lack pits in their walls. Xylem fibers are also absent. The observations that spermine application can reverse the effects of *acl5* on xylem differentiation, and that ACL5 is expressed specifically in developing xylem elements in WT *Arabidopsis*, lend strong support to the idea that PA synthesis is required for normal xylem development.

Further evidence for a role of PAs in xylem differentiation comes from work with *Zinnia elegans*, in which tracheary element differentiation can be studied at the cellular level in isolated cell cultures. The addition of spermine to cultured *Z. elegans* cells that have been induced to form tracheary elements dramatically increases vessel size, and wall thickenings are more elaborate. It is likely that in both *Arabidopsis and Z. elegans* PAs play a role in delaying programmed cell death of tracheary elements, the final step in the differentiation of functional vessels and tracheids (see Chapter 18).

Key points Polyamines are low molecular weight organic molecules having at least two amino groups. Putrescine, the smallest of the PAs, is synthesized from the amino acids ornithine or arginine. The larger PAs such as spermidine and spermine are synthesized from *S*-adenosyl methionine and putrescine. PAs are catabolized by two classes of enzymes, diamine oxidases and polyamine oxidases. PAs have a broad spectrum of effects on plant growth and development including xylem differentiation, flowering, fruit growth and responses to pathogens. Many of the responses to PAs may be related to their effects on programmed cell death but the precise mechanism of their action is yet to be resolved.

10.11 Salicylic acid

Salicylic acid (SA) has long been recognized as a biologically important molecule produced by plants. Until recently its biological importance was related to the medicinal role of its derivative acetylsalicylic acid, more commonly known as aspirin. SA was initially isolated from the bark of *Salix alba* or willow, where it is abundant, hence its name. Its formal chemical name is 2-hydroxybenzenecarboxcylic acid and it has a molecular mass of 138 Da.

10.11.1 There are two biosynthetic pathways of salicylic acid biosynthesis in plants

There are two possible routes for the synthesis of SA in plants, one from trans-cinnamic acid (tCA), the other from chorismic acid (Figure 10.46). In the tCA pathway trans-cinnamic acid is synthesized from phenylalanine in a reaction catalyzed by phenylalanine ammonia lyase (**PAL**). tCA is converted to benzoic acid by the reactions of β-oxidation in which CoA is added to tCA; then trans-cinnamoyl-CoA undergoes three rounds of β-oxidation to produce benzoyl-CoA which is converted to benzoic acid. Benzoic acid is hydroxylated to form SA by a soluble cytochrome P450 enzyme. Support for the role of this pathway in SA synthesis comes from experiments that show that when *Nicotiana tabacum* plants are inoculated with tobacco mosaic virus (TMV), PAL activity is suppressed and there are much lower levels of SA.

The chorismate pathway for SA synthesis occurs in two steps (Figure 10.46). The first step is catalyzed by isochorismate synthase, encoded by the ***ICS1/SID2*** gene; the second is catalyzed by isochorismate pyruvate lyase (**IPL**). The *ics1/sid2* mutant of *Arabidopsis* fails to accumulate SA under various inductive conditions, suggesting that the chorismate pathway predominates in this species. Observations on other species, however, show that SA is synthesized via the tCA pathway. Tobacco plants experiencing abiotic stress from ozone exposure synthesize SA from benzoic acid in the absence of *ICS* expression. In *Arabidopsis*, on the other hand, *ICS* is expressed in response to ozone and SA accumulates. These data suggest that the pathway for SA biosynthesis in plants is species-specific and perhaps treatment-specific.

SA is converted in vivo to methyl salicylate (MeSA), and glucosyl esters and glucosides can be produced (Figure 10.46). Methylation of SA on the carboxyl group

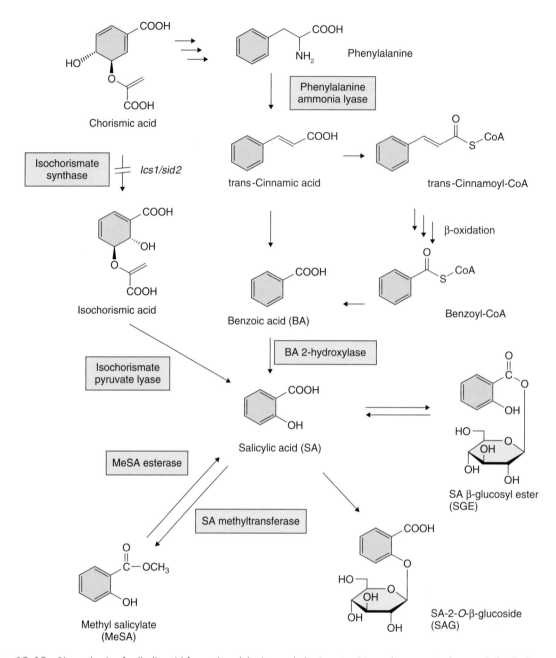

Figure 10.46 Biosynthesis of salicylic acid from phenylalanine and chorismate. SA can be converted to methyl salicylate or be conjugated to glucose.

to form MeSA is catalyzed by a carboxyl methyltransferase that uses SAM as the methyl donor. MeSA can be converted back to SA by MeSA esterase.

10.11.2 Salicylic acid induces flowering and heat production in some plants

Salicylic acid can retard senescence of petals, induce flowering in duckweeds and bring about **thermogenesis** (heat production) in arum lilies from the family Araceae (see Figure 7.18). Adding aspirin to a vase of cut flowers is a widely used remedy to delay their senescence. In this case, the mechanism of SA action is a consequence of its effects on the conversion of ACC to ethylene. SA inhibits ACO, which lowers ethylene levels and thus delays senescence. Duckweeds (small aquatic angiosperms of the family Araceae that includes *Lemna gibba, Spirodela polyrrhiza* and *Wolffia microscopia*) require a series of long days to flower but can be induced to flower in short days if treated with SA. It is unlikely that SA is an endogenous flowering hormone in duckweeds, since it is relatively abundant in plants grown under both long and short days.

Arum spadix thermogenesis is an interesting biological phenomenon and SA is thought to be the signal that induces this process. The spadix produces amines that become volatile as the temperature of the spadix tissue increases by 10 °C or more, the result of thermal energy released by uncoupled electron flow through the alternative oxidase pathway in mitochondria (see Chapter 7). The pungent volatilized amines attract pollinators to the flower. The thermogenic-inducing principle was initially called **calorigen** and was identified as SA in the late1980s. SA application to the spadix can bring about thermogenicity, accompanied by an increase in endogenous SA levels—an example of positive feedback.

10.11.3 Salicylic acid is involved in localized and systemic disease resistance

In response to pathogen infection many plants develop resistance to the disease by limiting the spread of the pathogen (see Chapter 15). Not only is localized spread of the pathogen limited but parts of the plant remote from the site of infection also develop resistance to the pathogen, a phenomenon called **systemic acquired resistance** (**SAR**). SAR is also associated with the production of a suite of proteins called pathogenesis-related (**PR**) proteins. SA is an important participant in this mechanism of resistance to pathogenic infection.

Evidence that SA was involved in SAR came from experiments showing that SA treatment of leaves of *Nicotiana tabacum* cv. Xanthi, which are susceptible to TMV, induces the synthesis of PA proteins and resistance to TMV infection. Exposure to SA has now been shown to induce SAR in a variety of pathogen-infected plants. Further proof that SA was involved in SAR came from experiments showing that infection of TMV-resistant *N. tabacum* plants with TMV causes a 40-fold increase in SA levels in inoculated leaves and a ten-fold increase in leaves remote from the infection site. The increase in SA concentrations in TMV-resistant plants is accompanied by enhanced expression of genes that encode PA proteins in inoculated and uninoculated leaves. *N. tabacum* plants that are susceptible to TMV do not show an increase in SA levels or in PA protein gene expression when infected with the virus. Results from these and similar experiments led to the conclusion that SA plays a key role in signal transduction leading to the development of SAR.

MeSA has a function in SAR that appears to differ between species. Experiments show that it is essential for SAR in *N. tabacum*, where it plays a role in regulating the

pool of available SA. There is, however no evidence that MeSA participates in SA homeostasis or SAR in *Arabidopsis*.

Key points Salicylic acid is involved plant responses to abiotic stress and in plant defense against pathogens. It can prevent local pathogen infection as well as mediate systemic acquired resistance (SAR), promoting resistance at sites remote from the point of infection. SA is synthesized via two pathways that depend on the nature of the elicitor. Plants exposed to abiotic stress synthesize SA from chorismate, whereas plants responding to the presence of pathogens make SA from trans-cinnamic acid. SA made during SAR is methylated and methyl SA is though to be the active form in some plant species. SA can also promote flowering in duckweeds, and in plants such as arum lilies it can trigger thermogenesis, a process that generates heat by uncoupled electron flow in mitochondria. Heat volatilizes pungent amines, which attract insect pollinators.

10.12 Nitric oxide

Nitric oxide (NO) is a gas produced by all living organisms. Investigations in the 1980s into the biochemistry of nitrate (NO_3^-) metabolism by plants led to the discovery that NO can be synthesized during nitrate assimilation when NO_3^- is reduced to NH_4^+. Bacteria, on the other hand, generate NO during the process of denitrification when NH_4^+ ions are oxidized to N_2 (see Chapter 13). Animals produce NO from arginine using the enzyme NO synthase. NO synthases have not been found definitively in plants and, as we will discuss below, plants can use a variety of other enzymes and pathways to produce NO, including the enzyme nitrate reductase that plays a key role in the conversion of NO_3^- to NH_4.

In mammals, NO is a regulator of vasodilation, a property that has generated a great deal of research interest and led to the development of drugs such as Viagra. NO also elicits a range of responses in plants. These include regulating stomatal closure, lateral root initiation, flowering, seed germination and responses to a wide variety of abiotic stresses including osmotic stress, salinity, flooding and temperature extremes. NO is also involved in plant defense against pathogen attack by stimulating the production of H_2O_2. Treatment of plant tissues with ABA and IAA results in increased NO production, and in stomatal guard cells H_2O_2 stimulates the synthesis of NO.

In contrast to mammals, there has been no unequivocal demonstration of a plant NO synthase activity that uses arginine as the substrate for the synthesis of NO. In plants the synthesis of NO from NO_3^- or nitrite (NO_2^-) catalyzed by nitrate reductase is now well established (see Chapter 13). During **normoxic** conditions, i.e. in the presence of normal atmospheric oxygen concentrations, nitrate reductase is the source of NO produced enzymatically by plants. *Arabidopsis* has two genes encoding nitrate reductase, **NIA1** and **NIA2**, and *nia1/nia2* double mutants have a dramatic reduction in the synthesis of NO by stomatal guard cells and in the ability of these guard cells to close stomata in response to ABA. While NIA2 is the more abundant enzyme in *Arabidopsis*, mutations in *NIA1* show that production of NO by NIA1 is more important in regulating stomatal closure.

Other biosynthetic pathways in plants can give rise to significant amounts of NO, especially under conditions of limited oxygen availability, when substrates such as NO_2^- accept electrons from cytochrome oxidase, cytochrome c, cytochrome P450 or other donors (Equation 10.1).

Equation 10.1 Nitric oxide formation by reduction of nitrite

$$NO_2^- + 2e^- + 2H^+ \rightarrow NO + H_2O$$

Non-enzymatic mechanisms for NO synthesis are also believed to operate in animals and plants. The chemistry of NO production from NO_2^- is well established. Under mildly acidic conditions nitrite is protonated to form nitrous acid (Equation 10.2). Nitrous acid can then be reduced to NO by a range of biological reductants, including ascorbate and phenolic components of the plant cell wall as shown in Equation 10.3.

Equation 10.2 Protonation of nitrite

$$NO_2^- + H^+ \rightarrow HNO_2$$

Equation 10.3 Reduction of nitrous acid

$$HNO_2 + 2RH \rightarrow NO + H_2O + 2R$$

Plants have several acidic compartments that would support the production of NO from nitrite, especially the cell wall where the activity of the plasma membrane H^+-ATPase acidifies the apoplast and where phenolic components of the wall can provide the reductant to reduce nitrite to NO.

Key points Nitric oxide is a reactive gas that affects many aspects of plant growth and development, including stomatal guard cell function, lateral root initiation, flowering, germination and responses to abiotic stresses. NO has been known to be produced by plants since the 1980s but its role as a hormone has only recently been explored. NO can be synthesized from nitrate, in a reaction catalyzed by the enzyme nitrate reductase, as well as by other enzyme-catalyzed reactions, but in contrast to the situation in animals there is as yet no evidence that plants make NO from arginine using an NO synthase. A significant amount of NO can be produced non-enzymatically from nitrite in the acidic conditions found in plant cell walls.

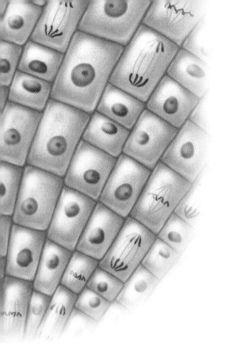

Chapter 11

The cell cycle and meristems

11.1 Introduction to cell division and meristems

In plants as in all other living organisms, every cell is a product of the division of an existing cell. Before division, the nuclear genome of the cell is replicated exactly, so that the new cell is genetically identical to its progenitor. It is crucial that the division process is regulated very precisely, because if cells divide at inappropriate times or positions, disorganized growth occurs, leading in some cases to tumor-like proliferation. The key elements of the mechanism for regulating cell division were present very early in eukaryotic evolution, and are highly conserved across all eukaryotes, but the plant cell cycle also has some unique features.

There are two types of cell cycle in eukaryotes: the vegetative, or **mitotic**, cell cycle, which in plants gives rise to all the roots, stems, leaves and other non-reproductive tissues; and its special variant the **meiotic** cell cycle, which occurs only in reproductive tissues and, in plants, produces spores (see Figure 1.6). This chapter will address both these types of cell cycle. It will provide an overview of the process of DNA replication and cell division, with an emphasis on the features that are unique to plants. Cell division takes place in meristems, regions of the plant producing new cells that can then differentiate into any one of a wide

range of cell types; regulation of the cell cycle in meristems will also be discussed.

11.1.1 The mitotic cell cycle consists of four phases: M, G1, S and G2

The two processes required for cell division to take place are **DNA replication**, and **segregation of the chromosomes** between the two daughter cells. These states are mutually incompatible: if the genome is to be passed on accurately to the daughter cell, DNA cannot be replicated while chromosomes are segregating. There must therefore be very tight regulation of the timing of these two processes. This is achieved by the organization of the cell cycle into four phases, which the cell must pass through in order, with irreversible transitions between each phase and the next. These four phases are known as **M phase**, which consists of mitosis followed by cytokinesis; gap 1 or post-mitotic interphase (**G1 phase**); DNA synthesis phase (**S phase**); and gap 2 or post-synthetic phase (**G2 phase**) (Figure 11.1). G1, S and G2 together are also known as **interphase**. During the M phase, the cell divides, partitioning its replicated genome between its two daughter cells. The M phase is usually described in terms of six stages, each of which shows a characteristic conformation of the chromosomes. These

The Molecular Life of Plants, First Edition. Russell Jones, Helen Ougham, Howard Thomas and Susan Waaland.
© 2013 John Wiley & Sons, Ltd. Published 2013 by John Wiley & Sons, Ltd.

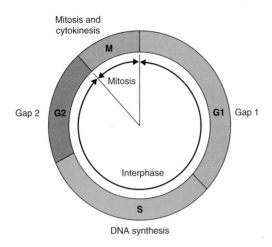

Figure 11.1   Phases of the cell cycle. Mitosis is the phase during which chromosomes condense and sister chromatids separate, followed by nuclear division. For the rest of the cell cycle, collectively known as interphase, the chromosomes are diffuse. Interphase consists of the G1 gap phase, S (DNA synthesis) phase and the second gap phase, G2.

stages, prophase, prometaphase, metaphase, anaphase, telophase and cytokinesis, are illustrated in Figure 11.2. The first five of these six stages, from prophase to telophase, constitute mitosis; cytokinesis, in which the two nuclei are segregated into separate daughter cells, completes the M phase. G1 is the phase during which cells grow by synthesizing proteins, lipids, carbohydrates and other essential molecules. In preparation for division, the DNA is replicated during the S phase, and, following the G2 gap, during which further biosynthesis of macromolecules occurs, the cell enters the next M phase.

Transitions from one phase to the next are regulated by a network of proteins that include specialized protein kinases, phosphatases and proteases. These act as switches to ensure that the correct stepwise progression occurs, so that DNA is replicated and the duplicated genetic material is properly partitioned between the two daughter cells. They also couple stages in the cell cycle with environmental conditions. The mode of regulation is such that, in a given environment, most cells divide

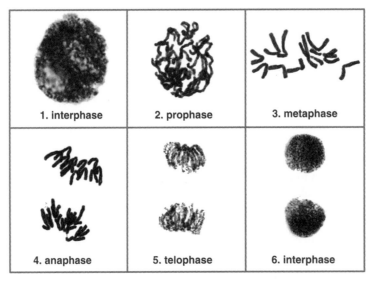

A

Figure 11.2   Mitosis in plant cells. (A) Photomicrographs of mitosis in root cells of an angiosperm with six pairs of chromosomes (2n = 2x = 12). The chromosomes are stained with Feulgen's solution, which reacts with DNA to form a colored product. (B) The first stage of mitosis is prophase, during which the nuclear membrane disappears, the chromosomes become condensed and the spindle forms. During prometaphase and metaphase the chromosomes attach to spindle fibers and move to become aligned at the metaphase plate, which is midway between the poles of the spindle. At anaphase, sister chromatids separate and are pulled towards the opposite poles. During telophase, new nuclear envelopes form around the chromosomes in the areas that will become the two daughter cells and chromosomes start to decondense. Cytokinesis begins during telophase; the cell plate (phragmoplast), which is unique to cell division in plants, starts to form between the daughter nuclei. When it reaches the sides of the cell, it fuses with the side walls of the parent cell and by the end of cytokinesis the daughter cells, each containing an intact nucleus and its own plasma membrane, are separated by a cell wall.

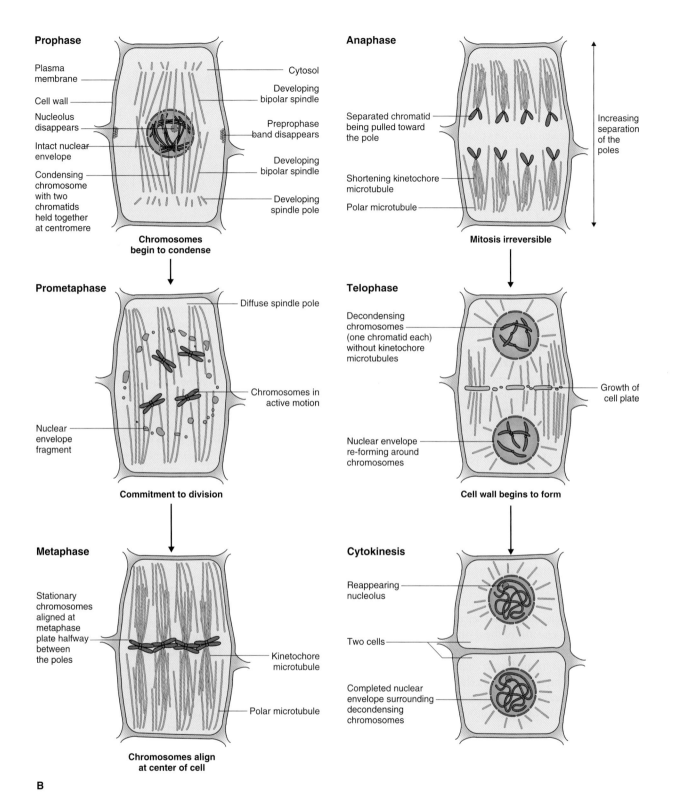

Prophase

Plasma membrane
Cell wall
Nucleolus disappears
Intact nuclear envelope
Condensing chromosome with two chromatids held together at centromere

Cytosol
Developing bipolar spindle
Preprophase band disappears
Developing bipolar spindle
Developing spindle pole

Chromosomes begin to condense

Prometaphase

Diffuse spindle pole

Chromosomes in active motion

Nuclear envelope fragment

Commitment to division

Metaphase

Stationary chromosomes aligned at metaphase plate halfway between the poles

Kinetochore microtubule

Polar microtubule

Chromosomes align at center of cell

Anaphase

Separated chromatid being pulled toward the pole

Increasing separation of the poles

Shortening kinetochore microtubule
Polar microtubule

Mitosis irreversible

Telophase

Decondensing chromosomes (one chromatid each) without kinetochore microtubules

Growth of cell plate

Nuclear envelope re-forming around chromosomes

Cell wall begins to form

Cytokinesis

Reappearing nucleolus

Two cells

Completed nuclear envelope surrounding decondensing chromosomes

B

Figure 11.2 (*continued*).

when they reach a particular threshold size. However, division normally cannot be triggered by growth to a certain size alone; cells must also be stimulated by other signals. The mechanisms that regulate the timing of cell division have another important role: that of quality control. At specific points in the cell cycle, called **checkpoints**, the cell monitors progress to ensure that incompletely replicated or damaged DNA is not passed on to the progeny cells.

11.1.2 Rigid plant cell walls impose some unique features upon the plant cell cycle

At both the DNA sequence and the function level, many of the components of the cell division cycle are highly conserved across eukaryotes, including plants. These components include the enzymes that carry out DNA replication; the cytoskeletal structures that control chromosome movements during mitosis; the protein kinases that control the major transitions between the phases of the cell cycle; and many components of the **ubiquitin–proteasome system of protein degradation** (**UbPS**; see Chapter 5). However, plant cells have evolved some differences in the way they control cell division. This is probably, at least in part, because of structural features that are unique to plant cells. One of these is the rigid plant cell wall. Late in the cell division process in animals, when the two daughter nuclei are moving apart, there is progressive constriction of the plasma membrane at the contractile ring between the two nascent daughter cells, which ultimately leads to the formation of two complete new cells. In plants, the cell wall prevents this constriction from occurring, so the two daughter nuclei are instead separated by a **cell plate** that grows at the equator of the mother cell (Figure 11.2B; see also Section 4.2.5 and Figure 4.7). This cell plate, which is formed from Golgi-derived vesicles carrying plasma membrane and cell wall components, subsequently fuses with the plasma membrane and the side walls of the parent cell resulting in two daughter cells.

Another important difference between the cell division processes in plants and animals is again due to the rigid plant cell walls. Adjacent cells in multicellular plants are firmly attached to one another by the middle lamella of the cell wall. Because of this, cells in multicellular plants cannot move, so organogenesis must depend on cell division and cell expansion at the sites where new organs are to form. Cell division takes place primarily in **meristems**, regions specialized for the production of new cells that initially are capable of becoming any one of a range of cell types; it is only as the

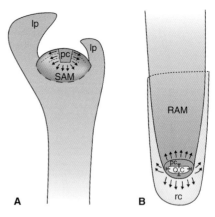

Figure 11.3   Apical meristems. Plants can form new organs and tissues throughout their life span. This is possible because they contain (A) shoot apical meristems (SAMs) and (B) root apical meristems (RAMs). Both types of meristem consist of undifferentiated cells that divide to provide new cells to the growing organs. Small populations of proliferative cells (pc) within the meristems are essential for the meristems to perpetuate themselves. Daughter cells produced by these proliferative cells form cell lines with distinct fates: one daughter cell will maintain the proliferative cells, while the other gives rise to a file of growing cells (arrows) which will eventually differentiate into specialized cell types. In the shoot apex, leaf primordia (lp), which also consist of dividing cells, are formed close to the SAM. In the root, which is covered by a protective root cap (rc), the apical meristem is just behind the cap and produces files of cells contributing to the growth of the root itself. At the boundary where the RAM and root cap meet, there is a small group of non-dividing cells, the quiescent center (QC). These cells emit signals that maintain the cells immediately around them in a proliferative, non-differentiating state.

cells are displaced from the meristem that they differentiate (Figure 11.3; see also Chapter 1).

Lastly, in contrast to mammals and many other animals, plant growth and development occurs mainly postembryonically. During plant embryogenesis, the root–shoot axis is established and some leaf primordia are formed, but plant growth takes place mostly after germination, and requires proliferation of cells in the meristems to produce new leaves, stems, roots and other structures (see Chapter 12). The regulation of cell division in plants must reflect all the structural and developmental features discussed in this section.

11.1.3 Meiosis is a specialized form of the cell cycle that gives rise to haploid cells

The **meiotic cell cycle** shares many features with the mitotic cycle: it has a DNA synthesis phase and a

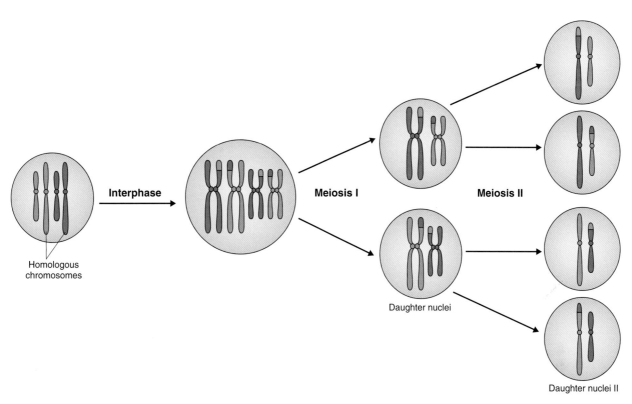

Figure 11.4 The essential features of meiosis. After DNA replication, homologous chromosomes, each consisting of two sister chromatids, pair and exchange some genetic material. There are then two rounds of cell division. During the first division homologs separate producing two haploid rather than diploid cells; during the second division sister chromatids separate producing four haploid cells. The daughter cells of meiosis are genetically different from the mother cell and from each other.

nuclear and cell division stage, separated by two gap phases. Many genes have regulatory roles in both types of cell cycle. However, because its function is to produce haploid cells from diploid cells instead of daughter cells that are identical to the parent cell, the meiotic cycle differs from the mitotic cycle in three important ways. Firstly, there are two divisions, not one, so that an individual diploid cell gives rise to four new cells (Figure 11.4) which are haploid rather than diploid. Another important difference is that during prophase of **meiosis I**, pairs of homologous chromosomes, which have already undergone DNA replication so that each consists of two sister chromosomes, are brought together and held in place along their entire lengths. In this configuration, the chromosomes undergo a process called **crossing over**, in which DNA is exchanged between the two members of the pair of homologous chromosomes. This is one of the means by which a eukaryotic organism brings about **recombination**, or reassortment, of the genetic material that it inherited from its parents, which ensures variation between its offspring. A final difference is that homologs segregate during meiosis I so the daughter cells of this division are haploid. The two cell

Key points For cell division to occur DNA must be replicated and the chromosomes must be segregated between the two daughter cells. To achieve this, the cell cycle is regulated very tightly. During mitosis (M phase) the chromosomes condense and sister chromatids separate. Mitosis is normally followed by cytokinesis in which two new daughter cells are formed. There is then a gap phase (G1), after which DNA synthesis (S phase) and a second gap phase (G2) prepare the cell for the next round of mitosis. Although the mechanism and components of the mitotic cell cycle are highly conserved across all eukaryotes, the presence of the rigid cell wall imposes some constraints that are unique to plants. Meiosis is a specialized version of the cell cycle with two rounds of division; it produces four haploid cells rather than two cells that are identical to the parent cell. During meiosis the DNA of the two parental chromosome sets can recombine, so that the progeny have new assortments of genetic material.

Table 11.1 The important differences between mitosis and meiosis.

Mitosis	Meiosis
In somatic cells	In cells in the sexual cycle
One division of the mother cell results in two daughter cells	Two divisions of the mother cell result in four haploid daughter cells
Mother cell can have any ploidy	Mother cell is always diploid (2n)
Number of chromosomes per nucleus remains the same after division	Number of chromosomes is halved in daughter cells (2n → 1n)
Division is preceded by S phase in which the amount of DNA is duplicated	Only meiosis I is preceded by S phase
Normally no pairing of homologous chromosomes occurs	During prophase I, pairing of all homologous chromosomes takes place
No exchange of DNA (crossing over) between chromosomes	During prophase I there is at least one crossing over event per homologous pair of chromosomes
Centromeres split during anaphase	Centromeres separate during anaphase II, but not during anaphase I
Genotype of the daughter cells is identical to that of the mother cells and to each other	Daughter cells differ in genotype from the mother cell and from each other

division stages, **meiosis I** and **meiosis II**, function to partition the newly recombined chromatids between the four daughter cells (Figure 11.4). Table 11.1 summarizes the similarities and differences between mitosis, meiosis I and meiosis II.

11.2 Molecular components of the cell cycle: kinases, cyclins, phosphatases and inhibitors

Coordination of the cell cycle requires the action of a large number of proteins, The key players include: kinases, which phosphorylate other proteins; cyclins, which are required to activate catalysis by many of the kinases; phosphatases, which remove phosphate groups from proteins; and kinase inhibitors. This section will introduce the most important of these proteins and describe their functions (Table 11.2).

Many of the components of the cell cycle regulatory machinery are highly conserved across all eukaryotes, but there are some, unique to plants, that will also be discussed.

11.2.1 Specific kinase complexes push the cell through the cell cycle

Progression through the cell cycle, in plants as in other organisms, is controlled by changes in the activity of specific **cyclin-dependent kinases** (**CDKs**). These kinases function in complexes, each of which is composed of two subunits: the catalytic CDK itself and a **cyclin** (**CYC**), a class of protein whose role is to activate catalysis. In most cases the CDK alone is inactive, and must associate with a cyclin as the first step in its activation. Unlike many single-celled eukaryotes, which have only a single CDK, plants and other multicellular eukaryotes have multiple CDKs. All eukaryotic cells also have multiple classes of cyclin proteins, each with a role at a specific point in the cell cycle. Different CDKs interact with different classes of cyclins. It is these individual CDK–cyclin interactions

Table 11.2 Kinases, cyclins and inhibitors of the mitotic cell cycle.

Protein	Characteristics
Cyclin-dependent kinases (CDKs): cell cycle transitions	
CDKA type	PSTAIRE domain
	Act at transitions G1 to S and G2 to M
CDKB1 type	PPTALRE domain
	Progression into/through M
	Transcripts accumulate in S, G2 and M phases, activity high in M
	Suppress exit from cell cycle
CDKB2 type	PPTTLRE domain
	Transcripts/protein accumulate in G2 and M
CDK-activating kinases (CAKs)	
CDKD	Phosphorylates CDKs and RNA pol II
	Requires CycH
CDKF	Phosphorylates CDKD
	Does not require Cyc
Cyclins (CYCs): required for CDK activity	
CYCA	Control the S to M phase transition
	Contain D-box motif
CYCB	Active both at the G2 to M transition and within M phase
	Contain D-box motif
CYCD	Control the G1 to S transition
	Growth-promoting hormones and sucrose trigger accumulation; in germinating seeds, expressed prior to the start of cell division
CYCH	Control the activity of CDKDs
CDK inhibitors (CDKIs)	
KRP	Inhibit activity of CDKA–CycD complexes
SIM, SMR	Inhibit activity of CDKA-CycD complexes

that determine the activity of the resulting enzyme complexes at given steps in the cell cycle, hence the association of CDKs with specific cyclins is considered to be an important regulatory mechanism in determining progression through the different phases of the cell cycle. However, activity of CDKs is influenced not only by cyclins but also by the phosphorylation status of a CDK, and by its association with other molecules including **CDK inhibitors** and docking factors (Figure 11.5).

11.2.2 CDK–cyclin complexes can phosphorylate protein substrates to regulate cell cycle progression and other cell processes

Once a eukaryotic CDK is complexed with a cyclin subunit, it can phosphorylate protein substrates on serine or threonine residues within a recognition motif which consists of the amino acid sequence (serine or threonine)–proline–X–(arginine or lysine), where X can be any amino acid. CDK molecules have a two-lobed structure in which the catalytic residues are located in a cleft between the N-terminal (upper) and C-terminal (lower) lobes (Figure 11.5). The amino acids required for ATP binding are close to the N terminus of the CDK protein, and are followed by a **PSTAIRE domain** (named after the single-letter codes for a sequence of seven amino acids), which is present in CDKs of all eukaryotes;

the PSTAIRE domain is required for cyclin binding. Although the first such domain that was identified contained the amino acid sequence PSTAIRE, several variants of this sequence motif have since been identified, in which one or more of the seven amino acids is different. These variations are used to classify CDKs into subgroups. CDKs that have similar or identical PSTAIRE domains generally interact with closely related cyclins.

All eukaryotes so far investigated have at least one CDK with the original PSTAIRE motif; plant CDKs in this category are known as **A-type CDKs** or **CDKAs**. They are important for the transitions from G1 to S phase and G2 to M phase, and plants with low CDK activity show a reduced rate of cell division.

One class of CDKs, the **B-type CDKs** (**CDKBs**), is unique to plants. In this class, either two or three of the amino acids in the PSTAIRE motif are substituted, giving either PPTALRE or PPTTLRE. The resulting CDKs are classified into two subgroups, CDKB1 and CDKB2. The expression patterns of the genes of these subgroups during the cell cycle differ slightly. *CDKB1* transcripts

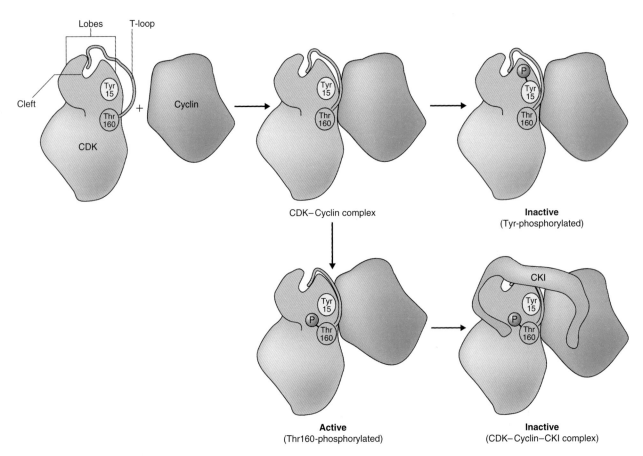

Figure 11.5 Schematic models of a CDK (cyclin-dependent kinase) and different forms of its complex with a cyclin. The CDK must associate with the cyclin in order to have protein kinase activity, and for full activity the threonine at position 160 must also be phosphorylated. In contrast, when the tyrosine residue at position 15 is phosphorylated, the activity of the CDK is inhibited. Inhibition can also be brought about by the binding of a CDK inhibitor (CKI, yellow), even though this does not alter the phosphorylation status of the CDK. The cleft between its two lobes contains the active site of the CDK.

accumulate in the S, G2 and M phases, while *CDKB2* expression occurs only during the G2 and M phases. The accumulation of each CDKB protein is closely matched to the patterns of transcription of its corresponding gene, and CDKB-associated kinase activity is at its maximum during mitosis. In particular, CDKB1 activity is essential for progression through mitosis, as shown by the observation that plants with reduced CDKB1 activity arrest at the point of transition from G2 to M. CDKB1 has additional roles in the cell cycle, indicated by its expression in all but the G1 phase. Kinase activity associated with CDKB1 has been shown to suppress the initiation of **endoreduplication** (see Section 11.4.3). Plants with reduced expression of the gene encoding CDKB2 have disorganized meristems.

At the primary protein sequence level, the amino acids required for CDK catalytic activity are located throughout the molecule. It is the folding of CDK proteins into their correct three-dimensional structures that organizes them into the appropriate spatial configuration for interaction with cyclin subunits and subsequently with their substrates. The entrance of CDK substrates to the catalytic site can be obstructed by a flexible domain, called the **T-loop** (see Figure 11.5). For substrate binding to occur, the T-loop needs to move out of the entrance to the active site. This open, accessible configuration is stabilized by the phosphorylation of a threonine residue in the T-loop by **CDK-activating kinases** (**CAKs**), which themselves are CDKs (see Figure 11.5).

Plants have two classes of CAKs: **CDKD**, which is functionally related to the CAKs of vertebrates and other animals, and **CDKF**, which is unique to plants. These classes are distinct in both their substrate specificity and their cyclin dependence or independence (Figure 11.6). In addition to phosphorylating CDKs, CDKDs also phosphorylate the C-terminal domain of the large subunit of RNA polymerase II, thereby coupling basal rates of DNA transcription to progression through the cell cycle. CDKD requires association with a cyclin, in this case a **CYCH**, for activity. In contrast, CDKF activation does not require any association with cyclins. Interestingly, there is a functional association between CDKD and CDKF: CDKF operates as a CAK-activating kinase phosphorylating and thereby activating CDKD. In turn CDKF itself is phosphorylated at a conserved site, amino acid Thr290, by a further upstream kinase which activates CDKF. This complex phosphorylation cascade is believed to link the activation of CDKs with developmental pathways and environmental inputs.

In addition to the well-characterized A-type, B-type, D-type and F-type CDKs, plants contain two classes of more distantly related CDKs, known as C-type (**CDKC**) and E-type (**CDKE**) CDKs. Neither class appears to have

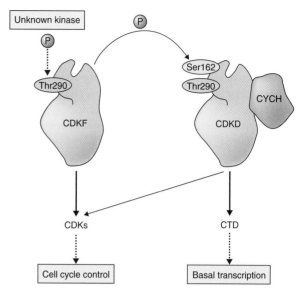

Figure 11.6 Roles of CDKF, CDKD and cyclin H (CYCH) in cell cycle control and basal transcription. CDKF is activated by phosphorylation of Thr290. Active CDKF is responsible for phosphorylation and activation of CDKD, which forms a stable complex with CYCH that phosphorylates both the C-terminal domain (CTD) of RNA polymerase II and several CDKs.

a role in the control of the cell cycle. CDKCs phosphorylate the C-terminal domain of the RNA polymerase II, presumably playing a role in DNA transcription. They also co-localize in the cell with splicing factors, and appear to link transcription with splicing of introns from nascent mRNA, ensuring that the mRNA is correctly spliced immediately after it is synthesized. The only known function of CDKE is during floral organ development, where it appears to control cell fate by a mechanism that is not well understood.

11.2.3 Binding of cyclins determines three-dimensional structure, specificity and subcellular localization of cell-dependent kinases

Binding of cyclins to CDKs has multiple consequences. First, it induces structural changes that are needed for recognition of the CDK's protein substrate including the repositioning of the T-loop to allow the substrate access to the catalytic site (see Figure 11.5). Phosphorylation by a CAK then stabilizes this open configuration as described in Section 11.2.2. Second, the identity of the cyclin that participates in the cyclin–CDK complex confers substrate specificity upon the complex. And third, cyclins have a role in targeting CDKs to specific subcellular compartments at appropriate stages during the cell cycle. All eukaryotes contain multiple cyclins, but plants have many more than have been identified so far in other organisms. For example, the genome of *Arabidopsis* has genes capable of encoding up to 50 different cyclins. Plant cyclins (CYCs) are classified into eight types, A, B, C, D, H, L, P and T, using nomenclature based on their functional similarity to the mammalian equivalents. A domain called the **cyclin box**, which interacts with the kinase subunit, is conserved across all eukaryotes, and it is variations in this domain which are used as the basis for classification. Of the eight classes of plant cyclin, four, the A-, B-, D- and H-types, have so far been shown to have roles in cell cycle progression. The H-type cyclins control the activity of CDKDs, while each of the other three classes is believed to control one or more of the transitions between phases of the cell cycle. D-type cyclins control the G1 to S transition; A-type cyclins control the S to M phase transition; and B-type cyclins function both at the G2 to M transition and within the M phase.

The **D-type cyclins** (**CYCDs**) are found in the majority of eukaryotes, but their sequences vary considerably. In mammals, the accumulation of D-type cyclins is stimulated by growth-promoting hormones. While the mechanisms of growth stimulation in plants are very different from those in animals, nevertheless certain plant hormones (e.g. cytokinins, brassinosteroids, gibberellins) and sucrose cause accumulation of plant CYCDs. This observation suggests that D-type cyclins have a role in coordinating the plant's hormonal and nutritional status with the progression of the cell cycle. In germinating seeds, D-type cyclin genes are expressed prior to the start of cell division, and knocking out CYCD expression causes a delay in activation of the cell cycle in the embryonic root meristem (Figure 11.7).

The **A-type cyclins** (**CYCAs**), which regulate progression through the S phase, and the **B-type cyclins** (**CYCBs**), all contain a protein domain, known as the **mitotic destruction box** or **D-box**, which is required to

ensure their destruction late in mitosis. In addition, plant *CYCB* genes contain a conserved cis-acting element, the **M-specific activator** (**MSA**) element, which is required to ensure that *CYCB* genes are expressed specifically at the G2 to M transition, in accordance with their role in triggering this transition. Plant B-type cyclins have been shown to initiate entry into mitosis even in very different organisms, such as *Xenopus* oocytes.

11.2.4 Kinases, phosphatases and specific inhibitors all have roles in regulating the activity of CDK–cyclin complexes

Specialized kinases and phosphatases regulate the activities of CDK–cyclin complexes in all eukaryotes. In yeast and mammals, the activity of CDK–cyclin complexes can be switched off or on by the actions of **WEE1 kinase** and **CDC25** (cell division cycle 25) **phosphatase** respectively. WEE1 phosphorylates CDKs, thus inactivating them. (The WEE1 kinase is so named because, in yeast, mutations in the *WEE1* gene lead to premature cell division and the production of daughter cells that are smaller than usual, or 'wee' in Scottish dialect.) Phosphorylation of the tyrosine residue Tyr15 in CDKs by **WEE1 kinase** causes inhibition of CDK activity in both yeast and vertebrates; vertebrates additionally phosphorylate the Thr14 residue. This phosphorylation is reversed by **CDC25** (Figure 11.8). There is evidence that plants also downregulate CDKs using phosphorylation (see Figure 11.5). For example, Tyr phosphorylation in CDKA has been detected under stress conditions, including DNA damage, where cell cycle arrest is required. The observation that overexpression of the plant gene encoding WEE1 kinase inhibits cell cycle progression suggests that WEE1 is responsible for this Tyr phosphorylation in plants as in other organisms. On the other hand, there is no obvious homolog to CDC25 in any higher plant genome so far sequenced, suggesting

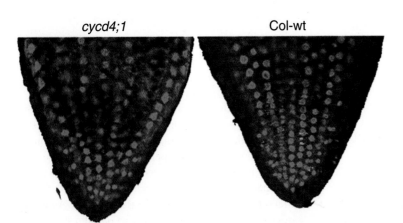

Figure 11.7 Photomicrograph of root apices of *Arabidopsis* seedlings 14 hours after the start of germination. Mitotic spindles have been stained pink. The right-hand panel shows a control root tip (Columbia wild-type, Col-WT) in which many mitotic spindles are visible; the left-hand panel shows a root from a mutant plant (*cycd4;1*) in which the *CYCD* gene has been inactivated, resulting in fewer cells undergoing mitosis.

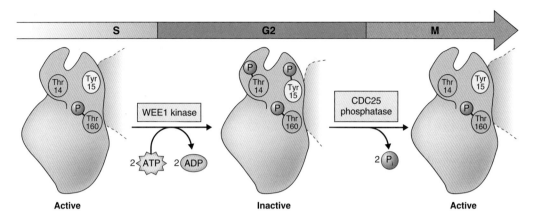

Figure 11.8   Regulation of protein kinase activity of a yeast (*Schizosaccharomyces pombe*) CDK by WEE1 kinase and CDC25 phosphatase. When WEE1 kinase phosphorylates two adjacent amino acid residues, Thr14 and Tyr15, the protein kinase activity of the CDK is inhibited. CDC25 removes the phosphate residues and restores protein kinase activity to the CDK. Plants do not appear to have a CDC25 homolog.

that Tyr CDK dephosphorylation may be carried out by a different, and as yet unidentified, phosphatase, possibly one unique to plants.

In addition to phosphorylation and dephosphorylation, CDK activity can be fine-tuned by **CDK inhibitors** (**CKIs**). When inhibition is required, CKIs associate with the activated CDK–cyclin complex and prevent it from phosphorylating its protein substrates. Inhibition of CDK activity by CKIs is reversible and does not change the phosphorylation status of the CDK–cyclin complex (see Figure 11.5). Plants contain two classes of CKIs. While both classes interact with CDKA–CYCD complexes, experimental evidence suggests that they are triggered by different stimuli.

The first class of CKIs in plants has a highly conserved C-terminal domain that is required for binding CDKs and cyclins. This domain is similar to a region at the N terminus of the mammalian CIP/KIP class of CKIs, so plant CKIs in this class are often referred to as **Kip-related proteins** or **KRPs**. Elevating KRP above normal levels suppresses CDK activity and leads to a strong inhibition of cell division, dwarfism and other changes in plant morphology (Figure 11.9A, B). The KRPs mainly inhibit CDKA–CYCD complexes. As discussed above, CYCDs have a role in plant responses to growth hormones; treatment of plants with growth-promoting hormones increases CYCD accumulation. In contrast, treatment of plants with abscisic acid, a hormone that inhibits cell division, induces transcription of the *KRP1* gene (also known as *ICK1*), suggesting that the KRP1 protein probably plays a role in mediating the cell cycle arrest in these plants.

The second class of plant CKIs is known as **SIM** and **SMR** proteins. The first gene in the family to be identified was that encoding the SIAMESE (SIM) protein in *Arabidopsis*. The *SIM* gene was found to suppress cell

division in trichomes (Figure 11.9C, D). SIM and the SIAMESE-RELATED (SMR) proteins all have a six-amino acid domain in common with KRPs, and this domain is required for the binding of cyclins and for inhibiting CDK activity. Like the KRPs, SIM and SMR proteins inhibit CDKA–CYCD complexes. The observation that the genes encoding SIMs and SMRs are strongly upregulated under a range of stress conditions suggests that they may regulate the cell cycle in response to biotic and abiotic stresses.

11.2.5 Proteolysis by the ubiquitin–proteasome system ensures that the cell cycle is irreversible

Proteolysis of specific proteins at key points in the cell cycle ensures that the cycle can only move in one direction. For example, towards the end of the G1 phase, degradation of CKIs in the KRP family (see Section 11.2.4) leads to an increase in CDK activity, particularly that of CDKAs. This increased activity is essential if the cell cycle is to progress to the S phase. Proteolysis during the cell cycle is carried out by the ubiquitin–proteasome system (UbPS; see Section 5.9), in which target proteins are tagged with the polypeptide ubiquitin, identifying them for degradation by the 26S proteasome. For this tagging to take place, the protein targets and ubiquitin-carrying enzymes must be brought together through the action of **ubiquitin ligases** (**E3s**). Two related E3 complexes have been shown to be important in cell cycle control; they are the **SKP1/CULLIN/ F-box (SCF) related complex**, which operates at the G1 to S transition, and the **anaphase-promoting complex** (**APC**), which acts in the M phase.

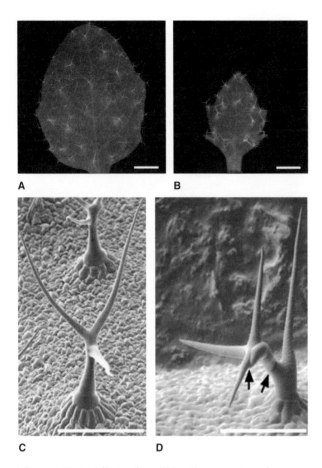

A   B

C   D

Figure 11.9   Effects of modifying the expression of genes encoding CDK inhibitors (CKIs). (A, B) Rosette leaves from (A) an *Arabidopsis* wild-type (Columbia) plant and (B) a transgenic plant expressing the *Arabidopsis KRP1* gene under the control of a promoter which is specific for expression in the cell line that gives rise to stomata. This abnormal expression of *KRP1* in stomatal guard cells results in smaller leaves and an altered leaf morphology because leaf cell numbers are reduced. (C, D) Knocking out the *SIM* gene in *Arabidopsis* causes cell division within trichomes, where it is normally suppressed. (C) Scanning electron micrograph of a wild-type trichome which consists of just one cell. (D) Scanning electron micrograph of a multicellular *sim* mutant trichome. Arrows indicate junctions between adjacent cells of the trichome. Scale bars = 1 mm (A), 100 μm (B) and 200 μm (C, D).

The SCF complex and its role in hormonal signal transduction are described in Chapter 10. Like most, though not all, SCFs, those that have roles in cell cycle regulation only recognize phosphorylated proteins as substrates for ubiquitination. In mammals, D-type cyclins are degraded in a SCF-dependent manner, and it is likely that the same is true for plant CYCDs. Other cell cycle regulators discussed in this chapter, including E2Fc and several of the CKIs, are also ubiquitinated by SCF-type E3 ligases.

In contrast to the SCF complex, the APC can recognize proteins that are not phosphorylated. It

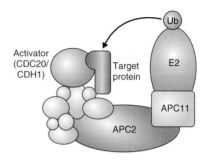

Figure 11.10   The anaphase-promoting complex (APC; an E3 ubiquitin ligase), a large protein complex that contains 11 different subunits, is important in regulating the G1 to S transition in the cell cycle. E2 ubiquitin-conjugating enzymes interact with the APC by associating with its APC11 subunit. Transfer of ubiquitin (small circle) to a lysine residue on a target protein substrate requires that the substrate is recognized by the APC via an activator protein, CDC20 or CCS52. The interaction between the APC and the substrates is mediated through the presence of specific amino acid motifs (such as the D-box or KEN box) in the substrate protein sequences. Once a protein has been polyubiquitinated, it can be recognized by the 26S proteasome and undergo proteolytic degradation.

consists of at least 11 proteins (Figure 11.10), and like SCF it can exist in several forms depending on which adaptor proteins are associated with it. Adaptor proteins of the CDC20 and CDH1 families contribute to the activation and the substrate specificity of the APC complex. The *Arabidopsis* genome has five CDC20-encoding genes. It also contains genes encoding three CDH1-related proteins: CCS52A1, CCS52A2 and

Key points   The cell cycle is a complex process in which many proteins have roles. One group of key players is the cyclin-dependent kinases (CDKs). Most CDKs require the presence of cyclins in order to gain activity; cyclin binding determines the three-dimensional structure and substrate specificity of the CDK, and often its subcellular localization. CDK–cyclin complexes can phosphorylate protein substrates, an important mechanism for regulating progression through the cell cycle. The CDK–cyclin complexes are themselves regulated by many proteins, including kinases, phosphatases and specific inhibitors. The irreversibility of the cell cycle is ensured by proteolytic removal of components once they have fulfilled their roles; this is mediated by the ubiquitin–proteasome system. Two ubiquitin E3 ligases are especially important in cell cycle control: the SCF-related complex operates at the G1 to S transition, and the anaphase-promoting complex (APC) acts during the M phase.

CCS52B. The *CCS52B* gene is expressed from G2/M to M, whereas *CCS52A1* and *CCS52A2* are expressed from the late M phase until early in G1. These observations suggest that the reason for multiple genes of this family of APC activators is that they carry out functions at consecutive stages in the plant cell cycle. The CCS52 proteins also interact with different subsets of the mitotic cyclins. The APC recognizes, among others, proteins containing D-box sequences, such as A- and B-type cyclins, which must be removed to ensure exit from mitosis. In these cyclins the presence of the D-box results in their destruction via proteolysis.

11.3 Control of progress through the cell cycle

This section describes the progress of the cell through the mitotic cell cycle, taking as its starting point the transition from the G1 gap phase into S phase, when DNA is replicated. The roles of the kinases, cyclins and other regulatory proteins, which were introduced in Section 11.2, are described here in the context of the other components of the cell cycle machinery. Figure 11.11 summarizes the stages of the cell cycle and the key molecular events occurring at each stage.

11.3.1 Transition from G1 to S phase is controlled by the interaction between CDK–CYCD complexes and the RBR/E2F pathway

Regulation of entry into the S phase from G1 is carried out by a pathway which is remarkably highly conserved across all eukaryotes, the **RBR/E2F pathway**. The entry into S is initiated by the synthesis of D-type cyclins. These cyclins form complexes with CDKs and phosphorylate a protein known in mammals as **pRB** (retinoblastoma tumor-suppressor protein) and in plants as **RBR** (RB-related). The RB/RBR protein regulates cell proliferation by associating with members of the **E2F transcription factor** family of proteins (Figure 11.12). Transcription factors in this family activate genes that encode proteins that are essential for DNA replication. Thus E2F transcription factors are crucial for the cell's decision to proceed past the G1 to S restriction point and enter the DNA synthesis phase. When RBR binds to an E2F transcription factor, the **transcriptional activation domain** of the E2F protein is masked, inactivating it. Under certain conditions, RBR also stimulates the recruitment of DNA-modifying

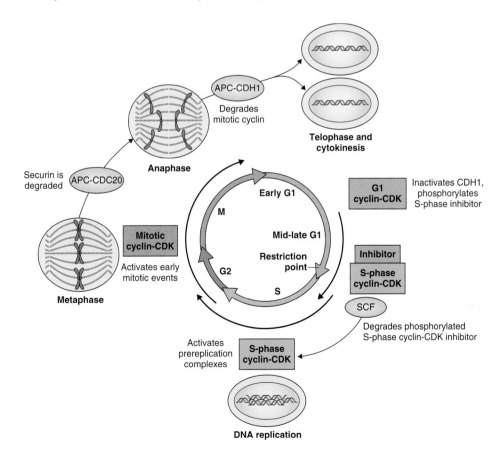

Figure 11.11 Stages of the mitotic cell cycle and the key molecular events at each stage.

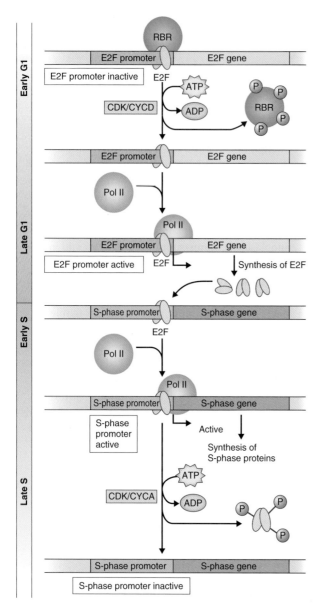

Figure 11.12   Regulation of the G1 to S transition by RBR and E2F transcription factors. In early G1, binding of RBR to the E2F promoter prevents transcription of the *E2F* gene. Once the cell has passed through the restriction point, the CDK–CYCD complex phosphorylates RBR, and the phosphorylated RBR dissociates from E2F allowing the *E2F* gene to be transcribed by RNA polymerase II (Pol II) and for E2F protein to accumulate. E2F binds to the promoters of S-phase-specific genes to stimulate their transcription, leading to the synthesis of S-phase proteins. Later in S phase, phosphorylation of E2F by the CDK–CYCA complex inactivates E2F-dependent S-phase transcription.

proteins that can induce **chromatin condensation** and repress the promoter activity of the genes that are targets of the E2F transcription factors. In late G1, when RBR undergoes phosphorylation by CDK–CYCD complexes,

its inhibitory function is suppressed, freeing E2Fs in a transcriptionally active form. These E2Fs can then promote transcription of genes encoding enzymes required for DNA replication. The genome of *Arabidopsis* contains one gene (*RBR1*) encoding an RB-related protein and three genes for E2F transcription factors (*E2Fa*, *E2Fb*, *E2Fc*). RBR also has an important role in non-dividing cells, since it is required for differentiation of male and female gametophytes (discussed in Chapter 16); RBR-deficient cells show a delay in cell differentiation.

Conservation of this mechanism is so high that even the DNA cis-acting element (TTTCCCGC) in the promoter region of E2F target genes is identical among plants, mammals and other eukaryotes. All E2Fs contain a single conserved DNA-binding domain; in most cases this is followed by the so-called DP (dimerization partner) domain (Figure 11.13A). This region allows the E2Fs to dimerize with members of the DP protein family, and the heterodimeric proteins so formed contain a second DNA-binding domain. The new domain formed by this dimerization is required for sequence-specific binding to the promoter regions of many E2F target

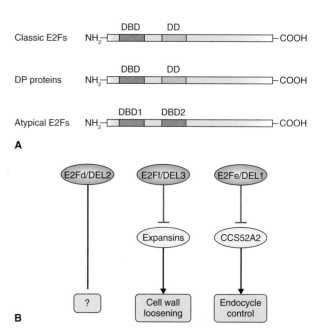

Figure 11.13   Typical and atypical E2F transcription factors. (A) Typical E2F proteins and the related DP proteins contain both a DNA-binding domain (DBD) and a dimerization domain (DD). Atypical E2F proteins lack a DD, but contain two DBDs. Typical E2F must dimerize with a DP protein in order to bind DNA, whereas the presence of the two DBDs enables the atypical E2Fs to bind DNA as monomers. (B) Overview of the biological roles of atypical E2F proteins (E2Fd/DELs) so far identified in *Arabidopsis*. Expansins are cell wall loosening enzymes; CCS52A2 is an activator protein that regulates the ubiquitin ligase activity and substrate specificity of the APC, and controls progression into the endocycle.

genes. E2Fa and E2Fb also contain transcriptional activation domains, DNA sequences which promote the assembly of RNA polymerase II complexes at the TATA box of genes and increase transcription of those genes. Overexpression of these E2Fa and E2Fb genes induces cell proliferation. In contrast, E2Fc lacks an obvious activation domain, and it functions as a negative regulator of E2F-responsive genes: when overexpressed it inhibits cell division, whereas plants lacking functional E2Fc protein show excessive cell proliferation. At the G1 to S transition, E2Fc undergoes phosphorylation by CDKs, which targets it for destruction in a SCF-dependent manner. E2Fc removal combined with RBR inactivation by phosphorylation results in increased expression of a number of genes required for DNA synthesis, including replication factors such as DNA polymerases and enzymes of nucleotide biosynthesis. Recently, a new class of E2F-related proteins has been identified in plants. These proteins can bind E2F target genes without the participation of DP proteins because they themselves contain a duplicated DNA-binding domain (Figure 11.13A). These atypical E2F proteins, which were first discovered in *Arabidopsis*, are known either as **DP-E2F-like** (**DEL**) proteins or as **E2Fd-E2Ff**. They lack transcriptional activation domains, and are postulated to act as repressors. They have not as yet been shown to play a role in the timing of the onset of S phase; however, results obtained using plants in which the genes have been inactivated suggests that they may have a function in coordinating the relationship between cell division and post-mitotic cell differentiation. One member of the family, E2Ff/DEL3, inhibits cell elongation in dividing cells by transcriptional repression of genes encoding cell wall-modifying enzymes such as **expansin**, while another member, E2Fe/DEL1, prevents cells that are undergoing division from entering the endocycle prematurely (see below) (Figure 11.13B). Although they were initially discovered in plants, DEL-related proteins are also found in animals, where they play an important role in the regulation of cell proliferation and apoptotic cell death.

11.3.2 S-phase progression is controlled by many proteins

In all eukaryotes, replication of DNA is initiated at multiple origins in the genome. The mechanism consists of four steps: recognition of an origin by an **origin recognition complex** (**ORC**); assembly of a **pre-replication complex** (**preRC**); activation of **DNA helicase**; and loading of the replication machinery onto the DNA at the origin (Figure 11.14). **Origins of replication** are determined not by primary DNA sequences but by structures dictated by epigenetic

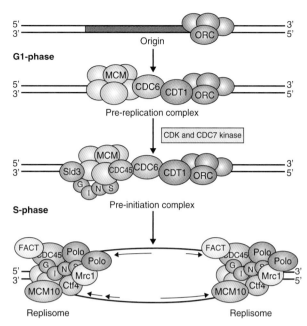

Figure 11.14 Assembly of the pre-replication complex. The replication proteins are associated with the origin recognition complex (ORC) in a carefully regulated series of steps, to avoid premature initiation of DNA replication. CDC6 and CDT1 proteins are the first to bind, followed by MCM proteins and the CDC45 protein, thus forming the pre-replication complex. This complex is activated by protein phosphorylation, carried out by CDKs and CDC7; this is the point in the cell cycle at which a cell becomes committed to a round of cell division, and defines the start of S phase. Next, the replisome complex, which contains many different replication proteins, including the DNA polymerases, is loaded, and the replication fork can be established.

determinants (see Chapter 3), with histone acetylation and histone methylation both playing roles. DNA synthesis from different origins is initiated at different points during the S phase; origins of DNA replication are classified into early, intermediate and late types, but it is not yet clear what determines into which category a given origin will fall.

The control of DNA replication is complex; many proteins interact either directly or indirectly with the origins of replication, and these interactions begin as early as the end of the previous S phase. Successful initiation and completion of DNA replication also require many other proteins which do not themselves directly regulate replication. The pathway for biosynthesis of nucleoside triphosphates is stimulated shortly before S phase, to ensure an adequate supply of the substrates for DNA synthesis. Expression of histone genes and synthesis of the corresponding proteins are also upregulated in early S phase, so that the newly replicated DNA can properly assemble into chromatin.

The first step leading to replication is the assembly of a pre-replication complex (Figure 11.14). During the S to early G1 phases, an **ORC** binds to the DNA. ORC is a complex consisting of the six proteins **ORC1** to **ORC6** in equal amounts, and the point where it binds to the DNA becomes a docking platform for other proteins. Next, during early G1, the **minichromosome maintenance (MCM) helicase**, including a complex of six proteins, is loaded onto the ORC. The protein CDC6, an ATP-binding protein, is recruited onto the ORC and is essential for loading MCM proteins onto the DNA. The related MCM proteins form a doughnut-shaped structure around the DNA helix. This complex will unwind the DNA, initially at the replication origin and subsequently, as the rest of the replication machinery travels along the DNA, it will continue to unwinding the DNA at the replication forks. The ORC and MCM, together with CDC6 and CDT1, bound to the DNA constitute the **preRC**.

At this point the DNA is prepared for replication, but its actual onset at the start of the S phase requires the assembly of a **replisome** (a complex containing many enzymes including DNA polymerases) at the origin of replication, where the preRC is already present. Loading of the replisome creates a larger complex, the **pre-initiation complex** (**preIC**). This opens the DNA helix, stabilizes the resulting single-stranded DNA, and gives DNA polymerases access so that they can copy the DNA, incorporating successive nucleoside triphosphates into the new strand. It is important that helicase activity does not lead to unreplicated single-strand DNA, so helicase activation is tightly coupled to replisome loading.

Once replication has been initiated, the MCM helicase complex starts to unwind the DNA, initially at the replication origin. As the rest of the replication machinery travels along the DNA, the helicase continues to unwind the DNA, proceeding in two directions from each origin, producing two Y-shaped DNA structures, called **replication forks**. Each individual replicating unit on the DNA is known as a **replicon**. Eventually, as replication proceeds along the DNA in both directions, replicons 'meet' and fuse, so that at the end of the S phase the complete genome has been duplicated. Both assembly of the replisome and normal progression of replication require the presence of a complex of four small proteins: Sld5, Psf1, Psf2 and Psf3. The complex is called the **GINS complex**, after the Japanese words for 5, 1, 2 and 3: go–ichi–ni–san.

11.3.3 DNA replication is strictly controlled during the cell cycle

During the mitotic cell cycle, initiation of DNA synthesis can only occur during the S phase; it is inhibited during the G2, M and G1 phases. This prevents two potential problems. First, if DNA were synthesized during late M or G1 the synthesis would be occurring when the cells should only be dividing or growing. Second, DNA synthesis in G2 and early M would lead to a change in ploidy (DNA content and genome copy number) and would also interfere with chromosome segregation. Therefore, although the preRC is present on the DNA from the end of the preceding S phase, it is only when the preIC is formed that DNA replication can begin, thus ensuring that replication occurs only once during each round of the cell cycle. The exception is the special situation of endoreduplication, discussed in Section 11.4.3, which is a variant of the cell cycle that does not include mitosis or cell division.

At the transition from metaphase to anaphase, the APC is activated and destroys several proteins, including the mitotic cyclins, resulting in a dramatic fall in CDK activity. It is this reduced activity that allows CDC6 to accumulate and interact with ORCs, since CDC6 would otherwise be phosphorylated by CDK and marked for destruction by the UbPS. Thus CDC6–ORC complexes, which are required for recruitment of the MCM complex onto the replication origin, can form only when mitosis is complete and CDK activity drops, but before the S phase begins (see Figure 11.14).

A critical point in the cell cycle occurs at the so-called **restriction point** in the late G1 phase. Here the cell must make the irreversible decision whether or not to divide. Once this point has been passed and the decision taken, the cell is committed to the S phase. Expression of S-phase genes, such as those that encode histones, is stimulated, and the S phase itself is then initiated through the activation of preRCs on the DNA by phosphorylation. This requires the action of CDKs together with the **Dbf4-dependent kinase** (**DDK4**), a heteromeric protein kinase complex (a catalytic CDC7 subunit and a regulatory dumbbell-forming protein 4 (DBF4) subunit) attached to origins of replication. DDK4 phosphorylation is believed to cause a conformational change in the MCM5 protein. This triggers the helicase, and also promotes binding of the CDC45 protein which, with the GINS complex, brings about loading of the replisome. The GINS complex, when phosphorylated by CDKs, causes binding of the DNA polymerase subunit B, which then recruits the replisome with CDC45. Thus phosphorylation of MCM by DDK4 is a key event in switching on DNA replication. In order to prevent reactivation of origins of replication, and ensure that DNA is copied only once during each S phase, CDC6 and other replication proteins undergo CDK-dependent phosphorylation, targeting them for destruction.

The replication of nuclear DNA during the S phase can be considered to be the critical event of the whole cell cycle. However, as we have seen, its occurrence depends

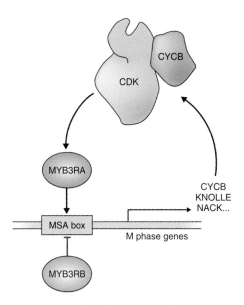

Figure 11.15 Binding of MYB3R proteins to the M-specific activator (MSA) element. The MSA element is necessary, and sufficient on its own, to drive transcription of genes that are expressed specifically at the G2 to M transition. *MYB3RA* genes display a transcription profile that is dependent on the phase of the cell cycle, whereas levels of MYB3RB protein remain constant throughout the cell cycle, and MYB3RB cannot activate MSA-containing promoters but instead inhibits transcription of the corresponding genes. The extent to which MYB3RA activates transcription of genes with MSA-containing promoters depends on the cell cycle phase and also on the extent to which its C-terminal domain is phosphorylated by CDKs. Thus the cyclins, whose synthesis has been induced by MYB3RAs, form complexes with CDKs that in turn superactivate MYB3RA activity.

on processes that occur during the M and G1 phases, during which protein complexes that mediate the initiation of DNA synthesis are assembled on the genome. It is the presence of these complexes that confers upon the cell competence to initiate DNA replication once the S phase is entered. During the replication process itself, assembly of these protein complexes is inhibited, just as DNA replication is inhibited during G2, M and G1 phases.

11.3.4 The MSA element in plant B-type cell-dependent kinases is important for the G2 to M transition

Once DNA replication is complete, the cell progresses to G2, where it is prepared for entry into mitosis. This requires synthesis of the M-phase-specific cyclins, which takes place in late G2. Then, at the G2 to M transition,

there is a rapid increase in mitotic CDK activity, which initiates mitosis and cytokinesis, beginning with the condensation of the chromosomes (see Figure 11.2B). In plants, evidence indicates that the CDKs that regulate the transition from G2 to M are CDKA and CDKB, which are both active at this point in the cell cycle but inactive later on. Both A- and B-type cyclin genes are also strongly transcribed during the G2 and M phases, and their products are considered to be important for mitotic events.

Plant B-type *CDK* genes have a cis-acting element, the **MSA (M-specific activator) element**, which is also found in other genes that are expressed during mitosis such as those encoding kinesins, the motor proteins required for the formation of the mitotic spindle and therefore for chromosome separation. The MSA element is both necessary and sufficient to direct gene expression specific to the G2 to M transition. It binds two types of proteins, designated **MYB3RA** and **MYB3RB**, that are structurally similar to animal c-Myb proteins, a class of transcription factors that regulates cell differentiation and, if overexpressed, can cause cancers and autoimmune diseases. The transcription of *MYB3RA* genes is dependent on cell cycle phase, whereas *MYB3RB* gene expression remains constant throughout the cell cycle. When MYBR3A proteins bind to a MSA element, they promote gene expression; however, when MYBR3B proteins bind to the MSA element, they cannot activate it. A model has been proposed in which MYB3RA and MYB3RB proteins act antagonistically to regulate the expression of genes specific to the G2 to M transition (Figure 11.15). The cyclins, whose synthesis is promoted by MYB3RA protein binding to the MSA element, form a complex with CDKs that leads to increased activation of MYB3RA activity in a positive feedback loop. This probably facilitates the expression of mitosis-specific genes in a sharp peak at this point in the cell cycle.

11.3.5 Condensation of replicated chromosomes marks the beginning of the M phase

As we have seen, although mitosis is suppressed during the G1, S and G2 phases, the process that promotes mitosis begins as early as the S phase. Prophase of M starts with the initiation of **chromosome condensation** (Figure 11.16), followed by the disassembly of the nuclear envelope, which normally maintains the nuclear matrix in a separate compartment from the cytosol. Chromosome condensation results in a much higher DNA packing density than is present throughout the rest of the cell cycle, and the chromatin structure is completely remodeled. Histone H1 becomes

Key points The transition from G1 to S phase is regulated by a highly conserved pathway. Entry into S phase is initiated by synthesis of D-type cyclins, which form complexes with CDKs and phosphorylate a protein called RBR. This protein associates with members of the family of E2F transcription factors, so that they can activate genes encoding proteins that are essential for DNA replication and thus for the cell to enter the DNA synthesis phase. The replication of DNA is initiated at multiple origins in the genome. The initiation process consists of four steps: recognition of an origin by an origin recognition complex (ORC); assembly of a pre-replication complex (preRC); activation of DNA helicase; and loading of the replication machinery onto the DNA at the origin. DNA synthesis then proceeds from the multiple origins. During the cell cycle, DNA synthesis is completely inhibited throughout all stages other than the S phase. The transition from G2 to M phase is regulated by A- and B-type CDKs, which are active at this point in the cell cycle but inactive at later stages. Plant B-type CDK genes have a cis-acting element, the MSA (M-specific activator) element, which is also found in other genes expressed during mitosis. The MSA element is both necessary and sufficient to activate the expression of genes that are specific to the G2 to M transition.

phosphorylated and is no longer able to bind to the linker DNA between pairs of nucleosomes (see Chapter 3). Initially, in early prophase, the chromosomes (each consisting of two **sister chromatids**) condense and begin to shorten. As further condensation occurs during late prophase, prometaphase and early metaphase, the **DNA packing density** increases still further when **topoisomerases** introduce supercoiling, and the sister chromatids become distinguishable as separate structures.

The complete sequence of chromosome condensation, chromatid cohesion and subsequent chromatid separation requires the participation of several classes of protein. The condensing chromosome assembles on a central axis of **cohesins, condensins** and other scaffold proteins. These are responsible for packaging the chromatin into compact chromosomes in a configuration that allows accurate segregation of the replicated DNA. Cohesins, synthesized during the S phase, interact with both sister chromatids to establish their tight linkage and to maintain it until the beginning of anaphase, when the chromatids begin to separate. Cohesion occurs along the whole chromosome, at both the centromere and the chromosome arms. Condensins

mediate chromosome shortening. Condensation and cohesion of sister chromatids are tightly coupled, and both cohesins and condensins are essential for normal plant development.

At the end of prophase, as the nuclear envelope breaks down, spindle microtubules begin to interact with chromosomes. The spindle is a specialized component of the cytoskeleton which assembles prior to cell division, in both mitosis and meiosis, in order to carry out the separation of the chromosomes into the daughter cells. Spindle fibers attach to the replicated chromosomes via their **kinetochores**, large protein complexes that are attached to the chromosomes at their centromeres (Figure 11.16). Kinetochore assembly is coupled with the completion of the DNA replication. The molecular composition of the kinetochore complex is still poorly understood. Kinetochore microtubules are responsible for moving chromosomes to the metaphase plate and subsequently pulling sister chromatids to the spindle poles in anaphase.

11.3.6 Chromatid separation and exit from mitosis is mediated by phosphorylation of the anaphase-promoting complex by cell-dependent kinases and proteolysis of securin

At metaphase, replicated chromosomes are aligned at the midpoint of the spindle, the metaphase plate. Although they are still attached to each other, the sister chromatids are connected to opposite ends of the cell by kinetochore microtubules. At this point, they are competent for chromatid separation and once the link between sister chromatids is severed, chromosome segregation can take place. The separation of sister chromatids is brought about by an enzyme, **separase**, that breaks down the cohesins which link them. To prevent premature separation of the chromatids, an inhibitory protein, **securin**, binds to the separase protein. Securin not only inhibits separase action early in metaphase, it ensures that the separase is correctly targeted to the sites where it will act to promote chromatid separation in late metaphase.

To ensure that separation of sister chromatids takes place in an ordered manner, the cell monitors whether the kinetochores of each pair of sister chromatids are correctly connected to the opposite poles of the spindle by the spindle microtubules. Until this has been achieved, the separase inhibitor securin continues to suppress the separation of sister chromatids. At the end

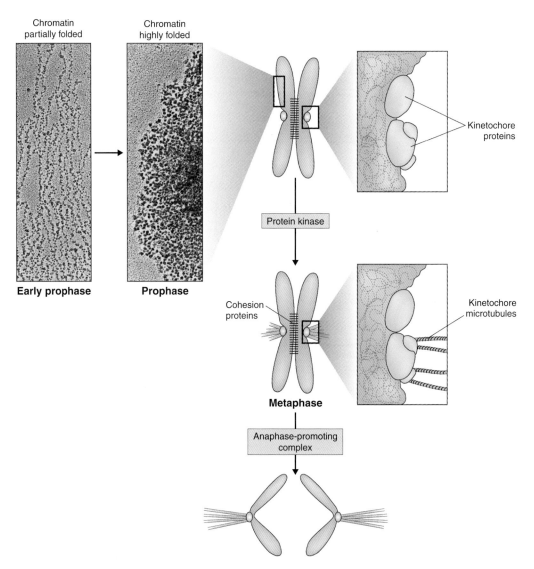

Chromatin
partially folded

Chromatin
highly folded

Early prophase   **Prophase**

Kinetochore
proteins

Protein kinase

Cohesion
proteins

Kinetochore
microtubules

Metaphase

Anaphase-promoting
complex

Figure 11.16 Chromosome condensation and the kinetochore complex. Except during mitosis and meiosis, the DNA molecule is wound around histones and packaged into higher-order chromatin structures. This protects it from shearing while still allowing transcription factors and RNA polymerase access to the DNA so that gene expression can occur. However, this organization of chromatin in non-dividing cells is still too diffuse to permit sister chromosomes to segregate in mitosis without tangling. Chromosome condensation therefore takes place during prophase (electron micrographs on left) in preparation for chromosome segregation. Kinetochores are protein complexes that bind to centromeres of the condensed chromosomes and act as attachment points for the spindle microtubules that will pull the sister chromatids apart and draw them to opposite poles of the spindle. For separation to occur, the cohesion proteins that join sister chromatids must be destroyed by proteolysis; this is mediated by the anaphase-promoting complex, which is activated by protein phosphorylation.

of metaphase, when all the protein complexes and other structural elements are in place, a trigger event initiates anaphase. This point is analogous to the primed preRC at the transition from G1 to S phase, and the trigger event is the same: protein phosphorylation. In this case, though, the metaphase to anaphase transition is brought about by the phosphorylation of the APC by CDKs.

Phosphorylation activates the APC, and this results in the ubiquitination of securin, which in turn leads to its degradation by the 26S proteasome. Proteolytic removal of securin activates separase, which can now break the

linkages between the sister chromatids, allowing them to be drawn to opposite poles of the spindle (see Figure 11.16). Activation of the APC has another function, too; it causes destruction of the mitotic cyclins and marks the exit from mitosis at telophase. At the same time, mitotic CDK activity declines, and this relieves the repression of G1 cyclin synthesis, allowing D-type cyclins to accumulate. The activity of G1 CDKs gradually increases, allowing new replication complexes to assemble in preparation for the next turn of the cell cycle.

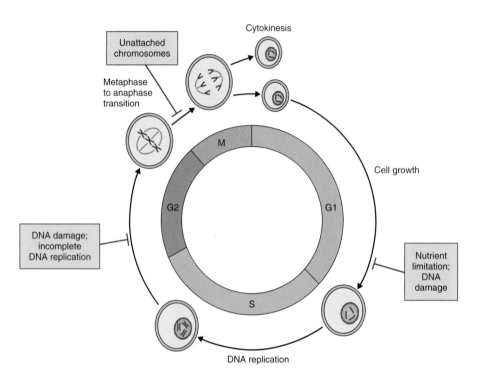

Cytokinesis

Unattached chromosomes

Metaphase to anaphase transition

M

G2

G1

Cell growth

DNA damage; incomplete DNA replication

Nutrient limitation; DNA damage

S

DNA replication

Figure 11.17 Monitoring of the cell cycle by checkpoint controls (red bars). If DNA has been damaged or incompletely replicated, or if chromosomes are incorrectly aligned and therefore not attached at metaphase, progression of the cell cycle is arrested before any irreversible steps are taken.

Telophase, the final stage in mitosis, is characterized by the formation of two new daughter nuclei. Nuclear envelopes are formed around the two sets of chromosomes at either end of the cell. Cytokinesis, during which the cell plate forms producing two separate daughter cells, often takes place while telophase is still in progress.

11.3.7 DNA damage and incomplete cell cycle progression are policed by checkpoint controls

Many environmental factors can disrupt the smooth progress of the cell cycle. For example, DNA damage can be caused by ionizing radiation and some chemicals. Other environmental stresses can result in a lag in completing DNA synthesis, or failure of the chromosomes to attach to the microtubules. If the cell were to continue through the phases of the cell cycle, the outcome would be daughter cells with defective DNA, abnormal chromosome numbers, or other problems. Repair mechanisms have evolved to correct these problems, but they can only work effectively if cell cycle progression is arrested. This arrest can occur at any of three **checkpoints**, depending on the timing and nature of the damage. Each of these checkpoints is positioned immediately before an irreversible point of commitment in the cell cycle, also known as a **restriction point**; they are: commitment to replicate DNA at the G1 to S

transition, commitment to enter mitosis on completion of DNA synthesis, and commitment to separate sister chromatids and initiate cell division (Figure 11.17).

The way in which plants control the latter two, mitosis-associated, checkpoints is still unclear. However, there is evidence that the checkpoint preceding the G1 to S transition is at least partly conserved between plants and other eukaryotes. If a cell's DNA is damaged, the cell needs to launch two mechanisms of response in order to survive. The DNA repair machinery must be activated, and progression of the cell cycle must be delayed or arrested so that the DNA damage is not propagated into new cells through mitosis. Two structurally related kinases are pivotal to both these responses. Named after homologs in mammals where they were first identified, they are the **ataxia telangiectasia mutated** (**ATM**) and **ATM-and-rad3-related** (**ATR**) proteins.

These proteins, which are highly conserved across all eukaryotes, respond to different types of DNA damage. The kinase activity of the ATM protein is activated by double-strand breaks (**DSBs**) in the DNA, whereas the ATR kinase is primarily activated by single-strand breaks or stalling of replication forks (Figure 11.18A). Mutations in the genes encoding either of these proteins make plants abnormally sensitive to chemicals that cause DNA damage, because failure to arrest the cell cycle when DNA is damaged means that the plants continue to replicate the damaged DNA and pass it on through each round of cell division. In plants, *atm* mutants show developmental abnormalities when exposed to γ-radiation or chemicals such as methyl methanesulfonate, which cause DSBs. In contrast, *atr*

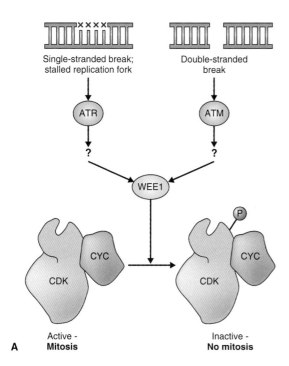

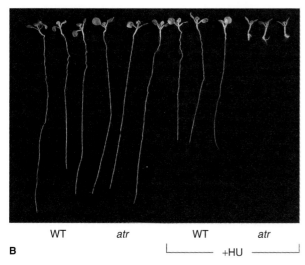

Figure 11.18 DNA damage control. (A) Monitoring of single- and double-strand breaks in DNA by ATR and ATM, respectively. When ATR or ATM signals the presence of broken DNA, transcription of the WEE1 kinase gene is induced by a pathway that is as yet unknown (indicated by question marks). WEE1 then phosphorylates CDK/CYC complexes, inactivating them, and this prevents cells with DNA damage from entering mitosis. (B) Hydroxyurea effects on *atr* mutants compared to wild-type (WT) *Arabidopsis* plants. The ATR protein is activated by compounds, such as hydroxyurea (HU), that slow down or stall progress of the replication fork in DNA synthesis; it keeps cells from progressing through the cell cycle if their DNA is incompletely replicated. Plants with a mutation in the gene encoding ATR are abnormally sensitive to hydroxyurea in comparison with wild-type plants.

mutants exhibit hypersensitivity to drugs such as **hydroxyurea** and **aphidicolin** that slow down or stall progression of the replication fork (Figure 11.18B). Mutations affecting either *ATM* or *ATR* have adverse effects on normal plant development, especially under environmental stress; double *atm atr* mutants show seriously impaired development.

DNA damage must also induce arrest of the cell cycle, and this is mediated by inactivation of CDK activity. In animals, this occurs when ATM and ATR activate the tumor-suppressor protein **p53** and the two checkpoint kinases **CHK1** and **CHK2**. Activation of these proteins leads to induction of transcription of genes encoding CKIs and to inactivation of a phosphatase, **CDC25 phosphatase**, which acts as a timer for mitosis in non-plant species. Plants, however, do not possess p53 or functionally homologous CHK1, CHK2 or CDC25-like proteins, so ATM and ATR must bring about cell cycle arrest in a different manner. It has been shown that WEE1 kinase is a key protein for cell cycle arrest in plants at times of DNA damage, though it is not necessarily the only mediator of the checkpoint. ATM and ATR rapidly bring about induction of *WEE1* gene expression and WEE1 protein synthesis. As we have previously seen (Figure 11.8), WEE1 phosphorylates CDKs, and this prevents the cell from entering mitosis (Figure 11.18A). The details of the signaling cascade that allows ATM/ATR to increase WEE1 activity are still unknown. Plants lacking a functional *WEE1* gene cannot arrest

Key points During the mitotic cell cycle, newly replicated chromosomes are segregated into two daughter cells. Once DNA replication is complete, each chromosome is attached to a specialized component of the cytoskeleton, the spindle, by a large protein complex called the kinetochore that forms at its centromere. Sister chromatids are separated by proteolysis catalyzed by separase, which breaks down the proteins that have hitherto held them together, and they are then able to move to opposite poles of the spindle. A cell plate forms across the equator of the mother cell, followed by reconstitution of plasma membranes and cell walls to produce two daughter cells. Checkpoint controls prevent cells from replicating their DNA if it has been damaged, and prevent cells from dividing if the cell cycle has not proceeded normally. Two kinases, ATM and ATR, have important roles in these checkpoints; ATM is activated by double-stranded breaks in DNA whereas ATR is activated by single-stranded breaks and by stalling of replication forks.

their cell cycles when DNA damage occurs, so they progress into mitosis with damaged DNA or an incompletely replicated genome. The resulting cells are often not viable, and as a consequence growth is arrested.

11.4 Cell cycle control during development

In contrast to animals, plants can continue to form new organs and tissues throughout their life span. They can do this because they have regions of localized cell division known as meristems, which are origins of tissues and organs and also replenish the stock of dividing cells. Cells in meristems are sometimes called 'stem cells', terminology that has been imported from a highly active area of current biomedical research. The term refers to the ability of these cells to go through numerous cycles of mitosis and remain undifferentiated while still retaining the potential to differentiate into a specialized cell type. Because of the scope for confusion over the precise meaning of 'stem cells' in the context of plant development, we will not use the term in this book.

11.4.1 Cell division is tightly regulated in meristems and during organogenesis

Meristems (first introduced in Chapter 1) are categorized as: apical, located at the growing tips of roots, shoots and branches; lateral, principally cambium, responsible for secondary growth; and intercalary, sites of growth located in internodes or, characteristically, at the base of the expanding leaf in grasses and other monocots. Chapter 12 will describe the roles of meristems in the development and growth of plant organs; in this chapter we will focus on the structure and molecular regulation of the meristems themselves. The **shoot apical meristems (SAMs)** and **root apical meristems (RAMs)** (see Figure 11.3) are responsible for producing new cells of the shoots and the roots, respectively. The apical meristems must also maintain themselves so that future new tissues can be produced from them. To achieve this, an apical meristem has at its center a population of cells in which the cumulative mutation rate is low because a very low rate of cell division minimizes the accumulation of DNA damage. The consequence of this is that even in very long-lived plants such as bristlecone pine, where a meristem may survive for 4000–5000 years, cells continue to divide accurately and without excessive DNA damage.

In the shoot, the initiation and development of new organs takes place on the sides of the SAM, in zones where cell division activity is locally increased to produce the new cells that are needed (Figure 11.19A). This process can be disrupted by mutation in any one of several genes that are essential for normal functioning of the SAM. These genes fall into one of three main categories (Figure 11.19B). Genes in the first category are required for the establishment and maintenance of the slowly dividing central zone cells, which are needed for the meristem to renew itself. The second category regulates cell differentiation, and the third controls localized cell division in organ primordia. Mutations affecting genes in any of these classes affect all aspects of normal shoot meristem activity, indicating that the individual activities of domains in the meristem in which these genes are expressed are not autonomous.

The first category of genes, those required for the establishment and maintenance of meristems, includes several different **homeodomain-containing** transcription factors. The first gene in this class to be identified was *Knotted* in maize, described in Chapter 3. Inappropriate activation of Knotted or the *Arabidopsis* homolog KNAT1 causes ectopic (literally 'out of place') production of shoot meristems, for example in the middle of leaves. Another gene in *Arabidopsis*, **SHOOT MERISTEM-LESS (STM)**, is necessary for the center of the meristem to renew itself as a region of undifferentiated cells. Lack of a functional *STM* gene results in disappearance of the meristem (Figure 11.20B), as does the absence of a functional version of another homeobox gene, **WUSCHEL (WUS)** (Figure 11.20C, E). Expression of *WUS* is negatively regulated by genes in the second class, those that promote cell differentiation in organ primordia. Mutations in these genes cause the meristem to increase in size because the cells continue to progress through the mitotic cycle, giving rise to increasing numbers of meristematic cells, rather than differentiating. An example of a gene in this category is **CLAVATA (CLV)** in *Arabidopsis*, which encodes a protein kinase. Mutations affecting *CLV* gene function result in extremely enlarged meristems (Figure 11.20D). It is not clear at this time how the homeodomain proteins and the factors which interact with them control cell cycle activity in meristems.

Genes in the third class required for meristem function regulate local proliferation of cells in developing organs. The product of the **PHANTASTICA (PHAN)** gene in *Antirrhinum majus* controls cell proliferation along the upper (adaxial) lamina, which generates the leaf blade. Recessive mutations in the *PHAN* gene result in a radially symmetrical, needle-like leaf that has a midrib but no mesophyll cells. In addition to impaired leaf development, *phan* mutants suppress the proliferation and production of new organs in the shoot apex, indicating that there is some mechanism of feedback signaling from developing organs which is important for maintaining the function of the shoot meristem. *ASYMMETRIC LEAVES 1* is the ortholog of *PHAN* in *Arabidopsis*. The roles of meristems in generating plant organs will be discussed in more detail in Chapter 12.

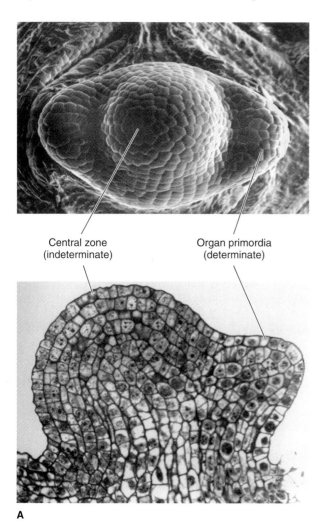

Central zone
(indeterminate)

Organ primordia
(determinate)

A

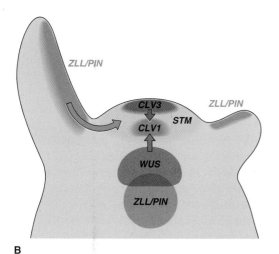

B

11.4.2 Many plant cells remain totipotent throughout the plant's life cycle

As every gardener knows, plants may readily be propagated by taking cuttings, by splitting the whole plant, by dividing a perennating organ such as a bulb or corm into pieces, by layering (inducing a shoot attached to the parent plant to grow roots and become capable of independent existence) or by grafting. The products of such procedures are clones, plants that are genetically identical to those from which they are derived. Clonal propagation is possible because fully differentiated tissues and organs are able to regenerate whole plants, reflecting a fundamental property of most, perhaps all, viable plant cell types, referred to as **totipotency**–the capacity of an individual cell to proliferate and produce all the differentiated cells of the whole organism. In many cases, under appropriate conditions, single plant cells are able to give rise to an entire plant. This is most commonly observed when plant cells are isolated and maintained on laboratory media. Perhaps surprisingly, even highly differentiated and specialized plant cell types can exhibit totipotency. For example, plants can be regenerated from stomatal guard cells of sugar beet (*Beta vulgaris*). Regeneration of a plant can take place by **embryogenesis** or **organogenesis**. During somatic embryogenesis, a whole embryo, like that found in a seed, develops from a single cultured plant cell. In organogenesis, in contrast, the cell gives rise first to shoots, and roots then develop from the shoots. Plants also use organogenesis to multiply asexually: in many species roots can generate shoots, in other cases shoots generate roots (Figure 11.21, also see Figure 10.4A).

Totipotency, the ability of individual plant cells to undergo somatic embryogenesis and pattern formation,

Figure 11.19   Organization of the shoot apical meristem (SAM). (A) Two different views of a SAM. The scanning electron micrograph taken from above (upper) shows primordia that will develop into organs (leaves), positioned around the periphery of the meristem. The micrograph of a longitudinal section (lower) shows the layers of tissue in the SAM and in a developing new organ. (B) The regulatory genes that are required for the establishment and maintenance of SAM are expressed in specific domains. The gene encoding WUS, which is expressed at the base of the meristem, is required for expression in a domain above it of the gene encoding STM. The genes encoding CLV1 and CLV3 are expressed in the indeterminate central zone of the meristem, and their products are believed to interact to determine the size of the central zone. The ZLL/PIN gene products are required for maintaining SAM activity.

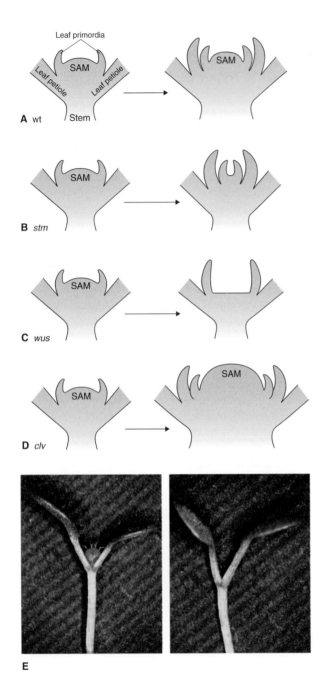

E

Figure 11.20 Effects of mutations in three genes (*SHOOT MERISTEM-LESS* (*STM*), *WUSCHEL* (*WUS*) and *CLAVATA* (*CLV*)) that are essential for normal functioning of the shoot apical meristem. The left panels represent an early shoot meristem; the right panels represent the meristem after further growth. (A) Wild-type (wt) shoot apical or lateral meristem. (B) Adventitious meristem in an *stm* mutant. (C) Adventitious meristem in a *wus* mutant. (D) Shoot apical or lateral meristem in a *clv* mutant. (E) Photograph of wild-type (left) and *wus* mutant (right) plants.

Figure 11.21 Promotion of root formation on a cutting from a begonia plant by application of a hormone rooting compound containing auxin.

without needing to go through flowering and sexual reproduction, is not observed in all species, or in all cell types within a species where it does occur, and in most cases cells need to be isolated and cultured in the lab before it can occur. However, there are plants (such as *Kalanchoe diagremontiana*) where somatic embryogenesis is seen in intact plants as part of normal development (Figure 11.22A). The common name of this succulent species, 'Mother of thousands', graphically describes the way in which totipotent cells along the margin of the mature leaf differentiate into plantlets that eventually drop off and commence independent existence. In this case, the plant uses somatic embryogenesis as an alternative, or in addition, to flowering in order to propagate itself. Some plant species may have retained this asexual mechanism from an earlier stage of plant evolution that predated the appearance of flowering plants. Related to this kind of propagation is pseudovivipary or vegetative apomixis, found in some species of onion (*Allium*; Figure 11.22B).

Plant regeneration, in any species, requires three processes: cellular **dedifferentiation**; re-entry into the cell cycle and cell proliferation; and finally induction of **redifferentiation**, the conversion of cells into the specialized types that make up root and shoot tissues. Dedifferentiation and entry into cell cycle are closely linked, which makes it difficult to analyze them separately at the molecular level. Experimental evidence indicates that the transition from fully differentiated cells to a dedifferentiated state is accompanied by a change in chromatin structure, mediated by post-transcriptional modification of histones in nucleosomes (see Chapter 3). At the onset of cell division, there is an increase in the proportion of euchromatin, the part of the genome that is accessible for transcription, relative to heterochromatin. The RBR/E2F pathway, which controls the transition from G1 to S phase (see Section 11.3.1), is also considered to have a role in the dedifferentiation

A   **B**

Figure 11.22   Examples of asexual reproduction. (A) Somatic embryogenesis in *Kalanchoe diagremontiana* ('Mother of thousands') which produces 'plantlets' along its leaf margins. (B) Pseudovivipary in *Allium cepa* (Egyptian onion).

process, since during dedifferentiation the chromatin structure of E2F target genes changes. The role of RBR in controlling a cell's differentiation status is supported by the observation that RBR deficiency delays cell differentiation, whereas induced overexpression of RBR promotes differentiation of meristematic cells.

11.4.3 Endopolyploidy is common in differentiated plant cells

At every stage in plant development some cells exit the mitotic cell cycle, stop dividing and begin to differentiate into specialized cells. Sometimes, exit from the cell cycle does not mean that DNA synthesis stops. Instead, a cell can enter an alternative cycle known as **endoreduplication**. Many organisms, including arthropods and mammals as well as plants, show endoreduplication, and it can occur in a wide range of cell types. It is characterized by repeated rounds of DNA replication without intervening cell division, so that at each new round the DNA content of the nucleus is doubled. In some cases the effects can be dramatic. For example, during embryogenesis, cells in the suspensor of *Phaseolus coccineus* undergo 12 additional cycles of DNA replication with no cell division, ending up with about 4000 times more DNA per cell than when they exited the mitotic cycle.

There are several different ways in which ploidy can be increased. When DNA replication is followed by mitosis, but without cytokinesis, the process gives rise to cells with multiple nuclei. However, when DNA is repeatedly replicated without nuclear division, so that all the replicated DNA remains within the original nucleus,

Key points   New cells of shoots and roots are produced by the shoot apical meristem (SAM) and root apical meristem (RAM) respectively. Cell division in meristems must be tightly regulated so that the meristems can maintain themselves as well as producing the cells from which organs develop. In the shoot, new organs are initiated and develop on the sides of the SAM. Three classes of genes are essential for normal functioning of the SAM: genes in the first category are needed to establish and maintain the slowly-dividing central zone cells from which the meristem renews itself. Genes in the second category regulate differentiation of cells into specialized types, and those in third control localized cell division in primordia of leaves and other organs. Many plant cells are totipotent; that is, they retain the capacity to proliferate and produce all the differentiated cells that make up the organism. Regeneration requires that cells dedifferentiate, re-enter the cell cycle and resume division; this must be followed by the induction of redifferentiation and conversion of cells into the specialized types that make up root and shoot tissues.

endoreduplication leads to **endopolyploidy** or **polyteny**. In endopolyploid cells, after DNA replication the sister chromatids separate and then return to the decondensed interphase state; however the nuclear envelope does not break down as it does in the normal mitotic cell cycle. The result is an increased number of individual chromosomes within the original nuclear envelope. In polyteny, on the other hand, DNA synthesis is not followed by chromosome condensation/decondensation

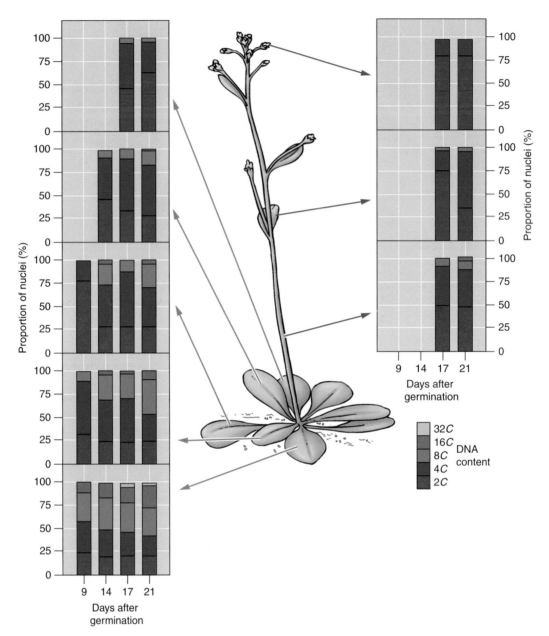

Figure 11.23 Developmental control of nuclear DNA content in *Arabidopsis*. Analysis of nuclei isolated from tissues at different developmental stages shows a strong correlation between tissue age and polyploid DNA content, expressed in multiples of C (the mass of DNA present in the haploid genome).

steps or by sister chromatid segregation. Under these circumstances 'giant' multistranded chromosomes are produced, and the number of chromosomes within the nuclear envelope remains unchanged.

In some plants, endoreduplication can occur in any vegetative tissue type, as seen in *Arabidopsis* (Figure 11.23). However, it is often tissue-specific, and is particularly widespread in storage tissues such as the maize endosperm and the fruit of tomatoes, and in the larger epidermal cells of grasses. There is still some debate about the physiological significance of endoreduplication. There is often a correlation between

DNA ploidy level and the size of cells (Figure 11.24), so endoreduplication may provide an increased gene dosage which is needed if cells are to grow beyond a certain size. For example, cells in storage tissues are often enlarged to accommodate accumulations of protein, carbohydrate or lipid storage bodies. Endoreduplication may also, or instead, be an evolutionary strategy to allow genome sizes to remain small in meristematic cells, while increasing in vegetative cells to protect against the effects of recessive mutations. Endoreduplication is common in plant species, like *Arabidopsis*, that have rapid life cycles. For such species endoreduplication, rather than repeated

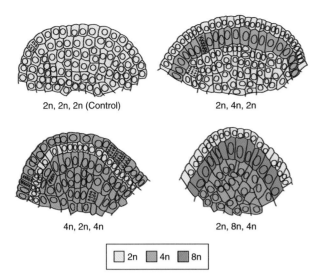

2n, 2n, 2n (Control)   2n, 4n, 2n

4n, 2n, 4n   2n, 8n, 4n

☐ 2n   ▨ 4n   ▪ 8n

Figure 11.24   The relationship between ploidy and cell size. Using grafting techniques, it is possible to produce plant chimeras (individuals composed of cells with different genotypes). Analysis of plants with chimeric meristems shows that the volume of a cell is determined not by its position in the meristem but by ploidy (expressed as multiples of n, where n represents one copy of the haploid genome). There is a strong positive correlation between increased ploidy and increased cell size.

cell division, may be a cost-effective way of expanding a surface (such as a leaf) by producing large cells with increased DNA content rather than many small, diploid cells. Finally, endoreduplication generally makes plants better able to adapt to a wide range of environmental conditions. An increased genome copy number may make it possible to accelerate the rate of gene evolution and genome reorganization, by promoting epigenetic regulation of gene expression through selective gene silencing, which can occur if closely related genes are present in a genome.

The endoreduplication cycle (**endocycle**) can be considered as a mitotic cell cycle with DNA replication and gap phases but without the M phase. Because cells that undergo an endocycle need to duplicate their DNA, just as dividing cells do, they share the components required for DNA synthesis, and the need to activate the origins of replication. It is therefore not surprising that the regulation and progression of the endocycle is influenced by many proteins (such as CDC6 and CDT1) that control the assembly of the preRC complex (see Figure 11.14), or that in transgenic plants, overexpression of E2F transcription factors, which promote DNA synthesis, causes increased levels of endoreduplication. This suggests that the RBR/E2F pathway (see Figure 11.12) is important for the onset of an endocycle.

The elimination of the M phase from the endoreduplication cycle requires that activity of the CDK–cyclin complexes that are specific to G2 to M transition must be suppressed. This is achieved both by repression of transcription of the *CDK* genes and by proteolytic degradation of the CDKs that are components of these complexes: the A-type and B-type CDKs. Plants with reduced CDKB1 activity enter the endocycle more rapidly than normal, and evidence suggests that increased activity of the CCS52A-activated APC probably leads to selective destruction of the CDKB1-associated cyclins by the 26S proteasome. CKIs are also believed to reduce CDK activity, particularly that of CDKAs.

11.4.4 Plant cells must replicate and maintain three genomes

So far, we have focused on replication of the nuclear genome in this discussion of the cell cycle. However, eukaryotic cells have other genomes, and unlike animals and fungi, plant cells have three different genomes: those of the nucleus, the plastids and the mitochondria. Each genome exists in a separate subcellular compartment, and while there is usually only one nucleus per plant cell, there are multiple mitochondria and plastids. Moreover, the copy number of mitochondrial and plastid genomes within each organelle is variable. The organellar genomes are replicated mainly in meristematic tissues and organ primordia, where there can be 20–100 copies of the genome in each mitochondrion and 50–150 copies of each plastid genome. In these rapidly dividing cells the number of organelles is relatively low. Following mitosis, however, replication of organellar genomes gradually ceases although the organelles themselves continue to divide. This ultimately leads to larger numbers of each organelle per cell, with a reduced genome copy per organelle (Figure 11.25).

Dividing plant cells maintain a population of plastids, and plastid numbers in different cell types are regulated; this indicates that there is a mechanism that coordinates division of the plastids and their distribution between daughter cells during the cell division process. In species with a single plastid per cell, such as certain algae, this regulation is precise; clearly each newly divided cell must end up with one, and only one, plastid. Regulation is probably less strict in flowering plants, where every meristematic cell has multiple plastids. The replication factor **CDT1** has a role in coordinating nuclear DNA replication with plastid division. In *Arabidopsis*, the CDT1 protein is targeted to both the nucleus and plastids; and plants with reduced CDT1 levels show abnormalities in plastid number and size. However, the

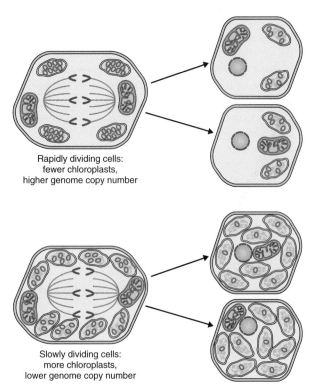

Rapidly dividing cells:
fewer chloroplasts,
higher genome copy number

Slowly dividing cells:
more chloroplasts,
lower genome copy number

Figure 11.25   Chloroplast DNA in rapidly and slowly dividing cells. There are relatively few chloroplasts in rapidly dividing cells, but each of these chloroplasts has a high copy number of the circular plastid genome. When cells exit from the shoot apical meristem and gradually cease to proliferate, the plastids continue to divide one or two more times, but no longer replicate their DNA. The result is that slowly dividing cells contain larger numbers of plastids, each with a lower genome copy number than in rapidly dividing cells. Note that sizes of organelles shown in this figure are not to scale.

molecular modes of action of this replication factor in controlling plastid division are as yet unknown.

11.5 The meiotic cell cycle

The discussion so far has covered the processes of the mitotic cell cycle which occur throughout the life of a plant; we now turn to the process of meiosis which occurs only during sexual reproduction. The meiotic cell cycle, like the mitotic cycle, includes two gap phases, G1 and G2, separated by a DNA synthesis or S phase. However, the M phase of the mitotic cell cycle, which results in two genetically identical daughter cells, is replaced in the meiotic cell cycle by two sequential cell division steps, meiosis I and meiosis II, which result in four haploid cells, each of which is usually genetically different from the other three. Both meiosis I and

Key points Plant cells, like those in many other organisms, can undergo a process called endoreduplication, in which repeated rounds of DNA replication occur without intervening cell division, so that at each round the DNA content of the cell is doubled. This can lead, in extreme cases, to cells with more than 4000 times the amount of DNA that they had when they first entered the mitotic cell cycle. In plants it is most common for DNA to be repeatedly replicated without nuclear division, so that all replicated DNA remains within the original nucleus. If after DNA replication the sister chromatids separate and decondense as in the normal cell cycle, the cell is described as endopolyploid. Endoreduplication can occur in any vegetative tissue type, but it is particularly common in storage tissues such as maize endosperm and tomato fruit, and in the larger epidermal cells of grasses. The endoreduplication cycle can be considered as a mitotic cell cycle with DNA replication and gap phases but without the M phase. Unlike other eukaryotes, plants must replicate and maintain three genomes: those of the nucleus, mitochondrion and plastid. The division of plastids is coordinated with cell division so that the number of plastids in daughter cells is controlled.

meiosis II can be divided into the same phases as mitosis: prophase, prometaphase, metaphase, anaphase, telophase and cytokinesis. However, the events of meiosis I differ significantly from those of mitosis (see Figures 11.2B and 11.4). In meiosis I, homologous chromosomes pair, recombination usually occurs, and homologs segregate; as a result, the chromosome number is reduced from 2n (diploid) in the parent cells to 1n (haploid) in the daughter cells. In contrast, the events of meiosis II are very similar to mitosis. In the following sections we will examine the unique processes that occur during meiosis I.

11.5.1 During meiotic prophase I homologous chromosomes pair and recombination usually occurs

The events of prophase I of meiosis are unique and complex. Homologous chromosomes pair in preparation for their separation during anaphase I and exchange genetic material in the process of crossing over, which

Prophase of meiosis I

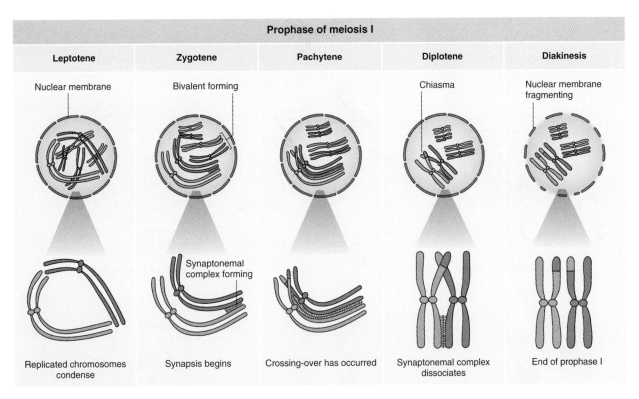

Leptotene	Zygotene	Pachytene	Diplotene	Diakinesis

Figure 11.26 The important events during prophase of meiosis I. During this stage of meiosis, chromosomes condense, homologs synapse and crossing over occurs between sister chromatids of paired homologs.

results in recombination of genes. Geneticists recognize five stages of prophase I (Figure 11.26). During the **leptotene** stage, the chromosomes, previously replicated during the S phase, condense. In the **zygotene** stage, homologous chromosomes pair and align, gene for gene, along their whole lengths, to form a structure called a **bivalent** (or **tetrad**) that consists of four chromatids, two from each chromosome in the homologous pair. This pairing and alignment, or **synapsis**, is mediated by a structure called the **synaptonemal complex**. The process of crossing over to exchange genetic material can then occur. Each point along a bivalent where crossing over has taken place is a **chiasma** (from the Greek letter chi, χ, which it resembles). The chiasmata hold homologs together at this stage in meiosis. By the **pachytene** stage, crossing over is complete but the four chromatids remain associated by the synaptonemal complex, which breaks down during the **diplotene** stage, allowing the bivalent to begin the process of separation, which will be completed at anaphase of meiosis I. By the last stage of prophase in meiosis I, **diakinesis**, the synaptonemal complex has disappeared.

The remainder of meiosis I has some features in common with the M phase of the mitotic cell cycle: during prometaphase, the nuclear membrane fragments and spindle fibers attach to centromeres, then the chromosomes line up along the metaphase plate prior to being drawn apart to opposite spindle poles, nuclear

envelopes reform and cytokinesis produces two daughter cells (Figure 11.27). There are two key differences between mitotic M phase and meiosis I. First, in meiosis I, the chromosomes line up on the metaphase plate not singly but as the paired homologs that were linked as bivalents during prophase. Because they have exchanged genetic material, the chromosomes that segregate to the daughter cells at the end of meiosis I are genetic mixtures, as Figure 11.27 illustrates. Second, meiosis I produces haploid cells with half the ploidy of the parent cell, whereas mitosis gives rise to cells with the same ploidy as the parent cell.

11.5.2 Synapsis is the process of pairing homologous chromosomes

In meiosis, the sister chromatids, formed by DNA replication during the S phase, are held together by a multiprotein, doughnut-shaped cohesin complex. This **sister chromatid cohesion (SCC) complex** is required for the subsequent steps in meiosis to proceed correctly. In animals and fungi, the meiotic cohesin complex differs from that which forms during mitosis, but in plants the difference is less clear-cut and one of its essential components, the plant REC8 homolog, is expressed in somatic as well as meiotic tissues.

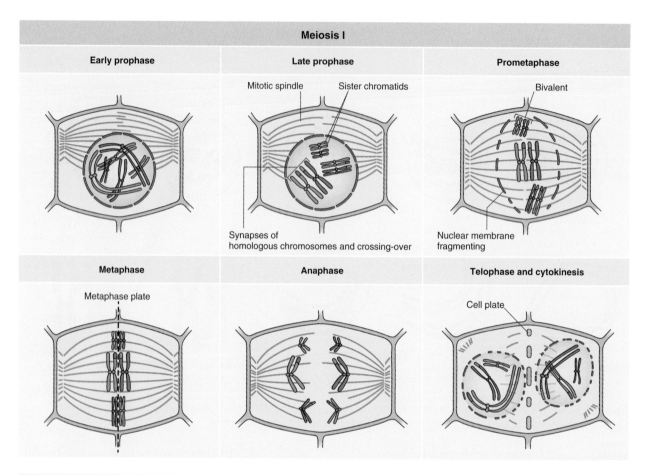

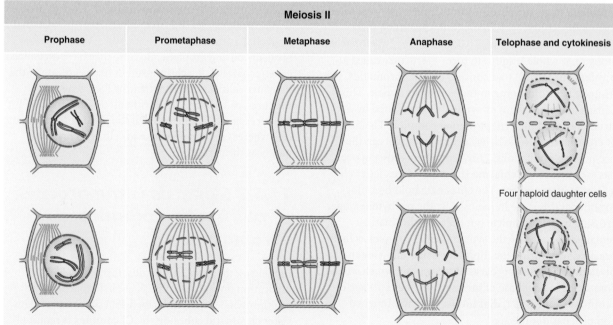

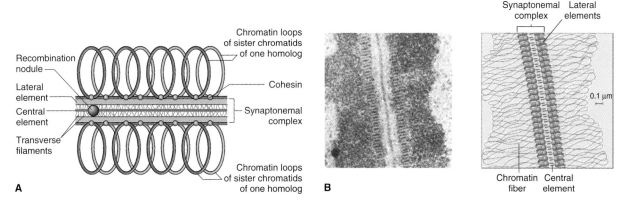

Figure 11.28   The synaptonemal complex that holds homologous chromosomes together during prophase of meiosis I. (A) Diagram of a synaptonemal complex. Each homolog consists of two sister chromatids. (B) Electron micrograph of a synaptonemal complex, and diagrammatic representation of its key features.

During the zygotene phase of meiosis I, each pair of homologous chromosomes forms a tripartite structure in which the whole lengths of the chromosomes are linked by the **synaptonemal complex** (**SC**). During leptotene, prior to formation of the SC, each chromosome develops a protein core, known as the **axial element**, to which the chromatin of the two chromatids is attached in a series of loops. Correct formation of the axial elements depends on sister chromatid cohesion. The SC consists of a central element, composed of protein, which is linked by transverse filaments to the axial elements of the two homologous chromosomes. When the homologous chromosomes synapse, the axial elements become the lateral elements of the SC (Figure 11.28). How the homologous chromosomes recognize one another is still an unsolved problem, but one phenomenon, observed during meiosis in plants, is believed to play a part. This is the so called 'bouquet', in which the telomeres of all the

chromosomes attach to the nuclear envelope and form a cluster (Figure 11.29). Synapsis is most commonly initiated near telomeres, and the bouquet may help bring the telomeres of homologous chromosomes into close proximity.

Although the SC is a feature of meiosis in most eukaryotes, including plants, it is not essential for recombination: some organisms, including the fungi *Saccharomyces pombe* and *Aspergillus nidulans*, carry out recombination without formation of an SC. Structurally the SC is highly conserved across all organisms that possess it, but this is not the case at the molecular level. The proteins of the central and lateral elements show very low sequence homology when compared among plants, fungi and animals, and even between closely related species. In *Arabidopsis*, the AtZIP1 protein, which is present only during prophase of meiosis I, is located in the central element of the SC. AtZIP1 is encoded by two

Figure 11.27   Stages of meiosis I and II that give rise to four daughter cells from each parent cell. During prophase I, the chromosomes condense. Each pair of homologous chromosomes undergoes synapsis, lining up to form a tetrad that consists of four chromatids. Crossing over occurs, resulting in recombination of DNA between the two homologous chromosomes. In prometaphase I, the nuclear membrane fragments and the tetrads begin to align at the metaphase plate. The alignment is complete during metaphase I, and in anaphase I the homologous chromosomes move to opposite poles of the spindle. Each consists of two chromatids, and where crossing over has occurred, the DNA of one chromatid in each chromosome will be partly derived from the other chromosome in the homologous pair. Telophase I and cytokinesis I follow, producing two daughter cells each with half the number of chromosomes of the parent cell; at this stage each chromosome still consists of two chromatids. The stages of meiosis II are similar to those of mitosis (see Figure 11.2). However, the daughter cells are not diploid but haploid, containing only one representative of each pair of homologous chromosomes, and each chromosome may or may not have undergone recombination with its homolog.

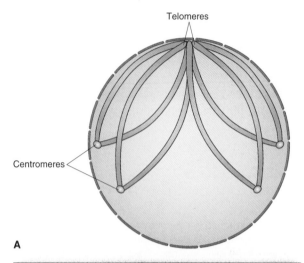

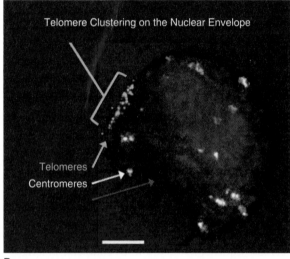

Figure 11.29 Clustering of chromosomes at the beginning of prophase I of meiosis. (A) Diagram showing the 'bouquet' configuration of chromosomes during meiosis, which is believed to help bring telomeres of homologous chromosomes into close proximity on the nuclear membrane and to facilitate synapsis. (B) In this pseudo-colored microscopic image of a maize nucleus in early prophase I of meiosis, the green spots represent telomeres, clustered together, and the white spots are centromeres.

genes, *AtZIP1a* and *AtZIP1b*. These appear to be functionally redundant since either can be knocked out without affecting the phenotype, whereas if both genes are knocked out, meiosis is delayed and in most cells pairing and synapsis of homologous chromosomes fails to occur. Nevertheless, when pairing does occur, crossing over takes place at the normal frequency, so the loss of the AtZIP1 protein does not prevent recombination events.

11.5.3 Recombination is most commonly initiated by double-stranded breaks in the DNA

The process of genetic recombination during meiosis is most commonly initiated by the **Spo11 protein**, which creates double-stranded breaks in the DNA of one of the two homologous chromosomes. Starting from a break point, the DNA strands are **resected** (trimmed back) from the 3′ ends to create regions of single-stranded DNA (Figure 11.30). Resection is carried out by the

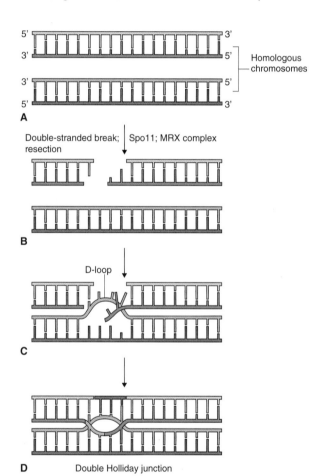

Figure 11.30 Initial stages in the process of crossing over. (A) The double-stranded DNA of one chromatid from each of the two paired homologous chromosomes. (B) Generation of a double-stranded break by Spo11 is followed by resection (cutting back) of each strand from the 5′ end by the MRX complex to produce regions of single-stranded DNA. (C) One strand invades the DNA of the other chromatid, in a process driven by RAD51 and DMC1. (D) DNA synthesis (dotted line) is followed by ligation, producing a double Holliday junction, which is stabilized.

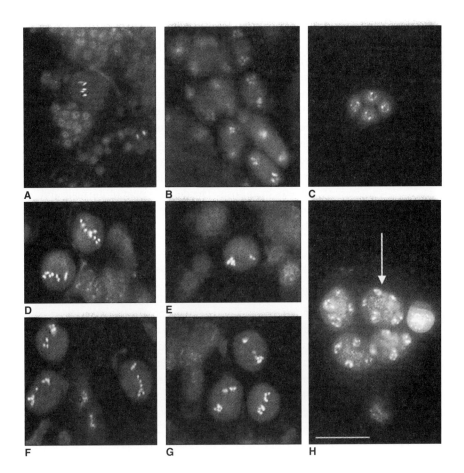

Figure 11.31 Mutations in the gene encoding DMC result in few bivalents forming during meiosis. Micrographs of cells stained with diamidino-2-phenylindole, a fluorescent reagent that binds to DNA. (A–C) Wild-type *Arabidopsis* microspore (pollen) mother cells. (A) Meiotic metaphase I, with five bivalents aligned. (B) Metaphase II, with two groups of five condensed chromosomes. (C) A tetrad at the end of telophase II, with four normal nuclei. (D–H) Microspore mother cells in the *dmc1* mutant. (D) A stage equivalent to metaphase I, with ten univalents scattered across the microspore mother cell and no bivalents. (E–G) Metaphase II equivalent, with ten univalents distributed randomly between two groups. (H) Arrow points to a tetrad during telophase II with more than four nuclei. Scale bar = 10 μm (applies to all parts).

MRX complex, which consists of three proteins: Mre11, Rad50 and Xrs2/Nbs1. The single-stranded DNA invades the double-stranded DNA of one chromatid of the homologous chromosome, in a step catalyzed by RAD51 (radiation-sensitive 51) and DMC1 (disrupted meiotic cDNA1) (Figure 11.30). These two proteins have single-stranded DNA-binding activity and DNA-dependent ATPase activity. RAD51 is active during mitosis as well as meiosis; it is believed to have a role in repairing double-stranded breaks in addition to being required for recombination during meiosis. DMC1 is required for bivalent formation and the segregation of chromosomes during meiosis; mutations in the *Arabidopsis AtDMC1* gene result in cells with mainly univalents and few if any bivalents visible in late prophase of meiosis I (Figure 11.31).

The DNA strand in the chromatid of the homolog that has been displaced by the invading single-stranded DNA from the chromatid of the other homolog forms the so-called **D-loop** (displacement loop) (Figure 11.30C). The process of DNA gap repair synthesis fills in the gaps, and during the process two structures called **Holliday junctions** are generated. These junctions need be resolved to allow the homologous

chromosomes, which have now exchanged some DNA between one of each of their chromatids, to separate. This can occur in one of two ways, which are illustrated in Figure 11.32. Sometimes the result is a so-called non-crossover event, in which the chromosomes largely retain their original identity, although even in this case there is a small region of recombination. Most frequently, though, resolution of the Holliday junctions leads to crossover and significant recombination between the two chromosomes. Chiasmata are the visible signs of Holliday junctions and recombination events.

11.5.4 Recombination is followed by two divisions to produce four haploid cells

When recombination is complete, the synaptonemal complex begins to break up, and by the end of diplotene it has largely disappeared. Microscopically it becomes easier to see that the bivalent is composed of four chromatids (for this reason, it is also sometimes referred to as a tetrad). Two of the four chromatids will normally

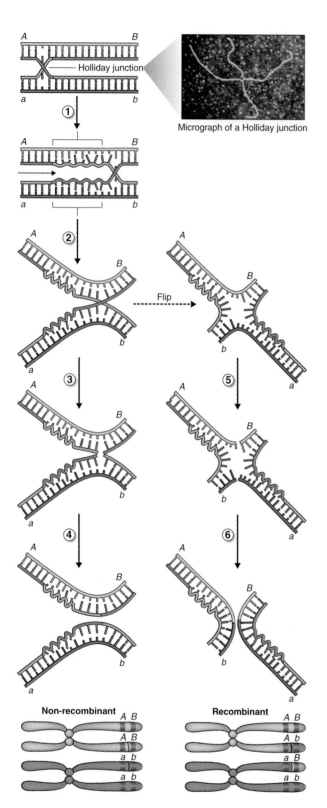

Micrograph of a Holliday junction

Flip

Non-recombinant

A B
A B
a b
a b

Recombinant

A B
A b
a B
a b

have undergone one or more recombination event, and thus consist of DNA from both of parents. Now the chromatids must be partitioned among the four daughter cells, which are the products of meiosis, and this is achieved by two successive divisions, as illustrated in Figure 11.27. The daughter cells are haploid, that is, they contain only one representative of each pair of homologous chromosomes. In plants, the haploid products of meiosis are spores that will divide by mitosis to produce gametophytes, which in turn will produce gametes by mitosis. The development of the male and female reproductive structures of plants is described in Chapter 16.

Key points In meiosis, unlike mitosis, the homologous chromosomes inherited from each parent are paired, and they undergo crossing over so that their genetic material is recombined in the daughter cells. The process of pairing is called synapsis, and it is mediated by a large protein complex, the synaptonemal complex. Recombination is usually initiated by the introduction of double-stranded breaks in the DNA of one chromatid of one of the paired homologs. The DNA is then trimmed back from the break points, and then one of the cut DNA strands invades the DNA of one chromatid of the homologous chromosome. DNA synthesis fills in the gaps created, producing a structure called a double Holliday junction. Resolution of a Holliday junction can result in crossing over to produce recombinant chromatids, though it can also result in a non-crossover event. After recombination, the cell undergoes two divisions to produce four haploid cells.

Figure 11.32   Resolution and crossing over at chiasmata during meiosis. A Holliday junction (an example is shown in the micrograph) can be resolved in two ways, only one of which leads to crossing over. (1) The Holliday junction migrates from left to right, creating two heteroduplex regions that have a few base mismatches. (2) In the next steps, the chromatids have been redrawn to make the figures more like a genuine Holliday junction, by bending ends A and B upwards and a and b downwards. (3) The strands that were originally nicked are broken. (4) The strands are connected to form chromosomes that are non-recombinant but do have short heteroduplex regions. (5) The strands that were *not* originally nicked are broken. (6) The strands are connected to form recombinant chromosomes with short heteroduplex regions.

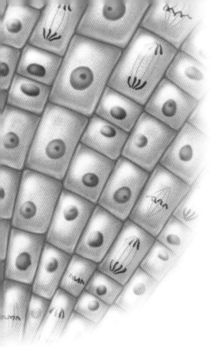

Chapter 12

Growth and development

12.1 Introduction to plant development

In this chapter we will examine the patterns of non-reproductive growth and differentiation during the plant life cycle. The development of reproductive structures is discussed in Chapter 16. In contrast to animals, plants grow and produce new organs throughout their lives. **Growth** in plants is a consequence of both an increase in cell number and an increase in cell size. It can result in plants that are as tall as 116 m and as much as 11 m in diameter. However, growth is not necessarily associated with increase in dry mass since plant cells may be up to 95% water by weight. **Differentiation**, the formation of cells, tissues and organs with specialized structures and functions, accompanies growth; it can be studied at the level of individual cells (**cell differentiation**), tissues (**histogenesis**) or entire organs (**organogenesis**). Together, growth and differentiation give rise to the overall form of a whole plant, a process called **morphogenesis**. As we discuss the cellular and molecular bases of development in plants it is important to remember that the cells of multicellular plants are interconnected by **plasmodesmata**, forming the integrated cytoplasmic network called the symplasm (see Chapters 4 and 14). **Ontogeny** is another term for morphogenesis and is often used in connection with

phylogeny (evolutionary relatedness) when comparing developmental processes across taxonomic groupings. Evolution provides clues about the developmental mechanisms that result in variations in organ form and function. The modern synthesis of phylogeny and ontogeny has been given the name **evo-devo** and seeks to establish how development originated, how the developmental repertoire evolved and how developmental processes became modified in evolution.

An example of the evo-devo approach applied to plant development is the comparative analysis of diploid and haploid stages in the life cycle. The **green alga** *Chlamydomonas reinhardtii* (see Figure 1.4B) alternates between diploid and haploid unicellular phases. Development of the diploid *Chlamydomonas* cell and promotion of meiosis to produce cells of the haploid phase is regulated by representatives of the TALE (three amino acid loop extension) class of **transcription factors**. TALE proteins, including those encoded by genes of the *KNOX* family (see Section 12.5), are known to control the development of sporophytic shoot meristems in angiosperms. Mutations of *KNOX* homologs in the **moss** *Physcomitrella patens* result in defective sporophyte development (Figure 12.1; for the life cycle of mosses see Figure 1.8). These genes are not expressed in the moss gametophyte. This evo-devo study thus establishes that the function of genes of the *KNOX* family of TALE transcription factors is **conserved** across the phylogeny

---

The Molecular Life of Plants, First Edition. Russell Jones, Helen Ougham, Howard Thomas and Susan Waaland.
© 2013 John Wiley & Sons, Ltd. Published 2013 by John Wiley & Sons, Ltd.

Figure 12.1   Expression of KNOX in the moss *Physcomitrella patens*. (A, B) The *KNOX* gene has been tagged so that its product can be visualized with a blue stain, showing (A) no detectable KNOX in gametophytes, and (B) intense expression of KNOX in the embryo sporophyte. (C) Wild-type gametophyte. (D) A gametophyte in which all three class 1 *KNOX* genes are inactivated by mutation (triple knockout) has a normal appearance. (E, F) Wild-type sporophyte morphology and structure in thin section. (G, H) Abnormal sporophyte morphology and structure of a triple mutant. c, columella; f, foot; op, operculum; se, seta; sp, sporangium with spores. Scale bars = 1 mm (A, C, D), 50 μm (B) and 100 μm (E–H).

of eukaryotic green plants, from algae to bryophytes to angiosperms, and is restricted to the development of structures in the diploid phase of the life cycle.

12.2 Cell origins and growth

Plants are distinguished by the localization of cell division to **meristems**, sources of cells that will become new tissues and organs (see Chapters 1 and 11). A further characteristic feature of plant development is

totipotency, the capacity of most living cell types to proliferate and produce all the differentiated cells of the whole organism. The meristematic origins of cells and the basis of totipotency are discussed in Chapter 11. The present chapter follows post-mitotic events in the meristem, from patterning through to the generation of the form and function of tissues and organs. In general, with the exception of certain **specialized cell types**, such as secretory cells and tracheary and sieve tube elements of the vascular tissues, growth and morphogenesis in plants take place without the loss of **genetic potential**. The extreme developmental plasticity of plant cells, tissues and organs is apparent in their capacity to dedifferentiate as well as redifferentiate. An example of this potential is provided by the re-greening senescent tobacco leaf as described in Chapter 18.

12.2.1 Apical meristems are organized into distinct regions

Apical meristems found at the tips of stems and roots add new cells to the lengths of these organs. These meristems give rise to: **periderm**, which produces epidermis; **procambium**, which generates vascular tissue; and **ground meristem**, the source of ground tissue. In addition, the shoot apical meristem gives rise to progenitors of leaves and branches.

The **shoot apical meristem** (**SAM**) has a series of identifiable zones (Figure 12.2A). The central zone (CZ) is the source of cells for the peripheral zone (PZ), which in turn contributes cells to the formation of lateral organs, and for the rib zone (RZ), which supplies cells for stem growth. The CZ also makes new cells to replenish its own stock. Cell layers in the SAM are referred to as L1 (outermost), L2, L3 and so on (Figure 12.2B). The outer **tunica** (comprising L1, and often L2) is distinguished from an inner **corpus** zone on the basis of the plane of cell division—at right angles to the surface of the apex (**anticlinal**) in the tunica, both perpendicular and parallel (**periclinal**) to the surface in the corpus. The outermost layer of the corpus is labeled L3. In some cases it is possible to assign fates to particular regions of the apex (though the plasticity of plant morphogenesis means that cells can be readily induced to switch lineage in response to experimental treatments or changing physiological conditions). Epidermal tissue usually develops from L1 (which is sometimes referred to as the protoderm). The corpus is the origin of ground tissue and also of the procambium, which differentiates into stem vascular tissue. As well as adding cells to stem, the

SAM produces **leaf primordia** that develop into new leaves (see Section 12.5.1) and the cells that will produce **lateral (axillary) bud primordia**, which have the potential to form new branches, in the axils of leaves. **Primordium** is the term that refers to an organ at its earliest stage of development. **Founder cells** are the cells of a meristem that give rise to a primordium.

The **root apical meristem** (**RAM**; Figure 12.3) differs from the SAM in a number of ways. First, it is not involved in organogenesis; branch roots are produced in mature tissue through the action of a **lateral meristem**, the **pericycle**, the cell layer between the endodermis and the central vascular tissue (see Section 12.4.2). Second, in addition to adding new cells basipetally to the length of the root, the RAM produces new cells acropetally to renew the **root cap**. The **quiescent center**, a reservoir of meristematic cells, is located near the apical end of the RAM. It is surrounded by a dome of rapidly dividing cells.

SAMs can be classified according to how long they continue to be active. A SAM that sustains the capacity to make new cells, tissues and organs is referred to as **indeterminate**. If a meristem gives rise to a terminal structure and ceases to replenish its reserve of proliferative cells, it is said to be **determinate**. The contrast between determinate and indeterminate growth is illustrated by the pattern of development of tomato varieties (Figure 12.4). The shoots of some lines (often called vining tomatoes) continue to grow at the apex while supporting the production of flowers and fruits at each node. In determinate bushy varieties, on the other hand, each shoot terminates in an inflorescence and fruit and grows no further. In many plants, the transition from the vegetative to the reproductive condition at the apex, described in detail in Chapter 16, is a switch from indeterminate to determinate growth. **Apical dormancy** (see Section 12.6.2 and Chapter 17) also constrains indeterminate growth, either by shutting down division and the onset of organogenesis, or by stockpiling primordia in an undifferentiated state within the bud.

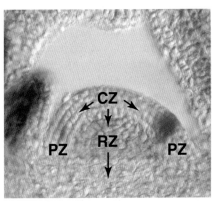

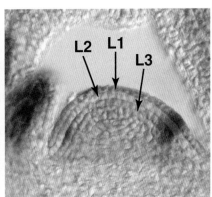

A   **B**

Figure 12.2   Pseudocolored photomicrograph of a section of the shoot apex of *Arabidopsis*. (A) The shoot apex is organized into distinct regions: central (CZ), rib (RZ) and peripheral (PZ) zones. (B) Cell layers: L1 and L2 are the tunica, L3 the corpus.

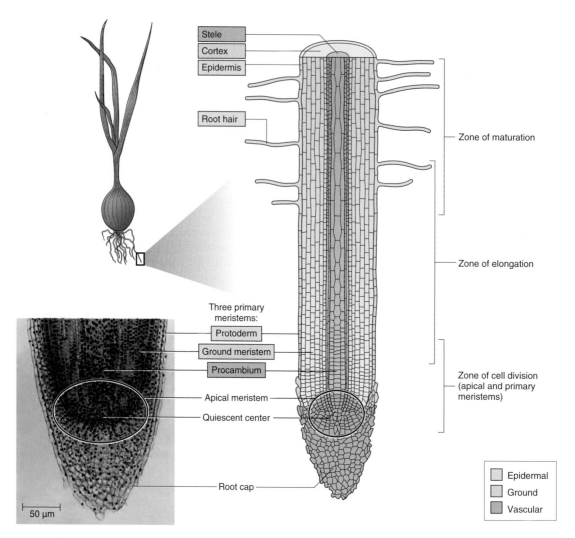

Figure 12.3   The root apical meristem and the cells and tissues derived from it. The micrograph shows a root apical meristem from onion (*Allium cepa*).

12.2.2 Morphogenesis is determined by polarity and differential growth

Morphogenesis is the three-dimensional extension of axes and lateral structures. The shape of such a structure is a consequence of regional differences in **rates** and **durations** of growth, which in turn reflects the pattern of constraints experienced during expansion into space. Figure 12.5 is a simple illustration of the role played by patterns of local growth rates in determining form. Based on photographic images of grape (*Vitis vinifera*) leaves taken over a period of 6 days, growth rates at points across the leaf blade were determined. Figure 12.5 shows maps of elemental growth rates superimposed on the outline of a normal leaf at day 6 compared with a malformed leaf of the same species. The abnormal shape of the latter is clearly

reflected in the asymmetrical distribution and irregular shapes of elemental growth rate contours (Figure 12.5B), whereas localized growth in the normal leaf retains its bilateral symmetry and regular ellipse-shaped isolines throughout lamina expansion (Figure 12.5A).

Growth rate and duration establish the time dimension for morphogenesis. An additional factor in the development of plant form and function is **selective cell death**—discussed in detail in Chapter 18. Unconstrained cell division and expansion will result in uniform three-dimensional growth. Imposing an asymmetrical limitation will establish **polarity**: that is, an imbalance in growth leading to top being differentiated from bottom or left from right. Polarized patterns of differentiation form under the influence of **morphogenetic fields**, which in turn originate as gradients of developmental regulators (**morphogens**), such as hormones, Ca^{2+} and transcription factors

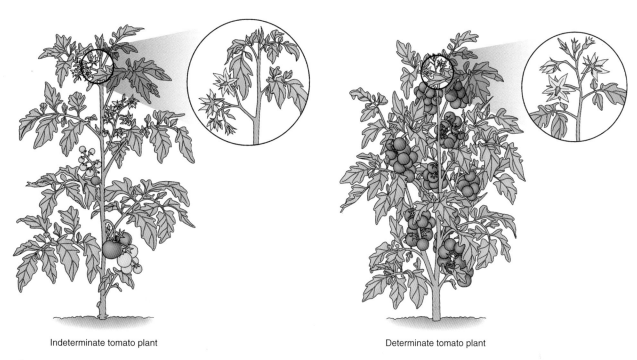

Indeterminate tomato plant   Determinate tomato plant

Figure 12.4   Indeterminate and determinate varieties of tomato. The apical meristem of indeterminate varieties continues to produce new vegetative tissues throughout the life of the plant, while the apical meristems of determinate varieties become floral meristems limiting growth of the plant.

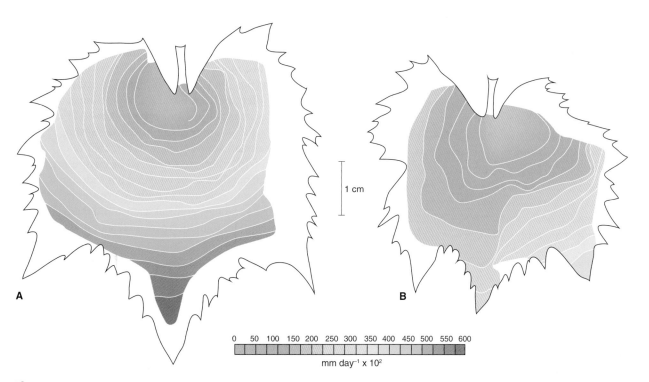

Figure 12.5   Distribution of elemental growth rates in the expanding lamina of a normal leaf of *Vitis vinifera* (A) compared with the pattern in a malformed leaf with marked asymmetry (B). Elemental growth rates are given as mm day^{-1} × 10^2; contours represent regions with the same rate, connected by isolines.

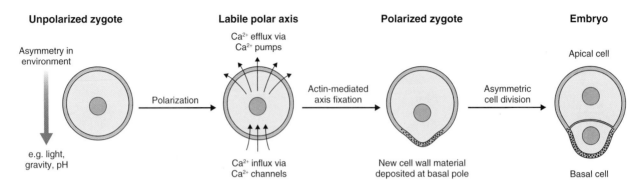

Figure 12.6   Steps in the establishment of apical-basal polarity in the embryo of the alga *Fucus* under the influence of a morphogenetic field.

(Figure 12.6). How are such fields generated and sustained or modified over time, and does the process of growth itself drive the formation of gradient profiles? Some answers to these questions of fundamental importance in developmental biology have come from studies of plant embryogenesis and organogenesis and are described in this chapter.

Key points Morphogenesis (ontogeny) is the process of growth and differentiation. Evo-devo is the integration of phylogeny (evolutionary relatedness) with ontogeny. Conservation of KNOX transcription factor function in sporophyte development is an example of a regulatory system revealed by an evo-devo study. Cells originate in meristems. The shoot apical meristem (SAM) is organized into distinct zones and gives rise to leaf and bud primordia. SAMs may be determinate or indeterminate. The root apical meristem (RAM) does not participate in organogenesis and is the source of the root axis and root cap. Morphogenesis is a process of differential growth which may be polarized under the influence of gradients of developmental regulators.

12.2.3 Plant growth is described by the universal S-shaped curve and is driven by water

Growth is essential for development, and differential growth is the basis of morphogenesis. Cells derived from the apical meristems increase

20–50-fold in length before growth ceases. Let us therefore examine the general features of plant growth. The pattern of growth in biological systems is normally **density-dependent**. It begins slowly when cell mass is small, reaches a maximal rate when density is optimal with respect to metabolic and environmental constraints, and declines towards an **asymptote** (upper size limit) as finite external and internal factors become increasingly influential. This behavior is expressed as the classic S-shaped (**sigmoidal**) growth curve typical of all levels of biological organization, from molecules to cells to tissues to organs to individuals to populations. The sigmoidal relationships between time and some measure of biomass (weight, volume, cell number and so forth) may be described mathematically by variations of the **logistic function** (Equation 12.1). The logistic function may be applied wherever the growth of a population at an initial stage is approximately exponential, followed by declining growth as the asymptote is approached, slowing to a halt at maturity. 'Population' here may refer to a cell population in culture, or organized in a tissue, organ or individual; but the logistic equation has wide applicability and is also routinely used in ecology, economics, chemistry, sociology, political science and many other disciplines to model increase to a limit over time.

Equation 12.1 Generalized archetypal logistic growth function

$$G = \alpha + \beta y^t$$

where G = biomass, t = time, α = the value of G at the asymptote and β refers to the initial state of the system. The parameter $y = e^{-c}$ represents exponential growth where e is the base of natural logarithms and c is the rate constant

Figure 12.7 presents an example of logistic-type curves fitted to growth data for five successive extending leaves of the grass *Lolium temulentum*. Leaf lengths were

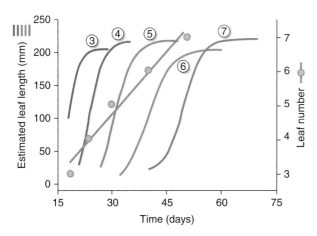

Figure 12.7 Fitted logistic growth curves and the corresponding plastochron index trend for leaves 3 to 7 of *Lolium temulentum*. The parameters of the equation fitted to the length/time data for each leaf allow a number of indices of growth to be calculated with statistical rigor, including the inflexion point of the curve. The mean plastochron (interval between appearance of successive leaves) can then be estimated from the linear regression of leaf number against time of inflexion point (-⊙-). The slope of the regression line gives a mean plastochron of 8.06 days.

measured daily and the logistic curve that is the best statistical fit to the length/time data for each leaf was determined by computer. By putting together coefficients of the best-fit logistic equations in various combinations, a range of biologically meaningful parameters quantifying growth can be derived, including mean and instantaneous relative and absolute growth rates, duration of growth, inflexion point of the S-shaped curve, and the plastochron index (see Section 12.5.1). The experiment in Figure 12.7 shows that *L. temulentum* produces a new leaf every 8.06 days.

Cell enlargement is constrained by the cell wall. In order to increase in size the protoplast of a cell must exert pressure on the cell wall and the wall must yield to that pressure. The pressure generated by protoplasts is called **turgor pressure** (see Chapter 14). The relationship between the relative rate of increase in cell volume, turgor pressure and wall extensibility is described by the **Lockhart equation** (Equation 12.2).

Equation 12.2 The Lockhart equation

$$1/V.dV/dt = w(P - Y)$$

where V = cell volume, t = time, w = volumetric extensibility of the cell wall, P = cell turgor pressure and Y = yield threshold pressure

Turgor pressure and **yield threshold** (the minimum turgor pressure that produces cell expansion) tend to be

more or less constant in growing cells, indicating that cell wall extensibility is the major determinant of growth rate. Extensible cell walls (characteristic of young growing tissues) are **plastic**, that is they yield to pressure by undergoing irreversible deformation, whereas the walls of mature cells tend to be rigid and capable only of **elastic** deformation.

12.2.4 Growing structures exhibit plastic growth at first and subsequently develop elastic properties as they mature and rigidify

Studies of the biomechanical properties of cell walls show that wall extensibility changes during cell expansion. The tools of **rheology**, the study of the deformation and flow of matter, have been applied to plant cell walls to determine their ability to extend in a plastic (irreversible) manner. Plastic extension is necessary for growth, but excessive plastic extension causes growth abnormalities, as the following example shows. The rheological properties of cell walls in growing and mature tissues of wild-type barley and an overgrowth mutant (gene name *slender, sln*) can be measured. Tissues are killed and weight is applied and then released. Mature, fully expanded tissue of either wild-type or *slender* undergoes a time-dependent change in length (**viscoelastic creep**) when a load is applied. This change is almost completely reversed when the load is removed (Figure 12.8). Thus the cell walls of fully expanded, non-growing tissue are essentially elastic. In contrast, a substantial fraction of the change in length brought about by a load applied to expanding tissue is not reversed when the load is removed, showing that extension is plastic. The results also show that the plasticity of growing cell walls of the *slender* mutant is appreciably greater than that of the wild-type genotype (Figure 12.8). We may conclude that the mutant has a defect in the biochemical process that stiffens the cell walls of tissue as it matures. The molecular basis of the *slender* phenotype is discussed further in Section 12.6.3.

Rheological measurements demonstrate that cell wall architecture changes during expansion. The increase in plasticity of walls is a result of **wall loosening**: that is, the controlled biochemical modification of the cell wall matrix that allows microfibrils to separate and newly synthesized polymers to be inserted in a coordinated fashion (Figure 12.9A). The mechanism of wall loosening and polymer deposition during cell expansion is explained by the **multinet growth hypothesis**. The primary wall of meristematic and parenchymatous cells

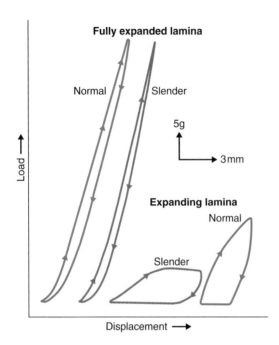

Figure 12.8 Load–displacement diagrams for growing and mature lamina tissues of wild-type and *slender* overgrowth mutants of barley. Upward arrows indicate stretching under load, downward arrows indicate relaxation when the load is removed. The area enclosed by the load–displacement shape indicates the work done to deform the system.

generally has five to ten layers. The multinet model pictures the insertion of new transversely-orientated microfibrils in strata on the inner surface of the wall, and displacement of the older microfibrils to the outer layers of the wall where they adopt a longitudinal alignment as the cell extends (Figure 12.9B). Wall extension is driven

by turgor, which exerts an outward pressure that separates the transversely-orientated microfibril bands, leading to expansion primarily along the longitudinal rather than the radial axis. Because the volume of the protoplast is large relative to that of the thin growing cell wall enclosing it, wall microfibrils can be under tension equivalent to several hundred times turgor pressure. This is the kind of hydrostatic power that allows plants to grow through concrete and undermine the foundations of buildings.

Plastic deformation of the cell wall requires breakage of chemical **cross-links** between the load-bearing wall elements. This has been intensively studied in stem and in coleoptile segments. **Auxin** treatment promotes cell elongation, an increase in wall extensibility and simultaneous wall loosening. Experiments have shown that auxin causes elongating tissue sections to **acidify** the medium in which they are bathed; additionally, if H^+ is substituted for auxin, it brings about a similar increase in growth. These observations led to the **acid-growth hypothesis**, which proposes that auxin acts by stimulating the plasma membrane proton pump and that the decrease in apoplastic pH activates a mechanism that cleaves the load-bearing bonds tethering cellulose microfibrils to other polysaccharides (Figure 12.10A). The consequence is separation of the microfibrils, cell wall loosening, uptake of water, and cell expansion in accordance with the multinet model.

The precise identity of the mechanism that breaks cross-linkages between cellulose and matrix polymers is not clear. There are hydrolytic enzymes that can cleave cross-linking glycans, but neither their pH responses nor their regulatory properties are consistent with a function

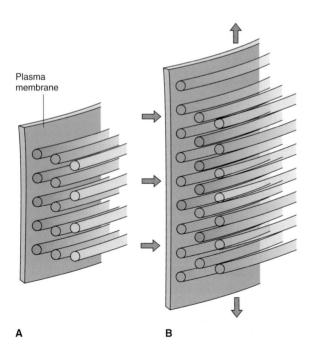

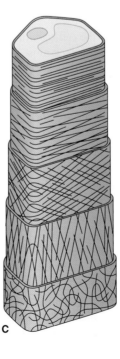

Figure 12.9 Orientation of microfibrils in growing cell walls. (A) Initial state of cell wall pre-growth. (B) Integration of wall loosening with incorporation of new wall polymers during cell expansion (vertical arrows). Horizontal arrows show the location of the addition of new microfibrils on the inner surface of the wall and integration into exterior strata. (C) Stratification of microfibrils according to the multinet growth model.

A   B   C

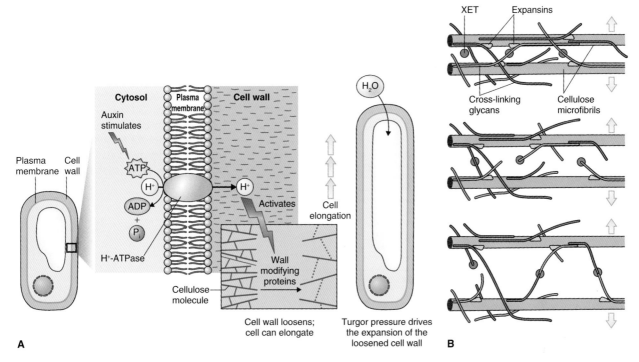

Figure 12.10   Cell wall microfibril separation and osmotically-driven cell expansion. (A) Acid growth is mediated by an auxin-stimulated proton pump in the plasma membrane, which decreases the pH of the cell wall solution, promoting the activity of wall-loosening proteins. (B) Loosening of cross-linking glycans, the consequence of the breakage of steric cellulose–glycan interactions by expansins, or by the action of a transglycosylase (XET), or both.

in wall loosening. One class of enzymes with many of the necessary characteristics, however, is the group encoded by the **xyloglucan endotransglucosylase/ hydrolase gene family**. An example is xyloglucan endo-β-transglucosylase (**XET**), which is able to cleave one xyloglucan chain and reattach it to the terminus of another. Studies using mutants defective in XET, or with their xyloglucan structure altered to a form for which XET has reduced affinity, have shown that XET has the properties of both a cell wall-loosening enzyme and a stiffening enzyme. In general, the search for an enzymatic mechanism responsible for cell wall loosening (Figure 12.10B) has not yet produced conclusive answers.

Expansins, a class of ubiquitous proteins that are encoded by two large multigene families, have been implicated in restructuring the cross-linking network in cell walls by a mechanism that does not involve detectable hydrolysis or transglycosylation. They act by disrupting hydrogen bonds between microfibrils and the load-bearing glycans to which cellulose is linked (Figure 12.10B), and can produce expansion when applied to stretched cell walls, to stem tissue killed by heat or freezing and thawing, or even to filter-paper. Furthermore, application of exogenous expansins to the flanks of meristems has been shown to result in bulging and in some cases the differentiation of leaf primordia (see Section 12.5.1). The **crystal structure** of a *Zea mays*

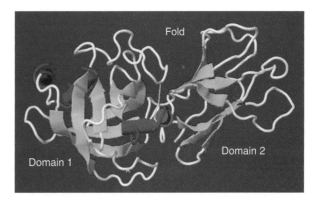

Figure 12.11   Crystal structure of the *Zea mays* β-expansin EXP B-1 showing sheets (yellow) and helices (pink) in two domains separated by a fold.

expansin (originally identified as a pollen allergen; see Chapter 16) is shown in Figure 12.11. The molecule has two domains with a high degree of secondary structure on either side of a fold. The surface of the molecule has a concentration of polar and aromatic amino acid residues and is thought to be able to bind to a region of branched polysaccharide up to ten residues long, and to bring about dissociation of glycan from the cellulose surface through a shift in the angle between the domains.

So far we have considered the requirement of tissue to undergo wall loosening so that growth can occur.

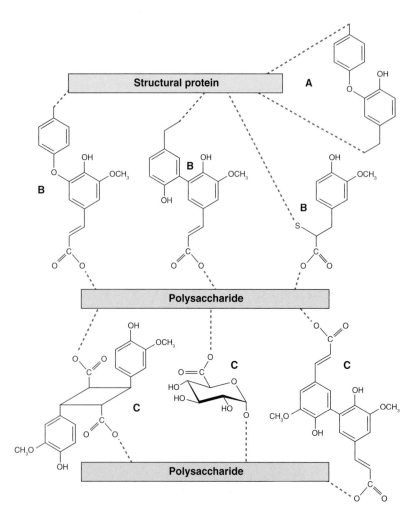

Figure 12.12
Polysaccharide–polysaccharide and polysaccharide–protein cross-bridges in cell walls. (A) Isodityrosine forms an intrapeptide linkage needed to stabilize the extensin rods. (B) Tyrosine, lysine (not shown) and sulfur-containing amino acids can form ether and aryl linkages with hydroxycinnamic acids esterified to polysaccharides. (C) Neighboring polysaccharides may contain cross-bridges esterified directly to sugars.

Rheological measurements, such as those illustrated in Figure 12.8, establish that underlying the transition from the expanding to the fully-grown state is a process of **wall stiffening**, locking the cell into shape. **Extensins**, hydroxyproline-rich glycoproteins (**HRGPs**) that cross-link other cell wall polymers (see Figure 4.6), become progressively less soluble as the wall stiffens. This suggests that they and other HRGP-like proteins may play a part in wall stiffening by participating in networks,

linking them to each other and to cellulose microfibrils. Structural proteins and polysaccharides can form **bridges** via tyrosine, lysine and S-containing amino acids, and through ether or ester-linked hydroxycinnamic acids and other **phenolic** compounds (Figure 12.12). **Isodityrosine** (Figure 12.12A) is a cross-linking product of the action of **peroxidase** on tyrosine residues in cell wall proteins (Equation 12.3), indicating a possible function for apoplastic forms of this

Equation 12.3 Formation of isodityrosine by peroxidase

$$2 \text{ Tyrosine} + H_2O_2 \longrightarrow \text{Isodityrosine} + 2H_2O$$

Tyrosine

Isodityrosine

enzyme in the locking mechanism. We can conclude that the transition from plasticity to elasticity during growth and maturation is a consequence both of the progressive cessation of wall-loosening activities and of metabolic and oxidative cross-linking events that stiffen the wall.

In addition to their intrinsic rheological properties, the inner cells of expanding axial and lateral structures are subject to physical and chemical constraints imposed by the **epidermis**. The role of the epidermis as a '**tensile skin**', which restricts the potential enlargement of the tissues it encloses, has received renewed attention recently. The biomechanical properties of epidermal cell walls and cuticles are consistent with such a constraining role. Conversely, hormonal signaling between epidermal and subepidermal tissues has been shown to exert a positive effect on shoot growth. Many studies since the early classic observations on the action of indole-3-acetic acid (IAA) have identified the outer cell layers of growing organs as the target for auxin-mediated enhancement of cell wall plasticity. **Brassinosteroids** (see Chapter 10) are another important hormonal influence. For example, brassinosteroid biosynthesis or response mutants of *Arabidopsis* are dwarfed; transgenic mutant lines expressing genes for brassinosteroid synthesis or perception specifically in the epidermis are normal in stature. Such biophysical and genetic studies have led to the conclusion that the balance between mechanical and chemical influences of and on the epidermis is a critical factor determining organ size and form.

12.2.5 Cells enlarge by tip growth or diffuse growth

Elongation of some specialized cells such as root hairs, pollen tubes and cotton fibers occurs by **tip growth** (Figure 12.13A). In tip growth, expansion is confined to the cell apex, which is generally dome-shaped, rich in secretory vesicles and lacking in microtubules. The dome is covered in a thin nascent cell wall and located above a subapical region that contains the nucleus and other organelles within a wall that becomes stiffer with increasing distance from the apex. Exocytosis of secretory vesicles inserts new plasma membrane and cell wall precursors into the site of growth at the apex. Arrays of microtubules assemble in the maturing subapical region, parallel to the axis of expansion, and control a gradient of cytoplasmic Ca^{2+} that focuses and stabilizes growth at the apical dome. Tip growth in pollen tubes is discussed in more detail in Section 16.5.6 and illustrated in Figures 16.39 and 16.40.

Elongating cells in the root and stem grow by **diffuse growth**, in which extension is uniformly distributed in the wall (Figure 12.13B). As we have seen in Chapter 4, cell walls are composed of cellulose microfibrils held

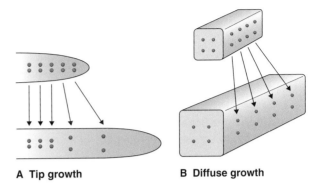

Figure 12.13 The two types of cell growth that occur in plants, illustrated using markers (dots) on the surfaces of hypothetical cells. (A) In tip growth, new cell wall and membrane components are inserted at the apex, resulting in a gradient of expansion, highest at the apex. Because expansion is confined to the apical dome, only the markers within this region separate (arrows). (B) In diffuse growth, expansion occurs over the entire surface of the cell, so the distance between the markers increases evenly. Note that the cell has elongated much more than it has widened.

together by cross-linking glycans and embedded in the pectin–glycoprotein matrix (see Figure 4.6). In cells that elongate by diffuse growth, cellulose microfibrils are orientated in hoops in transverse or slightly helical orientation. Lateral expansion of the cell is constrained by the orientation of cellulose microfibrils in the cell wall. Microfibrils in cells undergoing diffuse expansion tend to be orientated in parallel associations perpendicular to the axis of growth (see Figure 12.9A). The arrangement of cellulose microfibrils is thought to be related to the pattern of cortical microtubules, which may form channels in the plasma membrane that guide the movement of complexes of cellulose-synthesizing enzymes. As they travel, these complexes produce transversely-orientated cellulose microfibrils that become incorporated into the growing cell wall (see Figures 4.8 and 4.20). Treatment with colchicine, which disrupts cortical microtubules, causes cellulose microfibrils to be laid down in random arrays. In this case the cell wall yields equally in all directions and cells expand isodiametrically.

12.2.6 Cells of primary vascular tissues originate in stem and root apical meristems; cambium is the origin of secondary vascular tissues

During maturation of stems and roots, primary vascular tissues differentiate from procambial cells in the apical

Key points Growth typically follows a sigmoidal trend of size over time. Cell enlargement is driven by turgor pressure acting on the cell wall. The wall is plastic at first but becomes more elastic as the cell matures. During plastic growth, wall structure is loosened and new polymers are deposited. Acidification, auxin, glycan hydrolases and expansins promote cell wall loosening and enlargement. Growth is constrained and the wall becomes less plastic and more elastic as a consequence of cross-linking between microfibrils by phenolic bridges. The epidermis, which can exert a restrictive influence on the growth of internal tissues, is a site of action of growth-promoting hormones, notably brassinosteroids and auxin. Root hairs, pollen tubes and some other specialized cells enlarge by tip growth, but most tissues expand by diffuse growth. Orientation of cellulose microfibrils determines the extent of growth in different planes.

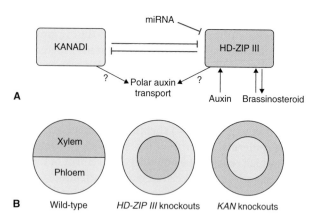

Figure 12.14 Regulation of vascular patterning by KAN and HD-ZIP III transcription factors. (A) The activities of *KAN* and *HD-ZIP III* family genes are mutually antagonistic. (B) Mutational knockout of *HD-ZIP III* or *KAN* genes results in a change in the arrangement of xylem and phloem in the vascular bundle from adaxial/abaxial to concentric organization.

meristem (see Section 12.2.1). Subsequently the primary stems and roots of many eudicots and all gymnosperms undergo **secondary growth** during which xylem and phloem develop from meristematic cells of the **cambium** (see Figure 1.40). Auxin has a decisive regulatory role in establishing the continuous procambial strands necessary for vascular differentiation. The hormone is also thought to be implicated in differentiating the outer (phloem) from the inner (xylem) region of the radially-distributed vascular bundles of the stem, and the adaxial from the abaxial sides of bundles in leaves. **Patterning** of vascular bundle organization is configured in the embryo and, like other morphogenetic processes, has been studied and modeled by analyzing mutants. In this way a number of networked **transcription factor** genes have been identified as critical for radial and adaxial/abaxial orientation. There is a mutually antagonistic relationship between the activity of the *KANADI* (*KAN*) group of *MYB* family transcription factor genes (named after myeloblastosis virus) and genes encoding type III *HD-ZIP* (HomeoDomain leucine Zipper) transcription factors (Figure 12.14A). The latter regulate vascular proliferation, interact with auxin and brassinosteroids, and have been shown to be under microRNA control (see Section 3.4.12). Mutational knockout of *HD-ZIP III* genes results in a radically altered pattern of vascular tissue, from the normal abaxial/adaxial arrangement of phloem and xylem, respectively, to one in which phloem surrounds xylem. Conversely, *kan* mutants have vascular

bundles in which xylem surrounds phloem (Figure 12.14B). It is thought that KAN and HD-ZIP III exert their influence on vascular patterning through changes in polar auxin transport. Further functions of these transcription factors in adaxial/abaxial patterning during leaf development are described in Section 12.5.4.

In addition to auxins, **cytokinins** have been identified as hormonal regulators of vascular patterning. In roots of mutants that have an impaired **cytokinin two-component histidine kinase** signaling pathway (see Chapter 10), protoxylem cell files in the developing vascular cylinders increase in number whereas other cell types are depleted. Cytokinins are also required for maintenance of the proliferative capacity of cambial cells during secondary growth.

A distinctive feature of tracheary elements in the xylem, and of the cells of mechanical tissue such as **sclerenchyma**, is their thickened secondary cell walls, in which the stiffening associated with cellulose–glycan cross-linking (see Figure 12.10) is further enhanced by impregnation with **lignins** (polymeric phenyl-propanoids). The monomeric phenolics, which form bridges in the primary cell wall (see Figure 12.12), are sometimes referred to as **monolignols**. The pathways of phenylpropanoid biosynthesis are described in Chapter 15. Polymerization of monolignols to form lignin in secondary walls occurs by coupling reactions involving oxygen radicals enzymatically generated from H_2O_2 by **peroxidases**, or from O_2 by **laccases**. The result is a lignin matrix cross-linked to cell wall polysaccharides via ester and ether bonds through hydroxycinnamic, ferulic and dehydroferulic acid bridges (Figure 12.15). The lignification phase of tracheary element differentiation

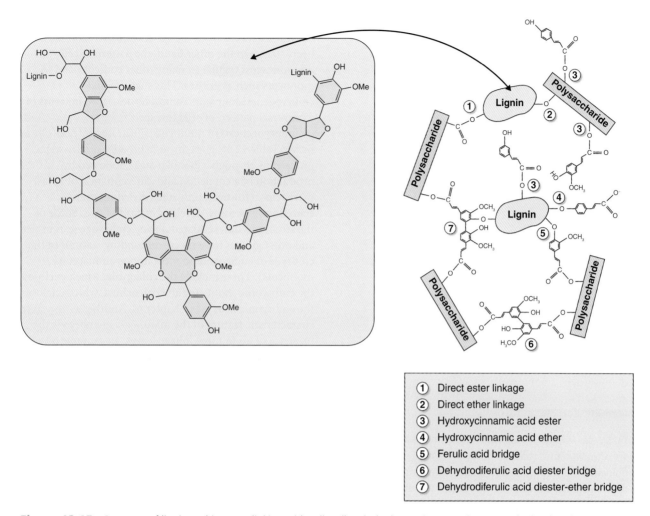

①	Direct ester linkage
②	Direct ether linkage
③	Hydroxycinnamic acid ester
④	Hydroxycinnamic acid ether
⑤	Ferulic acid bridge
⑥	Dehydrodiferulic acid diester bridge
⑦	Dehydrodiferulic acid diester-ether bridge

Figure 12.15   Structure of lignin and its cross-linking with cell wall carbohydrates via aromatic ester and ether bonds.

culminates in mass destruction of cytoplasm by a process of **programmed cell death**, resulting in a mechanically- and chemically-resistant tissue in the form of a rigid block of hollow tubes. Programmed death in tracheary element differentiation is discussed in detail in Chapter 18. Other inert macromolecules with barrier functions that plant cells accumulate include the waxy polymers **suberin** and **cutin**.

Phloem is the tissue in which long-distance transport of photosynthate occurs in vascular plants (see Chapter 14). The **sieve tube elements** in the phloem originate as **protophloem sieve element cells** that develop from procambium. Sieve tube elements are long and narrow and orientated parallel to the longitudinal axis of plant organs (see Figure 1.29). The transport of sugars is facilitated by the **sieve plates**, perforated walls at the apical and basal ends of the cell which allow large cytoplasmic connections between stacked sieve tube elements. A regulatory gene, *APL* (*ALTERED PHLOEM DEVELOPMENT*), is known from studies of a

developmental mutant in which protophloem differentiates into cells with a mixture of xylem and phloem characteristics. The mutation is lethal; seedlings develop only a few leaves before their development is arrested and they die. *APL*, which encodes a MYB-type transcription factor, is thought to be required for later steps of sieve element development and for spatially limiting the differentiation of xylem.

12.3 Embryogenesis

Angiosperm embryos most commonly develop from the zygote which, together with the triploid endosperm, is the product of the double fertilization event that occurs when the pollen tube reaches the ovule (see Chapters 1 and 16). Figure 12.16 depicts an embryo of the eudicot *Arabidopsis*, showing the two cotyledons and the root–shoot axis with the niches of proliferative cells

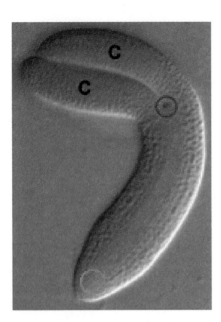

Figure 12.16 *Arabidopsis* embryo showing the two cotyledons (C) and proliferative cell niches in the root (green circle) and shoot (red circle) which will become the root and shoot apical meristems respectively.

Key points Procambial cells of the SAM and RAM are the source of primary vascular tissues. Xylem and phloem produced during secondary growth originate in the meristematic cells of the cambium. The organization of different cell types in vascular tissues is under the control of polar auxin transport, the cytokinin signaling pathway and MYB and HD-ZIP transcription factors. As they mature, the cell walls of xylem and sclerenchyma become lignified and cytoplasm is lost during a terminal phase of programmed cell death. A MYB-type transcription factor is implicated in development of the sieve tube elements of phloem.

destined to become the RAM and SAM. The earliest stages in the development of the *Arabidopsis* embryo from the zygote are shown in Figure 12.17. The initial division of the zygote is polarized and produces the basal cell, destined to become the suspensor, and the terminal cell which develops into the embryo proper. Embryo cells proliferate to form the **globular** stage, establishing ground meristem, protoderm and procambium. Two cotyledon primordia appear at the **heart-shaped** stage and the shoot and root apical meristems develop at the **torpedo** stage (see Figure 1.19). In *Arabidopsis* and other crucifers, the pattern of cell division during early embryogenesis is fixed and therefore it is possible to trace

cell fate in these species. Figure 12.17C maps the lineage of individual cells from the 16-cell embryo onto the early heart stage. On the other hand, the pattern of cell division is less regular in many other eudicots, and in monocots such as grasses. This suggests that embryogenesis in plants cannot be described in terms of a universal pattern of cell cleavage.

Embryogenesis in monocots, illustrated by rice (*Oryza sativa*; Table 12.1), goes through broadly similar steps, proceeding through a globular stage, followed by production of a cotyledon and differentiation of apical meristems. However, only one cotyledon, called a **scutellum** in grasses, develops instead of two. Some of the structures differentiated during embryogenesis in rice and other grasses have no corresponding features in the embryos of eudicots (and most non-graminaceous monocots). These include the **coleoptile**, which encloses the embryonic shoot, and the **coleorhiza**, which encloses the embryonic root (see Chapter 6). Another characteristic feature of the mature grass embryo is the presence of up to three foliage leaves, an example of precocious establishment of vegetative development. Incorporation of later stages of development into embryogenesis is an example of **heterochrony**, an adaptive mechanism seen widely in evolution.

12.3.1 Pattern formation is the result of polarity in the developmental fates of embryo cells

An important factor in the establishment of polarity in morphogenesis is **asymmetrical cell division**, that is, mitosis resulting in two daughter cells differing in size, structure, or both, and destined to experience different developmental fates. The first division of the *Arabidopsis* zygote into **basal** and **apical cells** is asymmetrical (Figure 12.17A)—a polarization event resembling that occurring in early embryogenesis of the alga *Fucus* (see Figure 12.6). Other examples of asymmetrical cell division leading to functional specialization include the separation of differentiation-competent daughter cells from undifferentiated, continuously mitotic cells in the **root meristem** (see Figure 12.3) and the formation of **stomata** during differentiation of leaf tissues (see Section 12.5.2).

In some cases there is good evidence that division is under the control of intrinsic regulatory systems that specify the formation of daughter cells differing in form and fate. Such intrinsic mechanisms may in turn be responsive to signals coming from surrounding cells. A number of different mechanisms are known to account for asymmetry in the products of eukaryotic cell division,

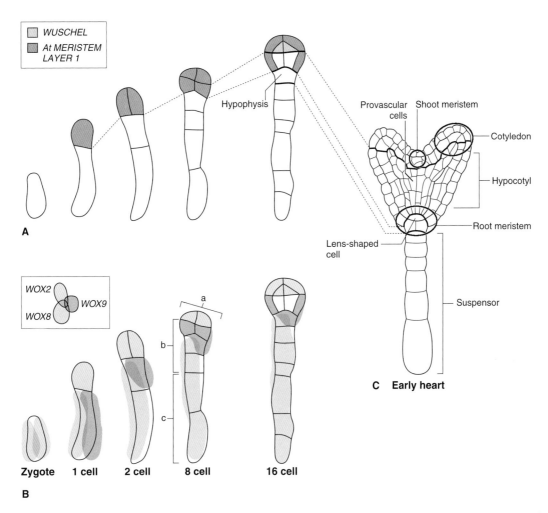

Figure 12.17   Molecular markers of cell fate in early development of the *Arabidopsis* embryo. (A) Expression patterns of *AtML1* and *WUS* transcription factors up to the 16-cell stage. (B) Cell-specific expression of *WOX* (*WUSCHEL HOMEOBOX*-like) family genes. Note the asymmetrical division of the zygote cell and the non-uniform distribution of markers between terminal and basal cells. Three domains of the eight-cell embryo are indicated by a, b and c. (C) Lineage relationships of cells at the heart stage to those of the 16-cell embryo.

though the precise cue that confers asymmetry in plant embryogenesis has yet to be identified. Genetic evidence suggests an essential role for a **signaling pathway** centered on a kinase encoded by the gene *YODA*, but details remain to be established. An important factor in the division mechanism that leads to non-equivalent daughter cells is orientation of the **mitotic spindle**, under the influence of actin cytoskeleton organization and associated motor proteins. Another way in which an unequal outcome may be imposed on mitosis is by localization of a regulatory factor in one region of the cell prior to division. Partitioning of this factor activates genes in, and thereby determines the fate of, one of the two daughter cells. The non-uniform distribution of homeobox transcription factors of the **WOX** (*WUSCHEL HOMEOBOX*-like) family can be related to the pattern of cell division and development in the early embryo of *Arabidopsis* (Figure 12.17A,B). *WOX2* and

WOX8, two of the 15 or so members of the *WOX* gene family, are cell markers of asymmetrical division of the zygote. *WOX2* and *WOX8* are initially coexpressed in the zygote, but become restricted to the apical and basal cells, respectively, following division. A third *WOX*, *WOX9*, is expressed only in the basal cell after division.

An important subject of continued debate is the question: to what extent does cell division drive subsequent organ growth and development? Early insights into the growth–differentiation relationship came from experiments on so-called **gamma plantlets**, embryos of cereal species that had been exposed to high doses of γ-rays. Gamma plantlets are incapable of DNA synthesis, mitosis or cell division but are morphologically more or less complete, producing the major highly differentiated cell types such as those of vascular and mesophyll tissues (Figure 12.18). They do not, however, initiate new organs, their meristematic

Table 12.1 Stages in the development of the rice embryo.

Stage	Days after pollination	Number of cells	Events	Structure in longitudinal section
Zygote	0	1	• Fertilization	
Early globular stage	1	1 to c. 25	• First division of fertilized egg • Rapid cell division	
Middle globular stage	2	c. 25 to c. 150	• Globular shaped • Relatively slow growth	
Late globular stage	3	c.150 to c.800	• Oblong-shaped • Onset of exponential growth, dual rhythmicity of growth and cell increase • Gradient of cell size along dorsoventral direction • Pattern formation, regionalization	
Shoot apical meristem and radicle formation	4	>800	• Onset of differentiation of coleoptile (black arrowhead), shoot apical meristem (arrow) and radicle (white arrowhead)	
First leaf formation	5–6		• Protrusion of first leaf primordium (arrow) • Enlargement of scutellum • Expression of *RAmy1A* in scutellar epithelium • Onset of juvenile vegetative stage	

Table 12.1 (*continued*)

Stage	Days after pollination	Number of cells	Events	Structure in longitudinal section
Second and third leaf formation	7–8		• Protrusion of second and third leaf primordia in alternate phyllotaxis	
Enlargement of organs	9–10		• Enlargement of organs and morphological completion • Expression of maturation-related genes such as *OsEM*, *Rab16A* and *REG2*	
Maturation	11–20		• Arrow indicates the shoot apical meristem • Dormancy develops from 21 DAP	

CO, coleoptile; EP, epiblast; RA, radicle; SC, scutellum.

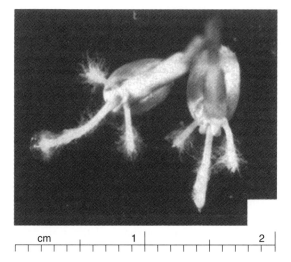

Figure 12.18 Gamma plantlets of wheat. These seedlings developed from seeds treated with high doses of γ-rays which made the plants unable to synthesize DNA or carry out mitosis and cell division.

regions show various abnormalities and they lack stomatal guard cells and trichome cells. Gamma plantlets demonstrate that embryo cells have considerable capacity for differentiation independent of any requirement for DNA replication and cell proliferation.

12.3.2 Axis formation during embryogenesis is the prelude to differentiation into root and shoot

Formation of the root and shoot axis during embryogenesis is fundamental to establishing the **body plan** of a plant. Axis formation is an early process in embryo patterning, involving the determination of symmetry and tissue identity, under the control of a regulatory network of hormones (particularly auxin) and transcription factors. The apical–basal axis of the eight-cell embryo consists of three **domains** (indicated by a, b and c in Figure 12.17B). The two apical cells generate the shoot apical meristem and most of the cotyledon tissue. The remaining cotyledon cells, hypocotyl and root apical initials are derived from the four cells of the central domain. The basal two cells will form the quiescent center and root cap initials. By the 16-cell stage, the embryo has become **concentrically organized**, with an outer protoderm, underlying subepidermal ground tissue and central vascular initials (Figure 12.17C). As the two cotyledons begin to differentiate (the onset of a left–right polarization), the **radial symmetry** of the globular embryo gives way to the **bilateral symmetry** of the heart-shaped stage. A line drawn between the shoot and root meristems traces the axis about which the radial symmetry of the stem and primary root will be maintained; but from heart to torpedo stage the cotyledons develop along a central-to-peripheral axis (Figure 12.19A), establishing an **adaxial** (facing towards the central axis) and **abaxial** (away-facing) surface.

In the embryo of the monocot rice, maize (*Zea mays*), and other grasses, there is a transition from the radially-symmetrical globular stage to a mode of organization in which the scutellum (equivalent to the single cotyledon) forms a bilaterally-symmetrical, shield-shaped mass situated between the coleoptile and the endosperm. The shoot meristem to root meristem axis is central with respect to the scutellum (Table 12.1). Thus the monocot embryo (exemplified by rice and maize) differs significantly from that of the eudicot *Arabidopsis* in having only one rather than two planes of bilateral symmetry (Figure 12.19B). This means that the functional role of a given gene for polarity and pattern regulation in eudicot embryogenesis must be divergent

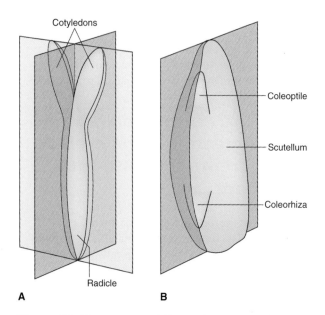

Figure 12.19   Symmetry of dicot and monocot embryos. (A) Dicot embryo, for example *Arabidopsis*, has two planes of bilateral symmetry. (B) Embryo of a monocot, such as *Zea mays*, has one axis of bilateral symmetry.

in timing and location from that of the homologous monocot gene.

12.3.3 Many genes with functions in pattern formation have been identified by studies of embryogenesis mutants

The genetic analysis of mutants allows development to be modeled as regulatory networks of interacting transcription factors. A large number of mutants aberrant in embryogenesis are known in *Arabidopsis* and a few in some other species such as maize and rice. Here we consider a group of **transcription factors** identified by genetic analysis whose expression profiles and interactions are consistent with critical roles in axis development and pattern formation. We have seen that *WOX2*, *WOX8* and *WOX9* are cell markers for patterning events in early embryogenesis (see Section 12.3.1). The gene *AtML1* (*Arabidopsis thaliana MERISTEM LAYER 1*) is required for differentiation of the protoderm in the SAM, but is not involved in root epidermis development. Figure 12.17 shows the spatial expression profiles of *WUSCHEL* (*WUS*; see Chapters 3 and 11), *AtML1* and members of the *WOX* gene family at different stages in early embryogenesis. A *WOX2*-like gene is expressed in the apical domain early in the development of the embryos of *Z. mays*, rice and other grasses and later becomes restricted to the outer L1 layer on the ventral side, where it pre-patterns the shoot apical meristem. On the other hand, cell-type expression profiles indicate that the homolog of *WUS* in grass embryos is primarily associated with the specification of new organs and does not operate in the feedback loop with the *CLAVATA* (*CLV*) gene that regulates the organization of the vegetative shoot apical meristem in *Arabidopsis*, as described in Chapter 3 (see Figure 3.32). Such observations point to the evolution of **divergent functions** for *WUS* between monocots and eudicots, correlated with changes in *CLV* signaling, possibly reflecting the contrasting organization and patterning of embryogenesis in the two classes of angiosperm.

Auxin plays a critical part in embryogenesis. This is illustrated by its influence on the regulatory transcription factor CUC during the transition to bilateral symmetry and cotyledon outgrowth in the *Arabidopsis* embryo (Figure 12.20). Genes of the **CUC** (*CUP-SHAPED COTYLEDON*) family initiate the expression of **STM** (*SHOOT MERISTEMLESS*) in the incipient shoot apical meristem. *STM*, a member of the *KNOX* gene family (see Sections 12.1 and 12.5), is a feedback suppressor of *CUC* expression and also downregulates *AS* (*ASYMMETRIC*

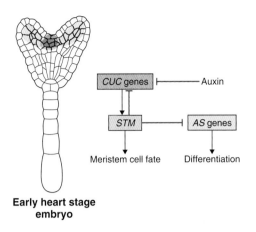

Early heart stage embryo

Figure 12.20 Interactions and tissue expression patterns of genes regulating the transition to bilateral symmetry and cotyledon outgrowth in the embryo of *Arabidopsis*.

LEAVES) genes, which promote the differentiation of leaf primordia. The *CUC–STM* interaction generates the boundary that separates the two cotyledons (Figure 12.20). Asymmetrical distribution of **auxin efflux carriers** prior to cotyledon outgrowth directs the hormone preferentially towards lateral domains where it inhibits *CUC* gene expression.

At a conservative estimate, more than 80 genes with functions in embryogenesis have been examined for tissue expression patterns and interactions. Of these, as described above, the vast majority are transcription factors and components of hormone or kinase-based signaling pathways. This has allowed a very detailed picture to be built up of the transcriptional and signal-transduction events underlying early morphogenesis. A major challenge for the future of plant developmental biology is to fill in the biochemical, cell biological and physiological details between genotype and phenotype—that is, between the switching circuitry that regulates developmental genes and the chemical and physical processes that build and animate biological structures.

12.4 Growth and differentiation of roots

The anatomy and morphology of roots across different plant species are somewhat less diverse than those of stems, probably because soil is more stable than the aerial environment, but they are still capable of considerable variation (see Figure 1.30). Comparative studies of fossils and phylogenies suggest that the characteristic structures common to most roots—for example root hairs and root caps—have a diversity of evolutionary origins. Endogenous initiation of lateral and adventitious roots

Key points Polarity is established at an early stage of embryogenesis, as a consequence of asymmetrical division of the zygote into basal and apical cells. Kinase signaling, orientation of the mitotic spindle and non-uniform distribution of transcription factors contribute to the asymmetry. Experiments on gamma plantlets, in which growth is disabled by exposure to high doses of radiation, show that morphogenesis can take place without cell division. Eudicot embryos are initially radially symmetrical and then develop two planes of bilateral symmetry as the cotyledons differentiate. The grass embryo, with its single cotyledon (scutellum), develops a single plane of bilateral symmetry. Identifying regulators of embryogenesis has been facilitated by analysis of mutants. Transcription factors of the WUS, WOX and CLV families are markers of embryonic cell patterning and components of regulatory networks that specify meristem organization. Auxin is the major hormonal regulator of embryo symmetry, cotyledon outgrowth and leaf primordium differentiation, acting via interaction between the transcription factors CUC and STM.

has also arisen repeatedly during evolution and is thought to be related to dominant patterns of auxin transport away from the stem apex and the requirement for vascular continuity. In the case of lateral roots, if they were initiated at the root apical meristem, there is a danger that they would be worn away as the root grew down through the soil. Endogenous initiation from the pericycle in a region in which elongation has ceased avoids this risk (Figure 12.21). External influences on root morphogenesis include nutrient availability and biotic interactions. Chemical signaling is of particular importance in pathological and symbiotic relationships of roots with other roots and with soil microorganisms. This section describes the development of root form and function and its regulation by internal and external factors.

12.4.1 Root architecture is important for functions that include support, nutrient acquisition, storage and associations with other organisms

The uptake of water and inorganic ions, anchoring and carbohydrate storage are major functions of roots (see

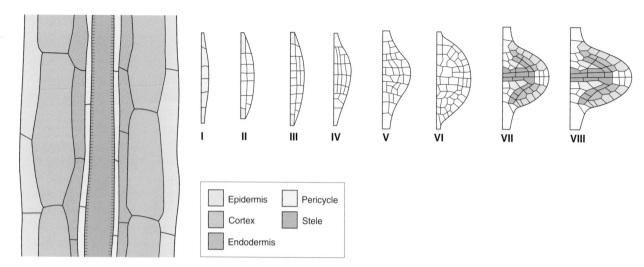

I   II   III   IV   V   VI   VII   VIII

	Epidermis		Pericycle
	Cortex		Stele
	Endodermis		

Figure 12.21   Eight stages of primordium development during lateral root morphogenesis. Lateral roots originate from pericycle cells deep within the primary root. (I) File of pericycle cells. (II–VI) Periclinal division of pericycle cells initiates a lateral root primordium. Further periclinal and anticlinal divisions increase the size of the primordium. (VII, VIII) As the primordium continues to grow, cell layers begin to differentiate.

Chapters 13, 14 and 17). In addition, roots may be specialized to carry out other functions including additional support (buttress roots of shallow-rooted tropical trees, contractile roots of *Lilium* bulbs, breathing roots of plants in swampy areas, grasping roots of ivy) and overwintering (tuberous roots of *Dahlia*, taproots of carrots and parsnips) (see Chapters 15 and 17). Over 4000 plant species are known to obtain some or all of their nutritional needs by parasitizing other plants. **Haustoria**, the roots of parasitic plants, penetrate and absorb nutrients from the root tissues of the host species (see Figure 15.42). The roots of many plants form symbiotic associations with other organisms. Of particular importance for plant nutrition are the nodule structures containing nitrogen-fixing prokaryotes and the **arbuscules** associated with interactions between roots and mycorrhizal fungi.

The architecture of a plant's root system is important for carrying out its functions. The **radicle** of the seedling together with the lateral (branch) roots it produces constitute the primary root system of the plant. In contrast to the stem, where new organ primordia are produced by the apical meristem, lateral root primordia are produced in mature tissue from the meristematic cells of the pericycle, which is adjacent to the vascular cylinder (see Figures 1.36 and 1.37); Figure 12.21 shows the stages in the formation of a lateral root primordium. Lateral roots extend from every side of the primary root. **Root hairs**, each of which is an outgrowth of a single epidermal cell, provide most of the surface through which roots absorb water and dissolved minerals (Figure 12.22; see also Figure 1.36). Stems and leaves may have the ability to produce **adventitious roots**. In this

case, adventitious root primordia originate from parenchyma cells that are adjacent to vascular bundles.

Further elaboration of the root system follows a different pathway in eudicots and in grasses, the most closely studied family of monocots. In eudicots, the primary root continues to grow and produce lateral roots; at its base it constitutes a **taproot**, which in some plant species can be a storage organ (see Chapter 17). Complex branching root systems result from laterals, giving rise to further laterals and so on. In grasses, adventitious roots, roots produced by the shoot, play a large part in the development of the plant's root system. Even before germination, **seminal root** primordia develop at the base of the embryonic stem, the **mesocotyl**. Early in the development of a grass seedling, the radicle emerges from the seed, and the seminal roots begin to grow (Figure 12.22). As the plant develops, the primary root, seminal roots and their laterals become less important, and adventitious roots that initiate at nodes on the shoot take over the roles of support, water and nutrient acquisition. The result is a **fibrous root system** that tends to be shallower than the taproot system of eudicots (see Figure 1.30).

As discussed above, the basic architecture of the root system of eudicots differs from that of grasses. Additionally there is great variation in root architecture within these two groups, and even within an individual species. The size and organization of a root system is dependent on the environment in which the root has developed (see Section 12.4.3 for example). Soil structure and compaction, and availability of water and nutrients, all play a part, and the other life forms present in the soil,

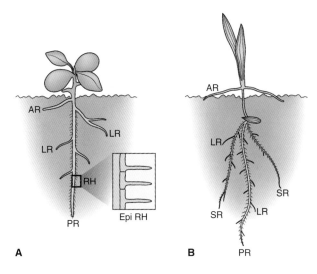

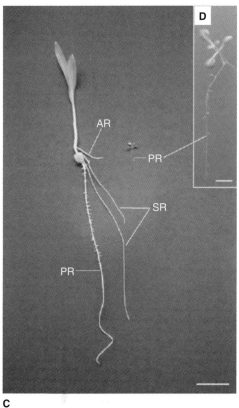

Figure 12.22 Typical root systems. (A) A dicot root, such as that of *Arabidopsis*, consists of a primary root (PR) from which lateral roots (LR) branch out as the seedling develops. The lateral roots subsequently undergo second- and third-order branching. Adventitious roots (AR) may also form. (B) A grass seedling root also possesses a primary root, but as the seedling develops, seminal roots (SR) originating at the top of the primary root and adventitious roots (AR) initiated from the base of the stem become more important. In both dicots and monocots, root hairs (RH) originate from single epidermal cells (Epi), and greatly increase the total surface area of the root system. (C, D) Fourteen-day-old seedlings of maize (C) and *Arabidopsis* (D) showing the seminal and adventitious roots in maize, whereas *Arabidopsis* has only a primary root.

including bacteria, fungi, invertebrates and burrowing vertebrates also influence root development.

A plant's ability to survive and be productive in stressful conditions, such as drought or mineral deficiency, will be affected by the size and structure of the root system. The development of modern crop varieties through domestication and selective breeding has led to many variations in root architecture, often selected inadvertently because of their advantageous effects on productivity. The initiation and growth of lateral roots is one of the most important factors in governing root system architecture. The next section will focus on the control of lateral root growth.

12.4.2 Lateral root initiation and growth are under complex genetic and hormonal control

Lateral roots in angiosperms and gymnosperms are initiated from the **pericycle**. In most eudicots, including *Arabidopsis*, lateral roots develop only from pericycle cells overlying the xylem tissue, whereas in other species, particularly grasses, the lateral roots arise from the pericycle that overlies the phloem, with cells of the endodermis also making a contribution. The formation of a lateral root begins when pericycle cells are stimulated to dedifferentiate; these cells then re-enter the cell cycle and undergo asymmetrical cell divisions, giving rise to a lateral root primordium. The cells of this primordium expand, driving its emergence through the outer layers of the primary root (see Figure 1.37). Once the lateral root has emerged, its root apical meristem is activated, and the lateral root grows by the same mechanism as that of the primary root that gave rise to it.

The initiation and growth of lateral roots are controlled by many genes and by the concentration and transport of plant hormones, particularly auxin. Some of the key genes regulating lateral root development are auxin-independent, however. The best characterized of these is the gene encoding the protein ABERRANT LATERAL ROOT FORMATION 4 (ALF4) in *Arabidopsis*. The *alf4* mutant completely lacks lateral roots, and it appears that the cell cycle in pericycle cells is blocked in the mutant plants. ALF4 is a nuclear-localized protein, but its exact function is not yet known. The two related proteins ARABIDILLO-1 and ARABIDILLO-2 also act in a hormone-independent manner and promote lateral root initiation; mutations in the genes encoding these two proteins cause a significant reduction in the number of lateral roots. ARABIDILLOs are **F-box proteins**, so they may be components of ubiquitin E3 ligases and participate in targeting certain proteins for degradation.

Auxins are important in many aspects of plant growth and development, including playing a key role in orchestrating root development in all plants so far studied. They are involved in the control of lateral root and adventitious root initiation as well as regulating root hair production. Hormone rooting compound, used by gardeners to promote rooting from cuttings of otherwise recalcitrant species, contains an auxin, usually **indole-3-butyric acid** (IBA), as the active ingredient. IBA is used because it is more chemically stable than its close relative indole-3-acetic acid (IAA), which is responsible for most auxin effects in plants (see Chapter 10).

Auxin signaling depends not only on the concentration of auxin in a given tissue, but also on the establishment of a concentration gradient. It has long been known that auxin synthesized in the above-ground part of the plant is transported to the roots; more recent work has revealed that some auxin is also synthesized in the roots themselves. Auxin is transported directionally within roots in a process known as polar auxin transport (see Chapter 10). Auxin moves towards the root tip in cells associated with the vascular cylinder (stele) and moves away from the tip in cells of the epidermis (Figure 12.23). Localized auxin concentrations are regulated both by diffusion across membranes and by the action of several auxin transport proteins; these include AUX1, which facilitates influx of auxin into cells, and PIN, which controls auxin efflux (see Figure 10.7). Taken together, the actions of these proteins establish local auxin maxima as well as concentration gradients, and it is in the areas of these maxima that lateral roots are initiated. Any mutation that prevents the establishment of normal auxin gradients disrupts the patterning of root development. The promotion of lateral root formation by auxin is inhibited by cytokinins (CKs), which act directly on the lateral root founder cells in the primary root to bring about this inhibition. There is evidence that cytokinin may interfere with PIN gene expression.

Other hormones have been implicated in the regulation of lateral root formation in certain plant species; for example, in *Arabidopsis*, both ethylene and brassinosteroids have been shown to promote lateral root initiation via an auxin-dependent pathway, while there is some evidence that ABA inhibits lateral rooting in this species. Figure 12.24 summarizes the current understanding of hormonal regulation of lateral root development in *Arabidopsis*. By contrast, ABA has been shown to stimulate lateral rooting in rice and in some legumes.

Auxin is required not only for the initiation of lateral roots but also for their continued growth, which takes place through a coordinated series of cell divisions. A key gene in the regulation of lateral root growth is *PUCHI*, which is expressed both in pericycle cells that will form

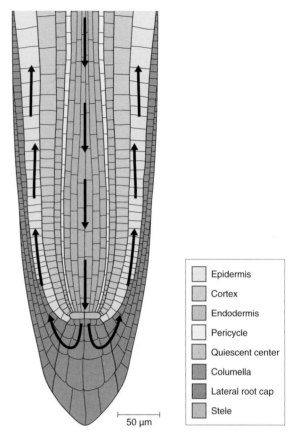

	Epidermis
	Cortex
	Endodermis
	Pericycle
	Quiescent center
	Columella
	Lateral root cap
	Stele

50 μm

Figure 12.23   Polar auxin transport in the *Arabidopsis* root tip. Auxin moves down from the shoot apex into the root cap in the vascular cylinder and then moves from the lateral root cap up through the epidermis.

the lateral root primordium and in the primordium itself. *PUCHI* encodes an APETALA2 (AP2) transcription factor that is upregulated by auxin. PUCHI protein acts to restrict the area within the lateral root primordium in which cells proliferate, thus ensuring properly controlled growth. As the lateral root primordium emerges, polar auxin transport creates a new maximum of auxin concentration at the primordium tip, and this new maximum regulates the expression of genes encoding several transcription factors that control meristem identity. These include the *Arabidopsis* genes *CLAVATA* (see Chapter 3) and *SCARECROW*; SCARECROW protein is required both to maintain the correct **radial pattern** of root cells in all roots, and to initiate lateral roots from the primary root. The maize equivalent of *SCARECROW*, *ZmSCR*, is important both for radial patterning in the tips of primary and lateral roots, and for establishment of the maize radicle during embryogenesis. Members of the *lob* domain family of transcription factors are involved in different aspects of root development depending on plant species. The *Arabidopsis lob* domain genes *LBD16*

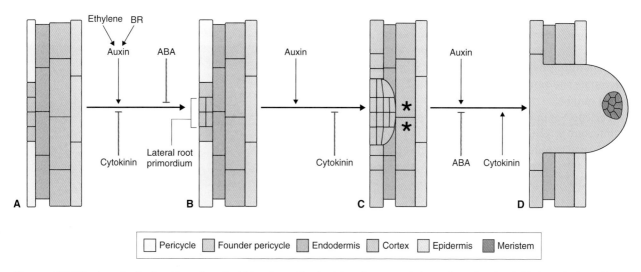

☐ Pericycle	▨ Founder pericycle	▨ Endodermis	▨ Cortex	☐ Epidermis	▨ Meristem

Figure 12.24 Longitudinal section of an *Arabidopsis* root basipetal to the zone of elongation, showing initiation of a lateral root and the stages at which plant hormones exert effects. (A, B) The process begins when a founder pericycle cell undergoes anticlinal divisions (perpendicular to the surface of the root), followed by periclinal divisions (parallel to the root surface). (C) Further cell divisions take place in a highly regulated manner, allowing the lateral root primordium to begin to emerge through the outer cell layers of the primary root; this causes cells (asterisked) to separate in these outer layers. (D) Once the new lateral root has emerged, its meristem is activated so that it can continue to grow. ABA, abscisic acid; BR, brassinosteroids.

and *LBD29* are both required for lateral root development, which is enhanced if either of the genes is overexpressed. The maize *lob* domain gene *rtcs* and the rice homolog *crl1/arl1* (two names for the same gene, identified in independent studies) are needed for adventitious root formation, illustrating how a gene family can evolve to perform different but related functions. Other genes modify the number of lateral roots produced; these include genes that encode the ARF (AUXIN RESPONSE FACTOR) family of transcription factors (see Chapter 10). In *Arabidopsis*, the *ARF* gene products ARF7 and ARF19 have been shown to activate the *lob* domain genes *LBD16* and *LBD29* discussed above.

> **Key points** Roots have a range of functions, including water and ion uptake, anchorage and support, storage and, in the case of haustoria, parasitism. Grasses have fibrous-type root architecture, whereas many eudicots have taproot systems. Lateral roots initiate in the pericycle, under the control of auxin and a range of auxin-dependent and -independent genes, some of which encode F-box proteins. Auxin movement in roots is mediated by transport proteins including PIN, LAX and AUX. Cytokinin is an inhibitor of lateral root development. Radial patterning of root tissues is maintained by auxin-regulated transcription factors.

12.4.3 Nutrients act as regulators of root development

Nutrient availability and the state of the plant's internal nutrient economy regulate the structure and function of the root system. The role of nutrients is discussed in more detail in Chapter 13. Here we summarize the effects of major soil nutrients.

The source of nitrogen present in the soil affects root architecture. High **nitrate** concentrations stimulate the growth of long, sparsely-branched primary roots. In contrast, primary roots grown with **ammonium** as the main source of nitrogen are usually short, thick and highly branched. If exogenous **glutamate** (Glu) is applied to primary roots, it inhibits their growth. This is thought to be due to an interaction with the auxin transport and signaling pathways described in Section 12.4.2. Nitrate antagonizes the effect of Glu via a pathway that requires the nitrate transporter protein NRT1 (see Chapter 13), but the precise mechanism of repression is unknown. Whichever form of nitrogen predominates, when the ratio of photosynthetically fixed carbon to nitrogen is high, the initiation of lateral roots is repressed. This repression requires the presence of another nitrate transporter, NRT2.1, which is also involved when lateral root formation is initiated following the transfer of plants from high to low nitrate concentrations. Recent results suggest that in addition to its role as a transporter, NRT2.1 can act as a fixed-nitrogen sensor that interacts with the auxin signaling pathway to control lateral root formation in response to the availability of fixed nitrogen. The term fixed nitrogen is used to distinguish forms of nitrogen that can be used directly by plants, e.g. nitrate and ammonium, from dinitrogen gas (N_2) which cannot.

The effect of nutrient availability on root system architecture is related in a complex fashion to other factors such as water supply, aeration and soil pH. For example, the external concentration of calcium required for maximal root growth in cotton is about 1 μM when the pH is 5.6, but this increases to more than 50 mM at pH 4.5. Nutrients regulate root development indirectly through influencing whole-plant growth according to the degree of deficiency or sufficiency, but they also have direct morphogenetic influence through modulation of the expression and function of specific genes.

In *Arabidopsis* and many other plants, P_i starvation inhibits the growth of primary roots. The effects of P_i starvation on lateral root growth are more complex since both positive and negative influences have been reported, and different plant species show different patterns of lateral root modification. For example, in P_i-starved *Arabidopsis*, lateral roots are produced more densely and they are more elongated than normal, resulting in a highly branched, shallow root system. P_i starvation in *Phaseolus vulgaris*, on the other hand, shifts the angle of lateral root growth from downward to outward; the result is again a shallower root system, but with no increase in the density or average length of lateral roots. The inhibition of primary root growth by P_i starvation is controlled by local P_i concentration at the tip of the root; this regulation requires the products of two genes, *LOW PHOSPHATE ROOT 1* (*LPR1*) and *LOW PHOSPHATE ROOT 2* (*LPR2*). Lateral root alterations are regulated by overall P_i status, and both auxin and ethylene have been implicated in such responses.

In certain plants, P_i starvation causes an interesting remodeling of root morphology. In white lupin (*Lupinus albus*), for example, P_i deficiency triggers the production of large numbers of short tertiary roots (Figure 12.25). These tertiary or **proteoid roots** form a tight cluster, which enables the plant to extract P_i and other nutrients more effectively from a small volume of soil (Figure 12.26). Proteoid roots are capable of releasing carboxylic acids, in particular citrate and malate, into the rhizosphere. These compounds act as chelators of metal ions including iron and aluminum, increasing the availability of the metals to the plant (see Chapter 13).

Days after emergence

5 6 7 8 10 12 14   22

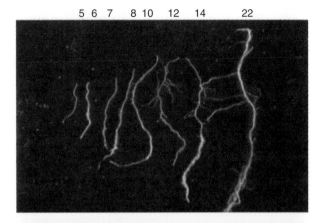

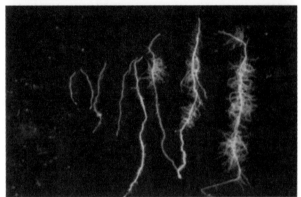

Figure 12.25   In white lupine (*Lupinus albus*), P_i deficiency (lower panel) triggers the production of large numbers of proteoid (tertiary) roots from secondary laterals. Compare with the upper panel, which shows roots developed when adequate P_i was present.

12.4.4 The formation of symbiotic associations with nitrogen-fixing bacteria modifies root development

Most plants are capable of growth in aseptic (sterile) conditions; symbiotic associations are not essential for survival provided mineral nutrient supplies are adequate. However, in the natural environment many plants form root associations with soil bacteria or fungi. In such cases the microorganisms increase the supply of essential nutrients to the plant while benefiting from the plant's ability to produce sugar by photosynthesis. Table 12.2 gives examples of symbiotic associations between nitrogen-fixing bacteria, termed **diazotrophs** (diazo = dinitrogen, troph = eater), and plants.

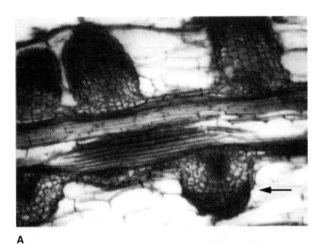

A

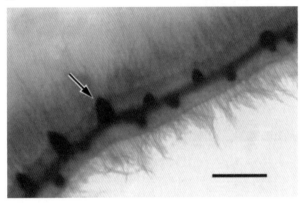

B

Figure 12.26 (A) Longitudinal section of a proteoid root from a P$_i$-deficient *Lupinus albus* plant, showing the dense cluster of emerging lateral roots. The solid arrow shows a tertiary root meristem. (B) Secondary lateral root excised from a P$_i$-deficient *L. albus* seedling 8 days after emergence. The root was cleared with sodium hypochlorite and stained with methylene blue to make it easier to see the tertiary lateral roots emerging through the epidermis (solid arrow). Root hairs are stained blue. Scale bar = 1 mm.

One of the best-known symbiotic relationships is that between plants in the **legume family** (Fabaceae) and soil bacteria in a group collectively known as rhizobia, which are capable of fixing nitrogen. In legumes, compatible rhizobial bacteria in the soil invade the plant's root and induce root cells to divide and produce **root nodules**, which can accommodate **bacteroids**, modified forms of the bacteria which lack cell walls (Figure 12.27). Within a root nodule, the rhizobial bacteroids fix atmospheric dinitrogen gas (N$_2$) into ammonia. This is an energy-demanding reaction, because the dinitrogen molecule has a triple bond which is difficult to break. The bacteroids derive the necessary energy from photosynthate produced by the plant. Plants cannot

themselves use gaseous nitrogen, but they can incorporate ammonia into amino acids and from there into proteins and other cell components. Thus legumes are able to grow in soils low in inorganic nitrogen. Many of the world's major food crops, including soybean (*Glycine max*), beans of the genera *Phaseolus* (haricot or kidney bean, navy bean, etc.), *Vicia* (broad bean) and *Vigna* (adzuki bean, mung bean), garbanzos (*Cicer arietinum*) and lentils (*Lens culinaris*), are legumes. The forage crops alfalfa (*Medicago sativa*) and clover (*Trifolium* spp.) are also legumes, as are peanuts (*Arachis hypogaea*) and mesquite (*Prosopis* spp.). Such species can grow with little nitrogenous fertilizer in comparison with crops such as cereals and leafy vegetables, though leguminous crops such as soybean that are cultivated intensively are often fertilized to maximize growth, and this decreases the amount of nitrogen fixed by their nodules.

The formation of nodules and establishment of the nitrogen-fixing symbiosis first requires that the host plant and the rhizobia recognize one another. This is achieved by the secretion from the plant roots of flavonoid compounds into the rhizosphere. Each rhizobial species recognizes specific flavonoids, and this is the basis of the specificity of the symbiosis. Once the bacteria have been attracted to the root, expression of their **nodulation-specific (*nod*) genes** is induced, and they synthesize lipo-chito-oligosaccharides, the so-called **Nod factors** (see Chapter 15). These molecules have an oligosaccharide backbone of *N*-acetyl-D-glucosamine units, with a fatty acyl group attached to each non-reducing sugar.

When a compatible *Rhizobium* species secretes its Nod factors adjacent to a plant root, nodule development is normally triggered. One characteristic early response to the presence of Nod factors is a transient, rapid increase in nuclear Ca^{2+} concentration in the root hair cells, known as **calcium spiking**. The rhizobia attach to root hairs, which undergo deformation within 6–8 hours. Rhizobia enter the root of the host plant either via the root hair or through cracks in the root epidermis. Root hair infection, which is more common, proceeds via the formation of infection threads from plant cell wall material (Figure 12.28). Cells in the root cortex and pericycle divide to form a nodule primordium and the infection threads inject modified rhizobia into cells of the developing nodule. The end result is a structure in which groups of the modified rhizobia, known as bacteroids, are enclosed within modified root cells, each surrounded by the root cell's plasma membrane and an extracellular matrix. Nodules can be determinate or indeterminate, depending on the host plant species. The differences between these two types concern the site of the initial cell

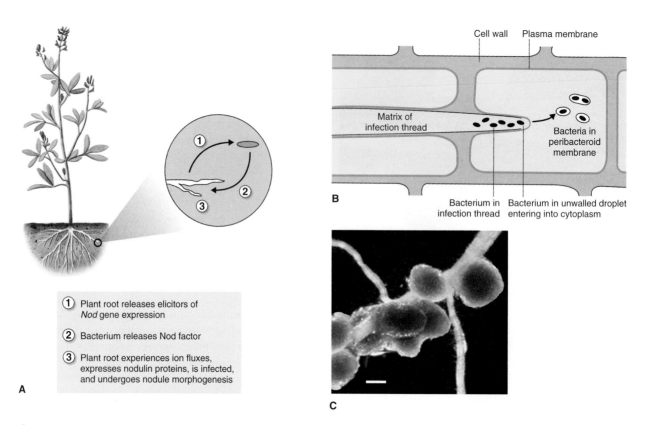

Figure 12.27   Stages in nodulation in roots of legume plants. (A) Interaction between the roots and bacteria. (B) The delivery of bacteria by infection thread to the cortical cell prior to nodule formation. (C) Pea root nodules. Scale bar = 100 μm.

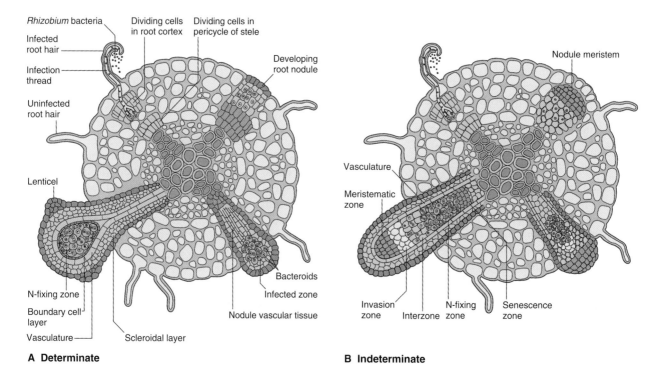

A Determinate   **B Indeterminate**

Figure 12.28   Root hair infection proceeds via the formation of infection threads from plant cell wall material. The end result is a root structure in which modified rhizobia are surrounded by a membrane and an extracellular matrix. (A) Determinate nodules like those of soybean (*Glycine max*) and bean (*Phaseolus vulgaris*) lack persistent meristems and are roughly spherical. (B) Indeterminate nodules such as those of alfalfa (*Medicago sativa*) and pea (*Pisum sativum*) have a persistent meristem and are cylindrical in shape.

Table 12.2 Examples of plants that form symbiotic associations with nitrogen-fixing prokaryotes.

Type of plant	Examples of species	Symbiont	Nature of association	Comments
Legumes	*Glycine max* (soybean)	*Bradyrhizobium japonicum*	Root nodules	There are five rhizobium genera (*Rhizobium*, *Bradyrhizobium*, *Sinorhizobium*, *Azorhizobium*, *Photorhizobium*). Biovars (bv) are spp. differentiated by host specificity
	Medicago sativa (alfalfa)	*Sinorhizobium meliloti*	Root nodules	
	Trifolium spp. (clover)	*Rhizobium leguminosarum* bv. *trifolii*	Root nodules	
	Pisum sativum (pea)	*Rhizobium leguminosarum* bv. *viciae*	Root nodules	
Non-legume trees and shrubs	*Parasponia* spp.	*Bradyrhizobium*	Root nodules	A rare example of a non-legume that forms root nodules with rhizobia
	Alnus spp. (alder)	*Frankia*	Actinorhizal root nodules	Nodulation by the actinomycete *Frankia* has some similarity both to rhizobium nodulation and to formation of mycorrhizal arbuscules
	Ceanothus (California lilac)	*Frankia*	Actinorhizal root nodules	
Members of Gunneraceae	*Gunnera* spp.	*Nostoc punctiforme*	Algal glands at the bases of petioles	Intracellular cyanobacterial symbiont
Grasses	*Saccharum* spp. (sugarcane)	*Gluconacetobacter diazotrophicus*	Endophytic colonization of xylem vessels and intercellular spaces	*G. diazotrophicus* has also been isolated from coffee and sweet potato plants
Aquatics	*Sesbania aculeata* (dhaincha)	*Azorhizobium*	Root and stem nodules	Used as a green manure for rice cropping in India
	Aeschenomene spp. (jointvetch)	*Photorhizobium*	Stem nodules	Symbiont is photosynthetic
	Azolla spp.	*Anabaena azollae*	Symbiont occupies specialized leaf cavities in the vegetative sporophyte	Aquatic fern used as an nitrogen fertilizer in East Asian lowland rice cultivation

divisions, the overall shape of the mature nodules, and whether or not a meristematic region is maintained. Indeterminate nodules, cylindrical in shape and with a persistent meristem, are found in species such as alfalfa (*M. sativa*), barrel medic (*M. truncatula*), white clover (*Trifolium repens*) and pea (*P. sativum*). In contrast, spherical, determinate nodules lacking persistent meristems occur in soybean (*G. max*), bean (*P. vulgaris*) and *Lotus japonicus*.

Initial recognition of Nod factors by plant roots is mediated by two **receptor-like kinases** (**RLKs**) which are present on epidermal cell plasma membranes. These receptors have an intracellular kinase domain, a transmembrane domain and an extracellular portion that has LysM domains (Figure 12.29). LysM (lysine motif) domains, which are about 40 amino acids long and are believed to have a peptidoglycan-binding function, are common in bacterial cell wall-degrading enzymes but are rather rare in eukaryotes. In each of the legume systems studied in detail, one of the RLK receptors has a typical serine/threonine kinase domain, while the other does not. It is thought therefore that the two LysM RLKs may assemble into a heterodimeric molecule which functions as the active receptor, with the functional kinase domain having a role in downstream signaling. The details of how the two-RLK system

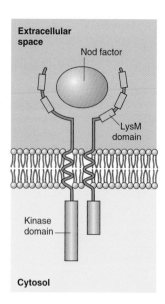

Figure 12.29   The initial recognition of Nod factors by plant roots is mediated by two receptor-like kinases present on epidermal cells. These receptors have an intracellular kinase domain, a transmembrane domain and an extracellular portion which has two or three LysM domains.

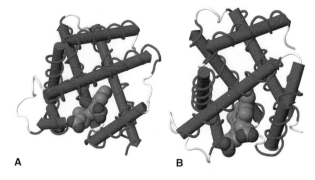

Figure 12.30   Leghemoglobin is structurally similar to the four subunits of hemoglobin, which transports oxygen in animal blood. (A) Leghemoglobin from yellow lupine (*Lupinus luteus*). (B) An α-subunit of human hemoglobin. α-Helices are shown as pink cylinders, heme group atoms are gray, red and blue.

interacts with other components of the complex signal transduction process have not been established.

The bacterial enzyme responsible for nitrogen fixation is **nitrogenase**, which catalyzes the conversion of atmospheric nitrogen to ammonia (see Equation 13.1). Nitrogenase must be protected from oxygen in order to function. In intact rhizobia, this protection is achieved by the bacterial outer membrane and cell wall. However, for bacteroids within a root nodule, an alternative means of protection is needed. This is provided by a **diffusion barrier** in the inner cortex of the nodule, which ensures that the majority of the oxygen entering the nodule does so through the nodule apex, generating a longitudinal gradient of oxygen concentration. At the centre of the nodule, where most of the nitrogen fixation takes place, the oxygen concentration is less than 50 nM. The molecule **leghemoglobin** is an important factor in regulating oxygen availability. Leghemoglobin is structurally similar to the hemoglobin that transports oxygen in animal blood (Figure 12.30), and like hemoglobin is red, giving root nodules their characteristic pink color. It functions by delivering to the bacteroids only that amount of oxygen which is needed to support respiration, so that excess oxygen, which would impair nitrogenase function, is avoided. Leghemoglobin genes, though closely related to other members of the plant globin family, are found only in legumes, where they normally occur as small gene

families; for example, *Lotus japonicus* has five leghemoglobin genes, three of which are expressed only in nodules.

The genes encoding nitrogenase and other enzymes of nitrogen fixation are under the control of regulatory networks that attune the process to development and to environmental and physiological conditions. The polypeptide subunits of the nitrogenase complex, and the other enzymes of nitrogen fixation, are the products of bacterial *nif* (*nitrogen fixation*) genes. Over 20 *nif* genes are known, and the functions of many of them are summarized in Table 12.3. The expression of *nif* genes is responsive to the concentrations of fixed nitrogen and external oxygen as well as to the redox status of the cell. Common regulatory principles and similar signaling pathways can be seen in different diazotrophs. The detailed mechanisms and relative influences of the various networks, however, vary with the species of microorganism and, in the case of endosymbionts, the physiological relationship with the host plant. In particular, the activities of regulatory cascades change markedly between the free-living and the symbiotic state.

Transcription of *nif* genes, including those that encode nitrogenase, is activated by NifA, a member of the enhancer-binding protein family of transcriptional regulators. NifA in endosymbiotic proteobacteria, including *Rhizobium*, senses oxygen by direct interaction. In free-living nitrogen-fixers it requires other proteins for oxygen sensitivity, as does FixK, a protein that regulates expression of the *fix* group of nitrogen-fixation genes. Figure 12.31 shows an example of the regulation of *nifA* and *fixK* gene expression in some rhizobial species by the two-component FixL–FixJ system. FixL, an oxygen-sensitive hemoprotein kinase,

Table 12.3 Known functions of *nif* genes. FeMoco is the prosthetic group of the nitrogen-fixing enzyme nitrogenase (see Chapter 13).

nif	Nitrogenase structural genes	Nitrogenase maturation	FeMoco biosynthesis	Electron donation	Regulatory
H	■				
D	■				
K	■				
Z		■			
M		■			
E			■		
N			■		
X			■		
Q			■		
S			■		
U			■		
V			■		
Y			■		
F				■	
J				■	
A					■
L					■

phosphorylates its partner, FixJ. In the phosphorylated form, FixJ activates transcription of genes for other regulatory proteins, including NifA and FixK, which in turn control expression of a number of *nif* and *fix* genes. When the oxygen concentration is high, FixL cannot phosphorylate FixJ, so it is only in the correct low-oxygen environment of a nodule that the bacteroids will express nitrogenase and the other proteins that are needed for the reduction of dinitrogen to ammonia.

While most nitrogen-fixing symbionts of higher plants are rhizobia (see Table 12.2), filamentous bacteria in the genus *Frankia* can also form root nodules specialized for nitrogen fixation. These symbioses are restricted to **actinorhizal** plants, almost all of which are trees and shrubs; they are predominantly species of temperate zones, such as alder (*Alnus* spp.) and bayberry (*Myrica* spp.). However, one actinorhizal genus, *Casuarina*, is native to the tropical climates of Australasia and the western Pacific and plants of this genus have been widely cultivated as ornamentals. They have established themselves as invasives in regions like Florida, where their ability to survive on low-nitrogen soil makes them effective competitors with native species. As in the plant–rhizobial symbiosis, the bacterial nitrogenase responsible for nitrogen fixation is oxygen-sensitive. Nodules on actinorhizal plants show a range of adaptations to protect the nitrogenase from excess oxygen while permitting bacterial respiration.

An interesting variation on the nitrogen-fixing symbiosis theme is presented by the association between the cyanobacterium *Anabaena azollae* and the seven species of aquatic ferns in the genus *Azolla*. Here, the photosynthetic bacteria are present not in the root nodules but in cavities within the leaves (Figure 12.32A). The majority of the ammonia produced from fixation of gaseous nitrogen by the nitrogenase of *Anabaena* is secreted by the bacteria and absorbed by the fern. The *Azolla–Anabaena* symbiosis has been used by rice farmers for hundreds of years. They inoculate rice paddies with *Azolla* and allow the ferns to spread, providing a rich source of fixed nitrogen for the developing crop (Figure 12.32B).

12.4.5 Mycorrhizal symbioses also modify root development

Mycorrhiza (mycor = fungus, rhiza = root) are symbioses between fungi and the roots of land plants. The fungal partner contributes by taking up minerals

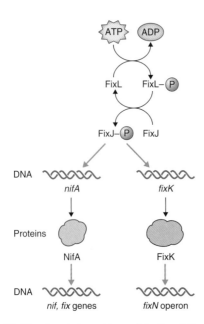

Figure 12.31 Regulation of bacterial *nif* and *fix* gene expression by the FixL–FixJ signal cascade. FixL is a heme-containing protein that phosphorylates the FixJ response regulator in the absence of oxygen. Phosphorylated FixJ binds to DNA and activates transcription of the genes of transcription factors NifA and FixK, which in turn promote transcription of genes needed for nitrogen fixation.

from the soil, particularly in nutrient-poor conditions; the plant partner contributes photosynthate as a source of carbon and energy for the fungus. Mycorrhizal associations allow plants to grow in drier, nutrient-poor soil. Mycorrhiza are classified as either **ectomycorrhiza**, in which the fungi are outside the plant cells, or **endomycorrhiza**, where part of the fungus penetrates the plant cell. In both cases, the fungal hyphae (filaments) extend out into the soil where they take up water and minerals, particularly P_i, and share them with the plant partner (see Section 13.3.3). The most common mycorrhizal symbioses are **arbuscular mycorrhiza**, in which the fungus, a member of the phylum Glomeromycota, forms tree-like structures ('arbuscule' means 'little tree') within the plant cells between the cell wall and the plasma membrane (Figure 12.33). Arbuscular mycorrhiza (AM) are found in at least 70% of land plant species.

Free-living fungi are induced to form AM by the presence in the soil of compounds called **strigolactones**, which are secreted by plant roots (see Chapter 10). The fungi in turn produce signaling molecules that induce symbiosis-specific responses in the plant (see Chapter 15). These molecules are collectively known as **Myc** (short for mycorrhiza) **factors**. In many cases they are poorly characterized, but some are diffusible, small, organic molecules that induce expression of symbiosis-related genes at the transcriptional level. Just

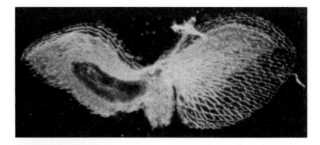

A

B

Figure 12.32 (A) In the symbiosis between aquatic ferns in the genus *Azolla* and the nitrogen-fixing cyanobacterium *Anabaena azollae*, the bacteria are present not in root nodules but in cavities within the leaves. In this picture, the darker-green area contains *Anabaena* cells. (B) The *Azolla*–rice–duck agricultural system used in rice paddies. The *Azolla*–*Anabaena* symbiosis provides fixed nitrogen to the rice plants; ducks graze on the *Azolla* and their feces are rich in other nutrients required by the rice.

as in nodulation, where bacterial Nod factors induce spikes of calcium concentration in the roots, plant roots have been observed to show calcium spiking as the hyphae of AM fungi approach them, but before contact is made. Formation of the mycorrhizal symbiosis requires that the plant can perceive **chitin**, a structural component of the cell wall in AM fungi. Chitin perception is mediated by receptor-like molecules containing the LysM domain, which is also present in Nod factor receptors, and the likelihood is that the chitin receptors and Nod factor receptors originated from the same ancestral proteins.

12.5 Growth and differentiation of leaves

This section discusses the origin and development of the angiosperm leaf. In the earliest multicellular green

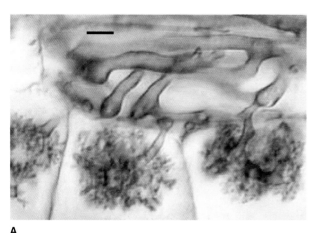

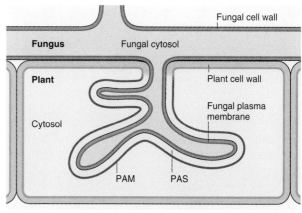

A   **B**

Figure 12.33   Arbuscular mycorrhiza. (A) The most common mycorrhizal symbioses are arbuscular mycorrhiza, in which the fungus forms tree-like structures. The figure shows arbuscules in the inner cortex of a root of *Asarum canadense*. Scale bar = 10 µm. (B) Schematic representation of the 'little tree' structure formed by the fungus within the plant cell. Each 'branch' of the fungus within a plant cell is surrounded by a membrane called the periarbuscular membrane (PAM), which is derived from the plant plasma membrane and is continuous with it. The PAM keeps the fungus out of the plant cytoplasm. The interface between the plant-derived PAM and the fungal plasma membrane is called the periarbuscular space (PAS).

Key points Root development is regulated by soil nutrients. Different forms of nitrogen, and the N:C ratio, influence root growth and development through nitrate transporters and auxin signaling. Phosphate starvation tends to inhibit primary root growth but has diverse effects on lateral roots. Auxin and ethylene have roles in responding to P status. Legumes and a range of other plant species can fix atmospheric nitrogen by forming symbiotic associations with soil bacteria, mostly representatives of various rhizobium genera. The bacteria occupy nodules on the root, which develop after the bacteria are recognized by the host and invade root cells. Flavonoid signals secreted from the root into the soil stimulate the bacterium to synthesize Nod factors that trigger a host kinase network leading to root nodule development. A physical barrier and the presence of leghemoglobin maintain the nodule occupied by the bacteroids at the low oxygen concentration required by the nitrogen-fixing enzyme nitrogenase. Nitrogenase and other enzymes of the N-fixation pathway are encoded by bacterial *nif* and *fix* genes, under the control of the transcriptional regulator NifA and the oxygen-sensitive FixK–FixJ system. The roots of most vascular plant species form symbiotic associations with mycorrhizal fungi, with beneficial effects on host nutrient uptake. Fungi are attracted by strigolactones secreted by roots and respond by synthesizing Myc factors that activate the host's symbiosis-related genes. In arbuscular mycorrhizal associations, the fungus invades root cells, forming arbuscules between the cell wall and plasma membrane.

plants, the green algae, photosynthetic cells are arranged in colonies or filaments, or in flattened cushion- or leaf-like **thalli**. The gametophyte of liverworts and ferns retains thallus morphology, whereas in the mosses and vascular plant sporophytes, photosynthesis is carried out primarily in lateral structures attached to the aerial axis. Evo-devo studies of foliar morphogenesis in representatives of the terrestrial plant phyla with true leaves have revealed common regulatory mechanisms.

A **leaf primordium** generally begins as a slight bulge on the flank of the SAM (Figure 12.34). In comparison to eudicots, the leaf initials of grasses and other monocots tend to be smaller relative to the size of the meristem as a whole. The SAM produces leaves at regular intervals and in predictable positions. The geometry of organ initiation on the SAM is the basis of **phyllotaxy**, the spatial pattern of leaf arrangement on the shoot. Figure 12.35 shows the major phyllotactic configurations observed in angiosperms. They range from instances of a single leaf per node, inserted on the stem **alternately** or in a **spiral**, through paired leaves arranged **opposite** each other at the node, to **whorls** of three or more leaves. Spiral phyllotaxy is the most common form and is characterized by a displacement angle of approximately $137°$ between one leaf and the next. The time interval between the formation of successive primordia is called the **plastochron** (see Figure 12.7).

Leaves are formed on the flanks of the SAM (Figure 12.34) in a three-stage process. The first phase is organogenesis (the initial establishment of organ identity). A group of cells (**founder cells**) is designated to form the leaf primordium. During the second stage the primordium becomes partitioned into the basic morphological domains that will form the component

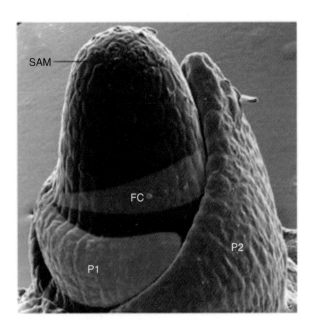

Figure 12.34 Scanning electron micrograph of a young maize apex showing the shoot apical meristem (SAM) with the two youngest leaf primordia, P1 and P2 (pseudocolored in green), initiating on its flanks. FC indicates the position of the founder-cell population that will give rise to the next leaf primordium.

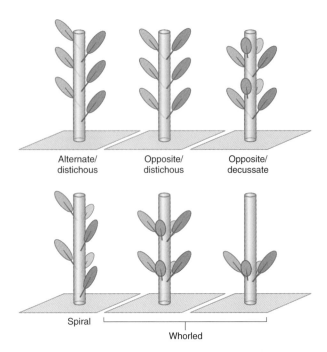

Figure 12.35 Types of phyllotaxy. Distichous refers to leaves arranged in two rows. In decussate phyllotaxy, four rows of leaves are arranged in opposite pairs. Two types of whorl are shown: one where leaves are arranged in successive nodes along the stem, the other where there is a single predominant whorl at the base of the shoot, as seen in certain rosette plants for example.

tissues of the growing and developing leaf. The final phase is a period of coordinated cell division, expansion and differentiation in which the morphological and anatomical potential of the primordium is expressed.

Development is a highly integrated process. A typical leaf will consist of up to 12 different cell types. The **epidermis** develops from the L1 layer of the meristem, while the **mesophyll** and **vascular** tissues usually develop from the L2 and L3 layers. While the origins of the internal, subepidermal cells of the leaf are distinct from those of the epidermis, morphogenesis of the leaf requires close developmental coordination between tissues of different origins within the primordium.

12.5.1 The vegetative shoot meristem produces leaf primordia at sites determined by a morphogenetic field

A leaf originates in the indeterminate apex of the vegetative shoot, but its growth is **determinate**, resulting in an organ of defined shape and size. The transition from the indeterminate to determinate state is critical for establishing organ identity. Members of the **KNOX gene family** are of particular importance, remaining in the *on* state in meristems and being turned *off* early in the development of simple leaves. The first plant KNOX-like homeobox gene to be identified was *Knotted1* (*Kn1*) in *Zea mays* (see Chapter 3). The *Arabidopsis* genome contains four class I *KNOX* genes, one of which (*STM*) has already been discussed as a regulator of cotyledon shape and outgrowth (see Section 12.3.3 and Figure 12.20). Downregulation of *KNOX* is associated with recruitment of founder cells from the flanks of the shoot apical meristem and formation of the leaf primordium. Genes of the *ASYMMETRIC LEAVES* (*AS*) family (see Section 12.3.3 and Figure 12.20) are expressed at the point of leaf initiation and act as repressors of *KNOX* gene expression.

Leaf initiation is under the negative influence of adjacent leaf primordia, and is sited in an available space on the meristem where the geometry of the morphogenetic field produces a local inhibitory minimum (Figure 12.36A). A number of factors contribute to the morphogenetic field. The initiation site is subject to **physical stress** from surrounding tissues. Cells at the site will begin to enlarge in response to upregulated expression of expansins (see Section 12.2.4) and an adequate supply of auxin via **polar transport**. It is thought that adjacent primordia are also sources of inhibitors of polar auxin transport (Figure 12.36B). The distribution pattern of **auxin transport proteins** in the meristem is consistent with auxin flux being directed to the site of incipient leaf formation.

Localized cell expansion appears to be sufficient to specify an initiation site and to activate the complete developmental pathway leading to formation of the

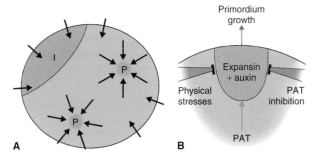

Figure 12.36 Determination of leaf initiation site (I) on the shoot apical meristem. (A) Apical dome viewed from above. A morphogenetic field is established by the influx of auxin (black arrows) from subtending tissues interacting with the directing of auxin transport to sites of previous leaf primordia (P). (B) Longitudinal section through a leaf initiation site. A new leaf primordium is thought to arise by the growth in volume of a set of cells that have sufficient expansin and auxin. Existing primordia influence the location of I by generating physical stresses close to themselves, and by releasing inhibitors of polar auxin transport (PAT). Beyond this inhibitory zone, PAT supplies sufficient auxin to support primordium initiation.

primordium and ultimately the fully differentiated leaf. However, as it stands, this model does not explain how different patterns of phyllotaxy can arise. Mutational analysis provides some clues. Normal maize plants have an alternate/distichous phyllotaxy (Figure 12.35). This pattern is the result of the SAM generating at each node a single leaf diametrically opposed to the previous leaf. A mutant genotype, *ABERRANT PHYLLOTAXY 1*, produces two leaves at each node. The corresponding gene, *ABPHYLL1*, has been cloned and found to encode a cytokinin signal transduction protein. There are indications that ABPHYLL1 is a positive regulator of auxin signaling in the SAM. It is likely that further studies

Key points Leaf primordia develop from founder cells on the flanks of the SAM. The spatial pattern of leaf initiation determines phyllotaxy, the arrangement of leaves on the stem. The plastochron is the time between the initiation of one leaf and the next. Up to 12 different cell types make up the fully developed leaf and each type is derived from a particular layer or layers within the SAM. The SAM may be indeterminate but leaf growth is determinate. The transition to determinacy during formation of the leaf primordium requires repression of *KNOX* by regulators of the AS family. The site of leaf initiation is specified by a combination of physical stresses across the SAM, polar transport of auxin, and upregulation of expansin synthesis. A combination of auxin and cytokinin signaling specifies the final pattern of phyllotaxy. Plastochron and phyllotaxy are regulated independently.

will support enhancement of the simple leaf initiation model shown in Figure 12.35 to include cross-talk between the auxin and cytokinin signaling pathways.

As in the **spatial** determination of leaf primordia, there is evidence that previously-formed leaves exert an inhibitory effect on the **timing** of initiation. Among the factors thought to be implicated in setting the plastochron is a negative regulatory gene encoding a specific member of the extensive family of **cytochrome P450** redox proteins. For the most part, phyllotaxy and plastochron are regulated independently of one another, although there is evidence of a role for cytokinin signaling in both processes.

12.5.2 During leaf epidermis development, three cell types are differentiated: pavement cells, trichomes and stomates

The **epidermis** is the interface between the organ and the aerial environment. As described in Section 1.8.1, epidermal cells form an interlocking mechanical system that combines the strength and flexibility necessary to allow morphogenesis. The epidermis is a barrier that is almost totally impermeable to water, and yet supports and regulates water and gas exchange. It protects the plant from pathogens and predators but in some organs—many flowers and fruits, for instance—epidermal cells are the source of chemical attractants for pollinating and dispersing animals. The varied functions of the epidermis are fulfilled by different specialized cell types that are distributed in a non-random pattern, which in turn reflects their foundation in, and development from, the primordium.

Epidermal identity is established very early in embryonic development. In monocot leaves, where the vascular bundles run in **parallel** along the axis of the lamina (Figure 12.37A), the unit of differentiation is a region of epidermal cells bounded by two adjacent lateral veins with an intermediate vein running through the middle. The founder cells of the **protoderm** are on the midvein side of this developmental compartment and undergo highly polarized cell division to create the incipient epidermal layer spanning the surface between the veins. The linear morphology of the lamina of grasses is generated by the extended activity of the **intercalary meristem** at the base of the leaf, with the consequence that expansion is predominantly in **one dimension** and the cells of epidermal and other tissues are arranged in regular, parallel files. The venation pattern of eudicot leaves, on the other hand, is **reticulate** (a network) rather than parallel (Figure 12.37B), expansion of the leaf surface is **two-dimensional** and patterning and cell lineages are correspondingly complex and arranged in patches or sectors rather than linear files (Figure 12.38).

Figure 12.37 (A) Parallel venation of a monocot leaf (daylily, *Hemerocallis*). (B) Reticulate venation pattern in leaf of a dicot (mint, *Mentha*).

Lateral veins (large, intermediate and small)

Midvein

Tertiary and lower order veins and veinlets

Secondary veins

1 cm

A

B

The leaf epidermis is normally derived from cells in the outer (L1) layer of the SAM, which forms the protoderm of the primordium. The major cell types of the epidermis are unspecialized **pavement cells**, **trichomes** (leaf hairs) and **guard cells** of the stomatal complex (see Figures 1.25 and 1.26). Mechanisms of differentiation of trichomes and stomata have common features in eudicots and monocots. For example, the distribution of stomata across the epidermis of eudicots like *Arabidopsis* appears at first sight to be random, whereas the stomata of grasses develop in distinct files; but in both cases stomatal differentiation is under the control of **bHLH** (basic helix-loop-helix) type transcription factors. Furthermore, there is evidence that the signaling pathways leading to trichome development are conserved across the angiosperms. A picture of tissue patterning in the protoderm determined by interacting transcriptional regulators has been built up based largely on mutant analysis in *Arabidopsis*.

The typical *Arabidopsis* trichome is a single cell with a distinctive tricorn (three-horned) structure (Figure 12.39A). The major players in the network of regulators specifying the site of trichome formation are transcription factors of the bHLH and MYB families, and a protein with a **WD40** repeat, TTG1 (TRANSPARENT TESTA GLABRA1). The designation WD40 refers to the

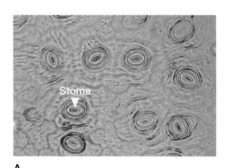

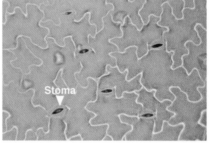

A

B

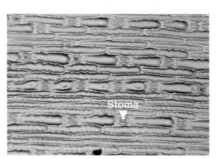

C

Figure 12.38 Epidermal cells of monocot and eudicot leaves. A replica of the epidermis is made by painting the leaf surface with nail polish. After drying, the layer of polish is removed using clear tape and examined under the microscope. (A) Cherry laurel (*Prunus laurocerasus*; eudicot). (B) Pea (*Pisum sativum*; eudicot). (C) Maize (*Zea mays*; monocot). Note that the stomata of the monocot leaf occur in rows that parallel the veins, while those in the eudicot leaves are scattered between veins.

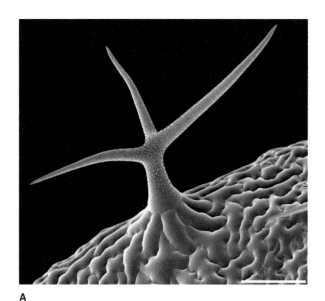

A

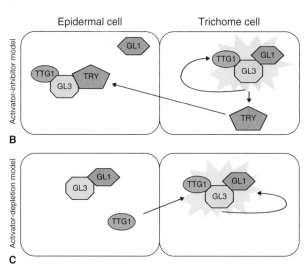

Figure 12.39 (A) Scanning electron micrograph of an *Arabidopsis* trichome. (B) Activator-inhibitor model of trichome patterning. (C) Activator-depletion model. The models are based on protein–protein interactions between TTG1 (WD40 repeat protein), GL3 (a bHLH transcription factor), GL1 (a MYB) and the MYB-like factor TRY.

40 amino acid motif with conserved tryptophan (W) and aspartate (D) residues typical of this widely distributed group of eukaryotic signal transduction components. Figure 12.39 also presents two possible models of trichome patterning based on protein–protein interactions between TTG1, GLABRA3 (GL3; a bHLH transcription factor), a MYB (GL1) and the MYB-like factor TRYPTICON (TRY). In the **activator-inhibitor model** (Figure 12.39B) TTG1, GL3 and GL1 form a self-activating complex that induces trichome differentiation. This trimeric complex activates its own inhibitor, TRY. TRY moves into the neighboring cell, replaces GL1 in the complex and inactivates it. The

activator–depletion model (Figure 12.39C) explains pattern formation in terms of the uneven distribution of TTG1 between incipient trichomes and neighboring cells. TTG1 is envisioned to move freely between cells and to bind to GL3, which is strongly expressed in trichomes. By becoming trapped in trichome initials, TTG1 is depleted in adjacent cells. A notable feature of both models is the cell-to-cell movement of regulatory proteins, thought to occur via plasmodesmata. Overlaying these protein–protein interactions is a complex network of further regulators and feedback loops that remain subjects of continuing research investigation.

The steps in the formation of a **stoma** are shown in Figure 12.40. A meristemoid mother cell divides asymmetrically (see Section 12.3.1) to form a small **meristemoid** cell which then undergoes several asymmetrical divisions before differentiation into a **guard mother cell (GMC)**. The GMC divides symmetrically to produce a pair of **guard cells**. Stomatal patterning, like that of trichomes, is regulated by protein–protein interactions involving bHLH transcription factors. One such factor is SPEECHLESS (SPCH), a bHLH protein that controls entry of the meristemoid mother cell into asymmetrical division. The meristemoid to guard mother cell transition is regulated by another putative bHLH factor, MUTE, and a third bHLH protein, FAMA, is required for the final differentiation step. ICE1/SCRM1 and SCRM2 are two further bHLH factors, which regulate stomatal differentiation and are believed to interact physically with SPCH, MUTE and FAMA (Figure 12.40). A number of genes have been implicated in regulatory pathways that determine the **spacing** of stomata across the epidermis. They include: two related *EPIDERMAL PATTERNING FACTOR* (*EPF*) genes; *TOO MANY MOUTHS* (*TMM*; Figure 12.41), which encodes a leucine-rich repeat receptor protein; and *STOMATAL*

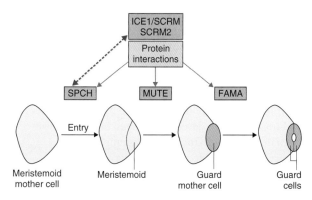

Figure 12.40 Stomatal differentiation and its regulation by bHLH transcription factors. Each transcription factor modulates a different step in the production of a stoma. Protein interactions are shown as blue arrows and the red dashed arrow indicates transcriptional feedback.

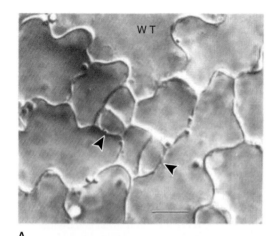

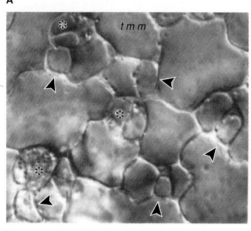

Figure 12.41 Abaxial epidermal cells of (A) a wild-type (WT) and (B) a *too many mouths* (*tmm*) mutant of *Arabidopsis*. Arrowheads indicate meristemoids, which occur at higher density in *tmm* than in WT. Asterisks identify developing stomata. Scale bar = 10 μm.

CARBON DIOXIDE) in regulating the response of stomatal development to CO_2 has been described. Stomatal density in plants with defective *HIC* increases by over 40% when they are grown at double ambient $[CO_2]$. *HIC* encodes an enzyme that functions in the formation of the **waxy cuticle**. Other genes for cuticular wax synthesis have also been implicated in the response of stomatal development to CO_2. These observations suggest that there is a more general role for the epidermal extracellular matrix and cuticle in the regulation of leaf structure, function and environmental responses. The nature of the systemic signal that links mature and developing leaves is uncertain, but the hormone abscisic acid (**ABA**), a well-established regulator of stomatal aperture in relation to plant water balance, has been proposed as a signaling component connecting stomatal density with transpiration.

12.5.3 Development of the internal structure of the leaf involves vascular and photosynthetic cell differentiation

The internal tissues of the leaf, vascular tissue and mesophyll cells, are derived from cells in the L2 and L3 layers of the apical meristem. Procambium forms vascular tissue and mesophyll cells develop from the ground meristem. The pattern of leaf vasculature varies widely among different plant species (e.g. Figure 12.37). Vein pattern is specified early in leaf development. Among the hypotheses that have been proposed to explain the organization of leaf venation is one that suggests that auxin flow is channeled through certain cells which become progressively more effective transporters of auxin and eventually differentiate into vascular tissue. Another proposes a morphogenetic field within which spacing and patterning of vascular bundles develop. Genetic analysis has identified *BRI1*, a gene that encodes a brassinosteroid receptor, as a component of the network that regulates vascular patterning. Differentiation of xylem and phloem tissues is described in Section 12.2.6.

Differentiation of the photosynthetic tissue of the leaf directly influences, and is influenced by, development of the photosynthetic apparatus, to the extent that genetic or environmental perturbation of the one often brings about reactions in the other. Detailed studies of this relationship have been made on *Zea mays* leaf development, with particular emphasis on the C_4 **pathway** of carbon fixation and the differentiation of photosynthetic tissues into **mesophyll** and **bundle sheath** (see Figure 9.39). The capacity for C_4

DENSITY AND DISTRIBUTION 1 (*SDD1*), a gene that encodes a protease. The TTM receptors are thought to exert negative regulation of stomatal development via a mitogen-activated protein kinase (**MAP kinase**) cascade of the kind shown in Figure 10.27.

Stomatal density (the number of stomata per unit area of epidermis) is sensitive to environmental influences, particularly light, CO_2 and water. The environmental stimulus is perceived by mature leaves and transmitted to developing leaves. Light responses are mediated by the **phytochrome** system. Recent observations have identified the **phytochrome-interacting factor PIF4** (a bHLH protein; see Chapter 8) as a critical component of the regulatory pathway. The density of stomata decreases as ambient $[CO_2]$ levels increase. Variation in stomatal density related to $[CO_2]$ in air has been used to document historical changes in $[CO_2]$ in ongoing studies of anthropogenic climate change (see Chapter 9). A role for the gene **HIC** (*HIGH*

photosynthesis is closely associated with vascular development. In the maize mutant **golden2** (**g2**), bundle sheath cell differentiation is perturbed, chloroplast development is arrested at the proplastid stage, and enzymes specific to bundle sheath cells fail to accumulate. Morphogenesis of mesophyll cells in *g2* is normal. *ZmGLK1*, a second **GOLDEN2-like** (**GLK**) gene in maize, is expressed in mesophyll cells. It is believed that *G2* and *GLK1* are critical components of the regulatory network that specifies bundle sheath and mesophyll cell differentiation in the leaves of C_4 species.

GLK genes encode transcription factors required for the expression of nuclear photosynthetic genes and for chloroplast development. They occur widely in land plants, generally as **partially redundant pairs** (*GLK1* and *GLK2*). C_3 plants such as *Arabidopsis* and the moss *Physcomitrella* differ from species with Kranz (C_4-specific) anatomy such as maize (see Chapter 9) in that their *GLK1* and *GLK2* homologs are not expressed in a cell-specific fashion. Expression of both genes is necessary for the transition of proplastid to chloroplast, is light-dependent, is detectable at all stages of leaf development, and is subject to circadian regulation (see Chapter 8). The characteristics of *GLK* genes are consistent with functions in coordinating, maintaining and fine-tuning the photosynthetic apparatus within individual cells.

The association between development of the photosynthetic cell and differentiation of its chloroplasts is evident from the phenotypes of a number of mutants in a range of species. For example the **dcl** (*defective chloroplast and leaf*) mutation in tomato and a similar but non-homologous mutation in *Antirrhinum* (**dag**, *differentiation and greening*) (Figure 12.42) lead to **aberrant chloroplast development** and misshapen palisade cells. DCL and DAG proteins are targeted to the plastid where they are believed to participate in the transition of proplastid to chloroplast. The nature of the link between DCL or DAG and cell differentiation is not clear. A third gene, **DOV1** (*Differential development Of Vascular-associated cells 1*), is thought to act at a later stage in development than *DCL* and *DAG*, since mesophyll chloroplasts of *dov1* mutants fail to develop but cell structure is otherwise normal.

During differentiation of photosynthetic cells, it is necessary that the genomes of the nucleus and plastid (and mitochondrion) act in a coordinated fashion. The nuclear genome is the master controller of organelle gene expression, but it is responsive to factors originating in organelles. This system of information sharing is known as **retrograde signaling**. Four pathways of retrograde signaling between the chloroplast and nucleus are known (Figure 12.43): (i) nuclear gene transcription induced by chloroplast-generated **reactive oxygen species** (ROS); (ii) nuclear gene regulation by the redox state of the

Wild-type   dag mutant

Figure 12.42   Abnormal leaf cell development in the *dag* mutant of *Antirrhinum*. Misshapen mesophyll cells in the mutant lead to gross changes in leaf morphology.

photosynthetic electron transport chain; (iii) accumulation of intermediates in **chlorophyll** biosynthesis; and (iv) inhibition of the expression of genes in the plastid genome in response to developmental factors or stress. The role of ROS and redox signaling in environmental regulation of gene expression is described in Chapter 15. Here we consider two signal-transduction pathways, operating in chloroplast development, in which **Mg–protoporphyrin IX** (**Mg-proto**), an intermediate in chlorophyll biosynthesis (see Chapter 8), is believed to play a critical role. Observations on the unicellular green alga *Chlamydomonas* and on angiosperm species have shown that perturbation around the chlorophyll–heme branch point of tetrapyrrole biosynthesis in the chloroplast (see Figure 8.23) generates a distress signal. Two signaling pathways are recognized. In one, Mg-proto export is facilitated by GENOMES UNCOUPLED 1 (**GUN1**) or a putative GUN1-dependent chloroplast protein (**GDCP**); this is followed by an interaction between Mg-proto and cytoplasmic factors and finally by signaling to the nucleus. In the second pathway, GUN1 or GDCP acts as a sensor for Mg-proto and other retrograde signals within the chloroplast and signals to the nucleus via unidentified factors (Figure 12.43). Several hundred nuclear genes have been shown to differ in expression between *gun* mutants and normal genotypes. The transcription factor **ABI4** (ABA insensitive 4) mediates between Mg-proto signaling and expression of nuclear genes encoding proteins of the photosynthetic apparatus, such as the light harvesting protein, LHCP2. ABI4 is also implicated in retrograde signaling between the mitochondrion and nucleus, suggesting a pivotal role for this regulator in the integration of ABA response and organelle development pathways.

Key points Epidermal cell identity is set in early embryogenesis. The parallel and reticulate arrangements of monocot and eudicot leaf epidermal cells, respectively, reflect the morphogenetic influence of the pattern of venation, which in turn is regulated by a network that includes a brassinosteroid receptor. Epidermal cells comprise pavement, trichome and guard cell types. Stomatal cell patterning is controlled primarily by bHLH transcription factors. The guard cells of a stoma arise through asymmetrical division of a guard mother cell. Stomatal density is influenced by light, water and CO_2, mediated by phytochrome, ABA and genes for cuticle development. Trichome differentiation is explained by an activator-inhibitor or activator-depletion model of regulatory interaction. Cells of the photosynthetic tissue of the C_4 species maize are developmentally sensitive to the proximity of vascular tissues, mediated by *G2* and *GLK* transcription factor genes. Nucleus–organelle regulatory interactions during mesophyll cell differentiation involve retrograde signaling pathways sensitive to redox conditions, intermediates in chlorophyll synthesis and the state of plastid gene expression. The transcription factor ABI4 has a central role in retrograde signaling.

12.5.4 Flattening, orientation and orientation and outgrowth of the lamina determine ultimate leaf size and shape

Leaves are broadly classified into **simple** and **compound**. In simple leaves the leaf blade is entire (undivided), while in compound leaves it is divided into leaflets (see Figure 1.32). The potential shape of a simple leaf is defined by **polarization events** in early organogenesis, which differentiate the adaxial (toward the axis) from the abaxial (away from the axis) surface, the proximal (nearest the node) from the distal (tip) region, and the different domains across the midvein–margin plane (Figure 12.44). Based on this ground plan, shape is expressed through cell division and expansion, accompanied by differentiation of epidermal and subepidermal tissues.

Analysis of mutants, largely in *Arabidopsis*, has identified a number of genes that regulate

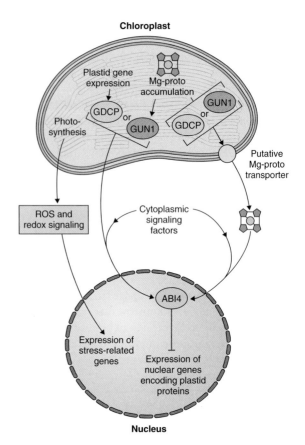

Figure 12.43 Chloroplast to nucleus retrograde signaling pathways. Sensitivity of nuclear gene expression to the state of the tetrapyrrole biosynthesis pathway is mediated by the size of the Mg–protoporphyrin IX (Mg-proto) pool, by GUN1 or GDCP, by unidentified cytoplasmic signaling components, and by the transcription factor ABI4. Reactive oxygen species (ROS) generated by the chloroplast, and the redox state of the photosynthetic electron transport chain, also regulate nuclear gene expression.

abaxial–adaxial flattening. As described in connection with vascular patterning in Section 12.2.6, the transcription factors *KANADI* and *HD-ZIP III* are mutually antagonistic (see Figure 12.14A). These gene families have more wide-ranging functions in morphogenesis. **Adaxial** identity is specified by *HD-ZIP III* genes and **abaxial** cell fate is specified by genes of the *KAN* family. KAN proteins repress expression of *HD-ZIP III* genes on the abaxial side, while HD-ZIP III proteins suppress adaxial *KAN* expression (Figure 12.45). Another class of gene expressed in abaxial but not adaxial domains is *YABBY*. YABBY and *KAN* interact, in some way that remains to be determined, to regulate outgrowth of the blade in eudicots such as *Arabidopsis* and *Antirrhinum*; but the polarization function of *YABBY* genes does not seem be conserved in maize. Genes of the *AS* family (see Section 12.3.3) act as *YABBY* repressors and are indirect positive regulators of *HD-ZIP*

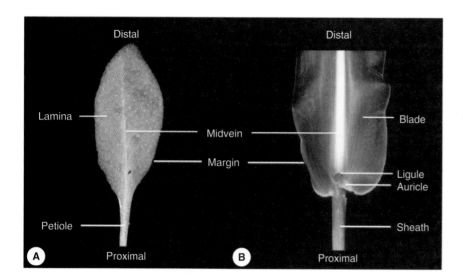

Figure 12.44 Adaxial side of the simple leaves of (A) *Arabidopsis* and (B) maize showing polarity features.

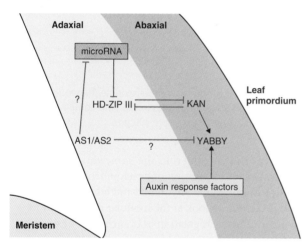

Figure 12.45 Adaxial/abaxial patterning in leaf development, regulated by interactions between *HD-ZIP III*, *KANADI*, *YABBY* and *AS* transcription factors. Adaxial identity is specified by *HD-ZIP III* genes; abaxial cell fate by genes of the *KAN* family.

III genes. *KAN* action in abaxial fate determination is also facilitated by auxin response factors (Figure 12.45).

After the polarity ground plan is in place, integrated division and expansion of cells across the length and width of the leaf establish final leaf size and shape. If dimensional control of initiation and termination during the period of cell growth is disrupted, the normally flat leaf surface becomes distorted. A number of genes with functions in lamina formation have been identified in studies of developmental mutants with crumpled leaves. They include **CINCINNATA** from *Antirrhinum* and the gene corresponding to a similar *Arabidopsis* phenotype, **JAW**, which appears to be part of a regulatory network involving microRNAs. Among the identified regulators of the degree of smoothness or serration of the leaf margin is the *Arabidopsis* gene **JAGGED**, as well as *CUC* and auxin transporter genes that also function in embryo

tissue patterning (see Section 12.3.3). *JAGGED* may also participate with other genes such as **LEAFY PETIOLE** in controlling differentiation into blades and stalks as specified by proximal–distal polarity.

Compound leaves are very widely distributed among the angiosperms, occurring in many unrelated eudicot taxa, and also among the monocots, in palms and some aroids (members of the *Arum* family) such as *Philodendron* and *Amorphophallus*. The compound character has arisen independently numerous times during plant evolution and, in the form of the **frond**, was already well established in the ferns and their allies. The initial suppression of *KNOX* associated with the transition from indeterminate apical growth to determinate development of the simple leaf (see Section 12.5.1) is reversed during development of the compound leaf of seed plants; in contrast, *KNOX* is never downregulated during differentiation of the fern frond. Figure 12.46 shows how interactions between *KNOX*, *CUC* and the gibberellin and auxin signaling pathways underlie early differentiation of the compound leaf in tomato. **Gibberellin** signaling promotes cell differentiation and leaflet outgrowth and is antagonized by *KNOX* expression. PIN proteins facilitate cell-to-cell auxin flow, resulting in local maxima of auxin response that both determine the positions of the **leaflet** primordia and downregulate *KNOX*. The auxin response in the interleaflet regions is suppressed by the product of the gene *ENTIRE* (*E*). *CUC* acts with the transcription factor *NAM* (*NO APICAL MERISTEM*) to define the boundary domain around the primordium, a requirement for **serration** of the leaflet margin.

Our understanding of the molecular basis of leaf development is typical of plant morphogenesis in general, in that it seems to consist of a bewildering array of transcription factors and other regulators interacting in networks of daunting complexity. In reality, the

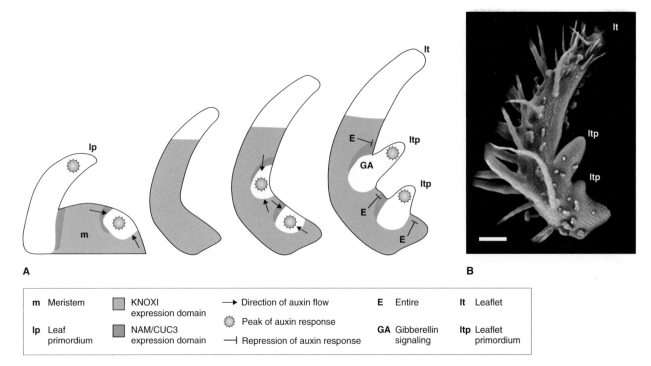

m	Meristem	KNOXI expression domain	→ Direction of auxin flow	**E** Entire	**lt** Leaflet
lp	Leaf primordium	NAM/CUC3 expression domain	✦ Peak of auxin response ⊣ Repression of auxin response	**GA** Gibberellin signaling	**ltp** Leaflet primordium

Figure 12.46 Regulation of compound leaf formation by transcription factors and hormones. (A) Developmental sequence showing the distribution of morphogenetic factors in meristem, leaf and leaflet primordia. (B) Developing tomato compound leaf with one distal leaflet and two lateral leaflet primordia that are initiated from tip to base. Scale bar = 100 μm.

number of major players is rather small, and as well as the issue of the biochemical mechanisms by which they exert their morphogenetic effects, it remains a fundamental question as to how such a comparatively limited genetic toolkit can produce such a rich variety of morphologies.

Key points Leaf shape is generated by polarized development, in which tip is differentiated from base, and lateral domains between the midvein and lamina margin are defined. Flattening of the leaf blade is under the control of transcription factors similar to those regulating vascular patterning. Other genes identified through analysis of mutants include regulators of lamina formation, serration of the leaf edge and differentiation of the petiole. The development of compound leaves is associated with incomplete suppression of *KNOX* expression and is under hormonal control by gibberellins and auxin.

12.6 Shoot architecture and stature

Shoot and root axes and branches establish the architectural framework on which the plant body is built.

As well as its structural role, the framework is a system for long-range integration of development through the routing of nutritional, hydraulic and hormonal signals. In this section we look at the modular nature of shoot structure and the regulation of branching. The practical importance of the control of plant architecture is illustrated with a discussion of 'Green Revolution' cereal varieties with reduced stature.

12.6.1 Plant structure is modular

The **body plan** of animals is fixed (maximal size, number of limbs, arrangement of internal organs and so forth), whereas in plants it is continuously expanding by the open-ended repetitive addition of structural units (Figure 12.47), as totipotency and the capacity for replication through vegetative propagation testify. The rich variety of plant forms is accounted for by variation in the spatial arrangement, timing of initiation, and development of **modules**. In some ways a plant is more like a **colonial organism** than an individual, behaving as an integrated population of parts. This view of plant structure considers morphogenesis of the individual to be the process of growth and development of the units from which it is built. The structural modules of shoots are termed **phytomers**. Typically, a phytomer comprises a node with an associated leaf and its axillary bud

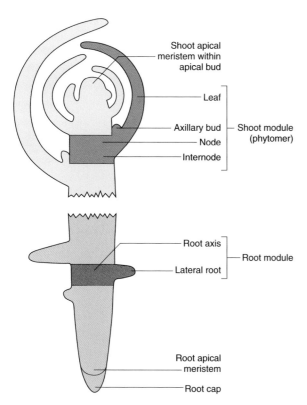

Figure 12.47 The plant body is built from structural modules. Root and shoot apical meristems produce the same structural modules repeatedly. Growth of axillary buds and lateral roots reiterate modular development. Stems grow by internode extension.

together with the internode below them (Figure 12.47). In general a phytomer follows the typical S-curve of initiation, growth and differentiation, maturity and finally senescence followed by death. Individual cells and tissues that make up a phytomer pass through a similar developmental sequence, as does the whole plant. Such a mode of organization, in which transformations are repeated at different scales (a property called **self-similarity**) is defined as **fractal**, a term from the mathematical discipline of topology. Mathematical modeling based on relatively simple **iterative** principles allows the morphogenesis of real and imaginary plants to be simulated. For example, Figure 12.48 is a visualization of the development of the herb *Capsella bursa-pastoris* (shepherd's purse), generated using **L-systems**, a mathematical formalism for modeling repetitive dynamic systems such as growth and differentiation of phytomers.

One way of **scaling up** from cell- and organ-level morphogenetic mechanisms to whole-plant development is to relate model parameters to particular physiological processes and the genes underlying them. We saw how a relatively simple logistic model of growth (see Equation 12.1 and Figure 12.7) can yield biologically meaningful outputs such as the plastochron, which in turn can be defined in terms of hormonal and genetic

regulatory networks (see Section 12.5.1). L-system representations and related approaches that simulate the origins and fates of phytomers give physiologists and molecular biologists insights into the rules of plant architecture and suggest new experimental approaches to the analysis of development. They show that morphogens operating at the whole-plant level must be **long-range signals**, and the vascular system is clearly a major communication channel, supplemented locally by the symplasm. Hormones are powerful mobile morphogens and, as discussed below in the example of branching regulation (see Section 12.6.2), their interactions make for a sensitive control mechanism. In addition to their nutritional functions, translocated inorganic compounds such as nitrate, and exported photosynthetic products such as sucrose, can be potent morphogenetic influences on sink tissues. There is also growing evidence for cell-to-cell and long-range movement of the products of regulatory gene expression, including both proteins and RNAs. Chapter 8 discusses in detail such a highly mobile morphogenetic protein, the so-called **florigen** responsible for signaling the transition from vegetative to reproductive development.

12.6.2 Branching is the result of interactions between apical and lateral growth

The pattern of branching varies greatly between species and often between varieties of the same species. These differences are in part genetically determined and related to programmed events at the SAM. But branching is often extremely responsive to environmental influences, an attribute that is essential for the fitness and survival of a sedentary organism. The source of this phenotypic plasticity resides in the structure of **axillary buds** and regulation of their growth. An axillary bud is situated in the angle between the stem and the adaxial side of an attached leaf, and its outgrowth results in the formation of a shoot branch. Each axillary bud contains an axillary meristem which is derived from cells that originate in the SAM and are located adjacent to the leaf primordium. Auxin is an essential regulator of primordium positioning. The axillary meristem produces a variable number of leaf and axillary bud primordia. A **dormant bud** forms when growth of the incipient axillary meristem is arrested after initiating a few unexpanded leaves, each of which in turn will be associated with a small axillary bud. Activation of such buds will produce a branch with a tertiary shoot apical meristem and so on, according to the principle of iterative development of self-similar modules described above.

The primary shoot apex exerts an inhibitory influence on growth of the axillary bud, an effect referred to as

Figure 12.48   Computer representation of the growth of *Capsella bursa-pastoris*. An iterative technique (L-systems modeling) was used to generate and transform successive phytomer units.

apical dominance. If the primary shoot tip is removed, apical dominance is relieved and axillary buds grow out into branches (see Figure 10.4B). This response is familiar to gardeners, who are accustomed to pinching out shoot tips to encourage bushy growth. Application of exogenous auxin to the site where the primary shoot tip was removed prevents outgrowth of axillary buds. Auxin itself does not accumulate in the suppressed axillary meristem, suggesting it acts via intermediary signals. In contrast to auxin, cytokinins inhibit apical dominance and promote axillary bud outgrowth. For example, apical dominance is suppressed both in transgenic plants overexpressing a gene for cytokinin synthesis, and in *Petunia* lines with the *shooting* (*sho*) mutation which results in elevated cytokinin content.

A novel second messenger that acts in the auxin regulatory pathway has been identified through studies of **shoot branching mutants** in a number of species: *more axillary growth* (***max***) in *Arabidopsis* (see Figure 10.39B); *ramosus* (***rms***) in pea; *decreased apical dominance* (***dad***) in petunia; and *high tillering dwarf* (***d***; Figure 10.39A) in rice. Genetic and biochemical analysis of these mutants established the existence of a long-range signaling pathway that involves the cleavage of carotenoids to produce **strigolactones**, a family of terpenoid derivatives originally identified in root exudates (see Section 10.8). Genes of the *MAX/RMS/DAD/D* families encode enzymes of strigolactone biosynthesis or targets in the network regulating the inhibition of branching in the bud.

Figure 12.49 presents a scheme for hormonal interactions regulating branch growth from axillary buds. Activation of axillary bud growth depends on the

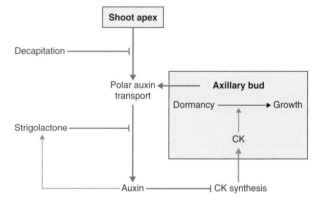

Figure 12.49   Shoot branching is regulated by a hormonal network that includes auxins, cytokinin (CK) and strigolactones acting through polar transport, inhibitory and promotive interactions, and source–sink relationships.

supply of cytokinin, carried in the transpiration stream, and the ability of the bud to export auxin along the polar transport pathway. Basipetal auxin transport from the shoot apex is negatively regulated by strigolactones moving up the plant, thereby reducing the sink strength of the stem for auxin. Auxin in turn is a positive regulator of strigolactone synthesis and an inhibitor of cytokinin synthesis in the stem. Decapitation removes the shoot apex as a source of auxin and releases from suppression genes for cytokinin synthesis. Newly-synthesized cytokinin moves to the dormant axillary bud and initiates sustained outgrowth. Eventually the activated SAM of the axillary shoot will produce auxin, cytokinin biosynthesis will close down, and a new cycle of apical dominance/axillary bud dormancy will be established.

Such a highly connected regulatory network of positive and negative interactions ensures integration of inputs from diverse sources external and internal to the plant.

Key points Plant architecture is modular. The structural module of the shoot is the phytomer. Development is fractal, the consequence of iterative formation and growth of modules. The arrangement of structural modules represents a framework for integrating whole-plant morphogenesis through long-range signaling. Branching occurs as a result of the interaction between environmental influences and programmed events at the SAM. Axillary buds are maintained in a dormant condition by apical dominance, the hormone-mediated suppression of growth by the primary shoot apex. The apex is a source of auxin, which inhibits axillary bud growth. Cytokinins promote outgrowth of axillary buds. Branching is controlled by auxins and cytokinins acting together in a transport and regulatory network with strigolactones.

12.6.3 Crop breeding has exploited genetic variation in stature to produce dwarf and semi-dwarf 'Green Revolution' cereals

Until the 20th century, domesticated cereals such as wheat and rice were tall, like the wild ancestors from which they were derived. Increasingly sophisticated plant-breeding methods were producing higher-yielding varieties of these crops, but yield came with a price. The long stems of the plants were not able to support the increased weight of seeds, and frequently collapsed, especially in strong winds and heavy rain. This phenomenon, known as **lodging**, resulted in significant losses of harvestable grain, and was the impetus that drove the development of modern, short-stemmed cereal varieties (see Figure 10.13C). The introduction of **semi-dwarf** varieties of rice and wheat, capable of producing high yields of grain without falling over in the field, was at the heart of the so-called 'Green Revolution' of the second half of the 20th century. The molecular basis of semi-dwarfism will be discussed later.

The fossil record shows that early land plants were tall. It is probable that light, water, carbon dioxide and in many cases soil minerals were abundantly available to them, and there was no particular reason for their growth to be constrained. However, for many groups of plants

the evolutionary trend has been a progressive decrease in stature, as they exploit new and more resource-limited ecological niches. Modern land plants exhibit a very wide range of stature, from redwood (*Sequoia sempervirens*) trees capable of attaining heights of 115 m, down to herb species with compressed internodes which do not exceed a centimeter in height. In general, the tallest plants are found in regions with high rainfall and close to sea level, an environment that probably most closely represents that experienced by the first land plants. The plant species found in dry, cold sites or at high altitudes never attain great height, even when transplanted to conditions where water, nutrients and light are not limiting.

As with many complex plant traits, height is controlled by multiple genes, but mutations in certain of these genes can have major effects on stature regardless of the plant's genetic makeup as a whole. Plant height was one of the seven characters studied by Gregor Mendel in the classic experiments that first shed light on genetics. Mendel crossed two lines of peas, one with long stems and one with short stems, and found that all the progeny were long-stemmed—the long-stem trait in peas is **dominant**. When these long-stemmed progeny plants were self-fertilized, the offspring consisted of long- and short-stemmed plants in a ratio of about 3:1, showing that in this case a single nuclear gene controlled the height character (Figure 12.50). The availability of height variation within a single crop species has been essential in the process of breeding modern, short-stemmed, high-yielding varieties.

The most important genes controlling plant height have functions related to the plant's ability to synthesize gibberellins, or to respond to these hormones. The biosynthetic pathway for this important group of diterpenoid carboxylic acid compounds is described in Chapter 10. In rice, semi-dwarf varieties result from a defect in gibberellin biosynthesis, while in wheat, gibberellin insensitivity is the basis for the dwarfing trait.

Many modern semi-dwarf varieties of rice (Figure 12.51) trace their ancestry back to a Chinese cultivar, Dee-geo-woo-gen, which was introduced into a breeding program in Taiwan in the 1950s and from there was taken up by the International Rice Research Institute (IRRI) in the Philippines. Other semi-dwarf rice cultivars, also giving high grain yields, have been produced by independent breeding programs in the USA, Japan and the People's Republic of China. However, molecular analysis has shown that in every one of these cases, the semi-dwarf trait results from mutation in a single gene, *sd1*, even though the parents in the crosses have been obtained independently.

In all these cultivars, normal plant height can be restored by applying gibberellic acid (GA) to the plants as they develop, indicating that they are deficient in some

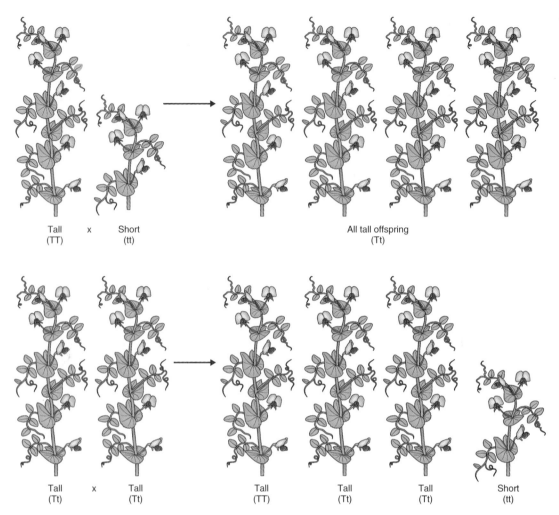

Figure 12.50   When Mendel crossed two lines of peas, one with long stems and one with short stems, he found that all the progeny were long-stemmed. The long-stem trait in peas is dominant, controlled by the allele designated *T*, whereas the short-stem trait, controlled by the allele *t*, is recessive. When these long-stemmed progeny plants, with the genotype *Tt*, were self-fertilized, the offspring consisted of long- and short-stemmed plants in a ratio of about 3:1. The result showed that in this case a single nuclear gene controlled the height character.

Figure 12.51   Two semi-dwarf rice cultivars and the corresponding tall isogenic lines. From left to right: Dee-geo-woo-gen (dwarf indica cultivar), Woo-gen (tall equivalent), Calrose 76 (dwarf japonica cultivar) and Calrose (tall equivalent). The Chinese cultivar Dee-geo-woo-gen was a progenitor of the so-called 'miracle rice' which was important in the Green Revolution.

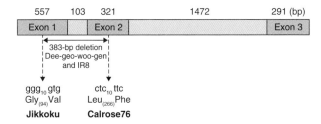

557   103   321   1472   291 (bp)

| Exon 1 | | Exon 2 | | | Exon 3 |

383-bp deletion
Dee-geo-woo-gen
and IR8

ggg$_{10}$gtg   ctc$_{10}$ttc
Gly$_{(94)}$Val   Leu$_{(266)}$Phe
Jikkoku   **Calrose76**

Figure 12.52 The rice *GA20ox2* gene consists of three exons and two introns. The mutations leading to short-stature phenotypes described in the text are either single nucleotide substitutions (dashed arrows) or the deletion present in Dee-geo-woo-gen (solid double-headed arrow).

component of GA production. The *sd1* gene has been isolated and shown to encode a **GA-20 oxidase** (**GA20ox**), one of the enzymes required for GA biosynthesis. Varieties with the *sd1* allele derived from Dee-geo-woo-gen contain a 383 base pair deletion in the *GA20ox* gene (known as *OsGA20ox2*, since it is the gene from rice, *Oryza sativa*). This deletion introduces an early stop codon (Figure 12.52), which probably results in a highly truncated and inactive form of the GA20ox enzyme. Other varieties, including Jikkoku and Reimie from Japan and the USA variety Calrose76, were bred from progenitors that had been treated with γ-radiation and contain point mutations in *OsGA20ox2* (Figure 12.52), which result in single amino acid substitutions. The latter dwarfing alleles are generally weaker than that resulting from the 383 base pair deletion, which indicates that the mutant enzymes might still have some GA biosynthetic activity. Gibberellin 20-oxidases, which catalyze the penultimate step in GA biosynthesis, are encoded by small gene families. Different members of these families can partially substitute for one another because of overlapping patterns of gene expression or because intermediates in the biosynthetic pathway can move between tissues, which may compensate for lack of expression of one member of the gene family in certain cell types. Because of this, mutations in the genes encoding GA20ox mutants are usually only moderately GA-deficient and have a **semi-dwarf phenotype**, whereas plants with severe GA deficiency caused by mutations in other genes are extremely dwarfed, are often sterile and can suffer other developmental abnormalities. Surprisingly, selection for semi-dwarfism in rice has always yielded mutations in *OsGA20ox2* rather than in any other GA biosynthesis gene. The most likely explanation is that mutations in other genes may have severe developmental consequences, or reduced rather than increased grain yield, so they have been discarded early in breeding programs.

The dwarfing genes used in bread wheat (*Triticum aestivum*) are derived from a Japanese variety, Daruma,

which was crossed with high-yielding American wheat to produce the variety **Norin 10**. Norin 10 is a progenitor of contemporary semi-dwarf wheats; it is adapted to many different environments. Well over half of all current commercial wheat varieties contain the Norin 10 dwarfing genes. There are two of these dwarfing genes. Unlike the recessive, loss-of-function *sd1* mutations in rice, the mutations in these wheat genes confer change of function and are semi-dominant. Each mutated gene causes some loss of height; when combined in the same plant their effect is additive and stature is further reduced (Figure 12.53). Bread wheat is a hexaploid species, deriving a set of chromosomes from each of its three wild diploid ancestors (see Figure 3.22). The three sets of chromosomes are identified as A, B and D. The **dwarfing alleles** are designated **Rht-B1b** (formerly *Rht1*) and **Rht-D1b** (*Rht2*); Rht stands for reduced height, and B1b/D1b mean that the alleles are on the short arms of chromosomes B1 and D1 respectively.

Rht-B1a
Rht-D1a   Rht-B1b   Rht-D1b   Rht-B1b
Rht-D1b

Figure 12.53 A series of otherwise isogenic wheat lines with different versions of the *Rht-B1* and *Rht-D1* genes. On the left, Rht-B1a/Rht-D1a plants are of normal (tall) stature. The Rht-B1b and Rht-D1b alleles each confer reduced height; the Rht-B1b/Rht-D1b combination shown on the right gives semi-dwarf plants. This is the same combination of alleles found in Norin 10, a progenitor of many contemporary semi-dwarf wheats.

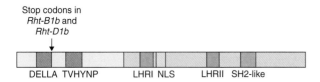

Stop codons in
Rht-B1b and
Rht-D1b

DELLA TVHYNP   LHRI NLS   LHRII SH2-like

Figure 12.54 The *Rht-B1b* and *Rht-D1b* mutations both introduce premature stop codons into the conserved DELLA region of the Rht/GAI protein.

The *Rht* genes encode members of the GRAS family of proteins, which are regulators of transcription. At the N-terminal end they contain two conserved regions, one of which includes a 27 amino acid motif called the **DELLA domain**. The wild-type Rht proteins are negative regulators of GA signaling, and GA acts by repressing their regulatory function. The Rht-like proteins are located in the nucleus. If they contain intact DELLA regions, they are rapidly degraded in the presence of GA. This degradation involves ubiquitin-mediated proteolysis by the ubiquitin–proteasome system described in Chapters 5 and 10.

The *Rht* dwarfing alleles cause a reduction in the plant's normal response to gibberellins. The *Rht-B1b* and *Rht-D1b* dwarfing alleles have DNA base substitutions which introduce premature stop codons into the DELLA region (Figure 12.54). The truncated proteins produced as a result of these stop codons act as constitutive repressors of growth; they are insensitive to GA. Thus once their N-terminal domains are altered, the Rht gene products can still act as repressors of growth, but can no longer themselves be repressed by GA, and the result is a semi-dwarf phenotype which is not reversed by applying GA to the plant.

Homologs of the *Rht* gene have been identified in other species. Mutations in the *Arabidopsis GAI* and maize *dwarf8* genes also result in gibberellin-insensitive dwarfs. In both these cases, as with the *Rht* mutations, the phenotype results from partial or complete deletion of one or both of the conserved N-terminal domains of the protein. In contrast, mutations that cause complete loss of Rht-like function produce overgrowth mutants such as *slender* in barley (see Section 12.2.4). In this case

Key points Stature (plant height) is an important trait in crop plants. Modern cereal varieties are shorter than the types traditionally grown and have advantages of lodging resistance, responsiveness to fertilizer and high yield. The Green Revolution was built on breeding short-straw wheat and rice. Alleles for tall stature are usually dominant to those for short stature. The semi-dwarf trait in rice is the phenotypic consequence of a defect in gibberellin biosynthesis. The gene responsible encodes GA-20 oxidase; alleles present in semi-dwarf rice varieties have base deletions or substitutions and fail to express functional enzyme. Dwarf wheat is gibberellin-insensitive. Most semi-dwarf bread wheat genotypes have *Rht* (*reduced height*) alleles from the variety Norin 10. Rht proteins negatively regulate GA signaling via the DELLA pathway. Base substitutions in *Rht* reduce the growth response to gibberellins. Complete loss of Rht-like function results in an overgrowth phenotype, as in the barley mutant *slender*.

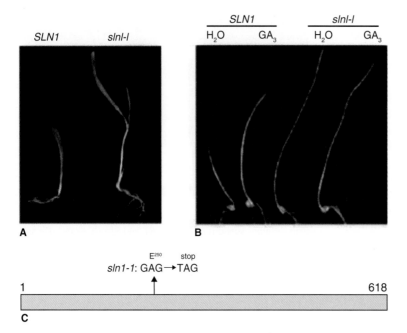

SLN1   sInI-I

SLN1   sInI-I
H$_2$O   GA$_3$ H$_2$O   GA$_3$

A   B

E^{250} stop
sln1-1: GAG → TAG

1   618

C

Figure 12.55 (A) Barley plants with the *slender* (*sln*) mutation are extremely elongated and spindly compared with wild-type (*SLN*), as these 5-day-old seedlings show. (B) The mutation mimics the effect of spraying the wild-type with gibberellin. (C) The mutation results from conversion of a GAG codon to the stop codon TAG. Numbers represent amino acid positions in the SLN1 protein (1 indicates the start Met, and 618 indicates the final Pro).

there is a base substitution in the middle of the *SLN1* gene that leads to plants with no detectable levels of the SLN1 protein, an Rht-like protein with a DELLA motif. Barley plants with this mutation grow as though sprayed with GA, becoming extremely elongated and spindly (Figure 12.55).

In grass species farmed for their grain, as has been shown above, selection for reduced stature is advantageous, because it decreases the likelihood of lodging, and causes the plant to allocate more of its fixed carbon to grain and less to stem. In industrialized western agriculture, the value of the non-grain part of the above-ground biomass has been low in comparison to that of the grain although, under less intensive agricultural regimes, stems and leaves have always been exploited as animal fodder, building materials, fuel and more. However, we are now moving into an era in which the fibrous stems of grasses and cereals are assuming major importance because of their potential as 'biorenewables'. It is clear that more research will be needed in order to maximize the yield of stem tissue while retaining the compositional traits required for the biorenewables industry, and minimizing the input of water and fertilizers.

Part V
Maturation

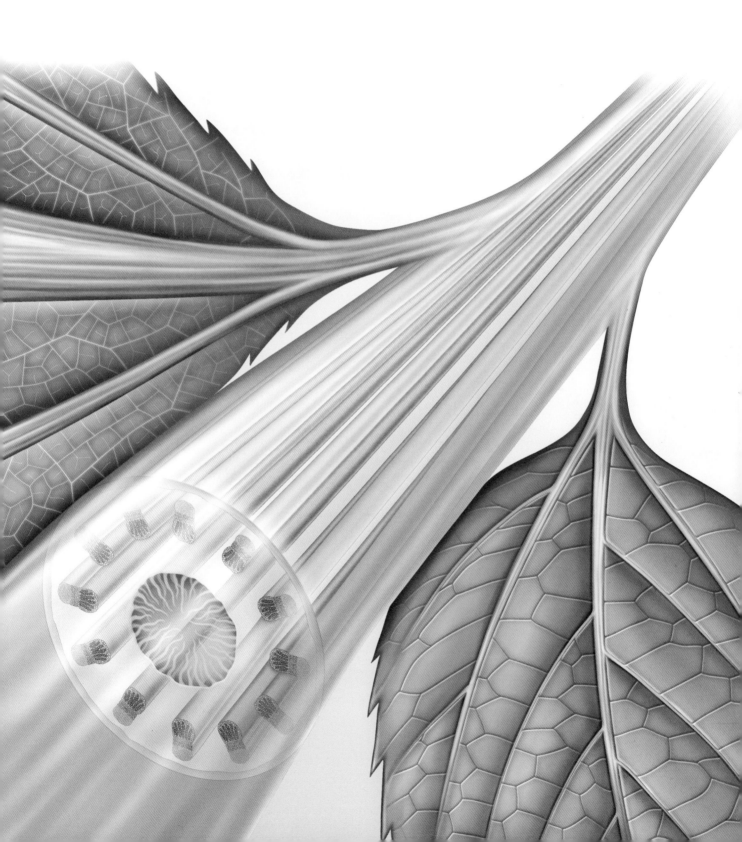

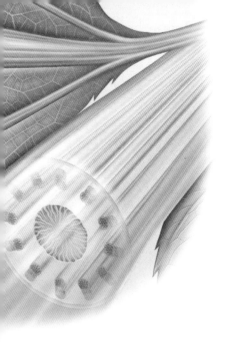

Chapter 13

Mineral nutrient acquisition and assimilation

13.1 Introduction to plant nutrition

Plants obtain all the elements that they need to survive, thrive and grow from carbon dioxide, water and dissolved inorganic ions (minerals). In Chapter 9 we discussed the acquisition of organic carbon compounds during the processes of photosynthesis. In this chapter we will investigate the uptake and role of essential inorganic ions. The value to agriculture of adding lime or wood ash to the soil has been known for more than 2000 years, but it was not until the mid-19th century that the mineral requirements of plants became the subject of systematic scientific study. The German chemist Justus von Liebig is generally credited with establishing that the essential elements for plant growth in addition to carbon (C), hydrogen (H) and oxygen (O) are nitrogen (N), phosphorus (P), sulfur (S), potassium (K), calcium (Ca), magnesium (Mg) and iron (Fe). During the 20th century other elements were identified as essential in minute quantities. These micronutrients include manganese (Mn), boron (B), zinc (Zn), copper (Cu), molybdenum (Mo) and nickel (Ni).

Table 13.1 lists the known macronutrients and micronutrients essential for plant growth and development. To be classified as 'essential', an element: (i) must be absolutely required for the plant to complete its life cycle; and (ii) must not be replaceable by another element. The concentrations of macronutrients in plant tissues are typically in the range 1 to 15 mg g^{-1} dry weight, while micronutrients may be one to several orders of magnitude less than this. The marked difference between macronutrient concentrations within the plant and the elemental composition of soils indicates that uptake and transport systems must be highly selective and capable of working against a concentration gradient. These features of plant nutrient acquisition have become the focus of current research that combines modern advances in genomics, membrane biophysics, structural biology and cell physiology to identify the underlying molecular structures and mechanisms.

13.1.1 Deficiency symptoms reflect the function and mobility of an element within the plant

If the supply of an essential element drops below the level required for optimal growth, the plant may develop

The Molecular Life of Plants, First Edition. Russell Jones, Helen Ougham, Howard Thomas and Susan Waaland.
© 2013 John Wiley & Sons, Ltd. Published 2013 by John Wiley & Sons, Ltd.

Table 13.1 Mineral elements required for healthy plant growth, development and survival.

Element (chemical symbol)	Abundance in Earth's crust (%)[a]	Typical range of concentrations in agricultural soils (mM)	Form in which element is taken up by plants	Concentration in plant tissues (molarity in fresh tissue)[b]	Physiological function	Deficiency symptoms
Macronutrients						
Nitrogen (N)	0.005	1.0–27.6 (NO_3^-), 5.0–6.1 (NH_4^+)	NO_3^-, NH_4^+, amino acids	71×10^{-3}	Used in structures of proteins, nucleic acids, cofactors, chlorophylls, alkaloids and their precursors; role in the synthesis of amino acids and many other primary and intermediate metabolites, including some hormones and other signaling molecules	Reduced synthesis of chlorophylls under N-limited conditions results in general chlorosis (yellowing) of older leaves and the rest of the plant is often light green
Potassium (K)	2.6	0.1–6.8 (K^+)	K^+	17×10^{-3}	Essential ion for protein synthesis; major solute functioning in water balance; stomatal function	Wilting and scorching of older leaves. Interveinal chlorosis begins at the base, scorching inward from leaf margins
Calcium (Ca)	3.6	1.7–19.6 (Ca^{2+})	Ca^{2+}	8.3×10^{-3}	Role in stability of cell walls, maintenance of membrane structure and permeability, and cell signaling	New leaves are distorted or irregularly shaped. Causes blossom-end rot
Magnesium (Mg)	2.0	0.3–10.3 (Mg^{2+})	Mg^{2+}	5.5×10^{-3}	Component of chlorophyll; essential ion for many enzyme reactions; role in the integrity of ribosomes and protein synthesis	Older foliage turns yellow at edge, often leaving a green arrowhead shape in the center of each leaf
Phosphorus (P)	0.12	0–0.09 ($H_2PO_4^-$)	$H_2PO_4^-$	4.3×10^{-3}	Essential nutrient for plants and animals in the form of phosphate ions; links nucleotide structural units of nucleic acids; central role in energy metabolism, as part of structures of ATP and ADP, phosphorylated intermediates and proteins; component of phospholipid cell membranes; role in phosphorylation of proteins in signaling cascades	Leaf tips look burnt, followed by older leaves turning a dark green or reddish-purple

Element (chemical symbol)	Abundance in Earth's crust (%)[a]	Typical range of concentrations in agricultural soils (mM)	Form in which element is taken up by plants	Concentration in plant tissues (molarity in fresh tissue)[b]	Physiological function	Deficiency symptoms
Sulfur (S)	0.05	0.3–3.5 (SO_4^{2-})	SO_4^{2-}	2.1×10^{-3}	Component of amino acids, proteins, coenzymes and prosthetic groups; component of metallothioneins, sulfolipids and antioxidants; role in stress responses through secondary metabolism	Younger leaves turn yellow first, sometimes followed by older leaves
Micronutrients						
Chlorine (Cl)	0.05	—	Cl^-	188×10^{-6}	Required for water-splitting step in photosynthesis	Wilting and intervein chlorosis of young leaves. Bronzing on the upper side of mature leaves
Boron (B)	0.05	—	$B(OH)_3$	123×10^{-6}	Cofactor in chlorophyll synthesis; role in cell wall function	Terminal buds die, witches' brooms form
Iron (Fe)	5.0	—	Fe^{3+}	120×10^{-6}	Component of cytochromes, iron-sulfur and other prosthetic groups	Yellowing occurs between the veins of young leaves
Manganese (Mn)	0.1	—	Mn^{2+}	61×10^{-6}	Active in formation of amino acids; required for water-splitting step in photosynthesis; component of superoxide dismutase	Yellowing occurs between the veins of young leaves. Reduction in size of plant parts (leaves, shoots, fruit) generally. Dead spots or patches
Zinc (Zn)	Trace	—	Zn^{2+}	20×10^{-6}	Component of RNA polymerase, alcohol dehydrogenase, carbonic anhydrase and superoxide dismutase; essential for chlorophyll biosynthesis	Shortened internodes. Terminal leaves may be rosetted, and yellowing occurs between the veins of the new leaves. Bronzing
Copper (Cu)	0.01	—	Cu^{2+}	6.2×10^{-6}	Component of plastocyanin, cytochrome oxidase, superoxide dismutase and lignin-biosynthetic enzymes	Leaves are dark green; plant is stunted

(continued overleaf)

Table 13.1 (continued).

Element (chemical symbol)	Abundance in Earth's crust (%)[a]	Typical range of concentrations in agricultural soils (mM)	Form in which element is taken up by plants	Concentration in plant tissues (molarity in fresh tissue)[b]	Physiological function	Deficiency symptoms
Molybdenum (Mo)	Trace	—	MoO_4^{2-}	70×10^{-9}	Essential for symbiotic relationship with N-fixing bacteria; cofactor in nitrate reduction; component of sulfite oxidase, xanthine dehydrogenase and aldehyde oxidase	General yellowing of older leaves (bottom of plant); the rest of the plant is often light green
Nickel (Ni)	Trace	—	Ni^{2+}	6×10^{-9}	Cofactor of urease	Necrotic lesions due to accumulation of toxic levels of urea
Nutrients required by some but not all plants						
Sodium (Na)	2.8	—	Na^+	$1.5\text{--}155 \times 10^{-3}$	Required by halophytes and plants with C_4 photosynthesis	
Cobalt (Co)	Trace	—	Co^{2+}	0.12×10^{-3}	Required by legumes for N fixation	
Silicon (Si)	27.8	—	$Si(OH)_4$	$2.6\text{--}255 \times 10^{-3}$	Not physiologically essential but certain grasses deposit Si as a deterrent to herbivores	

[a]Oxygen is the most abundant element in the Earth's crust (46.5%). Aluminum accounts for 8.1%. The nutrients Si, Fe, Ca, Na, K and Mg add up to 43.8% and the rest to 1.6%.
[b]Molarities calculated by assuming a 15:1 fresh weight:dry weight ratio. Concentrations presented are averages for bulk tissue, including both apoplastic and symplastic fractions. Concentrations in cytosol or individual subcellular compartments may be as much as orders of magnitude different from the overall figure.

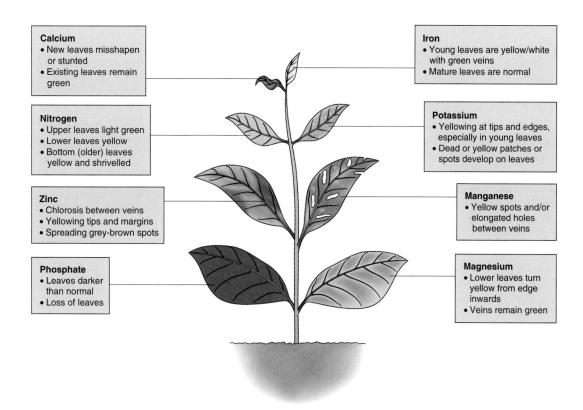

Figure 13.1 Signs of nutrient deficiency.

deficiency symptoms. Inadequate supply may be the result of low soil concentrations of the element, the presence of the element in a form that the plant cannot access or take up, or the influence of other factors such as soil pH, aeration, water status or high concentrations of antagonistic elements. A plant can reduce the severity of deficiency symptoms by sequestering the element in question at a time of abundance. For example, plants can store high concentrations (more than 20 mM in some cases) of nitrate in cell vacuoles. Thus, if the input of N from the soil becomes limiting, an immediate impact on metabolism can be offset by first draining these reservoirs.

The morphological pattern of deficiency symptoms is often dictated by the extent to which the limiting element is mobile within the plant (Figure 13.1). Inadequate supplies of a highly mobile element like P, for example, result in the development of deficiency symptoms in older organs first, as phosphate is salvaged from metabolites and macromolecules through catabolism and transported to younger tissues. By contrast, deficiency of a relatively immobile element such as Ca is apparent in the decreased growth and development of young organs. Different elements differ in their availability and interactions within the soil, in the physiology of storage and mobility, and in biochemical

function, and this results in deficiency responses that are often distinctive enough to be diagnostic. Table 13.1 gives characteristic deficiency symptoms for the major elements essential for plants. The availability of N, P and K often limits plant growth; these macronutrients are major ingredients in garden and agricultural fertilizers.

13.1.2 Other organs in addition to roots may function in nutrient acquisition

Roots are the principal nutrient-absorbing structures (see Chapters 1 and 12), but many plants are also able to obtain significant amounts of nutrients through other organs. For example, leaves are capable of nutrient uptake and foliar feeding has been used in horticulture and agriculture to enhance crop yield and quality. The leaves of about 600 plant species, representing a few families, notably the pitcher plants (Nepenthaceae) and the sundews and flytraps (Droseraceae), are modified into structures that capture and digest insects and other small animals. Compositional analyses of insects indicate they are rich sources of macronutrients: N, 99–121; P, 6–14.7; K, 1.5–31.8; Ca, 22.5; Mg, 0.94 g kg^{-1} dry weight. Such dietary supplements permit carnivorous plants to grow in nutrient-deficient bog, swamp and fenland environments where acid conditions limit the availability of many nutrients.

The roots of a plant make up a highly branched organ system with a large surface area available for the uptake of ions. The surface area of roots is further increased by the production of unicellular root hairs (see Chapter 1). The surface area of all the roots and root hairs of a plant may exceed the surface area of the above-ground shoot system. In addition, roots may function in anchoring a plant (for example buttress roots, contractile roots, breathing roots) and in reserve storage (root tubers, taproots). Over 4000 plant species are known to obtain some or all of their nutritional needs by parasitizing other plants. Haustoria, the roots of parasitic plants, penetrate and absorb nutrients from the root tissues of the host species. The roots of many plants form symbiotic associations with other organisms. Of particular importance for plant nutrition are the nodule structures that contain nitrogen-fixing prokaryotes and the arbuscules that are associated with interactions between roots and mycorrhizal fungi (see Chapter 12). The intimate relationship of roots, soil and microorganisms is called the **rhizosphere**.

Nutrient availability and the state of the plant's internal nutrient economy regulate the structure and function of the root system. Nutrient availability interacts in a complex fashion with other factors such as water supply, aeration and soil pH. For example, the external concentration of Ca required for maximal root growth in cotton (*Gossypium*) is about 1 μM when the pH is 5.6 but this increases to more than 50 mM at pH 4.5. Nutrients regulate root development indirectly through influencing whole-plant growth according to the degree of deficiency or sufficiency, but they also have direct morphogenetic influence through modulation of the expression and function of specific genes (see Chapter 12).

13.1.3 Technologies used to study mineral nutrition include hydroponics and rhizotrons

A major technical challenge for research into the mechanisms of plant nutrition is that the action takes place mostly underground. The rhizosphere is a complex system that is difficult to observe in situ. This has led to a number of experimental approaches to simplify plant–nutrient relations and make root development and function more accessible to analysis.

One widely used technique employs **rhizotrons**— 'windows' in the soil, in the form of glass or transparent plastic plates or tubes installed against soil profiles, through which roots may be observed. Recent developments based on this approach include collecting digital images at multiple wavelengths in the visible and infra-red ranges and applying computational techniques to resolve root architecture and dynamics. The use of chemical indicator dyes and fluorescent tags allows rhizosphere physiology to be probed non-invasively. Figure 13.2 illustrates how it is possible to discriminate between the roots of different species in a minirhizotron when the plants have been genetically transformed to express a fluorescent protein. In this example, the respective root systems of *Lolium multiflorum* and transgenic *Zea mays* growing in a mixed stand are not distinguishable to the naked eye, but can be easily identified under narrow spectrum light that excites the fluorescent protein.

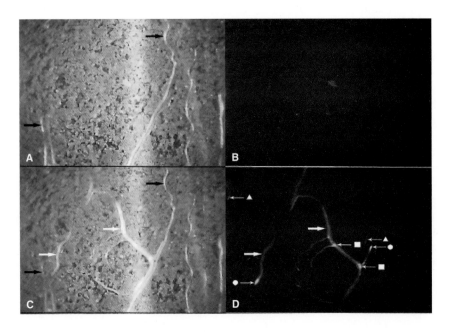

Figure 13.2 Roots of a mixed stand of *Zea mays* and *Lolium multiflorum* plants growing in a minirhizotron. The maize genotype was genetically transformed with a gene that expresses a green fluorescent protein. Conventional photographs of roots in visible light were taken (A) 37 days and (C) 48 days after the maize was sown. (B) and (D) are the corresponding fluorescence images. Roots of *Zea* (white arrows) and *Lolium* (black arrows) can be clearly distinguished. Fluorescence is particularly intense in root tips (circles) and branch points (squares).

Another method for studying root function and plant nutrition is **hydroponics**, the growing of plants in a soil-less, nutrient-enriched water solution. In some hydroponic systems, the roots are anchored in an inert medium such as sand, gravel or rockwool. Hydroponics is widely used for the production of commercial crops because it eliminates problems associated with soil-borne diseases, unfavorable physical and chemical characteristics of soil, water stress and fluctuations in nutrient supply and demand. There are many variations on the basic technique, some of which have become established as tools for research in plant nutrition. Figure 13.3 shows an example of a flowing nutrient facility, in which samples of nutrient solution are taken automatically on a 3-minute cycle; pH, nitrate, ammonium, potassium and further nutrients are measured and if they deviate from pre-set levels the appropriate chemicals are pumped into the culture unit. Plant nutrient uptakes are then calculated from the rates at which nutrients are re-supplied to the units. Growing plants in solution culture is a convenient way to access roots and their functions. It is important, however, to recognize that many, complex physical and biological interactions that occur in the rhizosphere are absent. This means that assumptions about soil-grown roots based on behavior under hydroponic conditions must be made cautiously.

13.1.4 The rhizosphere affects mineral availability to plants

Mineral ions must be dissolved in the soil solution before they can be taken up by roots. As we shall see, ions vary in their solubility. Solubility is affected by pH. For example, clay particles in the soil, which have a negative charge, attract and bind positively charged cations. Water percolating through the soil may wash away ions that are dissolved in the soil solution so clay particles behave as cation reserves. Protons can displace bound cations by cation exchange, releasing them into the soil solution (Figure 13.4). Plant roots may facilitate this exchange to acquire bound cations. Acid rain depletes soil nutrients because protons displace all other cations bound on clay.

The body of a plant can be roughly divided into two zones: the apoplast and the symplasm. The inter-connected cell walls of plant cells constitute the **apoplast**. The protoplasts of most of the cells in a plant are connected by thin cytoplasmic connections to form the **symplasm** (see Figure 14.6). Dissolved ions move into and across the root by two pathways: **apoplastic** and **symplasmic pathways** (Figure 13.5). To enter the symplasmic pathway ions must cross the plasma membranes of root cells using specific transporters, e.g. channels or co-transporters (see Chapter 5). This

A

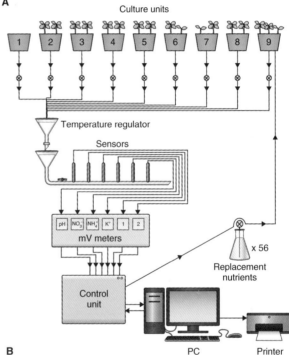

B

Figure 13.3 An automated hydroponics research system providing non-destructive continuous measurement of net uptake of NO_3^-, NH_4^+, K^+ and other ions at the whole-plant level, under defined environmental conditions and rhizosphere pH. (A) View of six culture units. (B) Schematic of control system. The resolution of uptake rates is from 10 minutes to 10 weeks and the system may be used for short-term (<24 hours) or long-term (several weeks) studies.

step excludes ions for which there are no transporters and allows the accumulation of ions in the cytosol to concentrations 100–1000-fold higher than their concentration in the soil solution. Ions can then diffuse across the root through the symplasm until they reach xylem parenchyma cells. Here they are unloaded into tracheary elements in the xylem and transported to cells in the rest of the plant (see Chapter 14). Ions may enter and leave the symplasmic pathway at any point along this

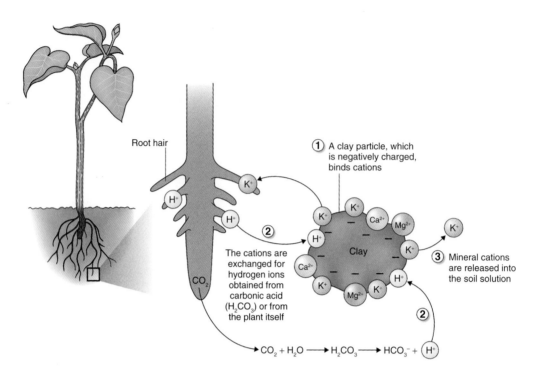

Figure 13.4   Cation exchange between clay particle and a root.

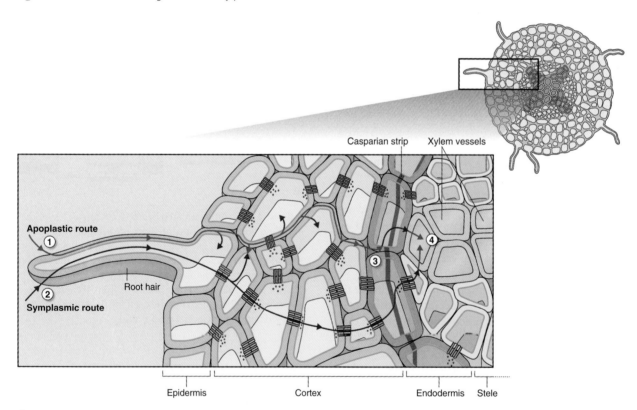

Figure 13.5   Apoplastic and symplasmic pathways for mineral transport across roots. (1) Apoplastic route: water and dissolved minerals from the soil diffuse into the hydrophilic walls of epidermal cells and travel through the apoplastic space of the cortex. (2) Symplasmic route: minerals that cross the root hair plasma membrane diffuse from cytoplasm to cytoplasm of adjacent cortical cells toward the stele. (3) The Casparian strip blocks apoplastic movement across the endodermis. To get from the cortex to the stele, minerals must move in the symplasm. (4) Water and minerals pass from the symplasmic space of endodermal and vascular parenchyma cells into xylem vessels for transport upward to the shoot.

pathway. Alternatively, ions may diffuse into the apoplast and diffuse across the root in cell wall solution until they reach the **endodermis** (Figure 13.5; see also Figure 1.31). The **Casparian strip**, a waxy layer in endodermal cell walls, forms a barrier in the apoplast that halts the diffusion of water and ions towards the vascular cylinder at the center of the root. To reach the xylem, ions in the apoplast must cross the plasma membrane of an endodermal cell and enter the symplasmic pathway.

At this point it is useful to summarize the basic principles of membrane transport (presented in detail in Chapter 5) that relate to the uptake and distribution of mineral ions. The plasma membrane of plant cells is maintained at a potential of -100 to -250 mV by the activity of the plasma membrane H^+-ATPase. Cations outside the cell can diffuse across the membrane through channels, moving down their electrochemical gradients. The movement of anions is against the electrochemical gradient and their passage across the membrane is via secondary active transport, usually mediated by the proton gradient maintained by the plasma membrane H^+-ATPase. Within the cell the vacuole has a major role in ion uptake and utilization. The tonoplast has a potential of about $+20$ to 30 mV, maintained by the action of vacuolar H^+-ATPase and H^+ pumping pyrophosphatase, and vacuolar sap is positively charged relative to the cytosol. Thus the electrochemical gradient favors the diffusion of anions between cytosol and vacuolar sap, whereas cation movement requires facilitation by co-transporters.

For the rest of this chapter, we will discuss each of the mineral nutrients starting with those required in the largest quantities followed by those that are needed in lower to trace amounts. For each element we will look at the form in which it is present in the soil and that in

Key points The macronutrient elements essential for plant growth are N, P, S, K, Ca and Mg. Fe, Mn, B, Zn, Cu, Mo and Ni are the essential micronutrients. Inadequate supply of a particular essential element produces characteristic deficiency symptoms. Plants acquire mineral nutrients from the soil mostly via roots, although some species (carnivorous plants, for example) also use other organs and nutrient sources. Because the root–soil zone (rhizosphere) is difficult to access, tools such as rhizotrons (windows for examining roots in situ) and hydroponics (soil-free cultivation) are often used for experimental observations of plant nutrition. Nutrients move from the rhizosphere into the plant where they are distributed via symplasmic (intracellular) or apoplastic (extracellular) routes.

which it is taken up by a plant and the mechanisms by which it enters root cells. Where applicable, we will discuss the way in which elements are incorporated into biological molecules, a process called **assimilation**.

13.2 Nitrogen

After carbon, hydrogen and oxygen, nitrogen is the most abundant element in living organisms. Most of this nitrogen comes from a pool fed by N recycled via decomposition from other organisms. New inputs to this pool are supplied by chemical reactions that accompany natural events (fire and electrical storms, for example) or human activity (such as the combustion of fossil fuels or the application of chemical fertilizers). During its uptake and release by living organisms, nitrogen passes through a series of reduced and oxidized states. Table 13.2 gives the oxidation status of the major forms of nitrogen.

The outer shell of the nitrogen atom has five electrons which are available to participate in bonding with other atoms. Elements more electronegative than nitrogen—that is, whose atoms have a higher affinity for electrons—will remove some or all of the five outer-shell electrons. Oxygen is such an element: up to three oxygen atoms may react with one nitrogen atom to make a series of nitrogen oxides (NO_x) up to the maximal oxidation state ($+5$) represented by the nitrate ion, NO_3^-. Nitrogen is reduced by removing electrons from less electronegative elements, notably hydrogen (as in ammonia gas, NH_3, and ammonium ion, NH_4^+) and carbon (as in the vast array of organic forms of nitrogen, including constituents of living cells). Nitrogen also bonds with itself in the atmospheric gas **dinitrogen, N_2**. The $N{\equiv}N$ triple bond is particularly stable, and it requires a great deal of energy as well as reducing power to reduce one molecule of dinitrogen to a biologically usable form. Nitrogen circulates on a global scale, moving between the oxidized, reduced and dinitrogen states in a perpetual biogeochemical cycle (Figure 13.6).

13.2.1 In the biosphere nitrogen cycles between inorganic and organic pools

Nitrogen is ubiquitous on Earth, but most of it is locked up in rocks and sediments (Table 13.3), where it is virtually inaccessible to living organisms except through weathering and other slow processes that take place over millions of years. Dinitrogen gas, which makes up about 80% of the planet's atmosphere, is unavailable to most eukaryotic organisms because of the intrinsic chemical

Table 13.2 Names and oxidation states of the major forms of nitrogen.

Compound	Oxidation state of N	Name	
N_2	0	Dinitrogen (nitrogen gas)	
$R\text{-}NH_x$	−1 to −3	Organic N ($x \geq 0$)	
NH_3	−3	Ammonia	Collectively NH_x
NH_4^+	−3	Ammonium ion	
NH_2OH	−1	Hydroxylamine	
N_2O	+1	Nitrous oxide	
NO	+2	Nitric oxide	
NO_2^-	+3	Nitrite	Collectively NO_x
NO_2	+4	Nitrogen dioxide	
NO_3^-	+5	Nitrate	

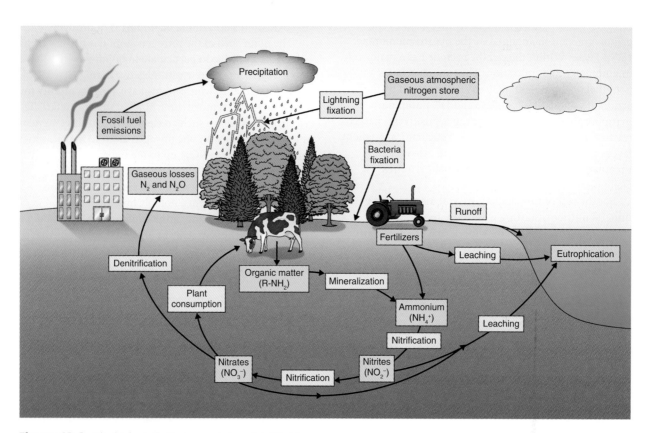

Figure 13.6   The biological nitrogen cycle in a simplified form.

stability of the N≡N triple bond. Atmospheric N_2, however, can be used by nitrogen-fixing prokaryotic organisms called **diazotrophs**. A small but significant amount of N_2 in the atmosphere is also converted to NO_x by electrical storms. NO_x may end up in the aquatic environment or soil where it is available to support plant growth (Figure 13.6). By contrast to inorganic forms,

once nitrogen has entered into organic combination ($R\text{-}NH_x$; see Table 13.2) it becomes highly labile and circulates with a turnover time measured in months to a few years (Table 13.3).

Plants can use both **nitrate** and **ammonium** as N sources. Once taken up, nitrate must be reduced to ammonium before it can be incorporated into organic

Table 13.3 Amounts and kinetics of different forms of nitrogen.

Nitrogen reservoir	Amount (kg)	Turnover rate[a]
Sediments and rocks	2×10^{20}	4×10^8 years
Atmosphere (mostly N_2)	4×10^{18}	4×10^7 years
Aquatic inorganics (mostly dissolved NO_x)	2×10^{16}	2×10^5 years
Terrestrial organics	3×10^{14}	1–40 years
Aquatic organics	3×10^{13}	1 month
Soil	2×10^{13}	Less than 1 year

[a] For any atom of N, the average length of time between its entering and leaving the reservoir.

molecules. Ammonium in the soil is derived from atmospheric N_2 by the process of nitrogen fixation (see Section 13.2.2) or by release from organic matter (R-NH_x) during decay, a process also called **mineralization** or **ammonification**. **Nitrifying bacteria** in the soil or in water convert NH_4^+ to NO_2^- and NO_3^-. Reduction of NO_3^- and NO_2^- to NH_4^+ and subsequent assimilation occurs in bacteria, fungi and plants. Nitrogen may be lost from the pool available for biological assimilation when **denitrifying bacteria** reduce soluble NO_x all the way back to N_2 and gaseous forms of NO_x. Nitrogen (particularly NO_3^-) may also be lost from this pool by diffusing into inaccessible domains within the soil. In addition, soluble forms of nitrogen may be leached from the land into aquatic environments (Figure 13.6). **Eutrophication**, high levels of available nitrogen and other nutrients in soils, lakes, rivers and seas, is associated with the degradation of natural communities, the reduction of biodiversity and, sometimes, as in the case of algal blooms, hazards to the food chain and human health.

In the following sections we will first discuss the process of dinitrogen fixation. Then we will follow the uptake of inorganic forms of fixed nitrogen, NH_4^+ and NO_3^-, by roots and finally look at how these ions are incorporated into organic molecules.

13.2.2 Nitrogen fixation converts dinitrogen gas into NH_3

Atmospheric N_2 gas is abundant, but is unusable by all but a few organisms. Before it enters into organic

Key points After C, H and O, N is the most abundant element in living organisms. Nitrogen exists in a number of reduced and oxidized forms, and cycles between organic and inorganic pools. Roots take up nitrate, NO_3^-, the most oxidized state of nitrogen. Nitrate must be reduced to the ammonium ion, NH_4^+, before it can be assimilated (enter organic combination). Roots can also take up ammonium, particularly in acid soils. Dinitrogen, N_2, the most abundant gas in the atmosphere, is so chemically inert that it can only be utilized biologically by diazotrophs, specialized nitrogen-fixing microorganisms that reduce N_2 to NH_4^+. In the global nitrogen cycle, ammonium is released from organic matter by the process of mineralization and may be converted back to nitrate by nitrifying bacteria. Denitrifying bacteria return organic N to gaseous oxides or atmospheric N_2. High levels of available nitrogen in the environment (eutrophication) can be ecologically harmful.

combination within cells, the nitrogen of N_2 must be fixed, that is, the very stable triple bond between the two atoms of dinitrogen gas must be opened and reduced to form NH_3. The $\Delta G^{\circ\prime}$ for nitrogen fixation overall is about -200 kJ mol^{-1}. The industrial-scale reduction of atmospheric dinitrogen uses the **Haber–Bosch** process in which N_2 and H_2 are combined under high temperature and pressure in the presence of an iron catalyst to make NH_3. This process is the basis of the manufacture of nitrogen fertilizer for agriculture. It is estimated that the 10^8 tons of fertilizer produced per year consume 1–2% of the global energy supply and sustain a third of the world's population. In contrast, **biological nitrogen fixation** occurs at ambient temperature and atmospheric pressure. It is limited to certain diazotrophic eubacterial species and a few representatives of the methanogenic archaea. Most diazotrophs are free-living but some, notably actinomycetes, cyanobacteria and α-proteobacteria, enter into symbiotic relationships with plants (see Table 12.2). The development of nitrogen-fixing root nodules and the regulation of genes for symbiosis and enzymes of N_2 assimilation are discussed in Chapter 12.

13.2.3 Biological nitrogen fixation is catalyzed by nitrogenase

The enzyme responsible for biological nitrogen fixation is **nitrogenase**, a complex of two enzymes: **dinitrogenase**

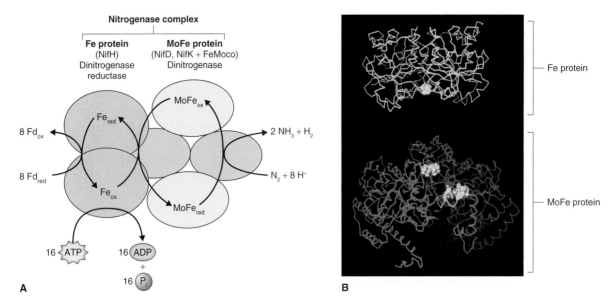

Figure 13.7 (A) The nitrogenase complex, showing the flow of reducing power and substrates in enzymatic nitrogen fixation. The Fe protein, encoded by *nifH*, accepts electrons from a carrier, e.g. ferredoxin (Fd), flavodoxin or another redox-active species of similar potential. The identity of the carrier varies, depending on the biological system involved. The Fe protein transfers single electrons at very low potential to the MoFe protein, accompanied by net hydrolysis of ATP. The MoFe protein, an $\alpha 2\beta 2$ heterotetramer of subunits encoded by *nifD* and *nifK*, accepts electrons and binds H^+ ions and N_2 gas in a stepwise cycle, ultimately leading to the production of H_2 and ammonia. (B) Model of the docking complex formed by the Fe protein (yellow) and an $\alpha\beta$ subunit pair of the MoFe protein (purple and red, respectively). Prosthetic groups of the enzyme subunits are shown as space-filling molecular structures.

reductase and **dinitrogenase** (Figure 13.7). The overall reaction catalyzed by nitrogenase (Equation 13.1) produces 2 moles of ammonia from one of dinitrogen and in the process uses 16 moles of ATP and generates 1 mole of dihydrogen (H_2). During this reaction, dinitrogenase (the **MoFe protein**) binds N_2 and hydrolyzes ATP, and dinitrogenase reductase (the **Fe protein**) binds ATP and supplies the high-energy electrons used to reduce N_2 to NH_3. The three-dimensional structures of the constituent enzymes and the stabilized nitrogenase complex have been determined by X-ray crystallography (Figure 13.7).

Equation 13.1 Overall reaction catalyzed by nitrogenase

$$N_2 + 16ATP + 8e^- + 8H^+ \rightarrow 2NH_3 + H_2 + 16ADP + 16P_i$$

Dinitrogenase reductase is a 64 kDa homodimer (that is, its structure consists of a pair of identical polypeptides) containing a single [Fe_4S_4] cluster. Dinitrogenase is a 240 kDa protein constructed from four polypeptides of more than one kind (a heterotetramer), in this case two α and two β subunits. The dinitrogenase holoenzyme (fully assembled protein with its prosthetic groups) contains two types of metal centers: the P cluster and the M cluster. The **P cluster** is an [Fe_8S_7] complex that, in the reduced state (P^N), consists of eight atoms of iron bonded to seven of sulfur (Figure 13.8A). Four of the Fe and four of the S atoms form a cuboidal fragment.

In the oxidized (P^{OX}) state, three further Fe atoms bond to three S and the remaining Fe links through oxygen to a serine residue in the polypeptide chain of the β subunit of the enzyme (Figure 13.8B). The dinitrogenase heterodimer contains a pair of identical P clusters. The **M cluster** is an **FeMo cofactor** (FeMoco; Figure 13.8C) and is the site of substrate reduction. Like the P cluster, there are two M centers per holoenzyme molecule and their structure changes with redox state, adopting at least three forms: native or semi-reduced (M^N), oxidized (M^{OX}) and reduced (M^R). It has an [Fe_4S_3] cuboid and an [Fe_3MoS_3] fragment, bridged by three sulfur atoms, and a molecule of the C_7 citric acid analog homocitrate.

All diazotrophs have a molybdenum–iron dinitrogenase system. Some free-living bacteria, such as *Azotobacter vinelandii* and *Rhodobacter capsulatus*, can synthesize alternative nitrogenases containing vanadium–iron or iron–iron cofactors in response to molybdenum depletion. However, endosymbionts of plants synthesize only the molybdenum–iron form of the enzyme. Dinitrogenase is capable of reducing not only dinitrogen but also several other compounds with multiple bonds, including acetylene (C_2H_2), cyanide (CN^-), nitrous oxide (N_2O) and azide (N_3^-). The enzymatic conversion of acetylene to ethylene is readily measured by gas chromatography and is the basis of a convenient assay for nitrogenase activity.

Nitrogenase is highly sensitive to inactivation by oxygen, as are some of the proteins that supply

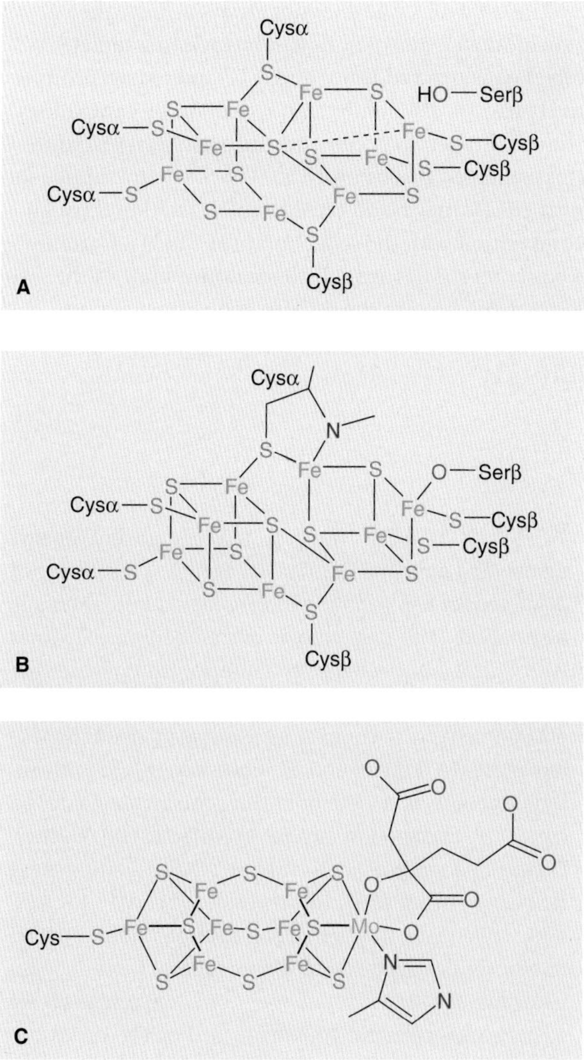

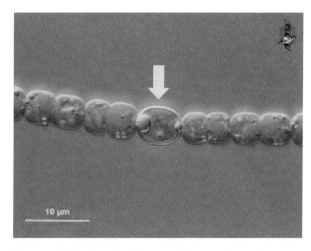

Figure 13.9   Part of a filament of the cyanobacterium *Anabaena* showing a N_2-fixing heterocyst (arrow).

Figure 13.8   The P and M prosthetic groups of dinitrogenase: (A) P^N, (B) P^{OX} and (C) FeMoco.

nitrogenase with reductant. Therefore, many nitrogen-fixing bacteria are obligate anaerobes (several species of *Clostridium*, for example). Others, such as the fast-respiring *Azotobacter*, are capable of maintaining an anaerobic environment by metabolically scavenging oxygen or by spatially separating nitrogen fixation from aerobic metabolism (as in *Anabaena* and other cyanobacteria that have specialized nitrogen-fixing cells called **heterocysts**; Figure 13.9).

Because the yield of ATP per unit carbon oxidized is much lower for fermentation and anaerobic respiration than for aerobic respiration, nitrogen-fixing anaerobes have a high demand for carbon substrate. As a consequence the endosymbiotic association (see Chapter 12) makes considerable demands on the plant host. It takes about 12 g of organic carbon to fix 1 g of nitrogen. At any given time the nitrogen-fixing root nodules of a legume may be consuming as much as 50%

of the carbon the plant acquires by photosynthesis. The penalty in terms of primary plant productivity explains why plants with endosymbiotic N_2-fixers, though important as sources of dietary protein, are not among the major crop plants that feed the world.

13.2.4 Dinitrogen fixation occurs via a catalytic cycle

The flow of electrons during dinitrogen fixation is shown in Figure 13.10. The oxidized form of the dinitrogenase reductase (Fe protein) accepts an electron from reduced ferredoxin (Fd) or flavodoxin, binds two molecules of MgATP and transfers the electron to dinitrogenase (MoFe protein). A single turn of the nitrogenase catalytic cycle requires eight such sequential transfers and hence the hydrolysis of 16 molecules of ATP to 16 molecules of ADP (see Equation 13.1). Electron transfer to dinitrogenase returns the Fe protein to the oxidized state ready for its next ATP-dependent redox cycle. The Fe protein undergoes marked conformational changes as ATP is bound and hydrolyzed.

The electron donor to dinitrogenase reductase is ferredoxin, or flavodoxin in some organisms (Figure 13.10). In many free-living diazotrophs, the enzyme pyruvate ferredoxin (flavodoxin) oxidoreductase can transfer electrons to oxidized ferredoxin

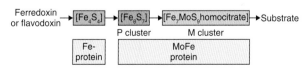

Figure 13.10   The sequence of electron transfer steps in the nitrogenase system. Fe-protein, dinitrogenase reductase; MoFe protein, dinitrogenase.

(flavodoxin) from α-keto acids such as pyruvate and α-ketoglutarate (Equation 13.2).

Equation 13.2 Pyruvate ferredoxin oxidoreductase

$$\text{Pyruvate} + \text{CoASH} + 2\text{Fd}_{ox} \rightleftharpoons$$
$$\text{acetyl-CoA} + \text{CO}_2 + 2\text{Fd}_{red} + 2\text{H}^+$$

Endosymbiotic rhizobia (see Table 12.2) may have alternative electron transport-coupled methods of generating the low-potential electrons required to reduce Fd. For example, Figure 13.11 shows the electron transport pathway from pyruvate to nitrogenase in *Azorhizobium caulinodans*, a diazotrophic α-proteobacterium that nodulates the tropical legume *Sesbania rostrata*. Here nitrogen fixation (which consumes ATP) and the cytochrome pathway (which produces ATP by oxidative phosphorylation) are in competition for electrons produced by the pyruvate dehydrogenase complex in the preparatory reaction (see Chapter 7).

Dinitrogenase uses the electrons donated by the Fe protein to convert N_2 and protons to NH_3 and H_2 in a reaction cycle with a number of intermediates. The enzyme (E) sequentially accepts up to four protons and four accompanying electrons from the Fe protein (Equation 13.3A). One molecule of N_2 reversibly displaces one molecule of H_2 from the EH complex (Equation 13.3B). The EN_2H_2 intermediate reacts with a further four protons and four electrons, releasing two molecules of NH_3 and regenerating the enzyme (Equation 13.3C).

Equation 13.3 Catalytic cycle of dinitrogenase

13.3A
$$E + 4H^+ + 4e^- \rightarrow EH_4$$

13.3B
$$EH_4 + N_2 \rightleftharpoons EN_2H_2 + H_2$$

13.3C
$$EN_2H_2 + 4H^+ + 4e^- \rightarrow 2NH_3 + E$$

The catalytic cycle does not always achieve the stoichiometry shown in Equation 13.3. Incomplete cycles consume reductant and ATP and reduce protons to H_2 gas. H_2 also inhibits N_2 reduction by competing with dinitrogen to occupy the nitrogenase active site. It is estimated that as much as 60% of energy supplied to nitrogenase may be diverted from N_2 fixation through competition with the reduction of H^+ to H_2. Certain *Rhizobium* species are able to recapture some of the lost energy through the activity of hydrogenase, an enzyme that recycles electrons by converting H_2 back to H^+.

Key points Nitrogen fixation is very energy demanding, with a $\Delta G^{o\prime}$ of -200 kJ mol^{-1}. A number of plant species can fix nitrogen by entering into symbiotic associations with diazotrophic microorganisms: legumes with *Rhizobium* bacteria, for example. Nitrogenase, the enzyme that reduces N_2 to NH_4^+ in a low-oxygen environment, is a complex consisting of dinitrogenase reductase (Fe protein) and dinitrogenase (MoFe protein). Dinitrogenase is a heterodimeric protein bound to an [Fe_8S_7] P cluster and an FeMo cofactor (M cluster). During the enzymatic N fixation cycle, dinitrogenase reductase hydrolyzes 16 molecules of ATP to ADP and sequentially transfers eight electrons from reduced ferredoxin or flavodoxin to dinitrogenase. Dinitrogenase uses the electrons to convert one molecule of N_2 to two molecules of ammonia. Incomplete cycles are energetically inefficient and produce dihydrogen (H_2) gas, which some *Rhizobium* species can convert back to protons via the enzyme hydrogenase.

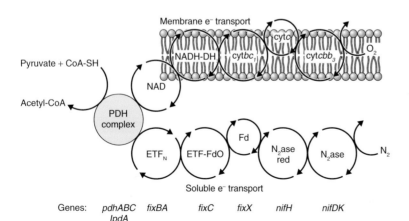

Genes: *pdhABC* *lpdA* *fixBA* *fixC* *fixX* *nifH* *nifDK*

Figure 13.11 Electron transfer pathways in support of N_2 fixation in *Azorhizobium caulinodans*. NAD$^+$ and electron-transferring flavoprotein (ETF$_N$) kinetically compete to reoxidize the pyruvate dehydrogenase (PDH) complex. Membrane electron transport (oxidative phosphorylation) yields ATP; soluble electron transport to N_2 consumes ATP. DH, dehydrogenase; ETF-FdO, ETF-ferredoxin oxidoreductase.

13.2.5 Uptake of ammonium into the symplasm occurs via specific membrane channels

We have seen that the major inorganic forms of nitrogen in the soil available to support plant growth are NO_3^- and NH_4^+. Roots take up these ions into the symplasm via specific transport proteins in the plasma membrane. In acidic soils where rates of nitrification are low, the availability of ammonium ions (NH_4^+) often greatly exceeds that of nitrate (NO_3^-). High ammonium levels will suppress the expression of genes for nitrate assimilation. The uptake of ammonium is multiphasic, indicating the presence of multiple transport systems. Measured K_m values for the transporters range from 10 to 70 μM with respect to NH_4^+ and to chemical analogs such as methylamine (CH_3NH_2). The structures and mechanism of one family of transporters, **AMT1**, have been inferred from studies of gene sequences and functional testing in *Xenopus* oocytes and other heterologous systems. The AMT1 proteins of plants are homologous to transporters from bacteria, mammals and other organisms and can complement NH_4^+ uptake mutations in yeast. They have multiple, putative membrane-spanning domains. There are five members of the *AMT1* family in the *Arabidopsis* genome; all five are expressed in roots, and two are expressed in shoots.

Electrophysiological measurements show that AMT1 proteins are NH_4^+ **channels** in the plasma membrane; NH_4^+ diffuses into the cytosol down an electrochemical gradient. A small proportion of cytosolic NH_4^+ may be deprotonated to NH_3 and moved across the tonoplast membrane into the vacuole (Figure 13.12) by an aquaporin-like tonoplast intrinsic protein (see Section 5.5.4). Excess ammonium as the exclusive N source can be toxic to the plant, resulting in reduced leaf expansion, impaired root growth and chlorosis (yellowing), probably as a consequence of acidification of the rhizosphere and disruption to acid–base balance and energy metabolism.

In nitrogen-fixing root nodules, ammonium is transported by a particular **nodulin** (a plant protein that is expressed in response to the formation of the symbiosis with diazotrophic bacteroids). **GmSAT1**, the ammonium transporter nodulin protein of soybean (*Glycine max*), has unusual features; for example, it has only one putative transmembrane domain, and its K_m for the ammonium analog methylamine is relatively high, about 5 mM, which is consistent with physiological studies of the host-derived membrane that surrounds the bacteroid.

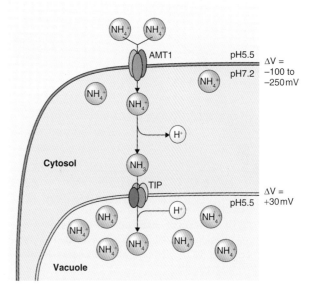

Figure 13.12   Uptake of NH_4^+ by the plasma membrane AMT1 transporter and transport across the vacuolar membrane by an aquaporin-like tonoplast intrinsic protein (TIP).

13.2.6 Roots take up nitrate in preference to other forms of nitrogen

Nitrate is more than a nutrient; it has potent morphogenetic effects. It acts as a signal both for modulating the expression of N metabolism genes and as part of large-scale mechanisms regulating growth rate, plant architecture, the interaction of carbon and nitrogen metabolism and responses to diverse environmental conditions. Given the choice, roots will normally take up NO_3^- in preference to NH_4^+, and will shut down the activity of AMT1 transporters. Re-supplying roots with NH_4NO_3 after a period of N depletion results in an increase in the endogenous pool of the N assimilation product glutamine, which in turn downregulates *AMT1* transcription. In the cytosol, nitrate is converted into ammonium, as a result of reactions catalyzed by nitrate reductase (NR) and nitrite reductase (NiR) (see Section 13.2.7) and then enters the organic N pool.

The uptake of soil nitrate by root cells occurs via **high (HATS)** and **low (LATS) affinity transport systems** located in the plasma membrane. HATS (sometimes also called system I) has a K_m for NO_3^- of 10–100 μM (Figure 13.13A). HATS has both a constitutive (**cHATS**) and a nitrate-inducible (**iHATS**) component. It has been suggested that the NO_3^--inducible element of HATS acts as a sensor for the availability of soil nitrate. LATS (system II) does not exhibit saturation (Michaelis–Menten type) uptake kinetics (Figure 13.13B). It operates up to and beyond 0.5 mM

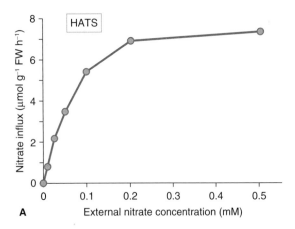
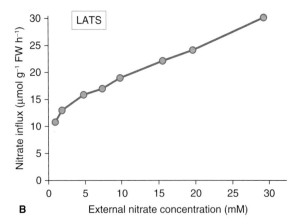

Figure 13.13   Kinetics of nitrate uptake. Nitrate influx was measured as a function of external nitrate concentration in barley roots that had been pretreated with 0.1 mM nitrate to induce the uptake system. (A) Uptake by the high-affinity system can be saturated at higher [NO$_3^-$], demonstrating Michaelis–Menten kinetics. (B) Uptake by the low-affinity system exhibits non-saturating kinetics. FW, fresh weight.

NO$_3^-$, at which concentrations HATS is saturated and the expression of its components is repressed. It is likely that plants in natural ecosystems, where soil nitrate availability is generally low, will tend to use HATS for uptake. LATS is more significant for agricultural species grown on well-fertilized soils and is consequently a focus of particular interest for research on nitrogen use efficiency, which is aimed at improving crop yield and quality and reducing eutrophication.

The transport systems deliver nitrate to the symplasm, where it may be reduced to ammonium, or may diffuse to the xylem to be transported to the shoot. Uptake of nitrate against the electrochemical gradient of the plasma membrane (−100 to −250 mV) is achieved by secondary active transport using an **electrogenic 2H$^+$/NO$_3^-$ symporter** which results in the import of a net positive charge (Figure 13.14). Evidence for this mechanism comes from observations of membrane depolarization when cells are exposed to nitrate, followed by repolarization as the plasma membrane H$^+$ pump reestablishes the electrochemical gradient. The two-phase electrophysiological behavior of cells in the presence of successively low and high NO$_3^-$ concentration indicates that this model provides a mechanism for active nitrate uptake by both LATS and HATS (Figure 13.15).

There are at least two families of **nitrate transporter genes**, **NRT1** and **NRT2**. Although *NRT1* and *NRT2* share little primary sequence homology, structural topologies of the corresponding proteins are similar, consisting of 12 transmembrane domains, distributed in two sets of six helices connected by a cytosolic loop. *NRT2* encodes the membrane-transporter and nitrate-sensor components of the inducible HATS. Activity of the NRT2 protein is controlled by feedback regulation of gene expression in response to reduced forms of N (ammonium, glutamine), membrane targeting through

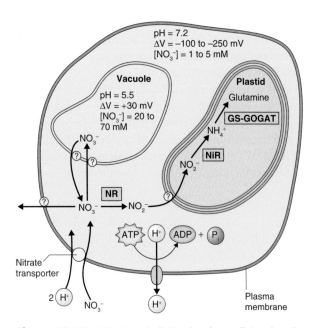

Figure 13.14   Nitrate assimilation by plant cells involves the transport of nitrate across the plasma membrane and then the reduction to ammonia in a two-step process. The plasma membrane H$^+$-ATPase maintains the electrochemical gradient that drives cellular uptake of nitrate. The values shown for electrical potentials and intracellular nitrate concentrations are typical but can vary significantly. GOGAT, glutamate synthase; GS, glutamine synthetase; NiR, nitrite reductase; NR, nitrate reductase.

interaction with a second component protein, and phosphorylation. There are seven members of the *NRT2* family in the *Arabidopsis* genome. *NRT2* is expressed predominantly in roots, with weak or no expression in the shoot. Levels of *NRT2* mRNA are highest in cells of the epidermis and endodermis near the root tip and decline sharply towards the base of the root axis.

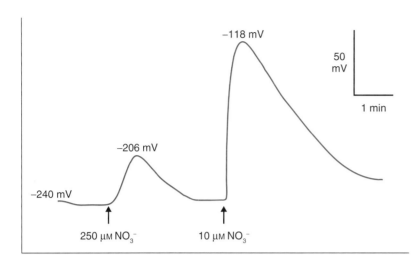

−118 mV

50 mV

1 min

−206 mV

−240 mV

250 μM NO$_3^-$ 10 μM NO$_3^-$

Figure 13.15 Membrane depolarization in response to nitrate. Potentials across the plasma membranes of root epidermal cells of *Arabidopsis* become more positive when plants are first exposed to nitrate. The first depolarization results primarily from uptake by the high-affinity system (measured at 250 μM nitrate), whereas the second depolarization reflects uptake by both the high- and low-affinity systems (measured at 10 mM nitrate). Plants were grown in the absence of nitrate before the measurements were taken.

Although the *NRT1* gene family is extensive—53 members have been identified in the *Arabidopsis* genome—the role of *NRT1*, and particularly its relationship to LATS, is complex and unresolved. The first *NRT1* was identified by screening *Arabidopsis* mutants for resistance to chlorate (ClO$_3^-$). Chlorate is a chemical analog of nitrate that can act as a substrate for nitrate transporters, and for nitrate reductase (NR) which reduces it to the toxic product chlorite (ClO$_2^-$). When a mutagenized population of *Arabidopsis* was treated with chlorate, a mutant, *chl1*, was isolated as a survivor (Figure 13.16). This mutant is deficient in nitrate uptake but has normal levels of NR activity. The *CHL1* gene of *Arabidopsis* is now called *AtNRT1.1*. Originally *AtNRT1.1* was thought to be a component of LATS, but its molecular and physiological properties do not meet expectations. Its expression is inducible by NO$_3^-$ and is localized to root tips; it is also found in stomatal guard cells where it has been shown to be sensitive to water stress. Some observations suggest that AtNRT1.1 protein may have a dual function, switching between high- and low-affinity transport ranges, possibly regulated by phosphorylation. Another member of the *NRT1* family in *Arabidopsis*, *AtNRT1.2*, is constitutively expressed in root epidermal cells and has a K$_m$ for NO$_3^-$ of around 6 mM, in the LATS range. Gene knockouts targeting *NRT1.1* or *NRT1.2* result in more than 45% decrease in LATS activity, but in neither case is the correlation of decreased transport activity with altered levels of gene transcripts strong, and uncertainty remains as to whether and to what extent these or other members of the *NRT1* family contribute to LATS. It is likely that additional or alternative candidates from different gene families remain to be characterized. In addition to their functions in NO$_3^-$ transport, NRT1 and NRT2 are important components of the signaling pathways that control developmental responses of roots to the plant's external and internal nitrogen relations (see Chapter 12).

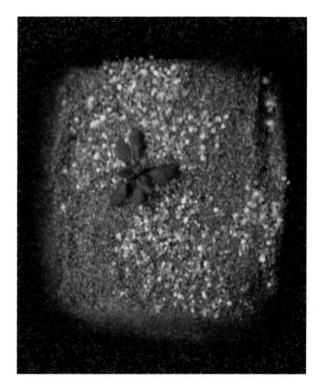

Figure 13.16 Nitrate-uptake mutants can be selected for by treating plants with chlorate. Wild-type plants take up chlorate and reduce it to the toxic product chlorite, which results in chlorosis. Nitrate-uptake mutants are unable to import chlorate and so remain green.

13.2.7 Nitrate reduction is the first step in nitrogen assimilation

Having entered a root cell, nitrate is available to the **assimilation pathway** which leads to organic nitrogen via nitrite and ammonia, with a change in the oxidation

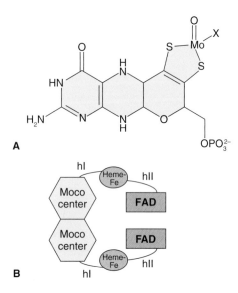

Figure 13.17 (A) Chemical structure of the molybdenum cofactor (Moco) of nitrate reductase. (B) Domain structure of nitrate reductase, showing the regions that bind Moco, heme and FAD, and the hinges hI and hII that connect them.

state of N from $+5$ to -3 (see Table 13.2). Most of the organic nitrogen in the biosphere has entered through the assimilatory nitrate reduction pathway of autotrophs. The reduction of NO_3^- to NH_4^+ requires eight electrons and is catalyzed by two enzymes, **nitrate reductase** (**NR**) and **nitrite reductase** (**NiR**). Nitrate reductase catalyzes the reduction of nitrate to nitrite, a reaction that involves the oxidation of one molecule of NADH or NADPH and the consumption of one proton per NO_3^- (Equation 13.4). NADH-specific NR is frequently present in land plants and algae and NADPH-specific forms of NR are unique to fungi. Forms able to use either NADH or NADPH are found in land plants, algae and (most commonly) fungi.

Equation 13.4 Nitrate reductase (NR)

$$NO_3^- + NAD(P)H + H^+ \rightarrow NO_2^- + NAD(P)^+ + H_2O$$

NR is an Mo-containing metalloenzyme, the active form of which is the homodimer. The structure of the Mo cofactor (**Moco**) consists of a molybdenum ion complexed with molybdopterin (Figure 13.17A). Each NR monomer is about 900 amino acids in length and consists of three regions (Figure 13.17B): the **Moco center** (which is subdivided into the N-terminal Moco-binding and C-terminal dimerization regions); the **cytochrome b_5-Fe heme domain**; and a C-terminal region associated with a **flavin adenine nucleotide (FAD) cofactor**. Protein structure studies show that the three functional domains are connected by two hinge regions,

hI and hII (Figure 13.17B). Moco, Fe heme and FAD function as redox centers in the sequence of electron transfer reactions that reduce NO_3^- to NO_2^-.

NR has a **three-stage catalytic cycle**: (i) a reductive half-reaction in which NAD(P)H reduces FAD; (ii) an electron transfer stage via the intermediate heme domain; and (iii) an oxidative half-reaction in which the Moco center transfers its electrons to nitrate and generates nitrite. This sequence of redox reactions is consistent with the observed progressive decrease in the negativity of midpoint potentials ($E^{\circ\prime}$) for FAD, heme and Moco from -272 to -160 to -10 mV respectively. The detailed reaction cycle in the Moco center that reduces nitrate to nitrite is shown in Figure 13.18. Molybdenum in the reduced form Mo(IV) (stage 1) reacts with NO_3^- in the active site (stage 2) to form OH^- and the reaction intermediate (stage 3). Oxidation of the Mo center to Mo(VI) breaks the O-N bond and releases NO_2^- (stage 4). Mo(IV) is regenerated by reduction with NAD(P)H (stages 5 and 6) and the cycle can continue.

The reduction of nitrite to ammonium during nitrogen assimilation is catalyzed by ferredoxin:nitrite oxidoreductase, commonly referred to as nitrite reductase. During the NiR reaction (Equation 13.5), six electrons are transferred from reduced ferredoxin ($\mathbf{Fd_{red}}$) to NO_2^-. Reduced ferredoxin is produced in plastids. In the colorless plastids of roots, the source of reductant is NADPH from the oxidative pentose phosphate pathway (see Chapter 7). In green tissues Fd_{red} is the product

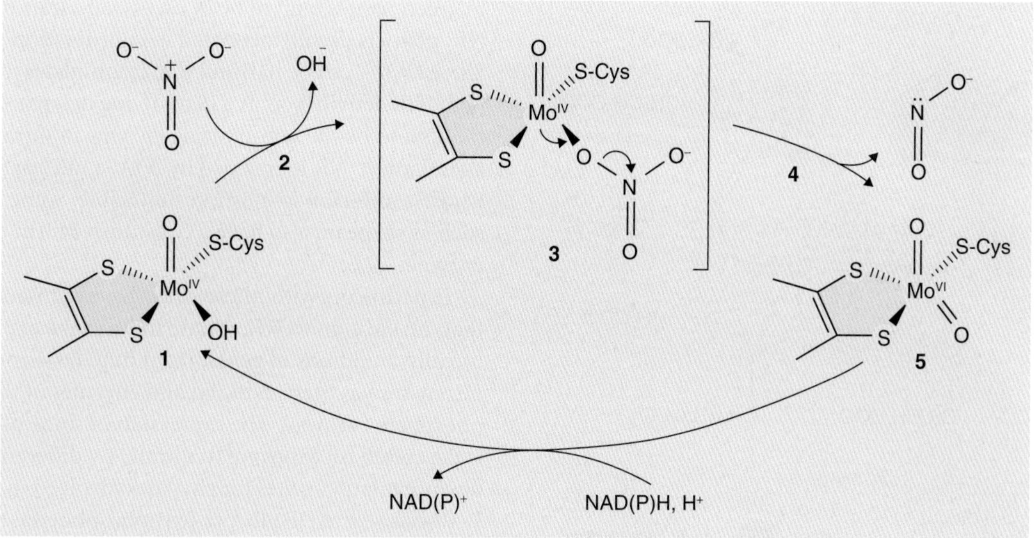

Figure 13.18 Proposed reaction cycle of nitrate reduction by nitrate reductase (NR). The reaction starts with the reduced Mo(IV) center (stage 1). Nitrate binds to the active site (stage 2), forming the reaction intermediate (stage 3) and releasing OH^-. Oxidation of the Mo center to Mo(VI) breaks the bond between the nitrate oxygen and nitrogen, releasing nitrite (stages 4 and 5). After completion of the reductive half-reaction, the Mo is regenerated [Mo(IV)] for the next cycle.

of non-cyclic photosynthetic electron transfer (see Chapter 9).

Equation 13.5 Nitrite reductase (NiR)

$$NO_2^- + 6Fd_{red} + 8H^+ \rightarrow NH_4^+ + 6Fd_{ox} + 2H_2O$$

The nitrite reductases of plants are soluble monomeric enzymes of approximately 65 kDa. The active site of NiR occupies the C-terminal half of the molecule and contains a **siroheme** (Figure 13.19A), coupled to an $[Fe_4S_4]$ center (Figure 13.19B). Four cysteines located in two clusters provide both the bridging ligand and the sulfur ligands for the $[Fe_4S_4]$ group. Site-directed mutagenesis to alter the amino acids at specific locations in the polypeptide chain has identified a number of ligands, including three lysine residues in the N-terminal region of the molecule, as important for ferredoxin binding. It is thought that the **NiR redox cycle** begins with the delivery of an electron to siroheme via the $[Fe_4S_4]$ center. Reduced (Fe^{2+}) siroheme rapidly binds nitrite. A second electron from reduced ferredoxin produces an Fe^{2+} siroheme–NO complex, with the iron–sulfur cluster in the $[Fe_4S_4]^{2+}$ state. Subsequent electron transport steps are unclear, though there is some evidence that hydroxylamine (NH_2OH, oxidation state -1; see Table 13.2) may be an intermediate. The pathway for the eight protons involved in the reaction is also undetermined. The final steps in nitrogen

assimilation, the incorporation of NH_4^+ into organic molecules via glutamine, are discussed in Section 13.2.9.

13.2.8 Nitrate reduction is regulated by controlling the synthesis and activity of nitrate reductase

Comparison of different species, and individuals of the same species growing in different environments, shows that plants employ diverse strategies for acquiring, reducing, storing and assimilating nitrate. Nitrate taken up may be reduced and assimilated in the root ($NO_3^- \rightarrow NH_4^+ \rightarrow$ glutamine) or it may be translocated as NO_3^- to the shoot in the xylem. Nitrate may also be stored in the vacuoles of root or shoot cells. Correspondingly, the distribution of NR and nitrate reduction varies between and within species, and even within different tissues in a particular organ. In some plants, for example cranberry (*Vaccinium macrocarpon*) and white clover (*Trifolium repens*), almost all the NR is localized in roots. In contrast the NR of other species, such as cocklebur (*Xanthium* sp.), is confined almost exclusively to leaf tissue. With increasing external NO_3^- concentration, the NR of cells at or near the root surface is often progressively supplemented by activity developing in deeper cortical tissues and the vascular

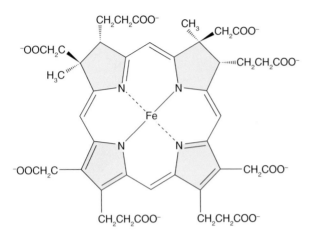

A   Siroheme

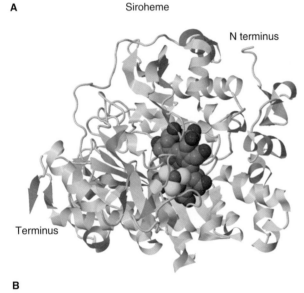

B

Figure 13.19   Nitrate reductase (NiR). (A) Structure of siroheme. (B) Model of spinach NiR, showing the protein backbone (silver), siroheme and iron–sulfur cluster (C atoms dark gray, O atoms red, N atoms blue, Fe atoms orange and S atoms yellow).

system. In C_4 plants (see Chapter 9), mesophyll cell photosynthesis has a greater capacity to generate reductant than that of bundle sheath cells, and in this case NR is found mostly in the mesophyll cells of leaves of maize (*Zea mays*) and other C_4 species.

NR is a highly regulated enzyme, sensitive to many different signals including nitrate abundance, ammonia, nitrogen metabolites (especially glutamine), CO_2, carbohydrates (especially sucrose), cytokinins and light. Long-term adjustments, over periods of hours to days, in NR activity are exerted at the level of gene expression while rapid alterations on the scale of minutes to hours are controlled by post-translational modulation of enzyme function and turnover. In contrast, levels of NiR

in green tissues tend to be in excess, and transcription of NiR genes is usually regulated in coordination with NR. Since NO_2^- can be harmful if it accumulates in the cell, high NiR activity acts as a detoxifying defense. NR is thought to be the rate-limiting enzyme in nitrate assimilation. NR is encoded by **NIA**, a nuclear gene whose expression is substrate-inducible. Some plants, such as soybean, also have a constitutive form of the enzyme.

Experiments with microarrays have demonstrated that, in addition to NR, nitrate induces the expression of, literally, hundreds of genes, including those encoding nitrate transporters, NiR, Fd and enzymes of the pentose phosphate pathway. The expression of some other genes is depressed on exposure to nitrate. By differentially downregulating genes for enzymes involved in starch synthesis (e.g. ADP glucose pyrophosphorylase; see Chapter 9) and upregulating genes for enzymes such as phosphoenolpyruvate (PEP) carboxylase (see Chapter 7), nitrate can bring about the redirection of carbon from carbohydrate storage into the production of amino acids and organic acids. The response to nitrate is detectable within minutes (Figure 13.20A) and is sensitive to concentrations down to micromolar levels. The induction of NR activity in the cytosol of green tissues is also under the influence of signals from the chloroplast. This ensures coordination with capacity of NiR in the plastid to prevent accumulation of excess NO_2^-.

NR induction requires light, acting not only via photosynthesis to provide reductant and carbon skeletons for nitrogen assimilation, but also through the signaling pathways of photomorphogenesis. NR induction in roots is supported by fixed carbon translocated from leaves. Etiolated shoot tissue will also induce NR synthesis in the dark if supplied with high concentrations of sugar. Once *NIA* is induced, the abundance of NR mRNA shows a pronounced **diel (day–night) rhythm** (Figure 13.20B) that is under the control of phytochrome and a downstream metabolite, probably the early product of nitrogen assimilation, glutamine (see Section 13.2.9). Over the daily cycle there is a reciprocal relationship between the concentration of glutamine and the abundance of NR mRNA in the leaves. Mutants with low NR have permanently low NR transcript and high glutamine levels. The diel cycle is in turn subject to the regulatory influence of **circadian clock genes** (see Chapter 8).

As well as modulation of NR activity through expression of its gene, the enzyme itself is controlled by phosphorylation and interaction with regulatory proteins (Figure 13.21). Such **post-translational mechanisms** work by directly modulating the catalytic activity of the enzyme and through the balance between synthesis and breakdown in protein turnover. They may operate over the long term, for example, bringing about

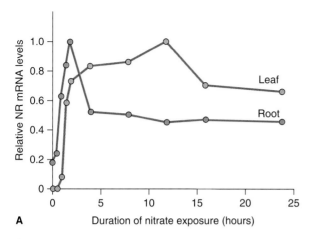
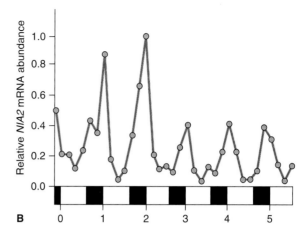

Figure 13.20   Regulation of nitrate reductase (NR) gene expression. (A) Transcription of genes encoding NR in roots and leaves, measured by the abundance of NR mRNA, is influenced by the availability of nitrate. Seven-day-old barley seedlings grown in the absence of nitrate were treated with 15 mM nitrate at time zero. RNA was extracted from roots and leaves at the indicated times, and relative NR mRNA levels were determined. (B) In plants grown in the presence of nitrate, *NIA2* mRNA concentrations demonstrate a diel cycle. RNA was extracted from leaves of tomato plants at the indicated times, and relative *NIA2* mRNA levels were determined.

a decrease in the amount of NR protein when plants are deprived of nitrogen or light for several days, even though the abundance of NR transcripts remains high. Regulation at the post-translational level may also be responsible for short-term and reversible responses such as the inhibition of NR activity in plants exposed to darkness or low CO_2 concentrations.

Inactivation of NR under such conditions is rapid (with a protein half-life of as little as 15 minutes) and requires ATP. The hinge1 region (hI), that joins the Moco and heme domains of the NR molecule (see Figure 13.17B), bears a regulatory Ser residue that undergoes ATP-dependent phosphorylation catalyzed by a specific enzyme, **NR-kinase**. Phosphorylation alone does not inhibit NR; an inactivator protein is also required. This was discovered to be a member of the **14-3-3 family** of highly conserved regulatory proteins that are known to function in cell cycle control and other processes (see Figure 5.12). The name 14-3-3 comes from the behavior of these proteins during purification. There are at least fifteen 14-3-3 genes or related pseudogenes in *Arabidopsis*. Dephosphorylation of NR by protein phosphatase 2A prevents 14-3-3 from binding to and inactivating the enzyme. Mechanisms of NR regulation through phosphorylation/dephosphorylation and interaction with 14-3-3 proteins are summarized in Figure 13.21.

The activity of NR-kinase is positively regulated by Ca^{2+} and inhibited by cytosolic hexose phosphates, levels of which are, in turn, directly attuned to photosynthesis and the export of fixed carbon from the chloroplast. Phospho-NR binds 14-3-3 in a Mg^{2+}- or Ca^{2+}-dependent fashion. It is believed that 14-3-3 attached to the hI hinge region inactivates NR by

blocking electron flow between the heme and Moco domains. Decreased enzymatic activity is accompanied by the loss of NR protein because the inactivated NR/14-3-3 complex is susceptible to attack by cytosolic proteases. The NR regulation system is also sensitive to adenine nucleotides, particularly 5′-AMP, which binds to 14-3-3 and disrupts the complex. The **kinase/14-3-3 post-translational regulatory mechanism** shown in Figure 13.21 permits the rapid and reversible inhibition of NR activity when conditions do not favor nitrate assimilation, such as a potentially harmful period of darkness.

13.2.9 Nitrogen enters into organic combination through the GS-GOGAT pathway

As we have seen, ammonium in cells may be the product of symbiotic nitrogen fixation, or the uptake and reduction of soil nitrate, or its release from organic combination during catabolism. It enters the amino acid pool and thence into metabolism in general through the operation of the **GS-GOGAT pathway** (Figure 13.22). **GS** is **glutamine synthetase; GOGAT** is **glutamate synthase**. The cycle supported by these key enzymes is the principal route of NH_4^+ assimilation in plants. GS catalyzes the ATP-dependent synthesis of glutamine from NH_4^+ and glutamate. Glutamine is one of the major forms in which organic nitrogen is translocated. Its amide group is available for donation to the amino acid pool through the activities of GOGAT and aminotransferases.

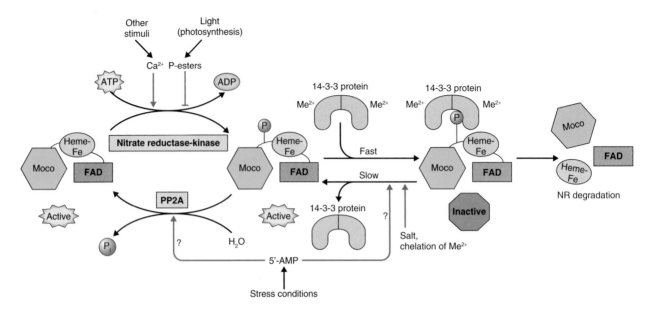

Figure 13.21   Proposed model for the regulation of nitrate reductase (NR) activity by phosphorylation/dephosphorylation and reversible binding of 14-3-3 protein. NR is phosphorylated by protein kinases on a serine residue in hinge 1. Phosphorylated NR is still active but is bound by 14-3-3 dimers, which inactivate NR. NR in 14-3-3 complexes is more rapidly degraded than free NR and is not an available substrate for phosphatases. If NR is released from 14-3-3 proteins, the regulatory phosphate can be removed by protein phosphatase PP2A. Me^{2+} equals Mg^{2+} or Ca^{2+}.

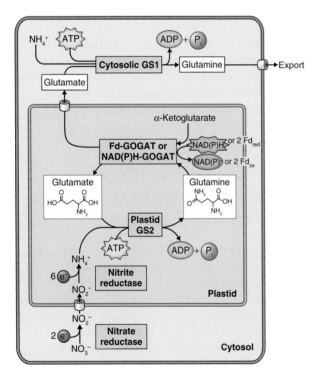

Figure 13.22   The glutamine synthetase–glutamate synthase (GS-GOGAT) pathway, the principal route of primary and secondary ammonium assimilation in plants. GS exists as a cytosolic (GS1) and a plastidic (GS2) isoform.

During the GS reaction, amidation of glutamate is accompanied by hydrolysis of ATP (Equation 13.6). The enzyme requires divalent cations such as Mg^{2+} and has a very high affinity for NH_4^+ (K_m 3–5 μM), allowing it to operate at the low ammonium concentrations present in living cells.

Equation 13.6 Glutamine synthetase (GS)

$$Glutamate + NH_4^+ + ATP \rightarrow glutamine + ADP + P_i + H^+$$

Active GS is a decamer, a protein with ten subunits. There are two isozymes of glutamine synthetase: a **cytosolic form**, **GS1**, and a **plastid-localized form, GS2**. There are up to five nuclear *GS1* genes; allelic variation in the *GS1* gene family may result in assembly of GS1 as a homo- or a heterodecamer. The protein crystal structure of cytosolic GS from *Zea mays* (Figure 13.23) shows that each decamer is composed of two face-to-face pentameric rings of subunits containing ten active sites, each site being formed between a pair of adjacent subunits. GS1 is more abundant in roots than in leaves, and is strongly expressed in phloem companion cells. This suggests that it plays a role in the primary assimilation of soil nitrogen and the synthesis of glutamine for long-distance transport. The plastid-localized isoenzyme, GS2, which is encoded by a single nuclear gene, is generally the predominant form in green tissues and functions both in primary assimilation and in the reassimilation of NH_4^+ released during

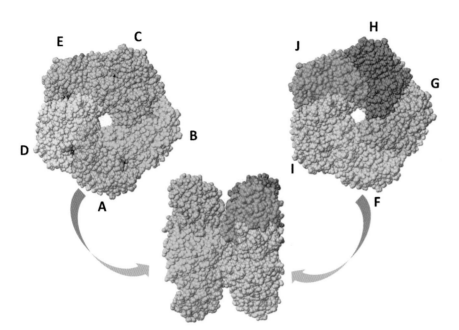

Figure 13.23 Model of the structure of the cytosolic glutamine synthetase of *Zea mays* determined by X-ray crystallography. The ten polypeptides of the holoenzyme (identified as subunits A to J) form two pentameric rings assembled into a face-to-face arrangement.

photorespiration (see Chapter 9). Based on DNA sequence comparisons, it is estimated that *GS1* and *GS2* are the products of an ancient gene duplication event.

GOGAT catalyzes the reductive transfer of the amide group of glutamine to the carbon skeleton of α-ketoglutarate, yielding two molecules of glutamate (Equation 13.7).

Equation 13.7 Glutamate synthase

13.7A Glutamate synthase (Fd-GOGAT)

$$\text{Glutamine} + \alpha\text{-ketoglutarate} + Fd_{red}$$
$$\rightarrow 2 \text{ glutamate} + Fd_{ox}$$

13.7B Glutamate synthase (NAD(P)H-GOGAT)

$$\text{Glutamine} + \alpha\text{-ketoglutarate} + NAD(P)H + H^+$$
$$\rightarrow 2 \text{ glutamate} + NAD(P)^+$$

The ferredoxin-dependent GOGAT (**Fd-GOGAT**) is found exclusively in photosynthetic organisms (Equation 13.7A). Plants and bacteria also have an **NAD(P)H-dependent GOGAT** activity (Equation 13.7B). The Fd-dependent enzyme is the predominant form of glutamate synthase in *Arabidopsis*, accounting for up to 96% of the total GOGAT activity in leaves and 68% in roots. In this species and in rice (*Oryza sativa*) there is a light-induced Fd-GOGAT that is predominantly expressed in leaves. A separate, constitutively expressed form is found in roots, where it probably functions in primary nitrogen assimilation. The enzyme in maize, by contrast, appears to be encoded by a single gene. Fd-GOGAT is localized in the chloroplasts of mesophyll

tissue and the plastids of root cells. There are reports of the presence of NAD(P)H-GOGAT in the vascular tissues of rice seedlings, where it probably functions in the reutilization of glutamine that has been exported from senescing tissues (see Chapter 18) and roots.

Before the discovery of the GS-GOGAT cycle, the route for assimilation of NH_4^+ in plants was thought to be via **glutamate dehydrogenase** (**GDH**). GDH is a ubiquitous enzyme that reversibly catalyzes the deamination of glutamate (Equation 13.8). The NADH-dependent form of GDH is found in the mitochondria and NADPH-dependent GDH is localized in the chloroplast. It is now thought that the main role of the enzyme is in glutamate catabolism.

Equation 13.8 Glutamate dehydrogenase (GDH)

$$\text{Glutamate} + NAD(P)^+ + H_2O \rightleftharpoons$$
$$NH_4^+ + \alpha\text{-ketoglutarate} + NAD(P)H$$

From glutamate, the product of ammonium assimilation via GS-GOGAT, nitrogen is made available for a wide range of metabolic processes through transamination and other reactions that make R-NH$_x$ (see Table 13.2) available for participation in biosynthesis and cell function.

13.3 Phosphorus

Elemental phosphorus does not occur in free form on Earth. Although P can react with other elements to make hydrides, halides, sulfides and metal phosphides, its

Key points Enzymatic reduction of NO_3^- to NH_4^+ via NO_2^- is catalyzed by nitrate reductase (NR) and nitrite reductase (NiR). NR, which reduces each mole of NO_3^- to NO_2^- with the oxidation of 1 mole of NAD(P)H and consumption of one H^+ in a three-stage catalytic cycle, is a homodimeric enzyme with an Mo-containing prosthetic group, a cytochrome b_5-Fe heme domain and an FAD cofactor. NiR, a monomeric enzyme containing siroheme and a $[Fe_4S_4]$ center, reduces NO_2^- to NH_4^+ with the transfer of six electrons from reduced ferredoxin. Expression of the NR gene, *NIA*, is subject to long-term regulation. In the short term, NR activity is controlled post-translationally, particularly through phosphorylation, interaction with 14-3-3-type regulatory proteins, and targeted proteolysis. Ammonium enters into organic combination via the glutamine synthetase–glutamate synthase (GS-GOGAT) pathway. GS is a decameric enzyme that catalyzes the ATP-dependent formation of glutamine from glutamate and ammonium. GOGAT transfers the amide group to α-ketoglutarate to form glutamate, with the reduction of ferredoxin (green tissues) or NAD(P)H (non-green tissues).

natural state is in combination with oxygen as **phosphate** (PO_4^{3-}, HPO_3^{2-}, $H_2PO_3^-$). Phosphates account for mineral deposits, inorganic and organic P in soils and water, and the various forms of P in living organisms. Phosphate (P_i) has critical functions in gene structure and function, as a component of nucleic acid molecules, and in bioenergetics in the form of the phosphoester and diphosphate bonds of high-energy compounds (see Chapter 2). It is both a substrate and a regulatory factor in photosynthesis and oxidative metabolism, participates in signal transduction pathways by way of covalent phosphorylation/dephosphorylation reactions and, in the form of phospholipids, plays a central role in membrane biochemistry.

Phosphates are salts of phosphoric acid H_3PO_4, which dissociates to yield up to three protons (Equation 13.9). Table 13.4 shows the pK_a values for each of these equilibria. For a typical acid $HA \rightleftharpoons H^+ + A^-$ the pK_a is

Table 13.4 pK_a values for phosphates.

Equilibrium	pK_a value
$H_3PO_4 \rightleftharpoons H_2PO_4^- + H^+$	$pK_{a1} = 2.15$
$H_2PO_4^- \rightleftharpoons HPO_4^{2-} + H^+$	$pK_{a2} = 7.20$
$HPO_4^{2-} \rightleftharpoons PO_4^{3-} + H^+$	$pK_{a3} = 12.37$

defined as $-\log_{10}$ of the dissociation constant K_a, which in turn is given by Equation 13.10.

Equation 13.9 Ionization of phosphoric acid

$$H_3PO_4 \rightleftharpoons H^+ + H_2PO_4^- \rightleftharpoons 2H^+ + HPO_4^{2-}$$
$$\rightleftharpoons 3H^+ + PO_4^{3-}$$

Equation 13.10 Dissociation constant K_a of a typical acid HA

$$K_a = [H^+][A^-]/[HA]$$

This tells us that at physiological pH (around 6–8) P_i in the cell is an equilibrium between the dihydro- and monohydro forms ($pK_{a2} = 7.2$; Table 13.4). With the exception of sodium, potassium, ammonium and lithium, most cations form virtually insoluble salts with phosphate. Thus in soils that are high in Fe and Al, much of the phosphate is unavailable for plant growth. Phosphate assimilation in plants differs from the reductive mode of nitrate and sulfate assimilation in that P remains in its oxidized state, entering into organic combination in the form of phosphate esters.

13.3.1 Phosphorus enters the biosphere as phosphate

Weathering, solubilization, sequestration, leaching and precipitation are stages in the phosphorus cycle. Unlike the nitrogen and sulfur cycles, the phosphorus biogeochemical cycle has no gas phase and, apart from small amounts in dusts (Aeolian P) and in acid rain as phosphoric acid, atmospheric P is of minuscule significance in the overall balance. By far the largest reservoir of phosphorus is in sedimentary rock (Table 13.5). Mineral P in sedimentary rocks such as apatite ($Ca_x(OH)_y(PO_4)_z$), and in biological sources such as bones and teeth, must be solubilized by weathering before it can be taken up by plants. Phosphates are also added to the soil in chemical fertilizers.

The uptake of soil phosphate by plants is the basis of a terrestrial food chain eventually leading to herbivores and carnivores. P is returned to the soil in the form of urine and feces, as well as by decomposition of plants and animals after death (Figure 13.24). Up to 80% of soil P may be in organic combination. In certain soils as much as half the organic P may be in the relatively unreactive form of **phytate** (myo-inositol hexaphosphates; see Chapter 6). Annual flux through the terrestrial biosphere is about 6×10^{10} kg. The flux through rivers into lakes and oceans is about a third of this (Table 13.5) and supports the aquatic food chain.

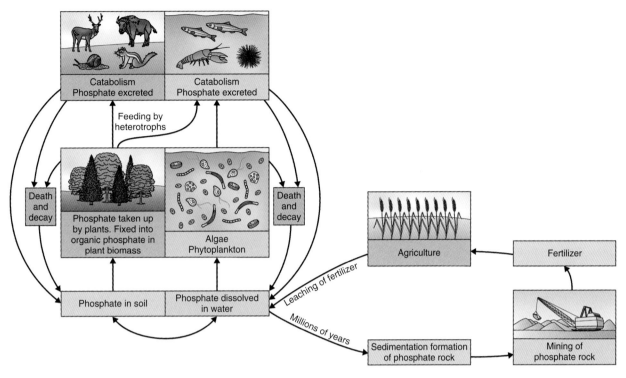

Figure 13.24 The phosphorus cycle.

Table 13.5 Global estimates of the amounts and kinetics of different forms of phosphorus (P).

Phosphorus reservoir	Amount (kg)	Flux (kg yr^{-1})
Sediments and rocks	4×10^{18}	
Mineable P	1×10^{13}	12×10^9
Terrestrial soils	2×10^{14} ⎫	Terrestrial cycle:
Terrestrial biota	3×10^{12} ⎭	60×10^9
Riverine input to the sea[a]		21×10^9
Aeolian P (dusts deposited by wind)		1×10^9
Seawater and biota	9×10^{13}	Marine cycle: 1×10^{12}

[a]Comprising 19×10^9 kg yr^{-1} bound to particles and 2×10^9 kg yr^{-1} free and reactive. About half is contributed by manmade eutrophication.

There is a general trend towards loss of phosphorus from the biosphere by the almost exclusively one-way discharge from rivers to the sea (Table 13.5) where it precipitates as insoluble calcium phosphate and may fall into the abyssal zone. At current flux rates, global phosphorus reserves are predicted to reach half-depletion midway through the present century. A small amount of marine P is returned to the land in the form of guano (the feces of seabirds). It has been estimated that the population of fish-eating birds on the islands off the coast of Peru, historically the principal source of fertilizer-grade material, produces about 11 000 tons of guano a year.

Plants take up P as soluble inorganic phosphate, designated P_i; organic P is not directly usable for plant growth. Soil organic matter releases P_i through the process of **mineralization**. The P accessible to support plant growth can be limited because soil phosphate is readily leached, precipitated, adsorbed by organic and mineral materials, made unavailable by drought or rapidly depleted from the rhizosphere by roots. Compared with those for other nutrients, diffusion coefficients for P_i in soil are typically very low, in the range 0.3–3.3×10^{-13} m^2 s^{-1}, and even lower when the soil is dry. To obtain otherwise inaccessible P from the soil, roots may modify their structure and function, or manipulate the chemistry of the rhizosphere, or enter into mutualistic associations with mycorrhizal fungi (see Section 13.3.3; also see Chapters 12 and 15).

13.3.2 Phosphate is actively accumulated by root cells

The concentration of P_i in root cells is in the millimolar range, whereas concentrations in soil are often 1 µM or less. In addition the membrane potential across the plasma membrane is very negative. Root cells must therefore import phosphate against an electrochemical gradient, requiring an input of energy equivalent to at least 1 mol ATP hydrolyzed per mole of $H_2PO_4^-$ taken up. Three types of P_i transporter have been identified in plants: **PHT1, PHT2** and **PHT3**. P_i uptake by roots is particularly dependent on high-affinity transporters of the PHT1 group. There are nine members of the *PHT1* gene family in *Arabidopsis*. Of these, *AtPHT1.1* and *AtPHT1.4* have been shown to encode two major transporters involved in P_i uptake by roots, while the other family members are thought to be expressed in different tissues. *PHT1* homologs have been identified in a wide range of species. Some members of the *PHT1* family function in P_i transport associated with mycorrhizal associations (see Section 13.3.3). PHT1-type transporters share common structural features and mechanisms. The PHT1 polypeptide has 12 membrane-spanning domains arranged in two (N- and C-terminal) halves connected with a hydrophilic loop (Figure 13.25A). Motifs in the primary structure indicate putative sites for post-translational modification by myristoylation (addition of myristic acid, a C_{14} fatty acid), phosphorylation and glycosylation. The signature amino acid sequence, GGDYPLSATIxSE, within transmembrane region 4 is conserved among all plant PHT1 transporters (Figure 13.25A).

Uptake of P_i via PHT1 is accompanied by a transient depolarization of the cell membrane followed by a repolarization (Figure 13.26). P_i uptake acidifies the cytoplasm, which stimulates the **plasma membrane H^+-ATPase** that repolarizes the membrane. Uptake is abolished by chemical treatments that collapse transmembrane proton gradients. Such observations show that PHT1 is a high-affinity **electrogenic P_i/H^+ symporter** (K_m 1–5 μM) that transports two or more protons per mole of $H_2PO_4^-$. Figure 13.25B shows a proposed six-stage mechanism for the transport of H^+ and $H_2PO_4^-$ by PHT1: (1) a transporter with a hydrophilic pore opening outside is protonated. The phosphate anion binds (2) and a conformational change leads to the opening of the pore inside (3). Phosphate is released (4), the transporter is deprotonated (5) and the outward face conformation is re-established (6). The magnitude of the change in membrane potential when P_i is added is small in P-sufficient plants and progressively larger with increasing P starvation (Figure 13.26), indicating that the activity of PHT1 is regulated according to the degree of P sufficiency or deficiency.

In addition to the high-affinity transporters of the PHT1 type, two groups of low-affinity transporters have been identified, which are mainly implicated in intracellular transfer of P_i across membranes. In *Arabidopsis*, expression of a chloroplast-associated PHT2

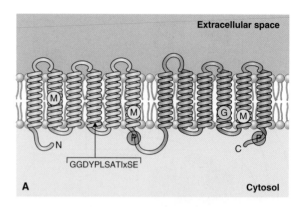
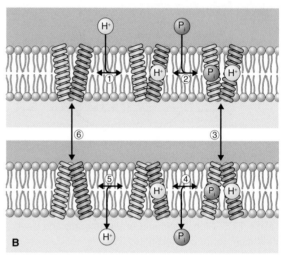

Figure 13.25 (A) Predicted transmembrane topology of PHT1 transporters. The amino acid sequence GGDYPLSATIxSE is conserved among all plant PHT1 transporters. M, G and P indicate myristoylation, glycosylation and phosphorylation sites respectively. (B) Putative mechanism of proton and phosphate symport through a PHT1 transporter. For an explanation of steps 1–6, see the text.

protein has been related to P_i distribution in leaves and cell wall metabolism in young tissues. Members of the PHT3 family are involved in mitochondrial phosphate transport. Other genes and proteins function in the uploading and distribution of P_i through the vascular system, and in vacuolar import and export of P_i, but these are less well characterized.

When the P supply is sufficient, more than 85% of the P_i in the plant may be stored in vacuoles. Under conditions of P deficiency, homeostasis is maintained by mobilizing stored vacuolar P_i. Until this reservoir is drained, the plant does not adjust its P import capacity. As phosphate starvation develops, the ability of a root to take up P_i is increased as a result of induced expression of high-affinity P_i transporters (Figure 13.27). P deficiency also induces the activities of several **acid phosphatases** (**APases**) and **ribonucleases** (**RNases**), thereby accelerating the salvage of P from old tissues and

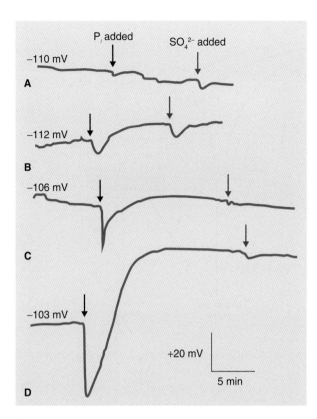

Figure 13.26 Phosphate-induced transients in transmembrane electrical potential of *Trifolium repens* roots. The solid arrows indicate the addition of 125 μM phosphate to media buffered at pH 5 and the resulting change in the root membrane potential. The red arrows indicate the addition of 63 μM sulfate as a control for the general effect of mineral anions on the membrane potential. (A) Plant grown on P_i for the entire 29-day period of the experiment. (B) P_i withheld for the final day of growth. (C) P_i withheld for the last 7 days of growth. (D) P_i withheld for the last 14 days of growth.

the movement of P_i to sites of new growth and development. Expression of *PHT1* in P-starved plants occurs in the root epidermis and the protein is localized to root epidermal cell plasma membranes. The response of *PHT1* gene expression to P limitation is complex, but at least in part it is related to the presence of a nucleotide sequence in the *PHT1* promoter region that recognizes the MYB-type transcription factor, **PHR1** (Figure 13.27). The PHR recognition sequence is also present in the promoters of a number of APase genes, suggesting that PHR is a component of a general regulatory mechanism invoked by P starvation (Figure 13.27). Other, less well-characterized transcription factors participate in the positive and negative regulation of genes for P transport and remobilization, and there is evidence for the operation of post-transcriptional mechanisms.

The P-starvation syndrome is a response not just to local conditions around the root, but to the P status of the whole plant. The nature of the regulatory signal that

connects the expression of P_i transporters and other P-sensitive genes to the internal P concentration has been examined in split-root experiments. Half of the root system of a tomato plant was deprived of P and the other half was exposed to sufficient P. *PHT* gene expression was found to be equal in the +P and −P roots, and intermediate between that of entire P-sufficient and P-starved roots. Such studies lead to the conclusion that transcriptional signals regulating the P-starvation response originate in, and reflect the P status of, the shoot. Thus P homeostasis at the level of the whole plant is maintained by the flexibly-regulated coordination of P_i acquisition in response to P_i supply and the state of the internal phosphorus economy.

13.3.3 Plants modify the rhizosphere and form mycorrhizal associations to improve phosphorus availability

We have seen that the availability to plants of soil P is often limited as a consequence of the low solubility of P_i in the presence of metals such as Fe and Al, oxides of which are common in clay soils, and Ca, abundant as $CaCO_3$ in limed or calcareous soils. P is also rendered unavailable when it is converted to organic forms by soil microbes. Plants have evolved a variety of strategies to mine otherwise unavailable reserves of soil P. The upregulation of P_i transporters in response to P depletion has been discussed above. Developmental changes to root architecture and function are considered in Chapter 12. Here we examine **rhizosphere processes** that release inaccessible inorganic and organic P, and the mycorrhizal associations formed between roots and some soil fungi to acquire P in non-rhizosphere soil. Different plant species, and genotypes within species, vary in the P-extraction strategies they deploy. A major priority for crop breeders is to use traditional and molecular approaches to improve phosphorus utilization efficiency by modifying the capacity of crop plants to mine soil P.

Several types of root exudate have been shown to enhance the uptake of soil P. In dry desert soils, some plants are able to release water from the roots at night, thereby improving P_i diffusion. But the commonest way in which plants modify the rhizosphere to increase P_i accessibility is through the secretion of **carboxylic acids** (citrate, malate, malonate) and **phosphatases** (Figure 13.28). The exudation of carboxylates is usually a response to P deficiency, but in some species such as *Cicer arietinum* (garbanzo, chickpea), the release of carboxylates appears to be constitutive. It is estimated that to meet its requirements for P, white lupine

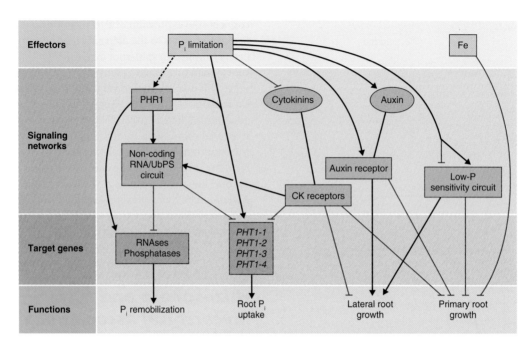

Figure 13.27 Overview of the P-dependent signaling cascades impacting on root P_i transport capacity, P_i remobilization and root system morphology. CK, cytokinin; UbPS, ubiquitin–proteasome system.

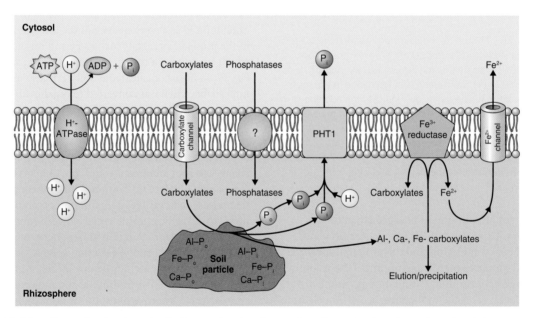

Figure 13.28 Effects of carboxylates (and other exudates) on inorganic and organic phosphorus (P_i, P_o) mobilization in soil. Carboxylates are released via an anion channel. It is not known how phosphatases are released. Carboxylates mobilize both P_i and P_o. Phosphatases hydrolyse P_o compounds, once these have been mobilized by carboxylates. Carboxylates will also mobilize a fraction of the cations that bind P, some of which (especially Fe) move to the root surface for uptake by the roots. Others move down the soil profile, where they are thought to form precipitates.

(*Lupinus albus*) releases up to 25% of its total fixed C as organic acids into the soil. Carboxylates are chelators, that is, they are polyanions that bind cations and prevent them from forming precipitates. By chelating metal cations that bind phosphate, carboxylates displace both inorganic and organic P from the soil matrix, making it

available to the plant. In addition, some chelated Fe may move to the root surface and be taken up by the plasma membrane Fe transport system (Figure 13.28; see Section 13.6.1). Carboxylate export occurs via an anion channel and is often, but not invariably, accompanied by acidification of the rhizosphere (Figure 13.28). Secretion of organic acids in response to P deficiency is associated with coordinated changes in root metabolism. The activities of tricarboxylic acid cycle enzymes such as **citrate synthase** and **malate dehydrogenase**, and of the glycolytic enzyme **PEP carboxylase**, are enhanced, as is expression of the genes encoding them.

Soil organic P is primarily in the form of esters of phosphate, including phytate; after these are rendered soluble by carboxylates, they must be hydrolyzed to become available for uptake by the plant. Acid phosphatases, released by roots in response to P starvation (Figure 13.28), can hydrolyze a range of organic P compounds. The roots of some species also release significant amounts of phytases.

More than 80% of all vascular plant species have the capacity to form mycorrhizal associations between their roots and fungal symbionts (see Chapter 12). Such symbioses benefit the plant predominantly by increasing the uptake of immobile nutrients, notably P_i, from the soil. P uptake by the fungal hypha occurs via high-affinity P_i/H^+ symporters that are functionally, topologically and phylogenetically related to the PHT transporter family of plants (see Figure 13.25). Transporters isolated from mycorrhizal *Glomus* spp. are expressed in the fungal plasma membrane and have a K_m for P_i of $8-18$ μM. Absorbed phosphate is incorporated into fungal energy metabolism, nucleic acids and phospholipids or condensed into polyphosphate, the storage form of P in fungi. After translocation within the fungal hyphae, which form arbuscules within root cells (see Figure 12.33), phosphate is transferred to the host plant by a process that has not been fully clarified. Associated with the activation of fungal P_i uptake, and possibly of the fungus-to-host P transfer system, is a major downregulation of the roots' high-affinity P_i transporters and of P-starvation-related genes such as APase, possibly reflecting improved plant P status.

13.4 Sulfur

Elemental sulfur occurs usually in areas of volcanic activity (where it was known by the old name brimstone), or where sulfate minerals have been reduced by anaerobic bacteria. Reduced inorganic sulfur takes the form of gaseous hydrogen sulfide, H_2S, and the sulfide anion S^{2-}. Oxidized, inorganic sulfur exists as gaseous sulfur dioxide, SO_2, and the anions sulfite, SO_3^{2-}, and sulfate, SO_4^{2-}. Among other inorganic S compounds are metabisulfites ($S_2O_5^{2-}$), thiosulfates ($S_2O_3^{2-}$), dithionites and dithionates ($S_2O_4^{2-}$, $S_2O_6^{2-}$) and thiocyanates (SCN^-). Estimates of the global reservoirs of sulfur are given in Table 13.6.

Organic S takes a variety of forms related to its role in vivo and to pathological or postmortem chemical transformations (Table 13.7). Functionally important sulfur moieties occur in amino acids and proteins, and in coenzymes and vitamins such as coenzyme A, *S*-adenosylmethionine, thiamine, biotin and

Key points Phosphorus exists in living organisms, and in mineral deposits, soils and water, as phosphate. Plants take up P as inorganic phosphate (P_i). At physiological pH, P_i in the cell is in equilibrium between the $H_2PO_4^-$ and HPO_4^{2-} ionic forms. P_i is released from organic matter into soil by mineralization. Uptake of P_i by roots is mainly via the PHT1 high-affinity electrogenic P_i/H^+ symporter. PHT2 and PHT3 are low-affinity transporters that transfer P_i across membranes within cells. P starvation depletes intracellular P_i stored in vacuoles before inducing expression of high-affinity transporters, acid phosphatases and RNases under the influence of the transcription factor PHR1. Root exudates containing carboxylic acids, phosphatases and phytases enable plants to access insoluble P_i in clay or calcareous soils. Symbioses with mycorrhizal fungi improve the uptake of immobile P_i. The transporters of mycorrhizal fungi are of the PHT-like high-affinity P_i/H^+ symporter type.

Table 13.6 Estimated global reservoirs of sulfur.[a]

Source	Amount (kg)
Sediments and rocks	10^{19}
Atmosphere	3×10^9
Inorganic S	
Aquatic	1×10^{18}
Terrestrial	NA
Organic S	
Aquatic	4×10^4
Terrestrial	$6-10 \times 10^{12}$

[a]Fluxes of sulfur are not as well understood as those of nitrogen and phosphorus. Estimates of total atmospheric flux range from 144 to 365×10^9 kg yr^{-1}. The annual flux from continents to oceans has been variously estimated to be in the range 8 to 100×10^9 kg and from oceans to continents as 4 to 20×10^9 kg. NA, not available.

Table 13.7 Structures and examples of sulfur-containing compounds formed by plants.

Compound	Generic structure	Examples
Thiols (mercaptans)	RSH	L-cysteine, coenzyme A
Sulfides or thioethers	R_1SR_2	Hydrogen sulfide (H_2S) L-methionine
Sulfoxides	R_1SOR_2	Allicin
Methylsulfonium compounds	$(CH_3)_2S^+R$	S-Adenosyl-L-methionine, S-methylmethionine, DMSP, dimethylsulfonic hydroxybutyrate
Sulfate esters	$$R-O-\overset{\displaystyle O}{\underset{\displaystyle O}{\overset{\|}{\underset{\|}{S}}}}-O^-$$	Phenol sulfates, polysaccharide sulfates
Sulfamates	$$R=N-O-\overset{\displaystyle O}{\underset{\displaystyle O}{\overset{\|}{\underset{\|}{S}}}}-O^-$$	Aryl sulfamates, mustard oil glycosides
Sulfonic acids	$$R-O-\overset{\displaystyle O}{\underset{\displaystyle O}{\overset{\|}{\underset{\|}{S}}}}-O^-$$	Glucose-6-sulfonate, cysteic acid, taurine, sulfoquinovosyl diacylglycerol

DMSP, dimethyl sulfoniopropionate.

methylmethionine. Other forms of organic sulfur in plants include sulfolipids, phytoalexins, redox compounds and many flavors and fragrances (either attractive or repulsive) such as the diallyls of onion and garlic (*Allium* spp.) and the glucosinolates of brassicas and other crucifers. Sulfur compounds of biological origin have been proposed to be of special significance for global climate change. According to the so-called CLAW hypothesis (named for the scientists who first advanced it in 1987: Charlson, Lovelock, Andreae and Warren), although burning of fossil fuels is the greatest contributor to atmospheric sulfur, marine algae are responsible for most of the biogenic component. As a defense against various environmental stresses, phytoplankton produce large quantities of the sulfur compound **dimethyl sulfoniopropionate** (**DMSP**) (Figure 13.29). DMSP released from planktonic algae is degraded by complex physical and microbial processes in the upper ocean to **dimethyl sulfide** (**DMS**), which is volatilized into the atmosphere and subsequently oxidized to dimethyl sulfoxide, sulfite and sulfate. Around 3×10^7 kg S per year is estimated to move from the ocean to the atmosphere by this pathway. The oxidation of DMS in the troposphere leads to the formation of sulfate aerosols that act as cloud condensation nuclei. Increased cloud cover in turn increases the Earth's **albedo** (the extent to which incoming solar radiation is reflected back into space) thereby, according to the CLAW hypothesis, representing a negative feedback loop that acts to stabilize the temperature of the Earth's atmosphere.

13.4.1 The sulfur cycle involves the interconversion of oxidized and reduced sulfur species

The **biogeochemical sulfur cycle** centers on the interconversion of oxidized and reduced sulfur species (Figure 13.30). Plants and microorganisms can assimilate sulfur by reducing sulfate and synthesizing the S-containing amino acid cysteine and other organic sulfur compounds. Microbial SO_4^{2-} reduction may also occur when facultatively anaerobic bacteria such as *Pseudomonas* and *Salmonella* use sulfate as a respiratory electron acceptor in place of oxygen and reduce it to sulfur. Sulfur is subsequently further reduced to H_2S by *Desulfovibrio*, *Desulfomonas* and other anaerobes. In nature, sulfate reduction takes place predominantly by this pathway.

In the oxidative phase of the sulfur cycle, SO_4^{2-} is regenerated from S and H_2S (Figure 13.30). Biological oxidation of reduced sulfur to sulfate is carried out by a range of organisms including chemoautotrophic bacteria,

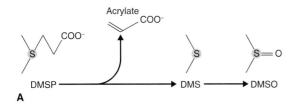

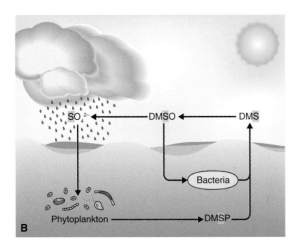

Figure 13.29 The phytoplankton–climate connection. (A) Formation of dimethyl sulfide (DMS) from dimethylsulfoniopropionate (DMSP) and its oxidation to dimethylsulfoxide (DMSO). (B) Phytoplankton-produced DMSP is broken down by bacteria to DMS and acrylate. DMS volatilizes and is oxidized to DMSO and to sulfate, which nucleates water droplets, leading to cloud formation. Sulfate is returned to the sea dissolved in rain. Because cloud cover reduces the growth of phytoplankton and is accompanied by atmospheric cooling, it has been proposed that phytoplankton serve as a homeostatic climate regulation mechanism.

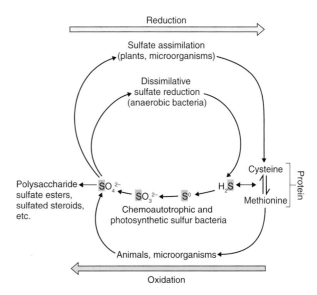

Figure 13.30 The biogeochemical sulfur cycle. Sulfate is reduced by sulfur-assimilating organisms, which use it for synthesis of cysteine and other organic sulfur compounds, and by dissimilators, anaerobic bacteria that use sulfate as a respiratory electron acceptor in place of oxygen. Many organisms oxidize reduced sulfur to sulfate. Chemoautotrophic bacteria extract electrons for energy, whereas phototrophic bacteria use the electrons for photosynthesis. Reduced sulfur is also oxidized geochemically when oxygen is present.

which extract electrons for energy, and phototrophic bacteria, which use the electrons for photosynthesis. Aerobes (e.g. *Thiobacillus*) and anaerobes such as *Chlorobium* oxidize sulfide to sulfur and then oxidize it further to sulfate. At neutral pH sulfide will also undergo spontaneous geochemical oxidation to gaseous sulfur dioxide, SO_2. Another important source of atmospheric SO_2 is the combustion of fossil fuels. Leaves are able to take up and assimilate SO_2 and this can be a significant source of S for plant growth in areas with high levels of air pollution. More generally, SO_2 re-enters the biological cycle in water and soil as the dissolved form SO_3^{2-}.

13.4.2 Plants acquire sulfur mainly as sulfate from the soil

Plants acquire sulfur mainly by the uptake of sulfate through the roots. Although S is not generally a growth-limiting nutrient because it is relatively abundant in the environment, its uptake and assimilation are tightly regulated to maintain coordination with sulfate supply, growth requirements and nitrogen assimilation. As with other anionic macronutrients, sulfate is actively accumulated by root cells (Figure 13.31). Although some reduction and assimilation of sulfate takes place in root plastids, most of the processing of, and demand for, S is in the shoot, where the chloroplasts of leaf cells are sites of light-driven assimilation of SO_4^{2-} into cysteine, glutathione and other metabolites (Figure 13.31). Plastids are known to contain the entire biosynthetic pathway leading from inorganic sulfate to cysteine. Like other mineral ions, sulfate is transported from the roots in the xylem. In leaves it enters mesophyll cells and is transported across the **chloroplast envelope**. Both in leaf and in root cells, some sulfate is transported across the **tonoplast** and stored in the vacuole (Figure 13.31).

Sulfate uptake against an electrochemical gradient at the **plasma membrane** is powered by the proton gradient generated by the plasma membrane H^+-ATPase. Sulfate is transported into the cytosol by an **electrogenic symport** that moves three H^+ per SO_4^{2-} (Figure 13.31). Sulfite, selenate, molybdate and chromate inhibit sulfate uptake by competing with sulfate for binding to the transporter. In contrast to the plasma membrane, the electrochemical gradient at the tonoplast favors the diffusion of sulfate into the vacuole and transfer across the tonoplast occurs through a **sulfate-specific channel**.

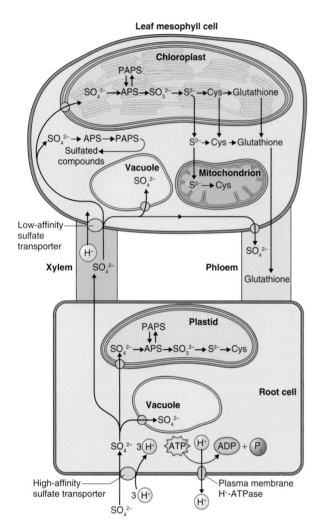

Figure 13.31 Overview of sulfur uptake, reduction and transport in plants. APS, 5'-adenosinephosphosulfate; PAPS, adenosine-3'-phosphate-5'-phosphosulfate.

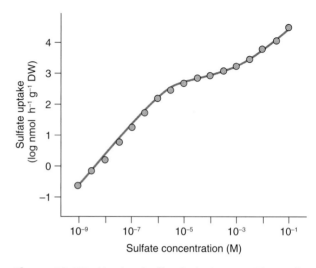

Figure 13.32 Uptake of sulfate by barley roots. The graph shows the multiphasic rate of sulfate uptake by barley roots incubated with a range of sulfate concentrations. DW, dry weight.

accumulation in the vacuole and transport into the chloroplast have not yet been identified. Observations on mutants indicate that SULTR1.2 is the major agent of sulfate uptake by *Arabidopsis* roots. Structure predictions based on the SULTR1.2 polypeptide sequence indicate that it has ten or 11 transmembrane helices (Figure 13.33).

The shape of a plot of the rate of sulfate uptake into roots against sulfate concentration is multiphasic (Figure 13.32), suggesting the presence of multiple transporters with differing affinities for sulfate. This interpretation is supported by genomic studies. For example, *Arabidopsis* has 14 genes encoding putative SO_4^{2-} transporters. Of these, two **high-affinity SO_4^{2-} transporters, SULTR1.1** and **SULTR1.2**, are mainly expressed in root epidermal and cortical cells, four **low-affinity transporters SULTR1.3, SULTR2.1, SULTR3.5** and **SULTR2.2**, appear to function in the vascular system, and **SULTR4.1** and **SULTR4.2** are tonoplast transporters facilitating the efflux of vacuolar sulfate. Genes encoding proteins mediating SO_4^{2-}

Key points Sulfur occurs in a reduced form as sulfide, as gaseous oxides, and as soluble sulfite (SO_3^{2-}) and sulfate (SO_4^{2-}) anions. Organic sulfur exists in amino acids, proteins, coenzymes, lipids, redox compounds, flavors and fragrances. The global S cycle involves microbial conversions between oxidized and reduced forms. Volatile sulfur compounds released from phytoplankton are thought to counteract climatic warming. Roots take up sulfate from the soil via electrogenic SO_4^{2-}/H+ symporters (SULTRs). Two high-affinity transporters are expressed in root epidermis and cortex. The structure of SULTR1.2, the major high-affinity transporter of *Arabidopsis* roots, is predicted to include ten or 11 transmembrane helices. Three low-affinity SULTRs are associated with vascular transport. Efflux of sulfate from the vacuole is mediated by two tonoplast transporters, whereas transfer into the vacuole occurs through a sulfate-specific channel.

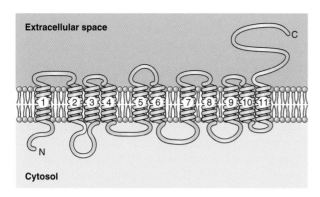

Figure 13.33   Model of the *Arabidopsis* sulfate transporter SULTR1.2.

13.4.3 The reduction of sulfate and its assimilation is catalyzed by a series of enzymes

The assimilation of sulfate involves the reduction of sulfate to sulfide and then the incorporation of sulfide into cysteine. The reduction of sulfate to sulfide requires one ATP, eight electrons and eight protons and yields AMP plus inorganic pyrophosphate (PP$_i$) (Equation 13.11). At 732 kJ mol^{-1}, this process is more energy-intensive than the assimilation of either nitrate or carbon (347 and 478 kJ mol^{-1}, respectively). The energy requirement for sulfate reduction in chloroplasts is largely met by ATP and reductant derived directly from photosynthesis. In non-photosynthetic plastids of root cells, sulfate assimilation may be powered by respiration and operation of the oxidative pentose phosphate pathway.

Equation 13.11 Reduction of sulfate to sulfide

$$SO_4^{2-} + ATP + 8e^- + 8H^+ \rightarrow S^{2-} + 4H_2O + AMP + PP_i$$

The first step in sulfate reduction is an activation step in which sulfate reacts with ATP to form **5′-adenosine phosphosulfate (APS)**, a reaction that is catalyzed by the enzyme **ATP sulfurylase** (Equation 13.12). This reversible reaction favors the direction of ATP formation, but removal of APS by APS reductase and of PP$_i$ by inorganic pyrophosphatase drives the reaction towards APS synthesis. The ATP sulfurylase of leaves exists as two isoforms with similar kinetic constants. The major isoform, representing up to 90% of total activity, is localized in the chloroplast; the minor isoform is cytosolic.

At least three genes encoding plastid ATP sulfurylase have been identified in *Arabidopsis*: **APS1**, **APS3** and **APS4**. It is proposed that the cytosolic isoform may be encoded by a fourth gene, **APS2**. ATP sulfurylases from spinach (*Spinacea oleracea*) and *Arabidopsis* have been reported to be homotetramers, but a recent study of the soybean enzyme suggested that the native form has a homodimeric structure, similar to the architecture of the enzyme from humans and a marine bacterium. The ATP sulfurylase polypeptide subunit has a molecular mass of 49–50 kDa. The holoenzyme binds MgATP and sulfate sequentially. Molybdate and selenate anions can compete for the sulfate binding site and inhibit the reaction.

Equation 13.12 ATP sulfurylase

$$SO_4^{2-} + MgATP \rightleftharpoons$$
$$MgPP_i + \text{5′-adenosine phosphosulfate (APS)}$$

APS is a high-energy compound by virtue of the phosphoric acid–sulfuric acid anhydride bond that potentiates the sulfuryl group for subsequent metabolic reactions (Figure 13.34). APS stands at the branchpoint of two pathways: **sulfate reduction** (leading to cysteine) and **sulfation** (leading to O-sulfated compounds, sulfonates and derivatives). In the sulfation pathway **APS kinase** phosphorylates APS to produce adenosine-3′-phosphate-5′-phosphosulfate (**PAPS**) (Equation 13.13). The enzyme contains six conserved cysteine residues per monomer, which are thought to be involved in redox regulation via the thioredoxin system (see Chapter 15). The *Arabidopsis* genome has four genes encoding, or

Equation 13.13 APS kinase

$$APS + ATP \rightarrow PAPS + ADP$$

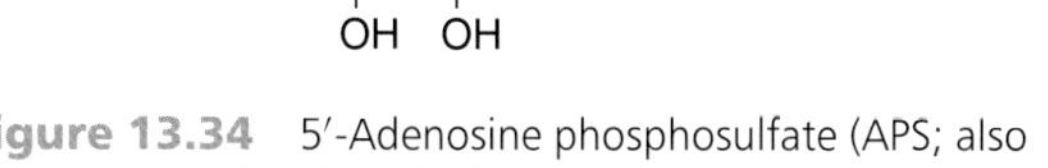

Figure 13.34   5′-Adenosine phosphosulfate (APS; also known as 5′-adenylylsulfate).

putatively encoding, APS kinase. The translation products of two of these, **Akn1** and **Akn2**, are predicted to be targeted to the plastid. PAPS in turn is a substrate for sulfotransferases, a multiprotein family responsible for sulfating a wide range of compounds, such as coumarins, glucosinolates, flavonoids, phenolic acids, steroids and sulfate esters (Table 13.7).

The sulfate reduction pathway from APS to cysteine involves several reactions. First, APS is reduced to **sulfite** (SO_3^{2-}) and then sulfite is reduced to **sulfide** (S^{2-}). Finally, sulfide reacts with *O*-acetylserine to form the thiol group of cysteine. The reduction of APS to sulfite is catalyzed by a thiol-dependent **APS reductase** which is unique to plants. This enzyme, which is localized in plastids, catalyzes the reaction between APS and reduced glutathione to produce sulfite, AMP and oxidized glutathione (Equation 13.14).

Equation 13.14 APS reductase

$$APS + 2 \text{ glutathione}_{red} \rightarrow SO_3^{2-} + \text{glutathione}_{ox}$$
$$+ AMP + 2H^+$$

Glutathione is a tripeptide consisting of glutamate, cysteine and glycine moieties (Figure 13.35A). It is one of an important group of compounds that mediate a wide range of redox reactions in vivo by cycling between dithiol (reduced, –SH HS–) and disulfide (oxidized, –S–S–) forms (see Chapter 15). Glutathione$_{red}$ is regenerated from glutathione$_{ox}$ by **NADPH glutathione reductase** (Figure 13.35B).

When the APS reductase protein is synthesized on cytosolic ribosomes it has an N-terminal transit peptide that directs the enzyme to the chloroplast and is cleaved during import. After cleavage, the mature enzyme has a

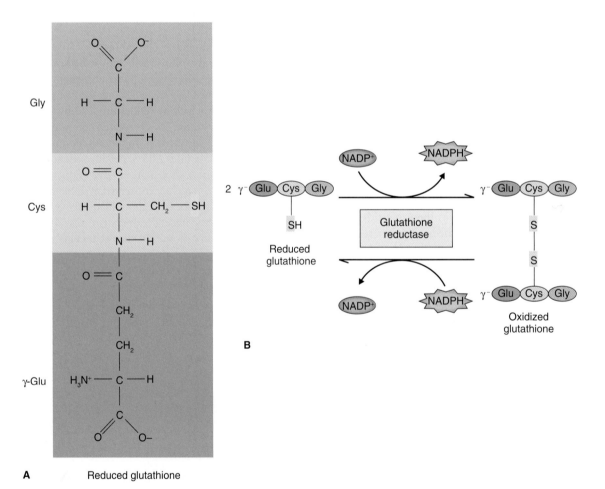

Figure 13.35 The tripeptide glutathione functions as a buffer of cellular redox potential, continually being interconverted between the reduced form, in which each glutathione molecule contains a single thiol group, and the oxidized form, in which two glutathione molecules join to form an intermolecular disulfide bond. (A) Structure of reduced glutathione. Note the unusual peptide linkage that joins the γ-carbonyl of glutamate and the amino group of cysteine. (B) Glutathione is maintained mostly in the reduced form by reaction with NADPH catalyzed by glutathione reductase, but under stress conditions, the proportion of oxidized glutathione in the cell can increase.

C-terminal redox domain and an N-terminal reductase domain (Figure 13.36). The C terminus shares a disulfide motif (CPFC, two cysteine residues separated by two other amino acids) with the redox-active proteins **glutaredoxin** and **thioredoxin**.

The next step in the sulfate reduction process is conversion of sulfite to sulfide by **sulfite reductase (SiR)**, a plastid-localized enzyme that uses ferredoxin to supply the six electrons required to reduce each mole of SO_3^{2-} (Equation 13.15). In green tissues ferredoxin is, in turn, directly reduced by non-cyclic electron transport via Fd-NADP$^+$ reductase. Ferredoxin is reduced indirectly in roots by NADPH from operation of the pentose phosphate pathway.

Equation 13.15 Sulfite reductase (SiR)

$$SO_3^{2-} + 6\ \text{ferredoxin}_{red} \rightarrow S^{2-} + 6\ \text{ferredoxin}_{ox}$$

The amino acid sequence of plant SiR has close homology to that of nitrite reductases (see Section 13.2.7). Also, like NiR, the C-terminal half of the SiR polypeptide binds prosthetic groups, consisting of one siroheme moiety (see Figure 13.19A) and one $[Fe_4S_4]$ cluster. The native enzyme of higher plants is a 64–71 kDa monomer encoded by one or two nuclear genes and is located in plastids. Purified SiR from spinach has a low but significant capacity to reduce nitrite, and such observations together with DNA sequence relationships indicate a common evolutionary origin within a superfamily of anion redox enzymes. Maintaining SiR activity in excess of APS reductase defends the plant against the accumulation of sulfite to toxic levels. Sulfite can also be detoxified by oxidation to sulfate, catalyzed by the molybdenum cofactor-containing enzyme sulfite oxidase.

13.4.4 Two enzymes catalyze the final steps of sulfate assimilation into cysteine

The final step in reductive sulfate assimilation is the condensation of *O*-acetylserine (**OAS**) and sulfide to produce cysteine. The synthesis of OAS from serine and acetyl-CoA is catalyzed by **serine acetyltransferase** (**SAT**; Equation 13.16). Possible sources of acetyl-CoA for this reaction are described in Chapter 7.

Equation 13.16 Serine acetyltransferase (SAT)

$$\text{Serine} + \text{acetyl-CoA} \rightarrow O\text{-acetylserine} + \text{CoASH}$$

O-acetylserine(thiol)lyase (**OASTL**), an enzyme that contains pyridoxal phosphate as a prosthetic group,

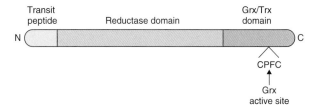

Figure 13.36 Domains of APS reductase. The N-terminal transit peptide (green) targets the protein to chloroplasts. The mature enzyme contains a reductase domain (blue) and a C-terminal domain (purple) with structural homology to proteins of the glutaredoxin/thioredoxin (Grx/Trx) family. It is thought that a redox interaction between glutathione$_{red}$ and the Grx/Trx region is the source of electrons for the reduction of APS in the reductase domain.

catalyzes the reaction of OAS and sulfide ion to form cysteine (Equation 13.17).

Equation 13.17 *O*-acetylserine(thiol)lyase (OASTL)

$$O\text{-acetylserine} + S^{2-} \rightarrow \text{cysteine} + \text{acetate}$$

Two dimers of OASTL and a homotetramer of SAT form a reversibly dissociable complex (Figure 13.37). In vivo only a small fraction of total OASTL associates with SAT in this way, but the complex has a critical role as a sensor of sulfur status. OAS concentration regulates the flux through this terminal step in sulfur assimilation by controlling the association state of the complex, which in turn determines the respective activities of the component enzymes. Sulfur-limiting conditions lead to high levels of OAS, to which the enzyme complex responds by dissociating, inactivating SAT and throttling back OAS synthesis. OASTL has maximal activity in the undissociated state and efficiently converts available sulfide to cysteine. When sulfur is abundant, free OAS is depleted, the complex reassociates, SAT is reactivated and OAS production is favored. While bacterial SATs are regulated by cysteine through negative feedback, many plant SATs are cysteine feedback-insensitive. However there are cysteine-sensitive isoforms that act to prevent runaway OAS synthesis when sulfide is plentiful. SAT and OASTL exist as cytosolic, chloroplast and mitochondrial forms. The corresponding genes have been identified, and are expressed in roots and leaves. Their transcription does not appear to respond to sulfate starvation.

Because cysteine, like glutathione (see Section 13.4.3), readily undergoes dithiol ⇌ disulfide interchange, it is important in cellular redox reactions. It also plays a role in higher-order protein structure, where two cysteine residues in the polypeptide chain may promote folding by forming a covalent disulfide bridge. Cysteine is a precursor for most cellular compounds that contain

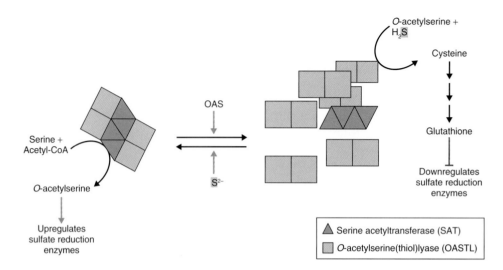

Figure 13.37 Regulation of the *O*-acetylserine(thiol)lyase (OASTL)–serine acetyltransferase (SAT) complex in cysteine synthesis. OASTL dimers (yellow boxes) are present in excess over SAT tetramers, indicated as blue triangles. The enzymes associate through specific interaction domains; SAT is active in the complexed state while OASTL is active when dissociated from the complex. Sulfide promotes formation of the complex, thereby stimulating *O*-acetylserine (OAS) formation. OAS positively regulates expression of proteins for sulfate assimilation. If OAS accumulates because of an insufficiency of sulfide, the complex is destabilized, thereby reducing OAS synthesis. OAS also reacts with sulfide to form cysteine catalyzed by free OASTL dimers. The resulting increased concentrations of cysteine and glutathione repress the expression of sulfate assimilation proteins.

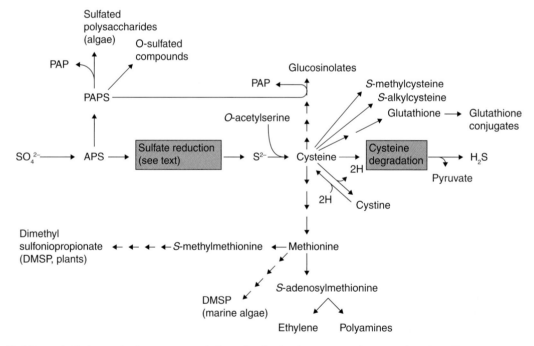

Figure 13.38 Assimilation and subsequent metabolism of sulfur in plants. APS, adenosine phosphosulfate; PAP, adenosine-3′,5′-diphosphate; PAPS, adenosine-3′-phosphate-5′-phosphosulfate.

sulfur, including methionine, glutathione, coenzymes, lipids and many natural products (see Table 13.7). A dietary supply of the sulfur-containing amino acids cysteine and methionine is essential for animals, including humans, which are unable to reduce sulfur.

Because crop plants, particularly legumes, are methionine-deficient, improving the flow of sulfur into protein amino acids is a continuing priority for plant breeding and biotechnology. Figure 13.38 is an outline of the major metabolic fates of assimilated sulfur.

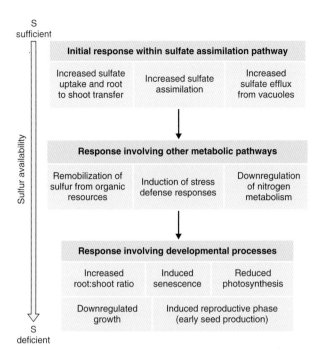

S
sufficient

Sulfur availability

S
deficient

Initial response within sulfate assimilation pathway

| Increased sulfate uptake and root to shoot transfer | Increased sulfate assimilation | Increased sulfate efflux from vacuoles |

Response involving other metabolic pathways

| Remobilization of sulfur from organic resources | Induction of stress defense responses | Downregulation of nitrogen metabolism |

Response involving developmental processes

| Increased root:shoot ratio | Induced senescence | Reduced photosynthesis |
| Downregulated growth | Induced reproductive phase (early seed production) | |

Figure 13.39 Model of the sequence of responses in plants grown under insufficient sulfur (S) supply. Mild S deficiency invokes changes at the level of S assimilation. The consequences of more severe deficiency become apparent in altered general metabolic activities. Under extreme S deficiency there are large-scale physiological adjustments.

13.4.5 Sulfur assimilation shares some features with nitrogen assimilation

A general overview of the regulation of sulfur assimilation in relation to sufficiency and deficiency is shown in Figure 13.39. The mechanisms regulating S and N assimilation have points of similarity and difference. Sulfate reduction takes place in plastids but, unlike nitrate reduction, it is not strongly regulated by light. It can occur in the absence of illumination both in roots and in etiolated shoots. Unlike nitrate reduction and carbon fixation, the rate of sulfur reduction does not show a day–night oscillation. Enzymes of sulfur assimilation are much more active in young leaves and root tips than in older tissues, reflecting the high demand for cysteine and methionine to support protein synthesis during growth and differentiation. Sulfate uptake activity is attuned to sulfur availability. Supplying sulfate represses uptake while S starvation stimulates sulfate transport and the gene transcription and enzymatic activity of APS reductase, particularly in roots. In contrast, enzymes of N assimilation are downregulated when nitrate is deficient. Leaves of mature plants are less responsive to sulfur deficiency, perhaps because vacuolar sulfate buffers leaf cells against short-term fluctuations in sulfur supply.

It is possible that glutathione, which is translocated downward in the phloem sap, acts in S signaling between the shoot and root. Short-term S starvation induces a cascade of metabolic and hormonal changes in *Arabidopsis* roots, including auxin synthesis and the participation of cytokinins and their receptors. *O*-acetylserine, a precursor of cysteine synthesis, also plays a prominent part in upregulating the expression of most genes that encode root SO_4^{2-} transporters, as well as those of the enzymes of S assimilation.

Two low-affinity sulfate transporters, SULTR3.5 and SULTR2.1, stimulate the retrieval of apoplastic SO_4^{2-} and its translocation to the shoot in S-starved plants. Expression of SULTR4.2, which functions in efflux from the vacuole to the cytosol in both roots and shoots, is also promoted by S starvation. A complex transcriptional system regulating SO_4^{2-} transport in response to S starvation is well characterized. There is evidence that a phosphorylation signaling cascade, similar to one that has been well studied in *Chlamydomonas* and fungi, operates in higher plants. The transcription factor **Sulfur Limitation 1** (**SLIM1**) has been shown to control upregulation of the genes *SULTR1.1*, *SULTR1.2* and *SULTR4.2* in response to S starvation. Site-directed mutagenesis of a putative phosphorylation site of SULTR1.2 interferes with the activity of the protein, indicating a functional role in the post-translational regulation of SO_4^{2-} transport activity.

Studies of the RNA populations in cells (**transcriptomics**) show that uptake of S (and N and P) by roots is tightly coordinated with photosynthesis in shoots, mediated by sugar signaling pathways (see Chapter 7). Furthermore, it is striking that the ratio of reduced sulfur to nitrogen in plants is strictly maintained at around 1:20, indicating a tight coordination of sulfur assimilation with nitrogen assimilation and growth rate. Variation in sulfate content among *Arabidopsis* ecotypes has been associated with a major genetic locus that shows a strong interaction with nitrogen availability in the soil. Mapping this locus allowed the underlying gene to be cloned and it was found to encode APS reductase. Thus this enzyme, whose activity is negatively related to sulfate accumulation in the plant, has an important function in the integration of S and N nutrition.

13.5 Cationic macronutrients: potassium, calcium and magnesium

The remaining macronutrients, potassium, calcium and magnesium, are acquired by plants as cations. Potassium is a macronutrient that is typically included in NPK

Key points Sulfate is assimilated into metabolism through reduction to sulfide (S^{2-}) and thence into cysteine, the precursor of most sulfur-containing metabolites. Sulfate is activated by a reaction with ATP to form the high-energy intermediate adenosine phosphosulfate (APS), catalyzed by ATP sulfurylase. Phosphorylation of APS by APS kinase produces adenosine-3'-phosphate-5'-phosphosulfate, which in turn is a substrate for sulfotransferases leading to the synthesis of a range of sulfated compounds. The first step in the synthesis of cysteine from APS is the reduction of APS to sulfite by glutathione-dependent APS reductase, a chloroplastic enzyme. Sulfite is reduced to sulfide in the chloroplast by sulfite reductase (SiR), in a reaction that transfers six electrons per mole of SO_3^{2-} from ferredoxin. SiR has a siroheme moiety and an $[Fe_4S_4]$ center, similar to NiR, with which it has a common evolutionary origin. Condensation of sulfide with *O*-acetylserine (OAS), catalyzed by OAS (thiol)lyase (OASTL), produces cysteine. OAS is synthesized from serine and acetyl-CoA by serine acetyltransferase (SAT). OASTL and SAT form a multi-subunit complex, the structure and activity of which are sensitive to sulfur status. Low or high sulfur availability also affects the transcription of genes for S transport and assimilation, through phosphorylation signaling cascades and the transcription factor Sulfur Limitation 1.

fertilizer as the third element after nitrogen and phosphorus. Averaged over many agronomic trials, wheat treated with NPK shows a 20% enhancement in yield over crops fed NP only. Most of the calcium taken up by plants is located in the apoplast and includes one fraction associated with the cell wall and another exchangeable at the plasma membrane. Deficiency symptoms (see Figure 13.1) show that apoplastic Ca^{2+} is essential for stable cell walls and membranes. Cytosolic Ca^{2+} is a fundamental mediator of signaling events, not just in plants but in eukaryotes in general, and its intracellular concentration is subject to close control. Magnesium and calcium represent the most important divalent cations for plant nutrition and some species require that the ratio of exchangeable Ca to Mg in the soil be within certain limits because excess of one ion will result in a deficiency of the other. The structural importance of Mg in the tetrapyrrole ring of chlorophyll (see Chapter 8) is evident from the chlorotic symptoms that result from an inadequate supply of this macronutrient (see Figure 13.1). Here we discuss the transport processes by which K, Ca and Mg are acquired by the plant and distributed within and between cells.

13.5.1 Potassium is the most abundant cation in plant tissues

At about 2.5%, potassium is one of the most plentiful elements in the lithosphere (see Table 13.1), and is the most abundant cation in plants, constituting up to 10% of dry matter. If this level falls below 1% of dry weight, most species will show deficiency symptoms, typically chlorotic (yellowing) and eventually necrotic (dying) interveinal areas in the oldest leaves and, under severe conditions, meristem death. Normally, cytoplasmic K concentrations are closely regulated at between 80 and 200 mM, while K levels in individual subcellular compartments may vary considerably. Fluctuations in K supply and demand are buffered by the vacuolar pool. Vacuolar potassium can be exchanged against sodium to maintain homeostatic potassium concentrations in the cytosol. Potassium is essential for a wide range of cellular and whole-plant functions, serving as an osmoticum, a diffusible counterion and an enzyme activator.

Concentrations of K in the soil range from 0.04% to 3%. Soil K exists in four different pools: soil solution (the form accessed by plants); exchangeable K; fixed K; and K sequestered in the molecular lattice structure of clay minerals. The dynamics of movement between these pools is a factor in determining the availability of K to plants. Ionic interactions with other macronutrients, notably nitrate, strongly influence the exchange of K between pools. Potassium is highly mobile in the soil solution as well as within a plant; the pathways of K movement are illustrated in Figure 13.40.

Movement of potassium from the soil into the symplasm in the root and its entry into different subcellular compartments require its transport across membranes. K^+ uptake has been among the most intensively studied transport processes in plants. Many of the approaches (particularly the application of electrophysiological and molecular techniques) and principles, described previously in connection with the movement of N, P and S into and within plants and their cells, have emerged from this research. Potassium transport consists of **low-** and **high-affinity systems** (**LATS** and **HATS** respectively; Figure 13.41). In molecular terms, low-affinity systems are commonly K^+ channels, whereas high-affinity components are co-transporters; however, both types of transporter may function in either HATS or LATS. As we saw in Chapter 5, transport through channels is passive while that mediated by co-transporters is secondarily active. In both cases, however, the activity of the plasma membrane proton pump is required to restore the membrane potential and/or proton gradient. In addition to K^+ channels, three major families of **potassium**

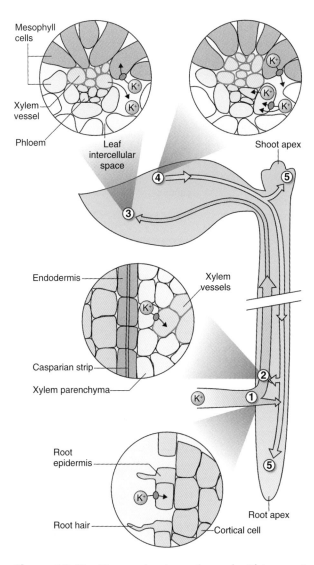

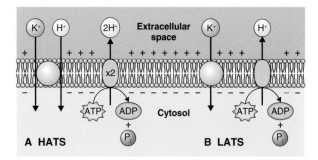

Figure 13.41 General mechanisms proposed for K+ influx into plant cells, via (A) the high-affinity transport system (HATS) and (B) the low-affinity transport system (LATS). In the HATS mechanism, the thermodynamically uphill flux of K+ is driven by the downhill flux of H+; charge balance is achieved by the outward pumping of two H+ by the plasma membrane proton ATPase. In the LATS mechanism, by contrast, uptake via an electrogenic K+ channel is electrically balanced by the ATP-driven efflux of one H+.

Figure 13.40 Diagram showing pathways for K+ transport into and within the plant. K+ is transported within the xylem (pink arrows) and phloem (blue arrows). The numbers represent important transport sites along the long-distance K+ transport pathway. For four of the five numbered sites, enlargements depict K+ transport at the cellular level. (1) K+ is taken up across the root cell plasma membrane (longitudinal view). (2) K+ is transported from living xylem parenchyma cells into non-living xylem vessels (cross-sectional view). (3) K+ is transported by the xylem to the shoot (leaf), moves from the xylem vessel to the apoplast surrounding the neighboring leaf cells, and is taken up into the leaf mesophyll cells (cross-sectional view). (4) K+ is loaded into the phloem of a fully expanded source leaf after K+ efflux from leaf cells. Transport into the sieve tube–companion cell complex can occur by a combination of apoplastic and symplasmic routes (cross-sectional view). (5) K+ moves through the phloem to the shoot and root apices, where it is unloaded for subsequent use.

co-transporters have been identified in the *Arabidopsis* genome: **KT/HAK/KUP** (K+/H+ symporters), **TRK/HKT** (K+/H+ or K+/Na+ symporters) and **CPA** (cation/H+ antiporters). Table 13.8 lists the major transporters that function in plant K nutrition, with their properties.

The KT/HAK/KUP proteins from *Arabidopsis* and rice are predicted to have between ten and 14 transmembrane domains. AtKUP1 is an example of such a HATS-type K+ transporter (Figure 13.42). The protein has a molecular mass of 79 kDa and is expressed in roots, leaves and flowers. Transport studies in yeast transformed with the *AtKUP1* gene suggest that the protein may have dual (high and low) K+ affinity. The *Arabidopsis* genome contains up to 12 members of the *KUP* family, with a range of locations and functions (Table 13.8).

Many of the **K+ channels** in plants, including AKT1, AKT2, SKOR and GORK (Table 3.8), are of the type called **Shaker**, named for the behavior of *Drosophila* mutants with defects in homologous genes. AKT1, for example, is a voltage-gated K+ channel that is expressed in all cell types of the root (see Figure 5.16). It accounts for a large portion of the low-affinity K+ uptake by potassium-starved *Arabidopsis* roots. There is evidence that AKT1 becomes important for high-affinity uptake into roots when other types of K+ transporters are inactive. In addition to their role in K uptake into roots, K+ channels are involved in loading and unloading K in both the xylem and the phloem (Table 13.8). Voltage-regulated K+ channels are important in turgor regulation in guard cells (see Chapter 14).

In response to low potassium status within the plant, expression of HATS-type transporters is stimulated and some K+ channels are upregulated. In addition, signaling cascades are activated, some of which are mediated by

Table 13.8 The major K⁺ transport proteins of *Arabidopsis* roots.

Protein	Type of channel or transporter	Localization in root	Putative function
AKT1	Shaker-type K⁺ channel	Root cap, epidermis, cortex, endodermis, stele	K⁺ uptake
ATKC1	Shaker-type K⁺ channel	Meristem, epidermis, cortex, endodermis	K⁺ uptake
GORK	Shaker-type K⁺ channel	Epidermis	K⁺ efflux, membrane repolarization, signaling, K⁺ sensing
SKOR	Shaker-type K⁺ channel	Pericycle and stellar parenchyma	Xylem loading
AKT2/AKT3	Bidirectional K⁺ channel	Phloem	Phloem loading and unloading
AtHAK5	K⁺/H⁺ symporter	Epidermis of main and lateral roots, stele of main roots	HATS component
TRH1 (AtKT3/AtKUP4)	K⁺/H⁺ symporter	Root cap	LATS component, root hair development, gravitropic responses
AtKUP1	K⁺/H⁺ symporter		HATS or dual high/low system component
AtKUP2	K⁺/H⁺ symporter	Root tip, root–hypocotyl junction	K⁺ transport, regulation of cell elongation
AtKUP3	K⁺/H⁺ symporter		LATS component
AtKUP12	K⁺/H⁺ symporter		HATS component
KEA5	Putative K⁺/H⁺ antiporter		
AtCHX17	Cation/H⁺ antiporter	Cortex and epidermis	K⁺ uptake

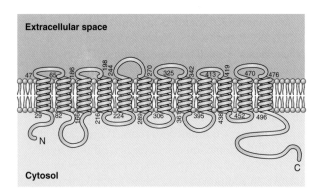

Figure 13.42   Structure of the high-affinity K⁺ transporter, AtKUP1.

reactive oxygen species (ROS) and hormones such as auxin, ethylene and jasmonic acid, using mechanisms reminiscent of reactions to stresses such as wounding (see Chapter 15).

13.5.2 Tightly regulated channels and transporters ensure cytosolic calcium is maintained at submicromolar concentrations

Calcium is a critical structural and regulatory nutrient in plants. Ca^{2+} concentrations in the rhizosphere solution are in the millimolar range, whereas cytosolic Ca^{2+} concentrations are submicromolar. In general Ca^{2+}

Key points Potassium, the major cationic macronutrient, is highly mobile in the soil and plant. Transport between soil and symplasm across membranes is facilitated by high-affinity (HATS) and low-affinity (LATS) systems. Potassium co-transporters include K^+/H^+ symporters, K^+/H^+ or K^+/Na^+ symporters and cation/H^+ antiporters. AtKUP1, an example of a HATS or dual HATS/LATS K^+/H^+ symporter, is a membrane protein with 12 transmembrane helices. AKT1, a voltage-gated Shaker-like K^+ channel, is responsible for much of the low-affinity K^+ uptake by potassium-starved roots. K^+ channels also participate in loading and unloading the vascular system and in stomatal turgor regulation. Expression of genes encoding some K^+ channels and transporters is sensitive to K availability, reactive oxygen species and hormones.

entry to roots occurs by diffusion. A number of tightly regulated Ca^{2+}-permeable cation channels in root cell plasma membranes have been described (Table 13.9). They include hyperpolarization-activated annexin-type Ca^{2+} channels, voltage-independent CNGC (cyclic nucleotide-gated channels) and glutamate receptor (GLR) cation channels, and depolarization-activated TPC1 (three pore channel 1). For example, Ca^{2+} influx into *Arabidopsis* root epidermal cells is mediated by cation channels with the following permeabilities to ion and ion combinations relative to Ca = 1.00: Ba (0.93) > Zn (0.51) > Ca/Na (0.19) > Ca/P (0.14). Cytosolic Ca^{2+} concentration is maintained by Ca^{2+}-ATPases and Ca^{2+}/H^+ antiporters, which export Ca^{2+} from the cytosol to the apoplast, compartments of the endomembrane system, plastids or vacuole. The low levels of cytosolic Ca are sensed by Ca-binding proteins including **calmodulin (CaM)**. CaM binds Ca at micromolar concentrations; after binding Ca it can interact with target proteins such as protein kinases and protein phosphatases. Vacuolar calcium is released

Table 13.9 Gene families implicated in the uptake and distribution of Mg and Ca between plant organs, cell types and subcellular compartments.

Element	Gene family	Transporter type	Membrane location	Putative cellular function
Mg	MRS2	Magnesium channel (named for Mitochondrial RNA Splicing)	PM or CP	Mg influx, chloroplast Mg uptake
	MHX	Magnesium–proton exchanger	TP	Vacuolar Mg accumulation
	TPC1	Three pore channel	TP	Vacuolar Mg release
Ca	Annexin	Hyperpolarization-activated annexin-type cation channel	PM	Ca influx
	CNGC	Cyclic nucleotide-gated channel	PM	Ca influx
	GLR	Calcium channel (named for GLutamate Receptor)	PM	Ca influx
	TPC1	Three pore channel	PM	Ca influx
	ECA/ACA	Ca^{2+}-ATPase	PM, EN, TP, plastid	Ca efflux from cytosol to apoplast, EN, ER, plastid and vacuole; xylem Ca loading
	CAX	Cation/H^+ antiporter	TP	Vacuolar Ca accumulation
	Annexin	Hyperpolarization-activated annexin-type cation channel	TP	Vacuolar Ca release
	TPC1	Three pore channel	TP	Vacuolar Ca release

CP, chloroplast envelope; EN, endomembrane system; PM, plasma membrane; TP, tonoplast.

through voltage- or ligand-gated Ca^{2+}-permeable cation channels in the tonoplast. The ER is also an important reservoir for maintaining cytosolic Ca^{2+} homeostasis, and **inositol trisphosphate** (**IP3**) plays a key role in opening channels that release Ca^{2+} from the ER into the cytosol.

Because the level of Ca in the cytosol is tightly regulated, there is not a gradient in Ca concentration in the symplasm, and movement of Ca across roots takes place largely by diffusion through the apoplast. The Casparian strip, which blocks diffusion in the apoplast, is a barrier to the movement of Ca^{2+} from the root cortex into the xylem. It is believed that calcium entry into the root xylem occurs largely in apical regions of the root before the endodermis differentiates and at points of lateral root initiation, where the Casparian strip is disrupted. Calcium is distributed within the plant, predominantly through the xylem, in the form of free Ca^{2+} or complexed with organic acids. Negatively charged groups of pectins and lignins in xylem cell walls tend to impede the movement of Ca^{2+} relative to mass flow. Significant amounts of calcium may be lost from the vascular system by lateral diffusion and precipitation as calcium oxalate. The generally low mobility of calcium can lead to deficiency symptoms in terminal organs, such as bitter pit in apples and blossom-end rot in tomato (see Table 13.1).

13.5.3 Channels in the plasma membrane deliver magnesium to the cytosol, and an antiporter mediates transfer from cytosol to vacuole

Magnesium ions are specifically required for the activity of many critical enzymes, including those of nucleic acid biosynthesis, photosynthesis and respiration. Mg also has a structural role in the tetrapyrrole ring of chlorophyll. The concentration of magnesium in soil solutions is 0.1–8.5 mM, and in the cytosol it is around 0.4 mM. Mg^{2+} entry to root cells appears to be predominantly through plasma membrane Mg^{2+} channels of the **MRS2** family. Influx and efflux of Mg^{2+} across the tonoplast membrane are thought to be via the **MHX Mg^{2+}/H$^+$ antiporter** and the **TPC1 Mg^{2+}-permeable cation channel** respectively (Table 13.9). Magnesium is transported in the xylem either as free Mg^{2+} or in chelated form.

Key points A high proportion of tissue calcium is apoplastic, where it stabilizes cell walls and membranes. Intercellular Ca^{2+} is essential for signaling processes. Calcium enters roots through tightly regulated Ca^{2+}-permeable cation channels. Ca^{2+}-ATPases and Ca^{2+}/H$^+$ antiporters move calcium from the cytosol to the apoplast or into subcellular compartments, and tonoplast-located Ca^{2+}-permeable cation channels release vacuolar calcium to the cytosol, thereby tightly controlling cytoplasmic Ca^{2+} concentrations. Calcium mobility is relatively restricted within the plant. Magnesium, another essential divalent cation, enters root cells through plasma membrane MRS2 Mg^{2+} channels, and vacuoles via the tonoplast MHX Mg^{2+}/H$^+$ antiporter. Magnesium efflux from the vacuole is mediated by the TPC1 Mg^{2+}-permeable cation channel. Long-range transport in the xylem occurs as free or chelated Mg^{2+}.

13.6 Micronutrients

Much more is known about how plants acquire and transport macronutrients than the mechanisms by which micronutrients (manganese, zinc, copper, boron, molybdenum, nickel) are taken up and utilized. Recent progress on Fe nutrition (see Section 13.6.1) has identified some of the genes and gene products in micronutrient physiology. Moreover, research interest is focusing on the low availability of Fe, Zn and Cu in many soils (which limits agricultural production), and particularly on excess levels of these and other minerals and their environmental importance as sources of heavy metal toxicity.

Zinc has also attracted special attention because of its important role in the structure and function of transcription factors, including the Zn finger, Zn cluster and RING finger domains, as well as its association with important enzymes such as superoxide dismutase. The cloning of the *IRT1* Fe transporter gene (see Section 13.6.1) led to the isolation of a number of **ZIPs** (ZRT-like (zinc regulated transporter-like), IRT-like protein), a family of genes that encode Zn^{2+} and micronutrient transporters. ZIP transporters are ubiquitous, having been identified in bacteria, fungi and mammals as well as plants. Most ZIP proteins have eight predicted

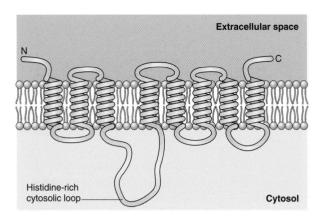

Figure 13.43 Predicted protein structure of the members of the ZIP family of micronutrient transporters.

transmembrane helices and in many cases there is a long loop region located between transmembrane domains 3 and 4 containing a histidine-rich sequence whose potential metal-binding ability suggests a function in zinc transport or its regulation (Figure 13.43).

Gene expression studies in yeast and experiments on roots indicate that the proteins encoded by **ZIP1, ZIP3** and **ZIP4** are high-affinity Zn^{2+} transporters that may also mediate the uptake of other divalent cations such as Cd and Cu. *ZIP1* and *ZIP3* are expressed in roots in response to Zn deficiency and *ZIP4* is expressed in roots and shoots. Overexpression of **ZNT1**, a homolog of *ZIP4*, is a factor in enhanced zinc uptake by the hyperaccumulator species *Thlaspi caerulescens* (see Section 13.6.4). Zinc is translocated and sequestered as complexes with citrate or nicotianamine or malate. A number of additional gene products have been associated with Zn transport, including YSL1, FER1, FER2 (see Section 13.6.1), PHT1–4 (see Section 13.3.2) and the stress-responsive enzymes of glutathione metabolism.

13.6.1 Iron is an essential component of biological electron transfer processes

Iron is essential for biological electron transfer processes through reversible redox reactions, cycling between Fe^{2+} and Fe^{3+}; for example, it is a component of cytochromes and ferredoxin. Inadequate Fe uptake leads to deficiency symptoms such as interveinal chlorosis in leaves and

reduced growth. Although Fe is the fourth most abundant element in the Earth's crust (see Table 13.1) its limited solubility results in concentrations of free Fe^{3+} and Fe^{2+} of less than 10^{-15} M in well-aerated soils at physiological pH. This means it is not readily available to plants. In addition, other minerals such as calcium may intensify Fe deficiency. One-third of the world's agricultural soils are calcareous and considered to be Fe-deficient, which leads to reductions in potential crop yield. The chemical properties of Fe also impose limitations on its accumulation by plants, because Fe^{2+} and Fe^{3+} catalyze the reduction of molecular oxygen to damaging **ROS** (reactive oxygen species). Once in the symplasm, Fe is maintained in a soluble, transportable form, and prevented from generating ROS by binding to chelators such as citrate and nicotianamine. Plants have evolved two strategies to take up Fe from the soil. In non-grasses, Fe starvation activates a process that involves both acidification of the soil solution and reduction of Fe^{3+} (Figure 13.44), whereas grass species employ a chelation-based strategy (Figure 13.45).

In aerobic conditions, soil Fe is largely in the form of ferric oxides. Under conditions of Fe deficiency, eudicots and non-grass monocots secrete protons into the

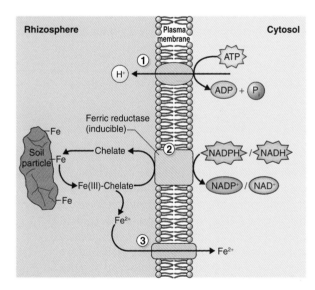

Figure 13.44 Iron uptake by eudicots and non-grass monocots. Key plasma membrane-localized components induced in roots by Fe deficiency are shown, including (1) AHA2, a member of a family of plasma membrane H^+-ATPases, (2) inducible ferric reductase oxidase 2 (FRO2) and (3) a high-affinity Fe^{2+} transporter (IRT1, iron regulated transporter 1).

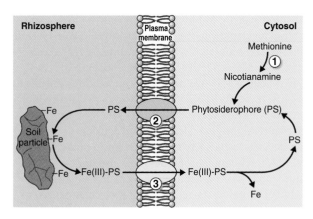

Figure 13.45   Iron uptake by grasses. Key components shown include (1) biosynthesis of phytosiderophore (PS), (2) inducible plasma membrane (PM) PS transporter and (3) Fe-PS transporter.

rhizosphere, lowering the pH of the soil solution and increasing the solubility of Fe^{3+} (Equation 13.18). Lowering the pH from 8 to 4 increases the concentration of Fe^{3+} in solution from about 10^{-20} to 10^{-9} M. The plasma membrane H^+-ATPase is thought to be involved this process. In *Arabidopsis*, this protein is encoded by the ***AHA*** gene family. Upregulation of one member of this family, *AHA2*, in response to low Fe has been shown to be mediated by an iron deficiency-induced transcription factor.

Equation 13.18 Solubilization of ferric ions by acidification

$$Fe(OH)_3 + 3H^+ \rightleftharpoons Fe^{3+} + 3H_2O$$

Before iron can be taken up by roots, it must be reduced from Fe^{3+} to the more soluble Fe^{2+}. An inducible plasma membrane **ferric reductase oxidase** (FRO) catalyzes the reduction of Fe^{3+} (Figure 13.44). It is encoded by the gene ***FRO2***, which is expressed in epidermal cells of Fe-deficient roots. The protein is predicted to have eight membrane-spanning domains, two histidine-coordinated heme groups and putative binding sites for FAD and NADPH. There are seven additional members of the *FRO* gene family of metal reductases in *Arabidopsis*, which are expressed elsewhere in roots, in vascular tissues and in shoots. Fe^{2+} is transported into root cells by **IRT1** (**iron regulated transporter 1**), a member of the extensive **ZIP metal transporter family** (Figure 13.44). IRT1 is expressed in the plasma membrane of epidermal cells in Fe-deficient roots. Studies of *Arabidopsis* mutants and of yeast cells transformed with the *IRT1* gene show that the protein is able to transport other divalent metals (Zn, Mn, Cd) as well as Fe.

Under Fe deficiency, grasses such as *Zea mays, Triticum* spp. and *Oryza sativa* use a different strategy for

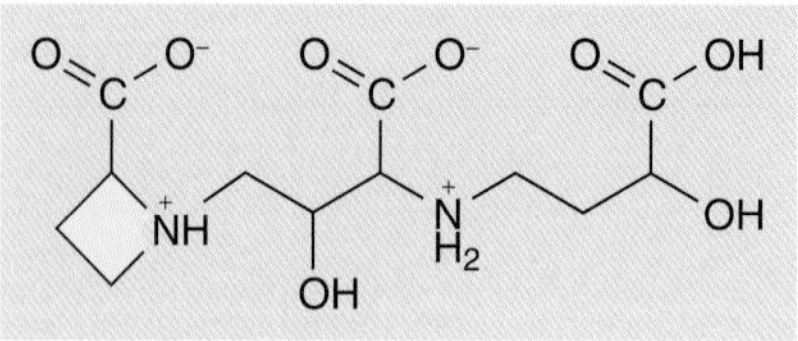

Figure 13.46   Mugineic acid.

obtaining iron. They release low molecular weight **phytosiderophores** (**PSs**) that are related to **mugineic acid** (Figure 13.46). PSs are chelators that bind Fe^{3+} in the rhizosphere and make it available for uptake by roots. The efficiency of this chelation-based response allows grasses to survive more extreme conditions of Fe deficiency than eudicots and other monocots. PSs released by grass roots effectively chelate Zn^{2+}, Cu^{2+}, Mn^{2+}, Ni^{2+} and Co^{2+} as well as iron and may have a general role in the uptake of these mineral micronutrients. Mugineic acid and related PS are biosynthesized from methionine via the intermediates *S*-adenosyl methionine (SAM; see Figure 13.38), nicotianamine and 2'-deoxymugineic acid (Equation 13.19).

Equation 13.19 Biosynthesis of mugineic acid.

3SAM → nicotianamine
(enzyme = nicotianamine synthase)

Nicotianamine → 2'-deoxymugineic acid
(enzyme = nicotianamine aminotransferase)

2'-Deoxymugineic acid → mugineic acid
(enzyme = dioxygenase)

The Fe(III)−PS complex is transported into the epidermal cells of Fe-deficient roots by a specific high-affinity transport protein in the plasma membrane (Figure 13.45). The gene ***YS1*** (*Yellow Stripe 1*) encodes the **Fe(III)-PS transporter**, an integral membrane protein with 12 putative transmembrane domains. The protein is a proton-coupled symporter for PS and mugineic acid−metal chelates. Once it has entered the root symplasm, Fe, in the form of chelation complexes, diffuses through plasmodesmata into the stele. Various chelation and oxidation/reduction steps, transport activities and protein associations solubilize, translocate, compartmentalize, remobilize, store and buffer Fe in vivo.

Some of the proteins that participate in Fe uptake from soil, or members of the same families, are expressed and may be functional in other cell types along the route from root to shoot. For example *YSL* (*Yellow Stripe-Like*) genes are expressed in a number of tissues, suggesting they play roles in Fe uptake in different parts of the plant. Fe homeostasis is required for building heme and Fe-S prosthetic groups and correctly assembling them onto apoproteins. Storage and buffering of iron at the subcellular level are essential to insure against iron scarcity and toxicity. Plastids sequester an important fraction of cellular iron in the form of **ferritin**, a protein molecule that can create a 'nanocage' enclosing up to 4500 atoms of Fe^{3+} in the form of an iron oxide. The *Arabidopsis* genome encodes four ferritins: **FER2** is confined to the seed, **FER1**, **FER3** and **FER4** are expressed in shoot tissue and FER1 is also found in roots. In mature seeds, stored iron is associated with vacuolar globoids.

Plants clearly have mechanisms for sensing Fe status and signaling Fe deficiency, but our understanding of them is incomplete. **FIT1** (Fe-induced deficiency transcription factor 1) is a transcription factor whose expression is upregulated by Fe deficiency. It has been shown to regulate about 40% of Fe-inducible genes, including *FRO2* and *IRT1*. FIT1 synthesis in turn is responsive to ethylene as well as Fe deficiency; this suggests that hormonal signaling may play a role in sensing Fe status.

Key points Iron, an essential micronutrient, is an abundant element but of limited availability in soils because of low solubility. In response to iron starvation, the plasma membrane H^+-ATPase in roots of non-grasses, encoded by the *AHA* gene family, is activated and acidifies the rhizosphere, thereby increasing the solubility of Fe^{3+}. Fe^{3+} is reduced to the more soluble Fe^{2+} by an inducible plasma membrane ferric reductase oxidase encoded by the gene *FRO2*. Fe^{2+} is transported into root cells by IRT1, a member of the extensive ZIP metal transporter family. Fe-starved grass roots release phytosiderophores (PSs), which are chelators that bind rhizosphere Fe^{3+} and make it available for uptake via the Fe(III)-PS transporter, a specific high-affinity proton-coupled symporter encoded by the gene *YS1*. Fe chelation complexes diffuse through the root symplasm to the stele. Iron is sequestered in plastids in the form of ferritin, an iron oxide-binding protein.

13.6.2 Several micronutrient elements are toxic in excess

A number of the essential micronutrient elements are heavy metals. While small quantities of Ni, Cu, Zn and so on are necessary for normal growth, exposure to supraoptimal levels produces toxicity reactions, although there is considerable variation for heavy metal tolerance between and within plant species. For example, maize plants are much less tolerant of excess Cu than common beans (*Phaseolus vulgaris*) are. The symptoms of Cu or Zn toxicity begin with inhibition of root growth and may progress to chlorotic reactions in the shoot, often associated with secondary iron deficiency. Excess Ni also results in chlorosis as a result of its competition with the uptake of other essential divalent cation nutrients, notably Ca^{2+}, Mg^{2+}, Fe^{2+} and Zn^{2+}.

Cu, Ni, Mn, Fe and Mo are all **transition metals**. A transition metal is defined as an element that forms one or more stable ions in which inner electron shells (the *d* orbitals) are incompletely filled. This gives transition metals distinctive chemical properties, which in turn allow them to do specialized jobs in living cells. In particular Cu, Fe, Mn and Mo exist in multiple redox states and therefore serve as critical cofactors for electron transport processes. The toxicity of such essential micronutrient metals when present in excess arises directly from their properties as transition elements. Exposure to toxic concentrations of heavy metal ions shifts the balance of free radical metabolism towards an accumulation of H_2O_2 (see Chapter 15). Redox-active transition metals such as Cu^+ and Fe^{2+} catalyze the conversion of H_2O_2 to the highly reactive hydroxyl radical ${}^\bullet OH$ via the **Fenton reaction** (Equation 13.20A). A subsequent reaction with superoxide radicals ($O_2^{\bullet -}$) regenerates reduced metal ions (Equation 13.20B).

Equation 13.20 Reactive oxygen species generation by transition metals

13.20A

$$H_2O_2 + Fe^{2+}/Cu^+ \rightarrow {}^\bullet OH + OH^- + Fe^{3+}/Cu^{2+}$$

13.20B

$$O_2^{\bullet -} + Fe^{3+}/Cu^{2+} \rightarrow Fe^{2+}/Cu^+ + O_2$$

The tendency of heavy metals in excess of micro-nutrient requirements to provoke the generation of reactive forms of oxygen links this type of toxicity with generalized responses to other stresses, such as pathogen and herbivore attack, that are mediated through ROS cascades and signaling pathways (see Chapter 15).

13.6.3 Aluminum is a non-nutrient mineral responsible for toxic reactions in many plants growing on acid soils

Of the non-nutrient elements with adverse effects on plant growth, Al is of greatest agricultural significance. At

8% (by volume), aluminum is the third most abundant element in the Earth's crust (after oxygen, 47%, and silicon, 28%). Low crop productivity on acid soils (pH <5) resulting from the solubilization of toxic forms of Al is a global agronomic problem. Dissolved aluminum at acid pH exists largely as the Al^{3+} ion, which is toxic to roots. Hydration products of Al, $Al(OH)_2^+$ and $Al(OH)^{2+}$, predominate at around neutral pH, and $Al(OH)_4^-$ is the major form of aluminum in alkaline solution. Breeding Al^{3+}-resistant crops, a priority for agricultural research and development, depends on exploiting knowledge of the underlying genetic and physiological mechanisms of Al toxicity in roots.

Al^{3+} in the soil complexes with organic acids, inorganic phosphate and sulfate. If it enters the cell, Al will also bind to these groups in proteins, nucleotides and other macromolecules. Two classes of Al-resistance mechanisms are recognized: (i) symplasmic tolerance, in which Al enters cells but metabolism is resilient and able to continue functioning normally; and (ii) Al exclusion, which defends the root apex against exposure to Al in the rhizosphere. One kind of exclusion mechanism that plants have evolved relies on the efflux of organic anions that protect roots by chelating Al^{3+} ions. The application of malate or citrate to the solution bathing the roots relieves Al toxicity because the complexes formed between Al and organic acid chelators do not readily cross the plasma membranes of root cells. Genes for exclusion-type Al resistance have been isolated from several species. They have been found to encode membrane proteins of the **Al^{3+}-activated malate transporter** (**ALMT**) and **multi-drug and toxin extrusion** (**MATE**) families.

The first plant Al^{3+}-resistance gene to be cloned was that encoding the ALMT-type malate transporter **TaALMT1** of *Triticum* spp. A wealth of evidence favors malate efflux regulated by TaALMT1 as the general mechanism of exclusion resistance to Al^{3+} toxicity in *Triticum*. For example, the position of this gene on the *Triticum* genetic map coincides with the single major genetic locus controlling Al^{3+} resistance in this species. Expressing the gene in heterologous transgenic systems such as tobacco (*Nicotiana tabacum*) confers increased Al^{3+} tolerance and enhances an Al^{3+}-activated efflux of malate. The response of resistant and sensitive cultivars of *Triticum* to micromolar levels of Al in the growth medium is shown in Figure 13.47. TaALMT1 is highly expressed in the wheat variety Atlas 66 but expression in variety Scout 66 is weak.

An example of a MATE-type gene is **HvAACT1** of barley, which has been identified and cloned by genetic mapping in Al-resistant and Al-sensitive cultivars. Genetically transformed *N. tabacum* plants overexpressing *HvAACT1* show enhanced citrate secretion and Al resistance compared with wild-type

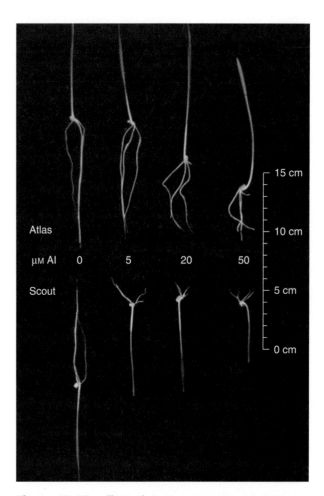

Figure 13.47 Effects of aluminum exposure on root growth of Al-resistant Atlas 66 and Al-sensitive Scout 66 cultivars of wheat. Seedlings were grown for 4 days in 0.6 mM $CaSO_4$ solutions containing 0, 5, 20 or 50 μM $AlCl_3$ (pH 4.5).

plants. The protein, which is localized in the plasma membranes of root epidermal cells, is constitutively expressed in the roots of Al-resistant barley. There is a good correlation between the expression of *HvAACT1* and citrate secretion in *Hordeum vulgare* cultivars differing in Al resistance.

Secondary structure predications for **ALMT** and **MATE** proteins (Figure 13.48) indicate molecular configurations with five to seven membrane-spanning domains; this is consistent with their role as transport proteins facilitating the efflux of Al-chelating organic acids. In ALMT proteins the membrane-spanning helices are at the amino terminal end of the protein (Figure 13.48A) whereas the predicted structure for HvAACT1 shows seven membrane-spanning domains in the C-terminal half of the molecule (Figure 13.48B). A proposed mechanism for the action of the constitutive transporters, such as the ALMT and MATE types of *Triticum* and *H. vulgare*, is presented in Figure 13.49A. Here, organic anion efflux is stimulated by direct

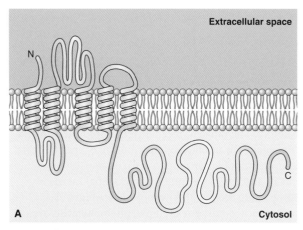

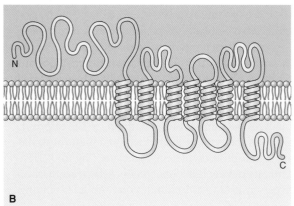

Figure 13.48 Predicted secondary structures of proteins associated with exclusion-type aluminum resistance. (A) ALMT1, an Al^{3+}-activated malate transporter from *Triticum*. (B) HvAACT1, a multidrug and toxin extrusion transporter from barley. In both cases, orientation of the structure across the membrane represents the most likely configuration based on model predictions.

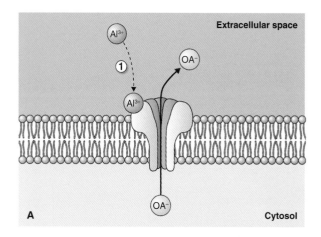

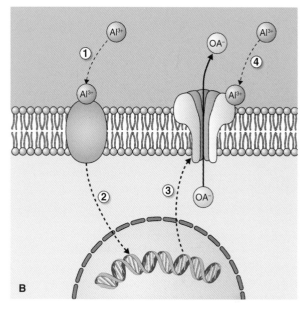

Figure 13.49 Hypothetical models for the Al^{3+}-activated efflux of organic anions by members of the ALMT and MATE families of proteins. (A) Pattern I. The protein is expressed constitutively in root apices with Al^{3+}-resistant genotypes showing greater expression than Al^{3+}-sensitive genotypes. Al^{3+} activates organic anion (OA⁻) efflux by interacting directly with the pre-existing proteins in the plasma membrane (arrow 1). (B) Pattern II. Al^{3+} first induces the expression of the proteins through a signal transduction pathway (arrows 2 and 3) possibly involving a specific Al^{3+} receptor (arrow 1) or non-specific stress responses. Al^{3+} then activates organic anion efflux by interacting with the newly synthesized proteins in the plasma membrane (arrow 4).

interaction between Al^{3+} and the pre-existing proteins in the plasma membrane. In many species, such as *Arabidopsis, Brassica* and *Sorghum* spp., expression of the genes for organic anion transporters is induced by Al^{3+} through a signal transduction pathway, via either a specific receptor or non-specific stress responses. Al^{3+} then interacts with the newly synthesized proteins at the plasma membrane and activates organic anion efflux (Figure 13.49B).

Most plants that are highly tolerant of harmful elements are excluders, restricting the accumulation of toxins by limiting their uptake and root-to-shoot translocation. However, some plants, called **hyperaccumulators**, are capable of building up and tolerating extremely high levels of trace elements in their tissues (in some cases exceeding 1% of leaf dry matter). Over 450 plant species have been identified as hyperaccumulators of trace metals (Zn, Ni, Mn, Cu, Co, Cd), metalloids (As) and non-metals (Se). Ni

hyperaccumulators account for 75% of this group. Hyperaccumulators are of considerable interest because of their potential use in **bioremediation**, the reclamation of contaminated land by removal of toxic metals. The hyperaccumulation character has evolved between and within species as an adaptation to the abundance of particular elements in the environment. For example, all populations of *Thlaspi caerulescens* (a member of the

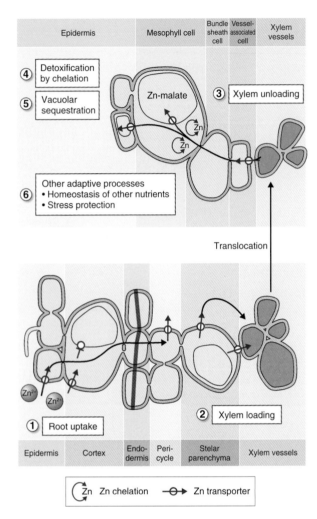

Figure 13.50 Overview of adaptations (highlighted in red) to hyperaccumulate metals, in particular zinc (Zn). Accumulation in the vacuoles of root cells may be blocked at the level of the tonoplast transporter. Zn ions transported from the root through the xylem may be sequestered in mesophyll cell vacuoles or detoxified by chelation.

Brassicaceae family) hyperaccumulate Zn, but when grown on the same substrate, populations from soils with high concentrations of trace metals generally accumulate less Zn than populations from soils with low concentrations. Comparative analysis of gene expression in relation to the physiology of Zn transport and storage indicates that *T. caerulescens* behaves as though it is Zn-deficient, even while it is accumulating high amounts of Zn. Figure 13.50 indicates the points in the system of uptake, translocation and sequestration where cell processes are altered in the hyperaccumulation syndrome.

13.6.4 Heavy metal homeostasis is mediated by metal-binding metabolites and proteins

The maintenance of metal homeostasis is essential for viability. The range of mechanisms employed by plants to regulate the uptake and distribution of specific metal ions includes the capacity to detoxify non-essential metals and excessive amounts of essential metals by complexing them with low molecular weight or polymeric ligands. For example, **phytochelatins** are metal-binding polymers of various sizes with the general structure $(\gamma\text{-Glu-Cys})_n\text{Gly}$ (n = 2 to 11), which are synthesized by plants in response to exposure to high concentrations of micronutrients such as Cu^{2+} or toxic cations such as Cd^{2+}. The precursor for phytochelatin synthesis is the tripeptide glutathione. Metal ions such as Cd^{2+} activate the enzyme **phytochelatin synthase** (γ-glutamylcysteine dipeptidyl transpeptidase), which catalyzes the transfer of a glutathione to the carboxyl group of the cysteine in another glutathione or to the analogous position in a phytochelatin (Figure 13.51). Mutants lacking functional phytochelatin synthase are hypersensitive to cadmium toxicity. An ATP-binding cassette (ABC)-type tonoplast transporter (see Chapter 5) transfers phytochelatin–cadmium complexes into the vacuole where reaction with additional cadmium and sulfide results in the formation of a CdS crystallite. Such a sequestration process is called **biomineralization**.

Metallothioneins (**MTs**) are low molecular mass (4–14 kDa) proteins with a high content of cysteine, arranged in metal-binding motifs (Cys-Cys, Cys-X-Cys or Cys-X-X-Cys where X represents any amino acid). These regions of the polypeptide are able to coordinate with bivalent metal ions through the sulfhydryl ligands of cysteine side-chains. MTs are found in organisms across the taxonomic range and have a variety of jobs, including binding Cd, Cu and other heavy metals, maintaining Zn homeostasis, controlling ROS and participating in signal transduction pathways. Based on the conserved positions of Cys residues among the range of genes encoding MTs and MT-like proteins in plants, four types of MT are recognized. **Type 1 MTs** are expressed predominantly in roots, **type 2 MTs** in leaves, **type 3 MTs** in fruits and **type 4** MTs in seeds. A model of the structure of a type 1 Cd-binding MT from durum wheat (*Triticum durum*) is shown in Figure 13.52. In many plant species, including *Thlaspi caerulescens*, MT gene expression is strongly induced by Cu treatment and, to a lesser degree, by Cd

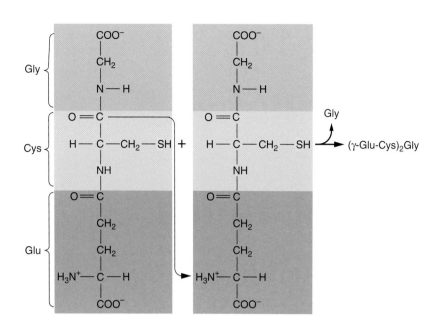

Figure 13.51   Synthesis of phytochelatin by γ-glutamylcysteine dipeptidyl transpeptidase (phytochelatin synthase).

molecules with roles in nutrition, homeostasis and detoxification include phytosiderophores, ferritins and metallochaperones.

Key points   ZIP transporters have high affinity for Zn^{2+} and may also mediate the uptake of other cations such as Cd and Cu, as well as Fe. Zn deficiency induces the expression of *ZIP*s in roots, and high expression of *ZIP*s is characteristic of species that hyperaccumulate Zn and other heavy metals. Zinc is translocated and stored in chelated form. Fe and the transition metal micronutrients, Cu, Ni, Mn and Mo, are potentially toxic in excess because they promote the generation of harmful reactive oxygen species. Aluminum is a non-nutrient element with agriculturally significant deleterious effects on growth. At acid pH soil aluminum is solubilized as the Al^{3+} ion which is toxic to roots. Some plants can resist Al toxicity by secreting chelators into the rhizosphere. Genes for this kind of resistance include those encoding Al^{3+}-activated malate transporter and multidrug and toxin extrusion proteins. Metal-binding proteins implicated in counteracting heavy metal toxicity include phytochelatins and metallothioneins.

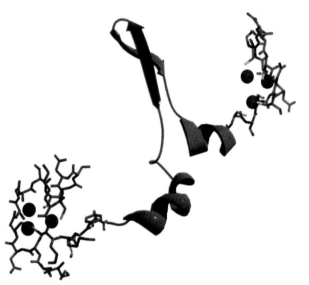

Figure 13.52   The predicted structure of a durum wheat metallothionein. Cadmium (purple spheres) binding metal centers at each pole of the dumbbell-shaped molecule are depicted in ball and stick representation with the extended hinge region highlighted in ribbon representation.

and Zn. Variation in Cu tolerance among *Arabidopsis* ecotypes and between populations of *Silene* spp. is correlated with MT gene expression. A type 2 MT has been implicated in ROS scavenging and responses to pathogen attack in *Oryza sativa*. Other metal-binding

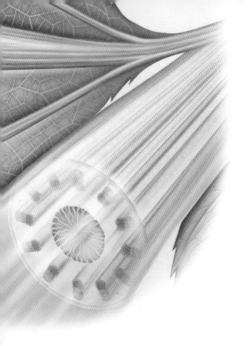

Chapter 14

Intercellular and long-distance transport

14.1 Introduction to transport of water and solutes

All living cells in a plant require water, dissolved ions and sugar to survive and grow. Most plants live with their roots in the soil and their leaves and stems above ground. Roots take up water and dissolved inorganic ions from the soil, and leaves and green stems produce sugar during photosynthesis. The challenge for the plant is to move water and minerals from the roots to the cells of stems and leaves and to transport sugar to non-photosynthetic parts. Often this transport must occur over long distances, from the tips of the deepest roots, in some cases tens of meters below ground, to the tops of the tallest stems—more than 100 m above ground in some trees. Transport must be fast enough, in some cases at rates as high as 4 mm s^{-1}, to replace water that evaporates from the above-ground surfaces of the plant. Plants are able to meet this challenge without any specialized pumping organ (Figure 14.1).

Transport of water, minerals and photosynthate within and between plant organs occurs in two transport systems: the **phloem** and the **xylem**. In each system, transport takes place within tubes formed by columns of specialized interconnected cells and is driven by bulk flow. We shall see that the contents of the tubes, the direction of transport and the source of the pressure gradient differ between these two systems. Transport in the phloem is referred to as **translocation** and transport in the xylem is a result of **transpirational** water loss through stomata.

Ideas about how water and dissolved solutes move over long distances in plants have not changed dramatically over the last century. The cohesion–tension hypothesis for water movement in the xylem, discussed in this chapter, was well established at the beginning of the 20th century and, in the 1920s, Munch proposed his pressure–flow hypothesis to explain movement in the phloem. Technical advances over the last 20 years or so have allowed researchers to measure the tensions that arise in the xylem during rapid transpiration. In order to do this micropipettes have been developed that can be inserted into functioning xylem; these have allowed tensions to be accurately determined. Measured tensions match the values predicted by the cohesion–tension hypothesis for water movement. Experimental evidence also substantially supports the pressure–flow hypothesis for water movement in the phloem and evidence in support of this has come from an interesting source. Researchers have used aphids, insects that feed on phloem contents, to measure the rate of flow in the phloem as well as the composition of phloem sap (see Section 14.5.2).

The Molecular Life of Plants, First Edition. Russell Jones, Helen Ougham, Howard Thomas and Susan Waaland.
© 2013 John Wiley & Sons, Ltd. Published 2013 by John Wiley & Sons, Ltd.

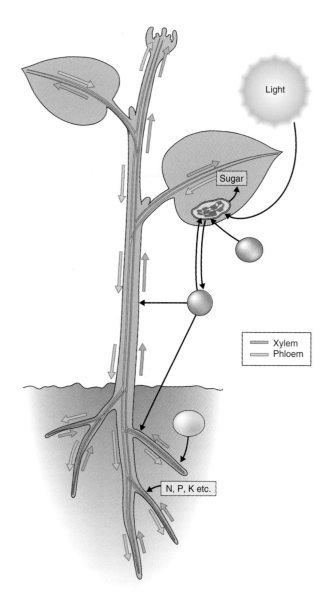

Figure 14.1 Overview of the long-distance transport system in plants showing the distribution of phloem and xylem and the pathway for uptake of water, solutes and gases.

Advances in molecular cloning have also played an important part in our understanding of water and solute movement. As discussed in Chapters 5 and 13, many specific solute transporters have been sequenced and the three-dimensional structure of several of these transport proteins is now known. For long-distance transport of water, the discovery of aquaporins about 20 years ago and the sequencing of members of this large protein family led to major conceptual advances in understanding how water moves over long distances, especially radially across the root. Advances in microscopy have also played important roles in elucidating the structure of plasmodesmata and, coupled with our knowledge of how viruses move, have greatly enhanced understanding of the function of plasmodesmata in transport.

In this chapter we will first discuss the principles that govern the movement of water into and out of cells. We will then examine cell to cell transport through plasmodesmata. Finally, we will investigate the mechanisms of long-distance transport of water, inorganic ions and photosynthate.

14.2 The concept of water potential

Before considering how water is taken up by plant cells we need to understand some of the physical properties of water and how these are influenced by forces such as pressure, gravity and dissolved solutes. Plant physiologists use the term **water potential** as a measure of the **chemical potential** of water. As we discussed in Chapter 5, chemical potential is a relative term: for water it relates the chemical potential of pure water under standard conditions to that of another aqueous solution. It is expressed in $J\ mol^{-1}$. Water potential is a measure of the free energy of water per unit volume; it is obtained by dividing the chemical potential of water by the partial molal volume of water ($= 18 \times 10^{-6}\ m^3\ mol^{-1}$). Water potential is expressed in pressure units, most commonly mega-Pascals (MPa), although other units of pressure are sometimes used. Table 14.1 shows how to convert between different, commonly used pressure units. Using this table, one can see that mountain bike tires are typically inflated to about 0.4 MPa and car tires to about 0.2 MPa. Plant cells by contrast have a pressure potential called **turgor pressure** that can reach as high as 1 MPa with values commonly around 0.4–0.6 MPa. By definition the water potential of pure water is 0 MPa.

Water potential is symbolized by the Greek letter psi, Ψ_w. In living cells three components contribute to Ψ_w: solute concentration, pressure and gravity. Their effects on water potential can be represented by Equation 14.1.

Equation 14.1 Components of water potential Ψ_w

$$\Psi_w = \Psi_s + \Psi_p + \Psi_g$$

Water potential = osmotic potential + pressure potential + gravitational potential

Table 14.1 Pressure unit equivalents.

1 mega-Pascal	$= 1\ J\ m^{-3}$
1 MPa	= 10 bars
1 MPa	= 9.87 atmospheres
1 MPa	= 145 pounds per square inch (psi)

14.2.1 Solutes lower the water potential

Dissolved solutes lower the chemical potential of water by effectively diluting water. This effect depends on the osmotic concentration of a solution. Osmotic concentration (osmol L^{-1}) is the sum of the concentrations of all solute particles in solution. We can calculate the osmotic concentration of a solute by multiplying its molal concentration by i, where i is the number of particles into which a solute dissociates. For sucrose and other non-ionizing solutes, I = 1, for salts such as KCl, I = 2, for $CaCl_2$, I = 3, and so on. For example, a 0.5 M solution of KCl is 1.0 osmolal. As most salts do not fully dissociate the values for i are only approximations, so for KCl, i is about 2, for $CaCl_2$ about 3, etc. Note that molal concentration refers to the number of moles of solute dissolved in 1 kg of water.

The **osmotic potential** (Ψ_s), also known as the **solute potential**, is a measure of the effect of dissolved solutes on water potential. A rough approximation of osmotic potential can be obtained from Equation 14.2, the **van't Hoff equation**:

Equation 14.2 The van't Hoff equation

$$\Psi_s \approx -RTc_s$$

where R is the gas constant (8.314 J mol^{-1} $°K^{-1}$), T the absolute temperature in °K, and c_s is the **osmolal concentration** of the solution.

From this equation one can see that the osmotic potential of a solution is always less than or equal to zero. A 1 osmolal, ideal solution at 20°C will have an osmotic potential of −2.44 MPa. Note that metabolites, and solutes such as sucrose and KCl, do not behave as ideal solutions, meaning that the van't Hoff equation provides only a rough estimate of a solution's osmotic potential.

14.2.2 Pressure can increase or decrease water potential

Increased pressure on a solution increases the chemical potential of water by effectively increasing its concentration while decreased pressure decreases its chemical potential. The effect of pressure on water potential is expressed as **pressure potential** (Ψ_p). Under standard conditions, Ψ_p is defined to be 0 MPa. Therefore, if the pressure is greater than 1 atmosphere (approximately 0.1 MPa), Ψ_p is positive, and if it is less than 1 atmosphere, Ψ_p is negative and referred to by plant physiologists as **tension**.

The opposing effects of pressure and tension on Ψ_w can best be demonstrated using a simple osmometer, which may be constructed by placing a **semipermeable** membrane in a glass U-tube as shown in Figure 14.2. If a 0.1 molal sucrose solution, having a Ψ_w of −0.24 MPa, is introduced into one arm of the U-tube and pure water ($\Psi_w = 0$ MPa) is placed into the other arm, water will move by **osmosis** across the semipermeable membrane and the water column in the osmometer will rise (Figure 14.2A). If a piston is placed into the U-tube as shown in Figure 14.2B and a pressure of 0.24 MPa is applied, net H_2O movement into the sucrose solution can stopped. The effect of pressure is to increase the Ψ_p of the sucrose solution. In this example, if 0.24 MPa of pressure is applied through the piston then Ψ_p will equal Ψ_s, $\Psi_w = 0$, and there will be no net water movement. If the pressure applied on sucrose solution in the U-tube is greater than its Ψ_s, water can be forced from sucrose to pure water. In the example in Figure 14.2C, a pressure of 0.3 MPa is applied to the sucrose solution causing water to flow into the U-tube arm containing pure H_2O because the Ψ of the sucrose solution is now 0.06 MPa.

If a piston is placed in the U-tube on top of the water column and pulled upward to generate a pressure of −0.30 MPa, water will move from sucrose to pure water. In this instance tension caused by raising the piston lowers the entropy of water, bringing about a negative Ψ_p and allowing for osmotic water flow from sucrose to water. It is tension that lifts or raises water up the stems of tall trees during the process of evapotranspiration.

Key points Water potential, denoted by Ψ_w, is the term used to describe the chemical potential of water, commonly expressed in units of mega-Pascals (MPa). It is a relative term as it compares the Ψ_w of pure water with that of a solution, such as water in the cytosol of plant cells. By definition the Ψ_w of pure water is 0. Several forces act on Ψ_w including the presence of solutes, pressures deviating from 1 atmosphere, and gravity. Solutes decrease Ψ_w and pressure above 1 atmosphere increases Ψ_w. Plants lack pumping organs yet move water and solutes over long distances, often exceeding 100 m in trees. Long-distance transport in plants occurs in tubes with low resistance and is driven by pressure gradients. In xylem, water and inorganic solutes move through large-diameter, dead tracheids and vessels, whereas in the phloem sugars and other organic molecules such as amino acids move in living sieve tubes. The flow of fluid in xylem tracheary elements is driven by negative pressure potential, called tension, that develops during rapid transpiration. In contrast, positive pressure drives flow in the phloem sieve tubes.

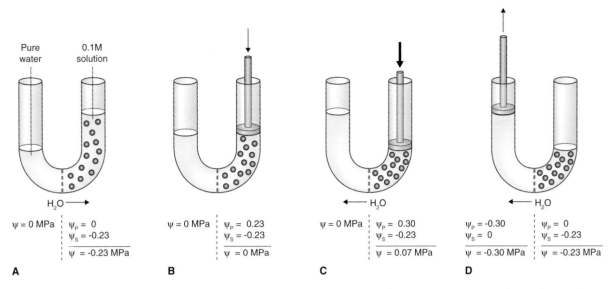

Figure 14.2   A simple osmometer illustrates the principles of osmosis. Water diffuses across semipermeable membranes from a region of high water potential (ψ_w) to one of a lower ψ_w. (A) Solutes decrease water potential. Water diffuses from pure water ($\psi_w = 0$) into 0.1 molal sucrose solution. Pressure can either increase or decrease water potential. (B, C) Pressures above atmospheric increase water potential. (D) Pressures below 1 atmosphere (tension) lower water potential. When the values of solute potential (ψ_s) and pressure potential (ψ_p) are known, ψ_w can be calculated: $\psi_w = \psi_s + \psi_p$.

14.2.3 Gravity increases water potential and is a large component of Ψ_w in trees

Gravity exerts an effect on water potential that is similar to the effects of pressure. As the water column increases in height, its water potential increases. The effect of gravity is expressed as gravitational potential, Ψ_g. The expression Ψ_g is the sum of three forces acting on water: the height of the water column, h; the density of water, ρ_w; and gravitational acceleration, g, as shown in Equation 14.3.

Equation 14.3 Forces contributing to gravitational potential Ψ_g

$$\Psi_g = \rho_w gh$$

At sea level and $20°C$, $\rho_w g \approx 0.01$ MPa m^{-2}. The effects of gravity on Ψ_w are considerable when dealing with water movement in tall trees. However, when the water relations of adjacent cells are considered, the gravitational component cancels out because the cells are at the same height. The expression for cellular water potential (Equation 14.1) can then be simplified to Equation 14.4.

Equation 14.4 Water potential in cells at the same height

$$\Psi_w* = \Psi_s + \Psi_p$$

14.3 Water uptake by plant cells

Water enters and leaves cells by osmosis, the diffusion of water across a semipermeable membrane. Water diffuses in and out of cells in response to gradients in Ψ_w, moving from regions of higher Ψ_w to regions of lower Ψ_w. In plant cells, Ψ_p is determined by turgor pressure. Turgor pressure builds up in a plant cell when water uptake causes the **protoplast** to push against the **rigid cell wall**; as we have seen (see Section 14.2.2), this pressure counteracts water uptake into cells. When cellular Ψ_p is greater than zero, a cell is stiff and is said to be **turgid** (Figure 14.3A); if Ψ_p drops to zero a cell becomes limp or **flaccid**. If a cell loses enough water so that its protoplast shrinks away from its wall, the cell is said to be **plasmolyzed** (Figure 14.3B). Turgor pressure is important in maintaining the structure of non-woody organs. If a plant loses too much water, its cells become limp and its leaves and stems **wilt**.

14.3.1 The permeability of biological membranes to water influences water uptake by plant cells

The driving force for water uptake, $\Delta\Psi_w$, is not the only parameter that determines how rapidly water moves into

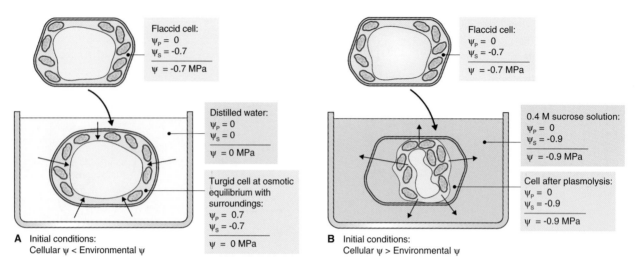

Figure 14.3   Movement of water into and out of cells is driven by gradients in water potential (ψ_w). Two identical plant cells are placed into different solutions. (A) When one of the cells is placed in pure water, water uptake leads to the development of a pressure potential (ψ_p) called turgor pressure. At equilibrium, water uptake ceases, the cell is fully turgid and the difference in ψ_w between the cell and the environment is zero. (B) When the other cell is placed in a solution whose ψ_s is lower than that of the cell, water is lost from the cell to the surrounding solution and the protoplast shrinks from the cell wall. In this case turgor pressure (ψ_p) is zero and the cell is said to be plasmolyzed.

cells. The **hydraulic conductivity**, Lp, of the plasma membrane can have a profound influence on how quickly cells equilibrate with their local environment. It is a measure of the ease with which water crosses the membrane and is expressed in units of volume of water per unit area per unit time per unit driving force ($m^3\ m^{-2}\ s^{-1}\ MPa^{-1}$). Equation 14.5 is a simplified expression that shows the relationship between J_v, the rate of water flow across a membrane (volume of water crossing the membrane per unit area of membrane per unit time, in $m^3\ m^{-2}\ s^{-1} = m-s^{-1}$), the hydraulic conductivity Lp and driving force $\Delta \psi_w$.

Equation 14.5 Rate of water flow across a membrane

$$J_v = Lp\Delta\psi_w$$

As we discussed in Chapter 5, water channels called aquaporins are present in cellular membranes. The opening and closing of these channels can be regulated, thus they can have a profound impact on the hydraulic conductivity of a membrane.

14.3.2 Diffusion and bulk flow drive movement of water and solute in plants

Two forces combine to drive the movement of water and solutes in plants: **diffusion** and **bulk flow**. Diffusion can account for the movement of molecules over short distances, for example, the movement of water across membranes by osmosis. The limitation of diffusion as an effective means of moving solutes over distances greater than a few millimeters was discussed in Section 5.2.1. The rate of diffusion decreases with the square of the distance traveled. Thus, while a solute could diffuse across a 10 μm cell in 2.5 seconds, it would take 50 hours to diffuse across an organ with a radius of 10 mm.

It is likely that cytoplasmic streaming is the most important contributor to the movement of water and solutes within plant cells. Cytoplasmic streaming is driven by the action of cytoskeletal motor proteins and is fueled by the hydrolysis of ATP, as we described in Chapter 4. This mode of intracellular transport is likely to be crucial not only for the functioning of giant algal cells such as those of *Nitella*, whose cells can be several millimeters in diameter and many centimeters in length (Figure 14.4), but also for ordinary parenchyma cells of vascular plants. Clearly, diffusion would not suffice to sustain the movement of water or solutes over such large distances.

In vascular plants, for transport over distances exceeding several millimeters, the movement of solutes occurs by bulk flow. Bulk flow is pressure-driven transport of a fluid and its solutes as a unit. This differs from diffusion in which each molecular species moves down its own concentration gradient. An example of bulk flow is the movement of water in the pipes of a house or blood flow in an animal circulatory system. Bulk flow of water through the plant, J_w, expressed as $m^3\ m^{-2}\ s^{-1}$, depends on a gradient in pressure as well as on the dimensions of the pathway and the viscosity of the

Figure 14.4   Habit photograph of *Nitella* with scale showing the relative size of the plant. Individual *Nitella* cells can exceed 10 cm in length and 2 mm in width. Single-celled branches form at each end of the cells.

3 cm

Key points The water potential, Ψ_w, of living plant cells is made up of a solute potential, Ψ_s, that is always less than zero and a pressure potential, Ψ_p, that is usually greater than zero. The pressure potential of living cells is also called turgor pressure; the cells of a wilted leaf have a pressure potential of zero. The effect of gravity on water potential, Ψ_g, depends on plant height; for plants that are only a few meters tall, Ψ_g is a relatively minor contributor to cellular Ψ_w. Osmosis is the diffusion of water through a semipermeable membrane from a region of high Ψ_w to one of lower Ψ_w. Biological membranes are examples of semipermeable membranes as they allow water to diffuse but generally not molecules that are charged or large. Movement of water into cells is governed by the difference in Ψ_w across the membrane ($\Delta\Psi_w$) and the permeability of the membrane to water. Despite the lipid content, biological membranes are permeable to water. The presence of aquaporin channels further increases their water permeability.

solution. These relationships were formalized by Poiseuille in Equation 14.6 where r is the radius of the conducting system, η the viscosity of the flowing solution, and ($\Delta\Psi_p/\Delta x$) the pressure gradient in the system.

Equation 14.6 The Poiseuille equation

$$J_w = (\pi r^4/8\eta)(\Delta\Psi_p/\Delta x)$$

It is clear from this equation that the radius of a tube dramatically affects flow of water. If the radius is doubled there is a 16-fold increase in flow rate; for example if r increases from 1 to 2, r^4 increases from 1 to 16. The smaller the radius of the transport system, the higher the resistance to flow. It follows that flow through a transport system that has a large radius and low resistance will be much higher than through a system where resistance is high. We shall see that long-distance transport in plants takes place in specialized cells that have large diameters and large intercellular connections. For example, sieve tubes in the phloem have cytoplasmic connections that are 30 times larger than plasmodesmata, resulting in much lower resistance to flow and faster rates of transport (Figure 14.5).

14.4 The role of plasmodesmata in solute and water transport

In Chapter 4 we saw that most of the cells of a plant are connected by fine cytoplasmic connections called **plasmodesmata** and that the plant body can be loosely divided into the **symplasm** and **apoplast** (Figure 14.6). The symplasm includes all the interconnected protoplasts in a plant; it forms a continuum between living cells in root tips and mesophyll cells of leaves. There are important **discontinuities** in the symplasm. In almost all plant species, **embryonic tissue** is **symplasmically isolated** from **maternal tissues**, and **stomatal guard cells** also lack plasmodesmata. The apoplast of a plant is the space outside the protoplasts and includes the network of cell walls and the contents of the water-conducting cells of the xylem.

Intercellular transport of water and solutes can occur through plasmodesmata, which provide high-resistance connections between almost all living cells in plants. The frequency of plasmodesmata varies widely from about 0.1 to 10 per square micrometer. Primary plasmodesmata form between cells as the cell plate grows during cell division. Strands of endoplasmic reticulum (ER) become trapped by the growing cell plate, establishing the sites at which plasmodesmata will form (see Chapter 4). Secondary plasmodesmata can form in

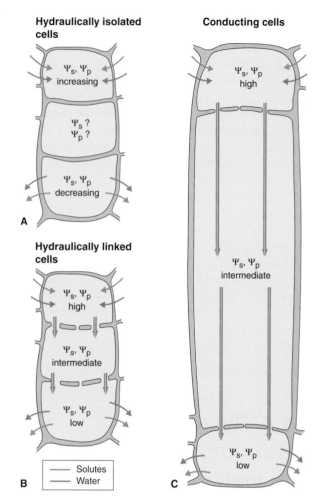

Hydraulically isolated cells

Ψ_s, Ψ_p increasing

Ψ_s? Ψ_p?

Ψ_s, Ψ_p decreasing

A

Hydraulically linked cells

Ψ_s, Ψ_p high

Ψ_s, Ψ_p intermediate

Ψ_s, Ψ_p low

— Solutes
— Water

B

Conducting cells

Ψ_s, Ψ_p high

Ψ_s, Ψ_p intermediate

Ψ_s, Ψ_p low

C

Figure 14.5 The path of water and solute movement in (A) cells that are hydraulically isolated, (B) cells that are connected by plasmodesmata, and (C) cells such as xylem vessels and phloem sieve elements that are connected by low-resistance cross walls. Cells in (A) are not interconnected by plasmodesmata and water moves from cell to cell by osmosis. Cells in (B) are connected by many small-diameter plasmodesmata through which water flows. Cells in (C), such as those of phloem and xylem, tend to have a larger diameter and cross walls with large perforations.

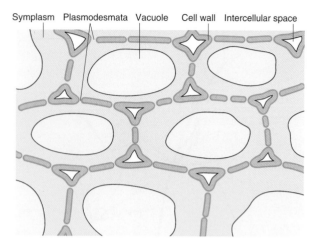

Symplasm Plasmodesmata Vacuole Cell wall Intercellular space

Figure 14.6 The apoplast (cell walls and intercellular air spaces) and symplasm. Plasmodesmata connect all of the cells in this diagram, meaning that they are all part of a single symplasm.

and the desmotubule is called the **central cavity**. The width of this central cavity has been estimated from electron microscope images at around 2–3 nm.

14.4.1 Plasmodesmata increase the flow of water and solutes between cells

Several experimental approaches have established conclusively that plasmodesmata increase the flow of water and solutes between cells. For example, measurements using microelectrodes show that cells that are linked by plasmodesmata are also **electrically coupled**. Using electrodes implanted into adjacent cells, it is possible to measure whether or not current injected into one cell is transmitted to an adjacent cell (Figure 14.9A). The extent to which current passes from cell to cell is called the **coupling ratio**, expressed as $\Delta E_1/\Delta E_2$, where ΔE_1 and ΔE_2 are the change in membrane potentials of cells 1 and 2 after injection of current as shown in Figure 14.9A. If cells 1 and 2 are devoid of plasmodesmata and therefore electrically insulated by adjacent plasma membranes, the coupling ratio would be 0, indicating no current flow. By contrast if both electrodes were placed in cell 1 the coupling ratio would be 1, indicating that there was no barrier to the flow of current between the electrodes. Measurements of the coupling ratio in rhizoids of the fern *Azolla* show coupling ratios as high as 0.4 and there is a striking positive correlation between the number of plasmodesmata and electrical coupling (Figure 14.9B).

pre-existing cell walls that originally lacked plasmodesmata. Examples of this can be found when graft unions are formed (Figure 14.7) and when host–parasite relationships are established, for example when mistletoes parasitize host plants.

One of the features common to all primary plasmodesmata is the presence of a **tube** formed by continuity of the plasma membrane between two adjacent cells (Figure 14.8). Enclosed within the tube is a **compressed strand of ER**, known as the **desmotubule**, in the centre of which is a **central rod**. A **cytoplasmic sleeve** lies between the desmotubule and the plasma membrane, and the space between the cytoplasmic sleeve

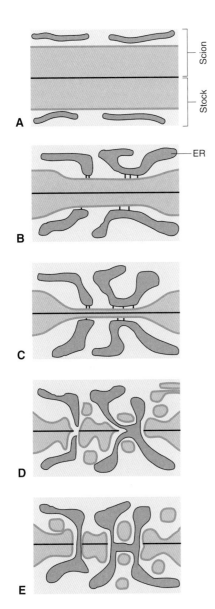

Figure 14.7   Formation of secondary plasmodesmata between adjacent cells during a graft union between stems cells from a stock plant and a scion branch. (A) Cells from the stock and scion stems are pressed together; initially, the cells are separated by debris from the cut surfaces. (B) First, in each adjacent cell, the endoplasmic reticulum (ER) fuses with the plasma membrane; (C) the adjacent cells walls thin; (D) adjacent plasma membranes and ER fuse; and (E) plasmodesmata are formed.

14.4.2 Fluorescent probes provide an estimate of the size exclusion limit of plasmodesmata

The injection of fluorescent dyes into cells has been a powerful tool for looking at plasmodesmatal function.

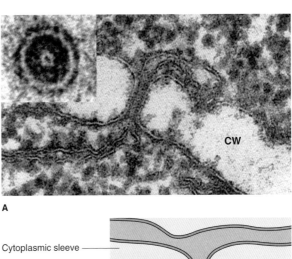

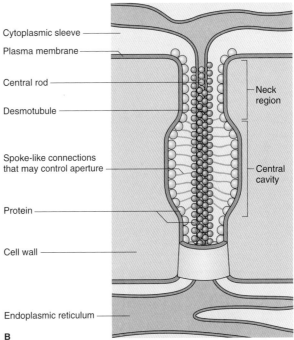

Figure 14.8   Structure of plasmodesmata. (A) Electron micrographs of a plasmodesma from *Azolla* root tip cells in cross-sectional (inset) and side view. (B) Diagram showing structural details of a typical plasmodesma in side view. CR, central rod; CS, cytoplasmic sleeve; CW, cell wall; DT, desmotubule; ER, endoplasmic reticulum; PM, plasma membrane; SP, spoke-like connections between the DT and PM that may control plasmodesmatal aperture. Blue and pale green circles represent plasmodesmata-specific proteins.

Experiments with fluorescently-labeled molecules of known molecular mass have allowed experimenters to determine the **size exclusion limit** (**SEL**) of plasmodesmata. Fluorescently-labeled molecules are injected into cells and fluorescence is monitored microscopically (Figure 14.10). Using this approach the SEL of most plasmodesmata was found to be around 800 Da, although there can be marked deviations from this value. Metabolic inhibitors, or stresses such as anoxia, can cause an increase in SEL from 800 Da to about

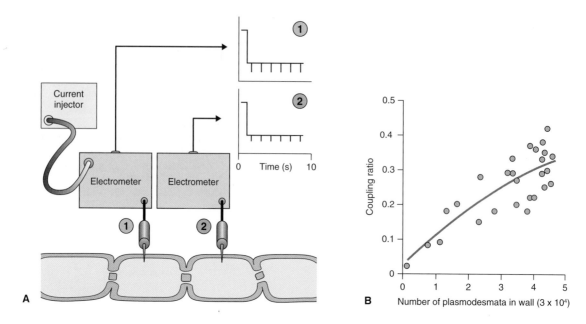

Figure 14.9 Plasmodesmata increase electrical coupling between cells. (A) Technique for measuring electrical coupling between cells. Electrodes are placed in the cytoplasm of adjacent cells. Passage of current between electrodes, called the coupling ratio, indicates the presence of intercellular connections. (B) Graph showing the relationship between current and the number of plasmodesmata. Symplasmically isolated cells show no electrical coupling, whereas cells connected by plasmodesmata show a passage of current. The extent of current is proportional to the number of plasmodesmata that interconnect the cells.

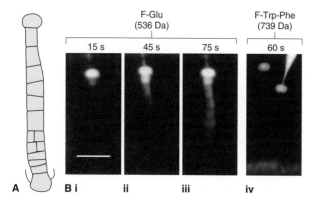

Figure 14.10 Microinjection of markers of differing sizes illustrates the size exclusion limit of plasmodesmata of *Abutilon* nectary trichomes. (A) Diagram of a trichome. (B) (i) Injection of a 536 Da fluorescent marker; (ii, iii) symplasmic movement of the dye; (iv) injection of a large (739 Da) marker shows no sign of intercellular movement via the symplasm.

5–10 kDa. On the other hand, elevation of cytosolic Ca^{2+}, from a resting concentration of around 100 nM to about 1 μM, reduces the SEL in stamen hair cells. Such experiments show that the SEL of plasmodesmata can vary in response to a wide range of conditions, allowing plants to regulate the flow of water and solutes between cells.

14.4.3 Endogenous macromolecules move from cell to cell via plasmodesmata

A surprising discovery made almost 10 years ago was that endogenous regulatory molecules, including RNAs and proteins, move from cell to cell via plasmodesmata. A transcription factor, KNOTTED1 (KN1), responsible for the abnormal development of leaves of the *Knotted1* (*Kn1*) mutant of maize (see Chapters 3 and 12), has been shown to move between cells in certain layers of the maize meristem. The KN1 protein has a molecular mass of 39.8 kDa and is present in the nuclei of all cell layers of the meristem of *Kn1* plants, including the epidermal layer (Figure 14.11). Surprisingly, mRNA encoding the KN1 protein is absent from epidermal cells of the meristem indicating that KN1 protein in the epidermis was synthesized elsewhere and transported to epidermal cells via plasmodesmata. Data indicating that KN1 protein mediates its own movement have come from experiments in which fluorescently-labeled KN1 protein was injected into leaves of tobacco and was able to move freely among the mesophyll cells. These experiments also showed that KN1 increased the SEL of leaf plasmodesmata from about 800 Da to between 20 and 40 kDa.

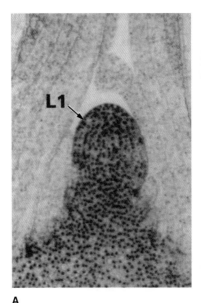
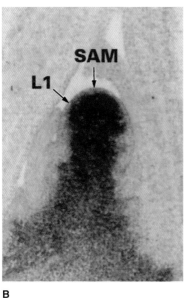

A   B

Figure 14.11   Plasmodesmata allow the movement of proteins as large as 40 kDa. (A) Immunolocalization of KN1 protein in *Zea mays* shows that the protein is found in all layers of the shoot apical meristem (SAM). (B) In situ hybridization shows that mRNA transcripts of the *Kn1* gene are expressed in all layers of the meristem except for the outermost epidermal layer (L1). This indicates that KN1 found in L1 is synthesized elsewhere and moves there through plasmodesmata. An interesting aspect of this phenomenon is that KN1 facilitates its own movement.

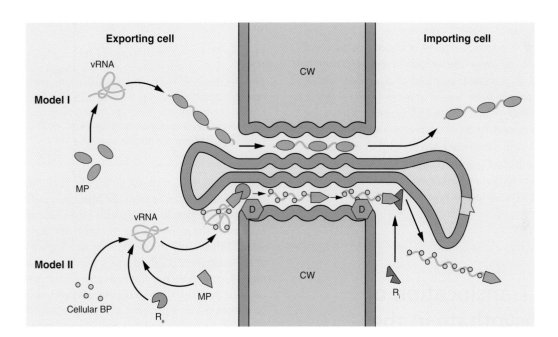

Figure 14.12   Models showing how movement proteins (MP) may facilitate the transport of viral RNA through plasmodesmata. One version, Model I, predicts that only one MP is required for RNA transport, whereas Model II invokes the participation of endogenous cellular proteins (cellular BP), putative receptor proteins (R_e, R_i) and docking proteins (D) as well as MP in RNA transport. CW, cell wall.

14.4.4 Viral RNA can move from cell to cell via plasmodesmata

It was suspected for many years that viruses moved from cell to cell in plants via plasmodesmata. The work with KN1 showing that proteins can facilitate their own movement provided a clue for the mechanism of viral RNA movement. We now know that a special class of **movement proteins** facilitates the transport of viral RNA through plasmodesmata (Figure 14.12).

In fact, movement proteins alone can increase the size exclusion limit of plasmodesmata, as shown by experiments in which injection of the tobacco mosaic virus (TMV) movement protein into tobacco leaves that lacked TMV caused the SEL of mesophyll cells to increase from 800 Da to 10 kDa.

> **Key points** Almost all cells of an individual plant are interconnected by plasmodesmata, cytoplasmic extensions lined by plasma membrane. Only stomatal guard cells and embryonic tissue are cytoplasmically isolated. Cells that are interconnected by plasmodesmata make up the symplast and all space outside the plasma membrane is called the apoplast. The apoplast includes cell walls, the contents of tracheary elements and intracellular spaces. Normally only relatively small molecules pass through plasmodesmata; however cells are able to regulate the size exclusion limit of plasmodesmata, allowing macromolecules including viruses and endogenous mRNA and proteins to move from cell to cell within the symplasm. KNOTTED1 (KN1) protein is an excellent example of a protein whose mRNA is absent from one of the cell layers in which KN1 is normally found. The mRNA encoding KN1 is not detectable in epidermal cell layers of the apical meristem but KN1 protein is abundant. KNI protein has the capacity to increase the size exclusion limit of plasmodesmata that interconnect the various cell layers in the meristem. This allows the protein to move symplasmically from its site of synthesis to epidermal cells.

14.5 Translocation of photosynthate in the phloem

The movement of dissolved sugars from parts of a plant that have net sugar production (**source regions**) to sites that have net sugar consumption (**sink regions**) is the primary function of the phloem. For example, photosynthetic cells in a mature leaf are a source of sugar while roots and seeds, which are converting sugar to starch, and expanding leaves, which are using sugar to support growth, are sinks. Translocation of newly synthesized sugars from source tissues in a photosynthesizing leaf to sink tissues of the plant is illustrated in Figure 14.13. The generally accepted mechanism for the transport of water and solutes in the

Figure 14.13   (A) Photograph of *Zebrina* plant labeled with $^{14}CO_2$ and (B) an autoradiogram of the labeled plant. One leaf of the plant (indicated by the arrow) was exposed to $^{14}CO_2$ and an autoradiogram was taken to display the location of labeled photosynthate. Note that label is prominent in the youngest leaves and meristem as well as the root system but is largely absent from the mature leaves.

phloem was first proposed in 1929 by Ernst Munch and is now widely referred to as the **Munch hypothesis**. Cells in a source region load sugars into the phloem, increasing the osmotic concentration of the sieve tubes; water diffuses into the sieve tubes, developing a high Ψ_p, at the source end. Cells in a sink region unload sugars and use or polymerize them, as a result decreasing osmotic concentration of the sieve tubes, causing water to diffuse out and lowering Ψ_p. The pressure difference between the source and sink end of sieve tubes drives the bulk flow of solution within sieve tubes from source to sink. A model that illustrates this mechanism is shown in Figure 14.14. Source–sink relationships are discussed further in Chapter 17.

14.5.1 Sieve elements and companion cells are unique cell types in the phloem of flowering plants

Two unique cell types make up the phloem of flowering plants: solute-conducting **sieve elements** and their associated **companion cells** (see Figure 14.29).

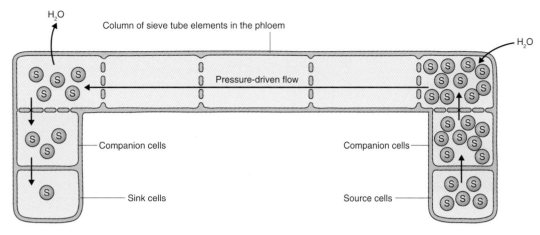

Figure 14.14 The principals of pressure-driven flow in the phloem. Sugar (S) loaded from source cells increases solute concentration. Water uptake leads to an increase in pressure potential in the phloem, and unloading of sugar at sink cells lowers solute concentration causing water to be lost and flow to occur.

At functional maturity, sieve elements (also known as sieve tube members) are elongate, living cells that lack a nucleus, ribosomes, microtubules, microfilaments and tonoplast and have perforated end walls referred to as **sieve plates**. Cytoplasmic connections pass through perforations in the sieve plate. Sieve elements of flowering plants are attached end-to-end at sieve plates to form long, low-resistance **sieve tubes** that can transport solutes both upward and downward in the plant (see Figure 1.29).

Developmental studies show that a sieve element and its associated companion cell are derived from a common **mother cell**. Unlike sieve elements, companion cells have nuclei, ribosomes and abundant mitochondria. They are regarded as the powerhouses and information centers for their associated sieve elements. Numerous plasmodesmata connect sieve elements to companion cells and the two cells are often referred to as the **sieve element/companion cell complex** (**SE/CC complex**) (Figure 14.15). With the exception of the SE/CC complex at the loading and unloading ends of the phloem, the phloem is symplasmically isolated from neighboring tissues because plasmodesmata are largely lacking.

14.5.2 Sieve elements contain high concentrations of solutes and have high turgor pressure

Because sieve elements transport solutes, especially organic solutes, they have high turgor pressures. Several ingenious methods have been devised to sample the contents of sieve elements and simultaneously measure their Ψ_p. Notable among these methods is the use of the aphid stylet. Aphids are insects that feed passively by

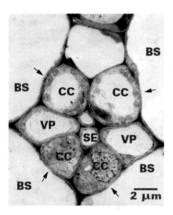

Figure 14.15 Electron micrograph showing a cross-section of phloem in a leaf including phloem sieve tube elements (SE), their adjacent companion cells (CC) and parenchyma cells (VP). The companion cells have abundant organelles, while the sieve tube elements lack organelles. Photosynthetic bundle sheath cells (BS) surround the phloem.

tapping into sieve tubes with their stylet (Figure 14.16A) and allowing turgor pressure to force phloem sap into their digestive system. To sample the phloem, the stylet of a feeding aphid can be severed and turgor pressure forces exudates from the cut end of the stylet (Figure 14.16B) where it can be collected and analyzed.

Table 14.2 shows an analysis of the contents of phloem exudates alongside those of exudates from the xylem obtained from two species. Sucrose is the predominant solute in the phloem but is absent from the xylem. The concentration of amino acids and potassium is also much higher in the phloem than the xylem. The general characteristics of phloem and xylem exudates are summarized in Table 14.3. In addition to emphasizing the difference in the concentration of organic solutes

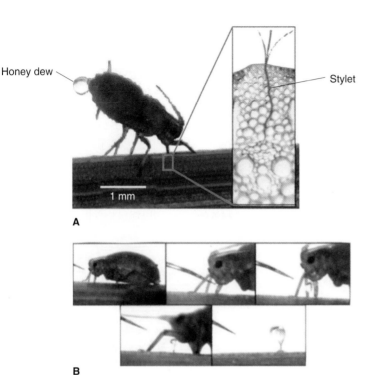

Honey dew

Stylet

1 mm

A

B

Figure 14.16   (A) Aphid feeding with stylet inserted into the phloem. Note the droplet of honeydew exuded from the rear of the insect. Inset shows micrograph with a stylet penetrating a phloem sieve element. (B) Series of images showing severing of the aphid stylet. The last image of the series shows a drop of phloem exudate exiting the severed end of the stylet.

Table 14.2 Comparison of xylem and phloem exudates from lupine and tree tobacco.

	Lupinus angustifolius		*Nicotiana glauca*	
	Xylem (mM)	Phloem (mM)	Xylem (mM)	Phloem (mM)
Sucrose	ND	490	ND	460
Amino acids	20	115	2.2	83
Potassium	4.6	47	5.2	94
Sodium	2.2	4.4	2.0	5.0
Phosphorus	NA	NA	2.2	14
Magnesium	0.33	5.8	1.4	4.3
Calcium	1.8	1.6	4.7	2.1
Iron	0.02	0.13	0.01	0.17
Zinc	0.01	0.08	0.02	0.24
Nitrate	0.50	Tr	NA	ND
pH	5.9	8.0	5.7	7.9

NA, not available; ND, not detectable; Tr, trace.

between phloem and xylem, which is reflected in the large differences in Ψ_s, Table 14.3 shows the contrast in pH between the two transport systems. The high pH of phloem sap reflects the high activity of the plasma membrane H^+ pump that transports H^+ into the apoplast to support sucrose uptake into the SE/CC complex (see Section 14.6.2).

14.5.3 Sieve elements have open sieve plates that allow pressure-driven solute flow

There has been considerable controversy about whether the structure of sieve elements is consistent with

Table 14.3 General characteristics of xylem and phloem exudates.

	Phloem exudates	Xylem exudates
Sugars	100–300 g l⁻¹	0 g l⁻¹
Amino acids (mostly Glu, Asp, Gln, Asn)	5–40 g l⁻¹	0.1–2 g l⁻¹
Inorganics	1–5 g l⁻¹	0.2–4 g l⁻¹
Total solutes	250–1200 mmol kg⁻¹ (Ψ_s c. −0.6 to −3 MPa)	10–100 mmol kg⁻¹ (Ψ_s c. −0.02 to −0.2 MPa)
pH	7.3–8.0	5.0–6.5

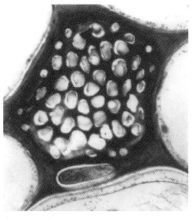

A   **B**

Figure 14.17 Electron micrographs showing (A) an unblocked sieve plate and (B) an open sieve tube element that is not occluded. The asterisk marks the open channel through which solute transport occurs. The tissue was prepared for microscopy by ultra-rapid freezing.

pressure-driven flow. Electron micrographs of sieve tube elements have often shown that cytoplasmic connections at the sieve plates appear to be blocked. Microscopists have concluded that this would preclude pressure-driven bulk flow of water and solutes through the phloem. However, this blockage has been shown to be an artifact of the way in which the tissue was prepared for microscopy. In this case, living tissue was cut into pieces before fixation. Because living sieve elements have high turgor pressure, when they are damaged or cut their cell contents surge to the point of the cut or wound, blocking the sieve plates of each sieve element. In vivo, this may be a mechanism to prevent leakage of sieve element contents when the phloem is damaged. Rapid freezing of individual vascular bundles in situ before preparation for microscopy prevents surging and has shown that neither the lumen of the sieve element nor the pores of the sieve plate are blocked by cellular debris (Figure 14.17).

Key points The transport of photosynthate in the phloem occurs through highly specialized sieve tube elements. New techniques for preserving living tissue for microscopy that involve ultra-cold, rapid freezing has shown that cytoplasmic connections through the sieve plates are free of blockages, which is essential for pressure-driven flow. The movement of solutes in the phloem is explained by the Munch pressure–flow hypothesis, which proposes that low-resistance tubes connect the loading end of the system, the source, to an unloading end, the sink. The loading of sugars into the source lowers the Ψ_w, causing water to flow into sieve elements by osmosis. Conversely, unloading of solute at the sink raises Ψ_w in the sieve tube and lowers it in sink cells, causing water to leave the unloading end of the phloem. Experiments with aphids, insects that feed directly on phloem contents, have established the occurrence of sugar gradients in phloem as well as the presence of positive pressure.

14.6 Phloem loading, translocation and unloading

The movement of sugars into the phloem at the source is referred to as **loading**, and the delivery of translocated sugars to the sink is termed **unloading**. Studies of the mechanisms of loading and unloading have focused on the relative contributions of symplasmic and apoplastic processes and have revealed marked species-to-species variations in strategies for moving photosynthate from sources to sinks.

14.6.1 At the source, phloem loading can occur from the apoplast or through the symplasm

Phloem loading has been almost exclusively studied in photosynthetic leaves. There are two mechanisms by which sugars may be loaded into the phloem. In plants with **apoplastic loading**, source cells release sugar into the apoplast; from there it is actively loaded into the SE/CC complex. In plants with **symplasmic loading**, there are uninterrupted symplasmic connections between photosynthesizing mesophyll cells and sieve tubes so that sugars diffuse via plasmodesmata from source cells to the SE/CC complex. Species with apoplastic loading have **ordinary companion cells**. These cells have many plasmodesmatal connections to their partner sieve elements, but few or none to surrounding cells (Figure 14.18A). The cell wall of an ordinary companion cell is often highly modified by the presence of ingrowths that are found in the cell wall facing away from the sieve element. These wall ingrowths have the effect of increasing the surface area of the plasma membrane to facilitate uptake of assimilates. This type of companion cell is referred to as having **transfer cell** characteristics. The plasmodesmata connecting the companion cell to the sieve element are complex and often highly branched on the companion cell side (Figure 14.18B). The absence of plasmodesmata between the SE/CC complex and neighboring cells ensures that they are symplasmically isolated and the SE/CC complex is referred to as having a **closed configuration**.

Species with **symplasmic loading** have **intermediary companion cells**. These cells have many plasmodesmata connecting them both to assimilate-producing cells and to sieve elements. Intermediary companion cells characterize the **open type** of SE/CC complex (Figure 14.18A). In leaves of plants that accumulate

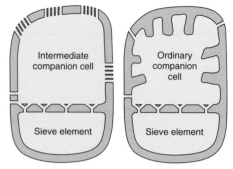

Number of plasmodesmata at the MC/CC interface	> 10 per µm²	< 0.1 per µm²
Uphill sugar gradient	Yes	Yes
Predominant transport sugar	Sucrose, galactosyl-oligosaccharides	Mainly sucrose

A

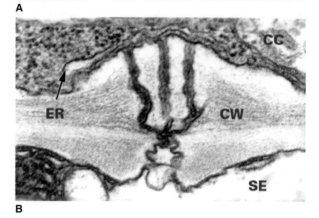

B

Figure 14.18 Symplasmic (left) and apoplastic (right) phloem loading. (A) Diagrams illustrating the types of companion cells commonly associated with sieve elements (SEs). On the left, intermediate companion cells (IC) are characteristic of symplasmic loaders and have many plasmodesmata that connect them to photosynthesizing cells. Sucrose and oligosaccharides are commonly loaded via intermediate companion cells. On the right, ordinary companion cells with cell wall ingrowths have very few plasmodesmata and load sucrose from the apoplast. CC, companion cell; MC, mesophyll cell. (B) Both types of companion cells are connected to SEs by branched plasmodesmata. CC, companion cell; CW, cell wall; ER, endoplasmic reticulum.

assimilates in the phloem by symplasmic loading, plasmodesmata form a continuum between the photosynthesizing cells and the SE/CC complex. As we shall see, however, the SEL of plasmodesmata between intermediary cells and sieve elements is larger than those between intermediary cells and sugar-producing cells.

14.6.2 Sucrose and other non-reducing sugars are translocated in the phloem

The mechanism of phloem loading determines the type of sugar that is transported in the phloem. In apoplastic loaders, where the SE/CC complex is symplasmically isolated, sucrose is transported almost exclusively; symplasmic loaders transport a range of sugars including glucose, fructose and larger oligosaccharides, such as raffinose, a trisaccharide of galactose. Most crop plants, including cereals and grasses, are apoplastic loaders, but trees and most tropical plants load assimilates via the symplasm (Figure 14.19).

During apoplastic loading of the SE/CC complex, sucrose synthesized in photosynthesizing leaf cells may be transported to the vicinity of the SE/CC complex via

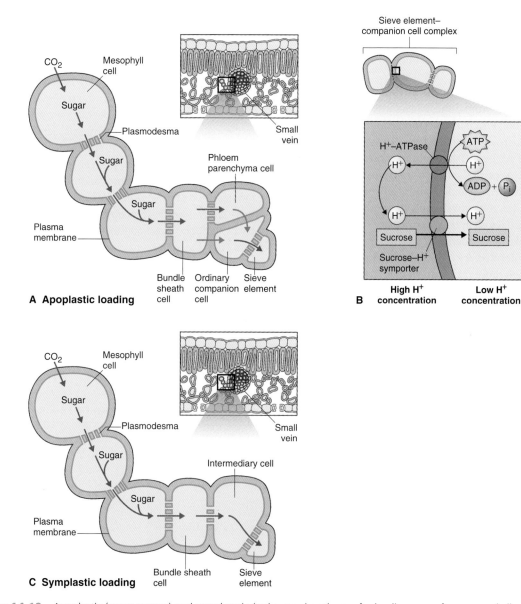

Figure 14.19 Apoplastic (green arrows) and symplasmic (red arrows) pathways for loading sugar from mesophyll cells into sieve tube elements. (A) In the apoplastic pathway sugar moves through plasmodesmata from mesophyll cells up to companion cells where it moves into the apoplast and is actively loaded into ordinary companion cells. (B) Active uptake of sucrose from the apoplast by a companion cell. The plasma membrane H^+-ATPase pumps protons out of the cell, creating a proton gradient. The energy of this gradient drives the uptake of sucrose via an H^+/sucrose symporter. (C) In the symplasmic pathway, sugar moves via plasmodesmata from mesophyll cells to intermediary cells and then into sieve tube elements.

the symplasm, but it then moves into the apoplast where it diffuses to the SE/CC complex; here it is taken up by secondary active transport (Figure 14.19A). Sucrose accumulation by the SE/CC complex involves plasma membrane H^+/sucrose co-transporters (Figure 14.19B). Experimental evidence shows that the uptake of sucrose from the apoplast is dependent on the activity of the plasma membrane H^+ pump, and treatments that reduce the pump's activity or alkalinize the apoplast reduce sucrose uptake. The observation that the phloem of apoplastic loaders transports sucrose almost exclusively also supports the idea that transport into the phloem is brought about by specific carriers. A variety of reducing and non-reducing sugars is found in the non-transporting cells of leaves, but only sucrose is found in the phloem.

This contrasts with the situation found in the phloem of plants that are symplasmic loaders where the phloem contains the oligosaccharides raffinose and stachyose in addition to sucrose (Figure 14.19C). The synthesis of raffinose and stachyose is thought to confer a degree of specificity to the process of sugar transport in symplasmic loaders. Raffinose and stachyose are tri- and tetrasaccharides, respectively, and are synthesized from sucrose and galactose in the intermediary companion cells of symplasmic loaders. The plasmodesmata connecting intermediary cells to sieve elements have a larger SEL than those that connect them to photosynthesizing cells. Sucrose (M_r 342) can diffuse through the smaller plasmodesmata into intermediary cells, but the larger size of raffinose (M_r 504) and stachyose (M_r 660) allows them only to move through the large SEL plasmodesmata to the sieve element (Figure 14.20). Oligosaccharide synthesis serves two

functions. It provides a mechanism to lower sucrose concentration in the SE/CC complex which then allows more sucrose to diffuse into the SE/CC complex down its concentration gradient. Also, the larger oligosaccharides are trapped and cannot diffuse back to the photosynthesizing cells. Analysis of the sugars in mesophyll cells and intermediary cells of the leaf supports this polymer-trapping model. Thus photosynthetic cells contain high concentration of sucrose but lack raffinose and stachyose.

Key points Sugars may be loaded into sieve elements for transport to sinks by one of two routes. One route is via the symplasm where photosynthesizing cells of leaves are connected directly via plasmodesmata to the loading end of the phloem sieve element. Polymer trapping is a mechanism that ensures that sugars travel only from the photosynthesizing cell to the sieve element. Sucrose is converted to larger oligosaccharides such as raffinose and stachyose in companion cells. Because of their larger size these oligosaccharides cannot diffuse back through plasmodesmata to photosynthesizing cells, but can diffuse to sieve tube elements through plasmodesmata with larger SELs. The other route of sugar transport to sieve elements is via the apoplast. In this case sucrose is released from photosynthesizing cells; it then diffuses in the apoplast to the companion cell/sieve element complex where it is taken up by sucrose/H^+ co-transporters.

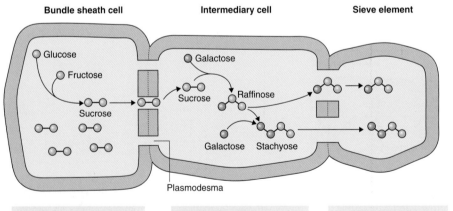

Figure 14.20   The principles of polymer trapping.

Sucrose, synthesized in the mesophyll, diffuses from the bundle sheath cells into the intermediary cells through the abundant plasmodesmata	In the intermediary cells, raffinose and stachyose are synthesized from sucrose and galactose, thus maintaining the diffusion gradient for sucrose. Because of their larger sizes, they are not able to diffuse back into the mesophyll	Raffinose and stachyose are able to diffuse into the sieve elements. As a result, the concentration of transport sugar rises in the intermediary cells and the sieve elements

14.6.3 Long-distance pressure–flow in the phloem is not energy-dependent

Because of lingering uncertainties about the existence of open pores in the sieve plates of the phloem, several alternative hypotheses have been proposed to explain solute translocation. One of these argues that cytoplasmic streaming is likely to move solutes through phloem from sieve element to sieve element; measured rates of translocation argue against such a mechanism. Solutes move in the phloem at rates that exceed 1 cm min^{-1} but cytoplasmic streaming rarely exceeds 1 mm min^{-1}. Also, while cytoplasmic streaming is an energy-dependent process, translocation along sieve tubes appears to be energy-independent. In one experiment a willow stem was cooled to 0°C and the movement of radioactively-labeled assimilates was compared with translocation at 23°C. If translocation required energy, one would expect that the rate of translocation would be much slower at 0°C than at 23°C. As Figure 14.21 shows, there was no difference in translocation rates between these two temperatures indicating that temperature-dependent processes are not required.

14.6.4 Phloem unloading involves a series of short-distance transport events

Sugars and other solutes transported in sieve tubes are unloaded into sink tissues where sugars may be used in respiration or converted to starch for storage. In a few plants, e.g. sugar beet (*Beta vulgaris*), sucrose is actively accumulated in the vacuoles of sink cells in the root. By removing and consuming sugar, phloem unloading increases Ψ_s and water diffuses out of the SE/CC complex, causing the pressure at the sink end of a sieve tube to be lower than that at the source end, a strict requirement of the Munch hypothesis. An interesting aspect of phloem unloading is that, unlike the loading process, it lacks specificity. In most cases solutes and water leave the SE/CC complex via the symplasm; therefore the unloading end of the phloem is not symplasmically isolated as is the loading end in apoplastic loaders. The plasmodesmata at the unloading end of the phloem have large SELs and consequently much higher conductivities than those in other parts of the plant. Evidence that supports symplasmic unloading includes the observation that the rate at which unloading occurs is generally at least an order of magnitude faster than what

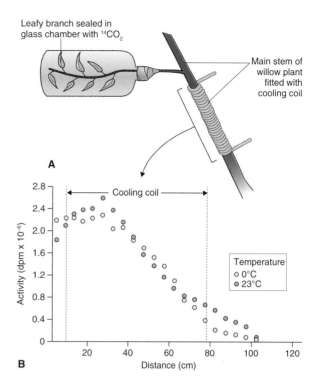

Figure 14.21 Effect of temperature on phloem transport. (A) Leafy willow branches were labeled with $^{14}CO_2$ and the stems below the labeled leaves were either maintained at 23 °C or cooled to 0 °C. (B) A plot of labeled sugar transport in the stem shows only a very minor impact of cooling on the rate of transport.

could be achieved by membrane transport. The large SELs of between 10 and 20 kDa make these high rates of symplasmic transport possible in plasmodesmata at the unloading SE/CC complex of several species.

14.7 Water movement in the xylem

The primary function of the xylem is to transport water and inorganic ions (minerals) from the roots to the stems and leaves. In contrast to the phloem, where a sugar solution is pushed along by an osmotically-generated pressure gradient, in the xylem water is pulled up by **tension** generated by the evaporation of water from leaf cell walls. Absorption of sunlight provides the energy that drives this process. Surface tension in the cell walls that line air spaces in leaves is transmitted via the apoplast to the water columns in the xylem and eventually extends to the root surface. The size of the forces that are involved should not be underestimated. For example, the pores in the cells walls of leaf parenchyma cells are as small as 5–7 nm in diameter but pressures in excess of 50 MPa

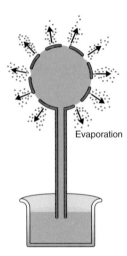

Figure 14.22 A porous ceramic globe attached to a thin glass tube that is immersed in water illustrates the principle of water transport by cohesion–tension. Evaporation of water from the fine pores of the ceramic globe pulls water up the glass tube, a phenomenon made possible by the ability of water molecules to form strong intermolecular hydrogen bonds.

would be required to remove water from them. Tensions in the xylem that develop from surface tension are equally impressive and can exceed −5 MPa. The ability of water molecules to form three hydrogen bonds (see Chapter 2) provides the **cohesive** strength to allow water to be pulled up the stems of even the tallest trees. That living tissues are not required for tension-driven water movement to take place can be illustrated by the working model shown in Figure 14.22. In this model, a porous ceramic globe attached to a narrow glass tube can mimic a living plant as far as evapotranspirational water movement is concerned. This hypothesis for the mechanism of water movement in the xylem is called the **cohesion–tension hypothesis**.

14.7.1 Water-conducting tissue of the xylem consists of low-resistance vessels and tracheids

Vessels and **tracheids**, known collectively as **tracheary elements**, make up the water-conducting cells of the xylem (see Figures 1.28 and 14.23). At functional maturity, water-conducting tracheary elements are dead and hollow, with thick, rigid secondary cell walls that are impregnated with lignin. Vessels are low-resistance tubes made up of **vessel elements** that have diameters as large as 70–200 μm and can be 1 mm or more in length. Vessel elements have secondary cell walls that are often

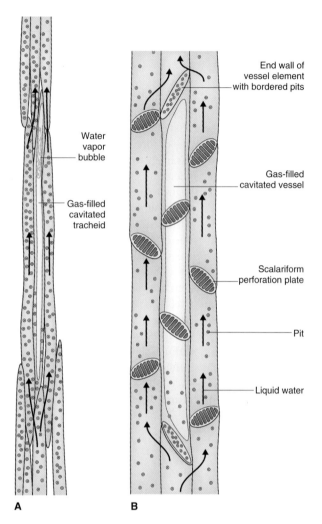

Figure 14.23 The path of water transport in tracheary elements. (A) Movement of water between adjacent tracheids occurs exclusively through pits in the cell wall. Pits are areas of the cell wall lacking secondary wall but where primary wall is still present. (B) Water movement along vessels occurs largely through unobstructed perforation plates in end walls, although as this diagram illustrates there may be lateral transport between vessels via pits. This figure also illustrates the phenomenon of cavitation—the introduction of air bubbles in the water column blocking water transport. Physical forces such as wind can cause air bubbles to form in tracheary elements. Because tracheids are interconnected by high-resistance pits, gas bubbles can be trapped in a single cell. The larger diameter of vessels elements and the presence of large perforations in their end walls (scalariform perforation plates), means that gas bubbles can expand to occupy a long series of continuous vessel elements. While this figure shows a vessel made up of three vessel elements, in nature a vessel may be made up of many vessel elements and be several meters in length.

laid down in annular, spiral or reticulate patterns. During maturation, the end walls of vessel elements dissolve forming **perforation plates**. The pattern of cell wall dissolution leaves perforations of various shapes. Being open-ended and having relatively large diameters makes vessel elements ideally suited for low-resistance water transport (Figure 14.23B). Vessels help define the flowering plants, because with a few exceptions, vessels are lacking in gymnosperms and other groups of non-flowering plants.

Tracheids are present in all vascular plants; they are narrower and longer than vessel elements (Figure 14.23A). Tracheids are also stacked to form long tubes but they lack perforation plates on their oblique end walls. Instead, some areas of the cell wall lack lignin; these areas are called **pits**. Pits occur in pairs between adjacent tracheary elements; only porous primary cell walls, sometimes called **pit membranes**, separate the cells that share a **pit pair** (see Figure 1.28). Water movement between tracheids occurs exclusively through pits. Pits can be simple pores or more complex structures represented by the bordered pits found in tracheary elements of some gymnosperms. Their smaller diameter and lack of perforations mean that tracheids have a higher resistance to water flow than do vessels.

14.7.2 Transpiration provides the driving force for xylem transport

Transpiration is the evaporation of water from the above-ground surfaces of plants. This is the driving force for movement of water in the xylem. During periods of rapid **evapotranspiration** water can travel up the xylem at speeds in excess of 4 mm s^{-1}. Water loss from the minute pores in cell walls leads to the build up of tension in the xylem that can exceed -5 MPa. Because the concentration of solutes in the xylem is very low (see Tables 14.2 and 14.3), tension makes up the largest component of the Ψ_w in the xylem of most plants. In small herbaceous plant xylem $\Psi_w = \Psi_p$ because Ψ_s and Ψ_g are negligible. In tall trees, however, Ψ_g contributes to a significant increase in Ψ_w of tracheary elements; for each rise of 10 m in height Ψ_g adds 0.1 MPa to Ψ_w. In coast redwood (*Sequoia sempervirens*), where heights of 100 m are common, Ψ_g is very important, and a driving force of -1.0 MPa is required to overcome the effects of Ψ_g.

The frictional resistance of the xylem must also be overcome to move water a given distance; it has been estimated than for a 100 m tall tree this resistance is equal to about 2 MPa. When this value is added to the effects of gravity it becomes clear that there must be a pressure difference of at least 3 MPa to move water to the top of a tree of this height. In practice, xylem tension easily exceeds this value in most plants.

Xylem tension can be measured easily using a pressure bomb (Figure 14.24). The pressure bomb operates by a simple principle. When a twig, leaf or similar piece of tissue is cut off, the tension on xylem water columns is released and the water columns retreat from the cut surface. This is similar to what happens if one severs a stretched rubber band. The severed plant part is inserted into a strengthened chamber ('the bomb'), with the cut end of the tissue remaining outside the chamber at normal atmospheric pressure. Then pressure is applied to the tissue in the chamber until the water in the xylem just returns to the cut surface. The pressure required to force water to the cut surface is a pressure equal, but opposite in sign, to the tension present in the xylem before it was cut from the intact plant. Measurements of xylem tension give average values of around -3 MPa for crop plants, -4 MPa for trees and -10 MPa for desert plants.

14.7.3 Under special circumstances sucrose may be transported from roots to shoots within the xylem

As we have seen, the concentration of sugars in xylem sap is usually very low (see Tables 14.2 and 14.3), and transport in xylem occurs by cohesion–tension. However, in the spring, many trees of the genus *Acer* (maples) leak a sugary 'sap' when a hole is drilled into the woody part of the stem (xylem). The best-known species in this regard is sugar maple, *Acer saccharrum*, whose sap typically contains 2–3% sucrose and is used to make maple syrup. These observations raise two questions: why does the xylem stream contain sugar and what is the mechanism by which positive pressure is generated within the xylem? During the winter, trees store starch in their roots. In the spring, when the tree is ready to leaf out, starch is converted into sucrose that diffuses into the apoplast and, in these species, it leaks into tracheary elements in the xylem.

The mechanism by which sugary sap moves up the xylem in sugar maple has been controversial, and only recently have plant physiologists agreed on how sap flow is likely to occur. Clues to the mechanism come from anecdotal accounts of conditions that must prevail for sap flow to occur, notably freezing nights followed by above-freezing days in springtime. By measuring sap movement in isolated segments of maple stems, it has been shown that water is pulled up in the stem at night and pushed up during the day. Increased pressure during the day will cause fluid to flow from a hole bored into the xylem. Why does sap rise in xylem at night and why does pressure

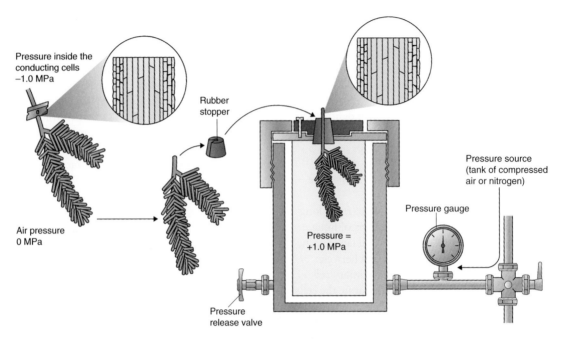

Figure 14.24 A pressure bomb allows measurement of tension in the xylem. Small branches are cut and inserted into the bomb chamber. When the branch is cut water columns in the xylem that are under tension recede. Application of pressure from a gas cylinder forces water back to the cut end of the stem. The pressure required to return the water to the cut end is equal and opposite in sign to the tension in the xylem.

in xylem increase during the day? The explanation is a simple physical one that is based on the fact the process requires nighttime temperatures below freezing and daytime temperature above freezing. At night freezing temperatures cause ice crystals to form and accumulate in the cell walls of stem tissues, while the roots and soil do not freeze. As ice forms, liquid water is drawn up into the stem from the root to replace the water removed by freezing. During the day the ice in the stem melts, causing an increase in its volume (water has maximum volume at $4°C$). An increase in the volume of living cells in the stem (e.g. xylem parenchyma) as they take up water causes pressure to rise, pushing sap up the tree.

14.7.4 Cavitation in tracheary elements interferes with water transport

Because of the high tension in the xylem during transpiration, the water is stretched taut and perturbations can cause **cavitation**, the introduction of bubbles. Cavitation causes a break in the water column. For example, violent winds can cause cavitation, as can freezing and thawing. Cavitation occurs by '**air seeding**' of tracheary elements. Air is likely to enter tracheary

elements via pit membranes (see Figure 1.28), where pores in the wall are in the range of $0.1–0.4$ μm, considerably larger than the $5–7$ nm pores in leaf parenchyma cells that permit high tensions to be built up. In many tree species, especially those growing in cold temperate zones, freezing and thawing may render most of the tracheary elements non-functional by the end of winter. Columns of tracheids are more resistant to blockage by bubbles than are vessels (see Figure 14.23).

Cavitation can be monitored using acoustic methods: the sound of a breaking water column can be detected with sensitive listening equipment. Another way of determining the extent of cavitation is to excise a piece of stem and measure the rate of water flow through the cut segment under low pressure. The segment is then flushed with water at high pressure, causing blocking air bubbles to redissolve and the rate of water flow at low pressure is again measured. The difference between the two measurements of water flow provides an estimate of the extent of cavitation.

One might ask whether cavitation is widespread in plants and whether it is found in the xylem of herbaceous species. The answer to these questions is yes. Cavitation is thought likely to be a daily occurrence in many species and is exacerbated by high winds, freezing and thawing, and drought.

14.7.5 Tracheary elements can be refilled with water by root pressure

If cavitation is a common phenomenon how do plants cope? There are several mechanisms that are thought to bring about the refilling of tracheary elements with water and most involve an increase in the xylem pressure. One mechanism involves **root pressure**. Root pressure develops because solutes accumulating at the base of the xylem in roots cause osmotic water flow into the xylem. Root pressure requires a specific set of environmental conditions, usually warm soil and a humid atmosphere, and it occurs at night when there is no evapotranspiration. At night, root pressure brings about an increased pressure in the xylem that can drive air back into solution. In herbaceous species root pressure can force liquid water out of the leaf through specialized cells called **hydathodes**, found at the tips or margins of leaves. This process, called **guttation**, accounts for the dew found on lawns and other plants at dawn (Figure 14.25). The water potential parameters of the xylem associated with root pressure are opposite in sign to those found during transpiration; that is, Ψ_s is more negative and Ψ_p is positive in the xylem as a result of root pressure.

Root pressure may be absent or not sufficient to drive air into solution in the trunks of tall trees. In this case it is thought that the secretion of solutes into tracheary

elements by living xylem parenchyma cells lowers Ψ_w, allowing an increase in pressure within the xylem from osmotic water flow into the system. Increased pressure would bring about the absorption of gas, restoring the conductive function of tracheary elements.

Key points Tracheary elements of the xylem are conduits for the movement of water and dissolved minerals from the soil to atmosphere. Tracheary elements, reinforced by rigid secondary walls, are under tension when transpiration is occurring. These tensions are made possible by the capacity of water to form strong intermolecular bonds and are aided by the relatively narrow diameters of tracheids and vessels and the microscopic pores within cell walls. Because tensions in tracheary elements can be as low as −5 MPa, water columns are easily and frequently broken by physical forces such as wind or freezing. This leads to cavitation, the introduction of air into tracheary elements, which renders them useless for water transport. One way that air in cavitated tracheary elements can be redissolved is by root pressure. This occurs at night when stomata are closed. Solutes are transported into tracheary elements in the root where they accumulate to concentrations that allow water to diffuse in. This increases pressure in the xylem water column and allows air bubbles to redissolve, which restores the integrity of the water column.

Figure 14.25 Guttation in strawberry. Under favorable conditions (at night when stomata are closed, with high relative humidity and warm, well-watered soil) osmosis causes water accumulation in the xylem and the build up of pressure that forces water from the margins of leaves through specialized hydathodes. Positive pressures that develop in the xylem are though to be one mechanism to redissolve air bubbles formed during cavitation.

14.8 The path of water from soil to atmosphere

Transport of water from the soil to the aerial parts of the plant is an integrated process requiring physical and physiological continuity between root, xylem and shoot. Water and solutes move in one direction, root to shoot, by bulk flow, pulled up by the tension that is generated by evapotranspiration through stomata.

14.8.1 There are two pathways by which water enters the root

Roots are well suited for water uptake. Root hairs greatly increase the ability of roots to take up water. For example, in rye it has been estimated that root hairs account for 60% of root surface area. Water can follow two basic pathways on its way into the vascular cylinder (**stele**). Water can move into root hairs by osmosis and travel symplasmically through plasmodesmata to the living xylem parenchyma cells in the stele; or it can diffuse through the apoplast until it reaches the **endodermis** (see Figure 13.5). Water can also move from one pathway to another at any point. At the endodermis the apoplastic pathway is blocked by the suberized region of endodermal wall called the **Casparian strip** (see Figures 1.31 and 13.5). Water movement across the endodermis must occur by osmosis, but once past this barrier, water can continue along a symplasmic or apoplastic pathway to the xylem. As we have already discussed, water will enter the tracheary elements because of xylem tension.

For water uptake to occur, soil water potential must be higher than that of root cells. This is the most likely scenario in well-watered soil, where the Ψ_s is around -0.02 to -0.04 MPa and the Ψ_w of the cell is usually in the range of -0.2 to -1.0 MPa. Saline soils can have much lower Ψ_s, in the range of -0.1 to -0.3 MPa. Low soil water content has the most dramatic effect on soil Ψ_w. As soil water content decreases, surface tension builds up on the fine particles in the soil resulting in a large negative Ψ_p. The soil Ψ_p can reach values lower than -1.0 MPa in poorly irrigated soil. It follows that poorly watered, saline soils can have a very negative Ψ_w thus severely limiting water availability to root cells.

14.8.2 The uptake of solutes and loading and unloading of the xylem are active processes

As Table 14.2 shows, solutes are present at millimolar concentrations in the xylem; for example, K^+ is present at concentrations of between 1 and 10 mM. Solutes such as K^+ follow the same paths from the soil to the xylem as does water. They enter the symplasm by moving through transporters in the plasma membrane and travel symplasmically to the stele, or they can move through the apoplast to the endodermis where they need transporters to cross the plasma membrane into the symplasm. Whereas the movement of water into tracheary elements of the stele follows a gradient in Ψ_w, the movement of solutes from the symplasm to the non-living tracheary elements presents physiologists with a dilemma. Because tracheary elements are dead, they have no symplasmic connections with the living cells of the root. Why then should living cells in the stele lose their solutes to dead tracheary elements? Several hypotheses have been proposed, and among the most recent is that xylem parenchyma cells release ions such as K^+, Na^+, Mg^{2+}, etc. into the xylem. This hypothesis is supported by the discovery of ion efflux channels in xylem parenchyma that could bring about xylem loading.

Unloading of solutes from the xylem into the leaf is also poorly understood. Solutes are generally not left behind in the apoplast when water evaporates from the cell walls lining the air spaces of the leaf. Rather, it is thought that solutes are taken up by specific membrane transporters into the living cells of the leaf and are distributed to other cells via plasmodesmata. Indirect support for this hypothesis comes from experiments with synthetic dyes. When these are administered to the leaf via the xylem, they accumulate at the ends of the tracheary elements, presumably because specific transporters for their uptake into the symplasm are lacking.

14.8.3 A number of structural and physiological features allow plants to control evapotranspiration from their shoots

The above-ground parts of plants also have coverings that minimize water loss. In herbaceous plants a waxy cuticle covers the epidermis, and modifications of the epidermis in the form of hairs and trichomes have also evolved to minimize water loss (Figure 14.26). Cuticles effectively seal the above-ground parts of plants. Most of the water that evaporates from a plant is lost through specialized pores in the epidermis called **stomata** or **stomates** (singular **stoma** or stomate; Figure 14.27). Stomata are formed by a pair of specialized guard cells. Although stomata are often abundant on the lower surfaces of leaves (Table 14.4), they are found in the epidermis of stems, fruits and floral tissues of herbaceous plants. In woody plants, stomata are largely confined to leaves. In most trees stomata are found only on the lower (abaxial) surfaces, while in water lily (*Lilium* spp.) stomata are only found on the upper (adaxial) sides of floating leaves. Many plants have stomata on both surfaces of their leaves; in these instances usually there are more stomata on the abaxial side. The density of stomata on a leaf may be influenced by environmental conditions during leaf development such as water availability and CO_2 concentration of the air.

14.8.4 Differences in water vapor concentration and resistances in the pathway drive evapotranspiration

Differences in **water vapor concentration** between the air spaces of leaves and atmosphere, and resistance to

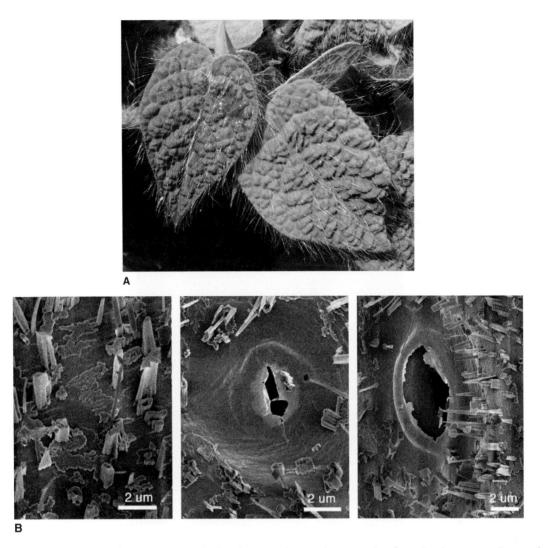

Figure 14.26   (A) Hairs and (B) wax deposits on leaf and stem surfaces reduce water loss from the above-ground parts of plants.

water vapor exchange determine the rate of water loss from leaves. There are several boundaries where differences in water vapor concentration and corresponding resistance are found. Water in tracheary elements evaporates from cell walls into leaf air spaces (Figure 14.28), and diffuses from leaf air spaces to the outside surface of the leaf, then to the atmosphere. The water potential gradient across these boundaries has been calculated (Table 14.5). Note that the largest difference in water potential occurs between leaf air spaces and leaf surface. The difference in water potential between the inner leaf air space and the air space just inside the stomatal pore is about −6 MPa; however, the difference between the air inside the stomatal pore and that just outside the pore is more than −63 MPa.

There are two boundaries that offer high resistance to water movement in leaves. One is at the stomatal pore, called the **leaf stomatal resistance**, and the other is at the

leaf surface/bulk atmosphere interface, the so-called **boundary layer resistance**. The number and size of stomatal pores will determine leaf stomatal resistance whereas the degree of mixing of water vapor at the surface of the leaf will affect the boundary layer resistance. Developmental programs determine the number of stomata in any given leaf; guard cells regulate the size of the stomatal pore. Boundary layer resistance is a function of the shells of water vapor that build up around an open stoma. These decrease the gradient in water vapor concentration C_{wv} between the inside of a leaf and the outside air. The anatomy of the leaf can affect the nature and extent of the boundary layer. Leaves with many hairs or trichomes have a greater boundary layer resistance than leaves lacking these modifications. Environmental conditions, especially air movement, affect the boundary layer.

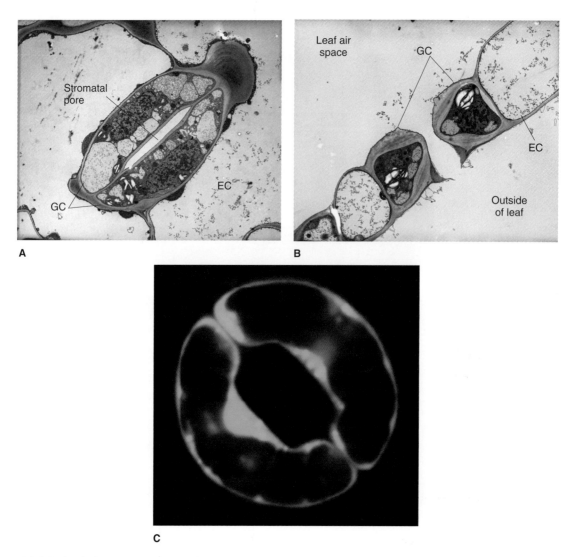

Figure 14.27 (A,B) Electron micrographs of eudicot stomata showing a view of a pair of guard cells (GC) and adjacent epidermal cells (EC) in (A) tangential section and (B) cross-section. (C) Light micrograph of a fluorescently-labeled pair of guard cells surrounding an open stomatal pore.

14.8.5 Stomatal guard cells are key regulators of water loss from leaves

Stomatal guard cells are pairs of highly specialized epidermal cells found in all vascular plants. They are symplasmically isolated from other leaf cells and unlike other epidermal cells they contain chloroplasts. They also have specialized wall thickenings that allow a pair of guard cells to open and close the stomatal pore. Two types of guard cells are found: the kidney-shaped guard cells found in most eudicots and many monocots (Figure 14.29A); and the dumbbell-shaped guard cells characteristic of grasses, sedges and palms (Figure 14.29B). Both types of guard cell have radially arranged cellulose microfibrils that reinforce their cell walls. Guard cells of grasses are found in a **stomatal complex** that includes a pair of **subsidiary cells** that flank the dumbbell-shaped guard cells. The bulbous ends of the guard cells of grasses have thin walls relative to the stem of the dumbbell, whose cell walls are reinforced by deposition of cellulose. In addition to radially arranged microfibrils, guard cell walls in dicots are also thickened along the side that is adjacent to the stomatal pore (Figure 14.29).

The presence of differentially reinforced cell walls allows guard cells to change shape when they take up or lose water, leading to the opening or closing of the stomatal pore (Figure 14.30). To open the pore, guard cells must accumulate solutes; this leads to the uptake of water and a subsequent increase in turgor pressure. In plants with kidney-shaped guard cells, an increase in turgor pressure

Table 14.4 Distribution of stomata.

Species	Stomatal distribution (% total)	
	Adaxial (upper) epidermis	Abaxial (lower) epidermis
Ferns		
Osmunda regalis (royal fern)	0	100
Phyllitis scolopendrium (hart's tongue fern)	0	100
Monocots		
Allium cepa (onion)	50	50
Hordeum vulgare (barley)	44	55
Zea mays (maize)	48	52
Eudicots		
Arabidopsis thaliana	35	65
Helianthus annuus (sunflower)	41	59
Nicotiana tabacum (tobacco)	21	79
Nymphaea alba (water lily)	100	0
Phaseolus vulgaris (green bean)	14	86
Pisum sativum (garden pea)	32	68
Prunus laurocerasus (cherry laurel)	0	100
Quercus palustris (pin oak)	0	100
Tilia americana (basswood)	0	100

Table 14.5 Values of water potential for the soil–plant–air continuum.[a]

Location	Ψ_w (MPa)
Soil at root surface	−0.50
Root xylem	−0.60
Stem xylem at 10 m	−0.80
Leaf mesophyll cell at 10 m	−0.80
Air inside leaf below stomata	−6.90
Air outside stomata	−70.0
Air above boundary layer	−95.0

[a]The value of Ψ_w in air is calculated from relative humidity (RT $\bar{V}_w^{-1}$ ln(%RH/100)).

causes guard cells to bend away from each other, forming the pore. In grass-type stomata, an increase in turgor pressure causes the thinner ends of the dumbbell-shaped cells to expand, pushing the guard cell pair apart.

Key points Most above-ground parts of plants are covered with an epidermis whose outer surface is covered with a waxy cuticle as well as hair-like extensions from epidermal cells. Collectively these modifications at the surface of the epidermis greatly reduce water loss so that the exchange of water vapor and gases occurs almost exclusively through the stomatal pore. Stomatal guard cells act as valves that regulate the opening and closing of stomatal pores, controlling the exchange of gases and water vapor between the air spaces in leaves and stems and the atmosphere. The stomatal pore may be completely open, partially open or tightly closed. Opening and closing of stomata is triggered by environmental conditions. Guard cells take up water and swell to open the stomatal pore; to close the pore, guard cells lose water and shrink. The influx or efflux of water is regulated by changes in the osmotic concentration in guard cells.

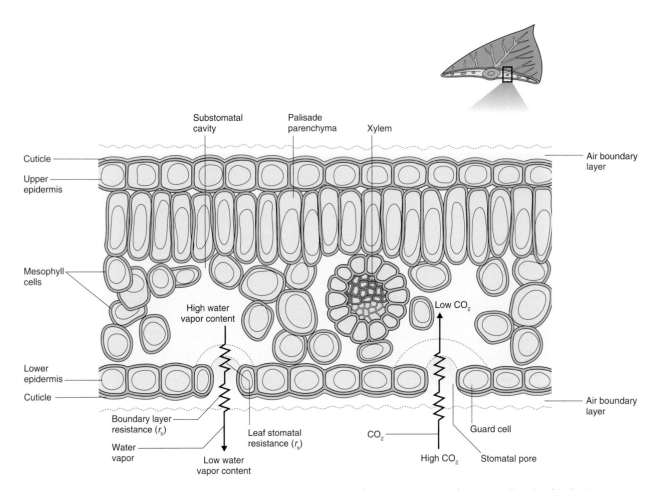

Figure 14.28   Pathways of CO_2 and water movement through the leaf. Water evaporates from the cell walls of leaf cells and, when stomata are open, water vapor diffuses to the atmosphere through the stomatal pore. CO_2 follows a similar path from the atmosphere to photosynthetic cells of the leaf. Stomata offer resistance to the diffusion of gases, the more open the stoma the lower the resistance to gas diffusion. Water vapor that escapes through stomata accumulates at the upper and lower surfaces of the leaf forming a boundary layer, seen by the dotted line. The boundary layer slows diffusion of water vapor from the leaf to the atmosphere.

14.8.6 Stomata open and close in response to a variety of environmental factors

Guard cells open and close stomata in response to a number of different environmental signals. For example, in most plants, stomata are open in the light and closed in the dark. There are two distinct responses to light in guard cells, both of which cause an increase in the osmotic concentration of guard cells leading to stomatal opening. Stomatal opening early in the day is associated with a peak of K^+ and its counter-ions (Figure 14.31) and is thought to be triggered principally by blue light. Later in the day, photosynthetically active light stimulates and maintains opening of stomata through the synthesis and accumulation of sucrose (Figure 14.31).

Blue light is absorbed by a photoreceptor in guard cell plasma membranes (see Chapter 8); this, in turn, stimulates the activity of the guard cell plasma membrane H^+ pump that causes acidification of the cell wall solution and a decrease in the plasma membrane potential (**hyperpolarization**). Hyperpolarization opens inwardly rectifying K^+ channels and leads to a 4–8-fold accumulation of K^+. The proton gradient is used to power the accumulation of anions by H^+/anion symporters. The resulting increase in osmotic concentration causes water to diffuse into the guard cells, leading to an increase in Ψ_p (Figure 14.32A). Evidence in support of the role of the H^+ pump in stomatal opening comes from two types of experiments. First, if the activity of the H^+ pump is inhibited, the effects of blue light on stomatal opening are blocked. Second, compounds that stimulate the H^+ pump trigger stomatal opening in the dark. Photosynthesis in guard cells may be required to

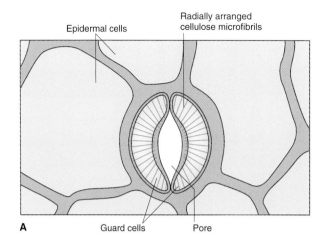

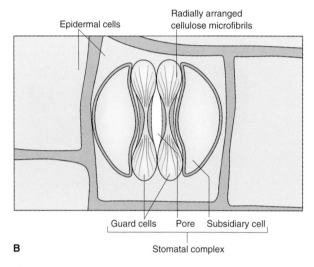

Figure 14.29 Orientation of cellulose microfibrils in the cell walls of guard cells. Microfibrils are radially arranged in the cell wall of both (A) kidney-shaped guard cells and (B) grass-type guard cells. In both types, the cell wall adjacent to the stomatal pore is thickened. In the guard cells of grasses the cell wall at the ends of the cells is relatively thin, resulting in dumbbell-shaped cells.

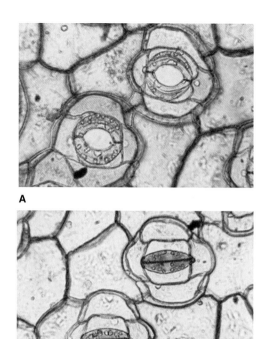

Figure 14.30 Abscisic acid (ABA)-induced stomatal closure. Epidermal strips from *Commelina* were incubated in buffered medium with and without 10 μM ABA. (A) Guard cells incubated in the absence of ABA are open, whereas (B) incubation in ABA for as little as 10 minutes brings about closure.

provide ATP for the operation of the plasma membrane H^+ pump as well as to increase the osmotic concentration in guard cells directly via the accumulation of sucrose. Inhibitors of photosynthesis are known to prevent the accumulation of sucrose and to block stomatal opening. It is likely that blue light acting via the H^+ pump and photosynthetically active light act cooperatively to increase osmotic concentration and therefore increase Ψ_p.

Humidity and low levels of soil moisture are also detected by guard cells. When the humidity is low, stomata close. Stomata also close when the soil is dry and root cells experience water stress. In this case, the hormone abscisic acid (ABA) plays a key role in communicating the water status of the roots to the

leaves. In response to water stress, ABA is synthesized by root cells; the hormone is quickly transported in the xylem to leaves where it induces stomatal closure (see Figure 14.30). ABA stimulates stomatal closing by causing Cl^- channels in the plasma membrane to open, allowing Cl^- to diffuse out of the cell. This causes the membrane potential to become less negative (**depolarize**); in addition ABA inhibits the activity of the plasma membrane H^+ pump, leading to further depolarization. Depolarization causes voltage-gated, outwardly rectifying K^+ channels to open and K^+ diffuses out of the cell. The net effect of Cl^- and K^+ loss is a lowering of the guard cell's osmotic concentration; as a consequence, water diffuses out of the guard cell, lowering Ψ_p and closing the stomatal pore (Figure 14.32B).

Intracellular CO_2 concentration is sensed by guard cells. A high level of CO_2 causes stomatal closure. The mechanism by which high CO_2 stimulates closure is similar to that of ABA. It triggers the opening of Cl^- channels, leading to the efflux of Cl^- and to membrane depolarization. Depolarization in turn triggers the opening of outwardly rectifying K^+ channels and subsequent K^+ efflux.

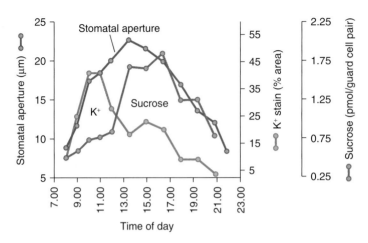

Figure 14.31 Daily changes in stomatal aperture and K^+ and the sucrose content of guard cells.

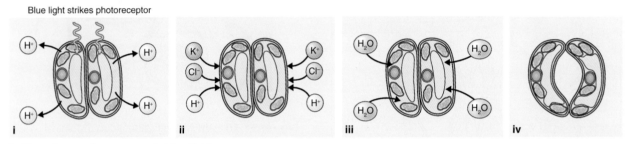

A Stomata open in response to blue light

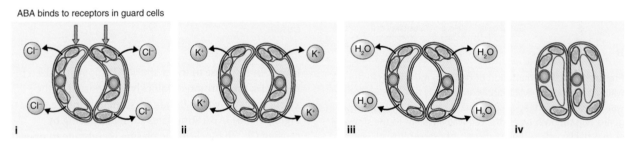

B Stomata close in response to ABA

Figure 14.32 Changes in osmotic potential are responsible for opening and closing stomata. (A) (i) Blue light photoreceptors activate the plasma membrane H^+ pump, (ii) which in turn activates K^+ and Cl^- channels. (iii) The increase in water potential as a result of solute accumulation causes water to enter cells by osmosis, increasing turgor pressure, and (iv) opening the pore. (B) (i) Abscisic acid (ABA) causes stomatal closure by triggering Cl^- efflux, and (ii) the resulting change in membrane potential opens outwardly rectifying K^+ channels. (iii) The resulting loss of solutes causes water to leave the guard cell by osmosis causing turgor pressure to fall and (iv) stomata to close.

14.8.7 The opening of stomata during the day represents a physiological compromise

Plants are faced with a dilemma when it comes to the regulation of gas exchange in leaves. Photosynthesis requires light and CO_2 and therefore the stomata of most plants open in daylight; however, at that time the driving force for water loss is greatest. By adjusting stomatal resistance, plants balance water loss with the requirement to carry out photosynthesis. The ultimate goal is to maximize CO_2 uptake while minimizing water loss. Regulating stomatal aperture in response to the demand of photosynthesis for CO_2 is one way of achieving this.

The concentration of CO_2 in the substomatal cavity can play a key role in balancing transpiration and photosynthesis. When the concentration of CO_2 in leaf air spaces is high, stomata tend to close, and the reverse happens as CO_2 levels fall. This observation suggests that guard cells regulate stomatal aperture in response to the demands of photosynthesis.

Plants have many adaptations to reduce transpirational water loss. For example, many plants close their stomata during the heat of the day when the rate of transpiration is high and reopen them later when temperatures drop and the humidity rises. Plants adapted for living in drier areas have special features that reduce surface area for water loss; these include having smaller stomatal pores, fewer stomata per leaf area and/or smaller leaves. Thicker cuticles, white surfaces and a proliferation of hairs all lead to lower rates of transpiration. Most of these adaptations to reduce water loss also result in a reduction in photosynthetic capacity. They are examples of the **photosynthesis/transpiration compromise** that plants must make to survive on dry land. Two groups, C_4 and CAM (crassulacean acid metabolism) plants, have biochemical as well as morphological adaptations that reduce transpiration; Chapter 9 discusses the relationship between transpiration and photosynthesis in these groups in further detail.

Key points Water and solutes travel upward from soil water in tracheary elements down a gradient in water potential. The driving force for water movement, the $\Delta\Psi_w$ between soil water and atmosphere, can be very large and even on humid days can exceed 10–20 MPa; on dry days this value can exceed 90 MPa. The existence of a driving force of such a magnitude requires that plants have a large resistance to the loss of water into the air. The biggest resistance to evaporative water loss is found in leaves and stems. It is a consequence of structural features that include the presence of a waxy cuticle on all epidermal cells and, in some species, trichomes. Water loss from the leaf surface is largely confined to stomata, and their number and location determine how readily water is lost by evapotranspiration.

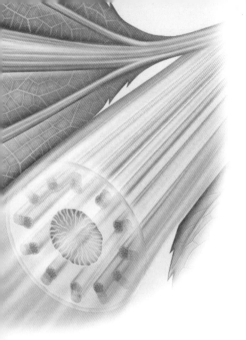

Chapter 15

Environmental interactions

15.1 Introduction to plant–environment interactions

Plants have evolved a wide range of characteristics to cope with a variable and sometimes hostile environment. Even though a plant is anchored in place, its open-ended, indeterminate mode of development (see Chapter 12) gives it the option of invading its environment as a means of overcoming limitations. For example, growing roots penetrate the soil and, by following gradients of water and nutrients, are able to maintain these inputs in response to demands from the rest of the plant. Similarly, **creeping plants** that produce rhizomes, stolons or runners—for instance, strawberries (*Fragaria* spp.; Figure 15.1)—effectively travel around in their environment, foraging for resources. Where light is limiting, such as beneath the forest canopy, shoots will extend and leaves will expand to escape from **shading** (see Chapter 12). Dispersal of seeds and other propagules is the means by which a plant, despite being rooted at a particular location, is able to send its genes elsewhere. So plants may be sessile, but they are far from immobile.

Nevertheless, for an individual that stays in one place, the environment is a constant test of **fitness**. The inescapable cycle of day and night is a case in point: as photoautotrophic organisms, plants must make relatively rapid physiological adjustments to an ever-changing light environment. In temperate regions, day length varies with the season and invokes longer-term changes in development (see Chapters 8 and 12). It is characteristic of plants that not only do they react to deviations from optimal environmental conditions; they also use such fluctuations as a source of information to trigger adaptive changes in structure and function. This chapter starts by considering general features of plant–environment interactions, then goes on to look at the most distinctive of these responses, namely the capacity to synthesize a vast range of phytochemical constituents with known or suspected roles in resisting or avoiding environmental pressures. We then proceed to describe specific examples of responses to non-biological external influences, followed by a survey of the mechanisms by which plants react to pathogenic organisms and other biological challenges.

15.2 General principles of plant–environment interactions

One approach to understanding the behavior of living organisms in response to the environment is to borrow

The Molecular Life of Plants, First Edition. Russell Jones, Helen Ougham, Howard Thomas and Susan Waaland.
© 2013 John Wiley & Sons, Ltd. Published 2013 by John Wiley & Sons, Ltd.

Figure 15.1 Strawberries move around, and exploit, their environment by sending out runners that form roots and new plantlets at intervals.

the concepts of **stress** and **strain** from engineering. Stress is defined as any factor that invokes a corresponding strain. In this context we should be careful not to think narrowly of *stress* in terms of *distress*, as a kind of analogy with human psychological and physical reactions to pressure. As we have seen, optimal conditions are not stress-free: for a plant, which cannot run away, stress and strain are a way of life. If the environment changes, the plant has no option but to change too. We have already met an example of the application of the stress–strain concept to a plant process in the discussion of the rheological properties of the growing cell wall (see Chapter 12), where the idea of **plastic** and **elastic** deformation was introduced. We can broaden this picture of stress–strain behavior to include the whole range of responses to environmental challenge. When a physiological reaction to the experience of an environmental perturbation is reversible on removal of the stress, the system is considered to have behaved elastically. Beyond the elasticity threshold, the physiological adjustment will be irreversible and the response is plastic. A third condition is reached if the stress exceeds the capacity of the system to react elastically or plastically in a physiologically coherent way. The result is **system failure**, a catastrophic outcome that leads to pathological changes and death.

Factors in the environment that influence plant growth, development and survival may be divided into **biotic** (those that arise directly or indirectly from interactions with other living organisms) and **abiotic** (originating in experiences of physical, chemical and energetic conditions). The major abiotic factors are light, temperature, water and nutrients (discussed in Section 15.6). Biotic influences include other plants,

pathogens, predators, pollinators and dispersers, as well as human interventions such as in selection and pollution.

15.2.1 Environmental factors may have both positive and negative effects

A simple model describes the interaction between environmental influences and the physiology of a plant. Imagine the case of an environmental factor F eliciting a biological response R. If F and R are measurable (for example F may be temperature, whereas R may be rate of growth), we can plot a graph to represent the relationship between stimulus and response (Figure 15.2). The curve typically rises from a minimum at a low value of F to a peak (the **optimum**), above which R either increases no further or (in the case of many environmental factors such as temperature) declines, ultimately to zero. The F–R curve is a combination of two opposing responses to increasing levels of the environmental factor. One, represented in Figure 15.2 as the R_s line, is the unconstrained stimulatory effect of exposure to progressively higher levels of F. For example, chemical

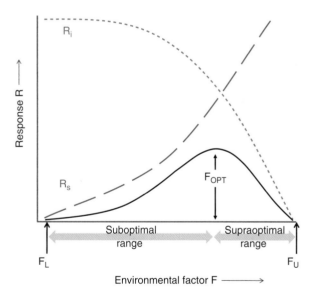

Figure 15.2 Generalized curves for the response, R, of a given biological process on exposure to different intensities of an environmental factor, F. The R_s and R_i lines are the positive (stimulatory) and negative (inhibitory) components, respectively, of the response to F. The combination of R_s and R_i is a curve (black line) that rises from a minimum at $F = F_L$ to a maximum (F_{OPT}) and declines again to an upper limit beyond which there is no further response (F_U).

reactions (the basis of most biological processes) will increase in a broadly exponential manner with increasing temperature. The R_s curve is counteracted by the tendency of biological processes to become progressively more disordered when overstimulated (Figure 15.2, R_i line). The combined curve is thus characterized by a **suboptimal** range, bounded by F_L (the value of F at which R is minimal) and F_{OPT}, and a **supraoptimal** range between F_{OPT} and F_U (Figure 15.2).

A typical example of response curves for different plants is shown in Figure 15.3. At low soil temperatures, dry matter is accumulated slowly. As the soil gets progressively warmer, dry matter productivity increases until it reaches a maximum at the **optimal growth temperature**. Increasing the temperature further results in decreasing production until a lethal temperature is reached and growth ceases completely. The production versus temperature plots in Figure 15.3 have similar features to the idealized curve in Figure 15.2. Temperatures below and above the optimum for growth are sub- and supraoptimal, respectively. Different species, especially from different habitats or geographical regions, have different temperature optima and growth–temperature ranges. Figure 15.3 shows that the optimum for wheat (*Triticum*), a temperate crop plant, is 18 °C, about 10 °C lower than that for soybean (*Glycine*), which is adapted to hotter, drier environments.

The lesson to take from this discussion of response curves is that, even at the optimum, plants experience **environmental constraint**. Living organisms are immensely complex combinations of individual components and processes, each of which has its own response curve for any given environmental variable. So although a *Zea mays* plant, for example, grows optimally at a soil temperature of 32 °C (Figure 15.3), for many constituent elements of its physiological machinery this temperature will be sub- or supraoptimal. Temperature is given here as an example of an environmental influence, but the principle is the same for any and every external factor. The optimum for a composite trait such as growth is the result of a great diversity of input–output actions and interactions among underlying processes.

15.2.2 Plants are equipped with mechanisms to avoid or tolerate stress

Organisms deal with non-optimal environmental conditions in different ways. They may adopt a strategy of **tolerance**, building resilient structures and physiologies able to survive by withstanding stress. Or they may take the **avoidance** route, confining growth to relatively favorable periods in a fluctuating environment, through life cycle strategies that minimize exposure of vulnerable stages to extreme conditions. The capacity to survive different degrees of stress varies greatly between species, and among genotypes of a single species. For example, many desert plants are **xerophytes**, that is, they have morphological and physiological characteristics that enable them to tolerate extreme drought. These include water-storing succulent tissues, CAM (crassulacean acid metabolism) photosynthesis and structural features that minimize dehydration (Figure 15.4A). Another drought-tolerance mechanism, adopted by species such as honey mesquite (*Prosopis glandulosa* var. *glandulosa*), is to develop deep **taproots** that improve access to groundwater and increase survival during long periods without rain (Figure 15.4B). In contrast, the strategy of **desert ephemerals** (short-lived

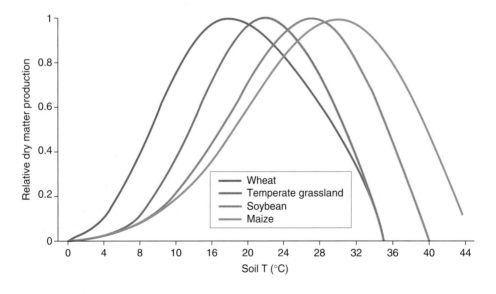

Figure 15.3 Plant growth responses to soil temperature in some agricultural crop systems. Growth is expressed as rate of dry matter accumulation, normalized with respect to maximal production = 1.

Figure 15.4 Examples of species that exhibit stress tolerance and stress avoidance. (A) The saguaro cactus (*Carnegiea giganteus*) is a highly drought-tolerant xerophyte. (B) The honey mesquite (*Prosopis glandulosa* var. *glandulosa*) is a drought-avoiding species with deep roots. (C) Mohave desert star (*Monoptilon bellioides*) is a desert ephemeral that only flowers after adequate winter/spring rain. (D) Spinach (*Spinacia oleracea*) is normally sensitive to osmotic stress, but has the capacity for physiological acclimation. (E) Black spruce (*Picea mariana*) is a cold-hardy tree that develops freezing tolerance by acclimation.

weedy plants of hot dry habitats) is to avoid drought by germinating and completing their life cycles while adequate water is available (Figure 15.4C).

15.2.3 Plants respond to the environment over the short term by acclimation, and on an evolutionary timescale by adaptation

Living systems are **homeostatic**. They tend to adjust to stress by minimizing strain and maintaining equilibrium. A homeostatic system has three modes of response to disturbance, the first two being: elastic, in which the system bounces back and resumes its former state; and plastic, in which it deforms and settles on a new stable configuration. When the limits of elastic and plastic resilience are exceeded, a catastrophic response results and the system becomes incoherent, entropy increases, and in the case of biology, death follows.

Homeostatic adjustment of an individual organism in response to changing environmental factors is termed **acclimation**. An example of acclimation is the reaction

of spinach plants to the presence of salt in the supply of water for growth (Figure 15.4D). These plants adjust their physiology by producing solutes that accumulate in the cytoplasm and, by sequestering salts in the vacuole, thereby maintain osmotic balance with the soil. Another example is found in trees of temperate regions. During the summer such trees generally cannot withstand freezing, but many species can acclimate in response to gradually decreasing temperatures in the fall. Eventually they may even be able to survive winter temperatures below $-50\,^{\circ}$C (Figure 15.4E). Acclimation is familiar to gardeners in the process known as **hardening off**, in which an otherwise vulnerable plant can be 'toughened up' by exposure to non-lethal levels of stress.

Constant or recurrent environmental challenge not only invokes acclimation; it may also exert **selective pressure**, driving the evolution of traits that increase fitness under stress. Such adjustments, which occur over many generations and across entire populations, are termed **adaptations**. In the case of cacti and similar xerophytes, for example, the adaptive traits, which include sunken stomata, light-reflective spines and CAM metabolism, are genetically determined, **constitutive** characters. These attributes have evolved for stress resistance but are expressed whether the plants are stressed or not. Table 15.1 summarizes the contrasting

Table 15.1 Adaptation and acclimation compared and contrasted.

	Adaptation	Acclimation
Individual or population level	Population	Individual
Caused by	Natural selection acting on allelic variation within populations.	Local environmental conditions acting on genetically-determined physiological responsiveness
Heritability	Genotypic	Generally non-heritable only; some instances of epigenetic transmission
Reversibility	Irreversible (except by further genotypic change and selection)	Reversible
Response of homeostasis to perturbation	Mostly plastic	Mostly elastic
Timescale	From generation time of the organism up to evolutionary	Short term (minutes/hours): metabolic and physiological adjustments of existing components without significant change in gene expression
		Long-term (up to weeks or months): altered patterns of gene expression, reallocation of resources, morphological change
Deployment in the life cycle	Strategic	Tactical

features of acclimation and adaptation. By acting over the short term in **individuals**, acclimation represents a **tactical** response to stresses that are often unscheduled. In contrast, adaptation is **strategic** in nature and is the basis of **population**-scale responses. These responses, such as winter dormancy and the loss of leaves in temperate forests, are often associated with predictable seasonal challenges. Figure 15.5 summarizes the relationships between stress, strain, acclimation and adaptation.

A critical difference between acclimation and adaptation is the role of the genome in each process (Table 15.1). Acclimation responses generally involve metabolic adjustments, which may or may not require the transcription of genes and translation of their products. Over the longer-term these metabolic changes lead to significant phenotypic changes in morphology and physiology. The traits acquired during acclimation are, in the vast majority of cases, not **heritable**. Nevertheless, there is evidence that **epigenetic** changes occur under some circumstances (see Chapter 3). A classic example is seen in flax (*Linum usitatissimum*), where plants grown with nitrogen-potassium fertilizer lacking phosphorus are only one-third to a quarter the size of those grown on NPK. This phenotypic difference is heritable, has been maintained over 50 generations, and remains even when plants are subsequently grown on NPK (Figure 15.6). The molecular basis of this so-called **genotrophic** effect is not known. Epigenetic effects apart, changes in gene expression, if they are

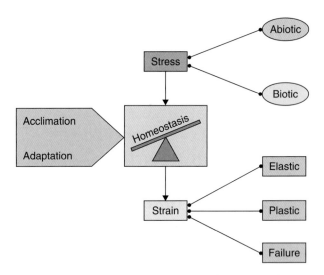

Figure 15.5 Biotic or abiotic stress invokes strain in the form of elastic, plastic or system failure responses. The extent to which homeostasis can be maintained is conditioned by the acclimatory or adaptive capacities of the organ, individual or population.

significant at all, play a reversible role in acclimation and are not associated with permanent changes to the genome. Adaptation, on the other hand, acts on **allelic variation** within a population, favoring those genotypes best fitted to survive the selective stress, and results in irreversible changes to the genome.

Key points Although they are sessile organisms, plants can adopt a range of developmental and adaptive strategies to escape from or tolerate a hostile environment. An environmental stress invokes a corresponding physiological strain. Response to stress may be elastic (the system returns to the pre-existing physiological state), plastic (change to a new state) or failure (collapse of homeostasis, often leading to death). Environmental stresses are classified into abiotic (physico-chemical in origin) or biotic (the result of organism–organism interactions). The response curve for a biological process under the influence of a given environmental factor typically consists of an optimum framed by sub- and supraoptimal ranges, and results from a combination of stimulation and inhibition. Tolerance and avoidance are strategies for surviving non-optimal conditions. Short-term, reversible, usually non-inherited adjustments that an organism makes when subject to environmental change are referred to as acclimation. Adaptation is a heritable, population-level response that becomes established over an evolutionary timescale.

15.2.4 Plants make a vast array of secondary metabolites, many of which are protective against biotic and abiotic challenges

Plants employ distinctive biochemical measures to deal with environmental stresses. The chemical intermediates and end-products that are directly involved in growth and development are referred to as **primary metabolites**. Primary metabolites are found in all plants, and include sugars, phytosterols, acyl lipids, nucleotides, amino acids and organic acids—the biochemical functions of which have been well established. But there is also an extraordinary variety of organic compounds that do not participate in primary metabolism and are differentially distributed within and between taxonomic groups within the plant kingdom. The specific functions of many of these so-called **secondary compounds** are unknown. It is becoming clear, however, that the principal roles of secondary metabolites are in acclimation and adaptation to environmental stresses, particularly those of biotic origin. It makes perfect sense for a sedentary, autotrophic organism to have evolved chemical defenses that react to, or prepare for, environmental challenges. In this chapter we will focus on three major groups of

Original parental line grown on: **NPK**   **NK**

Figure 15.6 Heritable phenotypes of flax (*Linum usitatissimum*) derived from a genetically uniform parental line that was grown on fertilizer with (NPK) or without (NK) phosphate. The photograph shows representative individuals of the 49th generation progeny grown on NPK. Such epigenetic variants resulting from responses to nutrient availability are termed genotrophs.

natural compounds with such functions in the molecular life of plants: **phenolics, alkaloids** and **terpenoids** (isoprenoids).

Phenolics (mostly phenylpropanoids), estimated to comprise at least 8000 different chemical structures, include low molecular weight phenolics, and condensation products such as tannins, lignins and flavonoids. The approximately 10 000 known alkaloids are protective, nitrogenous, secondary metabolites, derived principally from amino acids. Alkaloids are often pharmacologically active. The 25 000-plus terpenoid constituents of plants are built from five-carbon **isoprenoid** units. These include the **carotenoids** and related compounds, as well as a variety of toxins, antifeedants and attractants.

Secondary metabolites have evolved as a defense against pathogens, herbivores and other environmental hazards, thereby improving competitive ability and fitness. Secondary compounds are often toxic, antinutritional, allergenic, malodorous or bad-tasting and therefore detrimental for human consumption.

Domestication of wild species and selection for desirable **crop traits** usually results in the reduction or elimination of such natural products. Consequently, cultivated species frequently display weakened defenses against environmental (particularly biotic) challenges. The function of pesticides and other **crop protection** measures is to manage this vulnerability, but this incurs environmental and economic costs that significantly impact on agricultural efficiency. Understanding plant secondary metabolism is thus of importance for **food production**. In addition since many existing and potential **drugs** are derived from plant natural products, secondary metabolism is of fundamental pharmacological interest.

Key points Metabolites are classified as primary (indispensible chemical constituents that participate in core processes of growth and development and are found in all plants) and secondary (non-essential organic natural products, which are highly variable in structure and distribution between species). Many secondary metabolites have roles in adaptation and acclimation to environmental stresses. The major classes of secondary metabolites are: phenylpropanoids, which include phenolics, tannins, lignins and flavonoids; alkaloids, comprising compounds (many of which are highly bioactive) derived from amino acids; and terpenoids, the most numerous and structurally diverse metabolic products of plants. Because many secondary metabolites are antinutritional, the capacity to accumulate them has been bred out of crop species. This has generally weakened crop defenses against pests and diseases and led to the requirement for agrochemical treatments to maintain yield. Many of the most important drugs used in medicine are plant secondary metabolites or chemical derivatives of them.

15.3 Metabolic responses to stress I. Phenolics

Plant phenolics include a wide diversity of secondary metabolites. The basic chemical unit of phenolic compounds is a six-carbon **aromatic** (phenyl) ring to which one or more **hydroxyl groups** are attached (Figure 15.7A). The capacity to make a huge range of phenolics with a variety of structural and physiological roles is one of the characteristics that allowed the evolution of terrestrial plants from their aquatic ancestors. Phenolic cell wall components such as lignin stiffen cell walls and are essential for holding plants upright on land in the absence of the buoyant support provided by water. Terrestrial plants have also developed a wide variety of **non-structural phenolics**. These contribute to hardiness, colors, tastes, odors, defense against stresses and other attributes essential for survival. It is estimated that phenolics represent about 40% of the organic carbon circulating in the biosphere. The relative chemical inertness of macromolecular phenolic derivatives such as lignins and tannins makes the biodegradation of these products a rate-limiting step in the **recycling** of organic carbon into CO_2.

When an apple is damaged and exposed to air, the familiar browning reaction occurs rapidly. This is an example of a widespread response that occurs in plant tissues; it is the result of oxidation of phenolics. It frequently yields products that form inactivating complexes with proteins and nucleic acids and can make biochemical or molecular analysis difficult. For this reason, protocols for isolating plant proteins and nucleic acids generally include special precautions designed to minimize interference by phenolic compounds.

The chemical structures of most plant phenolics are related to their biosynthetic origins in phenylpropanoid (Figure 15.7B, D) and phenylpropanoid-acetate (Figure 15.7C, E) metabolism. Polymeric lignins, which reinforce the secondary cell walls of structural and transport tissues in vascular plants (see Figure 12.15), are products of the phenylpropanoid pathway. The **lignans** (Figure 15.7D) are di- and oligomeric phenylpropanoids structurally and biosynthetically related to lignin. They are widely distributed throughout the plant kingdom and function as defenses against biotic and abiotic stress in flowers, seeds, stems, bark, leaves and roots. **Suberin**, which is found in cork, bark and periderm tissues such as potato skin, contains alternating hydrophilic and hydrophobic layers of phenolic and aliphatic structural substances. Suberization establishes a barrier that protects against water loss and pathogen attack. The most diverse group of plant phenolics is the flavonoids (Figure 15.7E). Included among the estimated 4500 flavonoids present in plants are: **anthocyanins** (responsible for the pink/blue/red pigmentation in many vegetative tissues, flowers and fruits); **proanthocyanidins** or **condensed tannins** (antifeedants and wood protectants); and **isoflavonoids** (defensive products and signaling molecules). Plants are also protected against herbivores, microbial pathogens and competing neighbors by a range of miscellaneous phenolic-related

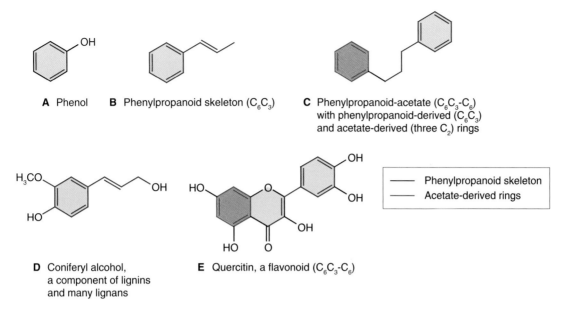

A Phenol   **B** Phenylpropanoid skeleton (C_6C_3)

C Phenylpropanoid-acetate (C_6C_3-C_6)
with phenylpropanoid-derived (C_6C_3)
and acetate-derived (three C_2) rings

D Coniferyl alcohol,
a component of lignins
and many lignans

E Quercitin, a flavonoid (C_6C_3-C_6)

— Phenylpropanoid skeleton
— Acetate-derived rings

Figure 15.7   Chemical structures of phenolics. (A) Phenol, the parent structure of phenylpropanoids. (B) The phenylpropanoid skeleton. (C) Structure of phenylpropanoid acetate, showing the origins of different parts of the molecule. (D) Coniferyl alcohol, a phenylpropanoid component of lignans. (E) Quercitin, a flavonol with a structure based on phenylpropanoid acetate.

products, including **coumarins, furanocoumarins** and **stilbenes**.

15.3.1 Phenylalanine and tyrosine are the metabolites that link primary metabolism to the secondary pathways of phenolic biosynthesis

Most phenolic and alkaloid secondary compounds in plants are derived from the **aromatic amino acids** phenylalanine, tyrosine and tryptophan. In addition to their role as monomers for protein synthesis, these primary metabolites are major points of contact between primary and secondary metabolism. Among the secondary metabolites that are derived from the aromatic amino acids are the indole hormones (e.g. indole-3-acetic acid (IAA); see Chapter 10), anthocyanin pigments, defensive phytoalexins, bioactive alkaloids and structural lignins. It is estimated that about 20% of the carbon fixed by plants flows through the common aromatic amino acid pathway, the largest proportion ending up in lignin.

Figure 15.8 is an outline of the reaction sequence leading from intermediates in primary carbon metabolism (phosphoenolpyruvate (PEP) and erythrose-4-phosphate) to phenylalanine, tyrosine and tryptophan, and onward to phenolics and alkaloids. The intermediate **shikimic acid**, a carboxylic acid with a

hydroxylated six-carbon ring, gives its name to the **shikimate pathway.** The shikimate pathway is a seven-step sequence localized in the chloroplast; its product **chorismic acid** is the final common intermediate in the synthesis of the three aromatic amino acids. The first reaction of this pathway is the formation of 3-deoxy-D-arabino-heptulosonate-7-phosphate (**DAHP**) by **DAHP synthase** (Equation 15.1), an enzyme that is regulated by tryptophan and Mn^{2+} activation.

Equation 15.1 DAHP synthase

$$PEP + erythrose\text{-}4\text{-}phosphate + H_2O \rightarrow DAHP + P_i$$

The next to last step in the shikimate pathway (Equation 15.2) is the target of the widely used herbicide **glyphosate** (Roundup; Figure 15.9) which is a competitive inhibitor of **EPSP synthase**. The lethal consequences of blocking this strategic reaction clearly demonstrate the central importance of the shikimate pathway in plant metabolism.

Equation 15.2 5-Enolpyruvylshikimate-3-phosphate (EPSP) synthase

$$Shikimate\text{-}3\text{-}phosphate + PEP \rightarrow EPSP + P_i$$

Chorismate, which is made from EPSP by the enzyme **chorismate synthase**, is the end-product of the shikimate pathway and the starting point for the reaction

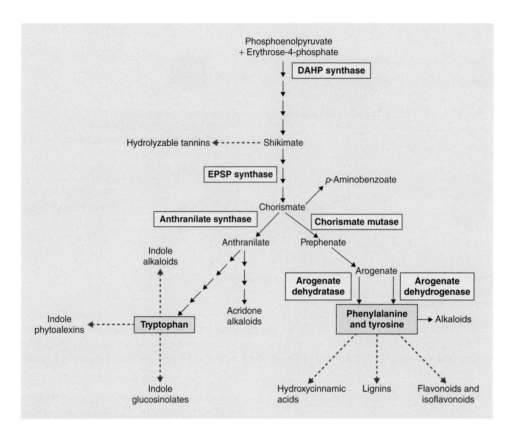

Figure 15.8 Biosynthesis of aromatic amino acids phenylalanine, tyrosine and tryptophan, the sources of phenolic and alkaloid natural products.

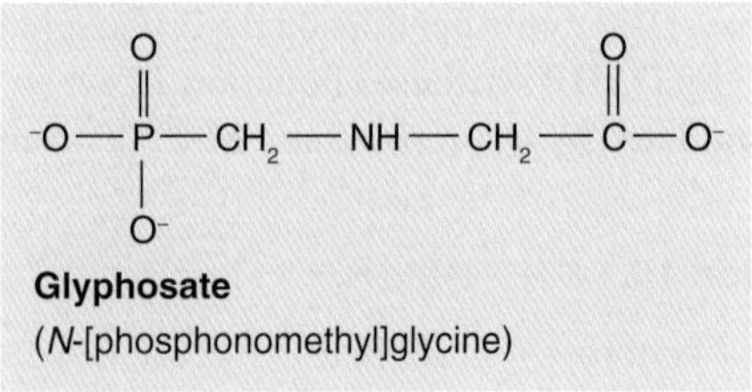

Glyphosate
(*N*-[phosphonomethyl]glycine)

Figure 15.9 The herbicide glyphosate is a competitive inhibitor of EPSP (5-enolpyruvylshikimate-3-phosphate) synthase, an enzyme in the phenylalanine and tyrosine synthesis pathway.

sequence that leads to the synthesis of phenylalanine and tyrosine (Figure 15.8). The enzyme **chorismate mutase** (**CM**) catalyzes the formation of prephenate from chorismate by an intramolecular rearrangement. CM1, an isoform located in the chloroplast, is a critical point for controlling entry into the **phenylalanine/tyrosine branch** of the aromatic amino acid synthesis pathway (Figure 15.8). Phenylalanine and tyrosine, the end-products of this pathway, modulate CM1 activity by **feedback inhibition** of the enzyme. Prephenate is transaminated to yield **arogenate**, the immediate precursor of phenylalanine and tyrosine. **Arogenate dehydratase** catalyzes the synthesis of phenylalanine

from arogenate in a decarboxylation reaction (Equation 15.3A) and **arogenate dehydrogenase** produces tyrosine from arogenate in an NADP$^+$-dependent reaction (Equation 15.3B). These two enzymes are regulated by end-product inhibition and act as sensitive control points in phenylpropanoid production.

Equation 15.3 Synthesis of phenylalanine and tyrosine from arogenate

15.3A Arogenate dehydratase

$$\text{Arogenate} \rightarrow \text{phenylalanine} + CO_2 + H_2O$$

15.3B Arogenate dehydrogenase

$$\text{Arogenate} + NADP^+ \rightarrow \text{tyrosine} + NADPH + H^+ + CO_2$$

Some phenolic compounds are generated through pathways other than the phenylalanine/tyrosine route. An example is **hydrolyzable tannins** (defense compounds found in leaves, fruits, pods and galls of a range of dicot species) which are derivatives of shikimate and carbohydrates. Hydrolyzable tannins are chemically distinct from condensed tannins, which are products of flavonoid metabolism, as described in Section 15.3.3.

15.3.2 Most phenolics are synthesized from phenylalanine or tyrosine via the phenylpropanoid pathway

In discussing the products of phenylalanine metabolism, it is useful to set out the basic structural features that underlie the chemical diversity of phenolic intermediates. The common molecular structure is the six-carbon aromatic ring. An unsaturated three-carbon (**propene**) side-chain is attached to the ring at the carbon atom designated 1 (Figure 15.10A). The other atoms of the aromatic ring are numbered clockwise with respect to C-1. Intermediate phenolic metabolites, derived from phenylalanine, vary in the atoms or groups attached to carbons 3, 4 and 5 of the aromatic ring (Figure 15.10A, groups R^1, R, R^2 respectively) and the terminal C atom of the propene side-chain (Figure 15.10A, R^3). The enzyme **phenylalanine (tyrosine) ammonia-lyase** (**PAL/TAL**) is the point of entry of aromatic amino acids into phenylpropanoid metabolism. PAL converts its preferred substrate, phenylalanine, to cinnamic acid, although the monocot enzyme is also able to utilize tyrosine to form 4-coumaric acid (Figure 15.10B). The ammonium ion, produced in this reaction, is **reassimilated** by the GS-GOGAT pathway (see Chapter 13). In some species, PAL is encoded by a single gene, but in other species it is the product of a multigene family. The cinnamate product of PAL is converted to 4-coumarate by the enzyme **cinnamate-4-hydroxylase** (**C4H**) (Figure 15.11). This oxygen-requiring, NADPH-dependent enzyme has a **cytochrome P450**

prosthetic group that specifically hydroxylates the aromatic ring in the 4- (that is, *para-*) position (Equation 15.4). Cytochrome P450-type enzymes carry out oxidations of organic compounds and are encoded by a large gene family (**CYP**); the *Arabidopsis* genome has more than 270 *CYP* or *CYP*-like sequences. C4H, like other CYP enzymes, is anchored in membranes of the endoplasmic reticulum (ER). There is evidence that PAL physically associates with C4H to form a loose complex that allows the product of the former enzyme to feed directly into the second, a process known as **metabolic channeling**.

Equation 15.4 Cinnamate-4-hydroxylase

$$\text{Cinnamate} + \text{NADPH} + \text{H}^+ + \text{O}_2 \rightleftharpoons \text{4-coumarate} + \text{NADP}^+ + \text{H}_2\text{O}$$

Cinnamoyl-CoA, the coenzyme A thioester synthesized from cinnamate, is the origin of the stilbene group of natural products (Figure 15.11) that are synthesized by plants in response to damage. Grapes, pines and legumes are rich sources of the **resveratrol** family of stilbenes, which function as **phytoalexins** (pathogen defense agents; see Section 15.7.1) and are of pharmaceutical interest as anticarcinogens and cardioprotectives. Hydroxylation of 4-coumarate at the 2-position yields 2,4-dihydroxycinnamate, which in turn gives rise to the **coumarins** (Figure 15.11), a group of fragrant but bitter-tasting products with appetite-suppressing properties. The anticoagulant warfarin (Coumadin), which is used as a rat poison and blood thinner, is a chemically-modified coumarin derivative.

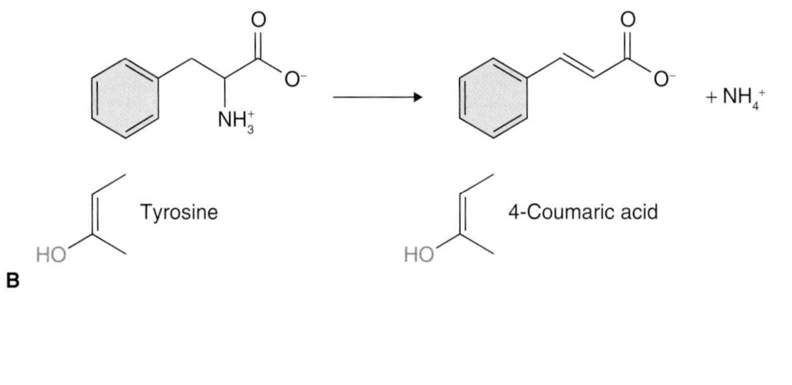

Figure 15.10 Chemical structures of phenolic intermediates derived from cinnamic acid. (A) Carbon atoms of the phenylpropanoid aromatic ring are numbered clockwise, beginning at the position of the propene side-chain. The convention in describing substitution patterns in aromatic rings is to refer to two substituents on adjacent carbon atoms as *ortho* (*o-*) with respect to each other; if there is one carbon atom between substituents the configuration is *meta* (*m-*), and *para* (*p-*) with two intervening carbon atoms. (B) Reactions catalyzed by phenylalanine (tyrosine) ammonia lyase. According to the convention described in (A), the hydroxylated derivative of cinnamic acid is referred to as either 4- or *p*-coumaric acid.

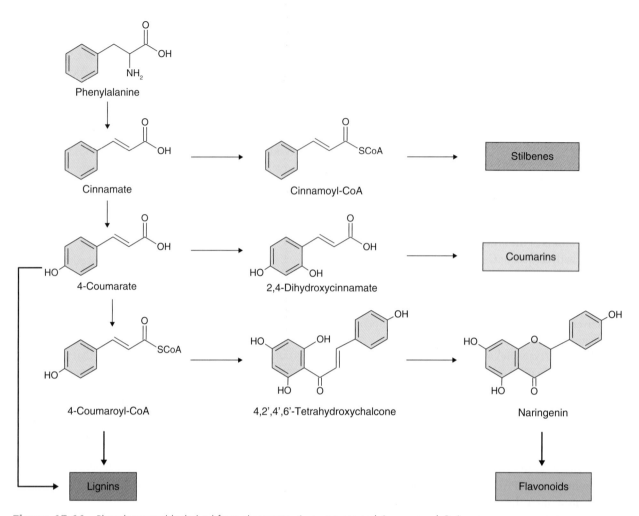

Figure 15.11 Phenylpropanoids derived from cinnamate, 4-coumarate and 4-coumaroyl-CoA.

15.3.3 The flavonoid pathway, leading to flavones, flavonols and anthocyanins, starts with chalcone synthase and chalcone isomerase

4-Coumaroyl-CoA is the metabolic origin of the **flavonoids** (Figure 15.11). Flavonoids can occur as monomers, dimers and higher oligomers, and are distributed throughout most plant tissues. In contrast to pigments such as carotenoids and chlorophylls, which are lipophilic and usually found in plastids, most flavonoids are water-soluble and are often located in **vacuoles**. The flux into the flavonoid synthesis pathway through 4-coumaroyl-CoA begins with two distinctive enzymes, **chalcone synthase** (**CHS**) and **chalcone isomerase** (**CHI**). The reaction catalyzed by CHS (Equation 15.5) consists of the condensation of three

molecules of malonyl-CoA, derived from acetate (see Figure 15.7E), with 4-coumaroyl-CoA to form 4,2′,4′,6′-tetrahydroxychalcone (Figure 15.11). Because of its importance in the pathway leading to floral pigmentation, the gene for CHS was one of the earliest targeted for biotechnological modification. Such interventions showed that flower color can be changed by altering *CHS* expression. CHS may act coordinately with chalcone reductase, an NADPH-dependent enzyme that generates isoliquiritigenin and leads to the isoflavonoid branch of phenylpropanoid metabolism (Figure 15.12).

Equation 15.5 Chalcone synthase (CHS)

4-Coumaroyl-CoA + 3malonyl-CoA

$\rightarrow$ 4,2′,4′,6′-tetrahydroxychalcone + 4CoASH + 3CO$_2$

Chalcone isomerase catalyzes a stereospecific ring closure isomerization step, usually to form **naringenin** (Figure 15.11). In isoflavone synthesis, CHI works with

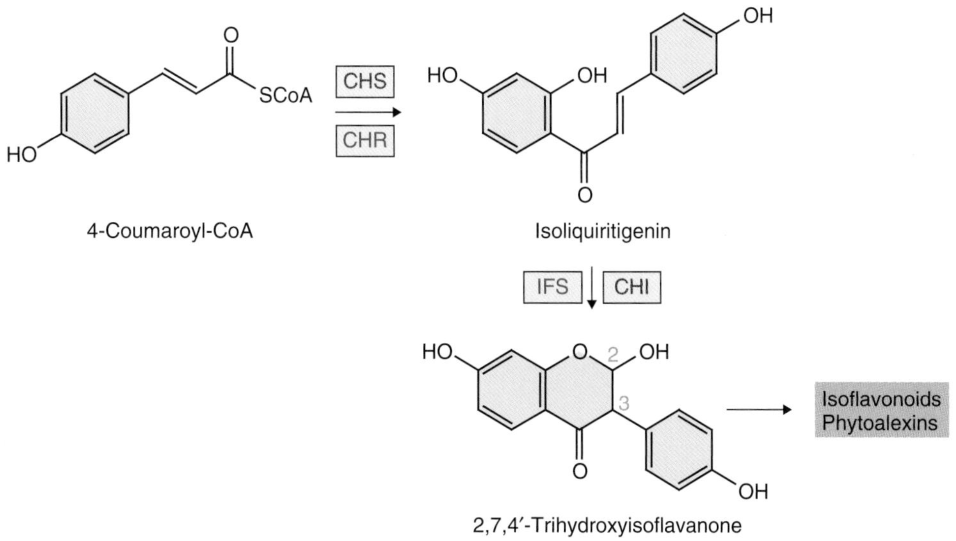

Figure 15.12 The pathway leading from 4-coumaroyl-CoA to isoflavonoids and phytoalexins. Enzymes in common with flavonoid biosynthesis are chalcone synthase (CHS) and chalcone isomerase (CHI). Committed enzymes of isoflavonoid synthesis are chalcone reductase (CHR) and isoflavone synthase (IFS). Note that the phenyl group in isoflavonoids is attached at position 3 on the oxane ring, whereas the bond in flavonoids derived from naringenin is at position 2 (compare Figure 15.11).

isoflavone synthase (an NADPH-dependent cytochrome P450 enzyme) to yield a product with the aromatic ring attached at position 3 (rather than 2 as in flavonoids) of the oxane ring (Figure 15.12). Further metabolism of isoflavonoids yields antipathogenic phytoalexins and insecticidal **rotenoids**.

Naringenin is the source of **flavones, flavonols, anthocyanins** and condensed tannins (Figure 15.13). The abundant flavone **apigenin** is reported to have a range of beneficial therapeutic properties; it is the reaction product of flavone synthase. Enzymatic hydroxylation of naringenin results in **dihydrokaempferol**, which in turn is the source of flavonols (through the flavonol synthase-catalyzed formation of **kaempferol**). A highly visible fate of dihydrokaempferol is to feed a **metabolic grid** leading to the synthesis of the vivid red, pink, mauve, violet, blue and purple anthocyanin pigments found in many petals, leaves, stems and fruits (Figure 15.13). **Dihydroquercitin** and **dihydromyricetin** are the products of hydroxylating dihydrokaempferol at the 3′ and 5′ positions. All three flavonols may be used as substrates by **dihydroflavonol-4-reductase** (DFR), an NADPH-dependent enzyme. The products of the DFR reaction are **leucoanthocyanidins** (e.g. leucopelargonidin, derived from dihydrokaempferol), which in turn can be converted to the colored anthocyanidins (e.g. pelargonidin) through the action of **anthocyanidin synthase**, an α-ketoglutarate-dependent dioxygenase (Figure 15.13). The final step in anthocyanin biosynthesis is glycosylation of anthocyanidin, which stabilizes the molecule. Further

modifications then include additional methylations, glycosylations and acylations of hydroxyl groups to produce the great range of anthocyanin colors present in the plant kingdom. The control of flower color is discussed further in Chapter 16, and the significance of foliar anthocyanins, responsible for many of the reds and purples of autumn leaves, is considered in Chapter 18.

Leucoanthocyanidins and anthocyanidins can also serve as precursors of **catechins** and condensed tannins (Figure 15.13), widely-distributed astringent defense compounds. Tea, cocoa, chocolate, wine and many fruits and vegetables are rich sources of catechins, which contribute to taste and nutritional effects. Tannins deter pathogens and herbivores by complexing with macromolecules and inactivating enzymes.

15.3.4 Lignin precursors are the products of a metabolic grid derived from 4-coumaric and cinnamic acids

Cinnamic and 4-coumaric acids are converted to a range of phenolic products through a series of enzymatic reactions including **aromatic hydroxylations, O-methylations, CoA ligations** and **NADPH-dependent reductions** (Figure 15.14). The result is a metabolic grid that supplies the precursors of lignin biosynthesis (**monolignols**), 4-coumaryl, coniferyl, 5-hydroxyconiferyl and sinapyl alcohols. Most of the enzymes of the grid are multifunctional with broad

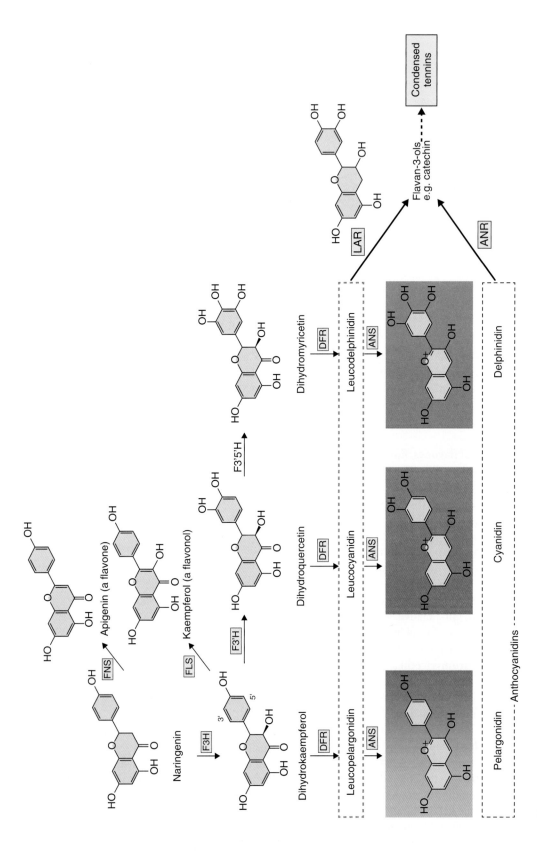

Figure 15.13 Conversion of naringenin to flavones, flavonols and intermediates in the metabolic grid leading to synthesis of anthocyanins and condensed tannins. Key to enzymes: ANR, anthocyanidin reductase; ANS, anthocyanidin synthase; DFR, dihydroflavonol-4-reductase; F3H, flavanone-3-hydroxylase; F3'H, flavonoid-3'-hydroxylase; F3'5'H, flavonoid-3',5'-hydroxylase; FLS, flavonol synthase; FNS, flavone synthase; LAR, leucoanthocyanidin reductase.

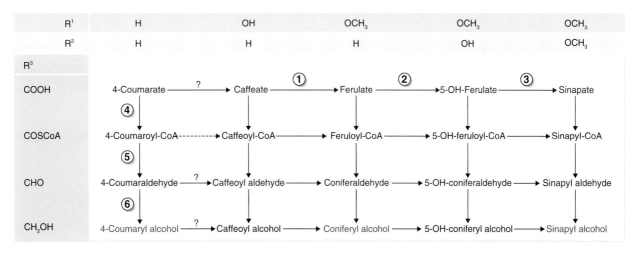

Figure 15.14 The grid of phenolic lignin precursors derived from 4-coumaric acid. R^1, R^2 and R^3 refer to the substitution pattern shown in Figure 15.10A. Monolignols are shown in red. Enzymes catalyzing horizontal reactions are: (1) O-methyl transferase, (2) ferulate-5-hydroxylase and (3) O-methyl transferase. Vertical reactions are catalyzed by: (4) 4-coumarate:coenzyme A ligase, (5) cinnamoyl-CoA reductase and (6) cinnamyl/sinapyl alcohol dehydrogenase. ? = identity of enzymes not certain. 4-Coumaroyl-CoA is converted to caffeoyl-CoA by three enzymatic steps.

substrate specificities and each catalyzes more than one reaction. **Ferulate-5-hydroxylase** (Figure 15.14, reaction 2) is an endomembrane-associated CYP-type enzyme. All the other enzymes of the grid are soluble, though some may be involved in metabolic channeling as described above for PAL and C4H.

Key points Phenylpropanoids are phenolics based on hydroxylated C_6 phenyl rings. Lignin is a structural phenolic. Non-structural phenolics, the most diverse class of which are the flavonoids, include tannins, pigments, volatiles and defense compounds. Most phenylpropanoids are metabolically derived from the shikimate pathway, which leads to phenylalanine, tyrosine and tryptophan. Cinnamate, the product of phenylalanine (tyrosine) ammonia-lyase, is the source of stilbenes and is also the substrate for cinnamate-4-hydroxylase. The product of this reaction, 4-coumarate, is the metabolic origin of lignins, coumarins and flavonoids. Coumarate feeds into the flavonoid pathway through chalcone synthase and chalcone isomerase. One branch of the pathway leads to isoflavonoids and phytoalexins via the intermediate isoliquiritigenin. Anthocyanins and condensed tannins, derived from the intermediate naringenin, are products of a second branch. Cinnamate and 4-coumarate are sources of monolignols (4-coumaryl, coniferyl, 5-hydroxyconiferyl and sinapyl alcohols), the precursors of lignin, which is synthesized in the cell wall from monolignol glucosides.

Lignin synthesis takes place in the cell wall. Before their export to the wall, monolignols synthesized in the cytosol are **glycosylated** by a reaction with uridine diphosphate (UDP) glucose, catalyzed by **UDP-glucose coniferyl alcohol glucosyltransferase** (Equation 15.6A). After the monolignol glucosides are transferred to the cell wall, they are hydrolyzed by **coniferin β-glucosidase** (Equation 15.6B). The monolignols released undergo polymerization, catalyzed by peroxidases and laccases, and are cross-linked to cell wall polysaccharides as described in Chapter 12.

Equation 15.6 Formation of lignin precursors
15.6A UDP-glucose coniferyl alcohol glucosyltransferase

UDP-glucose + coniferyl alcohol $\rightleftharpoons$ UDP + coniferin

15.6B Coniferin β-glucosidase

Coniferin + H_2O → coniferol + D-glucose

15.4 Metabolic responses to stress II. Alkaloids

From the earliest days of human history, alkaloids have been among the most potent and socially significant plant products. They include: **opium**, from the latex of the opium poppy (*Papaver somniferum*); **coniine**, the poison of hemlock (*Conium maculatum*; Figure 15.15A), famously used for the execution of the classical Greek

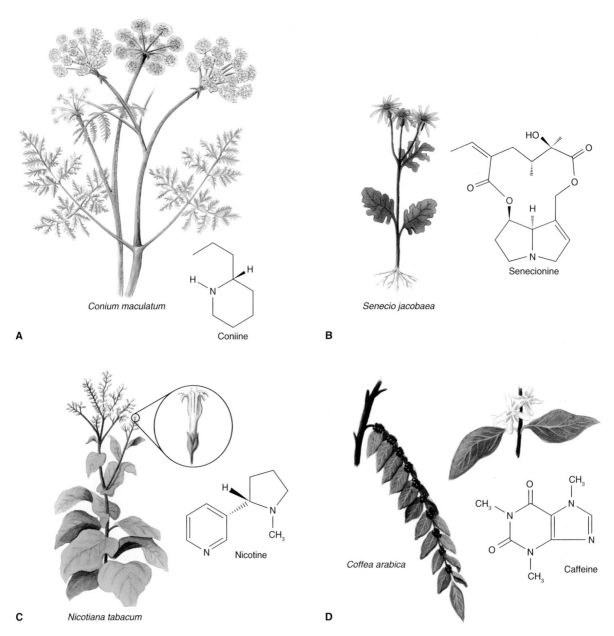

Figure 15.15   Examples of alkaloids and the plants that produce them, indicating the variety of chemical structures. (A) Coniine from poison hemlock (*Conium maculatum*). (B) Senecionine from ragwort (*Senecio jacobaea*). (C) Nicotine from tobacco (*Nicotiana tabacum*). (D) Caffeine from coffee (*Coffea arabica*).

philosopher Socrates; and **hyoscyamine**, which is the active principle in extracts of a number of solanaceous plants, and deadly in high concentrations but at trace levels is an effective drug treatment for a range of medical conditions. Alkaloid-containing plants were humankind's original medicines and even today the list of effective prescription drugs for the commonest ailments is dominated by pharmaceuticals of plant origin. Plant alkaloids are also chemical models on which the design of modern **synthetic drugs** is based. For example, the antimalarial drug **chloroquine** is an indole-derived alkaloid related to the natural product

quinine. Table 15.2 lists a number of the major plant-derived alkaloids used in modern medicine.

Alkaloids were originally characterized as pharmacologically active, basic compounds that contain nitrogen and are derived from plants. The term has since been extended to apply to many phytochemicals that do not strictly conform to this definition. It has also become clear that alkaloids are not confined to plants. Species of bacteria, fungi, sponges, arthropods, amphibians and mammals have been shown to accumulate alkaloids, which have significant ecological functions and in some cases are of plant origin. For example, larvae of the

Table 15.2 Physiologically active alkaloids used in modern medicine.

Alkaloid	Plant source	Use
Ajmaline	*Rauwolfia serpentina*	Antiarrhythmic that functions by inhibiting glucose uptake by heart tissue mitochondria
Atropine	*Atropa belladonna*	Anticholinergic antidote to nerve gas poisoning
Caffeine	*Coffea arabica*	Widely used central nervous system stimulant
Camptothecin	*Camptotheca acuminata*	Potent anticancer agent
Cocaine	*Erythroxylon coca*	Topical anesthetic, potent central nervous system stimulant, and adrenergic blocking agent; drug of abuse
Codeine	*Papaver somniferum*	Relatively non-addictive analgesic and antitussive
Coniine	*Conium maculatum*	First alkaloid to be synthesized; extremely toxic, causes paralysis of motor nerve endings, and used in homeopathy in small doses
Emetine	*Uragoga ipecacuanha*	Orally active emetic and amoebicide
Morphine	*Papaver somniferum*	Powerful narcotic analgesic and addictive drug of abuse
Nicotine	*Nicotiana tabacum*	Highly toxic, causes respiratory paralysis; horticultural insecticide; drug of abuse
Pilocarpine	*Pilocarpus jaborandi*	Peripheral stimulant of the parasympathetic system, used to treat glaucoma
Quinine	*Cinchona officinalis*	Traditional antimalarial, important in treating *Plasmodium falciparum* strains that are resistant to other antimalarials
Sanguinarine	*Eschscholzia californica*	Antibacterial showing antiplaque activity, used in toothpastes and oral rinses
Strychnine	*Strychnos nux-vomica*	Violent tetanic poison, used as rat poison; used in homeopathy
(+)-Tubocurarine	*Chondrodendron tomentosum*	Non-depolarizing muscle relaxant producing paralysis; used as an adjuvant to anesthesia
Vinblastine	*Catharanthus roseus*	Antineoplastic used to treat Hodgkin's disease and other lymphomas

cinnabar moth, *Tyria jacobaea*, gather alkaloid precursors from *Senecio jacobaea*, the host plant on which they feed (Figure 15.15B), and convert them into pheromones and defense compounds. Generally, however, plants deploy alkaloids as antibiotics, herbivore deterrents or other measures against biotic stresses. Alkaloids may be either **constitutive** (present all the time) or synthesized in response to physical or threatened trauma. **Nicotine**, for example (Figure 15.15C), is a potent insecticide and has long been used by humans to protect cultivated plants. Wild plants of tobacco (*Nicotiana* spp.) respond to herbivory by boosting the rate of endogenous nicotine biosynthesis. **Caffeine** (Figure 15.15D), found in seeds and leaves of cocoa (*Theobroma cacao*), coffee (*Coffea* spp.) and tea (*Camellia sinensis*), is another defense chemical, effective against insects at a fraction of the concentration found in fresh coffee beans or tea leaves. The pharmaceutical, nutritional or psychotropic properties of plant alkaloids can be directly attributed to their ecological function as chemical influences (often harmful) on the physiology of interacting animals and other organisms.

15.4.1 Tryptophan is the biosynthetic precursor of indole alkaloids

Most classes of alkaloid are derived directly or indirectly from amino acids (Table 15.3), either alone or in

Table 15.3 Precursors, structural groupings and examples of plant alkaloids.

Precursor	Alkaloid types	Example and source	Structure
Tyrosine	Isoquinoline	Mescaline (psychotropic product from various cacti)	
Tryptophan and anthranilate	Indole Quinoline Pyrroloindole Quinazoline Acridone	Eserine (cholinesterase inhibitor from *Physostigma venenosum*)	
Ornithine	Tropane Pyrrolizidine	Scopolamine (anticholinergic antifeedant induced by physical damage in species of the family Solanaceae)	
Lysine	Piperidine Quinolizidine Indolizidine	Lupanine (antiherbivory compound from species in the genus *Lupinus*)	
Nicotinic acid	Pyridine	Myosmine (carcinogenic constituent of tobacco and nuts)	
Histidine	Imidazole	Pilocarpine (narcotic from *Pilocarpus microphyllus*)	
Amination products	Acetate-derived Phenylalanine-derived Terpenoid Steroidal Purine	Ephedrine (stimulant of sympathetic nervous system, originally isolated from *Ephedra* spp.)	

combination with a terpenoid. Detailed biosynthetic pathways are known for relatively few alkaloids. Where information is available it often reveals exotic and complex chemical structures and mechanisms that testify to the great metabolic versatility of plants. By way of illustration, here we briefly consider the reaction sequence leading to one of the indole group of alkaloids, and in Section 15.4.2 we discuss a second example in which the product is an isoquinoline (Table 15.3).

The indole-type neurotoxin **strychnine** is a derivative of tryptophan. The first committed step in tryptophan formation is the synthesis of **anthranilate** from chorismate, catalyzed by **anthranilate synthase** (see Figure 15.8). Anthranilate synthase transfers an amine group from glutamine to chorismate and releases pyruvate (Figure 15.16). Two enzymatic steps convert anthranilate to indole-3-glycerol phosphate, the substrate for **tryptophan synthase** (**TS**). Tryptophan is

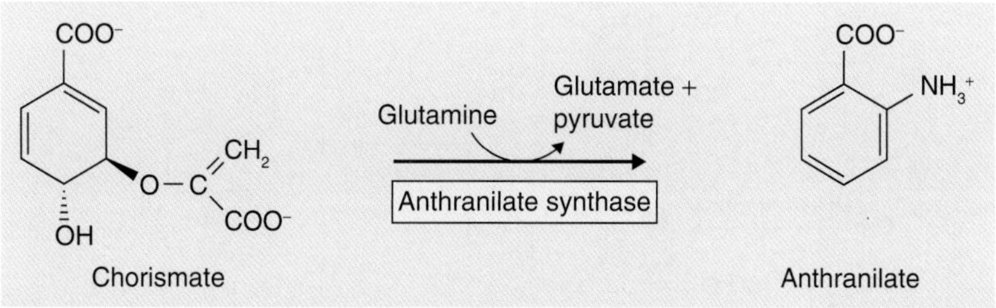

Figure 15.16 Conversion of chorismate to anthranilate by anthranilate synthase, first step in tryptophan biosynthesis.

the product of two reactions that are catalyzed by the α- and β-subunits of TS respectively (Equation 15.7).

Equation 15.7 Tryptophan synthase
15.7A Reaction catalyzed by α-subunit

Indole-3-glycerol phosphate →

indole + glyceraldehyde-3-phosphate

15.7B Reaction catalyzed by β-subunit

Indole + serine → tryptophan + H_2O

Strychnine biosynthesis begins with the decarboxylation of tryptophan by tryptophan decarboxylase to form **tryptamine** (Equation 15.8). **Strictosidine synthase** then catalyzes the stereospecific condensation of tryptamine with the terpenoid secologanin. The product is **strictosidine** (Figure 15.17), the starting point for species-specific enzymatic permutations that result in a multitude of diverse structures. Strychnine, one of the classic tools of poisoners in history and fiction, is just one of the many metabolic derivatives of strictosidine. Others include the

antimalarial quinine, and **vinblastine** and **vincristine**, tubulin-binding alkaloids from *Catharanthus* spp. that are widely used in cancer therapy.

Equation 15.8 Tryptophan decarboxylase

Tryptophan → tryptamine + CO_2

15.4.2 Morphine and related isoquinoline alkaloids are tyrosine derivatives

Tyrosine is the precursor of the **isoquinoline alkaloid, morphine** (Figure 15.18). One molecule of tyrosine is decarboxylated to form **tyramine**, which, in turn, is converted to 3,4-dihydroxyphenylamine (**dopamine**) by the action of a phenol oxidase. A second tyrosine is transaminated and decarboxylated to yield *p*-hydroxyphenylacetaldehyde. Dopamine and *p*-hydroxyphenylacetaldehyde are then stereoselectively condensed to form the first isoquinoline intermediate, (*S*)-norcoclaurine. A series of methylation and oxidation

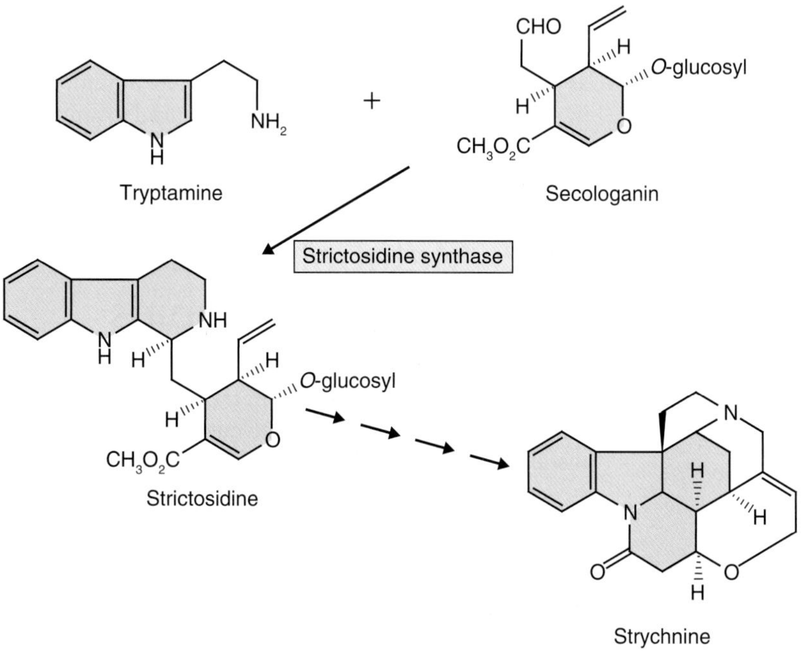

Figure 15.17 The indole alkaloid strychnine is biosynthesized from tryptophan via tryptamine (which is itself a biologically active alkaloid) and strictosidine.

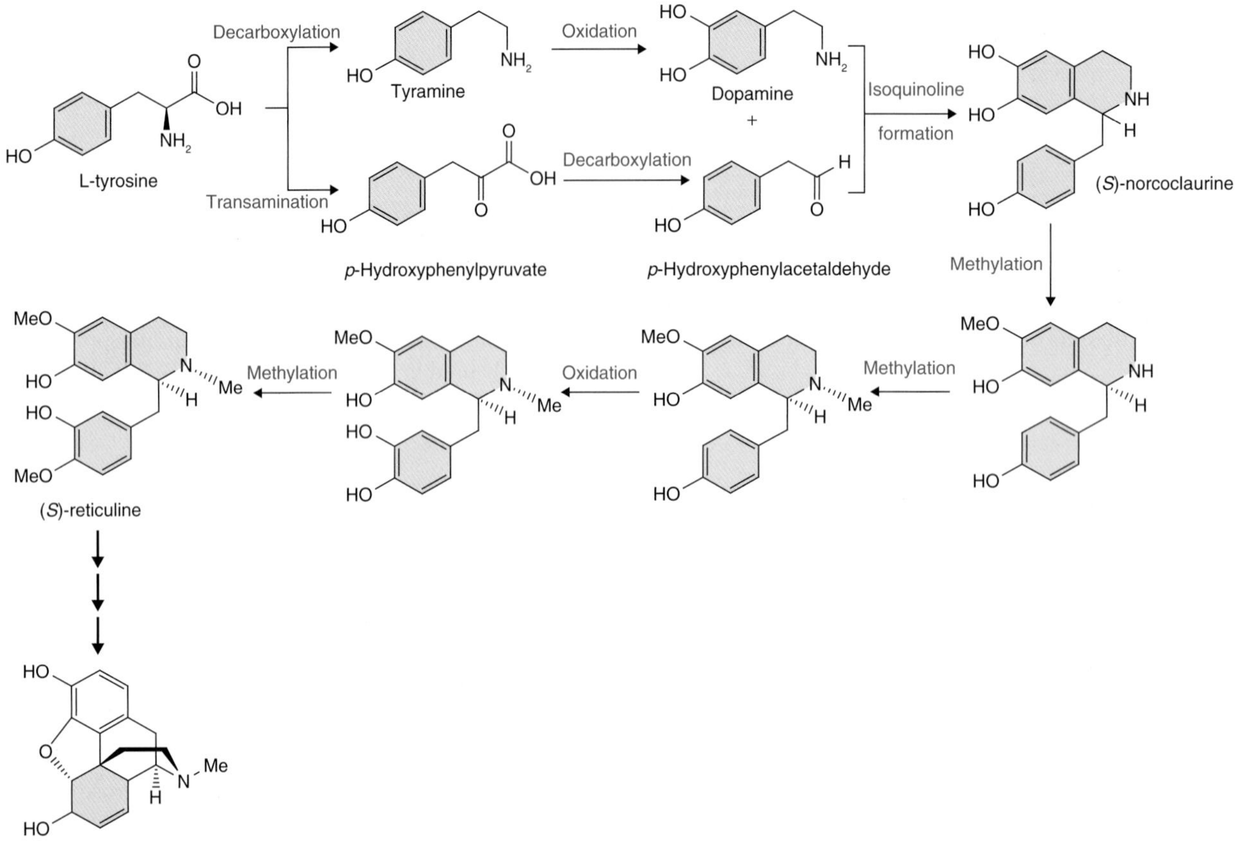

Figure 15.18   Outline of the morphine biosynthetic pathway.

Key points Alkaloid medicines, narcotics and poisons from plants have been used throughout human history. Some microorganisms and animals accumulate protective alkaloids, often acquired from the plants they feed on. Alkaloids may be present constitutively, or the plant may be induced to make them in response to attack by a predator or pathogen. The precursors of most alkaloids are amino acids, sometimes combined with terpenoids. The indole alkaloids, one of which is the neurotoxin strychnine, are synthesized from tryptophan, which in turn is derived from the shikimate pathway product chorismate. Strictosidine is a branch point intermediate in strychnine biosynthesis and is a precursor of a range of alkaloids, including quinine and the tubulin-binding drugs vinblastine and vincristine. Morphine and codeine are members of the isoquinoline class of alkaloids, which are derived from tyrosine. Important intermediates in this biosynthetic pathway are dopamine and (S)-reticuline.

reactions follow that produce (S)-**reticuline**, a branch point metabolite. (S)-reticuline, like strictosidine, is the precursor of an enormous array of structurally diverse products, that includes not only morphine and related compounds such as **codeine**, but also **berberine** (an active component of traditional medicines as well as a dyestuff) and **sanguinarine** (a product, found in a number of plant species, that is harmful to animal cells).

15.5 Metabolic responses to stress III. Terpenoids

Terpenoids take their name from **terpenes**, the volatile constituents of **turpentine** (a solvent produced by the distillation of pine tree resin). They are the most structurally varied group of plant natural products and are built from **isoprene** units, which consist of the branched five-carbon isopentane skeleton. Terpenoids are often called **isoprenoids**; however, isoprene itself is

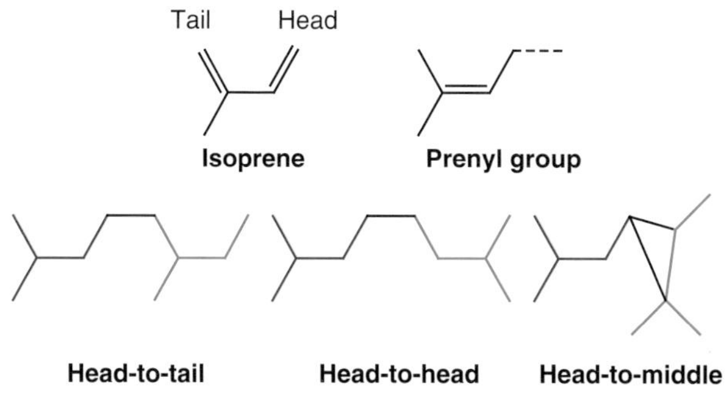

Figure 15.19 Terpenoids are built from prenyl units, based on the branched C_5 hydrocarbon isoprene. A great array of terpenoid structures can be generated by joining C_5 units in combinations of head–tail, head–head and head–middle configurations.

not an intermediate in terpenoid synthesis. The repetitive C_5 structural motif from which terpenoids are built is referred to as the **prenyl group** (Figure 15.19). The different classes of terpenoids are identified by the number of C_5 units (Table 15.4). Thus we have hemiterpenoids (C_5), monoterpenoids (C_{10}), sesquiterpenoids (C_{15}), diterpenoids (C_{20}), sesterpenoids (C_{25}), triterpenoids (C_{30}) and tetraterpenoids (C_{40}). Terpenoids consisting of more than eight isoprene units are referred to as polyterpenoids. Natural **rubber**, a polymer of isoprene consisting of 11 000 to 20 000 monomeric units, is an example (Table 15.4). Ubiquinone and plastoquinone, the electron carriers of respiration (see Figure 7.11) and photosynthesis (see Figure 9.15A), are also polyterpenoids. **Norterpenoids** are terpenoid derivatives that have lost one or more carbons and so are no longer C_5 multiples. For example, some **gibberellins** (see Chapter 10) are C_{19} structures, the result of the removal of one C from a diterpene precursor. Natural products of mixed biosynthetic origins that are partially derived from terpenoids are sometimes called **meroterpenoids**. Such compounds include **cytokinins** (see Chapter 10) and some proteins that are prenylated by the addition of lipophilic, membrane-anchoring 15- or 20-carbon isoprenoid side-chains.

Diterpenoids and triterpenoids include both primary and secondary metabolites. For example, the diterpene **kaurenoic acid** is an essential intermediate in the synthesis of gibberellins in all plants (see Chapter 10) and is therefore a participant in primary metabolism. On the other hand **abietic acid**, similar to kaurenoic acid in structure and biosynthetic origin (Figure 15.20), is a component of resin and largely restricted to members of the Fabaceae and Pinaceae families. Because it is not involved in primary metabolism, it is by definition a secondary compound. An important chemical group

derived from di- and triterpenoid precursors is the **steroids**. Steroids are widely distributed in plants, fungi and animals; they have essential roles in membrane structure and hormone signaling. Among the steroids of plants are **phytosterols**, natural constituents of vegetable oils with possible beneficial effects for human nutrition, and **brassinosteroids**, a class of plant hormone (see Chapter 10). Figure 15.21 shows some plant membrane sterols of triterpenoid origin. The distinctive structural feature is the sterane core, which comprises one cyclopentane (C_5) and three cyclohexane (C_6) rings.

The majority of tetraterpenoids are **carotenoids**, many of which (for instance, violaxanthin and lutein; see Figure 9.12) have essential roles in photobiology and oxygen metabolism across the taxonomic range of autotrophs and are therefore primary metabolites. Table 15.4 gives an example of a carotenoid secondary metabolite, **capsorubin**, which is responsible for the red color of the spice paprika. **Apocarotenoids**, the products of oxidative cleavage of tetraterpene carotenoids, are widely distributed in living organisms, and frequently perform important regulatory, receptor and signaling functions. The hormone **abscisic acid** (see Chapter 10) is the apocarotenoid product of asymmetrical cleavage of a C_{40} carotenoid.

15.5.1 Terpenoids are synthesized from IPP and DMAPP, frequently in specialized structures

The variety of terpenoids produced by plants is much wider than that produced by either microbes or animals. The accumulation, emission or secretion of large quantities of terpenoids by plants is almost always associated with the presence of specialized structures such as glandular trichomes, secretory cavities, resin ducts and blisters. These structures are usually non-photosynthetic and so are dependent on neighboring or remote tissues for the carbon and energy necessary to drive terpenoid synthesis. They are sources of defensive products, such as the diterpenoids of rosin from conifer species. The terpenoid essential oils emitted by the glandular epidermis of petals encourage insect pollination. Specialization of the epidermis is also associated with the formation and excretion of triterpenoid surface waxes, while lactiferous ducts produce certain triterpenoids and rubber. It is thought that, by separating bioactive and often harmful terpenoid products from the sites of metabolism in the major tissue types, a plant is able to avoid potential autotoxicity problems.

Table 15.4 Terpenoids grouped by number of carbon atoms, with examples and structures of members of each class.

Number of C atoms	Terpene class	Example	Structure
5	Hemiterpenoid	Isoprene, the structural unit of the terpenoids and a volatile product emitted by photosynthetic tissues	
10	Monoterpenoid	Limonene, a volatile inhibitor of seed germination from *Quercus ilex*	
15	Sesquiterpenoid	Farnesol, a natural insecticide and pheromone, found in essential oils of many species	
20	Diterpenoid	Taxadiene, precursor of the anticancer drug taxol derived from *Taxus* (yew)	
25	Sesterpenoid	Ophiobolane, a phytotoxic fungal terpenoid	
30	Triterpenoid	Lupane, member of a widely distributed family of cytotoxins	
40	Tetraterpenoid	Capsorubin, a colored component of paprika extract, from *Capsicum* spp.	

Table 15.4 (*continued*).

Number of C atoms	Terpene class	Example	Structure
>40	Polyterpenoid	Natural rubber is a polymer of isoprene (*n* = up to 20 000) found in the latex of the rubber tree, *Hevea*, and a number of other species	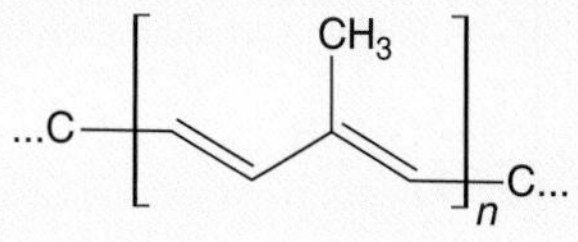

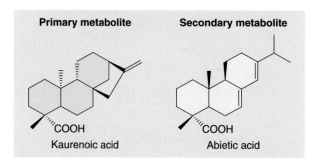

Primary metabolite   **Secondary metabolite**

Kaurenoic acid   Abietic acid

Figure 15.20 Examples of a primary and a secondary diterpenoid metabolite. Kaurenoic acid is a precursor of the gibberellin group of essential hormones, and is therefore classified as a primary metabolite. The resin component abietic acid is a secondary metabolite.

All terpenoids are biosynthesized from simple primary metabolites by a four-stage process. Stage 1 is the formation of **isopentenyl diphosphate** (**IPP**) and its isomer **dimethylallyl diphosphate** (**DMAPP**), the C_5 structural units from which all isoprenoids are built. During stage 2 a series of **prenyl diphosphate** homologs are made by repetitive additions of IPP. Stage 3 is the synthesis of terpenoid skeletons from prenyl diphosphates, catalyzed by specific **terpenoid synthases**. Secondary enzymatic modifications of terpenoid skeletons during stage 4 result

in the diversity of functional properties and chemical structures that characterize this family of natural products.

There are two pathways of IPP synthesis, one that occurs in the cytosol and a second that takes place in plastids. The synthesis of IPP in the cytosol (Figure 15.22) begins with a two-step condensation of three molecules of acetyl-CoA and a subsequent reduction to yield **mevalonic acid** (**MVA**), which gives this pathway its name. MVA is converted to IPP via two sequential, ATP-dependent phosphorylations and a decarboxylation. A critical regulatory reaction in the MVA pathway is that catalyzed by 3-hydroxy-3-methylglutaryl-CoA reductase (**HMGR**). HMGR is an ER-targeted enzyme and is encoded by a family of two to four genes, depending on species. Differential expression of *HMG* genes is thought to lead to tissue- and stress-specific biosynthesis of different types of isoprenoids.

Recently, an alternative pathway of IPP formation has been shown to operate in the plastids of land plants and algae, as well as in cyanobacteria and other eubacteria, and certain eukaryotic parasites. It begins with the formation of 1-deoxy-D-xylulose-5-phosphate (**DOXP**; Figure 15.23). In general, the **MVA pathway** of IPP synthesis is the origin of phytosterols, sesquiterpenes and triterpenoids, whereas the **DOXP pathway** supports synthesis of plastid isoprenoids, notably carotenoids,

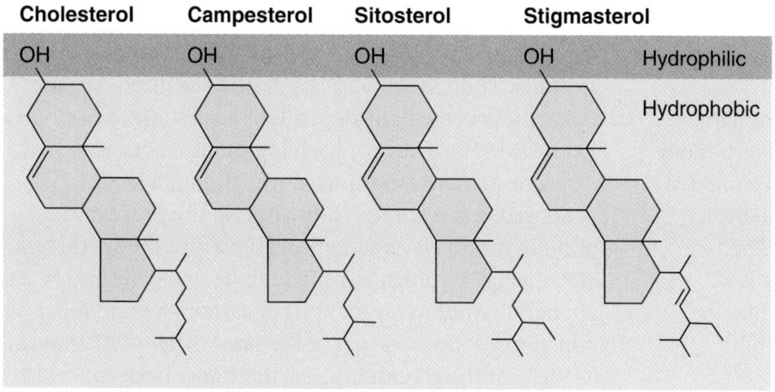

Cholesterol   **Campesterol**   **Sitosterol**   **Stigmasterol**

OH   OH   OH   OH   Hydrophilic

Hydrophobic

Figure 15.21 Steroids of plant membranes, showing amphipathic character. Cholesterol is a C_{27} molecule, campesterol is C_{28} and sitosterol and stigmasterol are C_{29}. These sterols are derived from the straight-chain C_{30} triterpenoid precursor by cyclization to create the 4-ring C_{17} sterane core, followed by removal of one or more carbons.

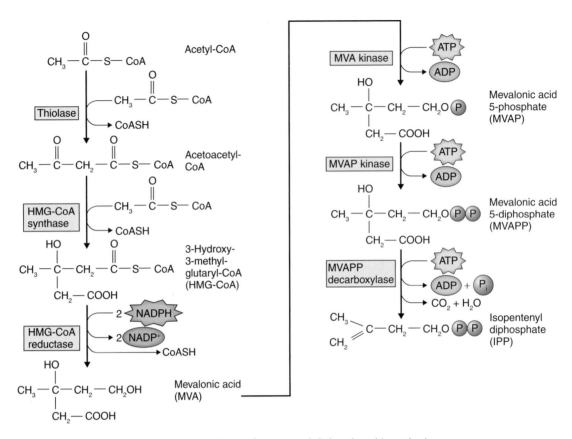

Figure 15.22   The mevalonic acid (cytosolic) pathway of isopentenyl diphosphate biosynthesis.

phytol, plastoquinones and tocopherols as well as the hormones gibberellin and abscisic acid.

15.5.2 Terpenoids of C_{10} and larger are made by condensation of IPP units on an initial DMAPP primer, catalyzed by prenyltransferases

Isoprenoid biosynthesis requires both IPP and its isomer DMAPP. IPP and DMAPP are used by **prenyltransferases** to form the C_{10}, C_{15}, C_{20} and larger prenyl diphosphates that serve as precursors in terpenoid biosynthetic pathways (Figure 15.24). Successive units of IPP are added to an initial DMAPP primer, usually via head-to-tail condensations (see Figure 15.19). The IPP-forming step of the plastid DOXP pathway (Figure 15.23) also yields DMAPP, though the precise mechanism is not known. On the other hand, the cytosolic MVA pathway produces only IPP

(Figure 15.22) and requires the action of the enzyme **IPP isomerase** to sustain equilibrium between IPP and DMAPP (Figure 15.24). IPP isomerase is of regulatory importance for the synthesis of the major terpenoid classes because the molar ratio of IPP to DMAPP needed to synthesize different terpenoid classes varies from 1:1 for monoterpenoids to 2:1 for sesquiterpenes and sterols, 3:1 for diterpenoids, carotenoids and phytol, and much higher for long-chain polyprenols and polyterpenoids.

Hemiterpenoids are derived from the IPP–DMAPP equilibrium. For example, DMAPP is the direct precursor of isoprene, a volatile hemiterpenoid released from photosynthetically active tissues. **Isoprene synthase** (Equation 15.9) is a plastid-localized enzyme that catalyzes the light-dependent formation of isoprene from DMAPP. The ecophysiological function, if any, of isoprene production is unknown, although it is proposed that isoprene is involved in the process of acclimation to elevated temperatures and the synthetase enzyme has an unusually high temperature optimum. At about 500 million metric tons of carbon, annual foliar emissions of isoprene are of the same order of magnitude as those of the greenhouse gas **methane**. Isoprene is also

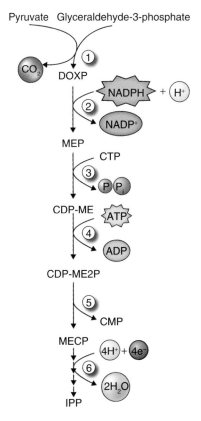

Figure 15.23 The DOXP pathway of IPP synthesis. Enzymes in the pathway are: (1) DOXP synthase, (2) DOXP reductoisomerase, (3) MEP cytidylyltransferase, (4) CDP-ME kinase, (5) MECP synthase and (6) synthase and reductase. CDP-ME, 4-diphosphocytidyl-2C-methyl-D-erythritol; CDP-ME2P, 4-diphosphocytidyl-2C-methyl-D-erythritol-2-phosphate; CMP, cytidine monophosphate; CTP, cytidine triphosphate; DOXP, 1-deoxy-D-xylulose-5-phosphate; IPP, isopentenyl diphosphate; MECP, 2C-methyl-D-erythritol-2,4-cyclodiphosphate; MEP, 4-diphosphocytidyl-2C-methyl-D-erythritol.

a principal reactant in the formation of **ozone** in the troposphere.

Equation 15.9 Isoprene synthase

Dimethylallyl diphosphate $\rightleftharpoons$ isoprene + PP$_i$

Geranyl diphosphate (**GPP**), farnesyl diphosphate (**FPP**) and geranylgeranyl diphosphate (**GGPP**) are the **diphosphate ester** precursors of mono-, sesqui- and diterpenoids, respectively, and are generated by the activities of prenyltransferase enzymes (Figure 15.24). The addition of one IPP moiety to the DMAPP primer yields the C_{10} intermediate GPP. Further reaction cycles result in FPP (C_{15}) and GGPP (C_{20}). Although prenyl

units are mostly added head to tail, there are many instances of exceptions to this pattern, involving head-to-head and head-to-middle patterns (see Figure 15.19). An example is the **pyrethrins**, monoterpenoids with head-to-middle configurations. These compounds were originally identified in members of the family Asteraceae and have been widely used as insecticides both in their native form and as synthetic derivatives. Most of the new allylic double bonds introduced by prenyl transferases are in the trans configuration, though occasionally cis double bonds are created. This is the case in rubber, for example, where cis bonds account for the polymer's elasticity. Prenyl transferase reactions can attach groups other than IPP and are responsible for adding prenyl side-chains to compounds such as proteins and other non-terpenoids.

Terpene synthases are families of enzymes that are responsible for the formation of the enormous diversity of carbon skeletons characteristic of terpenoids; prenyl diphosphate esters are the substrates for these enzymes. Terpenoid synthases that produce cyclic products, the commonest type of isoprenoid structure, are also referred to as **cyclases**. **Abietic acid** (see Figure 15.20), a common diterpenoid acid of **conifer resin**, is an example of a product of synthase/cyclase activities. Resin is important for wound-sealing and is familiar as its fossilized form, **amber**. In some cases a particular terpene synthase can generate more than one reaction product. For example, **pinene synthase** yields both α- and β-pinene, widely distributed monoterpene toxins that are active against bark beetles and their pathogenic fungal symbionts. There are **sesquiterpene synthases**, also involved in the production of conifer resin, that are known to be capable of individually making more than 25 different products.

15.5.3 Squalene and phytoene are precursors of phytosterols and carotenoids

The C_{30} triterpenoids are generated by the head-to-head joining of two C_{15} FPP chains to produce **squalene**. Similarly, the C_{40} tetraterpenoids are derived from **phytoene**, the product of combining two molecules of C_{20} GGPP in head-to-head fashion (Figure 15.24). **Squalene synthase** and **phytoene synthase** are prenyl transferases with similar reaction mechanisms that involve a complex series of rearrangements necessary to bring the head end (C-1) carbons of the two prenyl diphosphate precursors together.

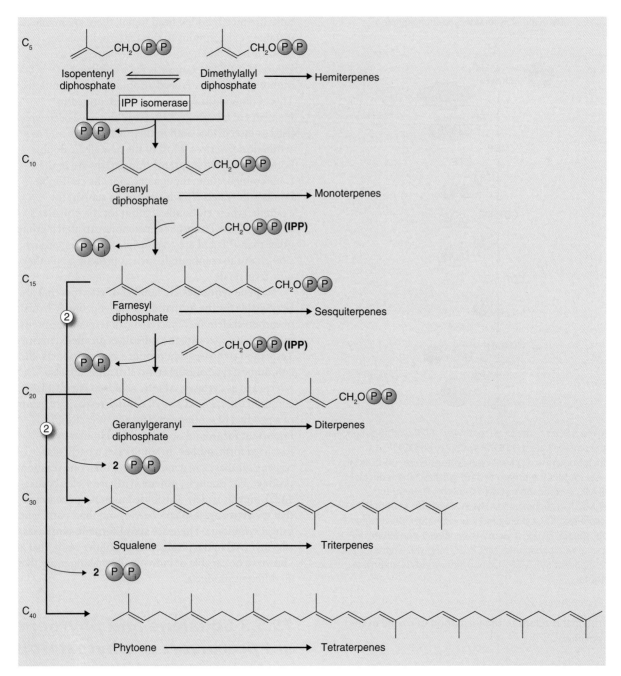

Figure 15.24   Biosynthesis of the major subclasses of terpenoids from isopentenyl diphosphate (IPP) and its isomer dimethylallyl diphosphate. Prenyl transferases catalyze the formation of monoterpenes (C_{10}), sesquiterpenes (C_{15}) and diterpenes (C_{20}) from the corresponding intermediates by sequential head-to-tail addition of C_5 units. Triterpenes (C_{30}) are formed from two C_{15} (farnesyl) units joined head to head, and tetraterpenes (C_{40}) are formed from two C_{20} (geranylgeranyl) units joined head to head.

Steroids are cyclization products of squalene. An NADPH-dependent **epoxidase** (Equation 15.10) oxidizes squalene to 2,3-epoxysqualene, which is then cyclized to form a range of intermediates in phytosterol biosynthesis. Figure 15.25 shows **cycloartenol** and **lanosterol**, major cyclization products which, through a series of methylation and demethylation reactions, give rise to membrane sterols such as cholesterol,

campesterol, sitosterol and stigmasterol (see Figure 15.21) and the brassinosteroid hormones (see Chapter 10).

Equation 15.10 Squalene epoxidase

$$\text{Squalene} + \text{NADPH} + \text{H}^+ + \text{O}_2 \rightarrow$$
$$2,3\text{-epoxysqualene} + \text{NADP}^+ + \text{H}_2\text{O}$$

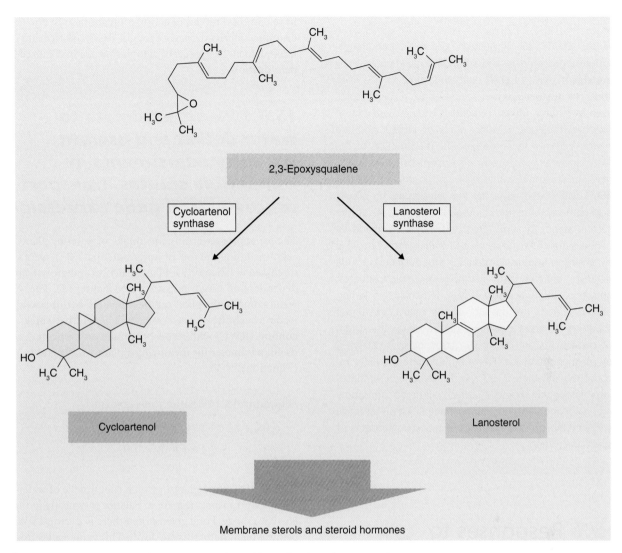

Figure 15.25   Precursors of phytosteroids, formed by cyclization of epoxysqualene.

Carotenoids are derived from phytoene via a series of desaturation and cyclization steps. Metabolic interconversions of carotenoids are important in protection against **photoinhibition** and oxidative damage, and are described further in Section 15.6. Carotenoid synthesis is also a central feature of **fruit development** in many species, contributing to color changes and other aspects of ripening, as discussed in Chapter 18.

15.5.4 Secondary modification of terpenoids results in a wide variety of bioactive compounds

The final step in the overall scheme of terpenoid biosynthesis involves modifications of the parent skeletons produced by terpenoid synthases and results in the rich array of different structures and functional properties found among the isoprenoids of the plant kingdom. Among these secondary transformations are oxidations, reductions, isomerizations and conjugations. Some terpenoid skeletons are highly decorated. For instance, the anticancer drug **taxol**, obtained from yew (*Taxus*) species, is a diterpenoid nucleus (taxadiene; see Table 15.4) that is extensively modified by a complex pattern of hydroxylations and acylations. **Saponins** and **cardenolides** are powerful bioactive compounds that are toxic to herbivores. In these molecules, the basic terpenoid structure is modified by **glycosylation**. The widely used cardioactive drug **digitoxigenin** is the aglycone (sugarless) moiety derived from foxglove (*Digitalis*) cardenolides.

Key points Terpenoids, also called isoprenoids, include both primary and secondary metabolites. They are chains of C_5 (prenyl) units and range in size from the C_5 hemiterpenoids to macromolecular polyterpenoids such as natural rubber. Terpenoids often accumulate in specialized glands, cavities, ducts or blisters and are synthesized in a four-step process. The products of the first stage are the C_5 precursors isopentenyl diphosphate (IPP) and its isomer dimethylallyl diphosphate (DMAPP). IPP is synthesized either by the cytoplasmic mevalonic acid pathway, or by the chloroplastic DOXP route. In the second stage, C_{10} and larger terpenoids are made by condensation of IPP units on a DMAPP primer, catalyzed by prenyl transferases. Stage 2 intermediates include geranyl, farnesyl and geranylgeranyl diphosphates. Squalene, the precursor of C_{30} terpenoids including phytosterols, is the product of the prenyl transferase squalene synthase. Synthesis of phytoene, the origin of carotenoids and other C_{40} terpenoids, is catalyzed by phytoene synthase. Prenyl diphosphate products are converted in stage 3 by specific terpenoid synthases into terpenoid skeletons, which undergo secondary enzymatic modifications during stage 4. Taxol, saponins and cardenolides are examples of bioactive products of stage 4 activities.

15.6 Responses to abiotic stresses

The experience of **abiotic stress** is not only normal for plants, it is the essential route by which information about the state of the environment is sensed and the physiological adjustments needed to insure survival are cued. Responses depend on the severity and duration of the stress, stage of development, tissue type and interactions between multiple stresses. Experiencing stress typically promotes changes in gene expression and metabolism. Reactions are frequently centered on altered patterns of secondary compounds as described in Section 15.2.4, and ultimately lead to localized or systemic modifications of physiology, development and life cycle. In the present section we will look at plant reactions to **physico-chemical variables** in the environment, particularly water (lack and excess), oxygen and its reactive derivatives, radiant energy (both heat and light) and mechanical influences. Although this discussion considers each type of factor individually, it is important to be aware of the extent to which different agents may

have common modes of action on plant tissues. For example, salinity damage, drought stress and freezing injury can all be aspects of cytoplasmic trauma caused by **dehydration**.

15.6.1 Plants acclimate to water deficit and osmotic stress by adjustments in compatible solutes, transport processes and gene expression

The principles governing the uptake of water by plant cells, including the role of water potential (Ψ_w), were discussed in Chapter 14. Plant structure and growth are dependent on turgor pressure (see Chapter 12). The water content of a plant can be measured by subtracting its dry weight from its wet weight. The **relative water content** (**RWC**) with respect to a fully turgid plant is a useful parameter for measuring a plant's water status (Equation 15.11).

> **Equation 15.11** Relative water content
>
> RWC = [(fresh weight − dry weight)/(turgid weight − dry weight)] × 100

Leaf RWC for a plant in which root uptake of water and shoot transpiration are in balance is typically in the 85–95% range. For a given organ there is a critical RWC, below which there is tissue death. This value varies from species to species but is often less than 50%. In general, when the water potential of the soil drops, the plant's RWC also declines. This may occur as a result of drought or osmotic limitation of water availability due to salinity. Most plants, however, have the capacity to adjust water potential by internally accumulating **osmotically active solutes**. This has the effect of decreasing Ψ_w and sustaining the influx of water along the steeper gradient in water potential between soil and cell (see Chapter 14), thereby preserving a high RWC and maintaining viability. Here we discuss the cell biology of osmotic adjustment, mechanisms of water flow into cells, the products and regulation of water stress-related gene expression, and the mechanisms by which water status is sensed and signaled.

In order to take up water, a plant must maintain a water potential gradient between root and soil. **Wilting** is an early symptom that the rate of transpiration by leaves and stems exceeds the rate of water uptake from the soil. Many plants that tolerate drought or saline conditions avoid the harmful effects of declining RWC by regulating their solute potentials. Osmotic adjustment is an acclimatory process that involves the synthesis of compounds which increase the osmotic concentration of the cell and protect it from dehydration damage. Comparative physiological studies show that some of these compounds function in acclimation not only in plants but also in bacteria and animals. In view of the fundamental importance of water for life, it is not surprising that cellular mechanisms of regulating water potential will show a degree of commonality in all living organisms.

In plants, the compounds used for osmotic adjustments in the central vacuole are different from those used in the remainder of the cell. In the vacuole, which occupies most of the volume of the cell, inorganic salts are usually the major osmoticum. Vacuolar enzymes tolerate high salt concentrations. In the cytosol and the rest of the organelles, high salt concentrations are lethal and small organic molecules are used instead. Compounds that function in osmotic adjustment in the cytosol and non-vacuole organelles are referred to as **compatible solutes** or **osmolytes**. They are members of a relatively small group of chemically diverse, water-soluble, organic compounds (Figure 15.26) that can be accumulated to high concentrations without significantly interfering with cellular metabolism (hence the name 'compatible'). For example, the cytosol and chloroplasts of salt-stressed spinach leaves (see Figure 15.4D) accumulate **glycine betaine** to more than 250 mM, whereas the concentration of this osmolyte in vacuolar sap is vanishingly small. The movement and differential distribution of osmolytes within and between cells are regulated by membrane-associated transporters. Accumulation of compatible osmolytes may have the additional benefit of protecting against other stresses. For example, **mannitol** and **proline** (Figure 15.26) are scavengers of hydroxyl radicals and could have a role in moderating the effects of reactive oxygen species (see Section 15.6.3).

Some compatible solutes, e.g. proline, occur widely throughout the plant kingdom. Others are more limited in their distribution; for example **β-alanine betaine** (Figure 15.26) is only found in a few species of the family Plumbaginaceae. Monomeric sugars derived from polymers, notably glucose from starch and fructose from fructans, can act as effective compatible solutes under stress conditions. Different mechanisms for solute accumulation are recognized. In some cases there is irreversible synthesis en masse. In others, buildup is achieved by shifting normal turnover toward synthesis and away from breakdown. Glucose derived from starch is an example of a solute released from an osmotically inactive polymer that can be readily repolymerized when the stress is relieved.

Mass synthesis of a compatible osmolyte in response to water limitation is illustrated by **glycine betaine** (**GB**) (Figure 15.26). Glycine betaine is a quaternary ammonium compound that acts as a compatible solute in a wide range of animals, bacteria and some **halophytic** (salt-tolerant) and drought-tolerant angiosperms. In angiosperms, it is abundant mainly in chloroplasts, where it maintains photosynthetic efficiency under osmotic stress by protecting the integrity of thylakoid membranes. Species such as maize (*Zea mays*), sugar beet (*Beta vulgaris*), barley (*Hordeum vulgare*) and spinach (*Spinacia oleracea*) are natural accumulators of GB in response to salt, drought or low-temperature stresses. GB is synthesized in chloroplasts from **choline**, which in turn is derived from the amino acid serine and the turnover of membrane phospholipid. Choline is dehydrogenated to **betaine aldehyde** by a monooxygenase that uses photosynthetically-reduced ferredoxin as a co-substrate. Betaine aldehyde is then converted to GB by betaine aldehyde dehydrogenase (Figure 15.27). The supply of choline is often the rate-determining step in GB biosynthesis. The activities of both enzymes, and the abundance of the mRNA transcripts of the genes that encode them, increase markedly under conditions of osmotic stress. Transcript levels decline when the stress is removed, but in the absence of a breakdown pathway, accumulation of GB itself is irreversible. Other routes of GB biosynthesis are known, but the one that originates with choline is common to all GB-accumulating plant species. Genetically engineering enzymes of GB biosynthesis into plants that normally fail to accumulate GB has been shown to improve tolerance of osmotic stress.

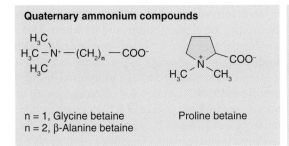
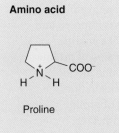
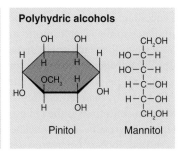

Figure 15.26   Structures of some common cell osmolytes.

The amino acid proline (see Figure 15.26) is an example of a compatible osmolyte that accumulates in the cytosol in response to a range of environmental stresses. Unlike GB, proline is rapidly catabolized when the stress is relieved; this provides breakdown products that are thought to contribute to the energy requirements of recovery and repair. In addition to its role as an osmolyte, proline supports the stabilization of proteins, membranes and other subcellular structures, acts as a **free radical scavenger**, and buffers cellular metabolism against fluctuations in redox and pH conditions. Moreover, expression of a number of salt stress-responsive genes is known to be under the control of promoters that possess **proline responsive elements**. There are two different biosynthetic pathways for proline in plants. The one that originates in the amino acid ornithine is less well understood than the major route from glutamate (Equation 15.12), which is also the pathway that operates in bacteria.

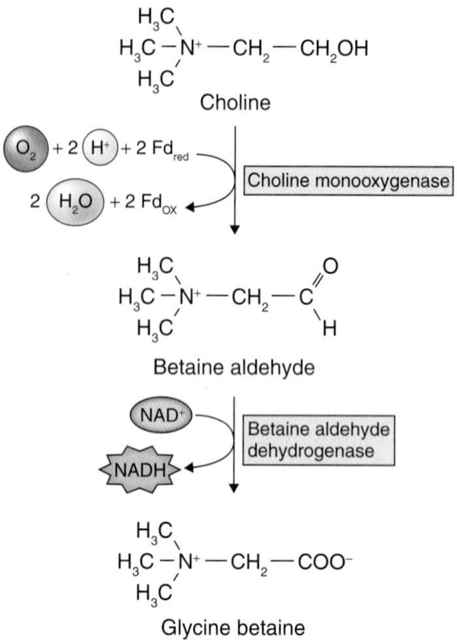

Figure 15.27 Biosynthesis of glycine betaine from choline.

Equation 15.12 Synthesis of proline from glutamate

15.12A Δ^1-pyrroline-5-carboxylate synthetase

Glutamate + ATP + NADPH + H$^+$

→ glutamate γ-semialdehyde + ADP

+ NADP$^+$ + P$_i$

15.12B Spontaneous cyclization of glutamate γ-semialdehyde

Glutamate γ-semialdehyde $\rightleftharpoons$

Δ^1-pyrroline-5-carboxylate + H$_2$O

15.12C Δ^1-pyrroline-5-carboxylate reductase

Δ^1-pyrroline-5-carboxylate + NADPH + H$^+$

→ proline + NADP$^+$

Formation of **glutamate γ-semialdehyde (GSA)** by the bifunctional enzyme Δ^1-pyrroline-5-carboxylate synthetase is the rate-limiting step and is subject to **allosteric feedback regulation** by proline. In coordination with tissue proline levels, transcripts of the gene encoding Δ^1-pyrroline-5-carboxylate synthetase accumulate rapidly when osmotic stress is imposed and decrease when the plant is rehydrated. The enzyme **proline dehydrogenase** catalyzes conversion of proline back to pyrroline-5-carboxylate. The abundance of proline dehydrogenase mRNA under dehydration/ rehydration conditions is the reciprocal of proline content and pyrroline-5-carboxylate synthetase transcription. By balancing stress-sensitive synthesis

and catabolism in this way, efficient control of proline content can be exerted through **metabolic turnover**.

A final group of compatible solutes considered here is the **polyhydric alcohols**, which include mannitol and **pinitol** (see Figure 15.26). Mannitol, a sugar alcohol, represents a substantial proportion of carbohydrates in some plant species, such as celery (*Apium graveolens*). It accumulates in response to osmotic stress as a result of decreased rates of consumption and reduced competition from **sucrose synthesis** for precursors. High concentrations of the cyclic sugar alcohol pinitol, found in members of the Pinaceae, Fabaceae and Caryophyllaceae families, are characteristic of halophytic and drought-adapted species. Pinitol is located in chloroplasts and the cytosol but is absent from vacuoles. Pinitol synthesis is **transcriptionally regulated**. Plants that have been genetically engineered to accumulate pinitol, mannitol or proline have been shown to have enhanced resistance to osmotic stresses.

The accumulation of compatible solutes in response to drought and salt stress enables a plant to acclimate to low water potentials. For osmotic adjustment to maintain integrity of cell structure and function, it must also control ion concentrations and transport. This is particularly significant for plants growing under **saline conditions**, in the presence of potentially toxic amounts of specific ions, notably Na$^+$. In addition to its direct osmotic effect, Na$^+$ can be toxic as a consequence of interference with the uptake of K$^+$, an essential mineral nutrient (see Chapter 13). No Na$^+$-selective ion channels have been found in plants. The major route of Na$^+$ influx across the plasma membrane is via **voltage-independent non-specific cation channels** (**VI-NSCCs**). Intracellular Ca^{2+} status has a direct influence on Na^+ flux since Ca^{2+} blocks VI-NSCC activity. Active export of Na^+ across the plasma membrane counteracts accumulation of Na^+ in the cytosol. Additionally, **Na^+/H^+ antiporters** in the tonoplast become active in response to salt stress and can move cytosolic Na^+ into the vacuole.

Acclimatory responses to drought and salinity, such as the accumulation of compatible solutes and redistribution of ions, are associated with alterations in gene expression under the control of regulatory networks sensitive to abiotic stresses. A comparison of global transcription patterns in water-deficient and -sufficient tissues reveals a range of stress-inducible genes, many of which have identifiable roles in acclimation. The protein products of such genes can be broadly classified into two groups (Figure 15.28). The **functional** group includes chaperones, antifreeze proteins, enzymes of osmolyte biosynthesis, aquaporins, sugar and proline transporters, detoxification enzymes, and various proteases. The **regulatory** group, comprising protein factors involved in further regulation of signal transduction and stress-responsive gene expression, includes transcription factors, protein kinases and phosphatases, enzymes of phospholipid metabolism, and components of the calmodulin system (see Chapter 13).

Among the functional gene products, **late embryogenesis-abundant (LEA) proteins** are of particular interest in the context of stress responses during the plant life cycle. Desiccation is a normal phase of embryo maturation, as the seed dries out in preparation for quiescence and dispersal. Desiccation-related induction of *LEA* genes was originally identified

in studies of seed maturation, but subsequently LEA proteins have also been observed to increase in abundance in vegetative tissues experiencing water deficit. The 400 or so LEA proteins are classified into up to seven families which differ in species of origin and structural motifs. LEAs are in turn members of a widespread group of proteins referred to as **hydrophilins**, characterized by high hydrophilicity and a glycine content of more than 6%. Hydrophilins occur throughout the archeal, eubacterial and eukaryotic domains and are believed to be components of a ubiquitous stress-response mechanism. Overexpression, in one species, of *LEA* genes from another species has been observed to improve drought and salinity resistance in plants, bacteria and yeast. There are, however, reports of instances of no or detrimental effects. It is not clear how LEAs work. A number of LEA proteins have been shown to protect against dehydration-induced **enzyme inactivation** in vitro. The common molecular motifs among the members of different LEA families indicate that most LEAs are intrinsically unstructured proteins, existing principally as random coils in solution. This may allow LEAs to prevent enzyme inactivation by closing in on their target enzymes and maintaining a water-rich environment that preserves protein integrity. LEAs may also form a tight hydrogen-bonding network in dehydrating cytoplasm that retains residual water and stabilizes cell structures.

As Figure 15.28 indicates, the hormone **abscisic acid** (ABA) has a role in plant responses to water limitation. Drought and high salinity trigger strong increases in endogenous ABA content, accompanied by major changes in gene expression and acclimatory and adaptive physiological responses. Mutants deficient in ABA biosynthesis are impaired in their responses to drought. Stomatal closure, mediated by ABA-triggered changes of ion fluxes in guard cells, is a fast response to water limitation (see Chapter 14). ABA-deficient mutants are abnormally **wilty** due to unrestrained transpiration. Drought sends an immediate **hydraulic signal** (possibly a rapid change in xylem tension) that is able to stimulate ABA biosynthesis over long distances. ABA is formed primarily in the vascular system and is rapidly distributed to neighboring tissues. ABA synthesized in response to water deficiency is swiftly broken down on rehydration. **Abscisic acid 8′-hydroxylase**, a key enzyme in ABA catabolism, is activated in vascular tissues and stomatal guard cells within minutes of exposing shoots to high humidity, resulting in the formation of inactive phaseic acid.

Several drought-sensitive genes are induced by exogenous ABA treatment, including those that encode enzymes of compatible solute metabolism. For example, *P5CS1*, the gene encoding the proline-synthesizing

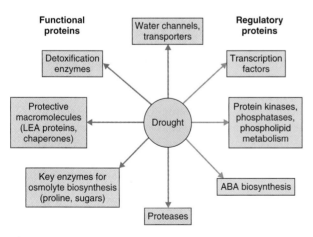

Figure 15.28 Drought stress-inducible genes classified into those whose products are responsible for executing stress responses (functional proteins) and those concerned with signal transduction and control of gene expression (regulatory proteins).

enzyme Δ^1-pyrroline-5-carboxylate synthetase (see Equation 15.12A), is regulated by an ABA-responsive subfamily of **bZIP** (basic leucine Zipper domain) **transcription factors**. A bZIP-binding **ABA responsive element** has also been identified in the gene for Em, a wheat LEA. Transcription factors of the MYC, MYB and NAC families have also been implicated in ABA regulation of responses to, and tolerance of, drought and salinity stress. ABA does not regulate all stress-induced genes, however. Accumulation of many carbohydrate-related compatible solutes, notably oligosaccharides of the **raffinose** family, occurs via an **ABA-independent pathway**. Genes in this pathway have a drought-responsive element in their promoter regions and are probably regulated by distinct or interacting signal transduction networks.

Key points Turgor is essential for plant structure and growth. Tissues die if water content falls below a critical level. Plants acclimate to inadequate water availability by accumulating osmotically active solutes. Compatible solutes (osmolytes) are localized in the cytosol and organelles. Specific transporters move osmolytes across membranes. Glycine betaine, a compatible osmolyte of plants and other organisms, is synthesized in response to water limitation. It is made and accumulated mainly in chloroplasts and protects thylakoid structure and function. Osmotic stress upregulates genes encoding glycine betaine biosynthesis. Proline, a cytosolic compatible osmolyte whose metabolic turnover is sensitive to water availability, stabilizes subcellular structures, scavenges free radicals and is a transcriptional regulator of stress-responsive genes. Some species accumulate the osmolytes mannitol and pinitol in response to osmotic stress. Saline conditions often result in osmotic stress and Na^+ toxicity. Acclimation to drought and salinity is associated with the expression of functional genes encoding a wide range of detoxification, enzymic, transporter and protective proteins. Among these are the late embryogenesis-abundant proteins, which stabilize cell structures against osmotic damage. Regulatory genes responsive to water stress encode a range of signaling components and transcription factors. Abscisic acid is a hormone with a major influence on the expression of a number of functional and regulatory genes, including those for compatible solute and protective protein synthesis. It also regulates ion fluxes during opening and closing of stomatal guard cells.

15.6.2 Flooding deprives plants of oxygen, affecting respiratory processes, gene expression and acclimatory changes in structure

From water deficiency we turn now to the problems of too much water. Plants, like most eukaryotes, are **obligate aerobes**. Because the diffusion coefficient of oxygen in air is about 10 000 times greater than that in water, flooding has the effect of blocking the entry of O_2 into the soil so that roots and underground stems cannot carry out respiration. Normal aerobic respiration yields up to 32 mol of ATP per mole of hexose, whereas respiration in the absence of oxygen produces only 2 mol ATP (see Chapter 7). To survive short-term flooding, plants must be able to make sufficient ATP, regenerate $NADP^+$ and NAD^+, and avoid accumulation of toxic metabolites. Flooding stimulates fermentative respiration, ethanol formation and the Pasteur effect (see Chapter 7). Many aquatic, wetland and shoreline species have evolved adaptations that allow them to survive or thrive under conditions of periodic or continuous immersion. However, tolerance of flooding is limited in vascular plants. In addition to its ecological significance in determining vegetation patterns, response to flooding is agriculturally important since waterlogged soil severely depresses **crop yields**.

Plant or cellular oxygen status can be defined as **normoxic** (normal O_2 levels), **hypoxic** (reduced O_2 levels) or **anoxic** (lacking O_2) (Table 15.5). Periods of oxygen deficit cause a shift from aerobic to anaerobic metabolism. Initially this is a direct metabolic response to O_2 deprivation, which is then followed by changes in gene expression. Over the longer term, acclimation to hypoxic or anoxic conditions can take the form of developmental responses involving modifications in growth behavior, morphology and anatomy. Injury caused by anoxia or hypoxia is a consequence of **acidification of cytoplasm**, which in turn results in greatly diminished protein synthesis, mitochondrial degradation, inhibition of cell division and elongation, disrupted ion transport, and root meristem cell death. Flooding-tolerant plants are able to avoid cytoplasmic acidosis and continue to make ATP during short-term inundation by stimulating **ethanolic fermentation**. Seedlings of maize (*Zea mays*), for example, can survive anoxia for up to 5 days before experiencing injury. Exposure to an episode of hypoxia can greatly enhance survival in a subsequent period of anoxia, a classic acclimatory or hardening response. Hypoxic pretreatment results in greater specific activity of

Table 15.5 Responses of respiratory metabolism to oxygen deprivation.

Oxygen status	Effect on metabolism
Normoxic (aerobic)	Aerobic respiration proceeds normally. Almost all ATP production results from oxidative phosphorylation
Hypoxic	The partial pressure of O_2 limits ATP production by oxidative phosphorylation. Glycolysis accounts for a larger proportion of ATP yield than under normoxic conditions. Metabolic and developmental changes are stimulated that result in adaptation to a low-oxygen environment
Anoxic (anaerobic)	ATP is produced only by way of glycolysis. Cells exhibit low ATP contents, diminished protein synthesis and impaired division and elongation. If anoxic conditions persist, many plant cells die

hexokinase, fructokinase, pyruvate kinase and other enzymes of glycolysis and ethanolic fermentation (see Chapter 7). This, in turn stimulates glycolytic flux, ATP production and the production of cytoplasm-acidifying lactate. The shift from aerobic metabolism to glycolytic fermentation is accompanied by rapid and dramatic changes in gene expression patterns and activities, which are regulated at both the transcriptional and the post-transcriptional level.

The mechanism by which plants sense oxygen deprivation is not known in detail, but there is clear evidence of interaction with metabolic and signaling networks associated with **reactive oxygen species** (**ROS**; see Section 15.6.3). Oxygen deprivation represses the expression of most genes, but upregulates an important subset, including genes encoding enzymes of sucrose and starch degradation, glycolysis and ethanol fermentation. Isoforms of alcohol dehydrogenase are differentially expressed in aerobic and anoxic cells because of sequence differences in their promoters. The promoters of many hypoxia-stimulated genes include the so-called **anaerobic response element** that interacts with hypoxia-responsive transcription regulators of the **ERF** (**ethylene responsive factor**) family.

Wetland plants, which are adapted to long-term flooding, possess anatomical, morphological and physiological features that permit survival in waterlogged soils. A particularly striking structural response to flooding and long-term anoxia is the differentiation of **aerenchyma**, continuous columnar intracellular spaces developed in root cortical tissues which facilitate the transport of O_2 from aerial structures to submerged roots. The production of aerenchyma, which is under the control of ethylene, is an example of **programmed cell death**; this is discussed in further detail in Chapter 18. Other adaptive features are **adventitious roots**, a thickened root **hypodermis** that reduces O_2 loss to the anaerobic soil, and **lenticels** (openings in the periderm of the stem that allow gas exchange). In addition some plants produce **pneumatophores**, shallow, negatively gravitropic roots that grow out of the water.

Rice (*Oryza sativa*) is the most important staple food for much of the world's population. Yield of rice per hectare has more than doubled since the 1960s, but further doubling will be necessary to meet projected nutritional requirements by the middle of the present century. More than 30% by area of Asian and 40% of African rice is grown in paddy fields with water depths from 15 cm to more than 50 cm. High-yielding rice varieties cannot survive deep water and prolonged submergence. Rice varieties that are adapted to deep water and those that are submergence-tolerant endure flooding but tend to be less productive.

Deep-water rice survives by outgrowing slowly rising floodwaters, which can reach 4 m in depth. The contrasting strategy of submergence-tolerant rice is to suppress elongation and adopt a quiescent state (Figure 15.29). The control of stature by gibberellin and 'Green Revolution' genes of the *Rht* family is described in detail in Chapter 12. Elongation growth in deep-water rice is under the control of *SNORKEL* and *SUBMERGENCE* ethylene response factor genes. Transcription factors encoded by *SNORKEL1* and *SNORKEL2* (*SK1* and *SK2*) are activated by ethylene and promote gibberellic acid (GA)-stimulated extension growth. Ethylene accumulated in submerged tissues also inhibits the synthesis, and promotes the breakdown, of ABA, a negative regulator of GA action. In the quiescence strategy the ethylene-induced action of the SUBMERGENCE (SUB1A-1) transcription factor on *SLENDER RICE-1* (*SLR1*) and *SLR LIKE-1* (*SLRL1*) expression suppresses GA-mediated growth. Transfer of *SK* or *SUB* genes into high-yielding varieties that are normally flooding-intolerant has the prospect of contributing significantly to improving the quality and quantity of rice produced on marginal land.

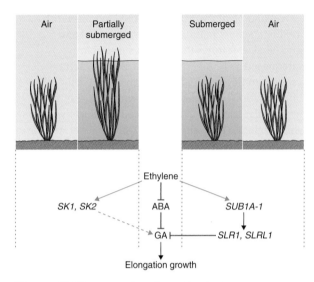

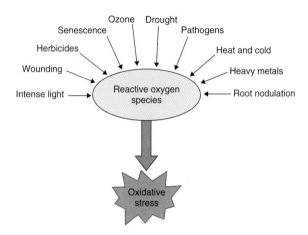

Figure 15.29   Strategies adopted by rice in response to flooding. The escape strategy of deep-water rice (shown on the left) is mediated by two *SNORKEL* genes, *SK1* and *SK2*. In submergence-tolerant varieties (shown on the right), quiescence is the result of interaction between the ethylene-sensitive *SUBMERGENCE* gene *SUB1A-1* and the stature-determining genes *SLR1* and *SLRL1*. Abscisic acid (ABA) inhibits gibberellic acid (GA)-mediated growth. Ethylene accumulated under submergence conditions promotes elongation by downregulating ABA synthesis and stimulating ABA degradation.

Figure 15.30   Oxidative stress mediated by reactive oxygen species underlies the entire range of plant responses to unfavorable abiotic and biotic factors in the environment.

15.6.3 Reactive oxygen species, common factors in plant responses to a range of stresses, regulate, and are regulated by, antioxidant systems

Molecular oxygen (O_2) accounts for about 20% by volume of the Earth's atmosphere. Living organisms have been exposed to atmospheric oxygen since the first oxygenic photoautotrophs evolved 2.7 billion years ago. Because many partially reduced oxygen derivatives are highly reactive and potentially toxic to cells, evolution has equipped all aerobic organisms with efficient mechanisms for sensing, scavenging and exploiting **reactive oxygen species** (**ROS**). ROS have multiple physiological roles in plants. Reactive oxygen is a substrate in a number of important biochemical reactions, for example the formation of lignin. ROS also participate in signaling networks, redox control of metabolism and development, and responses to biotic stress. Most significantly for the present discussion,

many environmental factors promote the formation of active oxygen species that are highly destructive to lipids, nucleic acids and proteins, resulting in cell damage or death. Oxidative stress mediated by ROS is a common factor in adverse reactions to air pollution, oxidant-forming herbicides, heavy metals, drought, heat and cold, wounding, and ultraviolet and excess visible light (Figure 15.30).

The various forms of reactive oxygen are given in Table 15.6. **Singlet oxygen** is the product of photosensitization and is discussed in the context of responses to light in Section 15.6.6. Here we consider the biological effects and metabolism of the **superoxide anion** ($O_2^{\bullet-}$), **hydrogen peroxide** (H_2O_2) and **ozone** (O_3). Interconversion, scavenging and elimination of these reactive molecules are the task of defense systems, comprising **antioxidant** enzymes and non-enzymatic antioxidant compounds that are present in various subcellular compartments. Figure 15.31 summarizes the activities and components of the major antioxidant systems of plants. The central importance of these systems is emphasized by the improvement in oxidative stress tolerance observed in plants selected or genetically engineered for increased synthesis of antioxidants and antioxidant enzymes.

Superoxide is the product of electron transport chains in the chloroplast and mitochondrion as well as a range of enzymes, including NAD(P)H oxidases, xanthine oxidase, lipoxygenase and cytochrome P450 monooxygenases. Superoxide anions are also produced in the chloroplast by the so-called **Mehler reaction**, when electrons are transferred directly from photosystem I to oxygen.

Hydrogen peroxide is formed from superoxide in a type of reaction known as **dismutation**, in which a

Table 15.6 Forms of reactive oxygen.

Compound	Shorthand notation(s)	Sources
Molecular oxygen (triplet ground state)	O_2; $^3\Sigma$	Most common form of dioxygen gas
Singlet oxygen (first excited singlet state)	$^1O_2$; $^1\Delta$	UV irradiation, photoinhibition, photosystem II e⁻ transfer reactions (chloroplasts)
Superoxide anion	$O_2^{\bullet-}$	Mitochondrial e⁻ transfer reactions, Mehler reaction in chloroplasts (reduction of O_2 by iron–sulfur center F_x of photosystem I), glyoxysomal photorespiration, peroxisome activity, plasma membrane, oxidation of paraquat, nitrogen fixation, defense against pathogens, reaction of O_3 and OH^- in apoplastic space
Hydrogen peroxide	H_2O_2	Photorespiration, β-oxidation, proton-induced decomposition of $O_2^{\bullet-}$, defense against pathogens
Hydroxyl radical	$HO^{\bullet-}$	Decomposition of O_3 in presence of protons in apoplastic space, defense against pathogens
Perhydroxyl radical	$HO_2^{\bullet-}$	Reaction of O_3 and OH^- in apoplastic space
Ozone	O_3	Electrical discharge or UV radiation in stratosphere, reactions involving combustion products of fossil fuels and UV radiation in troposphere

chemical species is simultaneously reduced and oxidized so as to form two different products. The biochemical reaction between $O_2^{\bullet-}$ and H^+ to produce H_2O_2 and O_2 is catalyzed by the enzyme **superoxide dismutase** (**SOD**; Figure 15.31). Different SODs are distinguished by their metal cofactors and subcellular location. **Cu/Zn SOD** occurs in cytosol, peroxisomes and plastids. It is also found in root nodules. **Mn SOD** is mitochondrial, and **Fe SOD** is located in the plastid. H_2O_2 is also a normal product of photosynthesis and photorespiration (see Chapter 9) as well as of a number of enzymes including glucose oxidases, amino acid oxidases, cell wall-bound peroxidases and plasma membrane NADPH oxidases. Enzymatic detoxification of H_2O_2 occurs via **catalase** or through a redox cycle catalyzed by **ascorbate peroxidase, monodehydroascorbate reductase, dehydroascorbate reductase** and **glutathione reductase** (Figure 15.31). **Ascorbate** (vitamin C), one of the reactants in this cycle, is a soluble, sugar acid antioxidant distributed throughout cytosol, plastids, vacuoles and apoplast. Another reactant, **glutathione** (see Chapter 13), is a tripeptide thiol antioxidant found in cytosol, mitochondria and plastids. H_2O_2, unlike other reactive forms of oxygen, is not a labile, short-lived radical and is a normal, measurable constituent of unstressed plant tissues. Under some circumstances, however, such as the presence of transition metal ions (**Fenton reaction**; see

Chapter 13), H_2O_2 can be a source of the extremely harmful hydroxyl ($HO^{\bullet-}$) and perhydroxyl ($HO_2^{\bullet-}$) radicals.

Ozone is a pollutant that causes injury to living organisms by oxidative stress mediated by ROS. Ozone is a normal component of the **stratosphere**, where it carries out the essential job of shielding the Earth from harmful ultraviolet radiation. However, living organisms are increasingly exposed to damaging levels of ozone in the lower atmosphere (**troposphere**) that are generated from reactions of O_2 with anthropogenic hydrocarbons and oxides of nitrogen (NO_x) and sulfur (SO_x) under the influence of solar ultraviolet radiation. Plant responses to ozone include impaired photosynthesis, leaf injury, reduced growth of shoots and roots and accelerated senescence, all of which contribute to significant reductions in yields from crop species.

Because of the potential for injury that results from the uncontrolled propagation of oxygen-containing radicals, the steady-state level of ROS in cells needs to be tightly managed. More than 150 genes have been identified in the **ROS regulatory network**, encoding a range of ROS-scavenging and ROS-producing proteins with a high degree of functional redundancy (a protein is functionally redundant if its activity can be substituted by that of one or more of the other proteins in the network). Plant cells sense ROS through receptor

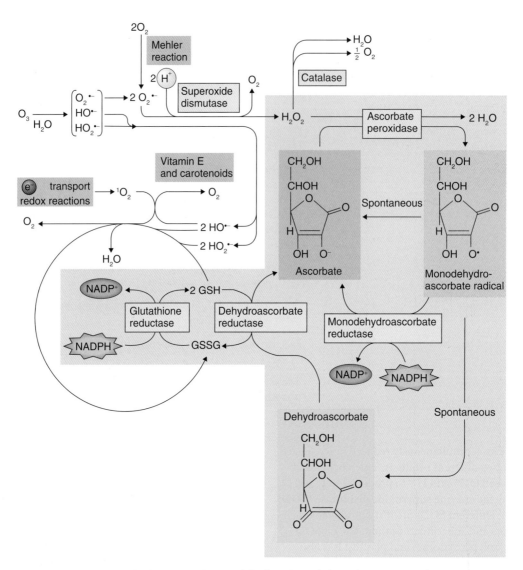

Figure 15.31 Reactive oxygen species interconversions and the functions of the major enzyme and non-enzymatic antioxidants. The ascorbate–glutathione cycle is highlighted. Superoxide radicals are eliminated by conversion to hydrogen peroxide (H_2O_2), a reaction catalyzed by superoxide dismutase. Catalase converts H_2O_2 to O_2 and H_2O. H_2O_2 may also be reduced to water by reaction with ascorbate. Ascorbate is regenerated by enzymatic reduction of monodehydroascorbate in the plastids. Alternatively, monodehydroascorbate may spontaneously dismutate to dehydroascorbate. Dehydroascorbate reductase then catalyzes the reaction of dehydroascorbate with reduced glutathione (GSH) to produce ascorbate and oxidized glutathione (GSSG). GSSG is reduced to GSH in an NADPH-dependent reaction catalyzed by glutathione reductase. Hydroxyl ions and singlet oxygen are eliminated in the glutathione pathway; damage by singlet oxygen is also diminished by non-enzymatic antioxidants.

proteins and redox-sensitive transcription factors and by direct inhibition of signal-transducing phosphorylation cascades. Most of the proposed mechanisms for **sensing cell redox status** are based on the reversible formation and breakage of **disulfide bridges** between cysteine residues in polypeptide chains. Equation 15.13 shows the making and breaking of an S-S bridge between polypeptides P_1 and P_2 under the influence of the redox mediator A. The resulting change in protein conformation and/or enzyme activity triggers the signal transduction network, which in turn leads to changes in gene expression and the corresponding acclimatory or injury response.

Equation 15.13 Redox sensitivity of the disulfide bridge

$$P_1\text{-Cys-SH} + \text{HS-Cys-}P_2 + A_{ox} \rightleftharpoons P_1\text{-Cys-S-S-Cys-}P_2 + A_{red}$$

Prominent among the genes whose expression is enhanced by oxidative stresses are those that encode enzymes of antioxidant synthesis and metabolism. Table 15.7 lists some example enzymes and illustrates the

Table 15.7 Antioxidants and antioxidant enzymes and the stress conditions that stimulate an increase in their levels or activities.

Antioxidant or antioxidant enzyme	Stress conditions
Ascorbate peroxidase	Drought, high CO_2, high light intensity, ozone, paraquat
Catalase	Chilling, aging
Glutathione	Chilling, drought, γ-irradiation, heat stress, high CO_2, ozone, SO_2
Glutathione reductase	Chilling, drought, high CO_2, ozone, paraquat
Superoxide dismutase	Chilling, high CO_2, high light, increased O_2, ozone, paraquat, SO_2

degree of cross-talk between oxidative and other stresses. This interaction between stress pathways is mediated in part by plant hormones such as abscisic acid, salicylic acid, jasmonic acid and ethylene (see Chapter 10). The regulation of catalase during aging and senescence, as described in Chapter 18, is an example of such an interaction.

15.6.4 Cold stresses are experienced through similar sensitivity, tolerance and acclimatory mechanisms to those of other abiotic challenges

The nature of the stress experienced by plants exposed to suboptimal temperatures is dependent on the physical states of water and cell membranes. Cooling to below the optimal temperature range causes metabolism, growth and development to slow. Initially this is a direct thermodynamic effect on rates of chemical reactions. Longer periods of exposure to low temperatures have more complex physiological consequences that depend on the temperature range, the adaptive features of the species and the extent to which it is able to acclimate. In discussing plant responses to low temperature, it is important to distinguish between **chilling** (temperatures in the range $0–15\,°C$, where water is in the liquid state) and **freezing** (below the temperature of ice formation). Plants of temperate regions are generally able to survive chilling temperatures. Species adapted to warmer

Key points Flooding leads to hypoxia or anoxia and a shift from aerobic to anaerobic metabolism, which may result in injury. Plants able to acclimate to flooding do so by activating ethanolic fermentation and may express anaerobiosis-sensitive genes for physiological responses including the development of aerenchyma. Transcription factors of the SNORKEL and SUBMERGENCE families, acting through the gibberellin-mediated growth mechanism, allow deep-water- and submergence-tolerant rice varieties to adapt to flooding. Metabolism of reactive oxygen species (ROS) is sensitive to many environmental stresses. ROS function as enzyme substrates and products, signaling components, redox regulators and harmful oxidizers. Major components of the antioxidant systems that interconvert and scavenge superoxide anion, hydrogen peroxide and the pollutant ozone are: superoxide dismutases; catalase; and the redox regulators ascorbate and glutathione and the enzymes that metabolize them. Receptor proteins and transcription factors sense ROS status and trigger signal transduction networks and acclimatory changes in gene expression.

climates are often **chilling-sensitive** and will show signs of damage after quite short exposures to temperatures below a defined lower limit which, for some tropical plants, could be well above $10\,°C$. Response to freezing is quantitative and statistical; that is, death in a genetically uniform population occurs over a temperature range rather than at a single threshold value. A widely used index of freezing sensitivity is LT_{50}, the temperature that is lethal for 50% of the population. LT_{50} varies with species and genotype, and can be altered by acclimation at low, non-lethal temperatures. Thus non-acclimated plants of winter varieties of **rye** (*Secale cereale*)—among the most freezing-tolerant crop species—may have an LT_{50} of $-5\,°C$, but exposure to a period of chilling temperatures can lower this to $-30\,°C$. Some trees of northern latitudes may experience winter temperatures of $-70\,°C$ and acclimated tissues of such species have been shown to have correspondingly low LT_{50} values.

Many of the mechanisms underlying responses to low-temperature stresses are shared with those related to water limitation (see Section 15.6.1) and the influence of ROS (Section 15.6.3). Freezing injury, for example, is largely a result of the formation of ice crystals, which in turn lead to dehydration, ionic imbalance and damage to membrane structure and function (Figure 15.32). Conditions that support acclimation to water stress frequently improve freezing tolerance and vice versa. Compatible solutes (see Section 15.6.1) are effective

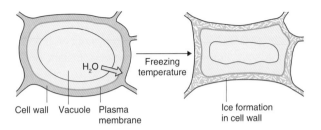

Figure 15.32 Freezing damage to plant cells. Ice formation in the wall withdraws water from the cell and results in dehydration injury, which shares a number of physiological features with drought stress responses.

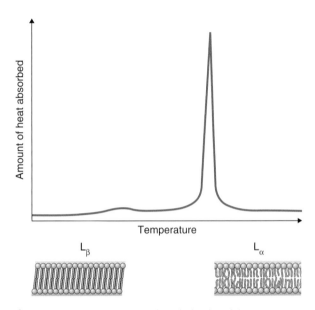

Figure 15.33 Membrane phospholipid undergoes a reversible state transition from gel (L_β) to liquid crystalline (L_α) at a critical temperature. The change of state, which is strongly endothermic, is detected by calorimetric measurements as a peak of heat absorption.

cryoprotectants (that is, chemicals that prevent low-temperature damage) and acclimation to freezing is commonly associated with the accumulation of carbohydrates, amino acids and amines. Freezing sensitivity and susceptibility to pathogens is a further example of cross-acclimation between stresses. A number of pathogenesis-related (PR) proteins (see Section 15.7.1) have been shown to have antifreeze properties.

The primary cellular basis of chilling injury differs from that of freezing damage. Much research has been aimed at linking the threshold below which chilling-sensitive species display symptoms of damage to a so-called **phase transition** in the state of membrane lipids, from **liquid crystalline** to **gel** (Figure 15.33). Gel phase lipids disrupt membrane fluidity, integrity and function and lead to disturbed metabolism, injury and death. The critical phase transition temperature for a given lipid is related to the degree of **saturation** of its constituent fatty acids (see Chapter 2). For example, pure **phosphatidylcholine** in which the glycerol moiety is esterified with two **stearic acid** residues (saturated C_{18} fatty acid) has a gel–liquid crystalline transition temperature in water of 58 °C. Replacing one of the stearates with the 18:1 unsaturated fatty acid **oleic acid** reduces the phase transition temperature to 3 °C. The relationship between chilling sensitivity and degree of saturation of membrane lipids is complex, however, and only broad correlations have been established. The precise cellular mechanism of injury in chilling-sensitive species remains to be established.

Membrane fluidity is one possible way in which plant cells can **sense cold stress**. Metabolite concentrations, protein conformation and redox status are other temperature-sensitive factors through which chilling may be perceived and transduced into changes in *COR* (*COLD RESPONSIVE*) gene expression. Recent evidence also implicates chromosomal DNA–histone interactions (**nucleosomes**; see Chapter 3) in the coordinated activation and deactivation of genes in response to shifts in temperature. Among the *COR*s are genes encoding transcription factors of the ERF family, genes of the ABA-responsive and ABA-independent drought

response pathways (see Section 15.6.1) and genes for antioxidant metabolism (see Table 15.7). A distinctive and visible reaction to chilling (and other stresses) in many plant species is the accumulation of **red anthocyanins**, indicative of activation of phenylpropanoid and other secondary metabolic pathways. The function of such pigments in stressed tissue is not certain, but there is evidence for a role in defense against light and/or oxidative damage.

15.6.5 Plants respond to high-temperature stress by making heat shock proteins

In contrast to the diversity of responses to suboptimal temperatures, most plants react uniformly to heat stress by invoking a physiological syndrome that shares basic features with that of organisms across the taxonomic range, from bacteria to vertebrates. Gene expression is altered, leading to the accumulation of **heat shock proteins** (HSPs; Figure 15.34), a family of proteins whose structure is conserved among different organisms. HSPs generally function as **molecular chaperones**; that is, they interact with mature or newly synthesized proteins to prevent aggregation or promote disaggregation, to assist folding or refolding of polypeptide chains, to facilitate protein import and translocation, and to participate in signal transduction and transcriptional activation (see Chapter 5).

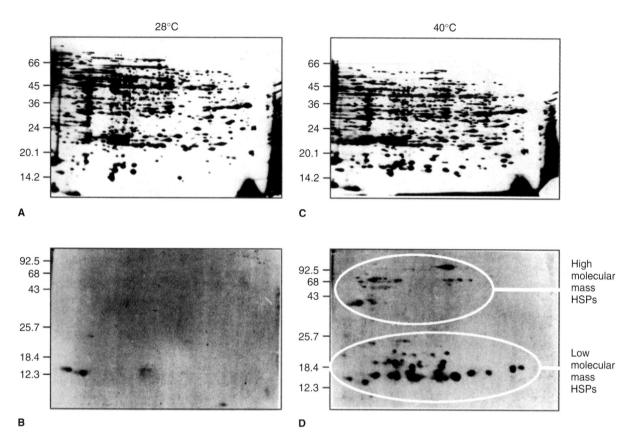

28°C   40°C

A   **C**

B   **D**

High molecular mass HSPs

Low molecular mass HSPs

Figure 15.34 Heat shock proteins (HSPs) accumulated by soybean seedlings in response to high temperature (40 °C compared with 28 °C control), separated by two-dimensional gel electrophoresis. (A, C) Silver-stained total proteins. (B, D) Newly synthesized proteins in gels (A) and (C), visualized by labeling with radioactive amino acid and fluorography. The vertical number scale alongside each gel represents polypeptide molecular weight (kDa) based on mobilities of standard proteins.

The heat shock response can often be detected in tissues experiencing temperatures as little as 5 °C above the growth optimum. It defends against damage to organelles, cytoskeleton and membrane function. Plants can acclimate to heat stress if exposed to a period of non-lethal supraoptimal temperatures. It is normal for plants under field conditions to experience such circumstances regularly—for example in sunny, dry conditions when the irradiance load is high and stomata are partially closed—and to go through cycles of acclimation and de-acclimation as environmental conditions fluctuate.

HSPs are classified into five groups according to molecular size: HSP100, HSP90, HSP70, HSP60 and small (sm) HSPs. The **HSP100** group consists of 100–140 kDa proteins. Included in this group are members of the **Clp** family of proteases, most of which are located in plastids or mitochondria, where they are thought to function in ATP-dependent protein disaggregation and turnover. Representatives of the **HSP90** (80–94 kDa) and **HSP70** (69–71 kDa) groups are molecular chaperones that are, in many cases, ATP-dependent and are essential for normal cell

function. They include constitutive members and those induced by cold stress as well as heat. They are present in several subcellular compartments—cytosol, ER, mitochondria and plastids. Members of the **HSP60** group (57–60 kDa), which are characteristically found in the mitochondrial matrix and chloroplast stroma, are abundant even at normal temperatures. This protein family includes **chaperonin 60**, a nuclear-encoded chloroplast protein required for assembling the subunits of **rubisco** into the functioning holoenzyme (see Chapter 9). The diversity of low molecular size HSPs (**smHSPs**, 15–30 kDa) is a unique feature of the plant heat shock response (Figure 15.34). They are thought not to be required for normal cell function, but to participate in the development of thermotolerance by promoting ATP-independent protein stabilization and preventing aggregation.

The promoter regions of genes encoding HSPs have the **heat shock element** (**HSE**), a conserved DNA sequence that is recognized by members of the **heat shock transcription factor** (HSF) family. The promoters of several of the *Hsf* genes also contain the HSE motif. Heat stress is required to convert monomeric HSF proteins to the active HSE-binding **oligomeric form**. Three HSF classes are recognized, A, B and C. Based on expression patterns of *HsfA*, *HsfB* and *HsfC* gene family members in *Arabidopsis* (Figure 15.35), tomato and rice, together with functional and comparative analyses, a picture is emerging of the roles and interactions of heat shock factors in development and stress responses. *HsfA2*, *HsfA7* and *HsfB1* are strongly upregulated in heat-stressed roots and shoots (Figure 15.35). Transcription of *HsfB1* is also stimulated in roots by the full range of abiotic stresses. The transcription factor HsfA1a (encoded by a member of the *HsfA* gene family) has been shown to act as a master regulator of induced thermotolerance in tomato, by triggering the expression of heat shock genes including *HsfA2* and *HsfB1*. The *HsfA2* gene encodes the dominant HSF in thermotolerant cells, and *HsfA3* functions in cross-talk with drought stress signaling. *HsfA4c* is expressed more or less uniformly in all tissues and all stress treatments (Figure 15.35). This kind of genomics study reveals the richness and interconnectedness of regulatory processes governing stress responses and plant development.

15.6.6 Plants have photochemical, acclimatory and adaptive mechanisms that defend against potentially harmful excess light

Plants are dependent on light for photomorphogenesis (see Chapter 8) and photosynthesis (see Chapter 9). However, light can be harmful if the absorbed photon energy is greater than the capacity of the plant to use it. Figure 8.4 showed how absorption of a photon raises the energy level of a pigment molecule from the ground state to an excited state. Excessive excitation of chlorophylls can result in the formation of singlet oxygen ($^1O_2$; see Table 15.6). Chlorophyll acts as a **photosensitizer** (S), providing the energy required to convert the stable triplet state O_2 into $^1O_2$. As shown in Equation 15.14A, a photosensitizer is a pigment that, like chlorophyll, absorbs light and flips from the excited singlet state ($^{*1}S$) into the excited triplet state ($^{*3}S$). Singlet oxygen is the product of reaction between $^{*3}S$ and O_2 (Equation 15.14B). Decreased efficiency of photosynthesis as a result of oxidative injury caused by $^1O_2$ and its reactive products is termed **photoinhibition**. The location and function of the **D1 protein** of the photosystem II (PSII) reaction center (see Chapter 9) make it particularly vulnerable to photooxidative damage. The requirement for constant synthesis to replace it accounts for the unusually high rate of **turnover** of this protein. Under

Key points Low temperatures influence plant growth and development through their effects on the physical state of water and cell membranes. Plants from environments prone to freezing can often acclimate to sub-zero temperatures. Freezing injury, which often involves extracellular ice formation and withdrawal of water from the cytosol, has features in common with osmotic stress. Compatible solutes are effective cryoprotectants. Plants adapted to warm climates are susceptible to being damaged or even killed by chilling (above-freezing temperatures). There is evidence that the threshold for chilling injury is associated with the temperature at which membrane lipids undergo phase transition from liquid crystalline to gel. Membrane fluidity is one possible temperature-sensing mechanism in acclimation. Others include concentrations of critical metabolites, protein conformation, redox status and the state of chromatin. Hormone regulated genes and genes for antioxidant and secondary metabolism are among those responsive to chilling. High-temperature stress invokes the heat shock response, which is highly conserved across all organisms. Cells make heat shock proteins (HSPs), many of which protect against protein aggregation and damage by acting as molecular chaperones. There are five size classes of HSP, ranging from the small HSPs (<30 kDa), which are unusually numerous and diverse in plants, up to >100 kDa. HSP gene expression is regulated by members of the heat shock transcription factor family, some of which are also active in responses to other abiotic and biotic stresses.

extreme conditions unrestrained production of singlet oxygen can be lethal. Plants are equipped with a range of mechanisms that protect against (i.e. **quench**) photosensitization and oxyradical damage.

Equation 15.14 Photosensitizer-mediated formation of singlet oxygen

15.14A Excitation of singlet state S and spin-flip to the excited triplet state

$$^1S + h\upsilon \rightarrow {}^{*1}S \rightarrow {}^{*3}S$$

15.14B Reaction between excited triplet photosensitizer and triplet state oxygen

$$^{*3}S + O_2 \rightarrow {}^1S + {}^1O_2$$

Triplet state chlorophyll can also cause **free radical** damage by directly reacting with the **fatty acids** of lipids. An important function of chloroplast carotenoids is to

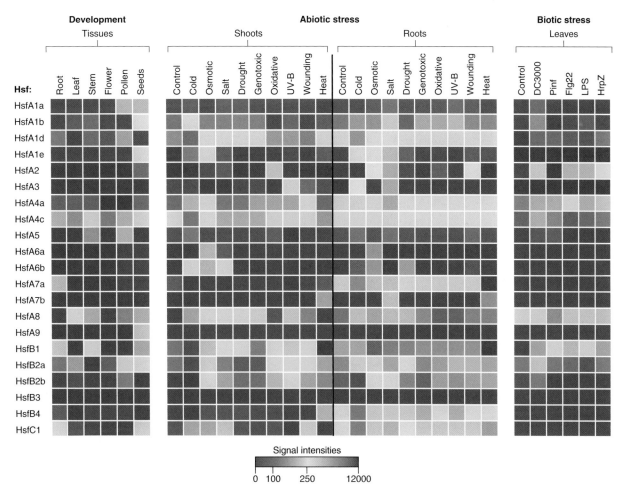

Figure 15.35 Relative abundances of mRNA transcripts of *Hsf* (heat shock transcription factor) genes in response to a range of abiotic and biotic stress treatments. Transcript levels (signal intensities) are presented in the form of a 'heat map', on a color scale between low (blue) and high (red). Abiotic stresses were applied to hydroponically grown *Arabidopsis* seedlings. Biotic stresses consist of infiltration with *Pseudomonas syringae* pv. tomato DC3000 (DC3000), or *Phytophthora infestans* (Pinf), or elicitation with flagellin (Flg22), lipopolysaccharide (LPS) or harpinZ (HrpZ).

act as quenchers. Carotenoids are able to accept excitation energy from triplet chlorophyll (and excited singlet state chlorophyll in some cases), thereby blocking the generation of singlet oxygen and free radical cascades. Carotenoid biosynthesis (Figure 15.36) also provides a route for dissipating excess excitation energy as heat by pH-dependent quenching of chlorophyll fluorescence via the **xanthophyll cycle**. The biosynthetic pathway from the C_{40} tetraterpenoid precursor phytoene (see Figure 15.24) branches at the intermediate **lycopene**. One branch leads to the most abundant chloroplast xanthophyll **lutein** via α-carotene; the other leads from **β-carotene** to the xanthophyll cycle and **neoxanthin** (Figure 15.36). Under conditions of low irradiance the carotenoid **violaxanthin** functions as a photosynthetic antenna pigment by transferring energy to chlorophyll a. In high light the thylakoid lumen becomes more acidic, driven by the water splitting reaction of PSII. Acidification of the lumen activates the enzyme

violaxanthin de-epoxidase, which rapidly converts violaxanthin to **antheraxanthin** and thence to **zeaxanthin**, a quencher of chlorophyll excited singlet states. In this way the combination of zeaxanthin, antheraxanthin and low pH in the thylakoid lumen facilitates **non-photochemical quenching**—the harmless dissipation as heat of excess excitation energy directly within the light-harvesting antennae. **Zeaxanthin epoxidase** recycles zeaxanthin to violaxanthin under reduced light flux (Figure 15.36). The essential protective functions of carotenoids in photosynthetic tissues explain why mutations or chemical treatments that block carotenoid biosynthesis usually result in the formation of lethal concentrations of singlet oxygen under high light intensity.

Fluctuating light quantity and quality is a normal feature of irradiance in many environments: for example **shading** and **sunflecks** are commonly experienced by forest floor plants. When sunlight passes through a leaf

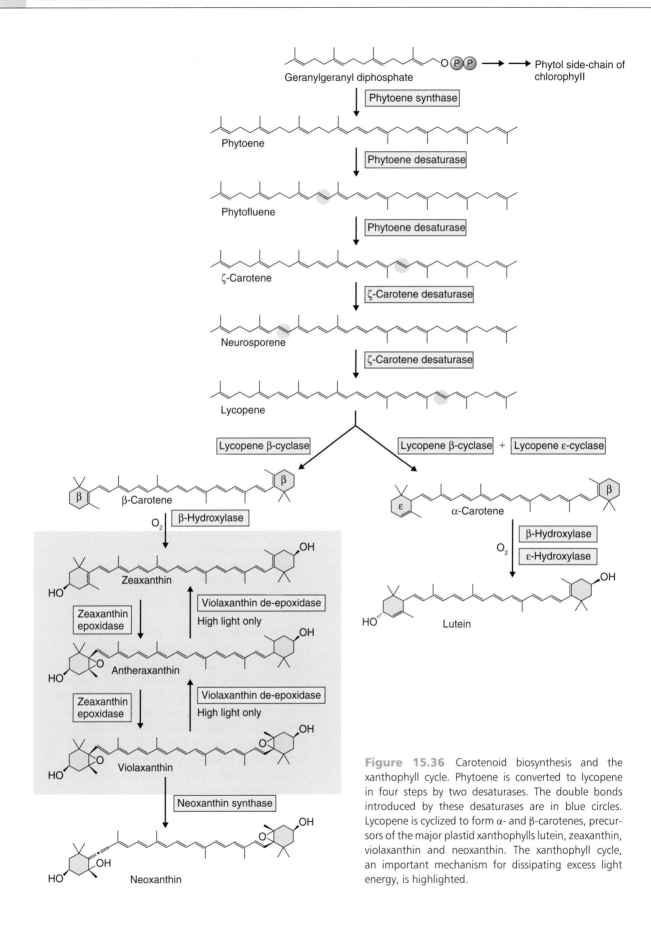

Figure 15.36 Carotenoid biosynthesis and the xanthophyll cycle. Phytoene is converted to lycopene in four steps by two desaturases. The double bonds introduced by these desaturases are in blue circles. Lycopene is cyclized to form α- and β-carotenes, precursors of the major plastid xanthophylls lutein, zeaxanthin, violaxanthin and neoxanthin. The xanthophyll cycle, an important mechanism for dissipating excess light energy, is highlighted.

canopy, the ratio of red to far-red wavelengths is greatly reduced (see Figure 8.22). Photosystem I (PSI) is able to use far-red light (>660 nm) more efficiently than PSII, whereas at red wavelengths (<660 nm) PSII drives photosynthesis more actively than does PSI. In red-depleted shade light there may be imbalanced excitation of the two photosystems and inefficiencies in electron transport. So-called **shade plants** have a number of adaptations to deal with such imbalances. These include an altered ratio of PSII to PSI centers in the thylakoid membranes. Some species accustomed to full-sun environments are able to acclimate by adjusting this ratio in response to shading. For example, the PSII:PSI ratio for chloroplasts of pea (*Pisum sativum*) plants grown in PSII light (550–660 nm) is 1.2; in sunlight it is 1.8; and in PSI light (>660 nm) it is 2.3. The distribution of light energy between PSI and PSII is also regulated minute-to-minute by a mechanism called **state transition**. When excess light is captured by PSII (which is located primarily in grana stacks; see Table 9.1), a subpopulation of PSII light-harvesting units, LHCII trimers, is phosphorylated by a redox-activated kinase. These then dissociate from PSII and move to the intergranal region of the thylakoid membrane and associate with PSI. The process is reversible and represents a sensitive way of balancing delivery of light energy captured by LHCII to PSII and PSI.

Overexcitation of the photosynthetic apparatus can result from production of NADPH and ATP in excess of that required by CO_2 fixation. This can be balanced by a system that links the redox state of ferredoxin, which is reduced by the light energy capture reactions, to that of **thioredoxin** (Figure 15.37) which in turn regulates the activity of several Calvin–Benson cycle enzymes (see Figure 9.29). **Ferredoxin–thioredoxin reductase** catalyzes the reaction between reduced ferredoxin and thioredoxin. Reduced thioredoxin activates target enzymes by reducing regulatory intramolecular disulfide bonds. **Calvin–Benson cycle** enzymes regulated in this way include fructose-1,6-bisphosphatase, sedoheptulose-1,7-bisphosphatase, phosphoribulokinase, $NADP^+$-glyceraldehyde-3-phosphate dehydrogenase and rubisco activase. The Calvin–Benson cycle is also directly regulated by light via changes in pH and Mg^{2+} concentration. Furthermore, reduced thioredoxin formed in the light stimulates the ATP synthase associated with the light energy capture reactions and the C_4 enzyme $NADP^+$-malate dehydrogenase, and inhibits the oxidative pentose phosphate pathway.

Key points Excessive light can be harmful to plants. Chlorophyll is a potential photosensitizer pigment. Injury caused by singlet oxygen derived from overexcitation of chlorophyll is referred to as photoinhibition. The high turnover rate of the D1 protein of photosystem II (PSII) reflects its constant exposure to photoinhibitory damage. High-energy triplet state chlorophyll is also a cause of free radical injury to membrane lipids. Chloroplast carotenoids function as quenchers of chlorophyll photosensitization by accepting excitation energy from chlorophyll. Excess excitation energy is also dissipated as heat through the xanthophyll cycle by non-photochemical quenching, in which the pH of the thylakoid lumen is responsive to light flux and regulates the de-epoxidation of violaxanthin to form the quencher zeaxanthin. Light transmitted through foliage is depleted in the red wavelengths preferentially utilized by PSII. Plants are able to control the distribution of light energy between PSI and PSII in response to short-term shading and sunflecking by a redox-regulated phosphorylation of light-harvesting complexes. Over a longer period, shade-adapted plants are able to develop chloroplasts with altered relative numbers of PSII and PSI centers. Thioredoxin and ferredoxin mediate redox regulation of Calvin–Benson cycle enzymes in response to the excitation state of the thylakoid reactions.

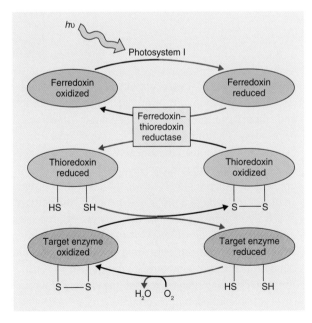

Figure 15.37 Light regulation of enzyme structure through the ferredoxin–thioredoxin system. Red arrows show reduction reactions, black arrows oxidations.

15.6.7 Gravity and touch are directional mechanical stresses that invoke tropic responses

Environmental influences on plant growth and development may not be exerted uniformly on all sides. Tropisms, directional growth responses to directional stimuli, are adaptations that contribute to plant survival by aiming growth towards resources or away from potentially harmful environments. Tropic responses occur under the influence of gradients of light, temperature, water, oxygen and mineral nutrient availability (Figure 15.38). The mechanisms of light perception and signaling in phototropism are discussed in Chapter 8. Here we focus on mechanical factors, in particular the influences of gravity (**gravitropism**) and touch (**thigmotropism**).

Roots are positively gravitropic, growing in the direction of gravitational pull, whereas shoots are negatively gravitropic. In general, redirection of growth during a tropic response is the consequence of asymmetrical elongation of cells on either side of an organ, in response to a lateral gradient in auxin concentration. Auxin accumulates in the downward-facing sides of roots and stems. Auxin promotes cell elongation in shoot cells leading to negative gravitropism, while it inhibits root cell elongation resulting in positive gravitropism in this organ. Polar auxin transport is mediated by localized influx (e.g. AUX1) and efflux (PINs, ABC transporters) transport carriers (see Chapters 10 and 12). In the root, gravity-sensing cells are located in the root cap while in the stem they are found toward the outside of the vascular tissue. In these sensory cells, gravity causes starch-filled **amyloplasts** in the cytoplasm to sediment downwards. Within the root cap, this initiates, by an as yet unknown mechanism, a biochemical signaling cascade leading to auxin redistribution (Figure 15.39). The promotion of asymmetrical growth by a lateral gradient of auxin is also the mechanism by which light perception is transduced into the phototropic response.

In some plants, stems, roots and/or tendrils grow around objects that they encounter (Figures 15.38 and 15.40). This response is known as thigmotropism. The identity of the touch sensor has not been established; however it is believed to activate a **Ca²⁺-related signaling network** leading ultimately to the thigmotropic response. Tropisms are integrated with other acclimatory and adaptive responses and are attuned to signaling and metabolic pathways that equip plants to deal with the diversity of abiotic stresses.

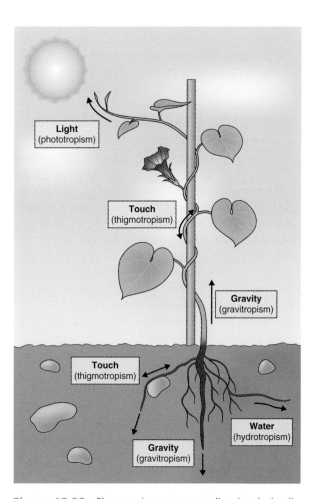

Figure 15.38 Plant tropic responses to directional stimuli.

Key points Tropic growth is an asymmetrical response to a non-uniform environmental stimulus. Roots and shoots are positively and negatively gravitropic, respectively. Phototropism refers to growth towards a directional light source. Unequal growth on the two sides of the responsive organ is generally the result of a lateral gradient of cell elongation under the control of auxin. Spatial distribution of auxin is determined by the pattern of activity of uptake and efflux transporters. Roots sense gravity through sedimentation of starch granules in the amyloplasts of root cap cells. The stems and tendrils of climbing plants are thigmotropic, responding to touch. The touch sensor in thigmotropism, the identity of which is not yet established, triggers growth via a Ca²⁺-signaling network.

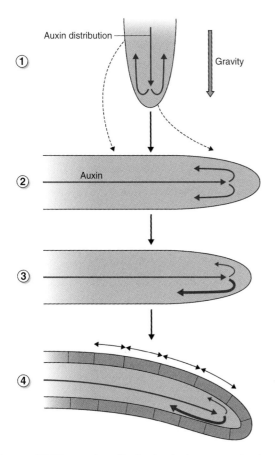

Figure 15.40 Thigmotropic growth in the stem of *Phaseolus vulgaris*. Note how the stem twines itself around the support.

Figure 15.39 Auxin redistribution in the root under the influence of gravity. (1) Normal distribution of auxin in root before exposure to unilateral gravitational stimulus. (2) Root tip moved from vertical to horizontal position. (3) Active redistribution of auxin to the lower side by gravity-sensing cells. (4) Asymmetrical distribution of auxin leads to bending as a consequence of growth inhibition on the lower side and stimulation on the upper.

15.7 Responses to biotic stresses

Plants form relationships with other living organisms from across the taxonomic range. These biotic interactions may be: **parasite/host** (one partner thrives at the expense of the other); **mutualistic** (mutually beneficial associations that improve access to resources or environmental resilience); **predator/prey** (relating to position in the food chain); **allelopathic** (deploying products of secondary metabolism to enhance competitiveness); or **coevolutionary** (exploiting other organisms through complementary adaptations). The different types of association are not mutually exclusive and, in many cases, the mechanisms underlying the interaction and the part the plant plays in it are **generic**.

Thus wounding and the formation of symbiotic root nodules converge on the metabolic and regulatory networks of ROS-mediated oxidative stress (see Figure 15.31), and defenses against pathogens include aspects of the heat shock syndrome (see Figure 15.35) and programmed cell death (see Chapter 18). In this section we examine how the basic principles of secondary metabolism and the adaptive and acclimatory features of abiotic stress response apply to the interplay between plants and other organisms.

15.7.1 Plants deploy constitutive and induced defenses against potential pathogens

Plants, like all living organisms, are constantly exposed to potential pathogens. A **plant pathogen** is defined as an organism that must grow on or inside the plant to complete a part or all of its life cycle and in so doing has a negative effect on its host. Each type of plant pathogen invades its host in a characteristic way. Some get in via stomata or lenticels, or through previously wounded tissue, whereas others directly penetrate the epidermis by mechanical or enzymatic attack. Pathogens are classified

Table 15.8 Strategies utilized by plant pathogens.

	Necrotrophic	**Biotrophic**	**Hemibiotrophic**
Attack strategy	Secreted cell wall-degrading enzymes, toxins, or both	Intimate intracellular contact with plant cells	Initial biotrophic phase, then necrotrophic phase
Specific features of interaction	Plant tissue killed and then colonized by the pathogen. Extensive tissue maceration	Plant cells remain alive throughout the infection. Minimal plant cell damage	Plant cells alive only in the initial stages of the infection. Extensive plant tissue damage at late stages
Host range	Broad	Narrow; often only a single species of plant is attacked	Intermediate
Examples	Rotting bacteria (e.g. *Erwinia* spp.); rotting fungi (e.g. *Botrytis cinerea*)	Fungal mildews and rusts; viruses and endoparasitic nematodes; *Pseudomonas* spp. bacteria	*Phytophthora infestans* (causal agent of potato late blight disease)

according to the strategy they adopt to use the infected plant as a substrate (Table 15.8). **Necrotrophs** kill the cells of the host plant; **biotrophs** cause minimal cellular damage and require living host tissue; and **hemibiotrophs** initially keep cells alive but kill them at later stages of the infection. A pathogen strain that invokes the symptoms of disease is referred to as **virulent**.

Necrotrophic fungi and bacteria injure a host by secreting cell wall-degrading enzymes, or toxins, or both. The fungi *Pythium* and *Botrytis* and bacteria of the genus *Erwinia* destroy cell walls and extract nutrients from the contents thus released. The fungus *Fusicoccum amygdali* is an example of a necrotrophic pathogen that secretes a toxin, in this case **fusicoccin**, which induces stomatal opening by stimulating the activity of the plasma membrane-localized H^+-ATPase. This leads to wilting and cell death in many plant species.

Biotrophic fungi, such as downy and powdery **mildews**, are highly specialized pathogens that interact with viable host cells, often through a feeding structure or **haustorium**. Many biotrophs secrete cytokinins into the host tissue. This prevents host cell senescence at the infection site and results in the formation of so-called **green islands**. Among the biotrophic bacterial pathogens are *Pseudomonas* spp., which have a group of genes, collectively referred to as the **hypersensitive resistant protein (hrp) cluster**, that are absolutely required for bacterial pathogenesis. *Rhizobium radiobacter* (formerly, and widely still, known as *Agrobacterium tumefaciens*), the cause of crown-gall disease in thousands of dicotyledonous plant species, is a biotrophic bacterium of major importance to biotechnology. *R. radiobacter* genetically transforms plant cells with a DNA fragment called T-DNA (transferred DNA), which is stably integrated into the plant genome and promotes tumor formation by altering host hormonal metabolism. *R. radiobacter* T-DNA is the most widely used vector for

stably integrating foreign DNA into the plant genome and for genetically transforming cells of higher plants with recombinant DNA molecules.

Viral pathogens are biotrophic, causing symptoms of tissue yellowing (**chlorosis**) or browning (**necrosis**), mosaic patterns (chlorotic mottling) and stunting in the host. Most plant viruses are single-stranded RNA viruses that replicate in the host cytoplasm and move between cells through **plasmodesmata** (see Chapter 14). Transmission from host to host is often via an **invertebrate vector** organism.

Among the hemibiotrophic pathogens are some of the world's most devastating plant disease organisms. For example, the Irish famine of 1846 and 1847, which resulted in the emigration of more than 1 million people from Ireland to the United States and other countries, was triggered by a catastrophic infection of ***Phytophthora infestans***, an oomycete or water mold that causes late blight disease of potato.

Measures adopted by plants against disease-causing microorganisms include **constitutive defenses** in the form of antimicrobial secondary metabolites; these are located in specific cellular compartments from which they are released following cell damage. Preformed antimicrobial secondary products are sometimes referred to as **phytoanticipins**, to distinguish them from the **phytoalexins**, secondary metabolites synthesized de novo following infection. Interestingly, a given phytoalexin in one plant species may behave as a phytoanticipin in another. An example of a phytoanticipin is **avenacin**, a triterpenoid saponin from oat (*Avena*) that prevents infection by *Gaeumannomyces graminis*, a major pathogen of cereal roots.

In contrast to constitutive defenses, **induced defense responses** require detection of the pathogen by the plant and a rapid activation of defense-related genes. The **resistance/avirulence** gene-for-gene system (see

Chapter 18) is one mechanism by which pathogens are recognized. Another is the interaction between **elicitors** (molecules of pathogen origin) and host receptors. Examples of **bacterial elicitors** are **flagellins** (flagellar proteins), **harpins** (secreted proteins) and **lipopolysaccharides** (see Figure 15.35). Fungal elicitors include cell wall polysaccharides, polypeptides, glycoproteins and lipid molecules. A plant–pathogen interaction that triggers the full defense syndrome is termed **incompatible**. It stimulates a diverse set of cellular mechanisms (Figure 15.41). The outcome is often localized cell death (**hypersensitive response, HR**), which impairs pathogen spread. HR is discussed in further detail, as an example of programmed cell death, in Chapter 18. It should be noted that even in **compatible** interactions, where the host fails to recognize the invader and there is no induced defense response, the pathogen must still overcome constitutive defenses before successful infection can occur.

An early induced defense response is superoxide production by a plasma membrane-associated **NADPH oxidase**. Metabolic fates and signaling roles of $O_2^{\bullet-}$ and H_2O_2 are described in Figure 15.31 and Section 15.6.3. An early consequence of pathogen recognition is the rapid synthesis of the signaling molecule **nitric oxide**, which acts synergistically with ROS to activate host defenses. ROS also participate in lignin synthesis that, together with the synthesis and cross-linking of **hydroxyproline-rich glycoproteins** and the production of **polygalacturonase inhibiting proteins**, renders the plant cell wall more resistant to microbial penetration and enzymatic degradation. Initiation of resistance involves the diversion of primary precursors into secondary metabolism. For example, the synthesis of phenylpropanoid phytoalexins is the result of coordinated activation, by transcription factors of the bHLH, MYB and WRKY families, of genes that encode the pathway leading from phenylalanine to isoflavonoids (see Figures 15.11 and 15.13).

Prominent among local and systemic defense responses is the expression of **pathogenesis-related** (**PR**) proteins, a diverse group of enzymes, antifungal agents and secondary signaling components. They include glucanases and chitinases that attack fungal cell walls, endoproteinases and ribonucleases that destroy pathogen macromolecules, and lipoxygenases that generate antimicrobial volatiles and jasmonate-related signaling molecules. Transcriptional activation of many PR genes is regulated by salicylic acid- and ethylene-mediated signal transduction cascades.

Plants form non-pathogenic (**endophytic**) relationships with a variety of fungi. In most cases the benefits to the participants in the interaction are unclear, but some endophytic associations have known ecological and agricultural significance. For example, *Neotyphodium* spp., fungal endophytes of forage and turfgrasses of the genera *Lolium* and *Festuca*, have been shown to improve (by mechanisms yet to be established) host tolerance of abiotic stresses and resistance to insect and mammalian herbivores.

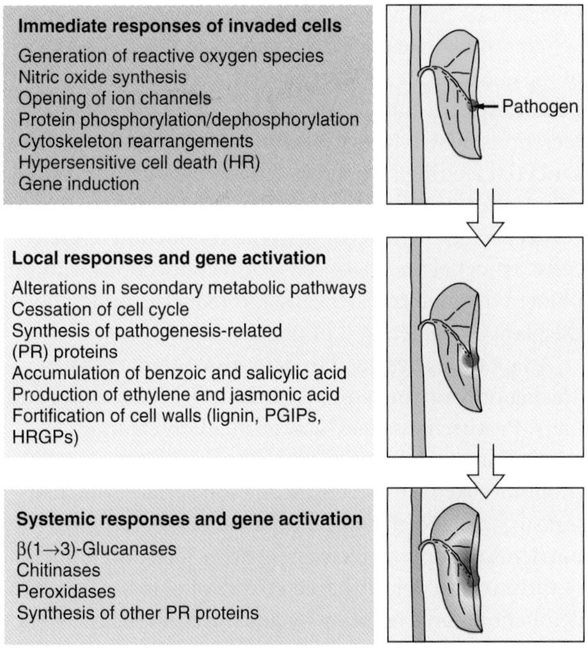

15.7.2 Plants compete by conducting chemical warfare with allelopathic secondary compounds

We have seen that roots are the source of compounds exuded into the rhizosphere that facilitate nutrient acquisition (see Chapter 13) and serve as signals to potential microbial symbionts (see Chapters 10 and 12). These functions of root exudates are only part of the story, however. Among the multiplicity of compounds found in **root exudates** are amino acids, organic acids, sugars, phenolics, high molecular mass polysaccharide mucilages and proteins. It has been estimated that from 5% to 20% of all carbon fixed in photosynthesis is secreted into the rhizosphere. To devote such a high level of resource to root exudation implies that secretion is important for plant survival.

An ecologically significant function of compounds secreted into the rhizosphere is to influence, in most cases negatively, the growth of neighboring plants—a

Immediate responses of invaded cells

Generation of reactive oxygen species
Nitric oxide synthesis
Opening of ion channels
Protein phosphorylation/dephosphorylation
Cytoskeleton rearrangements
Hypersensitive cell death (HR)
Gene induction

— Pathogen

Local responses and gene activation

Alterations in secondary metabolic pathways
Cessation of cell cycle
Synthesis of pathogenesis-related (PR) proteins
Accumulation of benzoic and salicylic acid
Production of ethylene and jasmonic acid
Fortification of cell walls (lignin, PGIPs, HRGPs)

Systemic responses and gene activation

β(1→3)-Glucanases
Chitinases
Peroxidases
Synthesis of other PR proteins

Figure 15.41 Immediate, local and systemic induced responses to pathogen invasion.

Key points The relationship between plants and their pathogens is an example of parasite–host biotic interactions. A virulent pathogen is one that causes disease symptoms. Necrotrophic pathogens injure or kill host cells by secreting toxins or cell wall-degrading enzymes. Powdery mildews are specialized biotrophic fungi that keep host cells alive by secreting cytokinins and extract nutrients through haustoria. Many biotrophs secrete cytokinins into host tissue. Pathogenesis in biotrophic bacteria of the genus *Pseudomonas* requires genes of the hypersensitive resistant protein (hrp) cluster. The crown-gall disease organism *Rhizobium radiobacter* (*Agrobacterium*) is exploited in biotechnology for genetic transformation of plants. Most pathogenic plant viruses are biotrophs of the single-stranded RNA type. Potato blight (*Phytophthora infestans*) is a hemibiotroph that initially keeps host tissue alive and then changes to the necrotrophic mode. Plants use antimicrobial secondary metabolites as constitutive defenses against disease-causing microorganisms. When they detect invading pathogens, plants induce additional defense mechanisms; these defenses can include localized cell death (the hypersensitive response), strengthening of the cell wall, and synthesis of pathogenesis-related proteins such as chitinases, nucleases and proteases.

phenomenon known as **allelopathy**. Allelopathically-active compounds (**allelochemicals**) are usually injurious and thus deter competitors from stealing soil resources. Many destructive or highly invasive species are very allelopathic. An example is **knapweed** (*Centaurea maculosa*), a competitive weed that can dominate large areas of landscape. Knapweed roots exude **catechins**, phytotoxic phenylpropanoids derived from leucoanthocyanidins and anthocyanidins (see Section 15.3.3). While allelochemicals can be an effective defense against competitors, they can also be a signal to parasites. Parasitic plants often find their hosts by using secondary metabolites secreted by host roots as signals to initiate the development of invasive roots (**haustoria**) (Figure 15.42). Flavonoids, *p*-hydroxy acids, quinones and cytokinins secreted by host roots have been shown to induce haustorium formation. *Striga* (Figure 15.42), a parasitic genus that causes major losses of maize, sorghum, millet and rice, is estimated to infest two-thirds of the area devoted to growing cereal crops in Africa. *Striga* gives its name to the **strigolactones**, important signaling molecules involved in many aspects of plant growth and development (see, for example, Chapters 10 and 12). These molecules also function as

rhizosphere allelochemicals that are implicated in host–parasite recognition, as well as the formation of mycorrhizal associations (see Chapter 12).

15.7.3 A range of generic local and systemic stress responses are invoked by herbivory, predation and wounding

Plants are under constant attack from **invertebrates**, which produce physical and physiological symptoms in a plant as a result of their feeding, reproductive or sheltering behaviors. **Chewing insects** can be responsible for spectacular plant tissue damage. Plagues of locusts, for example, owe their fearsome reputation to their capacity to defoliate an entire crop covering several hectares within a day or so. Tissue damage caused by chewing insects frequently permits secondary infection by necrotrophic fungi and bacteria. On the other hand, **sap-sucking insects**, such as aphids or thrips, drain the contents of phloem sieve elements using a specialized mouth part, the stylet (see Figure 14.16). These insects rarely cause extensive physical injury, although heavy infestations can cause chronic shortages of photosynthate which lead to severe reductions in growth potential. While feeding, many sap-sucking insect pests also transmit viruses, including some agriculturally significant biotrophic pathogens such as **potato virus X**.

Parasitic **nematodes** are among the most damaging invertebrate herbivores. In most cases they infect root systems, often causing large changes to the metabolism of the entire plant and to root architecture. **Cyst nematodes** are endoparasites that release glandular secretions, which trigger fusion of host cells to form **syncytial feeding structures**. Endoparasitic **root-knot nematodes** stimulate DNA endoreduplication and cortical cell growth, leading to the production of **giant cells**. Syncytial and giant cells form connections with the phloem through transfer cells and reduce productivity of the plant by depleting it of photosynthate.

Adaptations designed to deter tissue damage by grazing or browsing animals include prickles, spines and hairs. **Pasture** grasses exhibit particularly striking adaptations to defoliation stress, such as intercalary shoot meristems located near the soil surface, silicified leaf surfaces and efficient mechanisms for recycling nutrients to support recovery growth. The development of the morphological characteristics of plants adapted to defoliation and trampling by grazing animals is complemented by the evolution of features that fit the grazers for a diet of vegetation—for example, teeth designed for biting and chewing foliage, digestive systems

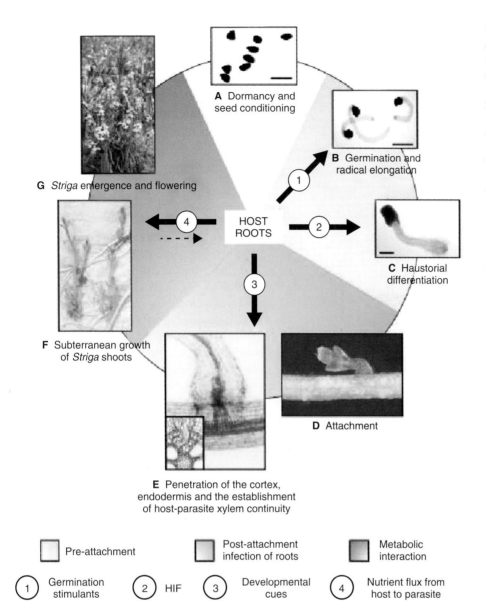

A Dormancy and seed conditioning

B Germination and radical elongation

G *Striga* emergence and flowering

HOST ROOTS

C Haustorial differentiation

F Subterranean growth of *Striga* shoots

D Attachment

E Penetration of the cortex, endodermis and the establishment of host-parasite xylem continuity

Pre-attachment

Post-attachment infection of roots

Metabolic interaction

1 Germination stimulants

2 HIF

3 Developmental cues

4 Nutrient flux from host to parasite

Figure 15.42 The life cycle of *Striga*. (A) Each *Striga* plant can produce 100 000 seeds. (B) Seeds germinate in response to germination stimulants present in host root exudates. (C) The elongating *Striga* radical perceives haustorial initiation factors (HIFs) secreted by the host root and forms a functional attachment organ. (D) Sticky hairs attach the haustorium to the host root. (E) Haustorium cells form a wedge and intrude into host xylem vessels (but not phloem). (F) Following establishment of xylem continuity with the host, there is further differentiation of the haustorium, followed by the formation of cotyledons and leaf initiation. (G) The *Striga* shoot emerges above ground and flowers and sets seed approximately 6 weeks later. Scale bar = 250 µm (A–C).

capable of processing high-bulk, low-nutrient content feed, and so forth. The relationship between herbage plants and ruminants is a **coevolutionary** one.

Chewing by herbivores and feeding by endoparasites trigger local and systemic **wound responses** in plant tissues. Wounding releases oligosaccharides from damaged cell walls, which activate a number of stress response genes, including those encoding pathogenesis-related proteins, heat shock factors (see Figure 15.35) and proteinase inhibitors (**PIs**). PIs are small defense proteins that interfere with the invertebrate digestive system by inhibiting serine, cysteine and aspartyl proteinases and thereby reduce the uptake of essential amino acids and retard the herbivore's growth and development. Many of the immediate biochemical changes in injured cells are shared with those of other abiotic and biotic stresses (see Figures 15.30 and 15.41):

production of ROS, induction of secondary metabolism and accumulation of phenolics, tannins and phytoalexins, and increased synthesis of ethylene, jasmonates and salicylic acid.

As a consequence of the export of signal molecules from a wound site via the vascular system, wounding often stimulates systemic responses in the whole plant. Wounding promotes proteolytic processing of the 200-amino acid precursor **prosystemin** to form the 18-amino acid peptide hormone **systemin**. Systemin engages with a receptor in the plasma membrane and activates the oxylipin pathway of jasmonic acid (JA) synthesis (see Chapter 10), as well as the transcription of genes encoding a PI (*PIN2*). JA is believed to move through the phloem to remote undamaged tissues where it induces synthesis of PIN2 and other systemic wound response proteins (Figure 15.43).

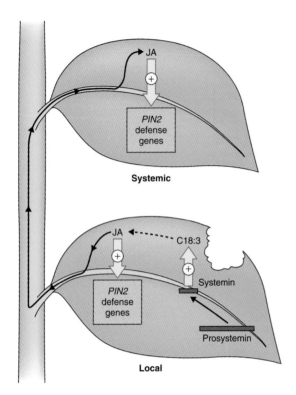

Figure 15.43 Local and systemic responses to wounding mediated by systemin and jasmonic acid (JA). Prosystemin is proteolytically processed to the peptide hormone systemin, which activates synthesis of JA from linolenic acid (C18:3). JA stimulates the transcription of genes encoding proteinase inhibitor (*PIN2*) and other wound response proteins locally and in remote tissue to which JA is transported.

In addition to inducing physiological reactions within the plant, wounding also stimulates the production of **volatile compounds** that can act as an alarm signal to other parts of the same plant or even to other individuals in the community. Among the volatiles emitted by plants are methanol, acetone, formaldehyde and other short-chain carbonyl compounds, plus a host of terpenoids (including isoprene—see Section 15.5.2), phenylpropanoids and fatty acid derivatives. Plant-to-plant signaling by volatiles is a relatively new field of research and details of mechanisms and ecological significance remain to be discovered.

Key points Plants secrete into the rhizosphere compounds that can influence (usually negatively) the growth of other plants around them. This phenomenon is known as allelopathy and is especially common among highly invasive species, such as knapweed. Secondary metabolites secreted from roots are also used by parasitic plants to locate suitable hosts. Parasitism by nematodes and herbivory by grazing animals can inflict considerable damage on plants. Some species defend themselves by producing spines or hairs. Most plants have a range of local and systemic stress responses that are invoked by herbivory, predation and wounding; many of these reactions have features in common with abiotic stress responses, such as ROS production and accumulation of phenolics, tannins and phytoalexins. The synthesis of protease inhibitors, which interfere with invertebrate digestive systems, is a frequent response to herbivore attack. Wounding can also induce the release of volatile compounds that act as a signal to other parts of the plant or to nearby plants.

Part VI
Renewal

Chapter 16

Flowering and sexual reproduction

16.1 Introduction to flowering

Charles Darwin called the sudden appearance and rapid spread of flowering plants during evolution an 'abominable mystery'. No less mysterious is why the sexual mode of biological reproduction should have arisen in the first place, particularly in plants, where the property of cellular totipotency allows efficient asexual propagation. But sexual reproduction by means of flowers clearly confers fitness benefits and competitive advantages: the first flowering plants emerged 125 million years ago and within the geologically brief timespan of 60 million years angiosperms had established themselves as the dominant form of vegetation, accounting for more than 80% of the species in most ecosystems.

As with the principles determining body plan and morphogenesis (see Chapter 12), the reproductive strategies of seed plants and higher animals differ in fundamental aspects (Figure 16.1). During early embryogenesis in animals, **somatic cells** (destined to give rise to all body tissues other than those producing eggs and sperm) are differentiated from **germ line cells** (committed to becoming the gametes) (Figure 16.1A). The decision about whether a given cell takes the somatic or germ line route to maturity depends on intrinsic factors, such as cytoplasmic polarity, and on signaling

from neighboring cells. Normally, once a cell is committed to the germ line or soma, neither it nor the cells derived from it can divert into the alternative developmental fate. As a consequence, the animal reproductive system is not exposed to the developmental signals that subsequently induce formation of the major somatic tissues and organs, and mutations can be inherited only if they affect germ line cells.

There is no such specification of germ line and soma cells in plants, however (Figure 16.1B). The shoot apex switches from the vegetative to reproductive mode of differentiation in response to developmental and environmental signals. The reproductive system of the flower is derived from cells with a history of exposure to internal and environmental influences experienced through many rounds of cell division during plant maturation. The alternation of generations between the diploid sporophyte and haploid gametophyte (see Chapter 1) is an efficient mechanism for selecting against potentially deleterious heritable somatic mutations occurring in meristem cells before gametogenesis (formation of eggs and sperm).

This chapter discusses the environmental and developmental factors determining the transition from the vegetative to the reproductive condition. The patterning of flower primordia and the differentiation of floral organs are considered and finally the formation of male and female gametes and their union during fertilization is discussed.

The Molecular Life of Plants, First Edition. Russell Jones, Helen Ougham, Howard Thomas and Susan Waaland.
© 2013 John Wiley & Sons, Ltd. Published 2013 by John Wiley & Sons, Ltd.

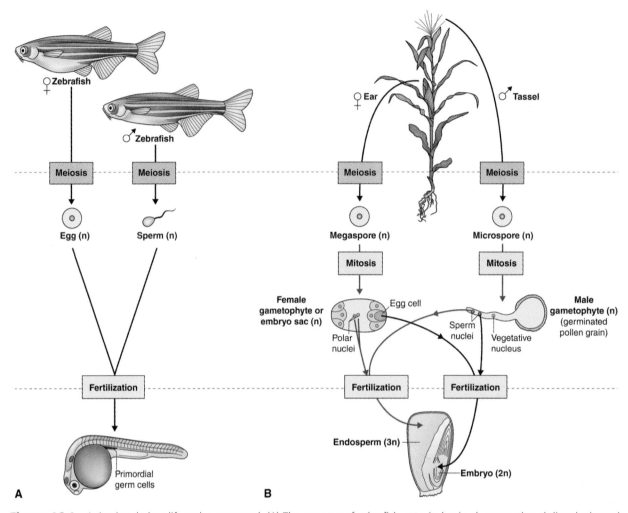

Figure 16.1   Animal and plant life cycles compared. (A) The gametes of zebrafish, a typical animal, are produced directly through meiosis of germ line cells, and gene expression is mostly restricted to diploid cells. Primordial germ line cells are differentiated early in embryogenesis. (B) The life cycle of a typical flowering plant (maize) comprises alternating sporophyte and gametophyte generations, and meiotic cells are not derived from a predefined germ line.

16.2 Induction of flowering

To maximize reproductive success, flowering must take place at the right time. Most plants are **cross-pollinated** and will synchronize flower production with others of the same species and with the availability of pollinators. In latitudes where the annual cycle alternates between favorable and unfavorable seasons, flowering is usually subject to close environmental control so that it coincides with the best conditions for seed development and fruit dispersal. Flowering is a three-stage process. First, the plant must be competent to respond to the factors that invoke the inductive signal. Next, the signal must be received by the leaves (see Section 8.5.4) and a message sent to the shoot apex, which undergoes the vegetative to reproductive transition. Then the reprogrammed apex switches from the production of

vegetative laterals to pathways of patterning and morphogenesis that lead to the differentiation of floral organs. The best understood factors stimulating the action of the inductive signal are light (**photoperiodism**) and temperature (**vernalization**).

16.2.1 Floral induction requires both the perceptive organ and the shoot apex to acquire competence during plant maturation

For flowering to occur, the plant must be competent. Competence resides in either the leaf, which perceives the inductive stimulus, or the shoot apex, or both. Figure 16.2 summarizes the possible combinations of

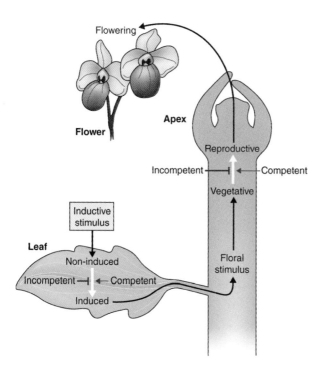

Figure 16.2 Points at which acquisition of competence is necessary for flowering to occur.

foliar and apical competence and their outcomes in response to the inductive stimulus. Apical incompetence is characteristic of the shoots of young trees, which are usually unable to make the transition into flowering even when grafted onto mature plants. The reverse is the case in some species such as *Kalanchoe* (mother of thousands; see Figure 11.22A), in which the juvenile apex is competent and will form flowers when grafted onto a mature stock. The regulation of competence to flower is complex and varies across the taxonomic range, but an important part of the mechanism is the transition from the juvenile to mature phase of whole-plant development.

Distinct periods of juvenility and maturity are identifiable in the life spans of many species. The maturation process is referred to as **heteroblasty** or phase change. Four phases or maturation stages are recognized in the life cycle: embryonic; post-embryonic juvenile; adult vegetative; and adult reproductive. Each phase is associated with a characteristic package of morphological and physiological features. Ivy (*Hedera helix*) is a well-studied example of phase change. Table 16.1 compares the characteristics of juvenile ivy with those of the adult phase plant. As well as competence to flower, many aspects of morphology change with maturity in this species, including leaf form and phyllotaxy, growth habit, shoot determinacy and root development. If tissue taken from a juvenile ivy plant is cultured, plants regenerated from it have a stable juvenile phenotype. Similarly, mature tissue yields regenerants with mature

characteristics. The difference between the phases has a strong epigenetic component (see Section 3.5).

Gibberellins (GAs) are important internal factors responsible for regulating phase change. Their effect is not the same in all species, however; for example, adult-phase ivy reverts to juvenility when sprayed with GA, whereas the juvenile–adult transition is delayed in *Arabidopsis* mutants with deficiencies in GA synthesis or perception. GA is also required for bolting, the process of stem elongation that precedes flowering in *Arabidopsis* and other rosette plants. Environmental variables, particularly light intensity and ambient temperature, significantly influence the timing of progress from juvenile to adult development. Figure 16.3 shows the phenotypes of phase-change mutants of maize (*Zea mays*), which produce shorter plants with more shoots than wild-type plants. The leaves of juvenile maize plants are short, hairless and covered in epicuticular wax. Adult-phase leaves are long and narrow with hairs but no wax. The analysis of mutants has identified the *Corngrass* (*Cg*) gene and genes of the *Teopod* (*Tp*) family as promoters of the adult condition and suppressors of the juvenile phase of vegetative development in this species. The signal that triggers phase transition is perceived directly in individual leaf primordia rather than by the shoot apical meristem. Mutations in *Cg* and *Tp* genes are associated with overexpression of the miR156 family of microRNAs (see Chapter 3). The miR156 locus has also been shown to be important in the architecture and maturation of rice, and variations in this locus are proposed to have been important in the evolution and domestication of cereal crops in general. Among the target genes silenced by miR156 are transcription factors of the extensive **SPL** (*SQUAMOSA promoter-binding-like*) family, which have a wide range of developmental functions including meristem identity, leaf shape and tissue patterning, and floral induction. The role of small RNAs in phase change, flowering and other morphogenetic processes is a fast-developing story which is expected to yield many new insights into the regulation of the development of plants and their parts.

16.2.2 Determinacy of the reproductive apex affects plant morphology and annual/perennial growth habit

The shoot apex may be continuously meristematic (**indeterminate**) or may differentiate into a terminal organ and cease to make new structures (**determinate**). The early events of organogenesis at the shoot apex, whether the outcome is a leaf (see Chapter 12) or a floral structure, involve the formation of a determinate

Table 16.1 Distinguishing characteristics of juvenile and adult ivy.

Juvenile characters	Adult characters
Three- or five-lobed palmate leaves	Entire, ovate leaves
Alternate phyllotaxy	Spiral phyllotaxy
Plastochron 4.2 days	Plastochron 3.2 days
Shoot apex relatively narrow with large cells	Wide apex with small cells
High rate of internode growth	Low rate of internode growth
Anthocyanin pigmentation of young leaves and stems	No anthocyanin pigmentation
Stems hairy	Stems smooth
Climbing and spreading growth habit	Upright or horizontal growth habit
Shoots show unlimited growth and lack terminal buds	Shoots show limited growth terminated by buds with scales
Absence of flowering	Presence of flowering
Rooting ability of cuttings good	Cuttings root poorly
Adventitious roots present	Adventitious roots absent

Figure 16.3 Phase-change mutants of *Zea mays*. The effects of mutations in genes of the *Teopod* family (*Tp1*, *Tp2*) and the *Corngrass* locus (*Cg*) are compared with the wild-type phenotype (Wt).

Key points Of all the plants on Earth, only the angiosperms produce flowers. There are three requirements for flowering to occur. Firstly, the plant must have reached a state in which it is capable of responding to an inductive signal. Secondly, the signal must be perceived in the leaves and transmitted to the shoot apical meristem. Finally, the meristem must respond by transforming from vegetative proliferation to the production of floral parts. The best understood factors that trigger flowering are day length and temperature. Many species have distinct juvenile and adult phases, and only adult-phase plants are competent to flower. Gibberellins are important in regulating the phase change from juvenile to adult. The shoot apex may be indeterminate—that is, continuously meristematic—or it may differentiate into a terminal, determinate organ and cease to make new structures. In annual species, all indeterminate vegetative shoot apices become determinate floral apices. In perennial species, the vegetative apex becomes temporarily determinate in order to produce a resting bud. Determinacy of a floral apex is usually irreversible, but reversion sometimes occurs in response to environmental conditions.

primordium. The pattern of apical and primordium determinacy is decisive not only for vegetative and floral morphology but also for plant habit and life cycle (see, for example, Figure 12.4). In annual species, all indeterminate vegetative shoot apices become determinate floral apices and the entire plant dies once the seeds have been dispersed. In perennial species, temporary determinacy of the vegetative apex results in the formation of a resting bud. **Dormancy**, the inactive condition adopted by plants to survive unfavorable

seasons, is a state of arrested apical growth, and the position of resting buds in the plant is the basis of one of the widely used systems of classifying plant life forms (see Chapter 17).

Generally, determinacy of a terminal floral apex is irreversible, but there are numerous examples of reversion, in which the meristem is kick-started into further activity, leading to the abnormal production of vegetative or floral structures. Reversion may be an environmental response. For example, florally-induced soybean (*Glycine max*) that is transferred from inductive short days to non-inductive long days has been observed to revert to the vegetative condition, making compound leaves consisting of three leaflets instead of foliar bracts (modified simple leaves associated with floral parts) and initiating vegetative shoot growth after aborting further development of flower buds. In some species, such as garden balsam (*Impatiens balsamina*), the tendency to reversion (Figure 16.4) has a strong genetic component; the extent and nature of reversion behavior in parents and offspring varies between different varieties. Mutations in genes that determine and maintain the reproductive state often result in a form of floral reversion in *Arabidopsis*.

16.2.3 Different inductive pathways lead to flowering

Several environmental signals may be involved in the induction of flowering. These include photoperiod, light quality and temperature. Photocontrol of flowering is exerted through day length and light quality; the role of light in controlling flowering is discussed in Chapter 8 (see Figure 8.33). Exposure to low temperature is also critical for the acquisition of competence to respond to photoinductive conditions in many long-day (LD) winter

A   **B**   **C**

Figure 16.4 Flowering, reversion and reflowering responses of *Impatiens* after inductive photoperiod treatment. (A) Top view of a terminal flower. (B) Top view of a terminal floral apex that has reverted to the production of leaves. (C) Reflowering phenotype observed in some of the progeny.

annual, biennial and perennial species from temperate regions. The cold requirement is called **vernalization** and acts as a kind of time-computing mechanism that measures the passage of winter and ensures that flowering does not begin until the favorable conditions of spring arrive. The analysis of a class of delayed-flowering mutants that are insensitive to photoperiod has identified another regulatory pathway of floral induction in *Arabidopsis*. This so-called **autonomous pathway** is independent of day length and represents the route by which endogenous developmental cues feed into the network of flowering genes and their interactions.

Figure 16.5 shows how the photoperiod pathway, light quality pathway, vernalization pathway and autonomous pathway form a regulatory network that converges on the activities of a set of genes that integrate the floral stimulus and trigger the vegetative to reproductive transition. Gibberellins play a number of roles in this transition. These include functions in competence (Section 16.2.1) and the development of floral organs (see Section 16.3) as well as in the promotion of bolting and flowering in *Arabidopsis* and many other LD and/or biennial species. Flowering in perennial species, on the other hand, tends to be insensitive to, or even inhibited by, GA. There is good evidence that GA is a mobile signal that transmits the photoperiodic floral stimulus in the LD grass *Lolium temulentum*; its action is independent from that of FT, the phloem-mobile protein that relays the flowering induction signal from leaf to shoot apex (see Section 8.5.4).

Expression of both *SOC1* (*SUPPRESSOR OF OVEREXPRESSION OF CONSTANS1*) and *LFY* in *Arabidopsis* is promoted by GA via the DELLA-mediated signaling mechanism (see Chapter 10). *SOC1* is thus regulated in a multifactorial manner and integrates the autonomous, vernalization and GA pathways. The GA and photoinductive pathways converge on *SOC1* and *LFY* (Figure 16.6).

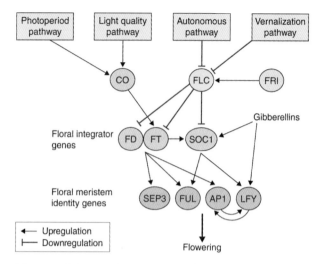

Figure 16.6 Regulatory network of the products of flowering genes in *Arabidopsis*. Interactions that promote flowering are shown as black lines, floral repression as red lines.

Floral meristem identity genes stand at the point of conjunction between the different inductive pathways. The genes **APETALA1** (**AP1**), **LEAFY** (**LFY**), **SEPALLATA3** (**SEP3**) and **FRUITFUL** (**FUL**), are all named for the *Arabidopsis* mutants identified by their abnormal floral phenotypes (see Section 16.3). Activation of floral identity genes during the switch to reproductive development involves either derepression of the target gene (as in the case of *AP1*) or reorganization of sites of expression within the apex (*LFY* is an example) before floral primordia develop (Figure 16.7). As we shall see, during the vegetative to reproductive transition, floral identity genes have distinct but overlapping functions and cooperate to modulate each other's activity (Figure 16.6).

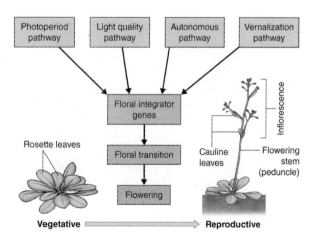

Figure 16.5 Pathways that regulate flowering in *Arabidopsis*.

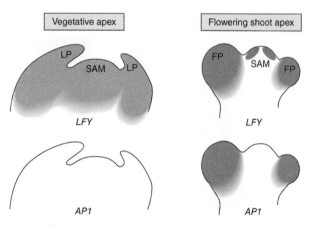

Figure 16.7 Patterns of expression of selected floral identity genes in vegetative and reproductive apices of *Arabidopsis*. FP, flower primordium; LP, leaf primordium; SAM, shoot apical meristem.

16.2.4 Regulation of floral induction by vernalization is epigenetic in nature

Many strains of *Arabidopsis*, including those most commonly used for laboratory experimentation, are summer annuals and have no vernalization requirement. Studies of natural variation, however, have identified **ecotypes** (genotypes adapted to particular habitats) that, like winter annuals, delay flowering unless they are exposed to a period of chilling. Genetic analysis has established that this vernalization requirement is conferred by *FLOWERING LOCUS C* (**FLC**) and a second dominant gene, *FRIGIDA* (**FRI**). FLC is a MADS-box transcription factor that functions as a repressor of flowering, in part by blocking expression of *FT*. The *FLC* gene is expressed in all the vegetative parts of the plant, especially the vegetative shoot apex and the roots, but *FLC* mRNA is not detectable in young floral meristems, indicating that expression of *FLC* is downregulated in the apex after the transition to flowering. FLC is the point of convergence of the autonomous and vernalization pathways (see Figure 16.6). Mutations in the autonomous pathway delay flowering, an effect that is suppressed in *flc* null mutants. Genes in the autonomous pathway that block *FLC* expression have a range of functions in development, many of which are not directly concerned with flowering. *FRI* encodes a nuclear protein, unique to plants, that upregulates *FLC* expression, thereby repressing transcription of *FT* and other floral integrator genes (see Figure 16.6). Most of the natural variation in the vernalization requirement of *Arabidopsis* ecotypes is due to allelic **variation** at the *FRI* and *FLC* loci.

Vernalization is an example of an **epigenetic switch**. Even after chilling treatment has been discontinued, vernalization-mediated change in competence to flower is retained through a large number of mitotic divisions in the apical meristem. The switch takes the form of modifications to the chromatin of the *FLC* locus, resulting in mitotically-stable repression of *FLC* gene transcription. In *Arabidopsis*, low temperature induces the synthesis of **VIN3**, the product of the *VERNALIZATION INSENSITIVE 3* gene. VIN3 is recruited to the so-called polycomb repressive complex 2 (**PRC2**) in association with other proteins including VRN2 and VRN1, encoded by *VERNALIZATION* (**VRN**) genes. Polycomb takes its name from the phenotype of a homeotic mutant of *Drosophila*. Proteins of the polycomb group silence gene expression in animals and plants by remodeling chromatin. The repressed state is perpetuated through cell divisions. PRC2 components increase during vernalization and modify histone methylation, which results in repression

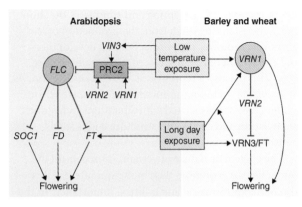

Figure 16.8 Comparison of vernalization mechanisms in *Arabidopsis* and cereals. During vernalization in *Arabidopsis*, PRC2 in association with VRN2 and VRN1 recruits VIN3 and initiates mitotically-stable repression of FLC. In wheat and barley, *VRN1* and *VRN2* are structurally and functionally distinct from the *Arabidopsis* genes of the same name. In this case, vernalization stimulates expression of *VRN1*, which in turn promotes flowering. Vernalization acts in concert with long days, signaling through the FT/VRN3 photoperiod pathway, to ensure that flowering and seed formation occurs in spring and summer.

of *FLC* transcription and consequent derepression of flowering (Figure 16.8).

Winter cereals are planted in fall and have a vernalization requirement which is satisfied by exposure to cold temperatures before seedling growth resumes in the spring. Most temperate perennial grasses also require vernalization to become florally competent to respond to LDs. As their name suggests, spring cereals are planted after winter is over and will flower without vernalization. The vernalization pathways in *Arabidopsis* and winter cereals show points of similarity and difference (Figure 16.8). Flowering in *Triticum* spp. is regulated at three *VRN* loci. Confusingly, the genes designated *VRN* in grasses are not structurally related to the genes of the same name in *Arabidopsis*. *VRN3* of wheat and barley (*Hordeum*) has high homology with *FT*. The *VRN2* region of the cereal genome encodes two similar zinc finger proteins which act as repressors of *VRN3/FT*. The spring growth habit in wheat and barley is associated with loss of the *VRN3/FT* repressor *VRN2*, a genetic locus with no clear homolog in *Arabidopsis*. *VRN1* is a MADS box gene with multiple functions. Plants with deletions at this locus remain permanently vegetative. *VRN1* is thought to be the primary target for the vernalization signal in grasses. In contrast to *FLC* of *Arabidopsis* (a structurally unrelated MADS box gene), *VRN1* in grasses is induced rather than repressed by low temperature. *VRN1*, *VRN2* and *VRN3/FT* interact to regulate vernalization and flowering (Figure 16.8). *VRN1*, which is expressed at low levels during winter, becomes strongly upregulated by VRN3 in shoot apices as day length increases during spring and stimulates the shoot apical meristem to switch from the production of leaf primordia to flower primordia. Further development and elongation of the inflorescence generally require LDs. VRN1 is also a repressor of *VRN2* transcription. As in *Arabidopsis*, chilling-induced floral competence in winter cereals is associated with alterations in histone methylation and chromatin structure and is stably transmitted during mitosis.

The epigenetic nature of vernalization control in winter cereals is of more than biological interest: it is also of major historical and political importance. In the 1930s, the influential Ukrainian agriculturalist Trofim Lysenko denounced the principles of Mendelian genetics, using vernalization as evidence for the inheritance of acquired characters. Lysenko's doctrine, enshrined in Soviet Communist Party orthodoxy (and propagated through its own journal, the *Bulletin of Vernalization*), dominated biology during the era of Josef Stalin and was responsible for the persecution of many prominent geneticists and agronomists, thereby contributing to the backwardness of Soviet agriculture and to frequent and widespread crop failure. It is no exaggeration to state that the genetic regulation of cereal vernalization was a significant influence on the course of 20th century history.

Key points Photoperiod and light quality are important in regulating the induction of flowering in many plants. Temperate species also often require a period of exposure to cold in order to become competent to flower; this cold requirement is called vernalization. Vernalization is an epigenetic switch, brought about by modification of the chromatin around the *FLC* gene. An autonomous pathway, which is independent of light regime and temperature, integrates endogenous development cues to ensure that flowering is not triggered inappropriately. The autonomous, photoperiod, light quality and vernalization pathways operate together to control expression of key genes that trigger the switch from vegetative to reproductive development.

16.3 Development of floral organs

The transition of shoot apex identity from the vegetative to the reproductive state described in Section 16.2 is only the first step in flowering. To make a functional flower, the identities of organ primordia in the apex must be established. Familiarity with the terminology used to describe floral morphology is necessary for understanding how flowers develop (see Section 1.7.1). The structure of the mature *Arabidopsis* flower is typical of that of many advanced angiosperms. It comprises four types of floral organs in an arrangement of four concentric whorls. The outermost whorl of four **sepals** (collectively called the **calyx**) and the next whorl of four **petals** (the **corolla**) enclose the male and female reproductive organs, comprising, respectively, a whorl of six **stamens** surrounding the central **gynoecium** that contains two **carpel**s (see Figure 1.17). Collectively, the calyx and corolla comprise the **perianth**.

Once *Arabidopsis* has been induced to flower it bolts; that is, the rosette produces a shoot carrying the floral apex and cauline leaves—cauline means growing on a stem rather than in a rosette (see Figure 16.5). This main flowering stem (peduncle) is indeterminate and bears the primary inflorescence, the first inflorescence to form. An inflorescence is defined as a cluster or group of flowers on a stem. Successive flowers that make up an inflorescence are arranged at an angle of $130°$ to $150°$ to each other in a spiral around the center of the shoot apex, with the oldest flowers at the base and the youngest at the top of the meristem. Axillary buds of cauline leaves develop into additional secondary inflorescences, bearing flowers along their length. Flower initiation on the primary shoot meristem begins shortly before secondary inflorescences form. A flower primordium first appears as a bulge (buttress) on the shoot apical meristem, and organ primordia as bulges on a flower primordium. Under experimental conditions, the sequence of events giving rise to a normal *Arabidopsis* flower is highly predictable. Table 16.2 is a timetable of developmental landmarks and Figure 16.9 shows a corresponding scanning electron microscope sequence of the morphology of the floral apex and developing flower and floral organ primordia.

Arabidopsis has been the subject of detailed genetic and molecular dissection of floral morphogenesis. Analysis of flower development in this species is based on the availability of **homeotic mutants**. In a homeotic mutant an organ or tissue develops in an abnormal location, often replacing the structure that would normally occur in that position. Such mutants have been intensively studied in insects, notably *Drosophila*, where the body is metameric, that is, organized into segments bearing different organs. In the *Drosophila* mutant *Antennapedia*, for example, mis-expression of the corresponding homeotic gene (*Antp*) results in primordia on the head segment developing into legs instead of antennae (see Figure 3.30). The body plan of plants is similarly metameric and subject to homeotic mutation, as exemplified by genes of the *KNOX* family

Table 16.2 Stages of flower development in *Arabidopsis thaliana*.

Stage	Landmark event at beginning of stage	Duration (hours)[a]	Age of flower at end of stage (days)
1	Flower buttress arises	24	1
2	Flower primordium forms	30	2.25
3	Sepal primordia arise	18	3
4	Sepals overlie flower meristem	18	3.75
5	Petal and stamen primordia arise	6	4
6	Sepals enclose bud	30	5.25
7	Long stamen primordia are stalked at base	24	6.25
8	Locules appear in long stamens	24	7.25
9	Petal primordia are stalked at base	60	9.75
10	Petals are level with short stamens	12	10.25
11	Stigmatic papillae appear	30	11.5
12	Petals are level with long stamens	42	13.25

[a]Estimated to the nearest 6 hours.

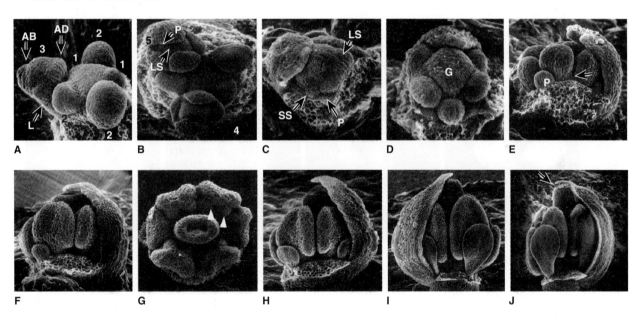

Figure 16.9 Scanning electron micrographs of shoot apical meristem (A, B) and young flower primordia (C–J) of *Arabidopsis*. Stages refer to the sequence of events summarized in Table 16.2. (A) Side view showing stage 1, 2 and 3 primordia. (B) Top view showing stage 4 and 5 primordia. (C) Stage 5 flower, with two sepals removed. (D) Side view of a stage 6 flower, with sepals removed. (E) Side view of a stage 7 flower. (F, G) Side and top views of stage 8. Arrowheads point to locules of stamens. (H) Early Stage 9. (I) Early stage 10. (J) Early stage 11. AB, abaxial sepal primordium; AD, adaxial sepal primordium; G, gynoecium; L, lateral sepal primordium; LS, long stamen; P, petal; SS, short stamen.

(see Chapters 3 and 12). Homeotic mutations affecting flowers were among the first recognized by geneticists. Familiar examples include the so-called double flowers of many commercial varieties of roses, camellias, carnations and other blooms, in which some or all of the stamens in a flower are replaced by petals. Genetic analysis of

homeotic mutants has been the key to building a model of the regulation of floral organ identity in *Arabidopsis*.

The following discussion examines the picture of flower development that emerges from work on *Arabidopsis* mutants in which the normal course of floral morphogenesis is disrupted. We go on to consider how the model for the specification of flower structures established in *Arabidopsis* applies, with appropriate adjustments, across the range of angiosperms. Finally we describe how floral symmetry, inflorescence architecture and flower color are determined.

16.3.1 Specification of floral structures in Arabidopsis is explained by the ABC model of gene expression

Analysis of *Arabidopsis* homeotic mutants that have abnormal identities of organs (sepals, petals, stamens, carpels) in their respective whorls supports a theory of flower development based on the interactions of a small number of regulatory genes. Some *Arabidopsis* mutants with mis-specified floral organs are shown in Figure 16.10. The flower of the **apetala2 (ap2)** mutant has stamens in the third whorl and carpels in the fourth whorl, like wild-type flowers. However, carpels replace sepals in the first whorl, and the second whorl may contain stamens or lack organs altogether. The flower of another mutant, **pistillata (pi)**, has sepals in the first and second whorls and carpels in the third and fourth. An **agamous (ag)** flower consists of perianth parts only, with

sepals in the first whorl, petals in the second and third whorls, and reiterations of this pattern in interior whorls. Observations of these and other homeotic mutants have been used to develop a model of floral organ determination that proposes three overlapping fields of gene activity, designated A, B and C, working in a combinatorial fashion to specify organ identity in each of the whorls of the flower primordium. The A field is conceived as covering whorls 1 and 2, the C field covers whorls 3 and 4, and B overlaps A and C in whorls 2 and 3, respectively. An important feature of the model is the antagonism between A and C, such that if expression of either is missing, the other will be expressed across the entire floral meristem (Figure 16.11). According to the model, in the wild-type flower production of a sepal in the normal position, whorl 1, requires class A gene expression. A combined with B specifies development of a petal in whorl 2. B and C together determine stamens in whorl 3, and class C gene expression alone establishes the identity of carpels in whorl 4 (Figure 16.11).

The ABC concept provides an explanation for the corresponding mutant phenotypes. The wild-type alleles of the genes shown in Figure 16.10 are classified according to the model thus: *AP2* is an A-type gene, *PI* is a B-type and *AG* is a C-type gene. The combination of functional *AP2*, *PI* and *AG* specifies a normal flower with sepals, petals, stamens and carpels occupying whorls 1 to

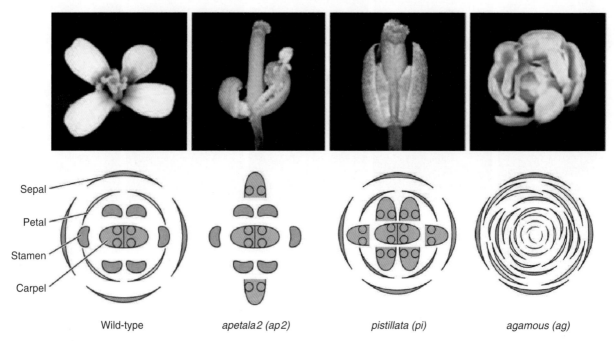

Sepal			
Petal			
Stamen			
Carpel			
Wild-type	apetala2 (ap2)	pistillata (pi)	agamous (ag)

Figure 16.10 Schematic diagrams and photographs of the phenotypes of floral organ identity mutants *apetala2* (*ap2*), *pistillata* (*pi*) and *agamous* (*ag*), in which the identities of organs in the whorls of the *Arabidopsis* flower are disturbed. In *ap2*, sepals and petals are absent, carpels replace petals in whorl 1 and the second whorl may contain stamens or, as in the example shown, may lack organs altogether. In *pi*, stamens and petals are absent, sepals replace petals in whorl 2 and carpels replace stamens in whorl 3. In *ag*, carpels and stamens are absent.

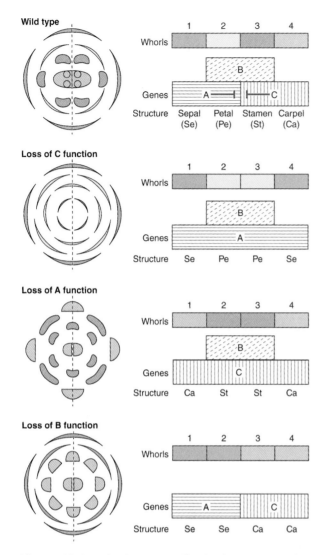

Figure 16.11 Floral organ specification by the ABC model. Field A acting alone specifies sepals; fields A+B act together to specify petals; fields B+C act together to specify stamens; and field C acting alone specifies carpels. Antagonism between the A and C fields is indicated by red T-ended lines. When one of these fields is missing, the other expands to cover all whorls. Floral diagrams and distribution of organs between whorls are shown for wild-type and loss-of-function mutants.

4 sequentially. Figure 16.11 shows the consequence of mutations in *AG* (loss of C function), in *AP2* (loss of A function) and in *PI* (loss of B function).

AG, *AP2* and *PI* are classified as homeotic, but unlike *KNOX* and many other homeotic genes, they do not encode homeodomain proteins (see Chapter 3). *AG* and *PI* are members of the MADS-box family of transcriptional regulators. Figure 16.12 is a comparison of plant MADS-box sequences, showing a high degree of conservation. MADS-box genes in the fern *Ceratopteris* are expressed in gametophytes and sporophytes and

show a high degree of homology with the MADS-box genes of the eudicots *Antirrhinum* and *Arabidopsis*. It is believed that MADS-box genes present in the last common ancestor of contemporary vascular plants underwent multiple duplications and diversifications and were recruited into novel developmental networks during the evolution of floral organs. *AP2* does not encode a MADS-box protein but belongs to the ERF (ethylene responsive factor) family of transcription regulators.

16.3.2 The original ABC model has been enhanced to include E class factors

Since the ABC model of floral organ identity determination was originally proposed, further studies of homeotic mutants have led to its elaboration to include additional morphogenetic fields, designated D and E. Class D genes, which act in carpels to confer ovule identity, will be further discussed in Section 16.4.4. Here we consider *Arabidopsis SEPALLATA* (*SEP*), a set of four MADS-box genes required to specify petals, stamens and carpels. The pattern of expression of *sep* mutations has led to their corresponding wild-type loci being assigned to class E in the extended flower development model. Flowers of *sep1-sep2-sep3* triple mutants consist of sepals only. A fourth *SEP* gene, *SEP4*, works with *SEP1*, *SEP2* and *SEP3* to specify sepal identity, and also contributes to the development of petals, stamens and carpels. In quadruple mutants (*sep1-sep2-sep3-sep4*) all floral organs are converted into leaf-like structures. Based on such observations the ABCE model has been proposed. In this model A + E function is needed for sepals, A + B + E function for petals, B + C + E function for stamens, and C + E function for carpels.

Genes of the ABCE model are expressed in the floral apex at characteristic locations that change as the flower primordium differentiates. Figure 16.13 shows how expression patterns become modified between early and later developmental stages (stages 3 and 6; see Table 16.2 and Figure 16.9). At stage 3, only sepal initials can be identified. Primordia of all four organ types are present at stage 6. *AP2* (class A) is expressed in all four whorls of the flower. *AP1* (also class A) is not detectable in the vegetative apex and is expressed specifically in the outer two whorls of the flower primordium (Figure 16.13). From an early stage, expression of the class B gene *PI* is strongly associated with cells destined to form petals and stamens. *AG* (class C) is expressed in the inner two floral whorls. Of the *SEP* genes (class E), *SEP1* and *SEP2* are expressed throughout the flower, *SEP3* is localized to the

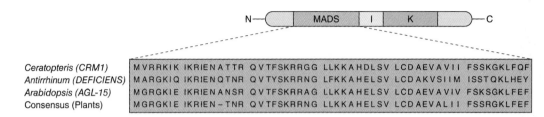

Figure 16.12   General structure of a MADS-box protein. The I and K domains and a region at the C terminus are sites at which MADS factors bind to each other in complexes necessary for transcriptional regulator function. Comparisons of amino acid sequences for selected MADS domains, from *Ceratopteris* (a fern), *Arabidopsis* and *Antirrhinum*, and the consensus sequence for plant MADS-box genes are shown.

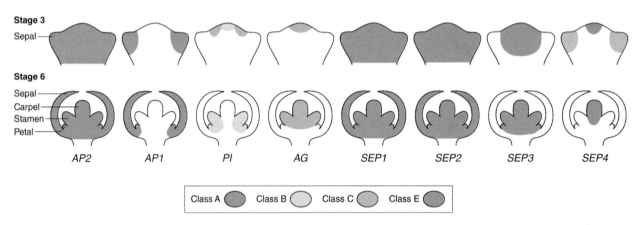

Figure 16.13   Patterns of expression of class A (*AP1, AP2*), B (*PI*), C (*AG*) and E (*SEP1–4*) genes in meristems at two stages of floral development (see Table 16.2).

inner three whorls and *SEP4* is prominent in the fourth whorl and detectable at low levels in sepal primordia at stage 3.

The proteins that are encoded by floral identity MADS-box genes interact at their I, K and C-terminal domains (Figure 16.12) to form different tetramers which bind with so-called CArG DNA sequences in the regulatory regions of target genes. The name of these sequences comes from CC(A/T)rich$_6$GG, meaning that each consists of two C nucleotides, followed by a region of six nucleotides which are all, or mostly, As and Ts, and finally two Gs. Figure 16.14 relates each organ type to the composition of the corresponding MADS-box tetramer that specifies it. Experimental studies of protein–protein and protein–DNA binding provide direct support for the 'floral quartet' mechanism underlying the ABCE model.

There is a certain degree of overlap between genes in the floral induction network (see Figure 16.6) and regulators of flower organ identity. For example, *AP1* functions as a class A homeotic gene. Interaction with FT-FD and LFY activates *AP1*, whereas A–C antagonism integral to the ABC model means that AG is a suppressor of *AP1* expression. The class E gene *SEP3* is directly responsive to FT-FD.

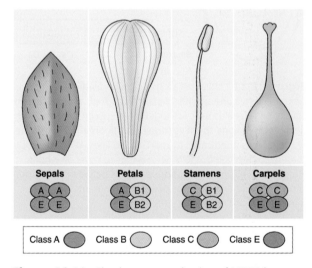

Figure 16.14   Floral quartet mechanism of MADS-box factor interactions specifying flower identity. Sepals and carpels are determined by a complex of class E homodimers with homodimers of class A (sepals) or class C (carpels) factors. Class B function is dependent on heterodimer formation, for example between AP3 (= B1) and PI (= B2). Petals and stamens are specified by the class B heterodimer complexed with class E and class A (petals) or class C (stamens) monomers.

16.3.3 The ABCE model, or modifications of it, applies to floral differentiation across the range of angiosperms

The ABCE concept was largely based on genetic studies in *Arabidopsis*, but work using other species contributed significantly to aspects of the model and to its establishment as the general scheme that can be applied to floral development across the range of angiosperms. One such species is snapdragon, *Antirrhinum majus*, the mature flowers of which are morphologically very different from those of *Arabidopsis* (Figure 16.15). *Antirrhinum* flowers are **bilaterally symmetrical** (**zygomorphic**) whereas those of *Arabidopsis* are **radially symmetrical** (**actinomorphic**). The four petals of the *Arabidopsis* flower are identical and each is bilaterally symmetrical, whereas the *Antirrhinum* corolla consists of a single, bilaterally symmetrical dorsal (adaxial) petal, two asymmetrical lateral petals and two asymmetrical ventral (abaxial) petals which are fused at their bases to form a corolla tube (Figure 16.15). Note that the terms adaxial and dorsal are often used interchangeably in the literature on the orientation of plant organs, as are abaxial and ventral. Species with zygomorphic flowers are thought to have evolved from actinomorphic ancestors on several independent occasions during angiosperm diversification. Coevolution with specialized pollinators gave rise to bilateral floral symmetry, which facilitates efficient and specific pollen exchange. Many hypotheses seek to explain this trend. For example, it is suggested that floral characteristics such as the morphology and color of the corolla and the quantity and quality of the reward to the pollinator form phenotypic clusters that match perceptional or behavioral preferences in particular taxonomic groups of pollinators. The genetic regulation of floral symmetry is discussed in Section 16.3.5.

Despite the differences in floral structure, the organ-specifying genes and mutants on which the ABCE model is based are closely similar in *Antirrhinum* and *Arabidopsis* (Table 16.3, Figure 16.16). Indeed, some of the *Arabidopsis* genes for floral meristem identity were cloned using sequences first isolated from *Antirrhinum*. Comparative studies of ABCE-related DNA sequences has established that conservation of genes and

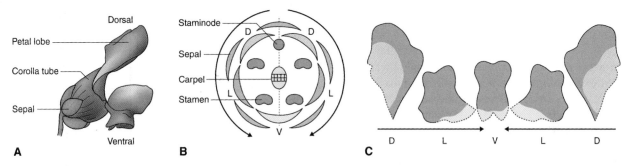

Figure 16.15 *Antirrhinum majus* floral anatomy. (A) The whole flower. (B) Floral diagram. The flower has a single plane of symmetry along the dorsoventral axis (dotted line). (C) Petal lobes. Each flower has five petals: a pair of individually asymmetrical dorsals (D), a pair of asymmetrical laterals (L) and one bilaterally symmetrical ventral (V)—each highlighted in different colors. The arrows running from dorsal to ventral in (B) indicate its relationship to the flattened representation shown in (C), where the corresponding arrows are straight. The staminode is a rudimentary sterile stamen.

Table 16.3 Orthologs of ABCE floral genes in *Arabidopsis, Antirrhinum* and *Zea*.

Species	Class A	Class B	Class C	Class E
Arabidopsis	*APETALA1 (AP1, AP2)*	*PISTILLATA (PI1), APETALA3 (AP3)*	*AGAMOUS (AG)*	*SEPALLATA (SEP1, SEP2, SEP3, SEP4)*
Antirrhinum majus	*SQUAMOSA (SQUA), LIPPLESS (LIP1, LIP2)*	*DEFICIENS (DEF), GLOBOSA (GLO)*	*FARINELLI (FAR)*	*DEFH49, DEFH200, DEFH72, AmSEP3b*
Zea mays	*ZAP1, GLOSSY15 (GL15)*	*SILKY1 (SI1), ZMM16, ZMM18, ZMM29*	*ZAG1, ZMM2*	*ZMM3, ZMM8, ZMM14, ZMM24, ZMM31, ZMM6, ZMM27*

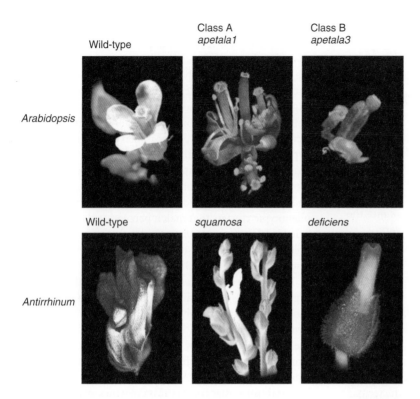

Wild-type   Class A *apetala1*   Class B *apetala3*

Arabidopsis

Wild-type   *squamosa*   *deficiens*

Antirrhinum

Figure 16.16 Floral morphologies of *Arabidopsis* and *Antirrhinum* mutants in orthologous class A (*apetala1, squamosa*) and class B (*apetala3, deficiens*) genes.

mechanisms extends across the taxonomic range, including the so-called basal angiosperms, which represent survivors of the earliest lineages of flowering plants. They include plants such as magnolia (order Magnoliales), avocado (*Persea americana*; Laurales), black pepper (*Piper nigrum*; Piperales) and water lily (any of several members of the family Nymphaeaceae).

Monocots are sometimes included among the basal angiosperms (see Figure 1.16). The principle of interacting morphogenetic fields applies to the specification of monocot floral organs, but the ABCE model has to be modified to account for some distinctive structural features in this group. Many non-grass monocot flowers, such as lily (*Lilium* spp.) and tulip (*Tulipa* spp.), have a variant of the four-whorl structure in which sepals and petals are replaced by petaloid organs called tepals, arranged in an outer and inner whorl of three each, surrounding six stamens and three fused carpels. The organ identity of tepals is explained by a model in which the B field extends to cover both outer whorls, so that A + B function specifies both outer and inner tepals, B + C specifies stamens and C alone specifies carpels (Figure 16.17).

Grasses are wind-pollinated and have correspondingly specialized flower structures in which sepals are highly modified and petals are absent (see Figure 6.6). A, B and C class genes have been isolated from a number of grass species (Table 16.3 lists orthologs from maize). Studies of homeotic mutants in maize and rice have confirmed that class B function is conserved in these species. The specification of carpels and the function of class C

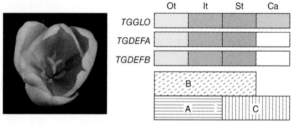

	Ot	It	St	Ca
TGGLO				
TGDEFA				
TGDEFB				

Figure 16.17 Specification of floral organs in *Tulipa gesneriana*. *TGGLO*, *TGDEFA* and *TGDEFB* are orthologs of the *Antirrhinum* class B genes *GLO*, *DEFA* and *DEFB*. The model is consistent with the floral quartet concept illustrated in Figure 16.14. *DEF*- and *GLO*-type genes must form a heterodimer to have class B function. The outer tepals (Ot) and inner tepals (It) require expression of class A+B genes, stamens (St) B+C genes and carpels (Ca) C genes only. Note the lack of B function in carpels, where there is expression of *GLO* but not *DEFA* or *DEFB*.

MADS-box genes are less clear. Although class E orthologs have been identified in many monocots (at least seven from maize for example; Table 16.3), their function is not yet confirmed.

While most angiosperms have flowers with both stamens and carpels, a number of species have separate staminate (male) and pistillate (female) flowers. Maize is an example of a **monoecious** species bearing separate male and female flowers on the same plant. In **dioecious** species such as asparagus and holly (*Ilex* spp.), male and female flowers are carried on different plants. The sex of the flower in monoecious and dioecious plants is

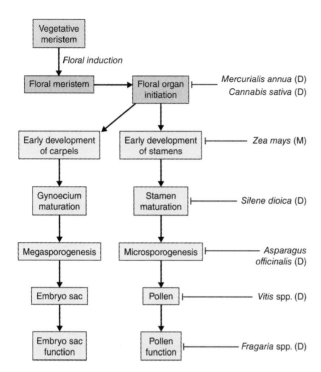

Figure 16.18   Steps in floral development blocked during the sex determination process for representative monoecious (M) and dioecious (D) plants. Note that *Fragaria* spp. form a polyploid series with 2n = 14, 28, 42 and 56. All diploid species are hermaphrodites; wild polyploid species are dioecious.

determined by mechanisms in which initiation, specification or differentiation of stamens or carpels is blocked at a point that varies from species to species (Figure 16.18). Maize flower development and the role of cell death in sex determination are described in Chapter 18.

16.3.4 Floral symmetry is determined by the interplay between TCP- and MYB-type transcription factors

As we have seen, the flowers of *Antirrhinum* are zygomorphic. Studies of *Antirrhinum* mutants in which the zygomorphic symmetry is disturbed have identified four key genes regulating dorsal/ventral development. *CYCLOIDEA* (*CYC*) and *DICHOTOMA* (**DICH**) promote dorsal identity and encode proteins belonging to the TCP family. TCP factors, which are exclusive to flowering plants, are named after the genes *TB1* (from maize), *CYC* (from *Antirrhinum*) and *PCF* (from rice), and are related to basic helix-loop-helix-type DNA-binding proteins. CYC and DICH interact with two members of the MYB gene family, *RADIALIS*

(**RAD**), and *DIVARICATA* (**DIV**). Expression of the two TCP proteins in dorsal regions of the floral meristem leads to localized activation of *RAD* transcription. *RAD* and *DIV* act antagonistically so activation of *RAD* on the ventral side of the meristem leads to inactivation of *DIV* there. Figure 16.19 interprets the dorsoventral and lateral patterning of the *Antirrhinum* flower in terms of interplay between the pairs of *TCP* and *MYB* transcription factors.

The decisive role of *CYC* homologs in the development of zygomorphy has been demonstrated for several species. In *Lotus japonicus* and *Pisum sativum*, for example, disruption of wild-type patterns of dorsoventral

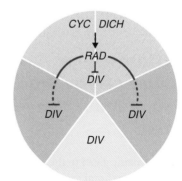

Figure 16.19   Model for interactions between genes controlling dorsoventral asymmetry in *Antirrhinum*. Expression of *CYC* and *DICH* in the dorsal domain (upper two sections) leads to activation of *RAD* (solid arrow), which in turn inhibits *DIV* activity in both dorsal regions (solid line) and in lateral regions (dotted line).

symmetry by ectopic (abnormally positioned) or reduced expression of *CYC*-like genes results in dorsalized or ventralized flower phenotypes, respectively. Evidence for involvement of *CYC* in the evolutionary transition from radially to bilaterally symmetrical flowers is seen in the zygomorphic candytuft (*Iberis*). In this genus, flowers have two dorsal petals that are smaller than the two ventral petals. In wild-type *Iberis* expression of a *CYC*-like gene in later stages of dorsal petal development is associated with a reduction in their size relative to that of ventral petals. This is in contrast to *Antirrhinum*, where the dorsal petals are larger than the ventral. In radially symmetrical mutants of *Iberis*, this pattern of *CYC* expression is absent. Expressing *Iberis CYC*-like genes in its actinomorphic close relative *Arabidopsis* has the effect of reducing petal size.

In addition to their role in floral symmetry, *TCP* genes also have a role in floral organ abortion. For example, the stamen in the adaxial position of the *Antirrhinum* flower is sterile (a staminode; see Figure 16.15B). *CYC* expression is necessary for dorsal stamen abortion. The desert ghost flower (*Mohavea confertiflora*), a relative of *Antirrhinum*, has both adaxial and lateral staminodes, and this correlates with expansion of *CYC*-like gene expression into the lateral domain. Genes of the *TCP* family have undergone extensive duplication, structural divergence and functional diversification during plant evolution and have been co-opted on multiple occasions into a variety of morphogenetic programs—and not only for floral organ patterning. *TCP* genes are known to be transiently expressed in a range of different developing tissues, including flower and shoot meristems and leaf and floral organ primordia. It is becoming clear that they are active in determining organ shape, size, number and curvature. For example the *Antirrhinum* gene *CINCINNATA* (*CIN*) encodes a TCP-type transcription factor responsible for regulating lobe formation both in petals and in leaves.

16.3.5 Inflorescence architecture can be modeled using veg, a meristem identity parameter

Inflorescence is the name given to the arrangement of flowers on the peduncle (floral axis). The terminology applied to the different forms of inflorescence architecture gives us some of the most beautiful words in the language (corymb, umbel, spadix, glomerule, thyrse are examples) but can be extremely complex and confusing. We can, however, group the diversity of forms into a small number of basic classes (Figure 16.20). The

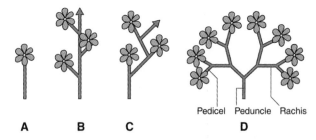

Figure 16.20   Four basic inflorescence architectures. (A) Single flower. (B) Raceme with sympodial growth in which flowers develop from axillary meristems. (C) Cyme with monopodial growth in which flowers develop from apical meristems. (D) Panicle with a branched series of axes with terminal flowers.

single flower is usually terminal. The **raceme** consists of a **monopodial** (that is, continuous) main stem bearing lateral flowers. The inflorescence of *Arabidopsis* is a raceme. The main stem of the **cyme** terminates in a flower, and growth is **sympodial**, that is, it continues from lateral shoots that also end in flowers, resulting in a flowering axis built of separate shoots. This gives the cyme a zigzag shape, though this is frequently evened out during development so that it comes to resemble the straight stem of the raceme. The **panicle** is a branching series of axes, each terminating in a flower. The stem within the inflorescence that bears the flowers or more branches within the inflorescence is called the **rachis** and the stalk of each single flower is the **pedicel**.

The inflorescence in species of the aster family (Asteraceae) is a highly modified raceme. Rachis and pedicels are so reduced that the minute individual flowers (florets) are compressed into a capitulum, a composite head resembling a single bloom. The individual 'flower' of *Helianthus* (sunflower), a typical example of a member of the Asteraceae, is in fact an entire inflorescence (Figure 16.21). Florets, each of which consists of perianth, stamens and carpels, are arranged on a cushion-like **receptacle** and are surrounded by a whorl of bracts. The inflorescence may, as in *Helianthus*, be differentiated into a central disk consisting of bisexual florets and the outer petal-like ray, in which the florets are usually female. In accordance with the fractal principle of plant architecture (see Chapter 12), groups of individual capitula are themselves clustered into second-order 'inflorescences of inflorescences' in many species. Since the Asteraceae is the most numerous of all plant families, the capitulum is evidently a highly successful floral configuration for promoting sexual propagation.

A scheme that accounts for different inflorescence architectures, called the **transient model**, has been used to simulate the development of racemes, cymes and panicles. It defines meristem identities in terms of

Figure 16.21 Inflorescence of sunflower (*Helianthus annuus*; Asteraceae), showing ray and disk florets.

'**vegetativeness**' (***veg***) a quantifiable property that is influenced by factors such as plant age or the state of regulatory gene networks. Above a certain threshold value of *veg*, the meristem produces a vegetative **phytomer** consisting of an internode and lateral meristem subtended by a leaf. Below the threshold the

meristem becomes floral. Below the vegetative/floral threshold, there is a second *veg* threshold that determines whether the floral meristem is determinate or indeterminate. If *veg* falls below the second threshold, the meristem is determinate, above it the meristem is indeterminate. After a period of branching, a simultaneous uniform decrease in *veg* in all meristems renders them determinate, resulting in the formation of a panicle. If *veg* reaches the threshold for floral determination in lateral meristems before apical meristems, a raceme is generated, whereas the reverse results in a cyme.

Inflorescence development has been simulated according to the transient model using two *Arabidopsis* architectural genes as determinants of *veg*. **LFY**, a floral identity gene that interacts directly with the class A gene *AP1* (see Figure 16.6), acts to reduce *veg*. The effect of the gene *TERMINALFLOWER 1* (**TFL1**) is to increase *veg*; as a result, *tfl1* mutants produce inflorescences with short axes that terminate in flowers. A meristem with high expression of *TFL1* and low *LFY* has a high *veg* score and is said to be in state A. High *LFY* and low *TFL1* (low *veg*) corresponds to state B. Figure 16.22A combines these interactions into a model in which *TFL1* and *LFY* are mutually antagonistic, are time-dependent and have opposite effects on *veg*. In state B, *TFL1* is suppressed and growth promotes the state B to A transition. Running a computer simulation based on this model, varying time and also values for *veg* through the effects of introducing mutations in *TFL1* and *LFY*, the branched raceme-type architecture of the wild-type *Arabidopsis* inflorescence is generated as well as mutant and

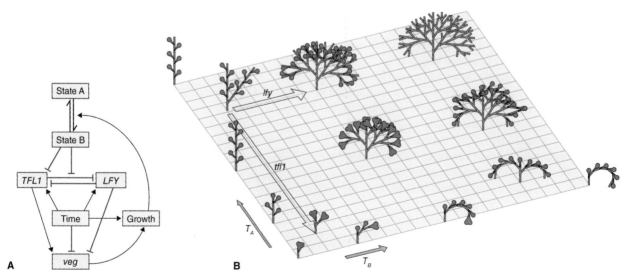

Figure 16.22 (A) Representation of the transient model of inflorescence architecture, showing interactions between growth and time, meristem state (state A and B) and genes (*TFL1*, *LFY*) determining vegetativeness (*veg*). (B) Different inflorescence phenotypes generated in a computer simulation based on the transient model. The T$_A$ and T$_B$ axes represent times at which flowers begin to form. Open arrows point away from the wild-type architecture of *Arabidopsis* and indicate the effect of *tfl1* and *lfy* mutations under inductive conditions.

overexpression phenotypes (Figure 16.22B). By altering the values of the various model parameters, cyme- and panicle-like inflorescence architectures can be created.

Grass inflorescences have distinctive structures. Each floral unit is called a **spikelet**, contains up to 40 individual florets and may be determinate or indeterminate depending on species. In rice, each spikelet meristem produces a single fertile floret and spikelets are arranged into panicles. Both male and female flowers of maize, the other grass species in which genetic regulation of floral architecture has been researched, have two florets per spikelet, only one of which is fertile. The functions of the *Arabidopsis LFY* gene are conserved in its maize homologs *ZFL1* and *ZFL2*. There is a rice homolog, *RFL*, but its main roles seem to be in panicle branching and the control of tillering (branch growth). Homologs of *CLAVATA* (*CLV*), a gene that regulates patterning of vegetative meristem growth in *Arabidopsis* (see Chapter 12), play a distinctive part in the architecture of rice and maize inflorescence development by controlling the number of branches produced by the apical meristem of the inflorescence. Functional studies and modeling of factors determining inflorescence structure in monocots and dicots are revealing both common and phylogenetically-specific regulatory features.

16.3.6 Colors of flower parts are due to betalain, anthocyanin or carotenoid pigments

Before the evolution of the angiosperms the color world of vegetation would have consisted largely of greens, yellows and browns, much as it does in modern conifer-dominated forest ecosystems. A rapid increase in the profusion of natural pigmentation accompanied the explosion in angiosperm evolution during the Cretaceous and was likely to have been an adaptive response to a combination of abiotic and biotic factors. For example, coevolution of angiosperms with insects and other animals was an important factor in the development of coloration in floral parts and in propagating structures such as seeds and fruits. Pigments in plant organs act as signaling molecules, advertising to pollinators, dispersers, or herbivores and other predators. Flower pigments are diverse in chemical structure, cellular localization and genetic regulation. Three major groups of pigments—betalains, carotenoids and anthocyanins—are responsible for the attractive colors of blooms (Figure 16.23). **Carotenoids** are tetraterpene products of the isoprenoid biosynthetic

pathway described in Chapter 15 (see Figure 15.36). **Anthocyanins** are synthesized as part of the flavonoid pathway, also discussed in Chapter 15 (see Figure 15.13). **Betalains** are red, violet, yellow or orange water-soluble derivatives of the aromatic amino acid tyrosine, and the early stages of their biosynthesis are shared with those leading to alkaloids such as morphine (Figure 16.24; see also Figure 15.18). They are restricted to some, but not all, families within the order Caryophyllales. Curiously, betalains are absent from species that accumulate anthocyanins and vice versa, but why this should be is unclear.

Floral betalains are glycosylated at the level of both cyclo-DOPA (cyclo-dihydroxyphenylalanine) and betanidin (Figure 16.24) and are located in vacuoles of petal cells. Anthocyanin pigments are also water-soluble glycosides stored in the vacuole. The glycosylation enzyme of maize, anthocyanidin 3-*O*-glucosyltransferase, is encoded by the *Bronze1* (*Bz1*) gene (see Figure 3.6). The carotenoids of colored flowers, like those of fruits, are hydrophobic and concentrated in chromoplasts, often in the form of globules, crystals or fibrils (see Chapter 18). The interaction between carotenoid-based flower coloration and the visual system of animals is an example of extreme convergence in evolution. The earliest prokaryotic photoautotrophs harvested light energy through carotenoid–protein receptor structures and metabolic systems. These show remarkable conservation of gene and protein sequence through to both the carotenoid structures and functions of angiosperm organs and the visual photoreceptors of the animals that interact with them. In contrast with the extremely ancient origins and functional organization

Key points The symmetrical organization of flowers is controlled by interactions between genes in two families of transcription factor, the *TCP* family and *MYB*-type factors. TCP gene products also play a part in the developmental abortion of floral organs and in other aspects of plant development. The arrangement of flowers on the peduncle (floral axis) is called the inflorescence. Plants exhibit a very wide range of inflorescence architectures. Models of inflorescence development and structure are based on *veg*, a quantifiable parameter that specifies the state of the meristem and determines whether it will remain vegetative or become an indeterminate or determinate floral organ. Many floral parts are brightly colored and act as signals to potential pollinating and seed-dispersing insects, birds and other animals. Depending on species, this color is due to pigments in one or more of three major groups: betalains, anthocyanins and carotenoids.

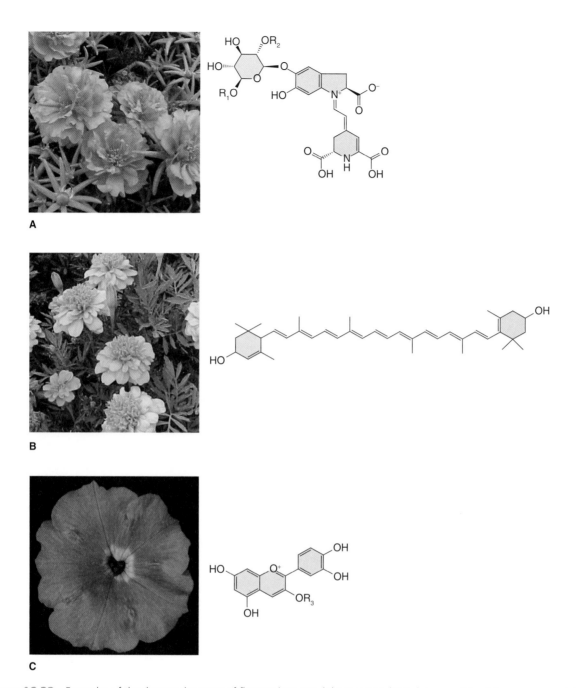

Figure 16.23 Examples of the three major types of flower pigments. (A) Moss-rose (*Portulaca grandiflora*) flowers accumulate primarily the betalain pigment betanin (R1 = R2 = H). (B) The carotenoid pigment lutein gives the marigold (*Tagetes patula*) flower its yellow-orange color. (C) *Petunia hybrida* flower pigment is an anthocyanidin, cyanidin.

of carotenoids, the vacuolar pigments of flowers, fruits and leaves are relatively late innovations of land plants, and their diversification was a key factor in angiosperm radiation.

A number of genes determining the patterns and shades of flower colors have been identified in *Antirrhinum*. The intensity and regional distribution of the magenta-colored anthocyanin of *Antirrhinum* petals is under the control of *DELILA* (**DEL**), *ROSEA* (**ROS**),

VENOSA (**VEN**) and *MUTABILIS* (**MUT**). Mutations in these regulatory genes do not abolish pigmentation but change the pattern of pigmentation within the flower. *DEL* determines corolla tube color. *MUT* affects pigmentation in the petal lobes. *ROS* regulates the pattern and intensity of color in both lobes and tube. *VEN* controls stripiness by specifying pigmentation of the epidermis overlying the veins in both lobes and tube. *DEL* encodes a basic helix-loop-helix (bHLH) protein

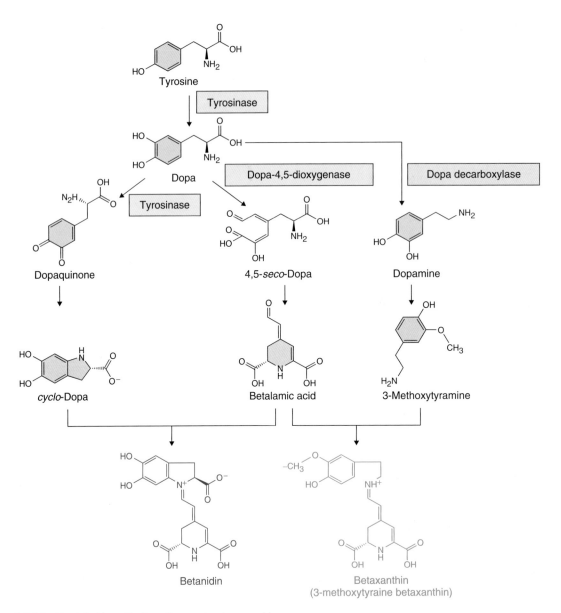

Figure 16.24 Betalain biosynthetic pathway, structures and known enzymes.

and *ROS* and *VEN* encode MYBs. Each of the transcription factors activates expression of genes for anthocyanin biosynthesis with its own characteristic specificity. The pigmentation of different parts of the *Antirrhinum* flower is determined by interactions between pairs of genes: *ROS1*, *ROS2* or *VE* with *MUT* or *DEL*, as shown in Figure 16.25. The range of anthocyanin pigmentation pattern across different *Antirrhinum* species can be attributed largely to variations in the activity of the *ROS* and *VEN* loci. This kind of region-specific activity in transcriptional regulators of pigment biosynthesis is one way of coloring flowers in sectors and stripes. Another common mechanism is transposon insertion and excision, as described in Chapter 3 for *Petunia* (see Figure 3.6).

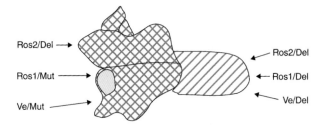

Figure 16.25 Summary of the interactions of genes encoding regulatory proteins in the different regions of the *Antirrhinum* flower. Red diagonal stripes sloping right indicate the area of *MUT* expression; red diagonal stripes sloping left indicate the area of *DEL* expression. *ROS1* and *VE* can interact with both *MUT* and *DEL* in the lobes, whereas *ROS2* can interact only with *DEL*.

16.4 Development of the male and female gametophytes

In order for the plant to complete its life cycle, it must form gametes: sperm and egg cells (Figure 16.26). In flowering plants, the male gametophyte—the structure that produces a male gamete—is the pollen grain, which is produced in the anther. The female gametophyte is the **embryo sac** produced within an **ovule** in the ovary of a carpel. When pollen is released from an anther, it is transferred to the stigma of the carpel by a process called **pollination**. As will be described in Section 16.5, when a pollen grain makes contact with stigmatic tissue, it extrudes a pollen tube which grows through the style until it reaches the embryo sac, where it releases its two sperm cells to carry out **fertilization**. At the molecular level, more is known about pollen grain formation than about the development of the embryo sac. This is, firstly, because a flower produces many more pollen grains than embryo sacs; for example a *Zea mays* plant can produce more than 25 million pollen grains, but only a few hundred embryo sacs, so that a typical maize plant bears one to two cobs with 200–400 kernels each. Secondly, because they are released from anthers, pollen grains are usually easy to isolate, whereas an embryo sac must be painstakingly dissected out from the ovule tissue that surrounds it in order to study the genes which are expressed exclusively in female gametes. In this and the subsequent section, we will describe the formation of male and female gametophytes of typical flowering plants.

16.4.1 The male gametophyte is the pollen grain, which forms in the anther

At both the cytological and the molecular level, the development of the plant's male reproductive tissue is highly conserved among all angiosperms. The sequence of events leading to the production of a male gametophyte, the pollen grain in the anther of the flower, is illustrated in Figure 16.27. The first stage in this process is the division of a diploid cell to give a **tapetal initial cell** and a **microspore mother cell**. These cells play very different roles in the production of pollen grains.

The microspore mother cell divides by meiosis (see Chapter 11) to produce four haploid cells, called **microspores**, which are initially surrounded by a wall that consists of the polysaccharide callose, $\beta(1 \rightarrow 3)$-linked glucan. This group of four cells, known as a tetrad, is released from the wall by the enzyme callase which is secreted by the tapetum. Once released from the callose

wall, each microspore undergoes an asymmetrical mitotic division to produce a pollen grain that consists of a large **vegetative cell** and a small **generative cell**. Subsequently, the generative cell will divide to yield two sperm cells. The timing of this division varies between plant species. In *Arabidopsis*, for example, it occurs while the pollen is still on the anther, but in the majority of angiosperms the pollen grain is shed from the anther while it is still at the two-cell (vegetative and generative) stage, and division of the generative cell takes place only after pollination when pollen tube growth begins (see Section 16.5). In either case, the final product is a pollen grain consisting of two small sperm cells surrounded by the cytoplasm of the vegetative cell (Figure 16.27). Nutrient reserves, proteins and transcripts required for pollen tube growth are accumulated in the vegetative cell during pollen maturation, so that once contact is made between pollen and the stigmatic surface, the pollen tube can grow rapidly.

The tapetal initial cells undergo mitotic divisions to form the tapetum, a tissue that lines the locule of the anther. The tapetum has two important functions. First, it produces the callase that is required to release the microspores from the wall of callose that surrounds them. Second, it manufactures many of the compounds, including structural polymers and pigments, that make up the outer wall of the pollen grain. Late in pollen development, the cells of the tapetum disintegrate and their cytoplasmic contents are deposited on the pollen coat. As we shall see in Section 16.5, the composition of the pollen coat is important for interactions between the pollen and the female tissue.

16.4.2 Many genes are expressed in the anther and nowhere else in the plant

Certain genes are expressed only in anthers and not at any other stage in the plant's life cycle. In *Arabidopsis*, for example, there are over 700 of these anther-specific genes. Some of them are members of multigene families, such as those encoding the cytoskeletal proteins tubulin and actin, other members of which are expressed at different developmental stages. However, there are also genes that encode types of protein unique to pollen. Most pollen-specific genes are expressed only in the vegetative cell, not in the smaller generative cell that will give rise to the sperm. During the first, asymmetrical division of the microspore, the vegetative cell receives most of the cytoplasm and thus of the ribosomes and other components of the translational apparatus. The chromatin of the generative cell nucleus is highly condensed and seems to be largely transcriptionally inactive.

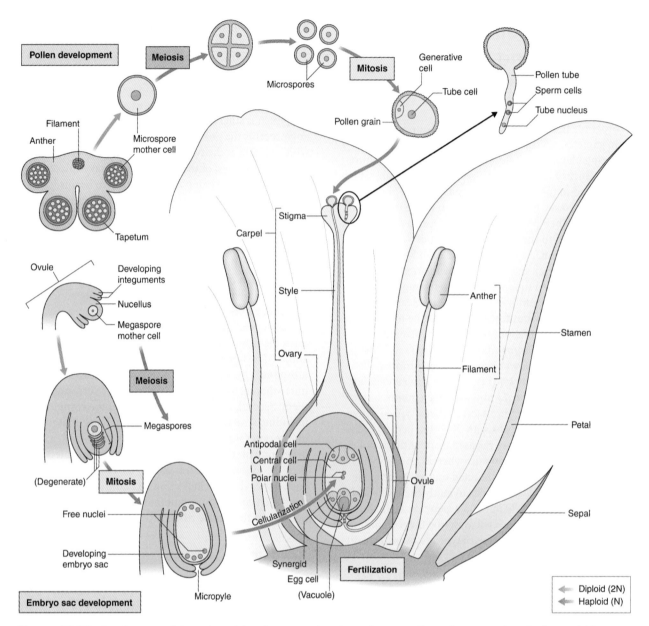

Figure 16.26   Development of the male and female gametophyte in angiosperms. The male gametophyte (pollen grain) is formed in the anther, and the female gametophyte (embryo sac) develops in the ovule.

The cell wall that surrounds a mature pollen grain is a complex structure. It has an inner, or **intine**, wall which completely encloses the grain, and an outer, or **exine**, wall, which has apertures at intervals. Following pollination, a pollen grain germinates on the surface of the stigma and the pollen tube emerges at one of the apertures in the exine wall, growing by tip growth and extending the intine cell wall (Figure 16.28). The intine wall differs from other plant cell walls in that its main component is callose, although it also contains cellulose and arabinans like other cell walls.

The exine wall of the pollen grain in many species shows an elaborate pattern of spines, ridges and other decorations and is almost indestructible. Because the patterns are often unique to particular species (Figure 16.29), they can be useful in identifying the plants associated with a particular location. This attribute has been applied both in paleobotany, to determine which plant species occupied a habitat in the distant past, and in forensic science, to identify crime scenes. The extraordinary durability of pollen resides in the main structural component of the exine wall,

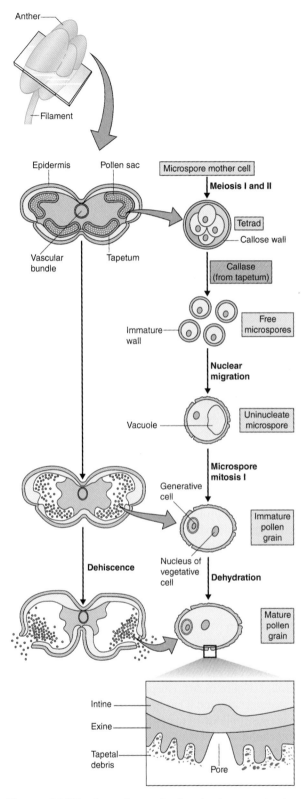

Figure 16.27 The pathway of pollen development that is typical of most angiosperms. Meiosis occurs in the anther and yields a tetrad of microspores. Free microspores are released after callase degrades the cell wall that has held the tetrad together. The first mitotic division is asymmetrical, forming a large vegetative cell and a small generative cell. Partially desiccated pollen is released when the anther dehisces.

sporopollenin, a highly decay-resistant phenolic polymer that allows the exine to retain its structure for millions of years and under harsh conditions.

The genetic control of exine patterns is interesting. If plants of two species that are very closely related but have different pollen grain morphologies are crossed, all the progeny plants have identical pollen grains, but their exine patterns can include characteristics from both parents (for example, the spine length from one parent and the spine density from the other). This suggests that different, and unlinked, genes control different elements of exine patterns, because the progeny plants inherit some from each parent. However, it also implies that pollen morphology is controlled not by the genetic makeup of the pollen grain itself (otherwise the pollen grains from a single parent plant would vary) but by the plant prior to the formation of the pollen grain. It is now known that the exine is made up of materials derived from three groups of cells: the microspore mother cells, the microspores themselves, and the tapetum. Genes that encode enzymes of the biosynthetic pathways for waxes and callose, as well as those needed for sporopollenin synthesis, have been shown to be important for normal development of the exine and for pollen fertility.

16.4.3 Mutations in genes that are active in the sporophyte can lead to male sterility

Several classes of mutations are known which lead to plants that are unable to act as male parents. Some male-sterile mutants have defects that affect an aspect of meiosis, and thus cannot form pollen. Others have abnormal tapetal cells, while still others appear to have normal tapetal tissues and pollen, but cannot release the pollen properly because of defects in anther structure. Although some genes underlying male sterility have been cloned, in most cases the functions of their products are poorly understood.

One class of male-sterile mutants deserves special mention because of their importance in plant breeding. They are the so-called **cytoplasmic male-sterile (CMS)** mutants. In such plants, which have been identified in a wide range of angiosperms including *Zea mays*, *Oryza sativa* and *Helianthus* and *Phaseolus* species, male sterility is transmitted through the female parent. In all the cases so far studied, it has been found that it is the mitochondria, which a plant inherits only from its female parent, that are responsible for the male sterility

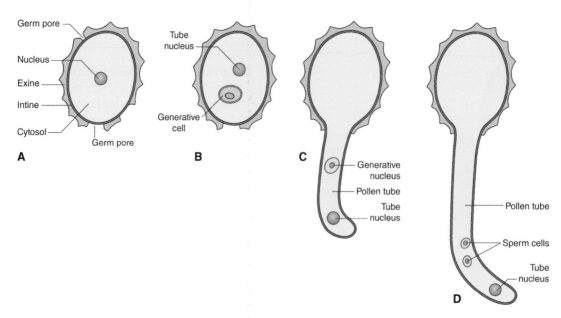

Figure 16.28 (A–D) When pollen germinates, a pollen tube emerges at one of the apertures in the exine surrounding the pollen grain. This occurs through tip growth, which extends the intine cell wall so that it continues to enclose the developing pollen tube.

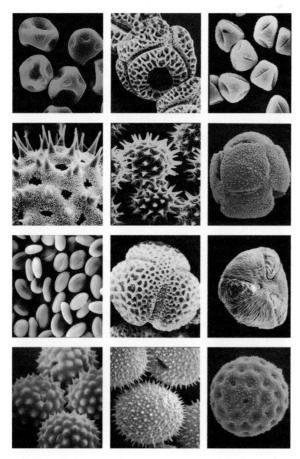

Figure 16.29 Pseudocolored scanning electron micrographs of pollen from diverse families, showing attractive sculpturing patterns in the exine wall. Left to right, top row: plantain, leucospermum, buttercup. Second row: iris, ox-eye daisy, geranium. Third row: nettle, ragweed, sycamore. Fourth row: water lily, hollyhock, love grass.

phenotype. In CMS plants, an abnormal protein is expressed in the mitochondria in all cells of the plant, but it is only during male gametophyte development that there is a phenotype, suggesting that mitochondria play a particularly important role in pollen formation. If the expression of the aberrant mitochondrially-encoded protein is reduced, the plant's fertility is restored. The nuclear genome in many plant species includes restorer genes, which can suppress expression of these toxic mitochondrial proteins.

Cytoplasmic male sterility in maize has been studied in detail. In one of the most intensively studied cases, the CMS-T line of maize has been shown to produce an abnormal mitochondrial protein called URF13 that causes complete male sterility. If, however, the CMS-T mutation is introduced into a genetic background in which the two restorer nuclear genes *Rf1* and *Rf2* are both expressed, male fertility is restored to normal. *Rf1* expression greatly reduces the amount of URF13 produced, but *Rf2* must also be expressed to confer fertility (Figure 16.30). The product of *Rf2* is an aldehyde dehydrogenase, and it has been suggested that URF13 may cause buildup of toxic aldehydes in the mitochondria of developing tapetal cells. A further example of cytoplasmic male sterility with a different molecular basis, the CMS-S system, will be described in Chapter 18. Plant breeders use cytoplasmic male sterility to produce F_1 hybrids in many crop plants: a CMS plant, which cannot self-fertilize, is used as the female parent and a plant with the nuclear restorer genes is used as the male. This restores fertility to the resulting F_1 progeny, which contain the desirable characteristics of both parent plants.

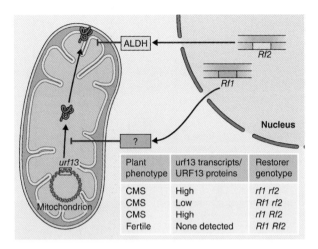

Plant phenotype	urf13 transcripts/ URF13 proteins	Restorer genotype
CMS	High	*rf1 rf2*
CMS	Low	*Rf1 rf2*
CMS	High	*rf1 Rf2*
Fertile	None detected	*Rf1 Rf2*

Figure 16.30 The cytoplasmic male sterility T (CMS-T) system in maize. A schematic illustration of the roles played by the mitochondrial-encoded toxic protein URF13 and the nuclear-encoded restorer proteins Rf1 and Rf2. Both Rf1 and Rf2 are required to restore fertility. Rf1 greatly reduces the accumulation of *urf13* transcripts, although how this occurs is not yet known. *Rf2* encodes an aldehyde dehydrogenase (ALDH) but has no effect on the abundance of *urf13* transcripts. The ALDH may remove a toxic compound produced by the URF13 protein.

16.4.4 The female gametophyte, or embryo sac, is produced by one meiotic division followed by several mitotic divisions

The female reproductive cells of seed plants are contained in the ovule, which consists of an outer layer (integument) enclosing the **nucellus** (megasporangium). Within the nucellus, meiotic division of the **megaspore mother cell** (MMC) gives rise to the female gametophyte (embryo sac), just as meiosis results in production of the male gametophyte in an anther (see Figure 16.26). The sequence of cell division events that give rise to the embryo sac varies between plant species, but more than 70% of flowering plants follow a pattern normally known as *Polygonum*-type development, named after the genus of dicots in which it was first identified. In these species, when an MMC divides by meiosis, three of the four haploid products die and the fourth, which is much larger than the other three, is the **megaspore** (Figure 16.31). The megaspore then undergoes three mitotic divisions, producing eight genetically identical daughter nuclei. Cytokinesis follows, producing a mature

Key points A plant must form gametes (sperm and egg cells) in order to complete its life cycle. The male gametophyte—the structure that produces a male gamete—is the pollen grain, which is formed in the anther. The female gametophyte, which is formed within an ovule in the ovary of a carpel, is the embryo sac. Pollination is the process in which pollen grains are transferred to the stigma of the carpel. When a pollen grain makes contact with stigmatic tissue, unless a self-incompatibility reaction prevents it, a pollen tube extruded from the pollen grain grows through the style until it reaches the embryo sac. The two sperm cells carried by the pollen grain are then able to fertilize the embryo sac. Hundreds of genes are expressed only in anthers and not at any other stage in the plant's life cycle. Gene expression in pollen occurs mainly in the vegetative cell, not in the generative cell that will give rise to the sperm. A mature pollen grain is surrounded by a complex, two-layer cell wall. The outer layer (the exine) is almost indestructible and in many species it has a complex pattern of spines, ridges and other decorations which can be used for identification purposes. There are several classes of gene which, if mutated, lead to plants that cannot act as male parents; that is, they are male sterile. Some have aberrant meiosis and cannot form pollen, while others have abnormal tapetum or anther structure. In cytoplasmic male sterility, the mitochondria inherited by the plant from the female parent produce an aberrant protein that results in failure to form normal pollen. Cytoplasmic male sterility is important in plant breeding because it is useful for hybrid seed production.

embryo sac which has seven cells: six uninucleate haploid cells (three antipodal cells, two synergids and the egg cell) and a central cell that contains two of the haploid nuclei. As will be discussed in Section 16.5.7, when the pollen tube delivers two sperm nuclei to the embryo sac, **double fertilization** occurs. The egg cell fuses with one sperm nucleus to form a **diploid zygote** which will give rise to an embryo progeny plant. The central cell fuses with the second sperm nucleus to produce the **triploid primary endosperm** cell which gives rise to the endosperm that surrounds the embryo in the seed and provides it with nutrients to support germination.

As with male gametophyte development, there are several genes that are expressed only during the steps that lead to the formation of the female gametophyte.

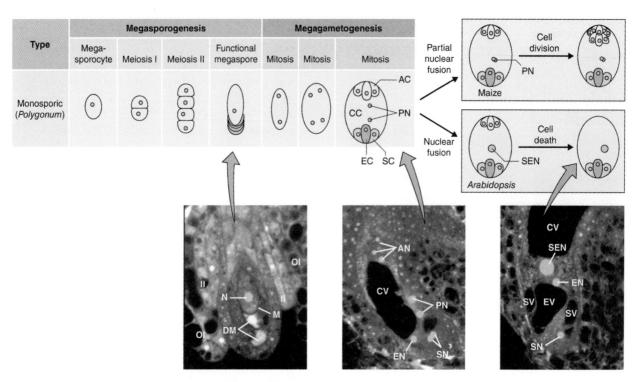

Figure 16.31 The *Polygonum* pattern of female gametophyte development. This pattern of cell division is typical of more than 70% of angiosperms. The process can be divided into two stages: megasporogenesis, during which meiosis occurs, and megagametogenesis, during which the surviving haploid megaspore divides mitotically to produce an embryo sac (female gametophyte). In *Arabidopsis* and some other species, the polar nuclei (PN) fuse to form a secondary endosperm nucleus (SEN). Elimination of antipodals in *Arabidopsis* is an example of programmed cell death (see Chapter 18). AC, antipodal cell; AN, antipodal cell nucleus; CC, central cell; CV, central cell vacuole; DM, degenerate megaspore; EC, egg cell; EN, egg nucleus; EV, egg vacuole; II, inner integument; M, megaspore; N, nucleus; OI, outer integument; SC, synergid cell; SN, synergid nucleus; SV, synergid vacuole.

Mutations in these genes cause aberrant development of the ovule and lead to female sterility. Some encode proteins that regulate the number of cell divisions, or the process of cell specialization in the embryo sac. Figure 16.32 shows the results of mutations in two such genes in *Arabidopsis*. In one mutant, development is arrested at the megaspore stage, while in the other the two polar nuclei do not fuse. Mutations in other genes can affect development of the ovule as a whole, causing ovules to be transformed into other structures and preventing normal embryo sac formation. For example, in *Arabidopsis*, four closely related genes of the MADS-box family of transcription factors, *AG*, *SEEDSTICK* (**STK**) and *SHATTERPROOF 1* and *2* (**SHP1** and **SHP2**), are required to confer the correct identity upon ovule tissues. In flowers of plants with the *stk-shp1-shp2* triple mutation, ovules are converted into organs resembling carpels or leaves. These ovule identity genes are sometimes included in an extension of the ABCE model of floral organ patterning (see Section 16.3.3), under the name of class D function genes. Similar phenotypes result from mutations in genes of the *SEP* family of genes, which encode another group of MADS-box transcription factors (Figure 16.33).

Key points The sequence of events giving rise to an embryo sac varies between plant species, but in more than 70% of flowering plants, after a megaspore mother cell has undergone division by meiosis, three of the four haploid products die. The fourth, which is the largest, is the megaspore, and it undergoes three mitotic divisions to produce eight genetically identical daughter nuclei. These develop into a mature embryo sac which consists of seven cells. Six are haploid cells with single nuclei (three antipodal cells, two synergids and the egg cell) while the seventh, the central cell, contains two of the haploid nuclei.

16.5 Pollination and fertilization

The goal of sexual reproduction is fertilization, the fusion of egg and sperm. Before fertilization can occur in

fem2: Never progresses beyond
megasporogenesis

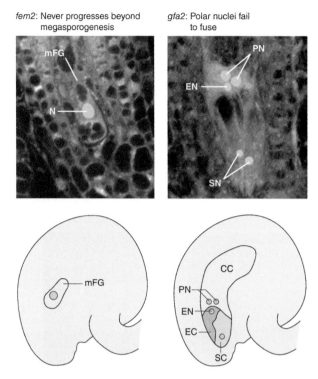

gfa2: Polar nuclei fail
to fuse

Figure 16.32 Micrographs of *Arabidopsis* mutants with defects in megagametogenesis. The *fem2* mutant stops development at the one-cell stage. In the *gfa* mutant the polar nuclei fail to fuse. mFG, mutant female gametophyte; other abbreviations as in Figure 16.31.

flowering plants, a pollen grain must make contact with a receptive stigma. It must then germinate producing a pollen tube, which grows through the tissues of the style towards the embryo sac. Fertilization to produce a zygote, the beginning of a new diploid sporophyte generation, can then occur. Each point in the sequence of events from pollen touchdown on the stigmatic surface to the initiation of embryogenesis requires that the appropriate cell–cell signaling interaction occurs on cue.

Following pollination the female partner must allow invasion of its tissues by the foreign cytoplasm of the male gametophyte; this step is mediated by highly specific molecular mechanisms for distinguishing self from non-self. Here we discuss the signals and regulatory networks that promote (and in some cases block) the union between male and female gametes which completes the cycle of alternation of generations.

16.5.1 Hydration and germination of the pollen grain require specific interaction between the pollen coat and stigmatic surface

During pollen maturation the developing grain becomes dehydrated. The pollen coat protects the grain from excess desiccation while it is in transit from the anther to the stigma; it also is involved in the early stages of pollen interaction with the stigma by contributing to pollen–stigma adhesion and facilitating rehydration. The outer layer of the pollen coat, which is generated by the tapetum during gametogenesis, is a lipid–protein matrix that represents the point of interaction between pollen and stigma. When a grain makes contact with the stigma, rehydration is initiated as a prelude to growth of the pollen tube and delivery of sperm to an embryo sac within an ovule (Figure 16.34A–C). Rehydration is a highly regulated step, at which self-pollen in many self-incompatible species is rejected and interspecific pollination is blocked (see Section 16.5.3). **Self-incompatible** species are those in which pollen cannot fertilize eggs on the same plant in which it was produced. Depending on plant species and the self-incompatibility mechanism it uses, successful production of an embryo may be blocked at any of

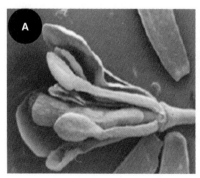

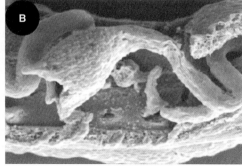

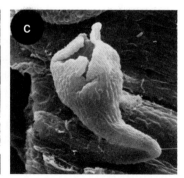

Figure 16.33 Phenotypes of *SEP1/sep1-sep2-sep3* and *stk-shp1-shp2* mutant plants. (A) A flower of a *SEP1/sep1-sep2-sep3* plant. Flower development is normal in this mutant. (B) Ovules of a *SEP1/sep1-sep2-sep3* mutant plant. Ovule development is affected severely; ovules are transformed into carpel- and leaf-like structures. (C) Ovules of a *stk-shp1-shp2* plant. Ovule development is affected severely. The phenotype is similar to that shown in (B).

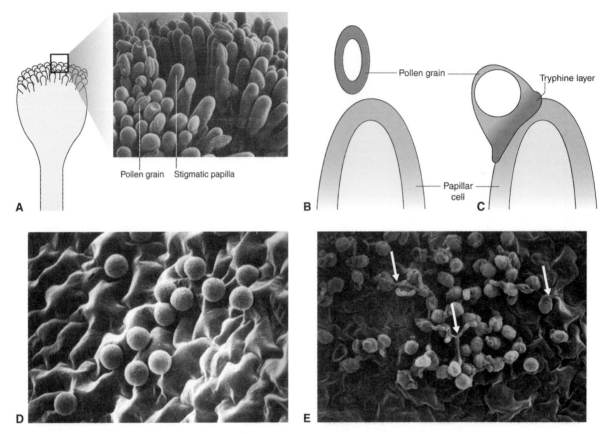

Figure 16.34   Interactions between pollen grains and the stigma in *Arabidopsis*. (A) Stigmatic surface and pollen grains. On making contact with the stigma, compatible pollen grains adhere, hydrate and grow a pollen tube. (B) Before contact, pollen grains have an elongated shape and are relatively desiccated. (C) When captured by the papillar cell, the outermost layer of the pollen grain (the tryphine layer in the case of brassicaceous species) is deposited on the papillar cell surface, an adhesion foot forms at the cell–cell interface and the grain rehydrates. (D) Pollen grains do not normally respond to contact with non-stigmatic tissue, such as the leaf epidermis of wild-type *Arabidopsis* shown here. (E) Pollen grains can rehydrate and germinate (arrows) on leaf surfaces of the *Arabidopsis fiddlehead* mutant.

several stages, including pollen germination, growth of the pollen tube, fertilization and development of the embryo.

Rehydration of pollen grains is facilitated by release of water from the stigmatic surface. The mechanism of this water release has not been fully determined, but a number of pollen coat and stigmatic components have been implicated. These include aquaporin-like genes, which are relatively highly expressed in stigmatic papillae. Secondary metabolites such as flavonols can also facilitate the initiation of pollen tube growth in some plants. The interaction between pollen and stigma is abnormal in certain mutants defective in lipid metabolism. Pollen grains of the *Arabidopsis* mutant *eceriferum* (*cer*), for example, which are depleted in pollen coat lipids, fail to rehydrate unless exposed to high humidity or when triacylglycerides are added to the stigma. The pollen coats of *Arabidopsis* and related members of the family Brassicaceae also contain lipases and glycine-rich lipid-binding proteins called oleosins.

Mutations in the gene that encodes the most abundant oleosin, GRP17, result in delayed pollen rehydration and reduce its ability to compete with wild-type pollen.

Further evidence that lipids are important in pollen grain germination comes from observations on *fiddlehead* (*fdh*), an *Arabidopsis* mutant with a modified epidermis that results in anomalous organ adhesion and fusion. Normally, pollen grains will not germinate on non-stigmatic tissues such as the epidermis of wild-type leaves (Figure 16.34D); but leaf surfaces of the *fdh* mutant can support pollen hydration and germination (Figure 16.34E). Plants with the *fdh* mutation retain the capacity for self- and non-self-recognition, however, since even the abnormal interaction with leaf epidermis occurs only with pollen from *Arabidopsis* or close relatives. The *fdh* mutant is deficient in β-ketoacyl-CoA synthase, an enzyme of long-chain fatty acid biosynthesis. As a consequence the high molecular mass lipids in *fdh* leaf cells differ from those in wild-type leaves, and epidermal cell permeability is altered. The range of genetic and biochemical studies of pollen rehydration in *Arabidopsis*, some of which we have discussed here, leads to the conclusion that plants normally restrict water flow across the cuticle in all tissues except the stigma, and that the stigmatic surface may be unique in its capacity interactively to change its water permeability on contact with pollen through alterations in long-chain fatty acid synthesis.

16.5.2 Pollen allergens, the cause of hay fever, have a range of functions in fertilization

An unwelcome consequence of the complexity of pollen coat structure is that exposure to pollen can stimulate the immune systems of sensitive individuals. The result is allergic rhinitis, the debilitating condition commonly known as hay fever, which reportedly affects almost 500 million people worldwide. Pollen allergens cross-link with antibodies in the mucous membranes of susceptible humans, leading to the release of histamines, leukotrienes and other inflammatory mediators and consequently to the development of hay fever symptoms including widening of the blood vessels, redness, swelling, increased mucus secretion, itching and sneezing.

Of course, allergenic pollen proteins exist not to make people sneeze but to facilitate pollen germination and growth. The route from stigma through the style to the embryo sac (the transmitting tract) is lined with an extracellular matrix consisting of cell wall carbohydrates, pectins and glycoproteins, with which pollen allergens interact. Pollen allergens are structurally and functionally diverse. They include proteins such as pectin-degrading enzymes, disease-resistance proteins, Ca^{2+}-binding proteins and unrelated structural proteins. Grass pollen is particularly allergenic: Table 16.4 lists the individual allergen classes in pollen from timothy grass (*Phleum pratense*), showing a typical range of sizes, immunological effects and protein functions. The major allergen, **Phl p 1**, is the *P. pratense* homolog of the *Zea mays* EXP B-1 allergen (see Figure 12.11). Allergens of this group are expressed at a low level prior to pollen mitosis and maximally in mature pollen. They function as β-expansins, relaxing the network of wall polysaccharides, thereby permitting turgor-driven cell enlargement (see Chapter 12). They are thought to facilitate invasion of the pollen tube into pistil tissues, by loosening the cell walls of the grass stigma and style.

Phl p 11 (Table 16.4) is a member of the OleI family, named for the major allergen of olive (*Olea europaea*) pollen, which is the commonest cause of hay fever in Mediterranean countries. Homologs have been described from *Arabidopsis*, tomato and maize. Allergens of this group accumulate from the early microspore stage onwards within the pollen wall and tapetum. The downregulation of *Phl p 11* results in a gametophytic lethal phenotype in which pollen matures correctly, but fails to rehydrate properly or to grow an effective tube.

Table 16.4 Immunological and biochemical characteristics of the *Phleum pratense* (timothy grass) pollen allergen groups.

Allergen	Allergenicity (% reactivity with IgE antibody)	Molecular mass (kDa)	Function
Phl p 1	95	35	β-Expansin
Phl p 2	50	11	
Phl p 3	50	11	
Phl p 4	70	50	
Phl p 5a	90	38	
Phl p 5b	90	32	RNase
Phl p 6	60	13	
Phl p 7	7	9	Ca^{2+}-binding protein
Phl p 11	40	20	Trypsin inhibitor
Phl p 12	10	14	Profilin
Phl p 13	60	60	Polygalacturonase

Comparison of protein structural motifs suggests that members of this group of allergens are related to inhibitors of trypsin-type proteases, although the relevance of this to their function in pollination is unclear. **Phl p 7** (Table 16.4) is a small, weakly allergenic Ca^{2+}-binding protein that is highly expressed in the cytosol of mature pollen, in the wall of the grain during hydration and on the surface and tip of the pollen tube. Such proteins are thought to function as calcium-sensitive signal molecules (see Section 16.5.6).

Phl p 12 is related to a family of small, highly conserved proteins called **profilins**. Profilin was first identified in the pollen of birch (*Betula*), the major cause of allergic responses in northern latitudes, where it is estimated to affect 15–20% of hay fever sufferers. The three-dimensional structure of the protein is shown in Figure 16.35. Profilins bind to the cytoskeleton component **actin**. Actin plays an important role in pollen tube growth, which suggests that profilin functions in pollination to support the process that delivers sperm cells to the embryo sac.

Phl p 13 (Table 16.4), a polygalacturonase, is one of the allergens that function in vivo in the enzymatic modification of pectins. In addition to polygalact-

uronases, pollen proteins with amino acid sequence similarity to pectate lyases have been identified as major allergens in a range of species, including tomato, common ragweed (*Ambrosia artemisiifolia*) and Japanese cedar (*Cryptomeria japonica*). Pectin is a major component of the middle lamella, which glues adjacent cells together (see Chapter 4). Pectin-modifying enzymes are employed by pathogenic microorganisms to soften plant tissues; it is likely that allergens with such activity play a role in the penetration of the style during pollen tube growth.

> **Key points** Pollination is followed by fertilization, the union of egg and sperm. On contact with the stigma, the pollen grain is rehydrated in a sequence of regulated events that includes expression of aquaporins and changes in lipid metabolism. Pollen allergens are a diverse group of proteins with a variety of functions in fertilization. They include cell wall modifiers, disease-resistance factors and Ca^{2+}-binding proteins. Profilin, the major allergen of birch pollen, interacts with the cytoskeleton during pollen tube growth and delivery of sperm cells to the embryo sac prior to fertilization.

16.5.3 Incompatibility mechanisms prevent self-pollination and promote outbreeding

Over 85% of angiosperms are hermaphrodites (that is, they are bisexual, bearing flowers each of which has both male and female reproductive organs). A further 5% or so are plants that have unisexual flowers and are monoecious, like maize (see Figure 16.3), where separate male and female flowers occur on the same plant. Therefore, for the great majority of plants, it is highly likely that pollen will land on a stigma from the same individual, a process called **self-pollination**. Self-pollination has the disadvantage of promoting inbreeding, which may decrease the fitness of the progeny. Several mechanisms have evolved to minimize or prevent fertilization of the egg cell by self-pollen. For example, there may be asynchrony between maturation of the stamens and pistils within a given hermaphroditic flower, or between the male and female flowers in monoecious individuals. Pollen produced either too

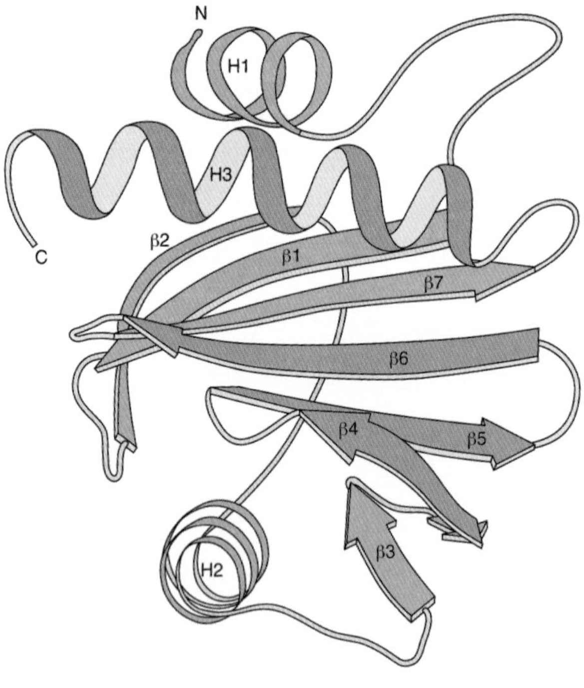

Figure 16.35 Ribbon representation of the X-ray crystallography structure of profilin from birch pollen, showing α-helices (H) and β-sheets (β). The protein has a core comprising a six-stranded antiparallel β-sheet, and an actin-binding domain composed predominantly of the C-terminal α-helix (H3) and the bottom three strands of the central sheet (β4, β5 and β6).

early or too late in relation to when the female tissue is receptive to pollen tube growth will fail to achieve fertilization.

Here we discuss genetic strategies for promoting outcrossing. Such **self-incompatibility** mechanisms ensure that pollen of an individual plant fails on the stigmas of the same plant, but pollen from other plants of the same species is successful. Self-incompatibility has arisen independently several times in evolution, and the molecular basis differs between different plant families. Self-incompatibility is usually controlled by an **S locus**, which consists of multiple genes (**determinants**) that are expressed either in the pollen grain (male) or in the pistil (female). Both the male and female determinants have many alleles and are inherited as a single segregating unit. The variants of this gene complex are called **S haplotypes**. Pollen is recognized as **self** (incompatible) or **non-self** (compatible) according to the proteins encoded by the different alleles of determinant genes. An incompatible response occurs if the partners carry determinants from the same S haplotype; but if the male and female carry male and female determinants from different S haplotypes, pollination and fertilization are permitted. There are two major types of self-incompatibility mechanisms, gametophytic and sporophytic, distinguished by the relationship between pollen and pistil tissue (Table 16.5). **Gametophytic self-incompatibility** (GSI) is determined by the haploid pollen genotype at the S locus. In contrast, **sporophytic self-incompatibility** (SSI) is determined by the diploid genotype of the male parent at the S locus. In a GSI interaction an incompatible pollen tube often initiates growth through the style before it arrests, while in SSI, the pollen grain fails to germinate (Figure 16.36).

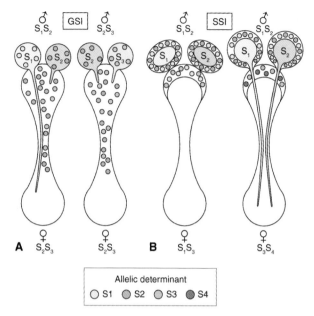

Figure 16.36 Self-incompatibility mechanisms. (A) Gametophytic self-incompatibility (GSI). Pollen is successful only when the haploid genotype of the pollen does not match the genotype of the female at the S locus. Both the S_2 and S_3 pollen tubes are arrested after some growth through an S_2S_3 style. Only the S_1 pollen tube can continue to grow through the style. (B) Sporophytic self-incompatibility (SSI). Pollen success depends on determinants donated by its parent sporophyte. The pollen grains can germinate and grow tubes only when the determinants of the diploid (sporophyte) phase do not match the genotype of the female. Even though the haplotype of the S_2 pollen on the style on the left is genotypically different from the S_1S_3 style, it still cannot grow a pollen tube because the sporophyte-derived determinants on its coat include S_1, leading to an incompatible response. In contrast, neither the S_1 nor the S_2 pollen on the style on the right has any determinants that match the S_3S_4 style, so both can grow pollen tubes.

16.5.4 In gametophytic self-incompatibility growth of the pollen tube is arrested by ribonucleases or programmed cell death

Gametophytic self-incompatibility systems are the commonest, having been identified in more than 60 plant families. There are at least two distinct GSI systems. Incompatibility in members of the Solanaceae, Rosaceae and Scrophulariaceae is based on destruction of the RNA in the tube of self-pollen by **cytotoxic ribonucleases** (RNases). In species of the poppy family (Papaveraceae), the incompatible reaction is a **programmed cell death** process that is described in further detail in Chapter 18.

Table 16.5 Types of self-incompatibility (SI) system. The SI system of Brassicaceae is of the sporophytic type (SSI). The system in Solanaceae, Rosaceae and Scrophulariaceae is gametophytic self-incompatibility (GSI). The identities and functions of male and female determinants are described in the text. The GSI system in Papaveraceae is based on a programmed cell death mechanism.

Family	Type of SI	Male determinant	Female determinant
Brassicaceae	SSI	SP11/SCR	SRK
Solanaceae, Rosaceae, Scrophulariaceae	GSI	SLF/SFB	S-RNase
Papaveraceae	GSI	(Unknown)	S-protein

Studies of GSI in the Solanaceae established a correlation between specific, abundant glycoproteins of molecular mass about 30 kDa in stylar extracts and the presence of particular S alleles. Subsequent analysis showed that these female determinant proteins are RNases. **S-RNase** gene expression occurs exclusively in the pistil. The protein is localized mostly in the upper segment of the style, the site of inhibition of self-pollen tube growth. Mutations that eliminate stylar S-RNase activity result in a self-compatible plant. The S haplotype specificity determinant of S-RNases has been found to reside in their protein backbone and not in their glycan side-chains. In genetic engineering experiments in which the N-glycosylation site of *Petunia* S_3-RNase was knocked out, plants retained the capacity to reject S_3 pollen. The genes encoding S-RNases can be highly polymorphic; S-RNases encoded by different alleles share amino acid identities ranging from 38% to 98%. Figure 16.37 is a model of a typical S-RNase of the Rosaceae, the S_3-RNase of *Pyrus pyrifolia*. A loop in the N-terminal half of the molecule, referred to as a hypervariable region, is the location of most of the allelic diversity in amino acid sequence between different female determinants in the S locus.

The nature and mode of action of male determinants in the GSI system of the Solanaceae, Rosaceae and Scrophulariaceae are not well understood. Genes encoding novel F-box proteins, components of one type of Cullin/RING E3 ubiquitin ligases (see Figure 5.36) are associated with the S locus in species from these families. Such proteins are referred to as **SLF** (S-locus F-box) or **SLB** (S-haplotype-specific F-box). The molecular mechanisms by which SLF/SLB and S-RNases interact and specifically inhibit self-pollen growth are not yet known. One simple model proposes that S-RNases pass non-specifically from the style to the pollen tube. Interaction between SLF/SLB and incompatible RNases protects the latter from degradation and they are therefore able to break down RNA in the pollen tube, halting its growth and preventing fertilization. RNases from a compatible source, however, are unprotected, or associate with an unidentified inhibitor, resulting in the survival of RNA in the growing tube. Other, more complex, schemes have also been suggested. Additional components of the system remain to be characterized and further research will be required before the full picture becomes clear.

16.5.5 Sporophytic self-incompatibility in the Brassicaceae is mediated by receptor kinases in the female and peptide ligands in the pollen coat

In sporophytic self-incompatibility, the diploid S genotype of the plant that produced the pollen grain determines whether the pollen tube is able to grow through the style (Figure 16.36B). The interaction in SSI systems is generally highly localized at the stigma surface and incompatibility takes the form of a blockage in pollen rehydration or pollen tube emergence. A single papillar cell can discriminate between genetically different pollen grains, inhibiting a self-type grain while allowing the development of non-self pollen. The response is typically rapid, occurring within minutes of pollen–stigma contact. Unlike GSI, SSI does not involve cytotoxic RNases or cell death. In this case incompatible pollen grains that have not formed pollen tubes remain viable for a time after landing on an incompatible stigma, and can develop pollen tubes when transferred to a compatible stigma.

The most fully understood SSI mechanism occurs in species of the family Brassicaceae; it is characterized by interaction between a stigma epidermal receptor kinase and the corresponding peptide ligand in the pollen coat.

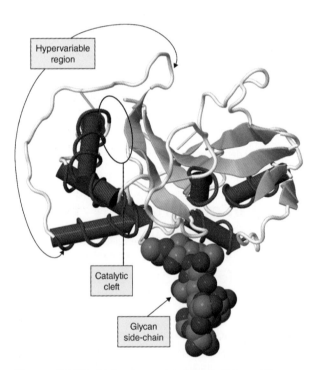

Figure 16.37 Molecular model of the S_3-RNase of *Pyrus pyrifolia* (Japanese pear). Its structure consists of eight helices and seven sheets. The hypervariable region (indicated) is a polypeptide loop where female determinant allelic variation at the S locus is expressed as divergent amino acid sequence. A glycan side-chain, consisting of mannose and *N*-acetyl glucosamine residues, is shown attached to one of two glycosylation sites. α-Helices are pink, β-sheets yellow.

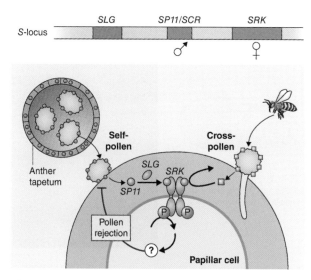

Figure 16.38 Model of the mechanism of sporophytic self-incompatibility in members of the family Brassicaceae. The S locus includes *SP11* (male determinant) and *SRK* (female determinant) genes. SP11 from the pollen coat binds to SRK and is thought to trigger a phosphorylation cascade leading to pollen rejection.

Key points Self-incompatibility is the means by which a plant prevents fertilization of its egg by its own pollen and thereby promotes outbreeding. Genetic control of self-incompatibility generally resides in multiple genes, called determinants, that constitute the S locus. Gametophytic self-incompatibility (GSI) is specified by the haploid pollen genotype and involves either destruction of pollen tube RNA by cytotoxic ribonucleases or, in poppy species, a type of programmed cell death. Sporophytic self-incompatibility (SSI) is determined by the male parent's diploid genotype at the S locus. An incompatible reaction between pollen and stigma in SSI results in an interaction between a pollen coat peptide ligand specified by the male S locus determinant and a kinase encoded by the female S locus determinant. This triggers a signaling cascade that results in blockage of pollen rehydration or tube growth.

The S locus in Brassicaceae consists of three genes: two female determinants, **SRK** and **SLG**, and one male determinant, **SP11** (Figure 16.38). *SRK* (*S-locus Receptor Kinase*) encodes a protein that spans the plasma membrane of a stigmatic papilla cell and has an extracellular domain, a transmembrane domain that anchors the protein in the plasma membrane, and a cytoplasmic domain with **serine/threonine kinase** activity. The male determinant, SP11 (S-locus protein 11), which is also designated SCR (S-locus cysteine-rich protein), is a small peptide that is expressed predominantly in the anther tapetum and accumulates in the exine layer during pollen maturation. Allelic variation gives rise to over 70% amino acid divergence among SRKs and 35% among SP11s. SLG (S-locus glycoprotein), which localizes to the papillar cell wall, is not essential for self/non-self recognition but enhances the incompatibility reaction in some S haplotypes.

When a self-pollen grain makes contact with the stigma, SP11 protein penetrates the papilla cell wall and binds to the extracellular domain of SRK in an S haplotype-specific manner (Figure 16.38). This induces autophosphorylation of SRK and triggers a signaling cascade that results in the rejection of the pollen. This mechanism is supported by the analysis of mutants and transgenic plants. For example, *Arabidopsis thaliana*, which is normally self-fertile, can be rendered self-incompatible by transformation with an *SRK-SP11* gene pair from its SSI-type relative *Arabidopsis lyrata*. Knowledge of the signaling cascade downstream of SRK is incomplete (Figure 16.38).

16.5.6 The growing pollen tube is actively guided toward the embryo sac

Following touchdown, rehydration and germination, the vegetative cell of the compatible male gametophyte produces a pollen tube that grows through the transmitting tract of the pistil toward an ovule. It enters the ovule at the micropyle and continues on to the embryo sac. Elongation of the pollen tube by tip growth can achieve rates as fast as $3~\mu m~s^{-1}$. Cytoplasm is concentrated near the tip as it grows (Figure 16.39A), and the part of the tube closest to the grain is blocked off by deposition of callose plugs (Figure 16.39A, B). Vesicles containing cell wall material are transported via the cytoskeleton to the tip of the pollen tube where they deliver their contents to the cell wall by exocytosis.

An important question is what guides the pollen tube on its journey to the waiting embryo sac. The route and rate of pollen tube growth are thought to be determined by signals from the embryo sac and the style, which also act as an energy source for the pollen tube. The chemical influences exuded by the style have been studied in Easter lily (*Lilium longiflorum*), which is experimentally convenient because, unusually, the style is hollow in this species, allowing access to pollen tubes as they grow. Gradients in Ca^{2+} concentration, which are mediated by Ca^{2+}-binding allergen proteins (see Section 16.5.2), appear to play an important role in determining the path of pollen tube growth. The direction of growth can be

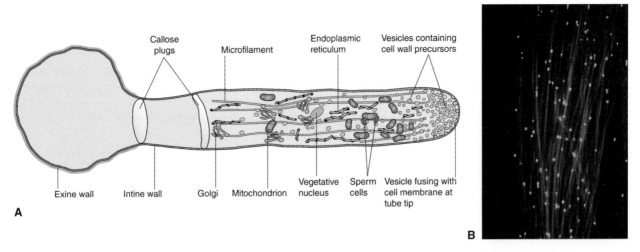

Figure 16.39 (A) A pollen tube elongating by tip growth. Callose plugs limit the volume occupied by the cytoplasm to regions close to the growing tip. (B) Fluorescence micrograph of pollen tubes growing through a pistil. The callose plugs that block off the pollen tube at regular intervals are visualized by staining with aniline blue.

changed if the Ca^{2+} gradient experienced by the tip is manipulated, for example by the application of chelators.

Fluxes of protons (H^+ ions) have been implicated in growth. There is a pH gradient in the tip region of pollen tubes; the most apical region has a pH of 6.8, while the region at the base of the vesicular zone (Figure 16.39A) is alkaline (pH 7.5). Proton fluxes in the tip, probably driven by plasma membrane H^+-ATPase activity, control a current loop in the apical dome, which in turn directly influences tube growth. Evidence of the link between protons and growth is provided by high-resolution studies that show regular oscillations in tube growth rate (between 100 and 500 nm s^{-1} in *Lilium*, with a periodicity of about 20–50 seconds) and an anticipatory oscillation in pH with the same cycle time. Similar rhythms in ATP supply have also been observed.

In addition to the roles of Ca^{2+} and H^+, regulation of pollen tube growth is exerted through signaling pathways involving **phosphoinositides** and small **G (guanosine triphosphate-binding) proteins**. Phosphoinositides such as phosphatidyl inositol-4,5-bisphosphate (PIP_2) are membrane lipids. PIP_2 associates with profilin (see Figure 16.35) and other actin-binding proteins. G proteins include **ROPs** (Rho family GTPases of plants) which represent a subfamily of molecular switches, unique to plants, with functions in developmental and signaling events. ROP and PIP_2 are localized to the plasma membrane of the pollen tube apex and contribute to cell polarity. Figure 16.40 is a model of the tip of the growing pollen tube showing ROP and PIP_2 and the participation of Ca^{2+}, H^+ and OH^- fluxes, respiratory ATP formation from starch stored in pollen amyloplasts, and interactions between vesicles, endoplasmic reticulum (ER) and actin.

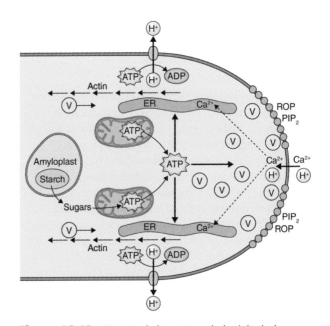

Figure 16.40 Structural elements and physiological processes in the apex that participate in growth of the pollen tube. Starch from amyloplasts is hydrolyzed to sugars, which are metabolized by mitochondria to generate ATP. ATP is used by the plasma membrane H^+-ATPase to support proton efflux. ATP is also required for actin polymerization and transport along microfilaments, for Ca uptake by the endoplasmic reticulum (ER), and for vesicle (V) trafficking and fusion in the apex. Rho family GTPases of plants (ROP) and phosphatidyl inositol-4,5-bisphosphate (PIP_2), are signaling molecules localized to the apical plasma membrane, where they contribute to specification of cell polarity.

As the pollen tube reaches the end of the transmitting tract, guidance is taken over by signals from the embryo sac (Figure 16.41). Pollen tubes are not attracted to embryo sacs that are developmentally aberrant, or have

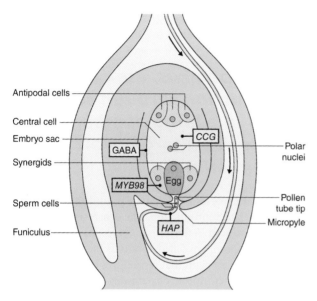

Figure 16.41 Factors that contribute to guiding the pollen tube from the end of the transmitting tract toward the micropyle. The mature embryo sac consists of an egg, two synergids, a central cell (which is shown with two polar nuclei) and zero to many antipodal cells. Both the male and female gametophyte contribute to proper guidance of the pollen tube to the micropylar opening of the ovule. GABA (γ-aminobutyric acid) can function as a chemical attractant. *MYB98* and *CCG* encode transcription factors expressed in synergids and the central cell, respectively, that ensure correct guidance of the pollen tube through the micropyle. *HAP* expresses a guidance recognition factor in the pollen tube.

already been fertilized, or have been heat-killed. There is evidence that **GABA** (γ-aminobutyric acid) secreted by the embryo sac can act as a chemical attractant in pollen tube guidance. Analysis of *Arabidopsis* mutants has identified a number of gametophytic genes whose expression is required for the correct routing of the pollen tube to the embryo sac. For example, synergid cells, located on either side of the egg, are believed to have a critical role in attracting the pollen tube to and through the micropyle. A transcription factor, MYB98, is specifically expressed in synergids. In *myb98* mutants, synergid ultrastructure is disrupted and ovules are unable to attract the pollen tube to the micropyle. The central cell may also play a role in pollen tube guidance; the gene *CENTRAL CELL GUIDANCE* (*CCG*) is thought to encode a transcriptional regulator, which must be functional in the central cell in order to ensure the pollen tube is properly guided through the micropyle. Pollen-borne genes are also important in the process. Pollen carrying mutations in genes of the *HAPLESS* (*HAP*) family produces tubes that meander aimlessly over the ovule surface, a behavior that suggests that a mechanism required for the recognition of guidance cues from the embryo sac has been disrupted. Figure 16.41

summarizes the components of the gametophytic guidance system.

16.5.7 Double fertilization completes the alternation of generations

When the pollen tube reaches the embryo sac, its tip bursts and releases its two sperm into one of the two synergid cells, which degenerates at the time of sperm release or shortly before it (Figure 16.42). In the *Arabidopsis* mutants *feronia* (*fer*) and *sirene* (*sir*), the pollen tube continues to grow, even though the synergid has degenerated, indicating that a signal from the embryo sac that normally halts pollen tube growth is missing in these genotypes. Generally, each ovule attracts only one pollen tube. Fertilization by extra pollen tubes is called **polyspermy** or **heterofertilization**. In *Arabidopsis* the normal rate of heterofertilization is less than 1%, but multiple pollen tubes are able to gain access to ovules in *fer* and *sir* mutants. This suggests that pollen tube reception is part of the mechanism that repulses additional pollen tubes and prevents polyspermy. *FERONIA* has been cloned and shown to encode a plasma membrane-localized, receptor-like serine-threonine kinase. A proposed signaling pathway suggests an interaction between FER and an (unidentified) ligand from the pollen tube in which activated FER causes the synergid to send another signal back to the pollen tube, inducing both growth arrest and bursting.

During double fertilization, one sperm fuses with the haploid egg cell to form a diploid zygote, thus initiating a new sporophyte generation. The second sperm fertilizes the central cell; its nucleus fuses with the two central cell nuclei, thus giving rise to a triploid primary endosperm cell (Figure 16.42). In some plants, the two sperm cells differ in size or shape or in the numbers and types of organelles they carry. An extreme example is white leadwort (*Plumbago zeylanica*). In this species the smaller sperm, which has more plastids, always fuses with the egg cell. Preferential fertilization related to unequal quantities of heritable organelles in the sperm is referred to as **cytoplasmic heterospermy**. Alternatively, differentiation between sperm cells can be the result of differences in nuclear content. Such a system, termed **nuclear heterospermy**, has been described for *Zea mays*. **B chromosomes** are extra chromosomes that are not homologous to the karyotype (the A chromosome set) and are found across the range of eukaryotes from plants to fungi to animals, although not all species have B chromosomes and not all genotypes within a species contain them. Unlike the A chromosomes, B chromosomes in some organisms frequently exhibit

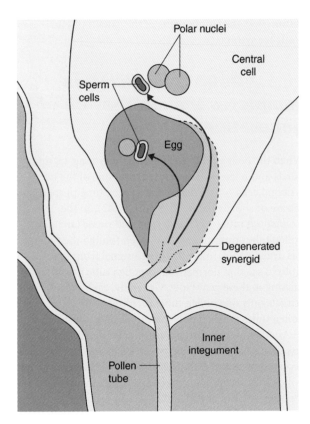

Figure 16.42 Double fertilization in maize. Two sperm are released from the pollen tube into a degenerated synergid. One fuses with the egg and the second with the central cell.

non-disjunction ('not coming apart'), in which a pair of chromosomes fails to separate correctly during meiosis I, meiosis II or mitosis. In lines of maize carrying B chromosomes, non-disjunction at the second pollen mitosis results in one sperm cell receiving two or more B chromosomes while the other receives none. Sperm cells with B chromosomes fertilize egg cells more frequently than those without (in a ratio of up to 3:1). In both nuclear and cytoplasmic heterospermy, selectivity is believed to be the result of signaling between egg and sperm, but the molecular mechanisms remain to be discovered.

Fertilization is the point at which organelle genomes are transmitted to the next generation, a phenomenon called cytoplasmic inheritance. The predominant type of cytoplasmic inheritance is **uniparental maternal**, in which the genotypes of plastids and mitochondria in the progeny are derived exclusively from the female parent. **Uniparental paternal** inheritance of plastids is found in a significant minority of flowering plants. Inheritance of mitochondria via the paternal line has been described for alfalfa (*Medicago sativa*), *Populus* and *Brassica napus*. The plastids of the egg cell are usually of variable size and shape, contain some starch grains and lamella-like structures, and tend to cluster around the nucleus.

Plastid number per cell ranges from more than 700 in *Plumbago* to as few as eight in *Daucus* (carrot) species.

16.5.8 Apomixis, asexual reproduction through seeds, occurs in a large number of taxa and is a target trait for crop breeding

In some plant species, seeds can be produced asexually as well as by sexual reproduction. The processes by which seeds are produced in the absence of fertilization of an egg cell by a sperm are called **apomixis** (also termed **agamospermy**); a plant that reproduces by apomixis is called an **apomict**. Apomixis pathways are classified as gametophytic or sporophytic according to the source of the cells that develop into an embryo. The apomictic habit has evolved independently several times and occurs in more than 400 taxa. **Sporophytic apomixis**, also called **adventitious embryony**, is so-named because embryos arise spontaneously from cells of ovule integuments or from nucellar cells that are adjacent to a haploid embryo sac. It is common among citrus species. The maturation and survival of adventitious embryos is dependent on the development of endosperm derived from normal double fertilization. This kind of apomixis is essentially a form of **somatic embryogenesis**, equivalent to the developmental pathway undertaken by cells in tissue culture that have been induced to regenerate into plants.

Three families, the Asteraceae, Rosaceae and Poaceae, which collectively constitute only 10% of flowering plant species, account for about 75% of **gametophytic apomicts**. Often, but not always, apomixis is associated with self-incompatibility, perenniality and dehiscent fruits (i.e. those that split open at maturity). Gametophytic apomixis originates with a diploid embryo sac, which develops by mitosis of a diploid cell with apomictic potential rather than from a haploid megaspore (Figure 16.43). Endosperm development may be either spontaneous (autonomous) or induced by fertilization with a pollen sperm nucleus (pseudogamous).

Two pathways of gametophytic apomixis, diplospory and apospory, are recognized, based on the cell type that gives rise to the diploid embryo sac. In **diplospory** (Figure 16.43B), the megaspore mother cell may initiate meiosis but abort it at an early stage and switch to mitotic development of the embryo sac, a condition referred to as **meiotic diplospory**. Dandelion (*Taraxacum*) is an example of such a diplosporous apomict. An alternative mechanism is **mitotic diplospory**, where the MMC proceeds directly to form the unreduced embryo sac by mitosis without an initial meiotic phase. The second

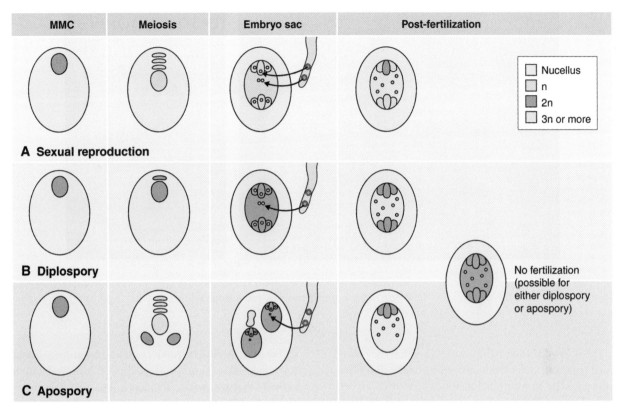

MMC	Meiosis	Embryo sac	Post-fertilization

A Sexual reproduction

B Diplospory

C Apospory

Nucellus
n
2n
3n or more

No fertilization
(possible for
either diplospory
or apospory)

Figure 16.43　Mechanisms of gametophytic apomixis compared with normal sexual fertilization. In each case, female reproduction occurs in the nucellus of the ovule, where a single cell typically becomes the megaspore mother cell (MMC). (A) Sexual fertilization. Meiosis of the 2n MMC gives rise to the four haploid cells of the reduced embryo sac. Double fertilization results in diploid embryo and triploid endosperm. (B) Diplosporous apomixis. In this type, there is no meiotic reduction of the MMC. The embryo is 2n without fertilization and only the sperm cell nucleus destined to form endosperm penetrates the embryo sac. (C) In apospory, the MMC undergoes meiotic reduction, but one or more embryo sacs form from 2n aposporous initials adjacent to the MMC or megaspore. Endosperm may result from the fusion of one sperm nucleus with the polar nucleus. In apospory and diplospory, endosperm may also form in the absence of a pollen nucleus.

form of gametophytic apomixis is **apospory**, exemplified by members of the hawkweed genus (*Hieracium*). In apospory the embryo sac is derived from one or more somatic ovule cells, called aposporous initials. Diploid embryo sacs can differentiate from aposporous initials at various times during ovule development and may coexist with meiotically haploid embryo sacs in the same ovule or may continue to develop while the haploid sexual embryo sac degenerates (Figure 16.43C).

Apomixis has been called an evolutionary dead end or blind alley because of its postulated association with genetic uniformity, low adaptive potential and the accumulation of deleterious mutations. On the other hand apomixis, like other forms of asexual reproduction (discussed in detail in Section 11.4.2 and in Chapter 17), is an abundant and widely distributed attribute that clearly confers adaptive and evolutionary advantages in dynamic natural populations. This makes the genetic regulation of apomixis a subject of ecological interest.

Apomixis has also attracted agronomic and biotechnological attention because introducing the trait

into major crop species has the prospect of revolutionizing crop breeding. Many of the most productive and stress-resistant agricultural and horticultural crop varieties are hybrids. The superior performance of the progeny of a cross between two inbred parents is called **heterosis** or hybrid vigor. Conversely, the consequence of self-pollinating hybrids over several generations is **inbreeding depression**, that is, progressive reduction in heterozygosity and vigor. An example of the exploitation of heterosis in a major crop is maize. The discoverer of heterosis in maize was Charles Darwin, who noted that the progeny of cross-pollinated maize were 25% taller than the progeny of inbred parents. Hybrid varieties were first developed by maize breeders early in the 20th century. Since that time their use has increased, together with crop yields, until today hybrids are planted in about 95% of maize acreage in the United States, and two-thirds worldwide. Figure 16.44 shows that the heterotic effect is apparent not only in grain yield at maturity but also from the earliest phase of seedling development.

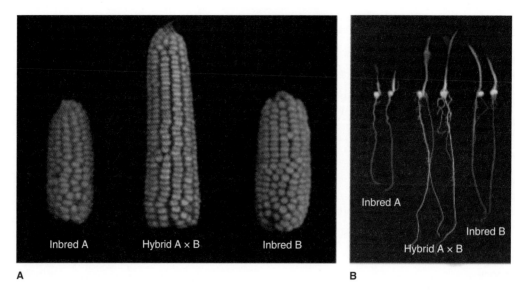

Figure 16.44   Heterosis in maize. (A) Grain yield in the hybrid between A and B exceeds that of either inbred parent. (B) Hybrid vigor is expressed from the earliest stages of seedling growth.

The downside to F_1 hybrid varieties is that they do not breed true, and suffer from inbreeding depression if propagated by conventional methods of seed multiplication. It is therefore necessary to recreate the hybrid each season by crossing inbred parents, an expensive and time-consuming business. This is where interest in apomixis comes in, because it offers the prospect of creating hybrids that propagate asexually via seeds and in which desirable traits are therefore transmitted faithfully to the progeny. The genus *Tripsacum* (gamagrass), which is related to *Zea*, includes apomictic species. *Tripsacum dactyloides* diploids are sexual, but tetraploid genotypes are pseudogamous diplosporous apomicts. Diplospory in *Tripsacum* is inherited as a single dominant locus. *T. dactyloides* will cross with maize and there have been numerous attempts to transfer the apomictic locus into maize by exploiting the interfertility between the two species. So far success has been limited, at least in part because of low rates of recombination between the genomes of the two species, but backcross programs to integrate the apomixis locus into the maize genetic background continue. Some authorities believe the organization of genes for apomixis is too complex for intergeneric transfer of the trait by conventional crossing ever to be possible.

An alternative strategy is to intervene in the process of meiosis in the MMC by genetic engineering. An example of this approach is a recent study that identified an *Arabidopsis* gene named ***OSD1*** (*OMISSION OF SECOND DIVISION 1*). The products of meiosis in *osd1* mutants are dyad rather than tetrad spores. In triple mutants consisting of *osd1* with a gene that eliminates recombination and another that modifies chromatid

segregation, meiosis is totally replaced by mitosis. Such so-called MiMe plants produce diploid male and female gametes that are genetically identical to their parent, and ploidy doubles at each generation. As more detailed molecular understanding of meiosis and apomixis is acquired, new tools and approaches such as MiMe plants are becoming available to help the biotechnologist and plant breeder reach the goal of being able to manipulate the reproductive systems of crops at will.

16.6 Seed and fruit development

Pollination and double fertilization are followed by the development of major structures of the seed: the embryo, endosperm and seed coat. The seed coat originates in ovule integuments, the maternal tissues that provide protection and facilitate nutrient transfer to the developing embryo they surround. A fruit (see Figure 1.22) is defined as a structure derived from an ovary and bearing or containing seeds. It may be simple (originating as a single ovary), aggregate (from several separate ovaries of a single flower) or multiple (derived from an inflorescence). The winged fruit of *Acer* spp. is an example of a simple type. Raspberries (*Rubus idaeus*) are aggregate fruits. Hop (*Humulus*) is a multiple fruit. The structures of some fruits include tissues other than those of the gynoecium, in which case they are called false fruits or pseudocarps. Strawberry (*Fragaria*), in which the red fleshy part is derived from the floral receptacle, is a false fruit (see Figure 6.5A).

Key points The pollen tube elongates by tip growth, and is actively guided by signals from the style and the embryo sac, mediated by Ca²⁺ gradients, Ca²⁺-binding allergens and proton fluxes. Pollen tube development is also regulated by phosphoinositide and G protein signaling pathways. The embryo sac produces chemical attractants that guide the pollen tube toward the point of fertilization. Transcription factors expressed in synergids, central cells and the pollen tube itself are essential for delivering the sperm cells to the embryo sac. Fertilization by more than one pollen tube is prevented by signals from the embryo sac that halt further tube growth. Double fertilization results in a 2n zygote and 3n endosperm. Fertilization transmits organelle genomes to the next generation, primarily from the maternal line, though some instances of paternal transmission are known. Asexual reproduction through seeds is called apomixis. In sporophytic apomixis, embryos are derived from diploid ovule cells adjacent to the haploid embryo sac. Gametophytic apomixis is the result of embryogenesis from a diploid embryo sac. Apomixis is of interest to crop breeders as a means of counteracting inbreeding depression and promoting hybrid vigor (heterosis), but there are formidable obstacles to be overcome before it can be a practical tool.

The fates of the different cell types destined to form embryo, endosperm, seed coat and maternally-derived fruit tissues are under the control of networks of gene

activity. The regulation of embryogenesis is described in detail in Section 12.3. In the period immediately after fertilization, endosperm proliferates and nourishes the early embryo. In many plants the endosperm is absorbed by the developing embryo and is absent from the mature seed, as in, for example, garden pea (*Pisum sativum*) and soybean (*Glycine max*). In other species, notably the grasses, endosperm is persistent and represents the major repository of the stored reserves that support germination and seedling growth. Here we discuss seed development in a typical non-endospermous species, the legume soybean. Subsequently, endosperm development in the cereal grain is considered. Finally we look at some general aspects of fruit tissue differentiation and its control.

16.6.1 Genomics analysis reveals tissue specificities and changes with time in gene expression patterns during seed development

A seed is an anatomically complex structure, comprising a number of tissues with a variety of developmental origins. One way of making sense of this complexity is to take a **genomics** approach, in which expression patterns of a global sample of genes from the species under examination are determined in different seed tissues over time. We can illustrate this approach with a study of soybean.

Figure 16.45 shows the major events in soybean seed development over the period from flowering and

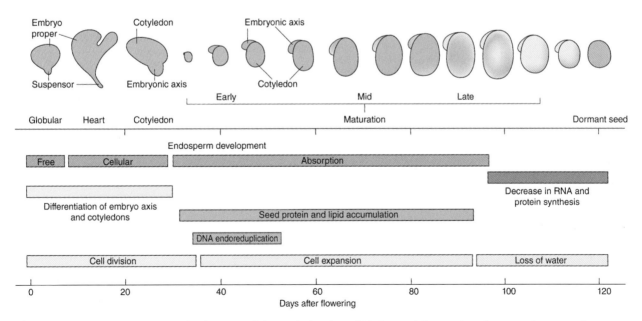

Figure 16.45   Events during the development of the seed of soybean (*Glycine max*) from early embryogenesis to maturity.

fertilization to maturity. The fully developed soybean seed is non-endospermous. Endosperm is present over the period of cell division and embryogenesis; but as cells expand, endosperm is absorbed and has disappeared by the time of reducing water content as the seed approaches maturity. During the period of desiccation the general level of transcription and translation declines, but some genes are upregulated, particularly those for late embryogenesis-abundant (LEA) and other proteins with protective functions in water-limited tissues (see Section 15.6.1). Protein, lipid and starch reserves are laid down in the cotyledons during the middle phase of seed development (see Section 17.3). DNA of the embryo at

early maturation undergoes **endoreduplication**, a process of genome replication that increases the number of gene copies per cell and enables a high rate of transcription to be supported (see Section 11.4.3).

More than 30 000 unique sequences from the soybean genome, arrayed on a gene chip, were probed with DNA copies (cDNAs) of the mRNAs expressed in soybean seed tissues at different times during maturation. Measuring the degree to which each gene hybridizes with its complementary cDNA quantifies the abundance of each mRNA and allows the pattern of transcription of the corresponding genes to be profiled in time and space. Figure 16.46 is an example of the output from such a

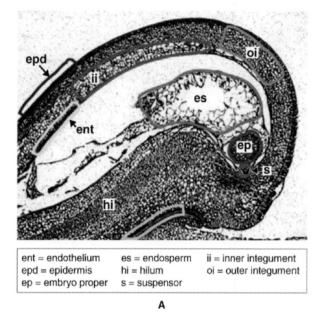

ent = endothelium	es = endosperm	ii = inner integument
epd = epidermis	hi = hilum	oi = outer integument
ep = embryo proper	s = suspensor	

A

Figure 16.46 Profiling gene expression in the developing soybean seed. (A) Light micrograph of a section of a seed at the globular stage of embryo development. (B) Functional categories of genes expressed in the suspensor.

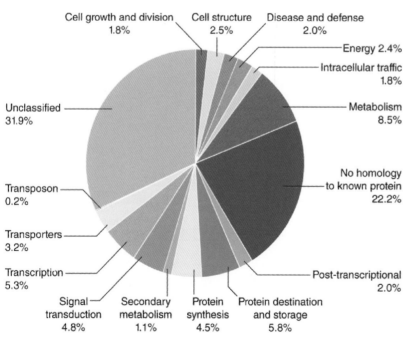

B

study. The micrograph (Figure 16.46A) identifies the different tissues of the seed at the globular stage of the embryo. Figure 16.46B is an example of the functional classification of the genes transcribed in a particular tissue—in this case, the **suspensor**. Notice that more than half the expressed genes have not yet been assigned a function or identity. This degree of ignorance is typical of the current state of knowledge of genomics and emphasizes that much work remains to be done before we have anything like a complete picture of development and the genes that control it. Of the classes of known genes expressed in the suspensor, those related to general metabolism, protein storage, transcription and signal transduction are predominant. It is estimated that to make a globular-stage soybean embryo requires the expression of genes corresponding to at least 20 000 different transcript sequences, of which 74 are found to be confined to the suspensor and no other tissue. An example of such a region-specific gene is that encoding a transcription factor of the *NAM* family (see Sections 12.5.4 and 18.3.5).

Genomics technology changes rapidly. Global transcription profiling using DNA microarrayed on gene chips is being replaced by high-throughput DNA sequencing methodology. The ever-increasing pace of data generation from such studies requires more bioinformatics, systems biology and computer capacity to turn the flood of information into understanding.

16.6.2 The development of nuclear endosperm comprises phases of syncytium formation, cellularization, endoreduplication and programmed cell death

The **primary endosperm cell** is the triploid fusion product of the two nuclei of the embryo sac central cell with the haploid nucleus of the second male gamete. Endosperm tissue development from the primary cell may take one of three forms. During **cellular endosperm** formation, cell walls are laid down at the same time as nuclei divide. Cellular endosperm formation occurs in about 25% of angiosperm families. In **nuclear endosperm** formation there is repeated free nuclear division without cell wall formation, resulting in a **syncytium** (a tissue consisting of cytoplasm containing many nuclei but not differentiated into separate cells); cell walls may form subsequently. Nuclear endosperm is the commonest type, occurring in 56% of angiosperms. Coconut (*Cocos nucifera*) is an example of a plant with nuclear endosperm formation (see Figure 1.21B). As the

seed matures, the syncytial endosperm begins to cellularize, forming the coconut 'meat'; the coconut 'water' remains a syncytium. **Helobial endosperm** is an intermediate type, in which the first mitotic division is followed by cytokinesis, resulting in the formation of two unequal cells. Subsequent divisions of these cells are of the free nuclear type; cell walls may form later in development, after completion of free nuclear division. Helobial endosperm formation is found in 19% of families, comprising mostly aquatic and other non-gramineous monocots. Here we discuss the formation of nuclear endosperm in cereals where it produces the storage tissue of grains, on which world food supply depends.

Early events in the formation of maize endosperm are shown in Figure 16.47A. After fertilization the primary endosperm nucleus rapidly enters a period of intense mitotic activity. Several rounds of more or less synchronized division without the formation of the cell plate and cytokinesis (see Section 11.1.2) produce a syncytium. Mitotic activity in early endosperm development is much greater than that in the embryo. As the proliferation rate of endosperm nuclei declines towards the level of embryo cell division, the syncytial phase is succeeded by a period of **cellularization** (Figure 16.47A). Microtubules radiating from the nuclear surface define nuclear–cytoplasmic domains. The pattern of cellularization takes the form of centripetal growth of cell files extending to the center of the endosperm cavity.

Endosperm cell fate is specified by positional signals. The mature endosperm comprises two types of storage tissue: **starchy endosperm** and **aleurone** (see Section 6.2.3). Some endosperm cells near vascular tissue and immediately surrounding the embryo have nutrient transfer functions. The cytoplasm of these **transfer cells** is typically dense, with many small, spherical mitochondria. A number of genes have been described with presumed functions in transfer cell specification and development. Mutation of *EMPTY PERICARP4*, a maize gene encoding a protein that regulates mitochondrial gene expression, results in a defective transfer cell layer and endosperm. Genes of the **BETL** (*Basal Endosperm Transfer Layer*) and **EBE** (*Embryo-sac Basal-endosperm-layer Embryo-surrounding-region*) families are preferentially expressed in maize transfer cells under the transcriptional control of the MYB-related protein **ZmMRP-1**. The basal layer of transfer cells in the maize mutant *globby-1* is abnormal as a consequence of errors in the timing of patterning events. There is evidence that signals from maternal sporophytic tissue exert a controlling influence on the development of basal transfer cell layers.

The maize mutant **dek1** (*defective kernel1*) lacks aleurone but has normal transfer cells, indicating

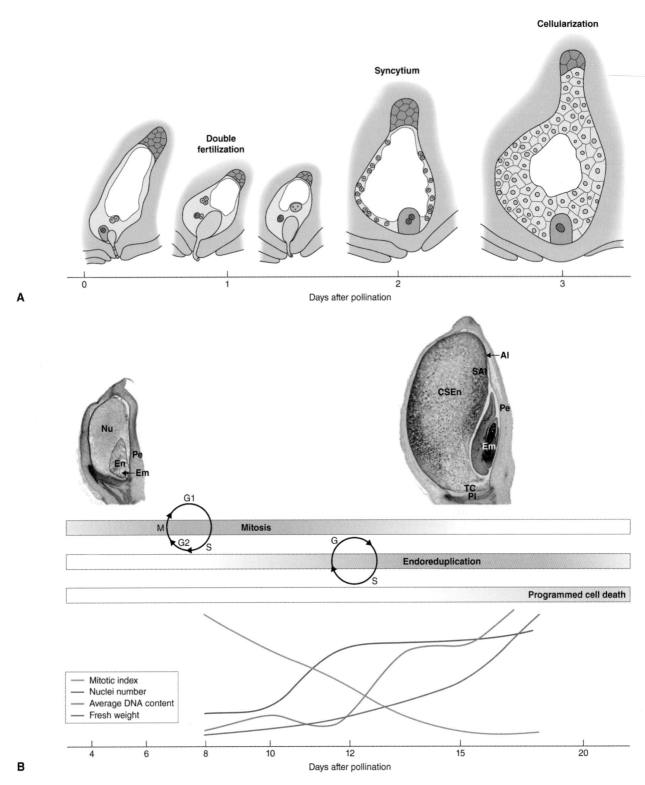

Figure 16.47 Phases in maize endosperm development. (A) The first 3 days after pollination: double fertilization, syncytium formation and cellularization of the endosperm. The pollen tube and sperm nuclei are shown in blue; the polar nuclei in the central cell of the female gametophyte, and the endosperm nuclei and tissue derived from them are in yellow; and the egg cell nucleus and embryo nuclei and tissue are in red/pink. (B) From 4 to 20 days after pollination: mitotic cell proliferation, endoreduplication and the final phase of programmed cell death as the grain approaches maturity. Al, aleurone; CSEn, central starchy endosperm; Em, embryo; En, endosperm; Nu, nucellus; Pe, pericarp; Pl, placentochalazal region of pedicel; SAl, subaleurone layer; TC, transfer cells.

that differentiation of each cell type is regulated independently of the other. *Dek1* encodes a membrane-associated protease that is thought to be part of a mechanism for transmitting **positional information** during aleurone cell specification. As is evident from the phenotypes of the *waxy, sugary* and *shrunken* mutants of maize, the development of starchy endosperm is sensitive to modifications in the enzymology of starch synthesis and the structure of starch granules and amyloplasts (see Sections 6.3.1 and 17.3.1). Storage proteins, mainly prolamins in cereals (see Section 6.3.4), accumulate throughout the middle to late phases of endosperm development (Figure 16.47B) and their expression is regulated predominantly at the transcription level.

A distinctive feature of nuclear endosperm development is the changing nature of the **cell cycle** (Figure 16.47). During formation of the syncytium cytokinesis is suppressed. Most of the cells of the mature endosperm are the products of mitosis uncoupled from cell division. These cells exhibit a high degree of **endopolyploidy** (up to 96x) as a result of extensive endoreduplication, involving reiterated DNA replication without chromatin condensation, chromatid segregation or cytokinesis. The roles of a number of cell cycle regulators (see Chapter 11) have been examined during endosperm development. For example, it has been shown that downregulation of mitotic **cyclin-dependent kinases** (CDKs) and upregulation of S-phase CDKs occurs at the point where the mitotic cycle switches to the endoreduplication cell cycle.

Further characteristic attributes of endosperm development are discussed elsewhere. They include the phenomenon of imprinting and **epigenetic control** (see Section 3.5.4 and Figure 3.36) and the ultimate mummification of the starchy endosperm at maturity at the culmination of a phase of **programmed cell death** (see Section 18.4.4).

16.6.3 Differentiation of fruit tissues is associated with the activities of MADS-box transcription factors

The extensive molecular resources established for *Arabidopsis* have been exploited to analyze aspects of fruit development in this species. The fruits of *Arabidopsis* and other members of the family Brassicaceae are **siliques**, derived from two fused carpels (valves) separated by a septum bearing the seeds

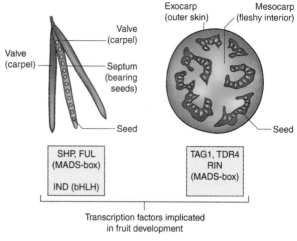

Figure 16.48　Structures, and transcription factors regulating development, of (A) *Arabidopsis* fruit and (B) tomato fruit.

(Figure 16.48A). Siliques are dry fruit that shed their seed by dehiscence when ripe. Valve and valve margin development are under the control of FRUITFUL (FUL) and SHATTERPROOF (SHP), tissue-specific MADS-box proteins, and INDEHISCENT (IND), a bHLH transcription factor. DNA profiling has identified more than 15 transcription factors associated with development of the **pericarp** tissues of the valve. These regulatory components seem to be well conserved across the range of species with dry dehiscent pod-type fruits, but details of how they network to direct fruit development remain to be established.

The evolutionary origins of fruits stretch back to more than 125 million years ago, in the Cretaceous period, when angiosperms began to appear with seeds enclosed in carpels. It is believed that the earliest fruits were dry and were the evolutionary predecessors of fleshy types, which date from about 65 million years ago. There is evidence of conservation of regulatory genes during this evolutionary progression. Tomato is an example of a **fleshy fruit** (a berry), consisting of a skin (exocarp) enclosing seeds embedded in mesocarp tissue (Figure 16.48B). Together the exocarp and mesocarp constitute the pericarp, the homolog of the tissue of the silique valve wall. It is not clear how fleshiness arises in evolution or development, but the effects of overexpressing the gene *TAG1* may give a clue. *TAG1* is the tomato homolog of the homeotic MADS-box gene *AGAMOUS* (*AG*; see Section 16.3.1). Overexpressing

TAG1 in tomato was observed to induce increased swelling and ripening of sepals. *TDR4* (*TOMATO DEFICIENS-RELATED4*) is a MADS-box gene with close sequence similarity to *FUL* and is thought to have a role in tomato pericarp development. Differentiating the structure of fleshy fruits is tightly integrated with **ripening** processes, and their regulatory networks overlap. One common factor is RIPENING INHIBITOR (RIN), a MADS-box protein that functions in the control of mesocarp softening. Chapter 18 further develops the story of RIN and other regulators of ripening in fleshy fruit (see Section 18.5).

Key points Double fertilization gives rise to the embryo and endosperm of the seed, and the seed coat develops from ovule integuments. Seeds are borne in or on a fruit, which is derived from the ovary. In non-endospermous seeds, such as those of soybean, the endosperm is absorbed by the developing embryo and the cotyledons accumulate protein, lipid and starch reserves during maturation. Endoreduplication of DNA occurs early in embryogenesis. Desiccation as the seed approaches maturity is associated with the upregulation of genes with functions in adaptation to water stress. Genomics studies suggest that differentiation of the globular-stage soybean embryo requires expression of over 20 000 different transcript sequences. In cereals and other endospermous species, the endosperm persists as the major reserve tissue of the mature seed. Endosperm differentiation in maize is an example of nuclear endosperm formation (the other, less common, types are cellular and helobial endosperm). In the period after fertilization, mitosis without cell wall formation results in a syncytium. Subsequently there is a phase of cellularization and specification of different endosperm cell types—starchy endosperm, aleurone and transfer cells. Cell identity and tissue differentiation are determined by transcription factors, are sensitive to the operation of the pathways of starch synthesis and are subject to epigenetic control. Mature starchy endosperm cells are highly endopolyploid and are in a mummified state. Species with fleshy fruits evolved relatively recently from those with dry fruits, and regulatory mechanisms have been conserved. In both cases the activities of MADS-box transcription factors are implicated in the differentiation of fruit tissues.

Chapter 17

Development and dormancy of resting structures

17.1 Introduction to resting structures in the plant life cycle

The life cycles of most plants include periods of **dormancy** (also called quiescence or rest), during which the whole plant or specific parts of it cease to grow and develop, even though environmental conditions may be permissive. Resting structures include the products of both vegetative development (including terminal and lateral buds, cambial tissues and the underground storage structures known collectively as perennating organs) and the reproductive phase (flower buds, seeds and fruits) (Figure 17.1). Characteristics shared by resting structures include accumulation of reserves, cell cycle and meristem arrest, resistance to stress, and signal perception mechanisms that sense the environmental factors required to shut down and subsequently restart growth. Many resting structures are **propagules**, that is, plant parts that become detached and are actively or passively dispersed. Propagules ensure that genes passed to progeny are spread widely, thereby increasing the chances of survival while minimizing competition with the parent. This chapter examines the structural and

physiological characteristics of dormant organs and relates them to other aspects of the plant life cycle, including cell division, morphogenesis, environmental responses, reproductive development and senescence.

17.2 Forms and functions of resting organs

The basic shape of the plant growth curve is sigmoidal (see Section 12.2.3) and consists of an initial phase of cell proliferation, a phase of cell expansion driven largely by water influx, and a final phase during which cell walls become rigid, growth rate declines to zero and size reaches a maximum asymptotically. The maturation of many resting structures, such as seeds and underground vegetative storage organs, conforms to this general pattern. The terminal stage, however, is usually associated with continued accumulation of dry weight as reserves are laid down, while the proportion of tissue water declines and fresh weight approaches a plateau. This is illustrated in Figure 17.2, which shows weight and compositional changes during maturation of sycamore (*Acer pseudoplatanus*) embryos. In this species the liquid endosperm is absorbed early in embryo development

The Molecular Life of Plants, First Edition. Russell Jones, Helen Ougham, Howard Thomas and Susan Waaland.
© 2013 John Wiley & Sons, Ltd. Published 2013 by John Wiley & Sons, Ltd.

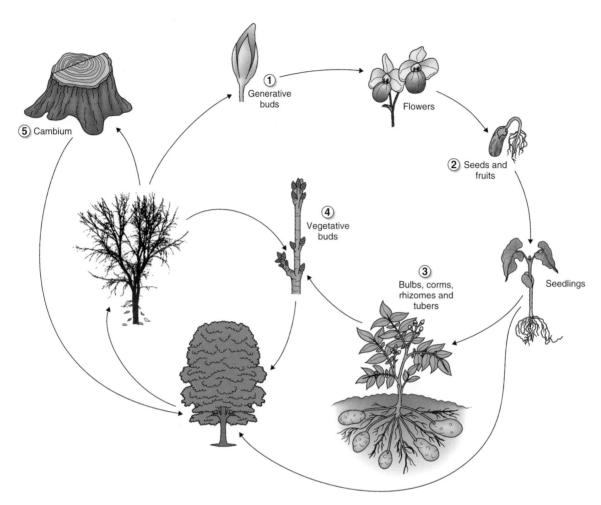

Figure 17.1 Resting structures in the plant life cycle. Organs and tissues showing arrested development can include: (1) buds containing floral organs, (2) seeds and fruits prior to germination, (3) bulbs, corms and other specialized reproductive structures, (4) apical and lateral vegetative buds and (5) meristematic tissues in the cambial layer of woody species.

and the fruit (samara; Figure 17.2A) consists of the winged pericarp (derived from the ovary wall) enclosing a single seed. The seed in turn is covered with a brown testa, within which are the embryonic axis and a pair of green cotyledons that accumulate reserves of each of the major types of storage compound—starch (Figure 17.2C), protein and lipid (Figure 17.2D).

In **resting buds** such as those containing the overwintering stem apices of woody species, growth is arrested at the cell expansion phase and the structure has the form of a telescoped shoot containing a number of organ initials and condensed internodes surrounded by a protective covering of bud scales. When the factor or factors constraining growth are relieved, there is resumption of development, either of the whole organ or from growth centers such as adventitious buds. In this section we examine the variety of resting structures found in plants, beginning with the products of sexual

reproduction and proceeding to consider perennating organs derived from modified stems, leaves and roots.

17.2.1 Dormancy of the embryo is conditioned by the associated storage tissue and seed coat

As described in Chapter 1, the 'seeds' of many species, including cereal grains and lettuce achenes, are, strictly speaking, single-seeded fruits because the ovary wall is fused to the seed coat (see Figure 6.5). Reserve materials to support germination and growth of the embryo may be stored in the perisperm, derived from maternal nucellar tissue as in coffee (*Caffea arabica*), in the endosperm or enlarged cotyledons, or in swollen

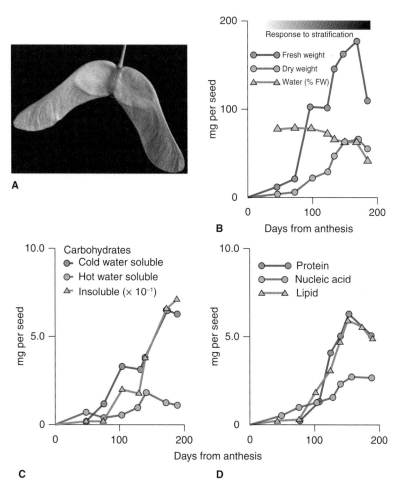

A

B

C

D

Figure 17.2 Changes in weight, chemical composition and dormancy during maturation of sycamore (*Acer pseudoplatanus*) embryos. (A) *A. pseudoplatanus* fruit (samara). (B) Fresh and dry weight per seed. The extent to which embryo dormancy could be overcome by a chilling treatment (stratification) is indicated by the intensity of the shaded bar. (C) Carbohydrate fractions. Cold water soluble = low molecular weight sugars; hot water soluble = larger oligosaccharides; insoluble = starch and cell wall polysaccharides. (D) Protein, nucleic acids and lipid.

hypocotyl, as in the case of Brazil nut (*Bertholletia excelsa*). Germination and development of the embryo are often constrained by the influence of surrounding tissues. The embryo and endosperm, particularly the aleurone, are sources of hormonal regulators as well as energy and nutrients (see Chapter 6). The seed coat, consisting of the **testa** (derived from the ovule integument), has important roles in the dormancy of many species. Dormancy of the intact seed may be broken with a chilling or light treatment, but removal of the seed coat will often eliminate the requirement. Lettuce (*Lactuca*) and birch (*Betula*) are examples of light-requiring species that will germinate in darkness if the seed coat is removed. The germination of freshly harvested grains of many cereals is poor, but gradually improves during dry storage. Removing the husk can often substitute for this so-called after-ripening requirement. Dormancy in a number of *Acer* species is broken by chilling intact seeds, but embryos freed from the testa are able to germinate without prior exposure to low temperature.

In many cases the seed coat inhibits germination for physical reasons. For example, seeds of *Cotoneaster* do not germinate until 2 years after shedding because only the imbibed seed is responsive to chilling and the hard seed coat does not become permeable to water until the following summer. Passage through the digestive tract of a seed-eating animal may also overcome inhibition by a hard seed coat. Another effect of the seed coat is to restrict gas exchange: removing or slitting the coat, or exposing the seed to high concentrations of oxygen, promotes germination of *Cucurbita* and *Betula*, for example. Restricting the oxygen supply can reimpose dormancy (**secondary dormancy**) on non-dormant seeds of a number of species including *Xanthium* (cocklebur), apple (*Malus*) and pear (*Pyrus*).

17.2.2 Terminal buds consist of leaf or flower primordia and unexpanded internodes enclosed in protective scales

Most woody plants in temperate regions form resting buds during the period of the year unfavorable for

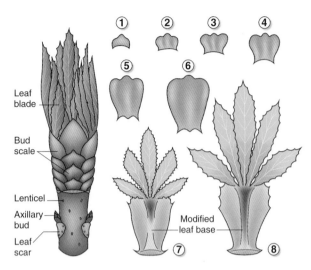

Figure 17.3   Dissection of the component parts of the expanding bud of horsechestnut (*Aesculus hippocastanum*), showing the transition from bud scales (1–6) to true leaves (7, 8).

growth. A typical resting bud is covered by protective scales derived either from stipules (outgrowths associated with leaf bases), as in *Betula* and *Quercus* (oak), or from modified leaves, such as those of *Viburnum*, *Acer*, *Malus* and *Fraxinus* (ash). Figure 17.3 presents the structure of the bud of horsechestnut (*Aesculus hippocastanum*), showing the transition between scale leaves and true leaves after dormancy has been broken and environmental conditions allow expansion to resume.

Overwintering buds generally contain a number of leaf primordia and compressed internodes. Bud break and shoot expansion in spring is usually followed by the production and growth of new primordia and the establishment of the summer canopy. In a few species, such as beech (*Fagus*) and oak, all the leaves that expand in spring and summer were pre-formed in the resting bud. Sometimes in these species a second flush of growth (known in folklore as lammas shoots) occurs from prematurely bursting terminal buds in late summer. The winter-dormant vegetative buds of *Salix gracilistyla* (rosegold pussy willow) are are formed about a month before the shortest day of winter and show 50% bud break 2 months later. Overwintering **generative buds** (that is, buds containing inflorescence primordia) become visible in mid to late June and the subsequent production of flowers (catkins) is completed in early April (Figure 17.4). In addition to members of the willow family (Salicaceae), woody species of the Betulaceae (birch), Fagaceae (beech) and Moraceae (mulberry) families develop both vegetative buds that produce leafy shoots and generative buds giving rise to catkin inflorescences.

The shoots of lime or basswood (*Tilia*), elm (*Ulmus*), chestnut (*Castanea*) and a number of other species do not form terminal buds. Instead, stem extension culminates in death and abscission of the apex, leaving the uppermost axillary bud to resume growth in the following season. Consequently the shoot system of these species has a distinctive, sympodial architecture (compare Figure 16.20).

17.2.3 Tree rings, the results of annual periods of vascular cambium growth and quiescence, are a historical record of environmental conditions

In the unfavorable season, both linear growth and increase in girth of stems are arrested. The yearly cycle of growth and quiescence in woody species gives rise to tree rings (Figure 17.5). When resting vascular cambium is activated, it first produces a zone of less dense xylem tissue with relatively large cells, called earlywood. Subsequently, as growth slows towards the end of the favorable season, cells become smaller and tissue becomes denser and is referred to as latewood. In temperate regions, each tree ring, consisting of a zone of earlywood and a zone of latewood, represents a single year's growth. As growth shuts down late in the season the parenchyma cells of the phloem, immediately beneath the outer bark layer, become sites of reserve storage.

Cambium activity and vascular development are sensitive to abiotic and biotic conditions such as water supply (Figure 17.5), temperature and prevalence of disease. As a direct consequence, the widths of rings represent a historical record of the tree's environment. **Dendrochronology**, the study of the dates and patterns of annual rings, based on statistical sampling of representative wood specimens, enables past ecologies to be reconstructed. Because it is often possible to establish dates for annual rings going back hundreds, and in a few cases thousands, of years, dendrochronology provides important data on past climate trends. The precision with which tree ring measurements can identify events in environmental history is illustrated by the record of North Alaskan summer temperatures (Figure 17.6). The aberrantly low value for the year 1783 (Figure 17.6B) corresponds to one of the largest volcanic eruptions ever recorded, that of the Craters of Laki in Iceland, which brought about global climatic disruption and killed an estimated 6 million people worldwide through crop failure and famine. The tree ring for 1780 (Figure 17.6A) contrasts appreciably with those on either side and is related to a year of major forest fires, fog and thick cloud

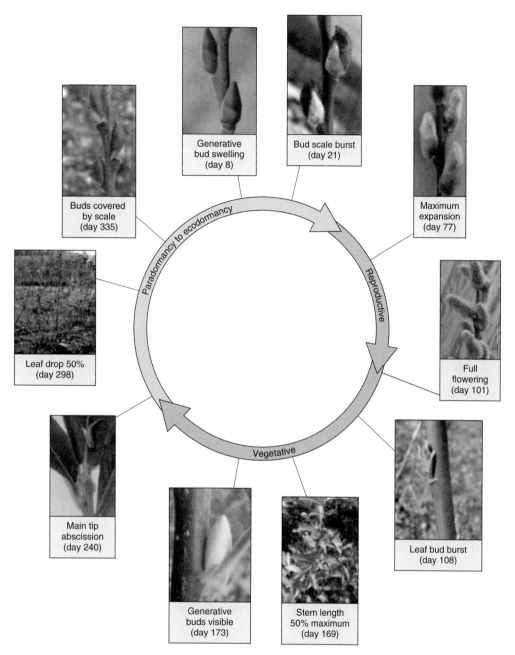

Figure 17.4   The annual cycle of the willow species *Salix gracilistyla* with the mean day of year when the selected stages were recorded in 2007 and 2008. The cycle is divided into reproductive and vegetative phases, as well as a period of dormancy preceding generative development.

cover over parts of North America. Other details of tree ring pattern can be similarly linked to recorded environmental variations.

17.2.4 Corms, rhizomes, stolons and tubers are modified stems

Modified stems may produce perennating structures. A **corm** is a swollen base of a stem consisting of a basal plate (from which adventitious roots develop), a bulky, uniformly solid mass of parenchymatous storage tissue, a thin covering tunic layer and an apical bud (which will form the aerial shoot). Examples of garden plants developing from corms include *Gladiolus*, *Crocus* (see Figure 1.33E) and *Freesia*. Among food plants, banana (*Musa*) forms corms, and the corm is the edible part of Chinese water chestnut (*Eleocharis dulcis*) and taro (*Colocasia esculenta*). Every year in the life of a corm-forming plant a new corm will develop on top of

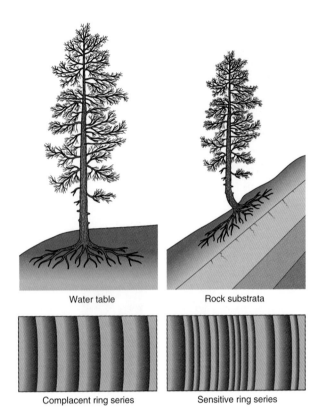

Water table   Rock substrata

Complacent ring series   Sensitive ring series

Figure 17.5 Comparison of annual ring size in two trees with differing annual water availability. The tree on the left received a consistent water supply, while the tree on the right received a variable annual water supply resulting in annual rings of variable width.

the old, which will usually shrivel away. The tendency for successive corms to emerge from the ground as they are built on the foundation of the previous year's organ is offset by the action of contractile roots that pull the corm deeper into the soil. In many species, small propagative structures called **cormels** may proliferate from the corm base.

Some plants such as sacred lotus (*Nelumbo nucifera*), quaking aspen (*Populus tremuloides*) and many grasses (e.g. bent grass, *Agrostis stolonifera*) produce underground stems called **rhizomes** (see Figure 1.33B). Unlike the erect stems of a shoot, which are negatively gravitropic, rhizomes grow horizontally, have scale-like leaves and are etiolated as a consequence of developing in darkness below the soil surface. Rhizomes may be relatively long and thin or, as is the case in ginger (*Zingiber officinale*), they may be compact and swollen, serving as food storage organs. Horizontal stems that grow along the surface of soil are called **stolons** or runners (see Figure 1.33A); these organs have much longer internodes than rhizomes. The spreading stem of creeping species like strawberry (*Fragaria*; see Figure 15.1) is a stolon.

Key points Dormancy is the suspension of growth and development under permissive environmental conditions. The dormancy phase in the plant life cycle can affect buds, cambium, underground perennating organs, seeds and fruits. Dormant structures are resistant to stress. Their growth curves are interrupted, meristem cell cycles are arrested and they generally store reserve substances that power the resumption of development once dormancy has been broken. Dormancy of seeds and single-seeded fruits is often imposed by the seed coat, which is a source of chemical inhibitors or constitutes a physical barrier to germination. Resting buds are telescoped shoots (vegetative or reproductive) covered by protective scales. Tree rings are formed by the annual cycle of activity and quiescence in vascular cambium, the meristematic tissue responsible for increase in stem girth. Because of the environmental sensitivity of cambial activity, the often extreme longevity of individual trees and the preservation of wood structure, tree rings represent a useful historical record of growth conditions and climate.

Many persistent weeds, for example couch grass or quack grass (*Agropyron repens*), owe their invasiveness to the production of rhizomes and the ability to regenerate plants from rhizome fragments left in the soil by tillage or mechanical weeding. The persistence and storage capacity of rhizomes are also essential factors in the agronomy of so-called second-generation energy crops such as the giant perennial grass *Miscanthus* (see Section 17.3.5).

The white potato (*Solanum tuberosum*) is an example of a **tuber**, a modified stem. Each potato tuber develops as a swelling at the tip of a horizontal stem variously referred to as a stolon or rhizome (see Figure 1.33C). Unlike a corm, a tuber lacks a basal plate from which roots develop. It has buds scattered over the surface in a spiral pattern, with an apical bud at the distal end relative to the point of attachment to the plant. The 'eyes' of a potato are axillary buds, each associated with a small curved scar, the remnant of a scale leaf. Each bud can develop into an upright aerial shoot; adventitious roots develop at the bases of these shoots. Unlike the tunic of a corm, potato skin is a layer of periderm that replaces the epidermis early in tuber development and consists of a protective barrier of phellem cells with suberized walls. The tuber of potato and most other tuberizing species exists for 1 year and decays after its reserves are withdrawn to support the next season's vegetative growth.

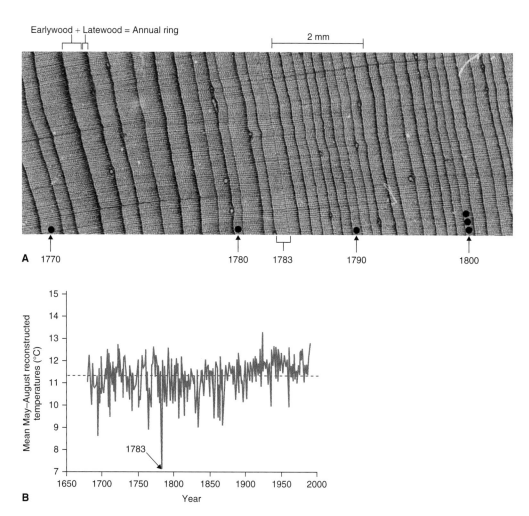

Figure 17.6 Annual rings are a record of the past environment to which the tree was exposed. (A) Tree rings in a representative cross-section of an Alaskan white spruce (*Picea glauca*), aligned with dates between 1770 and 1800. The ring corresponding to 1783 lacks a latewood zone. (B) Historical temperatures in northern Alaska reconstructed from tree ring data, showing the occurrence of an extreme event in 1783.

17.2.5 Bulbs are vegetative resting organs in monocots and each consists of swollen reserve-storing leaf bases surrounding a compressed shoot

Corms, tubers and other underground perennating organs are colloquially called 'bulbs' but a true **bulb** has certain specific characteristics that distinguish it from other modified shoots. A bulb consists of a bud surrounded by layers of swollen leaf bases, which store food (see Figure 1.33D). The modified leaves are inserted on a stem that forms a basal plate, from which adventitious roots emerge. Buds often develop into bulblets or offsets. Some species develop bulb-like

propagules (**bulbils**) in leaf axils on aerial stems (see Figure 11.22B). All true bulb-forming plants are monocots and the compressed construction of the organ with its lack of extensible internodes reflects the architectural principles on which the general morphology of this plant group is based. The most widely consumed bulb-forming food plants are onion, garlic, leek and other members of the family Alliaceae. Other families that develop true bulbs include the Liliaceae (e.g. lily, tulip) and Amaryllidaceae (e.g. *Amaryllis*, *Hippeastrum*, *Narcissus*). Perennial grasses generally survive the unfavorable season by losing foliage and storing reserves in resistant crowns consisting of bulb-like swollen leaf bases. As its name suggests, the resting structure of *Hordeum bulbosum*, a summer-dormant perennial grass closely related to cultivated barley, is a bulb. It forms from the lowest node of the flowering shoot in response to a change from short to long day lengths.

17.2.6 Tuberous roots and swollen taproots are forms of underground perennating storage organs

Cassava (*Manihot esculenta*), yam (*Dioscorea* spp.) and sweet potato (*Ipomoea batatas*) are examples of underground storage organs derived from modified lateral roots. Such so-called **tuberous roots** resemble, but are morphogenetically distinct from, stem tubers. The swollen secondary roots of sweet potato, for example, are internally and externally root-like in construction. They lack nodes and internodes, and produce adventitious roots and stems from the crown. Reserves (mostly starch) are stored in proliferated root parenchyma cells within which are embedded scattered small vascular strands, each surrounded by cambium. The storage roots of some species such as herbaceous peonies (*Paeonia* spp.) and forms of daylily (*Hemerocallis* spp.) are fleshy, but are not swollen to the extent that they could be considered to be tuberous.

In many species the **taproot** is enlarged and persistent, conical or spherical in shape, grows vertically downward and tapers to a point. Gardeners are aware that dandelion (*Taraxacum officinale*) and many other weeds are difficult to eradicate by uprooting because their taproots break, leaving fragments in the soil that rapidly resprout. The bulky tissues of swollen taproots largely consist of storage parenchyma which accumulates sucrose, starch and other carbohydrate reserves. Among the important root crop species grown for the nutritional qualities of their taproots are carrot, parsnip, radish, turnip and beet.

17.3 Synthesis and deposition of reserves

The capacity to lay down storage compounds in resting structures is one of the key traits evolved by vascular plants. By accumulating reserves, the plant evens out fluctuations in the supply of, and demand for, resources. Storing raw materials during periods of abundant carbon and nutrient assimilation is an investment against times when the environment becomes unfavorable. Cells that are packed with osmotically inactive storage polymers, which to a significant degree also displace intracellular water, are intrinsically resistant to dehydrating stresses (see Chapter 15). Rapid growth and development powered by mass mobilization of reserves gives the organs and individuals of the next generation a competitive edge.

Key points Underground resting organs are modified stems or roots. Corms are swollen stem bases, each with a basal plate, from which the roots grow, and a resting apical bud. Rhizomes and stolons are horizontally growing stems by means of which plants are able to invade their surroundings and propagate. In some cases stolons and rhizomes become enlarged and serve as storage organs. For example, the stem tubers of potato are developed from the swollen tips of stolons. A bulb is made up of enlarged, reserve-storing leaf bases surrounding a bud. Root tubers are derived from modified lateral roots and each generally consists of a toughened covering layer enclosing storage parenchyma within which small vascular bundles are scattered. Enlarged taproots serve as perennating organs in many species, including dandelion and other persistent weeds, and major root crops such as carrot, parsnip and beet.

If an organ is a net importer of nutrients (nitrogen, phosphorus, potassium, sulphur and other minerals) and of assimilates (carbon directly or indirectly derived from photosynthesis), it is referred to as a **sink**. Developing resting structures that accumulate storage compounds are strong sinks. Organs that supply the metabolism of a sink with precursors are referred to as **sources** (see Chapter 14). The development and composition of a resting structure are regulated by source–sink interactions. Sources and sinks communicate through the vascular system. Figure 17.7 shows how directional flows through phloem and xylem connect foliage and roots (sources) with developing leaves, fruits (sinks) and fleshy perennating organs (sinks). The **endosperm** of seeds and the **storage parenchyma** of corms, tubers and bulbs accumulate reserve carbohydrate and are supplied via the phloem with assimilated carbon mostly fixed by current photosynthesis (Figure 17.8). Reserve nitrogen is accumulated in the form of specific storage proteins. Some of the amino acid precursors of storage protein may be the products of newly-assimilated inorganic nitrogen, but generally speaking most amino acids imported by the sink are the phloem-borne recycled products of protein degradation occurring during senescence of source tissues (Figure 17.8; see Chapter 18).

Crop yield is essentially synonymous with sink size. A recurring question for agricultural research is: what determines yield—the capacity of the sink or the strength of the source? In fact there is no single answer. For a given crop, yield may be sink-limited under one set of conditions and source-limited under another. A related question is: how is sink demand communicated

a signaling role. Based on the examples of the flowering factor FT (see Chapters 8 and 16), and the antipathogenic signal systemin (see Chapter 15), we might even expect mobile proteins to be part of the source–sink communication system.

The following sections discuss the biochemistry, organization and regulation of storage compound synthesis in sink tissues during the development of resting structures, giving particular attention to starch and fructans, the lipids of oilseeds, the reserve proteins of seeds and vegetative organs and the relocation of mineral elements during dieback of above-ground biomass.

17.3.1 Starch is synthesized in plastids as semicrystalline granules by starch synthase and starch branching enzyme

Starch is the second most abundant plant carbohydrate polymer after cellulose, and the principal source of carbon and energy in the human diet. Cereal grains, potato (*Solanum tuberosum*), cassava, sweet potato, yam, banana (*Musa*) and legume seeds are the major sources of starch consumed by humans. Starch accounts for most of the dry weight of cereal grains and 'root' vegetables. For example, dry matter represents about 30% of the total weight of sweet potato root (compared with 5–15% for stalk and leaf) and 70–80% of root dry weight is starch.

Starch structure, consisting of **amylose** (linear polymer of glucose) and **amylopectin** (branched glucan), and starch biosynthesis from ADP-glucose were introduced in Chapters 6 and 9. Here we look in further detail at the enzymology and regulation of starch accumulation in storage tissues. Starch is made in **plastids**. Non-endospermous seeds with green embryos, such as peas (*Pisum*) and beans (*Vicia, Phaseolus*), store starch in the chloroplasts of cotyledons. In this case the processes that normally lay down starch as a dynamic repository of photosynthate in the mesophyll cells of leaves (see Chapter 9) have become adapted to produce starch grains, which may become so large they distort the structure of the thylakoid membrane system. The starch-storing chloroplasts in the cotyledons of seedlings of some species such as cucumber (*Cucumis sativus*) retain the capacity for photosynthesis. In germinated *C. sativus* seeds, the cotyledons emerge from the soil and become the first photosynthetic organs of the seedling (see Figure 8.1). Starch biosynthesis in endosperm, terminal buds and the storage parenchyma of perennating organs takes place in **amyloplasts**— colorless organelles that are part of the plastid developmental network (see Figures 4.23 and 4.24B).

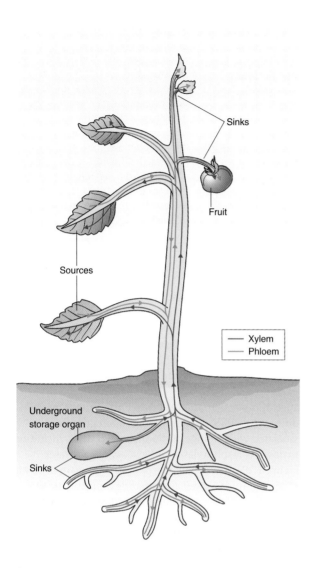

Figure 17.7 Relationship between sources (mature and senescing leaves, roots) and sinks (expanding leaves, fruits, underground storage organs) showing direction of flow in the xylem and phloem.

to the source, or source capacity signaled to a potential sink? Again, the answers are various and depend on species and circumstances. When the sink becomes full, biosynthesis may close down; precursors then back up in the translocation system, pool sizes of intermediates in source metabolism increase and assimilation is downregulated by feedback. In other cases events such as defoliation, shading or disease, which reduce source output, result in slowly developing and undersized sinks. The nature of the signal between source and sink is also diverse. Nutrients, and products of photosynthesis, such as nitrate and sucrose, are mobile integrators of source–sink physiology, acting both at the level of metabolism and through gene expression networks. Hormones moving through the vascular system also play

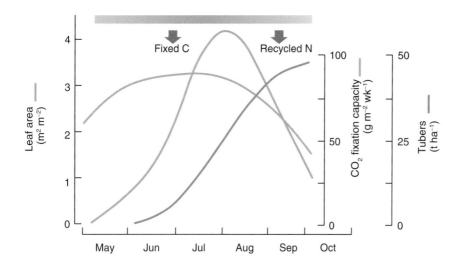

Figure 17.8 Relationship between photosynthesis and senescence of source leaves and yield of white potato tubers (sinks). Photosynthetic capacity, which supplies fixed carbon for the accumulation of tuber starch, begins to decline at the end of the summer. Senescence commences at the same time as the loss of leaf area (indicated at the top of the figure by the trend from green to yellow), and supplies remobilized amino acids (recycled N) to support the synthesis of tuber storage protein.

The enzymology of starch synthesis in amyloplasts is broadly similar to that in chloroplasts, but there are some significant points of difference (Figure 17.9). Sucrose transported to the developing endosperm or parenchyma cell from source leaves is unloaded from the phloem and undergoes a number of interconversions in the cytosol. Unlike chloroplasts, amyloplasts are heterotrophic and must import the carbon for starch biosynthesis from the cytosol. The amyloplast envelope has a glucose-6-phosphate transporter (see Figure 17.9). Synthesis of the linear $\alpha(1 \rightarrow 4)$-glucan chains of starch is catalyzed by **starch synthase** (Equation 17.1A), which extends a pre-existing amylose or amylopectin molecule by adding individual glucose molecules from ADP-glucose. Formation of the $\alpha(1 \rightarrow 6)$ branches in amylopectin is catalyzed by **starch branching enzyme** (Equation 17.1B), which works by cleaving an $\alpha(1 \rightarrow 4)$ linkage and linking the severed fragment $\alpha(1 \rightarrow 6)$ to a glucose about 20 residues downstream.

Equation 17.1 Enzymes of starch biosynthesis

17.1A Starch synthase

$$\text{ADP-glucose} + \alpha\text{-glucan}_{(n)} \rightarrow \alpha\text{-glucan}_{(n+1)} + \text{ADP}$$

17.1B Starch branching enzyme

$$\text{Linear } \alpha(1 \rightarrow 4)\text{-glucan}_{(n)} + \alpha(1 \rightarrow 4)\text{-glucan}_{(m)}$$
$$\rightarrow \text{branched } \alpha(1 \rightarrow 4)\text{-glucan}_{(n)}$$
$$|$$
$$\alpha(1 \rightarrow 6)$$
$$|$$
$$\alpha(1 \rightarrow 4)\text{-glucan}_{(m)}$$

Starch synthase (SS) is encoded by a family of five genes, classified as **GBSS** (**granule-bound starch synthase**), *SSI, SSII, SSIII* and *SSIV*. GBSS is tightly associated with the starch granule and makes amylose. The SS isoforms are soluble enzymes located in the plastid stroma or partly bound to the granule, and work with starch branching enzyme to synthesize amylopectin.

Genetic variants lacking specific SS isoforms produce amylopectins with different structures, indicating distinct properties and roles for each class of soluble synthase. There are two types of branching enzyme (**BE**), differing in the length of chain they transfer. By evolving multiple, specialized SS and BE enzymes, plants have developed the capacity to accumulate starches with a range of molecular architectures. Furthermore, in addition to GBSS, the developing starch granule is associated with a range of modifying enzymes, including amylases, phosphorylases and glycosyl transferases (Figure 17.9). These have the potential to restructure the growing granule as well as dismantle it during net degradation. This biochemical versatility is of major commercial significance, since starches with different properties have a wide range of industrial uses, including adhesives, coating agents, packaging materials, chemical feedstocks, bioenergy substrates and food additives.

The starch-synthesizing organs in various species differ in relative levels of SS and BE isoforms. SSIII accounts for about 80% of soluble SS activity in potato tubers, for example. SSII represents about 60% of synthase activity in pea embryos, whereas about 60% of maize endosperm SS activity is provided by SSI. Mutants deficient in GBSS activity accumulate starch that lacks amylose and are referred to as **waxy**. Table 6.5 lists mutants of several species with waxy seeds. Varieties of white potato are broadly divided into **floury** and waxy, based on the amylose:amylopectin ratio of tuber starch. Floury (high amylose) varieties are suitable for baking and mashing, whereas waxy (low amylose) potatoes have a firmer texture, retain their shape on cooking and so are favored for boiling whole and eating cold in salads (Figure 17.10A). A mutation affecting the synthesis of starch in pea embryos is of special significance in the history of genetics. Round/wrinkled seeds, together with green/yellow cotyledons (see Chapter 18) and tall/short stems (see Chapter 12) were three of the seven characters analyzed by Gregor Mendel in the studies that established

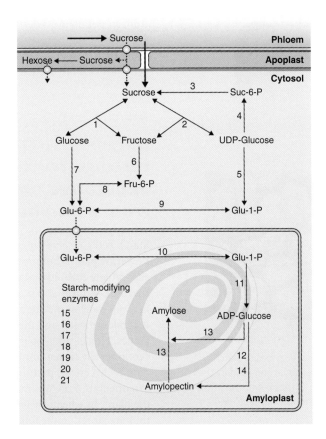

Figure 17.9   Pathway of starch biosynthesis in the storage parenchyma cell of a potato tuber. Enzymes: 1, invertase; 2, sucrose synthase; 3, sucrose-P phosphorylase; 4, sucrose-P synthase; 5, UDP-glucose pyrophosphorylase; 6, fructokinase; 7, hexokinase; 8, phosphoglucose isomerase; 9, cytosolic phosphoglucomutase; 10, plastidic phosphoglucomutase; 11, ADP-glucose pyrophosphorylase; 12, soluble starch synthase; 13, granule-bound starch synthase; 14, starch branching enzyme; 15, granule-associated phosphorylation factor R1; 16, isoamylase; 17, α-amylase; 18, β-amylase; 19, pullulanase; 20, starch phosphorylase; 21, maltosyl transfer enzyme (D-enzyme).

the laws of heredity. Wrinkled seeds (Figure 17.10B) are homozygous recessive at the *R* (*RUGOSUS*) locus and lack the BEI isoform of branching enzyme. Comparison of the DNA sequence of wild-type (*R*) and mutant (*r*) alleles revealed the insertion of a 0.8 kilobase **transposable element** in the latter, which results in an aberrant transcript and deficient BEI enzyme activity.

In many species, deficiencies in enzymes of starch biosynthesis are associated with a buildup of soluble sugars. An example is **sweetcorn** varieties, in which at least three distinct recessive genetic loci (*sugary1*, *shrunken2* and *sugary enhanced*) are responsible for high levels of endosperm sugar. Starch in wrinkled (*rr*) peas amounts to about 30% of dry weight, whereas the value for round types is about 50%, and the soluble sugar content of wrinkled is correspondingly greater—10% compared with 6%. The wrinkled phenotype arises in part because soluble sugars retain less water than starch grains and the seed shrinks in an irregular manner during the desiccation phase of maturation.

Soluble sugars accumulate in dormant starchy root and stem organs such as potatoes during post-harvest storage under chilling conditions (less than $10\,^{\circ}$C). This so-called low-temperature sweetening behavior is undesirable in potatoes used for the commercial manufacture of products such as chips or French fries. If levels of hexoses build up to more than about 2.6% of tuber dry weight, they may react with free amino acids during processing, resulting in dark-colored, bitter-tasting fries and chips and, in some cases, unacceptable levels of the suspected carcinogen acrylamide. At frying or oven temperatures, reducing sugars and amino acids undergo the **Maillard reaction**, beginning with the formation of *N*-glycosylamine (Equation 17.2). The initial phase is followed by a complex series of reactions and rearrangements, leading to the production of brown nitrogenous polymers and co-polymers called **melanoidins**, which are responsible for off flavors and aromas. Maillard reactions also underlie some desirable food sensations such as malty flavors, the golden appearance of bread crust and the aroma of roasted coffee.

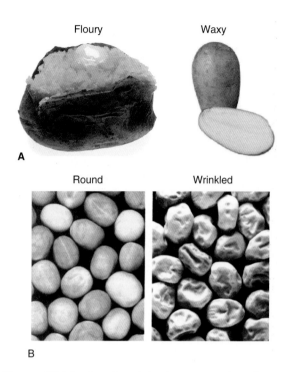

Floury   Waxy

A

Round   Wrinkled

B

Figure 17.10 Genetic variation in enzymes of starch biosynthesis has effects on the appearance and culinary uses of potatoes and peas. (A) Floury (high amylose) and waxy (low amylose) potatoes (*Solanum*). (B) Round (wild-type) and wrinkled (starch branching enzyme-deficient) pea (*Pisum*) seeds.

Equation 17.2 Initial phase of the Maillard reaction

$$R\text{-}CHO + R^1\text{-}NH_2 \rightarrow R\text{-}CHOH\text{-}HN\text{-}R^1$$
Sugar   amino
       acid

$$\rightleftharpoons \quad H_2O + R\text{-}CH\text{-}N\text{-}R^1$$

$$\rightleftharpoons \quad N\text{-glycosylamine}$$

rearranges

Low-temperature sweetening during dormancy is the result of the gradual degradation of starch and slow synthesis of sucrose using respiratory energy. Some enzymes of carbohydrate metabolism are particularly temperature-sensitive. For example the phosphofructokinase of potato tubers is cold-labile. Preferential inactivation of the enzyme during low-temperature storage leads to restriction of glycolysis, accumulation of hexose phosphates and increased sugar content. Chilling-related changes in transcription have also been observed, and genetic mapping of quantitative traits in populations derived from crosses between sweetening and non-sweetening genotypes identifies more than 20 loci, several of which are associated with genes for carbohydrate metabolism.

17.3.2 Fructans are storage polymers accumulated in the resting structures and other vegetative organs of species from a number of taxa

Starch is the reserve polysaccharide stored by most flowering plants, but more than 36 000 species accumulate carbohydrate reserves in the form of **fructans**—linear and branched polymers of fructose. Fructan-storing species include members of the order Asterales (dicots) and the monocot orders Cyperales, Poales, Liliales and Asparagales. Among the economically important fructan-storing plants are temperate cereals, pasture grasses, vegetables and ornamentals. Fructans may be accumulated throughout the plant. In wheat, barley and other temperate grasses they are stored in the foliage, stems and leaf bases, often in high concentrations, though the fructan content of mature grains is usually extremely low. Large quantities of fructan are laid down in the bulbs of *Allium* species (onions and their relatives), taproots of *Cichorium* (chicory) and tubers of *Helianthus tuberosus* (Jerusalem artichoke). Fructans are water-soluble and therefore osmotically active, and have a stabilizing effect on cell membranes, properties that make them effective antistress metabolites in addition to their storage role (see Chapter 15). Bacterial strains of the genera *Bacillus, Streptococcus, Pseudomonas, Erwinia, Actinomyces* and other taxa also synthesize fructans.

Fructans are polymers based on sucrose to which chains of fructose residues are added. Sucrose is a disaccharide (see Chapter 2) consisting of a glucopyranose moiety linked to a fructofuranose by a $\beta(1 \rightarrow 2)$ bond (Figure 17.11), and is the main storage carbohydrate in sugar beet and sugar cane. The trisaccharide **1-kestose**, which consists of a fructose residue linked $\beta(2 \rightarrow 1)$ to the fructosyl group of sucrose, is the parent molecule of the **inulin** series of fructan polymers. The storage polysaccharide of Jerusalem artichoke tubers is an inulin with a chain length of about 35 $\beta(2 \rightarrow 1)$-linked fructose units (Figure 17.12). Note that inulin (a carbohydrate) is not related in any way to the animal protein hormone insulin! A different fructan series, the **levans**, are chains

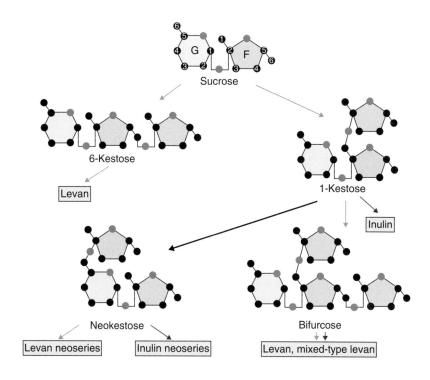

Figure 17.11 Types of fructan and their precursors, showing their structural and biosynthetic relationships to sucrose. Carbon atoms (black) of the glucose (G) and fructose (F) moieties of sucrose are numbered. Oxygen atoms are shown in red. Orange arrows indicate reactions catalyzed by sucrose:fructan 6-fructosyltransferase (6-SFT), green arrow sucrose:sucrose 1-fructosyltransferase (1-SST) and blue arrows fructan:fructan 1-fructosyltransferase (1-FFT). Neokestose is synthesized from sucrose and 1-kestose by fructan:fructan 6G-fructosyltransferase (6G-FFT). The oligosaccharides based on neokestose constitute a neoseries.

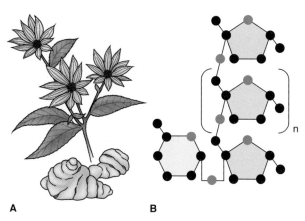

Figure 17.12 Fructan storage in Jerusalem artichoke (*Helianthus tuberosus*). (A) Shoots and tubers. (B) Structure of inulin, the linear β(2 → 1)-linked fructan of *H. tuberosus* tubers. The number of fructose units, n, is approximately 35.

of β(2 → 6)-linked fructose residues based on **6-kestose** (Figure 17.11). The high molecular weight fructans of graminaceous species and bacteria are predominantly levan types. The linear levan or inulin backbones of storage fructans often carry oligofructosyl branches. A third class of fructans, found in plants of the order Asparagales such as onion and asparagus, is based on **neokestose**, in which a glucose residue is enclosed by two fructose moieties. Extending the β(2 → 1)-linked fructose of neokestose gives rise to the inulin **neoseries**, whereas the addition of β(2 → 6) fructose units results

in the levan neoseries. The β(2 → 1)- and β(2 → 6)-linked tetrasaccharide bifurcose has been identified in barley and is the parent fructan for a range of mixed-linkage levans (Figure 17.11).

Fructans are synthesized and accumulated in the central vacuole of the cell. The synthesis of levan series fructans is catalyzed by sucrose:fructan 6-fructosyltransferase (**6-SFT**; Figure 17.11) (Equation 17.3A). The product of the first fructosyl transfer from sucrose (structure and linkage represented as G1 → 2F) is 6-kestose (G1 → 2F6 → 2F), which becomes the substrate for further transfers, building up polymers with a series of chain lengths (G1 → 2F6 → 2F6 → 2F6 → 2F6 → 2F6 → 2F...). Synthesis of 1-kestose from sucrose is catalyzed by sucrose:sucrose 1-fructosyltransferase (**1-SST**; Figure 17.11) (Equation 17.3B). The enzyme fructan:fructan 1-fructosyltransferase (**1-FFT**) transfers fructose residues to yield fructans of the inulin series (Equation 17.3C). Note that fructans, unlike starch and cell wall polysaccharides, are synthesized without substrate activation by reaction with high-energy nucleoside triphosphates. The thermodynamic requirement for fructosyl transfer is entirely met by the breakage of the glucose–fructose bond ($\Delta G^{\circ\prime} = -29.3$ kJ mol^{-1} compared with -30.5 kJ mol^{-1} for hydrolysis of ATP to ADP). A number of other enzymes participate in various aspects of fructan metabolism. The comparison of enzyme properties and DNA sequences of the corresponding genes strongly suggests that fructan synthesizing and degrading enzymes evolved from ancestral **invertases**, that is, enzymes that catalyze the reversible hydrolysis of the glucose–fructose bond of sucrose.

Equation 17.3 Enzymes of fructan biosynthesis

17.3A Sucrose:fructan 6-fructosyltransferase (6-SFT)

Sucrose (G1 → 2F) + sucrose (G1 → 2F)

$\rightleftharpoons$ glucose + 6-kestose (G1 → 2F6 → 2F)

17.3B Sucrose:sucrose 1-fructosyltransferase (1-SST)

Sucrose (G1 → 2F) + sucrose (G1 → 2F)

$\rightleftharpoons$ glucose + 1-kestose (G1 → 2F1 → 2F)

17.3C Fructan:fructan 1-fructosyltransferase (1-FFT)

G1 → 2F1 → 2F[1 → 2F1 → 2F]$_m$

+ G1 → 2F1 → 2F[1 → 2F1 → 2F]$_n$

$\rightleftharpoons$ G1 → 2F1 → 2F[1 → 2F1 → 2F]$_{m-1}$

+ G1 → 2F1 → 2F[1 → 2F1 → 2F]$_{n+1}$

17.3.3 Fatty acids are biosynthesized from acetyl-CoA in plastids and stored as triacylglycerols in oil bodies derived from the endoplasmic reticulum

The fats and oils stored by seeds are predominantly triacylglycerols, fatty acids esterified with glycerol (see Section 6.3.7; for structures of lipids, see Section 2.2.7). Among plants, carbon storage in the form of fats and oils is almost exclusively a feature of seeds and fruits. Few vegetative resting structures accumulate significant amounts of reserve lipid. An exception is the tuber of *Cyperus esculentus*, a perennial sedge native to warm temperate to subtropical regions of the northern hemisphere. The tubers are known as chufa or tigernuts (Figure 17.13) and have been used as food since the time of the ancient Egyptians. Chufa oil represents 20–36% tuber dry weight with a lipid composition similar to that of olive oil (Table 17.1), consisting of around 11% saturated (palmitic and stearic acids) and 88% unsaturated (oleic and linoleic acids) fatty acids (see Chapter 6). Chufa has been proposed as an oil crop for biofuel production. The following discussion of lipid biosynthesis focuses on oilseeds, but also makes reference to general features of fatty acid metabolism in other tissues.

Key points During their development, resting structures are strong sinks for carbon, nitrogen and other nutrients translocated from source organs. Source activity and the capacity of sinks to accept the products exported from sources determine yield in many crops. Starch and fructan are the major storage polysaccharides of resting organs. Starch is biosynthesized in plastids from monomeric precursors derived from translocated sucrose. Starch synthase catalyzes synthesis of the linear α(1 → 4)-linked glucan amylose. Starch branching enzyme is responsible for the α(1 → 6) branches of amylopectin. Deficiencies of specific isoforms of starch-synthesizing enzymes result in starch granules with different properties. Waxy variants of potato, and wrinkled pea seeds, are mutants low in amylose because they lack granule-bound starch synthase activity. Deficiencies in starch synthesis are associated with the buildup of soluble sugars, as occurs in sweetcorn. Accumulation of sugars during low-temperature storage of potatoes can lead to the formation of unpalatable and dangerous products on frying or roasting. Fructans are the soluble polysaccharide reserves of onion bulbs, chicory and Jerusalem artichoke roots, and the vegetative tissues of temperate grasses. Fructans are classified as inulins (chains of β(2 → 1)-linked fructose residues) or levans (β(2 → 6)-linked). They are biosynthesized from sucrose by the activity of fructosyl transferases and accumulate in the cell vacuole, where they can act as osmolytes.

Figure 17.13 Chufa (tigernuts, *Cyperus esculentus*), a rare example of tubers that store high levels of reserve lipid.

Acetyl-CoA, the precursor of fatty acid biosynthesis (Figure 17.14), acts as a donor of acetyl moieties in a series of reactions that build up the acyl chain two carbons at a time (which is why the vast majority of fatty

Table 17.1 Fatty acid composition of oils from tubers of chufa (*Cyperus esculentus*) compared with those of olive fruit (*Olea europaea*) and the fruit and seed of the oil palm (*Elaeis* sp.).

Fatty acid	Degree of saturation	Composition of oil (%)			
		Chufa	Olive	Palm fruit	Palm kernel
Lauric (12:0)	Saturated	—	—	—	48
Myristic (14:0)	Saturated	—	—	—	16
Palmitic (16:0)	Saturated	10	14	44	8
Stearic (18:0)	Saturated	1	3	5	3
Oleic (18:1$^{\Delta 9}$)	Monounsaturated	75	70	39	15
Linoleic (18:2$^{\Delta 9,12}$)	Polyunsaturated	12	12	11	2
Linolenic (18:3$^{\Delta 9,12,15}$)	Polyunsaturated	1	1	—	

acids have an even number of C atoms). The initial step is the ATP-dependent carboxylation of acetyl-CoA to form malonyl-CoA, catalyzed by **acetyl-CoA carboxylase** (**ACCase**). Acetyl-CoA and malonyl-CoA are converted to 16:0 and 18:0 fatty acids by **fatty acid synthase** (**FAS**), a multienzyme complex. In step 2 (Figure 17.14) a transacylase transfers the malonyl group of malonyl-CoA to **acyl-carrier protein** (**ACP**), an essential protein cofactor in the FAS complex. ACP is linked to the growing fatty acid chain, forming acyl-ACP, through a thioester (–S–CO–) bond. Subsequent reactions occur in a cyclic sequence. In the first turn of the cycle, a condensation-decarboxylation reaction between acetyl-CoA and malonyl-ACP yields 3-ketobutyryl-ACP. The enzyme responsible is isoenzyme III of **3-ketoacyl-ACP synthase** (**KAS III**; Figure 17.14, step 3). There follows an NADPH-dependent reduction (step 4), followed by a dehydration (step 5) and a further reduction (step 6), resulting in the formation of fully reduced acyl-ACP. Subsequent turns of the cycle follow a similar reaction sequence except that step 3 is catalyzed by isoenzyme I of KAS. Acyl-ACP carrying the fully saturated 16:0 acyl chain is the product of seven turns of the cycle. A further turn results in 18:0-ACP; in this case step 3 uses KAS II (Figure 17.14). The double bonds of unsaturated fatty acids are introduced by specific, mostly membrane-bound, desaturases. Finally the fatty acid is released from ACP by **thioesterase** (Equation 17.4)

Equation 17.4 Thioesterase

$$RCO\text{-}S\text{-}ACP + H_2O \rightarrow RCOOH + ACP\text{-}SH$$

The biosynthesis of reserve and membrane lipids is compartmentalized within the cell. Figure 17.15 summarizes the reactions and organelle interactions based on studies of lipid metabolism during the development of a number of oilseeds including *Arabidopsis*. Fatty acids are made in **plastids** where the major products, 16:0-ACP and 18:1$^{\Delta 9}$-ACP, are either used for plastid lipid synthesis, or hydrolyzed to free fatty acids by thioesterases in the stroma followed by conversion to acyl-CoAs in the outer plastid envelope and export. The acyl-CoA pool in the endoplasmic reticulum (ER) can be used for the synthesis of phospholipids in a network of transferase, desaturase and phosphatase reactions as shown in Figure 17.15. The triacylglycerols of reserve lipids are synthesized by acyl transfer to diacylglycerol from the acyl-CoA pool or from phosphatidyl choline and are accumulated in oil bodies (**oleosomes**; see Figure 4.18).

17.3.4 Seed and vegetative storage proteins are synthesized in response to the supply of sugars or nitrogen and accumulate in vesicles and vacuoles

Proteins stored by resting structures are a source of amino acids used to support biosynthesis and development when growth resumes. Characteristically, storage proteins are deposited en masse in membrane-limited cell compartments and, in the period between the phases of bulk synthesis and breakdown, they undergo little or no turnover. Amides (principally glutamine and asparagine) and amino acids, translocated from sources through the phloem, are the major precursors for storage protein synthesis, though some legume species such as soybean (*Glycine max*) and

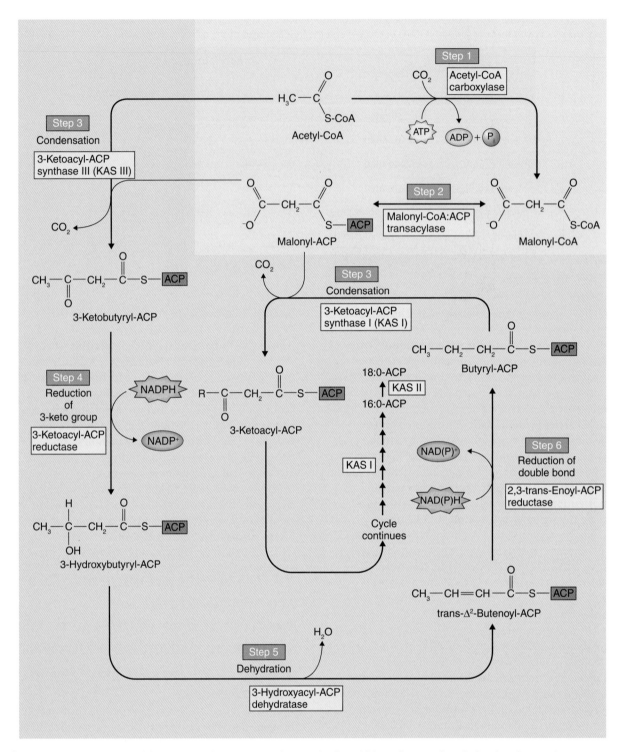

Figure 17.14 Overview of fatty acid synthesis. Fatty acids grow by the addition of two-carbon (C_2) units. The reactions highlighted in yellow show how malonyl-CoA enters the cycle; those on a pink background represent the cyclic reactions. Synthesis of a C_{16} fatty acid requires that the cycle be repeated seven times. During the first turn of the cycle, the condensation reaction (step 3) is catalyzed by ketoacyl-ACP synthase (KAS) III. For the next six turns of the cycle, the condensation reaction is catalyzed by isoform I of KAS. Finally, KAS II is used during the conversion of 16:0 to 18:0.

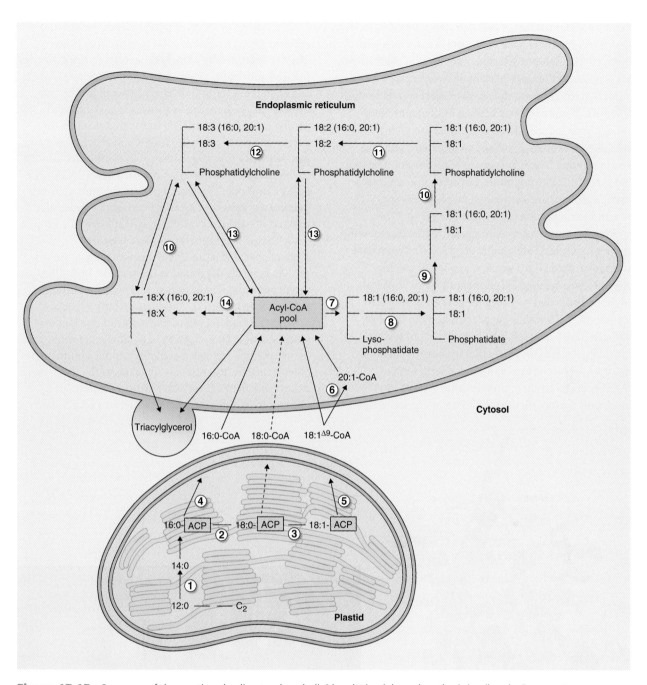

Figure 17.15 Summary of the reactions leading to phospholipid and triacylglycerol synthesis in oilseeds. Enzyme steps are numbered: 1, KAS I- and KAS III-dependent fatty acid synthase (FAS); 2, KAS II-dependent FAS; 3, stearoyl-ACP desaturase; 4, palmitoyl-ACP thioesterase; 5, oleoyl-ACP thioesterase; 6, oleate elongase (synthesis of gondoic acid found in *Arabidopsis* seed oil); 7, acyl-CoA:glycerol-3-phosphate acyltransferase; 8, acyl-CoA:lysophosphatate acyltransferase; 9, phosphatidate phosphatase, product diacylglycerol; 10, CDP-choline:diacylglycerol cholinephosphotransferase; 11, oleate desaturase; 12, linoleate desaturase; 13, acyl-CoA: sn-1 acyllysophosphatidylcholine acyltransferase; 14, same as in steps 7, 8 and 9 using fatty acids from the acyl-CoA pool product diacylglycerol.

cowpea (*Vigna unguiculata*) transport significant quantities of organic nitrogen in the form of **ureides**, derivatives of purine metabolism. When the translocated organic nitrogen compounds arrive at the sink, amide and amino transfers ensure that the composition of the amino acid pool from which reserve protein is synthesized is consistent with the proportions of the different amino acids in storage polypeptides. Plants provide two-thirds of the world supply of edible protein. About 50% of the world's food protein comes from cereal grains and a further 10% from pulses (the seeds of legume species), nuts and oilseeds.

The protein content of cereal grains is about 10–15% of dry weight. The capacity of legumes to fix nitrogen means that pulses are rich sources of protein, with an average content of up to 40% of dry weight (see Table 6.3). By contrast, the leaves of most species contain less than 5% protein. As discussed in Chapter 6, seed storage proteins are classified, according to their solubility properties, as albumins, globulins, glutelins and prolamins. They generally have no enzyme activities but can be enzyme inhibitors, lectins or thionins. Seed storage proteins are synthesized on rough ER and accumulate in vacuoles (see Figure 6.18).

Proteins are also stored in vegetative tissues, including resting stems, leaf bases and roots. The term **vegetative storage protein** (VSP) was first applied to specific glycoproteins synthesized abundantly in the leaves of soybean plants that had been de-podded during seed formation. VSPs with sequence similarity to those of soybean have been identified in a number of species, including *Arabidopsis* where they accumulate in floral tissues. VSP is now used as a general term for a heterogeneous group of proteins with reserve-like biochemical properties that build up in vegetative sinks.

The storage proteins of potato tubers, **patatins**, account for up to 40% of total soluble tuber protein. They are encoded by a single gene locus comprising two putative functional genes and 12 pseudo-genes. There is considerable structural diversity between products of these genes, and protein sequences vary significantly among potato cultivars. The translation products of patatin mRNAs carry a 23-amino acid N-terminal **pro-sequence** that targets the newly synthesized protein to the ER and is removed on import. Figure 17.16A is a molecular model of a typical patatin subunit. Mature patatins are dimers of subunits each with a molecular mass of 40–42 kDa. Nascent polypeptides in the ER pass through the Golgi apparatus where they are N-glycosylated (Figure 17.16A) before the mature protein is finally deposited in the vacuole. In addition to serving as sources of amino nitrogen for mobilization during tuber sprouting, patatins also have acyl hydrolase activity and are believed to play a role in resistance to biotic stresses. Transcription of patatin genes is induced by sucrose or a sucrose metabolite and the abundance of patatin mRNA progressively increases as the tuber swells and its sink demand for sugar grows (Figure 17.16B). Other starchy tuberous resting structures accumulate

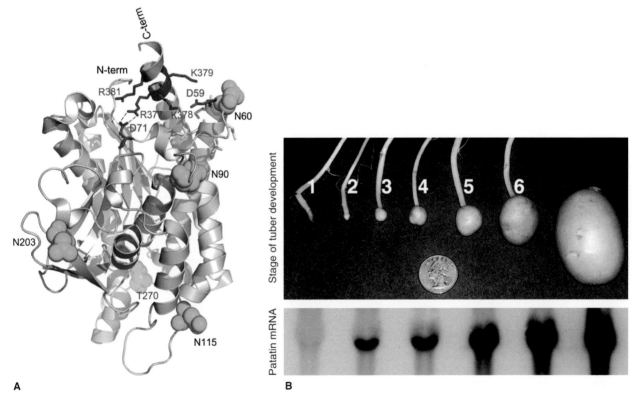

A   **B**

Figure 17.16   (A). Molecular structure of patatin. The C-terminal sequence of mature patatins contains four basic residues: Arg^{377}, Lys^{378}, Lys^{379} and Arg^{381} (blue side-chains). Arg^{377} binds strongly to Asp^{71} (red). Glycans (green) are attached to residues Asn^{60}, Asn^{90}, Asn^{115}, Asn^{203} and Thr^{270}. (B) Abundance of patatin mRNA (middle panel) in potato stolon tips at stages 1 to 6 of tuber development detected by hybridization with a patatin gene DNA sequence.

VSPs in a similar way. For example **sporamin**, a storage glycoprotein of sweet potato that accounts for up to 7% of root tuber dry weight, is also transcriptionally regulated by sugars and processed by the ER–Golgi–vacuole pathway. VSPs have been found in fructan-storing organs. For instance, a small family of VSPs with molecular mass around 17 kDa is accumulated in chicory taproots. In this case, there is evidence that expression of the corresponding genes is regulated by the supply of nitrogen from source tissues.

Deciduous trees recover nitrogen from senescing foliage and store it over winter as VSP in the parenchyma cells of the inner **bark tissue**. Bark storage protein (**BSP**) was originally identified as an important component of the annual nitrogen cycle in apple. Up to 90% of the nitrogen supporting spring growth in apple is supplied by remobilization of BSPs. BSPs have now been described in many species and are considered to be a near-universal factor in the nutrient economies of overwintering woody species. Since the genome of the genus *Populus* (poplar) has been sequenced and annotated, *Populus* spp. have become favored subjects for molecular studies of tree physiology, including BSP accumulation and mobilization. Figure 17.17 shows electron micrographs of bark tissues of *Populus deltoides* in winter and summer. Phloem parenchyma becomes densely packed with protein bodies in winter, when BSPs can account for 60–95% of total protein. Accumulation of BSPs in *Populus* is closely integrated with the mobilization events occurring during senescence of the foliage as described in Chapter 18. Decreasing day length triggers fall senescence and increases nitrogen translocation to the bark, which in turn activates BSP gene transcription. *Populus* BSPs are glycoproteins of molecular mass 32–38 kDa and are believed to be encoded by seven genetic loci. Transcripts of three of these genes are highly abundant in dormant cambium and dormant buds. BSPs encoded by the remaining four loci appear to function in short-term nitrogen storage in actively growing tissues.

Key points Seeds and fruits store lipids, mostly in the form of triacylglycerols. Chufa is an unusual example of a tuber that stores large quantities of oil. The C_{16} and C_{18} fatty acids of reserve lipids are biosynthesized in plastids from acetyl-CoA and malonyl-CoA in a multistep cycle catalyzed by the fatty acid synthetase complex. Unsaturated fatty acids are made by desaturases that introduce double bonds into hydrocarbon backbones of saturated fatty acids. Triacylglycerols, the products of acyl transfer to diacylglycerol, are stored in oil bodies. Nitrogen reserves are accumulated in the form of seed or vegetative storage proteins (VSPs). Patatins are the VSPs of potato tubers and sporamins are the VSPs of sweet potato root tubers. They are glycosylated vacuolar proteins that act as nitrogen sources and have stress-resistance properties. Deciduous trees withdraw nitrogen from foliage during senescence in the fall and store it as BSP, a class of VSP, in the phloem parenchyma of the inner bark tissue.

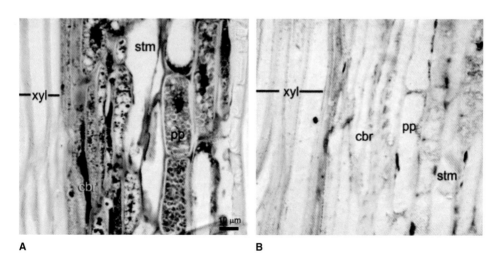

A   **B**

Figure 17.17 Seasonal variation in storage protein accumulation in stems of *Populus deltoides*. Images are light micrographs of longitudinal radial sections of stems sampled from trees in (A) January (winter) and (B) July (summer). Phloem parenchyma (pp) cells of the stem sampled in January contain numerous, small, densely-staining protein storage vacuoles, whereas phloem parenchyma cells of the stem sampled in July contain a large central vacuole and show protein staining only near the periphery of the cells. cbr, cambial region; stm, sieve tube member; xyl, xylem.

17.3.5 Mass transfer of mobile mineral elements to perennating structures occurs during dieback and ultimate desiccation of above-ground biomass

As well as supplying developing perennating organs with carbon from photosynthesis and reduced nitrogen from current assimilation and protein recycling (see Figure 17.8), source shoot tissues, before they die back, send mobile minerals for storage until growth resumes next season. An illustration of the practical importance of mineral cycling between shoots and rhizomes is provided by the highly productive giant grass *Miscanthus* (Figure 17.18). *Miscanthus* is being developed as a **second generation biofuel** crop. First generation biofuels are derived from sugar, starch or vegetable oil and are controversially in competition with food production for feedstocks. The sources for second generation biofuels include lignocellulose from non-food species such as *Miscanthus*.

The usefulness of *Miscanthus* biomass as a combustible or fermentable energy source depends on minimizing the amounts of residual elements other than C, H and O remaining after shoot senescence at the end of the growing season. Nutrient transfer from senescing green tissue to underground rhizomes (Figure 17.18) may occur with such efficiency that growth of the following season's biomass can be supported entirely by recycling the salvaged N, P, K and other nutrients from rhizomes back into leaves, thereby avoiding the need for the external application of any further fertilizer. The sustainability of perennial grasses as a source of renewable energy is in part a consequence of the timely and efficient way they move nutrients between shoots and rhizomes in their growth cycles.

A further important consequence of completely draining above-ground biomass of everything except lignocellulose is **dry-down**, large-scale desiccation which is necessary to maximize the economic yield of dry matter for harvest and transport. Dry-down following the mass transfer of nutrients to resting structures is also significant in cases where seed crop residues such as cereal straw are of practical value. Moreover, dry-down is desirable during the harvesting of grain maize because too much residual moisture in the shoot can clog the cutting mechanisms of combine harvesters.

Desiccation is a feature of the development of the resting structure itself. Seeds dry out during embryo maturation. During this process a range of stress-related

Figure 17.18 Biomass senescence and nutrient transfer to rhizomes in the perennial energy grass *Miscanthus*. Efficient relocation of minerals, and desiccation following shoot senescence, are desirable for economic recovery of lignocellulose for biofuel use.

genes are expressed, including many associated with drought tolerance (see Chapter 15) such as those encoding the appropriately named **late embryogenesis-abundant (LEA)** proteins. A similar spectrum of desiccation-related gene expression is seen during the development of other resting organs. For example, genes encoding LEA-like proteins are activated during differentiation and onset of dormancy in terminal buds, and the expression levels of genes for proteins of the LEA/dehydrin family of proteins are high in potato tubers. The phytohormone abscisic acid (ABA), a key component of networks regulating responses to water limitation (see Figure 15.29), is also influential during the development of dormant organs (see Section 17.5.4).

Key points The persistent underground organs of perennial species receive mobile minerals from shoots that senesce and die back at the end of the growing season. Efficient cycling of nutrients between above- and below-ground biomass reduces the need for minerals from the soil and is of economic importance in the cultivation of perennial biofuel crops such as *Miscanthus*. Associated with senescence and nutrient transfer is dry-down of shoot biomass, a desirable trait for a crop used to produce energy by fermentation or combustion. Desiccation also occurs during seed maturation, during which there is upregulation of genes associated with drought tolerance such as those encoding late embryogenesis-abundant (LEA) proteins and dehydrins. The expression of genes for LEA-like proteins is also enhanced during the formation of terminal resting buds and potato tubers.

17.4 Dormancy

Three types of dormancy have been recognized (Figure 17.19). **Ecodormancy** refers to quiescence due to limitations in environmental factors. An example of ecodormancy is the resting state of mature onion bulbs kept at low temperature. Although there is considerable variation between genotypes, most onion varieties have the capacity to sprout when transferred at any time from cool storage to warmer conditions. In this regard, onions resemble other overwintering monocots such as perennial grasses, which are inactive at chilling

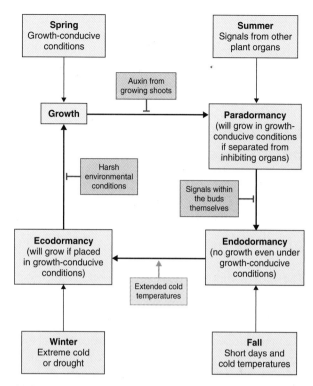

Figure 17.19 Different modes of dormancy in relation to seasonal conditions and growth.

temperatures but able to resume growth immediately the threshold temperature for leaf extension (usually about 5 °C) is exceeded (see Chapter 12). This is familiar to the gardeners of northern Europe and North America, who often find themselves having to mow the lawn during a mild spell in winter.

In **paradormancy** (sometimes called summer dormancy), growth is inhibited by the influence of another part of the plant. The suppression of lateral bud growth by apical dominance is an example of paradormancy and has been discussed in detail in Chapter 12.

Endodormancy is inhibition residing in the dormant structure itself (Figure 17.19). For the purposes of this chapter, we will draw the distinction between **developmental endodormancy**, where the quiescence of the structure is associated with physiological or structural immaturity, and **seasonal endodormancy**, in which the intrinsic dormancy program is executed in response to a sequence of environmental cues. During the seasonal cycle a quiescent structure may pass through some or all of the different modes of dormancy. Thus the resting condition of buds of woody species typically shifts from paradormancy in the summer to endodormancy during the winter to ecodormancy in the spring (as in the case of *Salix*; see Figure 17.4). Examples of the main types of endodormancy are examined in the following sections, and finally we compare and contrast the molecular mechanisms underlying the imposition and breakage of dormancy during development of the different categories of resting structure.

17.4.1 The relationship between embryo immaturity and capacity to germinate, which varies widely between species, is influenced by abscisic acid

The seeds of some species are dormant because their embryos are **immature** when shed. Examples include wood anemone (*Anemone nemorosa*) and ash (*Fraxinus excelsior*). Further embryo development and acquisition of competence to germinate requires adequate water and favorable temperatures. Under optimal conditions ash seeds take several months to complete maturation, whereas the immature seeds of marsh marigold (*Caltha palustris*) are able to germinate after about 10 days. The relationship between dormancy and stage of maturity varies widely across different species. For example certain plants, such as mangroves (*Rhizophora*), proceed directly from embryogenesis to the seedling state. On the other hand, sycamore embryos are dormant throughout maturation, and only become responsive to chilling or removal of the seed coat as they approach maximal weight and begin to dry out (see Figure 17.2B). In general, as discussed in further detail in Section 17.5.4, early and mid phases of maturation are dominated by the action of ABA, initially secreted by maternal tissues. Isolating developing embryos from the influence of ABA allows **precocious germination**. So-called viviparous mutants exhibit this behavior. Figure 17.20 shows a cob of the maize mutant ***vp-1*** (*viviparous-1*) bearing germinating immature grains. There are at least 15 *vp* mutants in maize, each affecting a different step in ABA biosynthesis or perception. *vp-12*, *vp-2*, *vp-5*, *vp-9* and *vp-7* block successive reactions in the pathway of β-carotene synthesis from geranylgeranyl diphosphate (see Figure 15.36), which leads ultimately to ABA (see Figure 10.33). *vp-1* interferes with ABA signal transduction and also has the pleiotropic effect of blocking anthocyanin synthesis, as the distinction between germinating non-pigmented grains and quiescent purple grains in Figure 17.20 demonstrates. Seeds of *ABA-insensitive* (*abi*) and *ABA-deficient* (*aba*) mutants of *Arabidopsis* also germinate precociously. The observations described here lead to the general conclusion that embryo maturation and embryo

Figure 17.20   Precocious germination of immature seed of the *vp-1* mutant of maize. Note the difference in pigment content between precocious and non-precocious grains.

dormancy are physiologically distinct processes under the control of separate though interacting regulatory systems.

17.4.2 Many organs need to experience a period of low temperature to break dormancy

Overwintering resting structures adopt the dormant condition as an adaptation against exposure to low-temperature stress (see Chapter 15), but the **chilling requirement** for breaking dormancy is also a way of measuring the progress of time from winter through spring. The optimal temperature for relieving seed dormancy in many species is around 5 °C. Apple (*Malus*) is typical (Figure 17.21A). Similarly, bud dormancy in woody species is most responsive to temperatures below 10 °C. Figure 17.21B shows that 12 °C is relatively ineffective in breaking the dormancy of an apple cultivar which is stimulated by 9 °C and, more strongly still, by 6 °C. Two environmental cues can induce endodormancy in overwintering apical buds or the resting buds of

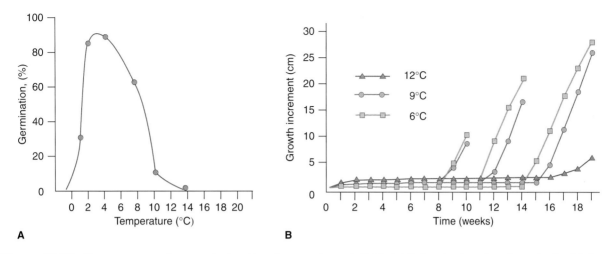

Figure 17.21 Temperature requirements for breaking dormancy in apple (*Malus*). (A) Germination of apple seeds after 85 days of stratification at a range of temperatures. (B) Release of bud dormancy in an apple rootstock cultivar, measured as growth capacity at 21 °C on transfer from the indicated temperatures at 6, 10 or 14 weeks.

tubers, bulbs, corms and other perennating organs: decreasing temperatures in the fall or shortening day lengths. In most temperate woody plants, including species of the model tree genus *Populus*, bud dormancy is typically controlled by **photoperiod**. The same is true of the development of potato tubers and many other underground resting structures. In a few species, notably apple and pear, bud dormancy is imposed by decreasing temperatures but not declining day lengths.

Treatment with gibberellin (GA) can substitute for the chilling requirement in dormant seeds of a number of species such as hazelnut (*Corylus*) and beech (*Fagus*). Dormant sycamore seeds are unresponsive to GA but cytokinin treatment substitutes for stratification. In a few species, such as sugar maple (*Acer saccharum*) and pear, either GA or cytokinin will break dormancy. GA treatment shortens the dormancy period of potato eyes and lateral buds of *Asparagus* rhizomes, and overcomes winter dormancy in buds of fig (*Ficus carica*) and peach (*Prunus persica*). The complementary roles of GA and ABA in growth and dormancy are discussed further in Section 17.5.4.

17.4.3 Weed seeds in the soil seed bank may be released from dormancy by exposure to light

Wherever soil is disturbed, whether by digging in the garden, plowing on the farm or felling in the forest, it is common to see a population of weed seedlings spring up, among which may be species that have not grown in that place for years or even centuries. The population of

> **Key points** Dormancy imposed by environmental limitation is termed ecodormancy. Paradormancy refers to inhibition of growth of one part of a plant by another. Endodormancy is an intrinsic incapacity to grow, because of either developmental immaturity or the requirement for a specific environmental cue. The seeds of many species are endodormant due to embryo immaturity and the inhibitory influence of ABA. Overwintering buds and seeds commonly require a period of chilling to break dormancy. Treatment with GA, or in a few species cytokinin, can overcome the need for chilling. Endodormancy in the apical buds of most woody species and the resting buds of many perennating organs is induced by decreasing day lengths after midsummer. Apple and pear are unusual in developing dormant apical buds in response to declining fall temperatures.

buried seeds is referred to as the **soil seed bank** and is an important reservoir of ecosystem biodiversity. Buried seeds may be considered to be in a dormant state, though whether this dormancy is of the eco- or endo- type depends on species, length of time in the soil and perhaps also the presence of allelochemicals (see Chapter 15). Exposure to light is known to be an absolute requirement for the germination of many weed species, including *Matricaria recutita* (chamomile), *Galinsoga* spp. (shaggy soldier) and *Veronica arvensis* (speedwell). Embryo (endo-) dormancy can rarely be overcome by light treatment. Many weed seeds are embryo-dormant at the time of shedding but over time in the soil become ecodormant, with light as the limiting variable. Seed bank populations have been observed to cycle between dormancy states under the influence of factors such as soil temperature. Experiments in which seeds were exposed to red and far-red wavelengths established that the light requirement is mediated by the phytochrome system as described in Chapters 6 and 8. *Arabidopsis* is, of course, a weed and its seeds display phytochrome-regulated germination (see Figure 8.13). Removing the seed coat of a light-sensitive species (or interfering with its structure, as happens in the various testa mutants of *Arabidopsis*) can abolish the light requirement, suggesting that an action of the phytochrome system in the intact seed is to stimulate a process that renders the coat permeable.

17.4.4 In ecosystems adapted to frequent fires, the chemical products of combustion act as dormancy-breaking signaling compounds

Fire is increasingly understood to be among the most important factors determining the distribution and properties of global ecosystems. Regular burning is a natural feature of the world's grasslands, savannas, mediterranean shrublands and boreal forests (Figure 17.22). Vegetation models predict that, without fire, vast areas of African and South American grasslands and savannas would potentially form forests under current climate conditions, and worldwide there would be a doubling of forested areas, at the expense mostly of C_4 plants but also of C_3 shrubs and grasses in cooler climates. The plants of inflammable ecosystems have a range of adaptations that allow them to tolerate or avoid regular fires, and in many cases periodic burning is a life

cycle requirement. In particular, the seeds of a number of species from fire-prone regions are dormant until they are exposed to **smoke** from burning vegetation. Stimulation of seed germination by smoke extracts has been shown for more than 170 species from 37 families and there are even commercial 'smoke-water' preparations available for horticultural use.

As a result of fractionating the constituents present in smoke, a new class of growth-promoting chemicals has been described, the **karrikins** (Figure 17.22, Table 17.2). The core structure of the bioactive karrikins is a pyran ring (consisting of five carbons and an oxygen; pyran appropriately takes its name from the Greek for fire) fused with a furan ring (four carbons and an oxygen) bearing a keto group. Karrikins are potent stimulants of germination, not only in species of ecosystems regularly subject to burning. *Arabidopsis* seeds are responsive to smoke-water, and KAR1 (Table 17.2) has been observed to induce expression of two key GA biosynthesis enzymes in dormant seed of this species. It is possible that karrikins work by interacting with GA/ABA signaling networks, but there are also indications that their activity is a consequence of their structural relatedness to the strigolactones (see Figure 10.38).

17.5 Regulation of development and dormancy of resting organs

This chapter has described the diversity of resting structures and the wide variety of physiologies they possess. Here we bring together developmental and regulatory principles to show that the mechanisms

Figure 17.22 Composite satellite image of African wildfires during 2002 with (inset) chemical structure of karrikins (see Table 17.2).

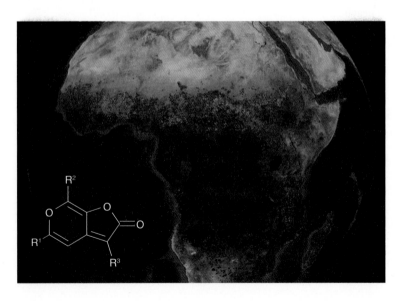

Table 17.2 Chemical structures of karrikins, growth-promoting chemicals derived from smoke. The core karrikin structure is shown in Figure 17.22.

Karrikin	Side group		
	R^1	R^2	R^3
KAR1	H	H	CH$_3$
KAR2	H	H	H
KAR3	CH$_3$	H	CH$_3$
KAR4	H	CH$_3$	CH$_3$

Key points The soil seed bank consists of buried dormant seeds that germinate when disturbance exposes them to light. The light response typically shows the red/far-red reversibility indicative of phytochrome control. In *Arabidopsis* (a seed bank weed under natural conditions) seed coat removal, or altered testa structure resulting from mutation, overcomes the light requirement. Seeds of species adapted to ecosystems subject to regular burning are often dormant until stimulated with compounds present in smoke. Among these are karrikins, bioactive germination promoters with chemical similarity to strigolactones.

plants deploy as they move into and out of the quiescent state have a high degree of commonality not only between different dormant structures but also with growth and differentiation processes occurring in other phases of the life cycle. Table 17.3 presents a developmental scheme for entry into and release from the dormant state and sets out parallel timelines for the different resting structures. The initial growth phase is followed by transition into dormancy, a maintenance phase and then termination of dormancy and transition back to growth. Aligned with the progression from the pre-dormant condition through dormancy to resumed growth are regulatory modules (Figure 17.23) that we have already encountered in the course of this book, emphasizing that the development of resting structures is essentially a set of variations on basic themes common to most aspects of plant growth and adaptation.

17.5.1 The cell cycle is arrested in dormant meristems

At the cellular level, dormancy may be understood in terms of arrested operation of the **mitotic cycle** in the

meristem of the quiescent structure. As described in Chapter 11, regulation of the cell cycle at the G1-phase restriction point, linked to protein kinase phosphorylation/dephosphorylation cascades, usually determines commitment to re-enter the cell cycle. Progression past the G1/S and G2/M mitotic checkpoints of the cell cycle is tightly regulated by **cyclin-dependent kinases** (CDKs). Buds of potato tubers are typical of endodormant structures because their meristem cells are arrested in G1 and exhibit reduced rates of DNA, RNA and protein synthesis. Paradormancy generally appears to occur via a signal transduction mechanism that arrests cells prior to the G1/S checkpoint. Release from cell cycle-related endodormancy is mediated by hormones and signaling mechanisms that control expression, assembly and activity of the CDK system. In *Populus*, transcription of most of the genes encoding **cyclins**, CDKs and other cell cycle regulators is discontinued in developing buds after 4 weeks of short days, about 1 week after bud scales first become visible and 2 weeks before full dormancy is established. Closing down the cell cycle reduces the requirement for synthesis of chromosomal proteins. This is reflected in low expression of **histone** genes during the dormancy phase and an increased activity in tissues prior to sprouting, as illustrated in Figure 17.24 for histone 2A of onion bulbs developed in the field and stored in cool conditions. In Figure 17.23, cell cycle arrest in response to declining expression of components of the cyclin/CDK/histone regulatory module is represented as a critical event in the transition to dormancy. Subsequent dormancy release involves the reactivation of cell cycle regulators and the resumption of meristematic activity. The resumption of DNA replication on release from dormancy is generally a late event and in some seeds, such as sycamore, replication does not fully resume until after radicle emergence.

17.5.2 Dormancy of apical buds is regulated by photoperiod in many temperate species

Seasonal endodormancy is usually a response to declining day lengths. Chapters 8 and 16 describe the **CO−FT** system, which mediates the photoperiodic induction of **flowering**. Overexpressing the poplar homologs of FT and CO in transgenic aspen (*Populus tremula*) results in plants that continue to grow and fail to form dormant buds when exposed to short days. Conversely, downregulating FT expression by RNA interference (see Chapter 3) leads to growth cessation and bud set irrespective of day length. In addition to its role in short-day induction of tree bud dormancy, the CO−FT module has been shown to regulate the tuberization response in potato and bulb development in

Table 17.3 Comparison of transitions between growth and dormancy phases in endo-, eco- and para-dormant resting organs.

	Resting organ	Growth phase	Transition phase	Dormancy established	Dormancy maintained	Dormancy released	Transition phase	Growth phase
Developmental endodormancy	Embryo	Embryogenesis	Growth	Maturation or dormancy		After-ripening	Imbibition	Germination
Seasonal endodormancy	Apical/axillary bud	Growth	Bud set		Endodormancy		Cell cycle resumes	Bud flush
	Cambium	Early growth	Late growth		Endodormancy		Cell cycle resumes	Growth
	Tuber	Stolon growth	Tuber		Tuber eye dormancy		Cell cycle resumes	Sprouting
	Seed/fruit	Embryogenesis, maturation	Embryo dormancy (frequently coat-imposed)				Cell cycle resumes	Germination
Ecodormancy	Bulb	Juvenility / Maturity	Bulb set	Rest period			Cell cycle resumes	Sprouting
Paradormancy	Axillary bud	Initiation	Growth	← (Brief or non-existent in *Arabidopsis*) →			Cell cycle resumes	Outgrowth

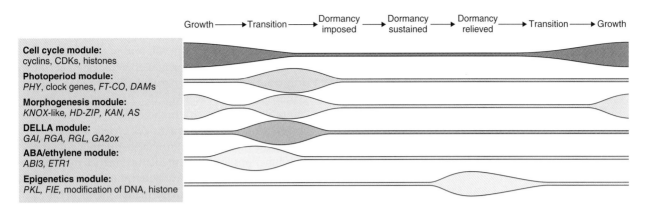

Figure 17.23   The growth–dormancy cycle in relation to the regulatory modules associated with progress between phases. Up- and downregulation are represented by the widths of the colored bands. ABA, abscisic acid; CDK, cyclin-dependent kinase.

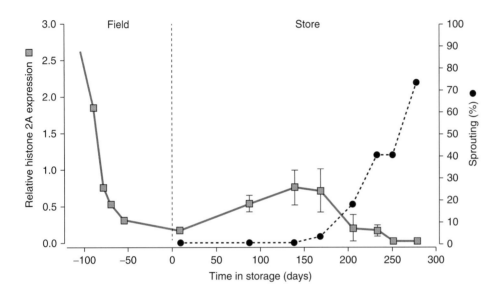

Figure 17.24   Expression of the gene encoding histone 2A in inner bulb tissue during field development, cold storage and sprouting of onion bulbs.

onion. In turn the CO–FT module interacts with the phytochrome and **circadian** systems to regulate entry into dormancy, as it does in the induction of flowering (Figure 17.25). Endodormancy induction and release in a number of species, including poplar, peach, apricot (*Prunus armeniaca*) and raspberry (*Rubus idaeus*), have been linked with the expression of a small family of **DORMANCY-ASSOCIATED MADS-BOX** (**DAM**) genes. These are induced under short days and act as suppressors of FT function (Figure 17.25). The mechanism of dormancy induction by temperature rather than photoperiod, as occurs in species of the Rosaceae such as apple and pear, is not well understood. *DAM* genes may be a point of interaction between the light and temperature pathways. The scheme in Figure 17.25 includes a DAM–CO/FT–phytochrome–circadian regulatory module.

17.5.3 Resting structures are formed by modification of vegetative development

The common features of dormancy and flowering extend beyond their inductive processes. The switch from vegetative development to the dormant condition that happens in buds is not simply a cessation of growth but, as in floral induction, it is a **morphogenetic** transition. Basic changes occur within the bud that may abruptly modify the course of development. For example, a leaf primordium laid down during the phase of spring development may become committed to form a bud scale during establishment of dormancy, whereas subsequent primordia retain the potential for foliar differentiation at bud break. In some cases a gradual transition

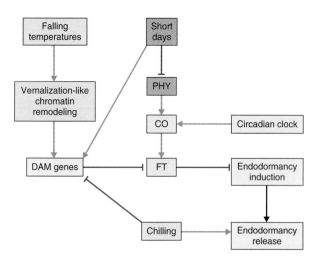

Figure 17.25 The regulatory network for the induction of dormancy by short days or declining temperatures, proposing a role for *DORMANCY-ASSOCIATED MADS-BOX* (*DAM*) genes. CO, Constans transcription factor; FT, flowering time transcription factor; PHY, phytochrome. Bar-ended lines indicate inhibition, arrows stimulation.

Key points Resting structures pass through a common sequence of phases: growth; transition into dormancy; dormancy maintenance; dormancy termination; and resumption of growth. The onset of dormancy is associated with arrest of the cell cycle in meristems. Endodormant buds of potato tubers are arrested in G1. Paradormancy is associated with arrest prior to the G1/S checkpoint. The expression of genes for cyclin-dependent kinases and chromosomal proteins decreases on entry into dormancy and is activated during the breaking of dormancy and prior to growth resumption. Daylength-determined entry into dormancy uses much of the regulatory machinery that controls photoperiodic induction of flowering, notably interaction between FT and CO, phytochrome and the circadian system. The CO–FT module has been shown to regulate bud set in *Populus*, tuberization in potato and bulb formation in onion. FT function in dormancy induction and release is suppressed by DAM proteins, whose genes are induced under short days.

development. Far from shutting down, there may even be a distinct increase in **meristematic activity** and accumulation of leaf primordia without internode expansion at an increasing rate up to the point at which dormancy sets in. This is particularly marked in gymnosperms such as *Abies* (fir), *Picea* (spruce), *Pinus* (pine) and *Larix* (larch). In these cases cell enlargement ceases before the meristematic cell cycle is arrested during the transition to dormancy. A number of genes with functions in meristem identity, meristem activity, organ development and leaf patterning are expressed differentially in relation to dormancy. Genes of the class 1 *KNOX* family—which sustain the meristematic condition and suppress commitment to leaf primordium initiation (see Chapter 12)—are upregulated during bud formation, as are HD-ZIP III, KANADI and AS transcription factors which regulate adaxial/abaxial patterning in leaf primordia (see Figure 12.45). Resumption of growth and development following release from dormancy is associated with reinstatement of the regulatory systems of morphogenesis.

17.5.4 Dormancy is controlled by the antagonistic actions of abscisic acid and gibberellin

Meristem cells in the dormant state are insensitive to growth-promoting signals. Internode extension is under GA control, and the persistence and resumption of growth during the transition into and out of the rest period, respectively, account for much of the effect of GA on dormancy. Day length and temperature regulation of GA synthesis and action are summarized in Figure 17.26. The transcription of genes encoding the **DELLA protein GAI** (GA-insensitive) is strongly upregulated in poplar apical buds immediately after exposure to short days (see Section 12.6.3). Another such DELLA protein, **RGA** (REPRESSOR OF GA1–3), increases on induction of cambium dormancy. The major DELLA regulating seed dormancy is **RGL** (RGA-like). DELLA proteins are growth suppressors. GA overcomes their effects by stimulating ubiquitin–proteasome-mediated DELLA breakdown (see Chapters 10 and 12). Chilling promotes DELLA stability by increasing the transcription of genes encoding DELLAs and the GA-inactivating enzyme GA-2 oxidase (Figure 17.26). Release from dormancy is the process of re-enabling growth. The cessation and resumption of growth during transition into and out of dormancy is shown in Figure 17.23 to be under control of a GA–DELLA module.

In poplar apical buds, ABA, which antagonizes the effects of GA in seed dormancy of some species, peaks in

between the two developmental fates is apparent in the intermediate forms of emerging leaves (see Figure 17.3). These structural shifts are clearly morphogenetic and not merely the consequence of non-specific inhibition of leaf

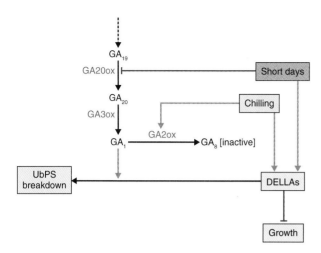

Figure 17.26 Control points in the gibberellin (GA) biosynthesis–DELLA system that determines the potential for growth during the transition into and out of dormancy. Enzymes are shown in blue, metabolic pathways in black, and regulatory interactions in green and red (bar-ended lines indicate inhibition, arrowheads indicate stimulation). UbPS, ubiquitin–proteasome system.

Figure 17.27 Slow shoot growth in a young peach tree after a mild winter. The chilling requirement for release from bud dormancy has been only partially satisfied.

concentration after growth cessation and before bud set, at a time when several genes that function in ABA signaling, such as *ABA INSENSITIVE-3* (**ABI3**), are upregulated. The case in favor of a role for ABA in maintaining dormancy is contradictory, however. It seems more likely that this hormone acts during development of the resting organ, by regulating the formation of buds or late embryogenesis and maturation of seeds, and is a negative regulator of growth after dormancy release. Similarly, upregulation of factors in ethylene signaling (including *ETHYLENE TRIPLE RESPONSE-1*, **ETR1**) is associated more with bud set than with dormancy maintenance in poplar. Figure 17.23 shows the action of ABI3 and ETR1 in the period preceding the transition to dormancy.

As Figure 17.25 suggests, the relation between light and cold induction of dormancy is comparable with that between the photoperiod and **vernalization** requirements for flowering (see Chapters 8 and 16). In both cases there is strong evidence for **epigenetic** regulation. For example, if the chilling requirement to break hazelnut dormancy is incompletely met, the seedlings that develop following germination are slow-growing dwarfs. This condition can persist for a long period but can be reversed by retrospectively meeting the chilling requirement (or by treatment with GA). Similarly, failure to completely satisfy the chilling requirements of dormant buds often results in poor shoot development in the following season (Figure 17.27). As in the case of vernalization (see Chapter 16), there is direct evidence that the repressed

state of chromatin in dormant tissues is mitotically transmitted; this condition can be relieved by low-temperature (and in some cases GA) treatments. Genes with functions in **chromatin** remodeling are among those strongly upregulated by short days early in bud set. They include *PICKLE* (**PKL**), a GA-responsive, general repressor of meristematic activity, and *FERTILIZATION-INDEPENDENT ENDOSPERM* (**FIE**), which is a regulator of the expression of *KNOTTED*-like genes. Further evidence for chromatin remodeling in endodormancy is provided by the observation that dormancy release in the eyes of potato tubers is associated with **DNA methylation** and histone multiacetylation. It is also well established that chromatin remodeling is required for appropriate expression of cell cycle-regulated genes. The general subject of the contribution of epigenetics and chromatin modification to the control of plant growth and development is in its infancy. Expanding knowledge will certainly give new insights into the nature and regulation of the dormant state.

The scheme illustrated in Figure 17.23 shows how environment and development cue the growth–dormancy cycle via the activities of a relatively small number of interacting regulatory systems. In turn, the entire dormancy syndrome is an adaptation integrated with the array of responses to environmental stresses that are described in Chapter 15.

Key points Dormant organs have specialized structures that develop in coordination with the transition to the dormant condition. For example, leaf morphogenesis is modified to form the scales of resting buds. During bud set, *KNOX* genes and genes regulating polarity and pattern formation in leaf primordia are upregulated. Day length and temperature control of entry into and exit from dormancy are mediated through the GA–DELLA–UbPS network. ABA acts as a GA antagonist but does not seem to function in dormancy maintenance. Similarities exist between the role of epigenetic regulation in dormancy induction and its role in vernalization. Transcription of genes with functions in chromatin remodeling is stimulated by short days early in bud development.

17.6 Adaptive and evolutionary significance of the resting phase

The development and physiology of structures that enter, maintain and emerge from the resting condition is closely attuned to the state of the environment. This gives plants the ability to 'travel' in time and space. Time travel by adopting a quiescent state in order to survive a prolonged period of unfavorable conditions is familiar from the images in motion pictures and literature of intergalactic exploration by astronauts in deep hibernation. Just as this is how humans are imagined as able to colonize the most hostile environments, so too it has enabled plants to spread across the face of the globe and occupy a vast diversity of ecological niches, including some of the most marginal and inhospitable of habitats. Dispersal is space travel and allows the immobile parent plant to broadcast its genes and regulate self-competition. The central significance of resting structures in plant architecture and life cycles is the basis of classification systems that relate form to ecological function. Here we discuss propagation through the dispersal of resting structures, the significance of growth and dormancy phases in the seasonal cycle, and the variety of life forms classified according to the

relationship between resting structures and whole-plant morphology. Finally, we consider the influence that the roots, shoots and fruits which currently feed the world have had on the course of human evolution.

17.6.1 Most resting structures are propagules

Plants are able to reproduce sexually, by seeds and fruits. They also reproduce asexually (vegetatively) in a number of ways, including by perennating organs such as the storage roots and shoots described in this chapter. The costs and benefits of sexual versus asexual reproduction are a subject much discussed in evolutionary biology. Generally speaking, the sexual mode of reproduction is costly in terms of resources but is advantageous for fitness of the species by promoting genetic diversity through meiosis and recombination. Vegetative reproduction is less expensive in terms of the commitment of material and energy, and is effective at rapidly propagating and maintaining a well-adapted genome. However, genetic uniformity means inflexibility and an increased risk of loss of fitness. The characteristics of seeds and fruits that facilitate their dispersal are a matter of everyday experience. The attractive colors and flavors of edible fruits originate in the inviting **signals** sent by plants to animal dispersers in the course of coevolution. Seeds with hooks, spines and other means of adhesion exploit animal vectors without rewarding them. Some fruits fly on wings, such as *Acer* samaras, or glide on parachutes, such as the achenes of dandelion (*Taraxacum*) (see Figure 1.22). Others, like those of squirting cucumber (*Ecballium elaterium*), are discharged explosively.

The morphological and physiological features of the rhizomes and stolons of creeping clonal species such as strawberry and clover may be considered to be adaptations that promote vegetative propagation and dispersal. A clonally-propagated colony is called a **genet** and consists of genetically identical individuals or **ramets**. Such populations are often very long-lived. Another example of dispersal of a vegetative propagule is the stress-resistant endodormant bud (**turion**) of aquatic species such as *Spirodela* (duckweed), *Potamogeton* (pondweed), *Hydrilla* (waterweed) and *Utricularia* (bladderwort). Buds are generally formed from modified shoot meristems under short days, become detached from the stolon of the mother frond following the formation of an abscission layer, and sink to the bottom of the water column where they overwinter before resuming growth when temperatures rise in the spring. During turion development cell expansion is reduced

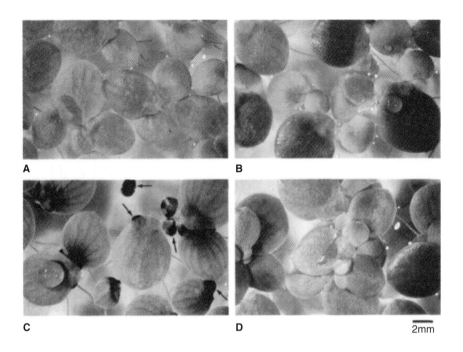

Figure 17.28 The effect of kinetin, a synthetic cytokinin, on ABA-induced turion formation in *Spirodella polyrrhiza*. (A) Control fronds. (B) Fronds treated with 20 μM kinetin. (C) Fronds treated with 250 nM ABA (arrows indicate turions and turion primordia). (D) Fronds treated with 250 nM ABA plus 20 μM kinetin. Fronds were photographed after 8 days of treatment.

2mm

and there is accumulation of starch and anthocyanin. The turion of *Spirodela polyrrhiza* (common duckmeat) has been studied as a model for bud dormancy. ABA stimulates and cytokinin inhibits turion formation (Figure 17.28). Among the genes activated by ABA treatment are ***tur1***, which encodes d-myo-inositol-3-phosphate synthase, the catalyst for the first committed step in the biosynthesis of the key signaling component **inositol**. ABA also induces a stress-related ABC transporter and a phosphatase in this species. Dormant and re-growing turions frequently experience anoxic conditions and express a number of genes related to fermentative metabolism. Vegetative propagation by turions is clearly effective, since a number of species using this strategy, such as *Hydrilla verticillata*, are classified as highly invasive aquatic weeds.

Key points Most vegetative, reserve-storing resting organs are propagules, means by which plants reproduce asexually. Ramets are genetically-identical individual plants propagated by rhizomes or stolons. They form a colony of clones, referred to as a genet. A number of species of water weed propagate by turions, endodormant buds that form under short days, disperse after detaching from the parent plant and grow into new individuals after overwintering. Turion formation is promoted by ABA and inhibited by cytokinin.

17.6.2 Phenology, the study of the timing of growth and quiescence phases in the annual life cycle, provides information on environmental change

Growth, reproduction and resting phases in the plant life cycle are synchronized with the seasons. **Phenology** is the study of the timing of periodic life cycle events over the course of a year, particularly in relation to climate. A typical phenological sequence is shown for *Salix gracilistyla* in Figure 17.4. Plant responses to daily weather tend to be cumulative. Thus the effect of temperature on growth or on the timing of a developmental event will increase with exposure time. This is often expressed as **thermal time**, or day-degrees, or heat units and is calculated as the degrees above a threshold temperature accumulated over a period (Equation 17.5). In dormant species, the effect of winter chilling is to reduce the thermal time to bud break. Figure 17.29 compares the thermal time responses of *Fagus* (beech), which has a high chilling requirement and flushes relatively late, with hawthorn (*Crataegus*) and black cottonwood (*Populus*), in which bud break occurs early. In responding to thermal time, plants act as integrators of environmental data, and this makes phenological observations an important source of information on climate trends.

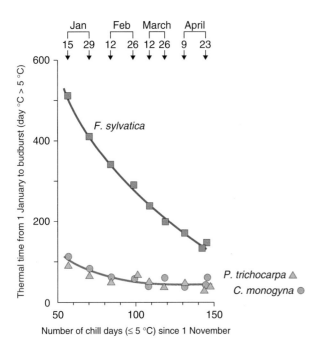

Figure 17.29   Relationship between thermal time (threshold temperature 5 °C) to bud burst and accumulated number of chill days (days below 5 °C) for late-flushing (*Fagus sylvatica*) and early-flushing (*Populus trichocarpa, Crataegus monogyna*) tree species. Replicate saplings were exposed to different lengths of time at winter temperatures by transferring them between a warm greenhouse and an outdoor field nursery. Mean dates of bud burst were recorded, and day-degrees experienced under each treatment were calculated according to Equation 17.5.

Equation 17.5 Calculation of thermal time (D) over the period j to k days

$$D = \sum_{j}^{k} T_n - T_t$$

where T_t is the threshold temperature and T_n the average temperature on day n

The timing of bud break after winter dormancy, observed as the first appearance of spring foliage, is directly related to thermal time and often shows the strongest response to year-to-year temperature change. Figure 17.30 presents the trend in date of annual bud break, based on data from more than 1400 stations across the northern hemisphere from 1955 to 2002, determined as the number of days by which it departs from the mean date over the period 1961 to 1990. The data are as variable as expected for such a large-scale set of observations, but the fitted line is statistically significant and shows that on average, across the hemisphere, spring is beginning about 1 day earlier each decade. Phenology,

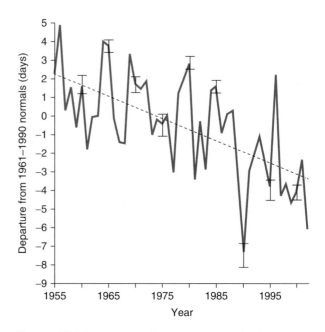

Figure 17.30   Date of leaf appearance in spring by year across the northern hemisphere, 1955–2002, shown as departure from the 1961–1990 mean. The linear regression line shows a trend towards earlier bud break at a rate of about 1 day per decade.

like tree rings (see Section 17.2.3), contributes important evidence to the debate about climate change.

17.6.3 Different life forms have evolved through changes in integration of component developmental processes

We saw in Chapter 12 that plant architecture is based on the reiteration of structural units. Variations on the theme of modular construction have given rise to a vast range of morphologies and life cycles in the modern flora. Many systems have been proposed to bring order to this diversity. The widely used **Raunkiaer** scheme is based on resting structures and classifies plant life forms according to the survival of apical meristems in the active or dormant state (Table 17.4). The various life forms are characterized by the extent to which shoot axes persist (**phanerophytes, chamaephytes**), retrench (**hemicryptophytes, cryptophytes**) or die outright (**therophytes**) at the end of the growing season. Annuals (such as *Arabidopsis* and most cereal crops) and many biennials are therophytes. The formation of resting structures, together with meristem determinacy (see Chapter 12) and the progressive programmed senescence

Table 17.4 The classification of plant life forms (after Raunkiaer).

Life form	Definition	Types included	Schematic representation
Phanerophytes	Generally tall plants visible throughout the year, carrying surviving buds or apices at least 25 cm up from the ground. Examples are trees, large shrubs and lianas	• Evergreens without bud covering • Evergreens with bud covering • Deciduous with bud covering • Less than 2 m high	
Chamaephytes	Low-growing plants visible all year round, bearing perennial buds between ground level and 25 cm up. Examples include shrubby tundra species	• Suffruticose chamaephytes, i.e. those bearing erect shoots which die back to the portion that bears the surviving buds • Passive chamaephytes with persistent weak shoots that trail on or near the ground • Active chamaephytes that trail on or near the ground because they are persistent and have horizontally directed growth • Cushion plants	
Hemicryptophytes	The surviving buds or shoot apices are situated at or just below the soil surface. Examples include perennial grasses, many forbs (non-grass herbs) and ferns	• Protohemicryptophytes with aerial shoots that bear normal foliage leaves, but of which the lower ones are less perfectly developed • Partial rosette plants bearing most of their leaves (and the largest) on short internodes near ground level • Rosette plants bearing all their foliage leaves in a basal rosette	
Cryptophytes	At the end of the growing season, they die back to bulbs, corms, rhizomes or similar underground (in some species, underwater) structures. For example lilies, onions, garlic, potatoes and similar forbs	• Geocryptophytes or geophytes, which include forms with: (i) rhizomes; (ii) bulbs; (iii) stem tubers; and (iv) root tubers • Marsh plants (helophytes) • Aquatic plants (hydrophytes)	

(continued overleaf)

Table 17.4 (*continued*).

Life form	Definition	Types included	Schematic representation
Therophytes	Plants that complete their life cycle from seed to seed and die within a season, or that germinate in fall and reproduce and die in the spring of the following year	• *Arabidopsis* is a therophyte	

and death of organs and tissues (see Chapter 18), determine a plant's position in the Raunkiaer classification described above, and represent the link between development of the whole and its constituent parts.

Understanding the genetic mechanisms underlying the diversity of life forms is a difficult challenge involving attempts to integrate molecular knowledge of developmental control with the functional and architectural principles of adaptation and evolution. It is too simplistic to talk of the 'genes for' the cryptophyte or phanerophyte condition, but genetic and molecular approaches are beginning to shed light on how the development of resting structures in the life cycle can be reorganized. For example, the rhizomatous character is a

desirable trait in some perennial crops. The productive temperate pasture legume white clover (*Trifolium repens*) vegetatively propagates largely by stolons. The related species, Caucasian clover (*T. ambiguum*), is rhizomatous, extremely persistent and drought-tolerant (Figure 17.31). Using interspecific hybridization aided by the identification of **linked DNA markers**, plant breeders have succeeded in transferring the capacity to make rhizomes into the white clover genetic background (Figure 17.31), thereby effectively converting a hemicryptophyte into a cryptophyte (Table 17.4). Similarly, the rhizomatous trait has been introgressed from perennial rice (*Oryza longistaminata*) into *O. sativa* and has been shown to be associated with two major genetic loci.

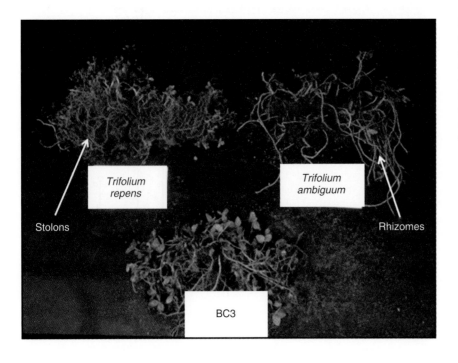

Figure 17.31 The stoloniferous species white clover (*Trifolium repens*), the rhizomatous species Caucasian clover (*T. ambiguum*), and the backcross progeny of an interspecific hybrid between them (BC3) in which the rhizomatous character has been introgressed into the white clover genetic background.

17.6.4 The consumption of plant resting organs has influenced the course of human evolution

In recent years information has emerged that identifies a critical role for the underground storage organs of plants in the early evolution of humans. Fossil evidence shows that **early hominids** used sticks and bones to dig up the starchy reserve organs of savanna plants. There are reports of modern savanna chimpanzees doing the same thing. Some authorities believe that tool use in humans can be traced back to this behavior. Consumption of such tough, fibrous foodstuffs may have influenced the structure of the teeth and jaws of early hominids. A starchy diet has also been shown to have a direct effect on the human genome. The *AMY1* locus, which encodes **human salivary amylase**, is a duplication hot-spot. Figure 17.32 shows that populations such as European–Americans and Hadza (hunter-gatherers of Tanzania) that subsist on a relatively starch-rich diet have significantly higher degrees of *AMY1* duplication than low-starch-consuming populations such as the

Yakut (nomadic hunters of Central Asia) and Mbuti (Congolese pygmy tribes). Higher *AMY1* copy numbers and protein levels are believed to improve the digestion of starchy foods and buffer against the fitness-reducing effects of intestinal disease. Soaking underground storage organs reduces the levels of antinutritional factors, and heating increases starch digestibility. It is conceivable that this is the origin of cooking. Another implication of the consumption of underground storage structures is that, having developed a taste for starchy food, early humans will have diversified their diet to include the grains of wild grasses, leading to the domestication of cereals and other crops and the evolution of agriculture. Thus we may trace the roots of human physiology, behavior, society and civilization back to the properties of the humble rhizome, tuber, corm, bulb and stolon.

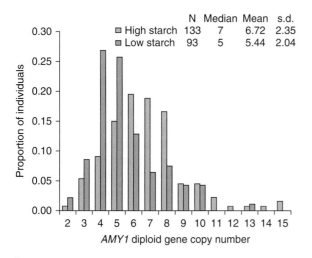

Figure 17.32 Relationship between diet and diploid copy number of the human salivary amylase gene *AMY1*. The histograms show frequency distributions of *AMY1* copy number variation in populations with traditional diets that are high or low in starch. N indicates the number of individuals screened in each group.

Key points Phenology is the study that accumulates, over a period of many years, records of the time points in the annual cycle at which significant events in plant development occur. Because the lengths of different phases in the plant life cycle are sensitive to environmental conditions, phenological data provide evidence of climate trends. Development is sensitive to the duration of exposure to temperature and may be represented on a thermal time scale. In dormant species with a chilling requirement, low winter temperatures reduce the thermal time to bud growth. Phenological studies estimate that the onset of spring in the northern hemisphere is commencing earlier at the rate of about 1 day per decade. The Raunkiaer system classifies different plant life forms according to the ways in which survival of apical meristems and resting structures are organized. Starchy vegetative storage organs have been part of the human diet since early in hominid evolution. Digging up roots with sticks or bones may be an early example of tool use. Cookery may be traced back to the habit of soaking and boiling to improve the palatability of such food sources. The tough, fibrous nature of storage roots and stems may have influenced the evolution of hominid teeth and jaws, and there is evidence from molecular studies of salivary amylase that the starchy diet represented by these foodstuffs has had a selective influence on the human genome.

Chapter 18

Senescence, ripening and cell death

18.1 Introduction to terminal events in the life of a plant and its parts

Selective death of plant tissues, individual cells and even whole organs is necessary for the normal development and survival of plants. The shedding of leaves from deciduous trees in temperate regions is one of the best-known examples, but all plants use controlled death and consequent elimination of parts as essential processes at many stages of their life cycles. **Programmed cell death** (**PCD**) is the general term given to any process by which plant protoplasm, and sometimes the associated cell wall, is selectively eliminated as part of a genetically determined event.

The purposeful nature of the cell death that takes place during development and adaptation tells us it must be a tightly regulated process, hence the use of the term 'programmed'. The PCD concept covers all the ways in which a cell can die in an organized way, but does not specify which mode of death is involved, or the mechanism. There are probably almost as many variants of cell death as there are types of plant cells and terminal events in the life cycles of plants and their parts. This chapter looks at some of the most important ways in which whole plants and their organs, tissues and cells initiate and carry out the processes that lead to PCD.

18.1.1 The different categories of cell death share some features

Programmed cell death is an active process in which metabolism proceeds in an orderly fashion and is dependent on sources of biological energy. PCD is under the control of specific genes, the functions of which can be shown, by mutation, transgenic modification or chemical intervention, to be necessary for terminal metabolic and cellular events to happen at the right time in the right place. Execution of the cell death program takes the form of **cascades** of biochemical activities, in which switching on a specific step in turn switches on other downstream actions, which in turn invoke yet more, thereby propagating the response throughout the entire cell.

Almost every stage in the life cycle of a plant, from germination through to development of the next generation of seeds, is influenced in some way by PCD (Figure 18.1). Interactions with other organisms (pathogens, pests, herbivores, pollinators, to mention a few), and responses to environmental stresses such as

The Molecular Life of Plants, First Edition. Russell Jones, Helen Ougham, Howard Thomas and Susan Waaland.
© 2013 John Wiley & Sons, Ltd. Published 2013 by John Wiley & Sons, Ltd.

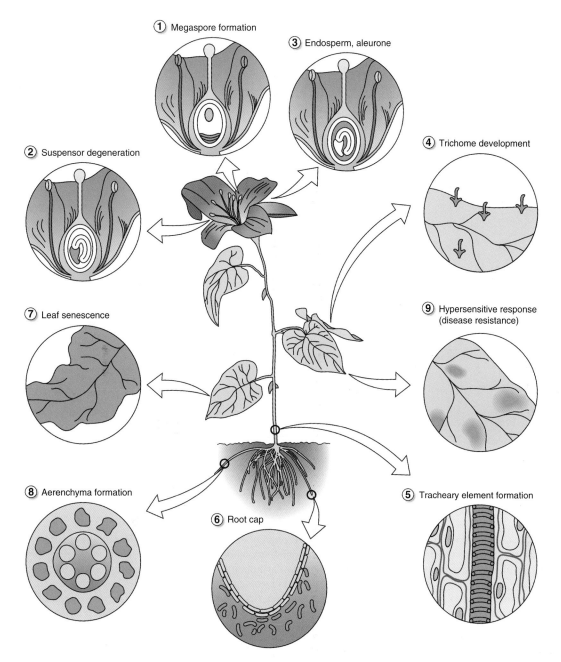

① Megaspore formation

③ Endosperm, aleurone

② Suspensor degeneration

④ Trichome development

⑦ Leaf senescence

⑨ Hypersensitive response
(disease resistance)

⑧ Aerenchyma formation

⑥ Root cap

⑤ Tracheary element formation

Figure 18.1 Programmed cell death (PCD) occurs in many plant cells and tissues and is intrinsic to numerous developmental and adaptive processes, including (1) gamete formation; (2) embryo development; (3) degeneration of tissues in the seed and fruit; (4–6) tissue and organ development; (7) senescence; and (8, 9) responses to environmental signals and pathogens.

drought, heat or nutrient deficiency, also often invoke the programmed death of certain cells. The most intensively studied form of PCD in animals is called **apoptosis**. It has been shown to be a highly regulated process which is under the control of specific well-characterized genes and executed via defined biochemical pathways (Figure 18.2). In most cases, plant PCD appears to involve different genes and pathways from apoptosis in animals.

18.1.2 Cell viability is maintained during the developmental program leading to death

Programmed cell death consists of two phases. In the first, cells are viable; their membranes and organelles remain intact, and organs such as leaves stay turgid. In

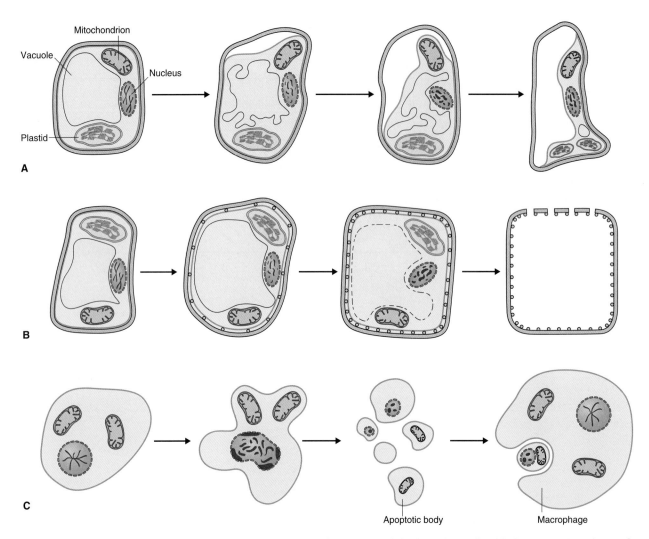

Figure 18.2   Comparison of cytological features of two kinds of programmed death in plant cells with the apoptotic pathway of animal cell death. (A) Hypersensitive cell death, a resistance response to pathogen attack. Condensation and cleavage of DNA in the nucleus precedes vacuole disruption and blebbing of the tonoplast and plasma membranes; the process ends with destruction of organelles, plasma membrane collapse and leakage of the dead cell's contents into the apoplast. (B) Tracheary element differentiation, an example of developmental cell death. Swelling and rupture of the vacuole happens as the cell walls undergo secondary thickening and restructuring. Nuclear DNA fragmentation occurs in the later stages, after vacuolar collapse. Finally autolysis eliminates the remaining cytoplasm, leaving an empty cell enclosed by a thickened, reticulated and perforated wall. (C) Apoptosis in animal cells. Chromatin condensation and fragmentation are early morphological events. The plasma membrane is disrupted and cell contents are repackaged into apoptotic bodies, which are finally engulfed by neighboring cells.

some cases, notably the **senescence** of green (mesophyll) cells in leaves, this phase is reversible: the mesophyll cells are still viable until almost all of the leaf's macromolecules have been recycled and exported to the rest of the plant. This phase can be protracted in some plant parts. The second phase is terminal; it is often rapid, and always irreversible.

18.1.3 Autolysis is a common form of cell death

An important general characteristic of many kinds of PCD in plants is that they are **autolytic**. That is, the cell digests itself, disposing of cell contents by processes that act on and through the very cytoplasm that is being eliminated. Digestion is carried out by **lytic enzymes** that hydrolyze macromolecules, cleaving them into oligomeric fragments and ultimately into monomers. The central vacuole of the plant cell is particularly rich in aggressive hydrolytic enzymes, including peptidases and nucleases, and has an essential role in many kinds of autolysis, either through an autophagy-type engulfing mechanism or by rupture of the tonoplast and release of vacuolar contents into the cytosol.

The hydrolysis of nucleic acids is catalyzed by **nucleases**. A distinction is made between an enzyme that cleaves nucleotide monomers sequentially from the ends of the substrate molecule (**exonuclease**) and a nuclease that hydrolyzes linkages between monomers within the polynucleotide chain (**endonuclease**). Nucleases may be specific for RNA (ribonuclease, **RNase**) or DNA (**DNase**) or be able to use either nucleic acid as a substrate.

The breakdown of proteins (**proteolysis**) is catalyzed by hydrolytic enzymes called **proteases** or proteinases (see Chapter 6). Particularly important in PCD are **endopeptidases**, proteases that cut internal peptide bonds in their substrates (see Figure 6.28). Endopeptidases are classified according to the nature of their active sites. For example, members of the **cysteine endopeptidase** family have a reactive cysteine residue in their catalytic center. Although cysteine endopeptidases are involved both in the apoptotic death of animal cells and in PCD in plants, DNA sequence comparisons show that genes encoding cysteine proteases of plants are unrelated, or only very distantly related, to those of animals. Cysteine proteases with a function in plant PCD are sometimes referred to as **metacaspases**. Members of the class of cysteine endopeptidases known as **vacuolar-processing enzymes** (**VPEs**) are examples of metacaspases. Other families of proteases classified by active site chemistry include **serine proteases** and **metalloproteases**. The **ubiquitin–proteasome system** (UbPS), discussed in Chapter 5, is also important in regulating cell functions through selective protein turnover and the removal of damaged or unwanted polypeptides.

The subcellular mechanisms of autolysis are diverse. Some plant cells dispose of their cellular contents by **autophagy** (literally 'eating oneself'; Figure 18.3). Portions of the cytosol, including intact organelles, are engulfed by vesicles called **autophagosomes**. These vesicles are taken up by the cell's large central vacuole, or, in some cases, fuse with lysosomes (small membrane-bounded lytic organelles). The contents released from autophagosomes, referred to as autophagic

bodies, are broken down by hydrolytic enzymes (Figure 18.3A). Chemical inhibitors of proteases block the operation of autophagic pathways (Figure 18.3B). Autophagosomes have been observed in senescing corolla cells of the short-lived flowers of Japanese morning glory (*Ipomoea tricolor*; Figure 18.3C) and during the hypersensitive response to pathogen attack (Figure 18.3D). Other instances of autolytic processes with autophagy-like features in plants include the death of nutrient-starved cell cultures and of tissues of germinating seeds, and terminal differentiation of vascular tissue.

The model for the mechanism of autophagy and its regulation has been established in detail for yeast, where the process is triggered by amino acid starvation. Homologs of most of the autophagy genes (*ATGs*) of yeast are present in *Arabidopsis* and other plants. Expression of *ATG* genes and the activities of ATG proteins are controlled by a network of regulatory **kinases**. Autophagy begins with a phase of vesicle induction, followed by vesicle expansion, docking and fusion with the tonoplast and finally digestion. The signaling pathways that direct the cell into autophagy converge on the expression and activity of ATG1 and ATG13 (Figure 18.4). ATG1 is phosphorylated by protein kinase A, a negative regulator of autophagy, which is itself regulated by the kinase TOR (Target Of

Key points The death of cells, tissues and organs occurs at all stages of the life cycle of a plant and is an essential part of plant development and adaptation. Programmed cell death (PCD) is an active, energy-requiring, genetically determined phase that propagates in an orderly manner and is followed by irreversible loss of viability. Most forms of PCD are autolytic: cell contents are eliminated by autodigestion, which is catalyzed by proteases, nucleases and other lytic enzymes. The central vacuole is a site or source of hydrolytic activity during PCD. Autophagy is a particular mechanism of autolysis shared by plants, yeast and other eukaryotic cells. A network of kinases, responsive to triggers such as nutrient starvation or developmental signals, regulates the expression of autophagy genes (*ATGs*), the assembly of autophagic vesicles that engulf portions of cytoplasm, and the fusion of vesicles with the central vacuole, where hydrolysis takes place.

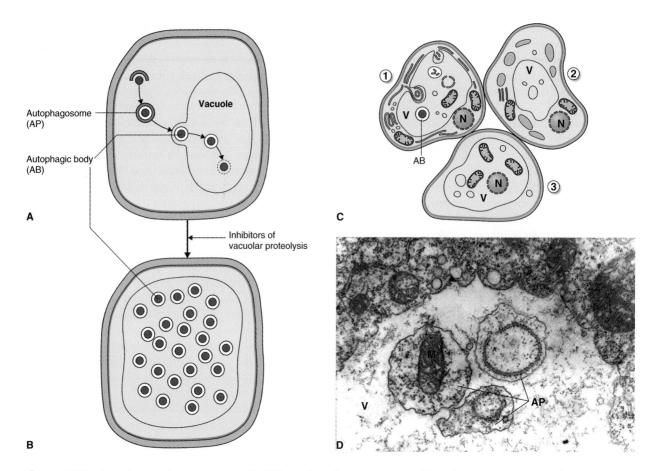

Figure 18.3 Autophagy pathways in plant cells. (A) Induction of autophagy in *Arabidopsis* and many other species results in the formation of a double-membrane autophagosome around a portion of cytoplasm. The outer membrane fuses with the tonoplast, and the inner membrane and contents enter the lumen of the vacuole and are degraded. (B) Treatment with inhibitors of vacuolar proteolysis results in the accumulation of autophagic bodies inside the vacuole. (C) Autophagy in dying corolla cells of the ephemeral flower of *Ipomoea tricolor*. Cell 1 is at an early stage, showing invaginations of the tonoplast and engulfing of portions of cytoplasm. Subsequently (cell 2) there is vacuolar shrinkage and swelling of cytoplasmic membrane systems. Cell 3 is approaching the terminal stage of autolysis which is initiated by tonoplast rupture. (D) Transmission electron micrograph of part of a sucrose-starved cell from a culture of the tobacco (*Nicotiana tabacum*) cell line BY-2 (bright yellow-2). The culture was treated with an inhibitor that prevents digestion of autophagosomes in the central vacuole. One of the autophagosomes shown has engulfed an entire mitochondrion. AB, autophagic body; AP, autophagosome; M, mitochondrion; N, nucleus; V, vacuole.

Rapamycin). Phosphorylation of ATG1 releases it into the cytosol from the pre-autophagosomal structure (**PAS**), a complex with ATG13. Expression of *ATG1* and *ATG13* is activated by the transcriptional regulator GCN4 (General Control Nonderepressible4). GCN4 is regulated by phosphorylated elongation initiation factor 2α, which in turn is activated by the starvation-induced kinases GCN2 and AMK (AMP-activated kinase). In addition to its role in *ATG* expression, AMK stimulates autophagy by phosphorylating and inactivating TOR. A vesicle is formed by coalescence of PASs into a cage that captures a portion of the cytoplasm. Expansion of the autophagosome vesicle is aided by many ATG proteins.

Ultimately the autophagosome is linked to cytoskeletal microtubules and targeted to the vacuole.

18.2 Cell death during growth and morphogenesis

Death of selected cells occurs throughout the life of the plant. From the earliest stages of differentiation in certain newly-formed tissue types there may be elimination of cell contents and even whole cells. **Vascular tissue**

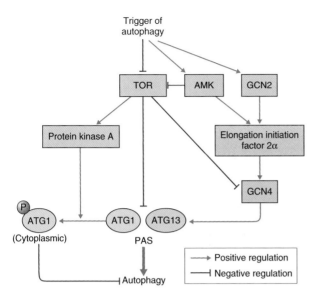

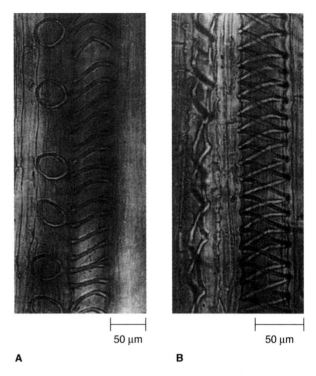

Figure 18.4 Regulation of autophagy. Multiple signaling pathways converge on the expression and activity of autophagy genes *ATG1* and *ATG13*. TOR (Target Of Rapamycin), protein kinase A, AMK (AMP-activated kinase) and GCN2 (General Control Nonderepressible2) are kinases operating in autophagy signaling pathways. Elongation initiation factor 2α and the transcription factor GCN4 regulate expression of *ATG1* and *ATG13*. PAS (pre-autophagosomal structure), the complex containing ATG1 and ATG13, is a component of the autophagosome.

Figure 18.5 Vessel elements from the primary xylem of castor bean (*Ricinus communis*) showing annular (A) and double helical (B) thickenings.

represents a well-studied example. Here tracheary elements, hollow cells with thickened walls, lose their protoplasm by a type of PCD (see Figure 18.2B). **Xylogenesis** (formation of tracheary elements), which begins during embryogenesis and continues throughout the life of the plant, is described in detail below. We also discuss other morphogenetic processes to which PCD contributes, including the formation of secretory glands and canals, the differentiation of protective surface structures such as spines and thorns, and the generation of perforations and indentations.

18.2.1 Cell death is an essential process in the formation of vascular and mechanical tissues

Xylem contains **tracheary elements** (**TEs**; Figure 18.5), hollow cells that form the system of continuous pipework through which the plant's water is transported

(see Section 1.8.3). During the last stages of differentiation, TEs undergo secondary cell **wall thickening**. This is followed by cell death and breakdown of protoplasm by autolysis. Finally, all that remains is the cell wall with its characteristic secondary thickenings.

TE differentiation has been extensively studied using an in vitro system derived from cultured cells of the flowering plant *Zinnia elegans*, which can be induced to redifferentiate synchronously into TEs. When mesophyll cells isolated from *Zinnia* leaves are incubated in a liquid medium containing auxin and cytokinin, up to 60% of them initially **dedifferentiate**, losing photosynthetic capacity. Then, without dividing, they redifferentiate into TEs (Figure 18.6). As they differentiate, these TEs develop thickening of the secondary cell wall—as do the TEs of xylem that forms in vivo. Shortly after thickening occurs, the tonoplast ruptures, followed by breakdown of the remaining organelles and the eventual complete loss of cell contents. The activity of many degradative enzymes, including DNases, RNases and proteases, increases in parallel with the loss of protoplasm.

TE differentiation in *Zinnia* occurs in three distinct stages over a period of about 4 days. The expression

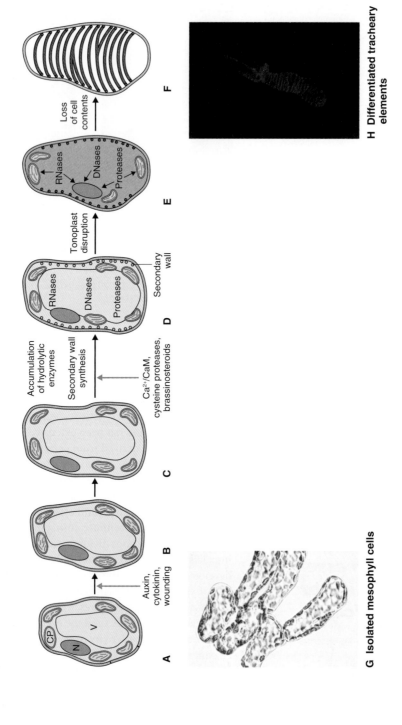

Figure 18.6 Programmed cell death during redifferentiation of cultured *Zinnia elegans* mesophyll cells into tracheary elements (TEs). (A) Cultured mesophyll cell. (B) Early stage of dedifferentiation in response to hormones and wounding. (C) TE precursor cell stage. (D) Immature TE with characteristic secondary cell wall thickenings. (E) Vacuole lysis and elimination of cell contents. (F) Mature, dead, empty TE. (G) Light microscope image of cells at stage (A). (H) Fluorescence microscope image of a mature TE cell stained to reveal cell wall lignin. CaM, calmodulin. CP, chloroplast; N, nucleus; V, vacuole.

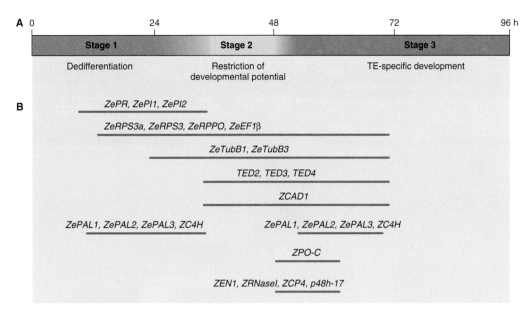

Figure 18.7 Stages of gene expression during *Zinnia* mesophyll redifferentiation. (A) TE differentiation is divided into three stages, totaling about 96 hours. (B) Periods over which TE differentiation-associated genes are expressed. The protease inhibitor genes *ZePI1* and *ZePI2* and the pathogenesis-related gene *ZePR* are part of the wound response. Ribosomal protein (*ZeRP*) and elongation factor (*ZeEF*) genes encode components of the protein synthesis machinery. *ZeTub* genes encode tubulins. TE differentiation-related (*TED*) transcripts appear when the cells commit to TE formation. Phenylalanine ammonia lyase (*ZePAL*), cinnamic acid 4-hydroxylase (*ZC4H*) and cinnamyl alcohol dehydrogenase (*ZCAD*) encode enzymes of phenylpropanoid metabolism. *ZPO-C* is a peroxidase gene, *ZEN1* and *ZRNaseI* encode nucleases, and *ZCP4* and *p48h-17* encode putative cysteine proteases.

patterns of many genes change during this process (Figure 18.7). Most of these genes have homologs in *Arabidopsis* and some have been functionally characterized by analysis of *Arabidopsis* mutants and transgenic plants. During stage 1, the dedifferentiation phase, genes which are upregulated include some of those involved in the **wound response** in plants (see Section 15.7.3) and those required for protein synthesis, including genes encoding ribosomal proteins (RPs) and elongation factors (EFs). At stage 2, when the cells become committed to redifferentiation as TEs, tracheary element differentiation-related (**TED**) transcripts are upregulated.

In stage 3, the secondary cell wall thickenings are laid down. Genes expressed during this phase include those encoding phenylalanine ammonia lyase (PAL), cinnamic acid 4-hydroxylase (C4H), cinnamyl alcohol dehydrogenase (CAD) and other enzymes of **phenylpropanoid metabolism** (see Chapter 15). Genes for the synthesis of structural components, for example cytoskeleton proteins such as tubulin, are also upregulated (Figure 18.7). After the secondary wall is established, the TE cell undergoes PCD. Autolysis occurs as the tonoplast ruptures and releases lytic vacuolar enzymes into the cytosol. Among the *TED*s expressed during late stage 2

and early stage 3 is *Zen1*, which encodes the nuclease responsible for the degradation of nuclear DNA during PCD in these cells. Also expressed at this time are genes encoding cysteine and serine proteases. The addition of inhibitors of cysteine protease to the culture medium before the onset of secondary wall thickening blocks further TE differentiation in *Zinnia*, indicating a central role for protein degradation in this type of PCD. Expression studies of the *Arabidopsis XYLEM CYSTEINE PROTEASE* genes *XCP1* and *XCP2* suggest that the enzymes they encode participate in clearing cell contents during xylem differentiation in vivo. In the final phase of TE development, the tonoplast ruptures and all recognizable organelles are finally erased. The last membrane structure to disappear is the plasma membrane.

Even after PCD is complete, **lignification** of the cell wall continues using precursors provided by adjacent cells. The result is a cell emptied of all cytoplasmic materials and open to the free movement of water through the xylem of which it is a component. Genetic studies in *Arabidopsis* show that cell death and cell wall thickening in TE differentiation are independent processes. For example, TEs of the mutant *gapped xylem* (*gpx*) undergo PCD but do not form secondary cell walls.

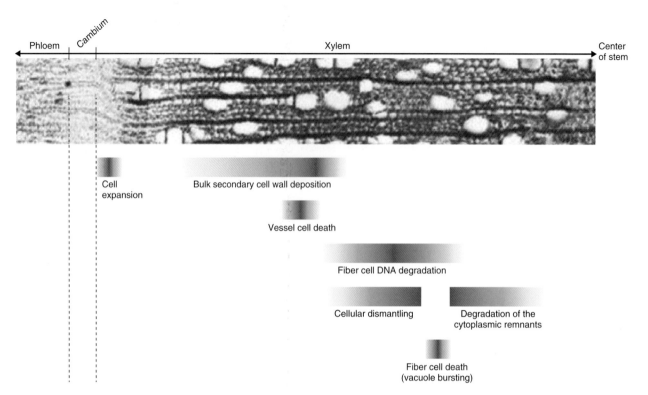

Figure 18.8   Events during xylem development in the woody stem tissues of a hybrid aspen (*Populus tremula* × *tremuloides*). Moving inward (left to right), xylem cells progressively further from the cambial layer represent a developmental gradient. Programmed cell death of vessel elements is complete by around the mid-point of the section. Development in xylem fibers is slower than in vessels, and cell death processes are detectable in tissue nearer the center of the stem.

The *Zinnia* system has been an important model for the study of **wood formation** in trees, a PCD process of major environmental impact and economic importance. Sequencing the genome of black cottonwood (*Populus trichocarpa*) has provided resources for the study of gene expression during secondary growth in trees of the genus *Populus*. Figure 18.8 shows a micrograph of a section of the woody tissue of hybrid aspen (*Populus tremula* × *tremuloides*). Sectioning shows discrete cell layers and allows the stage of xylem development to be related to distance from the cambium. Among the genes that are highly expressed in *Populus* woody tissue sections undergoing cell death are homologs of those encoding XCPs and a number of metacaspase-type proteases that have been implicated in the control of tracheary element PCD or other types of PCD in plants. Tree xylem contains both TEs and xylem fibers, and these cell types have significantly different cell death programs (Figure 18.8). Studies on *Populus* have shown that PCD in xylem fibers (which have much thicker walls than TEs) proceeds slowly, with only gradual degradation of the cytoplasmic contents. Several autophagy-related genes are upregulated in fibers, but not in the TEs, suggesting that cytoplasmic lysis in the fibers is an autophagic process.

Key points   PCD is an integral part of tracheary element (TE) development, clearing the protoplasm from the thick-walled cells of xylem, thereby creating the continuous pipework of water-conducting vascular tissue. Cells of *Zinnia* differentiate synchronously into TEs in culture, allowing the timetable and components of PCD to be determined. An initial dedifferentiation stage is followed by a phase of commitment to redifferentiation, during which there is upregulation of TE differentiation genes (*TEDs*), including those encoding enzymes catalyzing macromolecule breakdown and cytoplasmic lysis. Genes for cell wall thickening are activated and finally PCD proceeds to vacuole rupture and elimination of cell contents. The *Zinnia* system has proven to be a good model for xylogenesis in *Arabidopsis* and wood formation in trees such as *Populus* spp. Studies in *Populus* indicate that xylem fiber cells differentiate more slowly than TEs and follow an autophagy-like PCD pathway.

18.2.2 Lysigeny, schizogeny and abscission are responsible for the formation of tubes and cavities, and the shedding of organs

During animal embryonic development, topologically complex structures such as neural tubes are often formed by the migration of groups of cells (neurulation). In plants, however, the rigid cell wall locks most cells in position, making migration impossible. Instead, tubes and spaces often develop by means of localized, developmentally-regulated senescence and death of groups of cells. Similarly, plant shapes and adaptations for different purposes can be achieved by controlled cell death and hollowing out or shedding of parts. For example, the oil glands found on the surface of citrus peel develop when groups of cells below the epidermis undergo PCD. This forms a cavity that fills with essential oils (Figure 18.9A). The term **lysigeny** is used to describe the disintegration of cells to form glands, channels and secretory ducts. In some cases it is accompanied by the separation of cells, a process known as **schizogeny**. Citrus oil glands are products of schizolysigeny, a combination of lysigenous and schizogenous PCD. Mucilaginous canals found in the bud scales of *Tilia*

cordata (small-leaf linden, small-leaf lime) arise exclusively by lysigenous cell death (Figure 18.9B). The resinous secretory ducts in the phloem of plants of the sumac family (Anacardiaceae) develop schizogenously and are the source of urushiol, the severe allergen in the sap of poison oak and poison ivy (*Toxicodendron* spp.). Separation of adjoining cells is the mechanism underlying a number of prominent events in the plant life cycle (Figure 18.10), notably the shedding (**abscission**) of organs and the splitting of dry fruits to disperse seeds and anthers to release pollen (**dehiscence**). The middle lamella that joins plant cell walls together is composed primarily of pectin. During abscission and dehiscence, cell–cell adhesion is loosened by localized degradation of matrix polysaccharides. The process is related to that which occurs during the ripening of soft fruits (see Section 18.5.3). In the case of abscission, the region along which cell–cell adhesion will be dissolved is usually in a predetermined position called the **abscission zone**. Abscission is stimulated by ethylene and inhibited by auxin. Formation of the abscission zone is responsive to stress, abscisic acid and the degree of senescence of the organ to be shed. Under the control of abscission-related transcription factors and protein kinases, genes encoding cell wall-modifying proteins, particularly cellulase, polygalacturonase, peroxidase and expansin, are turned on in the abscission zone. Stress response factors are also upregulated, including metallothioneins and pathogenesis-related proteins, which function in defending the exposed surface from attack by pathogens and other environmental insults (see Chapter 15).

18.2.3 Organs may be shaped by selective death of cells and tissues

Selective cell death also controls other aspects of vegetative development in plants. Familiar leaf and stem structures, such as prickles, thorns and spines, are dead once they are mature. The **cacti** take this to extremes: in these plants, the green stem replaces leaves as the functional photosynthetic tissue, and the leaves are reduced to dead spines (see Figure 1.33F). Localized cell death as the leaf develops is responsible for the indentations and holes found in leaves of some plant species of the families Aponogetonaceae and Araceae; a well-known example in the latter family is the **Swiss cheese plant**, *Monstera* (Figure 18.11A). The **lace plant** (*Aponogeton madagascariensis*; Figure 18.11B) is an extreme case of perforation arising from patterning of PCD during leaf morphogenesis.

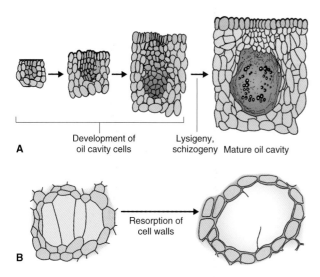

A Development of oil cavity cells | Lysigeny, schizogeny | Mature oil cavity

B Resorption of cell walls

Figure 18.9 Lysigenous and schizogenous cell death. (A) Formation of an oil gland in *Citrus* peel occurs by a combination of lysigeny, the death and dissolution of the protoplast and cell wall, and schizogeny, wall separation to produce an intercellular space. Cells surrounding the resulting cavity secrete essential oils into it. (B) The mucilaginous canals of *Tilia cordata* bud scales arise exclusively by lysigenous cell death.

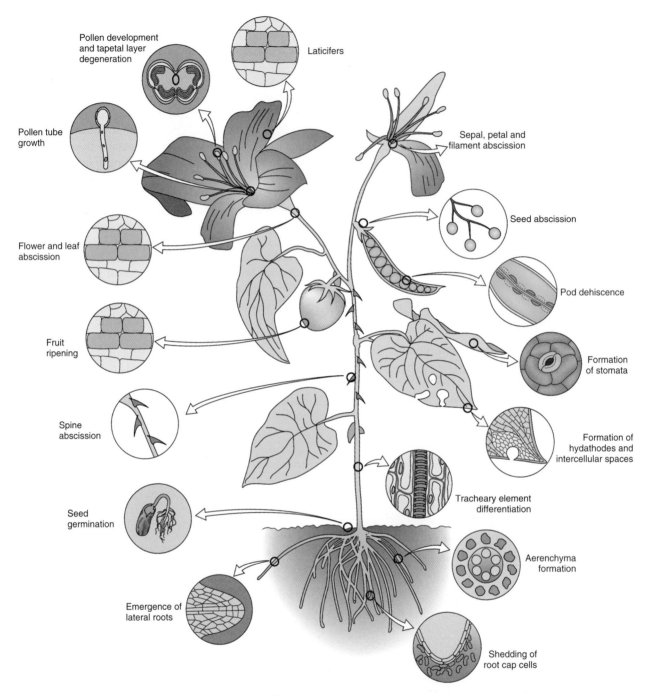

Figure 18.10 Sites of cell separation in plants. The illustration shows the range of developmental processes in which cell–cell contact is disrupted by changes in the intervening cell wall. They include shedding of tissues and organs, dehiscence, schizogenous formation of intercellular spaces and canals, penetrative growth and tissue softening.

18.3 Leaf senescence

In temperate parts of the world, senescence of green plant parts is one of the most visually striking of all biological processes. The autumnal color changes of trees, and the **ripening** of field crops, especially grain crops like wheat and maize, are defining features of late summer and fall

(Figure 18.12). Plant senescence follows a timetable that in some cases is governed by both day length and temperature, and is genetically controlled. Several genes have been identified as essential if the senescence program is to be correctly initiated and proceed normally. Execution of the program can also be strongly influenced by environmental factors, phase change (see Chapter 16) and hormonal status of the tissue.

A   **B**

Figure 18.11 Leaves with holes that develop from areas of localized programmed cell death. (A) Swiss cheese plant, *Monstera*. (B) Lace plant (*Aponogeton madagascariensis*).

Key points Ducts, glands and channels in plant tissues are formed by loss of cell contents (lysigeny) and/or cell separation (schizogeny). Mucilage-forming ducts in linden bud scales are lysigenous in origin. The intensely allergenic sap of poison oak and poison ivy is secreted into schizogenous ducts. Citrus peel bears oil glands that arise by a combination of schizogeny and lysigeny. Abscission of parts, such as leaves and flowers, and dehiscence of dry fruits and anthers, are cell separation events, the consequence of degradation of the middle lamellae of cell walls under the positive and negative regulation of ethylene and auxin, respectively. Localized PCD occurs during the differentiation of prickles and thorns, and also leads to the development of holes in leaves of species such as the Swiss cheese plant (*Monstera*) and the aquatic lace plant (*Aponogeton*).

All parts of the plant—leaves, stems, flowers and roots—undergo senescence and eventually death as the terminal phase of their development. Senescence usually takes place after organ growth is complete and the tissue has matured. It allows the plant to dispose of parts that are no longer required, and to reclaim valuable resources. These include many micronutrients, as well as nitrogen and phosphorus. Sometimes senescence follows maturity very rapidly, as in the case of certain flower petals. For example, the flower of Japanese morning glory (*Ipomoea*) is open for only 1 day before senescence is initiated (Figure 18.13). In other cases, a plant organ may be mature and functional for months or even years before the onset of senescence. It has been reported that the leaves of *Pinus longaeva* (**bristlecone pine**) have a lifespan of 45 years, and the desert plant *Welwitschia mirabilis* has large strap-like leaves (Figure 18.14) which, according to some estimates, survive for 1000 years or more.

A   **B**

Figure 18.12 Green tissues of a wheat field (A) senesce and turn yellow as the crop ripens (B).

0 h **4 h** **8 h** **12 h** **14 h**

Figure 18.13 Corolla senescence in the ephemeral flower of Japanese morning glory, *Ipomoea tricolor*. Flowers begin to bloom at daybreak, around 05:00 to 06:00 hours, and remain open until about 13:00. An early symptom of senescence is the change of corolla color from blue to purple. The corolla begins to curl at around 17:00 and has completely curled inward over the following 2 hours.

Figure 18.14 Each leaf of the extraordinary South African desert plant *Welwitschia mirabilis* may live for decades or even centuries.

18.3.1 Cell structures and metabolism undergo characteristic changes during senescence

During senescence, some metabolic pathways are activated, while others are switched off, in a highly regulated fashion. Figure 18.15 outlines steps in the senescence network of angiosperm leaves. In this scheme, triggers such as hormones or environmental conditions lead, via signal transduction cascades, to an **initiation phase** during which transcription of many genes is activated or suppressed. There is then a transition to a **reorganization** phase during which cell components

selectively redifferentiate and nutrients are **remobilized**. The final phase is **termination**, during which PCD takes place. It is only late in this terminal phase that subcellular membranes lose their integrity and the compartmentation of biochemical pathways breaks down.

Of the modifications that the major organelles undergo during senescence, those affecting the chloroplasts of mesophyll cells are the most dramatic. Until the terminal, death phase, these changes do not constitute deterioration; rather, they are due to redifferentiation. Leaf chloroplasts lose thylakoid membranes and form large lipid droplets called plastoglobuli as they redifferentiate into **gerontoplasts** (Figure 18.16). A corresponding set of organelle

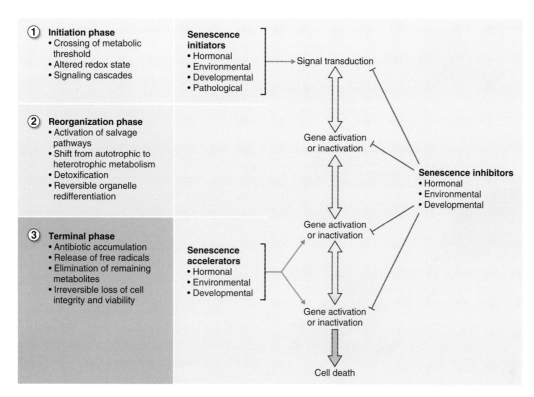

Figure 18.15   Steps in the process of leaf senescence from the initiating signal to cell death.

transformations occurs during the development of many kinds of colorful fruits and floral parts, during which chloroplasts become **chromoplasts**. Such interconversions are part of a complex developmental network connecting the different forms of plastids associated with the growth, maturation and terminal phases of plant cell differentiation (see Figure 4.23).

Other organelles may be affected as well. In senescing storage cotyledons, in the endosperm of many dicot seeds, and in the aleurone layer of cereal grains, the protein storage vacuoles undergo a major change. Instead of functioning as protein storage organelles, they now become large central vacuoles in which the integrity of the tonoplast is maintained (Figure 18.17). The specialized lipid bodies of cotyledons and endosperm are lost as these storage tissues senesce during germination. Peroxisomes are remodeled to support lipid metabolism and gluconeogenesis in senescing storage and photosynthetic tissues. All these changes in cell structure and compartmentation confirm that senescence is genuinely a programmed process, not a form of deterioration or necrosis.

As the cell's ultrastructure alters during senescence, primary and secondary metabolism are also reconfigured. Chlorophyll and macromolecules, especially storage proteins, are degraded. Salvaging metabolites and structural components, particularly those that are reserves of nitrogen and phosphorus, becomes a priority. Current photosynthesis is a declining

source of the energy required for remobilization and other metabolic processes in senescing cells. Increasingly, energy requirements are met by catabolism, associated with significant changes to respiratory and oxidative pathways (see Section 18.3.4). In many species, there is also synthesis or modification of **secondary metabolites**, influencing interactions with other organisms (see Chapter 15 and Section 18.6.3). Examples include the accumulation of antibiotic or antiherbivore compounds to protect potentially vulnerable senescing tissues against pathogens or pests, and the pigments that give ripening fruits and floral organs the bright colors needed to attract animal pollinators or seed-dispersers.

18.3.2 Leaves change color during senescence

In almost every case, the most striking symptom of leaf or flower senescence, fruit ripening or programmed death in plants is a change in color. Loss of green color during leaf senescence, sometimes accompanied by the development of orange or red pigmentation, is directly related to the regulation of nutrient mobilization from leaf cells, whether at the natural end of the leaf's life, or in response to a biotic or abiotic stress. Some pathways of pigment metabolism are specifically turned on, or upregulated, during senescence. The best characterized

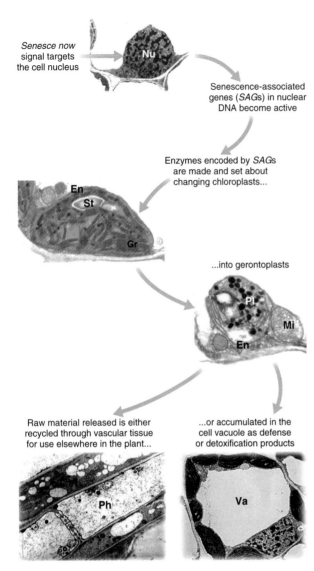

Senesce now signal targets the cell nucleus

Senescence-associated genes (*SAG*s) in nuclear DNA become active

Enzymes encoded by *SAG*s are made and set about changing chloroplasts...

...into gerontoplasts

Raw material released is either recycled through vascular tissue for use elsewhere in the plant...

...or accumulated in the cell vacuole as defense or detoxification products

Figure 18.16 Changes in cell organization during leaf senescence. Transcriptional and post-transcriptional events result in the differentiation of chloroplasts into gerontoplasts. Note that the plastid envelope (En) remains intact and that dismantling of grana (Gr) stacks and thylakoid membranes leads to the accumulation of plastoglobules (Pl) in the gerontoplast. Products of proteolysis, nucleic acid breakdown and other catabolic activities are exported via the phloem (Ph). Defense compounds and chlorophyll catabolites accumulate in the mesophyll cell vacuole (Va). Mi, mitochondrion; Nu, nucleus; St, starch granule.

of these are chlorophyll breakdown and the synthesis of anthocyanins. In other cases a color change does not result from de novo synthesis but is simply due to existing pigments that are unmasked as others are removed, or that are chemically modified leading to alterations in their spectral properties.

Whatever other pigmentation changes take place during leaf senescence, there is always loss of green

Key points During senescence, green tissues lose chlorophyll and become yellow, red, purple or brown. Ripening of field crops and the color changes of temperate forest trees in fall are examples of senescence. Senescence is genetically programmed and in many cases triggered by environmental signals. Nutrients are salvaged from senescing tissues and redistributed to parts of the plant that are growing or laying down reserves. The leaves of some long-lived plants can persist for decades before senescence is initiated, whereas more transient species have lateral organs with a life span of days or, in the case of some ephemeral flowers, hours. Senescence commences with an initiation phase, during which activation or downregulation of specific genes occurs on a large scale. There follows a period of reorganization during which the senescence program is executed, senescing cells are restructured and nutrients are remobilized. Chloroplasts become modified into gerontoplasts or chromoplasts. Vacuoles undergo marked structural and functional changes. Peroxisome metabolism becomes redirected towards lipid catabolism and in some cases gluconeogenesis. Pathways of hydrolytic, oxidative and secondary metabolism are activated in senescing cells. Senescence is succeeded by a terminal phase of PCD, culminating in loss of membrane integrity and cell viability.

color with time (Figure 18.18) as chlorophyll is degraded. The first step in this process is the release of chlorophyll from the pigment–protein complexes of the thylakoid membrane (Figure 18.19), for which the product of the *STAY GREEN* gene *SGR* is necessary (see Section 18.3.5). When the central magnesium atom of the chlorophyll ring is removed (by a poorly characterized activity called Mg-dechelating substance—Equation 18.1A), the color of the molecule shifts from leaf-green to the olive or khaki shade that is more characteristic of overcooked vegetables. The enzyme **pheophytinase** then cleaves off the **phytol** side-chain, leaving the much more polar molecule **pheophorbide** (Equation 18.1B).

Equation 18.1

18.1A Mg-dechelating substance

$$\text{Chlorophyll} \rightarrow \text{pheophytin} + \text{Mg}^{2+}$$

18.1B Pheophytinase

$$\text{Pheophytin} \rightarrow \text{pheophorbide} + \text{phytol}$$

The other product of pheophytinase action is phytol which, being highly hydrophobic, normally accumulates

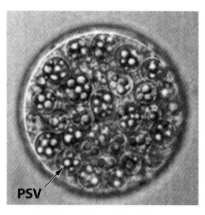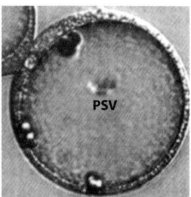

PSV

PSV

Figure 18.17 Changing vacuole morphology and function during programmed cell death in the barley aleurone cell. (A) During germination, cytoplasm becomes highly vacuolated as the small protein storage vacuoles (PSV) lose their reserve proteins. (B) The vacuoles derived from PSVs fuse to form one large central vacuole. Membrane integrity is maintained until death ensues.

14 21   37   53   7

Days after germination

Days after germination

A   B

Figure 18.18   (A) Stages in development of *Arabidopsis* plants from seedling (14 days after germination) to senescence (53 days). (B) Progressive yellowing of rosette leaves from 7 to 11 days after full expansion.

largely in the form of esters in the plastoglobules of gerontoplasts (see Figure 18.16). Chlorophyll b must be converted to chlorophyll a before it can be broken down by this catabolic pathway (Figure 18.19). This conversion is carried out by the enzyme **chlorophyll b reductase** which is activated during senescence. It catalyzes a two-step reaction with NADP and ferredoxin (Fd) as cofactors (Equation 18.2).

Equation 18.2 Chlorophyll b reductase

$$\text{Chlorophyll(ide) b} + \text{NADPH} + \text{H}^+$$
$$\rightarrow 7^1\text{-OH-chlorophyll(ide)} + \text{NADP}$$
$$7^1\text{-OH-chlorophyll(ide)} + \text{Fd}_{red}$$
$$\rightarrow \text{chlorophyll(ide) a} + \text{Fd}_{ox}$$

Up to and including pheophorbide, all the intermediates in the breakdown pathway are green. In the next, critical, steps, the ring is opened to yield a colorless straight-chain tetrapyrrole (Figure 18.19). Two enzymes participate in this step. The first is **pheophorbide a oxygenase** (PaO; Equation 18.3). As its full name implies, PaO uses pheophorbide a but not pheophorbide b as a substrate. The reaction catalyzed by PaO requires O_2 and involves iron, in a redox cycle driven by reduced ferredoxin (Fd_{red}).

Equation 18.3 Pheophorbide a oxygenase

$$\text{Pheophorbide a} + \text{Fd}_{red} + \text{O}_2 \rightarrow \text{RCC} + \text{Fd}_{ox}$$

The product of the PaO reaction is a red bilin compound (**red chlorophyll catabolite, RCC**). This is similar to the pigments excreted by some single-celled algae when they are transferred to a medium containing a carbon source so that they no longer need to photosynthesize and are starved of nitrogen. RCC does not normally accumulate in land plants but is instead immediately metabolized further by **RCC reductase** (Equation 18.4). This enzyme catalyzes the reduction of a double bond in RCC to produce a colorless tetrapyrrole that emits a strong blue fluorescence when excited with ultraviolet light; it is therefore known as **primary fluorescent chlorophyll catabolite (pFCC)**. Two different epimeric forms of pFCC are found in plants, depending on a single amino acid substitution (valine for phenylalanine) in the RCC reductase enzyme.

Equation 18.4 RCC reductase

$$\text{RCC} + \text{Fd}_{red} \rightarrow \text{pFCC} + \text{Fd}_{ox}$$

All the enzymes of chlorophyll catabolism described so far are found in the plastid (Figure 18.19). Chlorophyll b reductase, PaO and possibly Mg dechelatase are membrane-associated activities; RCC reductase is a soluble protein found in the stroma. Once pFCC has

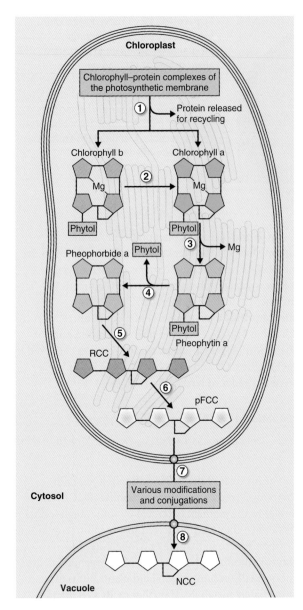

Figure 18.19 Pathway and subcellular organization of chlorophyll breakdown during senescence. Enzymes and activities are identified as follows: (1) SGR (*STAY GREEN* gene product); (2) chlorophyll b reductase; (3) dechelation reaction; (4) pheophytinase; (5) pheophorbide a oxygenase; (6) red chlorophyll catabolite (RCC) reductase; (7) ATP-dependent catabolite transporter; (8) ABC transporter. NCC, non-fluorescent chlorophyll catabolite; pFCC, fluorescent chlorophyll catabolite.

species. These terminal catabolites cross the tonoplast membrane via ABC transporters (see Chapter 5) and are deposited in the vacuole (Figure 18.19), after which they may undergo other minor modifications. In deciduous plants, the NCCs are still present when leaves are shed; thus the plant sacrifices all the nitrogen and carbon of the chlorophyll molecule in the NCCs. Though this appears wasteful, it is the price the plant pays for the opportunity to remobilize the much larger reserve of nitrogen that was previously locked into the proteins of the pigment complexes in the thylakoid membranes (see below).

18.3.3 During senescence macromolecules are broken down and nutrients are salvaged

One of the most important features of natural senescence is the **recycling** of macromolecules, including proteins, lipids and nucleic acids, for use in other, non-senescent parts of the plant. Why is this recycling process so important? Availability of light and carbon dioxide does not normally constrain plant growth, certainly not in an agricultural context, but in many situations water and mineral nutrients are limiting factors. This is especially the case for nitrogen and phosphorus, two of the most important elements in fertilizers. Plants have therefore evolved to be economical with N and P, and they manage this by recycling them, together with other macro- and micronutrients, from older organs (notably senescing leaves) as well as by reabsorption from differentiating cells (such as tracheary elements) that are undergoing PCD. A major function of senescence and cell death is to regulate the efficient salvage of the N and P locked up in the proteins and nucleic acids of mature tissues, and to insure that these elements are transported to growing plant parts or to storage tissues such as grains, oilseeds, roots, rhizomes or tubers.

In order that nutrients can be released for recycling, macromolecules must be broken down by hydrolytic enzymes, whose expression is upregulated during senescence. Amino acids released by protease action on the enzymes and structural proteins of plastids, other organelles, and cytosol during senescence may undergo further metabolism (Figure 18.20). In many species there is active synthesis of the **amides** glutamine—by the enzymes glutamate dehydrogenase and glutamine synthetase (see Chapter 13)—and asparagine, the product of transamidation from glutamine to aspartate. Amide nitrogen is the principal form in which recycled protein nitrogen is translocated from senescing tissues to developing sinks.

been synthesized it is exported from the gerontoplast, by an active process requiring an ATP-dependent **transporter** in the plastid envelope. It undergoes hydroxylation and other modifications—malonylation and glucosylation are common—to produce **non-fluorescent chlorophyll catabolites** (NCCs), which vary in structure and number depending on plant

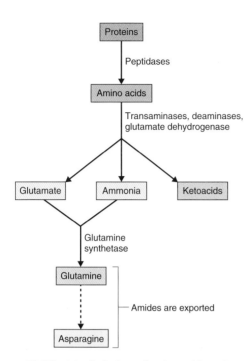

Figure 18.20 Metabolic fate of amino acid products of protein breakdown during leaf senescence. Amino acids released from protein by endo- and exopeptidases during senescence may be extensively metabolized before the nitrogen is exported from the leaf, principally in the form of the amides glutamine and asparagine.

In green cells of the leaf, most of the protein is located in chloroplasts, so these organelles contribute most of the nitrogen salvaged from senescing foliage. Of the chloroplast proteins (accounting for more than 60% of

leaf protein nitrogen), the most abundant are **rubisco**, which is located in the soluble stroma of the chloroplast, and the **light-harvesting chlorophyll-binding proteins** (**LHCPs**), found in the thylakoid membranes. The progress of leaf senescence, measured as loss of green color or as declining rate of photosynthesis, is generally found to be in step with decreasing amounts of rubisco, LHCP and other chloroplast proteins and with increasing proteolytic activity (Figure 18.21). However, despite the scale of the process, our knowledge about the location, regulation and biochemistry of plastid protein breakdown during senescence is still far from complete.

Several types of proteolytic enzymes have been identified within plastids, and some have been observed to be upregulated during senescence. Two classes of protease, which were originally described in bacteria and which include promising candidates for key roles in plastid protein breakdown, are the **Clp** (Caseinolytic protease) and **FtsH** (Filamentation temperature sensitive H) families. Clp proteases require ATP hydrolysis for activity. FtsHs are also ATP-dependent and have a Zn^{2+} requirement. Clp and FtsH proteases have a range of roles in the repair and adaptation of the photosynthetic apparatus. Two (Clp3 and ClpD) of the 15 or more plastid-localized Clps, and FtsH7 and FtsH8, two of the nine different FtsHs, have been shown to be highly expressed in senescent leaves, though it is not yet known whether they contribute to the net loss of rubisco, LHCP and other plastid proteins during senescence.

The genes associated with the initiation and execution of the leaf senescence program are generally referred to as **SAGs** (*Senescence Associated Genes*). Expression of **SAG12**, which encodes a protease, is widely used as a

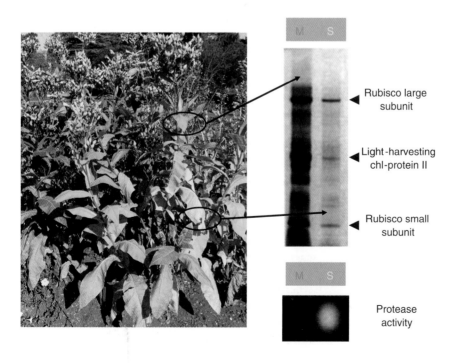

Figure 18.21 Proteins and proteolytic activity of mature green and senescent yellow tobacco (*Nicotiana*) leaves. Total proteins in extracts of equal weights of mature (M) and senescent (S) leaves were separated by electrophoresis. Staining reveals the large and small subunits of rubisco and the polypeptides of the light-harvesting complex of photosystem II to be the most abundant leaf proteins. Between maturity and the senescent state, there is large-scale breakdown of leaf proteins and export of the amino acid products. The proteolytic activity of senescing tissue greatly exceeds that of mature tissue, shown here by the extent to which equal volumes of leaf extract digest the protein gelatin in a gel plate assay. Proteolysis is visualized as a clear zone when the substrate is stained with a protein-specific dye.

marker for senescence. SAG12 and several other proteases expressed during leaf senescence are cysteine endopeptidases. They belong to the papain family of proteases and have similar three-dimensional structures and catalytic properties (see Figure 2.25). These enzymes do not have any of the sequence motifs that characterize proteins imported into the plastid. Instead, in many cases they are targeted to the vacuole or endoplasmic reticulum. There is some evidence that the proteases associated with senescence are found in the cytosol within vacuole-derived lytic vesicles. They are closely related to some of the proteases synthesized during seed germination, for example the **aleurains** and **oryzains** found in germinating barley and rice, respectively. It may be that the early stages of proteolysis in senescence take place within the plastid and subsequently the vacuole comes into play as gerontoplast differentiation gives way to the autolytic events of terminal PCD.

There are many parallels between nucleic acid degradation and its control during senescence, and the corresponding process of protein mobilization. In both cases, genes encoding catabolic enzymes are turned on, or upregulated, and hydrolytic activities increase, while amount of substrate—in this case nucleic acid, particularly RNA—declines. The locations of the enzymes known to be involved in nucleic acid remobilization (primarily in the cell vacuole) do not generally correspond to the sites where most of the degradation takes place.

The largest contribution to phosphorus remobilization is made by the RNA of **plastid ribosomes**. RNA is broken down by **ribonuclease (RNase)**, which rapidly converts its substrate to low molecular weight products. The *Arabidopsis* RNAse **BFN1** (Bifunctional Nuclease1) is one candidate for RNA mobilization during senescence, since it is strongly expressed in senescing leaves, stems (Figure 18.22A), petals and stamens. The enzyme is bifunctional in the sense that it is able to hydrolyze both RNA and single-stranded DNA. The *BFN1* gene is also active in differentiating xylem, in the abscission zones of leaves, flowers and fruit, and in developing anthers and seeds. It may therefore have a role not only in senescence but also in PCD at other stages in the plant life cycle. BFN1 is very similar to nucleases from other plants, fungi and protozoa, including the ZEN1 nuclease that is upregulated in PCD during TE development in *Zinnia* (Figure 18.22B; see Section 18.2.1).

Phosphorus is redistributed from senescing to young tissues in the form of inorganic phosphate. **Phosphatases**, which are highly active during senescence and PCD, release phosphate from the nucleotide products of nuclease attack on RNA and DNA. Nucleosides resulting from nuclease and phosphatase action are cleaved into

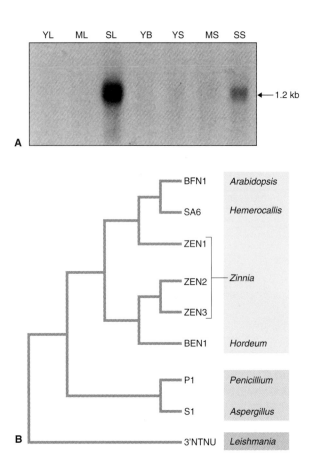

Figure 18.22 Bifunctional Nuclease1 (BFN1) of *Arabidopsis*. (A) The abundance of mRNA encoding BFN1 was determined by gel blot analysis of RNA from leaf and stem tissues of different ages, hybridized with a labeled DNA sequence from the *BFN1* gene. The size of the transcript detected (1.2 kb) corresponds to a protein of 35 kDa. YL, ML, SL, young, mature and senescent leaf; YB, young bolt; YS, MS, SS, young, mature and senescent stem. (B) Relationship tree showing amino acid sequence comparisons between BFN1 and nucleases of plant (green box), fungal (blue box) and protozoan (pink box) origin.

sugars, purines and pyrimidines, which are further catabolized, ultimately into ammonia and CO_2.

18.3.4 Energy and oxidative metabolism are modified during senescence

Senescence and PCD are active processes, requiring energy to power metabolic and transport activities. Photosynthesis meets the needs of the initial phase of senescence, but C fixation progressively declines as plastid proteins are broken up and recycled, and the substrates for respiration switch to the products of lipid and protein catabolism. The tricarboxylic acid (TCA)

Key points Chlorophyll breakdown is a defining feature of senescence and occurs via a metabolic pathway that becomes specifically turned on in mature cells of green tissue. The first step is the release of pigment from the chlorophyll–protein complexes of the thylakoid membrane, mediated by SGR, a factor encoded by the senescence-upregulated gene *STAY-GREEN*. Dissociation of the complex makes the protein nitrogen available for recycling. Loss of magnesium from the chlorophyll ring to yield pheophytin is followed by removal of the phytol side-chain by the enzyme pheophytinase. The product, pheophorbide, is oxidized to a colorless product, pFCC, in a ferredoxin-dependent two-step reaction via a red intermediate, RCC. These reactions are catalyzed by pheophorbide a oxygenase and RCC reductase, respectively. The end-product, NCC, accumulates in the vacuoles. To be broken down, chlorophyll b must first be converted to chlorophyll a by a reductase. Rubisco and the proteins of pigment–protein complexes together can account for more than 60% of the nitrogen salvaged during senescence. Proteases—possibly including the cysteine endopeptidase SAG12 and ATP-dependent enzymes of the Clp and FtsH families—release amino acids that are further metabolized to form amides, the principal form in which recycled nitrogen is translocated from senescing tissues. Nucleic acids are also hydrolyzed to nucleotides by nucleases during senescence. Phosphate released by phosphatases is available for export to meet the demand from growing tissues for phosphorus.

damage and even kill cells. The redox status of the cell controls, and is under the control of, specific enzyme systems and gene regulation mechanisms. As the tight regulation of oxidation and ROS propagation relaxes during the final stages of PCD, viability is lost and finally death occurs.

One ROS that is a component of the senescence signaling network is H_2O_2, which is a product of normal enzymatic reactions in peroxisomes, chloroplasts and other organelles. When rosette leaf senescence is triggered by bolting (extension of the flowering stem) in *Arabidopsis*, an increase in H_2O_2 is detectable very early before there is any measurable decrease in chlorophyll content (Figure 18.23). This is associated with a decline in the activity of **CAT2**, a peroxisomal **catalase** that scavenges H_2O_2 (Equation 18.5). At the same time there is a decrease in **APX1**, a cytosolic isoform of **ascorbate peroxidase** (Equation 18.6).

Equation 18.5 Catalase (CAT)

$$2H_2O_2 \rightarrow 2H_2O + O_2$$

Equation 18.6 Ascorbate peroxidase (APX)

$$Ascorbate + H_2O_2 \rightarrow dehydroascorbate + 2H_2O$$

Subsequently, as yellowing commences, a cytosolic catalase CAT3 is activated, APX1 increases again and H_2O_2 levels fall. Then, as senescence proceeds to completion, there is a steady buildup of H_2O_2 until cell

cycle is active in metabolizing the carbon skeletons of organic acids released by **transamination** and **deamination** of the amino acid products of proteolysis and amide formation (see Figure 18.20). In many species, enhanced activity of the glyoxylate cycle is associated with fatty acid degradation, sometimes leading to gluconeogenesis and the recycling of lipid carbon as exportable sugars (see Chapter 7). The state of energy metabolism in plastids, peroxisomes and mitochondria changes markedly during senescence, along with **cellular redox conditions**, with important consequences for the genetic and metabolic regulation of terminal events. **Reactive oxygen species** (**ROS**), discussed in detail in Chapter 15, play an important role in senescence. The amount of ROS generated during plant metabolism often increases with tissue age. Under conditions of tight biochemical control, ROS are components of **signaling pathways** that regulate senescence and death. If allowed to propagate without restraint, they have the potential to

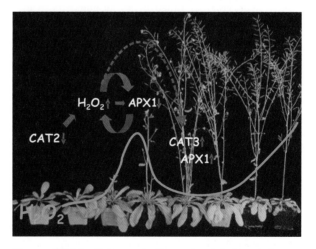

Figure 18.23 H_2O_2 production and metabolism during bolting of *Arabidopsis*. Downregulation of *CATALASE2* (*CAT2*) and the consequent decrease in CAT2 enzyme activity is thought to lead to an increase in H_2O_2 that in turn inactivates ascorbate peroxidase 1 (APX1). As APX1 activity declines, H_2O_2 levels increase further and ultimately induce *CAT3* expression. Increasing CAT3 activity removes H_2O_2 and restores APX1 activity.

death. The timing and interaction of CAT and APX activities are important in setting the balance between ROS production and the removal of ROS by **antioxidant** systems. This balance in turn determines the cellular redox state, with consequences for metabolism and gene expression.

An important regulatory gene in responses to and control of cell redox is that encoding the transcription factor **WRKY53**. WRKY53 is induced by H_2O_2 and autoregulates its own synthesis by feedback inhibition. It interacts with more than 60 genes of various kinds, including those for CAT1 and CAT3, the senescence-associated cysteine protease SAG12 and components of the salicylic acid and jasmonic acid signaling networks.

> **Key points** The energy requirements for senescence are satisfied by catabolism of the carbon skeletons of the products of lipid, carbohydrate and protein breakdown. This is often associated with increasing levels of reactive oxygen species (ROS) as cells age. ROS in general, and H_2O_2 in particular, have signaling functions during senescence and their accumulation is under the control of antioxidant metabolites and enzymes. Catalase and ascorbate peroxidase are enzymes that scavenge H_2O_2 and regulate its influence on gene expression and participation in oxidative metabolism. Among the genes induced by H_2O_2 is *WRKY53*, which encodes a transcription factor that in turn regulates its own gene and many other senescence-associated genes including those for catalase, SAG12 and hormone signaling networks.

18.3.5 Senescence is genetically regulated and under hormonal control

Genes that direct the **initiation** and **execution** of the leaf senescence program, *SAGs*, have been identified by studies of mutants and other heritable variations, by the analysis of gene expression patterns, and by the cloning and functional testing of DNA sequences. Mutations that interfere with normal senescence frequently reveal themselves by abnormal retention of chlorophyll and are referred to as **stay-green**. Some types of stay-green trait are beneficial in agriculture, because they are associated with extended photosynthetic productivity, enhanced crop quality and shelf-life, or better stress resistance. *Sorghum bicolor* is an example of a crop in which

stay-green is a valuable trait. But where the efficient recycling of N and P from foliage is desirable, for example in certain legumes whose seeds are harvested, stay-green has a negative effect. In nutrient-poor, natural ecosystems, low rates of internal nitrogen recycling are favored, which explains why **evergreen** shrubs and trees typical of such habitats have adopted the stay-green strategy as a fitness attribute.

Stay-green is a trait of historical significance. Gregor Mendel, the father of modern genetics, studied the inheritance of a number of characters in peas (*Pisum sativum*), including cotyledon senescence. Pea cotyledons normally turn yellow during seed maturation, as do pea leaves during senescence. Mendel investigated the inheritance of a mutant with mature seeds that retain greenness (Figure 18.24A). This stay-green gene, *SGR*, has been isolated by a combination of genetic mapping

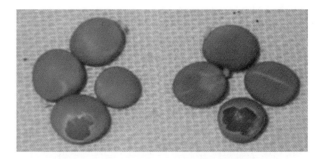

A

B

C

Figure 18.24 The stay-green trait. (A) Yellow and green pea seeds used by Mendel to establish the laws of inheritance. (B) Senescing leaf tissue of a normal yellowing genotype (left) and a stay-green mutant (right) of the pasture grass *Festuca pratensis*. (C) Wild-type (left) and stay-green phenotype created in *Arabidopsis* by knocking out expression of the gene *SGR* (right).

and functional analysis. By exploiting the availability of the complete DNA sequence of the **rice genome** and the genomics tools associated with it, the stay-green genes of pea and a number of grass species (e.g. Figure 18.24B) were shown to be homologs. Knocking out *SGR* expression using the RNA interference method creates a stay-green *Arabidopsis* (Figure 18.24C). The normal, non-mutant, form of this gene has a highly conserved structure and has been identified not only in angiosperms but also in other photosynthetic organisms, from mosses to unicellular green algae to cyanobacteria. It encodes a post-translational regulator of the disassembly of thylakoid photosystem complexes, making chlorophyll and protein available to their respective degradation pathways (see Sections 18.3.2 and 18.3.3).

The map-based cloning approach has also been successfully applied to the identification of an agriculturally important gene in cereal species that regulates senescence. Pasta (durum) wheat, *Triticum turgidum* ssp. *durum*, is the domesticated descendent of wild emmer (*T. turgidum* ssp. *dicoccoides*). The two species can be crossed, allowing the substitution of individual chromosomes of one species with the equivalents (**homeologous chromosomes**) of the other. Replacing chromosome 6 of durum wheat with emmer chromosome 6 results in durum genotypes with a higher grain protein content than non-substitution lines. Enhanced grain protein is the consequence of the expression of genetic loci on emmer chromosome 6 that determine rapid leaf senescence and mobilization of foliar nitrogen. Molecular mapping associated the high grain protein trait on this chromosome with an allele of the gene **NAM-B1**, which encodes a transcription factor closely similar to the *Arabidopsis* protein NO APICAL MERISTEM. The *NAM-B1* alleles of durum wheat (and of hexaploid bread wheat, *T. aestivum* ssp. *aestivum*) are non-functional. It is believed that **domestication** of cereals at the dawn of agriculture fixed the variant DNA sequences of non-functional *NAM-B1* alleles in the genetic backgrounds of wheat crop species. Because of this inadvertent manipulation of *NAM* expression in bread wheat, leaf senescence has been greatly delayed, resulting in deficiencies in the protein, Zn and Fe content of grain. These studies point to a central controlling function for *NAM* genes in cereal leaf senescence and the partitioning of nitrogen and minerals between the grain and crop residue (Figure 18.25). They dramatically illustrate the challenge for plant breeders in managing the trade-off between the advantages of delayed senescence (extended carbon fixation and higher grain yields) and the benefits of timely and efficient senescence (better recycling of nutrients and enhanced grain quality). Modern breeding can resolve such dilemmas by taking advantage of our increasing knowledge of the molecular basis of plant senescence combined with

	Transgenic	Non-transgenic control
Grain protein content %	13.27	19.08
Zn (ppm)	52.45	82.50
Fe (ppm)	37.40	60.83

Figure 18.25 Grain- and senescence-related traits of transgenic wheat plants segregating for the presence (transgenic) or absence (non-transgenic) of a construct suppressing the expression of the *NAM* genes of all three genomes. Hexaploid bread wheat (*Triticum aestivum* ssp. *aestivum*) was transformed with an RNAi construct that reduced expression of NAM transcripts by about 50%. The photo compares transgenic (left) and control (right) lines at 50 days after anthesis. Chlorophyll loss in the transformant was delayed by 24–30 days. The stay-green phenotype is associated with lower contents of protein, zinc and iron in the grain. ppm, parts per million.

high-precision tools for mapping and modifying specific genes, as in the examples of *NAM* and *SGR*.

Several types of stay-green mutation have been identified in *Arabidopsis*. One group of mutants with delayed senescence is called *ore* (the name comes from *oresara*, which means long-lived in Korean). The gene responsible for the phenotype of mutant *ORE9* has been cloned and found to encode an F-box protein that has a role in proteolysis mediated by the ubiquitin–proteasome system (UbPS). It is thought that the ORE9 protein may bring about initiation of senescence by triggering breakdown of a senescence repressor factor. In another *ore* mutant, *ore12-1*, leaf life span is extended as a result of interference with the signal transduction pathway for the antisenescence hormone cytokinin. *ORE12-1* encodes the **histidine kinase** AHK3. The *ore12-1* mutant is a gain-of-function variant which makes an altered form of the kinase that behaves as if it is permanently in the cytokinin-activated condition. AHK3 mediates phosphorylation of the **response regulator** ARR2 (see Chapter 10), essential for determining leaf longevity.

In contrast to cytokinin, ethylene induces premature leaf senescence in *Arabidopsis*. Leaf longevity is increased in the **ethylene-insensitive mutants** *etr1-1* and *ein2*. A leaf must reach a certain age before it becomes competent to initiate senescence in response to ethylene. A number of mutants have been described, designated *old* (*onset of leaf death*), in which the timing of the acquisition of sensitivity to ethylene is altered. Such studies of mutants reveal a complex signaling network that is regulated by ethylene and, in turn, regulates the ethylene response in *Arabidopsis* leaf senescence.

In addition to the genes identified by studies of mutants and the inheritance and mapping of genetic variation, *SAGs* have been defined and isolated by analyzing DNA **transcription patterns** in leaves before, during and after the initiation of senescence. More recently, high-throughput techniques for gene expression profiling such as DNA microarrays and sequencing have generated a wealth of information on gene expression during senescence of a number of model and crop plants. Plants analyzed include not only herbaceous species such as *Arabidopsis* and cereals but also the deciduous tree **aspen** (*Populus tremula*), in which autumnal leaf senescence proceeds in a coordinated fashion in the whole crown (Figure 18.26), simplifying gene expression profiling. Entry into senescence is always accompanied by extensive changes in gene expression. Many genes highly expressed in non-senescent leaves are turned off, notably nuclear and plastid genes encoding photosynthetic proteins and several relating to the synthesis of photosynthetic pigments (chlorophylls and carotenoids).

Over 800 senescence-upregulated genes have been described so far from *Arabidopsis*, and other species show activation of *SAGs* on a similar scale. Some SAGs are expressed early in senescence, others late; some appear in natural senescence of attached tissue but not in detached tissue; some respond to senescence modifiers such as salicylic acid, jasmonic acid or ethylene, and others do not. The general picture emerging from studies of *SAG* transcription profiles is that, although the overall patterns of gene expression during senescence seem similar in most plants, different senescence-inducing treatments in one species show quite unique gene expression profiles so there is no universal senescence gene expression pattern to be found.

SAGs may be classified according to the biochemical functions of the proteins they encode (Table 18.1). As might be expected, genes for protein processing and remobilization are prominent. They include SAG12, VPEs, Clp proteases, ubiquitin and enzymes of amino and keto acid metabolism. The complement of *SAGs* includes many genes encoding enzymes that catalyze β-oxidation of fatty acids, lipid and carbohydrate metabolism and gluconeogenesis. The presence in the list of numerous genes coding for **metallothioneins** is thought to reflect the need for protection against metal ion-mediated

Figure 18.26 Fall senescence in free-growing aspen. The pictures were taken at the same time of day from 7 September to 1 October at Umeå, Sweden.

Table 18.1 Senescence-associated genes (SAGs) classified according to potential functions.

Potential function	Proteins encoded
Protein degradation	Cysteine proteases Aspartic proteases Clp proteases Ubiquitin F-box protein
Protein processing	Vacuolar processing enzyme (VPE)
Nitrogen mobilization	Glutamine synthetase Aminotransferase Vegetative storage protein Branched chain α-ketoacid dehydrogenases
Lipid degradation	Lipase Acyl hydrolase Phospholipase D
Carbon mobilization	Isocitrate lyase, malate synthase Pyruvate orthophosphate dikinase Sugar transporter
Cell wall degradation	Endoxyloglucan transferase β-glucosidase
Phosphorus mobilization	RNase BFN1 bifunctional nuclease
Transport	Copper chaperone, RAN1 Sugar transporter
Transcriptional regulation	WRKY factors Leucine zipper proteins
Signaling pathways	Receptor kinase SARK Receptor kinase SIRK Calmodulin-binding protein
Antioxidants/metal binding	Catalase Metallothioneins Ferritin Blue copper-binding protein
Hormone biosynthesis	12 OPDA reductase, lipoxygenase, thiolase ACC synthase, ACC oxidase
Cell death	Cyclic nucleotide-gated ion channel
Defence-related gene products	PR1a, Chitinase Osmotin like, nitrilase β1.3 glucanase, etc. Hin 1 (harpin induced)

Table 18.1 (continued).

Potential function	Proteins encoded
Translation	Deoxyhypusine synthase Translation initiation factor 5A
Unknown function	Early light-induced protein Cytochrome P450s

ACC, 1-aminocyclopropane-1-carboxylic acid; OPDA, 12-oxo-phytodienoic acid.

oxidative damage or for ion storage and transport (see Chapter 13). Antifungal proteins, pathogenesis-related proteins and chitinases are among the components encoded by *SAGs* in a number of species. Genes for many **transcription factors**, including WRKY53 and members of the NAM family, also show large alterations in transcript abundance during senescence.

Growth regulators influence the initiation and progression of senescence and the expression of *SAGs*. We have seen that ethylene acts primarily as a promoter of senescence, whereas cytokinins are senescence antagonists. In an important experiment that combined knowledge of *SAG* expression and cytokinin function, researchers fused *ipt*, a gene encoding isopentenyl transferase, a limiting step in cytokinin biosynthesis, with the **promoter** region of the gene encoding the senescence-specific cysteine protease SAG12. Tobacco (*Nicotiana tabacum*) plants transformed with the *SAG12 promoter–ipt* fusion produce cytokinin in an **autoregulated** fashion (Figure 18.27). As the tissue starts

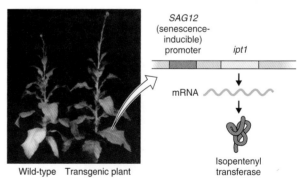

Figure 18.27 *Nicotiana* leaf senescence is delayed in plants genetically manipulated to autoregulate expression of a cytokinin biosynthesis gene. The bacterial gene *ipt1* encodes isopentenyl transferase, an enzyme that catalyzes a rate-limiting step in cytokinin biosynthesis. When fused to the promoter of *SAG12* and transformed into *Nicotiana*, ipt is specifically expressed in senescing tissue. Initiation of senescence in transgenic plants results in an autoregulatory loop in which increased cytokinin content blocks leaf senescence.

to senesce, the transgene is induced, and cytokinins are produced. The cytokinins inhibit senescence, which in turn decreases expression of the transgene. Thus, cytokinin is produced only in senescing tissue, and only in the amount needed to block senescence. Transformed plants differ from untransformed controls by exhibiting a significant delay in leaf senescence, an extension of photosynthetic activity and increased seed production. Remarkable improvements in drought tolerance of plants genetically modified in this way have also been observed. These characteristics have important practical implications for agriculture.

In some cases, cytokinins are able not only to inhibit senescence but even to reverse it (Figure 18.28). At flowering the lowest, oldest leaves of a mature plant of *N. rustica* are almost completely yellow. If the shoot is cut off just above the lowest node and the plant is kept in dim light, the leaves will gradually regain their green color. This process is greatly accelerated if the leaf is treated with cytokinin solution. During regreening,

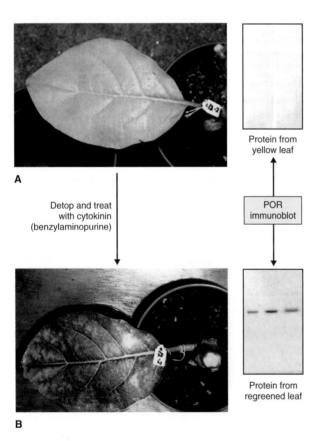

Protein from
yellow leaf

Detop and treat
with cytokinin
(benzylaminopurine)

POR
immunoblot

Protein from
regreened leaf

Figure 18.28 Regreening in *Nicotiana*. (A) Terminally senescent basal (oldest) leaf on a flowering *Nicotiana rustica* shoot. (B) The same leaf after removing the shoot at the node, treating with cytokinin and maintaining in dim light for 20 days. During regreening, gerontoplasts redifferentiate into functional chloroplasts, and proteins characteristic of chloroplast assembly in very young leaf cells, such as protochlorophyllide oxidoreductase (POR), reappear.

expression of *SAGs* is suppressed and genes for plastid assembly are turned on. Figure 18.28 shows the example of the chlorophyll biosynthesis enzyme protochlorophyllide oxidoreductase. The gerontoplasts of the yellow leaf redifferentiate into chloroplasts and photosynthetic activity returns. Not only is this a demonstration of the potency of cytokinin as an antisenescence factor, it shows that leaf senescence is potentially reversible at an advanced stage and thus fundamentally distinct from other types of PCD.

Key points The functions of a number of senescence-associated genes (*SAGs*) have been established by genetic and molecular analysis. Stay-greens are phenotypic consequences of mutations of *SAGs* that interfere with the characteristic color changes of senescence. In some cases stay-green traits are useful for crop productivity because they enhance the duration of photosynthesis, shelf-life or stress resistance, but they can result in undesirable inefficiencies in nutrient remobilization. *SGR*, a stay-green gene that functions in the breakdown of chlorophyll–protein complexes, has been isolated by a combination of comparative genetic mapping and functional testing in grasses, pea and *Arabidopsis*. A similar strategy has identified NAM, a transcription factor of wild and domesticated wheats that regulates the onset and progress of leaf senescence, the remobilization of nutrients including iron and zinc, and grain protein content. A number of stay-green mutants, collectively named *ore*, have been characterized in *Arabidopsis*. *ORE9* encodes an F-box protein that functions in the UbPS protein degradation pathway and *ORE12-1* is a component of the cytokinin signal transduction network. The senescence-inhibiting influence of cytokinin is dramatically illustrated by the stay-green phenotypes of plants genetically engineered to express a cytokinin synthesis gene under the autoregulatory control of a senescence-active promoter. In some circumstances cytokinin may also stimulate the reversal of yellowing and recovery of photosynthesis in leaves at an advanced stage of senescence. In *Arabidopsis* and other species, mutations interfering with ethylene metabolism and response also result in delayed leaf senescence. Genomics analyses have identified several hundred differentially expressed genes associated with the initiation and execution of senescence. Protein, lipid and carbohydrate remobilization, ion storage and transport, biotic and abiotic stress regulators and transcription factors are prominent among the functions encoded by these *SAGs*.

18.4 Programmed senescence and death in the development of reproductive structures and seeds

Senescence and other forms of PCD were well-established features of plant development and organization long before the **angiosperms** emerged and flowers, fruits and seeds evolved. Pre-existing genetic systems of regulated and selective senescence and death of cells, tissues and organs probably became recruited into programs of floral development from the time when the earliest flowering plants appeared. In angiosperms, distinctive kinds of PCD contribute to all stages of sexual reproduction, from the differentiation of male and female floral parts and gametes, through the development of structures attractive to pollinators, to embryo formation and seed maturation. Fruit ripening is a form of senescence that has been the subject of particularly intensive research and is considered separately, in Section 18.5.

18.4.1 Selective death of reproductive structures occurs during the development of unisexual flowers

Flowers, which evolved from modified shoots, first appeared over 125 million years ago. PCD of selected cells or groups of cells is a key factor in many aspects of floral development. In most plants that have unisexual flowers, at early stages of development those destined to become male flowers are indistinguishable from those which will become female flowers, since they all contain primordia for both male and female organs. But at a certain developmental stage, which varies with the species of plant, either the male or the female parts stop growing, and a cell death program is initiated that leads to their elimination. For example, the male inflorescence of *Zea mays* (the **tassel**; Figure 18.29A) is spatially separate from the female inflorescence (the **ear**; see Figure 16.1). In the early stages of tassel development, the young flowers contain primordia for both **stamens** (male) and **gynoecium** (female), but as the flower develops gynoecial cells stop growing and dividing. The nucleus and other organelles of these cells break down. It has been shown that the *TASSELSEED 2* (*TS2*) gene is required for this selective death of female organs in the tassel, since in plants carrying a mutation in the *TS2* gene, the arrest and degeneration of the gynoecia do not occur and the tassel develops female flowers (Figure 18.29B). In wild-type plants, the *TS2* gene is expressed in gynoecial cells in the tassel just before they start to undergo degeneration. In the ear, the gynoecia of female flowers do not undergo tasselseed-mediated cell death; the *TS2* gene is expressed in this tissue, but its action is suppressed by another gene, *SILKLESS 1* (*SK1*). Several other genes also interact with *TS2* and *SK1* to control the development of different tissues in maize flowers by selectively promoting or blocking cell death.

18.4.2 Petals and sepals undergo senescence

Macromolecule breakdown and modification of organelle structure and function are features of the senescence of **floral parts**. The different locations and modes of PCD in floral structures are summarized in Figure 18.30. In many species, for example carnation (*Dianthus caryophyllys*), wallflower (*Cheiranthus* sp.) and *Arabidopsis*, ethylene promotes floral senescence. In other species, however, normal flower senescence is essentially insensitive to ethylene. Lilies of the genus *Alstroemeria*, widely used as cut flowers, are among the

A

B

Figure 18.29 Control of development of unisexual flowers in the male inflorescence (tassel) of maize. (A) While the tassel contains flowers that were initially bisexual, PCD results in the death of female tissues, giving rise to a unisexual male inflorescence. (B) Female tissues of the *tasselseed2* mutant do not undergo PCD, and the resulting tassel flowers are mostly female.

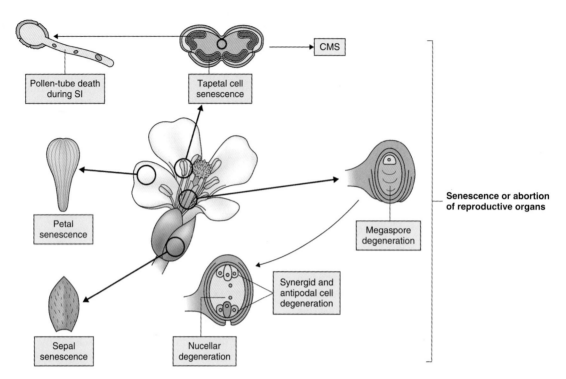

Figure 18.30   Programmed cell death events during the development and senescence of floral organs. CMS, cytoplasmic male sterility; SI, self-incompatibility.

best-known examples of ethylene-insensitive floral senescence. The vacuole plays an important part in programmed death of petal cells. Vacuolar metacaspases are upregulated during petal senescence. The loss of tonoplast integrity and the leakage of water, electrolytes and hydrolases from the vacuole lead to wilting, further

degradation of macromolecules and final loss of petal cell viability. Analysis of *SAG* expression indicates that about 25–30% of expressed genes are the same in senescing leaves and petals in *Arabidopsis*. When transcription during senescence of *Cheiranthus* petals was compared with leaf senescence in the same species, many upregulated genes were found to be shared by the two organ types: for example, remobilization-related genes, such as that encoding a SAG12-like cysteine protease. The expression of **defense genes**, such as those encoding chitinase and glutathione-*S*-transferase, remained constant or decreased with age in leaves, whereas they were strongly activated in petals. The enhancement of defenses against infection by disease-causing organisms during senescence of floral tissues safeguards the healthy development and dispersal of seeds.

Key points   Programmed cell death processes contribute to all stages of reproductive development. In species with unisexual flowers, all flowers initially possess the primordia of male and female parts but PCD selectively deletes one or other as floral differentiation proceeds. Elimination of gynoecial cells during development of the tassel (male flowers) of maize requires expression of the *TS2* gene. In female flowers *TS2* action and gynoecial PCD are suppressed by another gene, *SK1*. Corolla senescence often commences after, and in response to, pollination, and in many species, but not all, ethylene is a trigger. Macromolecule breakdown and vacuole-associated lysis occur during senescence of petals and other colored floral organs, and transcription analyses show considerable similarity between the *SAGs* expressed in petals and leaves—in particular genes encoding enzymes of remobilization and defense proteins.

18.4.3 Specific cells undergo senescence and death during gamete and embryo formation

Haploid tissues of many plants are also influenced by cell death programs. PCD is significant in the development of pollen grains (**microsporogenesis**; see Chapter 16). Microspores are surrounded by the **tapetum**, a layer of cells that provides nourishment for developing pollen grains. The tapetum degenerates during the later stages of pollen development by a PCD-like process. Microscopy studies have observed cell shrinkage, condensation of chromatin, swelling of the endoplasmic reticulum and the persistence of mitochondria. Fragmentation of DNA occurs and **cytochrome c** is released from mitochondria in tapetal cells towards the end of the unicellular stage of pollen development. Cytochrome c has a signaling role in some kinds of animal cell death and is suggested to be a factor in **cytoplasmic male sterility**, a mutant phenotype in which pollen abortion is determined by the mitochondrial genome (see Section 16.4.3). For normal pollen development, tapetum PCD must occur at the right time. Transgenic manipulation of early microsporogenesis to block the death of *Arabidopsis* tapetal cells results in pollen abortion. The development of viable pollen is also prevented if the tapetum undergoes PCD too early in microsporogenesis.

Many plant species are **self-incompatible**; that is, the plant is able to recognize and reject its own pollen, and hence cannot fertilize itself. This is an evolutionary strategy to ensure outcrossing and bring together novel combinations of genes in the progeny. Several very different self-incompatibility systems are known (see Section 16.5.3), some of them involving PCD processes. For example, *Papaver rhoeas*, the field poppy, has a pistil self-incompatibility locus that encodes a low molecular weight **S protein**. When incompatible pollen lands on a pistil, the S protein triggers a Ca^{2+}-dependent signaling network in the pollen. This causes a rapid cessation of pollen tube growth, depolymerization of the cytoskeletal protein actin, and activation of a mitogen-activated protein (MAP) kinase cascade. These events, in turn, promote PCD during which DNA is digested and metacaspases are activated.

During **megagametogenesis** (development of the egg cell), three of the four megaspores formed after meiosis of the megaspore mother cell undergo PCD, leaving one megaspore that will give rise to the female gametophyte (embryo sac), which in turn produces the egg. Following fertilization in most angiosperms the first mitotic division of the zygote gives rise to two cells, one that produces the embryo, the other developing into the **suspensor**. The suspensor may undergo a few rounds of mitosis, but eventually suspensor cells enter into PCD.

18.4.4 Programmed senescence and death occur during seed development and germination

The cereal endosperm contains two cell types, starchy endosperm and aleurone (see Chapters 6 and 16). Both undergo developmentally-regulated PCD, but in distinctly different ways. Localized PCD is also believed to play a part in the development of the embryo. The starchy endosperm of a mature seed is dead, but,

unusually, the contents of these cells are not broken down—instead they are preserved in a dehydrated, mummified state. When grain germinates, hydrolytic enzymes secreted by the scutellum of the embryo and aleurone layer cause degradation of the entire starchy endosperm. Thus breakdown of the cell contents takes place some time (perhaps many years) after the point at which the cells ceased to be viable. If this pattern of endosperm development is altered by mutation, the starchy endosperm cells die in a different way. In *Zea mays*, the **shrunken2** mutation causes these cells to undergo premature death and degradation (Figure 18.31). Autolysis of the starchy endosperm cells

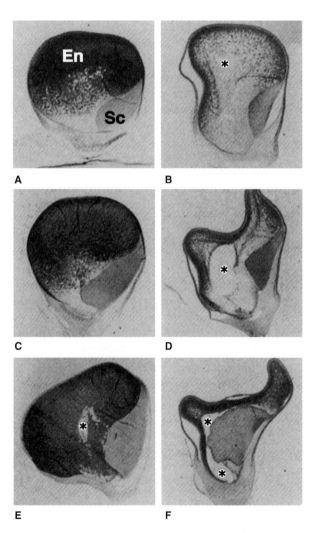

A   **B**   **C**   **D**   **E**   **F**

Figure 18.31   Endosperm (En) development in wild-type and *shrunken2* (*sh2*) maize genotypes. The starchy endosperm of the *sh2* mutant undergoes premature degradation and forms cavities (*). Shown are kernels of wild-type (left) and *sh2* mutant (right) plants at 28 days (A, B), 32 days (C, D) and 40 days (E, F) after pollination. The grains are stained with iodine-potassium iodide, which reacts with starch to form a dark blue complex. Sc, scutellum/embryo.

in *shrunken2* mutants causes deformation of the endosperm, leading to abnormal, shrunken kernels. Unlike starchy endosperm cells, those of the aleurone remain alive until germination is complete and all of the endosperm nutrient reserves have been mobilized. Only at this point do they undergo autolysis and die (see Figure 18.17). Gibberellin (GA) stimulates the initiation of PCD in the aleurone layer of barley and wheat grains, whereas abscisic acid (ABA) postpones PCD (Figure 18.32). Within hours of GA treatment, the **protein storage vacuoles** (**PSVs**) of aleurone cells, which hitherto had a neutral pH of about 7, become acidic (pH 5.5). They also accumulate lytic enzymes, including nucleases and several aspartic and cysteine proteases. In contrast, ABA-treated cells do not undergo PCD, and their PSVs maintain a near-neutral pH and do not accumulate these lytic enzyme activities. PSVs in GA-treated aleurone cells have much in common with vacuoles in senescing photosynthetic cells. Both kinds of organelle contain similar lytic enzymes, and the membranes that surround them stay intact throughout the period when macromolecules are degraded for export. This contrasts with, for example, the type of

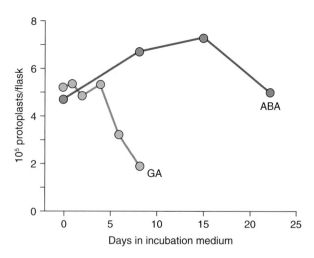

Figure 18.32 Programmed cell death in barley aleurone cells is promoted by gibberellic acid (GA) and delayed by abscisic acid (ABA). Aleurone cells were incubated in a medium containing 5 μM GA or 25 μM ABA; cell death was monitored by counting the number of live cells.

PCD seen in tracheary elements (Section 18.2.1), in which the surrounding membranes lose integrity.

18.5 Fruit ripening

The fruits we are familiar with in the human diet are structures that are unique to flowering plants (see Chapters 1 and 16). Ripening of fleshy fruits is typically associated with vivid color changes and the development of attractive textures, flavors and fragrances. The diversification of fruits with such characteristics at a time when the animals that act as dispersal agents were also undergoing rapid evolution strongly suggests that ripening processes have arisen by coevolutionary selection. We may expect, therefore, that the biochemistry and regulation of ripening will share features of the terminal processes found in leaves, from which they evolved, and will also show enhanced development of novel mechanisms for signaling to animal dispersers (including humans).

Fruit ripening is best understood as a terminal phase of development rather than a degenerative process, although there has been much research on the physiology and pathology of shelf-life and post-harvest deterioration. Unlike senescing leaves, fruits do not export significant amounts of products salvaged from the catabolism of macromolecules. The model species for investigations of fruit expansion, maturity, shelf-life and nutritional quality is tomato (*Solanum lycopersicum*). In the following discussion, tomato refers to the fruit of *S. lycopersicum* unless otherwise stated. The genomics

Key points Timely PCD during microsporogenesis is essential for the development of viable pollen. Tapetum degeneration, which nourishes the developing pollen grain, is accompanied by fragmentation of DNA and the release of mitochondrial cytochrome c, characteristics with some similarity to those of animal cell PCD. Some kinds of mutation in mitochondrial DNA cause cytoplasmic male sterility, a form of pollen abortion. PCD is a factor in the self-incompatibility system of field poppy, in which self-pollination is prevented by the termination of pollen tube growth and subsequent pollen cell PCD mediated by an S protein-triggered Ca²⁺-signaling network. Three of the four products of meiosis during egg cell formation, and cells of the suspensor following fertilization, are eliminated by PCD. Starchy endosperm cells of the developing cereal grain are dead at maturity, but this form of PCD may be likened to mummification because macromolecules are preserved rather than hydrolyzed. Mutations in endosperm starch synthesis interfere with PCD and result in abnormal kernel morphologies. Aleurone cells are viable at maturity but undergo gibberellin-stimulated PCD during germination, during which lytic enzymes accumulate and there are major changes in vacuole structure and function. Abscisic acid counteracts the PCD-promotive effect of gibberellin.

resources of *Arabidopsis* have also been (literally) fruitful in uncovering general mechanisms of molecular control in fruit formation and development.

18.5.1 A respiratory burst occurs during fruit ripening in some species

Harvested fruits are classified according to whether or not they display a ripening-associated respiratory burst, called the **climacteric**. Table 18.2 lists examples of climacteric and **non-climacteric** species. When harvested at full maturity, climacteric fruits can be ripened even when removed from the parent plant, but the detached fruits of non-climacteric species are generally incapable of proceeding to full ripeness.

The function of the respiratory climacteric is not fully understood. Non-climacteric fruits, for instance citrus (e.g. lemon, *Citrus limon*; orange, *C. sinensis*), *Fragaria* spp. or grape (*Vitus* spp.), ripen perfectly normally despite the absence of this physiological change. During the climacteric in fruits such as banana (*Musa* spp.), as much as 5% of carbohydrate may be lost as respiratory CO_2 (Figure 18.33). ATP formation through glycolysis and mitochondrial respiration, and the generation of carbon skeletons for transamination and other metabolic processes (see Chapter 7), are enhanced. The respiratory

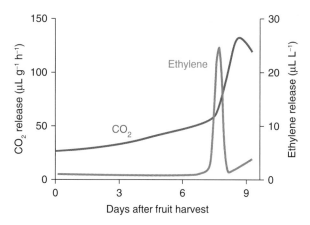

Figure 18.33 Ethylene and respiratory CO_2 production by ripening banana fruit. The respiratory burst (climacteric) is anticipated by a spike in ethylene emission.

climacteric is also associated with a burst of ethylene production (Figure 18.33). Ethylene is often called the '**ripening hormone**'. It regulates metabolism, including respiration, during senescence and ripening and also controls its own biosynthesis (see Section 18.5.5).

18.5.2 Fruits change color during ripening

Fruits become brightly colored during ripening (Figure 18.34). In many species, such as banana, pepper (*Capsicum*), tomato and citrus, fruits are green when immature and catabolize chlorophyll during ripening, by an essentially identical pathway to that of senescing leaves. The loss of chlorophyll unmasks underlying **carotenoids**, highly hydrophobic compounds that provide a yellow or orange background against which new pigments accumulate. The brightly colored carotenoids of ripe fruit are located in chromoplasts, specialized plastids derived from the chloroplasts of green immature fruit tissues (see Figures 4.23 and 4.24D). The carotenoids of chromoplasts form fibrils, crystals or globules and are associated with specific proteins called **fibrillins**. Fibrillin genes are strongly expressed during fruit ripening, leaf senescence and the development of floral parts, as well as in response to various environmental stresses.

The red carotenoid that appears during ripening of tomato, bell pepper (*Capsicum annuum*) and similar fruits is **lycopene**. Lycopene and other fruit carotenoids are synthesized by the isoprenoid pathway (see Chapter 15). **Phytoene synthase** (**PSY**) is the rate-limiting enzyme in the synthesis of lycopene by fruits. It condenses two molecules of the C_{20} precursor **geranylgeranyl diphosphate** (GGPP) into the C_{40} carotenoid phytoene (Equation 18.7), which in turn is

Table 18.2 Climacteric and non-climacteric fruits.

Climacteric fruits	Non-climacteric fruits
Apple	Cherry
Apricot	Cucumber
Banana	Grape
Guava	Grapefruit
Kiwifruit	Lemon
Mango	Lime
Papaya	Litchi
Passion fruit	Mandarin
Peach	Melon
Pear	Orange
Persimmon	Pineapple
Plum	Pomegranate
Sapodilla	Raspberry
Tomato	Strawberry

Figure 18.34 Colors of different kinds of bell pepper are the consequence of selection for arrest at different points in the chloroplast to chromoplast transition.

converted to lycopene by the enzymes **phytoene desaturase** (PDS; Equation 18.8) and plastoquinone-requiring ζ-carotene (zeta-carotene) desaturase (ZDS; Equation 18.9). The activities of GGPP synthase and PDS increase markedly during ripening of *Capsicum*.

Equation 18.7 Phytoene synthase (PSY) catalyzes a two-stage reaction

18.7A

2 Geranylgeranyl diphosphate (GGPP) →

PP_i + prephytoene diphosphate

18.7B

Prephytoene diphosphate → phytoene + PP_i

Equation 18.8 Phytoene desaturase (PDS)

Phytoene → ζ-carotene

Equation 18.9 Zeta-carotene desaturase (ZDS)

$\zeta\text{-Carotene} + 2PQH_2 + 2O_2 \rightleftharpoons \text{lycopene} + 4H_2O + 2PQ$

The second major group of pigments and secondary compounds that build up in ripening tissues comprises products of phenylpropanoid metabolism. Phenylpropanoid pathways originate with the amino acid phenylalanine and are complex, branched, metabolic sequences with several control points that lead to biosynthesis of a diverse group of phytochemicals, including phenolics, tannins and flavonoids (see Chapter 15). Unlike carotenoids, these are mainly water-soluble compounds and accumulate, not in chromoplasts, but in the central **vacuole. Anthocyanins** and proanthocyanins are **flavonoid** pigments that account for the colors of some ripe fruits, such as strawberry. Red grapes (*Vitus*), eggplants (*Solanum melongena*), blackcurrants (*Ribes nigrum*), cherries (*Prunus* spp.) and the skin of red apples (*Malus domesticus*) are rich in anthocyanins. Red and purple anthocyanins and yellow flavonoids are also responsible for the striking pigments of autumnal foliage in trees

such as maple species (*Acer* spp.). The development of anthocyanin coloration in fruits is the result of coordinated expression of the genes for flavonoid biosynthesis under the control of specific MYB transcription factors.

18.5.3 Fruit texture changes during ripening

Fleshy fruits become softer during ripening. Cell turgor pressure usually remains more or less constant but chemical changes in the cell walls of fleshy tissues result in the loss of firmness. **Hydrolytic enzymes** become activated during ripening and attack cell wall carbohydrates, de-esterifying and depolymerizing polysaccharides, changing the physical properties of the wall matrix and, in soft fruits like tomato and peach, loosening cell–cell adhesion. The basic structure of the cell wall in fruits conforms to that described in Chapter 4 in comprising a hemicellulose matrix with embedded cellulose microfibrils and pectinaceous middle lamella connecting adjacent cells.

Pectins, which may constitute more than 50% of the fruit wall, consist of homogalacturonan (HG) and rhamnogalacturonan (see Chapter 4). Pectins are more highly esterified in unripe than in ripe fruit, and pectin molecular weight and neutral sugar content are also greater. The softening of tomato flesh is associated with the loss of methyl esters of HG, the consequence of activity of the enzyme **pectin methylesterase** (**PME**), which removes methyl ester groups from the galacturonic acid residues of pectic polysaccharide backbones. De-esterified HG becomes available for hydrolysis by the enzyme **polygalacturonase** (**PG**). The tomato genome contains a small family of *PG* genes, of which one, encoding PG1, is implicated in ripening. Multiple isoforms of PG are detectable, arising by allelic variation and differences in protein glycosylation. A model of the molecular structure of polygalacturonase is shown in Figure 18.35. PG functions as an **endohydrolase**, progressively attacking glycosidic bonds within the unbranched HG backbone.

The enzymatic activity of PG, and the abundance of the mRNA products of PG gene transcription during

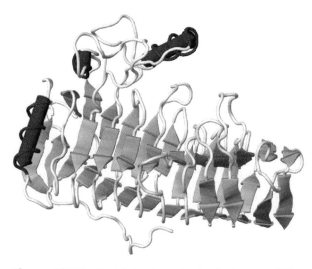

Figure 18.35 Crystal structure of polygalacturonase (PG) from *Erwinia carotovora* ssp. *carotovora*, a bacterial pathogen of plants. Crystal structures, and computed three-dimensional models of PGs such as this one, show a distinctive protein configuration consisting of a stack of parallel β-pleated strands (yellow arrows) coiled into a large helix. Pink rockets represent alpha helices. Based on high primary sequence homologies, plant polygalacturonases are predicted to have similar higher-order structures.

ripening, point to a central role in fruit softening. But when PG expression was suppressed in transgenic tomatoes by transformation with an **antisense** form of the PG gene, there was little or no reduction in pectin depolymerization or in fruit softening. Neither did antisense repression of PME activity bring about changes in tomato fruit texture, although the molecular weight of cell wall pectins was increased in these plants. Such manipulations have, however, improved shelf-life and post-harvest processing properties and for a limited time were the basis of marketed tomato varieties. Further evidence that the action of pectinolytic enzymes alone is not sufficient to bring about textural changes during fruit ripening comes from experiments with the tomato mutant *rin* (*ripening inhibitor*). The *rin* mutation effectively blocks the ripening process, resulting in fruits that remain green and firm and fail both to produce ethylene and to respond to it. The fruit of plants homozygous for the recessive *rin* allele are blocked in PG expression. If such a genotype is transgenically manipulated to express PG, enzyme activity is detected and pectin is depolymerized in vivo, but no significant fruit softening occurs.

Clearly, though PG and PME are important for the degradation of pectins, their induction during ripening is not the main determinant of tissue softening. Attention has been directed towards other cell wall-degrading enzymes that may be decisive. Among several **glycan-degrading enzymes** that increase in activity

during ripening are xyloglucan endotransglycosylase, endo-1,4-β-glucanases and expansins. They are believed to restructure the wall by altering the cross-linking network. It is likely that changes in fruit ripening require the cooperative action of a number of such wall-modifying enzymes.

18.5.4 Flavors and fragrances intensify during fruit ripening

Fruits generally become sweeter as they ripen. Sweetness is determined by the concentrations of the abundant sugars: sucrose and the derived hexoses. On a scale of sweetness based on sucrose = 1.0, fructose scores 1.2 whereas the value for glucose is 0.64. During domestication of the tomato by pre-Columbian civilizations in South America and its subsequent spread to Europe in the 16th century, sweetness was a highly valued trait. A legacy of this history is that sugar content in some varieties may be as high as 60% of the total dry weight. Fruits are major **terminal sinks** for carbohydrate exported from the photosynthetic tissues. Sucrose translocated into the developing fruit is unloaded from the transport system by a process in which the enzyme **invertase** plays a vital part (Equation 18.10).

Equation 18.10 Invertase (β-fructofuranosidase)

Sucrose + H_2O → glucose + fructose

Forms of invertase exist both inside (**symplasmic**) and outside (**apoplastic**) the cytoplasm. Figure 18.36A shows a structural model of the enzyme–substrate complex of vacuolar (symplasmic) invertase and Figure 18.36B a model of the cell wall-associated (apoplastic) form of the enzyme. It is believed that vacuolar invertases evolved from the apoplastic type at some time that pre-dated the divergence of monocots and eudicots. The two classes of invertase share a distinctive molecular structure consisting of a number of β-sheet regions arranged into so-called five-bladed β-propeller and β-sandwich domains (Figure 18.36). There is evidence that unloading of sucrose occurs symplasmically, early in tomato pericarp development. In later stages functional linkage of cell wall invertase to the activities of hexose transporters in the plasma membrane supports apoplastic unloading from the phloem and transfer to the developing fruit. Starchy storage organs such as potato tubers receive translocated sucrose by a similar mechanism (see Figure 17.9). The identification of invertase genes that coincide with genetic loci for various fruit traits on the tomato genetic map, coupled with experimental manipulation of the enzyme in transgenic plants, has demonstrated a critical

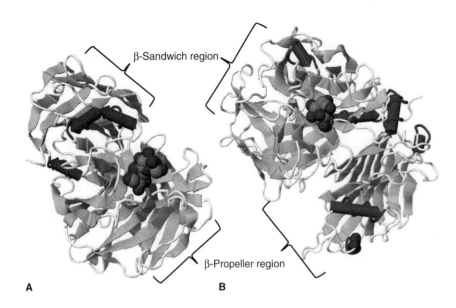

β-Sandwich region

β-Propeller region

A   **B**

Figure 18.36 Invertases. (A) One of the six identical polypeptide subunits of the β-fructofuranosidase (invertase) of the bacterium *Thermotoga maritima*, a model for the three-dimensional structure of plant symplasmic (vacuolar) invertases. The *T. maritima* enzyme is able to release fructose from sucrose and a range of related oligosaccharides, including raffinose (α-D-galactosyl sucrose), a molecule of which is shown occupying the active site. (B) Model of the enzyme–substrate complex of *Arabidopsis* cell wall (apoplastic) invertase with sucrose. Alpha helices are shown as pink rockets, and β-sheets as yellow arrows. C atoms of carbohydrate ligands are gray, and O atoms red.

role for apoplastic invertase in regulating sugar composition and fruit size in this species. Variability in sweetness reflects the relative proportions of sucrose and hexoses, which in turn is a consequence of differences in the relative activities of invertase and the other major sugar-hydrolyzing enzyme, **sucrose synthase** (Equation 18.11).

Equation 18.11 Sucrose synthase (UDP-glucose:fructose glucosyl transferase)

Sucrose + UDP ⇌ UDP-glucose + fructose

In many fruits there is a transient accumulation of starch in the early stages of development, followed by large-scale hydrolysis associated with sweetening (and, in climacteric fruits, the respiratory burst) as ripening proceeds. The activities of sucrose synthase, fructokinase and (particularly) **ADP-glucose pyrophosphorylase** (**AGP**; Equation 18.12) control the rate of starch accumulation. The genes for subunits of AGP co-locate with a number of genetic loci relating to tomato fruit sugar content, and antisense suppression of AGP in strawberry results in decreased starch and increased soluble sugar. Starch synthases and starch branching enzyme are bound to the starch granules of developing fruits (see Chapter 17). Starch degradation in fruits is not well understood, but enzymes likely to participate in starch hydrolysis, including starch phosphorylase and amylase, have been observed to increase in banana pulp during ripening, reaching their highest levels at around the climacteric.

Equation 18.12 ADP-glucose pyrophosphorylase (AGP)

Glucose-1-phosphate + ATP ⇌ PP$_i$ + ADP-glucose

Sourness (also known as tartness) is an important element in the complex mix of flavor compounds that give each fruit its distinctive taste and mouth feel. It is determined by the proportions of predominant organic acids, which are ranked relative to citric acid in the following order of sourness: citric (1.0) > malic (0.9) > tartaric (0.8). Amino acids, such as aspartate and glutamate, also contribute to sourness in some fruit. The metabolic origin of these compounds is glycolysis and the TCA cycle. The regulation of organic acid biochemistry and the basis of within- and between-species variation is not known in detail, but new approaches based on comprehensive metabolite profiling (**metabolomics**), together with recent developments in genetic mapping and sequencing of the tomato genome, are beginning to provide answers.

Another aspect of perception of the taste of fruit is **astringency**, which is largely determined by phenolic compounds, particularly proanthocyanidins or condensed tannins (see Chapter 15). The degree of astringency is related to the extent of polymerization and chemical modification of condensed tannins. Phenolics share a common metabolic origin in the phenylpropanoid pathway with pigments such as anthocyanins, and this is often reflected in coordinated regulation of taste and color during fruit ripening. For example, the branch point enzyme dihydroflavonol reductase is active early in strawberry development as condensed tannins accumulate, is subsequently downregulated, and then comes on again strongly as fruit color intensifies.

Together with color and flavor, the **fragrance** of many fruits is enhanced during ripening. This is the result of increased production of low molecular weight volatile compounds, notably esters, alcohols, aldehydes and ketones. Tomato fruits are estimated to contain some

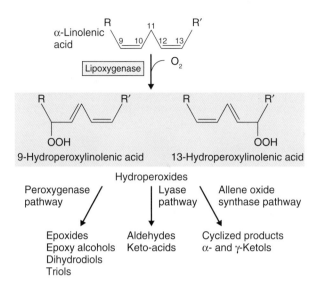

Figure 18.37 The enzyme lipoxygenase catalyses the oxidation of polyunsaturated fatty acids to form hydroperoxides. These are precursors of a range of volatiles and other products of diverging metabolic pathways.

400 different volatile compounds, prominent amongst which are cis-3-hexanal, cis-3-hexanol, hexanal, 3-methylbutanal, 6-methyl-5-hepten-2-one, 1-pentan-3-one, trans-2-hexanal, methyl salicylate, 2-isobutylthiazole and β-ionone. The biochemical origins of volatiles are diverse and trace back to the pathways of amino acid, fatty acid and carotenoid metabolism. An enzyme of particular interest is **lipoxygenase** (Figure 18.37), which catalyzes the hydroperoxidation of lipid precursors of some aroma compounds. The major sources of esters are acetyl CoA, the product of the pyruvate dehydrogenase complex, or acetaldehyde and ethanol via pyruvate decarboxylase and alcohol dehydrogenase (see Chapter 7). The ripening hormone ethylene is a volatile and it has been observed that genetic or chemical treatments aimed at suppressing ethylene production in climacteric fruits often result in decreased production of aroma compounds. Expression of the gene for **alcohol acyltransferase**, which has a role in the formation of esters, is strongly upregulated by ethylene, leading to a large increase in the activity of the enzyme.

As well as flavor and aroma factors, there is growing interest in nutritional compounds in ripe fruit that may have health benefits. They include essential chemicals that cannot be synthesized de novo in the human body. For example, **carotenoid deficiencies** may cause blindness, and carotenoids with a β-ring end group are required for the synthesis of **vitamin A**. Tomato fruits are the principal source of carotenoids in many Western diets. Another vitamin of high importance is **folic acid**: more than one-third of the daily intake of folate in the average diet comes from fruit and vegetable sources.

Fruits are also an important source of non-vitamin **antioxidants**. The biosynthesis and regulation of these minor but nutritionally important constituents in relation to fruit ripening are areas of active biotechnological research.

Finally, we should remember that although fruits have evolved to tempt animal dispersers, and have been selected through plant breeding to be attractive and appetizing to humans, plants (wild plants in particular) are generally equipped with an array of defenses that deter would-be predators. We see an example of this in tomato, where an allele of a gene called *malodorous*, which gives the fruit of wild species a disagreeable odor, has been shown to have been selected against during domestication.

18.5.5 Fruit ripening is subject to genetic and hormonal regulation

Our understanding of how fruit ripening is controlled is based to a large extent on analysis of tomato **mutants** (Figure 18.38), particularly those affecting ethylene synthesis and response. We have already mentioned *rin*, a mutant that fails to ripen in response to ethylene. The *RIN* gene has been cloned and found to encode a MADS-box transcription factor that is known to be also associated with floral development. *RIN* homologs are found in climacteric and non-climacteric species. Other transcription factor mutants with similar phenotypes to *rin* include *non-ripening* (*nor*) and *Colorless non-ripening* (**Cnr**). The CNR protein is a member of the SBP (*SQUAMOSA* promoter binding) family and *NOR* encodes a NAC domain factor.

It is thought that *NOR*, *RIN* and *CNR* are global regulators that function in the mechanism that directly triggers the ripening syndrome independently of ethylene. In ethylene-sensitive climacteric fruit like tomato, these factors also interact with pathways of ethylene synthesis and signal transduction. The first committed intermediate in ethylene biosynthesis, **1-aminocyclopropane-1-carboxylic acid** (**ACC**), is the product of **ACC synthase** (**ACS**) and is converted to ethylene by **ACC oxidase** (**ACO**) (see Chapter 10). Of the nine genes encoding ACS in tomato, four are differentially expressed during fruit ripening under the influence of *RIN*, with the products of different members of this gene family predominating pre- and post-climacteric. Similarly, three of the five genes encoding ACO are differentially transcribed in fruit ripening. Reducing *ACS* and *ACO* expression in antisense lines of tomato greatly suppresses ethylene output and results in plants with delayed leaf senescence and incomplete fruit ripening. The phenotype is

Key points During fruit ripening there is activation of metabolic pathways with similarities to those of senescing organs, resulting in color, texture, taste and odor changes that make fruits attractive for dispersers and for human consumption. A clear difference between ripening and senescence is the limited degree of salvage and export of materials from the breakdown of fruit macromolecules. Fruits are classified as climacteric (exhibiting a ripening-associated respiratory burst, often accompanying a spike in ethylene production) or non-climacteric. Mature climacteric fruits can be ripened when detached from the parent plant, a process that can be hastened by exposure to ethylene. Chlorophyll is lost by the same pathway that operates in yellowing leaves. In the fruits of many species, chloroplasts differentiate into chromoplasts. Carotenoids are unmasked, and new carotenoids synthesized via the isoprenoid pathway, resulting in the orange and red pigmentation characteristic of many of the fruits consumed by humans. The red carotenoid of tomato and bell pepper is lycopene. Another source of fruit color is the pigmented water-soluble products of phenylpropanoid metabolism, including the red and purple anthocyanins of grapes, cherries and strawberries. The softening of fleshy fruits is the result of activation of enzymes that hydrolyze cell wall polysaccharides. During tomato ripening, polygalacturonase enzyme activity and transcript abundance increase. The enzyme works cooperatively with a number of cell wall-modifying enzymes to bring about fruit softening. Fruits become sweet during ripening through the import of sugars and hydrolysis of starch. Apoplastic invertase has an important regulatory role in fruit sweetening. Other flavor components include organic acid products of respiratory pathways, and flavonols and other phenolics. Fruits become more fragrant as they ripen, producing a variety of volatile organic compounds (including ethylene). A number of volatiles are oxidation products of lipids, originating in reactions catalyzed by lipoxygenase. Fruits are important dietary sources of antioxidants and vitamins, such as carotenoids, vitamin A and folic acid.

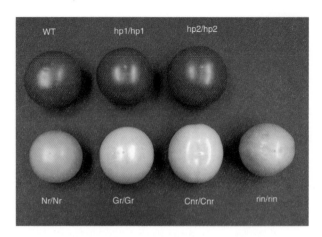

Figure 18.38 Wild-type and mutant tomato fruit. A ripe fruit of the normal tomato cultivar Ailsa Craig (WT) is shown 10 days after the onset of the color change from green to red, with equivalent-age fruits homozygous for the *high-pigment 1* (*hp1/hp1*), *high-pigment 2* (*hp2/hp2*), *Never-ripe* (*Nr/Nr*), *Green-ripe* (*Gr/Gr*), *Colorless non-ripening* (*Cnr/Cnr*) and *ripening-inhibitor* (*rin/rin*) mutations.

corrected by treatment with exogenous ethylene. Ripening-related expression of *ACS* and *ACO* genes has been demonstrated in a number of other species, including melon (*Cucumis melo*), *Malus*, *Musa* and persimmon (*Diospyros* sp.).

Fruits of the tomato mutant *Never-ripe* (**Nr**), as the name suggests, fail to ripen fully. *Nr* plants also show delayed leaf and flower senescence. *NR*, a member of the *ETR* (ethylene receptor) gene family, encodes an **ethylene receptor protein** which, if mutated, confers insensitivity to the hormone (see Chapter 10). *Green-ripe* (**Gr**) is another tomato mutant, in which ripening is incomplete as a consequence of selectively reduced ethylene responsiveness in fruit and floral tissues. As well as ethylene transduction, there is light-mediated regulation of tomato ripening. The carotenoid content of fruits of the recessive mutant *high pigment* (*hp*) is double that of wild-type fruits, as a consequence of a lesion in a gene functioning in responses to ultraviolet light. Similarly, overexpression of cryptochrome in tomato transgenics results in elevated levels of carotenoids and flavonoids.

Several mutations are known to block individual reactions in the metabolic program of ripening fruits. For instance, in the green-fruited tomato relative *Solanum pennellii* there is a null mutation in the gene encoding phytoene synthase. Another example is *gf* (*green flesh*), a tomato stay-green mutant. *GF* has been cloned and shown to be a homolog of *SGR*, the gene responsible for the stay-green phenotype in senescing leaves of grasses and developing cotyledons of *Pisum* seeds (see Section 18.3.2). In each case the mutant phenotype is insensitive to exogenous ethylene.

Based on the genetic and physiological characteristics of mutations affecting tomato fruits, it is possible to arrange genes and processes into a model of ripening that both accounts for experimental observations and defines targets for practical intervention to improve crop production and post-harvest quality (Figure 18.39).

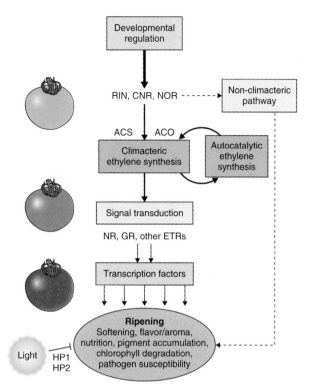

Figure 18.39   Model of the genetic regulation of tomato ripening. Transcription factors encoded by *RIN*, *CNR* and *NOR* genes regulate ripening by activating climacteric ethylene synthesis and are also thought to participate in an ethylene-independent (non-climacteric) pathway. Ethylene synthesis is sustained by autocatalysis. ACS and ACO are ACC (1-aminocyclopropane-1-carboxylic acid) synthase and ACC oxidase, respectively. NR, GR and other ethylene receptor proteins (ETRs) promote a cascade of gene activation that initiates and sustains ripening. Light regulates ripening through the activities of HP proteins.

Key points Mutations in a number of different genes result in altered ripening. The *rin*, *nor* and *Cnr* transcription factor mutants of tomato fail to ripen. *RIN* encodes a MADS-box factor that influences differential expression of genes encoding the ethylene synthesis enzymes ACC synthase and ACC oxidase before and after the climacteric. The mutants *Nr* and *Gr* are ethylene-insensitive. *NR* encodes an ethylene receptor protein. Carotenoid accumulation is controlled by ultraviolet light, mediated by the gene *HP*. Cryptochrome also functions in light regulation of ripening. The ethylene-insensitive *green flesh* (*gf*) mutant of tomato has a lesion in a homolog of the stay-green gene *SGR*, which is responsible for the breakdown of thylakoid chlorophyll–protein complexes.

18.6 Environmental influences on programmed senescence and death

Senescence is as responsive as any other physiological activity to sub- or supraoptimal environmental conditions. Seasonal or otherwise predictable environmental cues can trigger senescence as part of an adaptive **strategy**. Senescence is also a **tactic** deployed when an unpredictable stress (abiotic or biotic) is experienced. When the speed and severity of a developing environmental stress outruns the capacity of the tissue to invoke, coordinate and express the comparatively slow senescence program, cells divert more or less directly to the more rapid PCD pathways. General plant responses to environmental factors are discussed in Chapter 15. Here we look at how senescence and PCD in particular are influenced by the state of the environment.

18.6.1 Senescence varies with the seasons

Except in latitudes closest to the equator, the change in **day length** at different seasons is an important and reliable environmental input that can permit a plant to prepare for the likely stresses (heat, cold, drought and so on) to come (see Chapter 8). In plants of temperate regions, leaf senescence is a major developmental event. Proper timing of leaf senescence is vital if a plant is to balance its carbon and nitrogen demands: once the degradation of leaf photosynthetic pigments and proteins is initiated, photosynthetic carbon acquisition declines and eventually ceases, but nitrogen and other nutrients can be recovered from the senescing leaf. This process is most apparent in the highly coordinated—and often visually attractive—autumnal senescence seen in deciduous trees of temperate regions. Aspen has been used as a model to study senescence in the fall (see Figure 18.26), which in this species is initiated strictly according to **photoperiod**. A particular aspen tree always starts fall senescence at almost exactly the same date each year, independent of temperature or other environmental variables. Many stresses (such as attack by pathogens) can and do induce premature senescence of individual leaves or small groups of leaves, but when senescence in the fall is triggered by reducing day length, the whole crown of the aspen tree begins to senesce in a coordinated manner, according to a predictable timetable of cellular events (Figure 18.40).

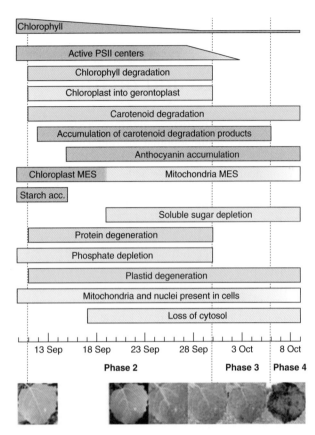

Figure 18.40 Timetable of events during autumnal senescence in aspen leaves. Senescence is divided into four phases. Phase 1 (the mature, pre-senescent stage; not shown here) is followed by phase 2, during which chloroplasts are converted to gerontoplasts, major pigmentation changes occur, N and P are mobilized and sugars are metabolized. During phase 2 the major energy source (MES) switches from chloroplasts to mitochondria. By phase 3 less than 5% of original chlorophyll remains; cell contents are severely depleted but metabolism continues and viability is sustained in some cells. Phase 4 is the stage at which cell death is complete and few structures are recognizable within residual cell walls. PSII, photosystem II.

Although initiation of senescence is determined by photoperiod, and therefore by date, once the process is underway, color changes are influenced by **temperature**. Chlorophyll degradation is enhanced at low temperatures, probably because of increased **photo-oxidative stress**. This means that the time from the onset of senescence until the leaves start to be visibly yellow could differ by up to 2 weeks, depending on how low the temperature falls during this period. Anthocyanin production (Figure 18.40) is also stimulated by the cold, and in years when the weeks immediately following senescence initiation are colder, decreased chlorophyll content and increased

Figure 18.41 Fall colors.

anthocyanin levels make the display of fall colors especially striking (Figure 18.41). The precise biological function of the anthocyanins accumulated in tree leaves during the fall remains something of a mystery. They may act as sunblockers or antioxidants, preventing light energy from damaging cells in which metabolic rate is constrained by suboptimal temperature. Additionally, or alternatively, there is evidence that fall colors act as visual warning signals to potentially predatory insects.

Photoperiodic control of the onset of senescence is especially valuable for trees growing at high latitudes. In northern forests, for example, nitrogen is often a limiting factor for growth. For many trees it is therefore preferable to sacrifice days or weeks of potential photosynthetic carbon gain in order not to lose large quantities of nitrogen and other nutrients if a sudden frost kills the leaf cells before remobilization has occurred. Climates at high latitudes often have rapid and inconsistent temperature changes in the fall, making photoperiod a more reliable predictor than temperature that winter is approaching. There is increasing evidence that phytochrome plays an important role in daylength perception and autumnal responses in *Populus*, and probably many other tree species. Transcription factors, including MADS-box homeotic genes, have been shown to play an important regulatory role in integrating autumnal senescence with growth, flowering and dormancy. It should be noted that not all temperate deciduous trees regulate senescence initiation according to photoperiod. *Malus*, pear (*Pyrus*) and some other woody members of the family Rosaceae are unusual in that, although they undergo foliar senescence, shed their leaves and spend winter in a dormant state, they are photoperiod-insensitive. It appears that in these species, chilling is the major trigger factor for the fall syndrome (see Chapter 15).

18.6.2 Programmed senescence and death are common responses to abiotic stresses

Because plants cannot move away from adverse environmental conditions, they need mechanisms to ensure their survival under circumstances that may be far from optimal for them. **Water** supply is an essential factor that is especially prone to shortage and, sometimes, excess. When plants encounter drought, flooding and other such variations in the environment, they often undergo senescence. Senescence-related genes are commonly among the genes upregulated in response to drought and other stresses. In addition to drought, temperature, atmospheric CO_2 content and **ultraviolet radiation** can all lead to increased expression of genes associated with senescence, and altered patterns of tissue senescence and death. One well-studied example of the way plants deploy PCD in response to environmental stress is the development of **aerenchyma**, tissue with large air pockets. When plant roots are subjected to flooding, they often undergo hypoxia or anoxia (see Chapter 15). Under these circumstances, whole cells are removed from mature roots by lysigeny (Figure 18.42). This creates channels through which the root can obtain oxygen by diffusion of air from the shoot. In a few plant species, aerenchyma is formed not by lysigeny but by schizogeny—that is, the air spaces are formed by cell separation rather than through loss of cells (see Figure 18.9). Most research on aerenchyma has been carried out on crop plants, particularly rice and maize in which flooding and poor aeration can seriously reduce yield. However, aerenchyma formation also occurs in species like *Arabidopsis*, which has been useful in understanding the genetic basis of aerenchyma development.

The trigger for aerenchyma formation is oxygen deficiency. Studies on the roots of *Zea mays* under oxygen-limiting conditions have shown that aerenchyma forms as a result of ethylene-regulated PCD in a specific group of root cortical cells. An entire cortical cell, including its cell wall, is removed, producing a space that facilitates the passage of oxygen through the roots. This PCD process can take place very rapidly and involves DNA fragmentation and significant alterations in the expression of genes encoding enzymes of glycolysis, fermentation, nitrogen utilization, trehalose metabolism and alkaloid synthesis. Regulatory systems are also activated, including those for calcium signaling, for protein phosphorylation and dephosphorylation, and for ethylene synthesis and perception. Anoxia or low-oxygen treatment of *Z. mays* roots results in cell wall breakdown associated with large increases in cellulase (Figure 18.43A), responses that correlate with the levels of the key ethylene biosynthesis enzyme ACC synthase (Figure 18.43B).

18.6.3 Senescence and cell death are adaptive and pathological responses to biotic interactions

Programmed cell death is an important weapon deployed by plants when threatened by, or exploiting, other organisms. As we have seen in this and other chapters,

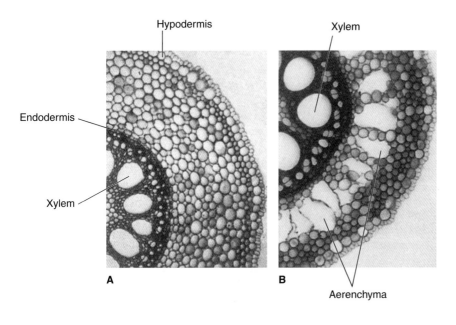

Figure 18.42 Aerenchyma formation in maize roots in response to low oxygen (hypoxia). Roots were grown under aerobic (A) and hypoxic (B) conditions. Root cortical cells respond to oxygen deprivation by undergoing lysigeny to form a network of continuous air spaces, allowing submerged roots to access atmospheric gases obtained by above-ground tissues.

Key points Senescence in many ecosystems of latitudes away from the equator is regulated by photoperiod. Autumnal senescence of temperate deciduous forests is generally initiated in response to decreasing day length. Apple and pear trees are exceptions in being induced to senesce by decreasing temperatures rather than day length. Studies on an established aspen tree growing in the field show that the onset of coordinated senescence in the foliage of the crown is predictably set by photoperiod. Phytochrome and MADS-box homeotic transcription factors mediate daylength regulation of the initiation of fall senescence. Temperature influences the timetable of subsequent events during execution of the senescence program. Low temperature promotes photo-oxidative bleaching of chlorophyll and the accumulation of anthocyanins. Plants commonly react to abiotic stresses by triggering PCD. Flooding is an example: low oxygen sensed by roots subject to flooding induces development of aerenchyma. Ethylene-signaling pathways regulate PCD in the root cortex, resulting in the development of lysigenous or schizogenous air spaces, thus improving oxygen flow. During aerenchyma differentiation, DNA is fragmented; genes for respiratory, carbohydrate, nitrogen and secondary metabolism are differentially activated; and cell wall breakdown is stimulated.

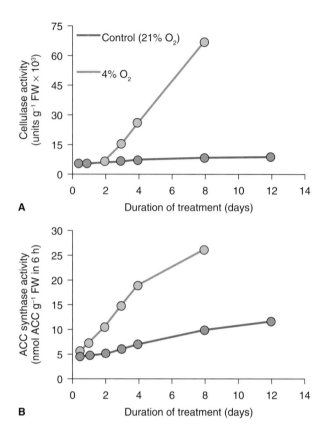

Figure 18.43   Effect of oxygen concentration on the activities of (A) cellulase and (B) ACC synthase, a rate-limiting enzyme in ethylene biosynthesis, accompanying aerenchyma formation in hypoxic maize roots. Both enzymes increase with the duration of the low-oxygen stress.

color changes associated with fall senescence, fruit ripening and floral display send invitations or warnings through the visual systems of potential pollinators, dispersers or predators. These interactions are the finely balanced outcome of coevolutionary adaptations dating back to the origins of flowering plants and beyond. Some pathogens, such as *Botrytis cinerea*, the highly virulent causal agent of gray mold disease of soft fruits, are **necrotrophs**: they kill host cells, often by producing lethal toxins (the sesquiterpenoid botrydial in the case of *Botrytis*), and live on the resulting dead tissue. By contrast, when subject to attack by **biotrophs** (pathological organisms that feed off tissue which must remain alive), the host plant will often invoke PCD to sacrifice parts in order to insure survival of the whole (see Chapter 15). The most prominent, and best understood, example of PCD used in this way is the **hypersensitive response** (**HR**) to fungal, bacterial and viral pathogens. The term *hypersensitivity* refers to the host plant's attempts to confine and neutralize the attempted infection. In spite of a great deal of detailed research on the HR mechanism, it remains unclear whether cell death is required for, or is a consequence of, resistance to pathogens.

The precise appearance of the HR response depends on the degree of resistance shown by the host (from extremely susceptible with no HR at one extreme to highly resistant with intense HR at the other) and on the infection strategy of the pathogen. Fungal and bacterial pathogens generally elicit **necrotic flecks**, localized areas of cell death (Figure 18.44), and single cell HR. HR cell death is usually rapid and, because it depends on the plant's own biochemistry, has features that identify it clearly as a type of PCD (see Figure 18.2A). Autophagy and the cell vacuole are important components of HR cell death. Vacuolar processing enzymes (VPEs) are thought to be particularly significant in the cell death process. For example, chemical inhibitors of VPEs can also inhibit the HR, and VPE-suppressed mutants or transgenics exhibit significant reductions in cell death and metacaspase activity when challenged with HR-eliciting pathogens. The in vivo substrates of VPEs are likely to be key triggers in the cell death process. Reactive oxygen and reactive nitrogen species, which are generated on a large scale during HR, are also critical for initiating and executing the cell death program. This is

Figure 18.44 Necrotic flecking as a result of yellow leaf blight of maize caused by the fungal pathogen *Mycosphaerella zeae-maydis*.

discussed in Chapter 15 in relation to the role of ROS in general responses to environmental stress.

Following pathogenic attack, the genetic program associated with the HR is likely to involve three types of genes: genes that will respond to triggers that initiate death; genes limiting the extent of death, which may suppress PCD in the absence of infection; and genes that integrate the death mechanism with other plant defenses. Identifying genes of each class has been a major objective for molecular plant pathologists.

The first genes identified as responsible for initiating the HR were the **R** (resistance) genes targeted by crop breeders, who have found them to be important sources of pathogen resistance. Work in the 1940s showed that to initiate an HR, the interaction of a single resistance gene product with a single product encoded in the pathogen by **avirulence** (**avr**) genes is all that is required. R−avr interactions may be considered, therefore, to be triggers of PCD. In more recent times researchers have sought to dissect further the HR process by generating and screening mutant lines that exhibit alterations in cell death.

Since PCD is a genetically-regulated process, it follows that mutations in genes contributing to PCD will lead to plants with altered forms of death. Alterations can take the form of abolition or reduction of cell death under circumstances where it would normally take place; or

premature and spontaneous cell death. Table 18.3 lists examples of mutations causing delayed or spontaneous death, and the molecular basis of the mutation in each case, where known.

Necrotic flecks resulting from spontaneous cell death have often been observed in crop plants, including maize (which has over 30 known examples) and rice, but *Arabidopsis* spontaneous death mutants are the most numerous and best characterized. Certain spontaneous lesion mutations have been exploited in agriculture, even where the precise molecular mechanism is not known, as they can confer increased pathogen resistance. For example, the Sekiguchi lesion (*sl*) mutation in rice confers resistance against the devastating fungal disease rice blast. The *mildew resistance locus o* (*mlo*) mutation of barley is widely used by breeders developing spring barley cultivars, even though it has necrotic flecks on its leaves, because it confers increased resistance against powdery mildew.

Few of the spontaneous death mutants identified in plants show abnormal developmental cell death pathways. Perhaps this is because, if the death processes leading to structures such as tracheary elements cannot proceed, the plant is not viable. However, there is one possible example in this category: the *Arabidopsis* mutant *vascular associated death1* (*vad1*), in which necrosis develops from the veins as leaves age (Figure 18.45). This mutation affects a membrane-associated protein which is also induced during a HR; thus *vad1* may represent a link between developmental cell death and that induced by certain pathogens.

Chlorosis, or yellowing of leaf tissue in a manner similar to senescence, is a common feature in plant diseases. Why would it be advantageous to a pathogen to initiate senescence? The bacterium *Pseudomonas syringae* pathovar (P. s. pv.) tomato, a pathogen of several dicot species, produces the toxin coronatine, which mimics the action of an endogenous class of plant hormones, **jasmonates**. Jasmonates contribute to the initiation of chlorosis, in addition to having a defense role. In bean (*Phaseolus vulgaris*), the chlorosis initiated by the pathogen P. s. pv. phaseolicola is due to a tetrapeptide toxin, phaseolotoxin, which blocks arginine biosynthesis. Another toxin, tab-toxin, initiates chlorosis in the wildfire infection of *Nicotiana tabacum*, caused by P. s. pv. tabaci. Tab-toxin is a dipeptide that inhibits glutamine synthase. This results in a build up of ammonia, which inhibits photosynthesis and eventually destroys the thylakoid membranes of the chloroplast. In each of these examples, the end result is that by initiating chlorosis, the pathogen establishes a new sink within its host plant, so that infected but still viable plant cells become a source of photoassimilates in the form of hexose sugars, and other nutrients, to support growth of the pathogen.

Table 18.3 Lesion mimic mutants and other cell death mutants of *Arabidopsis*, showing the range of genes, cell functions and signaling factors implicated in the corresponding PCD phenotypes.

Class		Allele	Gene product (where known)	Role (where known)	Main associated signals					
					Ca²⁺	ROS/¹O₂	Salicylic acid	Jasmonates	Ethylene	Sphingo-lipids
Reduced cell death (RCD) mutants		dnd1	AtCNGC2	Cyclic nucleotide-regulated ion channels			+	+	+	
		dnd2	AtCNGC4	-	+		+	+	+	
		executer1/2	Nuclear protein	Required for singlet oxygen-dependent death		+				
		Atrboh D/F	Plasma membrane component of NADPH oxidase	Generation of electron flow across membrane leading to ROS formation		+	+			
Spontaneous death (SD) mutants	Initiation class	acd5	Lipid kinase	-		+	+			+
		acd6	Ankyrin protein with transmembrane region				+			
		flu	Negative regulator of tetrapyrrole biosynthesis	Regulates Glu tRNA reductase, the first committed steps in tetrapyrrole biosynthesis		+				
		lsd2 to lsd7	-	-						
		cpr5	Transmembrane transporter	-			+			
		hrl1	-	-						
		cet1 to cet4	-	-				+		
	Propagative class	cpr22	Fusion of AtCNGC11/AtCNGC12	Cyclic nucleotide-regulated ion channels	+		+	+	+	
		lin2	Coproporphyrinogen III oxidase	Chlorophyll biosynthesis		+	+			
		acd1	Pheophorbide a oxygenase	Chlorophyll catabolism		+				
		acd2	Red chlorophyll catabolite reductase	Chlorophyll catabolism		+				
		lsd1	Zn finger protein	Transcriptional activator		+	+			
		acd11	-	Sphingosine transfer protein			+		+	
		vad1	Gram domain-containing protein	-			+			+

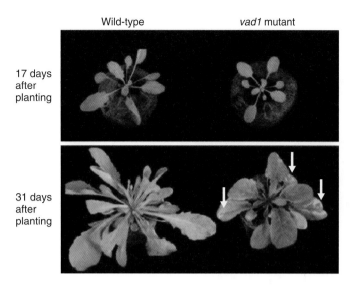

Figure 18.45 Phenotype of the *vascular associated death1* (*vad1*) mutant of *Arabidopsis*. The figure shows an early and late stage in the outward spread of necrotic regions (arrowed) from the veins as leaves age.

Key points Programmed cell death is significant in biotic interactions, both those of the beneficial kind represented by attracting pollinators and dispersers, and those of a harmful nature such as attack by pests and pathogens. Pathogenic organisms are classed as necrotrophs (which kill the host and feed off its dead tissues) and biotrophs (which require host tissues to remain alive). The hypersensitive response (HR) is a strategy whereby the host neutralizes infection by inducing PCD. HR in response to pathogenic fungi and viruses usually takes the form of single cell PCD and necrotic flecking. The mechanism of HR-associated PCD closely resembles autophagy in the role played by vacuolar lytic activities and in its sensitivity to chemical inhibitors. ROS and reactive nitrogen species are inducers of HR. Gene–gene interactions in the resistance–avirulence (R–avr) system are essential for triggering PCD during HR. The associated regulatory networks have been dissected by the use of mutants in which cell death is delayed or else occurs spontaneously. Among the agriculturally important spontaneous lesion mutants that confer enhanced disease resistance are *sl* in rice and *mlo* in barley. The *Arabidopsis* vascular differentiation mutant *vad1* is a possible example of a genetic variant defective in both developmental and HR-related PCD. Some pathogens, such as *Pseudomonas* spp., produce toxins that induce senescence-like yellowing of host leaf tissues. The potential vulnerability of senescing tissues to pathogen attack accounts for the prominence of defense genes among *SAGs*.

Leaf senescence is a vulnerable stage in the plant life cycle, when resources that are being remobilized for use elsewhere—for example, in grain filling or tuber development—could offer a feast to potential pathogens. This may explain why plants upregulate defense pathways as they initiate normal senescence (see Table 18.1), and why certain senescence hormones, particularly jasmonates and salicylates, are also mediators of resistance against pathogens.

18.6.4 The relationships between programmed senescence, death and aging are complex

Although not usually counted as an environmental influence, **time** has some of the properties of a stress factor. The term **aging** refers to biological changes over time. Human experience, and the biomedical discipline of **gerontology**, associate aging with questions of **longevity** and progressive deterioration of life functions. Plants are fascinating in this respect, since they seem to display some of the important age-related characteristics observed in animals and humans, but in an exaggerated form. For example, plants range from short-lived herbaceous types such as *Arabidopsis*, that go through a complete life cycle in a few weeks, to the oldest individual organisms on the planet (such as the Australasian clonal shrub *Lomatia tasmanica* (Figure 18.46), with an estimated age of more than 40 000 years). **Monocarpic** species flower once and die, a process reminiscent of the reproductive exhaustion associated with the death of so-called semelparous animal species such as the Pacific

Figure 18.46 *Lomatia tasmanica*. This species, known as King's lomatia or King's holly, exists in its native habitat in Tasmania as a single triploid clonal population. Although it flowers occasionally, it is apparently sterile and propagates by root suckering. It is claimed that the surviving clone is at least 43 600 years old. The photograph shows a cultivated specimen growing in the Hobart Botanical Garden.

salmon, mayfly and many kinds of octopus and squid. The apparent relationship between senescence, aging and life span is reinforced by pervasive allusions to fall leaves, fading flowers and ripening fruit as literary images describing the human condition.

It would seem obvious that there must be some link between the aging and longevity of an individual plant and the occurrence in its life cycle of programmed senescence and death of its parts. This connection is easier to make for monocarpic than for long-lived species. The life cycle strategy of a monocarp clearly consists of invoking the progressive senescence and death of the entire vegetative body to insure mass nutrient mobilization and transfer to the seeds, which survive to start the process over in the next generation. For trees and clonal species, the balance between the life span of the whole organism, on the one hand, and of its parts on the other, generally favors survival of the former. It remains an open question as to whether long-term accumulation of **genetic errors** and/or escalating stress associated with the increase in size and decline in **integration**, lead ultimately to the death of such plants. The answer would be of great biological and medical significance, and further bear out the central importance of programmed senescence and death in the molecular life of plants.

Key points The term aging refers to changes over time. Plants encompass the widest diversity of longevities of all multicellular organisms, with life spans ranging from weeks, in the case of ephemeral weeds like *Arabidopsis*, to millennia as exemplified by trees and clonal species. It has been suggested that monocarpic plants (which die after flowering once) suffer from a kind of reproductive exhaustion similar to that experienced by semelparous animals. Monocarpic senescence is one kind of survival strategy in which the whole of the vegetative body of the plant is sacrificed in the interests of seed production and propagation of the next generation. In long-lived plants that reproduce serially, the emphasis is on maintenance and survival of the individual. Such a strategy is accompanied by potential susceptibility to accumulated genetic errors and declining physiological integrity.

Acknowledgments, credits and sources

We would like to acknowledge those individuals, organizations, publishers and societies, who have granted permission to reproduce material in this book. Acknowledgements are listed by chapter:

Chapter 1
Figure 1.1, courtesy of Peggy Lemaux, University of California, Berkeley
Figure 1.2A & B, courtesy of Tracey Slotta @USDA-NRCS PLANTS Database
Figure 1.2C & D, courtesy of Steve Hurst @USDA-NRCS PLANTS Database
Figures 1.4, 1.14, 1.15, 1.26, 1.27A & C, 1.28B & D, 1.34, 1.35, 1.37, 1.38 and 1.41, all courtesy of J. Robert Waaland, Algamarine Ltd.
Figures 1.7, 1.10B–D, 1.21 and 1.39, all courtesy of Susan Waaland
Figure 1.27B, courtesy of Steven Ruzin, University of California, Berkeley

Chapter 2
Figures 2.16, 2.19 and 2.20, from Buchanan, Gruissem and Jones, *Biochemistry and Molecular Biology of Plants*, 2000, The American Society of Plant Biologists
Figure 2.25, source: RCSB PDB www.pdb.org
Figure 2.27, courtesy of J. Robert Waaland, Algamarine Ltd.
Figure 2.28, from Mieda, T. et al., 2004, *Plant Cell Physiology* 45:1271–9, Oxford University Press
Table 2.4, after Morris, *A Biologist's Physical Chemistry*, 2nd edition, 1967, Edward Arnold

Chapter 3
Figures 3.4, 3.14, 3.15, 3.17, 3.19, 3.23, 3.25, 3.34, 3.36, 3.37 and 3.39, from Buchanan, Gruissem and Jones, *Biochemistry and Molecular Biology of Plants*, 2000, The American Society of Plant Biologists
Figures 3.6A, 3.8B and 3.9A, source: Proceedings of the National Academy of Sciences of the United States of America, www.pnas.org
Figure 3.6B, source: www.thenakedscientists.com
Figure 3.9B, courtesy of Graham Moore, The John Innes Centre, UK

Figure 3.10, courtesy of Helen Ougham and Sid Thomas, Aberystwyth University, UK
Figure 3.12, source: http://plants.ensembl.org/index.html
Figure 3.16, from Freeman, *Biological Science*, 3rd edition, 2008, Pearson Education
Figure 3.20, courtesy of Neil Jones, Aberystwyth University, UK
Figure 3.22, source: www.wheatgenome.org
Figure 3.30, source: http://commons.wikimedia.org /wiki/File:Antennapedia.jpg
Figure 3.31, source: The MaizeGDB database, www.maizegdb.org.
Figure 3.35, from Napoli, C., Lemieux, C. and Jorgensen, R., 1990, *The Plant Cell* 2, 279–289, The American Society of Plant Biologists
Figure 3.46, source: RCSB PDB, www.pdb.org. Data originators Gabdoulkhakov, A.G., Savoshkina, Y., Krauspenhaar, R., Stoeva, S., Konareva, N., Kornilov, V., Kornev, A.N., Voelter, W., Nikonov, S.V., Betzel, C. and Mikhailov, A.M.
Tables 3.2, 3.3, 3.4 and 3.5, modified from Buchanan, Gruissem and Jones, *Biochemistry and Molecular Biology of Plants*, 2000, The American Society of Plant Biologists

Chapter 4
Figures 4.2, 4.5, 4.10, 4.15A, 4.18, 4.24A & D, 4.25, 4.26, 4.28, 4.36, 4.39 and 4.41, from Buchanan, Gruissem and Jones, *Biochemistry and Molecular Biology of Plants*, 2000, The American Society of Plant Biologists
Figure 4.3, from *Plant Cell Biology on DVD— Information for Students and a Resource for Teachers*, Gunning, B.E.S., 2009, Springer Verlag, www.plantcellbiologyonDVD.com. Micrograph courtesy of Adrienne Hardham, Australian National University
Figures 4.7 and 4.13A–C, from *Plant Cell Biology on DVD—Information for Students and a Resource for Teachers*, Gunning, B.E.S., 2009, Springer Verlag, www.plantcellbiologyonDVD.com
Figure 4.17, from *Plant Cell Biology on DVD— Information for Students and a Resource for Teachers*,

The Molecular Life of Plants, First Edition. Russell Jones, Helen Ougham, Howard Thomas and Susan Waaland.
© 2013 John Wiley & Sons, Ltd. Published 2013 by John Wiley & Sons, Ltd.

Gunning, B.E.S., 2009, Springer Verlag, www.plantcellbiologyonDVD.com. Micrograph A, courtesy of Karl Oparka; micrograph B, courtesy of Martin Steer

Figure 4.18B, courtesy of Peter Eastmond, University of Warwick, UK

Figure 4.21, from *Plant Cell Biology on DVD— Information for Students and a Resource for Teachers*, Gunning, B.E.S., 2009, Springer Verlag, www.plantcellbiologyonDVD.com. Micrograph B, courtesy of Ursula Meindl

Figures 4.22A–C, 4.24B & C and 4.27B, from *Plant Cell Biology on DVD—Information for Students and a Resource for Teachers*, Gunning, B.E.S., 2009, Springer Verlag, www.plantcellbiologyonDVD.com

Chapter 5

Figure 5.7B, from Bjørn, P. et al. 2007, *Nature* 450, 1111–14, Nature Publishing Group

Figure 5.13B, courtesy of C. Toyoshima, University of Tokyo, Japan

Figures 5.18 and 5.19, from Buchanan, Gruissem and Jones, *Biochemistry and Molecular Biology of Plants*, 2000, The American Society of Plant Biologists

Figure 5.23, from Maurel, C., 2007, *FEBS Letters* 581 (12): 2227–36, Elsevier

Figure 5.34, from *Plant Cell Biology on DVD— Information for Students and a Resource for Teachers*, Gunning, B.E.S., 2009, Springer Verlag, www.plantcellbiologyonDVD.com

Table 5.1, from Taiz, L. and Zeiger, E. *Plant Physiology*, 3rd edition, 2002, Sinauer Associates Inc. Data from Higinbotham et al., 1967

Chapter 6

Figure 6.2A & B, courtesy of Shinjiro Yamaguchi, published in Mikihiro Ogawa, et al., 2003, *The Plant Cell* 15: 1591–604, The American Society of Plant Biologists

Figures 6.3, 6.10, 6.12, 6.13, 6.13, 6.15, 6.19, 6.20, 6.30, 6.31 and 6.34 from Buchanan, Gruissem and Jones, *Biochemistry and Molecular Biology of Plants*, 2000, The American Society of Plant Biologists

Figure 6.11, courtesy of William Hurkman and Delilah Wood, University of California, Berkeley

Figure 6.16A, courtesy of Lacey Samuels, University of British Columbia, Canada

Figure 6.16B, courtesy of Paul Bethke, from Bethke et al., 2007, *Plant Physiology* 143: 1173–88, The American Society of Plant Biologists

Figure 6.17, courtesy of T. Okita, Washington State University

Figure 6.21, courtesy of Gerhard Leubner, originally published in Müller et al., 2006, *Plant and Cell Physiology* 47: 864–77, Oxford University Press

Figure 6.23, courtesy of Paul Bethke, from Bethke et al., 2007, *Plant Physiology* 143: 1173–88, The American Society of Plant Biologists

Figure 6.24, courtesy of Custom Life Science Images, © David McIntyre

Figure 6.27, courtesy of Paul Bethke, from Bethke et al., 2007, *Plant Physiology* 143: 1173–88, The American Society of Plant Biologists

Figure 6.29, courtesy of Bob Buchanan, University of California, Berkeley and J. Yin-Zhengzhou, National Engineering Research Centre for Wheat, Henan Agricultural University, China

Table 6.2, from Loren Cordain, 'Cereal grains: humanity's double-edged sword' in Simopoulos, A.P. (ed.), *Evolutionary Aspects of Nutrition and Health. Diet, Exercise, Genetics and Chronic Disease*, 1999, *World Review of Nutrition and Diet* 84: 19–73, Karger, Basel

Tables 6.3, 6.4, 6.7, 6.12, 6.14, 6.15 and 6.16, modified from Bewley and Black, *Seeds: Physiology of Development and Germination*, 2nd edition, 1994, Plenum Press

Table 6.5, source: International Starch Institute, Science Park Aarhus, Denmark, http://www.starch.dk/isi/starch/starch.asp

Table 6.8, source: Guy Inchbald, http://www.queenhill.demon.co.uk/seedoils/seedcontent.htm

Table 6.10, from Buchanan, Gruissem and Jones, *Biochemistry and Molecular Biology of Plants*, 2000, The American Society of Plant Biologists

Table 6.13, from Stevenson-Paulik et al., 2005, *Proceedings of the National Academy of Sciences* 102: 12612–17, National Academy of Sciences

Chapter 7

Figures 7.2, 7.3, 7.6, 7.7, 7.9, 7.10, 7.12, 7.13, 7.14, 7.15, 7.16, 7.17, 7.18, 7.21, 7.24 and 7.25, from Buchanan, Gruissem and Jones, *Biochemistry and Molecular Biology of Plants*, 2000, The American Society of Plant Biologists

Figure 7.30, output from Genevestigator meta-analysis tool, https://www.genevestigator.com/

Table 7.1, modified from Buchanan, Gruissem and Jones, *Biochemistry and Molecular Biology of Plants*, 2000, The American Society of Plant Biologists

Table 7.2, based on Rasmusson, A.G. and Escobar, M.A., 2007, *Physiologia Plantarum* 129: 57–67, Physiologia Plantarum

Chapter 8

Figure 8.1, courtesy of Susan Waaland

Figures 8.7, 8.14, 8.16, 8.20, 8.23 and 8.35, from Buchanan, Gruissem and Jones, *Biochemistry and Molecular Biology of Plants*, 2000, The American Society of Plant Biologists

Figure 8.10 B, output from Genevestigator meta-analysis tool, https://www.genevestigator.com/

Figure 8.12, from Lopez-Juez, E., Nagatani, A., Tomizawa, K.I., Deak, M., Kern, R., Kendrick, R.E. and Furuya, M., 1992, The cucumber long hypocotyl mutant lacks a light-stable PHYB-like phytochrome. *The Plant Cell* 4: 241–51, The American Society of Plant Biologists

Figure 8.21, adapted from Franklin and Whitelam, 2005, *Annals of Botany* 96 (2): 169–75, Oxford University Press

Figure 8.26, courtesy of Troy Paddock, National Renewable Energy Laboratory (NREL)

Figure 8.32, from Lagercrantz, U., 2009, *Journal of Experimental Botany* 60: 2501–15, Oxford University Press

Table 8.2, after Franklin, K.A. and Quail, P.H., 2010, *Journal of Experimental Botany* 61: 11–24, Oxford University Press

Table 8.3, after Banerjee, R. and Batschauer, A., 2005, *Planta* 220: 498–502, Springer Verlag

Chapter 9

Figures 9.3, 9.5, 9.7, 9.8, 9.11, 9.13, 9.14, 9.17, 9.20, 9.21, 9.24, 9.25, 9.26, 9.31, 9.32, 9.33, 9.34, 9.35, 9.36, 9.37, 9.39, 9.40 and 9.41, from Buchanan, Gruissem and Jones, *Biochemistry and Molecular Biology of Plants*, 2000, The American Society of Plant Biologists

Figure 9.19, source: RCSB PDB www.pdb.org

Figure 9.27, source: RCSB PDB www.pdb.org

Tables 9.1, 9.2, 9.3, 9.4, 9.5, 9.6, 9.7, 9.9 and 9.10, from Buchanan, Gruissem and Jones, *Biochemistry and Molecular Biology of Plants*, 2000, The American Society of Plant Biologists

Chapter 10

Figures 10.1, 10.3, 10.5, 10.7, 10.15, 10.20, 10.42, 10.45, 10.48 and 10.50 from Buchanan, Gruissem and Jones, *Biochemistry and Molecular Biology of Plants*, 2000, The American Society of Plant Biologists

Figure 10.4A, courtesy of Thomas G. Ranney, North Carolina State University

Figure 10.4B, courtesy of Vilem Reinohl, Mendel University, Brno, Czech Republic

Figure 10.8B & C, courtesy of J. Kleine-Vehn and J. Friml, Ghent University, Belgium

Figure 10.9, courtesy of Remko Offring, Leiden University, the Netherlands

Figure 10.13, courtesy of Yuji Kamiya, RIKEN, Japan and Nobutaka Takahashi, University of Tokyo, Japan

Figure 10.14A, courtesy of Tai-ping Sun, Duke University

Figure 10.14B, courtesy of Peter Hedden, Rothamstead Research, UK

Figure 10.14C, courtesy of Tina Barsby, National Institute of Agricultural Botany (NIAB), UK

Figures 10.19, 10.22B, 10.26, 10.33B and 10.43A, all courtesy of Shinjiro Yamaguchi, RIKEN, Japan

Figure 10.28, courtesy of Caren Chang, University of Maryland

Figure 10.34, courtesy of Zhi-yong Wang, Carnegie Institution at Stanford University

Figure 10.49, courtesy of Miguel Blazguez, originally published in Munoz et al., 2008, *Development*, 135: 2573, The Company of Biologists

Chapter 11

Figures 11.1, 11.2B, 11.12, 11.14, 11.16, 11.17, 11.19, 11.23, 11.24 and 11.25, from Buchanan, Gruissem and Jones, *Biochemistry and Molecular Biology of Plants*, 2000, The American Society of Plant Biologists

Figure 11.2A, courtesy of Neil Jones, Aberystwyth University, UK

Figure 11.7, source: *Proceedings of the National Academy of Sciences of the United States of America*, 2005, 102 (43): 15694–9, www.pnas.org,

Figure 11.9A & B, source: http://www.plantphysiol.org

Figure 11.9C & D, from Churchman et al., 2006, *The Plant Cell* 18: 3145–57, The American Society of Plant Biologists

Figure 11.18B, from Culligan, K., Tissier, A. and Britta, A., 2004, ATR regulates a G2-phase cell-cycle checkpoint in *Arabidopsis thaliana. The Plant Cell* 16: 1091–104, The American Society of Plant Biologists

Figure 11.20E, from Laux et al., 1996, *Development* 122: 87–96, The Company of Biologists

Figure 11.21, source: http://picasaweb.google.com/lh /photo/Xj-uQY8LOnE69RlAQQLe9A

Figure 11.22A, courtesy of Cal Lemke, University of Oklahoma

Figure 11.22B, source: http://s1.hubimg.com/u/ 1257364_f520

Figure 11.28B, source: http://bio3400.nicerweb.com /Locked/media/ch02/02_14-synaptonemal_complex

Figure 11.29B, courtesy of S.P. Murphy and H.W. Bass, Florida State University

Figure 11.31, from Couteau et al., 1999, *The Plant Cell* 11: 1623–34, The American Society of Plant Biologists

Figure 11.32, inset from Brooker, *Genetics: Analysis and Principles*, 4th edition, 2011, McGraw-Hill

Chapter 12

Figure 12.1, images from Keiko Sakakibara, Tomoaki Nishiyama, Hironori Deguchi and Mitsuyasu Hasebe, 2008, Class 1 KNOX genes are not involved in shoot development in the moss *Physcomitrella patens* but do function in sporophyte development. *Evolution and Development* 10: 555–66, John Wiley & Sons

Figure 12.2, courtesy of John Bowman, University of California, Davis

Figures 12.9, 12.12, 12.25, 12.26 and 12.27, from Buchanan, Gruissem and Jones, *Biochemistry and*

Molecular Biology of Plants, 2000, The American Society of Plant Biologists

Figure 12.16, from De Smet, I., Lau, S., Mayer, U. and Jurgens, G., 2010, Embryogenesis—the humble beginnings of plant life. *The Plant Journal* 61: 959–970, John Wiley & Sons

Figure 12.18, from Foard, D.E. and Haber, A.H., 1961, Anatomic studies of gamma-irradiated wheat growing without cell division. *American Journal of Botany* 48: 438–46, The Botanical Society of America

Figure 12.22C & D, source: Hochholdinger & Zimmermann, 2007

Figure 12.30, source: RCSB PDB www.pdb.org

Figure 12.32A, courtesy of T. Lumpkin

Figure 12.32B, source: www.asahi-net.or.jp/~it6i-wtnb /Aigamo-rice2

Figure 12.33A, courtesy of Mark Brundrett, University of Western Australia, http://mycorrhizas.info/vam.html

Figure 12.34, from Tsiantis, M. and Hay, A., 2003, Comparative plant development: the time of the leaf? *Nature Reviews Genetics* 4: 169–80, Nature Publishing Group

Figure 12.37, source: http://oak.cats.ohiou.edu/~braselto /readings/structure.html

Figure 12.42, from Chatterjee, M., Sparvoli, S., Edmunds, C., Garosi, P., Findlay, K. and Martin, C., 1996, DAG, a gene required for chloroplast differentiation and palisade development in *Antirrhinum majus*. *EMBO Journal* 15: 4194–420, Nature Publishing Group

Figure 12.44, from Byrne, M.E., 2005, Networks in leaf development. *Current Opinion in Plant Biology* 8: 59–66, Elsevier

Figure 12.48, from Prusinkiewicz, P., 1990, *The Algorithmic Beauty of Plants*, http://algorithmicbotany.org/papers/abop

Figure 12.51, from Hedden, P., 2003, The genes of the Green Revolution. *Trends in Genetics* 19: 5–9, Elsevier

Figure 12.53, source: Peter Hedden, Rothamstead Research, UK

Figure 12.55, from Fu et al., 2002, *The Plant Cell*, The American Society of Plant Biologists

Chapter 13

Figure 13.2, courtesy of M. Faget, ETH, Zürich, Switzerland

Figures 13.7, 13.13, 13.14, 13.15, 13.16, 13.17, 13.19, 13.20, 13.21, 13.22, 13.26, 13.29, 13.30, 13.31, 13.32, 13.34, 13.35, 13.36, 13.37, 13.38, 13.40, 13.42, 13.43, 13.44, 13.45, 13.47 and 13.51, from Buchanan, Gruissem and Jones, *Biochemistry and Molecular Biology of Plants*, 2000, The American Society of Plant Biologists

Figure 13.52, from Bilecen, K., Ozturk, U.H., Duru, A.D., Sutlu, T., Petoukhov, M.V., Svergun, D.I., Koch, M.H.J., Sezerman, U.O., Cakmak, I. and Sayers, Z., 2005, *Triticum durum* metallothionein. Isolation of the gene and structural characterization of the protein using solution scattering and molecular modeling. *Journal of Biological Chemistry* 280: 13701–11, The American Society for Biochemistry and Molecular Biology

Tables 13.2 and 13.7, from Buchanan, Gruissem and Jones, *Biochemistry and Molecular Biology of Plants*, 2000, The American Society of Plant Biologists

Table 13.8, from Ashley, M.K., Grant, M. and Grabov, A., 2006, *Journal of Experimental Botany* 57: 425–36, Oxford University Press

Table 13.9, from Karley, A.J. and White, P.J., 2009, *Current Opinion in Plant Biology* 12: 291–8, Elsevier

Chapter 14

Figure 14.4, courtesy of Jean-Pierre Metraux, University of Fribourg, Switzerland

Figures 14.5, 14.6, 14.7, 14.9, 14.10, 14.11, 14.12, 14.15, 14.16, 14.17, 14.18, 14.21, 14.22, 14.24 and 14.30, from Buchanan, Gruissem and Jones, *Biochemistry and Molecular Biology of Plants*, 2000, The American Society of Plant Biologists

Figure 14.8A, from *Plant Cell Biology on DVD—Information for Students and a Resource for Teachers*, Gunning, B.E.S., 2009, Springer Verlag, www.plantcellbiologyonDVD.com. Micrograph courtesy of Robyn Overall, University of Sydney, Australia

Figure 14.13, from Crafts and Yamaguchi, *The Auroradiography of Plant Materials*, Calif. Agr. Expt. Station Extension Serv. Manual 35, 1964

Figure 14.26A, courtesy of Susan Waaland

Figure 14.26B, courtesy of Lacey Samuels, University of British Columbia Canada, from Suh, M.C. et al., 2005, *Plant Physiology* 139: 1649–165, The American Society of Plant Biologists

Figures 14.27A & B, courtesy of David Robinson, University of Heidelberg, Germany

Figure 14.27C, courtesy of Julian Schroeder, University of California, San Diego

Tables 14.2 and 14.3, from Buchanan, Gruissem and Jones, *Biochemistry and Molecular Biology of Plants*, 2000, The American Society of Plant Biologists

Table 14.4, data from Willmer, *Stomata*, 1983, Prentice Hall

Table 14.5, modified from Nobel, *Biophysical Plant Physiology and Ecology*, 1st edition, 1983, Freeman

Chapter 15

Figure 15.1, source: http://www.kriyayoga.com /photography/photo_gallery/d/60773- 2/strawberry_runners-dsc02107-g1.jpg

Figure 15.3, data from USDA CENTURY Agroecosystem Version 4.0, http://www.nrel.colostate.edu/projects /century/MANUAL/html_manual/fig3-8b

Figures 15.4, 15.8, 15.9, 15.15, 15.16, 15.17, 15.19, 15.20, 15.21, 15.22, 15.24, 15.26, 15.27, 15.30, 15.31, 15.32, 15.33, 15.34, 15.36, 15.37 and 15.41, from Buchanan, Gruissem and Jones, *Biochemistry and Molecular Biology of Plants*, 2000, The American Society of Plant Biologists

Figure 15.35, data from von Koskull-Doring, P., Scharf, K-D. and Nover, L., 2007, *Trends in Plant Science* 12: 452–7, Elsevier

Figure 15.40, courtesy of Wendy Silk, University of California, Davis

Figure 15.42, from Scholes, D.J. and Press, M.C., 2008, Striga infestation of cereal crops. *Current Opinion in Plant Biology* 11 (2): 180–6, Elsevier

Tables 15.2, 15.5, 15.6, 15.7 and 15.8, from Buchanan, Gruissem and Jones, *Biochemistry and Molecular Biology of Plants*, 2000, The American Society of Plant Biologists

Chapter 16

Figures 16.7, 16.9, 16.11, 16.12, 16.16, 16.27, 16.29, 16.30, 16.31, 16.32, 16.34A, 16.34D, 16.34E, 16.35, 16.36, 16.39 and 16.42, from Buchanan, Gruissem and Jones, *Biochemistry and Molecular Biology of Plants*, 2000, The American Society of Plant Biologists

Figure 16.3, from Poethig, R.S., 2009, Small RNAs and developmental timing in plants. *Current Opinion in Genetics and Development* 19: 374–8, Elsevier

Figure 16.4, from Tooke, F., Ordidge, M., Chiurugwi, T. and Battey, N., 2005, Mechanisms and function of flower and inflorescence reversion. *Journal of Experimental Botany* 56: 2587–99, Oxford University Press

Figure 16.10, images of flowers from Krizek, B.A. and Fletcher, J.C., 2005, Molecular mechanisms of flower development: an armchair guide. *Nature Reviews in Genetics* 6: 688–98, Nature Publishing Group

Figure 16.23, from Grotewold, E., 2006, The genetics and biochemistry of floral pigments. *Annual Review of Plant Biology* 57: 761–80, Annual Reviews

Figure 16.33, from Favaro et al., 2003, MADS-box protein complexes control carpel and ovule development in *Arabidopsis*. *The Plant Cell* 15: 2603–11, The American Society of Plant Biologists

Figure 16.34B & C, from Lolle, S.J. and Pruitt, R.E., 1999, *Trends in Plant Science* 4: 14–20, Elsevier

Figure 16.37, source: RCSB PDB www.pdb.org

Figure 16.44, from Hochholdinger, F. and Hoecker, N., 2007, Towards the molecular basis of heterosis. *Trends in Plant Science* 12: 427–32, Elsevier

Figure 16.46, micrograph from Le et al., 2007, *Plant Physiology* 144: 562–74, The American Society of Plant Biologists

Figure 16.47, micrographs from Sabelli, P.A. and Larkins, B.A., 2009, *Plant Physiology* 149: 14–26, The American Society of Plant Biologists

Chapter 17

Figure 17.4, from Saska, M.M. and Kuzovkina, Y.A., 2010, *Annals of Applied Biology* 156: 431–7, John Wiley & Sons

Figures 17.7, 17.9, 17.14, 17.15, 17.20 and 17.32, from Buchanan, Gruissem and Jones, *Biochemistry and Molecular Biology of Plants*, 2000, The American Society of Plant Biologists

Figure 17.6, from Jacoby, G.C., Workman, K.W. and D'Arrigo, R.D., 1999, *Quaternary Science Reviews* 18: 1365–71, Elsevier

Figure 17.9, modified and simplified from Kloosterman, B., Vorst, O., Hall, R.D., Visser, R.G.F. and Bachem, C.W., 2005, Tuber on a chip: differential gene expression during potato tuber development. *Plant Biotechnology Journal* 3: 505–19, John Wiley & Sons

Figure 17.10, pea image courtesy of Neil Jones, Aberystwyth University, UK

Figure 17.13, source: Marco Schmidt, http://commons .wikimedia.org/wiki/User:Marco_Schmidt

Figure 17.16A, source: RCSB PDB www.pdb.org

Figure 17.16B, from Stupar, R.M., Beaubien, K.A., Jin, W., Song, J., Lee, M-K., Wu, C., Zhang, H-B, Han, B. and Jiang, J., 2006, *Genetics* 172: 1263–75, DOI, The Genetics Society of America

Figure 17.17, from Cooke, J.E.K. and Weih, W., 2005, Nitrogen storage and seasonal nitrogen cycling in *Populus*: bridging molecular physiology and ecophysiology. *New Phytologist* 167: 19–30, John Wiley & Sons. Courtesy of John Greenwood, University of Guelph, Canada

Figure 17.18, courtesy of John Clifton-Brown, Aberystwyth University, UK

Figure 17.22, source: NASA, www.nasa.gov

Figure 17.27, from Arora, R., Rowland, L.J. and Tanino, K., 2003, *HortScience* 38: 911–21, The American Society for Horticultural Science

Figure 17.28, from Chaloupkova, K. and Smart, C.C., 1994, *Plant Physiology* 105: 497–507, The American Society of Plant Biologists

Figure 17.31, courtesy of Athole Marshall, Aberystwyth University, UK

Chapter 18

Figures 18.1, 18.3C, 18.5, 18.6, 18.7, 18.8, 18.9, 18.15, 18.17, 18.18, 18.27, 18.28, 18.29, 18.31, 18.32 and 18.36, from Buchanan, Gruissem and Jones, *Biochemistry and Molecular Biology of Plants*, 2000, The American Society of Plant Biologists

Figure 18.3D, courtesy of David Robinson, University of Heidelberg, Germany

Figure 18.11B, courtesy of Adrian Dauphinee, Gunawardena Laboratory, Biology Department, Dalhousie University, Canada

Figure 18.13, from Shibuya, K., Yamada, T., Suzuki, T., Shimizu, K. and Ichimura, K., 2009, InPSR26, a putative membrane protein, regulates programmed cell death during petal senescence in Japanese morning glory. *Plant Physiology* 149: 816–24, The American Society of Plant Biologists.

Figure 18.14, courtesy of Thomas Schoch, License: CC-BY-SA 3.0

Figure 18.21, *Nicotania tabacum* image courtesy of Magnus Manske, 2009, http://commons.wikimedia .org/wiki/File:Nicotiana_tabacum_%27Tobacco %27_%28Solanaceae%29_plant

Figure 18.22, from Miguel, A., Perez-Amador, M.A. et al., 2000, Identification of BFN1, a bifunctional nuclease induced during leaf and stem senescence in *Arabidopsis. Plant Physiology* 122: 169–79, The American Society of Plant Biologists

Figure 18.23, from Zimmermann P., Heinlein, C., Orendi, G. and Zentgraf, U., 2006, Senescence-specific regulation of catalases in *Arabidopsis thaliana* (L.) Heynh. *Plant, Cell and Environment* 29: 1049–60, John Wiley & Sons

Figure 18.24, courtesy of Helen Ougham and Sid Thomas, Aberystwyth University, UK

Figure 18.25, adapted from Uauy, C., Distelfeld, A., Fahima, T., Blechl, A. and Dubcovsky, J., 2006, A NAC gene regulating senescence improves grain protein, zinc, and iron content in wheat. *Science* 314: 1298–301, American Association for the Advancement of Science

Figure 18.26, from Bhalerao, R., Keskitalo, J., Sterky, J.F., Erlandsson R., Björkbacka, H., Birve, S.J., Karlsson, J., Gardeström, P., Gustafsson, P., Lundeberg, J. and Jansson, S., 2003, *Plant Physiology* 131: 430–42, The American Society of Plant Biologists

Figure 18.28, images courtesy of Hilda Zavaleta-Mancera, Postgrado de Botánica, Colegio de Postgraduados en Ciencias Agricolas, Montecillo, Mexico

Figure 18.35, source: RCSB PDB www.pdb.org

Figure 18.39, from Giovannoni, J.J., 2007, Fruit ripening mutants yield insights into ripening control. *Current Opinion in Plant Biology* 10: 283–9, Elsevier

Figure 18.40, leaf images courtesy of Keskitalo, J., Bergquist, G., Gardeström, P. and Jansson S., 2005, *Plant Physiology* 139: 1635–48, The American Society of Plant Biologists

Figure 18.44, source: The MaizeGDB database, www.maizegdb.org

Figure 18.45, from Lorrain, Lin et al. 2004. Vascular associated death1, a novel GRAM domain-containing protein, is a regulator of cell death and defense responses in vascular tissues. *The Plant Cell* 16: 2217–32, The American Society of Plant Biologists

Figure 18.46, source: Shantavira, http://en.wikipedia.org /wiki/File:456509194_b4bab5b9e7_o.jpg

Table 18.1, from Buchanan-Wollaston, V., Earl, S., Harrison, E., Mathas, E., Navabpour, S., Page, T. and Pink, D., 2003, *Plant Biotechnology Journal* 1: 3–22, John Wiley & Sons

Table 18.2, from Prasanna, V., Prabha, T.N. and Tharanathan, R.N., 2007, *Critical Reviews in Food Science and Nutrition* 47: 1–19, Taylor & Francis

Table 18.3, from Buchanan, Gruissem and Jones, *Biochemistry and Molecular Biology of Plants*, 2nd edition, not yet published, John Wiley & Sons

Index

Page numbers in *italics* refer to figures, those in **bold** refer to tables

14-3-3 proteins 159, *160*, 475
26S proteasome 175, 178–9, *179*, 180

A-type cyclins (CYCAs) 380
a1-1, a2-1 genes *358*
aao3 gene 357
aba (ABA-deficient) genes 356, 357, 563, 650
ABA-independent pathway 564
ABA response elements (ABAREs) 359, 564
ABC, ABCE models of flower development 594–9, *595*, **597**, 610
ABC transporters 162, *162*, 333
ABI/abi (ABA-insensitive) genes 359, 650, 657
ABI4 transcription factor 441–2, *442*
Abies (fir) 15, 656
abietic acid 553, *555*, 557
abiotic stress responses 560, 564, 569, 572, 575, 576
  and abscisic acid 356, 358
  ACS gene expression 349
  to cold 569–70
  and cryoprotectants 570
  DELLA protein mediation 342
  to drought 560–4, *563*
  to excess light 572–3, 575
  to flooding 564–6, *566*
  gene expression 563, *563*
  to gravity and touch 576–7
  to heat 570–2
  to oxidative stress 566–9, *566*, **567**, *568*, **569**
  and salicylic acid 369
  and senescence 699–701, *700*, *701*, *702*
  to shade 575, *575*
ABPHYLL1 (aberrant phyllotaxy 1) gene 437
abscisic acid (ABA) 132, 329–30, *330*, 359, 553
  amylase synthesis, repression of 212
  catabolism 357–8, *358*
  dormancy 207–8, *208*, 655, 656–7
  drought response 563–4, *563*
  lateral root formation 426, *427*
  leaching prevention 204
  maturity 650

PCD delay 692, *692*
receptors 358–9, *359*
roles 356
signal transduction pathway 359, *359*
stomatal closure 165, *166*, 440, *531*, *532*
stress pathways 569
synthesis *340*, 356–8, *357*
water status communication 531
abscisic acid 8′-hydroxylase 563
abscission 348, 356, 673, *674*, 675
abscission zone 673
absorption spectrum 255
ACC (1-aminocyclopropane-1-carboxylic acid) 349, *350*, 368, 697
ACC oxidase (ACO) 349, *350*, 368, 697–8
ACC synthase (ACS) 349, *350*, 697–8
accessory light-harvesting pigments 294
accessory proteins 137, 142, *144*
acclimation 537–8, *538*, **538**, 561, 569
  see also adaptation
acd genes **704**
Acer pseudoplatanus (sycamore) 629–30, *631*
Acetabularia 75
acetyl-CoA 65, 315
  glyoxylate cycle 239, *240*
  IPP synthesis 555, *556*
  lipid synthesis *317*, 642–3, *644*
  respiration regulation 243, *243*
  sulfate assimilation 489
  TCA cycle 223–4, 227
  triacylglycerol breakdown 214, *215*, *216*, 239
acetyl-CoA carboxylase (ACCase) 643
acetylation 92, *92*
acid-growth hypothesis 412, *413*
acid phosphatases (APases) 480–1
acidification of cytoplasm 564
acids 45
ACO (ACC oxidase) genes 349, 697–8
ACS (ACC synthase) genes 349, 697–8
actin 138–9, 614, *614*, 618
actin filaments 137, 138, 138–9, *140*, 142
  assembly of 139–40, *141*
  exocytosis for tip growth 143–4, *144*
  motor proteins, interaction with *143*

  nucleotide hydrolysis 140, *140*, 142
  polarity 139, *140*
  treadmilling 140, *141*, 142
actinomorphic symmetry 597, 600
actinorhizal plants 433
action spectra 255, 261, *261*
activation energy 67, *68*
activator–depletion model 439, *439*
activator-inhibitor model 439, *439*
active transport 155–6, *156*
acyl-carrier protein (ACP) 643
acyl-CoA oxidase 214
acyl lipid synthesis 315
adaptation 537–8, *538*, **538**
  see also acclimation
ADC1/ADC2 (Arg decarboxylase) genes 366
adenine 58, 59, 342
adenosine diphosphate (ADP) 22, 23, 63, **63**, 64, 67
adenosine monophosphate (AMP) 23, 63, **63**, 64, 475, *476*
adenosine phosphosulfate (APS) 487, *487*
adenosyl methionine synthetase 349
Adh1 (alcohol dehydrogenase) gene 80, *80*
ADP-glucose 315
ADP-glucose pyrophosphorylase (AGP) 315, 696
adventitious embryony 620
adventitious roots 29, 331, 423, 424, *425*, 565
aerenchyma 565, *665*, *674*, 701, *701*, 702, *702*
aerobic respiration 136
Aesculus hippocastanum (horsechestnut) 632, *632*
after-ripening 205, 631
AG/ag (agamous) gene 97, 594, **597**
agamospermy 620
aging 705–6, *706*
agriculture 181–2, 540
AHA gene family 497, *498*, 499
AHP (histidine-containing phosphotransmitter) genes 346–7, *347*
AHP protein 346
'air seeding' 524

The Molecular Life of Plants, First Edition. Russell Jones, Helen Ougham, Howard Thomas and Susan Waaland.
© 2013 John Wiley & Sons, Ltd. Published 2013 by John Wiley & Sons, Ltd.